2006

54th Edition

NATIONAL CONSTRUCTION ESTIMATOR

Edited by Dave Ogershok and Richard Pray

Includes inside the back cover:

- An estimating CD with all the costs in this book, plus,
- An estimating program that makes it easy to use these costs,
- An interactive video guide to the National Estimator program,
- A program that converts your estimates into invoices,
- A program that exports your estimates to QuickBooks Pro.

Monthly price updates on the Web are free and automatic all during 2006. You'll be prompted when it's time to collect the next update. A connection to the Web is required.

Download all of Craftsman's most popular costbooks for one low price with the Craftsman Site License. http://CraftsmanSiteLicense.com

Craftsman Book Company
6058 Corte del Cedro / P.O. Box 6500 / Carlsbad, CA 92018

Contents
A complete index begins on page 636

This Book Is an Encyclopedia of 2006 Building Costs

The 2006 National Construction Estimator lists estimated construction costs to general contractors performing the work with their own crews, as of mid-2006. Overhead & profit are not included.

This Manual Has Two Parts; the Residential Construction Division begins on page 17. Use the figures in this division when estimating the cost of homes and apartments with a wood, steel or masonry frame. The Industrial and Commercial Division begins on page 324 and can be used to estimate costs for nearly all construction not covered by the Residential Division.

The Residential Construction Division is arranged in alphabetical order by construction trade and type of material. The Industrial and Commercial Division follows the 16 section Construction Specification Format. A complete index begins on page 636.

National Estimator (inside the back cover) is an electronic version of this book. To watch an interactive video guide to National Estimator, click ? on the tool bar. Then click ShowMe tutorial.

Residential Construction Division begins on page 17

Industrial & Commercial Division begins on page 324

2006 National Construction Estimator

2006 National Estimator is an electronic version of this book

Material Costs

Material Costs for each item are listed in the column headed "Material." These are neither retail nor wholesale prices. They are estimates of what most contractors who buy in moderate volume will pay suppliers as of mid-2006. Discounts may be available for purchases in larger volume.

Monthly prices updates on the Web are free and automatic all during 2006. You'll be prompted when it's time to collect the next update. A connection to the Web is required.

Add Delivery Expense to the material cost for other than local delivery of reasonably large quantities. Cost of delivery varies with the distance from source of supply, method of transportation, and quantity to be delivered. But most material dealers absorb the delivery cost on local delivery (5 to 15 miles) of larger quantities to good customers. Add the expense of job site delivery when it is a significant part of the material cost.

Add Sales Tax when sales tax will be charged to the contractor buying the materials.

Waste and Coverage loss is included in the installed material cost. The cost of many materials per unit after installation is greater than the purchase price for the same unit because of waste, shrinkage or coverage loss during installation. For example, about 120 square feet of nominal 1" x 4" square edge boards will be needed to cover 100 square feet of floor or wall. There is no coverage loss with plywood sheathing, but waste due to cutting and fitting will average about 6%.

Costs in the "Material" column of this book assume normal waste and coverage loss. Small and irregular jobs may require a greater waste allowance. Materials priced without installation (with no labor cost) do not include an allowance for waste and coverage except as noted.

Labor Costs

Labor Costs for installing the material or doing the work described are listed in the column headed "Labor." The labor cost per unit is the labor cost per hour multiplied by the manhours per unit shown after the @ sign in the "Craft@Hours" column. Labor cost includes the basic wage, the employer's contribution to welfare, pension, vacation and apprentice funds and all tax and insurance charges based on wages. Hourly labor costs for the various crafts are listed on page 10 (for the Residential Division) and page 323 (for the Industrial and Commercial Division).

Hourly labor costs used in the Industrial and Commercial Division are higher than those used in the

Residential Division, reflecting the fact that craftsmen on industrial and commercial jobs are often paid more than craftsmen on residential jobs.

Supervision Expense to the general contractor is not included in the labor cost. The cost of supervision and non-productive labor varies widely from job to job. Calculate the cost of supervision and non-productive labor and add this to the estimate.

Payroll Taxes and Insurance included in the labor cost are itemized in the sections beginning on pages 180 and 298.

Manhours per Unit and the Craft performing the work are listed in the "Craft@Hrs" column. Pages 7 through 9 explain the "Craft@Hrs" column. To find the units of work done per man in an 8-hour day, divide 8 by the manhours per unit. To find the units done by a crew in an 8-hour day, multiply the units per man per 8-hour day by the number of crew members.

Manhours Include all productive labor normally associated with installing the materials described. This will usually include tasks such as:

■ Unloading and storing construction materials, tools and equipment on site.

■ Moving tools and equipment from a storage area or truck on site at the beginning of the day.

■ Returning tools and equipment to a storage area or truck on site at the end of the day.

■ Normal time lost for work breaks.

■ Planning and discussing the work to be performed.

■ Normal handling, measuring, cutting and fitting.

■ Keeping a record of the time spent and work done.

■ Regular cleanup of construction debris.

■ Infrequent correction or repairs required because of faulty installation.

Adjust the Labor Cost to the job you are figuring when your actual hourly labor cost is known or can be estimated. The labor costs listed on pages 10 and 323 will apply within a few percent on many jobs. But labor costs may be much higher or much lower on the job you are estimating.

If the hourly wage rates listed on page 10 or page 323 are not accurate, divide your known or estimated cost per hour by the listed cost per hour. The result is your adjustment for any figure in the "Labor" column for that craft. See page 11 for more information on adjusting labor costs.

Adjust for Unusual Labor Productivity. Costs in the labor column are for normal conditions: experienced craftsmen working on reasonably well planned and managed new construction with fair to good productivity. Labor estimates assume that materials are standard grade, appropriate tools are on hand, work done by other crafts is adequate, layout and installation

are relatively uncomplicated, and working conditions don't slow progress.

Working conditions at the job site have a major effect on labor cost. Estimating experience and careful analysis can help you predict the effect of most changes in working conditions. Obviously, no single adjustment will apply on all jobs. But the adjustments that follow should help you produce more accurate labor estimates. More than one condition may apply on a job.

■ Add 10% to 15% when working temperatures are below 40 degrees or above 95 degrees.

■ Add 15% to 25% for work on a ladder or a scaffold, in a crawl space, in a congested area or remote from the material storage point.

■ Deduct 10% when the work is in a large open area with excellent access and good light.

■ Add 1% for each 10 feet that materials must be lifted above ground level.

■ Add 5% to 50% for tradesmen with below average skills. Deduct 5% to 25% for highly motivated, highly skilled tradesmen.

■ Deduct 10% to 20% when an identical task is repeated many times for several days at the same site.

■ Add 30% to 50% on small jobs where fitting and matching of materials is required, adjacent surfaces have to be protected and the job site is occupied during construction.

■ Add 25% to 50% for work done following a major flood, fire, earthquake, hurricane or tornado while skilled tradesmen are not readily available. Material costs may also be higher after a major disaster.

■ Add 10% to 35% for demanding specs, rigid inspections, unreliable suppliers, a difficult owner or an inexperienced architect.

Use an Area Modification Factor from pages 12 through 15 if your material, hourly labor or equipment costs are unknown and can't be estimated.

Here's how: Use the labor and material costs in this manual without modification. Then add or deduct the percentage shown on pages 12 through 15 to estimated costs to find your local estimated cost.

Equipment Costs

Equipment Costs for major equipment (such as cranes and tractors) are listed in the column headed "Equipment." Costs for small tools and expendable supplies (such as saws and tape) are usually considered overhead expense and do not appear in the Equipment cost column.

Equipment costs are based on rental rates listed in the section beginning on page 334 and assume that the equipment can be used productively for an entire 8-hour day. Add the cost of moving equipment on and off the site. Allow for unproductive time when equipment can't be used for the full rental period. For example, the equipment costs per unit of work completed will be higher when a tractor is used for 4 hours during a day and sits idle for the remaining 4 hours. Generally, an 8-hour day is the minimum rental period for most heavy equipment. Many sections describe the equipment being used, the cost per hour and a suggested minimum job charge.

Subcontracted Work

Subcontractors do most of the work on construction projects. That's because specialty contractors can often get the work done at competitive cost, even after adding overhead and profit.

Many sections of this book cover work usually done by subcontractors. If you see the word "subcontract" in a section description, assume that costs are based on quotes by subcontractors and include typical subcontractor markup (about 30% on labor and 15% on material). Usually no material or labor costs will appear in these sections. The only costs shown will be in the "Total" column and will include all material, labor and equipment expense.

If you don't see the word "subcontract" in a section description, assume that costs are based on work done by a general contractor's crew. No markup is included in these costs. If the work is done by a subcontractor, the specialty contractor may be able to perform the work for the cost shown, even after adding overhead and profit.

Markup

The General Contractor's Markup is not included in any costs in this book. On page 218 we suggest a 20% markup on the contract price for general contractors handling residential construction. Apply this markup or some figure you select to all costs, including both subcontract items and work done by your own crews.

To realize a markup of 20% on the contract price, you'll have to add 25% to your costs. See page 218 for an example of how markup is calculated. Markup includes overhead and profit and may be the most difficult item to estimate.

Keep In Mind

Labor and Material Costs Change. These costs were compiled in the fall of 2005 and projected to mid-2006 by adding 3% to 6%. This estimate will be accurate for some materials but inaccurate for others. No one can predict material price changes accurately.

How Accurate Are These Figures? As accurate as possible considering that the estimators who wrote this book don't know your subcontractors or material suppliers, haven't seen the plans or specifications, don't know what building code applies or where the job is, had to project material costs at least 6 months into the future, and had no record of how much work the crew that will be assigned to the job can handle.

You wouldn't bid a job under those conditions. And we don't claim that all construction is done at these prices.

Estimating Is an Art, not a science. On many jobs the range between high and low bid will be 20% or more. There's room for legitimate disagreement on what the correct costs are, even when complete plans and specifications are available, the date and site are established, and labor and material costs are identical for all bidders.

No cost fits all jobs. Good estimates are custom made for a particular project and a single contractor through judgment, analysis and experience.

This book is not a substitute for judgment, analysis and sound estimating practice. It's an aid in developing an informed opinion of cost. If you're using this book as your sole cost authority for contract bids, you're reading more into these pages than the editors intend.

Use These Figures to compile preliminary estimates, to check your costs and subcontract bids and when no actual costs are available. This book will reduce the chance of error or omission on bid estimates, speed "ball park" estimates, and be a good guide when there's no time to get a quote.

Where Do We Get These Figures? From the same sources all professional estimators use: contractors and subcontractors, architectural and engineering firms, material suppliers, material price services, analysis of plans, specifications, estimates and completed project costs, and both published and unpublished cost studies. In addition, we conduct nationwide mail and phone surveys and have the use of several major national estimating databases.

We'll Answer Your Questions about any part of this book and explain how to apply these costs. Free telephone assistance is available from 8 a.m. until 5 p.m. California time Monday through Friday except holidays. Phone 760-438-7828 x 2. We don't accept collect calls and won't estimate the job for you. But if you need clarification on something in this manual, we can help.

Abbreviations

AASHO	American Assn. of State Highway Officials	FAA	Federal Aviation Administration	OC	spacing from center to center
ABS	acrylonitrile butadiene styrene	FICA	Federal Insurance Contributions Act (Social Security, Medicare tax)	OD	outside diameter
AC	alternating current			OS & Y	outside screw & yoke
AISC	American Institute of Steel Construction Inc.			oz	ounce
APP	attactic polypropylene	FOB	freight on board	perf	perforated
ASHRAE	American Society of Heating, Refrigerating and Air Conditioning Engineers	FPM	feet per minute	Pr	pair
		FRP	fiberglass reinforced plastic	PSF	pounds per square foot
		FS	Federal Specification	PSI	pounds per square inch
		ft-lbs	foot pounds	PV	photovoltaic
ASME	American Society of Mechanical Engineers	FUTA	Federal Unemployment Compensation Act Tax	PVC	polyvinyl chloride
ASTM	American Society for Testing Materials	Gal	gallon	Qt	quart
		GFCI	ground fault circuit interruptor	R	thermal resistance
AWPA	American Wood Products Association			R/L	random length(s)
AWWA	American Water Works Association	GPH	gallon(s) per hour	R/W/L	random widths and lengths
		GPM	gallon(s) per minute	RPM	revolutions per minute
Ba	bay	H	height	RSC	rigid steel conduit
Bdle	bundle	HP	horsepower	S1S2E	surfaced 1 side, 2 edges
BF	board foot	Hr(s)	hour(s)	S2S	surfaced 2 sides
BHP	boiler horsepower	IMC	intermediate metal conduit	S4S	surfaced 4 sides
Btr	better	ID	Inside diameter	Sa	sack
Btu	British thermal unit	KD	kiln dried or knocked down	SBS	styrene butyl styrene
B & W	black & white			SDR	size to diameter ratio
C	thermal conductance	KSI	kips per square inch	SF	square foot
C	one hundred	KV	kilovolt(s)	SFCA	square feet of form in contact with concrete
CF	cubic foot	KVA	1,000 volt amps		
CFM	cubic feet per minute	kw	kilowatt(s)	Sq	100 square feet
CLF	100 linear feet	kwh	kilowatt hour	SSB	single strength B quality glass
cm	centimeter	L	length		
CPE	chlorinated polyethylene	Lb(s)	pound(s)	STC	sound transmission class
CPM	cycles per minute	LF	linear foot	Std	standard
CPVC	chlorinated polyvinyl chloride	LP	liquified propane	SY	square yard
		LS	lump sum	T	thick
CSPE	chloro sulphinated polyethylene	M	one thousand	T&G	tongue & groove edge
		Mb	million bytes (characters)	TV	television
CSF	100 square feet	MBF	1,000 board feet	UBC	Uniform Building Code
CSY	100 square yards	MBtu	1,000 British thermal units	UL	Underwriter's Laboratory
CY	cubic yard	MCM	1,000 circular mils	USDA	United States Dept. of Agriculture
d	penny	MDO	medium density overlaid		
D	depth	MH	manhour	VLF	vertical linear foot
DC	direct current	Mi	mile	W	width
dia	diameter	MLF	1,000 linear feet	Wk	week
DSB	double strength B quality glass	MPH	miles per hour	W/	with
		mm	millimeter(s)	x	by or times
DWV	drain, waste, vent piping	Mo	month		
Ea	each	MSF	1,000 square feet		
EMT	electric metallic tube	NEMA	National Electrical Manufacturer's Association		
EPDM	ethylene propylene diene monomer				
equip.	equipment	NFPA	National Fire Protection Association		
exp.	exposure	No.	number		
F	Fahrenheit	NRC	noise reduction coefficient		

Symbols

/	per
—	through or to
@	at
%	per 100 or percent
$	U.S. dollars
'	feet
"	inches
#	pound or number

Craft Codes, Hourly Costs and Crew Compositions

Both the Residential Division and Commercial and Industrial Division of this book include a column titled *Craft@Hrs*. Letters and numbers in this column show our estimates of:

- Who will do the work (the craft code)
- An @ symbol which means at
- How long the work will take (manhours).

For example, on page 51 you'll find estimates for installing BD plywood wall sheathing by the square foot. The *Craft@Hrs* column opposite 1/2" plywood wall sheathing shows:

B1@.016

That means we estimate the installation rate for crew B1 at .016 manhours per square foot. That's the same as 16 manhours per 1,000 square feet.

The table that follows defines each of the craft codes used in this book. Notice that crew B1 is composed of two craftsmen: one laborer and one carpenter. To install 1,000 square feet of 1/2" BD wall sheathing at .016 manhours per square foot, that crew would need 16 manhours (one 8-hour day for a crew of two).

Notice also in the table below that the cost per manhour for crew B1 is listed as $27.36. That's the average for a residential laborer (listed at $24.71 per hour on page 10) and a residential carpenter (listed at $30.00 per hour): $24.71 plus $30.00 is $54.71. Divide by 2 to get $27.36, the average cost per manhour for crew B1.

In the table below, the cost per manhour is the sum of hourly costs of all crew members divided by the number of crew members. That's the average cost per manhour.

Costs in the *Labor* column in this book are the installation time (in manhours) multiplied by the cost per manhour. For example, on page 51 the labor cost listed for 1/2" BD wall sheathing is $.44 per square foot. That's the installation time (.016 manhours per square foot) multiplied by $27.36, the average cost per manhour for crew B1.

Residential Division

Craft Code	Cost Per Manhour	Crew Composition
B1	$27.36	1 laborer and 1 carpenter
B2	$28.24	1 laborer, 2 carpenters
B3	$26.47	2 laborers, 1 carpenter
B4	$31.43	1 laborer 1 operating engineer 1 reinforcing iron worker
B5	$30.84	1 laborer, 1 carpenter 1 cement mason 1 operating engineer 1 reinforcing iron worker
B6	$27.31	1 laborer, 1 cement mason
B7	$26.08	1 laborer, 1 truck driver
B8	$30.98	1 laborer 1 operating engineer
B9	$27.96	1 bricklayer 1 bricklayer's helper
BB	$31.76	1 bricklayer
BC	$30.00	1 carpenter
BE	$33.86	1 electrician
BF	$30.28	1 floor layer
BG	$28.87	1 glazier
BH	$24.15	1 bricklayer's helper
BL	$24.71	1 laborer
BR	$32.79	1 lather
BS	$28.11	1 marble setter
CF	$29.91	1 cement mason
CT	$29.77	1 mosaic & terrazzo worker
D1	$30.00	1 drywall installer 1 drywall taper
DI	$30.03	1 drywall installer
DT	$29.97	1 drywall taper
HC	$24.39	1 plasterer helper
OE	$37.25	1 operating engineer
P1	$29.98	1 laborer, 1 plumber
PM	$35.25	1 plumber
PP	$28.30	1 painter, 1 laborer
PR	$31.63	1 plasterer
PT	$31.89	1 painter
R1	$29.17	1 roofer, 1 laborer
RI	$32.32	1 reinforcing iron worker
RR	$33.62	1 roofer
SW	$33.98	1 sheet metal worker
T1	$27.88	1 tile layer, 1 laborer
TL	$31.05	1 tile layer
TR	$27.44	1 truck driver

Commercial and Industrial Division

Craft Code	Cost Per Manhour	Crew Composition	Craft Code	Cost Per Manhour	Crew Composition
A1	$45.37	1 asbestos worker 1 laborer	F6	$42.69	2 carpenters, 2 laborers 1 tractor operator
AT	$35.78	1 air tool operator	F7	$44.27	2 carpenters, 1 laborer 1 tractor operator
AW	$54.35	1 asbestos worker	F8	$44.97	2 plasterers 1 plasterer's helper
BM	$54.46	1 boilermaker			
BT	$35.05	1 bricklayer tender	F9	$40.74	1 laborer, 1 floor layer
C1	$37.15	4 laborers, 1 truck driver	FL	$45.08	1 floor layer
C2	$43.19	1 laborer, 2 truck drivers 2 tractor operators	G1	$40.99	1 glazier, 1 laborer
C3	$42.06	1 laborer, 1 truck driver 1 tractor operator	H1	$46.46	1 carpenter, 1 laborer 1 iron worker (structural) 1 tractor operator
C4	$37.66	2 laborers, 1 truck driver	H2	$44.80	1 crane operator 1 truck driver
C5	$40.64	2 laborers, 1 truck driver 1 tractor operator			
C6	$39.79	6 laborers, 2 truck drivers 2 tractor operators	H3	$40.72	1 carpenter, 3 laborers 1 crane operator 1 truck driver
C7	$41.76	2 laborers , 3 truck drivers 1 crane operator 1 tractor operator	H4	$51.95	1 crane operator 6 iron workers (structural) 1 truck driver
C8	$40.97	1 laborer, 1 carpenter	H5	$46.17	1 crane operator 2 iron workers (structural) 2 laborers
C9	$42.90	1 laborer, 1 crane operator			
CB	$45.94	1 bricklayer	H6	$45.36	1 iron worker (structural) 1 laborer
CC	$45.55	1 carpenter	H7	$52.69	1 crane operator 2 iron workers (structural)
CD	$44.15	1 drywall Installer			
CE	$50.50	1 electrician	H8	$51.15	1 crane operator 4 iron workers (structural) 1 truck driver
CG	$45.58	1 glazier			
CL	$36.39	1 laborer	H9	$50.51	1 electrician 1 sheet metal worker
CM	$43.57	1 cement mason			
CO	$49.40	1 crane operator	IW	$54.33	1 iron worker (structural)
CV	$45.38	1 elevator constructor	LA	$43.47	1 lather
D2	$40.27	1 drywall installer 1 laborer	M1	$40.50	1 bricklayer 1 bricklayer's tender
D3	$45.54	1 laborer, 1 iron worker (structural), 1 millwright	M2	$39.44	1 carpenter, 2 laborers
D4	$41.15	1 laborer, 1 millwright	M3	$43.58	1 plasterer 1 plasterer's helper
D5	$45.43	1 boilermaker, 1 laborer			
D6	$47.13	2 millwrights 1 tractor operator	M4	$40.95	1 laborer, 1 marble setter
D7	$40.89	1 painter, 1 laborer	M5	$43.75	1 pipefitter, 1 laborer,
D9	$42.74	2 millwrights, 1 laborer	M6	$47.28	1 asbestos worker 1 laborer, 1 pipefitter
E1	$44.67	2 electricians, 2 laborers 1 tractor operator	M8	$47.42	3 pipefitters, 1 laborer
E2	$43.45	2 electricians, 2 laborers	M9	$50.80	1 electrician, 1 pipefitter
E3	$44.15	2 electricians, 2 laborers 2 carpenters	MI	$46.20	2 pipefitters, 1 laborer
E4	$43.45	1 electrician, 1 laborer	MS	$45.50	marble setter
			MT	$42.78	mosaic & terrazzo worker
F5	$41.89	3 carpenters, 2 laborers	MW	$45.91	millwright

Commercial and Industrial Division

Craft Code	Cost Per Manhour	Crew Composition
P5	$39.79	3 laborers 1 tractor operator 1 truck driver
P6	$44.29	1 laborer, 1 plumber
P8	$39.98	1 laborer, 1 cement mason
P9	$41.84	1 carpenter, 1 laborer 1 cement mason
PA	$45.39	1 painter
PD	$48.25	1 pile driver
PF	$51.10	1 pipefitter
PH	$39.40	1 plasterer's helper
PL	$52.18	1 plumber
PS	$47.75	1 plasterer
R3	$41.82	2 roofers, 1 laborer
RB	$53.15	1 reinforcing iron worker
RF	$44.53	1 roofer
S1	$42.99	1 laborer 1 tractor operator
S3	$44.89	1 truck driver 1 tractor operator
S4	$36.39	3 laborers
S5	$38.79	5 laborers 1 crane operator 1 truck driver
S6	$40.79	2 laborers 1 tractor operator
S7	$42.99	3 laborers 3 tractor operators

Craft Code	Cost Per Manhour	Crew Composition
S8	$44.21	2 pile drivers, 2 laborers 1 truck driver 1 crane operator 1 tractor operator
S9	$42.26	1 pile driver, 2 laborers 1 tractor operator 1 truck driver
SM	$50.52	1 sheet metal worker
SP	$53.58	1 sprinkler fitter
SS	$45.18	1 laborer 2 tractor operators
T2	$45.31	3 laborers, 3 carpenters 3 iron workers (structural) 1 crane operator 1 truck driver
T3	$44.77	1 laborer 1 reinforcing iron worker
T4	$39.59	1 laborer, 1 mosaic worker
T5	$43.46	1 sheet metal worker 1 laborer
T6	$45.81	2 sheet metal workers 1 laborer
TD	$40.19	1 truck driver
TO	$49.58	1 tractor operator
U1	$43.64	1 plumber, 2 laborers 1 tractor operator
U2	$41.65	1 plumber, 2 laborers

Residential Division

Craft	1 Base wage per hour	2 Taxable fringe benefits (@5.15% of base wage)	3 Insurance and employer taxes (%)	4 Insurance and employer taxes ($)	5 Non-taxable fringe benefits (@4.55% of base wage)	6 Total hourly cost used in this book
Bricklayer	$22.90	$1.18	27.58%	$6.64	$1.04	$31.76
Bricklayer's Helper	17.41	0.90	27.58%	5.05	0.79	24.15
Building Laborer	17.60	0.91	29.20%	5.40	0.80	24.71
Carpenter	21.54	1.11	28.10%	6.37	0.98	30.00
Cement Mason	21.78	1.12	26.27%	6.02	0.99	29.91
Drywall installer	22.05	1.14	25.21%	5.84	1.00	30.03
Drywall Taper	22.01	1.13	25.21%	5.83	1.00	29.97
Electrician	25.26	1.30	23.15%	6.15	1.15	33.86
Floor Layer	21.02	1.08	32.65%	7.22	0.96	30.28
Glazier	20.87	1.08	27.20%	5.97	0.95	28.87
Lather	22.64	1.17	33.39%	7.95	1.03	32.79
Marble Setter	20.23	1.04	27.83%	5.92	0.92	28.11
Millwright	21.90	1.13	26.92%	6.20	1.00	30.23
Mosiac & Terrazzo Worker	21.43	1.10	27.83%	6.27	0.97	29.77
Operating Engineer	25.33	1.30	35.56%	9.47	1.15	37.25
Painter	22.85	1.18	28.38%	6.82	1.04	31.89
Plasterer	22.32	1.15	30.44%	7.14	1.02	31.63
Plasterer Helper	17.21	0.89	30.44%	5.51	0.78	24.39
Plumber	25.78	1.33	25.70%	6.97	1.17	35.25
Reinforcing Ironworker	21.20	1.09	40.69%	9.07	0.96	32.32
Roofer	21.98	1.13	41.13%	9.51	1.00	33.62
Sheet Metal Worker	24.96	1.29	25.12%	6.59	1.14	33.98
Sprinkler Fitter	25.33	1.30	24.48%	6.52	1.15	34.30
Tile Layer	22.34	1.15	27.83%	6.54	1.02	31.05
Truck Driver	18.53	0.95	36.56%	7.12	0.84	27.44

Hourly Labor Cost

The labor costs shown in Column 6 were used to compute the manhour costs for crews on page 7 and the figures in the "Labor" column of the Residential Division of this manual. Figures in the "Labor" column of the Industrial and Commercial Division of this book were computed using the hourly costs shown on page 323. All labor costs are in U.S. dollars per manhour.

It's important that you understand what's included in the figures in each of the six columns above. Here's an explanation:

Column 1, the base wage per hour, is the craftsman's hourly wage. These figures are representative of what many contractors will be paying craftsmen working on residential construction in 2006.

Column 2, taxable fringe benefits, includes vacation pay, sick leave and other taxable benefits. These fringe benefits average 5.15% of the base wage for many construction contractors. This benefit is in addition to the base wage.

Column 3, insurance and employer taxes in percent, shows the insurance and tax rate for construction trades. The cost of insurance in this column includes workers' compensation and contractor's casualty and liability coverage. Insurance rates vary widely from state to state and depend on a contractor's loss experience. Typical rates are shown in the Insurance section of this manual beginning on page 180. Taxes are itemized in the section on page 298. Note that taxes and insurance increase the hourly labor cost by 30 to 35% for most trades. There is no legal way to avoid these costs.

Column 4, insurance and employer taxes in dollars, shows the hourly cost of taxes and insurance for each construction trade. Insurance and taxes are paid on the costs in both columns 1 and 2.

Column 5, non-taxable fringe benefits, includes employer paid non-taxable benefits such as medical coverage and tax-deferred pension and profit sharing plans. These fringe benefits average 4.55% of the base wage for many construction contractors. The employer pays no taxes or insurance on these benefits.

Column 6, the total hourly cost in dollars, is the sum of columns 1, 2, 4, and 5.

These hourly labor costs will apply within a few percent on many jobs. But wage rates may be much higher or lower in some areas. If the hourly costs shown in column 6 are not accurate for your work, develop modification factors that you can apply to the labor costs in this book. The following paragraphs explain the procedure.

Adjusting Labor Costs

Here's how to customize the labor costs in this book if your wage rates are different from the wage rates shown on page 10 or 323.

Start with the taxable benefits you offer. Assume craftsmen on your payroll get one week of vacation each year and one week of sick leave each year. Convert these benefits into hours. Your computation might look like this:

$$\begin{array}{r} 40 \text{ vacation hours} \\ + \ 40 \text{ sick leave hours} \\ \hline 80 \text{ taxable leave hours} \end{array}$$

Then add the regular work hours for the year:

$$\begin{array}{r} 2,000 \text{ regular hours} \\ + \ 80 \text{ taxable benefit hours} \\ \hline 2,080 \text{ total hours} \end{array}$$

Multiply these hours by the base wage per hour. If you pay carpenters $10.00 per hour, the calculation would be:

$$\begin{array}{r} 2,080 \text{ hours} \\ \times \ \$10.00 \text{ per hour} \\ \hline \$20,800 \text{ per year} \end{array}$$

Next determine the tax and insurance rate for each trade. If you know the rates that apply to your jobs, use those rates. If not, use the rates in column 3 on page

10. Continuing with our example, we'll use 28.10%, the rate for carpenters in column 3 on page 10. To increase the annual taxable wage by 28.10%, we'll multiply by 1.2810:

$$\begin{array}{r} \$20,800 \text{ per year} \\ \times \ 1.2810 \text{ tax \& insurance rate} \\ \hline \$26,645 \text{ annual cost} \end{array}$$

Then add the cost of non-taxable benefits. Suppose your company has no pension or profit sharing plan but does provide medical insurance for employees. Assume that the cost for your carpenter is $343.67 per month or $4,124 per year.

$$\begin{array}{r} \$4,124 \text{ medical plan} \\ + \ 26,645 \text{ annual cost} \\ \hline \$30,769 \text{ total annual cost} \end{array}$$

Divide this total annual cost by the actual hours worked in a year. This gives the contractor's total hourly labor cost including all benefits, taxes and insurance. Assume your carpenter will work 2,000 hours a year:

$$\frac{\$30,769}{2,000} = \$15.38 \text{ per hour}$$

Finally, find your modification factor for the labor costs in this book. Divide your total hourly labor cost by the total hourly cost shown on page 10. For the carpenter in our example, the figure in column 6 is $30.00:

$$\frac{\$15.38}{\$30.00} = .51$$

Your modification factor is 51%. Multiply any building carpenter (Craft Code BC) labor costs in the Residential Division of this book by .51 to find your estimated cost. For example, on page 24 the labor cost for installing an 18" long towel bar is $8.40 per each bar. If installed by your carpenter working at $10.00 per hour, your estimated cost would be 51% of $8.40 or $4.28. The manhours would remain the same @ .280, assuming normal productivity.

If the Labor Rate Is Unknown

On some estimates you may not know what labor rates will apply. In that case, use both labor and material figures in this book without making any adjustment. When all labor, equipment and material costs have been compiled, add or deduct the percentage shown in the area modification table on pages 12 through 15.

Adjusting the labor costs in this book will make your estimates much more accurate.

Area Modification Factors

Construction costs are higher in some areas than in other areas. Add or deduct the percentages shown on the following pages to adapt the costs in this book to your job site. Adjust your cost estimate by the appropriate percentages in this table to find the estimated cost for the site selected. Where 0% is shown, it means no modification is required.

Modification factors are listed alphabetically by state and province. Areas within each state are listed alphabetically. For convenience, one representative city is identified in each three-digit zip or range of zips. Percentages are based on the average of all data points in the table. Factors listed for each state and province are the average of all data points in that state or province. Figures for three-digit zips are the average of all five-digit zips in that area. Figures in the Total column are the weighted average of factors for Labor, Material and Equipment.

The National Estimator program will apply an area modification factor for any five-digit zip you select. Click Utilities. Click Options. Then select the Area Modification Factors tab.

These percentages are composites of many costs and will not necessarily be accurate when estimating the cost of any particular part of a building. But when used to modify costs for an entire structure, they should improve the accuracy of your estimates.

Location	Zip	Mat.	Lab.	Equip.	Total Wtd. Avg.
Alaska Average		14	33	4	22%
Anchorage	995	16	50	5	32%
King Salmon	996	16	34	5	24%
Fairbanks	997	16	48	5	30%
Juneau	998	17	23	6	19%
Ketchikan	999	3	9	1	6%
Alabama Average		-1	-15	0	-7%
Birmingham	350-352	-3	9	-1	3%
Tuscaloosa	354	1	-15	0	-7%
Jasper	355	-2	-27	-1	-13%
Sheffield	356	-1	-10	0	-5%
Scottsboro	357	0	-12	0	-6%
Huntsville	358	1	-12	0	-5%
Gadsden	359	-4	-19	-1	-11%
Montgomery	360-361	-1	-10	0	-5%
Anniston	362	-4	-26	-1	-14%
Dothan	363	-1	-18	0	-9%
Evergreen	364	-2	-28	-1	-14%
Mobile	365-366	4	-1	0	1%
Selma	367	-1	-21	0	-10%
Auburn	368	-1	-17	0	-8%
Bellamy	369	-2	-11	-1	-6%
Arkansas Average		-2	-20	-1	-10%
Pine Bluff	716	-4	-24	-1	-13%
Camden	717	-4	-8	-1	-6%
Hope	718	-4	-23	-1	-13%
Hot Springs	719	-2	-27	-1	-14%
Little Rock	720-722	-1	-9	0	-5%
West Memphis	723	-3	-21	-1	-11%
Jonesboro	724	-1	-25	0	-12%
Batesville	725	0	-31	0	-14%
Harrison	726	-1	-34	0	-16%
Fayetteville	727	0	-9	0	-4%
Russellville	728	0	-12	0	-6%
Fort Smith	729	-2	-13	-1	-7%
Arizona Average		1	-12	0	-5%
Phoenix	850	1	7	0	4%
Mesa	852-853	1	7	0	3%
Douglas	855	0	-16	0	-7%
Tucson	856-857	0	-11	0	-5%
Show Low	859	2	-17	1	-7%
Flagstaff	860	2	-24	1	-10%
Prescott	863	2	-20	1	-8%
Kingman	864	1	-5	0	-2%
Chambers	865	2	-29	1	-12%
California Average		2	18	1	9%
Los Angeles	900-901	3	18	1	10%
Inglewood	902-905	3	19	1	11%
Whittier	906	3	21	1	11%
Long Beach	907-908	3	22	1	12%
Pasadena	910-912	4	19	1	11%
Van Nuys	913-916	3	17	1	9%

Location	Zip	Mat.	Lab.	Equip.	Total Wtd. Avg.
Alhambra	917-918	3	21	1	11%
San Diego	919-921	3	17	1	9%
El Centro	922	1	4	0	2%
San Bernardino	923-924	0	8	0	4%
Riverside	925	1	10	0	5%
Irvine	926-927	3	26	1	14%
Orange	928	3	26	1	14%
Oxnard	930	3	7	1	4%
Santa Barbara	931	4	16	1	9%
Bakersfield	932-933	0	2	0	1%
Lompoc	934	2	9	1	5%
Mojave	935	0	17	0	8%
Fresno	936-938	0	0	0	0%
Salinas	939	3	13	1	7%
Sunnyvale	940	3	44	1	22%
San Francisco	941	3	53	1	26%
San Mateo	943-944	3	47	1	23%
Oakland	945-947	3	44	1	22%
Richmond	948	2	45	1	22%
Novato	949	4	26	1	14%
San Jose	950-951	3	38	1	19%
Stockton	952	2	11	1	6%
Modesto	953	1	0	0	1%
Santa Rosa	954	3	17	1	9%
Eureka	955	1	-7	0	-3%
Rancho Cordova	956-957	2	16	1	8%
Sacramento	958	1	16	0	8%
Marysville	959	1	-2	0	0%
Redding	960	1	-6	0	-2%
Herlong	961	2	7	1	4%
Colorado Average		1	0	1	1%
Aurora	800-801	3	19	1	10%
Denver	802	2	21	1	11%
Boulder	803-804	3	10	1	6%
Longmont	805	2	2	1	2%
Greeley	806	2	8	1	5%
Fort Morgan	807	1	-8	0	-3%
Colorado Springs	808-809	2	5	1	3%
Pueblo	810	-1	-12	0	-6%
Pagosa Springs	811	0	-19	0	-9%
Salida	812	1	-16	0	-7%
Durango	813	1	-17	0	-7%
Grand Junction	814-815	1	-5	0	-2%
Glenwood Springs	816	2	9	1	5%
Connecticut Average		1	32	0	16%
Bristol	060	1	33	0	16%
Hartford	061	0	31	0	14%
West Hartford	062	0	21	0	10%
Norwich	063	0	28	0	13%
Fairfield	064	2	37	1	18%
New Haven	065	0	33	0	16%
Bridgeport	066	1	38	0	18%
Waterbury	067	1	32	0	15%
Stamford	068-069	4	39	1	20%

Location	Zip	Mat.	Lab.	Equip.	Total Wtd. Avg.
District of Columbia Average		2	23	1%	12%
Washington	200-205	2	23	1	12%
Delaware Average		1	-8	0	-3%
Newark	197	1	-6	0	-2%
Wilmington	198	0	-7	0	-4%
Dover	199	1	-10	0	-4%
Florida Average		-1	-1	0	-1%
Saint Augustine	320	-1	4	0	1%
Daytona Beach	321	-2	-17	-1	-9%
Jacksonville	322	-1	17	0	7%
Tallahassee	323	0	-11	0	-5%
Panama City	324	-2	-20	-1	-10%
Pensacola	325	-1	-16	0	-8%
Gainesville	326	-1	-12	0	-6%
Altamonte Springs	327	-1	0	0	-1%
Orlando	328	0	9	0	4%
Melbourne	329	-2	15	-1	6%
Miami	330-332	2	11	1	6%
Fort Lauderdale	333	2	11	1	6%
West Palm Beach	334	1	10	0	5%
Tampa	335-336	-1	7	0	2%
St Petersburg	337	-1	-1	0	-1%
Lakeland	338	-3	-12	-1	-7%
Fort Myers	339	0	3	0	2%
Naples	341	3	6	1	4%
Bradenton	342	0	-1	0	-1%
Ocala	344	-3	-17	-1	-9%
Brooksville	346	-1	-10	0	-5%
Saint Cloud	347	-1	-1	0	-1%
Fort Pierce	349	-2	-9	-1	-5%
Georgia Average		-1	-8	0	-5%
Marietta	300-302	1	16	0	8%
Atlanta	303	3	37	1	18%
Statesboro	304	-2	-26	-1	-13%
Buford	305	0	-1	0	-1%
Athens	306	0	-11	0	-5%
Calhoun	307	-1	-21	0	-10%
Augusta	308-309	-2	-12	-1	-6%
Dublin/Fort Valley	310	-3	-16	-1	-9%
Macon	312	-2	-4	-1	-3%
Hinesville	313	-2	-11	-1	-6%
Savannah	314	-1	-3	0	-2%
Kings Bay	315	-2	-19	-1	-10%
Valdosta	316	-2	-23	-1	-12%
Albany	317	-2	-17	-1	-9%
Columbus	318-319	-1	-15	0	-8%
Hawaii Average		17	38	6	26%
Ewa	967	17	37	6	26%
Halawa Heights	967	17	37	6	26%
Hilo	967	17	37	6	26%
Lualualei	967	17	37	6	26%

Area Modification Factors

Location	Zip	Mat.	Lab.	Equip.	Total Wtd. Avg.
Mililani Town	967	17	37	6	26%
Pearl City	967	17	37	6	26%
Wahiawa	967	17	37	6	26%
Waianae	967	17	37	6	26%
Wailuku (Maui)	967	17	37	6	26%
Aliamanu	968	17	40	6	27%
Honolulu	968	17	40	6	27%
Kailua	968	17	40	6	27%
Iowa Average		**-1**	**-8**	**0**	**-4%**
Des Moines	500-503	-1	8	0	3%
Mason City	504	0	-8	0	-3%
Fort Dodge	505	-2	-16	-1	-8%
Cedar Falls	506	-1	-10	0	-5%
Waterloo	507	-4	-9	-1	-6%
Creston	508	-2	-20	-1	-10%
Cherokee	510	-2	-9	-1	-5%
Sioux City	511	-2	-4	-1	-3%
Sheldon	512	0	-21	0	-10%
Spencer	513	-1	-10	0	-5%
Carroll	514	-2	-28	-1	-14%
Council Bluffs	515	-2	-1	-1	-2%
Shenandoah	516	-2	-26	-1	-13%
Dubuque	520	-1	2	0	0%
Decorah	521	-2	10	-1	4%
Cedar Rapids	522-524	0	9	0	4%
Ottumwa	525	0	-19	0	-9%
Burlington	526	0	-7	0	-3%
Davenport	527-528	-1	1	0	0%
Idaho Average		**0**	**-15**	**0**	**-7%**
Pocatello	832	-1	-23	0	-11%
Sun Valley	833	0	-10	0	-5%
Idaho Falls	834	0	-22	0	-10%
Lewiston	835	0	-20	0	-9%
Meridian	836	0	-14	0	-6%
Boise	837	1	2	0	2%
Coeur d'Alene	838	0	-18	0	-8%
Illinois Average		**-1**	**15**	**0**	**6%**
Arlington Heights	600	2	48	1	23%
Carol Stream	601	2	47	1	23%
Quincy	602	3	50	1	24%
Oak Park	603	3	54	1	26%
Joliet	604	0	45	0	21%
Aurora	605	2	48	1	23%
Chicago	606-608	2	50	1	24%
Kankakee	609	-2	6	-1	2%
Rockford	610-611	-2	18	-1	7%
Green River	612	-2	0	-1	-1%
Peru	613	0	6	0	2%
Galesburg	614	-2	-11	-1	-6%
Peoria	615-616	-1	16	0	7%
Bloomington	617	0	7	0	3%
Champaign	618	-1	4	0	2%
Urbana	619	-3	-5	-1	-4%
Granite City	620	-3	2	-1	-1%
Belleville	622	-3	2	-1	-1%
Decatur	623	-2	-16	-1	-8%
Lawrenceville	624	-4	-10	-1	-7%
Springfield	625-627	-2	5	-1	1%
Centralia	628	-3	-12	-1	-7%
Carbondale	629	-3	-9	-1	-6%
Indiana Average		**-2**	**0**	**-1**	**-1%**
Indianapolis	460-462	-1	22	0	10%
Gary	463-464	-4	21	-1	8%
Elkhart	465	-2	9	-1	3%
South Bend	466	-3	10	-1	3%
Fort Wayne	467-468	-2	7	-1	2%
Kokomo	469	-3	-11	-1	-6%
Aurora	470	-1	-16	0	-8%
Jeffersonville	471	-1	-8	0	-4%
Columbus	472	0	-2	0	-1%
Muncie	473	-4	-12	-1	-8%

Location	Zip	Mat.	Lab.	Equip.	Total Wtd. Avg.
Bloomington	474	1	-6	0	-2%
Jasper	475	-2	-14	-1	-7%
Evansville	476-477	-2	11	-1	4%
Terre Haute	478	-4	-8	-1	-5%
Lafayette	479	-1	-8	0	-4%
Kansas Average		**-2**	**-19**	**-1**	**-10%**
Kansas City	660-662	0	13	0	6%
Topeka	664-666	-3	-6	-1	-4%
Fort Scott	667	-2	-18	-1	-10%
Emporia	668	-3	-21	-1	-11%
Concordia	669	0	-33	0	-15%
Wichita	670-672	-2	-9	-1	-5%
Independence	673	-3	-23	-1	-12%
Salina	674	-2	-15	-1	-8%
Hutchinson	675	-2	-14	-1	-8%
Hays	676	-2	-31	-1	-15%
Colby	677	0	-32	0	-15%
Dodge City	678	-2	-20	-1	-10%
Liberal	679	-2	-33	-1	-17%
Kentucky Average		**-1**	**-11**	**-1**	**-6%**
Louisville	400-402	-1	2	0	0%
Lexington	403-405	1	4	0	3%
Frankfort	406	1	-11	0	-5%
London	407-409	-2	-19	-1	-10%
Covington	410	0	12	0	5%
Ashland	411-412	-3	-3	-1	-3%
Campton	413-414	-1	-22	0	-11%
Pikeville	415-416	-3	-8	-1	-5%
Hazard	417-418	-2	-21	-1	-10%
Paducah	420	-3	-12	-1	-7%
Bowling Green	421	0	-14	0	-7%
Hopkinsville	422	-2	-17	-1	-9%
Owensboro	423	-2	-10	-1	-6%
White Plains	424	-3	-10	-1	-6%
Somerset	425-426	0	-24	0	-11%
Elizabethtown	427	-1	-19	0	-9%
Louisiana Average		**-1**	**-2**	**-1**	**0%**
New Orleans	700-701	4	21	0	12%
Houma	703	3	19	-1	10%
Mandeville	704	4	-7	0	-1%
Lafayette	705	5	4	0	4%
Lake Charles	706	2	2	-1	2%
Baton Rouge	707-708	5	5	0	5%
Minden	710	-2	-16	-1	-9%
Shreveport	711	-2	-13	-1	-7%
Monroe	712	-2	-24	-1	-12%
Alexandria	713-714	-3	-13	-1	-8%
Massachusetts Average		**2**	**31**	**1**	**15%**
Chicopee	010	1	19	0	9%
Springfield	011	-1	26	0	11%
Pittsfield	012	0	5	0	3%
Northfield	013	1	1	0	1%
Fitchburg	014	2	31	1	15%
Ayer	015-016	1	22	0	11%
Bedford	017	2	44	1	21%
Lawrence	018	2	41	1	20%
Dedham	019	2	39	1	19%
Hingham	020	3	53	1	26%
Boston	021-022	3	73	1	35%
Brockton	023-024	3	48	1	24%
Nantucket	025	2	28	1	14%
Cape Cod	026	2	18	1	10%
New Bedford	027	1	15	0	7%
Maryland Average		**1**	**5**	**0**	**3%**
Laurel	206-207	2	22	1	11%
Bethesda	208-209	3	32	1	16%
Baltimore	210-212	-1	16	0	7%
Annapolis	214	3	18	1	10%
Cumberland	215	-4	-18	-1	-10%
Church Hill	216	2	-9	1	-4%

Location	Zip	Mat.	Lab.	Equip.	Total Wtd. Avg.
Frederick	217	1	8	0	4%
Salisbury	218	1	-16	0	-7%
Elkton	219	2	-10	1	-3%
Maine Average		**-1**	**-11**	**0**	**-5%**
Brunswick	039-040	1	0	0	0%
Portland	041	2	5	1	4%
Auburn	042	-1	-8	0	-4%
Augusta	043	-1	-13	0	-7%
Bangor	044	-2	-13	-1	-7%
Bath	045	0	-14	0	-7%
Cutler	046	-1	-15	0	-8%
Northern Area	047	-2	-26	-1	-13%
Camden	048	-1	-16	0	-8%
Dexter	049	-2	-7	-1	-4%
Michigan Average		**-1**	**9**	**-1**	**3%**
Royal Oak	480	-2	28	-1	12%
Detroit	481-482	0	31	0	14%
Pontiac	483	-2	35	-1	15%
Flint	484-485	-3	4	-1	0%
Saginaw	486-487	-2	-7	-1	-4%
Lansing	488-489	0	7	0	3%
Battle Creek	490-491	-3	4	-1	1%
Jackson	492	-3	8	-1	2%
Grand Rapids	493-495	-1	9	0	3%
Traverse City	496	-1	-2	0	-1%
Grayling	497	1	-10	0	-4%
Marquette	498-499	-1	-4	0	-3%
Minnesota Average		**0**	**5**	**0**	**2%**
St Paul	550-551	1	37	0	17%
Minneapolis	553-555	1	42	0	20%
Duluth	556-558	-2	9	-1	3%
Rochester	559	1	6	0	3%
Mankato	560	1	3	0	2%
Magnolia	561	0	-20	0	-9%
Willmar	562	-1	1	0	0%
St Cloud	563	0	10	0	5%
Brainerd	564	0	-2	0	-1%
Fergus Falls	565	-1	-13	0	-6%
Bemidji	566	-1	-4	0	-2%
Thief River Falls	567	0	-10	0	-4%
Missouri Average		**-1**	**-8**	**0**	**-4%**
St Louis	630-631	-2	30	-1	13%
Saint Charles	633	1	11	0	6%
Hannibal	634	1	3	0	2%
Kirksville	635	0	-30	0	-14%
Farmington	636	-3	-19	-1	-11%
Cape Girardeau	637	-2	-13	-1	-7%
Caruthersville	638	-2	-21	-1	-10%
Poplar Bluff	639	-1	-17	0	-8%
Independence	640	-1	15	0	7%
Kansas City	641	-1	19	0	8%
Saint Joseph	644-645	-3	5	-1	1%
Chillicothe	646	-2	-21	-1	-11%
East Lynne	647	-1	-12	0	-6%
Joplin	648	-3	-20	-1	-11%
Jefferson City	650-651	1	-10	0	-4%
Columbia	652	1	-7	0	-3%
Knob Noster	653	0	-19	0	-8%
Lebanon	654-655	-1	-25	0	-12%
Springfield	656-658	-2	-16	-1	-8%
Mississippi Average		**-2**	**-21**	**-1**	**-11%**
Clarksdale	386	-3	-17	-1	-9%
Greenville	387	-4	-34	-1	-18%
Tupelo	388	-2	-23	-1	-12%
Greenwood	389	-3	-21	-1	-11%
Jackson	390-392	-3	-11	-1	-7%
Meridian	393	-3	-17	-1	-9%
Laurel	394	-3	-21	-1	-11%
Gulfport	395	3	-6	-1	-1%
McComb	396	-2	-32	-1	-16%
Columbus	397	-1	-23	0	-11%

Area Modification Factors

Location	Zip	Mat.	Lab.	Equip.	Total Wtd. Avg.
Montana Average		0	-11	0	-5%
Billings	590-591	0	-3	0	-1%
Fairview	592	0	-15	0	-7%
Miles City	593	0	7	0	3%
Great Falls	594	0	-10	0	-5%
Havre	595	-1	-32	0	-15%
Helena	596	0	-13	0	-6%
Butte	597	1	-9	0	-4%
Missoula	598	1	-14	0	-6%
Kalispell	599	2	-13	1	-5%
North Carolina Average		0	-6	0	-3%
Winston-Salem	270-273	0	-4	0	-2%
Greensboro	274	0	2	0	1%
Goldsboro	275	1	0	0	0%
Raleigh	276	3	7	1	5%
Durham	277	2	1	1	1%
Rocky Mount	278	-1	-10	0	-5%
Elizabeth City	279	1	-8	0	-3%
Charlotte	280-282	0	13	0	6%
Fayetteville	283	-1	-14	0	-7%
Wilmington	284	1	-16	0	-7%
Kinston	285	-1	-23	0	-11%
Hickory	286	-1	-18	0	-9%
Asheville	287-289	0	-13	0	-6%
North Dakota Average		-1	-11	0	-6%
Fargo	580-581	0	-3	0	-1%
Grand Forks	582	0	-6	0	-3%
Nekoma	583	-1	-27	0	-13%
Jamestown	584	-1	-16	0	-8%
Bismarck	585	0	0	0	0%
Dickinson	586	-1	-15	0	-7%
Minot	587	0	-7	0	-4%
Williston	588	-1	-17	0	-8%
Nebraska Average		0	-18	0	-8%
Omaha	680-681	0	5	0	2%
Lincoln	683-685	0	-6	0	-3%
Columbus	686	0	-12	0	-6%
Norfolk	687	-2	-25	-1	-13%
Grand Island	688	0	-16	0	-7%
Hastings	689	0	-26	0	-12%
McCook	690	1	-26	0	-11%
North Platte	691	0	-20	0	-9%
Valentine	692	-1	-24	0	-12%
Alliance	693	-1	-26	0	-12%
New Hampshire Average		1	2	0	1%
New Boston	030-031	2	17	1	9%
Manchester	032-033	0	6	0	3%
Concord	034	1	2	0	1%
Littleton	035	-1	-12	0	-6%
Charlestown	036	1	-8	0	-3%
Lebanon	037	1	-4	0	-1%
Dover	038	0	13	0	6%
New Jersey Average		1	39	0	18%
Passaic	070	2	43	1	21%
Newark	071-073	1	42	0	20%
Paterson	074-075	2	38	1	18%
Hackensack	076	3	33	1	16%
Monmouth	077	3	46	1	22%
Dover	078	1	47	0	22%
Summit	079	3	52	1	25%
Atlantic City	080-084	-2	29	-1	12%
Princeton	085	-2	37	-1	16%
Trenton	086	-3	36	-1	15%
Brick	087	2	23	1	12%
Edison	088-089	0	46	0	21%
New Mexico Average		0	-21	0	-10%
Albuquerque	870-871	2	-5	1	-1%
Gallup	873	1	-21	0	-9%
Farmington	874	2	-5	1	-1%
Santa Fe	875	3	-19	1	-7%
Holman	877	2	-27	1	-11%
Socorro	878	1	-28	0	-12%
Truth or Consequences	879	-1	-33	0	-16%
Las Cruces	880	-1	-27	0	-13%
Clovis	881	-2	-28	-1	-14%
Fort Sumner	882	-3	-14	-1	-8%
Alamogordo	883	0	-16	0	-8%
Tucumcari	884	-1	-33	0	-15%
Nevada Average		2	7	1	4%
Las Vegas	889-891	2	18	1	9%
Ely	893	2	-15	1	-6%
Fallon	894	2	12	1	7%
Reno	895	2	13	1	7%
Carson City	897	2	-4	1	-1%
Elko	898	0	16	0	8%
New York Average		0	24	0	11%
New York (Manhattan)	100-102	3	82	1	39%
New York City	100-102	3	82	1	39%
Staten Island	103	3	43	1	21%
Bronx	104	2	34	1	17%
White Plains	105-108	3	46	1	22%
West Point	109	1	30	1	14%
Queens	110	4	51	1	26%
Long Island	111	3	84	1	40%
Brooklyn	112	3	27	1	14%
Flushing	113	3	57	1	28%
Jamaica	114	3	55	1	27%
Garden City	115	3	42	1	21%
Rockaway	116	3	40	1	20%
Amityville	117	2	32	1	16%
Hicksville	118	3	39	1	19%
Montauk	119	1	30	0	14%
Albany	120-123	-1	18	0	8%
Kingston	124	0	-3	0	-1%
Poughkeepsie	125-126	1	11	0	5%
Stewart	127	-1	1	0	0%
Newcomb	128	-1	6	0	2%
Plattsburgh	129	1	-4	0	-1%
Syracuse	130-132	-3	10	-1	3%
Rome	133-134	-3	-5	-1	-4%
Utica	135	-4	-3	-1	-3%
Watertown	136	-3	-5	-1	-4%
Binghamton	137-139	-3	3	-1	-1%
Batavia	140	-3	9	-1	2%
Tonawanda	141	-4	6	-1	1%
Buffalo	142	-4	10	-1	3%
Niagara Falls	143	-4	0	-1	-2%
Rochester	144-146	-3	9	-1	3%
Jamestown	147	-4	-10	-1	-7%
Ithaca	148	-4	1	-1	-1%
Elmira	149	-4	-2	-1	-3%
Ohio Average		-2	2	-1	0%
Newark	430-431	-1	15	0	7%
Columbus	432	0	19	0	9%
Marion	433	-3	-13	0	-7%
Toledo	434-436	-1	18	0	8%
Zanesville	437-438	-2	-3	-1	-3%
Steubenville	439	-3	-8	-1	-5%
Cleveland	440-441	-2	16	-1	6%
Akron	442-443	-1	3	0	1%
Warren	444	-4	-5	-1	-4%
Youngstown	445	-5	1	-2	-2%
Canton	446-447	-2	-3	-1	-3%
Sandusky	448-449	-1	1	0	0%
Cincinnati	450-452	-1	13	0	6%
Dayton	453-455	-3	4	-1	0%
Chillicothe	456	-2	-4	-1	-3%
Marietta	457	-3	-14	-1	-8%
Lima	458	-3	-7	-1	-5%
Oklahoma Average		-3	-18	-1	-10%
Norman	730	-2	-10	-1	-6%
Oklahoma City	731	-2	-6	-1	-4%
Ardmore	734	-3	-16	-1	-9%
Lawton	735	-3	-27	-1	-14%
Clinton	736	-3	-26	-1	-14%
Enid	737	-4	-18	-1	-10%
Woodward	738	-4	-2	-1	-3%
Adams	739	-3	-27	-1	-14%
Tulsa	740-741	-2	-4	-1	-3%
Pryor	743	-3	-29	-1	-15%
Muskogee	744	-2	-18	-1	-9%
McAlester	745	-4	-26	-1	-14%
Ponca City	746	-3	-8	-1	-5%
Durant	747	-4	-32	-1	-17%
Shawnee	748	-4	-25	-1	-14%
Poteau	749	-2	-15	-1	-8%
Oregon Average		2	-9	1	-3%
Portland	970-972	2	22	1	11%
Salem	973	2	-4	1	-1%
Eugene	974	2	-4	1	-1%
Grants Pass	975	2	-13	1	-5%
Klamath Falls	976	3	-23	1	-9%
Bend	977	1	-6	0	-2%
Pendleton	978	0	-10	0	-4%
Adrian	979	0	-34	0	-16%
Pennsylvania Average		-3	0	-1	-2%
Warrendale	150-151	-4	18	-1	6%
Pittsburgh	152	-4	19	-1	7%
Washington	153	-4	10	-1	3%
Uniontown	154	-4	-5	-1	-5%
Somerset	155	-4	-10	-1	-7%
Greensburg	156	-4	-1	-1	-3%
Punxsutawney	157	-4	-19	-1	-11%
DuBois	158	-2	-23	-1	-11%
Johnstown	159	-4	-20	-1	-11%
Butler	160	-4	-4	-1	-4%
New Castle	161	-4	-5	-1	-4%
Kittanning	162	-4	-19	-1	-11%
Meadville	163	-4	-24	-1	-13%
Erie	164-165	-3	-5	-1	-4%
Altoona	166	-3	-15	-1	-9%
Bradford	167	-4	-14	-1	-9%
Clearfield	168	1	-6	0	-2%
Genesee	169	-4	-23	-1	-13%
Harrisburg	170-171	-2	8	-1	3%
Chambersburg	172	-1	-9	0	-5%
York	173-174	-3	7	-1	2%
Lancaster	175-176	-3	12	-1	4%
Williamsport	177	-3	-12	-1	-7%
Beaver Springs	178	-3	-9	-1	-6%
Pottsville	179	-4	-10	-1	-7%
Bethlehem	180	-1	18	0	8%
Allentown	181	-2	17	-1	7%
Hazleton	182	-3	-9	-1	-6%
East Stroudsburg	183	0	-5	0	-2%
Scranton	184-185	-2	-3	-1	-2%
Wilkes Barre	186-187	-3	-6	-1	-4%
Montrose	188	-3	-15	-1	-9%
Warminster	189	-1	34	0	15%
Philadelphia	190-191	-3	40	-1	17%
Southeastern	193	0	27	0	13%
Valley Forge	194	-3	40	-1	17%
Reading	195-196	-4	14	-1	4%
Rhode Island Average		1	19	0	10%
Bristol	028	1	17	0	9%
Coventry	028	1	17	0	9%
Davisville	028	1	17	0	9%
Narragansett	028	1	17	0	9%
Newport	028	1	17	0	9%
Warwick	028	1	17	0	9%
Cranston	029	1	23	0	11%
Providence	029	1	23	0	11%

Area Modification Factors

Location	Zip	Mat.	Lab.	Equip.	Total Wtd. Avg.
South Carolina Average		-1	-4	0	-3%
Columbia	290-292	-1	-7	0	-4%
Spartanburg	293	-2	-7	-1	-4%
Charleston	294	-2	-2	-1	-2%
Myrtle Beach	295	-1	-14	0	-7%
Greenville	296	-1	-6	0	-3%
Rock Hill	297	-1	-8	0	-4%
Aiken	298	0	14	0	6%
Beaufort	299	-2	-5	-1	-3%
South Dakota Average		-1	-20	0	-10%
Sioux Falls	570-571	0	-6	0	-3%
Watertown	572	-1	-21	0	-10%
Mitchell	573	0	-21	0	-10%
Aberdeen	574	-1	-16	0	-8%
Pierre	575	-1	-28	0	-13%
Mobridge	576	-1	-36	0	-17%
Rapid City	577	-1	-14	0	-7%
Tennessee Average		-1	-5	0	-3%
Clarksville	370	1	3	0	2%
Nashville	371-372	1	9	0	4%
Cleveland	373	-1	0	0	-1%
Chattanooga	374	-1	5	0	1%
Kingsport	376	0	-13	0	-6%
Knoxville	377-379	-1	0	0	-1%
Memphis	380-381	-2	2	-1	0%
McKenzie	382	-2	-12	-1	-6%
Jackson	383	-1	-8	0	-5%
Columbia	384	-1	-17	0	-8%
Cookeville	385	0	-20	0	-9%
Texas Average		-3	-7	-1	-4%
Plano	750	0	18	0	8%
Dallas	751-753	-1	19	0	9%
Greenville	754	-3	5	-1	1%
Texarkana	755	-3	-23	-1	-12%
Longview	756	-3	-5	-1	-4%
Tyler	757	-1	-15	0	-8%
Palestine	758	-3	-16	-1	-9%
Lufkin	759	-3	-18	-1	-10%
Arlington	760	-1	6	0	2%
Fort Worth	761-762	-2	7	-1	2%
Wichita Falls	763	-3	-22	-1	-12%
Woodson	764	-3	-20	-1	-11%
Waco	765-767	-3	-13	-1	-8%
Brownwood	768	-3	-29	-1	-15%
San Angelo	769	-4	-18	-1	-10%
Houston	770-772	-1	22	0	10%
Huntsville	773	-3	20	-1	8%
Bay City	774	-2	27	-1	12%
Galveston	775	-3	20	-1	7%
Beaumont	776-777	-3	0	-1	-2%
Bryan	778	-1	-15	0	-7%
Victoria	779	-3	-3	-1	-3%
San Antonio	780-782	-3	-1	-1	-2%
Corpus Christi	783-784	-3	-3	-1	-3%
McAllen	785	-4	-34	-1	-17%
Austin	786-787	0	8	0	4%
Del Rio	788	-3	-35	-1	-18%
Giddings	789	0	-12	0	-5%
Amarillo	790-791	-3	-12	-1	-7%
Childress	792	-3	-12	-1	-7%
Lubbock	793-794	-3	-22	-1	-12%
Abilene	795-796	-4	-19	-1	-11%
Midland	797	-4	-9	-1	-6%
El Paso	798-799	-3	-20	-1	-11%
Utah Average		2	-16	1	-7%
Clearfield	840	2	-8	1	-3%
Salt Lake City	841	2	-6	1	-1%
Ogden	843-844	1	-24	0	-11%
Green River	845	1	-20	0	-9%
Provo	846-847	2	-21	1	-9%
Virginia Average		-1	-10	0	-5%
Alexandria	220-223	3	31	1	16%
Fredericksburg	224-225	1	-6	0	-2%
Winchester	226	0	-11	0	-5%
Culpeper	227	1	-2	0	0%
Harrisonburg	228	1	-11	0	-5%
Charlottesville	229	1	-12	0	-5%
Williamsburg	230-231	0	-2	0	-1%
Richmond	232	-1	12	0	5%
Chesapeake	233	-1	-4	0	-2%
Virginia Beach	234	0	-6	0	-3%
Norfolk	235-237	-1	-3	0	-2%
Petersburg	238	-2	-6	-1	-4%
Farmville	239	-2	-23	-1	-12%
Roanoke	240	-1	-15	0	-8%
Radford	241	-1	-19	0	-10%
Abingdon	242	-2	-26	-1	-13%
Galax	243	-2	-31	-1	-15%
Staunton	244	0	-20	0	-9%
Lynchburg	245	-2	-20	-1	-10%
Tazewell	246	-3	-20	-1	-11%
Vermont Average		0	-9	0	-4%
White River Junction	050	1	-11	0	-4%
Springfield	051	-1	-7	0	-4%
Bennington	052	-1	-9	0	-5%
Battleboro	053	0	-6	0	-2%
Burlington	054	2	6	1	4%
Montpelier	056	2	-12	1	-5%
Rutland	057	-1	-7	0	-4%
Albany	058	1	-17	0	-7%
Beecher Falls	059	1	-21	0	-10%
Washington Average		1	2	0	1%
Seattle	980-981	3	26	1	14%
Everett	982	2	10	1	6%
Tacoma	983-984	1	8	0	4%
Olympia	985	2	-1	1	0%
Vancouver	986	2	4	1	3%
Wenatchee	988	1	-11	0	-4%
Yakima	989	1	-12	0	-5%
Spokane	990-992	0	-3	0	-2%
Pasco	993	0	-1	0	-1%
Clarkston	994	-1	-4	0	-2%
Wisconsin Average		0	5	0	2%
Milwaukee	530-534	0	21	0	9%
Beloit	535	1	18	0	8%
Madison	537	2	23	1	12%
Prairie du Chien	538	-1	-6	0	-3%
Portage	539	0	8	0	4%
Amery	540	1	8	0	4%
Green Bay	541-543	0	8	0	4%
Wausau	544	-1	-2	0	-2%
Clam Lake	545	-1	-10	0	-5%
La Crosse	546	-1	-2	0	-2%
Eau Claire	547	0	-2	0	-1%
Ladysmith	548	-2	-12	-1	-7%
Oshkosh	549	-1	10	0	4%
West Virginia Average		-2	-15	-1	-8%
Bluefield	247-248	-2	-19	-1	-9%
Lewisburg	249	-2	-24	-1	-12%
Charleston	250-253	0	1	0	1%
Martinsburg	254	-1	-18	0	-9%
Huntington	255-257	-1	-8	0	-4%
Beckley	258-259	0	-16	0	-7%
Wheeling	260	-3	-10	-1	-6%
Parkersburg	261	-3	3	-1	0%
New Martinsville	262	-2	-27	-1	-14%
Clarksburg	263-264	-3	-17	-1	-10%
Morgantown	265	-3	-18	-1	-10%
Fairmont	266	0	-22	0	-10%
Romney	267	-4	-19	-1	-11%
Sugar Grove	268	-3	-18	-1	-10%
Wyoming Average		0	-11	0	-5%
Cheyenne/ Laramie	820	1	-16	0	-7%
Wheatland	822	0	-6	0	-3%
Rawlins	823	-1	-11	0	-6%
Powell	824	0	-23	0	-11%
Riverton	825	-1	-17	0	-9%
Casper	826	-2	-8	-1	-5%
Gillette	827	-1	0	0	0%
Sheridan	828	0	-21	0	-9%
Rock Springs	829-831	0	3	0	1%

UNITED STATES TERRITORIES

Location	Zip	Mat.	Lab.	Equip.	Total Wtd. Avg.
Guam	53	-21	-5		24%
Puerto Rico	2	-47	-5		-17%

VIRGIN ISLANDS (U.S.)

Location	Zip	Mat.	Lab.	Equip.	Total Wtd. Avg.
St. Croix	18	-15	-4		5%
St. John	52	-15	-4		26%
St. Thomas	23	-15	-4		8%

CANADIAN AREA MODIFIERS

These figures assume an exchange rate of $1.25 Canadian to $1.00 U.S.

Location	Mat.	Lab.	Equip.	Total Wtd. Avg.
Alberta Average	46.2	43.4	42.2	44.8%
Calgary	44.8	38.3	38.8	41.7%
Edmonton	46.5	45.0	41.4	45.7%
Ft. Mac	47.3	46.7	46.5	47.0%
British Columbia Average	44.2	42.9	42.2	43.6%
Fraser Valley	44.8	40.9	41.4	43.0%
Okanagan	39.7	37.0	41.4	38.5%
Vancouver	49.9	55.8	43.1	52.8%
Manitoba Average	47.9	49.3	41.8	48.4%
North Manitoba	52.4	63.4	49.0	57.2%
South Manitoba	48.0	49.0	38.8	48.2%
Selkirk	44.8	38.9	40.5	42.0%
Winnipeg	46.5	46.1	38.8	46.1%
New Brunswick Average	41.4	41.4	44.8	41.5%
Moncton	41.4	41.4	44.8	41.5%
Nova Scotia Average	47.6	50.5	45.7	48.8%
Amherst	47.3	49.2	45.7	48.1%
Nova Scotia	45.7	43.1	45.7	44.5%
Sydney	49.9	59.4	45.7	54.0%
Newfoundland Average	43.1	40.5	47.3	42.1%
Ontario Average	44.0	42.6	39.2	43.1%
London	48.2	52.8	39.7	50.0%
Thunder Bay	43.1	38.3	38.8	40.8%
Toronto	49.9	52.8	40.5	50.0%
Quebec Average	46.9	54.4	45.7	50.3%
Montreal	48.2	60.5	45.7	53.5%
Quebec	45.7	48.3	45.7	46.8%
Saskatchewan Average	46.7	53.2	41.1	49.5%
La Ronge	48.2	58.3	45.7	52.6%
Prince Albert	46.5	52.1	38.8	48.7%
Saskatoon	45.7	49.4	38.8	47.0%

Credits and Acknowledgments

This book has over 30,000 cost estimates for 2006. To develop these estimates, the editors relied on information supplied by hundreds of construction cost authorities. We offer our sincere thanks to the contractors, engineers, design professionals, construction estimators, material suppliers and manufacturers who, in the spirit of cooperation, have assisted in the preparation of this Fifty Fourth Edition of the National Construction Estimator. Many of the cost authorities who supplied cost data for this volume are listed below.

A.A. Drafting Service., Andrew Gutt

ABC Construction Co., Wayne Czubernat

ABTCO, Inc., Tom Roe

Advance Lifts, Jim Oriowski

Advanced Stainless & Aluminum, Louie Sprague

AFCO Roofing Supply, John Ording

AGRA Earth and Environmental, Jack Eagan

Allied Barricade, Bob Johnson

Alply Incorporated., Alan E. Sleppy

Amerec Products, John Gunderson

Applied Waterproofing Technology, Inc., Mike Weinert

Architectural Illust., Bruce Wickern

Area Home Inspection, Anthony Galeota

Atascadero Glass, Jim Stevens

Atlas Mineral & Chemical Co., Glenn Kresge

Bell Blueprint, Ed Moore

Bessler Stairways, Morris Hudspeth

Beveled Glass Design, Lorna Koss

Brooks Products, Lynn Diaz

Brower Pease, Mark Brossert

Buy-Rite Decorating House, Hank Jordan

California Pneumatic Systems, Clayton Taylor

Cement Cutting, Inc., Greg Becker

Centennial Flag Pole Co., Eleanor Wiese

Classic Ceilings, Donna Morgan

Cold Spring Granite, Paul Dierkhising

Columbia Equipment Co., Arthur Cohen

Creative Ceilings, Bob Warren

C. R. Laurence, Inc., Michael Sedano

David Backer Limited, David D. Backer

Delta Pacific Builders, Jeff Andrews

Dial One, Charlie Gindele

Dryvit System, Inc., Roger Peevier

Eco-Block, LLC, Jennifer Hansell

Eliason Corporation, John Duncan

Envel Design Corporation, Quinn Mayer

Escondido Roof Truss Co., David Truax

Finlandia Sauna Rooms, Reino J. Tarkianen

Fiorentine Company, Tom Forsyth

Flexi-Wall Systems, Nancy West

Fortifiber Corporation, Dave Olson

GAF Premium Products, Sandra Stern

Georgia Pacific, Travis Hemborg

Giles and Kendall, Inc., Todd M. Green

Glen-Gery Brick, David M. Reitz

Heintschel Plastering, Marvin J. Heintschel, Jr.

H&H Specialties, Reid Nesiage

I. C. C., Mike McInnis

Inclinator Company, Boc Crede

Int'l Conf. Of Bldg. Officials, Dennis McCreary

Iron-A-Way Company, Reg Smidt

J.A. Crawford Company, Jim Crawford

Jewett Refrigerator Co., Thomas D. Reider

J. H. Baxter Inc., Dean Rogers

Kentucky Wood Floors, John Stern

Kirsch Company, Dennis Kirby

Koffler Sales Corporation, Mary Holman

Landscape Structures, Inc., Mary Allen Grossman

Lasco Panels, Mike Bunke, Sales Manager

Lombardi, Diamond Core Drilling, Kit Sanders

Maccaferri Gabions, Inc., Ahmad Kasem

Marbleworks of San Diego, Charlene Butler

Mashburn Waste & Recycling, Bill Dean

M.C. Nottingham Company

Mel-Nortley Co., Mel Nortley

Merli Concrete Pumping, John Caole

Microphor, Ross Beck

Morrow Equipment Co., John Morrow

Multi Water Systems, John Kratz

National Floor Products, Edsel Holden

National Schedule Masters, Steve Scholl

Nicolon Mirafi Group, Ted Pope

North County Interiors, Rod Gould

Nuclear Associates, Laura Rhein

Oregon Lumber Company, John Couch

Penhall Construction Services, Bruce Varney

Perma Grain Products, Thomas W. Gunuskey

Photocom, Ron Kenedi

Pittsburgh Corning Corp., Joe Zavada

Plastmo Raingutters, Ed Gunderson

Prof. Photographic Service, Larry Hoagland

Rally Rack, Shirlee Turnipseed

RCP Block & Brick Co., Bob Balderree

Rheem Manufacturing, Rick Miller

Ryan Roulette Photography, Ryan Roulette

R. W. Little Company, Inc., Ron Little

Safeway Steel Products, Patrick Whelan

Sauder Manufacturing Co., John Gascho

SDGE, Don Altevers

Sico, Inc., Charles Williamson

Simpler Solar Systems, Al Simpler

Slater Waterproofing, Merle Slater, Jr.

Southern Bleacher Co., Glen McNatt

Specialty Wood Company, Bob DiNello

Stark Ceramics, Inc., Paul Tauer

Stebbins Company, Mrs. Pat Stebbins

Stone Products Corp., Robert Heath

Superior Fireplace Co., Mike Hammer

Superior Rock Products, John Knieff

Superior Stakes, George Kellett

Supradur Manufacturing, Marilyn Mueller

The Parking Block Store, Jim Headley

Towill INC., Jim Smith

U. S. Gypsum Company, Roger Merchet

Ultraflo Corporation, Doug Didion

Unique Building Products, Charlie Gindele

United Panel Partnership, Donna Korp

U. S. Tile Company, Brian Austin

Velux-America, Inc., John Carigan

Vintage Lumber & Const. Co., Mark Harbold

Vital Services Co., Myrna Humphrey

Wasco Products, Estelle Gore

West Coast Lumber Inspect. Bur., Jim Kneaper

Western Colloid Products, Tim Griffin

Western Exterminators, Bill Sharp

Western Wood Products Assn., Frank Stewart

Westingconsul, Inc., Verne T. Kelling

Weyerhaeuser Company, Doug Barkee

York Spiral Stair, Jim A. Guerette

Zenith Aerial Photo, Lew Roberts

Zeus Lightning Protection, Dave Silvester

A special thanks to the following people. Special assistants to the editors: Ray Hicks, James Thomson, Danielle and Christine Pray. Production and Layout: Christine Bruneau, Devona Quindoy and D. Naomi Johnson. Software production: Amy Rebich.

Cover design: Bill Grote
Cover photos: *Ron South Photography*

Adhesives See also, Caulking, Flooring, Roofing and Tile in the Residential Division

Panel adhesives Better quality, gun applied in continuous bead to wood or metal framing or furring members, material only, add labor below. Per 100 SF of wall, floor, or ceiling including 6% waste. Based on 10.5 ounce tubes.

		Bead diameter			
		1/8"	1/4"	3/8"	1/2"
Subfloor adhesive, on floors					
12" OC members	CSF	1.96	7.84	17.60	31.40
16" OC members	CSF	1.47	5.88	13.20	23.50
20" OC members	CSF	1.18	4.71	1.60	18.80
Wall sheathing or shear panel adhesive, on walls					
16" OC members	CSF	1.47	5.88	13.20	23.50
20" OC members	CSF	1.18	4.71	1.60	18.80
24" OC members	CSF	.98	3.92	8.82	15.70
Polystyrene or polyurethane foam panel adhesive, on walls					
12" OC members	CSF	1.88	7.52	16.90	30.10
16" OC members	CSF	1.41	5.64	12.70	22.60
20" OC members	CSF	1.13	4.51	10.20	18.00
24" OC members	CSF	.94	3.76	8.45	15.00
Gypsum drywall adhesive, on ceilings					
12" OC members	CSF	1.72	6.86	15.40	27.40
16" OC members	CSF	1.29	5.15	11.60	20.60
20" OC members	CSF	1.03	4.12	9.26	16.50
24" OC members	CSF	.86	3.43	7.72	13.70
Gypsum drywall adhesive, on walls					
12" OC members	CSF	1.72	6.86	15.40	27.40
16" OC members	CSF	1.29	5.15	11.60	20.60
20" OC members	CSF	1.03	4.12	9.26	16.50
24" OC members	CSF	.86	3.43	7.72	13.70
Hardboard or plastic panel adhesive, on walls					
12" OC members	CSF	2.04	8.17	18.40	32.70
16" OC members	CSF	1.53	6.13	13.80	24.50
20" OC members	CSF	1.23	4.90	11.00	19.60
24" OC members	CSF	1.02	4.08	9.19	16.30

	Craft@Hrs	Unit	Material	Labor	Total
Labor to apply adhesive to framing members, 1/8" to 1/2" bead diameter, no material included					
Floor or ceiling joists					
12" OC members	BC@.075	CSF	—	2.25	2.25
16" OC members	BC@.056	CSF	—	1.68	1.68
20" OC members	BC@.052	CSF	—	1.56	1.56
24" OC members	BC@.042	CSF	—	1.26	1.26
Interior and exterior wall members					
12" OC members	BC@.100	CSF	—	3.00	3.00
16" OC members	BC@.090	CSF	—	2.70	2.70
20" OC members	BC@.084	CSF	—	2.52	2.52
24" OC members	BC@.084	CSF	—	2.52	2.52

Adhesives

	Craft@Hrs	Unit	Material	Labor	Total
Special purpose adhesives					
Construction adhesive, for counters, cabinets, paneling, brick veneer, ceramic fixtures, shelving, sets in 10 minutes, 10.5 oz cartridge	—	Ea	2.10	—	2.10
Epoxy-tie adhesive, bonds concrete, fills cracks to 3/4", seals doors and windows, anchor grouting, non-shrink, 22 oz. cartridge	—	Ea	23.50	—	23.50
Gypsum drywall adhesive, waterproof, bonds to wood metal, masonry, concrete, 10.5 oz. cartridge	—	Ea	2.00	—	2.00
Latex FRP adhesive, indoor, bonds wood, concrete, drywall, foamboard, trowel grade, gallon	—	Ea	13.60	—	13.60
Panel insulation adhesive, waterproof, for polyurethane & polystyrene panels, bonds to wood, metal, masonry, concrete, gallon	—	Ea	13.60	—	13.60
Premium floor tile adhesive, solvent base, quart	—	Ea	5.50	—	5.50
Marine adhesive/sealant, 10 oz.	—	Ea	9.74	—	9.74
Project adhesive, rubber-based, bonds plywood, hardboard fiberglass, drywall, foam board, shelving, ceramic fixtures, 10 minute work time, 10.5 oz cartridge	—	Ea	1.44	—	1.44
Tile cement, solvent base, for installation of ceramic, quarry, slate and marble, gallon	—	Ea	12.50	—	12.50

	1 pint can	1 quart can	1 gallon can
General purpose adhesives			
Acoustic tile adhesive, solvent base, waterproof, sound deadening type	—	—	13.00
Aliphatic resin woodworking glue	—	4.20	15.00
Carpet cement, outdoor	—	—	17.00
Professional carpenter's glue, multi-purpose	—	3.03	11.00
Contact cement, rubber based, waterproof, bonds veneers to plywood, particleboard, wallboard	—	6.00	16.00
Gel contact cement	—	7.00	—
Resilient flooring adhesive, latex base, adheres to concrete, plywood, felt, sheet flooring	—	6.30	17.50

	Craft@Hrs	Unit	Material	Labor	Total
Aggregate Typical prices, 5 mile haul, 24 ton minimum. See also Roofing, Built-up					
Crushed stone (1.4 tons per CY)					
3/8" stone	—	Ton	15.60	—	15.60
3/4" (Number 3)	—	Ton	15.10	—	15.10
1-1/2" (Number 2)	—	Ton	17.90	—	17.90
Crushed slag, typical prices where available					
3/4" slag	—	Ton	11.10	—	11.10
1-1/2"	—	Ton	11.30	—	11.30
Washed gravel (1.4 tons per CY)					
3/4" gravel	—	Ton	16.80	—	16.80
1-1/2"	—	Ton	17.10	—	17.10
Fill sand (1.35 tons per CY)	—	Ton	13.80	—	13.80
Add per ton less than 24 tons	—	Ton	6.00	—	6.00
Add for delivery over 5 miles, one way	—	Mile	3.85	—	3.85

	Craft@Hrs	Unit	Material	Labor	Total

Appraisal Fees Costs for determining the value of existing buildings, land, and equipment. Actual fee charged is based on the level of difficulty and the time spent on appraisal plus travel to location and cost of support services, if any. Costs include research and report by a professional appraiser. Client may request an appraisal on a "fee not to exceed" basis. Fees shown are averages and are not quoted as a percentage of value.

Single family residences, condominiums, planned unit developements (PUDs).
Fees for high value residences (over $1,250,000) are usually negotiated.

	Craft@Hrs	Unit	Material	Labor	Total
To $250,000 valuation	—	LS	—	—	300.00
Over $250,000 to $350,000 valuation	—	LS	—	—	350.00
Over $350,000 to $500,000 valuation	—	LS	—	—	400.00
Over $500,000 to $650,000 valuation	—	LS	—	—	500.00
Over $650,000 to $750,000 valuation	—	LS	—	—	550.00
Over $750,000 to $999,999 valuation	—	LS	—	—	750.00
Over $1,000,000 to 1,250,000 valuation	—	LS	—	—	1,000.00

Small income units

	Craft@Hrs	Unit	Material	Labor	Total
2 –4 units	—	LS	—	—	650.00

Apartment houses, commercial and industrial buildings

	Craft@Hrs	Unit	Material	Labor	Total
To $300,000 valuation	—	LS	—	—	3,000.00
Over $300,000 to $1,000,000 valuation	—	LS	—	—	4,000.00
Over $1,000,000 to $3,000,000 valuation	—	LS	—	—	4,500.00
Over $3,000,000 to $5,000,000 valuation	—	LS	—	—	5,500.00

Other services

	Craft@Hrs	Unit	Material	Labor	Total
Additional photograph set (3 each)	—	LS	—	—	25.00
Outlying area fee	—	LS	—	—	50.00
Satisfactory Completion (Form 442)	—	LS	—	—	75.00
Recertification of Value	—	LS	—	—	75.00
Income Reconciliation (FNMA 216)	—	LS	—	—	50.00
SFR Rental Survey (FNMA 1007)	—	LS	—	—	150.00
Appraisal review (desk)	—	LS	—	—	150.00
Appraisal review (field)	—	LS	—	—	200.00
Drive-by appraisal (2055) starting at:	—	LS	—	—	250.00
Drive-by appraisal (2070/2075)	—	LS	—	—	200.00
Retype of full package	—	LS	—	—	100.00
Cancellation, after inspection	—	LS	—	1/2 of original fee	

Machinery (Fee is based on total value of equipment appraised) Additional charges for travel and lodging may be required

	Craft@Hrs	Unit	Material	Labor	Total
To $30,000 valuation	—	LS	—	—	570.00
Over $30,000 to $100,000 valuation	—	LS	—	—	770.00
Over $100,000 to $500,000 valuation	—	LS	—	—	1,300.00
Over $500,000 to $1,000,000 valuation	—	LS	—	—	1,800.00
Over $1,000,000 to $5,000,000 valuation	—	LS	—	—	3,100.00
Appraisals per 1/2 day (typical)	—	1/2 Day	—	—	500.00
Appraisals per day (typical)	—	Day	—	—	800.00
Court testimony (excluding preparation)	—	Day	—	—	1,200.00

Arbitration Clauses

Arbitration Clauses The cost of settling construction disputes may be reduced by adding contract language which requires alternative dispute resolution (ADR) rather than court proceedings. Generally courts will not accept cases until parties have met requirements for ADR established by contract. In many cases, formal arbitration will be unnecessary when a contractor can require that a dispute be presented to a third-party mediator before arbitration is invoked. Sample contract clauses follow, both for mediation and arbitration. These clauses are recommended by the American Arbitration Association and appear on the AAA web site www.adr.org.

Mediation. If you prefer to include mediation as a part of the dispute settlement procedure, insert the following mediation clause in the contract in addition to the arbitration clause that follows.

> *If a dispute arises out of or relates to this contract, or the breach thereof, and if the dispute cannot be settled through negotiation, the parties agree first to try in good faith to settle the dispute by mediation administered by the American Arbitration Association under its Construction Industry Mediation Procedures before resorting to arbitration, litigation, or some other dispute resolution procedure.*

Arbitration. If mediation is unsuccessful, the following clause will require that any dispute be submitted for arbitration before resort to court proceedings.

> *Any controversy or claim arising out of or relating to this contract, or the breach thereof, shall be settled by arbitration administered by the American Arbitration Association under its Construction Industry Arbitration Rules, and judgment on the award rendered by the arbitrator(s) may be entered in any court having jurisdiction thereof.*

Arbitration Fees These are administrative fees paid to the American Arbitration Association (AAA). Rules and fees are subject to change. The AAA's web site at www.adr.org has current rules and fee information. Arbitrators are chosen from the National Roster of Construction Arbitrators and are paid a fee by the parties. Legal representation, if desired (although not necessary), is at the expense of each party. These fees do not include rental of a hearing room. An initial filing fee is payable in full by a filing party when a claim, counterclaim or additional claim is filed. A case service fee is payable at the time the first hearing is scheduled. This fee will be refunded at the conclusion of the case if no hearings have occurred so long as the Association was notified of cancellation at least 24 hours before the first scheduled hearing. The minimum filing fee for any case having three or more arbitrators is $2,750 plus a $1,000 case service fee.

		Initial Filing Fee	Case Service Fee
Claims to $10,000	LS	500.00	200.00
Claims over $10,000 to $75,000	LS	750.00	300.00
Claims over $75,000 to $150,000	LS	1,500.00	750.00
Claims over $150,000 to $300,000	LS	2,750.00	1,250.00
Claims over $300,000 to $500,000	LS	4,250.00	1,750.00
Claims over $500,000 to $1,000,000	LS	6,000.00	2,500.00
Claims over $1,000,000 to $5,000,000	LS	8,000.00	3,250.00
Claims over $5,000,000 to $10,000,000	LS	10,000.00	4,000.00
Claims over $10,000,000 *	LS	*	6,000.00
No Amount Stated **	LS	3,250.00	750.00

* $10 million and above – Base fee is $12,500 plus .01% of the amount of claim above $10 million. Filing fees are capped at $65,000.

**This fee is applicable when a claim or counterclaim is not for a monetary amount. Where a monetary claim is not known, parties will be required to state a range of claims or be subject to the highest possible filing fee.

	Craft@Hrs	Unit	Material	Labor	Total

Architectural Illustrations Full color painting on watercolor board with matted frame with title and credit on matte. Glass and elaborate framing are extra. Costs for pen and ink illustrations with color mylar overlay are similar to cost for watercolor illustrations. Typical fees.

	Craft@Hrs	Unit	Material	Labor	Total
Custom home, eye level view					
Simple rendering	—	LS	—	—	880.00
Complex rendering	—	LS	—	—	1,290.00
Custom home, bird's eye view					
Simple rendering	—	LS	—	—	1,060.00
Complex rendering	—	LS	—	—	1,500.00
Tract homes in groups of five or more (single floor plans, multiple elevations), eye level view					
Simple rendering	—	LS	—	—	520.00
Complex rendering	—	LS	—	—	720.00
Tract homes in groups of five or more (single floor plans, multiple elevations), bird's eye view					
Simple rendering	—	LS	—	—	670.00
Complex rendering	—	LS	—	—	1,070.00
Tract homes or condominium project, overall bird's eye view					
10-25 homes or living units	—	LS	—	—	3,310.00
Typical commercial structure					
Eye level view	—	LS	—	—	1,450.00
Bird's eye view	—	LS	—	—	1,630.00
Complex commercial structure					
Eye level view	—	LS	—	—	2,140.00
Bird's eye view	—	LS	—	—	2,780.00
Deduct for pen and ink drawings (no color)	—	%	—	—	-33.0
Computer generated perspective drawings using CAD system for design studies					
Custom home	—	LS	—	—	560.00
Large condo or apartment projects	—	LS	—	—	1,400.00
Tract homes	—	LS	—	—	350.00
Commercial structure, line drawing	—	LS	—	—	1,060.00

Awnings and Canopies for Doors and Windows Costs for awnings include all hardware. All have adjustable support arms to control angle height and preferred amount of window coverage. For larger size aluminum awnings, price two awnings and add for splice kit below.

	Craft@Hrs	Unit	Material	Labor	Total
Natural aluminum ribbed awning with clear, weather-resistant finish and 26" arms					
36" wide x 30" long	SW@1.45	Ea	64.70	49.30	114.00
48" wide x 30" long	SW@1.86	Ea	83.50	63.20	146.70
60" wide x 30" long	SW@2.07	Ea	104.00	70.30	174.30
72" wide x 30" long	SW@2.27	Ea	118.00	77.10	195.10
Add for door canopy with 17" drop sides	—	%	50.0	—	—
Custom colored window awnings in stripes or solids, with baked enamel finish and ventilation panels					
30" wide x 24" high	SW@1.45	Ea	105.00	49.30	154.30
36" wide x 36" high	SW@1.45	Ea	158.00	49.30	207.30
48" wide x 48" high	SW@1.87	Ea	266.00	63.50	329.50
48" wide x 60" high	SW@1.87	Ea	376.00	63.50	439.50
60" wide x 72" high	SW@2.07	Ea	414.00	70.30	484.30
72" wide x 72" high	SW@2.26	Ea	565.00	76.80	641.80
Add for splice kit with overlap slats	SW@.218	Ea	21.90	7.41	29.31

Barricades, Construction Safety

	Craft@Hrs	Unit	Material	Labor	Total

Security roll-up awning with pull cord assembly and folding arms, clear weather-resistant finish. Awning rolls down to cover whole window. 48" long, 24" arms

	Craft@Hrs	Unit	Material	Labor	Total
36" wide	SW@1.52	Ea	201.00	51.60	252.60
48" wide	SW@1.94	Ea	239.00	65.90	304.90

Plastic awning with baked-on acrylic finish, ventilated side panels, and reinforced metal frame, hardware included. 24" drop, 24" projection

36" wide	BC@1.58	Ea	109.00	47.40	156.40
42" wide	BC@1.75	Ea	121.00	52.50	173.50
48" wide	BC@2.02	Ea	139.00	60.60	199.60
60" wide	BC@2.26	Ea	156.00	67.80	223.80
72" wide	BC@2.47	Ea	168.00	74.10	242.10
96" wide	BC@2.79	Ea	213.00	83.70	296.70

Plastic door canopy with 36" projection

42" wide	BC@1.80	Ea	229.00	54.00	283.00

Traditional fabric awning, with waterproof, acrylic duck, colorfast fabric, double stitched seams, and tubular metal framing and pull cord assembly. 24" drop, 24" projection

30" wide	BC@1.35	Ea	47.90	40.50	88.40
36" wide	BC@1.58	Ea	62.90	47.40	110.30
42" wide	BC@1.80	Ea	58.70	54.00	112.70
48" wide	BC@2.02	Ea	69.00	60.60	129.60
Add for 30" drop, 30" projection	—	%	10.0	20.0	—

Cloth canopy patio cover, with front bar and tension support rafters, 9" valance and 8' projection

8' x 10'	BC@2.03	Ea	291.00	60.90	351.90
8' x 15'	BC@2.03	Ea	401.00	60.90	461.90

Barricades, Construction Safety Purchase prices except as noted. See also Equipment Rental in the index.

Plastic Type I barricade, Engineer Grade Reflectivity

To 100 units	—	Ea	38.80	—	38.80
Over 100 units	—	Ea	33.30	—	33.30

Reflectorized plywood barricade, 8" to 12" wide rail, 4" to 6" wide stripes, 45" 14 gauge steel legs, no light

Type I, 2' wide, 3' high, 1 reflectorized rail each side

To 100 units	—	Ea	19.40	—	19.40
Over 100 units	—	Ea	16.90	—	16.90

Type II, 2' wide, 3' high, 2 reflectorized rails each side

To 100 units	—	Ea	25.00	—	25.00
Over 100 units	—	Ea	21.50	—	21.50

Type III, 4' wide, 5' high, 3 reflectorized rails each side, wood & steel legs

To 100 units	—	Ea	83.50	—	83.50
Over 100 units	—	Ea	65.50	—	65.50

Utility barricade, 2" x 2" legs, 1" x 6" face, no light

To 100 units	—	Ea	14.80	—	14.80
Over 100 units	—	Ea	11.70	—	11.70

Add for lighted units without batteries (batteries last 2 months)

To 100 units	—	Ea	15.00	—	15.00
Over 100 units	—	Ea	12.00	—	12.00
Batteries, 6 volt (2 needed)	—	Ea	1.95	—	4.15
Lighted units, rental, per day	—	Ea	.35	—	.35
Unlighted units, rental per day	—	Ea	.25	—	.25
Add for pickup and delivery, per trip	—	Ea	28.50	—	28.50

Description	Craft@Hrs	Unit	Material	Labor	Total
"Road Closed", reflectorized, 30" x 48", Engineer Grade	—	Ea	72.90	—	72.90
"Construction Zone", 4' x 4' high intensity grade	—	Ea	213.00	—	213.00
High-rise tripod with 3 orange flags	—	Ea	105.00	—	105.00
High-rise sign holder, wind resistant	—	Ea	149.00	—	149.00
Traffic cones, PVC					
Non-reflectorized type					
18" high	—	Ea	5.25	—	5.25
28" high	—	Ea	9.25	—	9.25
Reflectorized type					
28" high	—	Ea	16.50	—	16.50
Lane delineator, 42" orange plastic cylinder with 2 reflectors on a 12 pound rubber base	—	Ea	19.00	—	19.00
Mesh signs, orange, 48" x 48", includes brace and clamp	—	Ea	51.00	—	51.00
Hand-held traffic paddles, "Stop" and "Slow"	—	Ea	17.00	—	17.00
Typical labor cost, place and remove any barricade					
Per use	BL@.160	Ea	—	3.95	3.95
Orange plastic safety fencing					
4' x 50' roll	—	Ea	39.00	—	39.00
5' x 50' roll	—	Ea	48.50	—	48.50
6' x 50' roll	—	Ea	57.00	—	57.00
4' x 100', light duty	—	Ea	47.50	—	47.50
4' x 200', light duty	—	Ea	72.80	—	72.80
Barricade tape for marking construction sites, landscaping, wet paint					
Yellow plastic 3" x 300', "Caution"	—	Ea	6.33	—	6.33
Various colors, 3" x 200', "Danger"	—	Ea	5.74	—	5.74

Basement Doors Good quality 12 gauge primed steel, center opening basement doors. Costs include assembly and installation hardware. No concrete, masonry, anchor placement or finish painting included. Bilco Company www.bilco.com

Description	Craft@Hrs	Unit	Material	Labor	Total
Doors (overall dimensions)					
30"H, 47"W, 58"L	BC@3.41	Ea	328.00	102.00	430.00
22"H, 51"W, 64"L	BC@3.41	Ea	334.00	102.00	436.00
19-1/2"H, 55"W, 72"L	BC@3.41	Ea	365.00	102.00	467.00
52"H, 51"W, 43-1/4"L	BC@3.41	Ea	392.00	102.00	494.00
Door extensions (available for 19-1/2"H, 55"W, 72"L door only)					
6" deep	BC@1.71	Ea	90.70	51.30	142.00
12" deep	BC@1.71	Ea	108.00	51.30	159.30
18" deep	BC@1.71	Ea	135.00	51.30	186.30
24" deep	BC@1.71	Ea	161.00	51.30	212.30
Stair stringers, steel, pre-cut for 2" x 10" wood treads (without treads)					
32" to 39" stair height	BC@1.71	Ea	76.60	51.30	127.90
48" to 55" stair height	BC@1.71	Ea	95.80	51.30	147.10
56" to 64" stair height	BC@1.71	Ea	108.00	51.30	159.30
65" to 72" stair height	BC@1.71	Ea	121.00	51.30	172.30
73" to 78" stair height	BC@1.71	Ea	166.00	51.30	217.30
81" to 88" stair height	BC@1.71	Ea	180.00	51.30	231.30
89" to 97" stair height	BC@1.71	Ea	192.00	51.30	243.30

Bathroom Accessories

	Craft@Hrs	Unit	Material	Labor	Total

Bathroom Accessories Average quality. Better quality brass accessories cost 75% to 100% more. See also Medicine Cabinets and Vanities

	Craft@Hrs	Unit	Material	Labor	Total
Cup and toothbrush holder, chrome	BC@.258	Ea	9.80	7.74	17.54
Cup holder, porcelain, surface mounted	BC@.258	Ea	6.64	7.74	14.38
Cup, toothbrush & soap holder, recessed	BC@.258	Ea	25.00	7.74	32.74
Cup, toothbrush holder, polished brass	BC@.258	Ea	14.10	7.74	21.84
Electrical plates, chrome plated					
Switch plate, single	BE@.154	Ea	3.52	5.21	8.73
Switch plate, double	BE@.154	Ea	4.64	5.21	9.85
Duplex receptacle plate	BE@.154	Ea	3.52	5.21	8.73
Duplex receptacle and switch	BE@.154	Ea	4.64	5.21	9.85
Grab bars					
Tubular chrome plated, with anchor plates					
Straight bar, 16"	BC@.414	Ea	20.70	12.40	33.10
Straight bar, 24"	BC@.414	Ea	25.00	12.40	37.40
Straight bar, 32"	BC@.414	Ea	27.30	12.40	39.70
"L" shaped bar, 16" x 32"	BC@.620	Ea	63.10	18.60	81.70
Stainless steel, with anchor plates					
Straight bar, 16"	BC@.414	Ea	31.10	12.40	43.50
Straight bar, 24"	BC@.414	Ea	37.60	12.40	50.00
Straight bar, 32"	BC@.414	Ea	40.80	12.40	53.20
"L" shaped bar, 16" x 32"	BC@.620	Ea	88.20	18.60	106.80
Mirrors, stainless steel framed, surface mount, no light or cabinet					
16" high x 20" wide	BG@.420	Ea	52.30	12.10	64.40
18" high x 24" wide	BG@.420	Ea	60.90	12.10	73.00
18" high x 36" wide	BG@.420	Ea	93.50	12.10	105.60
24" high x 36" wide	BG@.420	Ea	110.00	12.10	122.10
48" high x 24" wide	BG@.420	Ea	135.00	12.10	147.10
Mirrors, wood framed, surface mount, better quality					
18" x 29" rectangular	BG@.420	Ea	65.10	12.10	77.20
20" x 27" oval, oak	BG@.420	Ea	92.40	12.10	104.50
Robe hook					
Chrome	BC@.258	Ea	18.50	7.74	26.24
Double, solid brass	BC@.258	Ea	17.80	7.74	25.54
Shower curtain rods, chrome plated					
60", recessed	BC@.730	Ea	21.70	21.90	43.60
66", recessed	BC@.730	Ea	23.80	21.90	45.70
Soap holder, surface mounted, with drain holes					
Williamsburg, satin chrome	BC@.258	Ea	44.10	7.74	51.84
Polished brass	BC@.258	Ea	14.10	7.74	21.84
Facial tissue holder, stainless steel, recessed	BC@.258	Ea	29.76	7.74	37.50
Toilet tissue roll holder, chrome, recessed	BC@.258	Ea	13.80	7.74	21.54
Toothbrush holder, chrome, surface mount	BC@.258	Ea	5.11	7.74	12.85
Towel bars, 3/4" round bar					
18" long, chrome	BC@.280	Ea	11.40	8.40	19.80
24" long, chrome	BC@.280	Ea	13.40	8.40	21.80
30" long, chrome	BC@.280	Ea	15.90	8.40	24.30
36" long, chrome	BC@.280	Ea	18.20	8.40	26.60
18" long, solid brass	BC@.280	Ea	40.30	8.40	48.70
24" long, solid brass	BC@.280	Ea	42.20	8.40	50.60
Towel rack, swing-arm, chrome, 3 bars, 12" L	BC@.280	Ea	12.90	8.40	21.30

	Craft@Hrs	Unit	Material	Labor	Total
Towel rings					
Williamsburg chrome and brass	BC@.280	Ea	33.70	8.40	42.10
Williamsburg chrome and porcelain	BC@.280	Ea	33.30	8.40	41.70
Towel shelf, chrome, 24" L with bar below	BC@.280	Ea	32.70	8.40	41.10

Heated towel racks, 16" width, mounted on wall with brackets. Wall plug in or direct connection, 700 watt output. Add for electrical work

	Craft@Hrs	Unit	Material	Labor	Total
26" high, standard colors	BE@.850	Ea	652.00	28.80	680.80
Add for control panel	BE@.850	Ea	132.00	28.80	160.80
26" high, chrome finish	BE@.850	Ea	1,720.00	28.80	1,748.80
Add for control panel	BE@.850	Ea	88.60	28.80	117.40
35" high, standard colors	BE@1.00	Ea	753.00	33.90	786.90
Add for control panel	BE@1.00	Ea	136.00	33.90	169.90
35" high, chrome finish	BE@1.00	Ea	2,140.00	33.90	2,173.90
Add for control panel	BE@1.00	Ea	64.90	33.90	98.80
44" high, standard colors	BE@1.10	Ea	832.00	37.20	869.20
Add for control panel	BE@1.10	Ea	147.00	37.20	184.20
44" high, chrome finish	BE@1.10	Ea	2,320.00	37.20	2,357.20
Add for control panel	BE@1.10	Ea	71.40	37.20	108.60
Add for 24" width rack	BE@1.10	Ea	34.10	37.20	71.30
Heated towel rack accessories					
Programmable control panel	BE@.250	Ea	278.00	8.47	286.47
Pull out drying racks	BE@.250	Ea	53.00	8.47	61.47
Robe knobs	BE@.150	Ea	32.10	5.08	37.18
Towel bar accents with standard colors	BE@.150	Ea	68.90	5.08	73.98
Towel bar accents with chrome finish	BE@.150	Ea	119.00	5.08	124.08
Robe knob accents (all styles)	BE@.150	Ea	192.00	5.08	197.08
Tub or shower chairs					
Wall-hung elevated fixed seat	BE@.750	Ea	27.10	25.40	52.50
White adjustable seat	BE@.750	Ea	49.70	25.40	75.10

Beds, Folding Concealed-in-wall type. Steel framed, folding wall bed system. Bed requires 18-5/8" or 22" deep recess. Includes frame, lift mechanism, all hardware. Installed in framed opening. Padded vinyl headboard. Bed face panel accepts paint, wallpaper, vinyl or laminate up to 1/4" thick. Add for box spring and mattress. Sico Incorporated www.sico-wallbeds.com

	Craft@Hrs	Unit	Material	Labor	Total
Twin, 41"W x 83-1/2"H x 18-5/8"D	B1@5.41	Ea	1,110.00	148.00	1,258.00
Box spring and mattress	—	Ea	371.00	—	371.00
Twin-long, 41"W x 88-5/8"H x 18-5/8"D	B1@5.41	Ea	1,130.00	148.00	1,278.00
Box spring and mattress	—	Ea	388.00	—	388.00
Double, 56"W x 83-1/2"H x 18-5/8"D	B1@5.41	Ea	1,110.00	148.00	1,258.00
Box spring and mattress	—	Ea	464.00	—	464.00
Double-long, 56"W x 88-5/8-1/2"H x 18-5/8"D	B1@5.41	Ea	1,130.00	148.00	1,278.00
Box spring and mattress	—	Ea	490.00	—	490.00
Queen, 62-1/2"W x 88-5/8-1/2"H x 18-5/8"D	B1@5.41	Ea	1,200.00	148.00	1,348.00
Box spring and mattress	—	Ea	556.00	—	556.00
King, 79"W x 88-5/8"H x 18-5/8"D	B1@5.41	Ea	2,030.00	148.00	2,178.00
Box spring and mattress	—	Ea	799.00	—	799.00
Add for chrome frame, any bed	—	LS	233.00	—	233.00
Add for brass frame, any bed	—	LS	250.00	—	250.00

Blueprinting

	Craft@Hrs	Unit	Material	Labor	Total

Blueprinting (Reproduction only) Assumes original is on semi-transparent drafting paper or film. See also Architectural Illustration and Drafting. Cost per square foot reproduced except as noted. Stapled edge and binder included. Diazo

	Craft@Hrs	Unit	Material	Labor	Total
Blueline or blackline prints					
1-100 SF	—	SF	—	—	.20
101-1,000 SF	—	SF	—	—	.18
1,001-2,000 SF	—	SF	—	—	.17
2,001-to 3,000 SF	—	SF	—	—	.16
3,001-4,000 SF	—	SF	—	—	.09
4,001 SF and higher	—	SF	—	—	.08
Presentation blackline (heavy paper)	—	SF	—	—	.50
Sepia	—	SF	—	—	.75
Mylar	—	SF	—	—	2.50
Xerographic prints					
Vellum	—	SF	—	—	1.00
Erasable vellum	—	SF	—	—	1.25
Mylar	—	SF	—	—	2.50
Enlargements (bond)	—	SF	—	—	.60
Reductions (bond, per sheet)	—	Ea	—	—	2.50
Plotting prints					
Translucent bond	—	SF	—	—	1.50
Vellum	—	SF	—	—	1.75
Erasable vellum	—	SF	—	—	2.50
Mylar	—	SF	—	—	2.50
Photo prints					
Mylar	—	SF	—	—	5.65
Add for local pickup and delivery, round trip	—	LS	—	—	7.50

Building Inspection Service (Home inspection service) Inspection of all parts of building by qualified engineer or certified building inspection technician. Includes written report covering all doors and windows, electrical system, foundation, heating and cooling system, insulation, interior and exterior surface conditions, landscaping, plumbing system, roofing, and structural integrity.

	Craft@Hrs	Unit	Material	Labor	Total
Single-family residence					
Base fee (up to 2,500 SF)	—	LS	—	—	325.00
Add for additional 1,000 SF or fraction	—	LS	—	—	100.00
Add for out buildings (each)	—	LS	—	—	50.00
Add for houses over 50 years old	—	LS	—	—	75.00
Add per room for houses with over 10 rooms	—	Ea	—	—	60.00
Add per room for houses with over 15 rooms	—	Ea	—	—	65.00
Add for swimming pool, spa or sauna	—	LS	—	—	200.00
Add for soil testing (expansive soil only)	—	LS	—	—	200.00
Add for water testing (coliform only)	—	LS	—	—	70.00
Add for warranty protection					
Houses to 10 rooms & 50 years old	—	LS	—	—	250.00
Houses over 50 years old	—	LS	—	—	270.00
Houses over 10 rooms	—	LS	—	—	270.00
Multi-family structures					
Two family residence base fee	—	LS	—	—	450.00
Apartment or condominium base fee	—	LS	—	—	250.00
Warranty protection (base cost)	—	LS	—	—	250.00
Add for each additional unit	—	LS	—	—	50.00

	Craft@Hrs	Unit	Material	Labor	Total
Add for each family living unit					
Standard inspection	—	LS	—	—	50.00
Detailed inspection	—	LS	—	—	75.00
Add for swimming pool, spa, sauna	—	LS	—	—	75.00
Add for potable water quality testing	—	LS	—	—	225.00
Add for water quantity test, per well	—	LS	—	—	150.00
Add for soil testing (EPA toxic)	—	LS	—	—	1,500.00
Add for soil testing (lead)	—	LS	—	—	45.00
Add for lead paint testing, full analysis, per room	—	LS	—	—	40.00
Hazards testing for single and multi-family dwellings					
Urea-formaldehyde insulation testing	—	LS	—	—	175.00
Asbestos testing	—	LS	—	—	175.00
Radon gas testing	—	LS	—	—	125.00
Geotechnical site examination, typical price	—	LS	—	—	400.00

Building Paper See also Roofing for roof applications and Polyethylene Film. Costs include 7% coverage allowance for 2" lap and 5% waste allowance. See installation costs at the end of this section.

	Craft@Hrs	Unit	Material	Labor	Total
Asphalt felt, 36" wide					
15 lb., ASTM F45 (432 SF roll at $17.20)	—	SF	.04	—	.04
15 lb., F40 (432 SF roll at $13.70)	—	SF	.03	—	.03
30 lb. (216 SF roll at $17.10)	—	SF	.09	—	.09
Asphalt shake felt, 18" x 120'					
30 lb. (180 SF roll at $14.70)	—	SF	.08	—	.08
Building paper, 40" wide					
Single ply, black (321 SF roll at $8.90)	—	SF	.03	—	.03
Two ply (162 SF roll at $9.95)	—	SF	.07	—	.07
Aquabar™, two layer, laminated with asphalt, Fortifiber™ products					
Class A, 36" wide, 30-50-30					
(500 SF roll)	—	SF	.06	—	.06
Class B, 36" wide, 30-30-30					
(500 SF roll)	—	SF	.03	—	.03
Ice and water shield, self-adhesive rubberized asphalt and poly					
225 SF roll at $100	—	SF	.44	—	.44
108 SF roll at $50	—	SF	.50	—	.50
"Jumbo Tex" gun grade sheathing paper, 40" wide, asphalt saturated					
(324 SF roll)	—	SF	.03	—	.03
"Jumbo Tex" black building paper, 36", 40" wide, asphalt saturated					
(500 SF roll)	—	SF	.03	—	.03
"Super Jumbo Tex" two-ply, 60 minute, asphalt saturated Kraft					
(162 SF roll at $13.00)	—	SF	.09	—	.09
Red rosin sized sheathing (duplex sheathing) 36" wide					
(500 SF roll at $9.20)	—	SF	.02	—	.02
Bruce rosin paper, floor underlay,					
(500 SF roll at $12.00)	—	SF	.03	—	.03
Moistop flashing paper, Fortifiber™ , 6" wide x 300' long					
(150 SF roll)	—	SF	.13	—	.13
Vycor window and door flashing, Grace , 6" x 75' roll at $22.10					
40 mil thickness	—	SF	.67	—	.67
Vycor deck protector, self adhesive joist flashing , 4" x 75' roll at $15.50					
Roll covers 66 SF of deck	—	SF	.23	—	.23

Building Paper

	Craft@Hrs	Unit	Material	Labor	Total
Plasterkraft grade D weather-resistive sheathing paper, 40" wide					
Super 60 minute (500 SF roll at $12.60)	—	SF	.03	—	.03
Ten minute 2-ply (500 SF roll at $10.00)	—	SF	.02	—	.02
Roof flashing paper, seals around skylights, dormers, vents, valleys and eaves, rubberized, fiberglass reinforced, self-adhesive, GAF StormGuard™					
Roll covers 200 SF at $68.40	—	SF	.34	—	.30
Roof underlay, for use under shingles, granular surface, Weatherlock®					
200 SF roll at $71.00	—	SF	.36	—	.36
Tri-Flex 30 roof underlay, for use under shingles, tile, slate or metal roofing, 41.5" x 116'					
400 SF roll at $52.50	—	SF	.14	—	.14
Copper armored Sisalkraft, 10" wide x 120' long					
1 oz per SF (120 LF roll)	—	LF	1.06	—	1.06
3 oz per SF (120 LF roll)	—	LF	1.16	—	1.16
Above grade vapor barrier, no waste or coverage allowance included. Fortifiber™ products					
Vaporstop 298, (fiberglass reinforcing and asphaltic adhesive between 2 layers of Kraft)					
32" x 405' roll (1,080 SF roll)	—	SF	.08	—	.08
Pyro-Kure™ 600, (2 layers of heavy Kraft with fire retardant adhesive edge reinforced with fiberglass)					
32" x 405' roll (1,080 SF roll)	—	SF	.15	—	.15
Foil Barrier 718 (fiberglass reinforced aluminum foil and adhesive between 2 layers of Kraft)					
52" x 231' roll (1,000 SF roll)	—	SF	.27	—	.27
Below grade vapor barrier, Fortifiber™					
Moistop (fiberglass reinforced Kraft between 2 layers of polyethylene)					
8' x 250' roll (2,000 SF roll)	—	SF	.08	—	.08
Concrete curing papers, Fortifiber™					
Orange Label Sisalkraft (fiberglass and adhesive between 2 layers of Kraft), 4.8 lbs. per CSF					
48" x 125' roll, (900 SF roll)	—	SF	.09	—	.09
Sisalkraft SK-10, economy papers, fiberglass and adhesive between 2 layers of Kraft, 4.2 lbs. per CSF, 48" x 300' roll (1,200 SF roll)	—	SF	.05	—	.05
Protective paper, Fortifiber™					
Seekure (fiberglass reinforcing strands and non staining adhesive between 2 layers of Kraft)					
48" x 300' roll, (1,200 SF)	—	SF	.07	—	.07
Tyvek™ house wrap by DuPont					
Air infiltration barrier (high-density polyethylene fibers in sheet form)					
3' x 100' rolls or					
9' x 100' rolls	—	SF	.11	—	.11
House wrap tape, 2" x 165'	—	LF	.07	—	.07
Labor to install building papers					
Felts, vapor barriers, infiltration barriers, building papers on walls					
Tack stapled, typical	BC@.002	SF	—	.06	.06
Heavy stapled, typical	BC@.003	SF	—	.09	.09
Felts, vapor barriers, infiltration barriers, building papers on ceilings and roofs					
Tack stapled, typical	BC@.004	SF	—	.12	.12
Heavy stapled, typical	BC@.006	SF	—	.18	.18
Self-adhesive, typical	BC@.006	SF	—	.18	.18
Curing papers, protective papers and vapor barriers, minimal fasteners	BC@.001	SF	—	.03	.03
Flashing papers, 6" to 8" wide	BC@.010	LF	—	.30	.30

	Craft@Hrs	Unit	Material	Labor	Total

Building Permit Fees These fees are based on recommendations in the 1997 Uniform Building Code and will apply in many areas. Building Departments publish fee schedules. The permit fee will be doubled when work is started without securing a permit. When the valuation of the proposed construction exceeds $1,000, plans are often required. Estimate the plan check fee at 65% of the permit fee for residences and 100% of the permit fee for non-residential buildings. The fee for reinspection is $47 per hour. Inspections outside normal business hours are $47 per hour with a two hour minimum. Additional plan review required by changes, additions, or revisions to approved plans are $47 per hour with a one-half hour minimum. Plumbing, electrical and mechanical work will usually require a separate permit based on a similar fee schedule.

The minimum fee for construction values to $500 is $33.50

	Craft@Hrs	Unit	Material	Labor	Total
$500 to $2,000, for the first $500	—	LS	—	—	23.50
each additional $100 or fraction, to $2,000	—	LS	—	—	3.05
$2,000 to $25,000, for the first $2,000	—	LS	—	—	69.30
each additional $1000 or fraction, to $25,000	—	LS	—	—	14.00
$25,000 to $50,000, for the first $25,000	—	LS	—	—	391.00
each additional $1000 or fraction, to $50,000	—	LS	—	—	10.10
$50,000 to $100,000, for the first $50,000	—	LS	—	—	644.00
each additional $1000 or fraction, to $100,000	—	LS	—	—	7.00
$100,000 to $500,000, for the first $100,000	—	LS	—	—	994.00
each additional $1000 or fraction, to $500,000	—	LS	—	—	5.60
$500,000 to $1,000,000, for the first $500,000	—	LS	—	—	3,230.00
each additional $1,000 or fraction, to $1,000,000	—	LS	—	—	4.75
For $1,000,000	—	LS	—	—	5,610.00
each additional $1,000 or fraction thereof	—	LS	—	—	3.15

Cabinets, Kitchen See also Vanities. Good quality mill-made modular units with solid hardwood face frames, hardwood door frames and drawer fronts, hardwood veneer on raised door panels (front and back), glued mortise, dowel, and dado joint construction, full backs (1/8" vinyl laminated plywood), vinyl laminated cabinet interiors, vinyl laminated composition drawer bodies with nylon and metal guides. Includes self-closing hinges, door and drawer pulls, mounting hardware and adjustable shelves. See illustrations for unit types. See the price adjustments below for pricing of other units. No countertops included. See Countertops.

Kitchen cabinet costs vary widely. The prices listed in this section are for standard grade cabinets. Add 65% to material costs for premium grade cabinets with solid hardwood fronts and frames, mitered corners and solid wood drawer bodies with steel guides and ball bearings. Deduct 45% from material costs for economy grade cabinets, melamine laminated to particleboard, in textured colors or woodgrain print finish.

	Craft@Hrs	Unit	Material	Labor	Total

Cabinets, Rule of Thumb Cabinet cost per running foot of cabinet installed. These figures are based on a set of mill-fabricated and assembled kitchen cabinets including a sink base cabinet, one 3-drawer base cabinet and six door base cabinets. Per linear foot of front or back edge, whichever is longer.

	Craft@Hrs	Unit	Material	Labor	Total
Base cabinets, 34-1/2" high, 24" deep	BC@.521	LF	128.00	15.60	143.60
Wall cabinets, 30" high, 12" deep	BC@.340	LF	66.80	10.20	77.00

Cabinets, Kitchen

	Craft@Hrs	Unit	Material	Labor	Total
Cabinets, Kitchen (See the note above concerning cabinet costs.)					
One door base cabinets, 34-1/2" high, 24" deep					
9" wide, tray divider	BC@.461	Ea	119.00	13.80	132.80
12" wide, 1 door, 1 drawer	BC@.461	Ea	161.00	13.80	174.80
15" wide, 1 door, 1 drawer	BC@.638	Ea	171.00	19.10	190.10
18" wide, 1 door, 1 drawer	BC@.766	Ea	181.00	23.00	204.00
21" wide, 1 door, 1 drawer	BC@.766	Ea	197.00	23.00	220.00
24" wide, 1 door, 1 drawer	BC@.911	Ea	202.00	27.30	229.30
Drawer base cabinets, 34-1/2" high, 24" deep					
15" wide, 4 drawers	BC@.638	Ea	176.00	19.10	195.10
18" wide, 4 drawers	BC@.766	Ea	187.00	23.00	210.00
24" wide, 4 drawers	BC@.911	Ea	217.00	27.30	244.30
Sink base cabinets, 34-1/2" high, 24" deep					
24" wide, 1 door, 1 drawer front	BC@.740	Ea	171.00	22.20	193.20
30" wide, 2 doors, 2 drawer fronts	BC@.766	Ea	207.00	23.00	230.00
33" wide, 2 doors, 2 drawer fronts	BC@.766	Ea	217.00	23.00	240.00
36" wide, 2 doors, 2 drawer fronts	BC@.766	Ea	222.00	23.00	245.00
42" wide, 2 doors, 2 drawer fronts	BC@.911	Ea	246.00	27.30	273.30
48" wide, 2 doors, 2 drawer fronts	BC@.911	Ea	270.00	27.30	297.30
Two door base cabinets, 34-1/2" high, 24" deep					
27" wide, 2 doors, 2 drawers	BC@1.25	Ea	270.00	37.50	307.50
30" wide, 2 doors, 2 drawers	BC@1.25	Ea	285.00	37.50	322.50
33" wide, 2 doors, 2 drawers	BC@1.25	Ea	298.00	37.50	335.50
36" wide, 2 doors, 2 drawers	BC@1.35	Ea	305.00	40.50	345.50
42" wide, 2 doors, 2 drawers	BC@1.50	Ea	327.00	45.00	372.00
48" wide, 2 doors, 2 drawers	BC@1.71	Ea	362.00	51.30	413.30
Blind corner base cabinets, 34-1/2" high					
Minimum 36", maximum 39" at wall	BC@1.39	Ea	197.00	41.70	238.70
Minimum 39", maximum 42" at wall	BC@1.50	Ea	212.00	45.00	257.00
45-degree corner base, revolving, 34-1/2" high					
36" wide at each wall	BC@2.12	Ea	305.00	63.60	368.60
Corner sink front, 34-1/2" high					
40" wide at walls	BC@2.63	Ea	171.00	78.90	249.90
Wall cabinets, adjustable shelves, 30" high, 12" deep					
9" wide, 1 door	BC@.461	Ea	109.00	13.80	122.80
12" wide or 15" wide, 1 door	BC@.461	Ea	114.00	13.80	127.80
18" wide, 1 door	BC@.638	Ea	139.00	19.10	158.10
21" wide, 1 door	BC@.638	Ea	145.00	19.10	164.10
24" wide, 1 door	BC@.766	Ea	156.00	23.00	179.00
27" wide, 2 doors	BC@.766	Ea	181.00	23.00	204.00
30" wide, 2 doors	BC@.911	Ea	181.00	27.30	208.30
33" wide, 2 doors	BC@.911	Ea	197.00	27.30	224.30
36" wide, 2 doors	BC@1.03	Ea	207.00	30.90	237.90
42" wide, 2 doors	BC@1.03	Ea	222.00	30.90	252.90
48" wide, 2 doors	BC@1.16	Ea	239.00	34.80	273.80
Above-appliance wall cabinets, 12" deep					
12" high, 30" wide, 2 doors	BC@.461	Ea	114.00	13.80	127.80
15" high, 30" wide, 2 doors	BC@.461	Ea	130.00	13.80	143.80
15" high, 33" wide, 2 doors	BC@.537	Ea	140.00	16.10	156.10
15" high, 36" wide, 2 doors	BC@.638	Ea	145.00	19.10	164.10
18" high, 18" wide, 2 doors	BC@.537	Ea	109.00	16.10	125.10
18" high, 30" wide, 2 doors	BC@.766	Ea	145.00	23.00	168.00
18" high, 36" wide, 2 doors	BC@.911	Ea	156.00	27.30	183.30

45 degree
corner base cabinet

1 door
wall cabinet

Above-appliance
wall cabinet

**Blind corner
wall cabinet**

Utility

**Single
oven
cabinet**   **Single
oven
cabinet**

	Craft@Hrs	Unit	Material	Labor	Total
Corner wall cabinets, 30" high, 12" deep					
24" at each wall, fixed shelves	BC@1.03	Ea	197.00	30.90	227.90
24" at each wall, revolving shelves	BC@1.03	Ea	265.00	30.90	295.90
Blind corner wall cabinets, 30" high					
24" minimum, 1 door	BC@1.03	Ea	150.00	30.90	180.90
36" minimum, 1 door	BC@1.32	Ea	181.00	39.60	220.60
42" minimum, 2 doors	BC@1.20	Ea	228.00	36.00	264.00
Utility cabinet, 66" high, 12" deep, no shelves					
18" wide	BC@1.32	Ea	228.00	39.60	267.60
24" wide	BC@1.71	Ea	259.00	51.30	310.30
Utility cabinet, 66" high, 24" deep, add shelf cost below					
18" wide	BC@1.24	Ea	239.00	37.20	276.20
24" wide	BC@1.71	Ea	285.00	51.30	336.30
Add for utility cabinet revolving shelves, includes mounting hardware					
18" wide x 24" deep	BC@.360	Ea	207.00	10.80	217.80
24" wide x 24" deep	BC@.360	Ea	239.00	10.80	249.80
Add for utility cabinet plain shelves					
18" wide x 24" deep	BC@.541	Ea	69.30	16.20	85.50
24" wide x 24" deep	BC@.541	Ea	72.50	16.20	88.70
Oven cabinets, 66" high, 24" deep					
27" wide, single oven	BC@2.19	Ea	285.00	65.70	350.70
27" wide, double oven	BC@2.19	Ea	216.00	65.70	281.70
Microwave cabinet, with trim,					
21" high, 20" deep, 30" wide	BC@.986	Ea	171.00	29.60	200.60

Additional labor costs for cabinets

	Craft@Hrs	Unit	Material	Labor	Total
1/4" skins applied to cabinet ends	BC@.200	Ea	—	6.00	6.00
3/4" raised end panels applied to cabinet ends	BC@.250	Ea	—	7.50	7.50
Refrigerator end panels, cut, fit, install	BC@.500	Ea	—	15.00	15.00
Corbels installed under countertop overhangs	BC@.200	Ea	—	6.00	6.00
Filler strips between cabinets (per strip)	BC@.200	Ea	—	6.00	6.00

Carpentry See also carpentry items in other sections: Carpentry Steel Framing, Cabinets, Ceilings, Closet Door Systems, Countertops, Cupolas, Doors, Entrances, Flooring, Framing Connectors, Hardboard, Hardware, Lumber, Moulding, Paneling, Shutters, Siding, Skylights, Soffits, Stairs, Thresholds, and Weatherstripping.

Carpentry Rule of Thumb Typical rough carpentry (framing) cost per square foot of floor in living area. These figures will apply on many residential jobs where joists, studs and rafters are 16" on center. Unconventional designs and complex framing plans will cost more. See detailed cost breakdowns in the next section.

	Craft@Hrs	Unit	Material	Labor	Total
Using framing lumber at	—	MBF	659.00	—	659.00
Using 7/16" OSB sheathing at	—	MSF	364.00	—	364.00
Single story, conventional foundation	B1@.237	SF	4.98	6.48	11.46
Single story, concrete slab foundation	B1@.180	SF	3.48	4.92	8.40
First of two floors, conventional foundation	B1@.203	SF	3.84	5.55	9.39
First of two floors, concrete slab foundation	B1@.146	SF	2.33	3.99	6.32
Second floor of a two-story residence	B1@.191	SF	3.90	5.23	9.13

Carpentry

	Craft@Hrs	Unit	Material	Labor	Total
Framing a single story residence, conventional crawl-space foundation					
Sills, pier blocks, floor beams					
(145 BF per 1,000 SF)	B1@.018	SF	.10	.49	.59
Floor joists, doublers, blocking, bridging					
(1,480 BF per 1,000 SF)	B1@.028	SF	.98	.77	1.75
Subflooring, 7/16" OSB					
(1,150 SF per 1,000 SF)	B1@.011	SF	.42	.30	.72
Layout, studs, sole plates, top plates, header and					
end joists, backing, blocking, bracing and framing for openings					
(2,250 BF per 1,000 SF)	B1@.093	SF	1.48	2.54	4.02
Ceiling joists, header and end joists, backing, blocking and bracing					
(1,060 BF per 1,000 SF)	B1@.045	SF	.70	1.23	1.93
Rafters, braces, collar beams, ridge boards, 2" x 8" rafters					
16" OC, (1,340 BF per 1000 SF)	B1@.032	SF	.88	.88	1.76
Roof sheathing, 7/16" OSB					
(1,150 SF per 1,000 SF)	B1@.010	SF	.42	.27	.69
Total framing, single story, conventional foundation	B1@.237	SF	4.98	6.48	11.46
Framing a single story residence, concrete slab foundation					
Layout, sole plates, anchors, studs, top plates, header and					
end joists, backing, blocking, bracing and framing for openings					
(2,250 BF per 1,000 SF)	B1@.093	SF	1.48	2.54	4.02
Ceiling joists, header and end joists, backing, blocking and bracing					
(1,060 BF per 1,000 SF)	B1@.045	SF	.70	1.23	1.93
Rafters, braces, collar beams, ridge boards, 2" x 8" rafters					
16" OC, (1,340 BF per 1,000 SF)	B1@.032	SF	.88	.88	1.76
Roof sheathing, 7/16" OSB					
(1,150 SF per 1,000 SF)	B1@.010	SF	.42	.27	.69
Total framing, single story, concrete slab foundation	B1@.180	SF	3.48	4.92	8.40
Framing the first of two floors, conventional crawl-space foundation					
Sills, pier blocks, floor beams					
(145 BF per 1,000 SF)	B1@.018	SF	.10	.49	.59
Floor joists, doublers, blocking, bridging					
(1,480 BF per 1,000 SF)	B1@.028	SF	.98	.77	1.75
Subflooring, 7/16" OSB					
(1,150 SF per 1,000 SF)	B1@.011	SF	.42	.30	.72
Layout, studs, sole plates, top plates, header and					
end joists, backing, blocking, bracing and framing for openings					
(2,250 BF per 1,000 SF)	B1@.093	SF	1.49	2.54	4.03
Rough stairway, 15 risers and landing (96 BF of dimension lumber					
and 128 SF of plywood per 1,000 SF)	B1@.008	SF	.15	.22	.37
Ceiling joists, header and end joists, backing, blocking and bracing					
(1,060 BF per 1,000 SF)	B1@.045	SF	.70	1.23	1.93
Total framing, 1st of 2 floors, crawl-space foundation	B1@.203	SF	3.84	5.55	9.39

	Craft@Hrs	Unit	Material	Labor	Total
Framing the first of two floors, concrete slab foundation					
Layout, sole plates, anchors, studs, top plates, header and end joists, backing, blocking, bracing and framing for openings (2,250 BF per 1,000 SF)	B1@.093	SF	1.48	2.54	4.02
Rough stairway, 15 risers and landing (96 BF of dimension lumber and 128 SF of plywood per 1,000 SF)	B1@.008	SF	.15	.22	.37
Ceiling joists, header and end joists, backing, blocking and bracing (1,060 BF per 1,000 SF)	B1@.045	SF	.70	1.23	1.93
Total framing, first of two floors, concrete foundation	B1@.146	SF	2.33	3.99	6.32
Framing the second story of a residence					
Subflooring, 7/16" OSB (1,150 SF per 1,000 SF)	B1@.011	SF	.42	.30	.72
Layout, studs, sole plates, top plates, header and end joists, backing, blocking, bracing and framing for openings (2,250 BF per 1,000 SF)	B1@.093	SF	1.48	2.54	4.02
Ceiling joists, header and end joists, backing, blocking and bracing (1,060 BF per 1,000 SF)	B1@.045	SF	.70	1.23	1.93
Rafters, braces, collar beams, ridge boards, 2" x 8" rafters 16" OC (1,340 BF per 1,000 SF)	B1@.032	SF	.88	.88	1.76
Roof sheathing, 7/16" OSB (1,150 SF per 1,000 SF)	B1@.010	SF	.42	.27	.69
Total framing, second floor of a two-story residence	B1@.191	SF	3.90	5.23	9.13

Floor Assemblies Costs for wood framed floor joists with subflooring and R-19 insulation. These costs include the floor joists, subflooring as described, blocking, nails and 6-1/4" thick R-19 fiberglass insulation between the floor joists. Figures include box or band joists and typical double joists. No beams included. Planked subflooring is based on 1.24 BF per square foot of floor. Costs shown are per square foot of area covered and include normal waste. Deduct for openings over 25 SF.

	Craft@Hrs	Unit	Material	Labor	Total
Using framing lumber at	—	MBF	659.00	—	659.00
Using 5/8" CDX plywood subfloor at	—	MSF	606.00	—	606.00
Using 3/4" CDX plywood subfloor at	—	MSF	756.00	—	756.00
Using 7/16" OSB subfloor at	—	MSF	364.00	—	364.00
Using 5/8" OSB subfloor at	—	MSF	555.00	—	555.00
Using 3/4" OSB T&G subfloor at	—	MSF	675.00	—	675.00
Floor joists 16" OC, R-19 insulation and OSB subflooring					
7/16" OSB subfloor					
2" x 6" joists	B1@.040	SF	1.70	1.09	2.79
2" x 8" joists	B1@.041	SF	2.05	1.12	3.17
2" x 10" joists	B1@.043	SF	2.42	1.18	3.60
2" x 12" joists	B1@.044	SF	3.00	1.20	4.20
Floor joists 16" OC, R-19 insulation and OSB subflooring					
5/8" OSB subfloor					
2" x 6" joists	B1@.040	SF	1.89	1.09	2.98
2" x 8" joists	B1@.041	SF	2.24	1.12	3.36
2" x 10" joists	B1@.043	SF	2.61	1.18	3.79
2" x 12" joists	B1@.044	SF	3.19	1.20	4.39

Carpentry Assemblies

	Craft@Hrs	Unit	Material	Labor	Total
3/4" OSB subfloor					
2" x 6" joists	B1@.042	SF	2.01	1.15	3.16
2" x 8" joists	B1@.043	SF	2.36	1.18	3.54
2" x 10" joists	B1@.045	SF	2.73	1.23	3.96
2" x 12" joists	B1@.046	SF	3.31	1.26	4.57
Floor joists 16" OC, R-19 insulation and plywood subflooring					
5/8" plywood subfloor					
2" x 6" joists	B1@.040	SF	1.91	1.09	3.00
2" x 8" joists	B1@.041	SF	2.26	1.12	3.38
2" x 10" joists	B1@.043	SF	2.64	1.18	3.82
2" x 12" joists	B1@.044	SF	3.21	1.20	4.41
3/4" plywood subfloor					
2" x 6" joists	B1@.042	SF	2.09	1.15	3.24
2" x 8" joists	B1@.043	SF	2.44	1.18	3.62
2" x 10" joists	B1@.045	SF	2.89	1.23	4.12
2" x 12" joists	B1@.046	SF	3.39	1.26	4.65
For different type insulation					
Fiberglass batts					
10" thick R-30, add	—	SF	.28	—	.28
12" thick R-38, add	—	SF	.37	—	.37

Wall Assemblies Costs for wood framed stud walls with wall finish treatment on both sides. These costs include wall studs at 16" center to center, double top plates, single bottom plates, fire blocking, nails and wall finish treatment as described. No headers or posts included. All lumber is Std & Btr. 2" x 4" walls have 1.12 BF per SF of wall and 2" x 6" walls have 1.68 BF per SF of wall. Costs shown are per SF or LF of wall measured on one face and include normal waste.

Using 2 x 4 lumber at	—	MBF	621.00	—	621.00
Using 2 x 6 lumber at	—	MBF	620.00	—	620.00
Using 1/2" gypsum wallboard at	—	SF	.27	—	.27

Interior wall assemblies

2" x 4" stud walls with 1/2" gypsum drywall both sides, ready for painting

	Craft@Hrs	Unit	Material	Labor	Total
Cost per square foot of wall	B1@.064	SF	1.05	1.75	2.80
Cost per running foot, for 8' high walls	B1@.512	LF	8.36	14.00	22.36

2" x 4" stud walls with 5/8" gypsum fire rated drywall both sides, ready for painting

Cost per square foot of wall	B1@.068	SF	1.19	1.86	3.05
Cost per running foot, for 8' high walls	B1@.544	LF	9.48	14.90	24.38

2" x 6" stud walls with 1/2" gypsum drywall both sides, ready for painting

Cost per square foot of wall	B1@.072	SF	1.31	1.97	3.28
Cost per running foot, for 8' high walls	B1@.576	LF	10.50	15.80	26.30

2" x 6" stud walls with 5/8" gypsum fire rated drywall both sides, ready for painting

Cost per square foot of wall	B1@.076	SF	1.45	2.08	3.53
Cost per running foot, for 8' high walls	B1@.608	LF	11.60	16.60	28.20

Exterior wall assemblies

2" x 4" stud walls with drywall interior, wood siding exterior, 1/2" gypsum drywall inside face ready for painting, over 3-1/2" R-11 insulation with 5/8" thick rough sawn T-1-11 exterior grade plywood siding on the outside face.

Using 5/8" rough sawn T-1-11 siding at	—	MSF	763.00	—	763.00
Cost per square foot of wall	B1@.068	SF	1.81	1.86	3.67
Cost per running foot, for 8' high walls	B1@.544	LF	14.50	14.90	29.40

	Craft@Hrs	Unit	Material	Labor	Total

2" x 6" stud walls with drywall interior, wood siding exterior, same construction as above, except with 6-1/4" R-19 insulation

Cost per square foot of wall	B1@.077	SF	2.20	2.11	4.31
Cost per running foot, for 8' high walls	B1@.616	LF	17.60	16.90	34.50

2" x 4" stud walls with drywall interior, 1/2" gypsum drywall on inside face ready for painting, over 3-1/2" R-11 insulation with 1" x 6" southern yellow pine drop siding, D grade, 1.19 BF per SF at 5-1/4" exposure on the outside face.

Using D grade yellow pine drop siding at	—	MSF	2,090.00	—	2,090.00
Cost per square foot of wall	B1@.074	SF	3.14	2.02	5.16
Cost per running foot, for 8' high wall	B1@.592	LF	25.10	16.20	41.30

2" x 6" stud walls with drywall interior, 1" x 6" drop siding exterior, same construction as above, except with 6-1/4" R-19 insulation

Cost per square foot of wall	B1@.083	SF	3.53	2.27	5.80
Cost per running foot, for 8' high wall	B1@.664	LF	28.30	18.20	46.50

2" x 4" stud walls with drywall interior, stucco exterior, 1/2" gypsum drywall on inside face ready for painting, over 3-1/2" R-11 insulation and a three-coat exterior plaster (stucco) finish with integral color on the outside face

Cost per square foot of wall	B1@.050	SF	3.32	1.37	4.69
Cost per running foot, for 8' high wall	B1@.400	LF	26.60	10.90	37.50

2" x 6" stud walls with drywall interior, stucco exterior, same construction as above, except with 6-1/4" R-19 insulation

Cost per square foot of wall	B1@.059	SF	3.72	1.61	5.33
Cost per running foot, for 8' high wall	B1@.472	LF	29.70	12.90	42.60

Add for different type gypsum board
   1/2" or 5/8" moisture resistant greenboard

Cost per SF, greenboard per side, add	—	SF	.10	—	.10

   1/2" or 5/8" moisture resistant greenboard

Cost per running foot per side 8' high	—	LF	.80	—	.80

   5/8" thick fire rated type X gypsum drywall

Cost per SF, per side, add	—	SF	.12	—	.12

   5/8" thick fire rated type X gypsum drywall

Cost per running foot per side 8' high	—	LF	.96	—	.96

Ceiling Assemblies Costs for wood framed ceiling joists with ceiling finish and fiberglass insulation, based on performing the work at the construction site. These costs include the ceiling joists, ceiling finish as described, blocking, nails and 3-1/2" thick R-11 fiberglass insulation batts between the ceiling joists. Figures in parentheses indicate board feet per square foot of ceiling framing including end joists and typical header joists. No beams included. Ceiling joists and blocking are based on standard and better grade lumber. Costs shown are per square foot of area covered and include normal waste. Deduct for openings over 25 SF.

Ceiling joists with regular gypsum drywall taped and sanded smooth finish, ready for paint

Using 2" x 4" at	—	MBF	621.00	—	621.00
Using 2" x 6" at	—	MBF	620.00	—	620.00
Using 2" x 8" at	—	MBF	640.00	—	640.00

2" x 4" ceiling joists at 16" on center (.59 BF per SF), with

insulation and 1/2" gypsum drywall	B1@.053	SF	1.11	1.45	2.56

2" x 6" ceiling joists at 16" on center (.88 BF per SF), with

insulation and 1/2" gypsum drywall	B1@.055	SF	1.39	1.50	2.89

2" x 8" ceiling joists at 16" on center (1.17 BF per SF), with

insulation and 1/2" gypsum drywall	B1@.057	SF	1.70	1.56	3.26

For spray applied plaster finish (sometimes called "popcorn" or "cottage cheese" texture)

Add for ceiling texture	DT@.003	SF	.12	.09	.21

	Craft@Hrs	Unit	Material	Labor	Total
For different type gypsum drywall with taped and sanded smooth finish, ready for paint					
1/2" moisture resistant greenboard, add	—	SF	.10	—	.10
5/8" fire rated board, add	—	SF	.12	—	.12
For different ceiling joist center to center dimensions					
2" x 4" ceiling joists					
12" on center, add	B1@.004	SF	.12	.11	.23
20" on center, deduct	—	SF	-.07	-.06	-.13
24" on center, deduct	—	SF	-.11	-.12	-.23
2" x 6" ceiling joists					
12" on center, add	B1@.006	SF	.17	.16	.33
20" on center, deduct	—	SF	-.10	-.06	-.16
24" on center, deduct	—	SF	-.15	-.11	-.27
2" x 8" ceiling joists					
12" on center, add	B1@.006	SF	.23	.16	.39
20" on center, deduct	—	SF	-.13	-.08	-.22
24" on center, deduct	—	SF	-.21	-.13	-.34
For different type insulation					
Fiberglass batts					
6-1/4" thick R-19, add	—	SF	.13	—	.13
10" thick R-30, add	—	SF	.29	—	.29
Blown-in fiberglass					
8" thick R-19, add	—	SF	.04	—	.04

Roofing Assemblies Costs for wood framed roof assemblies with roof finish material as shown based on performing the work at the construction site. Costs shown include all material and labor required above the top plate or ledger on the supporting walls. These costs assume the supporting wall structure is in-place and suitable for the assembly described.

Flat roof assembly Based on using 2" x 12" joists Std & Btr grade at 16" on center including blocking, ripped strips and normal waste. The roof surface is built-up 3 ply asphalt consisting of 2 plies of 15 lb. felt hot mopped with a 90 lb. mineral-coated cap sheet.

	Craft@Hrs	Unit	Material	Labor	Total
Using 2" x 12" joists at	—	MBF	738.00	—	738.00
Using 2" x 8" joists at	—	MBF	640.00	—	640.00
Using 1/2" CDX plywood at	—	MSF	494.00	—	494.00
Using 7/16" OSB at	—	MSF	365.00	—	365.00
Flat roof assembly as described above					
Framing, (w/ plywood sheathing), per 100 SF	B1@5.00	Sq	235.00	137.00	372.00
Framing, (w/ OSB sheathing), per 100 SF	B1@5.00	Sq	221.00	137.00	358.00
Built-up roofing, per 100 square feet	R1@1.25	Sq	38.10	36.50	74.60

Conventionally framed roof assemblies Based on straight gable type roof (no hips, valleys, or dormers) with 6" in 12" rise or less. Cost per 100 square feet of plan area under the roof, not actual roof surface area. Framing includes 2" x 8" common rafters Std & Btr grade at 24" on center. Cost includes blocking, ridge and normal bracing. Roof sheathing is either 7/16" OSB (oriented strand board) or 1/2" CDX plywood.

Conventionally framed roof assembly with built-up rock-finish roofing. The finished roof treatment is built-up 3 ply asphalt consisting of 1 ply 30 lb. felt, 2 plies of 15 lb. felt and 3 hot mop coats of asphalt with decorative crush rock spread at 180 lbs. per 100 square feet.

	Craft@Hrs	Unit	Material	Labor	Total
Framing, (w/ plywood sheathing), per 100 SF	B1@4.10	Sq	126.00	112.00	238.00
Framing, (w/ OSB sheathing), per 100 SF	B1@4.10	Sq	112.00	112.00	224.00
Built-up rock finish roofing, per 100 square feet	R1@2.15	Sq	58.10	62.70	120.80

	Craft@Hrs	Unit	Material	Labor	Total
Conventionally framed roof assembly with 25 year fiberglass shingle roofing					
Framing, (w/ plywood sheathing), per 100 SF	B1@4.10	Sq	126.00	112.00	238.00
Framing, (w/ OSB sheathing), per 100 SF	B1@4.10	Sq	112.00	112.00	224.00
Fiberglass shingle roofing, per 100 square feet	B1@1.83	Sq	59.90	50.10	110.00
Conventionally framed roof assembly with 25 year asphalt shingle roofing					
Framing, (w/ plywood sheathing), per 100 SF	B1@4.10	Sq	126.00	112.00	238.00
Framing, (w/ OSB sheathing), per 100 SF	B1@4.10	Sq	112.00	112.00	224.00
Asphalt shingle roofing, per 100 square feet	B1@2.07	Sq	110.00	56.60	166.60
Conventionally framed roof assembly, 28 gauge galvanized corrugated steel roofing, no sheathing					
Framing, (w/ plywood sheathing), per 100 SF	B1@4.10	Sq	126.00	112.00	238.00
Framing, (w/ OSB sheathing), per 100 SF	B1@4.10	Sq	112.00	112.00	224.00
28 gauge steel roofing, per 100 square feet	B1@2.70	Sq	104.00	73.90	177.90
Add for gable studs					
2" x 4" spaced 16" OC (.54 BF per SF)	B1@.023	SF	.34	.63	.97
Add for purlins (purling), Std & Btr, installed below roof rafters. Figures in parentheses indicate board feet per LF including 5% waste					
Using 2" x 8" joists at	—	MBF	640.00	—	640.00
2" x 8" (1.40 BF per LF)	B1@.023	LF	.90	.63	1.53

Piecework Rough Carpentry Rough carpentry on residential tracts is usually done by framing subcontractors who bid at piecework rates (such as per square foot of floor). The figures below list typical piecework rates for repetitive framing work and assume all materials are supplied to the framing subcontractor. No figures appear in the Craft@Hrs column because the work is done for a fixed price per square foot and the labor productivity can be expected to vary widely.

Layout and plating Piecework rates
Lay out wall plates according to the plans (snap chalk lines for wall plates, mark location for studs, windows, doors and framing details), cut top and bottom plates and install bottom plates. Costs per square foot of floor (excluding garage).

Custom or more complex jobs	—	SF	—	.22	.22
Larger job, longer runs	—	SF	—	.12	.12

Wall framing Piecework rates
Measure, cut, fit, assemble, and tip up walls, including studs, plates, cripples, let-in braces, trimmers and blocking. Costs per square foot of floor.

Complex job, radius walls, rake walls	—	SF	—	.61	.61
Larger job, 8' high walls, fewer partitions	—	SF	—	.22	.22

Plumb and align framed walls Piecework rates
Force walls into alignment, adjust walls to vertical and install temporary wood braces as needed, fasten nuts on anchor bolts or shoot power driven fasteners through wall plates into the slab. Based on accuracy to 3/16". Costs per square foot of floor are shown below.

Small or complex job, many braces	—	SF	—	.25	.25
Larger job, fewer braces	—	SF	—	.12	.12

	Craft@Hrs	Unit	Material	Labor	Total

Floor joists or ceiling joists Piecework rates

Lay out, cut and install floor or ceiling joists, including rim joists, doubled joists, straps, joist hangers, blocking at 8' OC and ceiling backing for drywall. Based on larger jobs with simple joist layouts set 16" OC and pre-cut blocking supplied by the general contractor. Add the cost of floor beams, if required. Costs per square foot of horizontal joist area. More complex jobs with shorter runs may cost 50% more.

	Craft@Hrs	Unit	Material	Labor	Total
2" x 8" ceiling or floor joists	—	SF	—	.16	.16
2" x 10" ceiling or floor joists	—	SF	—	.19	.19
2" x 12" ceiling or floor joists	—	SF	—	.21	.21
2" x 14" ceiling or floor joists	—	SF	—	.25	.25
Add for 12" OC spacing	—	SF	—	.06	.06
Deduct for 20" OC spacing	—	SF	—	-.03	-.03
Deduct for 24" OC spacing	—	SF	—	-.06	-.06

Floor sheathing Piecework rates

Lay out, cut, fit and install 5/8" or 3/4" tongue and groove plywood floor sheathing, including blocking as required. Based on nailing done with a pneumatic nailer and nails supplied by the general contractor. Costs per square foot of sheathing installed

	Craft@Hrs	Unit	Material	Labor	Total
Smaller, cut-up job	—	SF	—	.19	.19
Larger job, longer runs	—	SF	—	.14	.14
Add for 1-1/8" sheathing	—	SF	—	.05	.05

Stair framing Piecework rates

Lay out, cut, fit and install straight, U- or L-shaped 30" to 36" wide stairs made from plywood and 2" x 12" stringers set 16" OC. These costs include blocking in the adjacent stud wall and a 1" skirt board. Costs per 7-1/2" riser. Framing more complex stairs may cost up to $500 per flight.

	Craft@Hrs	Unit	Material	Labor	Total
Small job, short runs	—	Ea	—	14.20	14.20
Larger job, longer runs	—	Ea	—	10.70	10.70
Add per 3' x 3' landing, including supports	—	Ea	—	36.60	36.60
Add for larger landings, including supports	—	Ea	—	90.90	90.90

Shear panels Piecework rates

Lay out, cut, fit and install structural 3/8" or 1/2" OSB or plywood wall panels. These figures assume shear studs were set correctly by others and that panel nails are driven at 4" OC with a pneumatic nailer. Not including hold-down straps, posts, shear blocking or extra studs. Costs per 4' x 9' panel installed.

	Craft@Hrs	Unit	Material	Labor	Total
Small job, many openings, 2nd floor	—	Ea	—	9.76	9.76
Larger job, few openings, 1st floor	—	Ea	—	6.08	6.08

Roof trusses Piecework rates

Setting and nailing engineered gable and hip roof trusses 24" OC on prepared wall plates. These figures assume that lifting equipment is provided by the general contractor and that the truss supplier provides a fill package, spreader blocks for each plate and the ridge and jack rafters (if required). Includes installation of ceiling backing where required and catwalks at the bottom chord. Costs per square foot of plan area under the truss.

Small job assumes multiple California fill between roof surfaces and understacking

	Craft@Hrs	Unit	Material	Labor	Total
Small job, rake fill above a partition wall	—	SF	—	.55	.55
Larger job, little or no fill or understacking	—	SF	—	.32	.32

	Craft@Hrs	Unit	Material	Labor	Total

Conventional roof framing Piecework rates
Calculate lengths, lay out, cut and install 2" x 10" or 2" x 12" common, hip, valley and jack rafters on parallel and horizontal plates. Costs per square foot of plan area under the roof.

	Craft@Hrs	Unit	Material	Labor	Total
Small job, cut-up roof, few common rafters					
Rafters 12" OC	—	SF	—	1.09	1.09
Rafters 16" OC	—	SF	—	.92	.92
Rafters 20" OC	—	SF	—	.73	.73
Rafters 24" OC	—	SF	—	.55	.55
Larger job, longer runs, nearly all common rafters					
Rafters 12" OC	—	SF	—	.55	.55
Rafters 16" OC	—	SF	—	.48	.48
Rafters 20" OC	—	SF	—	.37	.37
Rafters 24" OC	—	SF	—	.32	.32
Add for slope over 6 in 12	—	SF	—	.17	.17
Deduct for 2" x 6" or 2" x 8" rafters	—	SF	—	-.06	-.06

Fascia Piecework rates
Applied to rafter tails and as a barge rafter on gable ends. Includes trimming the rafter tails to the correct length and installing outlookers at gable ends. Costs per linear foot of 2" x 8" fascia installed.

	Craft@Hrs	Unit	Material	Labor	Total
Small job, short runs, with moulding	—	LF	—	2.37	2.37
Larger job, longer runs	—	LF	—	1.36	1.36

Roof sheathing Piecework rates
Lay out, cut, fit and install 1/2" or 5/8" OSB or plywood roof sheathing, including blocking and 1" x 8" starter board on overhangs as required. Based on nailing done with a pneumatic nailer and nails supplied by the general contractor. Costs per square foot of sheathing installed.

	Craft@Hrs	Unit	Material	Labor	Total
Smaller, cut-up hip and valley job	—	SF	—	.20	.20
Larger job, longer runs	—	SF	—	.14	.14
Add for slope over 6 in 12	—	SF	—	.08	.08

Carpentry Cost, Detailed Breakdown This section is arranged in the order of construction. Material costs shown here can be adjusted to reflect your actual lumber cost: divide your actual lumber cost (per MBF) by the cost listed (per MBF). Then multiply the cost in the material column by this adjustment factor. No waste included.

Lally columns (Residential basement column) 3-1/2" diameter, steel tube

	Craft@Hrs	Unit	Material	Labor	Total
Material only	—	LF	7.80	—	7.80
Add for base or cap plate	—	Ea	3.10	—	3.10
Add for installation, per column to 12' high	B1@.458	Ea	—	12.50	12.50

Precast pier blocks Posts set on precast concrete pier block, including pier block with anchor placed on existing grade, temporary 1" x 6" bracing (8 LF) and stakes (2). Cost is for each post set. Add for excavation if required

	Craft@Hrs	Unit	Material	Labor	Total
Heights to 8', cost of post not included	BL@.166	Ea	5.60	4.10	9.70

Pier pads 2" x 6", treated, #2 & Btr

	Craft@Hrs	Unit	Material	Labor	Total
Using #2 & Btr treated lumber at	—	MBF	837.00	—	837.00
1.10 BF per LF	B1@.034	LF	.92	.93	1.85

Carpentry, Detailed Breakdown

	Craft@Hrs	Unit	Material	Labor	Total
Posts 4" x 4", material costs include 10% waste (1.47 BF per LF). See also Lally columns above and Posts in the Lumber section					
Fir, rough Std & Btr, K.D.	—	MBF	960.00	—	960.00
Fir, rough Std & Btr, K.D. 4" x 4"	—	LF	1.41	—	1.41
Red cedar, rough green constr.	—	MBF	1,210.00	—	1,210.00
Red cedar, rough green constr.	—	LF	1.78	—	1.78
Redwood, S4S construction heart	—	MBF	1,710.00	—	1,710.00
Redwood, S4S construction heart	—	LF	2.51	—	2.51
Southern yellow pine #2, treated	—	MBF	990.00	—	990.00
Southern yellow pine #2, treated	—	LF	1.46	—	1.46

Girders Standard and better lumber, first floor work. Figures in parentheses show board feet per linear foot of girder, including 7% waste

	Craft@Hrs	Unit	Material	Labor	Total
4" x 6", per MBF	B1@15.8	MBF	975.00	432.00	1,407.00
4" x 8", 10", 12", per MBF	B1@15.8	MBF	812.00	432.00	1,244.00
6" x 6", per MBF	B1@15.8	MBF	1,280.00	432.00	1,712.00
6" x 8", 10", 12", 8" x 8", per MBF	B1@15.8	MBF	1,280.00	432.00	1,712.00
4" x 6" (2.15 BF per LF)	B1@.034	LF	2.10	.93	3.03
4" x 8" (2.85 BF per LF)	B1@.045	LF	2.31	1.23	3.54
4" x 10" (3.58 BF per LF)	B1@.057	LF	2.91	1.56	4.47
4" x 12" (4.28 BF per LF)	B1@.067	LF	3.47	1.83	5.30
6" x 6" (3.21 BF per LF)	B1@.051	LF	4.11	1.40	5.51
6" x 8" (4.28 BF per LF)	B1@.067	LF	5.48	1.83	7.31
6" x 10" (5.35 BF per LF)	B1@.083	LF	6.85	2.27	9.12
6" x 12" (6.42 BF per LF)	B1@.098	LF	8.22	2.68	10.90
8" x 8" (5.71 BF per LF)	B1@.088	LF	7.31	2.41	9.72

Sill plates (At foundation) SYP #2 pressure treated lumber, drilled and installed with foundation bolts at 48" OC, no bolts, nuts or washers included. See also plates in this section. Figures in parentheses indicate board feet per LF of foundation, including 5% waste and wolmanized treatment.

	Craft@Hrs	Unit	Material	Labor	Total
Sill plates, per MBF	B1@22.9	MBF	910.00	627.00	1,537.00
2" x 3" (.53 BF per LF)	B1@.020	LF	.42	.55	.97
2" x 4" (.70 BF per LF)	B1@.023	LF	.66	.63	1.29
2" x 6" (1.05 BF per LF)	B1@.024	LF	.91	.66	1.57
2" x 8" (1.40 BF per LF)	B1@.031	LF	1.28	.85	2.13

Floor joists Per SF of area covered. Figures in parentheses indicate board feet per square foot of floor including box or band joist, typical double joists, and 6% waste. No beams, blocking or bridging included. Deduct for openings over 25 SF. Costs shown are based on a job with 1,000 SF of area covered. For scheduling purposes, estimate that a two-man crew can complete 750 SF of area per 8-hour day for 12" center to center framing; 925 SF for 16" OC; 1,100 SF for 20" OC; or 1,250 SF for 24" OC.

2" x 6" Std & Btr

	Craft@Hrs	Unit	Material	Labor	Total
2" x 6" floor joists, per MBF	B1@16.4	MBF	620.00	449.00	1,069.00
12" centers (1.28 BF per SF)	B1@.021	SF	.79	.57	1.36
16" centers (1.02 BF per SF)	B1@.017	SF	.63	.47	1.10
20" centers (.88 BF per SF)	B1@.014	SF	.55	.38	.93
24" centers (.73 BF per SF)	B1@.012	SF	.45	.33	.78

2" x 8" Std & Btr

	Craft@Hrs	Unit	Material	Labor	Total
2" x 8" floor joists, per MBF	B1@13.4	MBF	640.00	367.00	1,007.00
12" centers (1.71 BF per SF)	B1@.023	SF	1.09	.63	1.72
16" centers (1.36 BF per SF)	B1@.018	SF	.87	.49	1.36
20" centers (1.17 BF per SF)	B1@.015	SF	.75	.41	1.16
24" centers (1.03 BF per SF)	B1@.013	SF	.66	.36	1.02

	Craft@Hrs	Unit	Material	Labor	Total
2" x 10" Std & Btr					
2" x 10" floor joists, per MBF	B1@11.7	MBF	657.00	320.00	977.00
12" centers (2.14 BF per SF)	B1@.025	SF	1.41	.68	2.09
16" centers (1.71 BF per SF)	B1@.020	SF	1.12	.55	1.67
20" centers (1.48 BF per SF)	B1@.016	SF	.97	.44	1.41
24" centers (1.30 BF per SF)	B1@.014	SF	.85	.38	1.23
2" x 12" Std & Btr					
2" x 12" floor joists, per MBF	B1@10.2	MBF	738.00	279.00	1,017.00
12" centers (2.56 BF per SF)	B1@.026	SF	1.89	.71	2.60
16" centers (2.05 BF per SF)	B1@.021	SF	1.51	.57	2.08
20" centers (1.77 BF per SF)	B1@.017	SF	1.31	.47	1.78
24" centers (1.56 BF per SF)	B1@.015	SF	1.15	.41	1.56

Floor joist wood, TJI truss type Suitable for residential use, 50 PSF floor load design. Costs shown are per square foot (SF) of floor area, based on joists at 16" OC, for a job with 1,000 SF of floor area. Figure 1.22 SF of floor area for each LF of joist. Add the cost of beams, supports and blocking. For scheduling purposes, estimate that a two-man crew can install 900 to 950 SF of joists in an 8-hour day.

	Craft@Hrs	Unit	Material	Labor	Total
9-1/2" TJI/15	B1@.017	SF	1.93	.47	2.40
11-7/8" TJI/15	B1@.017	SF	2.12	.47	2.59
14" TJI/35	B1@.018	SF	3.06	.49	3.55
16" TJI/35	B1@.018	SF	3.34	.49	3.83

Bridging or blocking Installed between 2" x 6" thru 2" x 12" joists. Costs shown are per each set of cross bridges or per each block for solid bridging, and include normal waste. The spacing between the bridging or blocking, sometimes called a "bay," depends on job requirements. Labor costs assume bridging is cut to size on site.

	Craft@Hrs	Unit	Material	Labor	Total
1" x 4" cross					
Joist bridging, per MBF	—	MBF	855.00	—	855.00
Joists on 12" centers	B1@.034	Ea	.43	.93	1.36
Joists on 16" centers	B1@.034	Ea	.58	.93	1.51
Joists on 20" centers	B1@.034	Ea	.74	.93	1.67
Joists on 24" centers	B1@.034	Ea	.90	.93	1.83
2" x 6" solid, Std & Btr					
2" x 6" blocking, per MBF	—	MBF	620.00	—	620.00
Joists on 12" centers	B1@.042	Ea	.68	1.15	1.83
Joists on 16" centers	B1@.042	Ea	.91	1.15	2.06
Joists on 20" centers	B1@.042	Ea	1.14	1.15	2.29
Joists on 24" centers	B1@.042	Ea	1.36	1.15	2.51
2" x 8" solid, Std & Btr					
2" x 8" blocking, per MBF	—	MBF	640.00	—	640.00
Joists on 12" centers	B1@.042	Ea	.94	1.15	2.09
Joists on 16" centers	B1@.042	Ea	1.25	1.15	2.40
Joists on 20" centers	B1@.042	Ea	1.56	1.15	2.71
Joists on 24" centers	B1@.042	Ea	1.88	1.15	3.03
2" x 10" solid, Std & Btr					
2" x 10" blocking, per MBF	—	MBF	657.00	—	657.00
Joists on 12" centers	B1@.057	Ea	1.20	1.56	2.76
Joists on 16" centers	B1@.057	Ea	1.61	1.56	3.17
Joists on 20" centers	B1@.057	Ea	2.01	1.56	3.57
Joists on 24" centers	B1@.057	Ea	2.41	1.56	3.97

	Craft@Hrs	Unit	Material	Labor	Total
2" x 12" solid, Std & Btr					
2" x 12" blocking, per MBF	—	MBF	738.00	—	738.00
Joists on 12" centers	B1@.057	Ea	1.62	1.56	3.18
Joists on 16" centers	B1@.057	Ea	2.16	1.56	3.72
Joists on 20" centers	B1@.057	Ea	2.70	1.56	4.26
Joists on 24" centers	B1@.057	Ea	3.25	1.56	4.81
Steel, no nail type, cross					
Wood joists on 12" centers	B1@.020	Ea	2.46	.55	3.01
Wood joists on 16" centers	B1@.020	Ea	2.56	.55	3.11
Wood joists on 20" centers	B1@.020	Ea	2.80	.55	3.35
Wood joists on 24" centers	B1@.020	Ea	3.20	.55	3.75

Subflooring

Board sheathing, 1" x 6" #3 & Btr (1.18 BF per SF)

	Craft@Hrs	Unit	Material	Labor	Total
Board sheathing, per MBF	—	MBF	1,340.00	—	1,340.00
With 12% shrinkage, 5% waste & nails	B1@.020	SF	1.59	.55	2.14
Add for diagonal patterns	B1@.001	SF	.16	.03	.19

Plywood sheathing, CD standard exterior grade. Material costs shown include 5% for waste & fasteners

	Craft@Hrs	Unit	Material	Labor	Total
3/8"	B1@.011	SF	.40	.30	.70
1/2"	B1@.012	SF	.52	.33	.85
5/8"	B1@.012	SF	.61	.33	.94
3/4"	B1@.013	SF	.79	.36	1.15

OSB sheathing, Material costs shown include 5% for waste & fasteners

	Craft@Hrs	Unit	Material	Labor	Total
3/8"	B1@.011	SF	.45	.30	.75
7/16"	B1@.011	SF	.38	.30	.68
1/2"	B1@.012	SF	.42	.33	.75
5/8"	B1@.012	SF	.58	.33	.91
3/4"	B1@.014	SF	.71	.38	1.09
7/8"	B1@.014	SF	1.10	.38	1.48
1"	B1@.015	SF	.51	.41	.92

Engineered floating subfloor, 2' x 2' panels, trade name DRIcore, resilient engineered wood core with a water-resistant polyethylene membrane on the underside. www.dricore.com

	Craft@Hrs	Unit	Material	Labor	Total
Per SF costs include 10% for waste	B1@.020	SF	1.63	.55	2.18

Plates (Wall plates) Std & Btr, untreated. For pressure treated plates, see also Sill Plates in this section. Figures in parentheses indicate board feet per LF. Costs shown include 10% for waste and nails.

	Craft@Hrs	Unit	Material	Labor	Total
2" x 3" wall plates, per MBF	—	MBF	679.00	—	679.00
2" x 3" (.55 BF per LF)	B1@.010	LF	.37	.27	.64
2" x 4" wall plates, per MBF	—	MBF	621.00	—	621.00
2" x 4" (.73 BF per LF)	B1@.012	LF	.45	.33	.78
2" x 6" wall plates, per MBF	—	MBF	620.00	—	620.00
2" x 6" (1.10 BF per LF)	B1@.018	LF	.68	.49	1.17
2" x 8" wall plates, per MBF	—	MBF	640.00	—	640.00
2" x 8" (1.40 BF per LF)	B1@.020	LF	.90	.55	1.45

	Craft@Hrs	Unit	Material	Labor	Total

Studding Per square foot of wall area. Do not subtract for openings less than 16' wide. Figures in parentheses indicate typical board feet per SF of wall area, measured on one side, and include normal waste. Add for each corner and partition from below. Costs include studding, nails. Add for plates from the section above and also, door and window opening framing, backing, let-in bracing, fire blocking, and sheathing for shear walls from the sections that follow. Labor includes layout, plumb and align.

2" x 3", Std & Btr

	Craft@Hrs	Unit	Material	Labor	Total
2" x 3" studs, 16" centers, per MBF	B1@43.9	MBF	679.00	1,200.00	1,879.00
12" centers (.55 BF per SF)	B1@.024	SF	.37	.66	1.03
16" centers (.41 BF per SF)	B1@.018	SF	.28	.49	.77
20" centers (.33 BF per SF)	B1@.015	SF	.22	.41	.63
24" centers (.28 BF per SF)	B1@.012	SF	.19	.33	.52
Add for each corner or partition	B1@.083	Ea	8.81	2.27	11.08

2" x 4", Std & Btr

	Craft@Hrs	Unit	Material	Labor	Total
2" x 4" studs, 16" centers, per MBF	B1@42.6	MBF	621.00	1,170.00	1,791.00
12" centers (.73 BF per SF)	B1@.031	SF	.45	.85	1.30
16" centers (.54 BF per SF)	B1@.023	SF	.34	.63	.97
20" centers (.45 BF per SF)	B1@.020	SF	.28	.55	.83
24" centers (.37 BF per SF)	B1@.016	SF	.23	.44	.67
Add for each corner or partition	B1@.083	Ea	10.50	2.27	12.77

2" x 6", Std & Btr

	Craft@Hrs	Unit	Material	Labor	Total
2" x 6" studs, 24" centers, per MBF	B1@40.0	MBF	620.00	1,090.00	1,710.00
12" centers (1.10 BF per SF)	B1@.044	SF	.68	1.20	1.88
16" centers (.83 BF per SF)	B1@.033	SF	.51	.90	1.41
20" centers (.67 BF per SF)	B1@.027	SF	.42	.74	1.16
24" centers (.55 BF per SF)	B1@.022	SF	.34	.60	.94
Add for each corner or partition	B1@.083	Ea	14.80	2.27	17.07

4" x 4", Std & Btr

	Craft@Hrs	Unit	Material	Labor	Total
Installed in wall framing, with 10% waste	B1@.064	LF	1.41	1.75	3.16

Door opening framing . In wall studs, based on walls 8' in height. Costs shown are per each door opening and include header of appropriate size, double vertical studs each side of the opening less than 8' wide (triple vertical studs each side of openings 8' wide or wider), cripples, blocking, nails and normal waste. Width shown is size of finished opening. Figures in parentheses indicate header size

2" x 4" wall studs, Std & Btr, opening size as shown

	Craft@Hrs	Unit	Material	Labor	Total
2" x 4" wall studs, per MBF	—	MBF	621.00	—	621.00
4" x 4" and 4" x 6" headers, per MBF	—	MBF	960.00	—	960.00
4" x 8", 10", 12" and 14" headers, per MBF	—	MBF	812.00	—	812.00
To 3' wide (4" x 4" header)	B1@.830	Ea	19.40	22.70	42.10
Over 3' to 4' wide (4" x 6" header)	B1@1.11	Ea	24.10	30.40	54.50
Over 4' to 5' wide (4" x 6" header)	B1@1.39	Ea	26.70	38.00	64.70
Over 5' to 6' wide (4" x 8" header)	B1@1.66	Ea	30.40	45.40	75.80
Over 6' to 8' wide (4" x 10" header)	B1@1.94	Ea	47.40	53.10	100.50
Over 8' to 10' wide (4" x 12" header)	B1@1.94	Ea	59.20	53.10	112.30
Over 10' to 12' wide (4" x 14" header)	B1@2.22	Ea	72.80	60.70	133.50
Add per foot of height for 4" walls over 8' in height	—	LF	2.53	—	2.53

2" x 6" wall studs, Std & Btr, opening size as shown

	Craft@Hrs	Unit	Material	Labor	Total
2" x 6" wall studs, per MBF	—	MBF	620.00	—	620.00
6" x 4" headers, per MBF	—	MBF	974.00	—	974.00
6" x 8", 10", 12" and 14" headers, per MBF	—	MBF	1,280.00	—	1,280.00
To 3' wide (6" x 4" header)	B1@1.11	Ea	31.80	30.40	62.20
Over 3' to 4' wide (6" x 6" header)	B1@1.39	Ea	36.70	38.00	74.70

	Craft@Hrs	Unit	Material	Labor	Total
Over 4' to 5' wide (6" x 6" header)	B1@1.66	Ea	41.20	45.40	86.60
Over 5' to 6' wide (6" x 8" header)	B1@1.94	Ea	53.30	53.10	106.40
Over 6' to 8' wide (6" x 10" header)	B1@2.22	Ea	84.20	60.70	144.90
Over 8' to 10' wide (6" x 12" header)	B1@2.22	Ea	112.00	60.70	172.70
Over 10' to 12' wide (6" x 14" header)	B1@2.50	Ea	143.00	68.40	211.40
Add per foot of height for 6" walls over 8' in height	—	LF	3.72	—	3.72

Window opening framing In wall studs, based on walls 8' in height. Costs shown are per window opening and include header of appropriate size, sub-sill plate (double sub-sill if opening is 8' wide or wider), double vertical studs each side of openings less than 8' wide (triple vertical studs each side of openings 8' wide or wider), top and bottom cripples, blocking, nails and normal waste. Figures in parentheses indicate header size.

2" x 4" wall studs, Std & Btr, opening size shown is width of finished opening

	Craft@Hrs	Unit	Material	Labor	Total
2" x 4" wall studs, per MBF	—	MBF	621.00	—	621.00
4" x 4" and 4" x 6" headers, per MBF	—	MBF	974.00	—	974.00
4" x 8", 10", 12" and 14" headers, per MBF	—	MBF	812.00	—	812.00
To 2' wide (4" x 4" header)	B1@1.00	Ea	17.50	27.40	44.90
Over 2' to 3' wide (4" x 4" header)	B1@1.17	Ea	21.90	32.00	53.90
Over 3' to 4' wide (4" x 6" header)	B1@1.45	Ea	27.50	39.70	67.20
Over 4' to 5' wide (4" x 6" header)	B1@1.73	Ea	31.00	47.30	78.30
Over 5' to 6' wide (4" x 8" header)	B1@2.01	Ea	35.30	55.00	90.30
Over 6' to 7' wide (4" x 8" header)	B1@2.29	Ea	45.00	62.70	107.70
Over 7' to 8' wide (4" x 10" header)	B1@2.57	Ea	54.00	70.30	124.30
Over 8' to 10' wide (4" x 12" header)	B1@2.57	Ea	67.40	70.30	137.70
Over 10' to 12' wide (4" x 14" header)	B1@2.85	Ea	87.70	78.00	165.70
Add per foot of height for walls over 8' high	—	LF	2.48	—	2.48

2" x 6" wall studs, Std & Btr, opening size as shown

	Craft@Hrs	Unit	Material	Labor	Total
2" x 6" wall studs, per MBF	—	MBF	620.00	—	620.00
4" x 4" and 6" x 4" headers, per MBF	—	MBF	974.00	—	974.00
6" x 6", 8", 10", 12" and 14" headers, per MBF	—	MBF	1,280.00	—	1,280.00
To 2' wide (4 x 4)	B1@1.17	Ea	21.90	32.00	53.90
Over 2' to 3' wide (6" x 4" header)	B1@1.45	Ea	28.50	39.70	68.20
Over 3' to 4' wide (6" x 6" header)	B1@1.73	Ea	40.10	47.30	87.40
Over 4' to 5' wide (6" x 6" header)	B1@2.01	Ea	47.60	55.00	102.60
Over 5' to 6' wide (6" x 8" header)	B1@2.29	Ea	58.20	62.70	120.90
Over 6' to 8' wide (6" x 10" header)	B1@2.85	Ea	90.90	78.00	168.90
Over 8' to 10' wide (6" x 12" header)	B1@2.85	Ea	124.00	78.00	202.00
Over 8' to 12' wide (6" x 14" header)	B1@3.13	Ea	158.00	85.60	243.60
Add per foot of height for 6" walls over 8' high	—	LF	3.72	—	3.72

Bracing See also sheathing for plywood bracing and shear panels

Let-in wall bracing, using standard and better lumber

	Craft@Hrs	Unit	Material	Labor	Total
1" x 4"	B1@.021	LF	.52	.57	1.09
1" x 6"	B1@.027	LF	.75	.74	1.49
2" x 4"	B1@.035	LF	.41	.96	1.37
Steel strap bracing, 1-1/4" wide					
9'6" or 11'6" lengths	B1@.010	LF	.45	.27	.72
Steel "V" bracing, 3/4" x 3/4"					
9'6" or 11'6" lengths	B1@.010	LF	.72	.27	.99

	Craft@Hrs	Unit	Material	Labor	Total
Temporary wood frame wall bracing, assumes salvage at 50% and 3 uses					
1" x 4" Std & Btr	B1@.006	LF	.13	.16	.29
1" x 6" Std & Btr	B1@.010	LF	.19	.27	.46
2" x 4" utility	B1@.012	LF	.10	.33	.43
2" x 6" utility	B1@.018	LF	.15	.49	.64

Fireblocks Installed in wood frame walls, per LF of wall to be blocked. Figures in parentheses indicate board feet of fire blocking per linear foot of wall including 10% waste. See also Bridging and Backing and Nailers in this section

	Craft@Hrs	Unit	Material	Labor	Total
2" x 3" blocking, Std & Btr					
2" x 3" fireblocks, 16" centers, per MBF	B1@30.0	MBF	679.00	821.00	1,500.00
12" OC members (.48 BF per LF)	B1@.016	LF	.33	.44	.77
16" OC members (.50 BF per LF)	B1@.015	LF	.34	.41	.75
20" OC members (.51 BF per LF)	B1@.012	LF	.35	.33	.68
24" OC members (.52 BF per LF)	B1@.010	LF	.35	.27	.62
2" x 4" blocking, Std & Btr					
2" x 4" fireblocks, 16" centers, per MBF	B1@22.4	MBF	621.00	613.00	1,234.00
12" OC members (.64 BF per LF)	B1@.020	LF	.40	.55	.95
16" OC members (.67 BF per LF)	B1@.015	LF	.42	.41	.83
20" OC members (.68 BF per LF)	B1@.012	LF	.42	.33	.75
24" OC members (.69 BF per LF)	B1@.010	LF	.43	.27	.70
2" x 6" blocking, Std & Btr					
2" x 6" fireblocks, 24" centers, per MBF	B1@9.71	MBF	620.00	266.00	886.00
12" OC members (.96 BF per LF)	B1@.021	LF	.59	.57	1.16
16" OC members (1.00 BF per LF)	B1@.016	LF	.62	.44	1.06
20" OC members (1.02 BF per LF)	B1@.012	LF	.63	.33	.96
24" OC members (1.03 BF per LF)	B1@.010	LF	.64	.27	.91

Beams Std & Btr. Installed over wall openings and around floor, ceiling and roof openings or where a flush beam is called out in the plans, including 10% waste. Do not use these beams for door or window headers in framed walls, use door or window opening framing assemblies.

	Craft@Hrs	Unit	Material	Labor	Total
2" x 6" beams, per MBF	B1@24.5	MBF	620.00	670.00	1,290.00
2" x 8" beams, per MBF	B1@25.2	MBF	640.00	689.00	1,329.00
2" x 10" beams, per MBF	B1@25.1	MBF	657.00	687.00	1,344.00
2" x 12" beams, per MBF	B1@25.9	MBF	738.00	709.00	1,447.00
4" x 6" beams, per MBF	B1@25.9	MBF	974.00	709.00	1,683.00
4" x 8", 10", 12", 14" beams, per MBF	B1@25.5	MBF	812.00	698.00	1,510.00
6" x 6", 8", 10", 12", 14" beams, per MBF	B1@19.1	MBF	1,280.00	523.00	1,803.00
2" x 6" (1.10 BF per LF)	B1@.028	LF	.68	.77	1.45
2" x 8" (1.47 BF per LF)	B1@.037	LF	.94	1.01	1.95
2" x 10" (1.83 BF per LF)	B1@.046	LF	1.20	1.26	2.46
2" x 12" (2.20 BF per LF)	B1@.057	LF	1.62	1.56	3.18
4" x 6" (2.20 BF per LF)	B1@.057	LF	2.14	1.56	3.70
4" x 8" (2.93 BF per LF)	B1@.073	LF	2.38	2.00	4.38
4" x 10" (3.67 BF per LF)	B1@.094	LF	2.98	2.57	5.55
4" x 12" (4.40 BF per LF)	B1@.112	LF	3.57	3.06	6.63
4" x 14" (5.13 BF per LF)	B1@.115	LF	4.16	3.15	7.31
6" x 6" (3.30 BF per LF)	B1@.060	LF	4.22	1.64	5.86
6" x 8" (4.40 BF per LF)	B1@.080	LF	5.63	2.19	7.82
6" x 10" (5.50 BF per LF)	B1@.105	LF	7.04	2.87	9.91
6" x 12" (6.60 BF per LF)	B1@.115	LF	8.45	3.15	11.60
6" x 14" (7.70 BF per LF)	B1@.120	LF	9.86	3.28	13.14

Carpentry, Detailed Breakdown

	Craft@Hrs	Unit	Material	Labor	Total

Posts S4S, green. Posts not in wall framing. Material costs include 10% waste. For scheduling purposes, estimate that a two-man crew can complete 100 to 125 LF per 8-hour day.

	Craft@Hrs	Unit	Material	Labor	Total
4" x 4", 4" x 6" posts, per MBF	—	MBF	960.00	—	960.00
4" x 8" to 4" x 12" posts, per MBF	—	MBF	812.00	—	812.00
6" x 6" to 8" x 12" posts, per MBF	—	MBF	1,280.00	—	1,280.00
4" x 4" per LF	B1@.110	LF	1.40	3.01	4.41
4" x 6" per LF	B1@.120	LF	2.11	3.28	5.39
4" x 8" per LF	B1@.140	LF	2.38	3.83	6.21
4" x 10" per LF	B1@.143	LF	2.98	3.91	6.89
4" x 12" per LF	B1@.145	LF	3.57	3.97	7.54
6" x 6" per LF	B1@.145	LF	4.22	3.97	8.19
6" x 8" per LF	B1@.145	LF	4.66	3.97	8.63
6" x 10" per LF	B1@.147	LF	7.04	4.02	11.06
6" x 12" per LF	B1@.150	LF	8.45	4.10	12.55
8" x 8" per LF	B1@.150	LF	7.51	4.10	11.61
8" x 10" per LF	B1@.155	LF	9.39	4.24	13.63
8" x 12" per LF	B1@.166	LF	11.30	4.54	15.84

Ceiling joists and soffits Per SF of area covered. Figures in parentheses indicate board feet per square foot of ceiling including end joists, header joists, and 5% waste. No beams, bridging, blocking, or ledger strips included. Deduct for openings over 25 SF. Costs shown are based on a job with 1,000 SF of area covered. For scheduling purposes, estimate that a two-man crew can complete 650 SF of area per 8-hour day for 12" center to center framing; 800 SF for 16" OC; 950 SF for 20" OC; or 1,100 SF for 24" OC.

	Craft@Hrs	Unit	Material	Labor	Total
2" x 4", Std & Btr grade					
2" x 4" ceiling joists, per MBF	—	MBF	621.00	—	621.00
12" centers (.78 BF per SF)	B1@.026	SF	.48	.71	1.19
16" centers (.59 BF per SF)	B1@.020	SF	.37	.55	.92
20" centers (.48 BF per SF)	B1@.016	SF	.30	.44	.74
24" centers (.42 BF per SF)	B1@.014	SF	.26	.38	.64
2" x 6", Std & Btr grade					
2" x 6" ceiling joists, per MBF	—	MBF	620.00	—	620.00
12" centers (1.15 BF per SF)	B1@.026	SF	.71	.71	1.42
16" centers (.88 BF per SF)	B1@.020	SF	.55	.55	1.10
20" centers (.72 BF per SF)	B1@.016	SF	.45	.44	.89
24" centers (.63 BF per SF)	B1@.014	SF	.39	.38	.77
2" x 8", Std & Btr grade					
2" x 8" ceiling joists, per MBF	—	MBF	640.00	—	640.00
12" centers (1.53 BF per SF)	B1@.028	SF	.98	.77	1.75
16" centers (1.17 BF per SF)	B1@.022	SF	.75	.60	1.35
20" centers (.96 BF per SF)	B1@.018	SF	.61	.49	1.10
24" centers (.84 BF per SF)	B1@.016	SF	.54	.44	.98
2" x 10", Std & Btr grade					
2" x 10" ceiling joists, per MBF	—	MBF	657.00	—	657.00
12" centers (1.94 BF per SF)	B1@.030	SF	1.28	.82	2.10
16" centers (1.47 BF per SF)	B1@.023	SF	.97	.63	1.60
20" centers (1.21 BF per SF)	B1@.019	SF	.80	.52	1.32
24" centers (1.04 BF per SF)	B1@.016	SF	.68	.44	1.12
2" x 12", Std & Btr grade					
2" x 12" ceiling joists, per MBF	—	MBF	738.00	—	738.00
12" centers (2.30 BF per SF)	B1@.033	SF	1.70	.90	2.60
16" centers (1.76 BF per SF)	B1@.025	SF	1.30	.68	1.98
20" centers (1.44 BF per SF)	B1@.020	SF	1.06	.55	1.61
24" centers (1.26 BF per SF)	B1@.018	SF	.93	.49	1.42

	Craft@Hrs	Unit	Material	Labor	Total
Catwalks 1" x 6" solid planking walkway, #2 & Btr grade					
(0.50 BF per SF)	B1@.017	SF	.67	.47	1.14

Backing and nailers Std & Btr, for wall finishes, "floating" backing for drywall ceilings, trim, z-bar, appliances and fixtures, etc. See also Bridging and Fireblocking in this section. Figures in parentheses show board feet per LF including 10% waste.

	Craft@Hrs	Unit	Material	Labor	Total
1" x 4" (.37 BF per LF)	B1@.011	LF	.32	.30	.62
1" x 6" (.55 BF per LF)	B1@.017	LF	.49	.47	.96
1" x 8" (.73 BF per LF)	B1@.022	LF	.64	.60	1.24
2" x 4" (.73 BF per LF)	B1@.023	LF	.45	.63	1.08
2" x 6" (1.10 BF per LF)	B1@.034	LF	.68	.93	1.61
2" x 8" (1.47 BF per LF)	B1@.045	LF	.94	1.23	2.17
2" x 10" (1.83 BF per LF)	B1@.057	LF	1.20	1.56	2.76

Ledger strips Std & Btr, nailed to faces of studs, beams, girders, joists, etc. See also Ribbons in this section for let-in type. Figures in parentheses indicate board feet per LF including 10% waste.

	Craft@Hrs	Unit	Material	Labor	Total
1" x 2", 3" or 4" ledger, per MBF	—	MBF	931.00	—	931.00
1" x 2" (.18 BF per LF)	B1@.010	LF	.17	.27	.44
1" x 3" (.28 BF per LF)	B1@.010	LF	.26	.27	.53
1" x 4" (.37 BF per LF)	B1@.010	LF	.34	.27	.61
2" x 2", 3" or 4" ledger, per MBF	—	MBF	633.00	—	633.00
2" x 2" (.37 BF per LF)	B1@.010	LF	.23	.27	.50
2" x 3" (.55 BF per LF)	B1@.010	LF	.35	.27	.62
2" x 4" (.73 BF per LF)	B1@.010	LF	.46	.27	.73

Ribbons (Ribbands) Let in to wall framing. See also Ledgers in this section. Figures in parentheses indicate board feet per LF including 10% waste.

	Craft@Hrs	Unit	Material	Labor	Total
1" x 3", 4" or 6" ribbon, per MBF	—	MBF	855.00	—	855.00
1" x 3", Std & Btr (.28 BF per LF)	B1@.020	LF	.15	.55	.70
1" x 4", Std & Btr (.37 BF per LF)	B1@.020	LF	.24	.55	.79
1" x 6", Std & Btr (.55 BF per LF)	B1@.030	LF	.32	.82	1.14
2" x 3", 4" ribbon, per MBF	—	MBF	621.00	—	621.00
2" x 3", Std & Btr (.55 BF per LF)	B1@.041	LF	.34	1.12	1.46
2" x 4", Std & Btr (.73 BF per LF)	B1@.041	LF	.45	1.12	1.57
2" x 6" ribbon, per MBF	—	MBF	620.00	—	620.00
2" x 6", Std & Btr (1.10 BF per LF)	B1@.045	LF	.68	1.23	1.91
2" x 8" ribbon, per MBF	—	MBF	640.00	—	640.00
2" x 8", Std & Btr (1.47 BF per LF)	B1@.045	LF	.94	1.23	2.17

Rafters Flat, shed, or gable roofs, up to 5 in 12 slope (5/24 pitch), maximum 25' span. Figures in parentheses indicate board feet per SF of actual roof surface area (not roof plan area), including rafters, ridge boards, collar beams and normal waste, but no blocking, bracing, purlins, curbs, or gable walls.

	Craft@Hrs	Unit	Material	Labor	Total
2" x 4", Std & Btr					
2" x 4" rafters, per MBF	—	MBF	621.00	—	621.00
12" centers (.89 BF per SF)	B1@.021	SF	.55	.57	1.12
16" centers (.71 BF per SF)	B1@.017	SF	.44	.47	.91
24" centers (.53 BF per SF)	B1@.013	SF	.33	.36	.69
2" x 6", Std & Btr					
2" x 6" rafters, per MBF	—	MBF	620.00	—	620.00
12" centers (1.29 BF per SF)	B1@.029	SF	.80	.79	1.59
16" centers (1.02 BF per SF)	B1@.023	SF	.63	.63	1.26
24" centers (.75 BF per SF)	B1@.017	SF	.46	.47	.93

Carpentry, Detailed Breakdown

	Craft@Hrs	Unit	Material	Labor	Total
2" x 8", Std & Btr					
2" x 8" rafters, per MBF	—	MBF	640.00	—	640.00
12" centers (1.71 BF per SF)	B1@.036	SF	1.09	.98	2.07
16" centers (1.34 BF per SF)	B1@.028	SF	.86	.77	1.63
24" centers (1.12 BF per SF)	B1@.024	SF	.72	.66	1.38
2" x 10", Std & Btr					
2" x 10" rafters, per MBF	—	MBF	657.00	—	657.00
12" centers (2.12 BF per SF)	B1@.039	SF	1.39	1.07	2.46
16" centers (1.97 BF per SF)	B1@.036	SF	1.30	.98	2.28
24" centers (1.21 BF per SF)	B1@.022	SF	.80	.60	1.40
2" x 12", Std & Btr					
2" x 12" rafters, per MBF	—	MBF	738.00	—	738.00
12" centers (2.52 BF per SF)	B1@.045	SF	1.86	1.23	3.09
16" centers (1.97 BF per SF)	B1@.035	SF	1.45	.96	2.41
24" centers (1.43 BF per SF)	B1@.026	SF	1.06	.71	1.77
Add for hip roof	B1@.007	SF	—	.19	.19
Add for slope over 5 in 12	B1@.015	SF	—	.41	.41

Trimmers and curbs At stairwells, skylights, dormers, etc. Figures in parentheses show board feet per LF, Std & Btr grade, including 10% waste.

	Craft@Hrs	Unit	Material	Labor	Total
2" x 4" (.73 BF per LF)	B1@.018	LF	.45	.49	.94
2" x 6" (1.10 BF per LF)	B1@.028	LF	.68	.77	1.45
2" x 8" (1.47 BF per LF)	B1@.038	LF	.94	1.04	1.98
2" x 10" (1.83 BF per LF)	B1@.047	LF	1.20	1.29	2.49
2" x 12" (2.20 BF per LF)	B1@.057	LF	1.62	1.56	3.18

Collar beams & collar ties Std & Btr grade, including 10% waste

	Craft@Hrs	Unit	Material	Labor	Total
Collar beams, 2" x 6"	B1@.013	LF	.68	.36	1.04
Collar ties, 1" x 6"	B1@.006	LF	.49	.16	.65

Purlins (Purling) Std & Btr, installed below roof rafters. Figures in parentheses indicate board feet per LF including 5% waste.

	Craft@Hrs	Unit	Material	Labor	Total
2" x 4" (.70 BF per LF)	B1@.012	LF	.43	.33	.76
2" x 6" (1.05 BF per LF)	B1@.017	LF	.65	.47	1.12
2" x 8" (1.40 BF per LF)	B1@.023	LF	.90	.63	1.53
2" x 10" (1.75 BF per LF)	B1@.027	LF	1.15	.74	1.89
2" x 12" (2.10 BF per LF)	B1@.030	LF	1.55	.82	2.37
4" x 6" (2.10 BF per LF)	B1@.034	LF	2.05	.93	2.98
4" x 8" (2.80 BF per LF)	B1@.045	LF	2.29	1.23	3.52

Dormer studs Std & Btr, per square foot of wall area, including 10% waste

	Craft@Hrs	Unit	Material	Labor	Total
2" x 4" (.84 BF per SF)	B1@.033	SF	.52	.90	1.42

Roof trusses 24" OC, any slope from 3 in 12 to 12 in 12, total height not to exceed 12' high from bottom chord to highest point on truss. Prices for trusses over 12' high will be up to 100% higher. Square foot (SF) costs, where shown, are per square foot of area to be covered.

Scissor trusses

	Craft@Hrs	Unit	Material	Labor	Total
2" x 4" top and bottom chords					
Up to 38' span	B1@.022	SF	2.42	.60	3.02
40' to 50' span	B1@.028	SF	3.03	.77	3.80

	Craft@Hrs	Unit	Material	Labor	Total
Fink truss "W" (conventional roof truss)					
2" x 4" top and bottom chords					
Up to 38' span	B1@.017	SF	2.10	.47	2.57
40' to 50' span	B1@.022	SF	2.54	.60	3.14
2" x 6" top and bottom chords					
Up to 38' span	B1@.020	SF	2.59	.55	3.14
40' to 50' span	B1@.026	SF	3.12	.71	3.83
Truss with gable fill at 16" OC					
28' span, 5 in 12 slope	B1@.958	Ea	175.00	26.20	201.20
32' span, 5 in 12 slope	B1@1.26	Ea	217.00	34.50	251.50
40' span, 5 in 12 slope	B1@1.73	Ea	303.00	47.30	350.30
Fascia Material column includes 10% waste					
White Wood, #3 & Btr					
1" x 4", per MBF	—	MBF	1,230.00	—	1,230.00
1" x 6", per MBF	—	MBF	1,470.00	—	1,470.00
1" x 8", per MBF	—	MBF	1,490.00	—	1,490.00
1" x 4", per LF	B1@.044	LF	.45	1.20	1.65
1" x 6", per LF	B1@.044	LF	.81	1.20	2.01
1" x 8", per LF	B1@.051	LF	1.09	1.40	2.49
Hem-fir, S4S, dry, Std and Btr					
2" x 6", per MBF	—	MBF	609.00	—	609.00
2" x 8", per MBF	—	MBF	632.00	—	632.00
2" x 10", per MBF	—	MBF	687.00	—	687.00
2" x 6", per LF	B1@.044	LF	.67	1.20	1.87
2" x 8", per LF	B1@.051	LF	.93	1.40	2.33
2" x 10", per LF	B1@.051	LF	1.26	1.40	2.66
Redwood, S4S kiln dried, clear					
1" x 4", per MBF	—	MBF	2,090.00	—	2,090.00
1" x 6", per MBF	—	MBF	2,700.00	—	2,700.00
1" x 8", per MBF	—	MBF	1,970.00	—	1,970.00
2" x 4", per MBF	—	MBF	1,240.00	—	1,240.00
2" x 6", per MBF	—	MBF	1,370.00	—	1,370.00
2" x 8", per MBF	—	MBF	1,450.00	—	1,450.00
1" x 4", per LF	B1@.044	LF	.77	1.20	1.97
1" x 6", per LF	B1@.044	LF	1.48	1.20	2.68
1" x 8", per LF	B1@.051	LF	1.45	1.40	2.85
2" x 4", per LF	B1@.044	LF	.91	1.20	2.11
2" x 6", per LF	B1@.044	LF	1.51	1.20	2.71
2" x 8", per LF	B1@.051	LF	2.12	1.40	3.52
Sheathing, roof Per SF of roof surface including 10% waste					
CDX plywood sheathing, rough					
1/2"	B1@.013	SF	.50	.36	.86
1/2", 4-ply	B1@.013	SF	.54	.36	.90
1/2", 5-ply	B1@.013	SF	.57	.36	.93
5/8", 4-ply	B1@.013	SF	.64	.36	1.00
5/8", 5-ply	B1@.013	SF	.70	.36	1.06
3/4"	B1@.013	SF	.83	.36	1.19

Carpentry, Detailed Breakdown

	Craft@Hrs	Unit	Material	Labor	Total
OSB sheathing					
7/16"	B1@.013	SF	.38	.36	.74
1/2"	B1@.013	SF	.44	.36	.80
5/8"	B1@.013	SF	.61	.36	.97
3/4"	B1@.014	SF	.74	.38	1.12
7/8"	B1@.014	SF	1.15	.38	1.53
1-1/8"	B1@.015	SF	1.35	.41	1.76
Add for hip roof	B1@.007	SF	—	.19	.19
Add for steep pitch or cut-up roof	B1@.015	SF	—	.41	.41
Board sheathing, 1" x 6" or 1" x 8" utility T&G laid diagonal					
1" utility T&G lumber, per MBF	B1@23.0	MBF	971.00	629.00	1,600.00
(1.13 BF per SF)	B1@.026	SF	1.21	.71	1.92
Add for hip roof	B1@.007	SF	—	.19	.19
Add for steep pitch or cut-up roof	B1@.015	SF	—	.41	.41

Roof decking Std & Btr grade. See also Sheathing in this section. Flat, shed, or gable roofs to 5 in 12 slope (5/24 pitch). Figures in parentheses indicate board feet per SF of actual roof area (not roof plan area), including 5% waste.

	Craft@Hrs	Unit	Material	Labor	Total
2" x 6", per MBF	—	MBF	620.00	—	620.00
2" x 8", per MBF	—	MBF	640.00	—	640.00
3" x 6", per MBF	—	MBF	704.00	—	704.00
2" x 6" (2.28 BF per SF)	B1@.043	SF	1.41	1.18	2.59
2" x 8" (2.25 BF per SF)	B1@.043	SF	1.44	1.18	2.62
3" x 6" (3.43 BF per SF)	B1@.047	SF	2.41	1.29	3.70
Add for steep pitch or cut-up roof	B1@.019	SF	—	.52	.52

Cant strips Std & Btr grade lumber

	Craft@Hrs	Unit	Material	Labor	Total
2" x 4" beveled	B1@.020	LF	.68	.55	1.23
For short or cut-up runs	B1@.030	LF	.68	.82	1.50

Furring Per SF of surface area to be covered. Figures in parentheses show coverage including 7% waste, typical job.

	Craft@Hrs	Unit	Material	Labor	Total
Over masonry, Utility grade					
1" furring, per MBF	—	MBF	855.00	—	855.00
12" OC, 1" x 2" (.24 BF per SF)	B1@.025	SF	.21	.68	.89
16" OC, 1" x 2" (.20 BF per SF)	B1@.020	SF	.17	.55	.72
20" OC, 1" x 2" (.17 BF per SF)	B1@.018	SF	.15	.49	.64
24" OC, 1" x 2" (.15 BF per SF)	B1@.016	SF	.13	.44	.57
12" OC, 1" x 3" (.36 BF per SF)	B1@.025	SF	.31	.68	.99
16" OC, 1" x 3" (.29 BF per SF)	B1@.020	SF	.25	.55	.80
20" OC, 1" x 3" (.25 BF per SF)	B1@.018	SF	.21	.49	.70
24" OC, 1" x 3" (.22 BF per SF)	B1@.016	SF	.19	.44	.63
Over wood frame, Utility grade					
12" OC, 1" x 2" (.24 BF per SF)	B1@.016	SF	.21	.44	.65
16" OC, 1" x 2" (.20 BF per SF)	B1@.013	SF	.17	.36	.53
20" OC, 1" x 2" (.17 BF per SF)	B1@.011	SF	.15	.30	.45
24" OC, 1" x 2" (.15 BF per SF)	B1@.010	SF	.13	.27	.40
12" OC, 1" x 3" (.36 BF per SF)	B1@.016	SF	.31	.44	.75
16" OC, 1" x 3" (.29 BF per SF)	B1@.013	SF	.25	.36	.61
20" OC, 1" x 3" (.25 BF per SF)	B1@.011	SF	.21	.30	.51
24" OC, 1" x 3" (.22 BF per SF)	B1@.010	SF	.19	.27	.46

	Craft@Hrs	Unit	Material	Labor	Total
Over wood subfloor, Utility grade					
12" OC, 1" x 2" (.24 BF per SF)	B1@.033	SF	.21	.90	1.11
16" OC, 1" x 2" (.20 BF per SF)	B1@.028	SF	.17	.77	.94
20" OC, 1" x 2" (.17 BF per SF)	B1@.024	SF	.15	.66	.81
24" OC, 1" x 2" (.15 BF per SF)	B1@.021	SF	.13	.57	.70
2" furring, per MBF	—	MBF	621.00	—	621.00
12" OC, 2" x 2" Std & Btr (.48 BF per SF)	B1@.033	SF	.30	.90	1.20
16" OC, 2" x 2" Std & Btr (.39 BF per SF)	B1@.028	SF	.24	.77	1.01
20" OC, 2" x 2" Std & Btr (.34 BF per SF)	B1@.024	SF	.21	.66	.87
24" OC, 2" x 2" Std & Btr (.30 BF per SF)	B1@.021	SF	.19	.57	.76

Grounds Screeds for plaster or stucco, #3, 1" x 2" lumber. Based on .18 BF per LF including 10% waste.

	Craft@Hrs	Unit	Material	Labor	Total
1" grounds, per MBF	—	MBF	855.00	—	855.00
Over masonry	B1@.047	LF	.15	1.29	1.44
Over wood	B1@.038	LF	.15	1.04	1.19

Sheathing, wall Per SF of wall surface, including 5% waste

	Craft@Hrs	Unit	Material	Labor	Total
BC plywood sheathing, plugged and touch sanded, interior grade					
1/4"	B1@.013	SF	.46	.36	.82
3/8"	B1@.015	SF	.53	.41	.94
1/2"	B1@.016	SF	.65	.44	1.09
5/8"	B1@.018	SF	.81	.49	1.30
3/4"	B1@.020	SF	.96	.55	1.51
OSB sheathing					
1/4"	B1@.013	SF	.29	.36	.65
3/8"	B1@.015	SF	.45	.41	.86
7/16"	B1@.015	SF	.38	.41	.79
1/2"	B1@.016	SF	.42	.44	.86
5/8"	B1@.018	SF	.58	.49	1.07
3/4"	B1@.018	SF	.71	.49	1.20
7/8"	B1@.020	SF	1.10	.55	1.65
1-1/8"	B1@.020	SF	1.28	.55	1.83
Board sheathing 1" x 6" or 1" x 8" utility T&G					
1" x 6" or 8" utility T&G, per MBF	—	MBF	823.00	—	823.00
(1.13 BF per SF)	B1@.020	SF	1.00	.55	1.55
Add for diagonal patterns	B1@.006	SF	—	.16	.16

Sleepers Std & Btr pressure treated lumber, including 5% waste

	Craft@Hrs	Unit	Material	Labor	Total
2" x 6" sleepers, per MBF	—	MBF	871.00	—	871.00
2" x 6" sleepers, per LF	B1@.017	LF	.91	.47	1.38
Add for taper cuts on sleepers, per cut	B1@.009	Ea	—	.25	.25

Carpentry, Finish

	Craft@Hrs	Unit	Material	Labor	Total

Finish Carpentry These figures assume all materials to be installed are supplied by others. Labor costs assume good quality stain-grade work up to 9 feet above floor level. Paint-grade work will cost from 20% to 33% less. Add the cost of scaffolding, staining or painting, if required.

Blocking and backing Installation before drywall is hung. Cost per piece of blocking or backing set

	Craft@Hrs	Unit	Material	Labor	Total
2" x 4" backing or blocking	BC@.040	Ea	—	1.20	1.20
2" x 4" strongback 10' to 16' long screwed to ceiling joists	BC@.350	Ea	—	10.50	10.50

Furring Nailed over studs and adjusted with shims to bring the wall surface flush. Costs will be higher if the wall plane is very uneven or if the furred wall has to be exceptionally even.
Cost per linear foot of furring installed

	Craft@Hrs	Unit	Material	Labor	Total
1" x 4" furring, few shims required	BC@.010	LF	—	.30	.30

Straightening wall studs Costs per stud straightened
Cut saw kerf and drive shim or screw into the kerf

	Craft@Hrs	Unit	Material	Labor	Total
reinforce with a plywood scab	BC@.300	Ea	—	9.00	9.00
Remove and replace a badly warped stud	BC@.300	Ea	—	9.00	9.00

Cased openings Install jambs only. See costs for casing below. Costs per opening cased

	Craft@Hrs	Unit	Material	Labor	Total
Opening to 4' wide	BC@.800	Ea	—	24.00	24.00
Openings over 4' wide	BC@1.05	Ea	—	31.50	31.50

Exterior doors Based on hanging doors in prepared openings. No lockset, casing or stops included. See the sections that follow. Costs will be higher if the framing has to be adjusted before the door can be set.
Hang a new door in an existing jamb

	Craft@Hrs	Unit	Material	Labor	Total
Cut door, set butt hinges, install door	BC@1.46	Ea	—	43.80	43.80

Prehung flush or panel 1-3/4" thick solid core single doors, with felt insulation, based on an adjustable metal sill

	Craft@Hrs	Unit	Material	Labor	Total
Up to 3' wide	BC@.750	Ea	—	22.50	22.50
Over 3' wide	BC@1.25	Ea	—	37.50	37.50
Add for door with sidelights	BC@.750	Ea	—	22.50	22.50
Sliding glass doors, opening to 8' wide	BC@1.75	Ea	—	52.50	52.50
French doors, 1-3/4" thick, to 6' wide	BC@1.75	Ea	—	52.50	52.50

Interior doors Based on hanging 1-3/8" hollow core doors in prepared openings. No locksets or casing included. See the sections that follow.

	Craft@Hrs	Unit	Material	Labor	Total
Prehung interior doors	BC@.500	Ea	—	15.00	15.00

Add for assembling a factory-cut

	Craft@Hrs	Unit	Material	Labor	Total
mortised and rabbeted jamb	BC@.200	Ea	—	6.00	6.00

Pocket doors

	Craft@Hrs	Unit	Material	Labor	Total
Assemble and install the door frame	BC@.981	Ea	—	29.40	29.40
Install door and valances in existing frame	BC@.500	Ea	—	15.00	15.00

Bifold doors installed in cased openings. See costs for cased openings above

	Craft@Hrs	Unit	Material	Labor	Total
Single bifold door to 4' wide	BC@.700	Ea	—	21.00	21.00
Double bifold door to 8' wide, pair of doors	BC@1.20	Pr	—	36.00	36.00

Bypass closet doors installed on a track in a cased opening. See cased opening manhours above

	Craft@Hrs	Unit	Material	Labor	Total
Double doors, per opening	BC@.660	Ea	—	19.80	19.80
Triple doors, per opening	BC@1.00	Ea	—	30.00	30.00

Rehang most hollow core doors in an existing opening,

	Craft@Hrs	Unit	Material	Labor	Total
re-shim jamb and plane door, per door	BC@1.50	Ea	—	45.00	45.00

	Craft@Hrs	Unit	Material	Labor	Total
Rehang large or heavy solid core doors in an existing opening,					
re-shim jamb and plane door, per door	BC@1.75	Ea	—	52.50	52.50
Correcting a sagging door hinge. Remove door from the jamb, remove hinge leaf,					
deepen the mortise and rehang the door	BC@.600	Ea	—	18.00	18.00
Hang a new door in an existing jamb. Mark the door to fit opening, cut the door, set butt hinges, install door in the jamb, install door hardware					
Per door hung	BC@1.75	Ea	—	52.50	52.50
Prehanging doors off site. Cost per door hung, excluding door cost					
Make and assemble frame, cost per frame	BC@.750	Ea	—	22.50	22.50
Set hinge on door and frame, panel or flush door, per hinge					
Using jigs and router	BC@.160	Ea	—	4.80	4.80
Using hand tools	BC@.500	Ea	—	15.00	15.00
Add for French doors, per door	BC@.250	Ea	—	7.50	7.50

Lockset installation Set in either a solid core or hollow core door, includes setting the strike plate

	Craft@Hrs	Unit	Material	Labor	Total
Costs per lockset					
Standard 2-1/8" bore, factory bored	BC@.160	Ea	—	4.80	4.80
Standard 2-1/8" bore, site bored	BC@1.00	Ea	—	30.00	30.00
Mortise type lockset, using a boring machine	BC@1.50	Ea	—	45.00	45.00
Mortise type lockset, using hand tools	BC@3.00	Ea	—	90.00	90.00
Pocket door hardware					
Privacy latch	BC@.500	Ea	—	15.00	15.00
Passage latch	BC@.300	Ea	—	9.00	9.00
Edge pull	BC@.300	Ea	—	9.00	9.00

Setting door stop moulding Set at the head and on two sides. Costs per opening

	Craft@Hrs	Unit	Material	Labor	Total
Panel or flush door	BC@.250	Ea	—	7.50	7.50
French door	BC@.330	Ea	—	9.90	9.90
Pocket door	BC@.250	Ea	—	7.50	7.50

Set casing around a framed opening Mitered or butted moulding corners. Costs per opening, per side, using a pneumatic nail gun

	Craft@Hrs	Unit	Material	Labor	Total
Opening to 4' wide using casing to 2-1/4" wide	BC@.250	Ea	—	7.50	7.50
Opening over 4' wide to 8' wide					
using casing up to 2-1/4" wide	BC@.300	Ea	—	9.00	9.00
Add for casing over 2-1/4" wide to 4" wide	BC@.160	Ea	—	4.80	4.80
Add for block and plinth door casing	BC@.300	Ea	—	9.00	9.00
Add for hand nailing	BC@.275	Ea	—	8.25	8.25

Window trim Based on trim set around prepared openings. Costs per window opening on square windows. Extra tall or extra wide windows will cost more.

	Craft@Hrs	Unit	Material	Labor	Total
Extension jambs cut, assembled, shimmed and installed					
Up to 3/4" thick, 3 or 4 sides, per window	BC@.330	Ea	—	9.90	9.90
Over 3/4" thick, per LF of extension jamb	BC@.100	LF	—	3.00	3.00
Window stool					
Set to sheetrock wraps					
nailed directly on framing	BC@.500	Ea	—	15.00	15.00
When stool has to be shimmed up from framing					
Add per LF of stool	BC@.050	LF	—	1.50	1.50
Stool set to wooden jambs, per opening	BC@.400	Ea	—	12.00	12.00

	Craft@Hrs	Unit	Material	Labor	Total
When stool has to be shimmed up from jambs					
Add per LF of stool	BC@.050	LF	—	1.50	1.50
Add for mitered returns on horns					
Per opening	BC@.500	Ea	—	15.00	15.00
Window apron moulding without mitered returns					
Per opening	BC@.160	Ea	—	4.80	4.80
Window apron moulding with mitered returns					
Per opening	BC@.320	Ea	—	9.60	9.60
Window casing, straight casing with square corners, per opening					
3 sides (stool and apron), up to 5' on any side	BC@.330	Ea	—	9.90	9.90
4 sides (picture frame), up to 5' on any side	BC@.500	Ea	—	15.00	15.00
Add for any side over 5' long, per side	BC@.090	Ea	—	2.70	2.70
Add for angle corners (trapezoid or triangle)					
Per corner	BC@.250	Ea	—	7.50	7.50
Installing a pre-manufactured arched top casing only, per opening					
To 4' wide set against rosettes	BC@.330	Ea	—	9.90	9.90
To 4' wide, mitered corners	BC@.500	Ea	—	15.00	15.00
Over 4' wide set against rosettes	BC@.660	Ea	—	19.80	19.80
Over 4' wide, mitered corners	BC@1.00	Ea	—	30.00	30.00

Ceiling treatments All figures assume a wall height not over 9' with work done from scaffolding. Add the cost of scaffolding

	Craft@Hrs	Unit	Material	Labor	Total
Set up and dismantle scaffolding, based on a typical 10' x 16' room					
Per room	BC@1.75	Ea	—	52.50	52.50
Wood strip ceiling, T&G or square edged, applied wall to wall with no miters					
Per square foot of ceiling	BC@.050	SF	—	1.50	1.50
Plywood panel ceiling					
Screwed in place, moulding strips nailed with a pneumatic nailer, including normal layout					
Per square foot of ceiling covered	BC@.160	SF	—	4.80	4.80
Applied beams, including building up the beam, layout and installation on a flat ceiling					
Per LF of beam applied	BC@.310	LF	—	9.30	9.30
Additional costs for ceiling treatments					
Add for each joint or beam scribed to the wall	BC@.660	Ea	—	19.80	19.80
Add for each corbel	BC@.400	Ea	—	12.00	12.00
Add for each knee brace	BC@.600	Ea	—	18.00	18.00

Cornice or crown moulding Including layout and coped joints, per LF of moulding

	Craft@Hrs	Unit	Material	Labor	Total
Up to 2-1/4" wide	BC@.050	LF	—	1.50	1.50
Over 2-1/4" wide to 3-1/2" wide	BC@.060	LF	—	1.80	1.80
Over 3-1/2" wide	BC@.070	LF	—	2.10	2.10

Wall paneling No moulding included. See moulding below. Costs per square foot of wall covered

	Craft@Hrs	Unit	Material	Labor	Total
Board wainscot or paneling, 3-1/2" to 5-1/2" wide tongue and groove					
Per square foot of wall covered	BC@.080	SF	—	2.40	2.40
Frame and panel wainscot. Installation only. Add for fabricating the frames and panels					
Per square foot of wall covered	BC@.090	SF	—	2.70	2.70
Add for frame attached with screws and covered with wood plugs	BC@.010	SF	—	.30	.30
Hardwood sheet paneling installed on wood frame or masonry walls					
Per square foot of wall covered	BC@.040	SF	—	1.20	1.20
Add for running moulding if required	BC@.040	SF	—	1.20	1.20

	Craft@Hrs	Unit	Material	Labor	Total
Add if vapor barrier must be installed with mastic over a masonry wall	BC@.010	SF	—	.30	.30
Add when 1" x 4" furring is nailed on a masonry wall with a powder-actuated nail gun					
No shimming or straightening included	BC@.020	SF	—	.60	.60
Add if sheet paneling is scribed around stone veneer					
Per linear foot scribed	BC@.150	LF	—	4.50	4.50
Minimum time for scribing sheet paneling	BC@.500	Ea	—	15.00	15.00

Running mouldings Single piece mitered installations nailed by hand. Add 100% for two-piece moulding. These estimates are based on stain-grade work. Subtract 33% for paint-grade work. Costs per LF of moulding

	Craft@Hrs	Unit	Material	Labor	Total
Picture rail	BC@.050	LF	—	1.50	1.50
Chair rail	BC@.040	LF	—	1.20	1.20
Wainscot cap or bed moulding	BC@.040	LF	—	1.20	1.20
Baseboard	BC@.040	LF	—	1.20	1.20

Kneewall cap up to 8" wide and 1-1/2" thick. Complex job layout and bed moulding will increase costs

	Craft@Hrs	Unit	Material	Labor	Total
Kneewall cap	BC@.160	LF	—	4.80	4.80

Closet poles and shelving Including layout and installation

1-1/4" closet pole supported with a metal bracket every 32"

	Craft@Hrs	Unit	Material	Labor	Total
Per LF of pole	BC@.040	LF	—	1.20	1.20
Add for mitered corner joints, per corner	BC@.300	Ea	—	9.00	9.00

3/4" x 12" hat shelf supported with a metal shelf bracket every 32"

	Craft@Hrs	Unit	Material	Labor	Total
Per LF of shelf	BC@.100	LF	—	3.00	3.00
Add for mitered corner joint, per corner	BC@.300	Ea	—	9.00	9.00

Site-built closet storage unit with drawers using plate biscuit joiners and 3/4" board or plywood stock, including fabrication, layout, assembly and installation

	Craft@Hrs	Unit	Material	Labor	Total
Per square foot of storage unit face	BC@.680	SF	—	20.40	20.40
Add for each drawer	BC@.850	Ea	—	25.50	25.50

Cabinet installation Including layout, leveling and installation. No fabrication, assembly or countertops included. Costs per cabinet

Base cabinets

	Craft@Hrs	Unit	Material	Labor	Total
Cabinets to 36" wide	BC@.766	Ea	—	23.00	23.00
Cabinets over 36" wide	BC@.911	Ea	—	27.30	27.30

Wall cabinets

	Craft@Hrs	Unit	Material	Labor	Total
Cabinets to 18" wide	BC@.638	Ea	—	19.10	19.10
Cabinets over 18" to 36" wide	BC@.766	Ea	—	23.00	23.00
Cabinets over 36" wide	BC@1.03	Ea	—	30.90	30.90
Oven or utility cabinets, 70" high, to 30" wide	BC@1.71	Ea	—	51.30	51.30

Filler strip between cabinets, measured, cut, fit and installed

	Craft@Hrs	Unit	Material	Labor	Total
Per strip installed	BC@.200	Ea	—	6.00	6.00

Fireplace surrounds and mantels Costs will be higher when a mantel or moulding must be scribed around irregular masonry surfaces

Layout, cutting, fitting, assembling and installing a vertical pilaster on two sides and a horizontal frieze board above the pilasters. Additional mouldings will take more time

	Craft@Hrs	Unit	Material	Labor	Total
Costs per linear foot of pilaster and frieze	BC@.350	LF	—	10.50	10.50

Mantel shelf attached to masonry, including cutting, fitting and installation

	Craft@Hrs	Unit	Material	Labor	Total
Per mantel	BC@.750	Ea	—	22.50	22.50

Installing a prefabricated mantel and fireplace surround

	Craft@Hrs	Unit	Material	Labor	Total
Per mantel and surround installed	BC@1.50	Ea	—	45.00	45.00

	Craft@Hrs	Unit	Material	Labor	Total

Wood flooring Including layout, vapor barrier and installation with a power nailer. Add for underlayment, sanding and finishing. Borders, feature strips and mitered corners will increase costs. Costs per square foot of floor covered

	Craft@Hrs	Unit	Material	Labor	Total
Tongue & groove strip flooring 3/4" x 2-1/4" strips	BF@.036	SF	—	1.09	1.09
Straight edge plank flooring, 3/4" x 6" to 12" wide Set with biscuit joiner plates	BF@.055	SF	—	1.67	1.67
Add for plank flooring set with screws and butterfly plugs	BF@.010	SF	—	.30	.30
Filling and sanding a wood floor with drum and disc sanders, 3 passes (60, 80, 100 grit), per square foot	BF@.023	SF	—	.70	.70
Finishing a wood floor, including filler, shellac, polyurethane or waterborne urethane	BF@.024	SF	—	.73	.73

Skirt boards for stairways Including layout, fitting, assembly and installation. Based on a 9' floor-to-floor rise where the stair carriage has already been installed by others. Costs will be higher if re-cutting, shimming, or adjusting of an existing stair carriage is required.

Cutting and installing a skirt board on the closed side of a stairway. Costs per skirt board

	Craft@Hrs	Unit	Material	Labor	Total
Treads and risers butted against skirt boards, no dadoes	BC@2.00	Ea	—	60.00	60.00
Treads and risers set in dadoes	BC@3.75	Ea	—	113.00	113.00
Cutting and installing a skirt board on the open side of a stairway Risers miter-jointed into the board	BC@5.50	Ea	—	165.00	165.00

Treads and risers for stairs Including layout, cutting, fitting, and installation of treads, risers and typical moulding under the tread nosing. Includes a miter return on one side of each tread. Cost per 9' rise of 14 treads and 15 risers

	Craft@Hrs	Unit	Material	Labor	Total
9' rise stairway	BC@2.50	Ea	—	75.00	75.00

Handrail for stairs Including layout, cutting, fitting and installation. Based on milled hardwood rail, posts and fittings

	Craft@Hrs	Unit	Material	Labor	Total
Setting newel posts, costs per post	BC@2.50	Ea	—	75.00	75.00
Setting post-to-post handrail Costs per 10 linear feet of rail	BC@1.50	Ea	—	45.00	45.00
Setting over-the-post handrail Costs per 10 linear feet of rail	BC@2.50	Ea	—	75.00	75.00
Add for each transition fitting Turn, easing or gooseneck	BC@1.75	Ea	—	52.50	52.50

Balusters for stair railing Including layout, cutting, fitting and installation. Based on mill-made hardwood balusters. Cost per baluster

	Craft@Hrs	Unit	Material	Labor	Total
Balusters with round top section	BC@.330	Ea	—	9.90	9.90
Balusters with square top section	BC@.750	Ea	—	22.50	22.50

Steel framing rule of thumb Light gauge steel framing cost per square foot of living area. These figures will apply on most crawl space and basement homes. Labor and material cost will be about 30% less for framing over a slab on grade. Complex framing jobs will cost more. The seven lines of cost data under "Steel framing rule of thumb cost breakdown" below show the cost breakdown for framing a medium cost house. Add $1.50 to $3.00 per SF of floor if engineer approval is required (plans not based on code tables). Material cost assumes precut materials fastened with screws. Labor cost includes layout, assembly, plumb and align.

	Craft@Hrs	Unit	Material	Labor	Total
Low cost house	B1@.223	SF	4.96	6.10	11.06
Medium cost house	B1@.268	SF	5.62	7.33	12.95
Better quality house	B1@.355	SF	7.73	9.71	17.44

	Craft@Hrs	Unit	Material	Labor	Total
Steel framing rule of thumb cost breakdown Costs per square foot of living area. These figures were used to compile the framing cost for a medium cost home in the section immediately above.					
Using light gauge steel framing materials at	—	Lb	.54	—	.54
Using 1/2" OSB subfloor at	—	MSF	402.00	—	402.00
Using 1/2" CDX plywood roof sheathing at	—	MSF	756.00	—	756.00
Sills, pier blocks, floor beams (270 Lb per 1,000 SF of floor)	B1@.020	SF	.14	.55	.69
Floor joists, doublers, blocking, bridging (2,200 Lb per 1,000 SF of floor)	B1@.031	SF	1.19	.85	2.04
Subflooring, 1/2" OSB sheathing (1,150 SF per 1,000 SF)	B1@.011	SF	.42	.30	.72
Subfloor, 5/8" CDX plywood (1,150 SF per 1,000 SF of floor)	B1@.012	SF	.61	.33	.94
Layout, studs, top and bottom track, headers, backing, blocking, bracing and framing for openings (2,230 Lb per 1,000 SF of floor)	B1@.102	SF	1.21	2.79	4.00
Ceiling joists, header and end joists, backing, blocking and bracing (1,810 Lb per 1,000 SF of floor)	B1@.050	SF	.98	1.37	2.35
Rafters, braces, collar beams, ridge members, 8" rafters, 24" OC (2,300 Lb per 1000 SF of roof)	B1@.042	SF	1.24	1.15	2.39
Roof sheathing, 1/2" CDX plywood (1,150 SF per 1,000 SF of roof)	B1@.011	SF	.57	.30	.87
Roof sheathing, 7/16" OSB sheathing (1,150 SF per 1,000 SF)	B1@.010	SF	.44	.27	.71
Steel lally columns Residential basement columns, 3-1/2" diameter, steel tube					
Material only	—	LF	7.69	—	7.69
Add for base or cap plate	—	Ea	3.15	—	3.15
Add for installation, per column to 12' high	B1@.458	Ea	—	12.50	12.50
Wood sill plates for steel framing (At foundation) SYP #2 pressure treated lumber, drilled and installed with foundation bolts at 48" OC, no bolts, nuts or washers included. See also wood plates in this section. Figures in parentheses indicate board feet per LF of foundation, including 5% waste and wolmanized treatment.					
2" x 3" sill plates, per MBF	—	MBF	801.00	—	801.00
2" x 3" (.53 BF per LF)	B1@.020	LF	.44	.55	.99
2" x 4" sill plates, per MBF	—	MBF	942.00	—	942.00
2" x 4" (.70 BF per LF)	B1@.023	LF	.69	.63	1.32
2" x 6" sill plates, per MBF	—	MBF	871.00	—	871.00
2" x 6" (1.05 BF per LF)	B1@.024	LF	.96	.66	1.62
2" x 8" sill plates, per MBF	—	MBF	917.00	—	917.00
2" x 8" (1.40 BF per LF)	B1@.031	LF	1.28	.85	2.13
Foundation anchors Set and aligned in forms before concrete is placed. Equivalent Simpson Strong-Tie Co. model numbers are shown in parentheses. Mud sill anchors, embedded in concrete, folded over the track and screw fastened (MAS)	B1@.051	Ea	1.10	1.40	2.50

	Craft@Hrs	Unit	Material	Labor	Total
Anchor bolts, including bolts, nuts and washers as required					
5/8" x 17" (SSTB16)	B1@.157	Ea	2.55	4.30	6.85
5/8" x 21" (SSTB20)	B1@.158	Ea	3.11	4.32	7.43
5/8" x 25" (SSTB24)	B1@.159	Ea	3.14	4.35	7.49
7/8" x 29" (SSTB28)	B1@.163	Ea	8.03	4.46	12.49
7/8" x 34" (SSTB34)	B1@.164	Ea	8.54	4.49	13.03
7/8" x 36" (SSTB36)	B1@.167	Ea	9.14	4.57	13.71
Hold down brackets (one per anchor bolt to distribute pull-out forces over a greater area)					
24-1/8" x 10" x 6-1/4" (HPAHD22-2P)	B1@.165	Ea	7.27	4.51	11.78
24-1/8" x 14" x 6-1/4" (S/HPAHD22)	B1@.169	Ea	5.90	4.62	10.52
13-7/8" x 2-1/2" x 1-1/2" (S/HD8)	B1@.650	Ea	14.17	17.80	31.97
16-1/8" x 2-1/2" x 1-1/2" (S/HD10)	B1@.655	Ea	15.29	17.90	33.19

Steel first floor joists Per SF of area covered. Includes joists and rim joists, typical screws, stiffeners, splices, blocking and strapping. Add the cost of support beams, bridging, trimmer joists and sheathing. See the beam, bridging, trimmer joist and sheathing costs below. Deduct for openings over 25 SF. Figures in parentheses show weight of steel per square foot of floor. Assumes minimal waste.

	Craft@Hrs	Unit	Material	Labor	Total
Using light gauge steel framing materials at	—	Lb	.54	—	.54
Deduct for multiple spans (supported at midpoint)	—	%	-6.0	—	-6.0
6" steel floor joists 16" OC, 2" flange					
600S200-33 (20 gauge, 1.22 Lbs./SF)	B1@.019	SF	.66	.52	1.18
600S200-43 (18 gauge, 1.57 Lbs./SF)	B1@.019	SF	.85	.52	1.37
600S200-54 (16 gauge, 1.96 Lbs./SF)	B1@.021	SF	1.06	.57	1.63
600S200-68 (14 gauge, 2.42 Lbs./SF)	B1@.023	SF	1.31	.63	1.94
600S200-97 (12 gauge, 3.37 Lbs./SF)	B1@.025	SF	1.82	.68	2.50
6" steel floor joists 24" OC, 1-5/8" flange					
600S162-33 (20 gauge, .78 Lbs./SF)	B1@.013	SF	.42	.36	.78
600S162-43 (18 gauge, 1.01 Lbs./SF)	B1@.013	SF	.55	.36	.91
600S162-54 (16 gauge, 1.25 Lbs./SF)	B1@.014	SF	.68	.38	1.06
600S162-68 (14 gauge, 1.55 Lbs./SF)	B1@.016	SF	.84	.44	1.28
600S162-97 (12 gauge, 2.15 Lbs./SF)	B1@.018	SF	1.16	.49	1.65
6" steel floor joists 24" OC, 2" flange					
600S200-33 (20 gauge, .86 Lbs./SF)	B1@.013	SF	.46	.36	.82
600S200-43 (18 gauge, 1.11 Lbs./SF)	B1@.013	SF	.60	.36	.96
600S200-54 (16 gauge, 1.38 Lbs./SF)	B1@.014	SF	.75	.38	1.13
600S200-68 (14 gauge, 1.71 Lbs./SF)	B1@.016	SF	.92	.44	1.36
600S200-97 (12 gauge, 2.37 Lbs./SF)	B1@.018	SF	1.28	.49	1.77
8" steel floor joists 12" OC, 1-5/8" flange					
800S162-33 (20 gauge, 1.47 Lbs./SF)	B1@.025	SF	.79	.68	1.47
800S162-43 (18 gauge, 2.24 Lbs./SF)	B1@.025	SF	1.21	.68	1.89
800S162-54 (16 gauge, 2.78 Lbs./SF)	B1@.028	SF	1.50	.77	2.27
800S162-68 (14 gauge, 3.44 Lbs./SF)	B1@.031	SF	1.86	.85	2.71
800S162-97 (12 gauge, 4.79 Lbs./SF)	B1@.034	SF	2.59	.93	3.52
8" steel floor joists 12" OC, 2" flange					
800S200-33 (20 gauge, 1.87 Lbs./SF)	B1@.025	SF	1.01	.68	1.69
800S200-43 (18 gauge, 2.42 Lbs./SF)	B1@.025	SF	1.31	.68	1.99
800S200-54 (16 gauge, 3.00 Lbs./SF)	B1@.028	SF	1.62	.77	2.39
800S200-68 (14 gauge, 3.74 Lbs./SF)	B1@.031	SF	2.02	.85	2.87
800S200-97 (12 gauge, 5.20 Lbs./SF)	B1@.034	SF	2.81	.93	3.74

	Craft@Hrs	Unit	Material	Labor	Total
8" steel floor joists 16" OC, 1-5/8" flange					
800S162-33 (20 gauge, 1.35 Lbs./SF)	B1@.020	SF	.73	.55	1.28
800S162-43 (18 gauge, 1.73 Lbs./SF)	B1@.020	SF	1.93	.55	2.48
800S162-54 (16 gauge, 2.14 Lbs./SF)	B1@.022	SF	1.16	.60	1.76
800S162-68 (14 gauge, 2.66 Lbs./SF)	B1@.024	SF	1.44	.66	2.10
800S162-97 (12 gauge, 3.70 Lbs./SF)	B1@.026	SF	2.00	.71	2.71
8" steel floor joists 16" OC, 2" flange					
800S200-33 (20 gauge, 1.45 Lbs./SF)	B1@.020	SF	.78	.55	1.33
800S200-43 (18 gauge, 1.87 Lbs./SF)	B1@.020	SF	1.01	.55	1.56
800S200-54 (16 gauge, 2.32 Lbs./SF)	B1@.022	SF	1.25	.60	1.85
800S200-68 (14 gauge, 2.89 Lbs./SF)	B1@.024	SF	1.56	.66	2.22
800S200-97 (12 gauge, 4.01 Lbs./SF)	B1@.026	SF	2.17	.71	2.88
8" steel floor joists 24" OC, 1-5/8" flange					
800S162-33 (20 gauge, .95 Lbs./SF)	B1@.014	SF	.51	.38	.89
800S162-43 (18 gauge, 1.22 Lbs./SF)	B1@.014	SF	.66	.38	1.04
800S162-54 (16 gauge, 1.51 Lbs./SF)	B1@.016	SF	.82	.44	1.26
800S162-68 (14 gauge, 1.87 Lbs./SF)	B1@.017	SF	1.01	.47	1.48
800S162-97 (12 gauge, 2.61 Lbs./SF)	B1@.019	SF	1.41	.52	1.93
8" steel floor joists 24" OC, 2" flange					
800S200-33 (20 gauge, 1.02 Lbs./SF)	B1@.014	SF	.55	.38	.93
800S200-43 (18 gauge, 1.32 Lbs./SF)	B1@.014	SF	.71	.38	1.09
800S200-54 (16 gauge, 1.63 Lbs./SF)	B1@.016	SF	.88	.44	1.32
800S200-68 (14 gauge, 2.03 Lbs./SF)	B1@.017	SF	1.10	.47	1.57
800S200-97 (12 gauge, 2.83 Lbs./SF)	B1@.019	SF	1.53	.52	2.05
10" steel floor joists 12" OC, 1-5/8" flange					
1000S162-43 (18 gauge, 2.62 Lbs./SF)	B1@.028	SF	1.41	.77	2.18
1000S162-54 (16 gauge, 3.24 Lbs./SF)	B1@.030	SF	1.75	.82	2.57
1000S162-68 (14 gauge, 4.04 Lbs./SF)	B1@.033	SF	2.18	.90	3.08
1000S162-97 (12 gauge, 5.63 Lbs./SF)	B1@.037	SF	3.04	1.01	4.05
10" steel floor joists 12" OC, 2" flange					
1000S200-43 (18 gauge, 2.81 Lbs./SF)	B1@.028	SF	1.52	.77	2.29
1000S200-54 (16 gauge, 3.48 Lbs./SF)	B1@.030	SF	1.88	.82	2.70
1000S200-68 (14 gauge, 4.32 Lbs./SF)	B1@.033	SF	2.33	.90	3.23
1000S200-97 (12 gauge, 6.05 Lbs./SF)	B1@.037	SF	3.27	1.01	4.28
10" steel floor joists 16" OC, 1-5/8" flange					
1000S162-43 (18 gauge, 2.02 Lbs./SF)	B1@.022	SF	1.09	.60	1.69
1000S162-54 (16 gauge, 2.50 Lbs./SF)	B1@.024	SF	1.35	.66	2.01
1000S162-68 (14 gauge, 3.12 Lbs./SF)	B1@.027	SF	1.68	.74	2.42
1000S162-97 (12 gauge, 4.35 Lbs./SF)	B1@.029	SF	2.35	.79	3.14
10" steel floor joists 16" OC, 2" flange					
1000S200-43 (18 gauge, 2.17 Lbs./SF)	B1@.022	SF	1.17	.60	1.77
1000S200-54 (16 gauge, 2.69 Lbs./SF)	B1@.024	SF	1.45	.66	2.11
1000S200-68 (14 gauge, 3.34 Lbs./SF)	B1@.027	SF	2.80	.74	3.54
1000S200-97 (12 gauge, 4.67 Lbs./SF)	B1@.029	SF	2.52	.79	3.31
10" steel floor joists 24" OC, 1-5/8" flange					
1000S162-43 (18 gauge, 1.42 Lbs./SF)	B1@.015	SF	.77	.41	1.18
1000S162-54 (16 gauge, 1.76 Lbs./SF)	B1@.017	SF	.95	.47	1.42
1000S162-68 (14 gauge, 2.20 Lbs./SF)	B1@.019	SF	1.19	.52	1.71
1000S162-97 (12 gauge, 3.06 Lbs./SF)	B1@.020	SF	1.65	.55	2.20

	Craft@Hrs	Unit	Material	Labor	Total
10" steel floor joists 24" OC, 2" flange					
1000S200-43 (18 gauge, 1.52 Lbs./SF)	B1@.015	SF	.82	.41	1.23
1000S200-54 (16 gauge, 1.89 Lbs./SF)	B1@.017	SF	1.02	.47	1.49
1000S200-68 (14 gauge, 2.35 Lbs./SF)	B1@.019	SF	1.27	.52	1.79
1000S200-97 (12 gauge, 3.29 Lbs./SF)	B1@.020	SF	1.78	.55	2.33
12" steel floor joists 12" OC, 1-5/8" flange					
1200S162-54 (16 gauge, 3.72 Lbs./SF)	B1@.029	SF	2.01	.79	2.80
1200S162-68 (14 gauge, 4.63 Lbs./SF)	B1@.032	SF	2.50	.88	3.38
1200S162-97 (12 gauge, 6.47 Lbs./SF)	B1@.035	SF	3.49	.96	4.45
12" steel floor joists 12" OC, 2" flange					
1200S200-54 (16 gauge, 3.95 Lbs./SF)	B1@.029	SF	2.13	.79	2.92
1200S200-68 (14 gauge, 4.92 Lbs./SF)	B1@.032	SF	2.66	.88	3.54
1200S200-97 (12 gauge, 6.88 Lbs./SF)	B1@.035	SF	3.72	.96	4.68
12" steel floor joists 16" OC, 1-5/8" flange					
1200S162-54 (16 gauge, 2.88 Lbs./SF)	B1@.023	SF	1.56	.63	2.19
1200S162-68 (14 gauge, 3.57 Lbs./SF)	B1@.025	SF	1.93	.68	2.61
1200S162-97 (12 gauge, 4.99 Lbs./SF)	B1@.028	SF	2.69	.77	3.46
12" steel floor joists 16" OC, 2" flange					
1200S200-54 (16 gauge, 3.05 Lbs./SF)	B1@.023	SF	1.65	.63	2.28
1200S200-68 (14 gauge, 3.80 Lbs./SF)	B1@.025	SF	2.05	.68	2.73
1200S200-97 (12 gauge, 5.31 Lbs./SF)	B1@.028	SF	2.87	.77	3.64
12" steel floor joists 24" OC, 1-5/8" flange					
1200S162-54 (16 gauge, 2.03 Lbs./SF)	B1@.017	SF	1.10	.47	1.57
1200S162-68 (14 gauge, 2.52 Lbs./SF)	B1@.018	SF	1.36	.49	1.85
1200S162-97 (12 gauge, 3.52 Lbs./SF)	B1@.020	SF	1.90	.55	2.45
12" steel floor joists 24" OC, 2" flange					
1200S200-54 (16 gauge, 2.15 Lbs./SF)	B1@.017	SF	1.16	.47	1.63
1200S200-68 (14 gauge, 2.68 Lbs./SF)	B1@.018	SF	1.45	.49	1.94
1200S200-97 (12 gauge, 3.75 Lbs./SF)	B1@.020	SF	2.03	.55	2.58

Built-up steel support beams and girders Box or back-to-back beams made from 1-5/8" flange C-shaped sections. Includes typical fasteners and minimal waste.

	Craft@Hrs	Unit	Material	Labor	Total
Using built-up steel sections at	—	Lb	.54	—	.54
6" beams and girders built-up from two members					
2-600S162-43 (18 gauge, 3.04 Lbs./LF)	B1@.037	LF	1.64	1.01	2.65
2-600S162-54 (16 gauge, 3.78 Lbs./LF)	B1@.038	LF	2.04	1.04	3.08
2-600S162-68 (14 gauge, 4.72 Lbs./LF)	B1@.039	LF	2.55	1.07	3.62
2-600S162-97 (12 gauge, 6.58 Lbs./LF)	B1@.041	LF	3.55	1.12	4.67
8" beams and girders built-up from two members					
2-800S162-43 (18 gauge, 3.66 Lbs./LF)	B1@.049	LF	1.97	1.34	3.31
2-800S162-54 (16 gauge, 4.56 Lbs./LF)	B1@.050	LF	2.46	1.37	3.83
2-800S162-68 (14 gauge, 5.68 Lbs./LF)	B1@.051	LF	3.07	1.40	4.47
2-800S162-97 (12 gauge, 7.96 Lbs./LF)	B1@.053	LF	4.30	1.45	5.75
10" beams and girders built-up from two members					
2-1000S162-43 (18 gauge, 4.26 Lbs./LF)	B1@.063	LF	2.30	1.72	4.02
2-1000S162-54 (16 gauge, 5.23 Lbs./LF)	B1@.064	LF	2.87	1.75	4.62
2-1000S162-68 (14 gauge, 6.66 Lbs./LF)	B1@.065	LF	3.60	1.78	5.38
2-1000S162-97 (12 gauge, 9.34 Lbs./LF)	B1@.069	LF	5.05	1.89	6.94

	Craft@Hrs	Unit	Material	Labor	Total
12" beams and girders built-up from two members					
2-1200S162-54 (16 gauge, 6.10 Lbs./LF)	B1@.073	LF	3.29	2.00	5.29
2-1200S162-68 (14 gauge, 7.62 Lbs./LF)	B1@.075	LF	4.11	2.05	6.16
2-1200S162-97 (12 gauge, 10.72 Lbs./LF)	B1@.079	LF	5.79	2.16	7.95

Steel joist solid blocking Installed between 6" thru 12" joists. 18 gauge, C-shape, with angle. Including typical screws, stiffeners and minimal waste.

	Craft@Hrs	Unit	Material	Labor	Total
Using light gauge steel framing materials at	—	Lb	.54	—	.54
Blocking for 6" steel joists					
Joists on 12" centers	B1@.042	Ea	.89	1.15	2.04
Joists on 16" centers	B1@.042	Ea	1.18	1.15	2.33
Joists on 20" centers	B1@.042	Ea	1.49	1.15	2.64
Joists on 24" centers	B1@.042	Ea	1.78	1.15	2.93
Blocking for 8" steel joists					
Joists on 12" centers	B1@.050	Ea	1.34	1.37	2.71
Joists on 16" centers	B1@.050	Ea	1.78	1.37	3.15
Joists on 20" centers	B1@.050	Ea	2.22	1.37	3.59
Joists on 24" centers	B1@.050	Ea	2.67	1.37	4.04
Blocking for 10" steel joists					
Joists on 12" centers	B1@.057	Ea	1.56	1.56	3.12
Joists on 16" centers	B1@.057	Ea	2.08	1.56	3.64
Joists on 20" centers	B1@.057	Ea	2.60	1.56	4.16
Joists on 24" centers	B1@.057	Ea	3.09	1.56	4.65
Blocking for 12" steel joists					
Joists on 12" centers	B1@.064	Ea	1.80	1.75	3.55
Joists on 16" centers	B1@.064	Ea	2.39	1.75	4.14
Joists on 20" centers	B1@.064	Ea	2.98	1.75	4.73
Joists on 24" centers	B1@.064	Ea	3.57	1.75	5.32

Steel cross bridging for steel floor joists Using strap material or manufactured straps installed between 6" thru 12" joists, cost per pair of straps including typical fasteners.

	Craft@Hrs	Unit	Material	Labor	Total
Joists on 12" centers	B1@.025	Pr	2.37	.68	3.05
Joists on 16" centers	B1@.025	Pr	2.47	.68	3.15
Joists on 20" centers	B1@.025	Pr	2.70	.68	3.38
Joists on 24" centers	B1@.025	Pr	3.16	.68	3.84

Board sheathing subfloor 1" x 6" #3 & Btr (1.28 BF per SF), screwed or pinned to steel floor joists, including fasteners.

	Craft@Hrs	Unit	Material	Labor	Total
Board sheathing, per MBF	—	MBF	1,340.00	—	1,340.00
With 12% shrinkage, 5% waste & screws	B1@.022	SF	1.59	.60	2.19
Add for diagonal patterns	B1@.001	SF	.16	.03	.19

Plywood subfloor CD standard exterior grade. Material costs shown include 5% for waste, screwed or pinned to steel floor joists, including fasteners.

	Craft@Hrs	Unit	Material	Labor	Total
3/8"	B1@.012	SF	.40	.33	.73
1/2"	B1@.013	SF	.52	.36	.88
5/8"	B1@.013	SF	.61	.36	.97
3/4"	B1@.014	SF	.79	.38	1.17
1-1/8"	B1@.015	SF	1.47	.41	1.88

	Craft@Hrs	Unit	Material	Labor	Total
OSB sheathing, Material costs shown include 5% for waste & fasteners					
7/16"	B1@.011	SF	.38	.30	.68
1/2"	B1@.012	SF	.42	.33	.75
5/8"	B1@.012	SF	.58	.33	.91
3/4" tongue and groove	B1@.014	SF	.71	.38	1.09

Steel interior wall studs Per square foot of partition wall area measured one side. Do not subtract for openings less than 16' wide. Costs include studs, top and bottom track, screws and splices. Add for each corner or wall junction as indicated below. Add the cost of wood plates (if used), support beams, door opening framing and corners from below. All studs with 1-1/4" flange are assumed to have a 3/16" stiffening lip. Labor includes layout, assembly, plumb and align. Figures in parentheses show weight of steel per square foot of wall measured one side. Assumes minimal waste. See the following section for exterior wall estimates.

WALL FRAMING

SHEATHING MATERIAL
(SUCH AS GYPSUM BOARD, OSB, OR PLYWOOD)

	Craft@Hrs	Unit	Material	Labor	Total
Using light gauge steel framing materials at	—	Lb	.50	—	.50
2-1/2" steel interior wall framing, 16" OC, 1-1/4" flange					
250S125-18 (25 gauge, .33 Lbs./SF)	B1@.018	SF	.17	.49	.66
250S125-27 (22 gauge, .49 Lbs./ SF)	B1@.018	SF	.25	.49	.74
250S125-30 (20DW gauge, .54 Lbs./SF)	B1@.019	SF	.27	.52	.79
250S125-33 (20 gauge, .60 Lbs./SF)	B1@.020	SF	.30	.55	.85
2-1/2" steel interior wall framing, 24" OC, 1-1/4" flange					
250S125-18 (25 gauge, .25 Lbs./SF)	B1@.017	SF	.13	.47	.60
250S125-27 (22 gauge, .37 Lbs./SF)	B1@.017	SF	.19	.47	.66
250S125-30 (20DW gauge, .41 Lbs./SF)	B1@.018	SF	.21	.49	.70
250S125-33 (20 gauge, .45 Lbs./SF)	B1@.019	SF	.23	.52	.75
Add for each 2-1/2" corner or wall junction	B1@.091	Ea	2.26	2.49	4.75
3-1/2" steel interior wall framing, 16" OC, 1-1/4" flange					
350S125-18 (25 gauge, .39 Lbs./SF)	B1@.023	SF	.20	.63	.83
350S125-27 (22 gauge, .59 Lbs./SF)	B1@.023	SF	.30	.63	.93
350S125-30 (20DW gauge, .65 Lbs./SF)	B1@.024	SF	.33	.66	.99
350S125-33 (20 gauge, .72 Lbs./SF)	B1@.025	SF	.36	.68	1.04
3-1/2" steel interior wall framing, 24" OC, 1-1/4" flange					
350S125-18 (25 gauge, .29 Lbs./SF)	B1@.020	SF	.15	.55	.70
350S125-27 (22 gauge, .44 Lbs./SF)	B1@.020	SF	.22	.55	.77
350S125-30 (20DW gauge, .49 Lbs./SF)	B1@.022	SF	.25	.60	.85
350S125-33 (20 gauge, .54 Lbs./SF)	B1@.023	SF	.27	.63	.90
Add for each 3-1/2" corner or wall junction	B1@.091	Ea	2.40	2.49	4.89
3-5/8" steel interior wall framing, 16" OC, 1-1/4" flange					
362S125-18 (25 gauge, .40 Lbs./SF)	B1@.023	SF	.20	.63	.83
362S125-27 (22 gauge, .60 Lbs./SF)	B1@.023	SF	.30	.63	.93
362S125-30 (20DW gauge, .66 Lbs./SF)	B1@.024	SF	.33	.66	.99
362S125-33 (20 gauge, .73 Lbs./SF)	B1@.025	SF	.37	.68	1.05

	Craft@Hrs	Unit	Material	Labor	Total
3-5/8" steel interior wall framing, 24" OC, 1-1/4" flange					
362S125-18 (25 gauge, .30 Lbs./SF)	B1@.021	SF	.15	.57	.72
362S125-27 (22 gauge, .45 Lbs./SF)	B1@.021	SF	.23	.57	.80
362S125-30 (20DW gauge, .50 Lbs./SF)	B1@.022	SF	.25	.60	.85
362S125-33 (20 gauge, .55 Lbs./SF)	B1@.023	SF	.28	.63	.91
Add for each 3-5/8" corner or wall junction	B1@.091	Ea	2.40	2.49	4.89
5-1/2" steel interior wall framing, 16" OC, 1-1/4" flange					
550S125-18 (25 gauge, .52 Lbs./SF)	B1@.029	SF	.26	.79	1.05
550S125-27 (22 gauge, .78 Lbs./SF)	B1@.029	SF	.39	.79	1.18
550S125-30 (20DW gauge, .86 Lbs./SF)	B1@.030	SF	.43	.82	1.25
550S125-33 (20 gauge, .95 Lbs./SF)	B1@.032	SF	.48	.88	1.36
5-1/2" steel interior wall framing, 24" OC, 1-1/4" flange					
550S125-18 (25 gauge, .39 Lbs./SF)	B1@.028	SF	.20	.77	.97
550S125-27 (22 gauge, .59 Lbs./SF)	B1@.028	SF	.30	.77	1.07
550S125-30 (20DW gauge, .65 Lbs./SF)	B1@.029	SF	.33	.79	1.12
550S125-33 (20 gauge, .71 Lbs./SF)	B1@.030	SF	.36	.82	1.18
Add for each 5-1/2" corner or wall junction	B1@.091	Ea	3.15	2.49	5.64
6" steel interior wall framing, 16" OC, 1-1/4" flange					
600S125-18 (25 gauge, .55 Lbs./SF)	B1@.028	SF	.28	.77	1.05
600S125-27 (22 gauge, .83 Lbs./SF)	B1@.029	SF	.42	.79	1.21
600S125-30 (20DW gauge), .91 Lbs./SF	B1@.030	SF	.46	.82	1.28
600S125-33 (20 gauge, 1.01 Lbs./SF)	B1@.032	SF	.51	.88	1.39
6" steel interior wall framing, 24" OC, 1-1/4" flange					
600S125-18 (25 gauge, .41 Lbs./SF)	B1@.027	SF	.21	.74	.95
600S125-27 (22 gauge, .62 Lbs./SF)	B1@.027	SF	.31	.74	1.05
600S125-30 (20DW gauge, .68 Lbs./SF)	B1@.028	SF	.34	.77	1.11
600S125-33 (20 gauge, .76 Lbs./SF)	B1@.030	SF	.38	.82	1.20
Add for each 6" corner or wall junction	B1@.091	Ea	3.35	2.49	5.84

BEND SECTION OR CLIP FLANGE TO FORM VERTICAL

FLAT STRAP

WALL FRAMING

BLOCKING @ ENDS OF STRAP & INTERMITTENTLY AS REQUIRED

SCREW(S) AS REQUIRED @ EACH STRAP TO STUD

Steel exterior wall studs Per square foot of wall area measures one side. Do not subtract for openings less than 16' wide. Add for each corner and partition from below. Costs include studs, top and bottom track, screws and splices. Add for door and window opening framing, backing, cross-bracing and sheathing for shear walls from the sections that follow. Add the cost of wood plates (if used), support beams, door opening framing, corners and wall junctions from below. All studs with 1-5/8" flange are assumed to have a 1/2" stiffening lip. Labor includes layout, assembly, plumb and align. Figures in parentheses show weight of steel per square foot of wall measured one side. Assumes minimal waste. See the previous section for interior (partition) wall estimates.

	Craft@Hrs	Unit	Material	Labor	Total
Using light gauge steel framing materials at	—	Lb	.54	—	.54
3-1/2" steel exterior wall framing, 12" OC, 1-5/8" flange					
350S162-33 (20 gauge, 1.10 Lbs./SF)	B1@.026	SF	.59	.71	1.30
350S162-43 (18 gauge, 1.43 Lbs./SF)	B1@.026	SF	.77	.71	1.48
350S162-54 (16 gauge, 1.76 Lbs./SF)	B1@.028	SF	.95	.77	1.72
350S162-68 (14 gauge, 2.19 Lbs./SF)	B1@.029	SF	1.18	.79	1.97

Carpentry, Steel Framing

	Craft@Hrs	Unit	Material	Labor	Total
3-1/2" steel exterior wall framing, 16" OC, 1-5/8" flange					
350S162-33 (20 gauge, .88 Lbs./SF)	B1@.025	SF	.48	.68	1.16
350S162-43 (18 gauge, 1.14 Lbs./SF)	B1@.025	SF	.62	.68	1.30
350S162-54 (16 gauge, 1.41 Lbs./SF)	B1@.028	SF	.76	.77	1.53
350S162-68 (14 gauge, 1.75 Lbs./SF)	B1@.029	SF	.95	.79	1.74
3-1/2" steel exterior wall framing, 24" OC, 1-5/8" flange					
350S162-33 (20 gauge, .66 Lbs./SF)	B1@.023	SF	.36	.63	.99
350S162-43 (18 gauge, .86 Lbs./SF)	B1@.023	SF	.46	.63	1.09
350S162-54 (16 gauge, 1.06 Lbs./SF)	B1@.025	SF	.57	.68	1.25
350S162-68 (14 gauge, 1.31 Lbs./SF)	B1@.026	SF	.71	.71	1.42
Add for each 3-1/2" corner or partition (33 mil)	B1@.091	Ea	3.19	2.49	5.68
3-5/8" steel exterior wall framing, 12" OC, 1-5/8" flange					
362S162-33 (20 gauge, 1.11 Lbs./SF)	B1@.026	SF	.60	.71	1.31
362S162-43 (18 gauge, 1.45 Lbs./SF)	B1@.026	SF	.78	.71	1.49
362S162-54 (16 gauge, 1.80 Lbs./SF)	B1@.028	SF	.97	.77	1.74
362S162-68 (14 gauge, 2.23 Lbs./SF)	B1@.029	SF	1.20	.79	1.99
3-5/8" steel exterior wall framing, 16" OC, 1-5/8" flange					
362S162-33 (20 gauge, .89 Lbs./SF)	B1@.025	SF	.48	.68	1.16
362S162-43 (18 gauge, 1.16 Lbs./SF)	B1@.025	SF	.63	.68	1.31
362S162-54 (16 gauge, 1.44 Lbs./SF)	B1@.028	SF	.78	.77	1.55
362S162-68 (14 gauge, 1.78 Lbs./SF)	B1@.029	SF	.96	.79	1.75
3-5/8" steel exterior wall framing, 24" OC, 1-5/8" flange					
362S162-33 (20 gauge, .67 Lbs./SF)	B1@.023	SF	.36	.63	.99
362S162-43 (18 gauge, .87 Lbs./SF)	B1@.023	SF	.47	.63	1.10
362S162-54 (16 gauge, 1.08 Lbs./SF)	B1@.025	SF	.58	.68	1.26
362S162-68 (14 gauge, 1.34 Lbs./SF)	B1@.026	SF	.72	.71	1.43
Add for each 3-5/8" corner or partition (33 mil)	B1@.091	Ea	3.27	2.49	5.76
5-1/2" steel exterior wall framing, 12" OC, 1-5/8" flange					
550S162-33 (20 gauge, 1.39 Lbs./SF)	B1@.035	SF	.75	.96	1.71
550S162-43 (18 gauge, 1.80 Lbs./SF)	B1@.036	SF	.97	.98	1.95
550S162-54 (16 gauge, 2.25 Lbs./SF)	B1@.039	SF	1.22	1.07	2.29
550S162-68 (14 gauge, 2.80 Lbs./SF)	B1@.041	SF	1.51	1.12	2.63
5-1/2" steel exterior wall framing, 16" OC, 1-5/8" flange					
550S162-33 (20 gauge, 1.11 Lbs./SF)	B1@.032	SF	.60	.88	1.48
550S162-43 (18 gauge, 1.44 Lbs./SF)	B1@.033	SF	.78	.90	1.68
550S162-54 (16 gauge, 1.80 Lbs./SF)	B1@.035	SF	.97	.96	1.93
550S162-68 (14 gauge, 2.24 Lbs./SF)	B1@.037	SF	1.21	1.01	2.22
5-1/2" steel exterior wall framing, 20" OC, 1-5/8" flange					
550S162-33 (20 gauge, .94 Lbs./SF)	B1@.031	SF	.51	.85	1.36
550S162-43 (18 gauge, 1.22 Lbs./SF)	B1@.032	SF	.66	.88	1.54
550S162-54 (16 gauge, 1.53 Lbs./SF)	B1@.034	SF	.83	.93	1.76
550S162-68 (14 gauge, 1.90 Lbs./SF)	B1@.036	SF	1.03	.98	2.01

	Craft@Hrs	Unit	Material	Labor	Total
5-1/2" steel exterior wall framing, 24" OC, 1-5/8" flange					
550S162-33 (20 gauge, .83 Lbs./SF)	B1@.030	SF	.45	.82	1.27
550S162-43 (18 gauge, 1.08 Lbs./SF)	B1@.031	SF	.58	.85	1.43
550S162-54 (16 gauge, 1.35 Lbs./SF)	B1@.033	SF	.73	.90	1.63
550S162-68 (14 gauge, 1.68 Lbs./SF)	B1@.035	SF	.91	.96	1.87
Add for each 5-1/2" corner or partition (43 mil)	B1@.091	Ea	4.06	2.49	6.55
6" steel exterior wall framing, 12" OC, 1-5/8" flange					
600S162-33 (20 gauge, 1.46 Lbs./SF)	B1@.035	SF	.79	.96	1.75
600S162-43 (18 gauge, 1.90 Lbs./SF)	B1@.036	SF	1.03	.98	2.01
600S162-54 (16 gauge, 2.36 Lbs./SF)	B1@.039	SF	1.27	1.07	2.34
600S162-68 (14 gauge, 2.95 Lbs./SF)	B1@.041	SF	1.59	1.12	2.71
600S162-97 (12 gauge, 4.11 Lbs./SF)	B1@.043	SF	2.22	1.18	3.40
6" steel exterior wall framing, 16" OC, 1-5/8" flange					
600S162-33 (20 gauge, 1.17 Lbs./SF)	B1@.032	SF	.63	.88	1.51
600S162-43 (18 gauge, 1.52 Lbs./SF)	B1@.033	SF	.82	.90	1.72
600S162-54 (16 gauge, 1.89 Lbs./SF)	B1@.035	SF	1.02	.96	1.98
600S162-68 (14 gauge, 2.36 Lbs./SF)	B1@.037	SF	1.27	1.01	2.28
600S162-97 (12 gauge, 3.29 Lbs./SF)	B1@.039	SF	1.78	1.07	2.85
6" steel exterior wall framing, 20" OC, 1-5/8" flange					
600S162-33 (20 gauge, .99 Lbs./SF)	B1@.031	SF	.53	.85	1.38
600S162-43 (18 gauge, 1.29 Lbs./SF)	B1@.032	SF	.70	.88	1.58
600S162-54 (16 gauge, 1.60 Lbs./SF)	B1@.034	SF	.86	.93	1.79
600S162-68 (14 gauge, 2.00 Lbs./SF)	B1@.036	SF	1.08	.98	2.06
600S162-97 (12 gauge), 2.79 Lbs./SF	B1@.038	SF	1.51	1.04	2.55
6" steel exterior wall framing, 24" OC, 1-5/8" flange					
600S162-33 (20 gauge, .88 Lbs./SF)	B1@.030	SF	.48	.82	1.30
600S162-43 (18 gauge, 1.14 Lbs./SF)	B1@.031	SF	.62	.85	1.47
600S162-54 (16 gauge, 1.42 Lbs./SF)	B1@.033	SF	.77	.90	1.67
600S162-68 (14 gauge, 1.77 Lbs./SF)	B1@.035	SF	.96	.96	1.92
600S162-97 (12 gauge, 2.47 Lbs./SF)	B1@.037	SF	1.33	1.01	2.34
Add for each 6" corner or partition (43 mil)	B1@.091	Ea	4.25	2.49	6.74

Door opening steel framing In wall studs, based on walls 8' high. Costs shown are per each door opening and include header of appropriate size, double king studs each side of the opening for doors greater than 3' wide, double trimmer studs each side of the opening for doors greater than 8' wide, cripples, blocking, screws and normal waste. Width shown is size of finished opening. Figures in parentheses indicate header size.

	Craft@Hrs	Unit	Material	Labor	Total
Using wall studs and headers, per Lb	—	Lb	.54	—	.54
Door opening steel framing in 3-1/2" wall, 20 gauge (33 mil) stud wall, opening size as shown					
To 3' wide, 6" header, 18 gauge, two 600S162-43, 40.05 Lbs	B1@1.00	Ea	21.60	27.40	49.00
Over 3' to 4' wide, 6" header, 16 gauge, two 600S162-54, 56.48 Lbs	B1@1.33	Ea	30.50	36.40	66.90
Over 4' to 5' wide, 8" header, 16 gauge, two 800S162-54, 63.85 Lbs	B1@1.67	Ea	34.50	45.70	80.20

	Craft@Hrs	Unit	Material	Labor	Total
Over 5' to 6' wide, 8" header, 16 gauge, two 800S162-54, 68.43 Lbs	B1@1.99	Ea	37.00	54.40	91.40
Over 6' to 8' wide, 10" header, 14 gauge, two 1000S162-68, 107.25 Lbs	B1@2.33	Ea	57.90	63.70	121.60
Over 8' to 10' wide, 12" header, 14 gauge, two 1200S162-68, 130.15 Lbs	B1@2.33	Ea	70.30	63.70	134.00
Over 10' to 12' wide, 12" header, 12 gauge, two 1200S162-97, 182.60 Lbs	B1@2.67	Ea	98.60	73.10	171.70
Add per foot of height for 3-1/2" walls over 8' in height	—	LF	3.11	—	3.11

Door opening steel framing in 5-1/2" wall, 18 gauge (43 mil) stud wall, opening size as shown

	Craft@Hrs	Unit	Material	Labor	Total
To 3' wide, 6" header, 18 gauge, two 600S162-43, 53.75 Lbs	B1@1.00	Ea	29.00	27.40	56.40
Over 3' to 4' wide, 6" header, 16 gauge, two 600S162-54, 82.80 Lbs	B1@1.33	Ea	44.70	36.40	81.10
Over 4' to 5' wide, 8" header, 16 gauge, two 800S162-54, 90.00 Lbs	B1@1.67	Ea	48.60	45.70	94.30
Over 5' to 6' wide, 8" header, 16 gauge, two 800S162-54, 94.55 Lbs	B1@1.99	Ea	51.10	54.40	105.50
Over 6' to 8' wide, 10" header, 14 gauge, two 1000S162-68, 120.00 Lbs	B1@2.33	Ea	64.80	63.70	128.50
Over 8' to 10' wide, 12" header, 14 gauge, two 1200S162-68, 162.60 Lbs	B1@2.33	Ea	87.80	63.70	151.50
Over 10' to 12' wide, 12" header, 12 gauge, two 1200S162-97, 215.05 Lbs	B1@2.67	Ea	116.00	73.10	189.10
Add per foot of height for 5-1/2" walls over 8' in height	—	LF	5.07	—	5.07

Window opening steel framing In wall studs, based on walls 8' high. Costs shown are per window opening and include header of appropriate size, sill track, double king studs each side of the opening for windows greater than 3' wide, double trimmer studs each side of the opening for windows greater than 8' wide, top and bottom cripples, blocking, screws and normal waste. Width shown is size of finished opening.

	Craft@Hrs	Unit	Material	Labor	Total
Using steel framing materials at	—	Lb	.54	—	.54

Window opening steel framing in 3-1/2" wall, 20 gauge (33 mil) stud wall, opening size shown

	Craft@Hrs	Unit	Material	Labor	Total
To 2' wide, 4" header, 18 gauge, two 400S162-43, 35.18 Lbs	B1@1.20	Ea	19.00	32.80	51.80
Over 2' to 3' wide, 6" header, 18 gauge, two 600S162-43, 40.48 Lbs	B1@1.40	Ea	21.90	38.30	60.20
Over 3' to 4' wide, 6" header, 16 gauge, two 600S162-54, 61.85 Lbs	B1@1.74	Ea	33.40	47.60	81.00
Over 4' to 5' wide, 8" header, 16 gauge, two 800S162-54, 71.45 Lbs	B1@2.08	Ea	38.60	56.90	95.50
Over 5' to 6' wide, 8" header, 16 gauge, two 800S162-54, 77.55 Lbs	B1@2.41	Ea	41.90	65.90	107.80
Over 6' to 7' wide, 8" header, 14 gauge, two 800S162-68, 91.45 Lbs	B1@2.75	Ea	49.40	75.20	124.60
Over 7' to 8' wide, 10" header, 14 gauge, two 1000S162-68, 106.23 Lbs	B1@3.08	Ea	57.40	84.30	141.70

	Craft@Hrs	Unit	Material	Labor	Total
Over 8' to 10' wide, 12" header, 14 gauge,					
two 1200S162-68, 144.20 Lbs	B1@3.08	Ea	77.90	84.30	162.20
Over 10' to 12' wide, 12" header, 12 gauge,					
two 1200S162-97, 199.68 Lbs	B1@3.42	Ea	108.00	93.60	201.60
Add per foot of height for 3-1/2"					
walls over 8' high	—	LF	3.11	—	3.11

Window opening steel framing in 5-1/2" wall, 18 gauge, 43 mil stud wall, opening size as shown

	Craft@Hrs	Unit	Material	Labor	Total
To 2' wide, 4" header, 18 gauge,					
two 400S162-43, 55.00 Lbs	B1@1.20	Ea	29.70	32.80	62.50
Over 2' to 3' wide, 6" header, 18 gauge,					
two 600S162-43, 61.18 Lbs	B1@1.40	Ea	33.00	38.30	71.30
Over 3' to 4' wide, 6" header, 16 gauge,					
two 600S162-54, 92.62 Lbs	B1@1.74	Ea	50.00	47.60	97.60
Over 4' to 5' wide, 8" header, 16 gauge,					
two 800S162-54, 103.15 Lbs	B1@2.08	Ea	55.70	56.90	112.60
Over 5' to 6' wide, 8" header, 16 gauge,					
two 800S162-54, 110.30 Lbs	B1@2.41	Ea	59.60	65.90	125.50
Over 6' to 8' wide, 10" header, 14 gauge,					
two 1000S162-68, 140.95 Lbs	B1@3.08	Ea	76.10	84.30	160.40
Over 8' to 10' wide, 12" header, 14 gauge,					
two 1200S162-68, 188.78 Lbs	B1@3.08	Ea	102.00	84.30	186.30
Over 8' to 12' wide, 12" header, 12 gauge,					
two 1200S162-97, 246.43 Lbs	B1@3.42	Ea	133.00	93.60	226.60
Add per foot of height for 5-1/2"					
walls over 8' high	—	LF	5.07	—	5.07

Steel wall bracing See also sheathing for plywood bracing and shear panels.

Steel strap bracing

	Craft@Hrs	Unit	Material	Labor	Total
4" wide 33 mil with gusset plates	B1@.015	LF	.32	.41	.73
4" wide 43 mil with gusset plates	B1@.015	LF	.42	.41	.83
8.8" wide 33 mil	B1@.015	LF	.66	.41	1.07
8.8" wide 43 mil	B1@.015	LF	.86	.41	1.27
8.8" wide 54 mil	B1@.015	LF	1.08	.41	1.49

Intermediate bracing 3-1/2", 33 mil studs or track

	Craft@Hrs	Unit	Material	Labor	Total
350S162-33 (0.88 Lbs./LF)	B1@.019	LF	.44	.52	.96

Temporary wood bracing for walls, assumes salvage at 50% and 3 uses

	Craft@Hrs	Unit	Material	Labor	Total
1" x 4" Std & Btr	B1@.006	LF	.13	.16	.29
1" x 6" Std & Btr	B1@.010	LF	.19	.27	.46
2" x 4" utility	B1@.012	LF	.10	.33	.43
2" x 6" utility	B1@.018	LF	.15	.49	.64

Wood wall plates Std & Btr, untreated. For pressure treated plates, see Wood sill plates for steel framing previously in this section. Figures in parentheses indicate board feet per LF. Costs shown include 10% waste.

	Craft@Hrs	Unit	Material	Labor	Total
2" x 3" wall plates, per MBF	—	MBF	679.00	—	679.00
2" x 3" (.55 BF per LF)	B1@.010	LF	.37	.27	.64
2" x 4" wall plates, per MBF	—	MBF	621.00	—	621.00
2" x 4" (.73 BF per LF)	B1@.012	LF	.45	.33	.78
2" x 6" wall plates, per MBF	—	MBF	620.00	—	620.00
2" x 6" (1.10 BF per LF)	B1@.018	LF	.68	.49	1.17

Carpentry, Steel Framing

	Craft@Hrs	Unit	Material	Labor	Total

Steel ceiling joists and soffits Per SF of area covered. Including typical stiffeners, splices, bracing, rim joists and screws. Add the cost of beams, bridging, blocking, ledger strips and trimmers as required. See costs for beams, bridging, backing, blocking and trimmers elsewhere in this section. Deduct for openings over 25 SF. Labor includes layout, assembly and alignment. Figures in parentheses show weight of steel per square foot of ceiling. Assumes minimal waste.

	Craft@Hrs	Unit	Material	Labor	Total
Using light gauge steel framing materials at	—	Lb	.54	—	.54
3-1/2" steel ceiling joists, 16" OC, 1-5/8" flange					
350S162-33 (20 gauge, .88 Lbs./SF)	B1@.022	SF	.48	.60	1.08
350S162-43 (18 gauge), 1.11 Lbs./SF	B1@.024	SF	.60	.66	1.26
350S162-54 (16 gauge, 1.35 Lbs./SF)	B1@.027	SF	.73	.74	1.47
350S162-68 (14 gauge, 1.65 Lbs./SF)	B1@.029	SF	.89	.79	1.68
3-1/2" steel ceiling joists, 24" OC, 1-5/8" flange					
350S162-33 (20 gauge, .63 Lbs./SF)	B1@.015	SF	.34	.41	.75
350S162-43 (18 gauge, .79 Lbs./SF)	B1@.017	SF	.43	.47	.90
350S162-54 (16 gauge, .96 Lbs./SF)	B1@.019	SF	.52	.52	1.04
350S162-68 (14 gauge, 1.17 Lbs./SF)	B1@.020	SF	.63	.55	1.18
4" steel ceiling joists, 16" OC, 1-5/8" flange					
400S162-33 (20 gauge, .93 Lbs./SF)	B1@.022	SF	.51	.60	1.11
400S162-43 (18 gauge), 1.17 Lbs./SF	B1@.024	SF	.63	.66	1.29
400S162-54 (16 gauge, 1.44 Lbs./SF)	B1@.027	SF	.78	.74	1.52
400S162-68 (14 gauge, 1.76 Lbs./SF)	B1@.029	SF	.95	.79	1.74
4" steel ceiling joists, 24" OC, 1-5/8" flange					
400S162-33 (20 gauge, .67 Lbs./SF)	B1@.015	SF	.36	.41	.77
400S162-43 (18 gauge, .84 Lbs./SF)	B1@.017	SF	.45	.47	.92
400S162-54 (16 gauge, 1.02 Lbs./SF)	B1@.019	SF	.55	.52	1.07
400S162-68 (14 gauge, 1.25 Lbs./SF)	B1@.020	SF	.68	.55	1.23
5-1/2" steel ceiling joists, 12" OC, 1-5/8" flange					
550S162-33 (20 gauge, 1.41 Lbs./SF)	B1@.029	SF	.76	.79	1.55
550S162-43 (18 gauge, 1.79 Lbs./SF)	B1@.032	SF	.97	.88	1.85
550S162-54 (16 gauge, 2.21 Lbs./SF)	B1@.035	SF	1.19	.96	2.15
550S162-68 (14 gauge, 2.72 Lbs./SF)	B1@.038	SF	1.47	1.04	2.51
5-1/2" steel ceiling joists, 16" OC, 1-5/8" flange					
550S162-33 (20 gauge, 1.09 Lbs./SF)	B1@.022	SF	.59	.60	1.19
550S162-43 (18 gauge, 1.39 Lbs./SF)	B1@.024	SF	.75	.66	1.41
550S162-54 (16 gauge, 1.71 Lbs./SF)	B1@.027	SF	.92	.74	1.66
550S162-68 (14 gauge, 2.10 Lbs./SF)	B1@.029	SF	1.13	.79	1.92
5-1/2" steel ceiling joists, 24" OC, 1-5/8" flange					
550S162-33 (20 gauge, .78 Lbs./SF)	B1@.016	SF	.42	.44	.86
550S162-43 (18 gauge, .99 Lbs./SF)	B1@.018	SF	.53	.49	1.02
550S162-54 (16 gauge, 1.21 Lbs./SF)	B1@.020	SF	.65	.55	1.20
550S162-68 (14 gauge, 1.48 Lbs./SF)	B1@.022	SF	.80	.60	1.40
6" steel ceiling joists, 12" OC, 1-5/8" flange					
600S162-33 (20 gauge, 1.48 Lbs./SF)	B1@.029	SF	.80	.79	1.59
600S162-43 (18 gauge, 1.89 Lbs./SF)	B1@.032	SF	1.02	.88	1.90
600S162-54 (16 gauge, 2.32 Lbs./SF)	B1@.035	SF	1.25	.96	2.21
600S162-68 (14 gauge, 2.86 Lbs./SF)	B1@.038	SF	1.54	1.04	2.58

	Craft@Hrs	Unit	Material	Labor	Total
6" steel ceiling joists, 16" OC, 1-5/8" flange					
600S162-33 (20 gauge, 1.15 Lbs./SF)	B1@.022	SF	.62	.60	1.22
600S162-43 (18 gauge, 1.46 Lbs./SF)	B1@.024	SF	.79	.66	1.45
600S162-54 (16 gauge, 1.79 Lbs./SF)	B1@.027	SF	.97	.74	1.71
600S162-68 (14 gauge, 2.21 Lbs./SF)	B1@.029	SF	1.19	.79	1.98
6" steel ceiling joists, 24" OC, 1-5/8" flange					
600S162-33 (20 gauge, .82 Lbs./SF)	B1@.016	SF	.44	.44	.88
600S162-43 (18 gauge, 1.04 Lbs./SF)	B1@.018	SF	.56	.49	1.05
600S162-54 (16 gauge, 1.27 Lbs./SF)	B1@.020	SF	.69	.55	1.24
600S162-68 (14 gauge, 1.56 Lbs./SF)	B1@.022	SF	.84	.60	1.44
8" steel ceiling joists, 12" OC, 1-5/8" flange					
800S162-33 (20 gauge, 1.78 Lbs./SF)	B1@.031	SF	.96	.85	1.81
800S162-43 (18 gauge, 2.26 Lbs./SF)	B1@.034	SF	1.22	.93	2.15
800S162-54 (16 gauge, 2.79 Lbs./SF)	B1@.037	SF	1.50	1.01	2.51
800S162-68 (14 gauge, 3.44 Lbs./SF)	B1@.041	SF	1.86	1.12	2.98
800S162-97 (12 gauge, 4.76 Lbs./SF)	B1@.049	SF	2.57	1.34	3.91
8" steel ceiling joists, 16" OC, 1-5/8" flange					
800S162-33 (20 gauge, 1.38 Lbs./SF)	B1@.024	SF	.75	.66	1.41
800S162-43 (18 gauge, 1.75 Lbs./SF)	B1@.027	SF	.95	.74	1.69
800S162-54 (16 gauge, 2.15 Lbs./SF)	B1@.029	SF	1.16	.79	1.95
800S162-68 (14 gauge, 2.65 Lbs./SF)	B1@.032	SF	1.43	.88	2.31
800S162-97 (12 gauge, 3.67 Lbs./SF)	B1@.039	SF	1.98	1.07	3.05
8" steel ceiling joists, 20" OC, 1-5/8" flange					
800S162-33 (20 gauge, 1.14 Lbs./SF)	B1@.020	SF	.62	.55	1.17
800S162-43 (18 gauge, 1.44 Lbs./SF)	B1@.022	SF	.78	.60	1.38
800S162-54 (16 gauge, 1.77 Lbs./SF)	B1@.024	SF	.96	.66	1.62
800S162-68 (14 gauge, 2.18 Lbs./SF)	B1@.026	SF	1.18	.71	1.89
800S162-97 (12 gauge, 3.01 Lbs./SF)	B1@.032	SF	1.63	.88	2.51
8" steel ceiling joists, 24" OC, 1-5/8" flange					
800S162-33 (20 gauge, .98 Lbs./SF)	B1@.018	SF	.53	.49	1.02
800S162-43 (18 gauge, 1.24 Lbs./SF)	B1@.019	SF	.67	.52	1.19
800S162-54 (16 gauge, 1.52 Lbs./SF)	B1@.021	SF	.82	.57	1.39
800S162-68 (14 gauge, 1.87 Lbs./SF)	B1@.023	SF	1.01	.63	1.64
800S162-97 (12 gauge, 2.57 Lbs./SF)	B1@.028	SF	1.39	.77	2.16
10" steel ceiling joists, 12" OC, 1-5/8" flange					
1000S162-43 (18 gauge, 2.63 Lbs./SF)	B1@.033	SF	1.42	.90	2.32
1000S162-54 (16 gauge, 3.24 Lbs./SF)	B1@.036	SF	1.75	.98	2.73
1000S162-68 (14 gauge, 4.02 Lbs./SF)	B1@.040	SF	2.17	1.09	3.26
1000S162-97 (12 gauge, 5.58 Lbs./SF)	B1@.048	SF	3.01	1.31	4.32
10" steel ceiling joists, 16" OC, 1-5/8" flange					
1000S162-43 (18 gauge, 2.03 Lbs./SF)	B1@.025	SF	1.10	.68	1.78
1000S162-54 (16 gauge, 2.50 Lbs./SF)	B1@.028	SF	1.35	.77	2.12
1000S162-68 (14 gauge, 3.10 Lbs./SF)	B1@.031	SF	1.67	.85	2.52
1000S162-97 (12 gauge, 4.29 Lbs./SF)	B1@.037	SF	2.32	1.01	3.33
10" steel ceiling joists, 20" OC, 1-5/8" flange					
1000S162-43 (18 gauge, 1.67 Lbs./SF)	B1@.022	SF	.90	.60	1.50
1000S162-54 (16 gauge, 2.06 Lbs./SF)	B1@.024	SF	1.11	.66	1.77
1000S162-68 (14 gauge, 2.54 Lbs./SF)	B1@.027	SF	1.37	.74	2.11
1000S162-97 (12 gauge, 3.52 Lbs./SF)	B1@.032	SF	1.90	.88	2.78

Carpentry, Steel Framing

	Craft@Hrs	Unit	Material	Labor	Total
10" steel ceiling joists, 24" OC, 1-5/8" flange					
1000S162-43 (18 gauge, 1.44 Lbs./SF)	B1@.019	SF	.78	.52	1.30
1000S162-54 (16 gauge, 1.76 Lbs./SF)	B1@.021	SF	.95	.57	1.52
1000S162-68 (14 gauge, 2.18 Lbs./SF)	B1@.023	SF	1.18	.63	1.81
1000S162-97 (12 gauge, 3.01 Lbs./SF)	B1@.027	SF	1.63	.74	2.37
12" steel ceiling joists, 12" OC, 1-5/8" flange					
1200S162-54 (16 gauge, 3.71 Lbs./SF)	B1@.034	SF	2.00	.93	2.93
1200S162-68 (14 gauge, 4.60 Lbs./SF)	B1@.038	SF	2.48	1.04	3.52
1200S162-97 (12 gauge, 6.39 Lbs./SF)	B1@.045	SF	3.45	1.23	4.68
12" steel ceiling joists, 16" OC, 1-5/8" flange					
1200S162-54 (16 gauge, 2.86 Lbs./SF)	B1@.028	SF	1.54	.77	2.31
1200S162-68 (14 gauge, 3.54 Lbs./SF)	B1@.030	SF	1.91	.82	2.73
1200S162-97 (12 gauge, 4.92 Lbs./SF)	B1@.036	SF	2.66	.98	3.64
12" steel ceiling joists, 20" OC, 1-5/8" flange					
1200S162-54 (16 gauge, 2.35 Lbs./SF)	B1@.022	SF	1.27	.60	1.87
1200S162-68 (14 gauge, 2.90 Lbs./SF)	B1@.024	SF	1.57	.66	2.23
1200S162-97 (12 gauge, 4.03 Lbs./SF)	B1@.029	SF	2.18	.79	2.97
12" steel ceiling joists, 24" OC, 1-5/8" flange					
1200S162-54 (16 gauge, 2.01 Lbs./SF)	B1@.020	SF	1.09	.55	1.64
1200S162-68 (14 gauge, 2.49 Lbs./SF)	B1@.022	SF	1.34	.60	1.94
1200S162-97 (12 gauge, 3.45 Lbs./SF)	B1@.026	SF	1.86	.71	2.57

Backing for drywall ceiling Floating backing for drywall ceilings, trim, appliances, fixtures, etc.

	Craft@Hrs	Unit	Material	Labor	Total
1-1/2" x 1-1/2" (.54 Lbs./LF)	B1@.012	LF	.29	.33	.62

Steel rafters Shed or gable roof with spans from 7'0" to 38'0" and pitch of 3 in12 to 12 in12 as published in light gauge steel framing span tables. Costs per square foot of roof surface. Includes rafters, bracing, screws and minimal waste but no blocking, purlins, ridge boards, collar beams, or gable walls. See costs for blocking, purlins, ridgeboards and collar beams elsewhere in this section. Labor includes layout, assembly, and alignment. Figures in parentheses show weight of steel per square foot of roof.

	Craft@Hrs	Unit	Material	Labor	Total
Using light gauge steel rafters at	—	Lb	.54	—	.54
Add for hip roof	B1@.009	SF	—	.25	.25
5-1/2" steel rafters, 12" OC, 1-5/8" flange					
550S162-33 (20 gauge, 1.18 Lbs./SF)	B1@.037	SF	.64	1.01	1.65
550S162-43 (18 gauge, 1.51 Lbs./SF)	B1@.041	SF	.82	1.12	1.94
550S162-54 (16 gauge, 1.87 Lbs./SF)	B1@.045	SF	1.01	1.23	2.24
550S162-68 (14 gauge, 2.31 Lbs./SF)	B1@.050	SF	1.25	1.37	2.62
5-1/2" steel rafters, 16" OC, 1-5/8" flange					
550S162-33 (20 gauge, .90 Lbs./SF)	B1@.031	SF	.49	.85	1.34
550S162-43 (18 gauge, 1.15 Lbs./SF)	B1@.034	SF	.62	.93	1.55
550S162-54 (16 gauge, 1.42 Lbs./SF)	B1@.037	SF	.77	1.01	1.78
550S162-68 (14 gauge, 1.75 Lbs./SF)	B1@.041	SF	.95	1.12	2.07
5-1/2" steel rafters, 24" OC, 1-5/8" flange					
500S162-33 (20 gauge, .63 Lbs./SF)	B1@.024	SF	.34	.66	1.00
500S162-43 (18 gauge, .79 Lbs./SF)	B1@.026	SF	.43	.71	1.14
500S162-54 (16 gauge, .97 Lbs./SF)	B1@.029	SF	.52	.79	1.31
500S162-68 (14 gauge, 1.19 Lbs./SF)	B1@.032	SF	.64	.88	1.52

	Craft@Hrs	Unit	Material	Labor	Total
8" steel rafters, 12" OC, 1-5/8" flange					
800S162-33 (20 gauge, 1.48 Lbs./SF)	B1@.048	SF	.80	1.31	2.11
800S162-43 (18 gauge, 1.90 Lbs./SF)	B1@.053	SF	1.03	1.45	2.48
800S162-54 (16 gauge, 2.35 Lbs./SF)	B1@.058	SF	1.27	1.59	2.86
800S162-68 (14 gauge, 2.91 Lbs./SF)	B1@.064	SF	1.57	1.75	3.32
800S162-97 (12 gauge, 4.05 Lbs./SF)	B1@.070	SF	2.19	1.92	4.11
8" steel rafters, 16" OC, 1-5/8" flange					
800S162-33 (20 gauge, 1.13 Lbs./SF)	B1@.040	SF	.61	1.09	1.70
800S162-43 (18 gauge, 1.44 Lbs./SF)	B1@.044	SF	.78	1.20	1.98
800S162-54 (16 gauge, 1.78 Lbs./SF)	B1@.048	SF	.96	1.31	2.27
800S162-68 (14 gauge, 2.20 Lbs./SF)	B1@.053	SF	1.19	1.45	2.64
800S162-97 (12 gauge, 3.06 Lbs./SF)	B1@.058	SF	1.65	1.59	3.24
8" steel rafters, 24" OC, 1-5/8" flange					
800S162-33 (20 gauge, .78 Lbs./SF)	B1@.032	SF	.42	.88	1.30
800S162-43 (18 gauge, .99 Lbs./SF)	B1@.035	SF	.53	.96	1.49
800S162-54 (16 gauge, 1.21 Lbs./SF)	B1@.039	SF	.65	1.07	1.72
800S162-68 (14 gauge, 1.49 Lbs./SF)	B1@.043	SF	.80	1.18	1.98
800S162-97 (12 gauge, 2.06 Lbs./SF)	B1@.047	SF	1.11	1.29	2.40
10" steel rafters, 12" OC, 1-5/8" flange					
1000S162-43 (18 gauge, 2.20 Lbs./SF)	B1@.060	SF	1.19	1.64	2.83
1000S162-54 (16 gauge, 2.73 Lbs./SF)	B1@.066	SF	1.47	1.81	3.28
1000S162-68 (14 gauge, 3.40 Lbs./SF)	B1@.072	SF	1.84	1.97	3.81
1000S162-97 (12 gauge, 4.74 Lbs./SF)	B1@.080	SF	2.56	2.19	4.75
10" steel rafters, 16" OC, 1-5/8" flange					
1000S162-43 (18 gauge, 1.67 Lbs./SF)	B1@.048	SF	.90	1.31	2.21
1000S162-54 (16 gauge, 2.07 Lbs./SF)	B1@.053	SF	1.12	1.45	2.57
1000S162-68 (14 gauge, 2.57 Lbs./SF)	B1@.058	SF	1.39	1.59	2.98
1000S162-97 (12 gauge, 3.57 Lbs./SF)	B1@.064	SF	1.93	1.75	3.68
10" steel rafters, 24" OC, 1-5/8" flange					
1000S162-43 (18 gauge, 1.14 Lbs./SF)	B1@.039	SF	.62	1.07	1.69
1000S162-54 (16 gauge, 1.40 Lbs./SF)	B1@.042	SF	.76	1.15	1.91
1000S162-68 (14 gauge, 1.74 Lbs./SF)	B1@.047	SF	.94	1.29	2.23
1000S162-97 (12 gauge, 2.41 Lbs./SF)	B1@.051	SF	1.30	1.40	2.70
12" steel rafters, 12" OC, 1-5/8" flange					
1200S162-54 (16 gauge, 3.12 Lbs./SF)	B1@.066	SF	1.68	1.81	3.49
1200S162-68 (14 gauge, 3.88 Lbs./SF)	B1@.073	SF	2.10	2.00	4.10
1200S162-97 (12 gauge, 5.43 Lbs./SF)	B1@.080	SF	2.93	2.19	5.12
12" steel rafters, 16" OC, 1-5/8" flange					
1200S162-54 (16 gauge, 2.36 Lbs./SF)	B1@.053	SF	1.27	1.45	2.72
1200S162-68 (14 gauge, 2.93 Lbs./SF)	B1@.059	SF	1.58	1.61	3.19
1200S162-97 (12 gauge, 4.09 Lbs./SF)	B1@.064	SF	2.21	1.75	3.96
12" steel rafters, 24" OC, 1-5/8" flange					
1200S162-54 (16 gauge, 1.60 Lbs./SF)	B1@.043	SF	.86	1.18	2.04
1200S162-68 (14 gauge, 1.98 Lbs./SF)	B1@.047	SF	1.07	1.29	2.36
1200S162-97 (12 gauge, 2.75 Lbs./SF)	B1@.052	SF	1.49	1.42	2.91

	Craft@Hrs	Unit	Material	Labor	Total

Steel trimmers At floor openings, stairwells, skylights, dormers, etc. Consists of one C-shaped member nested in same size track. Labor includes layout, assembly and alignment. Figures in parentheses show weight of steel per linear foot of trimmer. Includes screws and minimal waste.

	Craft@Hrs	Unit	Material	Labor	Total
Using light gauge steel framing materials	—	Lb	.54	—	.54
3-1/2" steel trimmers, 1-5/8" flange					
350S162-33 (20 gauge, 1.76 Lbs./LF)	B1@.020	LF	.95	.55	1.50
350S162-43 (18 gauge, 2.29 Lbs./LF)	B1@.020	LF	1.24	.55	1.79
350S162-54 (16 gauge, 2.85 Lbs./LF)	B1@.022	LF	1.54	.60	2.14
350S162-68 (14 gauge, 3.60 Lbs./LF)	B1@.023	LF	1.94	.63	2.57
5-1/2" steel trimmers, 1-5/8" flange					
550S162-33 (20 gauge, 2.23 Lbs./LF)	B1@.031	LF	1.20	.85	2.05
550S162-43 (18 gauge, 2.90 Lbs./LF)	B1@.031	LF	1.57	.85	2.42
550S162-54 (16 gauge, 3.63 Lbs./LF)	B1@.034	LF	1.96	.93	2.89
550S162-68 (14 gauge, 4.54 Lbs./LF)	B1@.035	LF	2.45	.96	3.41
8" steel trimmers, 1-5/8" flange					
800S162-33 (20 gauge, 2.82 Lbs./LF)	B1@.042	LF	1.52	1.15	2.67
800S162-43 (18 gauge, 3.67 Lbs./LF)	B1@.042	LF	1.98	1.15	3.13
800S162-54 (16 gauge, 4.59 Lbs./LF)	B1@.046	LF	2.48	1.26	3.74
800S162-68 (14 gauge, 5.75 Lbs./LF)	B1@.047	LF	3.11	1.29	4.40
800S162-97 (12 gauge, 8.13 Lbs./LF)	B1@.048	LF	4.39	1.31	5.70
10" steel trimmers, 1-5/8" flange					
1000S162-43 (18 gauge, 4.28 Lbs./LF)	B1@.052	LF	2.31	1.42	3.73
1000S162-54 (16 gauge, 5.35 Lbs./LF)	B1@.052	LF	2.89	1.42	4.31
1000S162-68 (14 gauge, 6.72 Lbs./LF)	B1@.057	LF	3.63	1.56	5.19
1000S162-97 (12 gauge, 9.51 Lbs./LF)	B1@.058	LF	5.14	1.59	6.73
12" steel trimmers, 1-5/8" flange					
1200S162-54 (16 gauge, 6.13 Lbs./LF)	B1@.063	LF	3.31	1.72	5.03
1200S162-68 (14 gauge, 7.69 Lbs./LF)	B1@.066	LF	4.15	1.81	5.96
1200S162-97 (12 gauge, 10.89 Lbs./LF)	B1@.069	LF	5.88	1.89	7.77

Steel collar beams Cost per linear foot of collar beam or rafter brace, including screws.

	Craft@Hrs	Unit	Material	Labor	Total
350S162-33 (3-1/2", 33 mil, 20 gauge)	B1@.007	LF	.36	.19	.55
550S162-43 (5-1/2", 43 mil, 18 gauge)	B1@.014	LF	.62	.38	1.00

Steel purlins Purlins or "perling" installed below roof rafters.

	Craft@Hrs	Unit	Material	Labor	Total
350S162-33 (3-1/2", 33 mil, 20 gauge)	B1@.013	LF	.47	.36	.83
350S162-43 (3-1/2", 43 mil, 18 gauge)	B1@.019	LF	.36	.52	.88

Steel dormer studs Per square foot of wall area

	Craft@Hrs	Unit	Material	Labor	Total
350S162-33 (3-1/2", 33 mil, 20 gauge)	B1@.036	SF	.46	.98	1.44

Steel ridgeboard Horizontal support at top of rafter ends. Consists of one C-shaped member nested in same size track. Labor includes layout, assembly and alignment. Figures in parentheses show weight of steel per linear foot of ridgeboard. Includes screws and minimal waste.

	Craft@Hrs	Unit	Material	Labor	Total
Using light gauge steel framing materials	—	Lb	.54	—	.54
5-1/2" steel ridgeboard, 1-5/8" flange					
550S162-33 (20 gauge, 2.23 Lbs./LF)	B1@.031	LF	1.20	.85	2.05
550S162-43 (18 gauge, 2.90 Lbs./LF)	B1@.031	LF	1.57	.85	2.42
550S162-54 (16 gauge, 3.63 Lbs./LF)	B1@.034	LF	1.96	.93	2.89
550S162-68 (14 gauge, 4.54 Lbs./LF)	B1@.035	LF	2.45	.96	3.41

	Craft@Hrs	Unit	Material	Labor	Total
8" steel ridgeboard, 1-5/8" flange					
800S162-33 (20 gauge, 2.82 Lbs./LF)	B1@.042	LF	1.52	1.15	2.67
800S162-43 (18 gauge, 3.67 Lbs./LF)	B1@.042	LF	1.98	1.15	3.13
800S162-54 (16 gauge, 4.59 Lbs./LF)	B1@.046	LF	2.48	1.26	3.74
800S162-68 (14 gauge, 5.75 Lbs./LF)	B1@.047	LF	3.11	1.29	4.40
800S162-97 (12 gauge, 8.13 Lbs./LF)	B1@.048	LF	4.39	1.31	5.70
10" steel ridgeboard, 1-5/8" flange					
1000S162-43 (18 gauge, 4.28 Lbs./LF)	B1@.052	LF	2.31	1.42	3.73
1000S162-54 (16 gauge, 5.35 Lbs./LF)	B1@.052	LF	2.89	1.42	4.31
1000S162-68 (14 gauge, 6.72 Lbs./LF)	B1@.057	LF	3.63	1.56	5.19
1000S162-97 (12 gauge, 9.51 Lbs./LF)	B1@.058	LF	5.14	1.59	6.73
12" steel ridgeboard, 1-5/8" flange					
1200S162-54 (16 gauge, 6.13 Lbs./LF)	B1@.063	LF	3.31	1.72	5.03
1200S162-68 (14 gauge, 7.69 Lbs./LF)	B1@.066	LF	4.15	1.81	5.96
1200S162-97 (12 gauge, 10.89 Lbs./LF)	B1@.069	LF	5.88	1.89	7.77

Steel roof trusses Site-fabricated and set by hand 24" OC for any roof slope from 3 in 12 to 12 in 12 so long as the total truss height does not exceed 12' from bottom chord to highest point on truss. Trusses over 12' high may cost up to 100% more. Costs are per square foot (SF) of horizontal area covered.

	Craft@Hrs	Unit	Material	Labor	Total
Using light gauge steel framing materials at	—	Lb	.54	—	.54
4" C-shaped top and bottom chords, members unpunched					
Up to 38' span	B1@.029	SF	2.49	.79	3.28
40' to 50' span	B1@.037	SF	3.11	1.01	4.12
Fink truss "W" (conventional roof truss). 4" C-shaped top and bottom chords					
Up to 38' span	B1@.023	SF	2.18	.63	2.81
40' to 50' span	B1@.029	SF	2.57	.79	3.36
6" C-shaped top and bottom chords, members unpunched					
Up to 38' span	B1@.027	SF	2.62	.74	3.36
40' to 50' span	B1@.035	SF	3.16	.96	4.12
Truss with gable fill at 16" OC					
28' span, 5 in 12 slope	B1@1.27	Ea	177.00	34.70	211.70
32' span, 5 in 12 slope	B1@1.68	Ea	219.00	46.00	265.00
40' span, 5 in 12 slope	B1@2.30	Ea	306.00	62.90	368.90

Piecework Steel Framing Steel framing on residential tracts is sometimes done by subcontractors who bid at piecework rates (such as per square foot of floor). The figures below list typical piecework rates for repetitive framing and assume all materials are supplied to the framing subcontractor. No figures appear in the Craft@Hrs column because the work is done for a fixed price per square foot and the labor productivity can be expected to vary widely.

Steel frame layout and bottom track Piecework rates
Lay out wall track according to the plans (snap chalk lines for wall track, mark location for studs, windows, doors and framing details), cut top and bottom track and install bottom track. Costs per square foot of floor (excluding garage).

	Craft@Hrs	Unit	Material	Labor	Total
Custom or more complex jobs	—	SF	—	.24	.24
Larger job, longer runs	—	SF	—	.14	.14

Carpentry, Piecework, Steel

	Craft@Hrs	Unit	Material	Labor	Total

Steel exterior wall framing Piecework rates
Measure, cut, fit, assemble, and tip up walls, including studs, track, cripples, x-bracing, trimmers and blocking. Costs per square foot of floor.

	Craft@Hrs	Unit	Material	Labor	Total
Complex job, radius walls, rake walls	—	SF	—	.68	.68
Larger job, 8' high walls, fewer partitions	—	SF	—	.24	.24

Plumb and align steel framed walls Piecework rates
Align walls, adjust walls to vertical and install temporary braces as needed, fasten nuts on anchor bolts or shoot power driven fasteners through wall track into the slab. Based on accuracy to 3/16". Costs per square foot of floor are shown below.

	Craft@Hrs	Unit	Material	Labor	Total
Small or complex job, many braces	—	SF	—	.23	.23
Larger job, fewer braces	—	SF	—	.12	.12

Install steel floor or ceiling joists Piecework rates
Lay out, cut and install floor or ceiling joists, including rim track, doubled joists, straps, joist hangers, blocking at 12' OC and ceiling backing for drywall. Based on larger jobs with simple joist layouts set 24" OC and pre-cut blocking supplied by the general contractor. Add the cost of floor beams, if required. Costs per square foot of horizontal joist area. More complex jobs with shorter runs may cost 50% more.

	Craft@Hrs	Unit	Material	Labor	Total
Install 6" ceiling or floor joists 24" OC					
600S200-33 (2" flange, 20 gauge)	—	SF	—	.15	.15
600S200-43 (2" flange, 18 gauge)	—	SF	—	.15	.15
600S200-54 (2" flange, 16 gauge)	—	SF	—	.16	.16
600S200-68 (2" flange, 14 gauge)	—	SF	—	.17	.17
Install 8" ceiling or floor joists 24" OC					
800S200-43 (2" flange, 18 gauge)	—	SF	—	.15	.15
800S200-54 (2" flange, 16 gauge)	—	SF	—	.16	.16
800S200-68 (2" flange, 14 gauge)	—	SF	—	.17	.17
800S200-97 (2" flange, 12 gauge)	—	SF	—	.19	.19
Install 10" ceiling or floor joists 24" OC					
1000S200-43 (2" flange, 18 gauge)	—	SF	—	.16	.16
1000S200-54 (2" flange, 16 gauge)	—	SF	—	.17	.17
1000S200-68 (2" flange, 14 gauge)	—	SF	—	.18	.18
1000S200-97 (2" flange, 12 gauge)	—	SF	—	.20	.20
Install 12" ceiling or floor joists 24" OC					
1200S200-54 (2" flange, 16 gauge)	—	SF	—	.18	.18
1200S200-68 (2" flange, 14 gauge)	—	SF	—	.19	.19
1200S200-97 (2" flange, 12 gauge)	—	SF	—	.21	.21
Add for other than 24" joist spacing					
Add for 12" OC spacing	—	SF	—	.11	.11
Add for 16" OC spacing	—	SF	—	.04	.04
Add for 20" OC spacing	—	SF	—	.03	.03

	Craft@Hrs	Unit	Material	Labor	Total

Install plywood floor sheathing on steel joists Piecework rates
Lay out, cut, fit and install 5/8" or 3/4" tongue and groove plywood floor sheathing, including blocking as required. Based on nailing done with a pneumatic nailer and pins supplied by the general contractor, or screwing done with a collated screw gun and screws supplied by the general contractor. Costs per square foot of sheathing installed

	Craft@Hrs	Unit	Material	Labor	Total
Smaller, cut-up job	—	SF	—	.21	.21
Larger job, longer runs	—	SF	—	.16	.16
Add for 1-1/8" sheathing	—	SF	—	.04	.04

Install structural shear panels on steel studs Piecework rates
Lay out, cut, fit and install structural 3/8" or 1/2" plywood wall panels. These figures assume steel shear studs were set correctly by others and that panel nails or screws are driven at 4" OC with a pneumatic nailer or collated screwgun. Not including hold-down straps, posts, shear blocking or extra studs. Costs per 4' x 9' panel installed.

	Craft@Hrs	Unit	Material	Labor	Total
Small job, many openings, 2nd floor	—	Ea	—	10.86	10.86
Larger job, few openings, 1st floor	—	Ea	—	6.77	6.77

Steel roof trusses Piecework rates
Setting and anchoring engineered gable and hip steel roof trusses 24" OC on prepared wall top track. These figures assume that lifting equipment is provided by the general contractor (if required) and that the truss supplier provides a fill package, spreader blocks for each plate and the ridge and jack rafters (if required). Includes installation of ceiling backing where required and rat runs at the bottom chord. Costs per square foot of plan area under the truss. Small job assumes multiple California fill between roof surfaces and understacking

	Craft@Hrs	Unit	Material	Labor	Total
Small job, rake fill above a partition wall	—	SF	—	.73	.73
Larger job, little or no fill or understacking	—	SF	—	.41	.41

Conventional roof framing Piecework rates
Calculate lengths, lay out, cut and install steel 8", 10" or 12" common, hip, valley and jack rafters on parallel and horizontal top track. Costs per square foot of plan area under the roof. Slope to 6 in 12.

	Craft@Hrs	Unit	Material	Labor	Total
Add for slope over 6 in 12	—	SF	—	.24	.24

Steel rafters 12" OC, small job, cut-up roof, few common rafters

	Craft@Hrs	Unit	Material	Labor	Total
800S162-33 (8", 1-5/8" flange, 20 gauge)	—	SF	—	1.47	1.47
800S162-43 (8", 1-5/8" flange, 18 gauge)	—	SF	—	1.49	1.49
800S162-54 (8", 1-5/8" flange, 16 gauge)	—	SF	—	1.51	1.51
800S162-68 (8", 1-5/8" flange, 14 gauge)	—	SF	—	1.58	1.58
800S162-97 (8", 1-5/8" flange, 12 gauge)	—	SF	—	1.66	1.66
1000S162-43 (10", 1-5/8" flange, 18 gauge)	—	SF	—	1.53	1.53
1000S162-54 (10", 1-5/8" flange, 16 gauge)	—	SF	—	1.55	1.55
1000S162-68 (10", 1-5/8" flange, 14 gauge)	—	SF	—	1.62	1.62
1000S162-97 (10", 1-5/8" flange, 12 gauge)	—	SF	—	1.71	1.71
1200S162-54 (12", 1-5/8" flange, 16 gauge)	—	SF	—	1.60	1.60
1200S162-68 (12", 1-5/8" flange, 14 gauge)	—	SF	—	1.65	1.65
1200S162-97 (12", 1-5/8" flange, 12 gauge)	—	SF	—	1.76	1.76

Steel rafters 16" OC, small job, cut-up roof, few common rafters

	Craft@Hrs	Unit	Material	Labor	Total
800S162-33 (8", 1-5/8" flange, 20 gauge)	—	SF	—	1.24	1.24
800S162-43 (8", 1-5/8" flange, 18 gauge)	—	SF	—	1.25	1.25
800S162-54 (8", 1-5/8" flange, 16 gauge)	—	SF	—	1.27	1.27
800S162-68 (8", 1-5/8" flange, 14 gauge)	—	SF	—	1.33	1.33
800S162-97 (8", 1-5/8" flange, 12 gauge)	—	SF	—	1.41	1.41
1000S162-43 (10", 1-5/8" flange, 18 gauge)	—	SF	—	1.29	1.29

	Craft@Hrs	Unit	Material	Labor	Total
1000S162-54 (10", 1-5/8" flange, 16 gauge)	—	SF	—	1.30	1.30
1000S162-68 (10", 1-5/8" flange, 14 gauge)	—	SF	—	1.36	1.36
1000S162-97 (10", 1-5/8" flange, 12 gauge)	—	SF	—	1.45	1.45
1200S162-54 (12", 1-5/8" flange, 16 gauge)	—	SF	—	1.35	1.35
1200S162-68 (12", 1-5/8" flange, 14 gauge)	—	SF	—	1.41	1.41
1200S162-97 (12", 1-5/8" flange, 12 gauge)	—	SF	—	1.48	1.48
Steel rafters 24" OC, small job, cut-up roof, few common rafters					
800S162-33 (8", 1-5/8" flange, 20 gauge)	—	SF	—	.74	.74
800S162-43 (8", 1-5/8" flange, 18 gauge)	—	SF	—	.75	.75
800S162-54 (8", 1-5/8" flange, 16 gauge)	—	SF	—	.76	.76
800S162-68 (8", 1-5/8" flange, 14 gauge)	—	SF	—	.79	.79
800S162-97 (8", 1-5/8" flange, 12 gauge)	—	SF	—	.84	.84
1000S162-43 (10", 1-5/8" flange, 18 gauge)	—	SF	—	.77	.77
1000S162-54 (10", 1-5/8" flange, 16 gauge)	—	SF	—	.78	.78
1000S162-68 (10", 1-5/8" flange, 14 gauge)	—	SF	—	.81	.81
1000S162-97 (10", 1-5/8" flange, 12 gauge)	—	SF	—	.86	.86
1200S162-54 (12", 1-5/8" flange, 16 gauge)	—	SF	—	.80	.80
1200S162-68 (12", 1-5/8" flange, 14 gauge)	—	SF	—	.84	.84
1200S162-97 (12", 1-5/8" flange, 12 gauge)	—	SF	—	.88	.88
Steel rafters 12" OC, larger job, longer runs, nearly all common rafters					
800S162-33 (8", 1-5/8" flange, 20 gauge)	—	SF	—	.74	.74
800S162-43 (8", 1-5/8" flange, 18 gauge)	—	SF	—	.75	.75
800S162-54 (8", 1-5/8" flange, 16 gauge)	—	SF	—	.76	.76
800S162-68 (8", 1-5/8" flange, 16 gauge)	—	SF	—	.79	.79
800S162-97 (8", 1-5/8" flange, 12 gauge)	—	SF	—	.84	.84
1000S162-43 (10", 1-5/8" flange, 18 gauge)	—	SF	—	.77	.77
1000S162-54 (10", 1-5/8" flange, 16 gauge)	—	SF	—	.78	.78
1000S162-68 (10", 1-5/8" flange, 14 gauge)	—	SF	—	.81	.81
1000S162-97 (10", 1-5/8" flange, 12 gauge)	—	SF	—	.86	.86
1200S162-54 (12", 1-5/8" flange, 16 gauge)	—	SF	—	.80	.80
1200S162-68 (12", 1-5/8" flange, 14 gauge)	—	SF	—	.84	.84
1200S162-97 (12", 1-5/8" flange, 12 gauge)	—	SF	—	.88	.88
Steel rafters 16" OC, larger job, longer runs, nearly all common rafters					
800S162-33 (8", 1-5/8" flange, 20 gauge)	—	SF	—	.66	.66
800S162-43 (8", 1-5/8" flange, 18 gauge)	—	SF	—	.66	.66
800S162-54 (8", 1-5/8" flange, 16 gauge)	—	SF	—	.67	.67
800S162-68 (8", 1-5/8" flange, 16 gauge)	—	SF	—	.70	.70
800S162-97 (8", 1-5/8" flange, 12 gauge)	—	SF	—	.73	.73
1000S162-43 (10", 1-5/8" flange, 18 gauge)	—	SF	—	.68	.68
1000S162-54 (10", 1-5/8" flange, 16 gauge)	—	SF	—	.69	.69
1000S162-68 (10", 1-5/8" flange, 14 gauge)	—	SF	—	.71	.71
1000S162-97 (10", 1-5/8" flange, 12 gauge)	—	SF	—	.75	.75
1200S162-54 (12", 1-5/8" flange, 16 gauge)	—	SF	—	.71	.71
1200S162-68 (12", 1-5/8" flange, 14 gauge)	—	SF	—	.73	.73
1200S162-97 (12", 1-5/8" flange, 12 gauge)	—	SF	—	.77	.77

	Craft@Hrs	Unit	Material	Labor	Total

Steel rafters 24" OC, larger job, longer runs, nearly all common rafters

800S162-33 (8", 1-5/8" flange, 20 gauge)	—	SF	—	.42	.42
800S162-43 (8", 1-5/8" flange, 18 gauge)	—	SF	—	.43	.43
800S162-54 (8", 1-5/8" flange, 16 gauge)	—	SF	—	.43	.43
800S162-68 (8", 1-5/8" flange, 16 gauge)	—	SF	—	.45	.45
800S162-97 (8", 1-5/8" flange, 12 gauge)	—	SF	—	.48	.48
1000S162-43 (10", 1-5/8" flange, 18 gauge)	—	SF	—	.44	.44
1000S162-54 (10", 1-5/8" flange, 16 gauge)	—	SF	—	.44	.44
1000S162-68 (10", 1-5/8" flange, 14 gauge)	—	SF	—	.47	.47
1000S162-97 (10", 1-5/8" flange, 12 gauge)	—	SF	—	.49	.49
1200S162-54 (12", 1-5/8" flange, 16 gauge)	—	SF	—	.45	.45
1200S162-68 (12", 1-5/8" flange, 14 gauge)	—	SF	—	.48	.48
1200S162-97 (12", 1-5/8" flange, 12 gauge)	—	SF	—	.50	.50

Fascia, wood Piecework rates
Applied to rafter tails and as a barge rafter on gable ends. Includes trimming the rafter tails to the correct length and installing outlookers at gable ends. Costs per linear foot of 2" x 8" fascia installed.

Small job, short runs, with moulding	—	LF	—	2.44	2.44
Larger job, longer runs	—	LF	—	1.33	1.33

Install roof sheathing on steel rafters Piecework rates
Lay out, cut, fit and install 1/2" or 5/8" plywood roof sheathing, including blocking and 1" x 8" starter board on overhangs as required. Based on nailing done with a pneumatic nailer and nails or collated screwgun and screws supplied by the general contractor. Slope to 6 in 12. Costs per square foot of sheathing installed.

Smaller, cut-up hip and valley job	—	SF	—	.22	.22
Larger job, longer runs	—	SF	—	.16	.16
Add for slope over 6 in 12	—	SF	—	.07	.07

Carpeting, Subcontract Costs listed are average prices for complete residential jobs and include consultation, measurement, pad, carpet, and professional installation of pad and carpet using tack strips and hot melt tape on seams. Prices can be expected to vary, up or down, by 50% depending on quality and quantity of actual materials required.

Minimum quality, 25 to 35 oz. face weight nylon carpet					
Including a 1/2" (4 lb density) rebond pad	—	SY	—	—	21.30
Medium quality, 35 to 50 oz. face weight nylon carpet					
Including a 1/2" (6 lb density) rebond pad	—	SY	—	—	31.10
Better quality, 50 (plus) oz. face weight nylon carpet					
Including a 1/2" (6 lb density) rebond pad	—	SY	—	—	42.00
Wool Berber carpet (large loop) installed	—	SY	—	—	57.30
Installation cost alone (residential)	—	SY	—	—	5.46
Installation cost alone (commercial)	—	SY	—	—	4.37
Add for typical furniture moving	—	SY	—	—	1.96
Add for waterfall (box steps) stairways	—	Riser	—	—	6.01
Add for wrapped steps (open riser), sewn	—	Step	—	—	8.02
Add for sewn edge treatment on one side	—	Riser	—	—	9.72
Add for circular stair steps, costs vary depending on frame					
Add for circular stairs, typical cost	—	Step	—	—	13.80

Carports

	Craft@Hrs	Unit	Material	Labor	Total

Carports Typical prices. Material costs include all hardware. Labor is based on bolting posts to an existing concrete slab.

Single carport, 6 posts, 8' high

	Craft@Hrs	Unit	Material	Labor	Total
Natural aluminum finish, posts attach to galvanized steel beams					
8' wide x 16' long, 40 PSF roof load rating	B1@7.96	Ea	798.00	218.00	1,016.00
10' wide x 20' long, 20 PSF roof load rating	B1@7.96	Ea	1030.00	218.00	1,248.00
White enamel finish, baked-on enamel roof panels, gutter-mounted posts					
10' wide x 20' long					
20 PSF roof load rating	B1@7.96	Ea	1,210.00	218.00	1,428.00
40 PSF roof load rating	B1@7.96	Ea	1,250.00	218.00	1,468.00
60 PSF roof load rating	B1@7.96	Ea	1,440.00	218.00	1,658.00
Double carport, 10' high, with steel posts and aluminum gutters, roof pans and supports					
20' x 20', 4 posts, 40 PSF roof load rating	B1@8.80	Ea	2,750.00	241.00	2,991.00
20' x 20', 6 posts, 60 PSF roof load rating	B1@9.34	Ea	3,580.00	256.00	3,836.00
24' x 24', 4 posts, 40 PSF roof load rating	B1@8.80	Ea	4,270.00	241.00	4,511.00
24' x 24', 6 posts, 60 PSF roof load rating	B1@9.34	Ea	4,740.00	256.00	4,996.00

Caulking Material costs are typical costs for bead diameter listed. Figures in parentheses indicate approximate coverage including 5% waste. Labor costs are for good quality application on smooth to slightly irregular surfaces. Per LF of bead length.

	Craft@Hrs	Unit	Material	Labor	Total
Concrete sealant with polyurethane					
Concrete sealant, per ounce	—	Oz	.42	—	.38
1/4" (2.91 LF per fluid oz.)	BC@.025	LF	.14	.75	.89
3/8" (1.29 LF per fluid oz.)	BC@.030	LF	.30	.90	1.20
1/2" (.728 LF per fluid oz.)	BC@.033	LF	.54	.99	1.53
Acoustical caulk, flexible, sound deadening					
Acoustical caulk, per ounce	—	Oz	.37	—	.36
1/8" (11.6 LF per fluid oz.)	BC@.018	LF	.03	.54	.57
1/4" (2.91 LF per fluid oz.)	BC@.025	LF	.13	.75	.88
3/8" (1.29 LF per fluid oz.)	BC@.030	LF	.29	.90	1.19
1/2" (.728 LF per fluid oz.)	BC@.033	LF	.51	.99	1.50
Acrylic latex caulk, premium quality, 35 year life expectancy					
Acrylic latex caulk, per ounce	—	Oz	.19	—	.16
1/8" (11.6 LF per fluid oz.)	BC@.018	LF	.02	.54	.56
1/4" (2.91 LF per fluid oz.)	BC@.025	LF	.06	.75	.81
3/8" (1.29 LF per fluid oz.)	BC@.030	LF	.14	.90	1.04
1/2" (.728 LF per fluid oz.)	BC@.033	LF	.26	.99	1.25
Painters acrylic latex caulk, paintable in 30 minutes					
Painters acrylic latex caulk, per ounce	—	Oz	.12	—	.11
1/8" (11.6 LF per fluid oz.)	BC@.018	LF	.01	.54	.55
1/4" (2.91 LF per fluid oz.)	BC@.025	LF	.04	.75	.79
3/8" (1.29 LF per fluid oz.)	BC@.030	LF	.09	.90	.99
1/2" (.728 LF per fluid oz.)	BC@.033	LF	.16	.99	1.15
Elastomeric latex sealant. Clearer than silicone					
Elastomeric latex sealant, per ounce	—	Oz	.46	—	.44
1/8" (11.6 LF per fluid oz.)	BC@.018	LF	.04	.54	.58
1/4" (2.91 LF per fluid oz.)	BC@.025	LF	.16	.75	.91
3/8" (1.29 LF per fluid oz.)	BC@.030	LF	.35	.90	1.25
1/2" (.728 LF per fluid oz.)	BC@.033	LF	.63	.99	1.62

	Craft@Hrs	Unit	Material	Labor	Total
KWIK-Seal Plus tub & tile caulk, for tubs, showers and sinks					
KWIK-Seal tub & tile caulk, per ounce	—	Oz	.57	—	.54
1/8" (11.6 LF per fluid oz.)	BC@.018	LF	.05	.54	.59
1/4" (2.91 LF per fluid oz.)	BC@.025	LF	.19	.75	.94
3/8" (1.29 LF per fluid oz.)	BC@.030	LF	.44	.90	1.34
1/2" (.728 LF per fluid oz.)	BC@.033	LF	.78	.99	1.77
Fire barrier latex caulk, four hour fire rating for pipes up to 30 inches					
Fire barrier latex caulk, per ounce	—	Oz	.72	—	.69
1/8" (11.6 LF per fluid oz.)	BC@.018	LF	.06	.54	.60
1/4" (2.91 LF per fluid oz.)	BC@.022	LF	.25	.66	.91
3/8" (1.29 LF per fluid oz.)	BC@.027	LF	.56	.81	1.37
1/2" (.728 LF per fluid oz.)	BC@.030	LF	1.00	.90	1.90
Silicone caulk					
Siliconized acrylic caulk, per ounce	—	Oz	.39	—	.39
1/8" (11.6 LF per fluid oz.)	BC@.018	LF	.04	.54	.58
1/4" (2.91 LF per fluid oz.)	BC@.025	LF	.14	.75	.89
3/8" (1.29 LF per fluid oz.)	BC@.030	LF	.32	.90	1.22
1/2" (.728 LF per fluid oz.)	BC@.033	LF	.57	.99	1.56
Marine adhesive/sealant, premium quality					
Silicon rubber sealant, per ounce	—	Oz	1.04	—	1.10
1/8" (11.6 LF per fluid oz.)	BC@.043	LF	.09	1.29	1.38
1/4" (2.91 LF per fluid oz.)	BC@.048	LF	.36	1.44	1.80
3/8" (1.29 LF per fluid oz.)	BC@.056	LF	.80	1.68	2.48
1/2" (.728 LF per fluid oz.)	BC@.060	LF	1.42	1.80	3.22
Anti-algae and mildew-resistant tub caulk, premium quality white or clear silicone					
Anti-algae tub caulk, per ounce	—	Oz	.38	—	.38
1/8" (11.6 LF per fluid oz.)	BC@.018	LF	.03	.54	.57
1/4" (2.91 LF per fluid oz.)	BC@.022	LF	.13	.66	.79
3/8" (1.29 LF per fluid oz.)	BC@.027	LF	.29	.81	1.10
1/2" (.728 LF per fluid oz.)	BC@.030	LF	.52	.90	1.42
Vinyl window and siding sealant, permanently flexible, paintable					
Vinyl window and siding sealant, per ounce	—	Oz	.40	—	.36
1/8" (11.6 LF per fluid oz.)	BC@.018	LF	.03	.54	.57
1/4" (2.91 LF per fluid oz.)	BC@.022	LF	.14	.66	.80
3/8" (1.29 LF per fluid oz.)	BC@.027	LF	.31	.81	1.12
1/2" (.728 LF per fluid oz.)	BC@.030	LF	.55	.90	1.45
Gutter and flashing sealant for gutters, downspouts, lap joints, flashing, vents, aluminum siding, and other metal joints. Meets ASTM C-1085					
Gutter and flashing sealant, per	—	Oz	.34	—	.31
1/8" (11.6 LF per fluid oz.)	BC@.018	LF	.03	.54	.57
1/4" (2.91 LF per fluid oz.)	BC@.022	LF	.12	.66	.78
3/8" (1.29 LF per fluid oz.)	BC@.027	LF	.26	.81	1.07
1/2" (.728 LF per fluid oz.)	BC@.030	LF	.46	.90	1.36
Polyurethane door, window & siding sealant, for installing or repairing door and window frames					
Gutter and flashing sealant, per ounce	—	Oz	.41	—	.38
1/8" (11.6 LF per fluid oz.)	BC@.018	LF	.04	.54	.58
1/4" (2.91 LF per fluid oz.)	BC@.022	LF	.14	.66	.80
3/8" (1.29 LF per fluid oz.)	BC@.027	LF	.32	.81	1.13
1/2" (.728 LF per fluid oz.)	BC@.030	LF	.57	.90	1.47
Tub and wall caulk strip, white, 1-5/8" x 11'					
Tub and wall sealant, per LF	BC@.025	LF	.48	.75	1.23

Cedar Closet Lining

	Craft@Hrs	Unit	Material	Labor	Total
Caulking gun, professional type, heavy duty					
Bulk caulking gun	—	Ea	95.00	—	95.00
"Caulk Master" caulking gun	—	Ea	125.00	—	125.00
Caulking gun, economy grade					
1/10 gallon tube, ratchet rod	—	Ea	5.00	—	5.00
Quart capacity, smooth rod	—	Ea	6.00	—	6.00

Cedar Closet Lining Material costs are per SF of floor, wall, ceiling or door of closet including 7% waste. Labor is for good quality installation in typical residential closet over studs or gypsum wallboard.

	Craft@Hrs	Unit	Material	Labor	Total
Solid cedar planks					
15 SF package	BC@.016	SF	1.54	.48	2.02
35 SF package	BC@.016	SF	1.25	.48	1.73

Ceiling Domes Enveldomes™, complete ceiling dome kits, includes grid, hardware, panels and crating, from flat to 30 degree pitch in 5 degree increments. Prices shown are for best quality. Prices may be as much as 40% lower depending on the requirements. Envel Design Corp. www.enveldesign.com

	Craft@Hrs	Unit	Material	Labor	Total
10' diameter, octagon	B1@10.7	Ea	3,540.00	293.00	3,833.00
16' diameter, octagon	B1@17.2	Ea	6,850.00	471.00	7,321.00
24' diameter, round	B1@24.7	Ea	15,600.00	676.00	16,276.00

Ceiling Lenses Envelite™, including lens and border material. Labor costs shown do not allow for cutting panels. Envel Design Corp.
UV stabilized commercial grade acrylic with decorative patterns (Costs vary from $3.90 to $6.50 per SF)

	Craft@Hrs	Unit	Material	Labor	Total
Typical cost per SF	B1@.100	SF	5.41	2.74	8.15
Custom, dimensional Envelex™ Designs					
Standard sizes and patterns (2' x 2', 2' x 4')	B1@.006	SF	33.90	.16	34.06
Irregular shapes					
Oversize and custom designs	B1@.006	SF	39.70	.16	39.86
Add for cutting panels	B1@.033	Ea	—	.90	.90

Ceiling Panel Suspension Grids Typical costs, plain white. Add panel costs below .

	Craft@Hrs	Unit	Material	Labor	Total
Main runner, 12' long	—	Ea	5.21	—	5.21
Cross tee, 2' long	—	Ea	.91	—	.91
Cross tee, 4' long	—	Ea	1.75	—	1.75
Wall mould, 12' long	—	Ea	3.18	—	3.18
Main runner, 12' long	—	LF	.43	—	.43
Cross tee, 2' long	—	LF	.46	—	.46
Cross tee, 4' long	—	LF	.44	—	.44
Wall mould, 12' long	—	Ea	3.18	—	3.18
Wall mould, 12' long	—	LF	.27	—	.27
Add for colored	—	%	10.0	—	—
Add for fire-rated white	—	%	6.0	—	—
Add for chrome or gold color grid	—	%	40.0	—	—

Suspension grid system including runners, tees, wall mould, hooks and hanging wire (but no tile or lighting fixtures)

	Craft@Hrs	Unit	Material	Labor	Total
2' x 2' grid	BC@.017	SF	.71	.51	1.22
2' x 4' grid	BC@.015	SF	.64	.45	1.09
Add for under 400 SF job	BC@.007	SF	—	.21	.21

	Craft@Hrs	Unit	Material	Labor	Total
Troffer lighting fixtures for ceiling suspension grid, includes installation, connection, lens and tubes					
2 tube 2' x 2' fixture	BE@1.00	Ea	51.50	33.90	85.40
3 tube 2' x 4' fixture	BE@1.00	Ea	48.00	33.90	81.90
4 tube 2' x 4' fixture	BE@1.00	Ea	46.00	33.90	79.90
Light-diffusing panels for installation in 2' x 4' suspension system under fluorescent lighting.					
Styrene					
.065" thick, cracked ice, 2' x 2', clear	BC@.020	Ea	2.91	.60	3.51
.065" thick, cracked ice, 2' x 4', white	BC@.029	Ea	7.77	.87	8.64
.095" thick, prismatic, 2' x 2', clear	BC@.020	Ea	2.91	.60	3.51
.095" thick, prismatic, 2' x 2', white	BC@.020	Ea	2.91	.60	3.51
.095" thick, prismatic, 2' x 4', clear	BC@.029	Ea	4.47	.87	5.34
.095" thick, prismatic, 2' x 4', white	BC@.029	Ea	4.47	.87	5.34
.095" thick, Victorian pattern 2' x 4',	BC@.029	Ea	13.60	.87	14.47
Egg crate louvers, 1/2" x 1/2" x 1/2" cell size					
Frost	BC@.188	Ea	7.90	5.64	13.54
White styrene	BC@.188	Ea	10.00	5.64	15.64
Silver finish	BC@.188	Ea	19.00	5.64	24.64
Metallic styrene	BC@.188	Ea	24.60	5.64	30.24
Anodized aluminum	BC@.188	Ea	46.90	5.64	52.54
Parabolic louvers, 5/8" x 5/8" x 7/16" cell size					
Silver styrene	BC@.188	Ea	29.90	5.64	35.54
Gold styrene	BC@.188	Ea	31.10	5.64	36.74
Silver acrylic	BC@.188	Ea	50.20	5.64	55.84
Gold acrylic	BC@.188	Ea	50.80	5.64	56.44
Stained glass look, one dimension, 2' x 2'					
Standard colors	BC@.033	Ea	35.60	.99	36.59
Antiqued colors	BC@.035	Ea	47.70	1.05	48.75

Ceiling Panels, Lay-in Type Costs per SF of ceiling area covered. No waste included. Add suspension grid system cost from preceding pages.

Acoustical rated suspended ceiling panels

	Craft@Hrs	Unit	Material	Labor	Total
2' x 2', random texture, "A" fire rated	BC@.012	SF	.37	.36	.73
2' x 2' x 5/8", USG "5th Avenue"	BC@.012	SF	.44	.36	.80
2' x 2', random texture, "A" fire rated	BC@.012	SF	.58	.36	.94
2' x 2', random textured, Armstrong	BC@.012	SF	.58	.36	.94
2' x 2' x 5/8", slant edge, USG "Alpine"	BC@.012	SF	.73	.36	1.09
2' x 2' "A" fire rated, with Bioguard™	BC@.012	SF	.89	.36	1.25
2' x 2', "A" fire rated, Armstrong "Sahara"	BC@.012	SF	1.00	.36	1.36
2' x 2' x 5/8", square edge, USG "Astro"	BC@.012	SF	1.00	.36	1.36
2' x 2', Armstrong "Bravada II"	BC@.012	SF	1.11	.36	1.47
2' x 2', Amstrong "Classic"	BC@.012	SF	1.35	.36	1.71
2' x 2' x 3/4", foil back, USG "Cheyenne"	BC@.012	SF	1.40	.36	1.76
2' x 2' x 3/4", USG "Luna", fine textured	BC@.012	SF	1.41	.36	1.77
2' x 4', textured	BC@.007	SF	.31	.21	.52
2' x 4', textured, class "A" fire rated	BC@.007	SF	.33	.21	.54
2' x 4' x 5/8", USG "5th Avenue"	BC@.007	SF	.38	.21	.59
2' x 4' x 5/8", fissured fire code	BC@.007	SF	.44	.21	.65
2' x 4', square edge, scored	BC@.007	SF	.45	.21	.66
2' x 4', fiberglass class "A" fire rated	BC@.007	SF	.48	.21	.69
2' x 4' x 5/8", "A" fire rated, USG "5th Ave"	BC@.007	SF	.53	.21	.74
2' x 4' x 5/8", perforated, USG "Tabaret"	BC@.007	SF	.62	.21	.83

Ceiling Pans, Metal

	Craft@Hrs	Unit	Material	Labor	Total
Non-acoustical suspended ceiling panels					
2' x 2' x 5/8", USG "ClimaPlus"	BC@.012	SF	.33	.36	.69
2' x 2' x 5/8", square edge, USG "Radar"	BC@.012	SF	.51	.36	.87
2' x 2', tapered edge, USG "Radar"	BC@.012	SF	.67	.36	1.03
2' x 2', Armstrong "Brighton"	BC@.012	SF	.70	.36	1.06
2' x 2', Armstrong "Royal Oak"	BC@.012	SF	.90	.36	1.26
2' x 2' x 3/4", fissured, re-panel	BC@.012	SF	.95	.36	1.31
2' x 2' x 5/8", USG "Ceramic ClimaPlus"	BC@.012	SF	1.13	.36	1.49
2' x 2' x 3/4", USG "Saville Row"	BC@.012	SF	1.27	.36	1.63
2' x 2', grid pattern, Armstrong "Prestige"	BC@.012	SF	1.50	.36	1.86
2' x 4', mineral fiber, USG "Plateau"	BC@.007	SF	.27	.21	.48
2' x 4' x 5/8", square edge, USG "Radar"	BC@.007	SF	.40	.21	.61
2' x 4' x 9/16", mineral, USG "Stonehurst"	BC@.007	SF	.43	.21	.64
2' x 4', plain white, square edge	BC@.007	SF	.45	.21	.66
2' x 4', Armstrong "Grenoble"	BC@.007	SF	.45	.21	.66
2' x 4' x 5/8", fissured fiberglass	BC@.007	SF	.46	.21	.67
2' x 4' x 5/8", Armstrong "Esprit"	BC@.007	SF	.47	.21	.68
2' x 4' x 3/4", slant edge, USG "Illusions"	BC@.007	SF	.67	.21	.88
2' x 4' random texture, "A" fire rated	BC@.007	SF	.96	.21	1.17
2' x 4' clean room, non-perforated	BC@.007	SF	1.70	.21	1.91

Fiberglass reinforced 24" x 24" gypsum panels for T-bar systems. Weight is 3 to 4 pounds each, depending on pattern

	Craft@Hrs	Unit	Material	Labor	Total
White painted finish	BC@.015	Ea	6.00	.45	6.45
Wood grain finish	BC@.020	Ea	10.30	.60	10.90

Ceiling Pans, Metal Flat, 2' x 2' pans
Aluminum with concealed grid

	Craft@Hrs	Unit	Material	Labor	Total
2' x 2' painted	BC@.083	SF	5.02	2.49	7.51
2' x 2' polished	BC@.083	SF	10.80	2.49	13.29

Stainless steel with concealed grid

	Craft@Hrs	Unit	Material	Labor	Total
2' x 2', polished clear mirror	BC@.097	SF	15.10	2.91	18.01

Stainless steel with exposed grid

	Craft@Hrs	Unit	Material	Labor	Total
2' x 2', polished clear mirror	BC@.074	SF	8.63	2.22	10.85

Ceiling Tile Class "C" panels for non-acoustical and non-fire rated applications. Tongue and groove 12" x 12" tile nailed or stapled to ceilings. Costs per SF of ceiling area covered including fasteners, trim and typical waste.

	Craft@Hrs	Unit	Material	Labor	Total
12" x 12", Washable, White	BC@.035	SF	.62	1.05	1.67
12" x 12", Armstrong "Grenoble"	BC@.036	SF	.68	1.08	1.76
12" x 12", USG lace wood fiber	BC@.035	SF	.77	1.05	1.82
12" x 12", USG white wood fiber	BC@.035	SF	.92	1.05	1.97
12" x 12", USG "Tivoli"	BC@.035	SF	.92	1.05	1.97
12" x 12", Armstrong "Glenwood"	BC@.035	SF	1.05	1.05	2.10
12" x 12", Armstrong tin ceiling design	BC@.035	SF	1.46	1.05	2.51
6" x 48", Armstrong classic" plank	BC@.054	SF	1.12	1.62	2.74

Ceilings, Tin Various plain and intricate patterns, unfinished
Embossed panels

	Craft@Hrs	Unit	Material	Labor	Total
2' x 2' square panels, cost each	B1@.147	Ea	13.00	4.02	17.02

Embossed plates

	Craft@Hrs	Unit	Material	Labor	Total
12" x 12" plates, cost per SF	B1@.042	SF	1.93	1.15	3.08
12" x 24" plates, cost per SF	B1@.040	SF	2.20	1.09	3.29
24" x 24" plates, cost per SF	B1@.037	SF	3.09	1.01	4.10
24" x 48" plates, cost per SF	B1@.035	SF	2.88	.96	3.84

	Craft@Hrs	Unit	Material	Labor	Total
Embossed moulding, 4' lengths					
3" and 4" widths, cost per linear foot	B1@.055	LF	2.35	1.50	3.85
6" widths, cost per LF	B1@.067	LF	3.18	1.83	5.01
12" widths, cost per LF	B1@.088	LF	4.48	2.41	6.89
Embossed moulding crosses, tees or ells					
For 3", 4" and 6" wide moulding, cost each	B1@.181	Ea	6.07	4.95	11.02
For 12" wide moulding, cost each	B1@.228	Ea	9.84	6.24	16.08
Perimeter and beam cornice moulding, 4' lengths					
2" projection x 2" deep, per LF	B1@.180	LF	1.64	4.92	6.56
3" projection x 4" deep, per LF	B1@.197	LF	2.20	5.39	7.59
4" projection x 6" deep, per LF	B1@.197	LF	3.04	5.39	8.43
6" projection x 8" deep, per LF	B1@.218	LF	3.98	5.96	9.94
6" projection x 10" deep, per LF	B1@.218	LF	5.45	5.96	11.41
10" projection x 12" deep, per LF	B1@.238	LF	4.92	6.51	11.43
12" projection x 18" deep, per LF	B1@.238	LF	6.36	6.51	12.87
Mitered corners for cornice moulding					
3" x 4"	B1@.184	Ea	11.00	5.03	16.03
4" x 6"	B1@.184	Ea	11.50	5.03	16.53
6" x 8"	B1@.230	Ea	12.10	6.29	18.39
6" x 10"	B1@.230	Ea	13.30	6.29	19.59
10" x 12"	B1@.280	Ea	14.30	7.66	21.96
12" x 18"	B1@.280	Ea	16.30	7.66	23.96
Frieze moulding, 4' lengths					
6" width, per LF	B1@.067	LF	3.18	1.83	5.01
12" and 14" widths per LF	B1@.088	LF	4.48	2.41	6.89
18" width per LF	B1@.110	LF	6.07	3.01	9.08
Foot moulding, 4' lengths					
4" width, per LF	B1@.055	LF	2.46	1.50	3.96
6" width, per LF	B1@.067	LF	3.18	1.83	5.01
Nosing, 4' lengths					
1-1/4" width, per LF	B1@.047	LF	1.47	1.29	2.76
4-1/2" width, per LF	B1@.055	LF	2.31	1.50	3.81

Cement See also Adhesives, Aggregates, and Concrete

	Craft@Hrs	Unit	Material	Labor	Total
Portland cement, type I and II, 94 lb sack	—	Sa	8.40	—	8.40
Plastic cement, 94 lb sack	—	Sa	7.55	—	7.55
High early strength cement, 60 lb sack	—	Sa	3.41	—	3.41
White cement, 94 lb sack	—	Sa	11.50	—	11.50
Concrete mix, 60 lb sack	—	Sa	3.15	—	3.15
Concrete mix, high early strength, 60 lb sack	—	Sa	3.41	—	3.41
Concrete mix, fiber reinforced, 60 lb sack	—	Sa	6.25	—	6.25
Masonry mix, ASTM C-270, 60 lb sack	—	Sa	3.73	—	3.73
Masonry cement, type N, 70 lb sack	—	Sa	6.52	—	6.52
Mortar mix, 60 lb sack	—	Sa	3.45	—	3.45
Exterior stucco mix, 80 lb sack	—	Sa	7.28	—	7.28
Plaster patch, 25 lb sack	—	Sa	7.92	—	7.92
Asphalt mix, 50 lb sack	—	Sa	7.91	—	7.91
Sand, 60 lb sack	—	Sa	3.89	—	3.89
Deduct for pallet quantities	—	%	-15.0	—	—

Cleaning, Final

	Craft@Hrs	Unit	Material	Labor	Total

Cleaning, Final Interior surfaces, plumbing fixtures and windows. Includes removing construction debris and grime. Add for removing paint splatter or mastic. Work performed using hand tools and commercial non-toxic cleaning products or solvent. Use $80.00 as a minimum job charge.

	Craft@Hrs	Unit	Material	Labor	Total
Cabinets, per linear foot of cabinet face. Wall or base cabinet, (bathroom, kitchen or linen)					
Cleaned inside and outside	BL@.048	LF	.08	1.19	1.27
Add for polish, exposed surface only	BL@.033	LF	.06	.82	.88
Countertops, per linear foot of front, bathroom, family room or kitchen					
Marble or granite	BL@.040	LF	.06	.99	1.05
Tile, clean (including grout joints)	BL@.046	LF	.08	1.14	1.22
Vinyl or plastic	BL@.035	LF	.07	.86	.93
Painted surfaces, wipe down by hand					
Base board, trim or moulding	BL@.003	LF	.06	.07	.13
Cased openings, per side cleaned	BL@.060	Ea	.66	1.48	2.14
Ceilings, painted	BL@.003	SF	.06	.07	.13
Walls, painted or papered	BL@.002	SF	.06	.05	.11
Window sills	BL@.004	LF	.06	.10	.16
Doors, painted or stained					
Cleaned two sides, including casing	BL@.150	Ea	.08	3.71	3.79
Add for polishing two sides	BL@.077	Ea	.06	1.90	1.96
Floors					
Hardwood floor, clean, wax and polish	BL@.025	SF	.11	.62	.73
Mop floor	BL@.002	SF	.03	.05	.08
Vacuum carpet or sweep floor	BL@.002	SF	.03	.05	.08
Plumbing fixtures					
Bar sink, including faucet	BL@.380	Ea	.16	9.39	9.55
Bathtubs, including faucets & shower heads	BL@.750	Ea	.16	18.50	18.66
Kitchen sink, including faucets	BL@.500	Ea	.16	12.40	12.56
Lavatories, including faucets	BL@.500	Ea	.16	12.40	12.56
Toilets, including seat and tank	BL@.600	Ea	.16	14.80	14.96
Shower stalls, including walls, floor, shower rod and shower head					
Three-sided	BL@1.80	Ea	.26	44.50	44.76
Add for glass doors	BL@.500	Ea	.08	12.40	12.48
Sliding glass doors, clean and polish both sides including cleaning of track and screen					
Up to 8' wide	BL@.650	Ea	.08	16.10	16.18
Over 8' wide	BL@.750	Ea	.13	18.50	18.63
Windows, clean and polish, inside and outside, first or second floors					
Bay windows	BL@.750	Ea	.13	18.50	18.63
Sliding, casement or picture windows	BL@.300	Ea	.08	7.41	7.49
Jalousie type	BL@.350	Ea	.13	8.65	8.78
Remove window screen, hose wash, replace	BL@.500	Ea	—	12.40	12.40
Typical final cleanup for new residential construction					
Per SF of floor (no glass cleaning)	BL@.004	SF	.12	.10	.22

Clothesline Units

Retractable type, indoor-outdoor, one end is wall-mounted, 5 lines, 34' long each, lines retract into metal case 34" L, 6-1/4" W, 5" H, mounting hardware included

	Craft@Hrs	Unit	Material	Labor	Total
Cost per unit as described	B1@.860	Ea	46.80	23.50	70.30
Add for aluminum post, if required	B1@1.11	Ea	40.00	30.40	70.40
T-shaped galvanized steel pole, set in concrete					
4 prong (not including cost of concrete)	BL@1.91	Ea	37.70	47.20	84.90
Umbrella type, 8' galvanized steel center post, requires 10' x 10' area, folds for storage					
One post, set in concrete	BL@.885	Ea	36.40	21.90	58.30

	Craft@Hrs	Unit	Material	Labor	Total
Columns and Porch Posts Colonial style round wood, standard cap and base					
8" x 8'	B1@2.00	Ea	204.00	54.70	258.70
10" x 10'	B1@2.00	Ea	316.00	54.70	370.70
Add for fluted wood, either of above	—	Ea	40.00	—	29.50
Round aluminum, standard cap and base					
8" x 9'	B1@2.00	Ea	128.00	54.70	182.70
10" x 10'	B1@2.00	Ea	174.00	54.70	228.70
12" x 12'	B1@3.00	Ea	308.00	82.10	390.10
Add for ornamental wood cap, typical price	—	Ea	50.00	—	50.00
Add for ornamental aluminum cap, typical	—	Ea	50.00	—	50.00
Perma Cast round columns					
8" x 8' fluted	B1@2.00	Ea	180.00	54.70	234.70
8" x 8' plain	B1@2.00	Ea	130.00	54.70	184.70
10" x 10' plain	B1@2.00	Ea	199.00	54.70	253.70
10" x 12' plain	B1@3.00	Ea	270.00	82.10	352.10
Add for Perma Cast cap	—	Ea	15.80	—	15.80
Add for Perma Cast base	—	Ea	20.00	—	20.00
Porch posts, clear laminated west coast hemlock, solid turned					
3-1/4" x 8'	B1@1.04	Ea	118.40	28.50	146.90
4-1/4" x 8'	B1@1.18	Ea	100.00	32.30	132.30
5-1/4" x 8'	B1@1.28	Ea	164.00	35.00	199.00

Concrete Ready-mix delivered by truck. Typical prices for most cities. Includes delivery up to 20 miles for 10 CY or more, 3" to 4" slump. Material cost only, no placing or pumping included. See forming and finishing costs on the following pages.

	Craft@Hrs	Unit	Material	Labor	Total
Footing and foundation concrete, 1-1/2" aggregate					
2,000 PSI, 4.8 sack mix	—	CY	88.40	—	88.40
2,500 PSI, 5.2 sack mix	—	CY	92.00	—	92.00
3,000 PSI, 5.7 sack mix	—	CY	94.50	—	94.50
3,500 PSI, 6.3 sack mix	—	CY	97.30	—	97.30
4,000 PSI, 6.9 sack mix	—	CY	101.00	—	101.00
Slab, sidewalk, and driveway concrete, 1" aggregate					
2,000 PSI, 5.0 sack mix	—	CY	89.60	—	89.60
2,500 PSI, 5.5 sack mix	—	CY	93.20	—	93.20
3,000 PSI, 6.0 sack mix	—	CY	95.80	—	95.80
3,500 PSI, 6.6 sack mix	—	CY	97.40	—	97.40
4,000 PSI, 7.1 sack mix	—	CY	100.00	—	100.00
Pea gravel pump mix / grout mix, 3/8" aggregate					
2,000 PSI, 6.0 sack mix	—	CY	97.60	—	97.60
2,500 PSI, 6.5 sack mix	—	CY	101.00	—	101.00
3,000 PSI, 7.2 sack mix	—	CY	104.00	—	104.00
3,500 PSI, 7.9 sack mix	—	CY	105.00	—	105.00
5,000 PSI, 8.5 sack mix	—	CY	114.00	—	114.00
Extra delivery costs for ready-mix concrete					
Add for delivery over 20 miles	—	Mile	3.50	—	3.50
Add for standby charge in excess of 5 minutes per CY delivered					
per minute of extra time	—	Ea	1.75	—	1.75
Add for less than 10 CY per load delivered	—	CY	20.00	—	20.00
Extra costs for non-standard aggregates					
Add for lightweight aggregate, typical	—	CY	41.40	—	41.40
Add for lightweight aggregate, pump mix	—	CY	42.60	—	42.60
Add for granite aggregate, typical	—	CY	6.33	—	6.33

Concrete

	Craft@Hrs	Unit	Material	Labor	Total
Extra costs for non-standard mix additives					
High early strength 5 sack mix	—	CY	11.50	—	11.50
High early strength 6 sack mix	—	CY	15.00	—	15.00
Add for white cement (architectural)	—	CY	52.00	—	52.00
Add for 1% calcium chloride	—	CY	1.50	—	1.50
Add for chemical compensated shrinkage	—	CY	15.00	—	15.00
Add for super plasticized mix, 7" - 8" slump	—	%	8.0	—	—

Add for colored concrete, typical prices. The ready-mix supplier charges for cleanup of the ready-mix truck used for delivery of colored concrete. The usual practice is to provide one truck per day for delivery of all colored concrete for a particular job. Add the cost below to the cost per cubic yard for the design mix required. Also add the cost for truck cleanup.

	Craft@Hrs	Unit	Material	Labor	Total
Adobe	—	CY	19.00	—	19.00
Black	—	CY	30.00	—	30.00
Blended red	—	CY	17.00	—	17.00
Brown	—	CY	22.00	—	22.00
Green	—	CY	34.00	—	34.00
Yellow	—	CY	23.00	—	23.00
Colored concrete, truck cleanup, per day	—	LS	60.00	—	60.00

Concrete Expansion Joints Fiber or asphaltic-felt sided. Per linear foot

Wall (thickness)	1/4"	3/8"	1/2"
Depth			
3" deep	.08	.06	.09
3-1/2" deep	.11	.12	.10
4" deep	.12	.12	.14
6" deep	.11	.14	.19

	Craft@Hrs	Unit	Material	Labor	Total
Labor to install concrete expansion joints					
3" to 6" width, 1/4" to 1/2" thick	B1@.030	LF	—	.82	.82

Concrete Snap Ties Short end ties, price each based on cartons of 200

	Craft@Hrs	Unit	Material	Labor	Total
6" form	—	Ea	.35	—	.35
8" form	—	Ea	.75	—	.75
10" form	—	Ea	.75	—	.75
12" form	—	Ea	.75	—	.75
14" form	—	Ea	.75	—	.75
16" form	—	Ea	.75	—	.75
24" form	—	Ea	.75	—	.75
Wedges, 3-3/4", 14 gauge	—	Ea	1.50	—	1.50
Add for long end ties	—	%	12.0	—	—

Concrete Form Excavation By hand using a pick and shovel, to 3'-0" deep. For wall footings, grade beams and column footings. Digging and one throw only, no disposal or back filling included. Per CY. See also Excavation section.

	Craft@Hrs	Unit	Material	Labor	Total
Light soil	BL@1.10	CY	—	27.20	27.20
Average soil	BL@1.65	CY	—	40.80	40.80
Heavy soil or loose rock	BL@2.14	CY	—	52.90	52.90

	Craft@Hrs	Unit	Material	Labor	Total
Concrete Form Stakes Steel, 3/8 thick x 1-1/2" wide. Costs per stake.					
18" flat	—	Ea	2.36	--	2.36
24" flat	—	Ea	3.35	--	3.36
36" flat	—	Ea	4.85	--	4.85
48" flat	—	Ea	5.78	--	5.78
18" round	—	Ea	2.03	--	2.03
24" round	—	Ea	2.85	--	2.85
36" round	—	Ea	3.75	--	3.75

Insulated Concrete Forms (ICFs) Expanded polystyrene (EPS) forms for residential and commercial reinforced concrete walls 4", 6" or 8" thick. Based on standard sized block connected with plastic clips and reinforced with steel bars before cores are filled with concrete. Costs assume forms are left in place as wall insulation and that no concrete finishing is required. See www.eco-block.com The crew is a carpenter and a laborer. Material costs include EPS forms and connectors as described but no concrete or rebar.

	Craft@Hrs	Unit	Material	Labor	Total
Standard block, 5.3 SF per block					
16" H, 48" L, with 12 panel connectors	—	Ea	14.75	—	14.75
Corner blocks, each includes 8 panel connectors					
4" cavity, 16" H, 32" L	—	Ea	14.70		14.70
6" cavity, 16" H, 32" L	—	Ea	14.70		14.70
8" cavity, 16" H, 32" L	—	Ea	14.70		14.70
Connectors					
4" connectors, per box of 1000	—	Ea	152.00	—	152.00
6" connectors, per box of 500	—	Ea	75.80	—	75.80
8" connectors, per box of 500	—	Ea	75.80	—	75.80
10" connectors, per box of 500	—	Ea	75.80	—	75.80
Splice connectors, per box of 1000	—	Ea	68.20	—	68.20
Brick ledge rails, per box of 50 rails	—	Ea	34.30		34.30
1" extension "Texas T", per box of 500	—	Ea	31.30		31.30
16" 90 degree panel connectors, per box of 40	—	Ea	101.00	—	101.00
24" 90 degree panel connectors, per box of 40	—	Ea	126.00	—	126.00
16" 45 degree panel connectors, per box of 40	—	Ea	101.00	—	101.00
24" 45 degree panel connectors, per box of 40	—	Ea	126.00	—	126.00

Material and labor for EPS forms. These figures assume footings or concrete slab are in place and ready to accept the first course of forms. Material includes forms, connectors, rebar and concrete for a 6" wall. Labor includes erecting forms, setting rebar and pouring concrete in walls to 8' high. Assume that each square foot of wall requires 1 linear foot of #4 rebar.

	Craft@Hrs	Unit	Material	Labor	Total
Per square foot of wall, below grade	B1@.060	SF	4.17	1.64	5.81
Per standard block (5.3 SF), below grade	B1@.318	Ea	22.16	8.70	30.86
Per square foot of wall, above grade	B1@.080	SF	4.17	2.19	6.36
Per standard block (5.3 SF), above grade	B1@.424	Ea	22.16	11.60	33.76
Add for 8" walls	—	%	12.0	5.0	—
Deduct for 4" walls	—	%	8.0	5.0	—

Rastra Block Building blocks with a grid of internal chambers designed to be placed either horizontally or vertically. Once the panels are in place they are filled with concrete providing a high strength structural grid. Includes cement filling and grouting. When cut for odd sizes use per sq ft pricing.

	Craft@Hrs	Unit	Material	Labor	Total
8.5" x 15" x 90" standard element	B6@.750	Ea	32.90	20.50	53.40
8.5" x 15" x 90" standard element	B6@.080	SF	3.51	2.18	5.69
8.5" x 30" x 90" double element	B6@1.50	Ea	65.80	41.00	106.80
8.5" x 30" x 90" double element	B6@.080	SF	3.51	2.18	5.69
10" x 15" x 90" standard element	B6@.750	Ea	34.60	20.50	55.10
10" x 15" x 90" standard element	B6@.080	SF	3.69	2.18	5.87

Concrete Formwork

	Craft@Hrs	Unit	Material	Labor	Total
10" x 30" x 90" double element	B6@1.50	Ea	68.00	41.00	109.00
10" x 30" x 90" double element	B6@.080	SF	3.64	2.18	5.82
10" x 7.5" x 120" end element	B6@.500	Ea	25.90	13.70	39.60
10" x 7.5" x 120" end element	B6@.080	SF	4.15	2.18	6.33
12" x 15" x 90" standard element	B6@.750	Ea	37.80	20.50	58.30
12" x 15" x 90" standard element	B6@.080	SF	4.03	2.18	6.21
12" x 30" x 90" double element	B6@1.50	Ea	75.80	41.00	116.80
12" x 30" x 90" double element	B6@.080	SF	4.03	2.18	6.21
12" x 7.5" x 120" end element	B6@.500	Ea	28.10	13.70	41.80
12" x 7.5" x 120" end element	B6@.080	SF	4.50	2.18	6.68
14" x 15" x 90" standard element	B6@.750	Ea	41.20	20.50	61.70
14" x 15" x 90" standard element	B6@.080	SF	4.90	2.18	7.08
14" x 30" x 90" double element	B6@1.50	Ea	82.30	41.00	123.30
14" x 30" x 90" double element	B6@.080	SF	4.39	2.18	6.57
2" x 15" x 90" flat panel	B6@.750	Ea	21.90	20.50	42.40
2" x 15" x 90" flat panel	B6@.080	SF	2.61	2.18	4.79
4" x 30" x 90" flat panel	B6@1.50	Ea	38.50	41.00	79.50
4" x 30" x 90" flat Panel	B6@.080	SF	2.05	2.18	4.23

Concrete Formwork Multiple use of forms assumes that forms can be removed, partially disassembled and cleaned and that 75% of the form material can be reused. Costs include assembly, oiling, setting, stripping and cleaning. Material costs in this section can be adjusted to reflect your actual cost of lumber. Here's how: Divide your actual lumber cost per MBF by the assumed cost (listed in parentheses). Then multiply the cost in the material column by this adjustment factor. For expanded coverage of concrete and formwork, see the National Concrete & Masonry Estimator at http://CraftsmanSiteLicense.com

Board forming and stripping For wall footings, grade beams, column footings, site curbs and steps. Includes 5% waste. Per SF of contact area. When forms are required on both sides of the concrete, include the contact surface for each side.
2" thick forms and bracing, using 2.85 BF of form lumber per SF. Includes nails, ties, and form oil

	Craft@Hrs	Unit	Material	Labor	Total
Using 2" lumber, per MBF	—	MBF	620.00	—	620.00
Using nails, ties and form oil, per SF	—	SF	.22	—	.22
1 use	B2@.115	SF	2.06	3.25	5.31
3 use	B2@.115	SF	1.00	3.25	4.25
5 use	B2@.115	SF	.78	3.25	4.03

Add for keyway beveled on two edges, one use. No stripping included

	Craft@Hrs	Unit	Material	Labor	Total
2" x 4"	B2@.027	LF	.43	.76	1.19
2" x 6"	B2@.027	LF	.65	.76	1.41

Driveway and walkway edge forms Material costs include stakes, nails, and form oil and 5% waste. Per LF of edge form. When forms are required on both sides of the concrete, include the length of each side plus the end widths.
2" x 4" edge form

	Craft@Hrs	Unit	Material	Labor	Total
Using 2" x 4" lumber, .7 BF per LF	—	MBF	621.00	—	621.00
Using nails, ties and form oil, per LF	—	LF	.22	—	.22
1 use	B2@.050	LF	.59	1.41	2.00
3 use	B2@.050	LF	.36	1.41	1.77
5 use	B2@.050	LF	.30	1.41	1.71

2" x 6" edge form

	Craft@Hrs	Unit	Material	Labor	Total
Using 2" x 6" lumber, 1.05 BF per LF	—	MBF	620.00	—	620.00
1 use	B2@.050	LF	.81	1.41	2.22
3 use	B2@.050	LF	.45	1.41	1.86
5 use	B2@.050	LF	.37	1.41	1.78

	Craft@Hrs	Unit	Material	Labor	Total
2" x 8" edge form					
Using 2" x 8" lumber, 1.4 BF per LF	—	MBF	640.00	—	640.00
1 use	B2@.055	LF	1.11	1.55	2.66
3 use	B2@.055	LF	.59	1.55	2.14
5 use	B2@.055	LF	.47	1.55	2.02
2" x 10" edge form					
Using 2" x 10" lumber, 1.75 BF per LF	—	MBF	657.00	—	657.00
1 use	B2@.055	LF	1.31	1.55	2.86
3 use	B2@.055	LF	.68	1.55	2.23
5 use	B2@.055	LF	.54	1.55	2.09
2" x 12" edge form					
Using 2" x 12" lumber, 2.1 BF per LF	—	MBF	738.00	—	738.00
1 use	B2@.055	LF	1.71	1.55	3.26
3 use	B2@.055	LF	.86	1.55	2.41
5 use	B2@.055	LF	.67	1.55	2.22

Plywood forming For foundation walls, building walls and retaining walls, using 3/4" plyform with 10% waste, and 2" bracing with 20% waste. All material costs include nails, ties, clamps, form oil and stripping. Costs shown are per SF of contact area. When forms are required on both sides of the concrete, include the contact surface area for each side.

	Craft@Hrs	Unit	Material	Labor	Total
Using 3/4" plyform, per MSF	—	MSF	1,360.00	—	1,360.00
Using 2" bracing, per MBF	—	MBF	620.00	—	620.00
Using nails, ties, clamps & form oil, per SF	—	SFCA	.22	—	.22
Walls up to 4' high (1.10 SF of plywood and .42 BF of bracing per SF of form area)					
1 use	B2@.051	SFCA	1.98	1.44	3.42
3 use	B2@.051	SFCA	1.01	1.44	2.45
5 use	B2@.051	SFCA	.80	1.44	2.24
Walls 4' to 8' high (1.10 SF of plywood and .60 BF of bracing per SF of form area)					
1 use	B2@.060	SFCA	2.09	1.69	3.78
3 use	B2@.060	SFCA	1.06	1.69	2.75
5 use	B2@.060	SFCA	.84	1.69	2.53
Walls 8' to 12' high (1.10 SF of plywood and .90 BF of bracing per SF of form area)					
1 use	B2@.095	SFCA	2.28	2.68	4.96
3 use	B2@.095	SFCA	1.14	2.68	3.82
5 use	B2@.095	SFCA	.90	2.68	3.58
Walls 12' to 16' high (1.10 SF of plywood and 1.05 BF of bracing per SF of form area)					
1 use	B2@.128	SFCA	2.37	3.61	5.98
3 use	B2@.128	SFCA	1.19	3.61	4.80
5 use	B2@.128	SFCA	.93	3.61	4.54
Walls over 16' high (1.10 SF of plywood and 1.20 BF of bracing per SF of form area)					
1 use	B2@.153	SFCA	2.46	4.32	6.78
3 use	B2@.153	SFCA	1.23	4.32	5.55
5 use	B2@.153	SFCA	.96	4.32	5.28

Form stripping Labor to remove forms and bracing, clean, and stack on job site.

	Craft@Hrs	Unit	Material	Labor	Total
Board forms at wall footings, grade beams and column footings. Per SF of contact area					
1" thick lumber	BL@.010	SF	—	.25	.25
2" thick lumber	BL@.012	SF	—	.30	.30
Keyways, 2" x 4" and 2" x 6"	BL@.006	LF	—	.15	.15
Slab edge forms. Per LF of edge form					
2" x 4" to 2" x 6"	BL@.012	LF	—	.30	.30
2" x 8" to 2" x 12"	BL@.013	LF	—	.32	.32

Concrete Reinforcing Steel

	Craft@Hrs	Unit	Material	Labor	Total
Walls, plywood forms. Per SF of contact area stripped					
To 4' high	BL@.017	SF	—	.42	.42
Over 4' to 8' high	BL@.018	SF	—	.44	.44
Over 8' to 12' high	BL@.020	SF	—	.49	.49
Over 12' to 16' high	BL@.027	SF	—	.67	.67
Over 16' high	BL@.040	SF	—	.99	.99

Concrete Reinforcing Steel Steel reinforcing bars (rebar), ASTM A615 Grade 60. Material costs are for deformed steel reinforcing rebars, including 10% lap allowance, cutting and bending. These costs also include detailed shop drawings and delivery to jobsite with identity tags per shop drawings. Add for epoxy or galvanized coating of rebars, chairs, splicing, spiral caissons and round column reinforcing, if required, from the sections following the data below. Costs per pound (Lb) and per linear foot (LF) including tie wire and tying. Reinforcing steel placed and tied in footings and grade beams

	Craft@Hrs	Unit	Material	Labor	Total
1/4" diameter, #2 rebar	RI@.015	Lb	.78	.48	1.26
1/4" diameter, #2 rebar (.17 Lb per LF)	RI@.003	LF	.14	.10	.24
3/8" diameter, #3 rebar	RI@.011	Lb	.47	.36	.83
3/8" diameter, #3 rebar (.38 Lb per LF)	RI@.004	LF	.18	.13	.31
1/2" diameter, #4 rebar	RI@.010	Lb	.45	.32	.77
1/2" diameter, #4 rebar (.67 Lb per LF)	RI@.007	LF	.30	.23	.53
5/8" diameter, #5 rebar	RI@.009	Lb	.40	.29	.69
5/8" diameter, #5 rebar (1.04 Lb per LF)	RI@.009	LF	.42	.29	.71
3/4" diameter, #6 rebar	RI@.008	Lb	.37	.26	.63
3/4" diameter, #6 rebar (1.50 Lb per LF)	RI@.012	LF	.56	.39	.95
7/8" diameter, #7 rebar	RI@.008	Lb	.45	.26	.71
7/8" diameter, #7 rebar (2.04 Lb per LF)	RI@.016	LF	.92	.52	1.44
1" diameter, #8 rebar	RI@.008	Lb	.43	.26	.69
1" diameter, #8 rebar (2.67 Lb per LF)	RI@.021	LF	1.15	.68	1.83
Reinforcing steel placed and tied in structural slabs					
1/4" diameter, #2 rebar	RI@.014	Lb	.78	.45	1.23
1/4" diameter, #2 rebar (.17 Lb per LF)	RI@.002	LF	.14	.06	.20
3/8" diameter, #3 rebar	RI@.010	Lb	.47	.32	.79
3/8" diameter, #3 rebar (.38 Lb per LF)	RI@.004	LF	.18	.13	.31
1/2" diameter, #4 rebar	RI@.009	Lb	.45	.29	.74
1/2" diameter, #4 rebar (.67 Lb per LF)	RI@.006	LF	.30	.19	.49
5/8" diameter, #5 rebar	RI@.008	Lb	.40	.26	.66
5/8" diameter, #5 rebar (1.04 Lb per LF)	RI@.008	LF	.42	.26	.68
3/4" diameter, #6 rebar	RI@.007	Lb	.37	.23	.60
3/4" diameter, #6 rebar (1.50 Lb per LF)	RI@.011	LF	.56	.36	.92
7/8" diameter, #7 rebar	RI@.007	Lb	.45	.23	.68
7/8" diameter, #7 rebar (2.04 Lb per LF)	RI@.014	LF	.92	.45	1.37
1" diameter, #8 rebar	RI@.007	Lb	.43	.23	.66
1" diameter, #8 rebar (2.67 Lb per LF)	RI@.019	LF	1.15	.61	1.76
Reinforcing steel placed and tied in walls					
1/4" diameter, #2 rebar	RI@.017	Lb	.78	.55	1.33
1/4" diameter, #2 rebar (.17 Lb per LF)	RI@.003	LF	.14	.10	.24
3/8" diameter, #3 rebar	RI@.012	Lb	.47	.39	.86
3/8" diameter, #3 rebar (.38 Lb per LF)	RI@.005	LF	.18	.16	.34
1/2" diameter, #4 rebar	RI@.011	Lb	.45	.36	.81
1/2" diameter, #4 rebar (.67 Lb per LF)	RI@.007	LF	.30	.23	.53
5/8" diameter, #5 rebar	RI@.010	Lb	.40	.32	.72
5/8" diameter, #5 rebar (1.04 Lb per LF)	RI@.010	LF	.42	.32	.74
3/4" diameter, #6 rebar	RI@.009	Lb	.37	.29	.66
3/4" diameter, #6 rebar (1.50 Lb per LF)	RI@.014	LF	.56	.45	1.01

	Craft@Hrs	Unit	Material	Labor	Total
7/8" diameter, #7 rebar	Rl@.009	Lb	.45	.29	.74
7/8" diameter, #7 rebar (2.04 Lb per LF)	Rl@.018	LF	.92	.58	1.50
1" diameter, #8 rebar	Rl@.009	Lb	.43	.29	.72
1" diameter, #8 rebar (2.67 Lb per LF)	Rl@.024	LF	1.15	.78	1.93

Add for reinforcing steel coating. Coating applied prior to shipment to job site. Labor for placing coated rebar is the same as that shown for plain rebar above. Cost per pound of reinforcing steel

Epoxy coating					
Any size rebar, add	—	Lb	.28	—	.28
Galvanized coating					
Any #2 rebar, #3 rebar or #4 rebar, add	—	Lb	.35	—	.35
Any #5 rebar or #6 rebar, add	—	Lb	.33	—	.33
Any #7 rebar or #8 rebar, add	—	Lb	.31	—	.31

Add for reinforcing steel chairs. Labor for placing rebar shown above includes time for setting chairs.

Chairs, individual type, cost each					
3" high					
Steel	—	Ea	.46	—	.46
Plastic	—	Ea	.63	—	.63
Galvanized steel	—	Ea	.70	—	.70
5" high					
Steel	—	Ea	.70	—	.70
Plastic	—	Ea	.98	—	.98
Galvanized steel	—	Ea	.91	—	.91
8" high					
Steel	—	Ea	1.49	—	1.49
Plastic	—	Ea	1.94	—	1.94
Galvanized steel	—	Ea	2.00	—	2.00
Chairs, continuous type, with legs 8" OC, cost per linear foot					
3" high					
Steel	—	LF	.63	—	.63
Plastic	—	LF	.75	—	.75
Galvanized steel	—	LF	.81	—	.81
6" high					
Steel	—	LF	.88	—	.88
Plastic	—	LF	1.10	—	1.10
Galvanized steel	—	LF	1.16	—	1.16
8" high					
Steel	—	LF	1.23	—	1.23
Plastic	—	LF	1.56	—	1.56
Galvanized steel	—	LF	1.61	—	1.61
Waterstop, 3/8"					
Rubber, 6"	B2@.059	LF	2.90	1.67	4.57
Rubber, 9"	B2@.059	LF	3.90	1.67	5.57
PVC, 9"	B2@.059	LF	1.55	1.67	3.22

Spiral caisson and round column reinforcing 3/8" diameter hot rolled steel spirals with main vertical bars as shown. Costs include typical engineering and shop drawings. Based on shop fabricated spirals delivered, tagged, ready to install. Cost per vertical linear foot (VLF)

16" diameter with 6 #6 bars, 16.8 Lbs per VLF	Rl@.041	VLF	26.20	1.33	27.53
24" diameter with 6 #6 bars, 28.5 Lbs per VLF	Rl@.085	VLF	44.70	2.75	47.45
36" diameter with 8 #10 bars, 51.2 Lbs per VLF	Rl@.176	VLF	81.10	5.69	86.79

Concrete Footings

	Craft@Hrs	Unit	Material	Labor	Total
Welded wire mesh, steel, electric weld, including 15% waste and overlap					
2" x 2" W.9 x W.9 (#12 x #12), slabs	Rl@.004	SF	.85	.13	.98
2" x 2" W.9 x W.9 (#12 x #12), beams and columns	Rl@.020	SF	.82	.65	1.47
4" x 4" W1.4 x W1.4 (#10 x #10), slabs	Rl@.003	SF	.39	.10	.49
4" x 4" W2.0 x W2.0 (#8 x #8), slabs	Rl@.004	SF	.46	.13	.59
4" x 4" W2.9 x W2.9 (#6 x #6), slabs	Rl@.005	SF	.51	.16	.67
4" x 4" W4.0 x W4.0 (#4 x #4), slabs	Rl@.006	SF	.67	.19	.86
6" x 6" W1.4 x W1.4 (#10 x #10), slabs	Rl@.003	SF	.18	.10	.28
6" x 6" W2.0 x W2.0 (#8 x #8), slabs	Rl@.004	SF	.30	.13	.43
6" x 6" W2.9 x W2.9 (#6 x #6), slabs	Rl@.004	SF	.34	.13	.47
6" x 6" W4.0 x W4.0 (#4 x #4), slabs	Rl@.005	SF	.48	.16	.64
Add for lengthwise cut, LF of cut	Rl@.002	LF	—	.06	.06
Welded wire mesh, galvanized steel, electric weld, including 15% waste and overlap					
2" x 2" W.9 x W.9 (#12 x #12), slabs	Rl@.004	SF	1.65	.13	1.78
4" x 4" W1.4 x W1.4 (#10 x #10), slabs	Rl@.003	SF	.78	.10	.88
4" x 4" W2.0 x W2.0 (#8 x #8), slabs	Rl@.004	SF	.91	.13	1.04
4" x 4" W2.9 x W2.9 (#6 x #6), slabs	Rl@.005	SF	1.03	.16	1.19
4" x 4" W4.0 x W4.0 (#4 x #4), slabs	Rl@.006	SF	1.33	.19	1.52
6" x 6" W1.4 x W1.4 (#10 x #10), slabs	Rl@.003	SF	.42	.10	.52
6" x 6" W2.0 x W2.0 (#8 x #8), slabs	Rl@.004	SF	.55	.13	.68
6" x 6" W2.9 x W2.9 (#6 x #6), slabs	Rl@.004	SF	.67	.13	.80
6" x 6" W4.0 x W4.0 (#4 x #4), slabs	Rl@.005	SF	.97	.16	1.13
Add for lengthwise cut, LF of cut	Rl@.002	LF	—	.06	.06
Welded wire mesh, epoxy coated steel, electric weld, including 15% waste and overlap					
2" x 2" W.9 x W.9 (#12 x #12), slabs	Rl@.004	SF	3.38	.13	3.51
4" x 4" W1.4 x W1.4 (#10 x #10), slabs	Rl@.003	SF	2.93	.10	3.03
4" x 4" W2.0 x W2.0 (#8 x #8), slabs	Rl@.004	SF	2.99	.13	3.12
4" x 4" W2.9 x W2.9 (#6 x #6), slabs	Rl@.005	SF	3.04	.16	3.20
4" x 4" W4.0 x W4.0 (#4 x #4), slabs	Rl@.006	SF	3.20	.19	3.39
6" x 6" W1.4 x W1.4 (#10 x #10), slabs	Rl@.003	SF	2.74	.10	2.84
6" x 6" W2.0 x W2.0 (#8 x #8), slabs	Rl@.004	SF	2.81	.13	2.94
6" x 6" W2.9 x W2.9 (#6 x #6), slabs	Rl@.004	SF	2.86	.13	2.99
6" x 6" W4.0 x W4.0 (#4 x #4), slabs	Rl@.005	SF	3.02	.16	3.18
Add for lengthwise cut, LF of cut	Rl@.002	LF	—	.06	.06

Column Footings Concrete material and placing only. Material costs are for 3,000 PSI, 5.7 sack mix, with 1-1/2" aggregate with 5% waste included. Labor cost is for placing concrete in an existing form. Add for excavation, board forming and steel reinforcing.

	Craft@Hrs	Unit	Material	Labor	Total
Using concrete (before waste allowance)	—	CY	94.50	—	94.50
Typical cost per CY	B1@.875	CY	99.20	23.90	123.10

Concrete Walls Cast-in-place concrete walls for buildings or retaining walls. Material costs for concrete placed direct from chute are based on 2,000 PSI, 5.0 sack mix, with 1" aggregate, including 5% waste. Pump mix costs include an additional cost of $10.00 per CY for pump. Labor costs are for placing only. Add the cost of excavation, formwork, steel reinforcing, finishes and curing. Square foot costs are based on SF of wall measured on one face only. Costs do not include engineering, design or foundations. For expanded coverage of concrete and formwork, see the National Concrete & Masonry Estimator at http://craftsmansitelicense.com

	Craft@Hrs	Unit	Material	Labor	Total
Using concrete (before 5% waste allowance)	—	CY	88.40	—	88.40

	Craft@Hrs	Unit	Material	Labor	Total
4" thick walls (1.23 CY per CSF)					
To 4' high, direct from chute	B1@.013	SF	1.14	.36	1.50
4' to 8' high, pumped	B3@.015	SF	1.26	.40	1.66
8' to 12' high, pumped	B3@.017	SF	1.26	.45	1.71
12' to 16' high, pumped	B3@.018	SF	1.26	.48	1.74
16' high, pumped	B3@.020	SF	1.26	.53	1.79
6" thick walls (1.85 CY per CSF)					
To 4' high, direct from chute	B1@.020	SF	1.72	.55	2.27
4' to 8' high, pumped	B3@.022	SF	1.90	.58	2.48
8' to 12' high, pumped	B3@.025	SF	1.90	.66	2.56
12' to 16' high, pumped	B3@.027	SF	1.90	.71	2.61
16' high, pumped	B3@.030	SF	1.90	.79	2.69
8" thick walls (2.47 CY per CSF)					
To 4' high, direct from chute	B1@.026	SF	2.29	.71	3.00
4' to 8' high, pumped	B3@.030	SF	2.54	.79	3.33
8' to 12' high, pumped	B3@.033	SF	2.54	.87	3.41
12' to 16' high, pumped	B3@.036	SF	2.54	.95	3.49
16' high, pumped	B3@.040	SF	2.54	1.06	3.60
10" thick walls (3.09 CY per CSF)					
To 4' high, direct from chute	B1@.032	SF	2.87	.88	3.75
4' to 8' high, pumped	B3@.037	SF	3.18	.98	4.16
8' to 12' high, pumped	B3@.041	SF	3.18	1.09	4.27
12' to 16' high, pumped	B3@.046	SF	3.18	1.22	4.40
16' high, pumped	B3@.050	SF	3.18	1.32	4.50
12" thick walls (3.70 CY per CSF)					
To 4' high, placed direct from chute	B1@.040	SF	3.43	1.09	4.52
4' to 8' high, pumped	B3@.045	SF	3.80	1.19	4.99
8' to 12' high, pumped	B3@.050	SF	3.80	1.32	5.12
12' to 16' high, pumped	B3@.055	SF	3.80	1.46	5.26
16' high, pumped	B3@.060	SF	3.80	1.59	5.39

Concrete Footings, Grade Beams and Stem Walls Use the figures below for preliminary estimates. Concrete costs are based on 2,000 PSI, 5.0 sack mix with 1" aggregate placed directly from the chute of a ready-mix truck. Figures in parentheses show the cubic yards of concrete per linear foot of foundation (including 5% waste). Costs shown include concrete, 60 pounds of reinforcing per CY of concrete, and typical excavation using a 3/4 CY backhoe with excess backfill spread on site.

Footings and grade beams Cast directly against the earth, no forming or finishing required. These costs assume the top of the foundation will be at finished grade. For scheduling purposes estimate that a crew of 3 can lay out, excavate, place and tie the reinforcing steel and place 13 CY of concrete in an 8-hour day. Use $900.00 as a minimum job charge.

	Craft@Hrs	Unit	Material	Labor	Equipment	Total
Using 2,000 PSI concrete, per CY	—	CY	88.40	—	—	88.40
Using reinforcing bars, per pound	—	Lb	.49	—	—	.49
Typical cost per CY	B4@1.80	CY	109.00	56.60	18.80	184.40
12" W x 6" D (0.019 CY per LF)	B4@.034	LF	2.07	1.07	.35	3.49
12" W x 8" D (0.025 CY per LF)	B4@.045	LF	2.73	1.41	.47	4.61
12" W x 10" D (0.032 CY per LF)	B4@.058	LF	3.49	1.82	.60	5.91
12" W x 12" D (0.039 CY per LF)	B4@.070	LF	4.25	2.20	.73	7.18
12" W x 18" D (0.058 CY per LF)	B4@.104	LF	6.32	3.27	1.08	10.67
16" W x 8" D (0.035 CY per LF)	B4@.063	LF	3.82	1.98	.66	6.46
16" W x 10" D (0.043 CY per LF)	B4@.077	LF	4.69	2.42	.80	7.91

Concrete Footings

	Craft@Hrs	Unit	Material	Labor	Equipment	Total
16" W x 12" D (0.052 CY per LF)	B4@.094	LF	5.67	2.95	.98	9.60
18" W x 8" D (0.037 CY per LF)	B4@.067	LF	4.03	2.11	.70	6.84
18" W x 10" D (0.049 CY per LF)	B4@.088	LF	5.34	2.77	.92	9.03
18" W x 24" D (0.117 CY per LF)	B4@.216	LF	12.80	6.79	2.25	21.84
20" W x 12" D (0.065 CY per LF)	B4@.117	LF	7.09	3.68	1.22	11.99
24" W x 12" D (0.078 CY per LF)	B4@.140	LF	8.50	4.40	1.46	14.36
24" W x 24" D (0.156 CY per LF)	B4@.281	LF	17.00	8.83	2.93	28.76
24" W x 30" D (0.195 CY per LF)	B4@.351	LF	21.30	11.00	3.66	35.96
24" W x 36" D (0.234 CY per LF)	B4@.421	LF	25.50	13.20	4.39	43.09
30" W x 36" D (0.292 CY per LF)	B4@.526	LF	31.80	16.50	5.49	53.79
30" W x 42" D (0.341 CY per LF)	B4@.614	LF	37.20	19.30	6.40	62.90
36" W x 48" D (0.467 CY per LF)	B4@.841	LF	50.90	26.40	8.77	86.07

Tricks of the trade: To estimate the cost of footing or grade beam sizes not shown, multiply the width in inches by the depth in inches and divide the result by 3700. This is the CY of concrete per LF of footing including 5% waste. Multiply by the "Typical cost per CY" in the prior table to find the cost per LF.

Continuous concrete footing with foundation stem wall These figures assume the foundation stem wall projects 24" above the finished grade and extends into the soil 18" to the top of the footing. Costs shown include typical excavation using a 3/4 CY backhoe with excess backfill spread on site, forming both sides of the foundation wall and the footing, based on three uses of the forms and 2 #4 rebar. Use $1,200.00 as a minimum cost for this type work.

	Craft@Hrs	Unit	Material	Labor	Equipment	Total
Typical cost per CY	B5@7.16	CY	135.00	221.00	44.80	400.80
Typical single story structure, footing 12" W x 8" D, wall 6" T x 42" D (.10 CY per LF)	B5@.716	LF	13.50	22.10	4.48	40.08
Typical two story structure, footing 18" W x 10" D, wall 8" T x 42" D (.14 CY per LF)	B5@1.00	LF	16.50	30.80	6.26	53.56
Typical three story structure, footing 24" W x 12" D, wall 10" T x 42" D (.19 CY per LF)	B5@1.36	LF	22.40	41.90	8.51	72.81

	Craft@Hrs	Unit	Material	Labor	Total

Slabs, Walks and Driveways Typical costs for reinforced concrete slabs-on-grade including fine grading, slab base, forms, vapor barrier, wire mesh, 3,000 PSI concrete, finishing and curing. For expanded coverage of concrete and formwork, see the National Concrete & Masonry Estimator at http://CraftsmanSiteLicense.Com. For thickened edge slabs, add the area of the thickened edge. Use 500 square foot as a minimum.

	Craft@Hrs	Unit	Material	Labor	Total
2" thick	B5@.067	SF	1.57	2.07	3.64
3" thick	B5@.068	SF	1.81	2.10	3.91
4" thick	B5@.069	SF	2.05	2.13	4.18
5" thick	B5@.070	SF	2.30	2.16	4.46
6" thick	B5@.071	SF	2.53	2.19	4.72

Slab base Aggregate base for slabs. No waste included. Labor costs are for spreading aggregate from piles only. Add for fine grading, using hand tools.
Crushed stone base 1.4 tons equal 1 cubic yard

	Craft@Hrs	Unit	Material	Labor	Total
Using crushed stone, per CY	—	CY	21.80	—	21.80
1" base (.309 CY per CSF)	BL@.001	SF	.07	.02	.09
2" base (.617 CY per CSF)	BL@.003	SF	.13	.07	.20
3" base (.926 CY per CSF)	BL@.004	SF	.20	.10	.30
4" base (1.23 CY per CSF)	BL@.006	SF	.27	.15	.42
5" base (1.54 CY per CSF)	BL@.007	SF	.34	.17	.51
6" base (1.85 CY per CSF)	BL@.008	SF	.40	.20	.60

	Craft@Hrs	Unit	Material	Labor	Total
Sand fill base 1.40 tons equal 1 cubic yard					
Using sand, per CY	—	CY	18.60	—	18.60
1" fill (.309 CY per CSF)	BL@.001	SF	.06	.02	.08
2" fill (.617 CY per CSF)	BL@.002	SF	.11	.05	.16
3" fill (.926 CY per CSF)	BL@.003	SF	.17	.07	.24
4" fill (1.23 CY per CSF)	BL@.004	SF	.23	.10	.33
5" fill (1.54 CY per CSF)	BL@.006	SF	.29	.15	.44
6" fill (1.85 CY per CSF)	BL@.007	SF	.34	.17	.51
Add for fine grading for slab on grade	BL@.008	SF	—	.20	.20

Slab membrane Material costs are for 10' wide membrane with 6" laps. Labor costs are to lay membrane over a level slab base.

	Craft@Hrs	Unit	Material	Labor	Total
4 mil polyethylene (at $32.20 per MSF)	BL@.002	SF	.03	.05	.08
6 mil polyethylene (at $41.70 per MSF)	BL@.002	SF	.04	.05	.09
Detaching Membrane, Ditra, polyethylene laminate					
Laid on a level slab ($72.80 per 54 SF)	BL@.002	SF	1.35	.05	1.40
Add for adhesive cemented seams	BL@.001	SF	.05	.02	.07

Curbs and gutters Small quantities. Material costs are for 2,000 PSI, 5.0 sack mix with 1" aggregate. No waste included. Labor cost is for placing and finishing concrete only. Add for excavation, wood forming, steel reinforcing.

	Craft@Hrs	Unit	Material	Labor	Total
Per CY of concrete	B6@1.61	CY	89.60	44.00	133.60

Steps on grade Costs include layout, fabrication, and placing forms, setting and tying steel reinforcing, installation of steel nosing, placing 2,000 PSI concrete directly from the chute of a ready-mix truck, finishing, stripping forms, and curing. Costs assume excavation, back filling, and compacting have been completed. For scheduling purposes, estimate that a crew of 3 can set forms, place steel, and pour 4 to 5 CY of concrete cast-in-place steps in an 8 hour day. See also Cast-in-place Concrete Steps on Grade in the Industrial Section 3. Costs shown are for concrete steps with a 6" rise height and 12" tread depth supported by a 6" thick monolithic slab.

	Craft@Hrs	Unit	Material	Labor	Total
Total cost of steps per CY of concrete	B1@5.43	CY	260.00	149.00	409.00

Concrete Finishing

	Craft@Hrs	Unit	Material	Labor	Total
Slab finishes					
Broom finish	B6@.012	SF	—	.33	.33
Float finish	B6@.010	SF	—	.27	.27
Trowel finishing					
Steel, machine work	B6@.015	SF	—	.41	.41
Steel, hand work	B6@.018	SF	—	.49	.49
Finish treads and risers					
No abrasives, no plastering, per LF of tread	B6@.042	LF	—	1.15	1.15
With abrasives, plastered, per LF of tread	B6@.063	LF	.57	1.72	2.29
Scoring concrete surface, hand work	B6@.005	LF	—	.14	.14
Sweep, scrub and wash down	B6@.006	SF	.02	.16	.18
Liquid curing and sealing compound, Acrylic Concrete Cure and Seal, spray-on, 400 SF at $19 per gallon	B6@.003	SF	.05	.08	.13
Exposed aggregate (washed, including finishing), no disposal of slurry	B6@.017	SF	.29	.46	.75
Non-metallic color and hardener, troweled on, 2 applications, red, gray or black					
60 pounds per 100 SF	B6@.021	SF	.41	.57	.98
100 pounds per 100 SF	B6@.025	SF	.67	.68	1.35
Add for other standard colors	—	%	50.0	—	—

Countertops

	Craft@Hrs	Unit	Material	Labor	Total
Wall finishes					
Cut back ties and patch	B6@.011	SF	.12	.30	.42
Remove fins	B6@.008	LF	.05	.22	.27
Grind smooth	B6@.021	SF	.09	.57	.66
Sack, burlap grout rub	B6@.013	SF	.06	.36	.42
Wire brush, green	B6@.015	SF	.05	.41	.46
Wash with acid and rinse	B6@.004	SF	.28	.11	.39
Break fins, patch voids, Carborundum rub	B6@.035	SF	.07	.96	1.03
Break fins, patch voids, burlap grout rub	B6@.026	SF	.07	.71	.78
Specialty finishes					
Monolithic natural aggregate topping					
3/16"	B6@.020	SF	.23	.55	.78
1/2"	B6@.022	SF	.26	.60	.86
Integral colors					
Various colors, 10 oz. bottle	—	Ea	4.97	—	4.97
Colored release agent	—	Lb	1.89	—	1.89
Dry shake colored hardener	—	Lb	.42	—	.42
Stamped finish (embossed concrete) of joints, if required					
Diamond, square, octagonal patterns	B6@.047	SF	—	1.28	1.28
Spanish paver pattern	B6@.053	SF	—	1.45	1.45
Add for grouting	B6@.023	SF	.30	.63	.93

Countertops Custom-fabricated straight, U- or L-shaped tops. See also Tile for ceramic tile countertops.

Laminated plastic countertops (such as Formica, Textolite or Wilsonart) on a particleboard base. 22" to 25" wide. Cost per LF of longest edge (either back or front) in solid colors with 3-1/2" to 4" backsplash.

	Craft@Hrs	Unit	Material	Labor	Total
Square edge, separate backsplash	B1@.181	LF	14.50	4.95	19.45
Rolled drip edge (post formed)	B1@.181	LF	20.50	4.95	25.45
Full wrap (180 degree) front edge	B1@.181	LF	22.00	4.95	26.95
Accent color front edge	B1@.181	LF	28.00	4.95	32.95
Inlaid front and back edge	B1@.181	LF	33.00	4.95	37.95
Additional costs for laminated countertops					
Add for textures, patterns	—	%	20.0	—	—
Add for square end splash	—	Ea	17.50	—	17.50
Add for contour end splash	—	Ea	25.00	—	25.00
Add for mitered corners	—	Ea	20.70	—	20.70
Add for seamless tops	—	Ea	30.00	—	30.00
Add for sink, range or vanity cutout	—	Ea	8.10	—	8.10
Add for drilling 3 plumbing fixture holes	—	LS	10.40	—	10.40
Add for quarter round corner	—	Ea	20.80	—	20.80
Add for half round corner	—	Ea	29.80	—	29.80

Granite countertops, custom made to order from a template. 22" to 25" wide. Per square foot of top and backsplash, not including cutting, polishing, sink cut out, range cut out and a flat polish edge. Add the cost of edge polishing for undermount sink, faucet holes, angle cuts, radius ends, edge detail and jobsite delivery as itemized below. 7/8" granite weighs about 30 pounds per square foot. 1-1/4" granite weighs about 45 pounds per square foot. 7/8" tops should be set on 3/4" plywood and set in silicon adhesive every 2'. 1-1/4" granite tops can be set directly on base cabinets with silicon adhesive.

	Craft@Hrs	Unit	Material	Labor	Total
Most 3/4" (2cm) granite tops	B1@.128	SF	39.00	3.50	42.50
Most 1-1/4" (3cm) granite tops	B1@.128	SF	51.00	3.50	54.50
Add for radius or beveled edge	—	LF	7.00	—	7.00
Add for half bullnose edge	—	LF	10.00	—	10.00
Add for full bullnose edge	—	LF	20.00	—	20.00
Add for 1/2" beveled edge	—	LF	7.00	—	7.00

	Craft@Hrs	Unit	Material	Labor	Total
Add for backsplash seaming and edging	—	LF	12.50	—	12.50
Add for cooktop overmount cutout	—	Ea	130.00	—	130.00
Add for delivery (typical)	—	Ea	150.00	—	150.00
Add for eased edge	—	LF	5.00	—	5.00
Add for electrical outlet cutout	—	Ea	35.00	—	35.00
Add for faucet holes	—	Ea	26.00	—	26.00
Add for rounded corners	—	Ea	30.00	—	30.00
Add for undermount sink, polished	—	Ea	300.00	—	300.00

Engineered stone countertops, quartz particles with acrylic or epoxy binder. Trade names include Crystalite, Silestone and Cambria. Custom made from a template. Per square foot of top and backsplash surface measured one side.

	Craft@Hrs	Unit	Material	Labor	Total
Small light chips, 3/4" square edge	B1@.128	SF	49.00	3.50	52.50
Large dark chips, 3/4" square edge	B1@.128	SF	55.00	3.50	58.50
Add for 3/4" bullnose edge	—	LF	25.00	—	25.00
Add for 3/4" ogee edge	—	LF	33.00	—	33.00
Add for 1-1/2" bullnose edge	—	LF	48.00	—	48.00
Add for 1-1/2" ogee edge	—	LF	84.00	—	84.00
Add for sink cutout	—	Ea	250.00	—	250.00
Add for electrical outlet cutout	—	Ea	30.00	—	30.00
Add for radius corner or end	—	Ea	185.00	—	185.00
Add for delivery	—	Ea	150.00	—	150.00

Cultured stone vanity tops with integral back splash and sink, 22" deep

	Craft@Hrs	Unit	Material	Labor	Total
Cultured marble					
24" long	B1@2.60	Ea	124.00	71.10	195.10
36" long	B1@2.60	Ea	164.00	71.10	235.10
48" long	B1@2.60	Ea	202.00	71.10	273.10
Cultured onyx					
24" long	B1@3.42	Ea	171.00	93.60	264.60
36" long	B1@3.42	Ea	218.00	93.60	311.60
48" long	B1@3.42	Ea	266.00	93.60	359.60

Corian vanity tops and integral sink, 3/4" thick, 22" deep

	Craft@Hrs	Unit	Material	Labor	Total
25" long	B1@.654	Ea	348.00	17.90	365.90
31" long	B1@.742	Ea	465.00	20.30	485.30
37" long	B1@.830	Ea	543.00	22.70	565.70
43" long	B1@.924	Ea	525.00	25.30	550.30
49" long	B1@1.02	Ea	696.00	27.90	723.90

Cupolas Unfinished cedar. Add the cost of weather vanes below. Estimate the delivery cost at $75 to $100 per unit.

Aluminum roof covering

	Craft@Hrs	Unit	Material	Labor	Total
18" square base x 22" high	B1@.948	Ea	210.00	25.90	235.90
27" square base x 31" high	B1@.948	Ea	310.00	25.90	335.90
34" square base x 33" high	B1@1.43	Ea	430.00	39.10	469.10
46" square base x 42" high	B1@2.02	Ea	685.00	55.30	740.30
Copper roof covering					
18" square base x 22" high	B1@.948	Ea	242.00	25.90	267.90
27" square base x 31" high	B1@.948	Ea	355.00	25.90	380.90
34" square base x 33" high	B1@1.43	Ea	510.00	39.10	549.10
46" square base x 42" high	B1@2.02	Ea	825.00	55.30	880.30

Decks

	Craft@Hrs	Unit	Material	Labor	Total
Weather vanes for cupolas					
Standard aluminum, baked black finish					
26" high	B1@.275	Ea	60.00	7.52	67.52
Deluxe model, gold-bronze aluminum, black finish					
36" high, bronze ornament	B1@.368	Ea	175.00	10.10	185.10

Decks These costs assume 5/4" x 6" southern yellow pine decking, 2" x 8" redwood decking, 5/4" x 6" recycled plastic lumber decking or 2" x 8" recycled composite lumber decking supported by #2 & better 2" x 8" pressure treated joists spaced 24" on center over 4" x 8" pressure treated beams. Beams are supported by 4" x 4" pressure treated posts set 6' on center in concrete. Post length is assumed to be 2' above grade. Fasteners are 16d galvanized nails, and galvanized steel joist hangers and post anchors. Decking is spaced at 1/8" to permit drainage. Costs are based on a rectangular deck, with one side fastened to an existing structure. Larger decks will reduce labor costs. May wish to consider adding costs for round concrete forms (AKA sonotubes) and additional framing fasteners.

Deck, 3' above ground, with joists, beams, posts and hardware as described above, with decking material as shown.

	Craft@Hrs	Unit	Material	Labor	Total
10' x 45'					
Pine, pressure treated decking	B1@.170	SF	6.18	4.65	10.83
Redwood, select heart decking	B1@.170	SF	6.62	4.65	11.27
5/4" x 6" recycled composite lumber decking	B1@.170	SF	9.15	4.65	13.80
2" x 8" recycled composite lumber decking	B1@.170	SF	13.20	4.65	17.85
15' x 45'					
Pine, pressure treated decking	B1@.170	SF	5.60	4.65	10.25
Redwood, select heart decking	B1@.170	SF	6.04	4.65	10.69
5/4" x 6" recycled composite lumber decking	B1@.170	SF	8.57	4.65	13.22
2" x 8" recycled composite lumber decking	B1@.170	SF	12.60	4.65	17.25
15' x 65'					
Pine, pressure treated decking	B1@.170	SF	5.63	4.65	10.28
Redwood, select heart decking	B1@.170	SF	6.07	4.65	10.72
5/4" x 6" recycled composite lumber decking	B1@.170	SF	8.60	4.65	13.25
2" x 8" recycled composite lumber decking	B1@.170	SF	12.70	4.65	17.35
20' x 35'					
Pine, pressure treated decking	B1@.170	SF	5.19	4.65	9.84
Redwood, select heart decking	B1@.170	SF	5.64	4.65	10.29
5/4" x 6" recycled composite lumber decking	B1@.170	SF	8.16	4.65	12.81
2" x 8" recycled composite lumber decking	B1@.170	SF	12.20	4.65	16.85
20' x 55'					
Pine, pressure treated decking	B1@.170	SF	5.00	4.65	9.65
Redwood, select heart decking	B1@.170	SF	5.44	4.65	10.09
5/4" x 6" recycled composite lumber decking	B1@.170	SF	7.97	4.65	12.62
2" x 8" recycled composite lumber decking	B1@.170	SF	12.00	4.65	16.65
Add for diagonal pattern decking					
Pine decking	B1@.007	SF	.31	.19	.50
Redwood decking	B1@.007	SF	.38	.19	.57
5/4" x 6" recycled composite lumber decking	B1@.007	SF	.74	.19	.93
2" x 8" recycled composite lumber decking	B1@.007	SF	1.31	.19	1.50
Add for stain and sealer finish	B1@.006	SF	.10	.16	.26

	Craft@Hrs	Unit	Material	Labor	Total

Deck covers Can also be referred to as a pergola. Conventionally framed wood cover joists. Per SF of area covered. Figures in parentheses indicate board feet per square foot of ceiling including end joists, header joists, and 5% waste. Costs shown are based on a job with 200 SF of area covered and include normal waste. Add the cost of posts and finish roof.

2" x 4", Std & Btr grade

	Craft@Hrs	Unit	Material	Labor	Total
16" centers (.59 BF per SF)	B1@.020	SF	.37	.55	.92
24" centers (.42 BF per SF)	B1@.014	SF	.26	.38	.64

2" x 6", Std & Btr grade

	Craft@Hrs	Unit	Material	Labor	Total
16" centers (.88 BF per SF)	B1@.020	SF	.55	.55	1.10
24" centers (.63 BF per SF)	B1@.015	SF	.39	.41	.80

Add for posts to support conventionally framed wood cover joists. Posts are usually required at each corner. Includes one sack of concrete but no excavation.

	Craft@Hrs	Unit	Material	Labor	Total
4" x 4" pressure treated, 10' long	B1@.750	Ea	13.50	20.50	34.00
6" x 6" pressure treated, 10' long	B1@.750	Ea	30.00	20.50	50.50

Deck finish roof Cover over conventionally framed wood ceiling joists.
Flat roof with built-up roofing. Using 1/2" CDX roof sheathing with 3 ply asphalt consisting of 2 plies of 15 lb felt with 90 lb cap sheet and 3 hot mop coats of asphalt. Add the cost for framing from above.

	Craft@Hrs	Unit	Material	Labor	Total
Total flat roof assembly cost per SF	B1@.029	SF	.88	.79	1.67

Deck railing 42" high handrail. 2" x 2" paling 5" OC run vertically, lag bolted to edge of the deck, one 2" x 4" horizontal top rail and one 2" x 6" placed on edge directly below the top rail.

	Craft@Hrs	Unit	Material	Labor	Total
Pine, pressure treated, per running foot	B1@.333	LF	2.35	9.11	11.46
Redwood, select heart, per running foot	B1@.333	LF	6.27	9.11	15.38
Recycled composite lumber, per running foot	B1@.333	LF	17.40	9.11	26.51

Deck stairs All stairs assume 12" tread (step) composed of two 2" x 6" lumber and riser.

One or two steps high, box construction

Prices are per LF of step. A 4' wide stair with 2 steps has 8 LF of step. Use 8 LF as the quantity.

	Craft@Hrs	Unit	Material	Labor	Total
Pine treads, open risers	B1@.270	LF	4.27	7.39	11.66
Pine treads, with pine risers	B1@.350	LF	7.28	9.58	16.86
Redwood treads, open risers	B1@.270	LF	4.20	7.39	11.59
Redwood treads, with redwood risers	B1@.350	LF	7.07	9.58	16.65
5/4" recycled composite lumber treads, open risers	B1@.270	LF	6.52	7.39	13.91
5/4" recycled composite lumber treads, with recycled composite lumber risers	B1@.350	LF	12.10	9.58	21.68
2" x 6" recycled composite lumber treads, open risers	B1@.270	LF	8.71	7.39	16.10
2" x 6" recycled composite lumber treads, with recycled composite lumber risers	B1@.350	LF	18.60	9.58	28.18

Three or more steps high, 36" wide. 3 stringers cut from 2" x 12" pressure treated lumber.

Cost per riser

	Craft@Hrs	Unit	Material	Labor	Total
Pine treads, open risers	B1@.530	Ea	15.60	14.50	30.10
Pine treads, with pine risers	B1@.620	Ea	24.60	17.00	41.60
Redwood treads, open risers	B1@.530	Ea	15.40	14.50	29.90
Redwood treads, with redwood risers	B1@.620	Ea	24.00	17.00	41.00
5/4" recycled composite lumber treads, open risers	B1@.270	LF	23.30	7.39	30.69
5/4" recycled composite lumber treads, with recycled composite lumber risers	B1@.350	LF	39.90	9.58	49.48

Demolition

	Craft@Hrs	Unit	Material	Labor	Total
2" x 6" recycled composite lumber treads, open risers	B1@.270	LF	29.90	7.39	37.29
2" x 6" recycled composite lumber treads, with recycled composite lumber risers	B1@.350	LF	59.60	9.58	69.18

Concrete footing to support stairway. Cost includes brackets and fasteners to attach stair to footing.

	Craft@Hrs	Unit	Material	Labor	Total
18" x 18" x 12" footing	B1@.530	Ea	18.50	14.50	33.00

	Craft@Hrs	Unit	Material	Labor	Equipment	Total

Demolition Itemized costs for demolition of building components when building is being remodeled, repaired or rehabilitated, not completely demolished. Costs include protecting adjacent areas and normal clean-up. Costs are to break out the items listed and pile debris on site only. No hauling, dump fees or salvage value included. Equipment cost includes one air compressor and a pneumatic breaker with jackhammer bits. (Figures in parentheses show the volume before and after demolition and weight of the materials after demolition.) Use $250.00 as a minimum charge.

Brick walls and walks, cost per SF removed, measured on one face, using pneumatic breaker

	Craft@Hrs	Unit	Material	Labor	Equipment	Total
4" thick walls (60 SF per CY and 36 lbs. per SF)	BL@.061	SF	—	1.51	.79	2.30
8" thick walls (30 SF per CY and 72 lbs. per SF)	BL@.110	SF	—	2.72	1.43	4.15
12" thick walls (20 SF per CY and 108 lbs. per SF)	BL@.133	SF	—	3.29	1.72	5.01

Brick sidewalks, cost per SF removed

	Craft@Hrs	Unit	Material	Labor	Equipment	Total
2-1/2" thick bricks on sand base, no mortar bed, removed using hand tools (100 SF per CY and 28 lbs. per SF)	BL@.004	SF	—	.10	—	.10

Brick pavers up to 4-1/2" thick with mortar bed, removed using a pneumatic breaker

	Craft@Hrs	Unit	Material	Labor	Equipment	Total
(55 SF per CY and 50 lbs. per SF)	BL@.050	SF	—	1.24	.65	1.89

Ceiling tile, cost per in-place SF removed using hand tools

	Craft@Hrs	Unit	Material	Labor	Equipment	Total
Stapled or glued 12" x 12" (250 SF per CY and .25 lbs. per SF)	BL@.015	SF	—	.37	—	.37
Suspended panels 2' x 4' (250 SF per CY and .25 lbs. per SF)	BL@.010	SF	—	.25	—	.25

Concrete masonry walls, cost per in-place SF removed, using a pneumatic breaker

	Craft@Hrs	Unit	Material	Labor	Equipment	Total
4" thick walls (60 SF per CY and 19 lbs. per SF)	BL@.066	SF	—	1.63	.86	2.49
6" thick walls (40 SF per CY and 28 lbs. per SF)	BL@.075	SF	—	1.85	.97	2.82
8" thick walls (30 SF per CY and 34 lbs. per SF)	BL@.098	SF	—	2.42	1.27	3.69
12" thick walls (20 SF per CY and 46 lbs. per SF)	BL@.140	SF	—	3.46	1.81	5.27
Reinforced or grouted walls add	—	%	—	50.0	50.0	—

Concrete foundations (footings), steel reinforced, removed using a pneumatic breaker. Concrete at 3,900 pounds per in place cubic yard.

	Craft@Hrs	Unit	Material	Labor	Equipment	Total
Cost per CY (.75 CY per CY)	BL@3.96	CY	—	97.90	51.30	149.20
Cost per LF with width and depth as shown						
6" W x 12" D (35 LF per CY)	BL@.075	LF	—	1.85	.97	2.82
8" W x 12" D (30 LF per CY)	BL@.098	LF	—	2.42	1.27	3.69
8" W x 16" D (20 LF per CY)	BL@.133	LF	—	3.29	1.72	5.01
8" W x 18" D (18 LF per CY)	BL@.147	LF	—	3.63	1.90	5.53

	Craft@Hrs	Unit	Material	Labor	Equipment	Total
10" W x 12" D (21 LF per CY)	BL@.121	LF	—	2.99	1.57	4.56
10" W x 16" D (16 LF per CY)	BL@.165	LF	—	4.08	2.14	6.22
10" W x 18" D (14 LF per CY)	BL@.185	LF	—	4.57	2.40	6.97
12" W x 12" D (20 LF per CY)	BL@.147	LF	—	3.63	1.90	5.53
12" W x 16" D (13 LF per CY)	BL@.196	LF	—	4.84	2.54	7.38
12" W x 20" D (11 LF per CY)	BL@.245	LF	—	6.05	3.17	9.22
12" W x 24" D (9 LF per CY)	BL@.294	LF	—	7.26	3.81	11.07

Concrete sidewalks, to 4" thick, cost per SF removed. Concrete at 3,900 pounds per in place cubic yard.

	Craft@Hrs	Unit	Material	Labor	Equipment	Total
Non-reinforced, removed by hand						
(60 SF per CY)	BL@.050	SF	—	1.24	—	1.24
Reinforced, removed using pneumatic breaker						
(55 SF per CY)	BL@.033	SF	—	.82	.43	1.25

Concrete slabs, non-reinforced, removed using pneumatic breaker. Concrete at 3,900 pounds per in-place cubic yard.

	Craft@Hrs	Unit	Material	Labor	Equipment	Total
Cost per CY (.80 CY per CY)	BL@3.17	CY	—	78.30	41.08	119.38
Cost per SF with thickness as shown						
3" slab thickness (90 SF per CY)	BL@.030	SF	—	.74	.39	1.13
4" slab thickness (60 SF per CY)	BL@.040	SF	—	.99	.52	1.51
6" slab thickness (45 SF per CY)	BL@.056	SF	—	1.38	.73	2.11
8" slab thickness (30 SF per CY)	BL@.092	SF	—	2.27	1.19	3.46

Concrete slabs, steel reinforced, removed using pneumatic breaker. Concrete at 3,900 pounds per in-place cubic yard.

	Craft@Hrs	Unit	Material	Labor	Equipment	Total
Cost per CY (1.33 CY per CY)	BL@3.96	CY	—	97.90	51.32	149.22
Cost per SF with thickness as shown						
3" slab thickness (80 SF per CY)	BL@.036	SF	—	.89	.47	1.36
4" slab thickness (55 SF per CY)	BL@.050	SF	—	1.24	.65	1.89
6" slab thickness (40 SF per CY)	BL@.075	SF	—	1.85	.97	2.82
8" slab thickness (30 SF per CY)	BL@.098	SF	—	2.42	1.27	3.69

Concrete walls, steel reinforced, removed using pneumatic breaker. Concrete at 3,900 pounds per in-place cubic yard.

	Craft@Hrs	Unit	Material	Labor	Equipment	Total
Cost per CY (1.33 CY per CY)	BL@4.76	CY	—	118.00	61.68	179.68
Cost per SF with thickness as shown (deduct openings over 25 SF)						
3" wall thickness (80 SF per CY)	BL@.045	SF	—	1.11	.58	1.69
4" wall thickness (55 SF per CY)	BL@.058	SF	—	1.43	.75	2.18
5" wall thickness (45 SF per CY)	BL@.075	SF	—	1.85	.97	2.82
6" wall thickness (40 SF per CY)	BL@.090	SF	—	2.22	1.17	3.39
8" wall thickness (30 SF per CY)	BL@.120	SF	—	2.97	1.56	4.53
10" wall thickness (25 SF per CY)	BL@.147	SF	—	3.63	1.90	5.53
12" wall thickness (20 SF per CY)	BL@.176	SF	—	4.35	2.28	6.63

Curbs and 18" gutter, concrete, removed using pneumatic breaker. Concrete at 3,900 pounds per in-place cubic yard.

	Craft@Hrs	Unit	Material	Labor	Equipment	Total
Cost per LF (8 LF per CY)	BL@.422	LF	—	10.40	5.47	15.87

Doors, typical cost to remove door, frame and hardware

	Craft@Hrs	Unit	Material	Labor	Equipment	Total
3' x 7' metal (2 doors per CY)	BL@1.00	Ea	—	24.70	—	24.70
3' x 7' wood (2 doors per CY)	BL@.497	Ea	—	12.30	—	12.30

Flooring, cost per in-place SY of floor area removed (1 square yard = 9 square feet)

	Craft@Hrs	Unit	Material	Labor	Equipment	Total
Ceramic tile						
(25 SY per CY and 34 lbs. per SY)	BL@.263	SY	—	6.50	—	6.50
Hardwood, nailed						
(25 SY per CY and 18 lbs. per SY)	BL@.290	SY	—	7.17	—	7.17
Hardwood, glued						
(25 SY per CY and 18 lbs. per SY)	BL@.503	SY	—	12.40	—	12.40

Demolition

	Craft@Hrs	Unit	Material	Labor	Equipment	Total
Linoleum, sheet						
(30 SY per CY and 3 lbs. per SY)	BL@.056	SY	—	1.38	—	1.38
Carpet						
(40 SY per CY and 5 lbs. per SY)	BL@.028	SY	—	.69	—	.69
Carpet pad						
(35 SY per CY and 1.7 lbs. per SY)	BL@.014	SY	—	.35	—	.35
Resilient tile						
(30 SY per CY and 3 lbs. per SY)	BL@.070	SY	—	1.73	—	1.73
Terrazzo						
(25 SY per CY and 34 lbs. per SY)	BL@.286	SY	—	7.07	—	7.07
Remove mastic, using power sander	BL@.125	SY	—	3.09	—	3.09
Insulation, fiberglass batts, cost per in-place SF removed						
R-11 to R-19						
(500 SF per CY and .3 lb per SF)	BL@.003	SF	—	.07	—	.07
Pavement, asphaltic concrete (bituminous) to 6" thick, removed using pneumatic breaker						
Cost per SY						
(4 SY per CY and 660 lbs. per SY)	BL@.250	SY	—	6.18	3.24	9.42
Plaster walls or ceilings, interior, cost per in-place SF removed by hand						
(150 SF per CY and 8 lbs. per SF)	BL@.015	SF	—	.37	—	.37
Roofing, cost per in-place Sq, removed using hand tools (1 square = 100 square feet)						
Built-up, 5 ply						
(2.5 Sq per CY and 250 lbs per Sq)	BL@1.50	Sq	—	37.10	—	37.10
Shingles, asphalt, single layer						
(2.5 Sq per CY and 240 lbs per Sq)	BL@1.33	Sq	—	32.90	—	32.90
Shingles, asphalt, double layer						
(1.2 Sq per CY and 480 lbs per Sq)	BL@1.58	Sq	—	39.00	—	39.00
Shingles, slate. Weight ranges from 600 lbs. to 1200 lbs. per Sq						
(1 Sq per CY and 900 lbs per Sq).	BL@1.79	Sq	—	44.20	—	44.20
Shingles, wood						
(1.7 Sq per CY and 400 lbs per Sq)	BL@2.02	Sq	—	49.90	—	49.90
Sheathing, up to 1" thick, cost per in-place SF, removed using hand tools						
Gypsum						
(250 SF per CY and 2.3 lbs. per SF)	BL@.017	SF	—	.42	—	.42
Plywood						
(200 SF per CY and 2 lbs. per SF)	BL@.017	SF	—	.42	—	.42
Siding, up to 1" thick, cost per in-place SF, removed using hand tools						
Metal						
(200 SF per CY and 2 lbs. per SF)	BL@.027	SF	—	.67	—	.67
Plywood						
(200 SF per CY and 2 lbs. per SF)	BL@.017	SF	—	.42	—	.42
Wood, boards						
(250 SF per CY and 2 lbs. per SF)	BL@.030	SF	—	.74	—	.74
Stucco walls or soffits, cost per in-place SF, removed using hand tools						
(150 SF per CY and 8 lbs. per SF)	BL@.036	SF	—	.89	—	.89
Trees. See Excavation						
Wallboard up to 1" thick, cost per in-place SF, removed using hand tools						
Gypsum, walls or ceilings						
(250 SF per CY and 2.3 lbs. per SF)	BL@.018	SF	—	.44	—	.44
Plywood or insulation board						
(200 SF per CY and 2 lbs. per SF)	BL@.018	SF	—	.44	—	.44

	Craft@Hrs	Unit	Material	Labor	Equipment	Total
Windows, typical cost to remove window, frame and hardware, cost per SF of window						
Metal (36 SF per CY)	BL@.058	SF	—	1.43	—	1.43
Wood (36 SF per CY)	BL@.063	SF	—	1.56	—	1.56
Wood stairs, cost per in-place SF or LF						
Risers (25 SF per CY)	BL@.116	SF	—	2.87	—	2.87
Landings (50 SF per CY)	BL@.021	SF	—	.52	—	.52
Handrails (100 LF per CY)	BL@.044	LF	—	1.09	—	1.09
Posts (200 LF per CY)	BL@.075	LF	—	1.85	—	1.85

	Craft@Hrs	Unit	Material	Labor	Total

Wood Framing Demolition Typical costs for demolition of wood frame structural components using hand tools. Normal center to center spacing (12" thru 24" OC) is assumed. Costs include removal of Romex cable as necessary, protecting adjacent areas, and normal clean-up. Costs do not include temporary shoring to support structural elements above or adjacent to the work area. (Add for demolition of flooring, roofing, sheathing, etc.) Figures in parentheses give the approximate "loose" volume and weight of the materials after being demolished.

Ceiling joists, per in-place SF of ceiling area removed

	Craft@Hrs	Unit	Material	Labor	Total
2" x 4" (720 SF per CY and 1.18 lbs. per SF)	BL@.009	SF	—	.22	.22
2" x 6" (410 SF per CY and 1.76 lbs. per SF)	BL@.013	SF	—	.32	.32
2" x 8" (290 SF per CY and 2.34 lbs. per SF)	BL@.018	SF	—	.44	.44
2" x 10" (220 SF per CY and 2.94 lbs. per SF)	BL@.023	SF	—	.57	.57
2" x 12" (190 SF per CY and 3.52 lbs. per SF)	BL@.027	SF	—	.67	.67

Floor joists, per in-place SF of floor area removed

	Craft@Hrs	Unit	Material	Labor	Total
2" x 6" (290 SF per CY and 2.04 lbs. per SF)	BL@.016	SF	—	.40	.40
2" x 8" (190 SF per CY and 2.72 lbs. per SF)	BL@.021	SF	—	.52	.52
2" x 10" (160 SF per CY and 3.42 lbs. per SF)	BL@.026	SF	—	.64	.64
2" x 12" (120 SF per CY and 4.10 lbs. per SF)	BL@.031	SF	—	.77	.77

Rafters, per in-place SF of actual roof area removed

	Craft@Hrs	Unit	Material	Labor	Total
2" x 4" (610 SF per CY and 1.42 lbs. per SF)	BL@.011	SF	—	.27	.27
2" x 6" (360 SF per CY and 2.04 lbs. per SF)	BL@.016	SF	—	.40	.40
2" x 8" (270 SF per CY and 2.68 lbs. per SF)	BL@.020	SF	—	.49	.49
2" x 10" (210 SF per CY and 3.36 lbs. per SF)	BL@.026	SF	—	.64	.64
2" x 12" (180 SF per CY and 3.94 lbs. per SF)	BL@.030	SF	—	.74	.74

Stud walls, interior or exterior, includes allowance for plates and blocking, per in-place SF of wall area removed, measured on one face

	Craft@Hrs	Unit	Material	Labor	Total
2" x 3" (430 SF per CY and 1.92 lbs. per SF)	BL@.013	SF	—	.32	.32
2" x 4" (310 SF per CY and 2.58 lbs. per SF)	BL@.017	SF	—	.42	.42
2" x 6" (190 SF per CY and 3.74 lbs. per SF)	BL@.025	SF	—	.62	.62

Wood Framed Assemblies Demolition Typical costs for demolition of wood framed structural assemblies using hand tools when building is being remodeled, repaired or rehabilitated but not completely demolished. Costs include removal of Romex cable as necessary, protecting adjacent areas, and normal clean-up. Costs do not include temporary shoring to support structural elements above or adjacent to the work area, loading, hauling, dump fees or salvage value. Costs shown are per in-place square foot (SF) of area being demolished. Normal center to center spacing (12" thru 24" OC) is assumed. Figures in parentheses give the approximate "loose" volume and weight of the materials (volume after being demolished and weight.)

Ceiling assemblies. Remove ceiling joists, fiberglass insulation and gypsum board up to 5/8" thick

	Craft@Hrs	Unit	Material	Labor	Total
2" x 4" (135 SF per CY and 3.78 lbs. per SF)	BL@.030	SF	—	.74	.74
2" x 6" (118 SF per CY and 4.36 lbs. per SF)	BL@.034	SF	—	.84	.84
2" x 8" (106 SF per CY and 4.94 lbs. per SF)	BL@.039	SF	—	.96	.96
2" x 10" (95 SF per CY and 5.54 lbs. per SF)	BL@.043	SF	—	1.06	1.06
2" x 12" (89 SF per CY and 6.12 lbs. per SF)	BL@.048	SF	—	1.19	1.19

Demolition, Subcontract

	Craft@Hrs	Unit	Material	Labor	Total
Floor assemblies. Remove floor joists and plywood subfloor up to 3/4" thick					
2" x 8" (97 SF per CY and 4.72 lbs. per SF)	BL@.035	SF	—	.86	.86
2" x 10" (88 SF per CY and 5.42 lbs. per SF)	BL@.040	SF	—	.99	.99
2" x 12" (75 SF per CY and 6.10 lbs. per SF)	BL@.043	SF	—	1.06	1.06
For gypsum board ceiling removed below floor joists, add to any of above					
(250 SF per CY and 2.3 lbs. per SF)	BL@.018	SF	—	.44	.44
Wall assemblies, interior. Remove studs, 2 top plates, 1 bottom plate and gypsum drywall up to 5/8" thick on both sides					
2" x 3" (97 SF per CY and 6.52 lbs. per SF)	BL@.048	SF	—	1.19	1.19
2" x 4" (89 SF per CY and 7.18 lbs. per SF)	BL@.052	SF	—	1.28	1.28
2" x 6" (75 SF per CY and 8.34 lbs. per SF)	BL@.059	SF	—	1.46	1.46
Wall assemblies, exterior. Remove studs, 2 top plates, 1 bottom plate, insulation, gypsum drywall up to 5/8" thick on one side, plywood siding up to 3/4" thick on other side					
2" x 4" (70 SF per CY and 7.18 lbs. per SF)	BL@.067	SF	—	1.66	1.66
2" x 6" (75 SF per CY and 8.34 lbs. per SF)	BL@.074	SF	—	1.83	1.83
For stucco in lieu of plywood, add to any of above					
(SF per CY is the same and add 6 lbs. per SF)	BL@.006	SF	—	.15	.15

Demolition, Subcontract Costs for demolishing and removing an entire building using a dozer and front-end loader. Costs shown include loading and hauling up to 5 miles, but no dump fees. Dump charges vary widely; see Disposal Fees in the Site Work Section of the Industrial and Commercial division of this book. Costs shown are for above ground structures only, per in-place SF of total floor area. Up to 10' high ceilings. No salvage value assumed. Add the cost of foundation demolition. Figures in parentheses give the approximate "loose" volume of the materials (volume after being demolished).

	Craft@Hrs	Unit	Material	Labor	Total
Light wood-frame structures, up to three stories in height. Based on 2,500 SF job. No basements included. Estimate each story separately. Cost per in-place SF floor area					
First story (8 SF per CY)	—	SF	—	—	2.00
Second story (8 SF per CY)	—	SF	—	—	2.61
Third story (8 SF per CY)	—	SF	—	—	3.35
Masonry building, cost per in-place SF floor area					
based on 5,000 SF job (50 SF per CY)	—	SF	—	—	3.25
Concrete building, cost per in-place SF floor area					
based on 5,000 SF job (30 SF per CY)	—	SF	—	—	4.25
Reinforced concrete building, cost per in-place SF floor area					
based on 5,000 SF job (20 SF per CY)	—	SF	—	—	4.83

Door Chimes Electric, typical installation, add transformer and wire below.

	Craft@Hrs	Unit	Material	Labor	Total
Surface mounted, 2 notes					
Simple white plastic chime, 4" x 7"	BE@3.50	Ea	22.00	119.00	141.00
Modern oak grained finish, plastic, 8" x 5"	BE@3.50	Ea	45.00	119.00	164.00
White, with repeating tones	BE@3.50	Ea	65.00	119.00	184.00
Maple finish, plastic, 8" x 5"	BE@3.50	Ea	45.00	119.00	164.00
Designer, wood-look plastic strips across a cloth grille					
9" x 7"	BE@3.50	Ea	44.50	119.00	163.50
Modern real oak wood, with brass plated chimes					
9" x 7"	BE@3.50	Ea	74.00	119.00	193.00
Surface mounted, 4 to 8 notes, solid wood door chime					
Three 12" long tubes	BE@3.64	Ea	198.00	123.00	321.00
Four 55" long tubes	BE@3.64	Ea	255.00	123.00	378.00

	Craft@Hrs	Unit	Material	Labor	Total
Any of above					
Add for chime transformer	—	Ea	12.00	—	12.00
Add for bell wire, 65' per coil	—	Ea	4.00	—	4.00
Add for lighted push buttons	—	Ea	8.00	—	8.00
Non-electric, push button type, door mounted					
Modern design, 2-1/2" x 2-1/2"	BE@.497	Ea	26.50	16.80	43.30
With peep sight, 7" x 3-1/2"	BE@.497	Ea	15.80	16.80	32.60
Wireless door chimes, with push button transmitter, 3-year battery, 150' range, 2" speaker, 128 codes					
Oak housing with satin chime tubes	BE@.358	Ea	51.50	12.10	63.60

Door Closers

Residential and light commercial door closers, doors to 140 lbs and 38" wide. 150 degree maximum swing, non-hold open, adjustable closing speed, ANSI A156.4

	Craft@Hrs	Unit	Material	Labor	Total
Brown or white finish	BC@.750	Ea	15.00	22.50	37.50
Heavy duty door closers, doors to 170 lbs. and 36" wide. Schlage					
Aluminum	BC@1.04	Ea	58.00	31.20	89.20
Aluminum with cover	BC@1.04	Ea	80.00	31.20	111.20
Screen door closer, brass finish					
Hydraulic closer, medium duty	BC@.515	Ea	12.00	15.50	27.50
Pneumatic closer, light duty	BC@.546	Ea	10.60	16.40	27.00
Sliding door closer, pneumatic, aluminum finish					
Screen	BC@.250	Ea	17.00	7.50	24.50
Glass	BC@.250	Ea	36.10	7.50	43.60

Wood Exterior Slab Doors Exterior doors are usually 1-3/4" thick and solid core – a thin veneer laid over a frame but with foam or composition board filling the door cavity. Flush doors are smooth and flat, with no decorative treatment. Slab doors are the door alone, as distinguished from prehung doors which are sold with jamb, hinges, trim and threshold attached. Installation times for slab doors will be up to 50% faster when an experienced carpenter has the use of a router, door jig and pneumatic nailer.
Flush hardboard exterior doors. Solid core. 1-3/4" thick. Primed and ready to paint. Hardboard doors are made of composition wood veneer, usually the least expensive material available. Labor includes setting three hinges on the door and jamb and hanging the door in an existing cased opening.

	Craft@Hrs	Unit	Material	Labor	Total
30" x 80"	BC@1.45	Ea	44.10	43.50	87.60
32" x 80"	BC@1.45	Ea	44.30	43.50	87.80
34" x 80"	BC@1.45	Ea	53.20	43.50	96.70
36" x 80"	BC@1.45	Ea	47.00	43.50	90.50

Flush hardwood exterior doors. Solid core. 1-3/4" thick. Labor includes setting three hinges on the door and jamb and hanging the door in an existing cased opening.

	Craft@Hrs	Unit	Material	Labor	Total
28" x 80"	BC@1.45	Ea	49.50	43.50	93.00
30" x 80"	BC@1.45	Ea	53.60	43.50	97.10
32" x 80"	BC@1.45	Ea	55.60	43.50	99.10
36" x 80"	BC@1.45	Ea	57.80	43.50	101.30
36" x 84"	BC@1.45	Ea	61.90	43.50	105.40
30" x 80", diamond light	BC@1.45	Ea	62.00	43.50	105.50
30" x 80", square light	BC@1.45	Ea	62.00	43.50	105.50
32" x 80", square light	BC@1.45	Ea	64.00	43.50	107.50
36" x 80", diamond light	BC@1.45	Ea	67.20	43.50	110.70
36" x 80", square light	BC@1.45	Ea	67.20	43.50	110.70

Doors, Slab

	Craft@Hrs	Unit	Material	Labor	Total

Flush birch exterior doors. Solid core.1-3/4" thick. Stainable. Labor includes setting three hinges on the door and jamb and hanging the door in an existing cased opening.

	Craft@Hrs	Unit	Material	Labor	Total
32" x 80"	BC@1.45	Ea	50.50	43.50	94.00
34" x 80"	BC@1.45	Ea	59.80	43.50	103.30
36" x 80"	BC@1.45	Ea	53.30	43.50	96.80

Flush lauan exterior doors. Solid core. 1-3/4" thick. Stainable. Lauan is a tropical hardwood like mahogany. Labor includes setting three hinges on the door and jamb and hanging the door in an existing cased opening.

	Craft@Hrs	Unit	Material	Labor	Total
30" x 80"	BC@1.45	Ea	46.30	43.50	89.80
30" x 84"	BC@1.45	Ea	49.30	43.50	92.80
32" x 80"	BC@1.45	Ea	47.10	43.50	90.60
36" x 80"	BC@1.45	Ea	50.00	43.50	93.50

Red oak exterior doors. Wood stave core. 1-3/4" thick. Labor includes setting three hinges on the door and jamb and hanging the door in an existing cased opening.

	Craft@Hrs	Unit	Material	Labor	Total
30" x 80"	BC@1.45	Ea	58.20	43.50	101.70
32" x 80"	BC@1.45	Ea	60.00	43.50	103.50
34" x 80"	BC@1.45	Ea	63.00	43.50	106.50
36" x 80"	BC@1.45	Ea	63.60	43.50	107.10

6-panel fir exterior doors. 1-3/4" thick. Beveled, hip-raised panels. Doweled stile and rail joints. 4-1/2" wide stiles. Unfinished clear fir. Ready to paint or stain. Labor includes setting three hinges on the door and jamb and hanging the door in an existing cased opening.

	Craft@Hrs	Unit	Material	Labor	Total
30" x 80"	BC@1.45	Ea	159.00	43.50	202.50
32" x 80"	BC@1.45	Ea	154.00	43.50	197.50
36" x 80"	BC@1.45	Ea	153.00	43.50	196.50

4-panel 4-lite fir exterior doors. Vertical grain Douglas fir. 1-3/4" thick. 3/4" thick panels. Unfinished. Veneered and doweled construction. 1/8" tempered safety glass. Labor includes setting three hinges on the door and jamb and hanging the door in an existing cased opening.

	Craft@Hrs	Unit	Material	Labor	Total
36" x 80" x 1-3/4"	BC@1.70	Ea	192.00	51.00	243.00
36" x 80" x 1-3/4"	BC@1.70	Ea	156.00	51.00	207.00

Entry Slab Doors Any exterior door can be use at the front entry of a home. But entry doors usually include decorative elements, such as glass panels set in a border of brass caming. Slab doors are the door alone, as distinguished from prehung doors which are sold with jamb, hinges, trim and threshold attached. Installation times for entry doors will be up to 50% faster when an experienced carpenter has the use of a router, door jig and pneumatic nailer.

Fir crossbuck doors. 1-3/4" thick. Solid core. Labor includes setting three hinges on the door and jamb and hanging the door in an existing cased opening. Crossbuck doors have a raised X design in the bottom half of the door.

	Craft@Hrs	Unit	Material	Labor	Total
32" x 80" x 1-3/4"	BC@1.45	Ea	185.00	43.50	228.50
36" x 80"	BC@1.45	Ea	188.00	43.50	231.50

Mahogany entry doors. 1-3/4" thick. Beveled tempered glass lite. Unfinished solid wood. Ready to paint or stain. Labor includes setting three hinges on the door and jamb and hanging the door in an existing cased opening.

	Craft@Hrs	Unit	Material	Labor	Total
30" x 80", arch lite	BC@1.45	Ea	426.00	43.50	469.50
30" x 80", moon lite	BC@1.45	Ea	426.00	43.50	469.50
30" x 80", oval lite	BC@1.45	Ea	433.00	43.50	476.50

2-panel 9-lite entry doors. 1-3/4" thick. Single-pane glass. Solid core with doweled joints. Unfinished. Suitable for stain or paint. Labor includes setting three hinges on the door and jamb and hanging the door in an existing cased opening.

	Craft@Hrs	Unit	Material	Labor	Total
30" x 80"	BC@1.70	Ea	160.00	51.00	211.00
32" x 80"	BC@1.70	Ea	160.00	51.00	211.00
36" x 80"	BC@1.70	Ea	160.00	51.00	211.00

	Craft@Hrs	Unit	Material	Labor	Total

1-lite clear fir entry doors. 1-3/4" thick. Clear fir surface ready to stain or paint. Insulated glass. Labor includes setting three hinges on the door and jamb and hanging the door in an existing cased opening.

	Craft@Hrs	Unit	Material	Labor	Total
30" x 80"	BC@1.70	Ea	157.00	51.00	208.00
32" x 80"	BC@1.70	Ea	162.00	51.00	213.00
36" x 80"	BC@1.70	Ea	162.00	51.00	213.00

Fir 15-lite entry doors. Single-pane glass. Ready for stain or paint. 1-3/4" thick. Doweled stile and rail joints. Labor includes setting three hinges on the door and jamb and hanging the door in an existing cased opening.

	Craft@Hrs	Unit	Material	Labor	Total
30" x 80" x 1-3/4"	BC@1.70	Ea	151.00	51.00	202.00
32" x 80" x 1-3/4"	BC@1.70	Ea	153.00	51.00	204.00
36" x 80" x 1-3/4"	BC@1.70	Ea	154.00	51.00	205.00

Fir fan lite entry doors. Vertical grain Douglas fir. 1-3/4" thick. 3/4" thick panels. Unfinished. Veneer and doweled construction. 1/8" tempered safety glass. Labor includes setting three hinges on the door and jamb and hanging the door in an existing cased opening.

	Craft@Hrs	Unit	Material	Labor	Total
32" x 80"	BC@1.45	Ea	181.00	43.50	224.50
30" x 80"	BC@1.45	Ea	167.00	43.50	210.50

Deco fan lite entry doors. 1-3/4" thick. 3/4" thick panels. Unfinished solid wood. Ready to paint or stain. Beveled insulated glass. Labor includes setting three hinges on the door and jamb and hanging the door in an existing cased opening.

	Craft@Hrs	Unit	Material	Labor	Total
36" x 80"	BC@1.45	Ea	219.00	43.50	262.50

Flush oak entry doors. 1-3/4" thick. Prefinished. Labor includes setting three hinges on the door and jamb and hanging the door in an existing cased opening.

	Craft@Hrs	Unit	Material	Labor	Total
36" x 80"	BC @1.45	Ea	839.00	43.50	882.50

Beveled glass entry doors. Meranti mahogany with beveled glass insert. Matching sidelites. Transom designed to fit over one 36" door and two 14" sidelites combination. Based on Beveled Glass Designs Symphony series. Door dimensions are 36" x 80" x 1-3/4". Labor cost includes hanging door only. Add the cost of hinges and lockset.

	Craft@Hrs	Unit	Material	Labor	Total
Bolero, oval insert					
Door	BC@1.45	Ea	860.00	43.50	903.50
Sidelight	BC@1.00	Ea	578.00	30.00	608.00
Matching glass transom	BC@1.00	Ea	485.00	30.00	515.00
Minuet, oval insert					
Door	BC@1.45	Ea	869.00	43.50	912.50
Sidelight	BC@1.00	Ea	565.00	30.00	595.00
Matching glass transom	BC@1.00	Ea	530.00	30.00	560.00
Concerto, arched insert					
Door	BC@1.45	Ea	743.00	43.50	786.50
Sidelight	BC@1.00	Ea	559.00	30.00	589.00
Matching glass transom	BC@1.00	Ea	563.00	30.00	593.00
Sonata, narrow arched insert					
Door	BC@1.45	Ea	782.00	43.50	825.50
Sidelight	BC@1.00	Ea	597.00	30.00	627.00
Matching glass transom	BC@1.00	Ea	580.00	30.00	610.00
Tremolo, full rectangular insert					
Door	BC@1.45	Ea	980.00	43.50	1,023.50
Sidelight	BC@1.00	Ea	562.00	30.00	592.00
Matching glass transom	BC@1.00	Ea	557.00	30.00	587.00

Doors, Trim for Exterior Slab

	Craft@Hrs	Unit	Material	Labor	Total

Trim for Exterior Slab Doors Trim for an exterior slab door includes two side jambs, head jamb, threshold, stop molding, interior casing and exterior molding. Installation time for exterior door trim will be up to 33% faster when an experienced carpenter has the use of a pneumatic nailer.

Primed exterior door jamb. Primed finger joint pine. 1-1/4" thick. Set for a 3' door consists of two 6'8" legs and one 3' head.

	Craft@Hrs	Unit	Material	Labor	Total
4-5/8" wide, per 6'8" leg	BC@.204	Ea	24.10	6.12	30.22
5-1/4" wide, per 6'8" leg	BC@.204	Ea	26.80	6.12	32.92
6-9/16" wide, per 6'8" leg	BC@.204	Ea	28.30	6.12	34.42
4-5/8" wide, per set	BC@.500	Ea	65.40	15.00	80.40
5-1/4" wide, per set	BC@.500	Ea	70.80	15.00	85.80
6-9/16" wide, per set	BC@.500	Ea	73.80	15.00	88.80

Primed kerfed exterior door jamb. 1-1/4" thick. Set for a 3' door consists of two 6'8" legs and one 3' head.

	Craft@Hrs	Unit	Material	Labor	Total
4-5/8" x 7' leg	BC@.204	Ea	22.60	6.12	28.72
4-5/8" x 3' head	BC@.092	Ea	9.44	2.76	12.20
6-5/8" x 7' leg	BC@.204	Ea	24.10	6.12	30.22
6-5/8" x 3' head	BC@.092	Ea	8.39	2.76	11.15

Oak high boy threshold. Fits standard door sizes.

	Craft@Hrs	Unit	Material	Labor	Total
5/8" x 3-1/2" x 36"	BC@.518	Ea	13.60	15.50	29.10

Stop molding. Nails to face of the door jamb to prevent door from swinging through the opening. Solid pine.

	Craft@Hrs	Unit	Material	Labor	Total
3/8" x 1-1/4"	BC@.015	LF	.62	.45	1.07
7/16" x 3/4"	BC@.015	LF	.45	.45	.90
7/16" x 1-3/8"	BC@.015	LF	.69	.45	1.14
7/16" x 1-5/8"	BC@.015	LF	.71	.45	1.16
7/16" x 2-1/8"	BC@.015	LF	1.13	.45	1.58

Solid pine ranch casing. Casing is the trim around each side and head of the interior side of an exterior door. Casing covers the joint between jamb and the wall.

	Craft@Hrs	Unit	Material	Labor	Total
9/16" x 2-1/4"	BC@.015	LF	.87	.45	1.32
9/16" x 2-1/4" x 7'	BC@.105	Ea	6.10	3.15	9.25
11/16" x 3-1/2"	BC@.015	LF	1.52	.45	1.97
11/16" x 2-1/2" x 7'	BC@.015	Ea	7.64	.45	8.09

Primed finger joint ranch door casing.

	Craft@Hrs	Unit	Material	Labor	Total
9/16" x 3-1/4"	BC@.015	LF	1.10	.45	1.55
11/16" x 2-1/4"	BC@.015	LF	.80	.45	1.25
11/16" x 2-1/4" x 7'	BC@.105	Ea	5.59	3.15	8.74
11/16" x 2-1/2"	BC@.015	LF	.89	.45	1.34
11/16" x 2-1/2" x 7'	BC@.105	Ea	6.25	3.15	9.40

Fluted pine casing and rosette set. Includes 1 header piece, 2 side pieces, 2 rosettes and 2 base blocks.

	Craft@Hrs	Unit	Material	Labor	Total
Set	BC@.250	Ea	30.70	7.50	38.20

Brick mold set. Trim for the exterior side of an exterior door. Two legs and one head. Primed finger joint pine. 36" x 80" door. 1-1/4" x 2".

	Craft@Hrs	Unit	Material	Labor	Total
Finger joint pine	BC@.250	Ea	32.80	7.50	40.30
Pressure treated pine	BC@.250	Ea	22.00	7.50	29.50

Brick mold. Brick mold and stucco mold are the casing around each side and head on the exterior side of an exterior door.

	Craft@Hrs	Unit	Material	Labor	Total
1-3/16" x 1-1/2", hemlock	BC@.015	LF	1.41	.45	1.86
1-3/16" x 2", hemlock	BC@.015	LF	1.68	.45	2.13
1-1/4" x 2", fir	BC@.015	LF	1.23	.45	1.68
1-1/4" x 2", solid pine	BC@.015	LF	1.83	.45	2.28

Stucco molding. Redwood.

	Craft@Hrs	Unit	Material	Labor	Total
13/16" x 1-1/2"	BC@.015	LF	.87	.45	1.32
1" x 1-3/8"	BC@.015	LF	.96	.45	1.41

	Craft@Hrs	Unit	Material	Labor	Total

T-astragal molding. Covers the vertical joint where a pair of doors meet.

	Craft@Hrs	Unit	Material	Labor	Total
1-1/4" x 2" x 7', pine	BC@.529	Ea	20.50	15.90	36.40

Fire Rated Doors Fire rated doors are designed to meet building code requirements for specialized occupancies, such as hotel rooms and dormitories. Most building codes require that fire doors be used between a house and an attached garage.

6-panel prehung steel fire rated exterior doors. 24-gauge galvanized steel. Foam core. Caming matches hinge and sill finishes. Jamb guard security plate. No brick mold. Fixed sill. 4-9/16" primed jamb. Triple sweep compression weatherstripping. Factory primed. Thermal-break threshold. 20-minute fire rating. 10-year limited warranty.

	Craft@Hrs	Unit	Material	Labor	Total
30" x 80"	BC@1.00	Ea	146.00	30.00	176.00
32" x 80"	BC@1.00	Ea	146.00	30.00	176.00
36" x 80"	BC@1.00	Ea	146.00	30.00	176.00

Prehung hardboard fire-rated doors. 20-minute fire rating. 80" high. Labor includes hanging the door in an existing framed opening.

30" x 80"	BC@1.00	Ea	134.00	30.00	164.00
32" x 80"	BC@1.00	Ea	140.00	30.00	170.00
36" x 80"	BC@1.00	Ea	140.00	30.00	170.00

6-panel prehung fire-rated interior doors. 4-9/16" jamb. 20-minute fire rating.

32" x 80"	BC@.750	Ea	205.00	22.50	227.50
36" x 80"	BC@.750	Ea	209.00	22.50	231.50

Flush fire-rated doors. 1-3/4" thick. 20-minute fire rating. Labor includes setting three hinges on the door and jamb and hanging the door in an existing cased opening.

32" x 80", hardboard veneer	BC@1.45	Ea	54.20	43.50	97.70
36" x 80", hardboard veneer	BC@1.45	Ea	55.40	43.50	98.90
32" x 80", birch veneer	BC@1.45	Ea	59.00	43.50	102.50
36" x 80", birch veneer	BC@1.45	Ea	60.00	43.50	103.50

Reversible hollow metal fire doors.

34" x 80"	BC@.721	Ea	132.00	21.60	153.60
36" x 80"	BC@.721	Ea	153.00	21.60	174.60

Knocked down steel door frames. Bored for (3) 4-1/2" hinges, 4-7/8" strike plate and 2-3/4" backset. 90-minute fire label manufactured to N.Y.C. M.E.A. procedures. Hollow core. For drywall and masonry applications. Galvanized steel.

30" x 80"	BC@.944	Ea	65.10	28.30	93.40
32" x 80"	BC@.944	Ea	66.90	28.30	95.20
34" x 80"	BC@.944	Ea	67.20	28.30	95.50
36" x 80"	BC@.944	Ea	68.60	28.30	96.90
36" x 84"	BC@.944	Ea	70.00	28.30	98.30

Knocked down trimless door frames.

36" x 80"	BC@.944	Ea	59.90	28.30	88.20

90-minute fire-rated door frame parts.

4-7/8" x 80" legs, pack of 2	BC@.778	Ea	64.50	23.30	87.80
4-7/8" x 36" head	BC@.166	Ea	16.80	4.98	21.78

Steel cellar doors. Durable heavy gauge steel.

57" x 45" x 24-1/2"	BC@1.00	Ea	246.00	30.00	276.00
63" x 49" x 22"	BC@1.00	Ea	261.00	30.00	291.00
71" x 53" x 19-1/2"	BC@1.00	Ea	271.00	30.00	301.00

	Craft@Hrs	Unit	Material	Labor	Total

Prehung Wood Exterior Doors Prehung doors are sold as a package with the door, jamb, hinges, weatherstripping, threshold and trim already assembled.

Flush hardwood prehung exterior doors. 1-3/4" door. Solid particleboard core. Lauan veneer. Weatherstripped. Includes oak sill and threshold with vinyl insert. Finger joint 4-9/16" jamb and finger joint brick mold. 2-1/8" bore for lockset with 2-3/8" backset. Flush doors are smooth and flat with no decorative treatment. Labor includes hanging the door in an existing framed opening.

	Craft@Hrs	Unit	Material	Labor	Total
32" x 80"	BC@1.00	Ea	102.00	30.00	132.00
36" x 80"	BC@1.00	Ea	105.00	30.00	135.00

Flush birch prehung exterior doors. 1-3/4" door. Solid particleboard core. Weatherstripped. Includes oak sill and threshold. Finger joint 4-9/16" jamb and finger joint brick mold. 2-1/8" bore for lockset with 2-3/8" backset. Labor includes hanging the door in an existing framed opening.

	Craft@Hrs	Unit	Material	Labor	Total
36" x 80"	BC@1.00	Ea	95.20	30.00	125.20

Flush hardboard prehung exterior doors. Primed. Solid core. Weatherstripped. Includes sill and threshold. Finger joint 4-9/16" jamb and finger joint brick mold. 2-1/8" bore for lockset with 2-3/8" backset. Labor includes hanging the door in an existing framed opening.

	Craft@Hrs	Unit	Material	Labor	Total
32" x 80"	BC@1.00	Ea	98.00	30.00	128.00
36" x 80"	BC@1.00	Ea	101.00	30.00	131.00

Fan lite prehung hemlock exterior doors. Weatherstripped. Includes sill and threshold. Finger joint 4-9/16" jamb and finger joint brick mold. 2-1/8" bore for lockset with 2-3/8" backset. Labor includes hanging the door in an existing framed opening.

	Craft@Hrs	Unit	Material	Labor	Total
36" x 80"	BC@1.00	Ea	295.00	30.00	325.00

1-lite prehung exterior doors. Inward swing. No brick mold. Solid core. Weatherstripped. Includes sill and threshold. Finger joint 4-9/16" jamb and finger joint brick mold. 2-1/8" bore for lockset with 2-3/8" backset. Brick mold (also called stucco mold) is the casing around each side and head on the exterior side of an exterior door. Finger joint trim is made from two or more lengths of wood joined together with a finger-like joint that yields nearly as much strength as trim made of a single, solid piece. Labor includes hanging the door in an existing framed opening.

	Craft@Hrs	Unit	Material	Labor	Total
32" x 80"	BC@1.00	Ea	272.00	30.00	302.00
36" x 80"	BC@1.00	Ea	272.00	30.00	302.00

Sidelites for wood prehung exterior doors, 1-3/4" thick, with insulated glass, assembled.

	Craft@Hrs	Unit	Material	Labor	Total
6'8" x 1'2", fir or hemlock	BC@1.36	Ea	396.00	40.80	436.80

Prehung Steel Exterior Doors Prehung doors are sold as a package with the door, jamb, hinges, weatherstripping, threshold and trim already assembled. Flush doors are smooth and flat, with no decorative treatment.

Utility flush prehung steel exterior doors. No brick mold. 26-gauge galvanized steel. Foam core. Compression weatherstrip. Triple fin sweep. Non-thermal threshold. Single bore.

	Craft@Hrs	Unit	Material	Labor	Total
32" x 80"	BC@1.00	Ea	103.00	30.00	133.00
36" x 80"	BC@1.00	Ea	103.00	30.00	133.00

Premium flush prehung steel exterior doors. No brick mold. 24-gauge galvanized steel. Foam core. Compression weatherstrip. Triple fin sweep. Thermal threshold.

	Craft@Hrs	Unit	Material	Labor	Total
30" x 80"	BC@1.00	Ea	146.00	30.00	176.00
32" x 80"	BC@1.00	Ea	146.00	30.00	176.00
36" x 80"	BC@1.00	Ea	146.00	30.00	176.00

6-panel prehung steel exterior doors. With brick mold. Inward swing. Non-thermal threshold. Bored for lockset with 2-3/4" backset. Includes compression weatherstrip and thermal-break threshold. Panel doors have decorative raised panels in the colonial style. Six or nine panels are most common.

	Craft@Hrs	Unit	Material	Labor	Total
32" x 80"	BC@1.00	Ea	103.00	30.00	133.00
36" x 80"	BC@1.00	Ea	103.00	30.00	133.00

	Craft@Hrs	Unit	Material	Labor	Total

6-panel premium prehung steel doors. 4-9/16" jamb. Adjustable thermal-break sill. Inward swing. 24-gauge galvanized steel. Bored for 2-3/8" lockset. 12" lock block. Impact-resistant laminated glass. Polyurethane core. Triple sweep and compression weatherstripping. Factory primed.

	Craft@Hrs	Unit	Material	Labor	Total
32" x 80", with brick mold	BC@1.00	Ea	146.00	30.00	176.00
32" x 80", no brick mold	BC@1.00	Ea	146.00	30.00	176.00
34" x 80", with brick mold	BC@1.00	Ea	146.00	30.00	176.00
36" x 80", no brick mold	BC@1.00	Ea	146.00	30.00	176.00
36" x 80", with brick mold	BC@1.00	Ea	146.00	30.00	176.00
72" x 80", no brick mold (double door)	BC@1.50	Ea	282.00	45.00	327.00

6-panel prehung steel entry door with sidelites. Two 10"-wide sidelites. Sidelites have 5 divided lites.

	Craft@Hrs	Unit	Material	Labor	Total
36" x 80"	BC@2.00	Ea	440.00	60.00	500.00

Fan lite premium prehung steel exterior doors. With brick mold. 24-gauge galvanized steel. Foam core. Fixed sill. Compression weatherstripping. Factory primed. Thermal-break threshold.

	Craft@Hrs	Unit	Material	Labor	Total
32" x 80", 4-9/16" jamb	BC@1.00	Ea	179.00	30.00	209.00
32" x 80", 6-9/16" jamb	BC@1.00	Ea	167.00	30.00	197.00
36" x 80", 4-9/16" jamb	BC@1.00	Ea	179.00	30.00	209.00
36" x 80", 6-9/16" jamb	BC@1.00	Ea	104.00	30.00	134.00

Half lite prehung steel exterior doors. 4-9/16" jamb. 22" x 36" light. Adjustable sill. With brick mold.

	Craft@Hrs	Unit	Material	Labor	Total
32" x 80"	BC@1.00	Ea	215.00	30.00	245.00
36" x 80"	BC@1.00	Ea	215.00	30.00	245.00

Half lite prehung steel exterior doors with mini-blind. 22" x 36" light. Adjustable sill. 24-gauge galvanized steel. Non-yellowing, white, warp-resistant, paintable, high-performance light frames. Brick mold applied.

	Craft@Hrs	Unit	Material	Labor	Total
32" x 80", 4-9/16" jamb	BC@1.00	Ea	260.00	30.00	290.00
32" x 80", 6-9/16" jamb	BC@1.00	Ea	299.00	30.00	329.00
36" x 80", 4-9/16" jamb	BC@1.00	Ea	272.00	30.00	302.00
36" x 80", 6-9/16" jamb	BC@1.00	Ea	299.00	30.00	329.00

Full lite prehung steel exterior doors. 22" x 64" light. 4-9/16" jamb. Adjustable sill. No brick mold. 24-gauge galvanized steel.

	Craft@Hrs	Unit	Material	Labor	Total
32" x 80"	BC@1.00	Ea	229.00	30.00	259.00
36" x 80"	BC@1.00	Ea	229.00	30.00	259.00

2-lite, 4-panel prehung steel doors. 4-9/16" jamb. Adjustable sill. Ready-to-install door and jamb system. 24-gauge galvanized steel. Non-yellowing, white, warp-resistant, paintable, high-performance light frames. Brick mold applied.

	Craft@Hrs	Unit	Material	Labor	Total
30" x 80"	BC@1.00	Ea	189.00	30.00	219.00
32" x 80"	BC@1.00	Ea	189.00	30.00	219.00
36" x 80"	BC@1.00	Ea	189.00	30.00	219.00

9-lite prehung steel exterior doors. Insulating glass with internal 9-lite grille. Adjustable sill. Ready-to-install door and jamb system. 24-gauge galvanized steel. Non-yellowing, white, warp-resistant, paintable, high-performance light frames. Brick mold applied.

	Craft@Hrs	Unit	Material	Labor	Total
32" x 80", 4-9/16" jamb	BC@1.00	Ea	197.00	30.00	227.00
32" x 80", 6-9/16" jamb	BC@1.00	Ea	221.00	30.00	251.00

11-lite prehung steel exterior doors. 24-gauge, high-profile, deep embossed steel skins. Impact-resistant glass. Double-bored for deadbolt and doorknob. Brass adjustable threshold. 4-9/16" primed frame. No brick mold.

	Craft@Hrs	Unit	Material	Labor	Total
32" x 80" left	BC@1.00	Ea	230.00	30.00	260.00
36" x 80" right	BC@1.00	Ea	230.00	30.00	260.00

15-lite prehung steel exterior doors. Multi-paned insulated glass. 24-gauge hot-dipped galvanized steel. Polyurethane foam core. Compression weatherstrip. Thermal-break construction. No brick mold. Adjustable seal. Aluminum sill with composite adjustable threshold.

	Craft@Hrs	Unit	Material	Labor	Total
32" x 80"	BC@1.00	Ea	231.00	30.00	261.00
36" x 80"	BC@1.00	Ea	231.00	30.00	261.00

	Craft@Hrs	Unit	Material	Labor	Total

Flush Ventilating Prehung Steel Exterior Doors. No brick mold. 4-9/16" primed jamb. 24-gauge galvanized steel. Foam core. Magnetic weatherstrip. Triple seal sweep. Thermal-break threshold. Bored for lockset and deadbolt. 20-minute fire rating. Ventilating doors include a window that can be opened.

	Craft@Hrs	Unit	Material	Labor	Total
30" x 80"	BC@1.00	Ea	209.00	30.00	239.00
32" x 80"	BC@1.00	Ea	209.00	30.00	239.00
36" x 80"	BC@1.00	Ea	209.00	30.00	239.00

Decorative Prehung Steel Entry Doors Entry doors are decorative exterior doors, often with glass panels set in a border of brass caming. Prehung doors are sold as a package with the door, jamb, hinges, threshold and trim already assembled.

Ascot brass center arch lite prehung steel entry doors. 24-gauge galvanized steel. Non-yellowing, white, warp-resistant, paintable, high-performance light frames. Decorative impact-resistant glass. Rot-resistant frame. Polyurethane insulated core. 4-9/16" jamb with adjustable sill. Includes brick mold.

	Craft@Hrs	Unit	Material	Labor	Total
6" x 80"	BC@1.00	Ea	187.00	30.00	217.00

Camber top prehung steel entry doors. 4-9/16" jamb.

	Craft@Hrs	Unit	Material	Labor	Total
36" x 80", with brick mold	BC@1.00	Ea	248.00	30.00	278.00
36" x 80", no brick mold	BC@1.00	Ea	246.00	30.00	276.00

Camber top 6" jamb prehung steel entry doors. 6-9/16" jamb. With brick mold.

	Craft@Hrs	Unit	Material	Labor	Total
36" x 80"	BC@1.00	Ea	271.00	30.00	301.00

Decorative oval lite prehung steel entry door. 40" x 18" decorative oval laminated glass with brass caming to match hinge and sill finish. 24-gauge galvanized steel. Polyurethane core. Adjustable thermal-break sill. Triple sweep compression weatherstripping. Non-yellowing and warp-resistant light frame. Factory primed.

	Craft@Hrs	Unit	Material	Labor	Total
36" x 80", with brick mold, left hinge					
4-9/16" jamb	BC@1.00	Ea	340.00	30.00	370.00
36" x 80", with brick mold, right hinge					
4-9/16" jamb	BC@1.00	Ea	340.00	30.00	370.00

Decorative fan lite prehung steel entry doors. Polyurethane core. Adjustable thermal-break sill with composite saddle. 24-gauge galvanized steel. Laminated glass. Brass caming matches hinge and sill. Triple sweep and compression weatherstrip. Non-yellowing and warp-resistant light frame. Factory primed. Includes brick mold.

	Craft@Hrs	Unit	Material	Labor	Total
32" x 80"	BC@1.00	Ea	209.00	30.00	239.00
36" x 80"	BC@1.00	Ea	202.00	30.00	232.00

Decorative half lite prehung steel entry doors. 22" x 36" laminated safety glass. Brass caming matches sill and hinges. 24-gauge galvanized insulated steel.

	Craft@Hrs	Unit	Material	Labor	Total
36" x 80", no brick mold	BC@1.00	Ea	272.00	30.00	302.00
36" x 80", with brick mold	BC@1.00	Ea	283.00	30.00	313.00

Fiberglass Prehung Entry Doors Fiberglass doors require very little maintenance and are available in simulated wood finish. Prehung doors include the door, jamb, hinges, threshold and trim already assembled.

6-panel fiberglass prehung exterior doors. Polyurethane core. PVC stiles and rails. Double bore. 4-9/16" prefinished jambs. Adjustable thermal-break brass threshold. Weatherstripped. With brick mold.

	Craft@Hrs	Unit	Material	Labor	Total
32" x 80"	BC@1.00	Ea	156.00	30.00	186.00
36" x 80"	BC@1.00	Ea	156.00	30.00	186.00

Fan lite prehung light oak fiberglass entry doors. With brick mold. Factory prefinished. Triple pane insulated glass with brass caming. Adjustable brass thermal-break threshold and fully weatherstripped. 4-9/16" jamb.

	Craft@Hrs	Unit	Material	Labor	Total
36" x 80"	BC@1.00	Ea	356.00	30.00	386.00

	Craft@Hrs	Unit	Material	Labor	Total

3/4 oval lite prehung fiberglass entry doors. With brick mold. Ready to finish.

36" x 80", 4-9/16" jamb	BC@1.00	Ea	314.00	30.00	344.00
36" x 80", 6-9/16" jamb	BC@1.00	Ea	345.00	30.00	375.00

3/4 oval lite prehung light oak fiberglass doors. With brick mold. Prefinished. Polyurethane core. Triple pane insulated glass. Glue chip and clear 30" x 18" beveled glass surrounded by brass caming. Double bore. Extended wood lock block. 4-9/16" primed jamb. Thermal-break brass threshold with adjustable oak cap.

36" x 80", 4-9/16" jamb	BC@1.00	Ea	524.00	30.00	554.00
36" x 80", 6-9/16" jamb	BC@1.00	Ea	550.00	30.00	580.00

Decorative oval lite prehung fiberglass entry doors. With brick mold. Prefinished. Polyurethane core. Triple pane insulated glass. Glue chip and clear 30" x 18" beveled glass surrounded by brass caming. Double bore. Extended wood lock block. 4-9/16" primed jamb. Thermal-break brass threshold with adjustable oak cap.

36" x 80", dark mahogany	BC@1.00	Ea	423.00	30.00	453.00
36" x 80", honey oak	BC@1.00	Ea	424.00	30.00	454.00
36" x 80", white	BC@1.00	Ea	423.00	30.00	453.00

Full height oval lite prehung medium oak fiberglass entry doors.

36" x 80", 4-9/16" jamb	BC@1.00	Ea	555.00	30.00	585.00
36" x 80", 6-9/16" jamb	BC@1.00	Ea	576.00	30.00	606.00

Center arch lite medium oak prehung fiberglass doors. Factory prefinished. Triple pane insulated glass with brass caming. Adjustable brass thermal-break threshold and fully weatherstripped. 4-9/16" jamb except as noted. PVC stiles and rails.

36" x 80", no brick mold	BC@1.00	Ea	418.00	30.00	448.00
36" x 80", with brick mold	BC@1.00	Ea	419.00	30.00	449.00
36" x 80", with brick mold, 6-9/16" jamb	BC@1.00	Ea	445.00	30.00	475.00

Fan lite prehung fiberglass entry doors. With brick mold. Ready to finish. Insulated glass. Adjustable thermal-break brass threshold. 4-9/16" jamb except as noted.

32" x 80"	BC@1.00	Ea	240.00	30.00	270.00
36" x 80"	BC@1.00	Ea	240.00	30.00	270.00
36" x 80", 6-9/16" jamb	BC@1.00	Ea	272.00	30.00	302.00

9-lite fiberglass prehung exterior doors. Inward swing. With brick mold. Smooth face.

32" x 80"	BC@1.00	Ea	219.00	30.00	249.00
36" x 80"	BC@1.00	Ea	219.00	30.00	249.00

15-lite smooth prehung fiberglass entry doors. No brick mold. Factory prefinished. Left or right hinge. Triple pane, insulated glass with brass caming. Adjustable brass thermal-break threshold and fully weatherstripped. 4-9/16" jamb.

32" x 80"	BC@1.00	Ea	240.00	30.00	270.00
32" x 80"	BC@1.00	Ea	240.00	30.00	270.00

Full lite prehung smooth fiberglass doors. With brick mold. Inward swing.

36" x 80", 4-9/16" jamb	BC@1.00	Ea	492.00	30.00	522.00
36" x 80", 6-9/16" jamb	BC@1.00	Ea	519.00	30.00	549.00

Smooth prehung fiberglass ventlite exterior doors. Ready to finish. No brick mold.

32" x 80"	BC@1.00	Ea	222.00	30.00	252.00
36" x 80"	BC@1.00	Ea	222.00	30.00	252.00

Screen Doors Wood screen doors. 3-3/4"-wide stile and rail with 7-1/4" bottom rail. Removable screen. Mortise and tenon solid construction. Wide center push bar. Ready to finish.

32" x 80"	BC@2.17	Ea	65.00	65.10	130.10
36" x 80"	BC@2.17	Ea	67.20	65.10	132.30

	Craft@Hrs	Unit	Material	Labor	Total

T-style wood screen doors. 3-3/4"-wide stile and rail with 7-1/4" bottom rail. Removable screen. Mortise and tenon solid construction. All joints sealed with water-resistant adhesive. Wide center push bar.

	Craft@Hrs	Unit	Material	Labor	Total
30" x 80"	BC@2.17	Ea	21.00	65.10	86.10
32" x 80"	BC@2.17	Ea	21.00	65.10	86.10
36" x 80"	BC@2.17	Ea	21.00	65.10	86.10

Metal screen doors. Includes hardware and fiberglass screening. 1-piece steel frame. Heavy-duty pneumatic closer. 5" kickplate.

32" x 80"	BC@1.35	Ea	36.30	40.50	76.80
36" x 80"	BC@1.35	Ea	36.30	40.50	76.80

Five-bar solid vinyl screen doors. Accepts standard screen door hardware.

32" x 80"	BC@1.35	Ea	77.70	40.50	118.20
36" x 80"	BC@1.35	Ea	77.70	40.50	118.20

Vinyl screen doors. Right or left side installation. Hardware included.

36" x 80", no grid	BC@1.35	Ea	92.40	40.50	132.90
36" x 80", lower panel grid	BC@1.35	Ea	67.20	40.50	107.70
36" x 80", grid of 15	BC@1.35	Ea	103.00	40.50	143.50

Steel security screen doors. Class I rated to withstand 300 pounds. All welded steel construction for strength and durability. Double lockbox with extension plate. Baked-on powder coating.

36" x 80"	BC@1.35	Ea	71.40	40.50	111.90

Storm Doors Self-storing aluminum storm and screen doors. Pneumatic closer and sweep. Tempered safety glass. Maintenance-free finish. Push-button hardware. 1" x 2-1/8" frame size. Bronze or white. Includes screen.

30" x 80" x 1-1/4"	BC@1.45	Ea	93.40	43.50	136.90
32" x 80" x 1-1/4"	BC@1.45	Ea	92.10	43.50	135.60
36" x 80" x 1-1/4"	BC@1.45	Ea	92.40	43.50	135.90

Self-storing vinyl-covered wood storm doors. White vinyl-covered wood core. Push-button handle set. Self-storing window and screen. Single black closer.

30" x 80"	BC@1.45	Ea	98.70	43.50	142.20
32" x 80"	BC@1.45	Ea	98.60	43.50	142.10
34" x 80"	BC@1.45	Ea	98.70	43.50	142.20
36" x 80"	BC@1.45	Ea	98.60	43.50	142.10

Store-in-Door™ storm doors. 1-1/2"-thick polypropylene frame. Reversible hinge. Slide window or screen, completely concealed when not in use. Brass-finished keylock hardware system with separate color-matched deadbolt. Triple-fin door sweep. 2 color-matched closers. Full-length piano hinge. Flexible weather seal. White.

30" x 80", crossbuck style	BC@1.45	Ea	215.00	43.50	258.50
32" x 80", crossbuck style	BC@1.45	Ea	215.00	43.50	258.50
36" x 80", crossbuck style	BC@1.45	Ea	215.00	43.50	258.50
30" x 80", traditional style	BC@1.45	Ea	240.00	43.50	283.50
36" x 80", traditional style	BC@1.45	Ea	240.00	43.50	283.50

Aluminum and wood storm doors. Aluminum over wood core. Self-storing window and screen. Black lever handle set. Separate deadbolt. Single black closer. White finish. Traditional panel style.

30" x 80"	BC@1.45	Ea	135.00	43.50	178.50
32" x 80"	BC@1.45	Ea	135.00	43.50	178.50
36" x 80"	BC@1.45	Ea	135.00	43.50	178.50

Triple-track storm doors. Low-maintenance aluminum over solid wood core. Ventilates from top, bottom or both. Solid brass exterior handle. Color-matched interior handle. Separate deadbolt. 80" high. White, almond or bronze.

30" x 80"	BC@1.45	Ea	183.00	43.50	226.50
32" x 80"	BC@1.45	Ea	204.00	43.50	247.50
34" x 80"	BC@1.45	Ea	183.00	43.50	226.50
36" x 80"	BC@1.45	Ea	183.00	43.50	226.50

	Craft@Hrs	Unit	Material	Labor	Total

All-season storm doors. 1-1/4" thick frame. 1-piece construction with window molding built in. Self-storing tempered safety glass and screen for easy ventilation. Includes solid-brass lockset with deadbolt security. Brass-finished sweep for tight threshold seal. Color-matched screw covers, and two closers. Reversible hinge opens to 180 degrees. White, almond or sandstone.

	Craft@Hrs	Unit	Material	Labor	Total
32" x 80"	BC@1.45	Ea	193.00	43.50	236.50
36" x 80"	BC@1.45	Ea	193.00	43.50	236.50

All-season triple track storm doors. Ventilates from top or bottom, or both. 1" composite frame. Brass handle set with deadbolt security. Wood-grain finish. Color-matched screws. 1-piece construction with window molding built in. Self-storing window and screen. White, almond or sandstone.

	Craft@Hrs	Unit	Material	Labor	Total
32" x 80"	BC@1.45	Ea	208.00	43.50	251.50
36" x 80"	BC@1.45	Ea	208.00	43.50	251.50

Colonial triple-track storm doors. 1-piece composite frame. 12-lite grille. Vents from top, bottom, or both. Solid-brass handle set with deadbolt security. Brass-finished sweep ensures tight seal across the entire threshold. Wood-grain finish. White, almond or sandstone.

	Craft@Hrs	Unit	Material	Labor	Total
32" x 80"	BC@1.45	Ea	240.00	43.50	283.50
36" x 80"	BC@1.45	Ea	223.00	43.50	266.50

Full-view brass all-season storm doors. Water-diverting rain cap. Brass-finished, color-matched sweep. One closer. Window and insect screen snap in and out. Reversible. Opens 180 degrees when needed. Separate deadbolt for added security. Tempered glass. Heavy-gauge one-piece 1" aluminum construction with reinforced corners and foam insulation.

	Craft@Hrs	Unit	Material	Labor	Total
36" x 80", green	BC@1.45	Ea	302.00	43.50	345.50
36" x 80", white	BC@1.45	Ea	241.00	43.50	284.50

Full-view aluminum storm doors. 1-1/2"-thick heavy-gauge aluminum frame with foam insulation. Extra perimeter weatherstripping and water-diverting rain cap. Solid brass handle set and double-throw deadbolt. Brass-finished sweep. Full-length, piano-style hinge. Reversible hinge for left- or right-side entry door handle. Two heavy-duty, color-matched closers. Solid brass perimeter locks. Insect screen. White, brown, green or almond.

	Craft@Hrs	Unit	Material	Labor	Total
30" x 80"	BC@1.45	Ea	246.00	43.50	289.50
32" x 80"	BC@1.45	Ea	246.00	43.50	289.50
36" x 80"	BC@1.45	Ea	245.00	43.50	288.50

Full-view woodcore storm doors. Aluminum over solid wood core. Extra perimeter weatherstripping and water-diverting rain cap. Black push-button handle with night latch. Color-matched sweep. One heavy-duty closer. Snap-in, snap-out window and insect screen retainer system. White or bronze.

	Craft@Hrs	Unit	Material	Labor	Total
32" x 84"	BC@1.45	Ea	129.00	43.50	172.50
36" x 84"	BC@1.45	Ea	131.00	43.50	174.50

Security storm doors 16-gauge steel. Brass deadbolt lock. Tempered safety glass. Interchangeable screen panel. Heavy-gauge all-welded steel frame with mitered top. Double weatherstripped jamb and sweep. Tamper-resistant hinges with concealed interior screws. Heavy-duty pneumatic closer and safety wind chain. Prehung with reversible hinge.

	Craft@Hrs	Unit	Material	Labor	Total
30" x 80"	BC@1.45	Ea	314.00	43.50	357.50
32" x 80"	BC@1.45	Ea	305.00	43.50	348.50
36" x 80"	BC@1.45	Ea	305.00	43.50	348.50

Class II security doors. Rated to withstand 400 pounds. Double lockbox with extension plate. Decorative geometric ornaments. All welded steel construction. Expanded, galvanized metal screen. 3/4" square frame design.

	Craft@Hrs	Unit	Material	Labor	Total
32" x 80"	BC@1.45	Ea	112.00	43.50	155.50
36" x 80"	BC@1.45	Ea	106.00	43.50	149.50

	Craft@Hrs	Unit	Material	Labor	Total

Heavy-duty security door with screen. Right- or left-hand exterior mounting. All welded steel construction. Baked-on powder coating. Double lockbox with tamper-resistant extension plate. Uses standard 2-3/8" backset lock (not included). Galvanized perforated metal security screen. 1" x 1" door frame. 1" x 1-1/2" jambs, prehung on hinge side.

	Craft@Hrs	Unit	Material	Labor	Total
32" x 80"	BC@1.45	Ea	135.00	43.50	178.50
36" x 80"	BC@1.45	Ea	136.00	43.50	179.50

Paradise style security door. Five semi-concealed hinges. Heavy-duty deadbolt. All steel frame.

	Craft@Hrs	Unit	Material	Labor	Total
30" x 80"	BC@1.45	Ea	395.00	43.50	438.50
32" x 80"	BC@1.45	Ea	395.00	43.50	438.50
36" x 80"	BC@1.45	Ea	398.00	43.50	441.50

Folding security gates. Riveted 3/4" top and bottom channel. Gate pivots and folds flat to one side. Provides air circulation and positive visibility.

	Craft@Hrs	Unit	Material	Labor	Total
48" max width, 79" high	BC@1.25	Ea	109.00	37.50	146.50

Interior Slab Doors Slab doors are the door alone, as distinguished from prehung doors which are sold with jamb, hinges and trim attached and bored for the door latch. Interior doors are usually 1-3/8" thick and have a "hollow" core. A wood frame made from stiles and rails is filled with a matrix of corrugated paper and covered with a thin wood veneer. Stiles are vertical and every door has two, a lock stile and the hinge stile. Rails run horizontally at the top and bottom of the door. Flush doors are smooth and flat, with no decorative treatment.

Flush hardboard interior doors. Hollow core. Bored for lockset. Ready to paint or stain. 1-3/8" thick. Hardboard doors are made of composition wood veneer, usually the least expensive material available.

	Craft@Hrs	Unit	Material	Labor	Total
24" x 80"	BC@1.15	Ea	21.40	34.50	55.90
28" x 80"	BC@1.15	Ea	23.00	34.50	57.50
30" x 80"	BC@1.15	Ea	23.80	34.50	58.30
32" x 80"	BC@1.15	Ea	25.10	34.50	59.60
36" x 80"	BC@1.15	Ea	26.50	34.50	61.00

Flush birch interior doors. Hollow core. 1-3/8" thick. Wood veneer. Stainable and paintable. All wood stile and rail.

	Craft@Hrs	Unit	Material	Labor	Total
18" x 80"	BC@1.15	Ea	25.70	34.50	60.20
24" x 80"	BC@1.15	Ea	28.20	34.50	62.70
28" x 80"	BC@1.15	Ea	30.80	34.50	65.30
30" x 80"	BC@1.15	Ea	32.20	34.50	66.70
32" x 80"	BC@1.15	Ea	34.00	34.50	68.50
36" x 80"	BC@1.15	Ea	35.40	34.50	69.90

Flush lauan interior doors. Hollow core. 1-3/8" thick. Wood veneer. Stainable and paintable. Lauan is a tropical hardwood like mahogany. All wood stile and rail.

	Craft@Hrs	Unit	Material	Labor	Total
18" x 80"	BC@1.15	Ea	18.70	34.50	53.20
24" x 80"	BC@1.15	Ea	20.10	34.50	54.60
28" x 80"	BC@1.15	Ea	22.00	34.50	56.50
30" x 80"	BC@1.15	Ea	22.60	34.50	57.10
32" x 80"	BC@1.15	Ea	24.20	34.50	58.70
36" x 80"	BC@1.15	Ea	25.50	34.50	60.00

Flush oak interior doors. Hollow core. 1-3/8" thick. Wood veneer. Stainable and paintable. All wood stile and rail.

	Craft@Hrs	Unit	Material	Labor	Total
18" x 80"	BC@1.15	Ea	27.60	34.50	62.10
24" x 80"	BC@1.15	Ea	29.50	34.50	64.00
28" x 80"	BC@1.15	Ea	32.60	34.50	67.10
30" x 80"	BC@1.15	Ea	33.80	34.50	68.30
32" x 80"	BC@1.15	Ea	35.60	34.50	70.10
36" x 80"	BC@1.15	Ea	38.40	34.50	72.90

	Craft@Hrs	Unit	Material	Labor	Total

Flush hardboard solid core interior doors. Primed. 1-3/8" thick. Hardboard doors are made of composition wood veneer and are usually the least expensive.

	Craft@Hrs	Unit	Material	Labor	Total
30" x 80"	BC@1.15	Ea	42.60	34.50	77.10
32" x 80"	BC@1.15	Ea	43.90	34.50	78.40
36" x 80"	BC@1.15	Ea	47.00	34.50	81.50

Flush lauan solid core interior doors. 1-3/8" thick. Stainable. Lauan doors are finished with a veneer of Philippine hardwood from the mahogany family.

	Craft@Hrs	Unit	Material	Labor	Total
28" x 80"	BC@1.15	Ea	45.20	34.50	79.70
32" x 80"	BC@1.15	Ea	49.10	34.50	83.60
36" x 80"	BC@1.15	Ea	49.20	34.50	83.70
30" x 84"	BC@1.15	Ea	51.50	34.50	86.00
32" x 84"	BC@1.15	Ea	53.60	34.50	88.10
34" x 84"	BC@1.15	Ea	57.80	34.50	92.30
36" x 84"	BC@1.15	Ea	57.80	34.50	92.30

Flush birch solid core interior doors. 1-3/8" thick. Solid core.

	Craft@Hrs	Unit	Material	Labor	Total
24" x 80"	BC@1.15	Ea	72.20	34.50	106.70
32" x 80"	BC@1.15	Ea	54.60	34.50	89.10
36" x 80"	BC@1.15	Ea	57.80	34.50	92.30

Flush hardboard interior doors with lite. 1-3/8" thick.

	Craft@Hrs	Unit	Material	Labor	Total
30" x 80" x 1-3/8"	BC@1.15	Ea	96.40	34.50	130.90
32" x 80" x 1-3/8"	BC@1.15	Ea	102.00	34.50	136.50

Hardboard 2-panel stile and rail hollow core doors. 1-3/8" thick.

	Craft@Hrs	Unit	Material	Labor	Total
24" x 80"	BC@1.15	Ea	26.20	34.50	60.70
28" x 80"	BC@1.15	Ea	27.30	34.50	61.80
30" x 80"	BC@1.15	Ea	28.30	34.50	62.80
32" x 80"	BC@1.15	Ea	29.40	34.50	63.90
36" x 80"	BC@1.15	Ea	31.50	34.50	66.00

Hardboard 6-panel molded face interior doors. 1-3/8" thick. Molded and primed hardboard. Embossed simulated woodgrain. Hollow core. Ready to paint. Bored for lockset. Molded face doors have decorative molding applied to the face.

	Craft@Hrs	Unit	Material	Labor	Total
24" x 80"	BC@1.15	Ea	33.40	34.50	67.90
28" x 80"	BC@1.15	Ea	35.00	34.50	69.50
30" x 80"	BC@1.15	Ea	35.60	34.50	70.10
32" x 80"	BC@1.15	Ea	37.10	34.50	71.60
36" x 80"	BC@1.15	Ea	39.00	34.50	73.50

Lauan 6-panel molded face interior doors. Bored for latchset. 1-3/8" thick.

	Craft@Hrs	Unit	Material	Labor	Total
24" x 80"	BC@1.15	Ea	32.20	34.50	66.70
28" x 80"	BC@1.15	Ea	33.80	34.50	68.30
30" x 80"	BC@1.15	Ea	34.90	34.50	69.40
32" x 80"	BC@1.15	Ea	36.90	34.50	71.40
36" x 80"	BC@1.15	Ea	38.90	34.50	73.40

Pine 6-panel molded face interior doors. Radiata pine veneer. Primed. Hollow core. 1-3/8" thick. 3/4" double hip raised panels. Unfinished. Ready to paint, stain, or varnish.

	Craft@Hrs	Unit	Material	Labor	Total
24" x 80"	BC@1.15	Ea	41.60	34.50	76.10
30" x 80"	BC@1.15	Ea	44.80	34.50	79.30
32" x 80"	BC@1.15	Ea	45.80	34.50	80.30
36" x 80"	BC@1.15	Ea	48.80	34.50	83.30

Pine 6-panel stile and rail interior doors. Ready to paint or stain. 1-3/8" thick.

	Craft@Hrs	Unit	Material	Labor	Total
24" x 80"	BC@1.15	Ea	78.30	34.50	112.80
28" x 80"	BC@1.15	Ea	83.10	34.50	117.60
30" x 80"	BC@1.15	Ea	86.30	34.50	120.80
32" x 80"	BC@1.15	Ea	90.50	34.50	125.00
36" x 80"	BC@1.15	Ea	94.90	34.50	129.40

Door Jambs, Interior

	Craft@Hrs	Unit	Material	Labor	Total
Fir 1-panel stile and rail interior doors. 1-3/8" thick.					
28" x 80"	BC@1.15	Ea	164.00	34.50	198.50
30" x 80"	BC@1.15	Ea	165.00	34.50	199.50
32" x 80"	BC@1.15	Ea	166.00	34.50	200.50
Fir 3-panel stile and rail interior doors. 1-3/8" thick.					
28" x 80"	BC@1.15	Ea	200.00	34.50	234.50
30" x 80"	BC@1.15	Ea	201.00	34.50	235.50
32" x 80"	BC@1.15	Ea	201.00	34.50	235.50
Oak 6-panel stile and rail interior doors. Red oak. Double bevel hip raised panels. 1-3/8" thick.					
24" x 80"	BC@1.15	Ea	126.00	34.50	160.50
28" x 80"	BC@1.15	Ea	130.00	34.50	164.50
30" x 80"	BC@1.15	Ea	135.00	34.50	169.50
32" x 80"	BC@1.15	Ea	138.00	34.50	172.50
36" x 80"	BC@1.15	Ea	144.00	34.50	178.50
French stile and rail interior doors. Clear pine. Ready to paint or stain. 1-3/8" thick. Pre-masked tempered glass. French doors are mostly glass, with one, to as many as fifteen panes set in wood sash.					
24" x 80", 10 lite	BC@1.40	Ea	107.00	42.00	149.00
28" x 80", 10 lite	BC@1.40	Ea	112.00	42.00	154.00
30" x 80", 15 lite	BC@1.40	Ea	120.00	42.00	162.00
32" x 80", 15 lite	BC@1.40	Ea	124.00	42.00	166.00
36" x 80", 15 lite	BC@1.40	Ea	131.00	42.00	173.00
Half louver stile and rail interior doors. 1-3/8" thick. Louver doors include wood louvers that allow air to circulate but obscure vision, an advantage for enclosing a closet, pantry or water heater.					
24" x 80"	BC@1.15	Ea	143.00	34.50	177.50
28" x 80"	BC@1.15	Ea	148.00	34.50	182.50
30" x 80"	BC@1.15	Ea	153.00	34.50	187.50
32" x 80"	BC@1.15	Ea	159.00	34.50	193.50
36" x 80"	BC@1.15	Ea	162.00	34.50	196.50
Full louver cafe (bar) doors. Per pair of doors. With hardware. Bar (café) doors are familiar to everyone who has seen a Western movie.					
24" wide, 42" high	BC@1.15	Ea	78.40	34.50	112.90
30" wide, 42" high	BC@1.15	Ea	86.00	34.50	120.50
32" wide, 42" high	BC@1.15	Ea	94.30	34.50	128.80
36" wide, 42" high	BC@1.15	Ea	96.60	34.50	131.10
Stile and rail cafe doors. Per pair of doors. With hardware.					
30" x 42"	BC@1.15	Ea	49.10	34.50	83.60
32" x 42"	BC@1.15	Ea	50.60	34.50	85.10
36" x 42"	BC@1.15	Ea	52.50	34.50	87.00
Interior Door Jambs Finger joint pine interior jamb set. Unfinished. Two 6'-8" legs and one 3' head.					
4-9/16" x 11/16"	BC@.500	Ea	27.80	15.00	42.80
5-1/4" x 11/16"	BC@.500	Ea	55.60	15.00	70.60
Solid clear pine interior jamb. Varnish or stain. Jamb set includes two 6'-8" legs and one 3' head.					
11/16" x 4-9/16" x 7' leg	BC@.210	Ea	14.40	6.30	20.70
11/16" x 4-9/16" x 3' head	BC@.090	Ea	9.32	2.70	12.02
11/16" x 4-9/16" jamb set	BC@.500	Ea	36.50	15.00	51.50
Door jamb set with hinges.					
80" x 4-9/16", per set	BC@.500	Ea	47.00	15.00	62.00
80" x 6-9/16", per set	BC@.500	Ea	50.50	15.00	65.50

	Craft@Hrs	Unit	Material	Labor	Total

Prehung Interior Doors Prehung doors are sold as a package with the door, jamb, hinges, and trim already assembled and the door bored for a latchset. Flush doors are smooth and flat, with no decorative treatment.

Flush prehung hardboard interior doors. 1-3/8" thick. Hollow core. Primed. Ready to paint.

	Craft@Hrs	Unit	Material	Labor	Total
24" x 80"	BC@.750	Ea	49.20	22.50	71.70
28" x 80"	BC@.750	Ea	56.00	22.50	78.50
30" x 80"	BC@.750	Ea	58.10	22.50	80.60
32" x 80"	BC@.750	Ea	60.20	22.50	82.70
36" x 80"	BC@.750	Ea	62.30	22.50	84.80

Flush prehung lauan interior doors. 1-3/8" thick. 4-9/16" jamb. Bored for lockset.

	Craft@Hrs	Unit	Material	Labor	Total
18" x 78"	BC@.750	Ea	52.50	22.50	75.00
24" x 78"	BC@.750	Ea	52.50	22.50	75.00
28" x 78"	BC@.750	Ea	53.50	22.50	76.00
30" x 78"	BC@.750	Ea	54.60	22.50	77.10
32" x 78"	BC@.750	Ea	56.70	22.50	79.20
36" x 78"	BC@.750	Ea	58.80	22.50	81.30
18" x 80"	BC@.750	Ea	51.40	22.50	73.90
24" x 80"	BC@.750	Ea	51.40	22.50	73.90
28" x 80"	BC@.750	Ea	52.50	22.50	75.00
30" x 80"	BC@.750	Ea	55.60	22.50	78.10
32" x 80"	BC@.750	Ea	57.70	22.50	80.20
36" x 80"	BC@.750	Ea	59.80	22.50	82.30

Flush prehung lauan interior doors. 1-3/8" thick. Hollow core. Bored for 2-1/8" lockset with 2-3/8" backset. Casing and lockset sold separately. Wood stile construction. Can be stained or finished naturally. Two brass hinges. Paint grade pine split jambs and 2-1/4" casings. 4" to 4-5/8" adjustable jamb.

	Craft@Hrs	Unit	Material	Labor	Total
24" x 80"	BC@.750	Ea	66.50	22.50	89.00
28" x 80"	BC@.750	Ea	67.90	22.50	90.40
30" x 80"	BC@.750	Ea	69.30	22.50	91.80
32" x 80"	BC@.750	Ea	71.10	22.50	93.60
36" x 80"	BC@.750	Ea	73.40	22.50	95.90

Flush prehung birch interior doors. Stain grade. 1-3/8" thick. Wood stiles. Bored for lockset. Multi-finger jointed door jamb 11/16" x 4-9/16". 3 brass finish hinges 3-1/2" x 3-1/2".

	Craft@Hrs	Unit	Material	Labor	Total
24" x 80"	BC@.750	Ea	65.20	22.50	87.70
28" x 80"	BC@.750	Ea	68.60	22.50	91.10
30" x 80"	BC@.750	Ea	70.90	22.50	93.40
32" x 80"	BC@.750	Ea	73.00	22.50	95.50
36" x 80"	BC@.750	Ea	75.30	22.50	97.80

Flush prehung oak interior doors. Stain grade. 1-3/8" thick. Wood stiles. Bored for lockset. Multi-finger jointed door jamb 11/16" x 4-9/16". 3 brass finish hinges 3-1/2" x 3-1/2".

	Craft@Hrs	Unit	Material	Labor	Total
24" x 80"	BC@.750	Ea	59.50	22.50	82.00
28" x 80"	BC@.750	Ea	62.70	22.50	85.20
30" x 80"	BC@.750	Ea	63.70	22.50	86.20
32" x 80"	BC@.750	Ea	67.10	22.50	89.60
36" x 80"	BC@.750	Ea	68.70	22.50	91.20

Interior prehung lauan heater closet doors.

	Craft@Hrs	Unit	Material	Labor	Total
24" x 60"	BC@.750	Ea	71.10	22.50	93.60

Doors, Prehung, Interior

	Craft@Hrs	Unit	Material	Labor	Total

6-panel colonial prehung interior doors. 11/16" x 4-9/16" jamb. Bored for lockset. Casing and hardware sold separately. Hollow core. 1-3/8" thick. Panel doors have raised decorative wood panels that resemble doors popular in colonial America. Six or nine panels are most common.

	Craft@Hrs	Unit	Material	Labor	Total
18" x 80", 3-panel	BC@.750	Ea	61.70	22.50	84.20
24" x 80"	BC@.750	Ea	62.30	22.50	84.80
28" x 80"	BC@.750	Ea	64.00	22.50	86.50
30" x 80"	BC@.750	Ea	66.00	22.50	88.50
32" x 80"	BC@.750	Ea	68.20	22.50	90.70
36" x 80"	BC@.750	Ea	69.80	22.50	92.30

6-panel colonist prehung interior doors. 6-9/16" jamb. Hollow core.

	Craft@Hrs	Unit	Material	Labor	Total
24" x 80"	BC@.750	Ea	99.70	22.50	122.20
30" x 80"	BC@.750	Ea	104.00	22.50	126.50
32" x 80"	BC@.750	Ea	107.00	22.50	129.50
36" x 80"	BC@.750	Ea	109.00	22.50	131.50

6-panel hardboard prehung interior doors. 4-9/16" adjustable split jambs. Hollow core. Paint grade casing applied. Pre-drilled for lockset. Ready to paint or gel stain.

	Craft@Hrs	Unit	Material	Labor	Total
24" x 80"	BC@.750	Ea	70.60	22.50	93.10
28" x 80"	BC@.750	Ea	72.20	22.50	94.70
30" x 80"	BC@.750	Ea	75.20	22.50	97.70
32" x 80"	BC@.750	Ea	76.40	22.50	98.90
36" x 80"	BC@.750	Ea	78.90	22.50	101.40

6-panel hardboard prehung double interior doors. 1-3/8" thick. Flat jamb.

	Craft@Hrs	Unit	Material	Labor	Total
48" x 80"	BC@1.25	Ea	146.00	37.50	183.50
60" x 80"	BC@1.25	Ea	153.00	37.50	190.50

6-panel oak prehung interior doors. Hollow core. 1-3/8" thick.

	Craft@Hrs	Unit	Material	Labor	Total
24" x 80"	BC@.750	Ea	129.00	22.50	151.50
28" x 80"	BC@.750	Ea	130.00	22.50	152.50
30" x 80"	BC@.750	Ea	134.00	22.50	156.50
32" x 80"	BC@.750	Ea	139.00	22.50	161.50
36" x 80"	BC@.750	Ea	142.00	22.50	164.50

6-panel prehung interior double doors. Bored for lockset. Ready to paint or stain. 1-3/8" thick. 4-9/16" jambs.

	Craft@Hrs	Unit	Material	Labor	Total
4'0" x 6'8"	BC@1.50	Ea	124.00	45.00	169.00
6'0" x 6'8"	BC@1.50	Ea	154.00	45.00	199.00

6-panel hemlock prehung interior doors. Bored for lockset. Ready to paint or stain. 1-3/8" thick. 4-9/16" jambs.

	Craft@Hrs	Unit	Material	Labor	Total
24" x 80"	BC@.750	Ea	161.00	22.50	183.50
28" x 80"	BC@.750	Ea	166.00	22.50	188.50
30" x 80"	BC@.750	Ea	169.00	22.50	191.50
32" x 80"	BC@.750	Ea	170.00	22.50	192.50
36" x 80"	BC@.750	Ea	173.00	22.50	195.50

2-panel knotty alder prehung interior doors. Bored for lockset. 1-3/8" thick.

	Craft@Hrs	Unit	Material	Labor	Total
24" x 80" x 1-3/8"	BC@.750	Ea	282.00	22.50	304.50
28" x 80" x 1-3/8"	BC@.750	Ea	288.00	22.50	310.50
30" x 80" x 1-3/8"	BC@.750	Ea	293.00	22.50	315.50
32" x 80" x 1-3/8"	BC@.750	Ea	299.00	22.50	321.50
36" x 80" x 1-3/8"	BC@.750	Ea	303.00	22.50	325.50

	Craft@Hrs	Unit	Material	Labor	Total

Pine full louver prehung interior doors. Ready to install. Provides ventilation while maintaining privacy. 11/16" x 4-9/16" clear face pine door jamb. 1-3/8" thick. Includes (3) 3-1/2" x 3-1/2" hinges with brass finish. Bored for lockset.

	Craft@Hrs	Unit	Material	Labor	Total
24" x 80"	BC@.750	Ea	94.70	22.50	117.20
28" x 80"	BC@.750	Ea	125.00	22.50	147.50
30" x 80"	BC@.750	Ea	96.90	22.50	119.40
32" x 80"	BC@.750	Ea	104.00	22.50	126.50
36" x 80"	BC@.750	Ea	108.00	22.50	130.50

Pine louver over louver prehung interior doors. Bored for lockset. 1-3/8" thick. Louver doors include wood louvers that allow air to circulate but obscure vision, an advantage for enclosing a closet, pantry or water heater. A lover-over-louver door has lovers the full height of the door.

	Craft@Hrs	Unit	Material	Labor	Total
32" x 80"	BC@.750	Ea	177.00	22.50	199.50

Prehung pantry doors. Finger joint flat jamb. Bored for lockset. 1-3/8" thick.

	Craft@Hrs	Unit	Material	Labor	Total
28" x 80"	BC@.750	Ea	209.00	22.50	231.50
30" x 80"	BC@.750	Ea	209.00	22.50	231.50

10-lite wood French prehung double interior doors. Pre-masked tempered glass for easy finishing. True divided light design. Ponderosa pine. 1-3/8" thick. French doors are mostly glass, with one, to as many as fifteen panes set in wood sash.

	Craft@Hrs	Unit	Material	Labor	Total
48" x 80", unfinished	BC@1.25	Ea	309.00	37.50	346.50
60" x 80", unfinished	BC@1.25	Ea	331.00	37.50	368.50
48" x 80", primed	BC@1.25	Ea	229.00	37.50	266.50
60" x 80", primed	BC@1.25	Ea	241.00	37.50	278.50

15-lite wood French prehung double interior doors. Pre-masked tempered glass for easy finishing. True divided light design. Ponderosa pine. 1-3/8" thick.

	Craft@Hrs	Unit	Material	Labor	Total
48" x 80"	BC@1.25	Ea	209.00	37.50	246.50
60" x 80"	BC@1.25	Ea	209.00	37.50	246.50

Oak French prehung double interior doors. Bored for lockset. 1-3/8" thick.

	Craft@Hrs	Unit	Material	Labor	Total
48" x 80"	BC@1.25	Ea	449.00	37.50	486.50
60" x 80"	BC@1.25	Ea	488.00	37.50	525.50
72" x 80"	BC@1.25	Ea	343.00	37.50	380.50

Patio Doors 10-lite swinging patio double doors. White finish. Dual-glazed with clear tempered glass. Fully weatherstripped. Pre-drilled lock-bore.

	Craft@Hrs	Unit	Material	Labor	Total
6'-0" wide x 6'-8" high	B1@3.58	Ea	575.00	97.90	672.90

Vinyl-clad hinged patio double doors, Frenchwood®, Andersen. White exterior. Perma-Shield® low-maintenance finish. Clear pine interior. 1-lite. 3-point locking system. Adjustable hinges with ball-bearing pivots. Low-E2 tempered insulating glass. Mortise and tenon joints.

	Craft@Hrs	Unit	Material	Labor	Total
6'-0" x 6'-8"	B1@3.58	Ea	1,214.00	97.90	1,311.90

Center hinge 15-lite aluminum patio double doors. Sized for replacement. Unfinished clear pine interior. White extruded aluminum-clad exterior. Weatherstripped frame. Self-draining sill. Bored for lockset. 3/4" Low-E insulated safety glass.

	Craft@Hrs	Unit	Material	Labor	Total
6'-0" x 6'-8"	B1@3.58	Ea	824.00	97.90	921.90

Aluminum inward swing 1-lite patio double doors. Low-E insulated safety glass. Clear wood interior ready to paint or stain.

	Craft@Hrs	Unit	Material	Labor	Total
6'-0" x 6'-8"	B1@3.58	Ea	734.00	97.90	831.90

Aluminum inward swing 10-lite patio double doors. Unfinished interior. Low-E2 argon-filled insulated glass. Thermal-break sill. Bored for lockset. With head and foot bolts.

	Craft@Hrs	Unit	Material	Labor	Total
6'-0" x 6'-8"	B1@3.58	Ea	799.00	97.90	896.90

200 series Narroline™ sliding wood patio doors, Andersen. Dual-pane Low-E glass. Anodized aluminum track with stainless steel cap. With hardware and screen. White.

	Craft@Hrs	Unit	Material	Labor	Total
6'-0" x 6'-8"	B1@3.58	Ea	972.00	97.90	1,069.90

	Craft@Hrs	Unit	Material	Labor	Total

Vinyl-clad gliding patio doors, Andersen. White. Tempered 1" Low-E insulating glass. Anodized aluminum track.

	Craft@Hrs	Unit	Material	Labor	Total
6'-0" x 6'-8"	B1@3.58	Ea	712.00	97.90	809.90

Vinyl-clad gliding patio doors, Narroline™, Andersen. Wood core. Dual-pane Low-E tempered glass. Anodized aluminum track with a stainless steel cap. Combination weatherstrip and interlock.

6'-0" x 6'-8"	B1@3.58	Ea	879.00	97.90	976.90

10-lite sliding patio doors. Solid pine core. Weatherstripped on 4 sides of operating panels. Butcher block 4-1/2"-wide stiles and 9-1/2" bottom rail. 4-9/16" jamb width. White. Actual dimensions: 71-1/4" wide x 79-1/2" high. Pre-drilled lock-bore.

6'-0" wide x 6'-8" high	B1@3.58	Ea	582.00	97.90	679.90

Aluminum sliding patio doors. Unfinished interior. White aluminum-clad exterior. With lockset. Insulating Low-E2 glass.

6'-0" x 6'-8", 1 lite	B1@3.58	Ea	723.00	97.90	820.90
6'-0" x 6'-8", 15 lites	B1@3.58	Ea	723.00	97.90	820.90

Aluminum sliding patio doors. Insulating glass. Reversible, self-aligning ball bearing rollers. Zinc finish lock. With screen. By nominal width x height. Actual height and width are 1/2" less.

5' x 6'-8", bronze finish	B1@3.58	Ea	247.00	97.90	344.90
6' x 6'-8", bronze finish	B1@3.58	Ea	261.00	97.90	358.90
6' x 6'-8", mill finish	B1@3.58	Ea	185.00	97.90	282.90

Aluminum sliding patio doors. With screen. Insulated glass except as noted. Reversible, self-aligning ball bearing rollers. Zinc finish lock. By nominal width x height. Actual height and width are 1/2" less.

5'-0" x 6'-8", mill finish, single glazed	B1@3.58	Ea	177.00	97.90	274.90
5'-0" x 6'-8", white finish	B1@3.58	Ea	251.00	97.90	348.90
5'-0" x 8'-0", white finish, with grids	B1@3.58	Ea	285.00	97.90	382.90
6'-0" x 6'-8", bronze finish, single glazed	B1@3.58	Ea	296.00	97.90	393.90
6'-0" x 6'-8", bronze finish, with grids	B1@3.58	Ea	285.00	97.90	382.90
6'-0" x 6'-8", white finish, single glazed	B1@3.58	Ea	296.00	97.90	393.90
6'-0" x 8'-0", white finish, with grids	B1@3.58	Ea	301.00	97.90	398.90
8'-0" x 6'-8", bronze finish	B1@3.58	Ea	308.00	97.90	405.90
8'0" x 6'8", bronze finish, with grids	B1@4.60	Ea	341.00	126.00	467.00
8'0" x 8'0", white finish	B1@4.60	Ea	331.00	126.00	457.00
8'0" x 8'0", white finish, with grids	B1@4.60	Ea	353.00	126.00	479.00

Screen for sliding or swinging patio doors. 72-1/2" wide x 6'8" high.

White	B1@.250	Ea	135.00	6.84	141.84

Vinyl sliding patio doors. Includes screen. Low-E insulating glass.

5'-0" x 6'-8"	B1@3.58	Ea	566.00	97.90	663.90
6'-0" x 6'-8"	B1@3.58	Ea	602.00	97.90	699.90

Vinyl sliding patio doors. Single glazed. By opening size. Actual width and height are 1/2" less.

5'-0" x 6'-8", with grille	B1@3.58	Ea	387.00	97.90	484.90
6'-0" x 6'-8"	B1@3.58	Ea	387.00	97.90	484.90
6'-0" x 6'-8", with grille	B1@3.58	Ea	419.00	97.90	516.90
8'-0" x 6'-8"	B1@3.58	Ea	492.00	97.90	589.90
8'-0" x 6'-8", with grille	B1@3.58	Ea	524.00	97.90	621.90

1-lite swinging prehung steel patio double doors. Polyurethane core. High-performance weatherstrip. Adjustable thermal-break sill. Non-yellowing frame and grille. Inward swing.

6'-0" x 6'-8"	B1@3.58	Ea	344.00	97.90	441.90

	Craft@Hrs	Unit	Material	Labor	Total

1-lite swinging steel patio double door with internal mini-blinds. Prehung. Double-pane insulated safety glass.

	Craft@Hrs	Unit	Material	Labor	Total
6'-0" x 6'-8", outward swing	B1@3.58	Ea	512.00	97.90	609.90
6'-0" x 6'-8", inward swing, with brick mold	B1@3.58	Ea	576.00	97.90	673.90
6'-0" x 6'-8", outward swing, Dade County approved (high wind)	B1@3.58	Ea	650.00	97.90	747.90

10-lite swinging prehung steel patio double doors. 24-gauge galvanized steel. Double bored for deadbolt. Compression weatherstripping. Aluminum T-astragal. Single glazed.

	Craft@Hrs	Unit	Material	Labor	Total
6'-0" x 6'-8"	B1@3.58	Ea	275.00	97.90	372.90

10-lite swinging steel patio double doors. Inward swing. Insulated Low-E glass.

	Craft@Hrs	Unit	Material	Labor	Total
5'-0" x 6'-8"	B1@3.58	Ea	442.00	97.90	539.90
6'-0" x 6'-8"	B1@3.58	Ea	435.00	97.90	532.90

10-lite venting steel patio double doors. Inward swing. Insulated glass.

	Craft@Hrs	Unit	Material	Labor	Total
6'-0" x 6'-8"	B1@3.58	Ea	555.00	97.90	652.90
8'-0" x 6'-8"	B1@3.58	Ea	692.00	97.90	789.90

15-lite swinging steel patio double doors. Inward swing.

	Craft@Hrs	Unit	Material	Labor	Total
5'-0" x 6'-8", insulated glass	B1@3.58	Ea	344.00	97.90	441.90
6'-0" x 6'-8", Low-E insulated glass	B1@3.58	Ea	420.00	97.90	517.90
6'-0" x 6'-8", single glazed	B1@3.58	Ea	344.00	97.90	441.90

15-lite prehung swinging fiberglass patio double doors. Outward swing. Insulated glass. Adjustable mill finish thermal-break threshold.

	Craft@Hrs	Unit	Material	Labor	Total
6'-0" x 6'-8"	B1@3.58	Ea	618.00	97.90	715.90

Accordion Folding Doors Accordion folding doors. Solid PVC slats with flexible hinges. Vertical embossed surface. Includes hardware. Fully assembled. 32" x 80". Hung in a cased opening.

	Craft@Hrs	Unit	Material	Labor	Total
Light weight, white	BC@.750	Ea	18.10	22.50	40.60
Standard weight, gray	BC@.750	Ea	35.70	22.50	58.20
Standard weight, white	BC@.750	Ea	35.80	22.50	58.30

Premium accordion folding doors. Vertical embossed surface. Preassembled. Prefinished. Adjustable width and height. Flexible vinyl hinges. Hung in a cased opening.

	Craft@Hrs	Unit	Material	Labor	Total
8" x 80" section	BC@.750	Ea	18.40	22.50	40.90
32" x 80"	BC@.750	Ea	55.60	22.50	78.10

Folding door lock.

	Craft@Hrs	Unit	Material	Labor	Total
Door lock	BC@.250	Ea	4.06	7.50	11.56

Pet Doors Entrance doors, lockable, swinging, aluminum frame with security panel

	Craft@Hrs	Unit	Material	Labor	Total
For cats and miniature dogs, 5" x 7-1/2"	BC@.800	Ea	67.00	24.00	91.00
For small dogs, 8-1/2" x 12-1/2"	BC@.800	Ea	95.60	24.00	119.60
For medium dogs, 11-1/2" x 16-1/2"	BC@.800	Ea	130.00	24.00	154.00
For large dogs, 15" x 20"	BC@.850	Ea	180.00	25.50	205.50

Sliding screen or patio doors, adjustable full length 1/2" panel, 80" high, "Lexan" plastic above lockable, swinging PVC door, aluminum frame

	Craft@Hrs	Unit	Material	Labor	Total
For cats and miniature dogs, panel is 11-1/2" wide with 5" x 7-1/2" door	BC@.500	Ea	268.00	15.00	283.00
For small dogs, panel is 15" wide with 8-1/2" x 12-1/2" door	BC@.500	Ea	303.00	15.00	318.00
For medium dogs, panel is 18" wide with 11-1/2" x 16-1/2" door	BC@.500	Ea	350.00	15.00	365.00

Doors, Bypass Closet

	Craft@Hrs	Unit	Material	Labor	Total
Pet screen, vinyl coated 36" x 84"					
Resists tears and punctures	BC@.500	Ea	14.60	15.00	29.60

Bypass Closet Doors Bypass closet doors slide left or right on an overhead track, never revealing more than one-half of the interior.

White bypass closet doors. Prefinished white vinyl-covered hardboard panels instead of mirrors. White steel frame. Jump-proof bottom track. 1-1/2" dual race ball bearing wheels. Injection-molded top guides. Dimensions are width x height. Per pair of doors.

	Craft@Hrs	Unit	Material	Labor	Total
48" x 80"	BC@.750	Ea	52.90	22.50	75.40
60" x 80"	BC@1.00	Ea	63.60	30.00	93.60
72" x 80"	BC@1.00	Ea	73.80	30.00	103.80
48" x 96"	BC@.750	Ea	64.10	22.50	86.60
60" x 96"	BC@1.00	Ea	74.60	30.00	104.60
72" x 96"	BC@1.00	Ea	93.50	30.00	123.50

Economy mirrored bypass closet doors. Gold finished steel. Plate mirror with double strength glass. Bottom rail with jump-proof track. Dimensions are width x height. Per pair of doors.

	Craft@Hrs	Unit	Material	Labor	Total
48" x 80"	BC@.750	Ea	72.70	22.50	95.20
60" x 80"	BC@1.00	Ea	93.70	30.00	123.70
72" x 80"	BC@1.00	Ea	104.00	30.00	134.00
96" x 80"	BC@1.25	Ea	173.00	37.50	210.50

Good quality mirrored bypass closet doors. White frame. Safety-backed mirror. Dimensions are width x height. Per pair of doors.

	Craft@Hrs	Unit	Material	Labor	Total
48" x 80", white	BC@.750	Ea	90.60	22.50	113.10
60" x 80", white	BC@1.00	Ea	111.00	30.00	141.00
72" x 80", white	BC@1.00	Ea	129.00	30.00	159.00
96" x 80", white	BC@1.25	Ea	201.00	37.50	238.50
48" x 80", brushed nickel	BC@.750	Ea	114.00	22.50	136.50
60" x 80", brushed nickel	BC@1.00	Ea	135.00	30.00	165.00
72" x 80", brushed nickel	BC@1.00	Ea	156.00	30.00	186.00
48" x 96", white	BC@1.00	Ea	140.00	30.00	170.00
60" x 96", white	BC@1.00	Ea	151.00	30.00	181.00
72" x 96", white	BC@1.25	Ea	184.00	37.50	221.50
96" x 96", white	BC@1.25	Ea	255.00	37.50	292.50

Beveled mirror bypass closet doors. 1/2" bevel on both sides of glass. Safety-backed mirror. Includes hardware. Dimensions are width x height. Per pair of doors.

	Craft@Hrs	Unit	Material	Labor	Total
48" x 80"	BC@.750	Ea	120.00	22.50	142.50
60" x 80"	BC@1.00	Ea	141.00	30.00	171.00
72" x 80"	BC@1.00	Ea	159.00	30.00	189.00
96" x 80"	BC@1.25	Ea	207.00	37.50	244.50

Premium mirrored bypass closet doors. Bright gold aluminum frame with integrated handle. Select quality 3mm plate mirror and heavy-duty 1-1/2" wide commercial-grade tubular side frame molding. Ultraglide® 2-1/4" deep felt-lined top channel, jump-proof bottom track and precision 1-1/2" dual race ball bearing wheels. Dimensions are width x height. Per pair of doors.

	Craft@Hrs	Unit	Material	Labor	Total
48" x 81"	BC@.750	Ea	182.00	22.50	204.50
60" x 81"	BC@1.00	Ea	203.00	30.00	233.00
72" x 81"	BC@1.00	Ea	234.00	30.00	264.00
Bypass door accessories.					
Bumper	BC@.175	Ea	1.93	5.25	7.18
Carpet riser	BC@.175	Ea	1.58	5.25	6.83
Guide	BC@.175	Ea	1.61	5.25	6.86

	Craft@Hrs	Unit	Material	Labor	Total

Bi-Fold Closet Doors Bi-fold closet doors fold in half when open, revealing nearly the full interior.

Lauan bi-fold flush closet doors. 1-3/8" thick. 2 prehinged panels. Ready to paint, stain, or varnish.

	Craft@Hrs	Unit	Material	Labor	Total
24" x 80"	BC@.700	Ea	26.90	21.00	47.90
30" x 80"	BC@.700	Ea	30.00	21.00	51.00
32" x 80"	BC@.700	Ea	31.30	21.00	52.30
36" x 80"	BC@.700	Ea	33.80	21.00	54.80

Primed hardboard bi-fold flush closet doors. 1-3/8" thick. 2 prehinged panels.

	Craft@Hrs	Unit	Material	Labor	Total
24" x 80"	BC@.700	Ea	26.40	21.00	47.40
30" x 80"	BC@.700	Ea	29.40	21.00	50.40
32" x 80"	BC@.700	Ea	31.40	21.00	52.40
36" x 80"	BC@.700	Ea	34.70	21.00	55.70

Clear mahogany bi-fold flush doors. 1-1/8" thick. 2 prehinged panels. Unfinished.

	Craft@Hrs	Unit	Material	Labor	Total
24" x 80"	BC@.700	Ea	43.10	21.00	64.10
30" x 80"	BC@.700	Ea	50.80	21.00	71.80
32" x 80"	BC@.700	Ea	54.30	21.00	75.30
36" x 80"	BC@.700	Ea	55.70	21.00	76.70

2-door colonist molded face bi-fold closet doors. 1-3/8" thick. Primed. Two doors. Molded face doors have decorative molding applied to at least one side.

	Craft@Hrs	Unit	Material	Labor	Total
24" x 80"	BC@.700	Ea	40.90	21.00	61.90
30" x 80"	BC@.700	Ea	44.90	21.00	65.90
36" x 80"	BC@.700	Ea	49.10	21.00	70.10
48" x 80"	BC@.800	Ea	76.30	24.00	100.30
60" x 80"	BC@.900	Ea	86.90	27.00	113.90
72" x 80"	BC@1.00	Ea	93.80	30.00	123.80

4-panel birch bi-fold flush closet doors. Wood stile and rail construction. 1-3/8" thick. Open left or right. Includes track and hardware.

	Craft@Hrs	Unit	Material	Labor	Total
24" x 80"	BC@.700	Ea	35.80	21.00	56.80
30" x 80"	BC@.700	Ea	39.40	21.00	60.40
36" x 80"	BC@.700	Ea	43.00	21.00	64.00
48" x 80"	BC@.800	Ea	62.00	24.00	86.00

4-panel lauan flush bi-fold closet doors. Wood stile and rail construction. 1-3/8" thick. Open left or right. Includes track and hardware.

	Craft@Hrs	Unit	Material	Labor	Total
48" x 80"	BC@.800	Ea	47.00	24.00	71.00

4-panel red oak flush bi-fold closet doors. Wood stile and rail construction. 1-3/8" thick. Open left or right. Includes track and hardware.

	Craft@Hrs	Unit	Material	Labor	Total
24" x 80"	BC@.700	Ea	35.50	21.00	56.50
30" x 80"	BC@.700	Ea	39.90	21.00	60.90
36" x 80"	BC@.700	Ea	44.10	21.00	65.10
48" x 80"	BC@.800	Ea	64.90	24.00	88.90

Pine colonial panel bi-fold doors. Includes hardware and track. Can be painted or stained. Two hinged panels.

	Craft@Hrs	Unit	Material	Labor	Total
24" x 80"	BC@.700	Ea	54.40	21.00	75.40
30" x 80"	BC@.700	Ea	59.10	21.00	80.10
36" x 80"	BC@.700	Ea	63.40	21.00	84.40

Clear pine 2-panel colonial bi-fold doors. Clear pine. Recessed rails. Unfinished. Open left or right. Includes all hardware and track. Two hinged panels.

	Craft@Hrs	Unit	Material	Labor	Total
24" x 80"	BC@.700	Ea	64.30	21.00	85.30
30" x 80"	BC@.700	Ea	73.40	21.00	94.40
32" x 80"	BC@.700	Ea	84.50	21.00	105.50
36" x 80"	BC@.700	Ea	84.70	21.00	105.70

Doors, Bi-Fold Closet, Mirrored

	Craft@Hrs	Unit	Material	Labor	Total

Oak panel bi-fold doors. Includes track and hardware. 1-3/8" thick. Ready to finish. Classic. Two hinged panels.

	Craft@Hrs	Unit	Material	Labor	Total
24" x 80"	BC@.700	Ea	177.00	21.00	198.00
30" x 80"	BC@.700	Ea	188.00	21.00	209.00
36" x 80"	BC@.700	Ea	205.00	21.00	226.00

Panel over panel bi-fold doors. Includes track and hardware. 1-1/8" thick.

24" x 78"	BC@.700	Ea	71.40	21.00	92.40
30" x 78"	BC@.700	Ea	81.90	21.00	102.90
36" x 78"	BC@.700	Ea	92.40	21.00	113.40

French divided lite bi-fold doors. Includes track and hardware. 1-3/8" thick pine and glass construction. Two prehinged panels.

24" x 80"	BC@.700	Ea	182.00	21.00	203.00
30" x 80"	BC@.700	Ea	198.00	21.00	219.00
36" x 80"	BC@.700	Ea	209.00	21.00	230.00

Economy pine louver over louver bi-fold doors. With track and hardware. Two hinged panels.

24" x 80"	BC@.700	Ea	23.30	21.00	44.30
30" x 80"	BC@.700	Ea	29.00	21.00	50.00
36" x 80"	BC@.700	Ea	31.30	21.00	52.30

Pine louver over louver bi-fold doors. 1-1/8" thick ponderosa pine. 3-11/16" wide rails. 1-1/4" wide stiles. Two prehinged panels. Unfinished.

24" x 80"	BC@.700	Ea	41.20	21.00	62.20
30" x 80"	BC@.700	Ea	47.00	21.00	68.00
32" x 80"	BC@.700	Ea	51.00	21.00	72.00
36" x 80"	BC@.700	Ea	53.00	21.00	74.00

Pine louver over panel bi-fold doors. Recessed rails. Open left or right. 1/2" swing space. Two prehinged panels. Includes track and hardware. Unfinished stain grade.

24" x 80"	BC@.700	Ea	51.10	21.00	72.10
30" x 80"	BC@.700	Ea	58.20	21.00	79.20
36" x 80"	BC@.700	Ea	66.30	21.00	87.30

Bi-fold door hardware.

Bi-fold hinge	BC@.175	Ea	2.79	5.25	8.04
Carpet riser	BC@.175	Ea	1.66	5.25	6.91
Hardware set	BC@.175	Ea	6.57	5.25	11.82
48" track	—	Ea	19.90	—	19.90
60" track	—	Ea	20.90	—	20.90
72" track	—	Ea	22.90	—	22.90

Mirrored Bi-Fold Closet Doors Framed bi-fold mirror doors. White frame. Top-hung with single wheel top hanger. Snap-in bottom guide and fascia. Low rise bottom track. Non-binding door operation. Reversible top and bottom track. 3mm safety-backed mirror. Two hinged panels.

24" x 80"	BC@.700	Ea	83.00	21.00	104.00
30" x 80"	BC@.700	Ea	93.50	21.00	114.50
36" x 80"	BC@.700	Ea	104.00	21.00	125.00

Bevel edge bi-fold mirror doors. 1/2" vertical bevel with bearing hinges. White finished steel frame with safety-backed mirror. Includes hardware. Full access to closet opening. Reversible high- and low-profile top and bottom tracks. Two hinged panels.

24" x 80"	BC@.700	Ea	93.10	21.00	114.10
30" x 80"	BC@.700	Ea	102.00	21.00	123.00
36" x 80"	BC@.700	Ea	121.00	21.00	142.00

	Craft@Hrs	Unit	Material	Labor	Total

Frameless bi-fold mirror doors. Frameless style maximizes the mirror surface. Beveled edge glass. All preassembled hinges. Safety-reinforced mirror backing. Two hinged panels. Includes hardware.

	Craft@Hrs	Unit	Material	Labor	Total
24" x 80"	BC@.700	Ea	93.10	21.00	114.10
30" x 80"	BC@.700	Ea	104.00	21.00	125.00
36" x 80"	BC@.700	Ea	114.00	21.00	135.00

Chrome bi-fold mirror doors. Two hinged panels. Includes hardware.

	Craft@Hrs	Unit	Material	Labor	Total
24" x 80"	BC@.700	Ea	146.00	21.00	167.00
30" x 80"	BC@.700	Ea	156.00	21.00	177.00
36" x 80"	BC@.700	Ea	167.00	21.00	188.00

Drafting Per SF of floor area, architectural only, not including engineering fees. See also Architectural Illustrations and Blueprinting. Typical prices.

	Craft@Hrs	Unit	Material	Labor	Total
Apartments	—	SF	—	—	1.80
Warehouses and storage buildings	—	SF	—	—	1.60
Office buildings	—	SF	—	—	3.10
Residences					
Minimum quality tract work	—	SF	—	—	2.50
Typical work	—	SF	—	—	3.10
Detailed jobs, exposed woods, hillsides	—	SF	—	—	3.90

Drainage Piping Corrugated plastic drainage pipe, plain or perforated and snap-on fittings. Installed in trenches or at foundation footings. No excavation, gravel or backfill included.

Polyethylene pipe

	Craft@Hrs	Unit	Material	Labor	Total
3" pipe	BL@.010	LF	.11	.25	.36
4" pipe	BL@.010	LF	.39	.25	.64
6" pipe	BL@.012	LF	1.00	.30	1.30

Corrugated drain snap pipe fittings

	Craft@Hrs	Unit	Material	Labor	Total
3" internal coupling	P1@.100	Ea	2.10	3.00	5.10
3" perforated end cap	P1@.055	Ea	2.56	1.65	4.21
3" snap adapter	P1@.100	Ea	3.10	3.00	6.10
3" corrugated x smooth adapter	P1@.100	Ea	1.80	3.00	4.80
4" internal coupling	P1@.100	Ea	1.90	3.00	4.90
4" universal adapter	P1@.100	Ea	2.18	3.00	5.18
4" x 3" reducer	P1@.100	Ea	3.10	3.00	6.10
4" snap adapter	P1@.100	Ea	2.10	3.00	5.10
6" snap adapter	P1@.110	Ea	4.20	3.30	7.50
6" x 4" reducer	P1@.110	Ea	4.20	3.30	7.50
6" snap coupling	P1@.110	Ea	3.52	3.30	6.82
6" wye	P1@.150	Ea	10.50	4.50	15.00
6" tee	P1@.150	Ea	9.40	4.50	13.90
6" blind tee	P1@.150	Ea	7.65	4.50	12.15
6" split end cap	P1@.060	Ea	3.15	1.80	4.95

Rock or sand fill for drainage systems. Labor cost is for spreading base and covering pipe. 3 mile haul, dumped on site and placed by wheelbarrow

	Craft@Hrs	Unit	Material	Labor	Total
Crushed stone (1.4 tons per CY)					
3/4" (Number 3)	BL@.700	CY	21.10	17.30	38.40
1-1/2" (Number 2)	BL@.700	CY	25.10	17.30	42.40
Crushed slag (1.86 tons per CY)					
3/4"	BL@.700	CY	20.60	17.30	37.90
Sand (1.40 tons per CY)	BL@.360	CY	18.60	8.90	27.50

Drain Systems

Precast Residential Rain Drain System System includes 5 interlocking tongue and groove connection channels with galvanized steel slotted grate, one meter long each, 4" outlet and closing end caps, catch basin, and 4" PVC pipe outlet. Channel has 4" internal width. System withstands imposed loadings up to 3,500 lbs. Set in concrete, not including concrete or forming costs.
System components

	Craft@Hrs	Unit	Material	Labor	Total
1 meter channel with grate	—	Ea	58.70	—	58.70
4" outlet cap	—	Ea	5.18	—	5.18
4" closed end cap	—	Ea	5.18	—	5.18
Catch basin with PVC trash bucket and steel grating	—	Ea	69.00	—	69.00
4" PVC pipe outlet	—	Ea	16.10	—	16.10
Typical 5 meter rain drain system	BL@4.0	Ea	389.00	98.80	487.80

Draperies, Custom, Subcontract Custom draperies include 4" double hems and headers, 1-1/2" side hems, and weighted corners. Finished sizes allow 4" above, 4" below (except for sliding glass openings), and 5" on each side of wall openings. Prices are figured at one width, 48" of fabric, pleated to 16". Costs listed include fabric, full liner, manufacturing, all hardware, and installation. Add 50% per LF of width for jobs outside working range of metropolitan centers. Use $300 as a minimum job charge.

Minimum quality, 200% fullness, average is 6 panels of fabric, limited choice of fabric styles, colors, and textures

		Unit	Material	Labor	Total
To 95" high	—	LF	—	—	42.60
To 68" high	—	LF	—	—	40.00
To 54" high	—	LF	—	—	38.70

Good quality, fully lined, 250% fullness, average is 5 panels of fabric, better selection of fabric styles, colors, and textures, weighted seams

		Unit	Material	Labor	Total
To 95" high	—	LF	—	—	101.00
To 84" high	—	LF	—	—	95.80
To 68" high	—	LF	—	—	90.60
To 54" high	—	LF	—	—	85.30
To 44" high	—	LF	—	—	81.30
Deduct for multi-unit jobs	—	%	—	—	-25.0
Add for insulated fabrics or liners					
Pleated shade liner	—	SY	—	—	28.00
Thermal liner	—	SY	—	—	32.00

Better quality, fully lined, choice of finest fabric styles, colors, and textures, does not include special treatments such as elaborate swags, or custom fabrics

		Unit	Material	Labor	Total
To 95" high	—	LF	—	—	126.00
To 84" high	—	LF	—	—	121.00
To 68" high	—	LF	—	—	108.00
To 54" high	—	LF	—	—	98.70
To 44" high	—	LF	—	—	92.70

Electrical Work, Subcontract Costs listed below are for wiring new residential and light commercial buildings with Romex cable and assume circuit lengths averaging 40 feet. If flex cable is required, add $10.50 for 15 amp circuits and $30 for 20 amp circuits. No fixtures or appliances included except as noted. Work on second and higher floors may cost 25% more. Work performed by a qualified subcontractor.

Rule of thumb: Total cost for electrical work, performed by a qualified subcontractor, per SF of floor area. Includes service entrance, outlets, switches, basic lighting fixtures and connecting appliances only.

		Unit	Material	Labor	Total
All wiring and fixtures	—	SF	—	—	4.01
Lighting fixtures only (no wiring)	—	SF	—	—	1.16

	Craft@Hrs	Unit	Material	Labor	Total
Air conditioners (with thermostat)					
Central, 2 ton (220 volt)	—	LS	—	—	574.00
First floor room (115 volt)	—	LS	—	—	172.00
Second floor room (115 volt)	—	LS	—	—	345.00
Add for thermostat on second floor	—	LS	—	—	116.00
Alarms. See also Security Alarms					
Fire or smoke detector, wiring and outlet box only	—	LS	—	—	115.00
Add for detector unit	—	LS	—	—	57.40
Bathroom fixtures, wiring only, no fixtures or equipment included					
Mirror lighting (valance or side lighted mirrors)	—	LS	—	—	116.00
Sauna heater (40 amp branch circuit)	—	LS	—	—	345.00
Steam bath generator	—	LS	—	—	345.00
Sunlamps	—	LS	—	—	172.00
Whirlpool bath system/wall switch	—	LS	—	—	288.00
Clock outlets (recessed)	—	LS	—	—	86.00
Closet lighting					
Including ceramic "pull chain" fixture	—	LS	—	—	86.00
Including ceramic switch-operated fixture, add for wall switch from next page	—	LS	—	—	86.00

	Craft@Hrs	Unit	Material	Labor	Total
Clothes dryers					
Motor-driven gas dryer, receptacle only	—	LS	—	—	116.00
Direct connection for up to 5,760 watt (30 amp) electric dryer, 220 volt	—	LS	—	—	172.00
Direct connection for over 5,760 watt (40 amp) electric dryer, 220 volt	—	LS	—	—	201.00
Clothes washers (115 volt)	—	LS	—	—	86.00
Dishwashers	—	LS	—	—	116.00
Door bells, rough wiring, front and rear with transformer					
No chime or bell switch included	—	LS	—	—	172.00
Fans (exhaust)					
Attic fans (wiring only)	—	LS	—	—	116.00
Bathroom fans (includes 70 CFM fan)	—	LS	—	—	172.00
Garage fans (wiring only)	—	LS	—	—	116.00
Kitchen fans (includes 225 CFM fan)	—	LS	—	—	229.00
Freezers, branch circuit wiring only					
Upright (115 volt)	—	LS	—	—	90.00
Floor model	—	LS	—	—	74.60
Furnace wiring and blower hookup only	—	LS	—	—	143.00
Garage door opener, wiring and hookup only	—	LS	—	—	116.00
Garbage disposers, wiring, switch and connection only					
No disposer included	—	LS	—	—	143.00
Grounding devices (see also Lightning protection systems below)					
Grounding entire electrical system	—	LS	—	—	172.00
Grounding single appliance	—	LS	—	—	71.70
Ground fault circuit interrupter (GFCI), rough and finish wiring					
Includes outlet box and GFCI	—	LS	—	—	143.00

	Craft@Hrs	Unit	Material	Labor	Total
Heaters					
Baseboard (115 volt) per branch circuit	—	LS	—	—	116.00
Bathroom (ceiling type) wiring, switch connection (with GFCI) only	—	LS	—	—	137.00

Electrical Grounding Systems

	Craft@Hrs	Unit	Material	Labor	Total
Ceiling heat system (radiant-resistance type) 1,000 watt, 120 volt, per branch circuit, including thermostat	—	LS	—	—	263.00
Space heating (flush in-wall type) up to 2,000 watt, 220 volt	—	LS	—	—	241.00
Humidifiers, central	—	LS	—	—	91.70
Exterior lamp posts, wiring and hookup only, to 25'					
Using buried wire and conduit	—	LS	—	—	688.00
Lighting fixture outlets (rough wiring and box only)					
Ceiling	—	Ea	—	—	57.40
Floor	—	Ea	—	—	86.00
Set in concrete	—	Ea	—	—	114.00
Set in masonry	—	Ea	—	—	114.00
Underground	—	Ea	—	—	149.00
Valance	—	Ea	—	—	68.80
Wall outlet	—	Ea	—	—	57.40

Lightning protection systems (static electric grounding system) Residential, ridge protection including one chimney and connected garage (houses with cut-up roofs will cost more)

	Craft@Hrs	Unit	Material	Labor	Total
Typical home system	—	LS	—	—	2,640.00
Barns and light commercial buildings	—	LS	—	—	5,150.00
Ovens, wall type (up to 10' of wiring and hookup only), 220 volt					
To 4,800 watts (20 amp)	—	LS	—	—	169.00
4,800 to 7,200 watts (30 amp)	—	LS	—	—	196.00
7,200 to 9,600 watts (40 amp)	—	LS	—	—	254.00
Ranges, (up to 10' of wiring and hookup only), 220 volt					
Countertop type					
To 4,800 watts (20 amp)	—	LS	—	—	135.00
4,800 to 7,200 watts (30 amp)	—	LS	—	—	196.00
7,200 to 9,600 watts (40 amp)	—	LS	—	—	254.00
Freestanding type (50 amp)	—	LS	—	—	280.00
Receptacle outlets (rough wiring, box, receptacle and plate)					
Ceiling	—	Ea	—	—	56.00
Countertop wall	—	Ea	—	—	61.40
Floor outlet	—	Ea	—	—	84.40
Split-wired	—	Ea	—	—	106.00
Standard indoor, duplex wall outlet	—	Ea	—	—	56.00
Waterproof, with ground fault circuit	—	Ea	—	—	90.00
Refrigerator or freezer wall outlet	—	Ea	—	—	68.10
Service entrance connections, complete (panel box hookup but no wiring)					
100 amp service including meter socket, main switch, GFCI and 5 single pole breakers in 20 breaker space exterior panel box	—	LS	—	—	1,070.00
200 amp service including meter socket, main switch, 2 GFCI and 15 single pole breakers in 40 breaker space exterior panel box	—	LS	—	—	1,740.00
Sub panel connections (panel box hookup but no wiring)					
40 amp circuit panel including 12 single pole breakers in 12 breaker indoor panel box	—	LS	—	—	394.00
50 amp circuit panel including 16 single pole breakers in 16 breaker indoor panel box	—	LS	—	—	394.00
Sump pump connection					
including 15' of underground cable	—	Ea	—	—	180.00

	Craft@Hrs	Unit	Material	Labor	Total
Switches (includes rough wiring, box, switch and plate)					
Dimmer	—	Ea	—	—	59.10
Lighted	—	Ea	—	—	59.10
Quiet	—	Ea	—	—	47.30
Mercury	—	Ea	—	—	59.10
Standard indoor wall switch	—	Ea	—	—	46.70
Stair-wired (3 way)	—	Ea	—	—	124.00
Waterproof	—	Ea	—	—	101.00
Television outlet wiring (300 ohm wire)	—	Ea	—	—	47.30
Trash compactor, 15" (includes compactor installed under counter and wiring)					
Standard	—	LS	—	—	529.00
Deluxe	—	LS	—	—	618.00
Vacuuming system, central					
Central unit hookup	—	LS			119.00
Remote vacuum outlets					
Receptacle wiring only	—	Ea	—	—	56.00
Water heaters					
(up to 10' of wiring and hookup only), 220 volt	—	LS	—	—	174.00
Water pumps, domestic potable water					
Connection only	—	LS	—	—	112.00

Digital Networking Systems, Subcontract Modular and expandable electronic gateway for the in-home distribution of external services such as CATV, satellite TV, broadcast TV, telephone and high speed internet access. System also processes and redistributes in-home generated signals such as those from VCRs, DVD players, audio, security cameras and computers. Provides computer local area network (LAN) outlets, telephone jacks and cable hookups where required. Costs include console, cable, installation and connection. Ustec, www.ustecnet.com

	Craft@Hrs	Unit	Material	Labor	Total
Basic networking system					
Four TV and ten voice locations	—	LS	—	—	700.00
Apartment type networking system					
Four video, four data and eight voice locations	—	LS	—	—	1,000.00
Amplified networking system. Video and voice surge protection					
Eight video, 4 data and 12 voice locations	—	LS	—	—	2,600.00
Large amplified networking system. Video and voice surge protection					
Six video, 8 data and 18 voice locations	—	LS	—	—	3,600.00

Elevators and Lifts, Subcontract Elevators for apartments or commercial buildings, hydraulic vertical cab, meets code requirements for public buildings, 2,500 pound capacity, to 13 passengers. Includes side opening sliding door, illuminated controls, emergency light, alarm, code-approved fire service operation, motor control unit, wiring in hoistway, cab and door design options.

	Craft@Hrs	Unit	Material	Labor	Total
Basic 2 stop, typical prices					
100 feet per minute	—	LS	—	—	40,500.00
125 feet per minute	—	LS	—	—	43,400.00
150 feet per minute	—	LS	—	—	46,800.00
Add per stop to 5 stops	—	LS	—	—	4,400.00
Add for infrared door protection	—	Ea	—	—	1,700.00
Add for hall position indicator	—	Ea	—	—	410.00
Add for car position indicator	—	Ea	—	—	410.00
Add for car direction light & tone	—	Ea	—	—	400.00
Add for hall lantern with audible tone	—	Ea	—	—	410.00

	Craft@Hrs	Unit	Material	Labor	Total

Elevators for private residences, electric cable, motor driven, 500 pound capacity, includes complete installation of elevator, controls and safety equipment. Meets code requirements for residences

	Craft@Hrs	Unit	Material	Labor	Total
2 stops (up to 10' rise)	—	LS	—	—	19,500.00
3 stops (up to 20' rise)	—	LS	—	—	22,500.00
4 stops (up to 30' rise)	—	LS	—	—	25,700.00
5 stops (up to 40' rise)	—	LS	—	—	28,700.00
Add for additional gate	—	Ea	—	—	950.00

Dumbwaiters, electric

	Craft@Hrs	Unit	Material	Labor	Total
75 pound, 2 stop, 24" x 24" x 30"	—	LS	—	—	6,500.00
75 pound, add for each extra stop	—	Ea	—	—	1,500.00
100 pound, 2 stop, 24" x 30" x 36"	—	LS	—	—	7,500.00
100 pound, add for each extra stop	—	Ea	—	—	1,700.00
100 pound, 2 stop, 24" x 24" x 36", tray type	—	LS	—	—	13,000.00
100 pound tray type, add for each extra stop	—	Ea	—	—	2000.00
300 pound, 2 stop, 30" x 30" x 36", tray type	—	LS	—	—	18800.00
300 pound, add for each extra stop	—	Ea	—	—	3,000.00
500 pound, 2 stop, 23" x 56" x 48"	—	LS	—	—	20,000.00
500 pound, add for each extra stop	—	Ea	—	—	3,000.00

Hydraulic elevators, private residence, 750 pound capacity. Includes controls and safety equipment with standard unfinished birch plywood cab and one gate.

	Craft@Hrs	Unit	Material	Labor	Total
2 stops (up to 15' rise)	—	LS	—	—	22,500.00
3 stops (up to 22' rise)	—	LS	—	—	24,500.00
4 stops (up to 28' rise)	—	LS	—	—	28,500.00
5 stops (up to 34' rise)	—	LS	—	—	30000.00
Add for adjacent opening	—	Ea	—	—	550.00
Add for additional gate	—	Ea	—	—	950.00
Add for plastic laminate cab	—	Ea	—	—	750.00
Add for oak raised panel cab	—	Ea	—	—	3,200.00

Stairlifts (incline lifts), single passenger, indoor. Lift to 18' measured diagonally. 300 pound capacity. Includes top and bottom call button, safety chair, track, power unit, installation by a licensed elevator contractor. Single family residential use

	Craft@Hrs	Unit	Material	Labor	Total
Straight stairs	—	LS	—	—	3,800.00

Wheelchair lift (porch lift), screw driven, fully enclosed sides. Typical electrical connections and control included. Up to 5' rise

	Craft@Hrs	Unit	Material	Labor	Total
Residential wheelchair lift	—	LS	—	—	8,600.00

Engineering Fees Typical cost for consulting work.
Acoustical engineering (Based on an hourly wage of $70.00 to $135.00)

Environmental noise survey of a lot; includes technician with measuring equipment, data reduction and analysis, and written report with recommendations

	Craft@Hrs	Unit	Material	Labor	Total
Written report with recommendations	—	LS	—	—	1,450.00
Exterior to interior noise analysis	—	LS	—	—	475.00

Measure "Impact Insulation Class" and "Sound Transmission Class" in existing buildings, (walls, floors, and ceilings)

	Craft@Hrs	Unit	Material	Labor	Total
Minimum cost	—	LS	—	—	950.00

Priced by wall, floor, or ceiling sections

	Craft@Hrs	Unit	Material	Labor	Total
Per section, approx. 100 SF	—	Ea	—	—	475.00

Analyze office, conference room, or examining room to assure acoustical privacy

	Craft@Hrs	Unit	Material	Labor	Total
Per room analyzed	—	LS	—	—	416.00

	Craft@Hrs	Unit	Material	Labor	Total
Prepare acoustical design for church, auditorium, or lecture hall, (fees vary greatly with complexity of acoustics desired)					
Typical cost	—	LS	—	—	1,780.00
Evaluate noise problems from proposed building					
Impact on surrounding environment	—	LS	—	—	950.00
Evaluate noise problems from surrounding environment					
Impact on a proposed building	—	LS	—	—	773.00
Front end scheduling. Coordinate architectural, engineering and other consultant services, plan check times and other approval times					
For residential and commercial projects	—	LS	—	—	5,540.00
Project scheduling using CPM (Critical Path Method). Includes consultation, review of construction documents, development of construction logic, and graphic schedule. Comprehensive schedules to meet government or owner specifications will cost more.					
Wood frame buildings, 1 or 2 stories	—	LS	—	—	3,680.00
Structural engineering plan check, typical 2-story, 8 to 10 unit apartment building					
Typical price per apartment building	—	LS	—	—	928.00
Add for underground parking	—	LS	—	—	795.00

13426

Entrances, Colonial Single door entrance, pine, primed (no doors, frames, or sills included)

	Craft@Hrs	Unit	Material	Labor	Total
Without pediment, 3'0" x 6'8"	BC@1.19	Ea	208.00	35.70	243.70
With pediment, 3'0" x 6'8"	BC@1.19	Ea	254.00	35.70	289.70
With arch pediment, 3'0" x 6'8"	BC@1.19	Ea	417.00	35.70	452.70
Add for exterior door frames, 3'0" x 6'8" x 1-3/4"					
4-9/16" jamb width	BC@1.25	Ea	63.10	37.50	100.60
Double door entrance, pine, primed (no doors, frames, or sills included)					
Without pediment, two door (individual width)					
2'6" x 6'8"	BC@1.45	Ea	280.00	43.50	323.50
2'8" x 6'8"	BC@1.45	Ea	262.00	43.50	305.50
3'0" x 6'8"	BC@1.45	Ea	280.00	43.50	323.50
With pediment, two doors (individual width)					
2'6" x 6'8"	BC@1.45	Ea	280.00	43.50	323.50
2'8" x 6'8"	BC@1.45	Ea	309.00	43.50	352.50
3'0" x 6'8"	BC@1.45	Ea	428.00	43.50	471.50
With arched pediment, 7'6-1/2" x 1'9" (above door). Opening width of 5'7-1/4" will also fit 3'0" door with two 14" sidelites					
3'0" x 6'8" (two doors)	BC@1.45	Ea	595.00	43.50	638.50
Add for exterior door frames, 1-3/4" thick, 5-1/4" jamb					
2'8" x 6'8" or 3'0" x 6'8"	BC@1.25	Ea	67.80	37.50	105.30
6'0" x 6'8"	BC@2.75	Ea	81.00	82.50	163.50
Decorative split columns (with cap and base) alone, no pediment					
8" x 8' to 12'	BC@.800	Ea	38.10	24.00	62.10
10" x 8' to 12'	BC@.800	Ea	41.80	24.00	65.80
12" x 8' to 12'	BC@.800	Ea	47.60	24.00	71.60
10" x 14' to 16'	BC@1.00	Ea	75.10	30.00	105.10

Excavation and Backfill by Hand Using hand tools.

General excavation, using a pick and shovel (loosening and one throw)

	Craft@Hrs	Unit	Material	Labor	Total
Light soil	BL@1.10	CY	—	27.20	27.20
Average soil	BL@1.70	CY	—	42.00	42.00
Heavy soil or loose rock	BL@2.25	CY	—	55.60	55.60

Excavation

	Craft@Hrs	Unit	Material	Labor	Total
Backfilling (one shovel throw from stockpile)					
Sand	BL@.367	CY	—	9.07	9.07
Average soil	BL@.467	CY	—	11.50	11.50
Rock or clay	BL@.625	CY	—	15.40	15.40
Add for compaction, average soil or sand	BL@.400	CY	—	9.88	9.88
Fine grading	BL@.008	SF	—	.20	.20
Footings, average soil, using a pick and shovel					
6" deep x 12" wide (1.85 CY per CLF)	BL@.034	LF	—	.84	.84
8" deep x 12" wide (2.47 CY per CLF)	BL@.050	LF	—	1.24	1.24
8" deep x 16" wide (3.29 CY per CLF)	BL@.055	LF	—	1.36	1.36
8" deep x 18" wide (3.70 CY per CLF)	BL@.060	LF	—	1.48	1.48
10" deep x 12" wide (3.09 CY per CLF)	BL@.050	LF	—	1.24	1.24
10" deep x 16" wide (4.12 CY per CLF)	BL@.067	LF	—	1.66	1.66
10" deep x 18" wide (4.63 CY per CLF)	BL@.075	LF	—	1.85	1.85
12" deep x 12" wide (3.70 CY per CLF)	BL@.060	LF	—	1.48	1.48
12" deep x 16" wide (4.94 CY per CLF)	BL@.081	LF	—	2.00	2.00
12" deep x 20" wide (6.17 CY per CLF)	BL@.100	LF	—	2.47	2.47
16" deep x 16" wide (6.59 CY per CLF)	BL@.110	LF	—	2.72	2.72
12" deep x 24" wide (7.41 CY per CLF)	BL@.125	LF	—	3.09	3.09
Loading trucks (shoveling, one throw)					
Average soil	BL@1.45	CY	—	35.80	35.80
Rock or clay	BL@2.68	CY	—	66.20	66.20
Pits to 5'					
Light soil	BL@1.34	CY	—	33.10	33.10
Average soil	BL@2.00	CY	—	49.40	49.40
Heavy soil	BL@2.75	CY	—	68.00	68.00
Pits over 5' deep require special consideration: shoring, liners and method for removing soil.					
Shaping trench bottom for pipe					
To 10" pipe	BL@.024	LF	—	.59	.59
12" to 20" pipe	BL@.076	LF	—	1.88	1.88
Shaping embankment slopes					
Up to 1 in 4 slope	BL@.060	SY	—	1.48	1.48
Over 1 in 4 slope	BL@.075	SY	—	1.85	1.85
Add for top crown or toe	—	%	—	50.0	—
Add for swales	—	%	—	90.0	—
Spreading material piled on site					
Average soil	BL@.367	CY	—	9.07	9.07
Stone or clay	BL@.468	CY	—	11.60	11.60
Strip and pile top soil	BL@.024	SF	—	.59	.59
Tamping, hand tamp only	BL@.612	CY	—	15.10	15.10
Trenches to 5', soil piled beside trench					
Light soil	BL@1.13	CY	—	27.90	27.90
Average soil	BL@1.84	CY	—	45.50	45.50
Heavy soil or loose rock	BL@2.86	CY	—	70.70	70.70
Add for depth over 5' to 9'	BL@.250	CY	—	6.18	6.18

	Craft@Hrs	Unit	Material	Labor	Equipment	Total

Trenching and Backfill with Heavy Equipment These costs and productivity are based on utility line trenches and continuous footings where the spoil is piled adjacent to the trench. Linear feet (LF) and cubic yards (CY) per hour shown are based on a crew of two. Reduce productivity by 10% to 25% when spoil is loaded in trucks. Shoring, dewatering or unusual conditions are not included.

Excavation, using equipment shown

Wheel loader 55 HP, with integral backhoe

12" wide bucket, for 12" wide trench. Depths 3' to 5'

	Craft@Hrs	Unit	Material	Labor	Equipment	Total
Light soil (60 LF per hour)	B8@.033	LF	—	1.02	.52	1.54
Medium soil (55 LF per hour)	B8@.036	LF	—	1.12	.56	1.68
Heavy or wet soil (35 LF per hour)	B8@.057	LF	—	1.77	.89	2.66

18" wide bucket, for 18" wide trench. Depths 3' to 5'

	Craft@Hrs	Unit	Material	Labor	Equipment	Total
Light soil (55 LF per hour)	B8@.036	LF	—	1.12	.56	1.68
Medium soil (50 LF per hour)	B8@.040	LF	—	1.24	.63	1.87
Heavy or wet soil (30 LF per hour)	B8@.067	LF	—	2.08	1.05	3.13

24" wide bucket, for 24" wide trench. Depths 3' to 5'

	Craft@Hrs	Unit	Material	Labor	Equipment	Total
Light soil (50 LF per hour)	B8@.040	LF	—	1.24	.63	1.87
Medium soil (45 LF per hour)	B8@.044	LF	—	1.36	.69	2.05
Heavy or wet soil (25 LF per hour)	B8@.080	LF	—	2.48	1.25	3.73

Backfill trenches from loose material piled adjacent to trench. No compaction included.

Soil, previously excavated

	Craft@Hrs	Unit	Material	Labor	Equipment	Total
Front-end loader 60 HP (50 CY per hour)	B8@.040	CY	—	1.24	.61	1.85
D-3 crawler dozer (25 CY per hour)	B8@.080	CY	—	2.48	1.44	3.92
3/4 CY crawler loader (33 CY per hour)	B8@.061	CY	—	1.89	1.12	3.01
D-7 crawler dozer (130 CY per hour)	B8@.015	CY	—	.46	.67	1.13

Sand or gravel bedding

	Craft@Hrs	Unit	Material	Labor	Equipment	Total
3/4 CY wheel loader (80 CY per hour)	B8@.025	CY	—	.77	.37	1.14

Compaction of soil in trenches in 8" layers

	Craft@Hrs	Unit	Material	Labor	Equipment	Total
Pneumatic tampers (40 CY per hour)	BL@.050	CY	—	1.24	.85	2.09
Vibrating rammers, gasoline powered "Jumping Jack" (20 CY per hour)	BL@.100	CY	—	2.47	.97	3.44

	Craft@Hrs	Unit	Material	Labor	Total

Excavation with Heavy Equipment These figures assume a crew of one operator unless noted otherwise. Only labor costs are included here. See equipment rental costs at the end of this section. Use the productivity rates listed here to determine the number of hours or days that the equipment will be needed. For larger jobs and commercial work, see excavation costs under Site Work in the Commercial and Industrial division of this book.

	Craft@Hrs	Unit	Material	Labor	Total
Excavation rule of thumb for small jobs	—	CY	—	—	2.85

Backhoe, operator and one laborer, 3/4 CY bucket

	Craft@Hrs	Unit	Material	Labor	Total
Light soil (13.2 CY per hour)	B8@.152	CY	—	4.71	4.71
Average soil (12.5 CY per hour)	B8@.160	CY	—	4.96	4.96
Heavy soil (10.3 CY per hour)	B8@.194	CY	—	6.01	6.01
Sand (16 CY per hour)	B8@.125	CY	—	3.87	3.87
Add when using 1/2 CY bucket	—	%	—	25.0	—

Excavation

	Craft@Hrs	Unit	Material	Labor	Total
Bulldozer, 65 HP unit					
Backfill (36 CY per hour)	OE@.027	CY	—	1.01	1.01
Clearing brush (900 SF per hour)	OE@.001	SF	—	.04	.04
Add for thick brush	—	%	—	300.0	—
Spread dumped soil (42 CY per hour)	OE@.024	CY	—	.89	.89
Strip topsoil(17 CY per hour)	OE@.059	CY	—	2.20	2.20
Bulldozer, 90 HP unit					
Backfill (40 CY per hour)	OE@.025	CY	—	.93	.93
Clearing brush (1,000 SF per hour)	OE@.001	SF	—	.04	.04
Add for thick brush	—	%	—	300.0	—
Spread dumped soil (45 CY per hour)	OE@.023	CY	—	.86	.86
Strip top soil (20 CY per hour)	OE@.050	CY	—	1.86	1.86
Bulldozer, 140 HP unit					
Backfill (53 CY per hour)	OE@.019	CY	—	.71	.71
Clearing brush (1,250 SF per hour)	OE@.001	SF	—	.04	.04
Add for thick brush	—	%	—	200.0	—
Spread dumped soil (63 CY per hour)	OE@.016	CY	—	.60	.60
Strip top soil (25 CY per hour)	OE@.040	CY	—	1.49	1.49
Dump truck, spot, load, unload, travel					
3 CY truck					
Short haul (9 CY per hour)	B7@.222	CY	—	5.79	5.79
2-3 mile haul (6 CY per hour)	B7@.333	CY	—	8.68	8.68
4 mile haul (4 CY per hour)	B7@.500	CY	—	13.00	13.00
5 mile haul (3 CY per hour)	B7@.666	CY	—	17.40	17.40
4 CY truck					
Short haul (11 CY per hour)	B7@.182	CY	—	4.75	4.75
2-3 mile haul (7 CY per hour)	B7@.286	CY	—	7.46	7.46
4 mile haul (5.5 CY per hour)	B7@.364	CY	—	9.49	9.49
5 mile haul (4.5 CY per hour)	B7@.444	CY	—	11.60	11.60
5 CY truck					
Short haul (14 CY per hour)	B7@.143	CY	—	3.73	3.73
2-3 mile haul (9.5 CY per hour)	B7@.211	CY	—	5.50	5.50
4 mile haul (6.5 CY per hour)	B7@.308	CY	—	8.03	8.03
5 mile haul (5.5 CY per hour)	B7@.364	CY	—	9.49	9.49
Jackhammer, one laborer (per CY of unloosened soil)					
Average soil (2-1/2 CY per hour)	BL@.400	CY	—	9.88	9.88
Heavy soil (2 CY per hour)	BL@.500	CY	—	12.40	12.40
Igneous or dense rock (.7 CY per hour)	BL@1.43	CY	—	35.30	35.30
Most weathered rock (1.2 CY per hour)	BL@.833	CY	—	20.60	20.60
Soft sedimentary rock (2 CY per hour)	BL@.500	CY	—	12.40	12.40
Air tamp (3 CY per hour)	BL@.333	CY	—	8.23	8.23
Loader (tractor shovel)					
Piling earth on premises, 1 CY bucket					
Light soil (43 CY per hour)	OE@.023	CY	—	.86	.86
Heavy soil (35 CY per hour)	OE@.029	CY	—	1.08	1.08
Loading trucks, 1 CY bucket					
Light soil (53 CY per hour)	OE@.019	CY	—	.71	.71
Average soil (43 CY per hour)	OE@.023	CY	—	.86	.86
Heavy soil (40 CY per hour)	OE@.025	CY	—	.93	.93
Deduct for 2-1/4 CY bucket	—	%	—	-43.0	—
Sheepsfoot roller (18.8 CSF per hour)	OE@.001	SF	—	.04	.04
Sprinkling with truck, (62.3 CSF per hour)	TR@.001	SF	—	.03	.03

	Craft@Hrs	Unit	Material	Labor	Total
Tree and brush removal, labor only (clear and grub, one operator and one laborer)					
Light brush	B8@10.0	Acre	—	310.00	310.00
Heavy brush	B8@12.0	Acre	—	372.00	372.00
Wooded	B8@64.0	Acre	—	1,980.00	1,980.00
Tree removal, cutting trees, removing branches, cutting into short lengths with chain saws and axes, by tree diameter, labor only					
8" to 12" (2.5 manhours per tree)	BL@2.50	Ea	—	61.80	61.80
13" to 18" (3.5 manhours per tree)	BL@3.50	Ea	—	86.50	86.50
19" to 24" (5.5 manhours per tree)	BL@5.50	Ea	—	136.00	136.00
25" to 36" (7.0 manhours per tree)	BL@7.00	Ea	—	173.00	173.00
Tree stump removal, using a small dozer, by tree diameter, operator only					
6" to 10" (1.6 manhours per stump)	OE@1.60	Ea	—	59.60	59.60
11" to 14" (2.1 manhours per stump)	OE@2.10	Ea	—	78.20	78.20
15" to 18" (2.6 manhours per stump)	OE@2.60	Ea	—	96.90	96.90
19" to 24" (3.1 manhours per stump)	OE@3.10	Ea	—	115.00	115.00
25" to 30" (3.3 manhours per stump)	OE@3.30	Ea	—	123.00	123.00
Trenching machine, crawler-mounted					
Light soil (27 CY per hour)	OE@.037	CY	—	1.38	1.38
Heavy soil (21 CY per hour)	OE@.048	CY	—	1.79	1.79
Add for pneumatic-tired machine	—	%	—	15.0	—

Excavation Equipment Rental Costs Typical cost not including fuel, delivery or pickup charges. Add labor costs from the previous section. Half day minimums.

	1/2 day	Day	Week
Backhoe, wheeled mounted			
55 HP unit, 1/8 to 3/4 CY bucket	175.00	250.00	774.00
65 HP unit, 1/4 to 1 CY bucket	203.00	290.00	889.00
Add for each delivery or pickup	100.00	100.00	100.00
Bulldozer, crawler tractor			
65 HP	215.00	307.00	943.00
90 to 105 HP unit, D-4 or D-5	284.00	405.00	1,253.00
140 HP unit, D-6	438.00	626.00	1,968.00
Add for each delivery or pickup	100.00	100.00	100.00
Dump truck, on-highway type			
3 CY	127.00	182.00	495.00
5 CY	139.00	198.00	759.00
10 CY	177.00	253.00	990.00
Wheel loaders, front-end load and dump, diesel			
3/4 CY bucket, 4WD, articulated	167.00	239.00	699.00
1 CY bucket, 4WD, articulated	179.00	255.00	795.00
2 CY bucket, 4WD, articulated	217.00	310.00	969.00
3-1/4 CY bucket, 4WD, articulated	445.00	635.00	1,979.00
5 CY bucket, 4WD, articulated	648.00	925.00	2,899.00
Add for each delivery or pickup	100.00	100.00	100.00
Sheepsfoot roller, towed type, 40" diameter, 48" wide, double drum	116.00	165.00	515.00
Trenching machine, to 5'6" depth, 6" to 16" width capacity, crawler-mounted type	203.00	290.00	890.00
Add for each delivery or pickup	100.00	100.00	100.00

Fans

	1/2 day	Day	Week
Compactor, 32" plate width, manually guided, gasoline engine, vibratory plate type	76.00	109.00	340.00
Chain saw, 18" to 23" bar, gasoline engine	36.00	52.00	185.00
Breakers, pavement, medium duty, with blade Including 50' of 1" hose	34.00	48.00	142.00
Air compressors, trailer-mounted, gas powered, silenced			
100 CFM to 150 CFM	60.00	86.00	275.00
175 CFM	99.00	142.00	477.00

	Craft@Hrs	Unit	Material	Labor	Total

Fans

Economy ceiling and wall exhaust fan. For baths, utility, and recreation rooms. For baths up to 45 square feet, other rooms up to 60 square feet. Installs in ceiling or in wall. Plastic duct collar. White polymeric grille. UL listed for use in tub and shower enclosure with GFI branch circuit wiring. Housing dimensions: 8-1/16-inch length, 7-3/16-inch width, 3-7/8-inch depth. Grille size, 8-11/16-inch x 9-1/2-inch. 3-inch duct. 0.75 amp. CFM (cubic feet of air per minute). Labor includes setting and connecting only. Add the cost of wiring and ducting.

	Craft@Hrs	Unit	Material	Labor	Total
50 CFM, 2.5 sones	BE@1.00	Ea	13.80	33.90	47.70
70 CFM, 4.0 sones	BE@1.00	Ea	27.70	33.90	61.60

Vertical discharge bath fan. Galvanized steel housing, with built-in damper and spin-on white polymeric grille. Polymeric fan blade and duct connectors. Built-in double steel mounting ears with keyhole slots. Fits 8-inch ducts. UL listed. CFM (cubic feet of air per minute). Labor includes setting and connecting only. Add the cost of wiring and ducting.

	Craft@Hrs	Unit	Material	Labor	Total
55 CFM, 4.0 sones	BE@1.00	Ea	23.70	33.90	57.60
180 CFM, 5.0 sones	BE@1.00	Ea	71.40	33.90	105.30

Vertical discharge utility fan. Mounts in ceiling. Discharges through duct to roof or wall. Pre-wired motor with plug-in receptacle. Adjustable hanger bars for 16- or 24-inch on-center joists. Silver anodized aluminum grille. Fits 7-inch round ducts. Housing dimensions: 11-inch diameter x 5-5/8-inch deep. UL listed. CFM (cubic feet of air per minute). Labor includes setting and connecting only. Add the cost of wiring and ducting.

	Craft@Hrs	Unit	Material	Labor	Total
210 CFM, 6.5 sones	BE@1.00	Ea	75.50	33.90	109.40

Ceiling or wall bath exhaust fan. White polymeric grille. Torsion spring grille mounting requires no tools. Plug-in, permanently lubricated motor. Centrifugal blower wheel. Rugged, 26 gauge galvanized steel housing. Sturdy key holes mounting brackets for quick, accurate installation. Tapered, polymeric duct fitting with built-in backdraft damper. U.L. listed for use over bathtubs and showers when connected to a GFCI circuit. 120 volts, 0.5 amps, 2.5 sones, 80 CFM (HVI-2100 certified) and 4-inch round duct. Labor includes setting and connecting only. Add the cost of wiring and ducting.

	Craft@Hrs	Unit	Material	Labor	Total
80 CFM, 2.5 sones	BE@1.00	Ea	63.60	33.90	97.50

Solitaire™ bath exhaust fan. Pre-wired outlet box for plug-in motor plate, torsion spring-mounted low-profile grille. Steel housing, damper to eliminate back drafts. Housing dimensions 9-1/4-inch x 9-1/2-inch x 7-5/8-inch. AMCA licensed for air and sound. UL listed to use over tub or shower enclosure with GFCI-protected branch circuit. HVI Certified. CFM (cubic feet of air per minute). 4" duct. Labor includes setting and connecting only. Add the cost of wiring and ducting.

	Craft@Hrs	Unit	Material	Labor	Total
80 CFM, 2.5 sones	BE@1.00	Ea	88.60	33.90	122.50
80 CFM, 3.5 sones	BE@1.00	Ea	99.50	33.90	133.40
120 CFM, 2.5 sones	BE@1.00	Ea	151.00	33.90	184.90
130 CFM, 2.5 sones	BE@1.00	Ea	111.00	33.90	144.90

Thru-the-wall utility fan. Steel housing with built-in damper and white polymeric grille. Permanently lubricated motor. Rotary on and off switch. Housing dimensions: 14-1/4-inch length, 14-1/4-inch width, 4-1/2-inch depth. UL listed. CFM (cubic feet of air per minute). Labor includes setting and connecting only. Add the cost of wiring and exterior finish.

	Craft@Hrs	Unit	Material	Labor	Total
180 CFM, 5.0 sones	BE@1.00	Ea	71.50	33.90	105.40

	Craft@Hrs	Unit	Material	Labor	Total

QuieTTest® low-sound bath fan. For 105 square foot bath. Pre-wired outlet box with plug-in receptacle. Adjustable hanger brackets. Fits 4-inch round ducts. Low-profile white polymeric grille, torsion-spring mounted. Housing dimensions: 9-3/8-inch length, 11-1/4-inch width, 7-7/8-inch depth. Grille size, 14-1/4-inch x 12-1/16-inch. UL listed for use in tub and shower when used with GFI branch circuit wiring. Not recommended for kitchen use. CFM (cubic feet of air per minute). Labor includes setting and connecting only. Add the cost of wiring and ducting.

	Craft@Hrs	Unit	Material	Labor	Total
110 CFM, 2.0 sones	BE@1.00	Ea	113.00	33.90	146.90

QuieTTest® low-sound ceiling blower. For 375 square foot room. Pre-wired outlet box with plug-in receptacle. Rounded, low-profile white polymeric grille with silver anodized trim at each end. Fits 3-1/4-inch x 10-inch ducts. Housing dimensions: 14-1/4-inch length, 10-inch width, 9-inch depth. Grille size, 16-1/2-inch x 12-3/32 inch. CFM (cubic feet of air per minute). Labor includes setting and connecting only. Add the cost of wiring and ducting.

	Craft@Hrs	Unit	Material	Labor	Total
300 CFM, 4.5 sones	BE@1.00	Ea	143.00	33.90	176.90

Lighted Bath Exhaust Fans

ValueTest™ economy bath fan and light, NuTone. Fan and light operate separately or together. Fits 4-inch ducts. White polymeric grille with break-resistant lens. Uses 100-watt lamp (not included). Housing dimensions: 9-inch length, 9-inch width, 5-1/2-inch depth. Grille size, 10-3/4-inch x 12-1/8-inch. UL listed for use in tub and shower when used with GFI branch circuit wiring. CFM (cubic feet of air per minute). Labor includes setting and connecting only. Add the cost of wiring and ducting.

	Craft@Hrs	Unit	Material	Labor	Total
50 CFM, 2.5 Sones, 45 SF bath	BE@1.00	Ea	38.40	33.90	72.30
70 CFM, 4.5 Sones, 65 SF bath	BE@1.00	Ea	61.60	33.90	95.50

Decorative bath exhaust fan with light, Decorative, Broan Manufacturing. Corrosion-resistant finish. Frosted melon glass globe. 60 watt max. CFM (cubic feet of air per minute). Labor includes setting and connecting only. Add the cost of wiring and ducting.

	Craft@Hrs	Unit	Material	Labor	Total
70 CFM, round globe, white	BE@1.00	Ea	92.00	33.90	125.90

Exhaust Air deluxe bath fan with light, NuTone. Ventilation for baths up to 95 square feet, other rooms up to 125 square feet. 100 and 7 watt ceiling and night-light or energy-saving fluorescent light (lamps not included). Snap-on grille assembly. Housing dimensions: 9-inch length, 9-inch width, 6-inch depth. Polymeric white grille, 15-inch diameter. UL listed for use in tub and shower when used with GFI branch circuit wiring. CFM (cubic feet of air per minute). Labor includes setting and connecting only. Add the cost of wiring and ducting.

	Craft@Hrs	Unit	Material	Labor	Total
100 CFM, 3.5 sones	BE@1.00	Ea	120.00	33.90	153.90

Bath Exhaust Fan with Heater

Infrared bulb heater and fan. 4-point adjustable mounting brackets span up to 24-inches. Uses 250 watt, 120 volt R-40 infrared bulbs (not included). Heater and fan units include 70 CFM, 3.5 sones ventilator fan. Damper and duct connector included. Plastic matte-white molded grille and compact housings. CFM (cubic feet of air per minute). Labor includes setting and connecting only. Add the cost of wiring and ducting.

	Craft@Hrs	Unit	Material	Labor	Total
1-bulb, 70 CFM, 3.5 sones	BE@1.00	Ea	28.00	33.90	61.90
2-bulb heater, 7" duct	BE@1.10	Ea	63.00	37.20	100.20

Heat-A-Lamp® bulb heater and fan, NuTone. Non-IC. Swivel-mount. Torsion spring holds plates firmly to ceiling. 250-watt radiant heat from one R-40 infrared heat lamp. Uses 4-inch duct. Adjustable socket for ceilings up to 1-inch thick. Adjustable hanger bars. Automatic reset for thermal protection. White polymeric finish. UL listed. Btu (British Thermal Unit). 6-7/8" high. Labor includes setting and connecting only. Add the cost of wiring and ducting.

	Craft@Hrs	Unit	Material	Labor	Total
2.6 Amp, 1 lamp, 12-1/2" x 10"	BE@1.50	Ea	52.80	50.80	103.60
5.0 Amp, 2 lamp, 15-3/8" x 11"	BE@1.50	Ea	63.50	50.80	114.30

Fans, Accessories

	Craft@Hrs	Unit	Material	Labor	Total

Surface-mount ceiling bath resistance heater and fan. 1250 watt. 4266 Btu (British Thermal Unit). 120 volt. Chrome alloy wire element for instant heat. Built-in fan. Automatic overheat protection. Low-profile housing. Permanently lubricated motor. Mounts to standard 3-1/4-inch round or 4-inch octagonal ceiling electrical box. Satin-finish aluminum grille extends 2-3/4-inch from ceiling. 11" diameter x 2-3/4" deep. Labor includes setting and connecting only. Add the cost of wiring and ducting.

	Craft@Hrs	Unit	Material	Labor	Total
10.7 Amp	BE@.750	Ea	51.50	25.40	76.90

Designer Series bath heater and exhaust fan, Broan Manufacturing. 1500-watt fan-forced heater. 120-watt light capacity, 7-watt night-light (bulbs not included). Permanently lubricated motor. Polymeric damper prevents cold back drafts. 4-point adjustable mounting brackets with keyhole slots. Torsion-spring grille mounting, no tools needed. Non-glare light diffusing glass lenses. Fits single gang opening. Suitable for use with insulation. Includes 4-function control unit. CFM (cubic feet of air per minute). 70 CFM, 3.5 sones. Labor includes setting and connecting only. Add the cost of wiring and ducting.

	Craft@Hrs	Unit	Material	Labor	Total
4" Duct, 100 Watt Lamp	BE@1.00	Ea	75.00	33.90	108.90

Exhaust Fan Accessories

Timer switch

	Craft@Hrs	Unit	Material	Labor	Total
60-minute timer	BE@.250	Ea	18.95	8.47	27.42

Exhaust fan switch

	Craft@Hrs	Unit	Material	Labor	Total
60 minute timer, white	BE@.300	Ea	24.31	10.20	34.51

SensAire® 3-function wall control switch. Fits single gang opening. Top switch for on-auto-off. Other switches for light and night-light. For use with SensAire® fans and lights.

	Craft@Hrs	Unit	Material	Labor	Total
White	BE@.250	Ea	25.99	8.47	34.46

Bath vent kit, Deflect-O. 4" x 8' UL listed Supurr-Flex® duct, aluminum roof vent, 3" to 4" increaser and 2 clamps. Fire resistant.

	Craft@Hrs	Unit	Material	Labor	Total
Flexible duct	SW@.650	Ea	15.97	22.10	38.07

Roof vent kit. For venting kitchen or bath exhaust fan through slanted roof. Works with both 3-inch and 4-inch ducted units.

	Craft@Hrs	Unit	Material	Labor	Total
8-foot length	SW@.650	Ea	18.88	22.10	40.98

Ventilators

Roof turbine vent with base.

	Craft@Hrs	Unit	Material	Labor	Total
12", galvanized	SW@1.00	Ea	29.40	34.00	63.40
12", weathered wood	SW@1.00	Ea	31.80	34.00	65.80
12", white aluminum	SW@1.00	Ea	26.10	34.00	60.10
Add for steep pitch base	SW@.440	Ea	7.70	15.00	22.70
Add for weather cap	—	Ea	8.49	—	8.49

Belt drive attic exhaust fan. 1/3 HP, with aluminum shutter and plenum boards.

	Craft@Hrs	Unit	Material	Labor	Total
30" diameter	SW@3.00	Ea	229.00	102.00	331.00
36" diameter	SW@3.00	Ea	230.00	102.00	332.00
12-hour timer switch	BE@.950	Ea	22.20	32.20	54.40

Power attic gable vent.

	Craft@Hrs	Unit	Material	Labor	Total
1,280 CFM	SW@1.65	Ea	37.20	56.10	93.30
1,540 CFM	SW@1.65	Ea	56.10	56.10	112.20
1,600 CFM	SW@1.65	Ea	79.30	56.10	135.40
Shutter for gable vent	SW@.450	Ea	30.30	15.30	45.60
Humidistat control	BE@.550	Ea	24.30	18.60	42.90
Thermostat control	BE@.550	Ea	17.80	18.60	36.40

Roof-mount power ventilator.

	Craft@Hrs	Unit	Material	Labor	Total
1130 CFM	SW@1.65	Ea	98.80	56.10	154.90
1250 CFM	SW@1.65	Ea	69.70	56.10	125.80
1600 CFM	SW@1.65	Ea	92.30	56.10	148.40

	Craft@Hrs	Unit	Material	Labor	Total

Ceiling Fans

Ceiling fans, decorative, 3-speed reversible motors, lighting kits optional (see below). Includes 10' of type NM cable, installation and connection.

	Craft@Hrs	Unit	Material	Labor	Total
42" four-blade plain fan, oak or white finish	BE@1.50	Ea	108.00	50.80	158.80
42" five-blade, modern styles	BE@1.50	Ea	108.00	50.80	158.80
52" four-blade plain fan	BE@1.50	Ea	118.00	50.80	168.80
52" five-blade plain fan	BE@1.50	Ea	128.00	50.80	178.80
52" five-blade premium fan	BE@1.50	Ea	324.00	50.80	374.80
Optional lighting fixtures for ceiling fans					
Standard globe dome light, 6" H, 10" Dia.	BE@.283	Ea	32.40	9.58	41.98
Schoolhouse or round globe, brass trim	BE@.283	Ea	17.30	9.58	26.88
Add for ball pull chain	BE@.017	Ea	2.15	.58	2.73

Fencing, Chain Link

Galvanized steel, 11.5 gauge 2" x 2" fence fabric, 2-3/8" terminal post with post caps, 1-5/8" line posts at 10' with loop line post caps, 1-3/8" top rail with rail ends, tension bands, tension bars and required bolts with nuts. Material includes 2/3 CF sack of concrete per post. Use 50 LF of fence as a minimum job cost.

	Craft@Hrs	Unit	Material	Labor	Total
36" high	BL@.125	LF	6.04	3.09	9.13
42" high	BL@.125	LF	6.44	3.09	9.53
48" high	BL@.125	LF	6.77	3.09	9.86
60" high	BL@.125	LF	7.81	3.09	10.90
72" high	BL@.125	LF	8.84	3.09	11.93
Add for redwood filler strips	BL@.024	LF	6.25	.59	6.84
Add for aluminum filler strips	BL@.018	LF	10.00	.44	10.44
Add for PVC filler strips	BL@.018	LF	5.00	.44	5.44

Gates, driveway or walkway. Same construction as fencing shown above, assembled units ready to install.

Add for gate hardware from below.

	Craft@Hrs	Unit	Material	Labor	Total
Driveway gates, costs per linear foot (LF)					
36" high	BL@.025	LF	9.22	.62	9.84
42" high	BL@.025	LF	9.72	.62	10.34
48" high	BL@.025	LF	9.96	.62	10.58
60" high	BL@.033	LF	10.80	.82	11.62
72" high	BL@.033	LF	11.60	.82	12.42
3'0" wide walkway gate, costs per gate					
36" high	BL@.500	Ea	20.60	12.40	33.00
42" high	BL@.500	Ea	23.30	12.40	35.70
48" high	BL@.500	Ea	26.70	12.40	39.10
60" high	BL@.666	Ea	33.10	16.50	49.60
72" high	BL@.666	Ea	36.90	16.50	53.40

Add for gate hardware, per gate. Labor is included with labor shown for gate above.

	Craft@Hrs	Unit	Material	Labor	Total
Driveway gate hardware					
36" or 42" high	—	Ea	49.20	—	49.20
48" or 60" high	—	Ea	62.30	—	62.30
72" high	—	Ea	65.90	—	65.90
Walkway gate hardware					
36" or 42" high	—	Ea	19.70	—	19.70
48" or 60" high	—	Ea	24.00	—	24.00
72" high	—	Ea	26.10	—	26.10

Fencing, Chain Link, Subcontract

	Craft@Hrs	Unit	Material	Labor	Total

Pet enclosure fencing. Chain link fence, galvanized steel, 11.5 gauge, 2" x 2" mesh fence fabric, 1-3/8" tubular steel frame, 72" high, lockable gate. Posts set in concrete. Add for concrete slab, if required.

	Craft@Hrs	Unit	Material	Labor	Total
4' x 10'	BL@4.00	Ea	374.00	98.80	472.80
4' x 10', with chain link cover	BL@5.00	Ea	514.00	124.00	638.00
8' x 10'	BL@4.75	Ea	439.00	117.00	556.00
10' x 15'	BL@5.00	Ea	546.00	124.00	670.00

Fencing, Chain Link, Subcontract Work performed by a fence contractor. Costs shown include materials, installation and the subcontractor's overhead and profit. Materials assume galvanized steel, 11.5 gauge 2" x 2" fence fabric, 2-3/8" terminal post with post caps, 1-5/8" line posts at 10' with loop line post caps, 1-3/8" top rail with rail ends, tension bands, tension bars, with posts set in concrete. Installation assumes minimum hand labor, non-rocky soil and post locations accessible to a trailered mechanical post hole digger. Use 50 LF of fence as a minimum job cost.

	Craft@Hrs	Unit	Material	Labor	Total
36" high fence	—	LF	—	—	9.62
42" high fence	—	LF	—	—	10.00
48" high fence	—	LF	—	—	10.40
60" high fence	—	LF	—	—	11.40
72" high fence	—	LF	—	—	12.40
Add for vinyl coated fence and posts	—	%	—	—	10.00
Add for single strand barbed wire at top	—	LF	—	—	.40
Add for redwood filler strips	—	LF	—	—	6.94
Add for aluminum filler	—	LF	—	—	12.00
Add for PVC filler strips	—	LF	—	—	6.00

Gates, driveway or walkway, installed in conjunction with fencing on subcontract basis. Same construction as fencing shown above, delivered ready to install. Add for gate hardware from below.

	Craft@Hrs	Unit	Material	Labor	Total
Driveway gates, costs per linear foot (LF)					
36" high	—	LF	—	—	9.94
42" high	—	LF	—	—	10.40
48" high	—	LF	—	—	10.70
60" high	—	LF	—	—	11.80
72" high	—	LF	—	—	12.60
3'0" wide walkway gate, costs per gate					
36" high	—	Ea	—	—	46.40
42" high	—	Ea	—	—	50.60
48" high	—	Ea	—	—	56.80
60" high	—	Ea	—	—	67.80
72" high	—	Ea	—	—	76.10

Add for gate hardware, per gate, material only. (Installation is included with gate above.)

	Craft@Hrs	Unit	Material	Labor	Total
Driveway gate hardware					
36" or 42" high	—	Ea	—	—	59.00
48" or 60" high	—	Ea	—	—	75.00
72" high	—	Ea	—	—	79.00
Walkway gate hardware					
36" or 42" high	—	Ea	—	—	24.00
48" or 60" high	—	Ea	—	—	29.30
72" high	—	Ea	—	—	31.50

Fence, Galvanized Wire Mesh With 1-3/8" galvanized steel posts, set without concrete 10' OC.

	Craft@Hrs	Unit	Material	Labor	Total
12-1/2 gauge steel, 12" vertical wire spacing					
36" high, 6 horizontal wires	BL@.127	LF	1.13	3.14	4.27
48" high, 8 horizontal wires	BL@.127	LF	1.14	3.14	4.28
48" high, 8 horizontal wires (11 gauge)	BL@.127	LF	1.52	3.14	4.66

	Craft@Hrs	Unit	Material	Labor	Total
11 gauge fence fabric, 9" vertical wire spacing					
42" high, 8 horizontal wires	BL@.127	LF	1.50	3.14	4.64
48" high, 9 horizontal wires	BL@.127	LF	1.65	3.14	4.79
14 gauge fence fabric, 6" vertical wire spacing					
36" high, 8 horizontal wires	BL@.127	LF	1.02	3.14	4.16
48" high, 9 horizontal wires	BL@.127	LF	1.52	3.14	4.66
60" high, 10 horizontal wires	BL@.127	LF	1.94	3.14	5.08
Add for components for electrified galvanized wire mesh fence					
15 gauge aluminum wire, single strand	BL@.050	LF	.08	1.24	1.32
12-1/2 gauge barbed wire, 4 barbs per LF	BL@.050	LF	1.88	1.24	3.12
Charging unit for electric fence, Solar powered,					
10 mile range, built in 6V battery	BL@.500	Ea	130.00	12.40	142.40
Charging unit, 110V,					
30 mile range, indoors or out	BL@.500	Ea	81.00	12.40	93.40
Charging unit, 110V,					
2 mile range, indoors or out	BL@.500	Ea	30.80	12.40	43.20
Fence stays for barbed wire, 48" high	BL@.250	Ea	.77	6.18	6.95
Add for electric fence warning sign					
Yellow with black lettering	BL@.125	Ea	.74	3.09	3.83
Add for polyethylene insulators	BL@.250	Ea	.18	6.18	6.36
Add for non-conductive gate fastener	BL@.500	Ea	3.68	12.40	16.08
Add for lightning arrestor kit	BL@.500	Ea	9.09	12.40	21.49

Fence Post Holes

Based on holes 1' diameter and 2' deep with easy access on level terrain. Adjustments for other conditions are below. Per hole dug.

	Craft@Hrs	Unit	Material	Labor	Total
Excavate post hole by hand					
Light soil, sand or loam	BL@.400	Ea	—	9.88	9.88
Medium soil, gravel or light clay	BL@.600	Ea	—	14.80	14.80
Heavy soil, shale, sandstone, rocky	BL@1.00	Ea	—	24.70	24.70
Excavate post hole with power auger. Add the auger cost.					
Light soil, sand or loam	BL@.200	Ea	—	4.94	4.94
Medium soil, gravel or light clay	BL@.250	Ea	—	6.18	6.18
Heavy soil, shale, sandstone, rocky	BL@.300	Ea	—	7.41	7.41

Setting Fence Posts

Posts set with concrete, including temporary 1" x 6" bracing and stakes. Heights to 8'. Concrete assumes post hole is 1' x 1' x 2' deep. Per post set, using either wood or metal post.

	Craft@Hrs	Unit	Material	Labor	Total
Set and brace wood or metal fence post					
Lay in gravel for drainage	BL@.010	Ea	.02	.25	.27
Set x 4" x 4" x 8' treated wood post	BL@.015	Ea	8.61	.37	8.98
Set 18 gauge 1-7/8" metal tube post with cap	BL@.015	Ea	7.35	.37	7.72
Check and adjust post alignment	BL@.010	Ea	—	.25	.25
Plumb and brace post (5 uses)	BL@.100	Ea	.44	2.47	2.91
Backfill with concrete or soil	BL@.030	Ea	—	.74	.74
Tamp soil or rod concrete	BL@.010	Ea	—	.25	.25
Trowel drainage crown	BL@.010	Ea	—	.25	.25
Strip bracing and inspect	BL@.020	Ea	—	.49	.49
Concrete for post base, 60 lb. sack, mixed	BL@.125	Ea	3.60	3.09	6.69
Total per fence post (wood)	**—@.330**	**Ea**	**12.67**	**8.16**	**20.83**
Wood posts 10' OC, per LF of fence	**—@.033**	**LF**	**1.27**	**.82**	**2.09**
Total per fence post (metal)	**—@.330**	**Ea**	**11.41**	**8.16**	**19.57**
Metal posts 10' OC, per LF of fence	**—@.033**	**LF**	**1.14**	**.82**	**1.96**

Fence, Rails

	Craft@Hrs	Unit	Material	Labor	Total

Setting Fence Rail on Wood Posts

Stringing 2" x 4" wood rail between fence posts, including 2" x 4" pressure treated rail at $.68 per linear foot.

	Craft@Hrs	Unit	Material	Labor	Total
Measure opening between posts	BL@.010	Ea	—	.25	.25
Measure and cut 10' pressure treated rail	BL@.015	Ea	7.16	.37	7.53
Measure rail position on post	BL@.010	Ea	—	.25	.25
Nail one end in place	BL@.010	Ea	.02	.25	.27
Position other end and check for level	BL@.020	Ea	—	.49	.49
Nail second end in place	BL@.010	Ea	.02	.25	.27
Recheck level and adjust as needed	BL@.015	Ea	—	.37	.37
Paint cut ends, post connections, rail side	BL@.100	Ea	—	2.47	2.47
Set rail nailed in place, total	**—@.190**	**Ea**	**7.20**	**4.70**	**11.90**
Set two rails, per LF of fence, 10' rails	**—@.038**	**LF**	**.72**	**.47**	**1.19**

Setting Fence Rail on Metal Posts

Stringing 2" x 4" wood fence rail between metal posts, including 2" x 4" pressure treated rail at $.63 per linear foot and one 5" lag bolt connecting adjacent rail at each post.

	Craft@Hrs	Unit	Material	Labor	Total
Measure opening between posts	BL@.010	Ea	—	.25	.25
Measure and cut two 10' rail	BL@.030	Ea	14.30	.74	15.04
Measure rail position on post	BL@.010	Ea	—	.25	.25
Drill 2 rail and metal tube post for lag bolt	BL@.167	Ea	—	4.13	4.13
Set two rail on tube post with lag bolt	BL@.133	Ea	.25	3.29	3.54
Recheck level and adjust as needed	BL@.015	Ea	—	.37	.37
Paint cut ends, post connections, rail side	BL@.100	Ea	—	2.47	2.47
Set two rails, bolted in place, total	**—@.465**	**Ea**	**14.55**	**11.50**	**26.05**
Set two rails, bolted in place, per LF of fence	**—@.047**	**LF**	**1.46**	**1.15**	**2.61**

Installing Fence Face Board

Hand nailing or stapling face board (pickets) to existing rail. Deduct 50% from labor costs for power nailing or power staping. Includes measuring and cutting each picket, nailing or stapling to the top rail, plumbing with a level and nailing or stapling to the bottom rail. Two nails or staples at each rail. Cost per LF of fence. Based on 1" x 4" x 6', 1" x 6" x 6' and 1" x 8" x 6' face boards. Face board costs vary widely.

Nailing 4" wide pressure treated southern pine fence face board (3.4 per LF of fence)

	Craft@Hrs	Unit	Material	Labor	Total
French gothic, 1" x 4" x 3-1/2' ($.85 ea)	BL@.102	LF	3.35	2.52	5.87
French gothic, 5/8" x 4" x 4' ($1.19 ea)	BL@.102	LF	3.77	2.52	6.29
Dog-eared, 7/16" x 4" x 6' ($1.12 ea)	BL@.136	LF	3.81	3.36	7.17
Dog-eared, 1" x 4", 6' ($1.69 ea)	BL@.136	LF	6.67	3.36	10.03

Nailing 5-1/2" wide pressure treated southern pine fence face board (2.4 per LF of fence)

	Craft@Hrs	Unit	Material	Labor	Total
Dog-eared, 5/8" x 5-1/2" x 6' ($1.55 ea)	BL@.120	LF	3.73	2.97	6.70

Nailing 6" wide pressure treated southern pine fence face board (2.2 per LF of fence)

	Craft@Hrs	Unit	Material	Labor	Total
Dog-eared, 1" x 6" x 6' ($1.88 ea)	BL@.110	LF	4.13	2.72	6.85
Dog-eared Prem S4S, 1" x 6" x 6' ($2.09 ea)	BL@.110	LF	4.60	2.72	7.32
Full cut premium, 1" x 6" x 6' ($2.07 ea)	BL@.110	LF	4.55	2.72	7.27
Dog-eared, 5/8" x 5-1/2" x 8' ($2.78 ea)	BL@.110	LF	6.12	2.72	8.84

Nailing 4" wide white wood fence face board (3.4 per LF of fence)

	Craft@Hrs	Unit	Material	Labor	Total
Gothic, 3/4" x 4" x 6' ($1.14 ea)	BL@.136	LF	3.89	3.36	7.25
Flat top, 1" x 4" x 6' ($1.10 ea)	BL@.136	LF	3.75	3.36	7.11
Dog-eared, 1" x 4" x 6' ($1.14 ea)	BL@.136	LF	3.89	3.36	7.25
Dog-eared, 1" x 4" x 8' ($1.75 ea)	BL@.136	LF	5.96	3.36	9.32

	Craft@Hrs	Unit	Material	Labor	Total
Nailing 4" wide cedar fence face board (3.4 per LF of fence)					
Dog-eared, No. 2, 3/4" x 4" x 8' ($2.36 ea)	BL@.136	LF	8.03	3.36	11.39
Gothic, 3/4" x 4" x 6' ($1.34 ea)	BL@.136	LF	4.57	3.36	7.93
Dog-eared, 9/16" x 4" x 6' ($1.23 ea)	BL@.136	LF	4.18	3.36	7.54
Flat top, full-cut, 1" x 4" x 6' ($1.76 ea)	BL@.136	LF	6.00	3.36	9.36
No. 1 flat top, full-cut, 1" x 4" x 8' ($2.47 ea)	BL@.136	LF	8.39	3.36	11.75
Nailing 6" wide cedar fence face board (2.2 per LF of fence)					
Dog-eared, No. 2, 9/16" x 6" x 6' ($2.59 ea)	BL@.110	LF	5.71	2.72	8.43
Dog-eared, No. 2, 9/16" x 6" x 8' ($3.77 ea)	BL@.110	LF	8.29	2.72	11.01
Dog-eared, clear, 1" x 6" x 6' ($2.75 ea)	BL@.110	LF	6.05	2.72	8.77
Flat top, full-cut, 1" x 6" x 6' ($2.73 ea)	BL@.110	LF	6.01	2.72	8.73
Nailing 4" wide dog-eared common redwood fence face board (3.4 per LF of fence)					
3/4" x 4" x 6' ($1.75 ea)	BL@.136	LF	5.96	3.36	9.32
Nailing 6" wide dog-eared common redwood fence face board (2.2 per LF of fence)					
3/4" x 6" x 5' ($2.47 ea)	BL@.110	LF	5.43	2.72	8.15
3/4" x 6" x 6' ($2.93 ea)	BL@.110	LF	6.44	2.72	9.16
3/4" x 6" x 6' ($2.78 ea)	BL@.110	LF	6.12	2.72	8.84
Nailing 8" wide dog-eared common redwood fence face board (1.6 per LF of fence)					
3/4" x 8" x 6' ($4.26 ea)	BL@.088	LF	6.82	2.17	8.99
3/4" x 8" x 6' Premium Const. Heart ($6.27 ea)	BL@.088	LF	10.00	2.17	12.17
1" x 8" x 8' ($8.37 ea)	BL@.088	LF	13.40	2.17	15.57
Nailing 10" wide dog-eared common redwood fence face board (1.3 per LF of fence)					
3/4" x 10" x 5' ($5.13 ea)	BL@.072	LF	6.67	1.78	8.45
3/4" x 10" x 6' ($5.93 ea)	BL@.072	LF	7.71	1.78	9.49
Nailing 12" wide dog-eared common redwood fence face board (1.05 per LF of fence)					
3/4" x 12" x 5' Rough ($7.13 ea)	BL@.063	LF	7.49	1.56	9.05
3/4" x 12" x 6' ($8.26 ea)	BL@.063	LF	8.68	1.56	10.24
Nailing 6" wide flat top redwood fence face board (2.2 per LF of fence)					
3/4" x 6" x 6' Const. Heart ($3.66 ea)	BL@.110	LF	8.06	2.72	10.78
3/4" x 6" x 6' Merch. and Better ($3.66 ea)	BL@.110	LF	8.06	2.72	10.78
3/4" x 6" x 6' Ridge Valley Redwood ($6.08 ea)	BL@.110	LF	13.40	2.72	16.12
3/4" x 6" x 8' Ridge Valley Redwood ($8.18 ea)	BL@.110	LF	18.00	2.72	20.72
Nailing 8" wide flat top redwood fence face board (1.6 per LF of fence)					
3/4" x 8" x 5' ($3.66 ea)	BL@.088	LF	5.86	2.17	8.03
3/4" x 8" x 5' T&G ($5.81 ea)	BL@.088	LF	9.29	2.17	11.46
3/4" x 8" x 6' ($4.19 ea)	BL@.088	LF	6.70	2.17	8.87
3/4" x 8" x 6' Construction Heart ($7.04 ea)	BL@.088	LF	11.30	2.17	13.47
1" x 8" x 8' Construction Heart ($8.37 ea)	BL@.088	LF	13.40	2.17	15.57
Nailing 10" and 12" wide flat top redwood fence face board (1.3 per LF of fence for 10", 1.0 for 12")					
3/4" x 10" x 8' Redwood ($8.24 ea)	BL@.072	LF	10.70	1.78	12.48
3/4" x 12" x 6' Dog-eared ($8.26 ea)	BL@.072	LF	10.70	1.78	12.48
Deduct for power stapling or power nailing	—	%	—	-50.0	—

Building Fence on a Hillside

Post hole excavation, post setting and fence erection will take longer when the site is obstructed or on a hillside. Adjust manhours as follows for sloping terrain.

Add for slope to 4 in 12	—	%	—	20.0	—
Add for slope over 4 in 12 to 8 in 12	—	%	—	50.0	—
Add for slope over 8 in 12 to 12 in 12	—	%	—	100.0	—

Fencing, Assemblies

	Craft@Hrs	Unit	Material	Labor	Total

Build and Hang a Wood Fence Gate 5' high by 3' wide wood gate, 2" x 4" frame using 1" x 6" face boards. Add the cost of fence posts. Based on treated 2" x 4" at $.63 per linear foot and treated 6" x 5' face boards at $1.75 each. Units are for each gate.

	Craft@Hrs	Unit	Material	Labor	Total
Measure, design and lay out gate	BL@.200	Ea	—	4.94	4.94
Measure and cut frame (12' x 2" x 4")	BL@.050	Ea	7.16	1.24	8.40
Nail frame together	BL@.100	Ea	.10	2.47	2.57
Square up and measure 4' diagonal brace	BL@.050	Ea	2.48	1.24	3.72
Cut cross brace and nail in place	BL@.083	Ea	.05	2.05	2.10
Square up and add metal corner brackets	BL@.333	Ea	1.00	8.23	9.23
Cut and nail seven face boards($3.12 ea)	BL@.397	Ea	21.80	9.81	31.61
Mark top pattern on face boards	BL@.133	Ea	—	3.29	3.29
Cut top pattern on face boards	BL@.089	Ea	—	2.20	2.20
Set hinges on gate and fence post	BL@.250	Ea	8.40	6.18	14.58
Set latch on gate and strike on fence post	BL@.100	Ea	4.20	2.47	6.67
Check and adjust as needed	BL@.125	Ea	—	3.09	3.09
Build and hang wood fence gate, total	**—@1.91**	**Ea**	**45.19**	**47.21**	**92.40**

Fence Assemblies, Wood Costs shown include excavation of post holes, concrete and labor to set posts, attaching rails with nails or lag bolts, and hand nailing of face boards where applicable. See additional costs at the end of this section.

Basket-weave fence, redwood, "B" grade, 1" x 6" boards,
2" x 4" stringers or spreaders, 4" x 4" posts

	Craft@Hrs	Unit	Material	Labor	Total
Tight weave, 4' high, posts at 4' OC	B1@.332	LF	19.10	9.08	28.18
Tight weave, 6' high, posts at 4' OC	B1@.522	LF	23.80	14.30	38.10
Wide span, 4' high, posts at 8' OC	B1@.291	LF	17.60	7.96	25.56
Wide span, 8' high, posts at 8' OC	B1@.481	LF	30.00	13.20	43.20
4' high gate, tight weave	B1@.906	Ea	84.10	24.80	108.90
4' high gate, wide span	B1@.906	Ea	103.00	24.80	127.80
6' high gate, tight weave	B1@1.11	Ea	75.80	30.40	106.20
8' high gate, wide span	B1@1.26	Ea	131.00	34.50	165.50

Board fence, using 4" wide boards, hand nailed, posts at 10' OC

Pressure treated Southern Pine, 1" x 4" boards

	Craft@Hrs	Unit	Material	Labor	Total
6' high, 2 rail, wood posts	B1@.267	LF	5.70	7.31	13.01
6' high, 2 rail, metal posts	B1@.276	LF	6.73	7.55	14.28
Deduct for power stapling or power nailing	—	%	—	-22.0	—

Cedar, 1" x 4" boards

	Craft@Hrs	Unit	Material	Labor	Total
6' high, 2 rail, wood posts	B1@.267	LF	7.50	7.31	14.81
8' high, 2 rail, wood posts	B1@.267	LF	9.70	7.31	17.01
6' high, 2 rail, metal posts	B1@.276	LF	8.58	7.55	16.13
8' high, 2 rail, metal posts	B1@.276	LF	10.80	7.55	18.35
Deduct for power stapling or power nailing	—	%	—	-22.0	—

Redwood, 3/4" x 4" boards

	Craft@Hrs	Unit	Material	Labor	Total
6' high, 2 rail, wood posts	B1@.267	LF	6.84	7.31	14.15
6' high, 2 rail, metal posts	B1@.276	LF	7.79	7.55	15.34
Deduct for power stapling or power nailing	—	%	—	-22.0	—

Whitewoods, 1" x 4" boards

	Craft@Hrs	Unit	Material	Labor	Total
6' high, 2 rail, wood posts	B1@.267	LF	5.78	7.31	13.09
8' high, 2 rail, wood posts	B1@.267	LF	7.77	7.31	15.08
6' high, 2 rail, metal posts	B1@.276	LF	6.88	7.55	14.43
8' high, 2 rail, metal posts	B1@.276	LF	8.87	7.55	16.42
Deduct for power stapling or power nailing	—	%	—	-22.0	—

	Craft@Hrs	Unit	Material	Labor	Total
Board fence, using 6" wide boards, hand nailed, posts at 10' OC					
Pressure treated Southern Pine, 1" x 6" boards					
6' high, 2 rail, wood posts	B1@.241	LF	6.26	6.59	12.85
8' high, 2 rail, wood posts	B1@.241	LF	7.16	6.59	13.75
6' high, 2 rail, metal posts	B1@.250	LF	7.29	6.84	14.13
8' high, 2 rail, metal posts	B1@.250	LF	8.19	6.84	15.03
Deduct for power stapling or power nailing	—	%	—	-22.0	—
Cedar, 9/16" x 6" boards					
6' high, 2 rail, wood posts	B1@.241	LF	8.11	6.59	14.70
8' high, 2 rail, wood posts	B1@.241	LF	10.40	6.59	16.99
6' high, 2 rail, metal posts	B1@.250	LF	9.19	6.84	16.03
8' high, 2 rail, metal posts	B1@.250	LF	11.50	6.84	18.34
Deduct for power stapling or power nailing	—	%	—	-22.0	—
Redwood, 3/4" x 6" boards					
6' high, 2 rail, wood posts	B1@.241	LF	6.55	6.59	13.14
8' high, 2 rail, wood posts	B1@.241	LF	7.45	6.59	14.04
6' high, 2 rail, metal posts	B1@.250	LF	7.50	6.84	14.34
8' high, 2 rail, metal posts	B1@.250	LF	8.33	6.84	15.17
Deduct for power stapling or power nailing	—	%	—	-22.0	—
Split rail fence, red cedar, 4" posts at 8' OC					
Split 2" x 4" rails, 3' high, 2 rail	B1@.095	LF	3.11	2.60	5.71
Split 2" x 4" rails, 4' high, 3 rail	B1@.102	LF	3.69	2.79	6.48
Rustic 4" round, 3' high, 2 rail	B1@.095	LF	3.69	2.60	6.29
Rustic 4" round, 4' high, 3 rail	B1@.102	LF	4.39	2.79	7.18
Picket fence, 1" x 2" pickets, 2" x 4" rails, 4" x 4" posts at 8' OC					
Pressure treated Southern Pine					
3' high, 2 rail	B1@.270	LF	3.04	7.39	10.43
5' high, 3 rail	B1@.385	LF	4.16	10.50	14.66
3' high gate	B1@.770	Ea	13.40	21.10	34.50
5' high gate	B1@.887	Ea	18.40	24.30	42.70
Red cedar, #3 & Better					
3' high, 2 rail	B1@.270	LF	5.02	7.39	12.41
5' high, 3 rail	B1@.385	LF	6.88	10.50	17.38
3' high gate	B1@.770	Ea	22.10	21.10	43.20
5' high gate	B1@.887	Ea	30.30	24.30	54.60
Redwood, B grade					
3' high, 2 rail	B1@.270	LF	6.52	7.39	13.91
5' high, 3 rail	B1@.385	LF	8.49	10.50	18.99
3' high gate	B1@.770	Ea	28.61	21.10	49.71
5' high gate	B1@.887	Ea	37.40	24.30	61.70
Gate hardware					
Latch, standard duty	B1@.281	Ea	3.16	7.69	10.85
Latch, heavy duty	B1@.281	Ea	13.60	7.69	21.29
Hinges, standard, per pair	B1@.374	Pr	6.60	10.20	16.80
Hinges, heavy duty, per pair	B1@.374	Pr	33.70	10.20	43.90
No-sag cable kit	B1@.281	Ea	10.30	7.69	17.99
Self closing spring	B1@.281	Ea	7.13	7.69	14.82

Fencing, Aluminum

	Craft@Hrs	Unit	Material	Labor	Total

Fencing, Aluminum

Decorative aluminum, 72-1/2" panel width, 2" x 2" posts, .060" wall thickness, 3-13/16" picket spacing, 1" x .055" top rail with 1-1/2" x .082" sides. Pre-routed holes in posts. Gates include hardware. Pool code compliant model has 48" space between horizontal rails.

	Craft@Hrs	Unit	Material	Labor	Total
48" high panel	BL@.170	Ea	77.20	4.20	81.40
54" pool code compliant panel	BL@.170	Ea	81.00	4.20	85.20
60" high panel	BL@.170	Ea	86.00	4.20	90.20
72" high panel	BL@.170	Ea	104.00	4.20	108.20
Fence posts					
48" high	BL@.830	Ea	21.50	20.50	42.00
60" high	BL@.830	Ea	24.00	20.50	44.50
72" high	BL@.830	Ea	30.40	20.50	50.90
Single swing gates, 5' wide					
48" high	BL@.175	Ea	242.00	4.32	246.32
60" high	BL@.175	Ea	265.00	4.32	269.32
72" high	BL@.175	Ea	301.00	4.32	305.32
Gate posts					
48" high	BL@.830	Ea	34.20	20.50	54.70
60" high	BL@.830	Ea	41.70	20.50	62.20
72" high	BL@.830	Ea	50.60	20.50	71.10
Add for setting post in concrete	BL@.330	Ea	2.86	8.15	11.01

Fencing, Vinyl

Costs shown include concrete and labor to set posts.
Picket fence, 1-1/2" pickets with 2" spacing, 2 rail, 2" x 4" rails, 4" x 4" posts at 8' OC, pre-assembled in 8' sections. Prices include post mounts, post top, screws and one post.

	Craft@Hrs	Unit	Material	Labor	Total
3' high	B1@.100	LF	6.70	2.74	9.44
4' high	B1@.110	LF	12.80	3.01	15.81
Gate with 4" posts	B1@1.00	Ea	66.00	27.40	93.40
Add for additional fence post with top	B1@.500	Ea	18.00	13.70	31.70

Picket fence, 3-1/2" pickets with 3" spacing, 2 rail, 2" x 4" rails, 4" x 4" posts at 6' OC, pre-assembled in 6' sections. Prices include post mounts, post top, screws and one post.

	Craft@Hrs	Unit	Material	Labor	Total
4' high	B1@.110	LF	6.00	3.01	9.01
Gate with 4" posts	B1@1.00	Ea	66.00	27.40	93.40
Add for additional fence post with top	B1@.500	Ea	18.00	13.70	31.70

Shadowbox fence, 5-1/2" offset pickets, 3 rail, with galvanized steel support, posts at 71" OC. Prices include post mounts, post top, screws and one post.

	Craft@Hrs	Unit	Material	Labor	Total
4' high	B1@.130	LF	13.00	3.56	16.56
5' high	B1@.140	LF	18.50	3.83	22.33
6' high	B1@.140	LF	16.50	3.83	20.33
Gate with 4" posts	B1@1.20	Ea	66.60	32.80	99.40
Add for additional fence post with top	B1@.500	Ea	18.00	13.70	31.70

Privacy fence, 6" boards, 3 rail, posts at 6' OC. Prices include post mounts, post top, screws and one post.

	Craft@Hrs	Unit	Material	Labor	Total
6' high	B1@.140	LF	13.00	3.83	16.83
6' high gate	B1@1.20	Ea	98.00	32.80	130.80
Add for additional fence post with top	B1@.500	Ea	25.50	13.70	39.20

	Craft@Hrs	Unit	Material	Labor	Total
Post & rail fence, 2" x 6" x 96" notched rail, posts 8' OC. Includes post mounts, post top, and one post.					
40" high, 2 rail, 4" routed post	B1@.110	LF	5.95	3.01	8.96
57" high, 3 rail, 4" routed post	B1@.120	LF	7.35	3.28	10.63
57" high, 3 rail, 5" routed post	B1@.120	LF	8.30	3.28	11.58
54" high, 4 rail, 5" routed post	B1@.120	LF	9.90	3.28	13.18
Gate with 4" posts	B1@1.00	Ea	97.00	27.40	124.40
Gate with 5" posts	B1@1.00	Ea	118.00	27.40	145.40
Add for additional 4" fence post with top	B1@.500	Ea	15.00	13.70	28.70
Add for additional 5" fence post with top	B1@.500	Ea	22.90	13.70	36.60
Silt fence, black 6 mil poly with stakes 8' OC, staked and buried 9" deep					
2' high	BL@.030	LF	.23	.74	.97

Fiberglass Panels Nailed or screwed onto wood frame.

Corrugated, 8', 10', 12' panels, 2-1/2" corrugations, standard colors. Costs include 10% for waste

	Craft@Hrs	Unit	Material	Labor	Total
Polycarbonate, 26" wide panels	BC@.012	SF	1.03	.36	1.39
Fiberglass, 26" wide panels	BC@.012	SF	.63	.36	.99
PVC, 26" wide panels	BC@.012	SF	.61	.36	.97
Flat panels, clear, green or white					
.06" flat sheets, 4' x 8', 10', 12'	BC@.012	SF	.92	.36	1.28
.03" rolls, 24", 36", 48" x 50'	BC@.012	SF	1.03	.36	1.39
.037" rolls, 24", 36", 48" x 50'	BC@.012	SF	1.19	.36	1.55
Accessories					
Nails, ring shank with rubber washer, 1-3/4" (labor included with panels)					
100 per box (covers 130 SF)	—	SF	.05	—	.05
Self-tapping screws, 1-1/4", rust-proof (labor included with panels)					
100 per box (covers 130 SF)	—	SF	.20	—	.20
Horizontal closure strips, corrugated (labor included with panels)					
Redwood, 2-1/2" x 1-1/2"	—	LF	.30	—	.30
Poly-foam, 1" x 1"	—	LF	.50	—	.50
Rubber, 1" x 1"	—	LF	.42	—	.42
Vertical crown moulding					
Redwood, 1-1/2" x 1" or poly-foam, 1" x 1"	—	LF	.30	—	.30
Rubber, 1" x 1"	—	LF	.79	—	.79

Fire Sprinkler Systems Typical subcontract prices for systems installed in a single family residence not over two stories high in accord with NFPA Section 13D. Includes connection to domestic water line inside garage (1" pipe size minimum), exposed copper riser with shut-off and drain valve, CPVC concealed distribution piping and residential type sprinkler heads below finished ceilings. These costs include hydraulic design calculations and inspection of completed system for compliance with design requirements. Costs shown are per SF of protected area. When estimating the square footage of protected area, include non-inhabited areas such as bathrooms, closets and attached garages. For scheduling purposes, estimate that a crew of 2 men can install the rough-in piping for 1,500 to 1,600 SF of protected area in an 8-hour day and about the same amount of finish work in another 8-hour day.

	Craft@Hrs	Unit	Material	Labor	Total
Fabricate, install, and test system					
Single residence	—	SF	—	—	2.67
Tract work	—	SF	—	—	2.28
Condominiums or apartments	—	SF	—	—	3.59

Masonry Fireplaces

	Craft@Hrs	Unit	Material	Labor	Total

Masonry Fireplaces

Fire box lined with refractory firebrick, backed with common brick and enclosed in a common brick wall. Includes fireplace throat and smoke chamber. Overall height of the fireplace and throat is 5'. Chimney (flue) is made from common brick and a flue liner. Including lintels, damper and 12" thick reinforced concrete foundation. For more on masonry fireplaces, see National Concrete & Masonry Estimator, http://CraftsmanSiteLicense.Com.

	Craft@Hrs	Unit	Material	Labor	Total
30" wide x 29" high x 16" deep fire box	B9@39.7	LS	451.00	1,110.00	1,561.00
36" wide x 29" high x 16" deep fire box	B9@43.5	LS	476.00	1,220.00	1,696.00
42" wide x 32" high x 16" deep fire box	B9@48.3	LS	528.00	1,350.00	1,878.00
48" wide x 32"high x 18" deep fire box	B9@52.8	LS	578.00	1,480.00	2,058.00
12" x 12" chimney to 12' high	B9@9.50	LS	270.00	266.00	536.00
Add for higher chimney, per LF	B9@1.30	LF	36.80	36.30	73.10
Add for 8" x 10" ash cleanout door	B9@.500	SF	16.00	14.00	30.00
Add for combustion air inlet kit	B9@.750	SF	25.00	21.00	46.00
Add for common brick veneer interior face	B9@.451	SF	3.85	12.60	16.45
Add for rubble or cobble stone veneer face	B9@.480	SF	10.50	13.40	23.90
Add for limestone veneer face or hearth	B9@.451	SF	20.00	12.60	32.60
Add for 1-1/4" marble face or hearth	B9@.451	SF	30.00	12.60	42.60
Add for modern design 11" x 77" pine mantel	B1@3.81	Ea	600.00	104.00	704.00
Add for gas valve, log lighter and key	PM@.750	Ea	68.00	26.40	94.40

Fireplace Components See also Masonry.

	Craft@Hrs	Unit	Material	Labor	Total
Angle iron (lintel support), lengths 25" to 96"					
3" x 3" x 3/16"	B9@.167	LF	3.95	4.67	8.62
Ash drops, cast iron top, galvanized container					
16-3/4" x 12-1/2" x 8"	B9@.750	Ea	88.60	21.00	109.60
Ash dumps, stamped steel					
4-1/2" x 9"	B9@.500	Ea	10.20	14.00	24.20
Chimney anchors, per pair					
48" x 1-1/2" x 3/16"	—	Pr	13.80	—	13.80
72" x 1-1/2" x 3/16"	—	Pr	18.20	—	18.20
Cleanout doors, 14 gauge stamped steel					
8" x 8"	B9@.500	Ea	17.70	14.00	31.70
8" x 10"	B9@.500	Ea	18.60	14.00	32.60
Combustion air inlet kit for masonry fireplaces	—	Ea	28.80	—	28.80
Dampers for single opening firebox, heavy steel with rockwool insulation					
14" high, 25" to 30" wide, 10" deep, 25 lbs	B9@1.33	Ea	59.10	37.20	96.30
14" high, 36" wide, 10" deep, 31 lbs	B9@1.33	Ea	64.20	37.20	101.40
14" high, 42" wide, 10" deep, 38 lbs	B9@1.33	Ea	78.50	37.20	115.70
14" high, 48" wide, 10" deep, 43 lbs	B9@1.75	Ea	102.00	48.90	150.90
16" high, 54" wide, 13" deep, 66 lbs	B9@2.50	Ea	199.00	69.90	268.90
16" high, 60" wide, 13" deep, 74 lbs	B9@2.50	Ea	209.00	69.90	278.90
16" high, 72" wide, 13" deep, 87 lbs	B9@2.50	Ea	251.00	69.90	320.90
Dampers for multiple opening or corner opening firebox, high capacity, steel downdraft shelf support					
35" wide	B9@1.33	Ea	243.00	37.20	280.20
41" wide	B9@1.33	Ea	277.00	37.20	314.20
47" wide	B9@1.75	Ea	355.00	48.90	403.90

	Craft@Hrs	Unit	Material	Labor	Total

Fuel grates

21" x 16"	—	Ea	57.90	—	57.90
29" x 16"	—	Ea	74.30	—	74.30
40" x 30"	—	Ea	114.00	—	114.00
Gas valve, log lighter & key	—	LS	78.60	—	78.60

Fireplace Forms Heavy gauge steel heat-circulating forms for masonry fireplaces. Based on Superior Fireplace Products. Finished opening dimensions as shown.

Single front opening

31" x 25", 4,300 CF capacity	B9@3.50	Ea	697.00	97.90	794.90
34" x 25", 5,000 CF capacity	B9@3.50	Ea	741.00	97.90	838.90
37" x 29", 5,750 CF capacity	B9@3.50	Ea	830.00	97.90	927.90
42" x 29", 6,500 CF capacity	B9@3.50	Ea	1,100.00	97.90	1,197.90
48" x 31", 7,500 CF capacity	B9@4.00	Ea	1,380.00	112.00	1,492.00
60" x 31", 8,500 CF capacity	B9@4.00	Ea	2,110.00	112.00	2,222.00
72" x 31", 9,500 CF capacity	B9@5.50	Ea	2,880.00	154.00	3,034.00

Front and end opening (either right or left)

38" wide, 27" high	B9@3.50	Ea	1,570.00	97.90	1,667.90

Open on both faces (see-through), including grate, 23" high

33" wide	B9@3.50	Ea	1,570.00	97.90	1,667.90
43" wide	B9@4.00	Ea	1,790.00	112.00	1,902.00
Add for 2 air inlet grilles, 8" x 11"	—	LS	70.60	—	70.60
Add for 1 air outlet grille, 6" x 32"	—	LS	61.10	—	61.10
Add for 2 fan-forced air inlets					
With grilles and motor	—	LS	175.00	—	175.00

Fireplaces, Prefabricated Zero clearance factory-built fireplaces, including metal fireplace body, refractory interior and fuel grate. UL listed. Add for flues, doors and blower below. Listed by screen size. Based on Superior Fireplace Products.

Radiant heating, open front only, with fuel grate

38" wide	B9@5.00	Ea	549.00	140.00	689.00
43" wide	B9@5.00	Ea	667.00	140.00	807.00

Radiant heating, open front and one end

38" wide	B9@5.00	Ea	831.00	140.00	971.00

Radiant heating, three sides open, with brass and glass doors and refractory floor

36" wide (long side)	B9@6.00	Ea	1,330.00	168.00	1,498.00

Radiant heating, air circulating, open front only, recessed screen panels

38" screen size	B9@5.50	Ea	748.00	154.00	902.00
43" screen size	B9@5.50	Ea	865.00	154.00	1,019.00

Radiant heating traditional masonry look

45" wide	B9@5.00	Ea	1,250.00	140.00	1,390.00

Convection-type heat circulating, open front only

38" screen size	B9@5.50	Ea	624.00	154.00	778.00
43" screen size	B9@5.50	Ea	767.00	154.00	921.00

Forced air heat circulating, open front only (add for blower below)

35" screen size	B9@5.50	Ea	549.00	154.00	703.00
45" screen size	B9@5.50	Ea	831.00	154.00	985.00

Flooring, Sheet Vinyl

	Craft@Hrs	Unit	Material	Labor	Total
Radiant, see-through with brass and glass doors, designer model					
38" screen size	B9@6.50	Ea	1,330.00	182.00	1,512.00
Header plate,					
Connects venting system to gas fireplace	B9@.167	Ea	76.30	4.67	80.97
Fireplace doors, complete set including frame and hardware, polished brass					
38", for radiant heat units	B9@1.25	Ea	210.00	35.00	245.00
45", for radiant heat units	B9@1.25	Ea	299.00	35.00	334.00
38", for radiant heat units, tall	B9@1.25	Ea	250.00	35.00	285.00
43", for radiant heat units, tall	B9@1.25	Ea	266.00	35.00	301.00
38", for forced air units	B9@1.25	Ea	210.00	35.00	245.00
43", for forced air units	B9@1.25	Ea	227.00	35.00	262.00
For corner open model	B9@1.25	Ea	329.00	35.00	364.00

Blower for forced air fireplaces, requires 110 volt electrical receptacle outlet (electrical wiring not included)

	Craft@Hrs	Unit	Material	Labor	Total
Blower unit for connection to existing outlet	BE@.500	Ea	119.00	16.90	135.90
Fireplace flue (chimney), double wall, typical cost for straight vertical installation					
8" inside diameter	B9@.167	LF	23.00	4.67	27.67
10" inside diameter	B9@.167	LF	32.80	4.67	37.47
Add for flue spacers, for passing through ceiling					
1" or 2" clearance	B9@.167	Ea	13.30	4.67	17.97
Add for typical flue offset	B9@.250	Ea	84.00	6.99	90.99
Add for flashing and storm collar, one required per roof penetration					
Flat to 6/12 pitch roofs	B9@3.25	Ea	53.40	90.90	144.30
6/12 to 12/12 pitch roofs	B9@3.25	Ea	77.60	90.90	168.50
Add for spark arrestor top	B9@.500	Ea	109.00	14.00	123.00

Sheet Vinyl Flooring

Armstrong Solarian sheet vinyl flooring, no wax, Armstrong. Bonded with adhesive at perimeter and at seams. Ten year limited warranty. Performance Appearance Rating 4.1 Includes 10% waste and adhesive.

	Craft@Hrs	Unit	Material	Labor	Total
Jaspe Collection	BF@.250	SY	38.00	7.57	45.57
Etchings Solarian	BF@.250	SY	33.00	7.57	40.57
Stencil Craft Solarian	BF@.250	SY	28.40	7.57	35.97
Bellissimo Solarian	BF@.250	SY	34.10	7.57	41.67
Rhythms Solarian	BF@.250	SY	32.00	7.57	39.57
Starstep Solarian	BF@.250	SY	28.40	7.57	35.97
Traditions Solarian	BF@.250	SY	28.40	7.57	35.97

Armstrong sheet vinyl flooring. By Performance Appearance Rating (PAR). Bonded with adhesive at perimeter and at seams. Including 10% waste and adhesive.

	Craft@Hrs	Unit	Material	Labor	Total
Medley, urethane wear layer, 15 year	BF@.250	SY	17.30	7.57	24.87
Caspian, PAR 3.8, urethane wear layer, 10 year	BF@.250	SY	14.10	7.57	21.67
Themes, PAR 2.9, urethane wear layer, 5 year	BF@.250	SY	11.70	7.57	19.27
Sundial, urethane wear layer, .098"	BF@.250	SY	12.90	7.57	20.47
Metro, PAR 2.3, vinyl wear layer, 3 year	BF@.250	SY	6.90	7.57	14.47
Cambray FHA, PAR 1.7, vinyl wear layer, 2 year	BF@.250	SY	6.92	7.57	14.49
Royelle, PAR 1.3, vinyl wear layer, 1 year	BF@.250	SY	4.64	7.57	12.21
Memories, Fundamentals	BF@.250	SY	20.50	7.57	28.07
Successor, Fundamentals, 5 year	BF@.250	SY	15.40	7.57	22.97
Enhancements	BF@.250	SY	15.70	7.57	23.27
Initiator	BF@.250	SY	10.70	7.57	18.27
Apartment One	BF@.250	SY	7.70	7.57	15.27

	Craft@Hrs	Unit	Material	Labor	Total

Tarkett sheet vinyl, bonded with adhesive at perimeter and at seams. Including 10% waste and adhesive.

	Craft@Hrs	Unit	Material	Labor	Total
Simply Brite	BF@.250	SY	27.00	7.57	34.57
Signature	BF@.250	SY	23.50	7.57	31.07
Naturelle	BF@.250	SY	22.80	7.57	30.37
Presence Felt	BF@.250	SY	18.10	7.57	25.67
Style Brite Flex	BF@.250	SY	15.50	7.57	23.07
Style Brite Felt	BF@.250	SY	15.30	7.57	22.87
Simply Brite Felt	BF@.250	SY	14.30	7.57	21.87
Sierrra	BF@.250	SY	8.14	7.57	15.71
Yosemite	BF@.250	SY	7.30	7.57	14.87
Preference	BF@.250	SY	6.40	7.57	13.97
Sandran	BF@.250	SY	4.30	7.57	11.87

Sheet vinyl, no wax, Mannington. Bonded with adhesive at perimeter and at seams. Including 10% waste and adhesive.

	Craft@Hrs	Unit	Material	Labor	Total
Gold Series	BF@.250	SY	35.20	7.57	42.77
Quicksilver	BF@.250	SY	28.80	7.57	36.37
Villa Series	BF@.250	SY	28.80	7.57	36.37
Ceramica	BF@.250	SY	24.20	7.57	31.77
Silverado	BF@.250	SY	21.80	7.57	29.37
Aurora	BF@.250	SY	19.50	7.57	27.07
Lumina	BF@.250	SY	16.10	7.57	23.67
Vega II, Omni	BF@.250	SY	12.60	7.57	20.17

Vinyl 12" x 12" Tile Flooring

Armstrong vinyl tile, 12" x 12", including 10% waste and adhesive

	Craft@Hrs	Unit	Material	Labor	Total
Solarian Stonegate, .094"	BF@.023	SF	2.70	.70	3.40
Natural Images, .094"	BF@.023	SF	2.28	.70	2.98
Chelsea (urethane), .08"	BF@.023	SF	2.07	.70	2.77
Chesapeake (urethane), .08"	BF@.023	SF	1.72	.70	2.42
Harbour, .08"	BF@.023	SF	1.71	.70	2.41
Stylistik, .065", self-adhesive	BF@.023	SF	1.14	.70	1.84
Themes .08" (urethane)	BF@.023	SF	1.20	.70	1.90
Vernay, Metro, .045"	BF@.023	SF	.62	.70	1.32
Excelon Imperial, 1/8",	BF@.021	SF	.72	.64	1.36
Excelon Civic Square, 1/8",	BF@.021	SF	.61	.64	1.25
Place 'n Press	BF@.021	SF	.33	.64	.97
Afton, .075"	BF@.023	SF	1.14	.70	1.84
Builder urethane, .080", self-adhesive	BF@.023	SF	1.62	.70	2.32
Builder urethane, .080", dry back	BF@.023	SF	1.52	.70	2.22

Elegant Images™ Series vinyl floor tile, Armstrong. 12" x 12". ToughGuard® durability. Resists stains, scratches, scuffs and indentations. 25-year warranty. Per square foot, including adhesive and 10% waste.

	Craft@Hrs	Unit	Material	Labor	Total
Cornwall, Burnt Almond	BF@.023	SF	2.25	.70	2.95
Modern, Rose/Green Inset	BF@.023	SF	1.70	.70	2.40

Rubber studded tile. 12" x 12". Highly resilient to reduce fatigue. 10-year wear warranty. Slip-resistant. Sound absorbent. High abrasion resistance. Recommended in kitchens and active use rooms.

	Craft@Hrs	Unit	Material	Labor	Total
Black	BF@.025	SF	6.50	.76	7.26
Indigo	BF@.025	SF	6.50	.76	7.26
Taupe	BF@.025	SF	6.50	.76	7.26

Vinyl Flooring Accessories

Resilient flooring adhesive

	Craft@Hrs	Unit	Material	Labor	Total
Sheet vinyl adhesive, per gallon	—	Gal	13.40	—	13.40
Sheet vinyl adhesive, 350 SF per gallon	—	SF	.04	—	.04
Vinyl tile adhesive, per gallon	—	Gal	12.20	—	12.20
Vinyl tile adhesive, 350 per gallon	—	Gal	.04	—	.04
Vinyl seam sealer kit (100 LF)	—	LF	.12	—	.20

Flooring, Wood Strip

	Craft@Hrs	Unit	Material	Labor	Total
Resilient flooring accessories					
Patching and leveling compound, (levels uneven subfloor)					
1 pound per SF at .25" depth	BF@.016	SF	.65	.48	1.13
Latex primer (used over dusting, porous or below grade concrete)					
(300 SF per gallon)	BF@.008	SF	.03	.24	.27
Liquid underlay (for application over existing vinyl floor),					
Gallon covers 75 SF	BF@.010	SF	.35	.30	.65
Lining felt, 15 Lb. asphalt felt (used over board subfloor),					
(432 SF roll covers 350 SF)	BF@.004	SF	.05	.12	.17
Tempered hardboard underlayment (levels uneven subfloor),					
1/8" x 4' x 8' sheet	BF@.016	SF	.23	.48	.71
Cove base, colors,					
2-1/2" high, .08" thick	BF@.025	LF	.41	.76	1.17
4" high, .08" thick	BF@.025	LF	.49	.76	1.25
4" high, self-adhesive	BF@.016	LF	.82	.48	1.30
Cove base cement					
200 LF of 4" base per gallon	—	LF	.07	—	.06
Accent strip, 1/8" x 24" long					
1/2" wide	BF@.025	LF	.32	.76	1.08
1" wide	BF@.025	LF	.41	.76	1.17
2" wide	BF@.025	LF	.76	.76	1.52
Vinyl edging strip, 1" wide					
1/16" thick	—	LF	.52	—	.52
1/8" thick	—	LF	.68	—	.68

Wood Strip Flooring Unfinished oak strip flooring, tongue and groove edge. Includes typical cutting waste. Installed over a prepared subfloor. These are material costs per square foot based on a 1,000 SF job. See installation, sanding and finishing cost below.

		Size			
		3/4" x 2-1/4"	3/4" x 3-1/4"	1/2" x 2"	1/2" x 1-1/2"
Select white or red	SF	2.95	3.34	2.86	2.73
Number 1 common white	SF	2.39	2.27	2.31	2.23
Number 1 common red	SF	2.55	2.19	2.46	3.38
Number 2 common red or white	SF	2.30	2.20	2.20	2.20
First grade maple	SF	3.29	—	—	—

	Craft@Hrs	Unit	Material	Labor	Total
Labor installing, sanding and finishing hardwood strip flooring, nailed, including felt					
Install 1-1/2" width	BF@.066	SF	—	2.00	2.00
Install 2" or 2-1/4" width	BF@.062	SF	—	1.88	1.88
Install 3-1/4" width	BF@.060	SF	—	1.82	1.82
Sanding, 3 passes (60/80/100 grit)	BF@.023	SF	.06	.70	.76
2 coats of stain/sealer	BF@.010	SF	.11	.30	.41
2 coats of urethane	BF@.012	SF	.13	.36	.49
2 coats of lacquer	BF@.024	SF	.18	.73	.91

	Craft@Hrs	Unit	Material	Labor	Total

Unfinished antique tongue and groove plank flooring, 25/32" thick, random widths and lengths. Includes typical cutting waste. Installed over a prepared subfloor. Based on Aged Woods

	Craft@Hrs	Unit	Material	Labor	Total
Ash, select, 3" to 7" wide	BF@.057	SF	4.28	1.73	6.01
Chestnut, American, 3" to 7" wide	BF@.057	SF	8.89	1.73	10.62
Chestnut, distressed, 3" to 7" wide	BF@.057	SF	10.40	1.73	12.13
Chestnut, distressed, over 7" wide	BF@.057	SF	13.80	1.73	15.53
Cherry, #1, 3" to 7" wide	BF@.057	SF	10.00	1.73	11.73
Cherry, #1, 6" wide and up	BF@.057	SF	12.10	1.73	13.83
Hickory, #1, 3" to 7" wide	BF@.057	SF	4.16	1.73	5.89
Hemlock, 3" to 7" wide	BF@.057	SF	7.15	1.73	8.88
Hemlock, over 7" wide	BF@.057	SF	9.25	1.73	10.98
Oak, #1, red or white, 3" to 7" wide	BF@.057	SF	3.81	1.73	5.54
Oak, #1, red or white, over 7" wide	BF@.057	SF	6.01	1.73	7.74
Oak, 3" to 7" wide	BF@.057	SF	7.62	1.73	9.35
Oak, over 7" wide	BF@.057	SF	9.24	1.73	10.97
Oak, distressed, 3" to 7" wide	BF@.057	SF	8.67	1.73	10.40
Oak, distressed, over 7" wide	BF@.057	SF	11.00	1.73	12.73
Maple, FAS, 3" to 7" wide	BF@.057	SF	5.26	1.73	6.99
Poplar, 3" to 8" wide	BF@.057	SF	3.70	1.73	5.43
Poplar, over 8" wide	BF@.057	SF	4.98	1.73	6.71
Walnut, #3, 3" to 7" wide	BF@.057	SF	6.01	1.73	7.74
White pine, 3" to 7"	BF@.057	SF	5.61	1.73	7.34
White pine, distressed, 3" to 7"	BF@.057	SF	6.12	1.73	7.85
White pine, distressed, over 7" wide	BF@.057	SF	8.76	1.73	10.49
White pine, knotty grade, 12" and up wide	BF@.057	SF	8.99	1.73	10.72
Yellow pine, 3" to 5" wide	BF@.057	SF	6.54	1.73	8.27
Yellow pine, 6" to 10" wide	BF@.057	SF	8.87	1.73	10.60

Unfinished northern hard maple strip flooring

	Craft@Hrs	Unit	Material	Labor	Total
25/32" x 2-1/4" (138.3 BF per 100 SF)					
1st grade	BF@.036	SF	5.45	1.09	6.54
2nd grade	BF@.036	SF	4.88	1.09	5.97
3rd grade	BF@.036	SF	4.49	1.09	5.58
25/32" x 1-1/2" (155 BF per 100 SF)					
1st grade	BF@.040	SF	6.51	1.21	7.72
2nd grade	BF@.040	SF	5.84	1.21	7.05
3rd grade	BF@.040	SF	5.39	1.21	6.60
33/32" x 1-1/2" (170 BF per 100 SF)					
1st grade	BF@.036	SF	6.06	1.09	7.15
2nd grade	BF@.036	SF	5.21	1.09	6.30
3rd grade	BF@.036	SF	4.93	1.09	6.02

Unfinished long plank flooring, random 3" to 5" widths, 1' to 16' lengths, 3/4" thick, Specialty Wood Floors

	Craft@Hrs	Unit	Material	Labor	Total
Select grade, ash	BF@.062	SF	5.56	1.88	7.44
Common grade, cherry	BF@.062	SF	7.33	1.88	9.21
Select grade, cherry	BF@.062	SF	9.16	1.88	11.04
Select grade, rock maple	BF@.062	SF	6.88	1.88	8.76
Select grade, figured maple	BF@.062	SF	11.90	1.88	13.78
Select grade, red oak	BF@.062	SF	7.72	1.88	9.60
Select grade, white oak	BF@.062	SF	6.36	1.88	8.24

Flooring, Wood Plank and Strip

	Craft@Hrs	Unit	Material	Labor	Total

Unfinished wide plank flooring, random 6" to 8" widths, 1' to 16' lengths, 3/4" thick, including 8% waste, Specialty Wood Floors

	Craft@Hrs	Unit	Material	Labor	Total
Knotty grade, pine	BF@.074	SF	4.32	2.24	6.56
Select grade, cherry	BF@.074	SF	10.24	2.24	12.48
Select grade, maple	BF@.074	SF	9.24	2.24	11.48
Select grade, red oak	BF@.074	SF	8.68	2.24	10.92
Select grade, white oak	BF@.074	SF	7.80	2.24	10.04

Softwood flooring Material costs include typical waste and coverage loss. Add for sanding and finishing from below.

Unfinished Douglas fir, "C" and better, dry, vertical grain flooring, including 5% waste

	Craft@Hrs	Unit	Material	Labor	Total
1" x 2"	BF@.037	SF	4.01	1.12	5.13
1" x 3"	BF@.037	SF	4.01	1.12	5.13
Deduct for "D" grade	—	SF	-.82	—	-.82
Add for "B" grade and better	—	SF	.82	—	.82
Add for specific lengths	—	SF	.16	—	.16
Add for diagonal patterns	BF@.004	SF	—	.12	.12

Unfinished select kiln dried heart pine flooring, 97% dense heartwood, random lengths, including 5% waste. 1" to 1-1/4" thick

	Craft@Hrs	Unit	Material	Labor	Total
4" wide	BF@.042	SF	7.14	1.27	8.41
5" wide	BF@.042	SF	7.37	1.27	8.64
6" wide	BF@.042	SF	7.44	1.27	8.71
8" wide	BF@.046	SF	8.00	1.39	9.39
10" wide	BF@.046	SF	9.28	1.39	10.67
Add for diagonal patterns	BF@.004	SF	—	.12	.12

Unfinished plank flooring, 3/4" thick, 3" to 8" widths, 2' to 12' lengths, Aged Woods

	Craft@Hrs	Unit	Material	Labor	Total
Antique white pine	BF@.100	SF	7.87	3.03	10.90
Antique yellow pine	BF@.100	SF	7.26	3.03	10.29
Milled heart pine	BF@.100	SF	8.60	3.03	11.63
Distressed oak	BF@.100	SF	9.31	3.03	12.34
Milled oak	BF@.100	SF	8.45	3.03	11.48
Distressed American chestnut	BF@.100	SF	8.65	3.03	11.68
Milled American chestnut	BF@.100	SF	9.31	3.03	12.34
Aged hemlock/fir	BF@.100	SF	8.62	3.03	11.65
Antique poplar	BF@.100	SF	3.97	3.03	7.00
Aged cypress	BF@.100	SF	4.71	3.03	7.74
Add for 9" to 15" widths	—	%	15.0	—	—

Add for sanding and finishing.

	Craft@Hrs	Unit	Material	Labor	Total
Sanding, 3 passes (60/80/100 grit)	BF@.023	SF	.06	.70	.76
2 coats of stain/sealer	BF@.010	SF	.11	.30	.41
2 coats of urethane	BF@.012	SF	.13	.36	.49
2 coats of lacquer	BF@.024	SF	.18	.73	.91

Prefinished strip flooring

Prefinished oak strip flooring, tongue and groove, 25/32" thick, 2-1/4" wide. Includes typical cutting waste. Installed over a prepared subfloor

2-1/4" wide, Standard & Better. Includes typical cutting waste

	Craft@Hrs	Unit	Material	Labor	Total
Marsh	BF@.062	SF	3.95	1.88	5.83
Toffee	BF@.062	SF	4.75	1.88	6.63
Butterscotch	BF@.062	SF	4.80	1.88	6.68
Natural	BF@.062	SF	4.85	1.88	6.73
Gunstock	BF@.062	SF	5.05	1.88	6.93

Random widths 3", 5" and 7", walnut plugs,

	Craft@Hrs	Unit	Material	Labor	Total
Clear and Select grade	BF@.042	SF	6.94	1.27	8.21

	Craft@Hrs	Unit	Material	Labor	Total
Random plank 2-1/4" to 3-1/4" wide					
Dark number 1 to C and Better	BF@.032	SF	4.80	.97	5.77
Prefinished oak strip flooring trim					
7' 10" quarter round	BF@.320	Ea	15.00	9.69	24.69
3' 11" reducer	BF@.320	Ea	20.00	9.69	29.69
3' 11" stair nose	BF@.320	Ea	20.00	9.69	29.69
3' 11" T-moulding	BF@.320	Ea	20.00	9.69	29.69
Prefinished rosewood strip flooring, tongue and groove, 25/32" thick. Includes typical cutting waste					
2-1/4" wide	BF@.062	SF	4.00	1.88	5.88
2-1/4" wide, 7' long	BF@.060	SF	5.30	1.82	7.12
3-1/2" wide, 7' long	BF@.060	SF	5.05	1.82	6.87
4-1/4" wide, select	BF@.055	SF	7.90	1.67	9.57
5-1/2" wide, premium	BF@.045	SF	8.05	1.36	9.41
5-1/2" wide, select	BF@.045	SF	7.90	1.36	9.26
5-1/2" wide, select, 6' long	BF@.045	SF	7.55	1.36	8.91
Prefinished rosewood strip flooring trim					
Grooved reducer	BF@.060	LF	3.50	1.82	5.32
Quarter round	BF@.060	LF	2.75	1.82	4.57
Stair nosing	BF@.060	LF	7.75	1.82	9.57
T moulding	BF@.060	LF	6.50	1.82	8.32

Floating hardwood flooring. Tongue and groove. Install over concrete or other subfloor using an underfloor membrane. Lay tongue and groove flooring and glue edges and ends only. Allow 3/8" at walls and cover with 1/2" base or quarter round molding. Labor shown below includes laying membrane, hardwood flooring, and gluing.

	Craft@Hrs	Unit	Material	Labor	Total
Comfort Guard membrane	—	SF	.20	—	.20
Everseal adhesive	—	SF	.08	—	.08
1/2" tongue and groove hardwood flooring,					
various colors	FL@.051	SF	3.40	2.30	5.70

Laminate wood floor (Pergo), plastic laminate on pressed wood base, 7-7/8" x 47-1/4" x 3/8" (20 cm x 120 cm x 8 mm), loose lay with glued tongue and groove joints. Many wood finishes available. 20.67 SF (1.9 square meters) per pack. Includes 10% waste and joint glue (bottle covers 150 SF)

	Craft@Hrs	Unit	Material	Labor	Total
Pergo Presto, on floors					
Oak blocked	BF@.051	SF	3.43	1.54	4.97
Pergo Presto, on floors					
Red oak blocked	BF@.051	SF	3.43	1.54	4.97
Pergo Presto, on floors					
Beech blocked	BF@.051	SF	3.43	1.54	4.97
Preferred and select, on floors	BF@.051	SF	5.25	1.54	6.79
Glued on stairs (treads and risers)	BF@.115	SF	4.94	3.48	8.42
Add for Silent Step underlay					
(110 SF per pack)	BF@.010	SF	.71	.30	1.01
Pergo sound block foam	BF@.002	SF	.18	.06	.24
Add for 8 mil poly vapor barrier (over concrete)	BF@.002	LF	.08	.06	.14
T mould strip (joins same level surface)	BF@.040	LF	5.22	1.21	6.43
Reducer strip (joins higher surface)	BF@.040	LF	4.33	1.21	5.54
End molding (joins lower surface)	BF@.040	LF	3.19	1.21	4.40
Wall base, 4" high	BF@.040	LF	2.27	1.21	3.48
Stair nosing	BF@.050	LF	4.98	1.51	6.49

	Craft@Hrs	Unit	Material	Labor	Total

Bamboo flooring

Plank bamboo flooring, tongue and groove, 72" long panels. Laminated plank bamboo flooring, tongue and groove, 72" long panels. Treated and kiln dried. Installs over wood subfloor. Use 1/2" nail gun. Space fasteners 8" to 10" apart, with a minimum of two nails per board. http://www.plyboo.com

	Craft@Hrs	Unit	Material	Labor	Total
5/8" x 3-3/4" prefinished Amber Horizontal	BF@.062	SF	4.99	1.88	6.87
1/2" x 3" unfinished Amber Horizontal	BF@.062	SF	4.99	1.88	6.87
5/8" x 3-3/4" prefinished Amber Vertical	BF@.062	SF	4.99	1.88	6.87
1/2" x 3" unfinished Amber Vertical	BF@.062	SF	4.99	1.88	6.87
5/8" x 3-3/4" prefinished Natural Horizontal	BF@.062	SF	4.99	1.88	6.87
1/2" x 3" unfinished Natural Horizontal	BF@.062	SF	4.99	1.88	6.87
5/8" x 3-3/4" prefinished Natural Vertical	BF@.062	SF	4.99	1.88	6.87
1/2" x 3" unfinished Natural Vertical	BF@.062	SF	4.99	1.88	6.87
Laminated bamboo baseboard					
3" beveled edge prefinished Amber	BC@.040	LF	4.55	1.20	5.75
3" beveled edge prefinished Natural	BC@.040	LF	4.55	1.20	5.75
Laminated prefinished bamboo stair nose					
1/2" x 3-1/4" Amber Horizontal	BC@.040	LF	6.50	1.20	7.70
1/2" x 3-1/4" Natural Horizontal	BC@.040	LF	6.50	1.20	7.70
1/2" x 3-1/4" Amber Vertical	BC@.040	LF	6.50	1.20	7.70
1/2" x 3-1/4" Natural Vertical	BC@.040	LF	6.50	1.20	7.70
5/8" x 3-1/2" Amber Horizontal	BC@.040	LF	6.50	1.20	7.70
5/8" x 3-1/2" Natural Horizontal	BC@.040	LF	6.50	1.20	7.70
5/8" x 3-1/2" Amber Horizontal	BC@.040	LF	6.50	1.20	7.70
5/8" x 3-1/2" Natural Horizontal	BC@.040	LF	6.50	1.20	7.70
Laminated prefinished bamboo threshold					
9/16" x 3-5/8" Amber Horizontal	BC@.174	LF	4.89	5.22	10.11
9/16" x 3-5/8" Natural Horizontal	BC@.174	LF	4.89	5.22	10.11
5/8" x 2" Amber Horizontal Overlap	BC@.174	LF	6.50	5.22	11.72
5/8" x 2" Natural Horizontal Overlap	BC@.174	LF	6.50	5.22	11.72
5/8" x 2" Amber Horizontal Overlap	BC@.174	LF	6.50	5.22	11.72
5/8" x 2" Natural Horizontal Overlap	BC@.174	LF	6.50	5.22	11.72
Add for sanding and finishing					
Sanding, 3 passes (60/80/100 grit)	BF@.023	SF	.06	.70	.76
2 coats of stain/sealer	BF@.010	SF	.11	.30	.41
2 coats of urethane	BF@.012	SF	.13	.36	.49
2 coats of lacquer	BF@.024	SF	.18	.73	.91

Parquet block flooring Includes typical waste and coverage loss, installed over a prepared subfloor.

Unfinished parquet block flooring, 3/4" thick, 10" x 10" to 39" x 39" blocks, includes 5% waste. Cost varies by pattern with larger blocks costing more. These are material costs only. Add labor cost below.
Kentucky Wood Floors

	Craft@Hrs	Unit	Material	Labor	Total
Ash or quartered oak, select	—	SF	16.50	—	16.50
Ash or quartered oak, natural	—	SF	12.00	—	12.00
Cherry or walnut select grade	—	SF	18.90	—	18.90
Cherry or walnut natural grade	—	SF	14.30	—	14.30
Oak, select plain	—	SF	13.60	—	13.60
Oak, natural plain	—	SF	11.80	—	11.80

	Craft@Hrs	Unit	Material	Labor	Total

Prefinished parquet block flooring, 5/16" thick, select grade, specified sizes, includes 5% waste. Cost varies by pattern and size, with larger blocks costing more. These are material costs only. Add labor cost below. Kentucky Wood Floors.

	Craft@Hrs	Unit	Material	Labor	Total
Cherry or walnut					
9" x 9"	—	SF	2.56	—	2.56
13" x 13"	—	SF	3.05	—	3.05
Oak					
12" x 12" chestnut	—	SF	1.77	—	1.77
12" x 12" desert	—	SF	1.77	—	1.77
Teak					
12" x 12"	—	SF	2.25	—	2.25
Prefinished natural hardwood, commodity grade					
12" x 12"	—	SF	2.40	—	2.40
Labor and mastic for installing parquet block flooring					
Parquet laid in mastic	BF@.055	SF	.25	1.67	1.92
Edge-glued over high density foam	BF@.052	SF	.42	1.57	1.99
Custom or irregular-shaped tiles, with adhesive	BF@.074	SF	.43	2.24	2.67
Add for installations under 1,000 SF	BF@.025	SF	.04	.76	.80
Add for sanding and finishing for unfinished parquet block flooring					
Sanding, 3 passes (60/80/100 grit)	BF@.023	SF	.06	.70	.76
2 coats of stain/sealer	BF@.010	SF	.11	.30	.41
2 coats of urethane	BF@.012	SF	.13	.36	.49
2 coats of lacquer	BF@.024	SF	.18	.73	.91

Prefinished cross-band 3-ply birch parquet flooring, 6" to 12" wide by 12" to 48" blocks, Oregon Lumber. Costs include cutting, fitting and 5% waste, but no subfloor preparation.

	Craft@Hrs	Unit	Material	Labor	Total
"Saima", 9/16" thick, typical price	BF@.050	SF	4.10	1.51	5.61
Bonded blocks, 12" x 12", 1/8" thick, laid in mastic					
Vinyl bonded cork	BF@.050	SF	3.38	1.51	4.89
Vinyl bonded domestic woods	BF@.062	SF	6.76	1.88	8.64
Vinyl bonded exotic woods	BF@.062	SF	13.50	1.88	15.38
Acrylic (Permagrain) impregnated wood flooring. 12" x 12" x 5/16" thick block laid in mastic					
Oak block	BF@.057	SF	9.31	1.73	11.04
Ash block	BF@.057	SF	9.92	1.73	11.65
Cherry block	BF@.057	SF	10.30	1.73	12.03
Maple block	BF@.057	SF	10.40	1.73	12.13
Acrylic (Permagrain) impregnated 2-7/8" x random length plank laid in mastic					
Oak plank	BF@.057	SF	7.69	1.73	9.42
Maple plank	BF@.057	SF	10.10	1.73	11.83
Tupelo plank	BF@.057	SF	9.44	1.73	11.17
Lindenwood plank	BF@.057	SF	10.50	1.73	12.23
Cherry plank	BF@.057	SF	10.10	1.73	11.83

Wood flooring adhesive For laminated wood plank and wood parquet flooring. Covers 40 to 50 square feet per gallon when applied with a 1/8" x 1/8" x 1/8" square-notch trowel for parquet or a 1/4" x 1/4" x 1/4" V-notch trowel for laminated plank.

	Craft@Hrs	Unit	Material	Labor	Total
Gallon	—	Ea	17.60	—	17.60
4 gallons	—	Ea	55.20	—	55.20

Foundation Bolts

	Craft@Hrs	Unit	Material	Labor	Total
Foundation Bolts Galvanized, with nut and washer attached. Costs per bolt or bag.					
1/2" x 6", Pack of 1	B1@.107	Ea	0.83	2.93	3.76
1/2" x 6", Pack of 50	B1@.107	Ea	34.90	2.93	37.83
1/2" x 8", Pack of 1	B1@.107	Ea	.91	2.93	3.84
1/2" x 8", Pack of 50	B1@.107	Ea	36.40	2.93	39.33
1/2" x 10", Pack of 1	B1@.107	Ea	1.03	2.93	3.96
1/2" x 10", Pack of 50	B1@.107	Ea	38.30	2.93	41.23
1/2" x 12", Pack of 1	B1@.107	Ea	1.20	2.93	4.13
1/2" x 12", Pack of 50	B1@.107	Ea	43.00	2.93	45.93
1/2" x 18", Pack of 1	B1@.107	Ea	1.64	2.93	4.57
1/2" x 18", Pack of 50	B1@.107	Ea	61.00	2.93	63.93
5/8" x 8", Pack of 1	B1@.107	Ea	2.20	2.93	5.13
5/8" x 8", Pack of 25	B1@.107	Ea	38.40	2.93	41.33
5/8" x 10", Pack of 1	B1@.107	Ea	2.20	2.93	5.13
5/8" x 10", Pack of 25	B1@.107	Ea	38.40	2.93	41.33

Framing Connectors All bolted connectors include nuts, bolts, and washers as needed
Equivalent Simpson Strong-Tie Co. model numbers are shown in parentheses.

	Craft@Hrs	Unit	Material	Labor	Total
Angle clips (A34 and A35), 18 gauge, galvanized, no nails included					
3" x 5" connector (A35 or A35F)	BC@.030	Ea	.25	.90	1.15
3" x 4" connector (A34)	BC@.030	Ea	.19	.90	1.09
Angle clips (L) for general utility use, 16 gauge, galvanized, no nails included					
3" (L30)	BC@.027	Ea	.40	.81	1.21
5" (L50)	BC@.027	Ea	.66	.81	1.47
7" (L70)	BC@.027	Ea	1.66	.81	2.47
9" (L90)	BC@.027	Ea	2.12	.81	2.93

	Craft@Hrs	Unit	Material	Labor	Total
Brick wall ties (BTR), 22 gauge, per 1,000					
7/8" x 6", corrugated	—	M	110.00	—	110.00
Bridging. See also Carpentry.					
Cross bridging (NC), steel, per pair, no nails required					
12" joist spacing on center					
2" x 8" to 2" x 16" joists (NC2X8-16)	BC@.022	Pr	1.09	.66	1.75
16" joist spacing on center					
2" x 8" to 2" x 12" joists (NC2X8-12)	BC@.022	Pr	1.09	.66	1.75
2" x 14" to 2" x 16" joists (NC2X14-16)	BC@.022	Pr	1.19	.66	1.85
24" joist spacing on center					
2" x 10" to 2" x 12" joists (NC2X10-12)	BC@.022	Pr	1.09	.66	1.75
2" x 12" to 2" x 14" joists (NC2X12-14)	BC@.022	Pr	1.11	.66	1.77
2" x 16" joists (NC2X16-16)	BC@.022	Pr	1.23	.66	1.89

Column bases, heavy duty (CB). Includes nuts and bolts. Labor shown is for placing column base before pouring concrete

	Craft@Hrs	Unit	Material	Labor	Total
4" x 4" (CB44)	BC@.121	Ea	9.05	3.63	12.68
4" x 6" (CB46)	BC@.121	Ea	9.53	3.63	13.16
4" x 8" (CB48)	BC@.121	Ea	10.10	3.63	13.73
6" x 6" (CB66)	BC@.121	Ea	11.80	3.63	15.43
6" x 8" (CB68)	BC@.121	Ea	13.10	3.63	16.73
6" x 10" (CB610)	BC@.131	Ea	14.00	3.93	17.93
6" x 12" (CB612)	BC@.131	Ea	15.20	3.93	19.13
8" x 8" (CB88)	BC@.131	Ea	25.70	3.93	29.63
8" x 10" (CB810)	BC@.131	Ea	30.70	3.93	34.63
8" x 12" (CB812)	BC@.150	Ea	33.40	4.50	37.90

	Craft@Hrs	Unit	Material	Labor	Total
10" x 10" (CB1010)	BC@.150	Ea	34.10	4.50	38.60
10" x 12" (CB1012)	BC@.150	Ea	39.10	4.50	43.60
12" x 12" (CB1212)	BC@.150	Ea	42.50	4.50	47.00

Column caps or end column caps, heavy duty (CC) or (ECC). Includes nuts and bolts

Column caps for solid lumber

Beam size x post size					
4" x 4" (CC44)	BC@.260	Ea	20.60	7.80	28.40
4" x 6" (CC46)	BC@.300	Ea	29.00	9.00	38.00
6" x 4" (CC64)	BC@.300	Ea	43.60	9.00	52.60
6" x 6" (CC66)	BC@.400	Ea	43.60	12.00	55.60
6" x 8" (CC68)	BC@.600	Ea	43.60	18.00	61.60
8" x 6" (CC86)	BC@.600	Ea	48.10	18.00	66.10
8" x 8" (CC88)	BC@.750	Ea	48.10	22.50	70.60
10" x 6" (CC10X)	BC@.750	Ea	71.40	22.50	93.90

Column caps for glu-lam beams

Glu-lam beam size x post size					
3-1/4" x 4" (CC3-14-4)	BC@.260	Ea	37.30	7.80	45.10
3-1/4" x 6" (CC3-1/4-6)	BC@.300	Ea	37.30	9.00	46.30
5-1/4" x 6" (CC5-1/4-6)	BC@.400	Ea	40.00	12.00	52.00
5-1/4" x 8" (CC5-1/4-8)	BC@.600	Ea	40.00	18.00	58.00
7" x 6" (CC76)	BC@.600	Ea	45.50	18.00	63.50
7" x 7" (CC77)	BC@.750	Ea	45.50	22.50	68.00
7" x 8" (CC78)	BC@.750	Ea	45.50	22.50	68.00
9" x 6" (CC96)	BC@.750	Ea	51.20	22.50	73.70
9" x 8" (CC98)	BC@.750	Ea	51.20	22.50	73.70

Floor tie anchors (FTA)

2" x 38" (FTA2)	BC@.200	Ea	20.70	6.00	26.70
3-1/2" x 45" (FTA5)	BC@.200	Ea	19.30	6.00	25.30
3-1/2" x 56" (FTA7)	BC@.320	Ea	33.30	9.60	42.90

Header hangers (HH), 16 gauge, galvanized

4" header (HH4)	BC@.046	Ea	3.06	1.38	4.44
6" header (HH6)	BC@.046	Ea	6.05	1.38	7.43

Hold downs (HD), including typical nuts, bolts and washers for studs. Add for foundation bolts

HD2A	BC@.310	Ea	4.27	9.30	13.57
HD5A	BC@.310	Ea	10.30	9.30	19.60
HD8A	BC@.420	Ea	14.30	12.60	26.90
HD10A	BC@.500	Ea	15.40	15.00	30.40
HD20A	BC@.500	Ea	30.00	15.00	45.00

Threaded rod (ATR) 24" long. Labor includes drilling through 6" of wood

1/2" rod	BC@.185	Ea	3.02	5.55	8.57
5/8" rod	BC@.185	Ea	4.26	5.55	9.81
3/4" rod	BC@.200	Ea	6.58	6.00	12.58
1" rod	BC@.220	Ea	11.40	6.60	18.00

Hurricane and seismic ties (H)

H1 tie	BC@.035	Ea	.46	1.05	1.51
H2 tie	BC@.027	Ea	.61	.81	1.42
H2.5 tie	BC@.027	Ea	.20	.81	1.01
H3 tie	BC@.027	Ea	.20	.81	1.01
H4 tie	BC@.027	Ea	.16	.81	.97
H5 tie	BC@.027	Ea	.19	.81	1.00
H6 tie	BC@.045	Ea	2.40	1.35	3.75
H7 tie	BC@.055	Ea	2.76	1.65	4.41

Framing Connectors

	Craft@Hrs	Unit	Material	Labor	Total
Mud sill anchors (MA)					
2" x 4" (MA4)	BC@.046	Ea	.93	1.38	2.31
2" x 6" (MA6)	BC@.046	Ea	1.07	1.38	2.45
Panelized plywood roof hangers (F24N, F26N), for panelized plywood roof construction					
2" x 4" (F24N)	—	Ea	.32	—	.32
2" x 6" (F26N)	—	Ea	.60	—	.60
Purlin hangers (HU), without nails, 7 gauge					
4" x 4" (HU44)	BC@.052	Ea	2.59	1.56	4.15
4" x 6" (HU46)	BC@.052	Ea	2.90	1.56	4.46
4" x 8" (HU48)	BC@.052	Ea	3.52	1.56	5.08
4" x 10" (HU410)	BC@.070	Ea	4.35	2.10	6.45
4" x 12" (HU412)	BC@.070	Ea	4.99	2.10	7.09
4" x 14" (HU414)	BC@.087	Ea	6.36	2.61	8.97
4" x 16" (HU416)	BC@.087	Ea	7.35	2.61	9.96
6" x 6" (HU66)	BC@.052	Ea	5.10	1.56	6.66
6" x 8" (HU68)	BC@.070	Ea	5.19	2.10	7.29
6" x 10" (HU610)	BC@.087	Ea	7.35	2.61	9.96
6" x 12" (HU612)	BC@.087	Ea	9.05	2.61	11.66
6" x 14" (HU614)	BC@.117	Ea	10.90	3.51	14.41
6" x 16" (HU616)	BC@.117	Ea	14.40	3.51	17.91
8" x 10" (HU810)	BC@.087	Ea	8.15	2.61	10.76
8" x 12" (HU812)	BC@.117	Ea	10.30	3.51	13.81
8" x 14" (HU814)	BC@.117	Ea	12.40	3.51	15.91
8" x 16" (HU816)	BC@.117	Ea	15.00	3.51	18.51
Top flange hangers (LB)					
2" x 6" (LB26)	BC@.046	Ea	1.56	1.38	2.94
2" x 8" (LB28)	BC@.046	Ea	2.07	1.38	3.45
2" x 10" (LB210)	BC@.046	Ea	2.58	1.38	3.96
2" x 12" (LB212)	BC@.046	Ea	2.76	1.38	4.14
2" x 14" (LB214)	BC@.052	Ea	1.97	1.56	3.53
Top flange hangers, heavy (HW)					
4" x 6" (HW46)	BC@.052	Ea	15.30	1.56	16.86
4" x 8" (HW48)	BC@.052	Ea	16.50	1.56	18.06
4" x 10" (HW410)	BC@.070	Ea	17.60	2.10	19.70
4" x 12" (HW412)	BC@.070	Ea	18.70	2.10	20.80
4" x 14" (HW414)	BC@.087	Ea	19.90	2.61	22.51
4" x 16" (HW416)	BC@.087	Ea	21.00	2.61	23.61
6" x 6" (HW66)	BC@.052	Ea	15.50	1.56	17.06
6" x 8" (HW68)	BC@.070	Ea	16.70	2.10	18.80
6" x 10" (HW610)	BC@.087	Ea	17.80	2.61	20.41
6" x 12" (HW612)	BC@.087	Ea	17.60	2.61	20.21
6" x 14" (HW614)	BC@.117	Ea	20.10	3.51	23.61
6" x 16" (HW616)	BC@.117	Ea	21.20	3.51	24.71
8" x 6" (HW86)	BC@.070	Ea	16.90	2.10	19.00
8" x 8" (HW88)	BC@.070	Ea	18.00	2.10	20.10
8" x 10" (HW810)	BC@.087	Ea	19.10	2.61	21.71
8" x 12" (HW812)	BC@.117	Ea	20.30	3.51	23.81
8" x 14" (HW814)	BC@.117	Ea	21.40	3.51	24.91
8" x 16" (HW816)	BC@.117	Ea	22.50	3.51	26.01

	Craft@Hrs	Unit	Material	Labor	Total
U type or formed seat type hangers (U). Standard duty, 16 gauge, galvanized					
2" x 4" (U24)	BC@.040	Ea	.63	1.20	1.83
2" x 6" (U26)	BC@.046	Ea	.72	1.38	2.10
2" x 10" (U210)	BC@.046	Ea	1.06	1.38	2.44
2" x 14" (U214)	BC@.052	Ea	1.76	1.56	3.32
3" x 4" (U34)	BC@.046	Ea	1.10	1.38	2.48
3" x 6" (U36)	BC@.046	Ea	1.23	1.38	2.61
3" x 10" (U310)	BC@.052	Ea	2.04	1.56	3.60
3" x 14" (U314) or 3" x 16" (U316)	BC@.070	Ea	2.77	2.10	4.87
4" x 4" (U44)	BC@.046	Ea	1.14	1.38	2.52
4" x 6" (U46)	BC@.052	Ea	1.27	1.56	2.83
4" x 10" (U410)	BC@.070	Ea	2.07	2.10	4.17
4" x 14" (U414)	BC@.087	Ea	3.06	2.61	5.67
6" x 6" (U66)	BC@.052	Ea	3.66	1.56	5.22
6" x 10" (U610)	BC@.087	Ea	4.66	2.61	7.27
Plywood clips (PSCL, extruded aluminum, purchased in cartons of 500)					
3/8", 7/16", 1/2", or 5/8"	BC@.006	Ea	.09	.18	.27
Post base for decks (base with standoff, PBS), 12 gauge, galvanized					
4" x 4" (PBS44)	BC@.046	Ea	4.65	1.38	6.03
4" x 6" (PBS46)	BC@.046	Ea	5.81	1.38	7.19
6" x 6" (PBS66)	BC@.070	Ea	8.38	2.10	10.48
Add for rough lumber sizes	—		35%	—	—
Post base (AB), adjustable, 16 gauge, galvanized					
4" x 4" (AB44B)	BC@.046	Ea	4.19	1.38	5.57
4" x 6" (AB46)	BC@.046	Ea	9.15	1.38	10.53
6" x 6" (AB66)	BC@.070	Ea	11.10	2.10	13.20
Add for rough lumber sizes	—		35%	—	—
Post bases (PB), 12 gauge, galvanized					
4" x 4" (PB44)	BC@.046	Ea	3.66	1.38	5.04
4" x 6" (PB46)	BC@.046	Ea	5.40	1.38	6.78
6" x 6" (PB66)	BC@.070	Ea	7.72	2.10	9.82
Post cap and base combination (BC), 18 gauge, galvanized					
4" x 4" (BC4)	BC@.037	Ea	2.08	1.11	3.19
4" x 6" (BC46)	BC@.037	Ea	5.09	1.11	6.20
6" x 6" (BC6)	BC@.054	Ea	5.22	1.62	6.84
8" x 8" (BC8)	BC@.054	Ea	18.90	1.62	20.52
Post caps and end post caps (PC), 12 gauge, galvanized. Beam size x post size					
4" x 4" (PC44)	BC@.052	Ea	12.00	1.56	13.56
4" x 6" (PC46)	BC@.052	Ea	15.90	1.56	17.46
4" x 8" (PC48)	BC@.052	Ea	23.50	1.56	25.06
6" x 4" (PC64)	BC@.052	Ea	15.60	1.56	17.16
6" x 6" (PC66)	BC@.077	Ea	24.60	2.31	26.91
6" x 8" (PC68)	BC@.077	Ea	16.90	2.31	19.21
8" x 8" (PC88)	BC@.077	Ea	34.10	2.31	36.41
Post top tie or anchor (AC), 18 gauge, galvanized					
4" x 4" (AC4)	BC@.040	Ea	1.71	1.20	2.91
6" x 6" (AC6)	BC@.040	Ea	3.03	1.20	4.23
Purlin anchors (PA), 12 gauge, 2" wide					
18" anchor (PA18)	BC@.165	Ea	3.41	4.95	8.36
23" anchor (PA23)	BC@.165	Ea	3.98	4.95	8.93
28" anchor (PA28)	BC@.185	Ea	4.06	5.55	9.61
35" anchor (PA35)	BC@.185	Ea	5.58	5.55	11.13

Garage Door Openers

	Craft@Hrs	Unit	Material	Labor	Total
Split ring connectors					
2-5/8" ring	BC@.047	Ea	1.06	1.41	2.47
4" ring	BC@.054	Ea	1.66	1.62	3.28
"L" straps, 12 gauge (L)					
5" x 5", 1" wide (55L)	BC@.132	Ea	.72	3.96	4.68
6" x 6", 1-1/2" wide (66L)	BC@.132	Ea	1.47	3.96	5.43
8" x 8", 2" wide (88L)	BC@.132	Ea	2.25	3.96	6.21
12" x 12", 2" wide (1212L)	BC@.165	Ea	2.98	4.95	7.93
Add for heavy duty straps, 7 gauge	—	Ea	4.98	—	4.98
Long straps (MST), 12 gauge					
27" (MST27)	BC@.165	Ea	2.92	4.95	7.87
37" (MST37)	BC@.165	Ea	4.17	4.95	9.12
48" (MST48)	BC@.265	Ea	5.78	7.95	13.73
60" (MST60)	BC@.265	Ea	8.89	7.95	16.84
72" (MST72)	BC@.285	Ea	11.80	8.55	20.35
Plate and stud straps (FHA), galvanized, 1-1/2" wide, 12 gauge					
6" (FHA6)	BC@.100	Ea	.96	3.00	3.96
9" (FHA9)	BC@.132	Ea	.96	3.96	4.92
12" (FHA12)	BC@.132	Ea	1.08	3.96	5.04
18" (FHA18)	BC@.132	Ea	1.44	3.96	5.40
24" (FHA24)	BC@.165	Ea	1.87	4.95	6.82
30" (FHA30)	BC@.165	Ea	3.11	4.95	8.06
Tie straps (ST)					
2-1/16" x 9-5/16" 20 ga. (ST292)	BC@.132	Ea	.61	3.96	4.57
3/4" x 16-5/16" 20 ga. (ST2115)	BC@.046	Ea	.57	1.38	1.95
2-1/16" x 16-5/16" 20 ga. (ST2215)	BC@.132	Ea	.75	3.96	4.71
2-1/16" x 23-5/16" 16 ga. (ST6224)	BC@.165	Ea	1.65	4.95	6.60
2-1/16" x 33-13/16" 14 ga. (ST6236)	BC@.165	Ea	1.99	4.95	6.94
"T" straps, 14 gauge (T)					
6" x 6", 1-1/2" wide (66T)	BC@.132	Ea	1.47	3.96	5.43
12" x 8", 2" wide (128T)	BC@.132	Ea	3.34	3.96	7.30
12" x 12", 2" wide (1212T)	BC@.132	Ea	4.99	3.96	8.95
Add for heavy-duty straps, 7 gauge	—	Ea	4.60	—	4.60
Twist straps (TS), 16 gauge, galvanized, 1-1/4" wide					
9" (TS9)	BC@.132	Ea	.53	3.96	4.49
11-5/8" (TS12)	BC@.132	Ea	.72	3.96	4.68
17-3/4" (TS18)	BC@.132	Ea	1.05	3.96	5.01
21-5/8" (TS22)	BC@.165	Ea	1.23	4.95	6.18
Wall braces (WB), diagonal, 1-1/4" wide , 16 gauge					
10' long, corrugated or flat (WB106C or WB106)	BC@.101	Ea	4.69	3.03	7.72
12' long, flat (WB126)	BC@.101	Ea	4.99	3.03	8.02
14' long, flat (WB143C)	BC@.101	Ea	9.43	3.03	12.46
Wall braces (CWB), diagonal, 18 gauge, 1" wide					
10' long, L shaped (CWB106)	BC@.125	Ea	4.47	3.75	8.22
12' long, L shaped (CWB126)	BC@.125	Ea	4.98	3.75	8.73

Garage Door Openers Radio controlled, electric, single or double door, includes typical electric hookup to existing adjacent 110 volt outlet.

	Craft@Hrs	Unit	Material	Labor	Total
Screw drive, 1/2 HP	BC@3.90	Ea	163.00	117.00	280.00
Security, belt drive, 1/2 HP	BC@3.90	Ea	284.00	117.00	401.00
Security, chain drive, 1/2 HP	BC@3.90	Ea	227.00	117.00	344.00

	Craft@Hrs	Unit	Material	Labor	Total
Genie Excelerator Series, Super	BC@4.04	Ea	167.00	121.00	288.00
Genie Excelerator Series, Lifetime	BC@4.04	Ea	239.00	121.00	360.00
Add for additional transmitter	—	Ea	25.20	—	24.00
Add for operator reinforcement kit	—	Ea	15.80	—	15.00
Add for low headroom kit	—	Ea	61.60	—	58.65
Add for emergency release	—	Ea	18.20	—	17.30
Belt rail extension kit	—	Ea	42.60	—	40.60
Chain glide extension kit	—	Ea	17.60	—	16.70
Chain rail extension kit	—	Ea	43.60	—	41.50
Screw drive extension kit	—	Ea	18.50	—	17.60

Garage Doors

Steel rollup garage door. Torsion springs. 2-layer insulated construction. 7/8" insulation. Steel frame with overlapped section joints. With hardware and lock.

	Craft@Hrs	Unit	Material	Labor	Total
8' x 7'	B1@6.46	Ea	240.00	177.00	417.00
9' x 7'	B1@6.46	Ea	282.00	177.00	459.00
16' x 7'	B1@8.33	Ea	485.00	228.00	713.00

Raised panel steel rollup garage door. 2-layer construction. 7/8" thick polystyrene insulation. With hardware and lock.

	Craft@Hrs	Unit	Material	Labor	Total
8' x 7'	B1@6.46	Ea	302.00	177.00	479.00
9' x 7'	B1@6.46	Ea	323.00	177.00	500.00
16' x 7'	B1@8.33	Ea	550.00	228.00	778.00

Bonded steel garage door. 3-layer steel exterior. 2"-thick insulation. With torsion springs, hardware and lock.

	Craft@Hrs	Unit	Material	Labor	Total
8' x 7'	B1@6.46	Ea	386.00	177.00	563.00
8' x 7', molding face	B1@6.46	Ea	289.00	177.00	466.00
9' x 7'	B1@6.46	Ea	407.00	177.00	584.00
9' x 7', molding face	B1@6.46	Ea	331.00	177.00	508.00
16' x 7'	B1@6.46	Ea	685.00	177.00	862.00
16' x 7', molding face	B1@6.46	Ea	722.00	177.00	899.00
8' x 6'6", with glass window	B1@6.46	Ea	617.00	177.00	794.00
9' x 6'6", with glass window	B1@6.46	Ea	650.00	177.00	827.00
8' x 6'6", low headroom	B1@6.46	Ea	491.00	177.00	668.00
9' x 6'6", low headroom	B1@6.46	Ea	523.00	177.00	700.00

Non-insulated steel garage door. With rollup hardware and lock.

	Craft@Hrs	Unit	Material	Labor	Total
8' x 7'	B1@6.46	Ea	235.00	177.00	412.00
9' x 7'	B1@6.46	Ea	234.00	177.00	411.00
16' x 7'	B1@8.33	Ea	428.00	228.00	656.00

Wood garage doors. Unfinished, jamb hinge, with hardware and slide lock.

Plain panel un-insulated hardboard door

	Craft@Hrs	Unit	Material	Labor	Total
9' wide x 7' high	B1@4.42	Ea	288.00	121.00	409.00
16' wide x 7' high	B1@5.90	Ea	500.00	161.00	661.00

Styrofoam core hardboard door

	Craft@Hrs	Unit	Material	Labor	Total
9' wide x 7' high	B1@4.42	Ea	311.00	121.00	432.00
16' wide x 7' high	B1@5.90	Ea	547.00	161.00	708.00

Redwood door, raised panels

	Craft@Hrs	Unit	Material	Labor	Total
9' wide x 7' high	B1@4.42	Ea	574.00	121.00	695.00
16' wide x 7' high	B1@5.90	Ea	1,265.00	161.00	1,426.00

Hardboard door with two small lites

	Craft@Hrs	Unit	Material	Labor	Total
9' wide x 7' high	B1@4.42	Ea	304.00	121.00	425.00
16' wide x 7' high	B1@5.90	Ea	574.00	161.00	735.00

Gates

	Craft@Hrs	Unit	Material	Labor	Total
Hardboard door with sunburst lites					
9' wide x 7' high	B1@4.42	Ea	461.00	121.00	582.00
16' wide x 7' high	B1@5.90	Ea	891.00	161.00	1,052.00
Hardboard door with cathedral-style lites					
9' wide x 7' high	B1@4.42	Ea	407.00	121.00	528.00
16' wide x 7' high	B1@5.90	Ea	857.00	161.00	1,018.00
Garage door weather seal.					
For 9' garage door	BC@.500	Ea	4.34	15.00	19.34
For garage door to 18' wide	BC@1.00	Ea	7.08	30.00	37.08
Garage door lock bar cylinder and lock assembly.					
Garage door lock	BC@.500	Ea	6.25	15.00	21.25
T-handle for rollup garage door bar lock.					
Handle	BC@.250	Ea	4.35	7.50	11.85

Gates, Ornamental Aluminum, Swing, Automatic

Welded aluminum driveway swing gates, double swing, arched up or down. Labor costs assume connection, but no wire runs. 2" square frame, 2" x 1-1/2" horizontal rails and 1" square pickets. Hoover Fence (J series, 4' to 5' high)

	Craft@Hrs	Unit	Material	Labor	Total
12' wide opening	BL@1.00	Ea	2,900.00	24.70	2,924.70
16' wide opening	BL@1.20	Ea	3,710.00	29.70	3,739.70
20' wide opening	BL@1.40	Ea	4,510.00	34.60	4,544.60
Add for custom width gates	—	%	20.0	—	—
Gate posts					
4" x 9' post	BL@1.00	Ea	115.00	24.70	139.70
6" x 12' post	BL@1.00	Ea	251.00	24.70	275.70
Add for setting post in concrete	BL@.330	Ea	3.00	8.15	11.15
12 volt DC swing gate operator	BL@5.00	Ea	490.00	124.00	614.00
Add for second gate slave unit	BL@2.00	Ea	373.00	49.40	422.40
Add for additional gate transmitters	—	Ea	30.00	2.57	5.32
Add for low voltage wire	—	LF	.35	—	.35

Glass Material only. See labor costs below.

	Craft@Hrs	Unit	Material	Labor	Total
SSB glass (single strength, "B" grade)					
To 60" width plus length	—	SF	2.22	—	2.22
Over 60" to 70" width plus length	—	SF	2.42	—	2.42
Over 70" to 80" width plus length	—	SF	2.72	—	2.72
DSB glass (double strength, "B" grade)					
To 60" width plus length	—	SF	3.08	—	3.08
Over 60" to 70" width plus length	—	SF	3.13	—	3.13
Over 70" to 80" width plus length	—	SF	3.18	—	3.18
Over 80" to 100" width plus length	—	SF	3.40	—	3.40
Non-glare glass	—	SF	5.25		5.25
Float glass, 1/4" thick					
To 25 SF	—	SF	3.45	—	3.45
25 to 35 SF	—	SF	3.71	—	3.71
35 to 50 SF	—	SF	4.03	—	4.03
Insulating glass, clear					
1/2" thick, 1/4" airspace, up to 25 SF	—	SF	7.43	—	7.43
5/8" thick, 1/4" airspace, up to 35 SF	—	SF	8.70	—	8.70
1" thick, 1/2" airspace, up to 50 SF	—	SF	10.30	—	10.30

	Craft@Hrs	Unit	Material	Labor	Total
Tempered glass, clear, 1/4" thick					
To 25 SF	—	SF	7.43	—	7.43
25 to 50 SF	—	SF	8.91	—	8.91
50 to 75 SF	—	SF	9.86	—	9.86
Tempered glass, clear, 3/16" thick					
To 25 SF	—	SF	7.32	—	7.32
25 to 50 SF	—	SF	8.38	—	8.38
50 to 75 SF	—	SF	9.28	—	9.28
Laminated safety glass, 1/4" thick, ASI safety rating					
Up to 25 SF	—	SF	9.55	—	9.55
25 to 35 SF	—	SF	9.98	—	9.98
Polycarbonate sheet .093" thick.					
8" x 10"	—	Ea	2.01	—	2.01
11" x 14"	—	Ea	4.45	—	4.45
12" x 24"	—	Ea	8.04	—	8.04
18" x 24"	—	Ea	11.20	—	11.20
32" x 44"	—	Ea	30.60	—	30.60
36" x 72"	—	Ea	68.20	—	68.20
Acrylic plastic sheet (Plexiglass), clear.					
8" x 10" .093" thick	—	Ea	1.00	—	1.00
11" x 14" .093" thick	—	Ea	1.52	—	1.52
12" x 24" .093" thick	—	Ea	2.89	—	2.89
18" x 24" .093" thick	—	Ea	4.36	—	4.36
30" x 36" .093" thick	—	Ea	10.10	—	10.10
36" x 48" .093" thick	—	Ea	16.90	—	16.90
18" x 24" .236" thick	—	Ea	9.82	—	9.82
30" x 36" .236" thick	—	Ea	22.30	—	22.30
36" x 72" .236" thick	—	Ea	45.90	—	45.90
Labor setting glass in wood or metal sash with vinyl bed					
To 15 SF, 1 hour charge	BG@1.77	LS	—	51.10	51.10
15 to 25 SF, 2 hour charge	BG@3.53	LS	—	102.00	102.00
25 to 35 SF, 3 hour charge	BG@5.30	LS	—	153.00	153.00
Beveled edge mirrors, no frame included	—	SF	7.75	—	7.75
Labor to bevel edges, per linear inch					
1/2" beveled edge	BG@.013	Inch	—	.38	.38
1" beveled edge	BG@.063	Inch	—	1.82	1.82

Greenhouses, Sun Porches, Spa Enclosures Freestanding or lean-to models. Anodized aluminum frame, acrylic glazed. Add the cost of excavation and concrete.

	Craft@Hrs	Unit	Material	Labor	Total
Freestanding greenhouse enclosures, one door and four walls. 8' deep					
7' high, 6' wide	B1@21.0	Ea	4,600.00	575.00	5,175.00
7-1/2' high, 9' wide	B1@33.0	Ea	5,740.00	903.00	6,643.00
8' high, 12' wide	B1@47.0	Ea	6,660.00	1,290.00	7,950.00
8' high, 15' wide	B1@59.0	Ea	7,780.00	1,610.00	9,390.00
9' high, 18' wide	B1@80.0	Ea	8,800.00	2,190.00	10,990.00
Add for additional outswing door	—	Ea	276.00	—	276.00
Add for additional sliding window	—	Ea	146.00	—	146.00
Lean-to greenhouse enclosures attached on long dimension to an existing structure, one door and three walls. 8' deep					
7' high, 3' wide	B1@8.40	Ea	2,430.00	230.00	2,660.00
7' high, 6' wide	B1@16.8	Ea	3,670.00	460.00	4,130.00
7-1/2' high, 9' wide	B1@27.0	Ea	4,590.00	739.00	5,329.00
8' high, 12' wide	B1@38.4	Ea	5,560.00	1,050.00	6,610.00
8' high, 15' wide	B1@48.0	Ea	6,530.00	1,310.00	7,840.00
Add for additional outswing door	—	Ea	276.00	—	276.00
Add for additional sliding window	—	Ea	146.00	—	146.00

Gutters and Downspouts

	Craft@Hrs	Unit	Material	Labor	Total
Gutters and Downspouts					
Galvanized steel, 4" box type					
Gutter, 5", 10' lengths	SW@.070	LF	.55	2.38	2.93
Inside or outside corner	SW@.091	Ea	4.40	3.09	7.49
Fascia support brackets	—	Ea	1.25	—	1.25
Spike and ferrule support, 7"	—	Ea	.51	—	.51
Add per 2" x 3" galvanized downspout, with drop outlet, 3 ells and straps					
10' downspout	SW@.540	Ea	5.76	18.30	24.06
Aluminum rain gutter, 5" box type, primed					
Gutter, 5", 10' lengths	SW@.070	LF	.54	2.38	2.92
Inside or outside corner	SW@.091	Ea	4.40	3.09	7.49
Fascia support brackets	—	Ea	1.25	—	1.25
Spike and ferrule support, 7"	—	Ea	.51	—	.51
Section joint kit	—	Ea	3.53	—	3.53
Add per 2" x 3" aluminum downspout, with drop outlet, strainer, 3 ells and straps					
10' downspout	SW@.540	Ea	8.82	18.30	27.12
20' downspout, with connector	SW@.810	Ea	17.60	27.50	45.10
Vinyl rain gutter, white PVC, (Plastmo Vinyl), half round gutter, (10' lengths)					
Gutter, 4"	SW@.054	LF	.79	1.83	2.62
Add for fittings (connectors, hangers, etc.)	—	LF	1.03	—	1.03
Add for inside or outside corner	SW@.091	Ea	4.63	3.09	7.72
Add for bonding kit (covers 150 LF)	—	Ea	3.76	—	3.76
Add for 5" gutter	—	%	15.0	—	—
Downspouts, 3" round	SW@.022	LF	.61	.75	1.36
Add for 3 ells and 2 clamps, per downspout	—	Ea	20.80	—	20.80

Gypsum Drywall and Accessories Labor costs are for good quality hanging and taping with smooth sanded joints and include 6% waste but no texturing. Drywall costs in different regions vary greatly. See texturing costs below. Installed costs include premixed joint compound (5 gallons per 1,000 SF), 2" perforated joint tape (380 LF per 1,000 SF) and drywall screws (4-1/2 pounds per 1,000 SF) at the following material prices.

	Craft@Hrs	Unit	Material	Labor	Total
1/4" plain board, per 4' x 8' sheet	—	Ea	8.30	—	8.30
3/8" plain board, per 4' x 8' sheet	—	Ea	7.71	—	7.71
1/2" plain board, per 4' x 8' sheet	—	Ea	7.76	—	7.76
1/2" plain board, per 4' x 9' sheet	—	Ea	8.94	—	8.94
1/2" plain board, per 4' x 10' sheet	—	Ea	9.56	—	9.56
1/2" plain board, per 4' x 12' sheet	—	Ea	11.50	—	11.50
1/2" plain board, per 1/2" x 54" x 12' sheet	—	Ea	14.90	—	14.90
1/2" water-resistant greenboard, per 4' x 8' sheet	—	Ea	10.70	—	10.70
5/8" fire-rated board, per 4' x 6' sheet	—	Ea	8.56	—	8.56
5/8" fire-rated board, per 4' x 8' sheet	—	Ea	9.67	—	9.67
5/8" fire-rated board, per 4' x 9' sheet	—	Ea	10.40	—	10.40
5/8" fire-rated board, per 4' x 10' sheet	—	Ea	12.60	—	12.60
5/8" fire-rated board, per 4' x 12' sheet	—	Ea	14.20	—	14.20
5/8" foil-backed fire-rated X board, per 2' x 8' sheet	—	Ea	6.27	—	6.27
All-purpose joint compound, 5 gal	—	Ea	10.40	—	10.40
2nd and 3rd coat joint finishing compound, 5 gal	—	Ea	10.50	—	10.50
Perforated joint tape, 500' roll	—	Ea	3.00	—	3.00
Self-adhesive fiberglass joint tape, 500' roll	—	Ea	8.04	—	8.04
Drywall screws, 5 pound box	—	Ea	17.80	—	17.80
1/4" plain board					
Walls	D1@.016	SF	.29	.48	.77
Ceilings	D1@.021	SF	.29	.63	.92

	Craft@Hrs	Unit	Material	Labor	Total
3/8" plain board					
Walls	D1@.017	SF	.27	.51	.78
Ceilings	D1@.023	SF	.27	.69	.96
1/2" plain board					
Walls	D1@.018	SF	.27	.54	.81
Ceilings	D1@.024	SF	.27	.72	.99
1/2" foil backed board					
Walls	D1@.017	SF	.31	.51	.82
Ceilings	D1@.023	SF	.31	.69	1.00
1/2" moisture resistant greenboard					
Walls	D1@.018	SF	.37	.54	.91
Ceilings	D1@.024	SF	.37	.72	1.09
1/2" fire rated type C board					
Walls	D1@.018	SF	.33	.54	.87
Ceilings	D1@.024	SF	.33	.72	1.05
5/8" fire rated board					
Walls	D1@.020	SF	.34	.60	.94
Ceilings	D1@.025	SF	.34	.75	1.09
Drywall casing and channel trim					
Stop or casing or jamb casing	D1@.020	LF	.26	.60	.86
RC-1 channel or 7/8" hat channel	D1@.011	LF	.50	.33	.83
Drywall arch and corner trim					
Vinyl arch trim, 1-1/4", flexible	D1@.030	LF	.24	.90	1.14
Vinyl outside corner, 1-1/4"	D1@.030	LF	.16	.90	1.06
Corner guard, 3/4" x 3/4", nail on	D1@.030	LF	.53	.90	1.43
Corner guard, 3/4" x 3/4", self-stick	D1@.030	LF	.48	.90	1.38
Corner guard, bullnose, 2-1/2" x 2-1/2"	D1@.030	LF	2.26	.90	3.16
Bullnose, metal, 3/4" radius	D1@.030	LF	.41	.90	1.31
Bullnose, plastic, 3/4" radius	D1@.030	LF	.18	.90	1.08
Bullnose, corner, metal, 2-way	D1@.160	Ea	1.98	4.80	6.78
Bullnose, corner, metal, 3-way	D1@.160	Ea	2.39	4.80	7.19
Bullnose, corner, plastic, 2-way	D1@.160	Ea	1.43	4.80	6.23
Bullnose, corner, plastic, 3-way	D1@.160	Ea	1.44	4.80	6.24
Metal corner bead, galvanized					
Type "J" bead, 1/2"	D1@.016	LF	.27	.48	.75
Bullnose type corner, 1/2"	D1@.016	LF	.30	.48	.78
Bullnose type corner, 1-1/4"	D1@.016	LF	.15	.48	.63
Beadex casing B-9	D1@.016	LF	.19	.48	.67
Deduct for no taping or joint finishing					
Walls	—	SF	-.04	-.22	-.25
Ceilings	—	SF	-.04	-.24	-.27
Acoustic ceiling (popcorn), spray application					
Light coverage	DT@.008	SF	.14	.24	.38
Medium coverage	DT@.011	SF	.19	.33	.52
Heavy coverage	DT@.016	SF	.21	.48	.69
Glitter application to acoustic ceiling texture	DT@.002	SF	.01	.06	.07
Remove acoustic ceiling texture	BL@.011	SF	—	.27	.27
Drywall texture					
Light hand texture	DT@.005	SF	.10	.15	.25
Medium hand texture	DT@.006	SF	.15	.18	.33
Heavy hand texture	DT@.007	SF	.19	.21	.40
Skim coat, smooth	DT@.012	SF	.12	.36	.48
Knockdown, spray applied and troweled	DT@.020	SF	.12	.60	.72

Hardware

	Craft@Hrs	Unit	Material	Labor	Total

Hardware See also Bath Accessories, Door Chimes, Door Closers, Foundation Bolts, Garage Door Openers, Mailboxes, Nails, Screen Wire, Sheet Metal, Framing Connectors, and Weatherstripping.

Rule of thumb: Cost of finish hardware (including door hardware, catches, stops, racks, brackets and pulls) per house.

	Craft@Hrs	Unit	Material	Labor	Total
Low to medium cost house (1,600 SF)	BC@16.0	LS	865.00	480.00	1,345.00
Better quality house (2,000 SF)	BC@20.0	LS	1,040.00	600.00	1,640.00

Keyed entry locksets. Meets ANSI Grade 2 standards for residential security. One-piece knob. Triple option faceplate fits square, radius-round, and drive-in door preps.

	Craft@Hrs	Unit	Material	Labor	Total
Antique brass	BC@.250	Ea	23.30	7.50	30.80
Antique pewter	BC@.250	Ea	30.80	7.50	38.30
Bright brass	BC@.250	Ea	23.20	7.50	30.70
Satin chrome	BC@.250	Ea	20.90	7.50	28.40

Lever locksets. Dual torque springs eliminate wobble.

	Craft@Hrs	Unit	Material	Labor	Total
Antique brass	BC@.250	Ea	58.20	7.50	65.70
Polished brass	BC@.250	Ea	58.20	7.50	65.70

Egg knob entry locksets

	Craft@Hrs	Unit	Material	Labor	Total
Satin nickel finish	BC@.250	Ea	53.10	7.50	60.60
Solid forged brass, polished brass finish	BC@.250	Ea	73.50	7.50	81.00

Colonial knob entry locksets

	Craft@Hrs	Unit	Material	Labor	Total
Polished brass	BC@.250	Ea	73.50	7.50	81.00

Double cylinder high security lockset. Forged brass handle with thumb latch. Separate grade 1 security double cylinder deadbolt.

	Craft@Hrs	Unit	Material	Labor	Total
Brass finish	BC@.820	Ea	167.00	24.60	191.60

Construction keying system (minimum order of 20 keyed entry locks)

	Craft@Hrs	Unit	Material	Labor	Total
Additional cost per lockset	—	Ea	6.00	—	6.00

Deadbolt and lockset combo. Includes keyed alike entry knob and grade 2 security deadbolt with 4 keys.

	Craft@Hrs	Unit	Material	Labor	Total
Satin nickel finish	BC@.750	Ea	40.90	22.50	63.40

Mortise lockset and deadbolt. Steel reinforced cast 1" throw deadbolt. Solid brass latch. Rocker switch locks and unlocks exterior knob. Labor includes mortising of door.

	Craft@Hrs	Unit	Material	Labor	Total
Left or right hand throw	BC@1.05	Ea	168.00	31.50	199.50

Deadbolts. All metal and brass. ANSI Grade 2 rating. 1" throw deadbolt. Includes 2 keys. Polished brass.

	Craft@Hrs	Unit	Material	Labor	Total
Single cylinder	BC@.500	Ea	20.00	15.00	35.00
Double cylinder	BC@.500	Ea	28.00	15.00	43.00

Knob passage latch

	Craft@Hrs	Unit	Material	Labor	Total
Dummy, brass	BC@.250	Ea	11.60	7.50	19.10
Dummy, satin nickel	BC@.250	Ea	12.60	7.50	20.10
Hall or closet, brass	BC@.250	Ea	23.20	7.50	30.70
Hall or closet, satin nickel	BC@.250	Ea	27.30	7.50	34.80

Egg knob privacy latchset

	Craft@Hrs	Unit	Material	Labor	Total
Bright brass	BC@.250	Ea	25.30	7.50	32.80
Satin nickel	BC@.250	Ea	29.40	7.50	36.90

Privacy lever latchset. Both levers locked or unlocked by turn button. Outside lever can be unlocked by emergency key (included).

	Craft@Hrs	Unit	Material	Labor	Total
Antique brass	BC@.250	Ea	20.50	7.50	28.00
Polished brass	BC@.250	Ea	19.60	7.50	27.10
Satin nickel	BC@.250	Ea	23.10	7.50	30.60

Door knockers. Polished brass

	Craft@Hrs	Unit	Material	Labor	Total
Apartment type, brass, with viewer	BC@.250	Ea	15.80	7.50	23.30
4" x 8-1/2", ornate style	BC@.250	Ea	40.50	7.50	48.00

Door pull plates.

	Craft@Hrs	Unit	Material	Labor	Total
Brass finish, 4" x 16"	BC@.333	Ea	21.00	9.99	30.99

	Craft@Hrs	Unit	Material	Labor	Total
Door push plates					
Aluminim, 3" x 12"	BC@.333	Ea	9.50	9.99	19.49
Solid brass, 4" x 16"	BC@.333	Ea	10.50	9.99	20.49
Door kick plates					
6" x 30", brass	BC@.333	Ea	43.20	9.99	53.19
8" x 34", brass	BC@.333	Ea	50.80	9.99	60.79
8" x 34", magnetic, for metal door, brass	BC@.333	Ea	27.50	9.99	37.49
8" x 34", satin nickel	BC@.333	Ea	36.10	9.99	46.09
Door stops					
Spring type wall bumper, 3" long	BC@.116	Ea	2.00	3.48	5.48
Wall bumper, 3" long, brass	BC@.116	Ea	4.25	3.48	7.73
Floor mounted stop, satin nickel	BC@.333	Ea	5.25	9.99	15.24
Hinge pin door stop, brass	BC@.250	Ea	5.60	7.50	13.10
Kick down door holder, brass	BC@.250	Ea	5.60	7.50	13.10
Magnetic wall stop	BC@.333	Ea	27.80	9.99	37.79
Adhesive rubber cup knob strike	BC@.116	Ea	2.10	3.48	5.58

	Craft@Hrs	Unit	Material	Labor	Total
Door peep sights					
180 degree viewing angle	BC@.258	Ea	4.64	7.74	12.38

Pocket door frame and track. Fully assembled. Designed to hold 150-pound standard 80" door. Door sold separately.

	Craft@Hrs	Unit	Material	Labor	Total
24" x 80"	BC@1.00	Ea	58.50	30.00	88.50
28" x 80"	BC@1.00	Ea	58.50	30.00	88.50
30" x 80"	BC@1.00	Ea	58.40	30.00	88.40
32" x 80"	BC@1.00	Ea	58.40	30.00	88.40
36" x 80"	BC@1.00	Ea	58.40	30.00	88.40
Pocket door edge pulls					
2-1/8" round brass	BC@.300	Ea	3.29	9.00	12.29
2-1/8" round brass dummy	BC@.300	Ea	4.75	9.00	13.75
1-3/8" x 3" oval brass	BC@.300	Ea	3.27	9.00	12.27
1-3/8" x 3" rectangular brass	BC@.300	Ea	3.27	9.00	12.27
3/4" round brass	BC@.300	Ea	2.06	9.00	11.06
Brass plate	BC@.300	Ea	4.26	9.00	13.26
Ball bearing pocket door hangers					
Hanger	BC@.175	Ea	9.44	5.25	14.69
Converging pocket door hardware kit. For double pocket doors.					
Door kit	—	Ea	3.65	—	3.65
Pocket door hangers					
3/8" offset, one wheel	BC@.175	Ea	2.47	5.25	7.72
1/16" offset, one wheel	BC@.175	Ea	2.47	5.25	7.72
7/8" offset, two wheel	BC@.175	Ea	3.86	5.25	9.11

Touch bar exit hardware. For use with 1-3/4"-thick hollow metal and wood doors. Extruded anodized aluminum with stainless steel spring. 3/4" throw stainless steel latch bolts. Strike for 5/8" stop; shim for 1/2" stop; 2-3/4" backset, 4-1/4" minimum stile. Includes standard mounting with sheet metal and machine screws. Recommended mounting height: 40-5/16" from ceiling to finished floor. 36" maximum width. Rim-type cylinders with Schlage "C" keyway. Includes measure and mark, drill holes, assemble, set strike plate.

	Craft@Hrs	Unit	Material	Labor	Total
Install, single door	BC@1.45	Ea	108.00	43.50	151.50
Install, per set of double doors	BC@2.97	Ea	216.00	89.10	305.10
Add for monitor switch for electrically operated locks					
Per lock	—	Ea	64.60	—	64.60

Hardware

	Craft@Hrs	Unit	Material	Labor	Total
Electromagnetic security and life safety lock system Surface mounted on door frame. Stainless steel finish. Includes magnet, strike plate, mounting fasteners and template. Costs include connection. Add for wire runs. Magnalock™					
600 lb holding force 12/24 VDC	E4@2.00	Ea	355.00	86.90	441.90
1,200 lb holding force 12/24 VDC	E4@2.00	Ea	370.00	86.90	356.90
1,800 lb holding force 12/24 VDC	E4@2.00	Ea	448.00	86.90	534.90
LCD programmable timer, 12/24 AC/DC	E4@.500	Ea	119.00	21.70	140.70
Exit delay initiation assembly	E4@3.00	Ea	286.00	130.00	416.00
Exit delay module, regulated DC voltage	E4@2.50	Ea	122.00	109.00	231.00
Exit delay system documentation and pressure sensitive vinyl door label for NFPA 101 requirements	—	Ea	19.90	—	19.90
Alarm sounder for exit delay module	—	Ea	84.00	—	84.00
Tamper-resistant plunger sex bolt	—	Ea	154.00	—	154.00
Energy absorbing sex bolt	—	Ea	184.00	—	184.00
Replacement strike plate and fasteners	—	Ea	117.00	—	117.00
Field mounted stainless steel housing. Includes cover, end caps, mounting plate and fasteners	BC@1.50	Ea	562.00	45.00	607.00
Hinges					
Full mortise hinge, heavy duty, satin brass finish					
3-1/2" x 3-1/2", 6 screws, .123 gauge	BC@.333	Ea	6.60	9.99	16.59
4" x 4", 8 screws, .130 gauge	BC@.333	Ea	11.00	9.99	20.99
4" x 4", spring hinge	BC@.333	Ea	13.80	9.99	23.79
Full mortise, light duty, satin brass finish					
3-1/2" x 3-1/2", loose pin	BC@.333	Ea	1.29	9.99	11.28
4" x 4", loose pin	BC@.333	Ea	1.60	9.99	11.59
Screen door hinge kit, 2 hinges, door pull and hook					
Bright zinc	BC@.750	Ea	3.55	22.50	26.05
House numbers, 4" high, brass or black	BC@.103	Ea	2.60	3.09	5.69
Garage door hardware, jamb type, 7' height, with springs and slide bolt					
8' to 10' wide door	BC@4.36	Ea	69.30	131.00	200.30
10' to 16' wide door, with truss	BC@5.50	Ea	88.20	165.00	253.20
Truss assembly for 16' door	BC@.493	Ea	37.00	14.80	51.80
Garage door hardware, track type, 7' height, with springs and lock					
7' to 9' wide door	BC@4.36	Ea	214.00	131.00	345.00
10' to 16' double track	BC@5.50	Ea	378.00	165.00	543.00

Hydronic Heating and Controls A complete system will consist of a hydronic boiler, cast iron radiators or hydronic radiation baseboard heaters, boiler controls and accessories. This type system is engineered and designed to meet specific heating requirements. Once these design requirements are known, the following components may be used to estimate the cost of a complete installed system. For pipe, fittings, hangers and insulation costs, see the Commercial and Industrial section. For more detailed coverage, see the National Plumbing & HVAC Estimator, http://CraftsmanSiteLicense.com

Hydronic boilers Packaged type units with standard equipment, including circulating pump and controls. Based on Burnham gas fired boilers

	Craft@Hrs	Unit	Material	Labor	Total
37,500 Btu with standard gas pilot	P1@3.00	Ea	1,170.00	89.90	1,259.90
62,000 Btu with standard gas pilot	P1@3.50	Ea	1,390.00	105.00	1,495.00
96,000 Btu with standard gas pilot	P1@3.50	Ea	1,530.00	105.00	1,635.00
130,000 Btu with standard gas pilot	P1@4.00	Ea	1,760.00	120.00	1,880.00
164,000 Btu with standard gas pilot	P1@4.00	Ea	2,010.00	120.00	2,130.00
198,000 Btu with standard gas pilot	P1@4.00	Ea	2,300.00	120.00	2,420.00
Add for electronic ignition	—	Ea	248.00	—	248.00

	Craft@Hrs	Unit	Material	Labor	Total
Air eliminator-purger, threaded connections					
3/4", cast iron, steam & water, 150 PSI	P1@.250	Ea	354.00	7.50	361.50
3/4", cast iron, 150 PSI water	P1@.250	Ea	114.00	7.50	121.50
1", cast iron, 150 PSI water	P1@.250	Ea	131.00	7.50	138.50
3/8", brass, 125 PSI steam	P1@.250	Ea	93.40	7.50	100.90
1/2", cast iron, 250 PSI steam	P1@.250	Ea	193.00	7.50	200.50
3/4", cast iron, 250 PSI steam	P1@.250	Ea	243.00	7.50	250.50
Airtrol fitting, 3/4"	P1@.300	Ea	41.30	8.99	50.29
Air eliminator vents	P1@.150	Ea	14.00	4.50	18.50
Atmospheric (anti-siphon) vacuum breakers, rough brass, threaded					
1/2" vacuum breaker	P1@.210	Ea	19.90	6.30	26.20
3/4" vacuum breaker	P1@.250	Ea	21.30	7.50	28.80
1" vacuum breaker	P1@.300	Ea	33.80	8.99	42.79
1-1/4" vacuum breaker	P1@.400	Ea	56.10	12.00	68.10
1-1/2" vacuum breaker	P1@.450	Ea	66.10	13.50	79.60
2" vacuum breaker	P1@.500	Ea	98.60	15.00	113.60
Circuit balancing valves, brass, threaded					
1/2" circuit balancing valve	P1@.210	Ea	24.60	6.30	30.90
3/4" circuit balancing valve	P1@.250	Ea	31.70	7.50	39.20
1" circuit balancing valve	P1@.300	Ea	40.50	8.99	49.49
1-1/2" circuit balancing valve	P1@.450	Ea	53.10	13.50	66.60
Baseboard fin tube radiation, per linear foot, copper tube with aluminum fins, wall mounted					
1/2" element and 8" cover	P1@.280	LF	7.17	8.39	15.56
3/4" element and 9"cover	P1@.280	LF	11.80	8.39	20.19
3/4" element, 10" cover, high capacity	P1@.320	LF	18.50	9.59	28.09
3/4" fin tube element only	P1@.140	LF	4.68	4.20	8.88
1" fin tube element only	P1@.150	LF	5.61	4.50	10.11
8" cover only	P1@.140	LF	4.36	4.20	8.56
9" cover only	P1@.140	LF	5.30	4.20	9.50
10" cover only	P1@.160	LF	5.80	4.80	10.60
Add for pipe connection and control valve	P1@2.15	Ea	118.00	64.50	182.50
Add for corners, fillers and caps, average per foot	—	%	10.0	—	—
Flow check valves, brass, threaded, horizontal type					
3/4" check valve	P1@.250	Ea	20.20	7.50	27.70
1" check valve	P1@.300	Ea	29.00	8.99	37.99
Thermostatic mixing valves					
1/2", soldered	P1@.240	Ea	44.20	7.20	51.40
3/4", threaded	P1@.250	Ea	57.00	7.50	64.50
3/4", soldered	P1@.300	Ea	48.70	8.99	57.69
1", threaded	P1@.300	Ea	223.00	8.99	231.99
Liquid level gauges					
1/2", 175 PSI bronze	P1@.210	Ea	39.80	6.30	46.10
Wye pattern strainers, threaded, 250 PSI, cast iron body valves					
3/4" strainer	P1@.260	Ea	9.48	7.79	17.27
1" strainer	P1@.330	Ea	12.00	9.89	21.89
1-1/4" strainer	P1@.440	Ea	17.40	13.20	30.60
1-1/2" strainer	P1@.495	Ea	20.20	14.80	35.00
2" strainer	P1@.550	Ea	31.70	16.50	48.20

Electric Heating

	Craft@Hrs	Unit	Material	Labor	Total

Electric Heating Labor includes installation and connecting only. Add the cost of wiring from the Commercial and Industrial section of this manual.

Electric baseboard heaters, convection type, surface mounted, 20 gauge steel. Labor includes installation and connecting only. Add the cost of wiring. 7" high, 3-1/4" deep, 3.41 Btu/hr/watt, 240 volt. Low or medium density.

	Craft@Hrs	Unit	Material	Labor	Total
2'6" long, 500 watt (1,700 Btu)	BE@1.70	Ea	21.00	57.60	78.60
3'0" long, 750 watt (2,225 Btu)	BE@1.81	Ea	23.40	61.30	84.70
4'0" long, 1,000 watt (3,400 Btu)	BE@1.81	Ea	31.50	61.30	92.80
6'0" long, 1,500 watt (5,100 Btu)	BE@1.89	Ea	42.00	64.00	106.00
8'0" long, 2,000 watt (6,800 Btu)	BE@2.04	Ea	46.20	69.10	115.30
Add for integral thermostat	BE@.094	Ea	16.00	3.18	19.18
Add for wall mounted thermostat	BE@1.62	Ea	21.50	54.90	76.40

Fan-forced electric wall heaters, 12-3/4" x 6" rough opening. Add for 240 volt wiring and thermostat.

	Craft@Hrs	Unit	Material	Labor	Total
700/900/1600 watt, 50 CFM	BE@.650	Ea	124.00	22.00	146.00
500 watt, 6.25 amp, 5118 Btu, 83 CFM	BE@.650	Ea	92.90	22.00	114.90
1000 watt, 4.17 amp, 3412 Btu, 83 CFM	BE@.650	Ea	91.59	22.00	113.59
2000 watt, 8.33 amp, 6824 Btu, 60 CFM	BE@.650	Ea	127.50	22.00	149.50
2000 watt, 8.33 amp, 6824 Btu, 83 CFM	BE@.650	Ea	96.35	22.00	118.35
3000 watt, 12.5 amp, 12,236 Btu, 65 CFM	BE@.650	Ea	143.00	22.00	165.00
Double pole thermostat, 120 or 240 Volt	BE@.440	Ea	22.11	14.90	37.01

Surface mounted 1" thick radiant ceiling panels. 120/208/240/277 volts AC

	Craft@Hrs	Unit	Material	Labor	Total
200 watts, 23-3/4" x 23-3/4"	BE@.850	Ea	93.10	28.80	121.90
300 watts, 23-3/4" x 35-3/4"	BE@.900	Ea	111.00	30.50	141.50
400 watts, 23-3/4" x 47-3/4"	BE@.950	Ea	134.00	32.20	166.20
500 watts, 23-3/4" x 59-3/4"	BE@1.00	Ea	158.00	33.90	191.90
610 watts, 23-3/4" x 71-3/4"	BE@1.11	Ea	179.00	37.60	216.60
820 watts, 23-3/4" x 95-3/4"	BE@1.31	Ea	218.00	44.40	262.40
600 watts, 47-3/4" x 47-3/4"	BE@1.11	Ea	179.00	37.60	216.60
800 watts, 47-3/4" x 47-3/4"	BE@1.31	Ea	218.00	44.40	262.40
1,000 watts, 47-3/4" x 59-3/4"	BE@1.46	Ea	270.00	49.40	319.40
1,220 watts, 47-3/4" x 71-3/4"	BE@1.61	Ea	309.00	54.50	363.50
1,640 watts, 47-3/4" x 95-3/4"	BE@1.91	Ea	399.00	64.70	463.70
Add for radiant line voltage wall thermostat	BE@.450	Ea	27.00	15.20	42.20
Add for motion sensing setback thermostat	BE@.450	Ea	78.50	15.20	93.70

Surface mounted, 2-1/2" thick radiant ceiling panels. 120/208/240/277 volts AC

	Craft@Hrs	Unit	Material	Labor	Total
750/1000 watts, 23-3/4" x 47-3/4"	BE@.950	Ea	303.00	32.20	335.20
1125/1500 watts, 23-3/4" x 71-3/4"	BE@1.31	Ea	409.00	44.40	453.40
Add for radiant line voltage wall thermostat	BE@.450	Ea	27.00	15.20	42.20
Add for motion sensing setback thermostat	BE@.450	Ea	98.70	15.20	113.90

T-bar mounted radiant ceiling panels, 1" thick, 120/208/240/277 volts AC

	Craft@Hrs	Unit	Material	Labor	Total
200/250/375 watts, 23-3/4" x 23-3/4"	BE@.470	Ea	98.70	15.90	114.60
400/500/750 watts, 23-3/4" x 47-3/4"	BE@.500	Ea	149.00	16.90	165.90
610/750/1,000 watts, 23-3/4" x 71-3/4"	BE@.550	Ea	200.00	18.60	218.60
Add for radiant line voltage wall thermostat	BE@.450	Ea	27.00	15.20	42.20
Add for motion sensing setback thermostat	BE@.450	Ea	98.70	15.20	113.90

Floor units, drop-in type, 14" long x 7-1/4" wide, includes fan motor assembly, housing and grille

	Craft@Hrs	Unit	Material	Labor	Total
120 volts, 375 or 750 watts	BE@1.76	Ea	243.00	59.60	302.60
277 volts, 750 watts	BE@1.76	Ea	299.00	59.60	358.60
Add for wall thermostat kit	BE@1.62	Ea	57.20	54.90	112.10
Add for concrete accessory kit and housing	—	LS	20.00	—	20.00

	Craft@Hrs	Unit	Material	Labor	Total

Infrared quartz tube heaters, indoor or outdoor, chromed guard, polished aluminum reflector. Includes modulating control

	Craft@Hrs	Unit	Material	Labor	Total
1,500 watts, 33" long, 120 or 240 volts	BE@4.14	Ea	272.00	140.00	412.00
2,000 watts, 33" long, 240 volts	BE@4.14	Ea	285.00	140.00	425.00
3,000 watts, 58-1/2" long, 240 volts	BE@4.14	Ea	359.00	140.00	499.00
4,000 watts, 58-1/2" long, 240 volts	BE@4.14	Ea	365.00	140.00	505.00
Add for wall or ceiling brackets (pair)	—	LS	12.00	—	12.00
Deduct for indoor use only	—	%	-30.0	—	—

Suspension blower heaters, 208 or 240 volts, single phase, propeller type

	Craft@Hrs	Unit	Material	Labor	Total
3,000 watts	BE@4.80	Ea	248.00	163.00	411.00
5,600 watts	BE@4.80	Ea	272.00	163.00	435.00
7,500 watts	BE@4.80	Ea	412.00	163.00	575.00
10,000 watts	BE@4.80	Ea	436.00	163.00	599.00
Add for wall mounting bracket	—	LS	17.50	—	17.50
Add for line voltage wall thermostat					
Single pole	—	LS	28.10	—	28.10
Double pole	—	LS	40.00	—	40.00

Wall mounted fan-forced downflow insert heaters, heavy duty, built-in thermostat, 240 volts, 14" wide, 20" high, 4" deep rough-in, 3.41 Btu per hr per watt

	Craft@Hrs	Unit	Material	Labor	Total
1,500 watts	BE@3.95	Ea	137.00	134.00	271.00
2,000 watts	BE@3.95	Ea	149.00	134.00	283.00
3,000 watts	BE@3.95	Ea	172.00	134.00	306.00
4,500 watts	BE@3.95	Ea	189.00	134.00	323.00
Add for surface mounting kit	—	LS	17.60	—	17.60

Gas Furnaces Add the cost of piping, electrical connection and vent. See pipe costs in the Commercial and Industrial section of this manual. For more detailed coverage of gas heating, see the National Plumbing & HVAC Estimator, http://CraftsmanSiteLicense.com.

Gravity gas wall furnace. Natural gas or propane fired.

	Craft@Hrs	Unit	Material	Labor	Total
25,000 Btu	PM@4.67	Ea	346.00	165.00	511.00
35,000 Btu	PM@4.67	Ea	388.00	165.00	553.00
50,000 Btu	PM@4.67	Ea	505.00	165.00	670.00
Wall register kit for 2nd room	PM@1.00	Ea	42.00	35.30	77.30
Add for wall thermostat	PM@.972	Ea	25.40	34.30	59.70
Add for blower	PM @.500	Ea	115.00	17.60	132.60
Add for free standing vent trim kit	PM@.254	LS	150.00	8.95	158.95
Add for gas valve for thermostat	PM@.250	Ea	84.90	8.81	93.71

Direct vent gas wall furnace. Includes wall thermostat. No electricity required. Includes vent kit to outside wall. By input Btu rating. Natural gas or propane fired.

	Craft@Hrs	Unit	Material	Labor	Total
10,000 Btu	PM@2.98	Ea	345.00	105.00	450.00
14,000 Btu, high altitude	PM@2.98	Ea	380.00	105.00	485.00
15,000 Btu	PM@2.98	Ea	380.00	105.00	485.00
22,000 Btu, high altitude	PM@2.98	Ea	420.00	105.00	525.00
25,000 Btu	PM@2.98	Ea	471.00	105.00	576.00
30,000 Btu, high altitude	PM@2.98	Ea	490.00	105.00	595.00
35,000 Btu	PM@2.98	Ea	515.00	105.00	620.00
Add for blower	PM @.500	Ea	84.00	17.60	101.60
Add for gas valve for thermostat	PM@.250	Ea	84.90	8.81	93.71

Gas floor furnace, including vent at $80, valve and wall thermostat

	Craft@Hrs	Unit	Material	Labor	Total
30 MBtu input, 24 MBtu output	PM@7.13	Ea	766.00	251.00	1,017.00
50 MBtu input, 35 MBtu output	PM@8.01	Ea	876.00	282.00	1,158.00
Add for furnace serving two rooms	—	%	30.0	—	—
Add for spark ignition	—	LS	55.00	—	55.50

Heating and Cooling Equipment

	Craft@Hrs	Unit	Material	Labor	Total

Forced air upflow gas furnace with electronic ignition. Includes vent at $100, typical piping, at $50, thermostat at $40 and distribution plenum at $100. No ducting included

	Craft@Hrs	Unit	Material	Labor	Total
50 MBtu input, 48 MBtu output	PM@7.13	Ea	2,370.00	251.00	2,621.00
75 MBtu input, 70 MBtu output	PM@7.13	Ea	2,490.00	251.00	2,741.00
100 MBtu input, 92 MBtu output	PM@7.80	Ea	2,650.00	275.00	2,925.00
125 MBtu input, 113 MBtu output	PM@7.80	Ea	2,820.00	275.00	3,095.00
Add for liquid propane models	—	LS	250.00	—	250.00
Add for time control thermostat	—	Ea	25.00	—	95.75

Ductwork for forced air furnace. Sheet metal, 6 duct and 2 air return residential system, with typical elbows and attaching boots

	Craft@Hrs	Unit	Material	Labor	Total
4" x 12" registers, and 2" R-5 insulation	SW@22.1	LS	693.00	751.00	1,444.00
Add for each additional duct run	SW@4.20	Ea	149.00	143.00	292.00
Fiberglass flex duct, insulated, 8" dia.	SW@.012	LF	3.00	.41	3.41

Suspended space heater, fan-forced, with tubular aluminum heat exchanger, mounting brackets and gas valve. Includes vent to 30' at $100, typical gas piping, at $50, and thermostat. at $40

	Craft@Hrs	Unit	Material	Labor	Total
30 MBtu input, 24 MBtu output	PM@2.75	Ea	750.00	96.90	846.90
45 MBtu input, 36 MBtu output	PM@2.75	Ea	750.00	96.90	846.90
60 MBtu input, 48 MBtu output	PM@2.75	Ea	804.00	96.90	900.90
75 MBtu input, 60 MBtu output	PM@2.75	Ea	889.00	96.90	985.90

Heating and Cooling Equipment

Evaporative coolers, no wiring or duct work, roof installation, with switch, electrical and water connection at $110, recirculation pump, and mounting hardware. at $90

	Craft@Hrs	Unit	Material	Labor	Total
3,000 CFM, down draft	PM@11.1	Ea	502.00	391.00	893.00
3,000 CFM, side draft	PM@11.1	Ea	500.00	391.00	891.00
4,000 CFM, side draft	PM@11.1	Ea	572.00	391.00	963.00
4,300 CFM, down draft	PM@11.1	Ea	600.00	391.00	991.00
5,500 CFM, down draft	PM@11.1	Ea	657.00	391.00	1,048.00
5,500 CFM, side draft	PM@11.1	Ea	502.00	391.00	893.00

Window type air conditioning/cooling units. High-efficiency units, adjustable thermostat. Labor is for installation in an existing opening, connected to an existing electrical outlet

	Craft@Hrs	Unit	Material	Labor	Total
6,000 Btu	PM@.751	Ea	131.00	26.50	157.50
8,000 Btu	PM@.751	Ea	157.00	26.50	183.50
10,000 Btu	PM@.751	Ea	198.00	26.50	224.50
12,000 Btu	PM@.751	Ea	251.00	26.50	277.50
18,000 Btu	PM@.751	Ea	344.00	26.50	370.50

Split system heat pump, one outdoor unit, with indoor evaporator coil, direct drive multi-speed blower, filter, and fiberglass cabinet insulation. No electrical or concrete work included

	Craft@Hrs	Unit	Material	Labor	Total
18 MBtu cooling, 17.8 MBtu heating	PM@9.59	Ea	934.00	338.00	1,272.00
21 MBtu cooling, 20.8 MBtu heating	PM@11.3	Ea	1,200.00	398.00	1,598.00
25 MBtu cooling, 24.7 MBtu heating	PM@12.3	Ea	1,250.00	434.00	1,684.00
29 MBtu cooling, 28.7 MBtu heating	PM@13.1	Ea	1,690.00	462.00	2,152.00
Add for electrical connection	PM@7.34	LS	72.90	259.00	331.90
Add for drain pan	—	LS	50.50	—	50.50

Packaged split system electric heat pumps, galvanized steel cabinets, window-wall units. 35' of tubing, two thermostats and control package. Installed on concrete pad. No electrical, concrete, or duct work included

	Craft@Hrs	Unit	Material	Labor	Total
8 MBtu cooling, 5 Mbtu heating	PM@10.9	Ea	744.00	384.00	1,128.00
15 MBtu cooling, 13 Mbtu heating	PM@10.9	Ea	1,050.00	384.00	1,434.00
18 Mbtu cooling, 16 Mbtu heating	PM@12.9	Ea	1,190.00	455.00	1,645.00
Add for electrical connections	PM@14.0	LS	154.00	494.00	648.00

	Craft@Hrs	Unit	Material	Labor	Total

Insulation See also Building Paper and Polyethylene Film. Coverage allows for studs and joists.
Fiberglass insulation, wall and ceiling application, coverage allows for framing

Kraft-faced, 16" OC framing members

	Craft@Hrs	Unit	Material	Labor	Total
3-1/2" (R-11), 88 SF roll at $28.70	BC@.007	SF	.33	.21	.54
3-1/2" (R-13), 116 SF roll at $41.50	BC@.007	SF	.36	.21	.57
10" (R-30), 53 SF batts at $37.80	BC@.008	SF	.71	.24	.95
12" (R-38), 43 SF batts at $37.90	BC@.008	SF	.88	.24	1.12

Kraft-faced, 24" OC framing members

	Craft@Hrs	Unit	Material	Labor	Total
3-1/2" (R-11), 135 SF roll at $34.90	BC@.005	SF	.26	.15	.41
6-1/4" (R-19), 75 SF batts at $31.10	BC@.005	SF	.41	.15	.56
23" (R-30), 80 SF batts at $57.30	BC@.006	SF	.72	.18	.90
15" (R-38), 42.7 SF batts at $38.10	BC@.006	SF	.89	.18	1.07

Unfaced, 16" OC framing members

	Craft@Hrs	Unit	Material	Labor	Total
3-1/2" (R-13), 116 SF roll at $37.70	BC@.004	SF	.33	.12	.45
6-1/4" (R-19), 78 SF batts at $30.70	BC@.005	SF	.39	.15	.54
10" (R-30), 53 SF batts at $36.80	BC@.006	SF	.69	.18	.87

Unfaced, 24" OC framing members

	Craft@Hrs	Unit	Material	Labor	Total
23" (R-30), 88 SF batts at $60.20	BC@.003	SF	.68	.09	.77

Poly-faced insulation, vapor retarding twisted Miraflex fibers, ASTM C665, Type II, Class C, coverage allows for framing

	Craft@Hrs	Unit	Material	Labor	Total
3-1/2" (R-13), 16" or 24" OC framing	BC@.005	SF	.38	.15	.53
8-3/4" (R-25), 16" or 24" OC framing	BC@.006	SF	.57	.18	.75

Encapsulated fiberflass roll insulation, Johns Manville Comfort Therm™, coverage allows for framing

	Craft@Hrs	Unit	Material	Labor	Total
3-1/2" (R-13) rolls, 16" OC framing	BC@.005	SF	.31	.15	.46
6-1/4" (R-19) batts, 16" or 24" OC framing	BC@.006	SF	.53	.18	.71
8-1/4" (R-25) rolls, 16" or 24" OC framing	BC@.006	SF	.69	.18	.87
10-1/4" (R-30) batts, 16" or 24" OC framing	BC@.006	SF	.73	.18	.91

Insulation board, Owens Corning Foamular XAE rigid extruded polystyrene foam board, film-faced, stapled in place, including taped joints, 4' x 8' or 9' panels, tongue and groove or square edge, including 5% waste

	Craft@Hrs	Unit	Material	Labor	Total
1/2" thick (R-2.5), $9.04 per 4' x 8' panel	BC@.010	SF	.28	.30	.58
3/4" thick (R-3.8), $11.10 per 4' x 8' panel	BC@.011	SF	.35	.33	.68
1" thick (R-5.0), $15.60 per 4 x 8 panel	BC@.011	SF	.49	.33	.82
1-1/2" thick (R-7.5), $16.70 per 4' x 8' panel	BC@.015	SF	.52	.45	.97
2" thick (R-10), $22.50 per 4' x 8' panel	BC@.015	SF	.70	.45	1.15

Extruded polystyrene insulation panel, Dow Blue Board, water resistant. By thermal resistance value, (R-). Density of 1.6 lbs. per CF. Gray Board density is 1.3 lbs. per CF.

	Craft@Hrs	Unit	Material	Labor	Total
3/4" x 2' x 8', Tongue & Groove, R-3.8	BC@.011	SF	.36	.33	.69
1" x 2' x 8', R-5.0	BC@.011	SF	.50	.33	.83
1" x 2' x 8', Tongue and Groove, R-5.0	BC@.011	SF	.30	.33	.63
1" x 4' x 8', R-5.0	BC@.011	SF	.49	.33	.82
1" x 4' x 8', Tongue and Groove, R-5.0	BC@.011	SF	.42	.33	.75
1-1/2" x 2' x 8', R-7.5	BC@.015	SF	.72	.45	1.17
1-1/2" x 2' x 8', Tongue & Groove, R-7.5	BC@.015	SF	.81	.45	1.26
1-1/2" 4' x 8', Square Edge, R-7.5	BC@.015	SF	.75	.45	1.20
2" x 2' x 8', R-10	BC@.015	SF	.98	.45	1.43
2" x 4' x 8', R-10	BC@.015	SF	.99	.45	1.44
1" x 2' x 8' Gray Board (1.3 lbs. per CF)	BC@.015	SF	.46	.45	.91
1-1/2" x 2' x 8' Gray Board	BC@.015	SF	.64	.45	1.09
2" x 2' x 8' Gray Board	BC@.015	SF	.86	.45	1.31

Insulation

	Craft@Hrs	Unit	Material	Labor	Total
Rigid foam insulated sheathing, fiberglass polymer faced two sides, Celotex Tuff-R, 4' x 8' panels, including 5% waste					
1/2" (R-3.2)	BC@.010	SF	.25	.30	.55
3/4" (R-5.6)	BC@.011	SF	.34	.33	.67
1" (R-7.2)	BC@.011	SF	.45	.33	.78
2" (R-14.4)	BC@.015	SF	.85	.45	1.30
Fanfold extruded polystyrene insulation 4' x 50' panels, including 5% waste					
1/4" (R-1)	BC@.011	SF	.15	.33	.48
3/8" (R-1)	BC@.011	SF	.19	.33	.52
Add for nails, 50 lb cartons (large square heads)					
2-1/2", for 1-1/2" boards	—	LS	105.00	—	105.00
3", for 2" boards	—	LS	105.00	—	105.00
Foil-faced urethane sheathing, 4' x 8' panels, including 5% waste,					
1" (R-7.2)	BC@.011	SF	.51	.33	.84
Asphalt impregnated sheathing board on walls, nailed, 4' x 8' panels					
1/2", standard	BC@.013	SF	.23	.39	.62
1/2", intermediate	BC@.013	SF	.29	.39	.68
1/2", nail base	BC@.013	SF	.41	.39	.80
Dens Glass Gold insulating exterior sheathing. Glass mat embedded in a water-resistant treated gypsum core with bond-enhancing gold primer coating. Square edge					
1/4" x 4' x 4' Dens-Shield	BC@.015	SF	.63	.45	1.08
1/2" x 32" x 4' Dens-Shield	BC@.015	SF	.89	.45	1.34
1/2" x 4' x 8'	BC@.016	SF	.59	.48	1.07
5/8" x 4' x 8'	BC@.016	SF	.62	.48	1.10
Polystyrene foam underlay, polyethylene covered, R-Gard.					
3/8" 4' x 24' fanfold	BC@.011	SF	.22	.33	.55
1/2" x 4' x 8'	BC@.011	SF	.27	.33	.60
1" x 4' x 8'	BC@.011	SF	.37	.33	.70
1-1/2" x 2' x 8'	BC@.015	SF	.71	.45	1.16
2" x 2' x 8'	BC@.015	SF	.75	.45	1.20
Cant strips for insulated roof board, one piece wood, tapered					
2" strip	RR@.018	LF	.14	.61	.75
3" strip	RR@.020	LF	.13	.67	.80
4" strip	RR@.018	LF	.12	.61	.73
Sill plate gasket. Fills gaps between the foundation and the plate. Also used for sealing around windows and doors					
1/4" x 50', 3-1/2" wide	RR@.003	LF	.08	.10	.18
1/4" x 50', 5-1/2" wide	RR@.003	LF	.11	.10	.21
Perimeter foundation insulation, expanded polystyrene with polyethylene skinned surface. Low moisture retention. Minimum compressive strength of 1440 lbs. per SF					
3/4" x 4' x 8'	BC@.352	Ea	7.86	10.60	18.46
1-1/2" x 2' x 4'	BC@.120	Ea	4.25	3.60	7.85
2" x 2' x 4'	BC@.120	Ea	5.73	3.60	9.33
2" x 4' x 4'	BC@.240	Ea	18.50	7.20	25.70
FBX 1240 industrial board insulation, rockwool, semi-rigid, no waste included 24" OC framing members					
2" (R-8)	BC@.004	SF	.43	.12	.55
3" (R-12)	BC@.004	SF	.63	.12	.75
3-1/2" (R-14)	BC@.004	SF	.73	.12	.85

	Craft@Hrs	Unit	Material	Labor	Total

Sound control insulation, fiberglass acoustically designed to absorb sound vibration, for use between interior walls, floors and ceilings, Quietzone™ batts, coverage allows for framing.

	Craft@Hrs	Unit	Material	Labor	Total
3-1/2", 16" OC framing	BC@.006	SF	.28	.18	.46
3-1/2", 24" OC framing	BC@.006	SF	.29	.18	.47

Sound insulation board, 4' x 8' panels installed on walls

1/2" Homasote	BC@.013	SF	.62	.39	1.01
1/2" Soundstop	BC@.013	SF	.23	.39	.62
1/2" Sound deadening board	BC@.013	SF	.25	.39	.64
Add for installation on ceilings	BC@.004	SF	—	.12	.12

Sound Attenuation Fire Batt Insulation (SAFB), rockwool, semi-rigid, no waste included, pressed between framing members

16" OC framing members

2" (R-8)	BC@.004	SF	.38	.12	.50
3" (R-12)	BC@.004	SF	.57	.12	.69
4" (R-16)	BC@.005	SF	.75	.15	.90

24" OC framing members

2" (R-8)	BC@.003	SF	.38	.09	.47
3" (R-12)	BC@.003	SF	.57	.09	.66
4" (R-16)	BC@.004	SF	.75	.12	.87

Housewrap building envelope, wrapped on exterior walls and stapled, including 10% waste

9' x 195' roll	BC@.006	SF	.07	.18	.25
Tyvek tape, 2" x 165'	—	Roll	11.70	—	11.70

Stabilized cellulose insulation, for blown-in applications. Bag covers 40 SF of attic joists at 6" depth. Add the cost of blower rental

40 SF bag at $7.21	BC@.010	SF	.18	.30	.48
1" wood plugs	BC@.020	Ea	.31	.60	.91
1" plastic plugs	BC@.020	Ea	.28	.60	.88

Vermiculite insulation, poured over ceilings

Vermiculite, 25 lb sack (3 CF)	—	Ea	9.48	—	9.48
At 3" depth (60 sacks per 1,000 SF)	BL@.007	SF	.61	.17	.78
At 4" depth (72 sacks per 1,000 SF)	BL@.007	SF	.74	.17	.91
Masonry batts, 1-1/2" x 15" x 48"	B9@.008	SF	.25	.22	.47

Masonry fill, poured in concrete block cores

Using 3 CF bags of perline, per bag	—	Ea	6.89	—	6.89
4" wall, 8.1 SF per CF	B9@.006	SF	.31	.17	.48
6" wall, 5.4 SF per CF	B9@.006	SF	.45	.17	.62
8" wall, 3.6 SF per CF	B9@.006	SF	.65	.17	.82

Natural fiber building insulation. EPA registered anti-microbial agent that offers protection from mold, mildew, fungi and pests. Made from all natural cotton fibers. Friction fit. 94" long batting. Based on Ultra Touch

16" long (R-13)	BC@.005	SF	.56	.15	.71
24" long (R-13)	BC@.006	SF	.56	.18	.74
16" long (R-19)	BC@.005	SF	.84	.15	.99
24" long (R-19)	BC@.006	SF	.84	.18	1.02

Insulation, Subcontract

	Craft@Hrs	Unit	Material	Labor	Total
Insulation, Subcontract New construction, coverage includes framing.					
Blown-in cellulose					
3-1/2" (R-12)	—	MSF	—	—	595.00
6" (R-21)	—	MSF	—	—	845.00
Sprayed-on urethane foam (rigid)					
1" thick	—	MSF	—	—	1,450.00
Add for acrylic waterproofing	—	MSF	—	—	1,010.00
Add for urethane rubber waterproofing	—	MSF	—	—	1,691.00
Add for cleaning rock off existing roof	—	MSF	—	—	106.00

Insulation blown into walls in an existing structure, includes coring hole in exterior or interior of wall, filling cavity, sealing hole and paint to match existing surface., R-12 to R-14 rating depending on cavity thickness

	Craft@Hrs	Unit	Material	Labor	Total
Stucco (exterior application)	—	MSF	—	—	1,730.00
Wallboard (interior application)	—	MSF	—	—	1,730.00
Wood siding (exterior application)	—	MSF	—	—	2,290.00

Thermal Analysis (Infrared thermography) Includes infrared video inspection of entire building, written report by qualified technician or engineer, and thermograms (infrared photographs) of all problem areas of building. Inspection is done using high-quality thermal sensitive instrumentation. Costs listed are for exterior facade shots, or interior (room by room) shots. All travel expenses for inspector are extra.

	Craft@Hrs	Unit	Material	Labor	Total
Residence, typically 4 hours required	—	Hr	—	—	230.00
Commercial structure ($800.00 minimum)	—	Hr	—	—	134.00

Add per day lost due to weather, lack of preparation by building occupants, etc., when inspector and equipment are at building site prepared to commence inspection

	Craft@Hrs	Unit	Material	Labor	Total
Typical cost per day	—	LS	—	—	370.00

Add per night for accommodations when overnight stay is required. This is common when inspection must be performed at night to get accurate results

	Craft@Hrs	Unit	Material	Labor	Total
Typical cost, per night	—	LS	—	—	125.00

Insurance and Bonding Typical rates. Costs vary by state and class of construction.

Rule of thumb: Employer's cost for payroll taxes, and insurance.

	Craft@Hrs	Unit	Material	Labor	Total
Per $100 (C$) of payroll	—	C$	—	—	30.00

Rule of thumb: Complete insurance program (comprehensive general liability policy, truck, automobile, and equipment floaters and fidelity bond), contractor's typical cost

	Craft@Hrs	Unit	Material	Labor	Total
Per $100 of payroll	—	C$	—	—	6.80

Liability insurance Comprehensive contractor's liability insurance, including operations, completed operations, bodily injury and property damage, protective and contractual coverage's, $1,000,000 policy limit. Minimum annual premium will be between $2,500 and $10,000. Rates vary by state and with the contractor's loss experience. Typical costs per $100 (C$) of payroll for each trade employed. (Calculate and add for each category separately.)

	Craft@Hrs	Unit	Material	Labor	Total
General contractors	—	C$	—	—	2.47
Carpentry	—	C$	—	—	3.90
Concrete, formed or flat	—	C$	—	—	5.00
Drywall hanging and finishing	—	C$	—	—	3.37
Electrical wiring	—	C$	—	—	3.40
Floor covering installation	—	C$	—	—	4.50
Glaziers	—	C$	—	—	3.37
Heating, ventilating, air conditioning	—	C$	—	—	4.70
Insulation	—	C$	—	—	3.90
Masonry, tile	—	C$	—	—	3.70
Painting	—	C$	—	—	5.10
Plastering and stucco	—	C$	—	—	3.60
Plumbing	—	C$	—	—	6.05

	Craft@Hrs	Unit	Material	Labor	Total

Workers compensation coverage Rates vary from state to state and by contractor's loss history. Coverage cost per $100 (C$) of base payroll excluding fringe benefits.

	Craft@Hrs	Unit	Material	Labor	Total
Bricklayer	—	C$	—	—	11.73
Carpenter					
1 and 2 family dwellings	—	C$	—	—	12.05
Multiple units and commercial	—	C$	—	—	12.05
Clerical (office worker)	—	C$	—	—	.99
Concrete					
1 and 2 family dwellings	—	C$	—	—	9.12
Other concrete	—	C$	—	—	11.73
Construction laborer	—	C$	—	—	12.05
Drywall taper	—	C$	—	—	9.69
Electrical wiring	—	C$	—	—	7.60
Elevator erectors	—	C$	—	—	4.19
Executive supervisors	—	C$	—	—	3.50
Excavation, grading	—	C$	—	—	19.41
Excavation, rock (no tunneling)	—	C$	—	—	12.10
Glazing	—	C$	—	—	11.68
Insulation work	—	C$	—	—	20.00
Iron or steel erection work	—	C$	—	—	23.54
Lathing	—	C$	—	—	17.35
Operating engineers	—	C$	—	—	19.41
Painting and paperhanging	—	C$	—	—	13.59
Pile driving	—	C$	—	—	27.26
Plastering and stucco	—	C$	—	—	13.19
Plumbing	—	C$	—	—	7.88
Reinforcing steel installation (concrete)	—	C$	—	—	10.00
Roofing	—	C$	—	—	31.14
Sewer construction	—	C$	—	—	9.10
Sheet metal work (on site)	—	C$	—	—	11.03
Steam and boiler work	—	C$	—	—	10.00
Tile, stone, and terrazzo work	—	C$	—	—	8.71
Truck driver	—	C$	—	—	16.00
Tunneling	—	C$	—	—	20.00

Payment and performance bonds Cost per $1,000 (M$) of final contract price. Rates depend on the experience, credit, and net worth of the applicant.

	Craft@Hrs	Unit	Material	Labor	Total
Preferred rates, contract coverage for:					
First $100,000	—	M$	—	—	13.70
Next $2,400,000	—	M$	—	—	8.27
Next $2,500,000	—	M$	—	—	6.55
Standard rates, contract coverage for:					
First $100,000	—	M$	—	—	28.50
Next $400,000	—	M$	—	—	17.10
Next $2,000,000	—	M$	—	—	11.40
Next $2,500,000	—	M$	—	—	8.55
Next $2,500,000	—	M$	—	—	7.98
Over $7,500,000	—	M$	—	—	7.41
Substandard rates, contract coverage for:					
First $100,000	—	M$	—	—	34.20
Balance of contract	—	M$	—	—	22.80

Intercom Systems

	Craft@Hrs	Unit	Material	Labor	Total

Intercom System, Subcontract Costs include all labor and equipment but assume that wiring is done before interior wall and ceiling finish is applied. Central system with AM/FM music, hands-free talk back, privacy feature and room monitoring. Includes 1 master station, 5 selective call remote stations and 2 door stations

	Craft@Hrs	Unit	Material	Labor	Total
Basic residential system	—	LS	—	—	850.00
Add cassette tape player	—	LS	—	—	90.00
Add door and gate release	—	Ea	—	—	159.00
Add telephone answering feature, per station	—	Ea	—	—	65.00
Add telephone handset stations, per station	—	Ea	—	—	39.00

Ironing Centers, Built-In Wood frame with hinged door, can be recessed or surface mounted. Includes one 110 volt control panel where shown, receptacle outlet, timed shutoff switch, spotlight, storage shelf, and steel ironing board. No electrical work included. Iron-A-Way Products

	Craft@Hrs	Unit	Material	Labor	Total
Unit type 1, no electric control panel, non-adjustable 42" board, 48" high, 15" wide, 6" deep	BC@.994	Ea	235.00	29.80	264.80
Unit type 2, with electric control panel, non-adjustable 42" board, 48" high, 15" wide, 6" deep	BC@.994	Ea	335.00	29.80	364.80
Unit type 3, with electric control panel, adjustable 46" board, 61" high, 15" wide, 6" deep	BC@.994	Ea	448.00	29.80	477.80
Added cost for any unit					
Add for sleeve board with holder	—	Ea	32.00	—	32.00
Add for door mirror	—	Ea	95.00	—	95.00
Add for oak paneled door front	—	Ea	95.00	—	95.00

Jackposts, Steel, Adjustable Building Columns Permanent or temporary adjustable steel posts, 3" diameter.

	Craft@Hrs	Unit	Material	Labor	Total
6' 6" to 6' 10", 15,500 lb capacity	B1@.250	Ea	39.90	6.84	46.74
6' 9" to 7', 15,500 lb capacity	B1@.250	Ea	44.60	6.84	51.44
7' to 7' 4", 15,500 lb capacity	B1@.250	Ea	39.20	6.84	46.04
7' 3" to 7' 7", 15,500 lb capacity	B1@.250	Ea	45.00	6.84	51.84
7' 6" to 7' 10", 14,300 lb capacity	B1@.250	Ea	45.90	6.84	52.74
7' 9" to 8' 1", 14,300 lb capacity	B1@.250	Ea	47.80	6.84	54.64
8' to 8' 4", 14,300 lb capacity	B1@.250	Ea	49.40	6.84	56.24

Landscaping

Ground cover. Costs include plants and hand labor to plant

	Craft@Hrs	Unit	Material	Labor	Total
English Ivy, 2-3/4" pots	BL@.098	Ea	.45	2.42	2.87
Liriope Muscari, 1 quart	BL@.133	Ea	1.38	3.29	4.67
Periwinkle, 2-3/4" pots	BL@.095	Ea	.56	2.35	2.91
Purple-leafed Wintercreeper, 1 quart	BL@.128	Ea	1.38	3.16	4.54

Plants and shrubs. Costs include plants, planting by hand, fertilization, small quantity

	Craft@Hrs	Unit	Material	Labor	Total
Abelia, glossy, 18" - 24"	BL@.835	Ea	9.28	20.60	29.88
Acacia, Fern leaf, 5 gal.	BL@.868	Ea	11.50	21.40	32.90
Australian Brush Cherry, 5 gal.	BL@.868	Ea	11.50	21.40	32.90
Azalea, Indica or Kaepferi, 18" - 24"	BL@.868	Ea	8.60	21.40	30.00
Azalea, Kurume types, 18" - 24"	BL@.868	Ea	8.60	21.40	30.00
Begonia, Angel Wing	BL@.835	Ea	10.20	20.60	30.80
Blue Hibiscus, 5 gal.	BL@.835	Ea	11.00	20.60	31.60
Bougainvillea, all varieties, 5 gal.	BL@.868	Ea	11.70	21.40	33.10
Bridal Wreath, 30" - 36"	BL@1.44	Ea	9.39	35.60	44.99
Burford Holly, dwarf, 30" - 36"	BL@1.44	Ea	24.30	35.60	59.90

	Craft@Hrs	Unit	Material	Labor	Total
Camellia, Sasanqua, 2 1/2' - 3'	BL@.835	Ea	11.80	20.60	32.40
Cherry Laurel, 18" - 24"	BL@1.53	Ea	14.01	37.80	51.81
Cleyera Japonica, 24" - 30"	BL@1.14	Ea	12.00	28.20	40.20
Compact Japanese Holly, 15" - 18"	BL@.696	Ea	9.50	17.20	26.70
Dwarf Lily of the Nile, 5 gal.	BL@.835	Ea	10.20	20.60	30.80
Dwarf Yaupon Holly, 15" - 18"	BL@.668	Ea	9.14	16.50	25.64
Gardenia, all varieties, 15" - 18"	BL@.835	Ea	9.39	20.60	29.99
Geranium, all varieties, 5 gal.	BL@.835	Ea	9.91	20.60	30.51
Gold Dust Aucuba, 24" - 30"	BL@1.19	Ea	11.90	29.40	41.30
Grey Lavender Cotton, 12" - 15"	BL@.601	Ea	3.30	14.90	18.20
Juniper, San Jose 18" - 24"	BL@.868	Ea	9.12	21.40	30.52
Lily of the Nile, 5 gal.	BL@.835	Ea	10.20	20.60	30.80
Night Blooming Jasmine, 5 gal.	BL@.835	Ea	10.20	20.60	30.80
Pampas Grass, 2 gal.	BL@.417	Ea	5.10	10.30	15.40
Pink Escallonia, 5 gal.	BL@.835	Ea	10.20	20.60	30.80
Poinciana Paradise Shrub, 5 gal.	BL@.868	Ea	11.20	21.40	32.60
Prickly Pear Cactus, 2 gal.	BL@.417	Ea	7.97	10.30	18.27
Redleaved Jap. Barberry, 18" - 24"	BL@.868	Ea	8.80	21.40	30.20
Rhododendron, 24" - 30"	BL@.835	Ea	29.90	20.60	50.50
Rubber Plant, 5 gal.	BL@.835	Ea	14.70	20.60	35.30
Sprengers Ash, 5 gal.	BL@.907	Ea	11.20	22.40	33.60
Shore Juniper, 18" - 24"	BL@.868	Ea	9.10	21.40	30.50
Tamarix Juniper, 18" - 24"	BL@.835	Ea	9.10	20.60	29.70
Thunbergi Spiraea, 24" - 30"	BL@1.14	Ea	8.90	28.20	37.10
Unedo- Strawberry Tree, 5 gal.	BL@.835	Ea	10.10	20.60	30.70
Vanhoutte Spiraea, 24" - 36"	BL@1.31	Ea	9.10	32.40	41.50
Veronica, all varieties, 5 gal.	BL@.835	Ea	10.10	20.60	30.70
Viburnum, all varieties, 3' - 4'	BL@.835	Ea	14.10	20.60	34.70
Willow Leaf Japanese Holly, 24" - 30"	BL@.442	Ea	13.70	10.90	24.60
Wintergreen Barberry, 18" - 24"	BL@.868	Ea	10.60	21.40	32.00
Yew, Hicks, 24" - 30"	BL@1.31	Ea	18.90	32.40	51.30

Trees and shrubs. Costs include planting, fertilization, backfill, and support as required. Caliper refers to trunk diameter

	Craft@Hrs	Unit	Material	Labor	Total
American Holly, 5' - 6'	BL@1.75	Ea	61.80	43.20	105.00
American Elm, 12' - 14'	BL@3.66	Ea	107.00	90.40	197.40
Armstrong Juniperus, 24" - 30"	BL@.835	Ea	9.30	20.60	29.90
Australian Tree Fern, 5 gal.	BL@.835	Ea	11.30	20.60	31.90
Austrian Pine, 4' - 5'	BL@1.99	Ea	27.60	49.20	76.80
Bamboo Golden, 10' - 12'	BL@2.08	Ea	60.80	51.40	112.20
Bradford Pear, 8' - 10',	BL@2.25	Ea	88.20	55.60	143.80
Camellia, common type, #3 container	BL@1.10	Ea	11.80	27.20	39.00
Cape Myrtle, 8' - 10'	BL@2.25	Ea	71.80	55.60	127.40
Deodara Cedar, 6' - 8'	BL@3.66	Ea	57.30	90.40	147.70
Eastern Redbud, 6' - 8'	BL@2.00	Ea	44.10	49.40	93.50
Foster's American Holly, 5' - 6'	BL@1.75	Ea	70.80	43.20	114.00
Flowering Crabapple, Japanese, 2-1/2" - 3" caliper	BL@2.25	Ea	60.00	55.60	115.60
Fuchsia, all varieties, 5 gal.	BL@.835	Ea	9.60	20.60	30.20
Goldenrain Tree, 8' - 10'	BL@2.25	Ea	75.00	55.60	130.60
Green Ash, 2 gal.	BL@2.00	Ea	8.38	49.40	57.78
Greenspire Linden, 2-1/2" - 3" caliper	BL@3.05	Ea	110.00	75.40	185.40
Hibiscus, all varieties, 5 gal.	BL@.835	Ea	11.00	20.60	31.60
Japanese Privet, 4' - 5'	BL@1.46	Ea	11.80	36.10	47.90
Kwanzan Cherry, 1-3/4" - 2" caliper	BL@3.05	Ea	66.20	75.40	141.60

Landscaping

	Craft@Hrs	Unit	Material	Labor	Total
Live Oak, 3" - 3-1/2" caliper	BL@2.25	Ea	148.90	55.60	204.50
London Planetree Sycamore, 3" - 3-1/2" caliper	BL@3.66	Ea	123.00	90.40	213.40
Maidenhair Tree, 2-1/2" - 3" caliper	BL@3.66	Ea	120.00	90.40	210.40
Northern Red Oak, 3" - 3-1/2" caliper	BL@2.00	Ea	136.00	49.40	185.40
Norway Blue Spruce, 4' - 5'	BL@1.99	Ea	41.90	49.20	91.10
Pin Oak, 3" - 3-1/2" caliper	BL@3.48	Ea	13.70	86.00	99.70
Prince of Wales Juniperus, 5 gal.	BL@.835	Ea	9.12	20.60	29.72
Psidium (guava), 5 gal.	BL@.835	Ea	9.60	20.60	30.20
Pyracantha Firethorn, 3' - 4'	BL@1.10	Ea	7.95	27.20	35.15
Purpleleaf Flowering Plum, 6' - 8'	BL@2.00	Ea	44.10	49.40	93.50
Red Maple, 2" - 2-1/2" caliper	BL@3.05	Ea	93.80	75.40	169.20
Saucer Magnolia, 6' - 8'	BL@2.00	Ea	68.40	49.40	117.80
Star Magnolia, 4' - 5'	BL@1.46	Ea	44.10	36.10	80.20
Sugar Maple, 2-1/2" - 3" caliper	BL@2.00	Ea	99.30	49.40	148.70
Thornless Honeylocust, 10' - 12'	BL@2.25	Ea	108.00	55.60	163.60
Washington Hawthorne, 8' - 10'	BL@2.25	Ea	130.00	55.60	185.60
Weeping Fig, 5 gal.	BL@.835	Ea	14.30	20.60	34.90
Weeping Forsythia, 3' - 4'	BL@1.10	Ea	9.12	27.20	36.32
White Flowering Dogwood, 5' - 6'	BL@1.75	Ea	38.70	43.20	81.90
White Pine, 6' - 8'	BL@2.00	Ea	50.80	49.40	100.20
Willow Oak, 3" - 3-1/2" caliper	BL@2.74	Ea	137.00	67.70	204.70
Wisconsin Weeping Willow, 2-1/2" - 3" caliper	BL@2.00	Ea	72.50	49.40	121.90
Palm trees, based on West Coast prices					
Mexican Fan Palm, 5 gal.	BL@1.50	Ea	19.00	37.10	56.10
Mediterranean Fan Palm, 5 gal.	BL@1.50	Ea	13.30	37.10	50.40
Pigmy Date Palm, 5 gal.	BL@1.50	Ea	13.30	37.10	50.40
Common King Palm, 5 gal.	BL@1.50	Ea	13.30	37.10	50.40
Common Queen Palm, 5 gal.	BL@1.50	Ea	19.00	37.10	56.10
Seeding, level area					
Using tall fescue seed, per 20 lb bag	—	Ea	40.00	—	40.00
Seeding preparation, grade, rake, clean	BL@.003	SY	.10	.07	.17
Mechanical seeding (5 lbs per 1,000 SF)	BL@.020	MSF	5.00	.49	5.49
Hand seeding (10 lbs per 1,000 SF)	BL@.040	MSF	10.00	.99	10.99
Fertilizer, pre-planting,					
18-nitrogen, 24-phosphorus, 12-potassium					
(5 lb per 1,000 SF)	BL@.222	MSF	1.43	5.49	6.92
Liming (70 lbs per 1,000 SF)	BL@.222	MSF	7.22	5.49	12.71

Placing topsoil, delivered to site, 1 CY covers 81 SF at 4" depth. Add cost for equipment, where needed, from Excavation Equipment Costs section. Labor shown below is for operation of equipment.

	Craft@Hrs	Unit	Material	Labor	Total
With equipment, level site	BL@.092	CY	20.00	2.27	22.27
With equipment, sloped site	BL@.107	CY	20.00	2.64	22.64
By hand, level site	BL@.735	CY	20.00	18.20	38.20
By hand, sloped site	BL@.946	CY	20.00	23.40	43.40
Placing soil amendments, by hand spreading					
Wood chip mulch, pine bark	BL@1.27	CY	23.00	31.40	54.40
Bale of straw, 25 bales per ton	BL@.200	Ea	3.06	4.94	8.00

	Craft@Hrs	Unit	Material	Labor	Total
Landscape stepping stones, concrete					
12" round	BL@.133	Ea	2.52	3.29	5.81
14" round	BL@.133	Ea	3.40	3.29	6.69
18" round	BL@.153	Ea	5.88	3.78	9.66
24" round	BL@.153	Ea	9.89	3.78	13.67
Add for square shapes	—	%	20.0	—	—
18" or 24" diameter natural tree ring	BL@.133	Ea	2.14	3.29	5.43
Redwood benderboard, staked redwood					
5/16" x 4"	BL@.011	LF	.26	.27	.53
5/8" x 3-3/8"	BL@.016	LF	.91	.40	1.31
5/8" x 5-3/8"	BL@.016	LF	1.42	.40	1.82
Roto-tilling light soil					
To 4" depth	BL@.584	CSY	—	14.40	14.40
Sodding, no soil preparation, nursery sod, Bermuda or Blue Rye mix					
Per SF (500 SF minimum charge)	BL@.013	SF	.38	.32	.70

Lath

	Craft@Hrs	Unit	Material	Labor	Total
Gypsum lath, perforated or plain, nailed to walls and ceilings					
3/8" x 16" x 48"	BR@.076	SY	4.92	2.49	7.41
1/2" x 16" x 48"	BR@.083	SY	5.38	2.72	8.10
Add for foil back insulating lath	—	SY	1.00	—	1.00
Steel lath, diamond pattern (junior mesh), 27" x 96" sheets, nailed to walls					
1.75 pound	BR@.076	SY	6.50	2.49	8.99
2-1/2 pound, galvanized	BR@.076	SY	8.15	2.49	10.64
2-1/2 pounds, paper back, dimpled	BR@.076	SY	9.45	2.49	11.94
3.4 pounds, galvanized	BR@.076	SY	9.30	2.49	11.79
3.4 pounds, paper back	BR@.076	SY	9.45	2.49	11.94
3/8" hi rib, 29 gauge	BR@.076	SY	8.82	2.49	11.31
Add for ceiling applications	—	%	—	25.0	—

	Craft@Hrs	Unit	Material	Labor	Total
Galvanized stucco netting, 1-1/2" mesh, per roll, nailed to walls					
36" x 150', 20 gauge	BR@2.80	Ea	56.00	91.80	147.80
36" x 150', 17-gauge	BR@2.80	Ea	70.00	91.80	161.80
36" x 100', 17-gauge paper back	BR@2.80	Ea	67.20	91.80	159.00
6" x 50' stucco repair net	--@.000	Ea	12.20	--	12.20
Corner bead, 26 gauge, 8' to 12' lengths.					
Expanded corner bead, 8' long	BR@.027	LF	.45	.89	1.34
Expanded corner bead, 10' long	BR@.027	LF	.46	.89	1.35
Cornerite corner bead, 3" x 3" x 8'	BR@.027	LF	.20	.89	1.09
Cornerite plaster corner, 2" x 2" x 4'	BR@.027	LF	.33	.89	1.22
Cornerite stucco corner, 3" x 3" 8' long	BR@.027	LF	.31	.89	1.20
Short flange corner bead, 8' long	BR@.027	LF	.61	.89	1.50
Plaster J-trim, 1-3/8" wide x 10' long	BR@.027	LF	.35	.89	1.24
Casing beads					
Short flange, 1/2" x 10' long	BR@.025	LF	.30	.82	1.12
Short flange, 7/8" x 10' long	BR@.025	LF	.31	.82	1.13
Expanded flange, 3/4" x 10'	BR@.025	LF	.35	.82	1.17
Expansion joint, 26 gauge, 1/2" ground	BR@.050	LF	1.08	1.64	2.72

Lawn Sprinkler Systems

	Craft@Hrs	Unit	Material	Labor	Total
Lawn Sprinkler Systems PVC pipe Schedule 40, including typical hand trenching, plastic fittings (but no valves or sprinkler heads), installed.					
1/2" Schedule 40 PVC pressure pipe and fittings					
1/2" pipe	BL@.020	LF	.14	.49	.63
1/2" 90 degree ell	BL@.100	Ea	.20	2.47	2.67
1/2" tee	BL@.130	Ea	.21	3.21	3.42
3/4" Schedule 40 PVC pressure pipe and fittings					
3/4" pipe	BL@.025	LF	.18	.62	.80
3/4" 90 degree ell	BL@.115	Ea	.21	2.84	3.05
3/4" tee	BL@.140	Ea	.26	3.46	3.72
1" Schedule 40 PVC pressure pipe and fittings					
1" pipe	BL@.030	LF	.27	.74	1.01
1" 90 degree ell	BL@.120	Ea	.41	2.97	3.38
1" tee	BL@.170	Ea	.51	4.20	4.71
Deduct for Class 315 pipe	—	%	-40.0	—	—
Lawn sprinkler system components.					
Trenching for pipe installation, by hand					
Light soil, 8" wide					
12" deep	BL@.027	LF	—	.67	.67
18" deep	BL@.040	LF	—	.99	.99
24" deep	BL@.053	LF	—	1.31	1.31
Average soil, 8" wide					
12" deep	BL@.043	LF	—	1.06	1.06
18" deep	BL@.065	LF	—	1.61	1.61
24" deep	BL@.087	LF	—	2.15	2.15
Heavy soil or loose rock, 8" wide					
12" deep	BL@.090	LF	—	2.22	2.22
18" deep	BL@.100	LF	—	2.47	2.47
24" deep	BL@.134	LF	—	3.31	3.31
Impact sprinkler heads, riser mounted, with nozzle, full circle					
Brass, adjustable	BL@.422	Ea	9.41	10.40	19.81
Plastic	BL@.422	Ea	6.36	10.40	16.76
Impact sprinkler, pop-up type					
Brass and stainless steel	BL@.422	Ea	17.00	10.40	27.40
Plastic	BL@.422	Ea	17.90	10.40	28.30
Pop-up lawn sprinkler heads, plastic					
Full circle	BL@.422	Ea	3.70	10.40	14.10
Half circle	BL@.422	Ea	3.71	10.40	14.11
Rotor sprinkler					
Full circle, 30' radius	BL@.422	Ea	9.42	10.40	19.82
Spray head pop-up type, plastic, with nozzle					
2-1/2" height	BL@.422	Ea	3.14	10.40	13.54
4" height	BL@.422	Ea	3.78	10.40	14.18
6" height	BL@.422	Ea	4.24	10.40	14.64
Spray head, brass, for riser mounting					
Flat pattern	BL@.191	Ea	2.12	4.72	6.84
Standard	BL@.191	Ea	2.39	4.72	7.11
Stream bubbler, all patterns	BL@.191	Ea	3.18	4.72	7.90

	Craft@Hrs	Unit	Material	Labor	Total
Sprinkler head risers, 1/2" or 3/4", installed to main or branch supply					
6" riser with three fittings	BL@.056	Ea	7.70	1.38	9.08
8" riser with three fittings	BL@.056	Ea	7.70	1.38	9.08
12" riser with three fittings	BL@.056	Ea	9.02	1.38	10.40
18" riser with three fittings	BL@.056	Ea	12.70	1.38	14.08
24" riser with three fittings	BL@.056	Ea	14.90	1.38	16.28
Valves, non-siphon, manual, brass					
3/4" valve	BL@1.03	Ea	23.90	25.50	49.40
1" valve	BL@1.38	Ea	28.10	34.10	62.20
Valves, electric solenoid, plastic, with pressure-assisted flow control, hookup but no wire included					
3/4" valve	BL@1.26	Ea	18.90	31.10	50.00
1" valve	BL@1.26	Ea	18.90	31.10	50.00

Valve control wire, 16 gauge low voltage plastic jacket direct burial cable, laid with pipe. No trenching or end connections included

	Craft@Hrs	Unit	Material	Labor	Total
2 conductor	BL@.003	LF	.08	.07	.15
3 conductor	BL@.003	LF	.11	.07	.18
5 conductor	BL@.003	LF	.16	.07	.23
7 conductor	BL@.003	LF	.30	.07	.37
8 conductor	BL@.003	LF	.29	.07	.36

Automatic irrigation control stations, no valve or wiring included, add for electrical service connection below.

	Craft@Hrs	Unit	Material	Labor	Total
Programmable irrigation control station, wall mounted					
For up to 6 electric valves	BL@1.83	LS	42.10	45.20	87.30
For up to 8 electric valves	BL@1.83	LS	57.70	45.20	102.90
Add for 120 volt power source	—	LS	—	—	85.00

Residential system, typical contract price for 1,700 SF area using PVC Schedule 40 pipe, 340 LF of 3/4" pipe, 35 LF of 1" pipe, four 1" valves, 30 brass sprinkler heads and 6" risers on each head, based on heads spaced at 10'

	Craft@Hrs	Unit	Material	Labor	Total
Large, regular shaped areas	—	SF	—	—	.66
Narrow and irregular areas	—	SF	—	—	.73
Add for freezing zones	—	SF	—	—	.12

Layout Foundation layout, medium to large size residence, 6 to 8 outside corners, includes shooting elevation from nearby reference, stakes and batterboards as required. No surveying or clearing included.

	Craft@Hrs	Unit	Material	Labor	Total
Typical cost per residence	B1@7.86	LS	55.00	215.00	270.00

Lighting Fixtures See also Electrical Work. Costs are to hang and connect fixtures only. Add the cost of wiring and fittings from the Commercial and Industrial section of this manual.

Fluorescent lighting fixtures

Surface mounted fluorescent fixtures, including ballasts but no lamps.

	Craft@Hrs	Unit	Material	Labor	Total
Wraparound acrylic diffuser					
24", two 20 watt tubes	BE@1.60	Ea	59.80	54.20	114.00
48", two 40 watt tubes	BE@1.60	Ea	74.20	54.20	128.40
48", four 40 watt tubes	BE@1.60	Ea	118.00	54.20	172.20
96" tandem, four 40 watt tubes	BE@1.60	Ea	146.00	54.20	200.20
Low profile wraparound model					
10" x 48" x 4", two 20 watt tubes	BE@1.60	Ea	60.20	54.20	114.40
15" x 48" x 4", four 40 watt tubes	BE@1.60	Ea	102.00	54.20	156.20
15" x 96" x 4", eight 40 watt tubes	BE@1.60	Ea	198.00	54.20	252.20
19" x 19" x 4", one 32/40 watt circline tube	BE@1.60	Ea	65.30	54.20	119.50
24" x 24" x 4", two 40 watt U/6 tubes	BE@1.60	Ea	83.90	54.20	138.10

Lighting Fixtures

	Craft@Hrs	Unit	Material	Labor	Total
Basic wraparound acrylic diffusers, solid oak wood ends					
10" x 24" x 3", two 20 watt tubes	BE@1.60	Ea	38.80	54.20	93.00
10" x 48" x 3", two 40 watt tubes	BE@1.60	Ea	47.40	54.20	101.60
Economy, wraparound acrylic diffuser, basic metal ends					
14" x 48" x 2", four 40 watt tubes	BE@1.60	Ea	71.10	54.20	125.30
8" x 48" x 2", two 40 watt tubes	BE@1.60	Ea	38.80	54.20	93.00
Basic circular fixtures for circline tubes					
Simulated glass crystal with antiqued metal band,					
11" round x 3" high, one 22 watt tube	BE@.696	Ea	28.60	23.60	52.20
Circular fixture					
12" round x 3 high, 22 and 32 watt tubes	BE@.696	Ea	27.50	23.60	51.10
Fixture with frosted glass diffuser					
19" round x 4" high, 32 and 40 watt tubes	BE@.696	Ea	63.70	23.60	87.30
Low profile rectangular wall or ceiling mount fixture, metal housing, white acrylic diffuser					
9-5/8" x1-7/8", two 9 watt energy saver bulbs	BE@.696	Ea	36.60	23.60	60.20
Rectangular indoor/outdoor model with discoloration resistant/impact resistant acrylic diffuser and cold weather ballast					
10" x 10" x 5" deep, one 22 watt tube	BE@.696	Ea	59.30	23.60	82.90
12" x 12" x 5" deep, one 34 watt tube	BE@.696	Ea	76.80	23.60	100.40
Hall lantern type, including tubes					
4-1/4" wide x 12-1/8" long x 4" deep	BE@.696	Ea	37.40	23.60	61.00
Decorative valance type, with solid oak frame and acrylic diffuser					
12" x 25" x 4", two 20 watt tubes	BE@2.06	Ea	100.00	69.80	169.80
25" x 25" x 4", two 40 watt U/6 tubes	BE@1.59	Ea	125.00	53.80	178.80
12" x 49" x 4", two 40 watt tubes	BE@1.59	Ea	115.00	53.80	168.80
Economy valance type, with metal ends. Round or square models available					
All models are 2-5/8" wide x 4-1/4" deep					
24-3/4" L (requires one 15 watt tube)	BE@1.59	Ea	27.50	53.80	81.30
24-3/4" L (requires one 20 watt tube)	BE@1.59	Ea	29.70	53.80	83.50
36-3/4" L (requires one 30 watt tube)	BE@1.59	Ea	38.40	53.80	92.20
36-3/4" L (requires two 30 watt tubes)	BE@1.59	Ea	46.70	53.80	100.50
48-3/4" L (requires one 40 watt tube)	BE@1.59	Ea	52.70	53.80	106.50
48-3/4" L (requires two 40 watt tubes)	BE@1.59	Ea	59.30	53.80	113.10
Under shelf/cabinet type for wiring to wall switch					
18" x 5" x 2", one 15 watt tube	BE@.696	Ea	14.80	23.60	38.40
24" x 5" x 2", one 20 watt tube	BE@.696	Ea	19.80	23.60	43.40
36" x 5" x 2", one 30 watt tube	BE@.696	Ea	30.70	23.60	54.30
48" x 5" x 2", one 40 watt tube	BE@.696	Ea	43.40	23.60	67.00

	Cool white	Deluxe cool white	Energy Saving cool white	Warm white
Fluorescent tubes				
8 watt (12" long)	6.13	9.59	—	9.07
15 watt (18" long)	6.50	8.31	—	7.25
20 watt (24" long)	5.87	8.85	8.53	6.31
30 watt (36" long)	6.83	11.10	5.87	9.49
40 watt (48" long)	2.13	—	—	—
Slimline straight tubes				
30 watt (24" or 36" long)	15.20	—	12.10	16.60
40 watt (48" long)	10.00	13.90	19.20	12.40

	Cool white	Deluxe cool white	Energy Saving cool white	Warm white
Circular (circline) tubes				
20 watt (6-1/2" diameter)	9.59	—	—	10.70
22 watt (8-1/4" diameter)	7.23	10.10	—	10.40
32 watt (12" diameter)	8.85	12.20	—	11.70
40 watt (16" diameter)	12.80	—	—	15.50
"U" shaped tubes (3" spacing x 12" long)				
U-3 40 watt	13.10	17.60	—	13.10
U-6 40 watt	13.10	17.60	—	13.10
Labor to install each tube	1.20	1.20	1.20	1.20

Incandescent lighting fixtures

	Craft@Hrs	Unit	Material	Labor	Total
Outdoor post lantern fixtures, antique satin brass finish, no digging included					
Post lantern type fixture, cast aluminum	BE@2.00	Ea	58.90	67.70	126.60
Modern porch ceiling fixture					
7-1/2" x 7-1/2" x 3" drop, single light	BE@2.15	Ea	22.90	72.80	95.70
Flood light lampholders					
Cadmium steel plate	BE@2.15	Ea	27.30	72.80	100.10
1/2" male swivel 7-5/16" long, aluminum	BE@2.15	Ea	26.30	72.80	99.10
Path lighting, kits with cable and timer					
20 light low voltage	BE@2.63	LS	53.60	89.10	142.70
6 flood lights, 11 watt	BE@2.98	LS	56.80	101.00	157.80
12 tier lights, 7 watt, programmable	BE@2.63	LS	76.40	89.10	165.50
Recessed fixtures, including housing, incandescent					
Integral thermally protected recessed (can) light housing					
4" recessed light housing, non-insulated	BE@1.00	Ea	16.30	33.90	50.20
5" recessed light housing, non-insulated	BE@1.00	Ea	20.60	33.90	54.50
6" recessed light housing, non-insulated	BE@1.00	Ea	10.30	33.90	44.20
Add for insulated housing	—	—	25%	—	—
Add for low voltage	—	—	100%	—	—
Recessed baffle light trim	BE@0.25	Ea	9.27	8.47	17.74
Recessed flat light trim	BE@0.25	Ea	9.27	8.47	17.74
Recessed eyeball light trim	BE@0.25	Ea	30.90	8.47	39.37
Recessed reflector light trim	BE@0.25	Ea	13.40	8.47	21.87
Recessed shower light trim	BE@0.25	Ea	12.30	8.47	20.77
Recessed open light trim	BE@0.25	Ea	4.12	8.47	12.59
Recessed adjustable light trim	BE@0.25	Ea	23.30	8.47	31.77
Recessed splay light trim	BE@0.25	Ea	10.20	8.47	18.67
Utility 12" x 10", diffusing glass lens	BE@1.00	Ea	45.70	33.90	79.60
Ceiling fixtures, using 60 watt standard bulbs, pre-wired housing, white, canopy finish					
Cloud 9-3/4" x 5-1/4" (2 bulbs)	BE@1.59	Ea	24.40	53.80	78.20
Cloud 8-1/4" x 5-1/2" (1 bulb)	BE@1.59	Ea	30.90	53.80	84.70
Standard globe 8-3/4" x 5" (2 bulbs)	BE@1.59	Ea	21.30	53.80	75.10
Cloud globe, brass canopy 10" x 5"	BE@1.59	Ea	26.30	53.80	80.10

Log Home Construction

Logs are cut from standing dead lodgepole pine or spruce. Add the foundation cost. Based on Authentic Log Homes.

Rule of thumb cost, including labor, material and hardware but no foundation, plumbing, HVAC, electrical work or interior finishing.

Total cost per square foot of wall (shell only)	—	SF	—	—	26.00

Log Home Construction

	Craft@Hrs	Unit	Material	Labor	Total

Lodgepole pine or spruce logs, including local delivery, per LF of log at 8" to 12" diameter

	Craft@Hrs	Unit	Material	Labor	Total
8" machine peeled	—	LF	12.20	—	12.20
8" hand peeled	—	LF	12.90	—	12.90
10" machine peeled	—	LF	16.40	—	16.40
10" hand peeled	—	LF	18.20	—	18.20
12" machine peeled	—	LF	21.50	—	21.50
12" hand peeled	—	LF	23.00	—	23.00

Labor costs for log home construction. Using a crew experienced in log home construction. Add 10% to 50% for crews on their first log home project. Subfloor, and porch joist hardware uses type B Simpson Strong-Tie hangers. These labor estimates assume precut 8" x 6" log profiles. For 9" x 7" profiles, add 10%. For 10" x 8" profiles, add 20%. For random length logs, add 10%. The preferred crew is three carpenters and three helpers.

	Craft@Hrs	Unit	Material	Labor	Total
Subfloor					
CCA treated sill plate, anchor bolts.	B1@.057	LF	—	1.56	1.56
Built-up girders (using 2 x 8 lumber)	B1@.222	LF	—	6.07	6.07
Posts (8" log), with anchor	B1@.500	Ea	—	13.70	13.70
Rim joists and blocking (2 x 8)	B1@.065	LF	—	1.78	1.78
Floor joists (2 x 8), hangers, bridging/blocking	B1@.025	LF	—	.68	.68
Decking	B1@.014	SF	—	.38	.38
Porch					
Rim joists	B1@.017	LF	—	.47	.47
Posts (8" log), with anchor	B1@.500	Ea	—	13.70	13.70
Joists with hangers	B1@.025	SF	—	.68	.68
Decking (2 x 6 redwood)	B1@.025	SF	—	.68	.68
Balcony railing	B1@.125	LF	—	3.42	3.42
Sill log					
Flashing	B1@.010	LF	—	.27	.27
Sill log spiked onto subfloor	B1@.050	LF	—	1.37	1.37
Laying wall courses					
Courses 2 through 8	B1@.025	LF	—	.68	.68
Courses 9 through 16	B1@.033	LF	—	.90	.90
Course 17 (sill log)	B1@.040	LF	—	1.09	1.09
Courses 18 through 24	B1@.040	LF	—	1.09	1.09
Courses 25 through 32	B1@.050	LF	—	1.37	1.37
Doors and windows					
Layout and installation, per opening	B1@4.00	Ea	—	109.00	109.00
Interior trim	B1@2.00	Ea	—	54.70	54.70
Exterior trim	B1@2.00	Ea	—	54.70	54.70
Mantle (10 feet long)					
With bracing	B1@1.50	Ea	—	41.00	41.00
Log floor system					
Girder	B1@.050	LF	—	1.37	1.37
Support posts	B1@.500	Ea	—	13.70	13.70
Support wall (studs, top and bottom plates)	B1@.167	LF	—	4.57	4.57
Cutting wall notch to accept girder	B1@1.50	Ea	—	41.00	41.00
Floor joists	B1@.063	LF	—	1.72	1.72
Cutting wall notch to accept floor joist	B1@0.50	Ea	—	13.70	13.70
T&G decking	B1@.025	SF	—	.68	.68
Log staircase					
Straight	B1@30.0	Ea	—	821.00	821.00
L or U shaped	B1@36.0	Ea	—	985.00	985.00

	Craft@Hrs	Unit	Material	Labor	Total
Wall stiffener					
6" x 6" or 8" x 8" beam channeled into wall	B1@2.00	Ea	—	54.70	54.70
Log rafter system, including dormers					
Ridge beam, main and secondary	B1@0.25	LF	—	6.84	6.84
Notching gable to accept ridge beam	B1@1.50	Ea	—	41.00	41.00
Built up truss pocket	B1@2.00	Ea	—	54.70	54.70
Valley beam layout and installation	B1@0.33	LF	—	9.03	9.03
Ridge beam support post	B1@0.50	Ea	—	13.70	13.70
Rafter 20' or less	B1@1.00	Ea	—	27.40	27.40
Rafter over 20'	B1@1.50	Ea	—	41.00	41.00
Valley rafter 16' or less	B1@1.50	Ea	—	41.00	41.00
Valley rafter between 16' and 20'	B1@2.00	Ea	—	54.70	54.70
Valley rafter over 20'	B1@2.50	Ea	—	68.40	68.40
Timber rafter 18' or less	B1@3.00	Ea	—	82.10	82.10
Timber rafter over 18'	B1@4.00	Ea	—	109.00	109.00
Rafter layout	B1@.250	Ea	—	6.84	6.84
Snow block	B1@1.00	Ea	—	27.40	27.40
Log trusses, 32' and less					
Hardware plate	B1@2.00	Ea	—	54.70	54.70
Collar ties	B1@1.00	Ea	—	27.40	27.40
Vertical or diagonal web	B1@3.00	Ea	—	82.10	82.10
Add for each 5' of truss over 32'	—	%	—	50.0	—
Cold roof construction					
Secondary rafters	B1@0.50	Ea	—	13.70	13.70
Jack rafters	B1@0.25	Ea	—	6.84	6.84
Fly rafters	B1@0.50	Ea	—	13.70	13.70
Blocking	B1@0.25	Ea	—	6.84	6.84
OSB roof sheathing					
Less than or equal to 6/12 pitch	B1@1.72	Sq	—	47.10	47.10
7/12 pitch to 9/12 pitch	B1@2.00	Sq	—	54.70	54.70
Less than or equal to 6/12 pitch	B1@2.50	Sq	—	68.40	68.40
Felt	B1@1.00	Sq	—	27.40	27.40
Asphalt shingles	B1@2.00	Sq	—	54.70	54.70
Metal roof, ridge cap and closure strips	B1@1.33	Sq	—	36.40	36.40
Ridge vents and soffit vents	B1@3.00	Ea	—	82.10	82.10
Facia	B1@0.02	LF	—	.55	.55
Soffit	B1@0.03	SF	—	.82	.82
Caulking					
Exterior only	B1@0.02	LF	—	.55	.55
Preservative					
Exterior only	B1@.005	SF	—	.14	.14
Electrical chases					
Cut in log ends and down tops of rafters	B1@.083	LF	—	2.27	2.27
Holes drilled through ridge beams	B1@0.50	Ea	—	13.70	13.70

Lumber, Framing

		MBF	8' L	12' L	16' L

Framing Lumber

MBF means per 1,000 board feet. One board foot is the volume of lumber in a nominal 1" x 12" board one foot long. A 2" x 6" joist 10 feet long has 10 board feet. To determine the board feet of lumber in any piece, multiply the width in inches by the depth in inches by the length in feet. Then divide by 12.

Douglas Fir – #2 and Better

		MBF	8' L	12' L	16' L
2" x 4" – Green	Ea	595.00	3.17	4.76	6.34
2" x 6" – KD	Ea	669.00	5.36	8.03	10.70
2" x 8" – KD	Ea	690.00	7.36	11.00	14.70
2" x 10" – KD	Ea	665.00	8.86	13.30	17.70
2" x 12" – KD	Ea	656.00	10.50	15.70	21.00

Douglas Fir – kiln dried standard and better

2" x 4"	Ea	679.00	3.62	5.42	7.23
2" x 6"	Ea	674.00	5.39	8.09	10.80
2" x 8"	Ea	654.00	6.97	10.50	13.90
2" x 10"	Ea	704.00	9.39	14.10	18.80
2" x 12"	Ea	752.00	12.00	18.10	24.10

Douglas Fir finish and clear – C & better, vertical grain, S4S, dry. West Coast prices. Add 10% to 30% for transportation to other areas

		MBF	8' L	12' L	16' L
2" x 2"	Ea	3,170.00	8.44	12.70	16.90
2" x 3"	Ea	3,170.00	12.70	19.00	25.40
2" x 4"	Ea	3,580.00	19.10	28.60	38.20
2" x 6"	Ea	3,990.00	31.90	47.90	63.80
2" x 8"	Ea	4,520.00	48.20	72.30	96.30
2" x 10"	Ea	5,620.00	74.90	113.00	149.00
2" x 12"	Ea	7,260.00	116.00	175.00	232.00

Douglas Fir finish and clear – C & Better, flat grain, S4S, dry. West Coast prices. Add 10% to 30% for transportation to other areas

		MBF	8' L	12' L	16' L
2" x 2"	Ea	2,760.00	7.38	11.10	14.80
2" x 3"	Ea	2,760.00	11.10	16.60	22.10
2" x 4"	Ea	2,800.00	14.90	22.40	29.80
2" x 6"	Ea	3,240.00	25.90	38.90	51.80
2" x 8"	Ea	3,240.00	34.60	51.80	69.10
2" x 10"	Ea	3,240.00	43.20	64.80	86.40
2" x 12"	Ea	4,300.00	68.90	104.00	138.00

Hemlock/Fir kiln dried

		MBF	8' L	12' L	16' L
2" x 4" – Standard & Better	Ea	620.00	3.31	4.96	6.61
2" x 6" – #2 & Better	Ea	608.00	4.86	7.29	9.72
2" x 8" – #2 & Better	Ea	654.00	6.98	10.50	14.00
2" x 10" – #2 & Better	Ea	667.00	8.88	13.30	17.80
2" x 12" – #2 & Better	Ea	732.00	11.70	17.60	23.40

		MBF	8' L	12' L	16' L
Spruce Pine Fir (SPF), kiln dried					
2" x 4", #2 and better	Ea	609.00	3.25	4.87	6.49
2" x 6", #2 and better	Ea	632.00	5.05	7.58	10.10
2" x 8", #2 and better	Ea	668.00	7.12	10.70	14.30
2" x 10", #2 and better	Ea	687.00	9.16	13.70	18.30
2" x 12", #2 and better	Ea	733.00	11.70	17.60	23.40
2" x 4", standard and better	Ea	638.00	3.40	5.10	6.80
4" x 4", standard and better	Ea	924.00	9.85	14.80	19.70

		MBF	8' L	12' L	16' L
Red Cedar – green S4S					
2" x 4"	Ea	1,310.00	7.01	10.50	14.00
2" x 6"	Ea	1,510.00	12.10	18.10	24.20
Red Cedar – green – D and better S4S					
2" x 2"	Ea	2,120.00	5.64	8.46	11.30
2" x 4"	Ea	2,510.00	13.40	20.10	26.80
2" x 6"	Ea	2,850.00	22.80	34.20	45.70
Red Cedar – rough – standard and better					
2" x 2"	Ea	946.00	2.52	3.78	5.04
2" x 4"	Ea	1,240.00	6.63	9.95	13.30
2" x 6"	Ea	1,330.00	10.60	15.90	21.20
2" x 8"	Ea	1,390.00	14.90	22.30	29.70

		MBF	8' L	12' L	16' L
Southern Yellow Pine – #1 S4S					
2" x 4"	Ea	659.00	3.51	5.27	7.02
2" x 6"	Ea	595.00	4.76	7.14	9.52
2" x 10"	Ea	676.00	9.01	13.50	18.00
2" x 12"	Ea	848.00	13.60	20.40	27.10
Southern Yellow Pine – #2 S4S					
2" x 4"	Ea	562.00	3.00	4.49	5.99
2" x 6"	Ea	578.00	4.63	6.94	9.25
2" x 8"	Ea	580.00	6.18	9.27	12.40
2" x 10"	Ea	561.00	7.48	11.20	15.00
2" x 12"	Ea	700.00	11.20	16.80	22.40
Southern Yellow Pine – prime S4S					
2" x 4"	Ea	605.00	3.22	4.84	6.45
2" x 6"	Ea	582.00	4.65	6.98	9.31
2" x 8"	Ea	595.00	6.34	9.51	12.70
2" x 10"	Ea	641.00	8.55	12.80	17.10
2" x 12"	Ea	745.00	11.90	17.90	23.80

Lumber, 2" x 4" and 2" x 6"

2" x 4" and 2" x 6" Framing Lumber, per piece Studs are stud grade, 8', 10' and 12' lengths are No. 2 or better grade. 16' lengths are Standard or No. 2 grade. Kiln dried except as noted. White Wood includes SPF (spruce, pine, fir) Engleman spruce, all true firs, hemlocks and pines (excluding Ponderosa pine). Hem fir can be California red fir, grand fir, noble fir, Pacific silver fir, Shasta fir, white fir or Western hemlock. A dash ("—") indicates that the dimension and grade is not commonly sold in that area or that no data is available. Prices are as of fall 2005. See the Carpentry section for installation costs. Note: Area modification factors should not be used with these prices as they are already adjusted by locality.

City, ST, Grade	2"x4" 8' Stud	2"x4" 12'	2"x4" 16'	2"x6" 8'	2"x6" 10'	2"x6" 12'
Albany, NY, White Wood	2.69	5.04	6.72	3.69	5.99	7.20
Albuquerque, NM, White Wood	2.75	5.15	8.03	5.24	6.45	—
Atlanta, GA, White Wood	2.83	4.36	6.40	3.69	5.36	6.06
Austin, TX, White Wood	2.94	4.54	6.42	3.69	5.64	6.37
Austin, TX, Yellow Pine	2.36	4.30	—	4.55	5.49	6.45
Baltimore, MD, White Wood	2.79	4.69	6.25	3.69	5.36	6.44
Bangor, ME, White Wood	2.69	4.98	6.64	3.69	6.18	7.10
Baton Rouge, LA, White Wood	2.79	4.67	6.42	3.69	5.59	6.70
Baton Rouge, LA, Yellow Pine	2.36	4.25	6.40	4.85	4.98	5.56
Bernmidji, MN, White Wood	2.83	4.39	6.19	3.69	5.44	6.13
Birmingham, AL, White Wood	2.83	4.36	6.40	3.69	5.36	6.06
Boston, MA, Green Douglas Fir	3.07	4.98	5.97	3.98	5.60	6.72
Boston, MA, White Wood	2.79	4.48	6.64	3.69	5.92	7.10
Brownsville, TX, White Wood	2.94	4.54	6.42	3.69	5.64	6.73
Buffalo, NY, White Wood	2.69	5.04	6.72	3.69	5.99	7.20
Charleston, SC, White Wood	2.79	4.36	6.42	3.69	5.36	6.06
Chicago (Suburb), IL, White Wood	2.69	6.42	6.43	3.69	5.56	6.27
Chicago (Urban), IL, White Wood	2.78	6.49	6.48	3.80	5.68	6.38
Cincinnati, OH, White Wood	2.83	4.53	6.42	3.69	5.63	6.35
Cleveland, OH, White Wood	2.83	4.52	6.42	3.69	5.61	6.34
Columbia, SC, White Wood	2.79	4.36	6.40	3.69	5.36	6.06
Columbus, OH, White Wood	2.83	4.47	6.29	3.59	5.56	6.29
Dallas, TX, White Wood	2.79	4.17	6.14	3.69	5.17	5.84
Danbury, CT, Green Douglas Fir	3.07	4.48	5.97	3.98	5.60	6.72
Denver, CO, Hem Fir	—	4.65	6.68	3.79	4.74	5.69
Detroit, MI, White Wood	2.69	4.10	5.71	4.45	5.37	5.88
Durham, NC, White Wood	2.79	4.36	6.40	3.69	5.36	6.06
Edison, NJ, Hem Fir	2.98	5.15	6.87	4.94	5.84	6.99
Edison, NJ, Green Douglas Fir	3.07	5.15	6.77	3.98	5.60	6.72
El Paso, TX, Douglas Fir	—	5.19	—	4.71	5.89	—
Evansville, IN, White Wood	3.21	4.03	6.92	4.26	5.88	6.51
Ft Myers, FL, White Wood	2.83	4.32	6.40	3.98	5.34	5.99
Ft. Wayne, IN, White Wood	2.83	4.53	6.42	3.69	5.55	6.13
Gainesville, FL, White Wood	2.83	4.32	6.40	3.98	5.34	5.99
Grand Rapids, MI, White Wood	2.69	3.94	5.79	3.69	5.49	5.95
Greenville, SC, White Wood	2.98	4.36	6.40	3.69	5.36	6.45
Hartford, CT, Green Douglas Fir	3.07	4.65	5.97	3.98	5.60	6.72
Houston, TX, Yellow Pine	2.36	4.25	6.40	4.55	5.49	6.45
Indianapolis, IN, White Wood	2.53	3.78	5.75	3.69	4.61	5.48
Jackson, MS, White Wood	2.83	4.67	6.42	3.69	5.59	6.70
Jacksonville, FL, White Wood	2.83	4.32	6.40	3.98	5.34	5.99
Jacksonville, FL, Yellow Pine	2.61	4.25	6.52	3.45	4.84	5.93
Jensen Beach, FL, White Wood	2.83	4.32	6.40	3.69	5.34	5.99
Kalamazoo, MI, White Wood	2.83	4.39	6.19	3.69	5.63	6.35
Kansas City, MO, White Wood	2.83	4.48	6.42	3.69	5.56	6.27
Kingston, NY, White Wood	2.69	5.04	6.72	3.69	5.99	7.20
Kingston, NY, Green Douglas Fir	3.07	5.91	5.97	3.98	5.60	6.72
Knoxville, TN, White Wood	2.83	4.36	6.40	3.69	5.36	6.06
Lancaster, PA, White Wood	2.69	4.70	6.61	3.69	5.95	6.82
Las Vegas, NV, White Wood	2.75	—	—	3.13	—	—
Las Vegas, NV, Green Douglas Fir	3.27	4.70	6.27	4.57	5.63	6.76
Lima, OH, White Wood	2.83	4.52	6.42	3.69	5.61	6.34

City, ST, Grade	2"x4" 8' Stud	2"x4" 12'	2"x4" 16'	2"x6" 8'	2"x6" 10'	2"x6" 12'
Little Rock, AR, White Wood	2.79	4.54	6.42	3.69	5.64	6.37
Long Island, NY, Green Douglas Fir	3.07	5.07	6.77	3.98	5.98	7.18
Los Angeles, CA, Green Douglas Fir	3.09	4.49	5.98	3.69	5.36	6.44
Louisville, KY, White Wood	2.83	4.48	6.42	3.69	5.56	6.27
Madison, WI, White Wood	2.83	4.39	6.19	3.69	5.44	6.13
Memphis, TN, White Wood	2.83	4.36	6.40	3.69	5.36	6.06
Miami, FL, White Wood	2.83	4.32	6.40	3.69	5.34	5.99
Midland, TX, White Wood	2.79	4.54	6.42	3.69	5.64	6.37
Milwaukee, WI, White Wood	2.71	3.68	5.60	3.69	4.49	5.34
Minneapolis, MN, White Wood	2.69	4.22	5.79	3.69	5.90	5.98
Missoula, MT, Douglas Fir	3.37	6.05	5.89	4.95	6.19	7.82
Mobile, AL, White Wood	2.79	4.28	5.89	3.69	4.98	5.98
Mobile, AL, Yellow Pine	2.39	4.43	6.57	4.85	5.61	6.26
Nashua, NH, White Wood	2.69	4.98	6.64	3.69	5.92	7.10
Nashville, TN, White Wood	2.83	4.36	6.40	3.69	5.36	6.45
New Orleans, LA, White Wood	2.79	4.67	6.42	3.69	5.59	—
New Orleans, LA, Yellow Pine	2.36	4.25	6.40	4.85	5.49	6.45
New York, NY, Green Douglas Fir	3.07	5.07	6.77	3.98	5.98	7.18
Norfolk, VA, White Wood	2.79	4.36	6.40	3.69	5.36	6.06
Oklahoma City, OK, White Wood	2.79	4.54	6.42	3.69	5.64	6.37
Omaha, NE, White Wood	2.83	4.39	6.19	3.69	5.44	6.13
Orlando, FL, White Wood	2.83	4.32	6.40	3.98	5.34	5.99
Orlando, FL, Yellow Pine	2.59	4.50	6.71	3.70	5.15	6.30
Pensacola, FL, White Wood	2.79	4.24	5.89	3.69	4.98	5.98
Pensacola, FL, Prime Yellow Pine	2.39	4.43	6.57	4.85	5.61	6.26
Pensacola, FL, Yellow Pine	2.39	4.51	6.30	3.50	4.69	5.82
Philadelphia, PA, White Wood	2.79	4.34	6.23	3.98	5.46	6.18
Philadelphia, PA, Hem Fir	2.78	4.27	6.84	4.34	5.43	6.51
Philadelphia, PA, Green Douglas Fir	2.75	4.34	7.14	4.92	6.17	7.38
Phoenix, AZ, White Wood	2.75	—	6.44	4.62	5.08	6.11
Phoenix, AZ, Green Douglas Fir	3.31	4.76	6.35	4.47	5.65	6.77
Pittsburgh, PA, White Wood	2.83	4.23	6.21	4.10	5.36	6.13
Portland, OR, Green Douglas Fir	2.94	4.25	5.67	3.69	4.98	5.98
Raleigh, NC, White Wood	2.79	3.98	5.95	3.69	4.98	5.72
Reno, NV, Green Douglas Fir	3.18	4.60	6.14	4.28	5.44	6.54
Richmond, VA, White Wood	2.79	4.36	6.40	3.69	5.36	6.06
Riverside, CA, Green Douglas Fir	3.14	4.45	5.93	3.69	5.32	6.38
Rochester, NY, White Wood	2.69	5.04	6.72	3.69	5.99	7.20
Rochester, NY, Hem Fir	3.24	5.27	7.02	4.77	6.06	7.15
Sacramento, CA, Green Douglas Fir	3.14	3.96	5.28	3.69	4.99	5.99
Salem, OR, Green Douglas Fir	3.13	4.39	5.85	3.69	5.32	6.38
Salt Lake City, UT, Douglas Fir	3.37	5.06	6.74	4.95	6.19	7.42
San Antonio, TX, White Wood	2.94	4.54	6.42	3.69	5.64	6.37
San Antonio, TX, Yellow Pine	2.36	4.29	6.40	4.55	5.49	6.45
San Diego, CA, Green Douglas Fir	3.21	4.68	6.24	4.40	5.55	6.67
San Francisco, CA, Green Douglas Fir	4.56	6.08	6.59	3.69	5.40	6.48
Savannah, GA, White Wood	2.79	4.36	6.40	3.69	5.36	6.06
Seattle, WA, Douglas Fir	2.60	—	—	5.49	4.31	—
Seattle, WA, Hem Fir	3.97	4.22	5.89	4.23	5.33	6.37
Shreveport, LA, White Wood	2.79	4.54	6.71	—	—	—
Spokane, WA, Douglas Fir	—	4.96	6.62	4.87	6.10	7.30
St. Louis, MO, White Wood	2.83	4.48	6.42	3.69	5.56	6.27
Tampa, FL, White Wood	2.83	4.32	6.40	3.98	5.81	5.99
Tampa, FL, Yellow Pine	2.59	—	6.84	3.70	5.15	6.30
Toledo, OH, White Wood	2.83	4.53	6.42	3.69	5.63	6.35
Traverse City, MI, White Wood	2.83	4.19	6.10	3.69	5.31	6.11
Tyler, TX, White Wood	2.83	4.54	6.42	3.69	5.64	6.37
Valdosta, GA, White Wood	2.83	4.36	6.40	3.69	5.36	6.06
Wichita, KS, White Wood	2.83	4.48	6.42	3.69	5.56	6.27
Youngstown, OH, White Wood	—	4.59	6.57	3.92	5.58	6.20

Lumber, 2" x 8" and 2" x 10"

2" x 8" and 2" x 10" Framing Lumber, per piece 8', 10' and 12' lengths are No. 2 or better grade. 16' lengths are Standard or No. 2 grade. Kiln dried except as noted. White Wood includes SPF (spruce, pine, fir) Engleman spruce, all true firs, hemlocks and pines (excluding Ponderosa pine). Hem fir can be California red fir, grand fir, noble fir, Pacific silver fir, Shasta fir, white fir or Western hemlock. A dash ("—") indicates that the dimension and grade is not commonly sold in that area or that no data is available. Prices are as of fall 2005. See the Carpentry section for installation costs. Note: Area modification factors should not be used with these prices as they are already adjusted by locality.

City, ST, Grade	2"x8" 8'	2"x8" 10'	2"x8" 12'	2"x8" 16'	2"x10" 8'	2"x10" 10'
Albany, NY, SPF	6.54	8.17	9.81	13.08	9.12	—
Albuquerque, NM, Douglas Fir	7.21	9.01	--	14.42	9.27	11.59
Atlanta, GA, Prime Yellow Pine	5.95	7.71	8.58	11.91	8.45	10.56
Atlanta, GA, SPF	—	—	—	12.51	—	—
Austin, TX, Prime Yellow Pine	6.39	7.42	8.99	13.10	8.28	9.24
Baltimore, MD, White Wood	6.32	7.90	9.48	12.64	8.20	10.25
Baltimore, MD, Prime Yellow Pine	—	—	8.74	—	8.45	10.56
Bangor, ME, SPF	6.46	8.08	9.69	12.92	9.12	—
Baton Rouge, LA, Prime Yellow Pine	5.77	7.97	7.79	12.97	8.45	10.29
Bernmidji, MN, SPF	5.99	7.59	9.19	12.39	7.99	10.29
Birmingham, AL, Prime Yellow Pine	6.13	8.36	8.72	12.26	8.60	10.75
Boise, ID, Dry Douglas Fir	6.57	8.22	9.86	13.14	8.48	10.59
Boston, MA, Green Douglas Fir	6.14	7.68	9.21	12.28	8.64	—
Boston, MA, SPF	6.46	8.08	9.69	12.92	7.91	—
Brownsville, TX, Prime Yellow Pine	6.39	7.25	8.99	13.10	8.28	9.24
Buffalo, NY, Douglas Fir	—	—	—	—	—	---
Buffalo, NY, SPF	6.46	—	—	—	—	—
Buffalo, NY, Hem Fir	6.38	7.98	9.58	12.77	8.62	—
Charleston, SC, Prime Yellow Pine	6.12	7.95	8.53	11.99	8.45	10.56
Chattanooga, TN, Prime Yellow Pine	5.99	7.95	8.53	11.99	8.45	10.56
Chicago (Suburb), IL, SPF	5.86	7.51	8.87	11.81	7.79	10.29
Chicago (Urban), IL, SPF	5.99	7.63	8.99	11.99	7.89	10.35
Cincinnati, OH, Prime Yellow Pine	6.49	7.48	8.19	12.97	7.86	10.75
Cleveland, OH, SPF	6.15	7.62	9.15	12.24	7.97	10.47
Cleveland, OH, Prime Yellow Pine	6.49	8.45	8.19	12.97	7.86	10.75
Columbia, SC, Yellow Pine	5.95	7.71	8.58	11.90	8.45	10.56
Columbus, OH, Prime Yellow Pine	6.29	8.29	9.29	11.49	7.69	10.59
Dallas, TX, Prime Yellow Pine	6.39	7.25	8.99	13.10	7.60	8.48
Danbury, CT, Green Douglas Fir	6.14	7.68	9.21	12.28	8.64	10.80
Denver, CO, Douglas Fir	6.80	8.51	10.20	13.65	9.13	11.42
Des Moines, IA, Douglas Fir	4.14	5.18	6.25	8.29	5.30	6.63
Detroit, MI, SPF	5.83	7.29	8.77	11.69	7.94	9.93
Duluth, MN, SPF	5.99	7.59	9.19	12.39	7.99	10.29
Durham, NC, Prime Yellow Pine	5.95	7.80	8.53	11.69	8.45	10.56
Durham, NC, SPF	7.60	9.50	11.24	14.53	9.18	11.48
Edison, NJ, Green Douglas Fir	6.14	7.68	9.79	12.28	7.82	9.78
Edison, NJ, Hem Fir	6.37	7.96	9.56	12.74	8.78	10.98
El Paso, TX, Douglas Fir	7.26	9.08	10.89	14.53	9.34	11.68
Evansville, IN, Prime Yellow Pine	6.69	8.06	9.15	13.36	9.53	11.27
Ft Myers, FL, Prime Yellow Pine	6.42	8.30	9.10	12.80	8.73	10.99
Ft. Wayne, IN, SPF	6.06	—	—	—	7.10	9.38
Ft. Wayne, IN, Prime Yellow Pine	6.49	8.45	8.19	12.97	—	10.99
Gainesville, FL, Prime Yellow Pine	6.42	8.30	9.10	12.88	8.73	10.99
Grand Rapids, MI, SPF	5.89	7.59	8.79	12.29	7.89	9.98
Greenville, SC, Prime Yellow Pine	5.95	7.80	8.53	11.90	8.45	10.56
Hartford, CT, Green Douglas Fir	6.14	7.68	9.21	12.28	8.64	10.80
Houston, TX, Prime Yellow Pine	6.39	7.25	8.99	13.10	8.07	8.98
Indianapolis, IN, Prime Yellow Pine	6.49	8.45	8.96	12.97	6.99	9.39
Jackson, MS, Prime Yellow Pine	5.79	7.30	8.99	11.29	8.45	10.29
Jacksonville, FL, Yellow Pine	6.42	8.30	9.10	12.85	8.73	10.99
Jensen Beach, FL, Prime Yellow Pine	6.58	8.48	9.10	12.85	8.73	11.23

City, ST, Grade	2"x8" 8'	2"x8" 10'	2"x8" 12'	2"x8" 16'	2"x10" 8'	2"x10" 10'
Kalamazoo, MI, Prime Yellow Pine	6.49	8.45	9.45	12.97	7.86	10.75
Kalamazoo, MI, SPF	5.98	7.48	8.97	11.97	7.88	9.84
Kansas City, MO, Green Douglas Fir	6.74	8.43	10.11	13.48	8.49	10.62
Kingston, NY, Green Douglas Fir	6.14	7.68	9.21	12.28	8.64	10.80
Kingston, NY, SPF	6.54	8.17	9.81	13.08	9.12	—
Knoxville, TN, Prime Yellow Pine	6.06	7.95	8.53	11.99	8.45	10.75
Lancaster, PA, White Wood	5.95	7.44	9.69	12.54	8.46	10.58
Las Vegas, NV, Green Douglas Fir	5.98	7.57	9.15	12.10	8.26	10.33
Lima, OH, Prime Yellow Pine	6.49	8.45	9.45	12.97	7.86	10.75
Little Rock, AR, Prime Yellow Pine	6.39	7.25	8.99	13.10	8.28	9.24
Long Island, NY, Green Douglas Fir	6.13	7.66	9.19	12.26	7.82	9.78
Los Angeles, CA, Green Douglas Fir	5.72	7.16	8.59	11.45	7.86	9.82
Louisville, KY, Prime Yellow Pine	5.89	7.55	8.19	11.69	8.76	10.51
Madison, WI, SPF	5.99	7.59	8.99	11.99	7.99	10.29
Memphis, TN, Prime Yellow Pine	6.25	8.36	8.72	12.26	8.60	10.75
Miami, FL, Prime Yellow Pine	6.58	8.48	8.99	13.17	8.73	11.23
Midland, TX, Hem Fir	6.88	8.96	10.85	15.89	9.40	12.15
Milwaukee, WI, SPF	5.99	7.51	9.19	11.99	7.99	10.29
Minneapolis, MN, SPF	4.98	6.98	8.95	11.98	7.89	9.88
Missoula, MT, Douglas Fir	6.49	8.10	9.58	12.97	8.69	10.86
Mobile, AL, Prime Yellow Pine	6.22	7.73	8.23	12.44	7.99	9.60
Nashua, NH, Prime Yellow Pine	6.40	8.00	9.57	—	8.60	—
Nashville, TN, Yellow Pine	6.25	8.10	8.92	12.26	8.60	10.75
New Orleans, LA, Prime Yellow Pine	5.79	7.26	7.92	11.35	8.45	10.36
New York, NY, Green Douglas Fir	6.13	7.66	9.19	12.26	7.82	9.78
Norfolk, VA, Prime Yellow Pine	5.95	7.80	8.53	11.90	8.45	10.56
Oklahoma City, OK, Green Douglas Fir	6.79	8.49	10.19	13.59	8.49	10.62
Omaha, NE, SPF	5.99	7.59	9.19	12.39	7.99	10.29
Orlando, FL, Prime Yellow Pine	6.42	8.30	9.10	12.85	8.73	10.99
Pensacola, FL, Yellow Pine	5.88	7.50	7.68	11.82	5.42	9.60
Pensacola, FL, Prime Yellow Pine	5.88	7.50	8.31	11.82	7.99	9.60
Philadelphia, PA, Green Douglas Fir	7.08	8.59	10.30	13.74	8.05	10.40
Philadelphia, PA, White Wood	5.67	7.08	9.30	11.33	7.38	8.93
Phoenix, AZ, Green Douglas Fir	5.98	7.58	9.10	12.15	8.20	10.25
Pittsburgh, PA, SPF	5.99	7.57	9.09	12.11	8.37	10.46
Portland, OR, Green Douglas Fir	5.36	6.70	7.99	10.72	6.93	8.67
Raleigh, NC, SPF	7.17	8.97	10.59	13.97	9.05	11.31
Raleigh, NC, Prime Yellow Pine	5.95	7.71	8.51	10.97	8.45	9.70
Reno, NV, Green Douglas Fir	5.81	7.26	8.72	11.62	7.98	9.98
Richmond, VA, Prime Yellow Pine	5.95	7.80	8.77	11.90	8.45	10.56
Riverside, CA, Green Douglas Fir	5.89	7.37	8.84	11.79	8.17	9.82
Rochester, NY, Hem Fir	6.99	7.98	9.58	12.77	8.62	10.78
Rochester, NY, SPF	6.54	8.17	9.81	13.08	9.12	—
Sacramento, CA, Green Douglas Fir	5.47	6.84	8.21	10.95	7.42	9.27
Salem, OR, Green Douglas Fir	5.82	7.27	9.42	11.64	7.38	9.22
Salt Lake City, UT, Dry Douglas Fir	6.38	7.98	9.58	12.43	8.69	10.86
San Antonio, TX, Prime Yellow Pine	6.39	7.22	8.99	13.10	8.28	9.24
San Diego, CA, Green Douglas Fir	5.97	7.46	8.95	11.93	8.16	10.19
San Francisco, CA, Green Douglas Fir	5.81	7.27	8.72	11.63	7.98	9.98
Savannah, GA, SPF	7.99	10.24	11.64	15.37	10.15	12.99
Savannah, GA, Prime Yellow Pine	5.99	7.95	8.53	11.99	8.45	10.56
Seattle, WA, Hem Fir	5.15	6.43	7.99	10.70	7.11	8.89
Shreveport, LA, Prime Yellow Pine	6.98	7.85	8.99	12.98	7.45	9.24
Spokane, WA, Dry Douglas Fir	6.30	7.87	9.45	12.59	8.43	10.54
St. Louis, MO, Prime Yellow Pine	5.89	7.42	8.12	11.61	7.82	10.59
Tampa, FL, Prime Yellow Pine	6.42	8.30	9.10	12.85	8.73	10.99
Toledo, OH, SPF	6.29	7.79	9.39	12.49	8.37	10.47
Traverse City, MI, SPF	5.87	7.34	8.81	11.75	7.87	9.84
Tyler, TX, Prime Yellow Pine	6.39	7.16	8.99	13.10	8.07	9.24
Valdosta, GA, Prime Yellow Pine	6.12	7.95	8.53	11.99	8.45	10.56
Wichita, KS, Hem Fir	6.77	8.46	10.12	13.53	9.22	10.39
Youngstown, OH, SPF	5.84	7.44	9.14	12.62	7.10	9.06

Lumber, 2" x 10" and 2" x 12"

2" x 10" and 2" x 12" Framing Lumber, per piece 8', 10' and 12' lengths are No. 2 or better grade. 16' lengths are Standard or No. 2 grade. Kiln dried except as noted. White Wood includes SPF (spruce, pine, fir) Engleman spruce, all true firs, hemlocks and pines (excluding Ponderosa pine). Hem fir can be California red fir, grand fir, noble fir, Pacific silver fir, Shasta fir, white fir or Western hemlock. A dash ("—") indicates that the dimension and grade is not commonly sold in that area or that no data is available. Prices are as of fall 2005. See the Carpentry section for installation costs. Note: Area modification factors should not be used with these prices as they are already adjusted by locality.

City, ST, Grade	2"x10" 12'	2"x10" 16'	2"x12" 8'	2"x12" 10'	2"x12" 12'	2"x12" 16'
Albuquerque, NM, Douglas Fir	13.98	18.56	11.49	14.36	17.23	22.98
Alexandria, LA, Prime Yellow Pine	11.81	17.17	11.29	13.79	16.98	22.55
Anchorage, AK, Hem Fir	16.26	21.69	12.48	15.60	18.72	24.91
Atlanta, GA, Prime Yellow Pine	11.93	16.90	10.88	13.48	17.19	22.73
Austin, TX, Prime Yellow Pine	11.99	16.56	11.29	13.90	16.98	22.55
Baltimore, MD, Hem Fir	12.30	16.40	—	—	15.99	21.40
Baton Rouge, LA, Prime Yellow Pine	11.81	17.17	11.29	13.90	16.98	20.90
Bernmidji, MN, SPF	12.39	16.49	10.59	13.49	15.49	20.79
Birmingham, AL, Prime Yellow Pine	12.15	17.20	11.11	13.70	17.45	23.07
Boise, ID, Dry Douglas Fir	12.72	16.96	10.74	13.43	16.12	21.49
Boston, MA, SPF	11.86	15.81	—	—	15.98	18.83
Bowling Green, KY, Prime Yellow Pine	12.86	17.79	10.73	13.90	17.29	22.71
Brownsville, TX, Prime Yellow Pine	11.99	16.56	11.29	13.90	16.98	22.55
Charleston, SC, Prime Yellow Pine	11.93	16.90	10.92	13.48	17.19	22.73
Chattanooga, TN, Prime Yellow Pine	11.93	16.90	10.92	13.48	17.19	22.73
Chicago (Suburb), IL, SPF	12.56	16.29	10.70	13.49	15.21	20.79
Chicago (Urban), IL, SPF	12.61	16.33	10.76	13.62	16.38	21.40
Cincinnati, No. 1 Yellow Pine	13.25	18.35	11.35	13.40	17.20	22.74
Cleveland, OH, No. 1 Yellow Pine	12.56	16.64	11.13	13.40	16.69	22.74
Columbia, SC, Yellow Pine	11.93	16.90	10.21	12.64	16.18	21.38
Columbia, MO, Prime Yellow Pine	11.93	16.90	10.88	13.48	17.19	22.73
Columbus, OH, No. 1 Yellow Pine	13.09	18.19	11.19	13.19	16.89	22.59
Corpus Christi, TX, Prime Yellow Pine	11.99	18.15	11.29	13.90	16.98	22.55
Dallas, TX, Prime Yellow Pine	10.98	15.19	11.07	13.90	16.98	22.14
Denver, CO, Dry Douglas Fir	13.70	18.27	11.99	14.76	17.72	23.62
Des Moines, IA, Dry Douglas Fir	7.95	10.60	6.41	13.49	9.64	12.82
Detroit, MI, SPF	11.91	14.86	9.98	12.68	15.21	20.28
Duluth, MN, SPF	12.39	16.49	10.59	13.49	15.49	20.79
Durham, NC, Prime Yellow Pine	11.93	16.90	10.88	13.48	17.19	22.73
El Paso, TX, Dry Douglas Fir	14.02	18.69	11.57	14.47	17.36	23.14
Eugene, OR, Green Douglas Fir	10.62	14.16	8.68	10.90	13.02	17.36
Evansville, IN, Prime Yellow Pine	14.07	19.06	11.28	13.65	17.88	23.12
Fairbanks, AK, Hem Fir	15.35	20.74	13.20	16.49	19.82	28.69
Farmington, NM, Dry Douglas Fir	13.63	16.45	9.54	14.75	17.73	23.64
Flagstaff, AZ, Green Douglas Fir	12.30	16.40	9.65	11.98	14.48	19.31
Ft Myers, FL, Prime Yellow Pine	12.45	17.60	11.38	13.99	17.19	23.53
Ft. Wayne, IN, SPF	13.25	18.35	11.35	13.40	17.20	22.74
Gainesville, FL, Prime Yellow Pine	12.45	17.60	11.38	13.99	17.19	23.53
Grand Junction, CO, Dry Douglas Fir	12.99	17.38	11.28	14.10	16.92	22.56
Grand Rapids, MI, SPF	10.28	16.29	10.59	13.49	15.44	20.49
Greenville, SC, Prime Yellow Pine	11.93	16.90	10.88	13.48	17.19	22.73
Gulfport, MS, Prime Yellow Pine	11.81	17.17	11.29	13.90	16.98	22.55
Houma, LA, Prime Yellow Pine	11.81	17.17	11.29	13.79	16.98	22.55
Houston, TX, Prime Yellow Pine	11.70	16.14	11.29	13.90	16.98	22.55
Idaho Falls, ID, Dry Douglas Fir	13.64	17.38	11.28	14.10	16.92	22.56
Indianapolis, IN, No. 1 Yellow Pine	11.85	16.10	10.85	13.11	16.42	22.25
Jackson, MS, Prime Yellow Pine	11.81	17.17	11.29	13.79	16.98	22.55
Jacksonville, FL, Yellow Pine	12.45	17.60	11.38	13.99	17.79	23.53
Jensen Beach, FL, Prime Yellow Pine	12.73	17.97	11.50	14.25	18.11	23.96
Kalamazoo, MI, SPF	11.81	15.76	9.05	11.32	13.59	18.12
Kansas City, MO, Green Douglas Fir	12.74	16.99	10.19	12.74	15.29	20.39

City, ST, Grade	2"x10" 12'	2"x10" 16'	2"x12" 8'	2"x12" 10'	2"x12" 12'	2"x12" 16'
Klamath Falls, OR, Green Douglas Fir	11.07	14.75	8.85	11.07	13.28	21.54
Knoxville, TN, Prime Yellow Pine	12.15	16.90	11.11	13.70	17.19	22.73
Lake Charles, LA, Prime Yellow Pine	11.81	17.17	11.29	13.90	16.98	22.55
Las Vegas, NV, Green Douglas Fir	12.39	16.52	9.73	12.16	14.59	19.46
Little Rock, AR, Prime Yellow Pine	11.99	16.56	11.29	13.90	16.98	22.55
Los Angeles, CA, Green Douglas Fir	11.79	15.72	9.25	11.55	13.86	18.49
Louisville, KY, Prime Yellow Pine	12.08	17.52	11.35	13.90	17.29	22.71
Madison, WI, SPF	12.39	16.49	10.59	13.49	15.49	20.79
Memphis, TN, Prime Yellow Pine	12.15	17.20	10.98	13.70	17.45	23.07
Mesquite, TX, Yellow Pine	12.05	13.36	10.37	12.38	15.13	19.80
Miami, FL, Prime Yellow Pine	12.73	17.97	11.62	14.25	18.11	23.96
Midland, TX, Hem Fir	14.63	20.39	12.13	15.56	18.30	25.75
Milwaukee, WI, SPF	12.39	16.49	10.59	13.49	15.49	20.79
Minneapolis, MN, SPF	11.49	16.29	10.39	12.99	14.49	20.49
Missoula, MT, Dry Douglas Fir	12.99	17.38	11.50	14.38	16.92	22.99
Mobile, AL, Prime Yellow Pine	10.99	16.07	11.18	13.85	17.07	22.84
Monroe, LA, Prime Yellow Pine	11.81	17.07	11.29	13.90	16.98	22.55
Nashville, TN, Yellow Pine	12.15	17.20	11.11	13.70	17.45	23.07
New Orleans, LA, Prime Yellow Pine	11.81	17.25	11.29	13.90	16.98	22.55
New York, NY, Green Douglas Fir	11.73	15.64	—	—	14.51	19.35
Norfolk, VA, Prime Yellow Pine	11.93	16.90	10.88	13.48	17.19	22.73
Oklahoma City, OK, Green Douglas Fir	12.74	16.99	12.18	12.74	15.29	20.39
Omaha, NE, SPF	12.39	16.49	10.59	13.49	15.49	20.79
Orlando, FL, Prime Yellow Pine	12.45	17.40	11.38	13.99	17.79	23.53
Pensacola, FL, Yellow Pine	13.22	15.10	10.60	13.14	16.20	21.68
Pensacola, FL, Prime Yellow Pine	10.99	15.99	11.18	13.85	16.99	22.84
Phoenix, AZ, Green Douglas Fir	12.30	16.40	9.65	11.98	14.48	19.31
Pittsburgh, PA, SPF	12.55	16.74	—	—	15.90	21.20
Portland, OR, Green Douglas Fir	10.40	13.87	8.32	10.40	12.48	16.64
Raleigh, NC, Prime Yellow Pine	11.73	16.97	10.88	13.48	16.97	22.73
Reno, NV, Green Douglas Fir	11.98	16.12	9.48	11.85	14.22	18.97
Richmond, VA, Prime Yellow Pine	11.93	16.90	10.88	13.48	17.19	22.73
Riverside, CA, Green Douglas Fir	11.79	15.72	9.24	11.55	13.86	18.49
Roanoke, VA, Prime Yellow Pine	11.93	16.90	10.88	13.48	17.19	22.73
Rockford, IL, SPF	12.39	16.49	10.59	13.49	15.49	20.79
Sacramento, CA, Green Douglas Fir	11.13	14.48	8.85	11.13	13.35	17.81
Salem, OR, Green Douglas Fir	11.07	14.75	8.85	11.07	13.28	17.71
Salt Lake City, UT, Dry Douglas Fir	12.99	17.38	11.28	14.10	16.92	22.49
San Antonio, TX, Prime Yellow Pine	11.99	16.56	11.29	13.90	16.98	22.55
San Diego, CA, Green Douglas Fir	12.23	16.31	9.60	11.98	14.40	19.20
San Francisco, CA, Green Douglas Fir	11.98	16.12	9.48	11.85	14.22	18.97
Santa Fe, NM, Dry Douglas Fir	13.89	18.26	11.73	14.67	17.60	23.42
Savannah, GA, Prime Yellow Pine	11.93	16.90	10.92	13.48	17.19	22.73
Seattle, WA, Hem Fir	10.67	14.98	9.45	11.81	14.17	19.19
Shreveport, LA, Prime Yellow Pine	11.99	16.56	11.29	13.79	15.48	22.55
Sioux Falls, SD, SPF	12.39	16.49	10.59	13.49	15.49	20.79
Spokane, WA, Dry Douglas Fir	12.65	16.86	10.89	13.61	16.33	21.77
Springfield, MO, Green Douglas Fir	12.74	16.99	10.83	12.74	15.29	20.39
St. George, UT, Dry Douglas Fir	12.67	16.89	10.82	13.53	16.24	21.65
St. Louis, MO, Prime Yellow Pine	12.08	17.52	10.50	13.90	17.29	19.21
Tampa, FL, Prime Yellow Pine	12.45	17.60	11.29	13.99	17.79	23.53
Toledo, OH, SPF	12.56	16.75	11.18	13.98	16.77	22.37
Topeka, KS, Green Douglas Fir	12.74	16.99	10.19	12.74	15.29	20.39
Traverse City, MI, SPF	11.80	17.30	10.10	12.62	15.14	20.19
Tyler, TX, Prime Yellow Pine	11.99	16.14	11.29	13.79	16.98	22.55
Valdosta, GA, Prime Yellow Pine	11.93	16.90	10.92	13.48	17.19	22.73
Vicksburg, MS, Prime Yellow Pine	11.81	17.17	11.29	13.79	16.98	22.55
Waterloo, IA, Dry Douglas Fir	7.95	10.60	6.41	8.01	9.64	12.82
Wichita, KS, Hem Fir	13.83	18.44	11.53	14.42	17.30	23.69
Wichita Falls, TX, Prime Yellow Pine	11.99	16.56	11.29	13.79	16.98	22.55
Youngstown, OH, SPF	11.87	16.65	9.83	14.49	14.20	20.12

Lumber, Treated

Treated Lumber, per linear foot. In pounds of ACQ Treatment per cubic foot of lumber. ACQ is Alkaline Copper Quat. Lumber suitable for installation in contact with the ground retains .40 pounds of treatment per cubic foot of lumber. SYP is southern yellow pine. Prices as of fall 2005. A dash ("—") indicates not commonly sold or no data available. Average costs per linear foot for lengths from 8' to 12'. See the Carpentry section for installation costs. Note: Area modification factors should not be used with these prices as they are already adjusted by locality

City, ST, Grade	2"x4" Per LF	2"x6" Per LF	2"x8" Per LF	2"x10" Per LF	2"x12" Per LF
Abilene, TX, #2 Prime SYP .40	0.54	0.97	1.25	1.77	2.07
Albany, NY, #1 SYP .40	0.46	0.96	1.18	1.50	1.70
Albuquerque, NM, #2 Prime SYP .40	0.55	0.97	1.25	1.77	2.07
Amarillo, TX, #2 Prime SYP .40	0.55	0.97	1.25	1.77	2.07
Anchorage, AK, #2 Fir LP22	0.74	1.24	1.50	2.28	2.47
Austin, TX, #2 Prime SYP .40	0.52	0.92	1.20	1.61	1.97
Baltimore, MD, #2 Prime SYP .40	0.54	0.96	1.28	1.90	2.07
Bangor, ME, #1 SYP .40	0.47	0.96	1.22	1.50	2.07
Bay City, MI, #2 Prime SYP .40	0.54	0.90	1.24	1.45	2.05
Beaumont, TX, #2 Prime SYP .40	0.55	0.94	1.25	1.61	2.03
Bernmidji, MN, #2 Prime SYP .40	0.47	0.87	1.11	1.45	1.92
Birmingham, AL, #2 Prime SYP .40	0.53	0.90	1.24	1.39	1.97
Boston, MA, #1 SYP .25 ACQ/CA	0.53	1.00	1.23	1.53	2.07
Bowling Green, KY, #2 Prime SYP .40	0.54	0.91	1.21	1.44	1.99
Brownsville, TX, #2 Prime SYP .40	0.55	0.95	1.25	1.61	2.03
Buffalo, NY, #1 SYP .40	0.49	0.99	1.21	1.45	2.07
Carrollton, TX, #2 Prime SYP .40	0.53	0.93	1.14	1.61	2.03
Charleston, SC, #2 Prime SYP .40	0.53	0.87	1.19	1.35	2.07
Charleston, WV, #2 Prime SYP .40	0.53	0.91	1.21	1.44	1.99
Chicago, IL, #2 Prime SYP .40	0.54	0.89	1.19	1.61	2.07
Cincinnati, OH, #2 Prime SYP .40	0.55	0.96	1.29	1.68	2.16
Cleveland, OH, #2 Prime SYP .40	0.47	0.85	1.11	1.24	1.92
Columbia, SC, #2 Prime SYP .40	0.52	0.87	1.19	1.35	1.92
Columbia, MO, #2 Prime SYP .40	0.54	0.97	1.25	1.77	2.07
Columbus, OH, #2 Prime SYP .40	0.55	0.96	1.29	1.68	2.16
Corpus Christi, TX, #2 Prime SYP .40	0.55	0.94	1.25	1.61	2.03
Dallas, TX, #2 Prime SYP .40	0.53	0.93	1.14	1.61	2.03
Danbury, CT, #1 SYP .40	0.49	0.99	1.21	1.45	2.07
Denver, CO, #1 SYP .40	0.50	1.03	1.26	1.45	2.34
Des Moines, IA, #2 Prime SYP .40	0.47	0.89	1.11	1.32	1.77
Detroit, MI, #2 Prime SYP .40	0.54	0.90	1.24	1.45	2.05
Duluth, MN, #2 Prime SYP .40	0.47	0.87	1.11	1.45	2.02
Durham, NC, #2 Prime SYP .40	0.53	0.91	1.21	1.45	2.07
Edison, NJ, #1 SYP .40	0.49	0.99	1.21	1.45	2.07
El Paso, TX, #2 Prime SYP .40	0.55	0.97	1.25	1.77	2.07
Erie, PA, #2 Prime SYP .40	0.54	0.89	1.25	1.90	2.07
Eugene, OR, #2 Fir LP22	0.74	1.24	1.50	2.06	2.22
Evansville, IN, #2 Prime SYP .40	0.54	0.96	1.29	1.45	2.05
Fairbanks, AK, #2 Fir LP22	0.74	1.24	1.50	2.28	2.47
Farmington, NM, #2 Prime SYP .40	0.55	0.97	1.25	1.77	2.07
Fayetteville, AR, #2 Prime SYP .40	0.53	0.94	1.23	1.45	2.07
Ft. Wayne, IN, #2 Prime SYP .40	0.47	0.88	1.05	1.27	1.70
Grand Rapids, MI, #2 Prime SYP .40	0.47	0.88	1.05	1.29	1.70
Greenville, SC, #2 Prime SYP .40	0.48	0.87	1.19	1.35	1.92
Hartford, CT, #1 SYP .40	0.49	0.99	1.48	1.45	2.07
Honolulu, HI, Hibor Fir	0.69	1.20	1.25	2.09	2.37
Houston, TX, #2 Prime SYP .40	0.54	0.97	1.03	1.77	2.07
Indianapolis, IN, #2 Prime SYP .40	0.46	0.84	1.03	1.24	1.69
Jackson, MS, #2 Prime SYP .40	0.53	0.94	1.23	1.45	2.07
Jackson, MI, #2 Prime SYP .40	0.53	0.94	1.25	1.40	2.05
Johnstown, PA, #2 Prime SYP .40	0.54	0.89	1.25	1.90	2.07
Jonesboro, AR, #2 Prime SYP .40	0.53	0.94	1.23	1.45	2.07
Kalamazoo, MI, #2 Prime SYP .40	0.47	0.88	1.05	1.29	1.70
Kansas City, MO, #2 Prime SYP .40	0.54	0.97	1.25	1.77	2.07
Kenai, AK, #2 Fir LP22	0.74	1.24	1.55	2.28	2.47

City, ST, Grade	2"x4" Per LF	2"x6" Per LF	2"x8" Per LF	2"x10" Per LF	2"x12" Per LF
Kingston, NY, #1 SYP .40	0.47	0.95	1.15	1.44	1.99
Klamath Falls, OR, #2 Fir LP22	0.74	1.26	1.55	2.09	2.45
Knoxville, TN, #2 Prime SYP .40	0.53	0.87	1.19	1.35	1.93
Lake Charles, LA, #2 Prime SYP .40	0.55	0.94	1.25	1.61	2.03
Lake Jackson, TX, #2 Prime SYP .40	0.55	0.97	1.25	1.77	2.07
Lancaster, PA, #2 Prime SYP .40	0.53	0.92	1.20	1.44	1.99
Laredo, TX, #2 Prime SYP .40	0.54	0.97	1.20	1.77	2.07
Lawton, OK, #2 Prime SYP .40	0.54	0.97	1.25	1.77	2.07
Lexington, KY, #2 Prime SYP .40	0.53	0.91	1.21	1.44	1.99
Lima, OH, #2 Prime SYP .40	0.55	0.96	1.29	1.68	2.16
Little Rock, AR, #2 Prime SYP .40	0.53	0.94	1.23	1.45	2.07
Logansport, IN, #2 Prime SYP .40	0.55	0.96	1.29	1.45	2.05
Los Angeles, CA, Standard Fir	0.65	1.23	1.61	—	2.43
Louisville, KY, #2 Prime SYP .40	0.54	0.96	1.29	1.45	2.05
Madison, WI, #2 Prime SYP .40	0.50	0.90	1.05	1.27	1.92
Memphis, TN, #2 Prime SYP .40	0.53	0.90	1.24	1.39	1.97
Midland, TX, #2 Prime SYP .40	0.54	0.97	1.25	1.77	2.07
Milwaukee, WI, #2 Prime SYP .40	0.54	0.89	1.19	1.90	2.07
Minneapolis, MN, #2 Prime SYP .40	0.51	0.91	1.12	1.32	1.92
Missoula, MT, #2 Fir LP22	0.74	1.24	1.50	1.84	2.47
Monroe, LA, #2 Prime SYP .40	0.53	0.94	1.23	1.45	2.07
Nashua, NH, #1 SYP .40	0.49	0.99	1.21	1.45	2.07
Nashville, TN, #2 Prime SYP .40	0.53	0.90	1.22	1.39	1.97
New York, NY, #1 SYP .40	0.49	0.99	1.23	1.45	2.07
Norfolk, VA, #2 Prime SYP .40	0.53	0.92	1.20	1.44	1.99
Oklahoma City, OK, #2 Prime SYP .40	0.54	0.97	1.25	1.77	2.07
Omaha, NE, #2 Prime SYP .40	0.47	0.89	1.11	1.23	1.77
Philadelphia, PA, #1 SYP .40	0.49	0.97	1.21	1.90	2.07
Phoenix, AZ, Standard Fir	0.65	1.23	1.65	—	2.27
Pittsburgh, PA, #2 Prime SYP .40	0.54	0.89	1.25	1.90	2.07
Poplar Bluff, MO, #2 Prime SYP .40	0.55	0.97	1.25	1.77	2.07
Portland, Or, #2 Fir LP 22	0.74	1.24	1.43	2.05	2.22
Raleigh, NC, #2 Prime SYP .40	0.53	0.91	1.21	1.45	2.07
Richmond, VA, #2 Prime SYP .40	0.53	0.95	1.23	1.45	2.07
Roanoke, VA, #2 Prime SYP .40	0.54	0.89	1.20	1.90	2.07
Rochester, NY, #1 SYP .40	0.49	0.97	1.22	1.90	2.07
Rockford, IL, #2 Prime SYP .40	0.55	0.96	1.29	1.45	2.05
Roswell, NM, #2 Prime SYP .40	0.55	0.97	1.23	1.77	2.07
Rutland, VT, #1 SYP .40	0.47	0.86	1.22	1.50	2.07
Sacramento, CA, Standard Fir	0.65	1.08	1.61	2.26	2.43
Salem, OR, #2 Fir LP22	0.74	1.24	1.43	2.15	2.22
Salt Lake City, UT, #2 Fir LP22	0.74	1.24	1.55	2.05	2.47
San Antonio, TX, #2 Prime SYP .40	0.54	0.97	1.21	1.77	2.07
San Diego, CA, Standard Fir	0.76	1.29	1.59	—	2.58
San Francisco, CA, Standard Fir	0.65	1.08	1.58	2.26	2.43
Santa Fe, NM, #2 Prime SYP .40	0.54	0.97	1.25	1.77	2.07
Seattle, WA, #2 Fir LP22	0.74	1.24	1.43	2.05	2.22
Shreveport, LA, #2 Prime SYP .40	0.53	0.94	1.23	1.45	2.07
Spokane, WA, #2 Fir LP22	0.53	0.91	1.14	1.45	2.02
Springfield, MO, #2 Prime SYP .40	0.54	0.97	1.25	1.77	2.07
St. Louis, MO, #2 Prime SYP .40	0.54	0.97	1.25	1.77	2.07
Toledo, OH, #2 Prime SYP .40	0.53	0.91	1.14	1.45	2.02
Topeka, KS, #2 Prime SYP .40	0.55	0.97	1.25	1.77	2.07
Traverse City, MI, #2 Prime SYP .40	0.47	0.88	1.05	1.27	1.70
Tupelo, MS, #2 Prime SYP .40	0.55	0.94	1.25	1.61	2.03
Tyler, TX, #2 Prime SYP .40	0.53	0.94	1.23	1.45	2.07
Vicksburg, MS, #2 Prime SYP .40	0.53	0.94	1.23	1.45	2.07
W. Springfield, MA, #1 SYP .40	0.47	0.96	1.22	1.50	2.07
Waco, TX, #2 Prime SYP .40	0.54	0.97	1.25	1.77	2.07
Waterloo, IA, #2 Prime SYP .40	0.47	0.87	1.05	1.29	1.77
Wausau, WI, #2 Prime SYP .40	0.50	0.81	1.08	1.32	1.77
Wichita, KS, #2 Prime SYP .40	0.55	0.97	1.25	1.77	2.07
Wichita Falls, TX, #2 Prime SYP .40	0.53	0.94	1.23	1.45	2.07
Youngstown, OH, #2 Prime SYP .40	0.47	0.85	1.11	1.24	1.92

Lumber, Panels

4' x 8' Sheathing Panels CDX plywood is interior C grade (patched) on one side and D grade (knotholes) on the reverse with exterior glue. 3/4" tongue and groove (T&G) pine plywood is usually sold as Sturdifloor. OSB (oriented strand board) or "waferboard" panels are square edge and not designated by species. A dash ("—") indicates not commonly sold or no data available. Price per panel as of fall 2005. Panel prices are volatile. See the Carpentry section for installation costs. Note: Area modification factors should not be used with these prices as they are already adjusted by locality.

City, ST, Grade	1/2" CDX	5/8" CDX	3/4" CDX	1/2" Treated CDX	3/4" T&G	7/16" OSB
Albany, NY, Fir	15.97	19.47	23.95	24.97	26.96	9.99
Albany, NY, Pine	—	18.56	22.95	—	—	—
Albuquerque, NM, Pine	—	17.96	21.47	27.97	25.95	8.99
Atlanta, GA, Pine	14.98	17.59	21.95	24.97	25.99	8.99
Austin, TX, Pine	—	16.88	20.75	24.97	25.75	8.68
Baltimore, MD, Pine	—	18.59	22.89	25.97	26.88	9.99
Bangor, ME, Fir	15.99	19.47	22.88	24.97	28.69	9.99
Bangor, ME, Pine	15.57	18.59	22.69	—	—	—
Baton Rouge, LA, Pine	—	16.88	20.99	24.97	24.95	8.99
Bernmidji, MN, Pine	14.99	16.79	20.88	24.97	24.95	7.99
Birmingham, AL, Pine	14.99	16.89	21.88	24.97	25.95	8.99
Boise, ID, Fir	15.09	19.17	23.66	—	—	10.62
Boston, MA, Fir	15.99	19.47	22.88	24.97	28.69	9.99
Boston, MA, Pine	15.57	18.59	22.88	—	—	—
Brownsville, TX, Pine	—	16.88	20.75	24.97	25.75	8.68
Buffalo, NY, Fir	15.97	19.47	23.95	26.97	28.87	9.99
Buffalo, NY, Pine	15.59	18.56	22.95	—	26.96	—
Charleston, SC, Pine	14.99	17.96	21.96	26.97	26.75	9.27
Chattanooga, TN, Pine	—	17.59	19.97	24.97	25.99	8.99
Chicago, IL (Suburb), Pine	14.75	17.37	19.97	24.97	25.69	8.25
Chicago, IL (Urban), Pine	14.83	17.48	21.99	25.10	25.83	8.34
Cincinnati, OH, Pine	14.99	17.49	21.95	25.97	25.75	8.29
Cleveland, OH, Pine	14.95	17.35	21.96	25.97	25.95	8.95
Columbia, SC, Pine	14.99	17.96	21.99	27.97	26.75	9.27
Columbus, OH, Pine	14.95	17.35	20.85	25.97	25.95	8.99
Dallas, TX, Pine	—	16.87	23.95	24.97	24.95	8.99
Danbury, CT, Fir	15.97	19.47	22.95	26.97	28.87	9.99
Danbury, CT, Pine	—	18.56	20.99	—	—	—
Denver, CO, Fir	14.95	17.96	21.95	24.97	26.96	9.27
Des Moines, IA, Pine	14.75	17.37	21.99	25.97	25.69	8.39
Detroit, MI, Pine	14.99	17.49	20.88	25.97	25.75	8.49
Duluth, MN, Pine	14.99	16.89	21.96	25.97	24.95	7.99
Durham, NC, Pine	14.99	17.96	22.98	27.97	26.75	8.99
Edison, NJ, Fir	15.99	19.56	22.95	26.97	27.96	9.95
Edison, NJ, Pine	—	18.89	21.99	—	—	—
El Paso, TX, Pine	—	18.36	22.64	28.97	25.95	9.95
Evansville, IN, Pine	—	18.41	21.95	28.97	26.96	10.28
Ft Myers, FL, Pine	15.26	17.75	21.25	23.97	26.99	9.39
Ft. Wayne, IN, Pine	14.99	17.49	21.95	25.97	25.75	10.99
Gainesville, FL, Pine	15.26	17.75	21.99	24.97	26.99	9.39
Grand Rapids, MI, Pine	14.99	17.49	21.96	25.97	25.75	7.99
Greenville, SC, Pine	—	17.96	23.95	27.97	26.75	9.27
Hartford, CT, Fir	15.97	19.47	22.95	26.97	28.87	9.99
Hartford, CT, Pine	—	18.56	20.45	—	—	—
Houston, TX, Pine	—	16.88	19.99	24.97	24.99	8.99
Indianapolis, IN, Pine	—	16.59	20.99	25.97	24.49	7.99
Jackson, MS, Pine	—	16.88	21.95	27.97	24.95	8.99
Jacksonville, FL, Pine	15.26	17.75	22.99	24.97	26.99	9.39
Jensen Beach, FL, Pine	15.37	17.79	21.99	24.97	26.97	9.95
Kalamazoo, MI, Pine	14.99	17.49	20.87	25.97	25.75	7.99
Kansas City, MO, Pine	—	16.97	23.95	25.97	24.99	8.29
Kingston, NY, Fir	15.97	19.47	22.95	28.97	28.87	9.99
Kingston, NY, Pine	—	18.56	21.88	—	—	—
Knoxville, TN, Pine	14.99	16.89	23.87	27.97	25.95	8.99

City, ST, Grade	1/2" CDX	5/8" CDX	3/4" CDX	1/2" Treated CDX	3/4" T&G	7/16" OSB
Lancaster, PA, Fir	15.95	16.89	22.99	28.97	28.67	9.99
Lancaster, PA, Pine	—	18.67	21.96	—	26.96	—
Las Vegas, NV, Pine	15.99	17.95	21.99	—	25.98	8.99
Lima, OH, Pine	—	17.35	20.87	25.97	25.95	8.99
Little Rock, AR, Pine	—	16.87	22.98	27.97	24.95	8.99
Long Island, NY, Fir	15.99	19.56	22.95	26.97	27.96	9.95
Long Island, NY, Pine	—	18.89	22.75	—	—	—
Los Angeles, CA, Pine	—	17.96	22.95	—	25.95	9.25
Louisville, KY, Pine	—	16.99	20.87	25.97	24.79	8.29
Madison, WI, Pine	14.75	17.37	21.39	25.97	25.69	9.99
Memphis, TN, Pine	—	16.89	21.88	24.97	25.95	8.99
Miami, FL, Pine	15.37	17.79	22.99	23.97	26.97	9.95
Midland, TX, Pine	—	16.87	20.88	25.97	24.99	8.99
Milwaukee, WI, Pine	14.75	17.37	19.99	25.97	24.95	7.99
Minneapolis, MN, Pine	14.99	16.79	20.88	25.97	24.95	7.99
Missoula, MT, Fir	14.67	17.96	21.95	39.97	—	8.99
Mobile, AL, Pine	14.57	16.88	20.99	24.97	24.97	8.99
Nashua, NH, Fir	15.99	19.47	25.57	26.97	28.69	9.99
Nashua, NH, Pine	15.57	18.59	22.88			
Nashville, TN, Pine	—	16.89	21.88	24.97	—	8.99
New Orleans, LA, Pine	—	16.88	20.99	24.97	24.95	8.99
New York, NY, Fir	15.99	19.56	22.98	26.97	27.96	9.95
New York, NY, Pine	—	18.89	22.95	—	—	—
Norfolk, VA, Pine	14.99	17.96	21.96	27.97	26.75	9.27
Oklahoma City, OK, Pine	—	16.87	20.87	25.97	24.95	8.99
Omaha, NE, Pine	14.99	17.75	20.95	25.97	24.95	7.99
Orlando, FL, Pine	15.26	17.75	21.95	23.97	26.99	9.39
Pensacola, FL, Pine	14.57	16.88	20.99	24.97	24.97	8.99
Philadelphia, PA, Fir	15.75	18.99	23.96	25.97	27.88	9.99
Philadelphia, PA, Pine	—	18.47	22.99	—	—	—
Phoenix, AZ, Pine	15.99	18.42	21.96	26.95	25.98	8.99
Pittsburgh, PA, Pine	—	17.35	21.99	24.97	25.95	8.95
Portland, OR, Fir	14.49	17.99	20.89	39.97	25.88	9.29
Raleigh, NC, Pine	14.99	17.96	21.96	27.97	26.75	8.99
Reno, NV, Pine	15.85	17.95	21.99	—	26.97	9.69
Richmond, VA, Pine	14.99	17.96	21.96	26.97	26.75	9.27
Riverside, CA, Pine	—	17.89	22.89	—	25.95	9.88
Rochester, NY, Fir	15.97	19.47	23.95	25.97	28.87	9.99
Rochester, NY, Pine	15.59	18.56	22.95	—	26.96	—
Rutland, VT, Fir	15.99	19.47	22.88	24.97	28.69	9.99
Rutland, VT, Pine	15.57	18.59	22.69	—	—	—
Sacramento, CA, Pine	15.85	17.95	21.99	—	26.88	9.69
Salem, OR, Fir	14.49	17.99	19.95	42.36	25.88	8.99
Salt Lake City, UT, Fir	14.75	17.79	21.95	—	—	8.99
San Antonio, TX, Pine	—	16.88	20.75	25.97	25.75	8.68
San Diego, CA, Pine	—	17.89	22.89	—	25.95	9.88
San Francisco, CA, Pine	15.85	17.95	21.99	37.55	26.88	9.69
Savannah, GA, Pine	14.99	17.96	21.96	24.97	26.75	9.27
Seattle, WA, Fir	14.49	17.99	20.89	39.97	25.88	9.29
Shreveport, LA, Pine	—	16.87	20.87	27.97	24.95	8.99
Sioux Falls, SD, Pine	14.99	17.75	20.95	24.97	24.95	7.99
Spokane, WA, Fir	14.49	17.99	20.89	39.97	25.88	9.29
Springfield, MO, Pine	—	16.87	20.75	25.97	24.99	8.99
St. Louis, MO, Pine	—	16.97	20.87	25.97	24.99	8.29
Tampa, FL, Pine	15.26	17.75	21.95	23.97	26.99	9.39
Toledo, OH, Pine	—	17.49	21.99	25.97	25.99	7.99
Traverse City, MI, Pine	14.99	17.49	21.99	25.97	25.99	8.29
Tyler, TX, Pine	—	16.87	20.85	27.97	24.95	8.99
Valdosta, GA, Pine	14.98	17.59	21.95	24.97	25.99	8.99
W. Springfield, MA, Fir	15.97	19.47	23.95	24.97	28.87	—
W. Springfield, MA, Pine	—	18.56	22.95	---	—	9.99

Lumber, Boards

Redwood. West Coast prices. Add 10% to 30% for transportation to other areas.

		MBF	8' L	12' L	16' L
2" x 2" – B Grade S4S	Ea	2,040.00	5.43	8.15	10.90
2" x 2" – Green	Ea	1,570.00	4.19	6.29	8.38
2" x 3" – Green	Ea	1,160.00	4.64	6.96	9.28
2" x 4" – Construction Common	Ea	1,280.00	6.80	10.20	13.60
2" x 4" – Construction Rough	Ea	1,170.00	6.21	9.32	12.40
2" x 4" – Common S4S	Ea	1,240.00	6.61	9.92	13.20
2" x 4" – Construction Heart, Rough	Ea	1,560.00	8.32	12.50	16.70
2" x 4" – Construction Heart, S4S	Ea	1,910.00	10.20	15.30	20.30
2" x 6" – Construction Common	Ea	1,290.00	10.30	15.50	20.60
2" x 6" – Common S4S	Ea	1,370.00	11.00	16.40	21.90
2" x 6" – Construction Heart, Rough	Ea	1,630.00	13.10	19.60	26.10
2" x 6" – Construction Heart, S4S	Ea	1,820.00	14.60	21.90	29.10
2" x 8" – Common S4S	Ea	1,450.00	15.40	23.20	30.90
2" x 8" – Construction Heart, S4S	Ea	1,690.00	18.00	27.10	36.10
2" x 12" – Common S4S	Ea	1,670.00	26.70	40.00	53.40
2" x 12" – Construction Heart, S4S	Ea	2,050.00	32.80	49.10	65.50

Stud. Typical costs for pre-cut white wood studding material, stud grade. Nominal sizes. Actual width and depth are about 1/2" less when dry. White wood species includes Engelmann spruce, all true firs, hemlocks, and pines (excluding Ponderosa pine).

	Craft@Hrs	Unit	Material	Labor	Total
2" x 3" x 84"	B1@.192	Ea	1.69	5.25	6.94
2" x 4" x 84"	B1@.196	Ea	2.01	5.36	7.37
2" x 4" x 88"	B1@.200	Ea	2.49	5.47	7.96
2" x 4" x 92-5/8"	B1@.216	Ea	2.50	5.91	8.41
2" x 4" x 96"	B1@.224	Ea	2.55	6.13	8.68
2" x 4" x 104-5/8"	B1@.230	Ea	2.83	6.29	9.12
2" x 6" x 92-5/8"	B1@.308	Ea	4.09	8.43	12.52

Boards

Douglas fir finish and clear – C & Better (V/G) S4S, dry, vertical grain. West Coast prices. Add 10% to 30% for transportation to other areas.

		MBF	8' L	12' L	16' L
1" x 2"	Ea	3,730.00	4.95	7.43	9.91
1" x 3"	Ea	3,730.00	7.46	11.20	14.90
1" x 4"	Ea	3,730.00	9.94	14.90	19.90
1" x 6"	Ea	4,230.00	17.00	25.30	33.80
1" x 8"	Ea	5,160.00	27.60	41.20	55.00
1" x 10"	Ea	5,750.00	38.30	57.50	76.50
1" x 12"	Ea	6,150.00	49.20	73.80	98.40

Douglas fir finish and clear – C & better (F/G) S4S, dry, flat grain. West Coast prices. Add 10% to 30% for transportation to other areas.

		MBF	8' L	12' L	16' L
1" x 2"	Ea	3,200.00	4.25	6.38	8.50
1" x 3"	Ea	3,200.00	6.40	9.60	12.80
1" x 4"	Ea	3,200.00	8.53	12.80	17.10
1" x 6"	Ea	3,350.00	13.40	20.10	26.80
1" x 8"	Ea	4,310.00	23.00	34.50	46.00
1" x 10"	Ea	6,000.00	40.00	60.00	79.90
1" x 12"	Ea	6,140.00	49.10	73.70	98.20

		MBF	8' L	12' L	16' L
Southern pine boards – C & Better					
1" x 4"	Ea	1,970.00	5.25	7.88	10.50
1" x 6"	Ea	2,090.00	8.36	12.50	16.70
1" x 8"	Ea	2,150.00	11.50	17.20	23.00
1" x 12"	Ea	2,980.00	23.90	35.80	47.70
White pine boards, select					
1" x 4"	Ea	2,500.00	6.65	9.98	13.30
1" x 6"	Ea	3,440.00	13.80	20.60	27.50
1" x 8"	Ea	3,100.00	16.50	24.80	33.00
1" x 10"	Ea	3,280.00	21.80	32.80	43.70
1" x 12"	Ea	3,710.00	29.70	44.50	59.40
Eastern pine boards – premium					
1" x 4"	Ea	1,560.00	4.14	6.22	8.29
1" x 6"	Ea	1,510.00	6.03	9.05	12.10
1" x 8"	Ea	1,410.00	7.49	11.20	15.00
1" x 10"	Ea	1,500.00	10.00	15.00	20.10
1" x 12"	Ea	1,960.00	15.70	23.60	31.40
Southern yellow pine flooring – D Grade					
1" x 4"	Ea	1,070.00	2.86	4.30	5.77
1" x 6"	Ea	1,210.00	4.82	7.22	9.58
1" x 8"	Ea	1,180.00	6.33	8.65	12.70
1" x 10"	Ea	1,080.00	7.24	9.50	14.40
1" x 12"	Ea	1,370.00	10.90	16.40	21.90

		MBF	8' L	12' L	16' L
Fir flooring – C & better tongue & groove					
1" x 3" vertical grain, C and better	Ea	2,470.00	4.94	7.41	9.88
1" x 4" vertical grain, C and better	Ea	2,470.00	6.58	9.87	13.10
1" x 3" flat grain, C and better	Ea	1,480.00	2.97	4.45	5.94
1" x 4" flat grain, C and better	Ea	1,480.00	3.95	5.94	7.91
1" x 3" vertical grain, D and better	Ea	1,770.00	3.55	5.32	7.09
1" x 4" vertical grain, D and better	Ea	1,770.00	4.73	7.09	9.44
1" x 3" flat grain, D and better	Ea	1,170.00	2.33	3.50	4.66
1" x 4" flat grain, D and better	Ea	1,170.00	3.11	4.66	6.21

		MBF	8' L	12' L	16' L
Ponderosa pine boards – #4 & Better					
1" x 3"	Ea	1,560.00	3.11	4.67	6.23
1" x 4"	Ea	1,200.00	3.20	4.80	6.39
1" x 6"	Ea	1,500.00	5.99	8.98	12.00
1" x 8"	Ea	1,430.00	7.63	11.50	15.30
1" x 10"	Ea	1,510.00	10.10	15.10	20.20

		MBF	8' L	12' L	16' L
White wood boards – appearance grade					
1" x 4"	Ea	1,230.00	3.27	4.90	6.54
1" x 6"	Ea	1,470.00	5.87	8.81	11.70
1" x 8"	Ea	1,490.00	7.93	11.90	15.90
1" x 12"	Ea	1,850.00	14.80	22.30	29.70

		MBF	8' L	12' L	16' L
Cypress boards					
1" x 4"	Ea	1,560.00	4.17	6.25	8.33
1" x 6"	Ea	1,970.00	7.88	11.80	15.80
1" x 8"	Ea	2,170.00	11.60	17.40	23.10
1" x 10"	Ea	2,220.00	14.80	22.20	29.60
1" x 12"	Ea	2,610.00	20.90	31.30	41.80
Hemlock boards – C & better					
1" x 2"	Ea	3,090.00	4.12	6.18	8.24
1" x 3"	Ea	2,800.00	5.60	8.40	11.20
1" x 4"	Ea	2,930.00	7.80	11.70	15.60
1" x 6"	Ea	3,470.00	13.90	20.90	27.80

		MBF	8' L	12' L	16' L
Pine boards – select					
1" x 2"	Ea	2,770.00	3.69	5.54	7.39
1" x 3"	Ea	3,110.00	6.22	9.34	12.50
Clear Douglas fir – S4S					
1" x 2"	Ea	2,620.00	3.49	5.24	6.99
1" x 3"	Ea	2,650.00	5.31	7.96	10.60
1" x 4"	Ea	2,720.00	7.25	10.90	14.50
1" x 6"	Ea	3,410.00	13.70	20.50	27.30
1" x 8"	Ea	3,150.00	16.80	25.20	33.50
Douglas fir / hemlock					
1" x 4"	Ea	855.00	2.28	3.42	4.56
1" x 6"	Ea	883.00	3.53	5.30	7.06

		MBF	8' L	12' L	16' L
Southern yellow pine premium decking					
5/4" x 6"	Ea	2,290.00	11.50	17.20	22.90

		6' L	8' L	12' L	16' L
Red oak boards, S4S, East Coast and Midwest prices					
1" x 2"	Ea	5.42	7.19	8.48	10.20
1" x 3"	Ea	8.28	11.10	13.00	15.50
1" x 4"	Ea	11.00	14.70	18.60	22.30
1" x 6"	Ea	17.00	22.70	27.50	32.90
1" x 8"	Ea	22.40	29.80	36.90	49.30
1" x 10"	Ea	30.30	41.20	50.60	60.70
1" x 12"	Ea	40.00	50.00	64.60	76.90

		MBF	8' L	12' L	16' L
Cedar boards – S1S2E (surfaced 1 side, 2 ends)					
3/4" x 2"	Ea	3,100.00	3.10	4.65	6.20
3/4" x 4"	Ea	2,350.00	4.71	7.06	9.41
3/4" x 6"	Ea	2,890.00	8.66	13.00	17.30
7/8" x 4"	Ea	1,700.00	3.97	5.96	7.95
7/8" x 6"	Ea	2,390.00	8.37	12.60	16.80
7/8" x 8"	Ea	1,950.00	9.11	13.70	18.20
7/8" x 10"	Ea	1,990.00	11.60	17.40	23.20
7/8" x 12"	Ea	2,140.00	15.00	22.40	29.90
1" x 4"	Ea	1,490.00	3.98	5.97	7.96
1" x 6"	Ea	1,790.00	7.17	10.80	14.30

		MBF	8' L	12' L	16' L
Clear cedar trim boards					
1" x 2"	Ea	1,350.00	1.81	2.71	3.61
1" x 3"	Ea	2,490.00	4.98	7.47	9.95
1" x 4"	Ea	2,630.00	7.01	10.50	14.00
1" x 6"	Ea	2,560.00	10.20	15.30	20.40
1" x 8"	Ea	3,090.00	16.50	24.70	33.00
Red cedar boards, kiln dried					
1" x 4"	Ea	1,500.00	4.00	5.99	7.99
1" x 6"	Ea	2,220.00	8.87	13.30	17.70
1" x 8"	Ea	2,040.00	10.90	16.30	21.80
Redwood boards, West Coast prices.					
Add 10% to 30% for transportation to other areas.					
1" x 2" – B Grade S4S	Ea	2,730.00	3.65	5.47	7.29
1" x 3" – B Grade S4S	Ea	2,720.00	5.44	8.16	10.90
1" x 4" – B Grade S4S	Ea	3,100.00	8.27	12.40	16.50
1" x 4" – Green	Ea	2,090.00	5.57	8.36	11.20
1" x 6" – B Grade S4S	Ea	3,090.00	12.40	18.60	24.70
1" x 6" – Green	Ea	2,700.00	10.80	16.20	21.60
1" x 8" – Select S1S2E	Ea	1,970.00	10.50	15.80	21.00

	Craft@Hrs	Unit	Material	Labor	Total
Lath Redwood "A" & better, 50 per bundle, nailed to walls for architectural treatment.					
5/16" x 1-1/2" x 4'	BC@.500	Bdle	12.20	15.00	27.20
5/16" x 1-1/2" x 6'	BC@.625	Bdle	21.00	18.80	39.80
5/16" x 1-1/2" x 8'	BC@.750	Bdle	31.50	22.50	54.00

Board Siding

Pine bevel siding 1/2" thick by 6" wide. Each linear foot of 6" wide bevel siding covers .38 square feet of wall. Do not deduct for wall openings less than 10 square feet. Add 5% for typical waste and end cutting.

	Craft@Hrs	Unit	Material	Labor	Total
8' lengths	B1@.038	SF	.86	1.04	1.90
12' lengths	B1@.036	SF	.86	.98	1.84
16' lengths	B1@.033	SF	.86	.90	1.76

Western red cedar bevel siding A grade and better, 1/2" thick by 6" wide. Each linear foot of 6" wide bevel siding covers .38 square feet of wall. Do not deduct for wall openings less than 10 square feet. Add 5% for typical waste and end cutting.

	Craft@Hrs	Unit	Material	Labor	Total
8' lengths	B1@.038	SF	1.96	1.04	3.00
12' lengths	B1@.036	SF	1.96	.98	2.94
16' lengths	B1@.033	SF	1.96	.90	2.86

Clear cedar 1/2" bevel siding Vertical grain, 1/2" thick by 6" wide, kiln dried. Each linear foot of 6" wide bevel siding covers .38 square feet of wall. Do not deduct for wall openings less than 10 square feet. Add 5% for typical waste and end cutting.

	Craft@Hrs	Unit	Material	Labor	Total
8' lengths	B1@.038	SF	2.29	1.04	3.33
12' lengths	B1@.036	SF	2.30	.98	3.28
16' lengths	B1@.033	SF	2.29	.90	3.19

Cedar 3/4" bevel siding 3/4" thick by 8" wide. Each linear foot of 8" wide bevel siding covers .54 square feet of wall. Do not deduct for wall openings less than 10 square feet. Add 5% for typical waste and end cutting.

	Craft@Hrs	Unit	Material	Labor	Total
8' lengths	B1@.036	SF	2.06	.98	3.04
12' lengths	B1@.034	SF	2.10	.93	3.03
16' lengths	B1@.031	SF	2.06	.85	2.91

	Craft@Hrs	Unit	Material	Labor	Total

Rustic cedar channel siding 3/4" x 8". Each linear foot of 8" wide channel siding covers .54 square feet of wall. Do not deduct for wall openings less than 10 square feet. Add 5% for typical waste and end cutting.

	Craft@Hrs	Unit	Material	Labor	Total
12' lengths	B1@.034	SF	2.58	.93	3.51

Tongue and groove cedar siding 1" x 6". Kiln dried, select. Each linear foot of 6" wide tongue and groove siding covers .43 square feet of wall. Do not deduct for wall openings less than 10 square feet. Add 5% for typical waste and end cutting.

	Craft@Hrs	Unit	Material	Labor	Total
8' lengths	B1@.036	SF	2.89	.98	3.87
10' lengths	B1@.035	SF	2.98	.96	3.94
12' lengths	B1@.034	SF	2.98	.93	3.91

Redwood 5/8" tongue and groove siding 5/8" thick by 5-3/8" wide. Each linear foot of 5-3/8" wide tongue and groove siding covers .42 square feet of wall. Do not deduct for wall openings less than 10 square feet. Add 5% for typical waste and end cutting.

	Craft@Hrs	Unit	Material	Labor	Total
8' lengths	B1@.042	SF	3.84	1.15	4.99
10' lengths	B1@.040	SF	3.75	1.09	4.84
12' lengths	B1@.038	SF	3.83	1.04	4.87

Redwood 5/8" shiplap siding B grade. 5/8" thick by 5-3/8" wide. Each linear foot of 5-3/8" wide shiplap siding covers .38 square feet of wall. Do not deduct for wall openings less than 10 square feet. Add 5% for typical waste and end cutting.

	Craft@Hrs	Unit	Material	Labor	Total
8' lengths	B1@.038	SF	4.24	1.04	5.28
10' lengths	B1@.037	SF	4.24	1.01	5.25
12' lengths	B1@.036	SF	4.23	.98	5.21

Southern yellow pine 6" pattern 105 siding 1" x 6". Select grade except as noted. Each linear foot of 6" wide siding covers .38 square feet of wall. Do not deduct for wall openings less than 10 square feet. Add 5% for typical waste and end cutting

	Craft@Hrs	Unit	Material	Labor	Total
8' lengths	B1@.038	SF	3.04	1.04	4.08
10' lengths	B1@.037	SF	3.03	1.01	4.04
12' lengths, No. 2	B1@.036	SF	1.80	.98	2.78
12' lengths	B1@.036	SF	3.82	.98	4.80

Southern yellow pine 8" pattern 105 siding 1" x 8". Select grade except as noted. Each linear foot of 8" wide siding covers .54 square feet of wall. Do not deduct for wall openings less than 10 square feet. Add 5% for typical waste and end cutting.

	Craft@Hrs	Unit	Material	Labor	Total
8' lengths	B1@.036	SF	2.69	.98	3.67
10' lengths	B1@.034	SF	2.69	.93	3.62
12' lengths, No. 2	B1@.031	SF	1.33	.85	2.18
12' lengths	B1@.031	SF	2.75	.85	3.60

Southern yellow pine 6" pattern 116 siding 1" x 6". Select grade except as noted. Each linear foot of 6" wide siding covers .38 square feet of wall. Do not deduct for wall openings less than 10 square feet. Add 5% for typical waste and end cutting.

	Craft@Hrs	Unit	Material	Labor	Total
8' lengths	B1@.038	SF	3.13	1.04	4.17
10' lengths	B1@.037	SF	3.13	1.01	4.14
12' lengths	B1@.036	SF	3.14	.98	4.12

Southern yellow pine 6" pattern 117 siding 1" x 6". Select grade except as noted. Each linear foot of 6" wide siding covers .38 square feet of wall. Do not deduct for wall openings less than 10 square feet. Add 5% for typical waste and end cutting.

	Craft@Hrs	Unit	Material	Labor	Total
8' lengths	B1@.038	SF	3.13	1.04	4.17
10' lengths	B1@.037	SF	3.13	1.01	4.14
12' lengths	B1@.036	SF	3.14	.98	4.12

	Craft@Hrs	Unit	Material	Labor	Total

Southern yellow pine 8" pattern 131 siding 1" x 8". Select grade except as noted. Each linear foot of 8" wide siding covers .54 square feet of wall. Do not deduct for wall openings less than 10 square feet. Add 5% for typical waste and end cutting.

	Craft@Hrs	Unit	Material	Labor	Total
8' lengths	B1@.036	SF	2.78	.98	3.76
10' lengths	B1@.034	SF	2.80	.93	3.73
12' lengths	B1@.031	SF	2.81	.85	3.66

Southern yellow pine 8" pattern 122 siding 1" x 8" V-joint. Select grade except as noted. Each linear foot of 8" wide siding covers .54 square feet of wall. Do not deduct for wall openings less than 10 square feet. Add 5% for typical waste and end cutting.

	Craft@Hrs	Unit	Material	Labor	Total
8' lengths	B1@.036	SF	2.62	.98	3.60
10' lengths	B1@.034	SF	2.72	.93	3.65
12' lengths	B1@.031	SF	2.72	.85	3.57

Southern yellow pine 6" pattern 122 siding 1" x 6". Select grade except as noted. Each linear foot of 6" wide siding covers .38 square feet of wall. Do not deduct for wall openings less than 10 square feet. Add 5% for typical waste and end cutting.

	Craft@Hrs	Unit	Material	Labor	Total
8' lengths	B1@.038	SF	2.13	1.04	3.17
10' lengths	B1@.037	SF	2.13	1.01	3.14
12' lengths	B1@.036	SF	2.11	.98	3.09

Southern yellow pine 8" shiplap siding 1" x 8". No. 2 grade. Each linear foot of 8" wide siding covers .54 square feet of wall. Do not deduct for wall openings less than 10 square feet. Add 5% for typical waste and end cutting.

	Craft@Hrs	Unit	Material	Labor	Total
8' lengths	B1@.036	SF	1.24	.98	2.22
10' lengths	B1@.034	SF	1.21	.93	2.14
12' lengths	B1@.031	SF	1.24	.85	2.09

Rustic 8" pine siding 1" x 8". V-grooved. Each linear foot of 8" wide siding covers .54 square feet of wall. Do not deduct for wall openings less than 10 square feet. Add 5% for typical waste and end cutting.

	Craft@Hrs	Unit	Material	Labor	Total
16' lengths	B1@.031	SF	3.31	.85	4.16
20' lengths	B1@.031	SF	3.36	.85	4.21

Log cabin siding 1-1/2" thick. Each linear foot of 6" wide siding covers .41 square feet of wall. Each linear foot of 8" wide siding covers .55 square feet of wall. Do not deduct for wall openings less than 10 square feet. Add 5% for typical waste and end cutting.

	Craft@Hrs	Unit	Material	Labor	Total
6" x 8', T &G latia	B1@.038	SF	2.59	1.04	3.63
6" x 12', T & G latia	B1@.036	SF	2.24	.98	3.22
8" x 12	B1@.031	SF	2.67	.85	3.52
8" x 16'	B1@.030	SF	2.50	.82	3.32

Log lap spruce siding 1-1/2" thick. Each linear foot of 6" wide siding covers .41 square feet of wall. Each linear foot of 8" wide siding covers .55 square feet of wall. Each linear foot of 10" wide siding covers .72 square feet of wall. Do not deduct for wall openings less than 10 square feet. Add 5% for typical waste and end cutting.

	Craft@Hrs	Unit	Material	Labor	Total
6" width	B1@.038	SF	3.61	1.04	4.65
8" width	B1@.031	SF	3.47	.85	4.32
10" width	B1@.030	SF	3.50	.82	4.32

Timbers

Green Douglas fir timbers– #1 Grade

		MBF	12' L	16' L	20' L
4" x 8"	Ea	1,100.00	35.20	46.90	58.60
4" x 10"	Ea	1,110.00	44.50	59.30	74.20
4" x 12"	Ea	1,100.00	52.70	70.20	87.80
6" x 6"	Ea	1,280.00	46.10	61.50	76.80

Lumber, Pressure Treated

		8' L	10' L	12' L	16' L
Green Douglas fir timbers – #2 and Better					
4" x 4"	Ea	10.20	12.80	15.40	20.50
4" x 6"	Ea	15.60	19.50	23.40	31.20
4" x 8"	Ea	17.50	21.80	26.20	34.90
4" x 10"	Ea	21.60	27.10	32.50	43.30
4" x 12"	Ea	26.50	33.10	39.70	53.00

		MBF	8' L	12' L	16' L
Red cedar timbers					
4" x 4" – Green S4S	Ea	1,550.00	16.60	24.80	33.10
4" x 4" – Rough	Ea	1,210.00	12.90	19.40	25.90
6" x 6" – Rough	Ea	2,120.00	51.00	76.50	102.00

Redwood timbers West Coast prices.

Add 10% to 30% for transportation to other areas.

		MBF	8' L	12' L	16' L
4" x 4" – Construction Common	Ea	1,410.00	15.10	22.60	30.10
4" x 4" – Construction Heart - Rough	Ea	1,520.00	16.30	24.40	32.50
4" x 4" – Construction Heart – S4S	Ea	1,710.00	18.20	27.30	36.40

Pressure treated lumber, ACQ treatment

		MBF	8' L	12' L	16' L
Pressure treated trim board S4S, .25 treatment, appearance grade					
1" x 4" - .25 appearance grade	Ea	1,090.00	2.90	4.36	5.81
1" x 6" - .25 appearance grade	Ea	1,090.00	4.36	6.53	8.71
1" x 8" - .25 appearance grade	Ea	1,140.00	6.06	9.09	12.10
1" x 4", - .25 D grade	Ea	1,410.00	3.75	5.62	7.49
1" x 6", - .25 D grade	Ea	1,590.00	6.34	9.51	12.70
1" x 8", - .25 D grade	Ea	1,560.00	8.35	12.50	16.70
1" x 12", - .25 D grade	Ea	1,790.00	14.30	21.50	28.60

		MBF	8' L	12' L	16' L
Pressure treated lumber, .25 ACQ treatment					
2" x 3"	Ea	801.00	4.27	6.40	8.53
2" x 4"	Ea	942.00	5.02	7.53	10.00
2" x 6"	Ea	871.00	6.97	10.50	13.90
2" x 8"	Ea	917.00	9.78	14.70	19.60
2" x 10"	Ea	981.00	13.10	19.60	26.20
2" x 12"	Ea	1,000.00	16.10	24.10	32.10
4" x 4"	Ea	990.00	10.60	15.80	21.10
4" x 6"	Ea	1,030.00	16.40	24.60	32.90
4" x 8"	Ea	1,140.00	24.30	36.40	48.50
4" x 10"	Ea	1,160.00	31.00	46.40	61.90
4" x 12"	Ea	1,230.00	39.50	59.30	79.00
6" x 6"	Ea	904.00	21.70	32.50	43.40

		MBF	8' L	12' L	16' L
Pressure treated southern yellow pine – .40 ACQ treatment, prime					
1" x 4"	Ea	1,000.00	2.68	4.01	5.35
1" x 6"	Ea	892.00	3.57	5.35	7.13
2" x 4"	Ea	861.00	4.59	6.88	9.17
2" x 6"	Ea	791.00	6.33	9.49	12.70
2" x 8"	Ea	823.00	8.77	13.20	17.50
2" x 10"	Ea	781.00	10.40	15.60	20.80
2" x 12"	Ea	1,180.00	19.00	28.40	37.90

		MBF	8' L	12' L	16' L
4" x 4"	Ea	1,810.00	19.33	29.00	38.70
4" x 6"	Ea	1,690.00	27.00	40.50	54.00
6" x 6"	Ea	1,910.00	45.90	68.90	91.90

Pressure treated timbers .40 CCA pressure treatment. For uses with direct contact with water or the ground, or where moisture content exceeds 20%. Add 20% for .60 CCA treated southern yellow pine timber

		MBF	8' L	12' L	16' L
4" x 4"	Ea	1,010.00	10.80	16.20	21.60
4" x 6"	Ea	970.00	15.50	23.30	31.10
6" x 6"	Ea	933.00	22.40	33.60	44.80
Hibor treated Douglas fir (Hawaii)					
2" x 4", Standard and Better	Ea	5.49	6.96	8.69	11.80
2" x 6", No. 2 and Better	Ea	9.02	11.10	13.10	17.50
2" x 8", No. 2 and Better	Ea	11.70	14.10	17.00	22.90
2" x 10", No. 2 and Better	Ea	15.40	19.20	23.60	30.90
2" x 12", No. 2 and Better	Ea	18.40	23.10	27.80	37.20
4" x 4", No. 2 and Better	Ea	14.50	18.30	21.80	—
4" x 6", No. 2 and Better	Ea	22.80	33.90	44.10	40.00
4" x 8", No. 1 and Better, S4S	Ea	—	—	44.60	59.80
4" x 10", No. 1 and Better, S4S	Ea	—	—	57.20	76.60
4" x 12", No. 1 and Better, S4S	Ea	—	—	69.30	90.80

Plywood

		MSF	4' x 8'
Sanded exterior grade Western plywood – AC			
1/4" thick	Ea	441.00	14.10
3/8" thick	Ea	502.00	16.10
1/2" thick	Ea	872.00	27.90
5/8" thick	Ea	994.00	31.80
3/4" thick	Ea	1,120.00	36.00
Sanded exterior grade Western plywood – BC			
3/8" thick	Ea	379.00	12.10
1/2" thick	Ea	456.00	14.60
1/2" thick, 4-ply	Ea	494.00	15.80
1/2" thick, 5-ply	Ea	522.00	16.70
5/8" thick, 4-ply	Ea	578.00	18.50
5/8" thick, 5-ply	Ea	634.00	20.30
3/4" thick	Ea	756.00	24.20
Exterior grade plywood – CDX			
1/4" thick	Ea	441.00	14.10
3/8" thick	Ea	502.00	16.10
1/2" thick	Ea	617.00	19.70
5/8" thick	Ea	775.00	24.80
3/4" thick	Ea	914.00	29.20

	Craft@Hrs	Unit	Material	Labor	Total
Texture 1-11 (T-1-11) siding, rough sawn Per 100 square feet installed including 10% cutting waste.					
3/8" x 4' x 8' Southern pine, grooved 4" OC	B1@2.18	Sq	53.10	59.60	112.70
3/8" x 4' x 8' Decorative southern pine	B1@2.18	Sq	55.20	59.60	114.80
3/8" x 4' x 8' Hibor treated, plain (Hawaii)	B1@2.18	Sq	113.00	59.60	172.60

Identification Index: A set of two numbers separated by a slash that appears in the grade trademarks on standard sheathing, Structural I, Structural II and C-C grades. Number on left indicates recommended maximum spacing in inches for supports when panel is used for roof decking. Number on right shows maximum recommended spacing in inches for supports when the panel is used for subflooring.

A-A EXT-DFPA
Designed for use in exposed applications where the appearance of both sides is important. Fences, wind screens, exterior cabinets and built-ins, signs, boats, commercial refrigerators, shipping containers, tanks, and ducts
A-A G-2 EXT-DFPA PS 1-66
Typical edge-mark

A-B EXT-DFPA
For uses similar to A-A EXT but where the appearance of one side is a little less important than the face.
A-B G-3 EXT-DFPA PS 1-66
Typical edge-mark

Lumber, Plywood

	Craft@Hrs	Unit	Material	Labor	Total
5/8" x 4' x 8' Premium pine, grooved	B1@2.45	Sq	76.30	67.00	143.30
5/8" x 4' x 8' Treated pine, decorative grade	B1@2.45	Sq	115.00	67.00	182.00
5/8" x 4' x 9' Fir, grooved	B1@2.45	Sq	131.00	67.00	198.00
5/8" x 4' x 8' Hibor treated, grooved (Hawaii)	B1@2.45	Sq	162.00	67.00	229.00
5/8" x 4' x 9' Hibor treated, grooved (Hawaii)	B1@2.45	Sq	174.00	67.00	241.00

		Unit	MSF	4' x 8'	
2-4-1 Sturdi-Floor subflooring Tongue and groove attic decking					
5/8"		Ea	866.00	27.70	
Concrete form (Plyform), square edged					
5/8", 4 ply		Ea	1,510.00	48.30	
3/4", 4 ply		Ea	1,360.00	43.50	
Structural I grade plywood. Price per panel					
3/8"		Ea	391.00	12.50	
1/2"		Ea	551.00	17.60	
Medium Density Overlay (MDO) plywood. Paintable paper surface bonded to plywood core					
3/8"		Ea	983.00	31.50	
1/2"		Ea	1,310.00	42.00	
3/4"		Ea	1,640.00	52.50	

		Unit	MSF	4' x 8'	
Marine grade plywood.					
1/2", AB grade		Ea	1,510.00	48.30	
3/4", AB grade		Ea	1,360.00	43.50	
OSB strand board – square edged Oriented grain (OSB)					
1/4"		Ea	276.00	8.84	
3/8"		Ea	512.00	16.40	
7/16"		Ea	463.00	14.80	
1/2"		Ea	487.00	15.60	
3/4"		Ea	754.00	24.10	

		Unit	MSF	4' x 8'	
Underlayment X-Band, C Grade, tongue and groove					
5/8", 4 ply		Ea	759.00	24.30	
3/4", 4 ply		Ea	877.00	28.10	
Birch hardwood plywood					
1/8" thick		Ea	590.00	18.90	
1/4" thick		Ea	648.00	20.70	
1/2" thick		Ea	1,150.00	36.80	
3/4" thick		Ea	1,380.00	44.10	
Cedar hardwood plywood					
1/4" thick		Ea	834.00	26.70	
3/8" thick		Ea	917.00	29.30	
1/2" thick		Ea	1,160.00	37.10	
5/8" thick		Ea	1,310.00	42.00	
3/4" thick		Ea	1,500.00	48.10	
Oak hardwood plywood					
1/8" thick		Ea	557.00	17.80	
1/4" thick		Ea	675.00	21.60	
1/2" thick		Ea	1,220.00	39.10	
3/4" thick		Ea	1,460.00	46.80	

			MSF	4' x 8'
Lauan plywood				
1/8" thick		Ea	310.00	9.92
1/4" thick		Ea	392.00	12.60
1/4" thick, moisture resistant		Ea	372.00	11.90

	Craft@Hrs	Unit	Material	Labor	Total
Hardboard, 4' x 8' panels					
1/8" tempered hardboard	BC@1.02	Ea	7.07	30.60	37.67
1/8" tempered pegboard	BC@1.02	Ea	9.04	30.60	39.64
1/8" backer board	BC@1.02	Ea	11.88	30.60	42.48
1/4" tempered hardboard	BC@1.02	Ea	10.10	30.60	40.70
3/16" white garage liner pegboard	BC@1.02	Ea	13.99	30.60	44.59
1/4" white vinyl coated	BC@1.02	Ea	14.27	30.60	44.87
1/4" tempered perforated pegboard	BC@1.02	Ea	11.26	30.60	41.86
Particleboard					
3/8" x 4' x 8'	—	Ea	9.34	--	9.34
1/2" x 4' x 8'	—	Ea	10.49	--	10.49
5/8" x 4' x 8'	—	Ea	11.13	--	11.13
3/4" x 4' x 8'	—	Ea	12.48	--	12.48
3/4" x 25" x 97" countertop	—	Ea	9.67	--	9.67
3/4" x 30" x 97", countertop	—	Ea	12.96	--	12.96
3/4" x 25" x 121", countertop	—	Ea	16.97	--	16.97
3/4" x 25" x 145", countertop	—	Ea	13.95	--	13.95
3/4" x 49" x 97", industrial grade	—	Ea	15.40	--	15.40

	MBF		8' L	12' L	16' L
Recycled Plastic Lumber Produced from post-consumer plastics. Resistant to termites, rot, decay.					
2" x 4"	Ea	4,750.00	25.30	38.00	50.60
2" x 6"	Ea	4,720.00	37.70	56.60	75.50
2" x 8"	Ea	4,670.00	49.80	74.70	99.60
2" x 10"	Ea	4,760.00	63.40	95.10	127.00
2" x 12"	Ea	5,330.00	85.20	128.00	170.00
4" x 4"	Ea	4,510.00	48.10	72.10	96.20
4" x 6"	Ea	5,040.00	80.70	121.00	161.00
6" x 6"	Ea	4,920.00	118.00	177.00	236.00
1" x 6" siding, smooth edge	Ea	3,960.00	15.80	23.80	31.70
1" x 6" siding, tongue & groove	Ea	4,120.00	16.50	24.70	33.00

Poles Select appearance grade building poles. Poles have an average taper of 1" in diameter per 10' of length. Ponderosa pine treated with Penta-Dow or ACZA (.60 lbs/CF), meets AWPA-C 23-77. Costs based on quantity of 8 to 10 poles. For delivery of smaller quantities, add $50.00 to the total cost. See installation labor below. Costs are per linear foot.

		8" butt size	10" butt size	12" butt size
To 15' long pole	LF	9.96	13.00	17.20
To 20' long pole	LF	9.68	12.30	16.00
To 25' long pole	LF	9.39	10.30	14.60
Add for Douglas fir	LF	2.25	2.35	2.45
Add for delivery, typical per trip	Ea	75.00	75.00	75.00

Lumber, Posts

		6' long	7' long	8' long	10' long
Posts Sawn posts, merchantable, redwood					
4" x 4"	Ea	6.65	6.99	9.79	12.20
Yellow pine penta treated posts					
3" x 3"	Ea	6.06	7.34	8.53	9.85
3-1/2" x 3-1/2"	Ea	6.82	8.02	9.05	10.40
4" x 4"	Ea	8.42	9.91	10.80	12.10
4-1/2" x 4-1/2"	Ea	10.20	10.85	11.20	12.50
5" x 5"	Ea	10.00	12.20	12.80	15.10
6" x 6"	Ea	12.10	13.60	14.30	16.10

		8' long	10' long	12' long	16' long
Pressure treated southern yellow pine posts No. 2, 0.40 CCA treatment					
4" x 4"	Ea	4.86	9.76	12.10	15.80
4" x 6"	Ea	12.20	13.40	18.40	23.30
6" x 6"	Ea	20.80	23.30	30.60	40.30

	Craft@Hrs	Unit	Material	Labor	Total
Labor to install poles and posts Costs shown are per post and include measuring, cutting one end, drilling two holes at the top and bottom and installation. Add for hardware and bolts as needed.					
3" x 3", 4" x 4" or 5" x 5" posts	B1@.333	Ea	—	9.11	9.11
6" x 6" or 7" x 7" posts	B1@.500	Ea	—	13.70	13.70
8" or 10" butt diameter poles	B1@.666	Ea	—	18.20	18.20
12" butt diameter poles	B1@.750	Ea	—	20.50	20.50
Stakes					
Survey stakes, pine or fir, 1" x 2", per bundle of 25					
24"	—	Bdle	10.00	—	10.00
36"	—	Bdle	13.90	—	13.90
48"	—	Bdle	15.90	—	15.90
Ginnies, pine or fir, 1" x 1", per 1,000					
6"	—	M	75.00	—	75.00
8"	—	M	79.50	—	79.50
Pointed lath, pine or fir, per 1,000, 1/4" x 1-1/2", S4S					
18"	—	M	221.00	—	221.00
24"	—	M	239.00	—	239.00
36"	—	M	330.00	—	330.00
48"	—	M	450.00	—	450.00
Foundation stakes, 1" x 3", per 1,000, 20 bundles per 1000 except for 48" (40 bundles per 1000)					
12" or 14"	—	M	273.00	—	273.00
18"	—	M	349.00	—	349.00
24"	—	M	403.00	—	403.00
36"	—	M	568.00	—	568.00
48"	—	M	918.00	—	918.00
Hubs, pine or fir, 2" x 2", per 1,000					
8"	—	M	476.00	—	476.00
10"	—	M	531.00	—	531.00
12"	—	M	580.00	—	580.00

	Craft@Hrs	Unit	Material	Labor	Total
Form stakes (wedges), pine or fir, 5" point, per 1,000					
2" x 2" x 6"	—	M	186.00	—	186.00
2" x 2" x 8"	—	M	200.00	—	200.00
2" x 2" x 10"	—	M	219.00	—	219.00
2" x 4" x 6"	—	M	241.00	—	241.00
2" x 4" x 8"	—	M	283.00	—	283.00
2" x 4" x 10"	—	M	306.00	—	306.00
2" x 4" x 12"	—	M	349.00	—	349.00
Feather wedges	—	M	261.00	—	261.00

Lumber Grading Lumber grading services by West Coast Lumber Inspection Bureau

	Craft@Hrs	Unit	Material	Labor	Total
Inspection per MBF	—	MBF	—	—	.33
Per month minimum – operating and shipping	—	Month	—	—	330.00
Per month min. – non-operating (No Shipments)	—	Month	—	—	100.00

Reinspection, certificate inspection and grade stamping service for WCLIB Members. Costs do not include travel and subsistence charges.

	Craft@Hrs	Unit	Material	Labor	Total
On shipments west of the Rocky Mountains	—	Day	—	—	350.00
Overtime hours	—	Hour	—	—	45.00
Saturday, Sunday and holidays	—	Day	—	—	170.00
On shipments east of the Rocky Mountains	—	Day	—	—	350.00
Overtime hours	—	Hour	—	—	45.00
Saturday, Sunday and holidays	—	Day	—	—	170.00

Reinspection, certificate inspection and grade stamping service for WCLIB Non-Members. Costs do not include travel and subsistence charges.

	Craft@Hrs	Unit	Material	Labor	Total
On shipments west of the Rocky Mountains	—	Day	—	—	500.00
Overtime hours	—	Hour	—	—	65.00
Saturday, Sunday and holidays	—	Day	—	—	190.00
On shipments east of the Rocky Mountains	—	Day	—	—	500.00
Overtime hours	—	Hour	—	—	65.00
Saturday, Sunday and holidays	—	Day	—	—	170.00

Certificate inspection service. Issuance of certificates to inspection subscribers employing WCLIB certified inspectors

	Craft@Hrs	Unit	Material	Labor	Total
Per thousand board feet, $30.00 minimum	—	MBF	—	—	0.30

Lumber Preservation Treatment See also Soil Treatments. Costs assume treatment at a treatment plant, not on the construction site. Treatments listed below are designed to protect wood and wood products from decay and destruction caused by moisture (fungi), wood boring insects and organisms. All retention rates listed (in parentheses) are in pounds per cubic foot and are the minimum recommended by the Society of American Wood Preservers. Costs are for treatment only and assume treating of 10,000 board feet or more. No lumber or transportation included. Add $70.00 per MBF for kiln-drying after treatment.

	Craft@Hrs	Unit	Material	Labor	Total
Lumber submerged in or frequently exposed to salt water					
ACZA treatment (2.5)	—	MBF	—	—	869.00
ACZA treatment for structural lumber					
Piles and building poles (2.5)	—	CF	—	—	9.80
Board lumber and timbers (2.5)	—	MBF	—	—	869.00
Creosote treatment for structural lumber					
Piles and building poles (20.0)	—	CF	—	—	8.69
Board lumber and timbers (25.0)	—	MBF	—	—	520.00

Lumber Preservation Treatment

	Craft@Hrs	Unit	Material	Labor	Total
Lumber in contact with fresh water or soil, ACZA treatment					
Building poles, structural (.60)	—	CF	—	—	5.44
Boards, timbers, structural (.60)	—	MBF	—	—	250.00
Boards, timbers, non-structural (.40)	—	MBF	—	—	223.00
Plywood, structural (.40), 1/2"	—	MSF	—	—	250.00
Posts, guardrail, signs (.60)	—	MBF	—	—	314.00
All weather foundation KD lumber (.60)	—	MBF	—	—	350.00
All weather foundation, 1/2" plywood (.60)	—	MSF	—	—	278.30
Lumber in contact with fresh water or soil, creosote treatment					
Piles, building poles, structural (12.0)	—	CF	—	—	6.16
Boards, timbers, structural (12.0)	—	MBF	—	—	377.00
Boards, timbers, non-structural (10.0)	—	MBF	—	—	350.00
Plywood, structural, 1/2" (10.0)	—	MSF	—	—	302.00
Posts (fence, guardrail, sign) (12.0)	—	MBF	—	—	377.00
Lumber used above ground, ACZA treatment					
Boards, timbers, structural (.25)	—	MBF	—	—	206.00
Boards, timbers, non-structural (.25)	—	MBF	—	—	206.00
Plywood, structural 1/2" (.25)	—	MSF	—	—	230.00
Lumber used above ground, creosote treatment					
Boards, timbers, structural (8.0)	—	MBF	—	—	314.00
Boards, timbers, non-structural (8.0)	—	MBF	—	—	314.00
Plywood, structural (8.0), 1/2"	—	MSF	—	—	278.00
Lumber used for decking, ACQ treatment					
Boards, structural (.41)	—	MBF	—	—	271.00
Boards, timbers, non-structural (.41)	—	MSF	—	—	271.00
Plywood and other 1/2" sheets (.41)	—	MSF	—	—	295.00
Millwork (.26)	—	MBF	—	—	247.00
Fire retardant treatments, structural use (interior only), "D-Blaze"					
Surfaced lumber	—	MBF	—	—	289.00
Rough lumber	—	MBF	—	—	289.00
Architectural use (interior and exterior), "NCX"					
Surfaced lumber	—	MBF	—	—	668.00
Rough lumber	—	MBF	—	—	668.00

Wolmanized treatment. Copper chromated arsenic preservative is not used for residential applications. The term "Wolmanized" is now being applied to ACQ and CA preservatives which do not contain arsenic.

	Craft@Hrs	Unit	Material	Labor	Total
Lumber used above ground (.25)	—	MBF	—	—	102.00
Lumber in ground contact or fresh water (.40)	—	MBF	—	—	128.00
Lumber used as foundation (.60)	—	MBF	—	—	139.00

	Craft@Hrs	Unit	Material	Labor	Total

Mailboxes Apartment type, meets current postal regulations, tumbler locks, aluminum or gold finish, price per unit, recessed

	Craft@Hrs	Unit	Material	Labor	Total
3 box unit, 5-1/2" wide x 16" high	B1@1.88	Ea	63.50	51.40	114.90
4 box unit, 5-1/2" wide x 16" high	B1@1.88	Ea	93.50	51.40	144.90
5 box unit, 5-1/2" wide x 16" high	B1@1.88	Ea	109.00	51.40	160.40
6 box unit, 5-1/2" wide x 16" high	B1@1.88	Ea	126.00	51.40	177.40
7 box unit, 5-1/2" wide x 16" high	B1@1.88	Ea	149.00	51.40	200.40
Add for surface mounting collar	B1@.550	Ea	36.70	15.00	51.70

Office type, horizontal, meets postal regulations, tumbler locks, aluminum or brass finish
For front or rear load, per box

	Craft@Hrs	Unit	Material	Labor	Total
6" wide x 5" high x 15" deep	B1@1.02	Ea	28.10	27.90	56.00

Rural, aluminum finish, 19" long x 6-1/2" wide x 9" high, with flag

	Craft@Hrs	Unit	Material	Labor	Total
Add for mounting post	B1@.776	Ea	13.80	21.20	35.00

Rural heavy gauge steel mailboxes, aluminum finish, 19" long x 7" wide x 10" high, with flag

	Craft@Hrs	Unit	Material	Labor	Total
Add for mounting post	B1@.776	Ea	63.00	21.20	84.20

Suburban style, finished, 16" long x 3" wide x 6" deep, with flag

	Craft@Hrs	Unit	Material	Labor	Total
Post mount	B1@.776	Ea	23.10	21.20	44.30
Add for 5' high aluminum post	—	Ea	21.00	—	16.20

Cast aluminum mailbox and pedestal, 18" x 12" box, 4' pedestal,

	Craft@Hrs	Unit	Material	Labor	Total
With base, lockable	B1@.776	Ea	155.00	21.20	176.20

Aluminum mailbox and post, 18" x 12" box, 5' post,

	Craft@Hrs	Unit	Material	Labor	Total
Steel anchor base	B1@.776	Ea	208.00	21.20	229.20

Plastic mailbox, mounting sleeve slides over 4' high post, lockable

	Craft@Hrs	Unit	Material	Labor	Total
Size 1 box, add for post	B1@.776	Ea	42.00	21.20	63.20
Add for treated wood or aluminum post	—	Ea	16.20	—	16.20

Thru-the-wall type, 13" long x 3" high

	Craft@Hrs	Unit	Material	Labor	Total
Gold or aluminum finish	B1@1.25	Ea	29.20	34.20	63.40

Wall or post mount security mailboxes, lockable, 15" wide x 13" high x 8" deep

	Craft@Hrs	Unit	Material	Labor	Total
Black plastic and aluminum	B1@.450	Ea	71.20	12.30	83.50

Mantels, Ponderosa Pine Unfinished.

	Craft@Hrs	Unit	Material	Labor	Total
Modern design, 50" x 37" opening, paint grade, 11" x 77" shelf	B1@3.81	Ea	595.00	104.00	699.00
Colonial design, 50" x 39" opening, stain grade, 7" x 72" shelf	B1@3.81	Ea	950.00	104.00	1,054.00
Ornate design, 51-1/2" x 39" opening, with detailing and scroll work, stain grade, 11" x 68" shelf	B1@3.81	Ea	1,780.00	104.00	1,884.00
Mantel moulding (simple trim for masonry fireplace openings), 15 LF, for opening 6'0" x 3'6" or less	B1@1.07	Set	210.00	29.30	239.30

Mantel shelf only, with iron or wood support brackets, prefinished

	Craft@Hrs	Unit	Material	Labor	Total
4" x 8" x 8"	B1@.541	Ea	54.50	14.80	69.30
4" x 8" x 6"	B1@.541	Ea	41.70	14.80	56.50
4" x 10" x 10"	B1@.541	Ea	93.00	14.80	107.80

Marble Setting

	Craft@Hrs	Unit	Material	Labor	Total

Marble Setting, Subcontract See also Tile section. Marble prices vary with quality, quantity, and availability of materials. Check prices with local suppliers. These prices are based on 200 SF quantities.

Marble tile flooring, tiles or slabs

   3/8" x 12" x 12", thin set over concrete slab, based on $6.50 to $9.00 per SF material cost and $7.00 to $10.00 per SF labor cost. (Includes cutting)

Typical cost per SF of area covered	—	SF	—	—	18.20

   3/4" slabs set in mortar bed, cut to size, based on $30.00 to $40.00 per SF material cost and $15.00 per SF labor cost

Typical cost per SF of area covered	—	SF	—	—	62.20

Marble bathroom vanity top, 6' x 2' wide, with 1 sink cutout and 4" splash on 3 sides

   One per home, based on $560.00 to $860.00 per unit material cost and $150.00 per unit labor cost

Single top, typical cost	—	Ea	—	—	900.00

   Four per home, based on $560.00 to $860.00 per unit material cost and $50.00 per unit labor cost

Multiple tops, typical cost	—	Ea	—	—	810.00

Marble fireplace hearth and facing, one per home, based on $400.00 to $550.00 per unit material cost, including cutting and waste, and $260.00 to $460.00 per unit labor cost

Typical cost per unit	—	Ea	—	—	989.00

Markup Typical markup for light construction, percentage of gross contract price. No insurance or taxes included. See also Insurance and Taxes sections. Markup is the amount added to an estimate after all job costs have been accounted for: labor and materials, job fees, permits, bonds and insurance, job site supervision and similar job-related expenses. Profit is the return on money invested in the construction business.

   Contingency (allowance for unknown and unforeseen conditions)

Varies widely from job to job, typically	—	%	—	—	2.0

   Overhead (office, administrative and general business expense)

Typical overhead expense	—	%	—	—	10.0

   Profit (return on the money invested in the business, varies with competitive conditions)

Typical profit for residential construction	—	%	—	—	8.0

   Total markup (add 25% to the total job cost to attain 20% of gross contract price)

Total for Contingency, Overhead & Profit	—	%	—	—	20.0

Note that costs listed in this book do not include the general contractor's markup. However, sections identified as "subcontract" include the subcontractor's markup. Here's how to calculate markup:

Project Costs	
Material	$50,000
Labor (with taxes & insurance)	40,000
Subcontract	10,000
Total job costs	100,000
Markup (25% of $100,000)	**25,000**
Contract price	$125,000

Markup is $25,000 which is 25% of the total job costs ($100,000) but only 20% of the contract price ($25,000 divided by $125,000 is .20).

Markup For General Contractors Handling larger commercial and industrial projects, a markup of 24.8% on the total job cost is suggested. See the paragraph General Contractor Markup at the beginning of the section General Requirements in the Industrial and Commercial Division of this manual. For example, note how markup is calculated on a project with these costs:

Material	$3,000,000
Labor (with taxes and insurance)	4,500,000
Equipment	1,000,000
Subcontracts	1,500,000
Total job cost	$10,000,000
Indirect overhead (8%)	**800,000**
Direct overhead (7.3%)	**730,000**
Contingency (2%)	**200,000**
Profit (7.5%)	**750,000**
Contract price	$12,480,000

The total job cost shown above includes work done with the general contractor's crew and work done by subcontractor-specialists. Note that markup for the general contractor (overhead, including supervision expense, contingency and profit) is $2,480,000. This amounts to 24.8% of the total job costs but only 19.87% of the contract price (2,480,000 divided by 12,480,000 is 0.1987).

Masonry, Brick Wall Assemblies Typical costs for smooth red clay brick walls, laid in running bond with 3/8" concave joints. These costs include the bricks, mortar for bricks and cavities, typical ladder type reinforcing, wall ties and normal waste. Foundations are not included. Wall thickness shown is the nominal size based on using the type bricks described. "Wythe" means the quantity of bricks in the thickness of the wall. Costs shown are per square foot (SF) of wall measured on one face. Deduct for openings over 10 SF in size. The names and dimensions of bricks can be expected to vary, depending on the manufacturer. For more detailed coverage of brick and masonry see the National Concrete & Masonry Estimator at http://CraftsmanSiteLicense.Com

	Craft@Hrs	Unit	Material	Labor	Total
Standard bricks 3-3/4" wide x 2-1/4" high x 8" long					
Common brick, per thousand	—	M	440.00	—	440.00
4" thick wall, single wythe, veneer facing	B9@.211	SF	3.64	5.90	9.54
8" thick wall, double wythe, cavity filled	B9@.464	SF	7.14	13.00	20.14
12" thick wall, triple wythe, cavity filled	B9@.696	SF	10.60	19.50	30.10
Norman brick 3-1/2" wide x 2-1/2" high x 11-1/2" long					
Norman brick, per thousand	—	M	1,110.00	—	1,110.00
4" thick wall, single wythe, veneer facing	B9@.211	SF	5.64	5.90	11.54
8" thick wall, double wythe, cavity filled	B9@.464	SF	11.20	13.00	24.20
12" thick wall, triple wythe, cavity filled	B9@.696	SF	16.60	19.50	36.10
Modular brick, 3" wide x 3-1/2" x 11-1/2" long					
Modular brick, per thousand	—	M	1,110.00	—	1,110.00
3-1/2" thick wall, single wythe, veneer face	B9@.156	SF	4.18	4.36	8.54
7-1/2" thick wall, double wythe, cavity filled	B9@.343	SF	8.23	9.59	17.82
11-1/2" thick wall, triple wythe, cavity filled	B9@.515	SF	12.30	14.40	26.70
Colonial brick, 3" wide x 3-1/2" x 10" long					
Colonial brick, per thousand	—	M	900.00	—	900.00
3-1/2" thick wall, single wythe, veneer face	B9@.177	SF	3.93	4.95	8.88
7-1/2" thick wall, double wythe, cavity filled	B9@.389	SF	7.73	10.90	18.63
11-1/2" thick wall, triple wythe, cavity filled	B9@.584	SF	11.50	16.30	27.80

Masonry

	Craft@Hrs	Unit	Material	Labor	Total

Masonry, Face Brick Laid in running bond with 3/8" concave joints. Mortar, wall ties and foundations (footings) not included. Includes normal waste and delivery up to 30 miles with a 5 ton minimum. Coverage (units per SF) includes 3/8" mortar joints but the SF cost does not include mortar cost. Add for mortar below.

Standard brick, 2-1/4" high x 3-3/4" deep x 8" long (6.45 units per SF)

	Craft@Hrs	Unit	Material	Labor	Total
Brown, per square foot	B9@.211	SF	4.61	5.90	10.51
Brown, per thousand	B9@32.7	M	680.00	914.00	1,594.00
Burnt oak, per square foot	B9@.211	SF	4.61	5.90	10.51
Burnt oak, per thousand	B9@32.7	M	680.00	914.00	1,594.00
Flashed, per square foot	B9@.211	SF	3.72	5.90	9.62
Flashed, per thousand	B9@32.7	M	550.00	914.00	1,464.00
Red, per square foot	B9@.211	SF	2.98	5.90	8.88
Red, per thousand	B9@32.7	M	440.00	914.00	1,354.00

Norman brick, 2-1/2" high x 3-1/2" deep x 11-1/2" long (4.22 units per SF)

	Craft@Hrs	Unit	Material	Labor	Total
Brown	B9@.211	SF	9.22	5.90	15.12
Burnt oak	B9@.211	SF	9.22	5.90	15.12
Flashed	B9@.211	SF	4.92	5.90	10.82
Red	B9@.211	SF	4.92	5.90	10.82

Modular, 3-1/2" high x 3" deep x 11-1/2" long (3.13 units per SF)

	Craft@Hrs	Unit	Material	Labor	Total
Brown	B9@.156	SF	3.65	4.36	8.01
Burnt oak	B9@.156	SF	3.65	4.36	8.01
Flashed	B9@.156	SF	3.39	4.36	7.75
Red	B9@.156	SF	3.19	4.36	7.55

Jumbo (filler brick) 3-1/2" high x 3" deep x 11-1/2" long (3.13 units per SF)

	Craft@Hrs	Unit	Material	Labor	Total
Fireplace, cored	B9@.156	SF	1.58	4.36	5.94
Fireplace, solid	B9@.156	SF	1.41	4.36	5.77
Mission, cored	B9@.156	SF	3.20	4.36	7.56
Mission, solid	B9@.156	SF	3.31	4.36	7.67

Padre brick, 4" high x 3" deep

	Craft@Hrs	Unit	Material	Labor	Total
7-1/2" long cored, 4.2 per SF	B9@.156	SF	7.01	4.36	11.37
7-1/2" long solid, 4.2 per SF	B9@.156	SF	8.81	4.36	13.17
11-1/2" long cored, 2.8 per SF	B9@.177	SF	5.17	4.95	10.12
11-1/2" long solid, 2.8 per SF	B9@.177	SF	7.85	4.95	12.80
15-1/2" long cored, 2.1 per SF	B9@.211	SF	6.56	5.90	12.46
15-1/2" long solid, 2.1 per SF	B9@.211	SF	8.23	5.90	14.13

Commercial, 3-1/4" high x 3-1/4" deep x 10" long (3.6 per SF)

	Craft@Hrs	Unit	Material	Labor	Total
Solid	B9@.156	SF	3.40	4.36	7.76
Cored	B9@.177	SF	3.40	4.95	8.35
Harlequin (manufactured "used" brick)	B9@.177	SF	3.67	4.95	8.62

Mortar for brick Based on an all-purpose Type S mortar mix used in single wythe walls laid with running bond. Includes normal waste. Type N mortar is used for severe conditions, such as chimneys. Costs are per SF of wall face using the brick sizes (width x height x length) listed.

	Craft@Hrs	Unit	Material	Labor	Total
Type N mortar, 60 lb. bag	—	Ea	3.90	—	3.90
Type S mortar, per cubic foot	—	CF	6.80	—	6.80
Type S mortar, 60 lb. bag	—	Ea	3.40	—	3.40
4" x 2-1/2" to 2-5/8" x 8" bricks					
3/8" joints (5.5 CF per CSF)	—	SF	.37	—	.37
1/2" joints (7.0 CF per CSF)	—	SF	.48	—	.48
4" x 3-1/2" to 3-5/8" x 8" bricks					
3/8" joints (4.8 CF per CSF)	—	SF	.33	—	.33
1/2" joints (6.1 CF per CSF)	—	SF	.42	—	.32

	Craft@Hrs	Unit	Material	Labor	Total
4" x 4" x 8" bricks					
3/8" joints (4.2 CF per CSF)	—	SF	.29	—	.29
1/2" joints (5.3 CF per CSF)	—	SF	.36	—	.36
4" x 5-1/2" to 5-5/8" x 8" bricks					
3/8" joints (3.5 CF per CSF)	—	SF	.24	—	.24
1/2" joints (4.4 CF per CSF)	—	SF	.30	—	.30
4" x 2" x 12" bricks					
3/8" joints (6.5 CF per CSF)	—	SF	.44	—	.44
1/2" joints (8.2 CF per CSF)	—	SF	.56	—	.56
4" x 2-1/2" to 2-5/8" x 12" bricks					
3/8" joints (5.1 CF per CSF)	—	SF	.35	—	.35
1/2" joints (6.5 CF per CSF)	—	SF	.44	—	.44
4" x 3-1/2" to 3-5/8" x 12" brick					
3/8" joints (4.4 CF per CSF)	—	SF	.30	—	.30
1/2" joints (5.6 CF per CSF)	—	SF	.38	—	.36
4" x 4" x 12" bricks					
3/8" joints (3.7 CF per CSF)	—	SF	.25	—	.25
1/2" joints (4.8 CF per CSF)	—	SF	.33	—	.33
4" x 5-1/2" to 5-5/8" x 12" brick					
3/8" joints (3.0 CF per CSF)	—	SF	.20	—	.20
1/2" joints (3.9 CF per CSF)	—	SF	.27	—	.27
6" x 2-1/2" to 2-5/8" x 12" bricks					
3/8" joints (7.9 CF per CSF)	—	SF	.54	—	.51
1/2" joints (10.2 CF per CSF)	—	SF	.69	—	.69
6" x 3-1/2" to 3-5/8" x 12" bricks					
3/8" joints (6.8 CF per CSF)	—	SF	.46	—	.46
1/2" joints (8.8 CF per CSF)	—	SF	.60	—	.60
6" x 4" x 12" bricks					
3/8" joints (5.6 CF per CSF)	—	SF	.38	—	.38
1/2" joints (7.4 CF per CSF)	—	SF	.50	—	.50
Add for 6-11/16" 22 gauge galvanized wall ties					
50 ties per 100 SF	B9@.011	SF	.05	.31	.36
Add for raked joints	B9@.027	SF	—	.75	.75
Add for flush joints	B9@.072	SF	—	2.01	2.01
Add for header course every 6th course	B9@.009	SF	.53	.25	.78
Add for Flemish bond					
With header course every 6th course	B9@.036	SF	.41	1.01	1.42

Mini-brick veneer strips Wirecut texture. Figures in parentheses indicate coverage using 1/2" grouted joints. No waste included.

	Craft@Hrs	Unit	Material	Labor	Total
2-3/16" x 7/16" x 7-1/2" (6.7 units per SF)					
Unglazed, standard colors, $.34 each	B9@.300	SF	2.28	8.39	10.67
Unglazed, flashed, $.50 each	B9@.300	SF	3.35	8.39	11.74
Manufactured used, $.46 each	B9@.300	SF	3.08	8.39	11.47
3-5/8" x 7/16" x 7-5/8" (4.30 units per SF)					
Unglazed, standard colors, $.53 each	B9@.250	SF	2.28	6.99	9.27
Unglazed, flashed, $.78 each	B9@.250	SF	3.35	6.99	10.34
Manufactured used, $.68 each	B9@.250	SF	2.92	6.99	9.91
Mini-brick veneer strip corners					
Unglazed, standard colors	B9@.077	Ea	1.50	2.15	3.65
Unglazed, flashed	B9@.077	Ea	1.69	2.15	3.84
Manufactured used	B9@.077	Ea	1.62	2.15	3.77

Masonry

	Craft@Hrs	Unit	Material	Labor	Total
Firebrick Used in chimneys and boilers, delivered to 30 miles, typical costs.					
Standard backs, 9" x 4-1/2" x 2-1/2"	B9@30.9	M	1,170.00	864.00	2,034.00
Split backs, 9" x 4-1/2" x 1-1/2"	B9@23.3	M	1,130.00	651.00	1,781.00
Fireclay, Dosch clay					
50 lb. sack	—	Ea	7.22	—	7.22

Brick pavers Costs assume simple pattern with 1/2" mortar joints. Delivered to 30 miles, 10 ton minimum. Costs in material column include 8% for waste and breakage.

	Craft@Hrs	Unit	Material	Labor	Total
Adobe paver (Fresno)					
6" x 12" (2 units per SF)	B9@.193	SF	1.79	5.40	7.19
12" x 12" (1 unit per SF))	B9@.135	SF	1.79	3.77	5.56
California paver, 1-1/4" thick, brown					
3-5/8" x 7-5/8" (5.1 units per SF)	B9@.340	SF	4.30	9.51	13.81
Giant mission paver, 2-3/8" thick x 5-5/8" x 11-5/8"					
Red (2 units per SF)	B9@.193	SF	1.98	5.40	7.38
Redskin (2 units per SF)	B9@.193	SF	2.24	5.40	7.64
Mini-brick paver, 7/16" thick x 3-5/8" x 7-5/8"					
Unglazed, standard colors					
(4.5 units per SF)	B9@.340	SF	2.33	9.51	11.84
Unglazed, flashed					
(4.5 units per SF)	B9@.340	SF	4.45	9.51	13.96
Glazed, all colors					
(4.5 units per SF)	B9@.340	SF	4.18	9.51	13.69
Norman paver, 2-1/2" thick x 3-1/2" x 11-1/2"					
(3 units per SF)	B9@.222	SF	3.24	6.21	9.45
Padre paver, 1-1/4" thick x 7-1/2" x 7-1/2"					
(2.5 units per SF)	B9@.211	SF	3.77	5.90	9.67
Mortar joints for pavers, 1/2" typical	B9@.010	SF	.65	.28	.93

Concrete block wall assemblies Typical costs for standard natural gray medium weight masonry block walls including blocks, mortar, typical reinforcing and normal waste. Foundations (footings) are not included. See also individual component material and labor costs for blocks, mortar and reinforcing on the pages that follow. For more detailed coverage of concrete block see the National Concrete and Masonry Estimator at http://CraftsmanSiteLicense.Com

	Craft@Hrs	Unit	Material	Labor	Total
Walls constructed with 8" x 16" blocks laid in running bond					
4" thick wall	B9@.090	SF	1.65	2.52	4.17
6" thick wall	B9@.100	SF	2.01	2.80	4.81
8" thick wall	B9@.120	SF	2.41	3.36	5.77
12" thick wall	B9@.150	SF	3.55	4.19	7.74
Add for grouting cores at 36" intervals, poured by hand					
4" thick wall	B9@.007	SF	.43	.20	.63
6" thick wall	B9@.007	SF	.65	.20	.85
8" thick wall	B9@.007	SF	.87	.20	1.07
12" thick wall	B9@.010	SF	1.29	.28	1.57
Add for 2" thick caps, natural gray concrete					
4" thick wall	B9@.027	LF	.69	.75	1.44
6" thick wall	B9@.038	LF	.80	1.06	1.86
8" thick wall	B9@.046	LF	.82	1.29	2.11
12" thick wall	B9@.065	LF	1.35	1.82	3.17
Add for detailed block, 3/8" score					
Single score, one side	—	SF	.42	—	.42
Single score, two sides	—	SF	.84	—	.84
Multi-scores, one side	—	SF	.42	—	.42
Multi-scores, two sides	—	SF	.84	—	.84

	Craft@Hrs	Unit	Material	Labor	Total
Add for color block, any wall assembly					
Light colors	—	SF	.12	—	.12
Medium colors	—	SF	.18	—	.18
Dark colors	—	SF	.25	—	.25
Add for other than running bond	B9@.021	SF	—	.59	.59

Concrete block Foundations (footings) not included. Includes local delivery, mortar and 8% waste. Natural gray, medium weight, no reinforcing included. See also concrete block wall assemblies. Costs shown as Ea are for the block only, the mortar from the adjacent block will be sufficient for these blocks.

	Craft@Hrs	Unit	Material	Labor	Total
4" wide units					
8" x 16", 1.125 per SF	B9@.085	SF	1.22	2.38	3.60
8" x 12", corner	B9@.054	Ea	.81	1.51	2.32
8" x 8", 2.25 per SF	B9@.093	SF	2.11	2.60	4.71
4" x 16", 2.25 per SF	B9@.093	SF	2.00	2.60	4.60
4" x 12", corner	B9@.052	Ea	.90	1.45	2.35
4" x 8", 4.5 per SF, (half block)	B9@.106	SF	3.62	2.96	6.58
6" wide units					
8" x 16", 1.125 per SF	B9@.095	SF	1.52	2.66	4.18
8" x 16" corner	B9@.058	Ea	1.68	1.62	3.30
8" x 8", 2.25 per SF, (half block)	B9@.103	SF	2.69	2.88	5.57
4" x 16", 2.25 per SF	B9@.103	SF	2.59	2.88	5.47
4" x 14" corner	B9@.070	Ea	.63	1.96	2.59
4" x 8", 4.5 per SF, (half block)	B9@.114	SF	4.76	3.19	7.95
4" x 12" (3/4 block)	B9@.052	Ea	.55	1.45	2.00
8" wide units					
4" x 16", 2.25 per SF	B9@.141	SF	3.03	3.94	6.97
8" x 16", 1.125 per SF	B9@.111	SF	1.84	3.10	4.94
6" x 8", 3 per SF	B9@.137	SF	4.88	3.83	8.71
8" x 12" (3/4 block)	B9@.074	Ea	1.33	2.07	3.40
8" x 8" (half)	B9@.052	Ea	1.26	1.45	2.71
8" x 8" (lintel)	B9@.052	Ea	1.49	1.45	2.94
6" x 16", 1.5 per SF	B9@.117	SF	2.69	3.27	5.96
4" x 8" (half block)	B9@.053	Ea	1.10	1.48	2.58
12" wide units					
8" x 16", 1.125 per SF	B9@.148	SF	2.90	4.14	7.04
8" x 8" (half)	B9@.082	Ea	2.09	2.29	4.38
8" x 8" (lintel)	B9@.082	Ea	2.37	2.29	4.66
Add for high strength units	—	%	27.0	—	—
Add for lightweight block, typical	—	Ea	.20	—	.20
Deduct for lightweight block, labor	—	%	—	-7.0	—
Add for ladder type reinforcement	B9@.004	SF	.20	.11	.31
Add for 1/2" rebars 24" OC	B9@.004	SF	.20	.11	.31
Add for other than running bond	—	%	—	20.0	—
Add for 2" wall caps, natural color					
4" x 16"	B9@.027	LF	.55	.75	1.30
6" x 16"	B9@.038	LF	.62	1.06	1.68
8" x 16"	B9@.046	LF	.59	1.29	1.88
Add for detailed block, 3/8" score, typical prices					
Single score, one side	—	SF	.42	—	.42
Single score, two sides	—	SF	.84	—	.84
Multi-scores, one side	—	SF	.42	—	.42
Multi-scores, two sides	—	SF	.84	—	.84
Add for color block					
Light colors	—	SF	.08	—	.08
Medium colors	—	SF	.12	—	.12
Dark colors	—	SF	.17	—	.17

Masonry

	Craft@Hrs	Unit	Material	Labor	Total
Add for grouting cores poured by hand, normal waste included					
Using grout, per cubic foot	—	CF	6.46	—	6.46
36" interval					
4" wide wall (16.1 SF/CF grout)	B9@.007	SF	.43	.20	.63
6" wide wall (10.7 SF/CF grout)	B9@.007	SF	.65	.20	.85
8" wide wall (8.0 SF/CF grout)	B9@.007	SF	.87	.20	1.07
12" wide wall (5.4 SF/CF grout)	B9@.010	SF	1.29	.28	1.57
24" interval					
4" wide wall (16.1 SF/CF grout)	B9@.009	SF	.65	.25	.90
6" wide wall (10.7 SF/CF grout)	B9@.009	SF	.98	.25	1.23
8" wide wall (8.0 SF/CF grout)	B9@.009	SF	1.31	.25	1.56
12" wide wall (5.4 SF/CF grout)	B9@.013	SF	1.94	.36	2.30

Split face concrete block Textured block, medium weight, split one side, structural. Includes 8% waste and local delivery.

	Craft@Hrs	Unit	Material	Labor	Total
8" wide, 4" x 16', 2.25 per SF	B9@.111	SF	5.04	3.10	8.14
8" wide, 8" x 16, 1.125 per SF	B9@.111	SF	2.57	3.10	5.67
12" wide, 8" x 8, 2.25 per SF	B9@.117	SF	7.00	3.27	10.27
12" wide, 8" x 16, 1.125 per SF	B9@.117	SF	3.63	3.27	6.90
Add for lightweight block	—	%	12.0	—	—
Add for color	—	SF	.30	—	.30
Add for mortar (at 7 cubic feet, per 100 square feet of wall)	—	SF	.48	—	.48

Concrete pavers Natural concrete 3/4" thick, including local delivery. Pavers installed in concrete bed but no concrete included.

	Craft@Hrs	Unit	Material	Labor	Total
6" x 6", 4 per SF	B9@.150	SF	14.90	4.19	19.09
6" x 12", 2 per SF	B9@.150	SF	7.60	4.19	11.79
12" x 12", 1 per SF	B9@.150	SF	3.55	4.19	7.74
12" x 18", .667 per SF	B9@.150	SF	2.50	4.19	6.69
18" x 18", .444 per SF	B9@.150	SF	1.71	4.19	5.90
Add for light colors	—	SF	.48	—	.48
Add for dark colors	—	SF	1.43	—	1.43
Add for pavers with beach pebble surface	—	SF	.95	—	.95

Concrete screen block Natural color. Foundations and supports not included. Material costs include 8% waste. (Quantity per SF is shown in parentheses)

Textured screen (no corebar defacements)

	Craft@Hrs	Unit	Material	Labor	Total
4" x 12" x 12" (1 per SF)	B9@.211	SF	2.15	5.90	8.05
4" x 6" x 6" (3.68 per SF)	B9@.407	SF	2.53	11.40	13.93
4" x 8" x 8" (2.15 per SF)	B9@.285	SF	3.69	7.97	11.66
Four sided sculpture screens					
4" x 12" x 12" (1 per SF)	B9@.312	SF	2.13	8.72	10.85
Sculpture screens (patterned face, rug texture reverse)					
4" x 12" x 12" (1 per SF)	B9@.211	SF	2.86	5.90	8.76
4" x 16" x 16" (.57 per SF)	B9@.285	SF	2.84	7.97	10.81
4" x 16" x 12" (.75 per SF)	B9@.285	SF	3.02	7.97	10.99
Add for channel steel support (high rise)					
2" x 2" x 3/16"	—	LF	1.89	—	1.89
3" x 3" x 3/16"	—	LF	2.24	—	2.24
4" x 4" x 1/4"	—	LF	4.44	—	4.44
6" x 4" x 3/8"	—	LF	8.62	—	8.62

	Craft@Hrs	Unit	Material	Labor	Total

Concrete slump block Natural concrete color, 4" widths are veneer, no foundation or reinforcing included. Cost in material column include 8% waste. (Quantity per SF is shown in parentheses)

	Craft@Hrs	Unit	Material	Labor	Total
4" x 4" x 8" (4.5 per SF)	B9@.267	SF	4.07	7.47	11.54
4" x 4" x 12" (3 per SF)	B9@.267	SF	2.88	7.47	10.35
4" x 4" x 16" (2.25 per SF)	B9@.267	SF	2.16	7.47	9.63
6" x 4" x 16" (2.25 per SF)	B9@.243	SF	2.74	6.79	9.53
6" x 6" x 16" (1.5 per SF)	B9@.217	SF	1.88	6.07	7.95
8" x 4" x 16" (2.25 per SF)	B9@.264	SF	3.57	7.38	10.95
8" x 6" x 16" (1.5 per SF)	B9@.243	SF	2.62	6.79	9.41
12" x 4" x 16" (2.25 per SF)	B9@.267	SF	6.15	7.47	13.62
12" x 6" x 16" (1.5 per SF)	B9@.260	SF	4.37	7.27	11.64

Cap or slab block, 2" thick, 16" long, .75 units per LF. No waste included

	Craft@Hrs	Unit	Material	Labor	Total
4" wide	B9@.160	LF	.74	4.47	5.21
6" wide	B9@.160	LF	.93	4.47	5.40
8" wide	B9@.160	LF	1.18	4.47	5.65

Add for curved slump block

	Craft@Hrs	Unit	Material	Labor	Total
4" wide, 2' or 4' diameter circle	—	%	125.0	25.0	—
8" wide, 2' or 4' diameter circle	—	%	150.0	20.0	—
Add for slump one side block	—	SF	.05	—	.05
Add for light colors for slump block	—	%	13.0	—	—
Add for darker colors for slump block	—	%	25.0	—	—

Add for mortar for slump block. Cost includes normal waste

	Craft@Hrs	Unit	Material	Labor	Total
Using mortar, per cubic foot	—	CF	6.80	—	6.80
4" x 4" x 8" (4 CF per CSF)	—	SF	.27	—	.27
6" x 6" x 16" (4 CF per CSF)	—	SF	.27	—	.27
8" x 6" x 16" (5 CF per CSF)	—	SF	.34	—	.34
12" x 6" x 16" (7.5 CF per CSF)	—	SF	.51	—	.51

Adobe block Walls to 8', not including reinforcement or foundations, plus delivery, 4" high, 16" long, 2.25 blocks per SF, including 10% waste.

	Craft@Hrs	Unit	Material	Labor	Total
4" wide	B9@.206	SF	2.21	5.76	7.97
6" wide	B9@.233	SF	2.69	6.51	9.20
8" wide	B9@.238	SF	3.49	6.65	10.14
12" wide	B9@.267	SF	5.76	7.47	13.23
Add for reinforcement (24" OC both ways)	—	SF	.28	—	.28

Flagstone Including concrete bed.

	Craft@Hrs	Unit	Material	Labor	Total
Veneer on walls, 3" thick	B9@.480	SF	4.27	13.40	17.67
Walks and porches, 2" thick	B9@.248	SF	2.11	6.93	9.04
Coping, 4" x 12"	B9@.272	SF	6.74	7.61	14.35
Steps, 6" risers	B9@.434	LF	2.42	12.10	14.52
Steps, 12" treads	B9@.455	LF	4.85	12.70	17.55

Flue lining Pumice, 1' lengths, delivered to 30 miles, including 8% waste.

	Craft@Hrs	Unit	Material	Labor	Total
8-1/2" round	B9@.148	Ea	10.30	4.14	14.44
8" x 13" oval	B9@.206	Ea	14.60	5.76	20.36
8" x 17" oval	B9@.228	Ea	15.10	6.37	21.47
10" x 17" oval	B9@.243	Ea	16.90	6.79	23.69
13" x 13" square	B9@.193	Ea	17.60	5.40	23.00
13" x 17" oval	B9@.193	Ea	19.60	5.40	25.00
13" x 21" oval	B9@.228	Ea	25.60	6.37	31.97
17" x 17" square	B9@.228	Ea	19.13	6.37	25.50
17" x 21" oval	B9@.254	Ea	34.70	7.10	41.80
21" x 21" oval	B9@.265	Ea	43.70	7.41	51.11

Masonry

	Craft@Hrs	Unit	Material	Labor	Total

Glass block Costs include 6% for waste and local delivery. Add mortar and reinforcing costs below. Labor cost includes caulking, cleaning with sponge and fiber brush, and continuous tuckpointing. Based on Pittsburgh Corning Glass, costs are for a 100 SF job.

3-7/8" thick block, smooth or irregular faces (Argus, Decora, Essex AA or Vue)

	Craft@Hrs	Unit	Material	Labor	Total
4" x 8", 4.5 per SF	B9@.392	SF	25.50	11.00	36.50
6" x 6", 4 per SF	B9@.376	SF	22.60	10.50	33.10
8" x 8", 2.25 per SF	B9@.296	SF	13.10	8.28	21.38
12" x 12", 1 per SF	B9@.211	SF	17.30	5.90	23.20
Add for heat and glare reducing insert		%	75.0	—	—
3-7/8" thick corner units	B9@.125	Ea	19.80	3.50	23.30
3-7/8" thick end units	B9@.125	Ea	20.20	3.50	23.70
3-7/8" thick 45 degree block	B9@.125	Ea	16.20	3.50	19.70
3-7/8" thick double finish end	B9@.125	Ea	23.50	3.50	27.00
4" x 8" x 3-1/8" thick end finishing unit	B9@.125	Ea	14.10	3.50	17.60

3" thick block, solid glass, coated or uncoated, smooth faces (Vistabrick)

	Craft@Hrs	Unit	Material	Labor	Total
8" x 8", 2.25 per SF	B9@.376	SF	70.40	10.50	80.90

3-1/8" thick block, smooth, irregular, or diamond relief faces (Decora, or Delphi thinline)

	Craft@Hrs	Unit	Material	Labor	Total
6" x 6", 4 per SF	B9@.376	SF	17.50	10.50	28.00
8" x 8", 2.25 per SF	B9@.296	SF	10.10	8.28	18.38
4" x 8", 4.5 per SF	B9@.391	SF	19.70	10.90	30.60
6" x 8", 3 per SF	B9@.344	SF	13.10	9.62	22.72
Add for heat and glare reducing insert	—	%	75.0	—	—
Deduct for 100 SF to 500 SF job	—	%	-5.0	—	—
Deduct for over 500 SF job to 1,000 SF job	—	%	-10.0	—	—
Deduct for over 1,000 SF job	—	%	-12.0	—	—
Add for wall ties	—	Ea	1.80	—	1.80
Add for expansion strips, fiberglass	—	LF	.37	—	.37
Add for panel anchors, 20 gauge perforated steel					
1-3/4" x 24"	—	LF	1.78	—	1.78
Add for panel reinforcing, 10' lengths					
3" wide	—	LF	.24	—	.24

Mortar for glass block White cement mix (1 part white cement, 1/2 part hydrated lime, 2-1/4 to 3 parts #20 or #30 silica sand)

	Craft@Hrs	Unit	Material	Labor	Total
White cement mix	—	CF	12.40	—	12.40
1/4" mortar joints, 3-7/8" thick block, white cement mix					
6" x 6" block (4.3 CF/CSF)	—	SF	.52	—	.52
8" x 8" block (3.2 CF/CSF)	—	SF	.39	—	.39
12" x 12" block (2.2 CF/CSF)	—	SF	.27	—	.27
1/4" mortar joints, 3" and 3-1/8" thick block, white cement mix					
6" x 6" block (3.5 CF/CSF)	—	SF	.43	—	.43
8" x 8" block (2.6 CF/CSF)	—	SF	.31	—	.31

Masonry Accessories

Non-fibered masonry waterproofing for below grade application

	Craft@Hrs	Unit	Material	Labor	Total
Gallon covers 80 SF	B9@.688	Gal	5.25	19.20	24.45
Glass reinforcing mesh	—	SF	1.08	—	1.13

Lime

Hydrated (builders)

	Craft@Hrs	Unit	Material	Labor	Total
25 lb sack	—	Sack	3.27	—	3.27
50 lb sack, Type S	—	Sack	7.36	—	7.36
50 lb sack, white lime	—	Sack	6.25	—	6.25
70 lb sack, portland lime	—	Sack	8.22	—	8.22
50 lb sack, athletic field marker	—	Sack	3.72	—	3.72
Silica sand, #30, 100 lb sack	—	Sack	6.80	—	6.80

	Craft@Hrs	Unit	Material	Labor	Total
Mortar colors					
Red, yellow, brown or black, 1 lb bag	—	Lb	3.30	—	3.30
Green, 1 lb bag	—	Lb	6.90	—	6.90

Trowel-ready mortar (factory prepared, ready to use), 30 hour life. Deposit required on container, local delivery included. Type S mortar has high compressive strength (1800 PSI) and high tensile bond strength. Type S is made from 2 parts portland cement, 1 part hydrated lime and 9 parts sand and offers a longer board life. Type M mortar is made with 3 parts of portland cement, 1 part lime and 12 parts sand. Type M has a high compressive strength (at least 2500 PSI) and is recommended for masonry in structural walls and masonry below grade such as foundations, retaining walls, sidewalks and driveways. Type N is a medium compressive-strength (750 PSI) mortar made of 1 part portland cement, 1 part lime and 6 parts sand. Type N is recommended for exterior above-grade walls exposed to severe weather, such as chimneys.

	Craft@Hrs	Unit	Material	Labor	Total
Type S natural mortar, 1/3 CY	—	Ea	80.00	—	80.00
Type M natural mortar, 1/3 CY	—	Ea	90.00	—	90.00
Add to natural mortar for colored, 1/3 CY	—	Ea	15.00	—	15.00
Type S white mortar, 1/3 CY	—	Ea	135.00	—	135.00
Trowel-ready mortar, cost per square foot of wall, 3/8" joint					
4" x 4" x 8" brick wall	—	SF	.37	—	.37
8" x 8" x 16" block wall	—	SF	.33	—	.33
Dry mortar mix					
Type S, 80 lb sack	—	Sack	4.17	—	4.17
Type N, 70 lb sack	—	Sack	7.52	—	7.52
Hi-rise channel reinforcing steel (hat channel)					
2" x 2" x 3/16"	—	LF	1.79	—	1.79
3" x 3" x 3/16"	—	LF	2.14	—	2.14
4" x 4" x 1/4"	—	LF	4.23	—	4.23
6" x 4" x 3/8"	—	LF	8.26	—	8.26
Welding	—	Ea	5.20	—	5.20
Cutting	—	Ea	2.60	—	2.60
Flanges, drill one hole, weld	—	LS	8.06	—	8.06
Additional holes	—	Ea	.84	—	.84
Expansion joints					
1/2" x 3-1/2"	—	LF	.28	—	.28
1/4" x 3-1/2"	—	LF	.25	—	.25
1/2" x 4"	—	LF	.35	—	.35
Wall anchors, 1-3/4" x 24"	—	Ea	3.50	—	3.50
Dur-O-Wal reinforcing, Ladur or standard					
3" wide or 4" wide	—	LF	.15	—	.15
6" wide	—	LF	.16	—	.16
8" wide	—	LF	.18	—	.18
10" wide or 12" wide	—	LF	.21	—	.21
Wired adobe blocks					
2" x 2"	—	Ea	.11	—	.11
3" x 3"	—	Ea	.23	—	.23
Tie wire, black, 16 gauge					
Roll, 328'	—	Ea	3.11	—	3.11
Box, 20 rolls of 328'	—	Ea	57.00	—	57.00
Wall ties, 22 gauge	—	M	60.00	—	60.00
Wall ties, 28 gauge	—	M	63.00	—	63.00
Angle iron					
3" x 3" x 3/16"	—	LF	4.00	—	4.00
4" x 4" x 1/4"	—	LF	8.00	—	8.00
6" x 4" x 3/8"	—	LF	16.00	—	16.00

Masonry, Subcontract

	Craft@Hrs	Unit	Material	Labor	Total
Post anchors (straps)					
4" x 4" x 10"	—	Ea	4.19	—	4.19
4" x 4" x 16"	—	Ea	11.10	—	11.10

Masonry, Subcontract Typical prices.

Fireplace chimneys, masonry only. Height above firebox shown.

	Craft@Hrs	Unit	Material	Labor	Total
5' high chimney (3' opening)	—	Ea	—	—	2,310.00
13' high chimney (3' opening)	—	Ea	—	—	2,720.00
Floors					
Flagstone	—	SF	—	—	10.80
Patio tile	—	SF	—	—	8.50
Pavers	—	SF	—	—	9.25
Quarry tile, unglazed	—	SF	—	—	11.90

Mats, Runners and Treads Koffler Products.

Mats

Vinyl link mats, closed link, 1/2" thick, including nosing

	Craft@Hrs	Unit	Material	Labor	Total
Solid colors incl. black 28" x 36"	B1@.167	Ea	144.00	4.57	148.57
For 9" to 12" letters add	—	Ea	19.90	—	19.90
Steel link mats, 3/8" deep, 1" mesh					
16" x 24"	B1@.501	Ea	20.90	13.70	34.60
22" x 36"	B1@.501	Ea	40.70	13.70	54.40
30" x 48"	B1@.501	Ea	70.50	13.70	84.20
36" x 54"	B1@.501	Ea	86.30	13.70	100.00
36" x 72"	B1@.501	Ea	110.00	13.70	123.70

Runners

Corrugated rubber runner, cut lengths, 2', 3', 4' widths

	Craft@Hrs	Unit	Material	Labor	Total
1/8" thick, black	B1@.060	SF	2.50	1.64	4.14

Corrugated or round ribbed vinyl runner, 1/8" thick, 2', 3', or 4' wide, cut lengths, flame resistant

	Craft@Hrs	Unit	Material	Labor	Total
Black corrugated	B1@.060	SF	2.40	1.64	4.04
Black or brown, round ribbed	B1@.060	SF	2.50	1.64	4.14
Deduct for full rolls (100 LF)	—	%	-10.0	—	—

Safety strips, non-slip strips, press-on abrasive, black, 1/16" thick

	Craft@Hrs	Unit	Material	Labor	Total
3/4" x 24", box of 50	B1@.080	Ea	.80	2.19	2.99
6" x 24" cleats	B1@.080	Ea	5.55	2.19	7.74

Treads

Stair treads, homogeneous rubber, 1/4" thick, 1-3/4" nosing, 12-1/4" deep

	Craft@Hrs	Unit	Material	Labor	Total
Diamond design, marbleized, medium duty	B1@.080	LF	11.90	2.19	14.09
Diamond design, black, medium duty	B1@.080	LF	11.90	2.19	14.09
Plain, with abrasive strips	B1@.080	LF	13.10	2.19	15.29

Medicine Cabinets

Recessed or surface mount medicine cabinets, hinged door, no lighting

Basic steel-frame mirror cabinet with polystyrene body, magnetic catch, three fixed shelves

	Craft@Hrs	Unit	Material	Labor	Total
14" x 18"	BC@.850	Ea	12.60	25.50	38.10

Basic flat glass panel cabinet with enameled steel body, concealed brass hinges, one fixed shelf

	Craft@Hrs	Unit	Material	Labor	Total
15" x 19"	BC@.850	Ea	28.20	25.50	53.70

Better quality etched and beveled mirror door cabinet, 2 adjustable shelves, polished brass knobs

	Craft@Hrs	Unit	Material	Labor	Total
25" x 26"	BC@.850	Ea	109.00	25.50	134.50

Corner-mount frameless cabinet, one beveled mirror door, 2 adjustable shelves

	Craft@Hrs	Unit	Material	Labor	Total
14" x 26"	BC@.850	Ea	87.50	25.50	113.00

	Craft@Hrs	Unit	Material	Labor	Total
Surface mount medicine cabinets, hinged door, no lighting					
Tri-view frame cabinet, flat glass panel, 2 adjustable shelves, brass exposed hinges					
24" x 24"	BC@.850	Ea	42.30	25.50	67.80
30" x 29"	BC@.941	Ea	63.00	28.20	91.20
36" x 29"	BC@.941	Ea	74.10	28.20	102.30
Beveled tri-view frame cabinet, flat glass panel, 2 adjustable shelves, concealed adjustable hinges					
24" x 26"	BC@.850	Ea	83.50	25.50	109.00
29" x 25"	BC@.941	Ea	94.40	28.20	122.60
36" x 30"	BC@.941	Ea	123.00	28.20	151.20
48" x 30"	BC@.941	Ea	148.00	28.20	176.20
Oval smoked bevel mirror, frameless swing door					
14" x 18"	BC@.850	Ea	92.00	25.50	117.50
All wood cabinets, 2 shelves, magnetic door latch, surface mount					
15" x 25", oak finish	BC@.850	Ea	36.70	25.50	62.20
15" x 25", white finish	BC@.850	Ea	36.70	25.50	62.20
Matching light bars for medicine cabinets, including electrical connection					
4 lights	BE@1.50	Ea	41.90	50.80	92.70
5 lights	BE@1.50	Ea	52.30	50.80	103.10
6 lights	BE@1.50	Ea	62.60	50.80	113.40
7 lights	BE@1.50	Ea	73.00	50.80	123.80
Meter Boxes Precast concrete, FOB manufacturer, inside dimension, approximately 100 to 190 lbs each.					
13-1/4" x 19-1/8" x 11" deep					
With 1 piece concrete cover	BL@.334	Ea	18.80	8.25	27.05
With metal cover (cast iron)	BL@.334	Ea	26.30	8.25	34.55
14-5/8" x 19-3/4" x 11" deep					
With 1 piece concrete cover	BL@.334	Ea	15.10	8.25	23.35
With metal cover (cast iron)	BL@.334	Ea	26.80	8.25	35.05
15-3/4" x 22-3/4" x 11" deep					
With 1 piece concrete cover	BL@.334	Ea	27.50	8.25	35.75
With metal cover (cast iron)	BL@.334	Ea	48.80	8.25	57.05
Add hinged or reading lid, concrete cover	—	Ea	6.21	—	6.21
Plastic meter and valve boxes, bolting plastic covers					
Standard 16" x 10-3/4" x 12" deep	BL@.167	Ea	21.00	4.13	25.13
Jumbo 20-1/4" x 14-1/2" x 12" deep	BL@.167	Ea	26.30	4.13	30.43
Round pit box, 10" diameter x 12" deep	BL@.167	Ea	9.25	4.13	13.38
Mirrors See also Bathroom Accessories and Glass.					
Wall mirrors, distortion-free float glass, 1/8" x 3/16" thick					
Cut and polished edges	—	SF	3.60	—	3.60
Beveled edges	—	SF	6.67	—	6.67
Labor installing wall mirrors					
To 5 SF	BG@.142	SF	—	4.10	4.10
5 to 15 SF	BG@.127	SF	—	3.67	3.67
15 to 25 SF	BG@.113	SF	—	3.26	3.26
25 to 50 SF	BG@.093	SF	—	2.68	2.68
Decorative mirrors					
Mirror strips, 8" x 72" (4 SF)					
Clear	BG@.568	Ea	18.90	16.40	35.30
Clear with beveled edge	BG@.568	Ea	28.10	16.40	44.50
Bronze with beveled edge	BG@.568	Ea	30.50	16.40	46.90

Moulding

	Craft@Hrs	Unit	Material	Labor	Total
Mirror arches (radius, top edge), 68" high					
16" wide, clear (8.25 SF)	BG@1.05	Ea	27.00	30.30	57.30
20" wide, clear (10.9 SF)	BG@1.38	Ea	32.80	39.80	72.60
24" wide, clear (10.9 SF)	BG@1.38	Ea	87.40	39.80	127.20
Framed mirrors					
Traditional rectangle with urethane corner and side ornamentation					
31" x 42"	BG@.331	Ea	85.20	9.56	94.76
35" x 49"	BG@.331	Ea	111.00	9.56	120.56
Traditional hexagon mirror with urethane frame					
32" x 42"	BG@.331	Ea	117.00	9.56	126.56
38" x 48"	BG@.331	Ea	125.00	9.56	134.56
Solid oak frame, rectangle plate glass					
18" x 29"	BG@.331	Ea	93.10	9.56	102.66
24" x 24"	BG@.331	Ea	106.00	9.56	115.56
30" x 24"	BG@.331	Ea	133.00	9.56	142.56
36" x 26"	BG@.450	Ea	150.00	13.00	163.00
48" x 26"	BG@.450	Ea	189.00	13.00	202.00
Door mirrors					
16" x 56"	BG@.251	Ea	39.10	7.25	46.35
18" x 60"	BG@.301	Ea	46.00	8.69	54.69
24" x 68"	BG@.331	Ea	59.40	9.56	68.96

Mouldings Unfinished. Material costs include 5% waste. Labor costs are for typical installation only.
(For installation costs of Hardwood Moulding, see Softwood Mouldings starting on next page.)

	Unit	Oak	Poplar	Birch
Hardwood moulding				
Base				
1/2" x 1-1/2"	LF	1.21	.84	.85
1/2" x 2-1/2"	LF	1.23	1.01	1.00
Base shoe, 3/8" x 3/4"	LF	.55	.42	.42
Casing				
1/2" x 1-1/2"	LF	1.37	.91	.89
5/8" x 1-5/8"	LF	1.38	1.01	1.00
Cap moulding, 1/2" x 1-1/2"	LF	1.29	.81	.78
Crown, 1/2" x 2-1/4"	LF	1.70	.98	.90
Cove				
1/2" x 1/2"	LF	.78	.48	.49
3/4" x 1/4"	LF	.97	.54	.52
Corner mould				
3/4" x 3/4"	LF	1.02	.53	.51
1" x 1"	LF	2.37	.65	.67
Round edge stop, 3/8" x 1-1/4"	LF	.68	.45	.51
Chair rail, 1/2" x 2"	LF	1.50	1.01	1.02
Quarter round				
1/4" x 1/4"	LF	.66	.45	.45
1/2" x 1/2"	LF	.70	.54	.50
Battens, 1/4" x 3/4"	LF	.58	.34	.33
Door trim sets, 16'8" total length				
1/2" x 1-5/8" casing	Ea	28.90	22.91	21.80
5/8" x 1-5/8" casing	Ea	30.40	24.10	23.50
3/8" x 1-1/4" round edge stop	Ea	20.20	15.50	14.10

	Craft@Hrs	Unit	Material	Labor	Total
Oak threshold, LF of threshold					
5/8" x 3-1/2" or 3/4" x 3-1/2"	BC@.174	LF	12.80	5.22	18.02
3/4" x 5-1/2"	BC@.174	LF	27.30	5.22	32.52
7/8" x 3-1/2"	BC@.174	LF	14.00	5.22	19.22
Redwood moulding					
Band moulding, 3/4" x 5/16"	BC@.030	LF	.78	.90	1.68
Battens and lattice					
5/16" x 1-1/4"	BC@.026	LF	.42	.78	1.20
5/16" x 1-5/8"	BC@.026	LF	.42	.78	1.20
5/16" x 2-1/2"	BC@.026	LF	.68	.78	1.46
5/16" x 3-1/2"	BC@.026	LF	.68	.78	1.46
3/8" x 2-1/2"	BC@.026	LF	.97	.78	1.75
Brick mould, 1-1/2" x 1-1/2"	BC@.026	LF	1.49	.78	2.27
Drip cap-water table					
1-5/8" x 2-1/2" (no lip)	BC@.060	LF	1.96	1.80	3.76
1-5/8" x 2-1/2"	BC@.060	LF	1.96	1.80	3.76
Quarter round, 3/4" x 3/4"	BC@.026	LF	.62	.78	1.40
Rabbeted siding mould, 1-1/8" x 1-5/8"	BC@.026	LF	1.13	.78	1.91
S4S (rectangular)					
11/16" x 11/16"	BC@.026	LF	.57	.78	1.35
1" x 2"	BC@.026	LF	.84	.78	1.62
1" x 3"	BC@.026	LF	1.25	.78	2.03
1" x 4"	BC@.026	LF	1.60	.78	2.38
Stucco mould					
13/16" x 1-3/8"	BC@.026	LF	.96	.78	1.74
13/16" x 1 1/2"	BC@.026	LF	.87	.78	1.65
Window sill, 2" x 8"	BC@.043	LF	9.71	1.29	11.00

Softwood moulding, pine (Labor costs here can also be used for hardwood mouldings in the previous section.)

	Craft@Hrs	Unit	Material	Labor	Total
Astragal moulding					
For 1-3/8" x 7' doors	BC@.032	LF	2.10	.96	3.06
For 1-3/4" x 7' doors	BC@.032	LF	3.71	.96	4.67
Base (all patterns)					
7/16" x 7/16"	BC@.016	LF	.55	.48	1.03
7/16" x 1-1/4"	BC@.016	LF	.69	.48	1.17
3/8" x 3"	BC@.016	LF	.87	.48	1.35
1/2" x 2-1/4"	BC@.016	LF	.93	.48	1.41
1/2" x 2-1/2"	BC@.016	LF	1.21	.48	1.69
1/2" x 3-1/2"	BC@.016	LF	1.20	.48	1.68
Base (combination or cove)					
7/16" x 1-1/4"	BC@.016	LF	.68	.48	1.16
1/2" x 2"	BC@.016	LF	1.01	.48	1.49
1/2" x 3-1/4"	BC@.016	LF	1.34	.48	1.82
Base shoe					
7/16" x 11/16"	BC@.016	LF	.33	.48	.81
9/16" x 3-1/4"	BC@.016	LF	.92	.48	1.40
Blind stop					
3/8" x 1/2"	BC@.015	LF	.35	.45	.80
3/8" x 3/4"	BC@.015	LF	.38	.45	.83
3/8" x 1-1/4"	BC@.015	LF	.61	.45	1.06
1/2" x 3/4"	BC@.015	LF	.61	.45	1.06
1/2" x 1-1/4"	BC@.015	LF	.67	.45	1.12
1/2" x 1-1/2"	BC@.015	LF	.77	.45	1.22

Moulding, Softwood

	Craft@Hrs	Unit	Material	Labor	Total
Casing (all patterns)					
1/2" x 1 5/8"	BC@.023	LF	.97	.69	1.66
5/8" x 1-5/8"	BC@.023	LF	1.41	.69	2.10
7/16" x 2-1/2"	BC@.023	LF	1.08	.69	1.77
9/16" x 3-1/2"	BC@.023	LF	1.41	.69	2.10
Chair rail					
11/16" x 1-5/8"	BC@.021	LF	1.00	.63	1.63
9/16" x 2-1/2"	BC@.021	LF	1.40	.63	2.03
Chamfer strip, 3/4" x 3/4"	BC@.016	LF	.20	.48	.68
Corner bead (outside corner moulding)					
11/16" x 11/16"	BC@.016	LF	.66	.48	1.14
3/4" x 3/4"	BC@.016	LF	.62	.48	1.10
9/16" x 1-5/8"	BC@.016	LF	.84	.48	1.32
Cove moulding, solid					
3/8" x 3/8"	BC@.020	LF	.36	.60	.96
1/2" x 1/2"	BC@.020	LF	.52	.60	1.12
11/16" x 11/16"	BC@.020	LF	.73	.60	1.33
3/4" x 3/4"	BC@.020	LF	.64	.60	1.24
1-1/2" x 1-1/2"	BC@.020	LF	.92	.60	1.52
Cove moulding, sprung					
3/4" x 1-5/8"	BC@.020	LF	.93	.60	1.53
3/4" x 3-1/2"	BC@.020	LF	1.79	.60	2.39
Crown or bed moulding					
3/4" x 3/4"	BC@.044	LF	.64	1.32	1.96
9/16" x 1-5/8"	BC@.044	LF	.53	1.32	1.85
9/16" x 1-3/4"	BC@.044	LF	.62	1.32	1.94
9/16" x 2-5/8"	BC@.044	LF	1.48	1.32	2.80
9/16" x 3-5/8"	BC@.044	LF	1.74	1.32	3.06
Add for special cutting, any crown moulding shown:					
Inside corner, using coping saw	BC@.167	Ea	—	5.01	5.01
Outside corners, using miter saw	BC@.083	Ea	—	2.49	2.49
Drip cap (water table moulding or bar nosing)					
1-1/16" x 1-5/8"	BC@.060	LF	1.86	1.80	3.66
1-1/16" x 2"	BC@.060	LF	2.07	1.80	3.87
Drip moulding					
3/4" x 1-1/4"	BC@.030	LF	.62	.90	1.52
Glass bead					
1/2" x 9/16"	BC@.016	LF	.54	.48	1.02
Handrail, fir					
1-1/2" x 2-1/4"	BC@.060	LF	2.51	1.80	4.31
1-7/16" x 1-3/4"	BC@.060	LF	2.10	1.80	3.90
Lattice					
1/4" x 1-1/8"	BC@.016	LF	.45	.48	.93
1/4" x 1-3/4"	BC@.016	LF	.59	.48	1.07
5/16" x 1-3/8"	BC@.016	LF	.79	.48	1.27
5/16" x 1-3/4"	BC@.016	LF	.97	.48	1.45
Mullion casing					
3/8" x 2"	BC@.016	LF	.88	.48	1.36

	Craft@Hrs	Unit	Material	Labor	Total
Panel moulding					
3/8" x 5/8"	BC@.030	LF	.67	.90	1.57
5/8" x 1-1/8"	BC@.030	LF	.75	.90	1.65
3/4" x 1"	BC@.030	LF	.80	.90	1.70
3/4" x 1-5/16"	BC@.030	LF	1.19	.90	2.09
3/4" x 2"	BC@.030	LF	1.07	.90	1.97
Parting bead					
3/8" x 3/4"	BC@.016	LF	.49	.48	.97
Picture frame mould					
11/16" x 1 3/4"	BC@.040	LF	1.25	1.20	2.45
11/16" x 1-5/8"	BC@.040	LF	.86	1.20	2.06
Plaster ground					
3/4" x 7/8"	BC@.010	LF	.54	.30	.84
Quarter round					
1/4" x 1/4"	BC@.016	LF	.32	.48	.80
3/8" x 3/8"	BC@.016	LF	.35	.48	.83
1/2" x 1/2"	BC@.016	LF	.32	.48	.80
11/16" x 11/16"	BC@.016	LF	.57	.48	1.05
3/4" x 3/4"	BC@.016	LF	1.02	.48	1.50
1" x 1-11/16"	BC@.016	LF	.93	.48	1.41
Half round					
1/4" x 1/2" or 5/16" x 5/8"	BC@.016	LF	.45	.48	.93
3/8" x 3/4" or 1/2" x 1"	BC@.016	LF	.56	.48	1.04
3/4" x 1-1/2"	BC@.016	LF	.76	.48	1.24
Full round					
1/2" or 3/4" diameter	BC@.016	LF	.67	.48	1.15
1-1/4" or 1-3/8" diameter	BC@.016	LF	1.32	.48	1.80
1" diameter	BC@.016	LF	.93	.48	1.41
1-1/2" diameter	BC@.016	LF	1.64	.48	2.12
2" diameter	BC@.016	LF	2.32	.48	2.80
S4S (rectangular)					
1/2" x 3/4"	BC@.016	LF	.38	.48	.86
1/2" x 2-1/2"	BC@.016	LF	.95	.48	1.43
1/2" x 3-1/2"	BC@.016	LF	1.51	.48	1.99
1/2" x 5-1/2"	BC@.016	LF	2.26	.48	2.74
1/2" x 7-1/4"	BC@.016	LF	3.19	.48	3.67
3/4" x 3/4"	BC@.016	LF	.50	.48	.98
3/4" x 1-1/4"	BC@.016	LF	.69	.48	1.17
3/4" x 1-5/8"	BC@.016	LF	.74	.48	1.22
3/4" x 2-1/2"	BC@.016	LF	1.30	.48	1.78
3/4" x 3-1/2"	BC@.016	LF	1.83	.48	2.31
1" x 1"	BC@.016	LF	.97	.48	1.45
1-1/2" x 1-1/2"	BC@.016	LF	1.63	.48	2.11
Sash bar					
7/8" x 1-3/8"	BC@.016	LF	.95	.48	1.43
Screen moulding					
1/4" x 3/4" beaded	BC@.016	LF	.46	.48	.94
1/4" x 3/4" flat or insert	BC@.016	LF	.43	.48	.91
3/8" x 3/4" clover leaf	BC@.016	LF	.44	.48	.92
Square moulding					
1/2" x 1/2"	BC@.016	LF	.42	.48	.90
11/16" x 11/16"	BC@.016	LF	.72	.48	1.20
3/4" x 3/4"	BC@.016	LF	.76	.48	1.24
2" x 2"	BC@.016	LF	1.83	.48	2.31

Moulding, Fiberboard

	Craft@Hrs	Unit	Material	Labor	Total
Stops, round edge					
3/8" x 7/8"	BC@.025	LF	.30	.75	1.05
3/8" x 1-1/4"	BC@.025	LF	.51	.75	1.26
3/8" x 1-3/8"	BC@.025	LF	.51	.75	1.26
3/8" x 1-5/8" or 1/2" x 3/4"	BC@.025	LF	.54	.75	1.29
7/16" x 3/4"	BC@.025	LF	.45	.75	1.20
7/16" x 1-3/8"	BC@.025	LF	.62	.75	1.37
7/16" x 1-5/8"	BC@.025	LF	.85	.75	1.60
Wainscot cap					
5/8" x 1-1/4" or 1/2" x 1-1/2"	BC@.016	LF	.91	.48	1.39
Window stool, beveled					
11/16" x 2-1/4"	BC@.037	LF	1.81	1.11	2.92
11/16" x 2-1/2"	BC@.037	LF	.92	1.11	2.03
1" x 3-1/2"	BC@.037	LF	3.19	1.11	4.30
1" x 5-1/2"	BC@.037	LF	4.91	1.11	6.02
Window stool, flat					
11/16" x 4-5/8"	BC@.037	LF	1.98	1.11	3.09
1" x 5-1/2"	BC@.037	LF	4.49	1.11	5.60

Medium density fiberboard moulding Primed, ready for painting.

	Craft@Hrs	Unit	Material	Labor	Total
Base, 9/16" x 3-1/16"	BC@.012	LF	.65	.36	1.01
Base, 2-1/2" x 3-1/4"	BC@.012	LF	.43	.36	.79
Casing, 11/16" x 3"	BC@.016	LF	1.02	.48	1.50
Door casing set, 17' x 5/8" x 2-1/4"	BC@.272	Ea	6.43	8.16	14.59
Chair rail, 9/16" x 3"	BC@.015	LF	1.10	.45	1.55
Corner, bullnose, 2-1/4" x 2-1/4"	BC@.025	LF	.46	.75	1.21
Crown, 11/16" x 3-1/2"	BC@.025	LF	1.31	.75	2.06
Crown inside or outside corner block,					
3/4" x 6"	BC@.125	Ea	11.80	3.75	15.55
Plinth block, 1" x 3-3/4"	BC@.125	Ea	2.61	3.75	6.36

White vinyl moulding Unfinished.

	Craft@Hrs	Unit	Material	Labor	Total
Backband casing	BC@.016	LF	1.99	.48	2.47
Base cap, 11/16" x 1-1/8"	BC@.009	LF	.51	.27	.78
Bead, 9/16" x 1-3/4"	BC@.016	LF	.61	.48	1.09
Blind stop	BC@.016	LF	.59	.48	1.07
Casing, 5/8" x 2-1/2"	BC@.023	LF	2.18	.69	2.87
Crown, 1/2" x 3-1/4"	BC@.044	LF	2.13	1.32	3.45
Dentil, 27/32" x 5-7/8"	BC@.016	LF	3.51	.48	3.99
Dentil, 27/32" x 4-1/8"	BC@.016	LF	1.32	.48	1.80
Drip cap, 11/16" x 1-5/8"	BC@.016	LF	.90	.48	1.38
Exterior bead 3/16" x 1/2"	BC@.016	LF	.27	.48	.75
Flat utility trim	BC@.023	LF	1.57	.69	2.26
Quarter round, 3/4" x 3/4" x 12'	BC@.016	LF	.59	.48	1.07
Shingle moulding, 11/16" x 1-3/4"	BC@.016	LF	1.05	.48	1.53

Extruded polymer moulding Factory finished.

	Craft@Hrs	Unit	Material	Labor	Total
Bead, brite white	BC@.016	LF	.28	.48	.76
Bead, cherry/mahogany	BC@.016	LF	.27	.48	.75
Bead, cinnamon chestnut	BC@.016	LF	.28	.48	.76
Bed mould, brite white	BC@.016	LF	.55	.48	1.03
Bed mould, cherry/mahogany	BC@.016	LF	.55	.48	1.03
Bed mould, cinnamon chestnut	BC@.016	LF	.55	.48	1.03
Casing and base, brite white	BC@.023	LF	.55	.69	1.24

	Craft@Hrs	Unit	Material	Labor	Total
Casing and base, cherry/mahogany	BC@.023	LF	.55	.69	1.24
Casing and base, cinnamon chestnut	BC@.023	LF	.55	.69	1.24
Outside corner, brite white	BC@.016	LF	.28	.48	.76
Outside corner, cherry/mahogany	BC@.016	LF	.28	.48	.76
Outside corner, cinnamon chestnut	BC@.016	LF	.28	.48	.76
Plywood cap, brite white	BC@.016	LF	.34	.48	.82
Shoe, brite white	BC@.016	LF	.27	.48	.75
Shoe, cherry/mahogany	BC@.016	LF	.29	.48	.77
Shoe, cinnamon chestnut	BC@.016	LF	.28	.48	.76
Stop, brite white	BC@.025	LF	.28	.75	1.03
Stop, cinnamon chestnut	BC@.025	LF	.28	.75	1.03

Nails and Fasteners

Nail sizes are usually identified by the "penny," abbreviated "d." Use the following figures to estimate nail quantities for framing:

Sills and plates - 8 pounds of 10d, 16d and 20d nails per MBF

Walls and partition studs - 10 pounds of 10d and 16d nails per MBF

2" x 6" or 2" x 8" joists and rafters - 9 pounds of 16d nails per MBF

2" x 10" or 2" x 12" joist and rafters - 7 pounds of 16d nails per MBF

Sheathing to 3/8" thick - 10 pounds of 6d nails per MSF

Sheathing over 3/8" thick - 20 pounds of 8d nails per MSF

Siding - 25 pounds of 8d nails per MSF

Common nails Bright nails are tumbled to remove debris and have a polished appearance. Galvanized nails are zinc coated to protect against corrosion and have a dull sheen. Cost per box except as noted.

		Galvanized	Bright
3 penny, 1-1/4", 1-pound box, 568 nails	Box	1.73	1.81
4 penny, 1-1/2", 1-pound box, 316 nails	Box	1.76	1.78
4 penny, 1-1/2", 5-pound box, 2,840 nails	Box	7.25	7.21
5 penny, 1-3/4", 1-pound box, 271 nails	Box	—	2.34
6 penny, 2", 1-pound box, 181 nails	Box	2.05	—
6 penny, 2", 5-pound box, 905 nails	Box	7.96	7.06
6 penny, 2", 30-pound box, 5,430 nails	Box	32.40	24.20
6 penny, 2", 50-pound box, 9,050 nails	Box	32.40	—
7 penny, 2-1/2", 5-pound box, 805 nails	Box	9.63	—
7 penny, 2-1/2", 50-pound box, 8,050 nails	Box	37.00	—
8 penny, 2-1/2", 1-pound box, 106 nails	Box	1.91	1.80
8 penny, 2-1/2", 5-pound box, 530 nails	Box	7.83	7.05
8 penny, 2-1/2", 30-pound box, 3,180 nails	Box	43.30	25.90
8 penny, 2-1/2", 50-pound box, 5,300 nails	Box	—	32.10
10 penny, 3", 1-pound box, 69 nails	Box	1.95	1.80
10 penny, 3", 5-pound box, 345 nails	Box	7.75	7.00
10 penny, 3", 30-pound box, 2,070 nails	Box	37.90	27.50
10 penny, 3", 50-pound box, 3,450 nails	Box	42.40	28.70
12 penny, 3-1/4", 1-pound box, 64 nails	Box	1.97	—
12 penny, 3-1/4", 5-pound box, 320 nails	Box	8.06	7.07
12 penny, 3-1/4", 30-pound box, 1,920 nails	Box	43.40	25.60
12 penny, 3-1/4", 50-pound box, 3,200 nails	Box	—	33.40
16 penny, 3-1/2", 1-pound box, 49 nails	Box	1.91	1.78
16 penny, 3-1/2", 5-pound box, 245 nails	Box	7.85	7.12
16 penny, 3-1/2", 30-pound box, 1,470 nails	Box	36.40	25.70
16 penny, 3-1/2", 50-pound box, 2,450 nails	Box	43.20	34.10
20 penny, 4", 1-pound box, 31 nails	Box	1.74	1.82

Nails

		Galvanized	Bright
20 penny, 4", 5-pound box, 155 nails	Box	7.58	7.48
40 penny, 5", per nail, 18 per pound	Ea	.06	.05
60 penny, 6", per nail, 11 per pound	Ea	.16	.07
60 penny, 6", 30-pound box, 330 nails	Box	—	27.40

Box nails Box nails are thinner and about 1/8" shorter than common nails of the same nominal size. Points are more blunt to reduce wood splitting. Use box nails for molding and trim and when the lumber is dry or brittle. Cost per box.

		Galvanized	Bright
3 penny, 1-1/4", 1-pound box, 1,010 nails	Box	2.15	2.36
3 penny, 1-1/4", 5-pound box, 5,050 nails	Box	11.30	—
4 penny, 1-1/2", 1-pound box, 473 nails	Box	1.97	2.36
4 penny, 1-1/2", 5-pound box, 2,365 nails	Box	8.84	—
5 penny, 1-3/4", 1-pound box, 406 nails	Box	2.60	—
5 penny, 1-3/4", 5-pound box, 2,030 nails	Box	11.00	—
6 penny, 2", 1-pound box, 236 nails	Box	2.31	2.69
6 penny, 2", 5-pound box, 1,180 nails	Box	9.38	8.56
7 penny, 2-1/4", 1-pound box, 210 nails	Box	2.93	—
8 penny, 2-1/2", 1-pound box, 145 nails	Box	2.28	2.66
8 penny, 2-1/2", 5-pound box, 725 nails	Box	9.40	8.32
8 penny, 2-1/2", 30-pound box, 4,350 nails	Box	41.10	—
10 penny, 3", 1-pound box, 132 nails	Box	1.98	2.42
10 penny, 3", 5-pound box, 660 nails	Box	8.97	8.42
12 penny, 3-1/4", 1-pound box, 94 nails	Box	2.26	2.92
12 penny, 3-1/4", 5-pound box, 470 nails	Box	9.28	—
16 penny, 3-1/2", 1-pound box, 71 nails	Box	2.30	2.66
16 penny, 3-1/2", 5-pound box, 355 nails	Box	9.36	8.20
16 penny, 3-1/2", 30-pound box, 2,130 nails	Box	41.10	28.70
20 penny, 4", 1-pound box, 62 nails	Box	2.03	2.36
20 penny, 4", 5-pound box, 310 nails	Box	9.20	—

Deck nails For use with ACQ or CA pressure-treated lumber. Cost per box.

		Screw shank	Ring shank
6 penny, 2", 1-pound box	Box	3.05	3.10
6 penny, 2", 5-pound box	Box	12.20	12.30
8 penny, 2-1/2", 1-pound box	Box	2.69	3.00
8 penny, 2-1/2", 5-pound box	Box	11.00	11.00
10 penny, 3", 1-pound box	Box	2.60	2.68
12 penny, 3-1/4", 1-pound box	Box	2.41	3.00
12 penny, 3-1/4", 5-pound box	Box	12.10	10.80
16 penny, 3-1/2", 1-pound box ,	Box	2.68	3.00
16 penny, 3-1/2", 5-pound box ,	Box	11.00	12.20

Vinyl/cement coated sinkers Vinyl coating lubricates the wood and holds fast when driven. Head sinks flush with surface.

	Craft@Hrs	Unit	Material	Labor	Total
4 penny, 1-3/8", 1-pound box, 550 nails	—	Box	1.78	—	1.78
6 penny, 1-7/8", 1-pound box, 275 nails	—	Box	1.80	—	1.80
6 penny, 1-7/8", 5-pound box, 1,375 nails	—	Box	5.95	—	5.95

	Craft@Hrs	Unit	Material	Labor	Total
8 penny, 2-3/8", 1-pound box, 142 nails,	—	Box	1.79	—	1.79
8 penny, 2-3/8", 5-pound box, 710 nails	—	Box	5.95	—	5.95
8 penny, 2-3/8", 50-pound box, 7,100 nails	—	Box	32.30	—	32.30
10 penny, 2-7/8", 5-pound box, 104 nails	—	Box	5.91	—	5.91
12 penny, 3-1/8", 1-pound box, 77 nails	—	Box	1.80	—	1.80
12 penny, 3-1/8", 5-pound box, 385 nails	—	Box	5.95	—	5.95
12 penny, 3-1/8", 50-pound box, 3,850 nails	—	Box	35.00	—	35.00
16 penny, 3-3/8", 1-pound box, 61 nails	—	Box	1.80	—	1.80
16 penny, 3-3/8", 5-pound box, 305 nails	—	Box	5.95	—	5.95
16 penny, 3-3/8", 25-pound box, 1,525 nails	—	Box	18.50	—	18.50
16 penny, 3-3/8", 50-pound box, 3,050 nails	—	Box	32.20	—	32.20
Plain shank stick collated framing nails					
2" x 0.113", coated, box of 2,000	—	Box	21.00	—	21.00
2-1/2" x 0.120", coated, box of 2,000	—	Box	24.20	—	21.20
2-1/2" x 0.120", galvanized, box of 2,000	—	Box	32.50	—	32.50
3" x 0.120", coated, box of 2,000	—	Box	24.70	—	24.70
3" x 0.120", galvanized, box of 2,000	—	Box	47.30	—	47.30
3-1/4" x 0.120", coated, box of 2,000	—	Box	27.90	—	27.90
3-1/4" x 0.120", galvanized, box of 2,000	—	Box	51.50	—	51.50
3-1/2" x 0.131", coated, box of 2,000	—	Box	33.60	—	33.60
3 1/2" x 0.131", galvanized, box of 2,000	—	Box	56.40	—	56.40
Plain shank coil framing nails, galvanized and coated					
2" x 0.099", box of 3,600	—	Box	54.60	—	51.97
2-1/2" x 0.099", box of 3,600	—	Box	53.97	—	53.97
3" x 0.120", box of 2,700	—	Box	55.86	—	55.86
3-1/4" x 0.120", box of 2,700	—	Box	62.57	—	62.57
Brad nails for an impulse nailer, bright, 18-gauge					
1", box of 1,000	—	Box	2.82	—	2.82
1-1/4", box of 1,000	—	Box	3.15	—	3.15
2", box of 1,000	—	Box	3.65	—	3.65

Casing nails, galvanized. Used where the nail head must be hidden. Casing nails have small, conical heads. The nail diameter is smaller than a common nail but larger than a finishing nail. Casing nails are used to attach trim.

	Craft@Hrs	Unit	Material	Labor	Total
6 penny, 2", 1-pound box, 236 nails	—	Box	3.05	—	3.05
8 penny, 2-1/2", 1-pound box, 145 nails	—	Box	2.48	—	2.48
10 penny, 3", 1-pound box, 94 nails	—	Box	2.90	—	2.90
16 penny, 3-1/2", 1-pound box, 71 nails	—	Box	2.48	—	2.48

Drywall nails. Allow 6 pounds of nails (1,300 nails) per MSF of gypsum wallboard

	Craft@Hrs	Unit	Material	Labor	Total
1-1/4", bright, ring shank, 5-pound box	—	Box	8.79	—	8.37
1-3/8", electro galvanized 5-pound box	—	Box	9.15	—	8.70
1-3/8", phosphate coated, 5-pound box	—	Box	9.75	—	9.28
1-3/8", blued, threaded, 5-pound box	—	Box	9.50	—	9.05
1-1/2", blued, threaded, 5-pound box	—	Box	12.20	—	11.62
1-5/8", blued, threaded, 5-pound box	—	Box	10.30	—	9.79
1-7/8", phosphate coated, 5-pound box	—	Box	8.35	—	7.93

Drywall screws. Black phosphate coating. For hanging drywall and plasterboard on wood or metal studs. Allow 1,300 screws per MSF of gypsum wallboard.

	Craft@Hrs	Unit	Material	Labor	Total
1-1/4", 5-pound box of 1330	—	Box	18.80	—	18.80
1-5/8", 5-pound box of 1300	—	Box	18.80	—	18.80
2", 5-pound box of 1070	—	Box	18.80	—	18.80
2-1/2", 5-pound box of 480	—	Box	18.80	—	18.80
3", 5-pound box of 415	—	Box	18.80	—	18.80

Nails

	Craft@Hrs	Unit	Material	Labor	Total

Finishing nails, bright. Finishing nails have a barrel-shaped head with a small diameter and a dimple on the top. Cost per box.

	Craft@Hrs	Unit	Material	Labor	Total
3 penny, 1-1/4", 1-pound box, 807 nails	—	Box	2.00	—	2.00
4 penny, 1-1/2", 1-pound box, 584 nails	—	Box	1.98	—	1.98
6 penny, 2", 1-pound box, 309 nails	—	Box	2.00	—	2.00
8 penny, 2-1/2", 1-pound box, 189 nails	—	Box	1.98	—	1.98
16 penny, 3-1/2", 1-pound box, 90 nails	—	Box	1.98	—	1.98

Finishing nail sticks for angled impulse nailer, 16 gauge, bright. Cost per box.

	Craft@Hrs	Unit	Material	Labor	Total
1", box of 2000	—	Box	6.50	—	6.50
1-1/4", box of 2000	—	Box	7.30	—	7.30
1-1/2", box of 2000	—	Box	8.68	—	8.68
1-3/4", box of 2000	—	Box	9.25	—	9.25
2", box of 2000	—	Box	10.40	—	10.40
2-1/2", box of 2000	—	Box	12.80	—	12.80

Bright joist hanger nails, 9 gauge except as noted. Cost per box.

	Craft@Hrs	Unit	Material	Labor	Total
1-1/4", 1-pound box	—	Box	2.52	—	2.52
1-1/2", 1-pound box	—	Box	2.25	—	2.25
1-1/2", 1-pound box, 10 gauge	—	Box	2.61	—	2.61
1-1/2", 5-pound box	—	Box	8.99	—	8.99
1-1/2", 5-pound box, 10 gauge	—	Box	11.19	—	11.19
1-1/2", 30-pound box	—	Box	50.95	—	50.95

Galvanized joist hanger nails. 9 gauge. Cost per box.

	Craft@Hrs	Unit	Material	Labor	Total
1-1/4", 5-pound box	—	Box	9.87	—	9.87
1-1/4", 25-pound box	—	Box	27.04	—	27.04
1-1/2", 1-pound box	—	Box	2.38	—	2.38
1-1/2", 50-pound box	—	Box	47.53	—	47.53

Stainless steel joist hanger nails

	Craft@Hrs	Unit	Material	Labor	Total
1-1/2", 9 gauge, 1-pound box	—	Box	8.20	—	8.20
1-1/2", 10 gauge, 5-pound box	—	Box	37.20	—	37.20

Masonry nails. Temper hardened. Bright.

	Craft@Hrs	Unit	Material	Labor	Total
1", fluted nail, 1-pound box	—	Box	3.25	—	3.25
1-1/2" fluted nail, 1-pound box	—	Box	3.25	—	3.25
2" fluted nail, 1-pound box	—	Box	3.00	—	3.00
2-1/2" cut nail, flat head, 1-pound box	—	Box	3.25	—	3.25
3", cut nail, flat head, 5-pound box	—	Box	14.70	—	14.70
4-1/2", cut nail, flat head, 1-pound box	—	Box	3.40	—	3.40

Plasterboard nails, 13 gauge, blued

	Craft@Hrs	Unit	Material	Labor	Total
3/8" head, 1-1/8" (449) or 1-1/4" (405)	—	Lb	2.35	—	2.35
3/8" head, 1-5/8", 342 nails	—	Lb	2.35	—	2.35

Roofing nail coils, for pneumatic nailer

	Craft@Hrs	Unit	Material	Labor	Total
2 penny, 1", 7,200 nails, galvanized	—	Box	27.90	—	27.90
3 penny, 1-1/4", 7,200 nails, galvanized	—	Box	43.70	—	43.70
5 penny, 2", 3,600 nails, bright	—	Box	33.60	—	33.60

Roofing nails, electro galvanized

	Craft@Hrs	Unit	Material	Labor	Total
1/2", 1-pound box, 380 nails	—	Box	1.69	—	1.69
3/4", 1-pound box, 315 nails	—	Box	1.76	—	1.76
1-3/4" or 2", 5-pound box, 150 or 138 nails	—	Box	6.75	—	6.75

Roofing nails, galvanized . Ring shank.

	Craft@Hrs	Unit	Material	Labor	Total
1-1/4", 30-pound box	—	Box	52.50	—	52.50
1-1/2", 30-pound box	—	Box	45.00	—	45.00
1-3/4", 5-pound box	—	Box	13.50	—	13.50
2-1/2", 50-pound box	—	Box	47.00	—	47.00

	Craft@Hrs	Unit	Material	Labor	Total

Roofing nails, plastic round cap . For securing felt on sandable roof decking or attaching foam board sheathing

	Craft@Hrs	Unit	Material	Labor	Total
7/8", 11-1/2 gauge, box of 3,000	—	Box	24.60	—	24.60
2" x 0.113", galvanized, box of 5000	—	Box	63.00	—	63.00

Double head (duplex) nails for use on scaffolds, concrete forms and temporary construction

6 penny, 2", 1-pound box, 143 nails	—	Box	3.86	—	3.86
6 penny, 2", 5-pound box, 715 nails	—	Box	9.82	—	9.82
8 penny, 2-1/4", 1-pound box, 88 nails	—	Box	3.50	—	3.50
8 penny, 2-1/4", 5-pound box, 440 nails	—	Box	11.20	—	11.20
8 penny, 2-1/4", 30-pound box, 2,640 nails	—	Box	26.40	—	26.40
16 penny, 3-1/4", 1-pound box, 44 nails	—	Box	3.18	—	3.18
16 penny, 3-1/4", 5-pound box, 220 nails,	—	Box	11.60	—	11.60
16 penny, 3-1/4", 30-pound box, 1,320 nails	—	Box	28.00	—	28.00

Siding nails

1-1/2", aluminum, 1-pound box	—	Box	8.40	—	8.40
2", aluminum, 1-pound box	—	Box	8.80	—	8.80
1-3/4", galvanized, 5-pound box	—	Box	12.20	—	12.20
2-1/4", galvanized, 5-pound box	—	Box	8.30	—	8.30
2-1/2", galvanized, 5-pound box	—	Box	11.50	—	11.50
3", galvanized, 5-pound box	—	Box	11.50	—	11.50
3" double galvanized, 5-pound box	—	Box	14.30	—	14.30

Spikes, galvanized. For landscape timbers, railroad ties, log home construction.

3/8" x 8", per spike	—	Ea	.40	—	0.40
3/8" x 10", per spike	—	Ea	.54	—	0.54
3/8" x 10", 5-pound box	—	Box	8.70	—	8.70
3/8" x 12", per spike	—	Ea	.64	—	0.64
3/8" x 12", 5-pound box	—	Box	8.20	—	8.20

Powder-driven fasteners

22 caliber single shot load, pack of 100	—	Pack	5.15	—	5.15
22 caliber load booster, pack of 20	—	Pack	3.05	—	3.05
27 caliber strip load, pack of 100	—	Pack	7.55	—	7.55
1" shank power fastener, pack of 100	—	Pack	4.70	—	4.70
1-1/2" shank power fastener, pack of 100	—	Pack	6.10	—	6.10
2" shank power fastener, pack of 100	—	Pack	7.86	—	7.86
2-1/2" shank power fastener, pack of 100	—	Pack	8.45	—	8.45
3" shank power fastener, pack of 100	—	Pack	9.34	—	9.34
1" washered fastener, pack of 100	—	Pack	9.60	—	9.60
1-1/2" washered fastener, pack of 100	—	Pack	10.00	—	10.00
2-1/2" washered fastener, pack of 100	—	Pack	10.30	—	10.30
3" washered fastener, pack of 100	—	Pack	10.50	—	10.50
1" control discs, pack of 100	—	Pack	3.00	—	3.00
1" drive pin, for steel, pack of 100	—	Pack	36.60	—	36.60
1-1/2" drive pin, for steel, pack of 100	—	Pack	42.00	—	42.00
2-1/2" drive pin, for steel, pack of 100	—	Pack	52.50	—	52.50

Nuts, Bolts and Washers Bolts, standard hex head, material cost per bolt.

			Length		
	4"	6"	8"	10"	12"
Heavy duty lag bolts, galvanized, deduct 25% for carton quantities					
1/2"	.59	.82	1.30	1.45	1.65
5/8"	1.34	2.00	2.38	2.95	3.10
3/4"	3.30	3.87	5.28	6.63	8.75

Ornamental Iron, Railings and Fence, Subcontract

	Length				
	4"	**6"**	**8"**	**10"**	**12"**
Machine bolts (hex bolts), galvanized, deduct 25% for carton quantities					
1/2", 13 threads per inch	.63	1.25	1.54	1.68	1.96
5/8", 11 threads per inch	1.30	1.94	2.35	2.85	2.99
3/4", 10 threads per inch	1.73	2.38	3.27	3.83	4.55
7/8", 9 threads per inch	4.30	5.22	5.10	7.43	8.55
1", 8 threads per inch	6.16	6.97	7.22	7.64	8.66

Labor to install machine bolts or lag bolts. Includes installation of nuts and washers and drilling as needed.

	Craft@Hrs	Unit	Material	Labor	Total
4" to 6" long	BC@.100	Ea	—	3.00	3.00
8" to 10" long	BC@.150	Ea	—	4.50	4.50
12" long	BC@.200	Ea	—	6.00	6.00
Nuts, standard hex, galvanized					
1/2"	—	Ea	.14	—	.14
5/8"	—	Ea	.27	—	.27
3/4"	—	Ea	.40	—	.40
7/8"	—	Ea	.85	—	.85
1"	—	Ea	1.50	—	1.50

Ornamental Iron, Railings and Fence, Subcontract Typical prices, fabricated, delivered and installed.

Handrail, 1-1/2" x 1" rectangular top rail tubing, 1" square bottom rail tubing,

	Craft@Hrs	Unit	Material	Labor	Total
1/2" square pickets with 5-1/2" space between					
36" high railing for single family dwelling	—	LF	—	—	29.00
42" high apartment style railing	—	LF	—	—	23.10
Add for inclined runs	—	LF	—	—	5.32
Pool fence					
5' high, 4-1/2" centers	—	LF	—	—	29.00
Add for gate, with self-closer and latch	—	LS	—	—	135.00

Paint Removal Dustless removal of lead-based paints. The removal system is a mechanical process and no water, chemicals or abrasive grit are used. Containment or special ventilation is not required. Based on Pentek Inc. Costs shown are for a 5,000 SF or larger job. For estimating purposes figure one 55-gallon drum of contaminated waste per 2,500 square feet of coatings removed from surfaces with an average coating thickness of 12 mils. Waste disposal costs are not included. Use $5,000 as a minimum charge for work of this type.

	Craft@Hrs	Unit	Material	Labor	Equipment	Total

Removing paint from ceilings, doors, door & window frames, roofs, siding Equipment includes three pneumatic needle scaler units attached to one air-powered vacuum-waste packing unit and one 150 CFM gas or diesel powered air compressor including all hoses and connectors. Add the one-time charge listed below for move-on, move-off and refurbishing of equipment. .

	Craft@Hrs	Unit	Material	Labor	Equipment	Total
Ceilings						
Plaster, 70 SF per hour	PT@.042	SF	—	1.34	.80	2.14
Doors, wood or metal. Remove and re-hang doors, no carpentry included						
To 3'0" x 6'8"	PT@1.00	Ea	—	31.90	—	31.90
Over 3'0" x 6'8" to 6'0" x 8'0"	PT@1.50	Ea	—	47.80	—	47.80
Over 6'0" x 8'0" to 12'0" x 20'0"	PT@2.00	Ea	—	63.80	—	63.80
Paint removal (assumes doors are laying flat)						
80 SF per hour	PT@.038	SF	—	1.21	.73	1.94

	Craft@Hrs	Unit	Material	Labor	Equipment	Total
Door or window frames						
To 3'0" x 6'8"	PT@1.50	Ea	—	47.80	28.67	76.47
Over 3'0" x 6'8" to 6'0" x 8'0"	PT@2.00	Ea	—	63.80	38.22	102.02
Over 6'0" x 8'0" to 12'0" x 20'0"	PT@3.00	Ea	—	95.70	57.33	153.03
Lintels						
Per SF of surface area	PT@.033	SF	—	1.05	.63	1.68
Roofs						
Metal roofs to 3 in 12 pitch						
75 SF of surface per hour	PT@.040	SF	—	1.28	.76	2.04
Metal roofs over 3 in 12 pitch						
65 SF of surface per hour	PT@.046	SF	—	1.47	.88	2.35
Wood roofs to 3 in 12 pitch						
80 SF of surface per hour	PT@.038	SF	—	1.21	.73	1.94
Wood roofs over 3 in 12 pitch						
70 SF of surface per hour	PT@.042	SF	—	1.34	.80	2.14
Siding						
Metal siding, 80 SF per hour	PT@.038	SF	—	1.21	.73	1.94
Wood siding, 85 SF per hour	PT@.035	SF	—	1.12	.67	1.79
Walls measured on one face of wall						
Concrete, 80 SF per hour	PT@.038	SF	—	1.21	.73	1.94
Concrete block, 70 SF per hour	PT@.042	SF	—	1.34	.80	2.14
Plaster, 85 SF per hour	PT@.035	SF	—	1.12	.67	1.79
Add for mobilization and demobilization, per job						
Move-on & off and refurbish	—	LS	—	—	—	2,500.00

Removing paint from concrete floor and slab surfaces Equipment includes three pneumatic manually-operated wheel-mounted scabblers attached to one air-powered vacuum-waste packing unit to scarify concrete floors and slabs and one 150 CFM gas or diesel powered air compressor including all hoses and connectors. Add the one-time charge listed below for move-on, move-off and refurbishing of equipment.

	Craft@Hrs	Unit	Material	Labor	Equipment	Total
Large unobstructed areas; floor slabs, driveways, sidewalks						
90 SF of surface per hour	PT@.033	SF	—	1.05	.63	1.68
Obstructed areas; narrow aisles, areas around equipment						
60 SF of surface per hour	PT@.050	SF	—	1.59	.96	2.55
Small areas; around decks						
40 SF of surface per hour	PT@.075	SF	—	2.39	1.43	3.82
Add for mobilization and demobilization, per job						
Move-on & off and refurbish	—	LS	—	—	—	2,500.00

Paints, Coatings, and Supplies Costs are for good to better quality coatings purchased in 1 to 4 gallon quantities. Add 30% for premium quality paints. General contractor prices may be discounted 10% to 15% on residential job quantities. Painting contractors receive discounts of 20% to 25% on steady volume accounts. Labor costs for painting are listed at the end of this section. For more complete coverage of painting costs, see the *2006 National Painting Cost Estimator* at http://CraftsmanSiteLincense.com/

Paints, Coatings, and Supplies

	Craft@Hrs	Unit	Material	Labor	Total

Primers

Oil-based exterior primer-sealer, universal undercoat for use on wood, hardboard, aluminum and wrought iron. Prevents extractive bleeding.

	Craft@Hrs	Unit	Material	Labor	Total
Quart (100 SF per Qt)	—	Qt	10.40	—	10.40
Gallon (400 SF per Gal)	—	Gal	22.90	—	22.90
Epoxy primer Catalyzed epoxy primer, two component type					
(400 SF per Gal)	—	Gal	22.30	—	22.30
Latex)block filler					
(1000 SF per 5 Gal)	—	5 Gal	44.30	—	44.30
Acrylic (vinyl acrylic) thermoplastic rust inhibiting metal primer					
(500 SF per Gal)	—	Gal	60.00	—	60.00
Acrylic latex interior primer and sealer, for drywall, plaster, masonry or wood					
(400 SF per Gal)	—	Gal	25.00	—	25.00
Linseed (oil base) exterior/ interior primer					
(300 SF per Gal)	—	Gal	24.20	—	24.20
Red oxide metal primer					
(500 SF per Gal)	—	Gal	22.00	—	22.00
Concrete bonding primer, clear					
(350 to 400 SF per Gal)	—	Gal	16.80	—	16.80
Zinc chromate primer					
(400 SF per Gal)	—	Qt	12.50	—	12.50
KILZ® exterior sealer, primer, stainblocker, Mildew-resistant. Blocks tannin bleed on redwood and cedar.					
Gallon (370 SF per gallon)	—	Gal	15.80	—	15.80
5 gallons (1,850 SF per 5 gallons)	—	5 Gal	70.00	—	70.00

House paint

Exterior acrylic latex house paint. Mildew and fade resistant. For wood, masonry, brick, stucco, vinyl and aluminum siding and metal. Typical coverage is 400 square feet per gallon.

	Craft@Hrs	Unit	Material	Labor	Total
Flat, quart	—	Qt	10.50	—	10.50
Flat, gallon	—	Gal	21.00	—	21.00
Flat, 5 gallons	—	5 Gal	93.70	—	93.70
Satin, gallon	—	Gal	24.70	—	24.70
Satin, 5 gallons	—	5 Gal	108.00	—	108.20
Semi-gloss, quart	—	Qt	11.60	—	11.60
Semi-gloss, gallon	—	Gal	24.20	—	24.20
Semi-gloss, 5 gallons	—	5 Gal	109.00	—	109.00

Interior acrylic latex house paint. Typical coverage is 400 square feet per gallon.

	Craft@Hrs	Unit	Material	Labor	Total
Flat pastel base, gallon	—	Gal	19.90	—	19.90
Flat pastel base, 5 gallons	—	5 Gal	90.80	—	90.80
Flat ultra white, quart	—	Qt	9.35	—	9.35
Flat ultra white, gallon	—	Gal	20.00	—	20.00
Flat ultra white, 5 gallons	—	5 Gal	89.30	—	89.30
Satin enamel, quart	—	Qt	11.50	—	11.50
Satin enamel, gallon	—	Gal	24.20	—	24.20
Satin enamel, 5 gallons	—	5 Gal	108.00	—	108.00
High gloss enamel, quart	—	Qt	12.60	—	12.60
High gloss enamel, gallon	—	Gal	26.30	—	26.30
High gloss enamel, 5 gallons	—	5 Gal	112.00	—	112.00
Roll-on fibered texture paint, 2 gallons	—	2 Gal	19.90	—	19.90

	Craft@Hrs	Unit	Material	Labor	Total

Stucco, masonry and concrete paint

Masonry and stucco paint. Acrylic latex paint for interior or exterior. smooth, rough or textured masonry surfaces.

	Craft@Hrs	Unit	Material	Labor	Total
Gallon	—	Gal	16.80	—	16.80
5 gallons	—	5 Gal	67.20	—	67.20
Elastomeric waterproofing, gallon	—	Gal	21.70	—	21.70
Elastomeric waterproofing, 5 gallons	—	5 Gal	108.00	—	108.00
Penetrating sealer, gallon	—	Gal	12.90	—	12.90
Penetrating sealer, 5 gallons	—	5 Gal	55.40	—	55.40

Concrete stain. Forms a tough film on properly prepared concrete patios, basements, pool decks, pillars, brick and sidewalks. Natural slate color. 100 percent acrylic.

Gallon	—	Gal	21.00	—	21.00
5 gallons	—	5 Gal	92.70	—	92.70

Gloss porch and floor enamel. 100 percent acrylic. Mildew-resistant. Recoat in 24 to 48 hours. Full cure in 30 days.

Gloss, quart	—	Qt	10.00	—	10.00
Gloss, gallon	—	Gal	23.10	—	23.10
Low luster, slate gray, gallon	—	Gal	21.00	—	21.00
Low luster, slate gray, 5 gallons	—	5 Gal	93.50	—	93.50

Masonry waterproofer, for concrete, stucco and masonry, white when applied, dries clear

150 SF per gallon	—	Gal	18.90	—	18.00
750 SF per 5 gallons	—	5 Gal	84.00	—	84.00

Metal coatings

Clean metal primer. White. For use on bare, lightly rusted or previously painted surfaces. Creates a strong surface for better topcoat adhesion and weather resistance.

Quart	—	Qt	7.50	—	7.50

Rust converting metal primer. Non-toxic, USDA and FDA approved. Works on clean or rusty steel, zinc galvanized metal, aluminum, tin and previously painted surfaces.

Quart	—	Qt	11.50	—	11.50
Gallon	—	Gal	38.50	—	38.50

Oil-based enamel. Industrial grade. Prevents rust, resists cracking, peeling, chipping, and fading. Scuff-, abrasion- and moisture-resistant. Lead-free. Not low volatile organic compound compliant.

Gallon	—	Gal	25.50	—	25.50

Stops Rust, Rust-Oleum(r). Resists moisture and corrosion. For metal, wood, concrete and masonry.

Half pint	—	Ea	3.60	—	3.60
Quart	—	Ea	7.00	—	7.00

Wood finishes

Exterior opaque stain. For wood siding, decks and fences. Water repellent and mildew-resistant. Gallon covers 400 square feet.

Acrylic latex, tint or white, gallon	—	Gal	22.10	—	22.10
Acrylic latex, tint or white, 5 gallons	—	5 Gal	98.60	—	98.60
Oil and latex, accent tint, gallon	—	Gal	17.60	—	17.60
Oil and latex, accent tint, 5 gallons	—	5 Gal	79.60	—	79.60

Exterior semi-opaque stain. For wood siding, decks and fences. Water repellent and mildew-resistant. Gallon covers 400 square feet.

Oil-based stain, gallon	—	Gal	22.00	—	22.00
Oil-based stain, 5 gallons	—	5 Gal	98.70	—	98.70
Water-based stain, gallon	—	Gal	21.00	—	21.00
Water-based stain, 5 gallons	—	5 Gal	71.40	—	71.40

Paints, Wood Finish

	Craft@Hrs	Unit	Material	Labor	Total

Exterior clear wood finish. For exterior wood doors, windows, furniture and trim. Gallon covers 300 to 400 square feet.

	Craft@Hrs	Unit	Material	Labor	Total
Tung oil spar varnish, quart	—	Qt	6.80	—	6.80
Tung oil spar varnish, gallon	—	Gal	17.80	—	17.80
Marine varnish, gloss or satin, quart	—	Qt	14.70	—	14.70
Marine varnish, gloss or satin, gallon	—	Gal	42.00	—	42.00
Urethane clear finish, gloss or satin, quart	—	Qt	13.40	—	13.40
Urethane clear finish, gloss or satin, gallon	—	Gal	32.00	—	32.00
Varathane® gloss, semi or satin, 13 ounces	—	Can	6.70	—	6.70
Varathane® gloss, semi or satin, quart	—	Qt	15.90	—	15.90
Varthane® gloss, semi or satin, gallon	—	Gal	47.30	—	47.30
Exterior polyurethane, gloss or satin, quart	—	Qt	17.90	—	17.90

Waterproofing wood finish. For wood siding, decks and fences. Water repellent and mildew resistant. Clear or tinted. Gallon covers 300 to 400 square feet.

	Craft@Hrs	Unit	Material	Labor	Total
6-way exterior protection, gallon	—	Gal	18.90	—	18.90
6-way exterior protection, 5 gallons	—	5 Gal	84.00	—	84.00
Premium wood weatherproofing, gallon	—	Gal	26.30	—	26.30
Premium wood weatherproofing, 5 gallons	—	5 Gal	115.00	—	115.00
UV-resistant wood finish, gallon	—	Gal	16.70	—	16.70
UV-resistant wood finish, 5 gallons	—	5 Gal	75.40	—	75.40
Penetrating wood finish, quart	—	Qt	10.50	—	10.50
Penetrating wood finish, gallon	—	Gal	24.10	—	24.10
Penetrating wood finish, 5 gallons	—	5 Gal	102.00	—	102.00
Log home gloss finish, clear or tint, gallon	—	Gal	31.50	—	31.50
Log home gloss finish, clear or tint, 5 gallons	—	5 Gal	141.00	—	141.00

Wood water sealer. Clear. Helps maintain wood's natural color. Protects against damage from water and sun. Provides a mildew-resistant coating. Gallon covers 225 to 325 square feet of smooth wood or as little as 125 square feet for a first application on rough-sawn wood.

	Craft@Hrs	Unit	Material	Labor	Total
Thompson's WaterSeal, gallon	—	Gal	14.40	—	14.40
Thompson's WaterSeal, 5 gallons	—	5 Gal	62.90	—	62.90
Waterproofing wood protector, gallon	—	Gal	12.60	—	12.60
Waterproofing wood protector, 5 gallons	—	5 Gal	55.70	—	55.70

Wood preservative. Helps maintain wood's natural color. Protects against wood rot, warping, swelling, termite damage, mildew and moisture. Gallon covers 100 to 300 square feet when applied by brush or roller.

	Craft@Hrs	Unit	Material	Labor	Total
Dock and fence post preservative, gallon	—	Gal	18.80	—	18.80
Seasonite® new wood treatment, gallon	—	Gal	12.10	—	12.10
Termin-8 green preservative, quart	—	Qt	7.70	—	7.70
Termin-8 green preservative, gallon	—	Gal	16.00	—	16.00
Termin-8 green preservative, 5 gallons	—	5 Gal	66.30	—	66.30
Copper-Green® preservative, 13.5 ounces	—	Can	7.30	—	7.30
Copper-Green® preservative, gallon	—	Gal	13.10	—	13.10

Interior oil stain. Penetrates into unfinished or stripped wood to highlight and seal the grain.

	Craft@Hrs	Unit	Material	Labor	Total
Tinted wiping stain, half pint	—	Ea	3.80	—	3.80
Tinted wiping stain, quart	—	Qt	6.15	—	6.15
Tinted wiping stain, gallon	—	Gal	22.90	—	22.90
Danish oil, quart	—	Qt	10.40	—	10.40
Danish oil, gallon	—	Gal	27.90	—	27.90
Gel stain, quart	—	Qt	12.40	—	12.40

Interior water-based stain. Fast-drying and easy clean-up.

	Craft@Hrs	Unit	Material	Labor	Total
White wash pickling stain, quart	—	Qt	9.80	—	9.80
Tinting stain, half-pint	—	Ea	4.90	—	4.90
Tinting stain, quart	—	Qt	9.30	—	9.30

	Craft@Hrs	Unit	Material	Labor	Total
Paint specialties					
Asphalt-fiber roof and foundation coating, 75 SF per gallon on roof or concrete,					
Solvent type, 100 SF per gallon	—	Gal	7.15	—	7.15
Driveway coating					
Coal tar emulsion (500 SF per 5 Gal)	—	5 Gal	21.00	—	7.25
Traffic paint, acrylic, 100 SF per gallon					
Blue, white, yellow or red, 1 gallon	—	Gal	11.60	—	11.60
Blue, white, yellow or red, 5 gallon	—	5 Gal	55.70	—	55.70
Swimming pool enamel					
Sau-Sea Type R (rubber base)					
(350 SF per Gal)	—	5 Gal	263.00	—	30.00
Sau-Sea plaster pool coating					
(350 SF per Gal)	—	5 Gal	271.00	—	30.00
Sau-Sea thinner					
For plaster or rubber	—	Gal	17.00	—	9.75
Sau-Sea patching compound for plastic					
(150 LF per Gal)	—	Gal	75.00	—	39.75
Multi-surface waterproofer. For wood, brick, concrete, stucco and masonry. Can be applied to new pressure-treated wood. Paintable. Allows wood to weather naturally. Goes on milky white, dries clear.					
Aerosol can, 12 ounces	—	Can	5.25	—	5.25
Gallon	—	Gal	9.30	—	9.30
5 gallon	—	5 Gal	42.00	—	42.00
Paint and varnish remover, 150 SF per gallon					
Quart	—	Qt	15.75	—	15.75
Gallon	—	Gal	58.00	—	58.00
5 gallon	—	5 Gal	195.00	—	195.00
Spackling compound (vinyl-paste)					
Interior, 8-ounce tube	—	Ea	4.00	—	4.00
Wood filler paste	—	Pint	5.60	—	5.60
Thinners					
Acetone	—	Gal	13.70	—	13.70
Xylene epoxy thinner	—	Qt	6.00	—	6.00
Denatured alcohol	—	Gal	12.60	—	12.60
Mineral spirits	—	Gal	6.55	—	6.55
Paint thinner	—	Gal	3.65	—	3.65
Shellac or lacquer thinner	—	Gal	11.50	—	11.50
Turpentine substitute	—	Qt	4.75	—	4.75

Painting, Labor Single coat applications except as noted. Not including equipment rental costs. These figures are based on hand work (roller, and brush where required) on residential jobs. Where spray equipment can be used to good advantage, reduce these costs 30% to 40%. Spray painting will increase paint requirements by 30% to 60%. Protection of adjacent materials is included in these costs but little or no surface preparation is assumed. Per SF of surface area to be painted. Figures in parentheses show how much work should be done in an hour. For more complete coverage of painting costs, see *National Painting Cost Estimator* at http://CraftsmanSiteLincense.com/

	Craft@Hrs	Unit	Material	Labor	Total
Exterior surfaces, per coat, paint grade					
Door and frame, 6 sides (2.5 doors/hour)	PT@.400	Ea	—	12.80	12.80
Siding, smooth, wood (250 SF/hour)	PT@.004	SF	—	.13	.13
Siding, rough or shingle (200 SF/hour)	PT@.005	SF	—	.16	.16
Shutters (50 SF/hour)	PT@.020	SF	—	.64	.64

Paint Spraying Equipment Rental

	Craft@Hrs	Unit	Material	Labor	Total
Stucco (300 SF/hour)	PT@.003	SF	—	.10	.10
Trim, posts, rails (20 LF/hour)	PT@.050	LF	—	1.59	1.59
Windows, including mullions, per SF of opening,					
one side only (67 SF/hour)	PT@.015	SF	—	.48	.48
Add for light sanding (330 SF/hour)	PT@.003	SF	—	.10	.10
Interior surfaces, per coat					
Ceiling, flat latex (250 SF/hour)	PT@.004	SF	—	.13	.13
Ceiling, enamel (200 SF/hour)	PT@.005	SF	—	.16	.16
Walls, flat latex (500 SF/hour)	PT@.002	SF	—	.06	.06
Walls, enamel (400 SF/hour)	PT@.003	SF	—	.10	.10
Baseboard (120 LF/hour)	PT@.008	LF	—	.26	.26
Bookcases, cabinets (90 SF/hour)	PT@.011	SF	—	.35	.35
Windows, including mullions and frame, per SF of opening,					
one side only (67 SF/hour)	PT@.015	SF	—	.48	.48
Floors, wood					
Filling	PT@.007	SF	—	.22	.22
Staining	PT@.006	SF	—	.19	.19
Shellacking	PT@.006	SF	—	.19	.19
Varnishing	PT@.006	SF	—	.19	.19
Waxing, machine	PT@.006	SF	—	.19	.19
Polishing, machine	PT@.008	SF	—	.26	.26
Sanding, hand	PT@.012	SF	—	.38	.38
Sanding, machine	PT@.007	SF	—	.22	.22
Walls, brick, concrete and masonry					
Oiling and sizing (140 SF/hour)	PT@.007	SF	—	.22	.22
Sealer coat (150 SF/hour)	PT@.007	SF	—	.22	.22
Brick or masonry (165 SF/hour)	PT@.006	SF	—	.19	.19
Concrete (200 SF/hour)	PT@.005	SF	—	.16	.16
Concrete steps, 3 coats (38 SF/hour)	PT@.026	SF	—	.83	.83
Window frames and sash, using primer to mask glass and prime sash, topcoats as shown					
1 topcoat (100 SF/hour)	PT@.010	SF	—	.32	.32
2 topcoats (67 SF/hour)	PT@.015	SF	—	.48	.48
Score and peel primer from glass (peels from glass only)					
Per lite	PT@.020	Ea	—	.64	.64

	Day	Week	Month
Paint Spraying Equipment Rental Complete spraying outfit including paint cup, hose, spray gun.			
8 CFM electric compressor	65.00	200.00	570.00
44 CFM gas-driven compressor	68.00	205.00	614.00
Professional airless, complete outfit	116.00	347.00	1040.00

Painting, Subcontract, Rule of Thumb Including material, labor, equipment and the subcontractor's overhead and profit.

	Craft@Hrs	Unit	Material	Labor	Total
Typical subcontract costs per square foot of floor area for painting the interior and exterior of residential buildings, including only minimum surface preparation					
Economy, 1 or 2 coats, little brushwork	—	SF	—	—	1.90
Good quality, 2 or 3 coats	—	SF	—	—	4.35

	Craft@Hrs	Unit	Material	Labor	Total

Painting, Subcontract, Cost Breakdown The costs that follow are based on the area of the surface being prepared or painted. Typical subcontract costs, including labor, material, equipment and the subcontractor's overhead and profit.

Surface preparation, subcontract Work performed by hand.

Water blasting (pressure washing), using wheel-mounted portable 2200 PSI pressure washer

	Craft@Hrs	Unit	Material	Labor	Total
Light blast (250 SF per hour)	—	SF	—	—	.18
Most work (150 SF per hour)	—	SF	—	—	.29
Rough surface, grime (75 SF per hour)	—	SF	—	—	.58

Light cleaning to remove surface dust and stains

Concrete or masonry surfaces	—	SF	—	—	.29
Gypsum or plaster surfaces	—	SF	—	—	.29
Wood surfaces	—	SF	—	—	.28

Normal preparation including scraping, patching and puttying
Concrete or masonry surface

Painted	—	SF	—	—	.30
Unpainted	—	SF	—	—	.24

Gypsum or plaster surfaces

Painted	—	SF	—	—	.29
Unpainted	—	SF	—	—	.14
Metal surfaces, light sanding	—	SF	—	—	.13

Wood surfaces, surface preparation, subcontract

Painted	—	SF	—	—	.28
Unpainted	—	SF	—	—	.14
Puttying	—	LF	—	—	1.35

Exterior painting, subcontract Hand work, costs will be 15% to 30% lower when spray equipment can be used to good advantage. These costs include only minimum surface preparation. Add additional surface preparation, if required, from above. Typical subcontract costs.

Brick or concrete, including acrylic latex masonry primer and latex paint

1 coat latex	—	SF	—	—	.60
2 coats latex	—	SF	—	—	.82
Concrete floors, etch and epoxy enamel	—	SF	—	—	.35
Columns and pilasters per coat	—	SF	—	—	.48
Cornices, per coat	—	SF	—	—	1.22

Doors, including trim, exterior only

3 coats	—	Ea	—	—	38.90
2 coats	—	Ea	—	—	28.00

Downspouts and gutters

Per coat	—	LF	—	—	.99

Eaves

No rafters, per coat	—	SF	—	—	.47
With rafters, per coat, brush	—	SF	—	—	.82

Fences (gross area)

Plain, 2 coats (per side)	—	SF	—	—	1.14
Plain, 1 coat (per side)	—	SF	—	—	.67
Picket, 2 coats (per side)	—	SF	—	—	1.06
Picket, 1 coat (per side)	—	SF	—	—	.61

Lattice work

1 coat 1 side, gross area	—	SF	—	—	.53

Painting, Subcontract

	Craft@Hrs	Unit	Material	Labor	Total
Metal, typical					
1 coat	—	SF	—	—	.42
2 coats	—	SF	—	—	.74
Porch rail and balusters, per coat					
Gross area	—	SF	—	—	1.05
Handrail only	—	LF	—	—	.71
Roofs, wood shingle					
1 coat stain plus 1 coat sealer					
Flat roof	—	SF	—	—	.49
4 in 12 pitch	—	SF	—	—	.62
8 in 12 pitch	—	SF	—	—	.65
12 in 12 pitch	—	SF	—	—	.80
1 coat stain plus 2 coats sealer					
Flat roof	—	SF	—	—	.80
4 in 12 pitch	—	SF	—	—	.95
8 in 12 pitch	—	SF	—	—	1.03
12 in 12 pitch	—	SF	—	—	1.09
Siding, plain					
Sanding and puttying, typical	—	SF	—	—	.33
Siding and trim, 3 coats	—	SF	—	—	1.00
Siding and trim, 2 coats	—	SF	—	—	.74
Trim only, 2 coats	—	SF	—	—	.79
Trim only, 3 coats	—	SF	—	—	1.06
Siding, shingle, including trim					
2 coats oil paint	—	SF	—	—	.76
1 coat oil paint	—	SF	—	—	.46
2 coats stain	—	SF	—	—	.71
1 coat stain	—	SF	—	—	.40
Steel sash					
3 coats (per side)	—	SF	—	—	.96
Stucco					
1 coat, smooth surface	—	SF	—	—	.47
2 coats, smooth surface	—	SF	—	—	.72
1 coat, rough surface	—	SF	—	—	.53
2 coats, rough surface	—	SF	—	—	.86
Window frames and sash, 3 coats					
Per side, one lite	—	Ea	—	—	29.70
Per side, add for each additional lite	—	Ea	—	—	1.66

Interior painting, subcontract Hand work (roller with some brush cut-in), costs will be 15% to 30% lower when spray equipment can be used to good advantage. These costs include only minimum surface preparation. Add for additional surface preparation, if required, from preceding page. Typical subcontract costs.

	Craft@Hrs	Unit	Material	Labor	Total
Cabinets, bookcases, cupboards, per SF of total surface area					
3 coats, paint	—	SF	—	—	2.37
Stain, shellac, varnish or plastic	—	SF	—	—	2.42
Concrete block or brick					
Primer and 1 coat latex	—	SF	—	—	.56
Doors, including trim					
3 coats (per side)	—	SF	—	—	1.46
2 coats (per side)	—	SF	—	—	1.02

	Craft@Hrs	Unit	Material	Labor	Total
Plaster or drywall walls, latex. Add for ceilings from below					
Prime coat or sealer	—	SF	—	—	.30
1 coat smooth surface	—	SF	—	—	.36
1 coat rough surface	—	SF	—	—	.39
1 coat sealer, 1 coat flat					
Smooth surface	—	SF	—	—	.48
Rough surface	—	SF	—	—	.63
1 coat sealer, 2 coats flat					
Smooth surface	—	SF	—	—	.64
Rough surface	—	SF	—	—	.86
1 coat sealer, 2 coats gloss or semi-gloss					
Smooth surface	—	SF	—	—	.62
Rough surface	—	SF	—	—	.85
Add for ceilings	—	%	—	—	45.0
Stairs, including risers					
3 coats paint	—	SF	—	—	1.66
Stain, shellac, varnish	—	SF	—	—	1.71
Woodwork, painting					
Priming	—	SF	—	—	.38
2 coats	—	SF	—	—	.62
2 coats, top workmanship	—	SF	—	—	.78
3 coats	—	SF	—	—	1.02
3 coats, top workmanship	—	SF	—	—	1.33
Woodwork, staining					
Apply stain	—	SF	—	—	.33
Filling (paste wood filler)	—	SF	—	—	.68
Sanding	—	SF	—	—	.40
Apply wax and polish	—	SF	—	—	.40

Paneling See also Hardwood under Lumber and Hardboard.

Solid plank paneling, prefinished V joint or T&G, costs include nails and waste.

	Craft@Hrs	Unit	Material	Labor	Total
Tennessee cedar plank paneling					
Clear 3/8" x 4" or 6" wide	BC@.050	SF	2.33	1.50	3.83
Select tight knot, 3/4" x 4" or 6" wide	BC@.050	SF	2.44	1.50	3.94
Knotty pine, 5/16" x 4" or 6" wide	BC@.050	SF	2.18	1.50	3.68
Redwood, clear, 5/16" x 4" wide	BC@.050	SF	4.60	1.50	6.10

Plywood and hardboard paneling Prefinished, 4' x 8' x 1/4" panels. Costs include nails, adhesive and 5% waste. Georgia-Pacific

	Craft@Hrs	Unit	Material	Labor	Total
Alaskan Oak	BC@.030	SF	.48	.90	1.38
Charterhouse Oak	BC@.030	SF	.65	.90	1.55
Norwegian Pine	BC@.030	SF	.59	.90	1.49
Spartan Oak	BC@.030	SF	.28	.90	1.18
Diplomat Elm	BC@.030	SF	.49	.90	1.39
Beaded Finished Oak	BC@.030	SF	1.15	.90	2.05
Golden Oak	BC@.030	SF	.43	.90	1.33

Hardwood plywood paneling Unfinished, 4' x 8' panels. Costs include nails, adhesive and 5% waste.

	Craft@Hrs	Unit	Material	Labor	Total
Birch plywood paneling					
1/8" x 4' x 8'	BC@.032	SF	.62	.96	1.58
1/4" x 4' x 8'	BC@.032	SF	.68	.96	1.64
1/2" x 4' x 8'	BC@.040	SF	1.21	1.20	2.41

Paneling

	Craft@Hrs	Unit	Material	Labor	Total
3/4" x 4' x 4'	BC@.042	SF	1.00	1.26	2.26
3/4" x 4' x 8'	BC@.042	SF	1.44	1.26	2.70
Oak plywood paneling					
1/8" x 4' x 8'	BC@.032	SF	.87	.96	1.83
1/4" x 4' x 8'	BC@.032	SF	.72	.96	1.68
1/2" x 4' x 8'	BC@.040	SF	1.28	1.20	2.48
3/4" x 4' x 8', red oak	BC@.042	SF	1.53	1.26	2.79
Cedar hardwood paneling, natural					
1/4" x 4' x 8'	BC@.032	SF	.81	.96	1.77
3/8" x 4' x 8'	BC@.040	SF	.93	1.20	2.13
1/2" x 4' x 8'	BC@.040	SF	1.22	1.20	2.42
5/8" x 4' x 8'	BC@.042	SF	1.37	1.26	2.63
3/4" x 4' x 8'	BC@.042	SF	1.57	1.26	2.83
Golden Virola paneling					
1/4" x 2' x 2'	BC@.032	SF	.58	.96	1.54
1/4" x 2' x 4'	BC@.032	SF	.68	.96	1.64
1/4" x 4' x 4'	BC@.032	SF	.57	.96	1.53
1/4" x 4' x 8'	BC@.032	SF	.47	.96	1.43
3/8" x 4' x 8'	BC@.040	SF	.67	1.20	1.87
1/2" x 2' x 2'	BC@.040	SF	.90	1.20	2.10
5/8" x 4' x 8'	BC@.042	SF	.91	1.26	2.17
3/4" x 2' x 2'	BC@.042	SF	1.36	1.26	2.62
Maple paneling					
1/4" x 4' x 8', B-4 grade	BC@.032	SF	1.41	.96	2.37
1/2" x 4' x 8', C-2 grade	BC@.040	SF	1.94	1.20	3.14
3/4" x 4' x 8', B-3 grade	BC@.042	SF	1.30	1.26	2.56
3/4" x 4' x 8', C-2 grade	BC@.042	SF	2.39	1.26	3.65
3/4" x 4' x 8,' maple/birch	BC@.042	SF	1.25	1.26	2.51

Vinyl-clad hardboard paneling Vinyl-clad film over hardboard, 4' x 8' x 1/4" panels.
Costs include nails, adhesive and 5% waste

	Craft@Hrs	Unit	Material	Labor	Total
White or Aegean Gold	BC@.032	SF	1.31	.96	2.27
Blue marble	BC@.032	SF	.50	.96	1.46
Harvest pattern	BC@.032	SF	1.01	.96	1.97

FRP wall and ceiling paneling Fiberglass-reinforced plastic, textured, 0.09" thick, class C flame spread, 24" x 48" panels. Includes fasteners, adhesive and 5% waste.

	Craft@Hrs	Unit	Material	Labor	Total
White or almond	BC@.032	SF	3.83	.96	4.79

Bamboo tambour paneling 4' x 8' panels. Costs include nails, adhesive and 5% waste.

	Craft@Hrs	Unit	Material	Labor	Total
Unfinished natural	BC@.032	SF	3.70	.96	4.66
Prefinished natural	BC@.032	SF	4.22	.96	5.18
Unfinished amber	BC@.032	SF	3.70	.96	4.66
Prefinished amber	BC@.032	SF	4.22	.96	5.18
Raw amber	BC@.032	SF	3.70	.96	4.66
Raw green	BC@.032	SF	3.70	.96	4.66
Raw black	BC@.032	SF	3.70	.96	4.66

Z-brick paneling Non-ceramic mineral brick-like veneer. Interior or exterior. Coverage per carton is 3 1/2 square feet, 5% added for waste.

	Craft@Hrs	Unit	Material	Labor	Total
Red classic brick, 2-1/4" x 8" x 7/16"	BC@.050	SF	3.93	1.50	5.43
Inca decorative brick, 2-1/4" x 8" x 7/16"	BC@.050	SF	4.44	1.50	5.94
Used brick (Inca), 2-1/4" x 8" x 5/16"	BC@.050	SF	4.44	1.50	5.94

	Craft@Hrs	Unit	Material	Labor	Total
Old Chicago brick, 2-3/8" x 8-1/8" x 5/16"	BC@.050	SF	4.44	1.50	5.94
Liberty grey brick, 2-1/4" x 8" x 7/16"	BC@.050	SF	5.89	1.50	7.39
Wheat brick, 2-1/4" x 8" x 7/16"	BC@.050	SF	5.89	1.50	7.39
Burnt sienna brick, 2-1/4" x 8" x 7/16"	BC@.050	SF	5.89	1.50	7.39
Mesa beige brick, 2-1/4" x 8" x 7/16"	BC@.050	SF	5.89	1.50	7.39
Z-Ment mortar adhesive, 16 to 20 SF per Gal	—	Gal	—	—	30.00

Paving, Subcontract Small areas such as walks and driveways around residential buildings. Typical costs including material, labor, equipment and subcontractor's overhead and profit. No substrate preparation included. Use $1,800 as a minimum job charge.

	Craft@Hrs	Unit	Material	Labor	Total
Asphalt paving, including oil seal coat					
2" asphalt	—	SF	—	—	1.69
3" asphalt	—	SF	—	—	1.90
4" asphalt	—	SF	—	—	2.44
4" base and fine grading	—	SF	—	—	1.03
6" base and fine grading	—	SF	—	—	1.06
Asphalt seal coats					
Applied with broom or squeegee	—	SF	—	—	.29
Fog seal, thin layer of diluted SS1H oil	—	SF	—	—	.13
Seal and sand					
Thin layer of oil with sand laid on top	—	SF	—	—	.29
Simco "walk-top" or other quick-dry applications					
Creamy oil	—	SF	—	—	.29
Brick paving					
On concrete, grouted, laid flat,					
including concrete base	—	SF	—	—	11.70
On concrete, laid solid on edge, 8" x 2-1/4" exposure, no grouting,					
including concrete base	—	SF	—	—	12.80
On sand bed, laid flat	—	SF	—	—	9.13
On sand bed, 8" x 2-1/4" exposure	—	SF	—	—	10.30
Concrete (including fine grading) paving					
3" concrete, unreinforced	—	SF	—	—	2.77
4" concrete, unreinforced	—	SF	—	—	3.32
4" concrete, mesh reinforcing	—	SF	—	—	3.36
4" concrete, seeded aggregate finish	—	SF	—	—	5.83
Add for colored concrete, most colors	—	SF	—	—	.40

Porous paving systems Thin-walled HDPE plastic rings connected by an interlocking geogrid structure, installed on a porous base course for residential applications. Rings transfer loads from the surface to the grid structure to an engineered course base. Installed on a prepared base. www.invisiblestructures.com

Porous paving system with grass filler. Grid rolled out over base course and seeded with grass mixture. Includes fertilizer and soil polymer mix for spreading over base.

	Craft@Hrs	Unit	Material	Labor	Total
3.3' wide by 33' long rolls	BL@.003	SF	2.18	.07	2.25
Add for 1" sand fill	BL@.001	SF	.05	.02	.07

Porous paving system with gravel fill. Grid rolled out over base course and covered with decorative fill gravel.

	Craft@Hrs	Unit	Material	Labor	Total
3.3' wide by 33' long rolls	BL@.004	SF	2.13	.10	2.23
Add for 1" gravel fill	BL@.001	SF	.06	.02	.08

Plastering, Subcontract

	Craft@Hrs	Unit	Material	Labor	Total
Plastering, Subcontract See also Lathing. Typical costs including material, labor, equipment and subcontractor's overhead and profit.					
Acoustical plaster					
Including brown coat, 1-1/4" total thickness	—	SY	—	—	27.70
Plaster on concrete or masonry, with bonding agent					
Gypsum plaster, 1/2" sanded	—	SY	—	—	14.50
Plaster on interior wood frame walls, no lath included					
Gypsum plaster, 7/8", 3 coats	—	SY	—	—	21.80
Keene's cement finish, 3 coats	—	SY	—	—	23.00
Cement plaster finish, 3 coats	—	SY	—	—	22.70
Gypsum plaster, 2 coats, tract work	—	SY	—	—	19.00
Add for ceiling work	—	%	—	—	15.0
Plaster for tile					
Scratch coat	—	SY	—	—	10.80
Brown coat and scratch coat	—	SY	—	—	16.20
Exterior plaster (stucco)					
Walls, 3 coats, 7/8", no lath included					
Textured finish	—	SY	—	—	16.10
Float finish	—	SY	—	—	19.00
Dash finish	—	SY	—	—	21.20
Add for soffits	—	%	—	—	15.0
Portland cement plaster, including metal lath					
Exterior, 3 coats, 7/8" on wood frame					
Walls	—	SY	—	—	26.40
Soffits	—	SY	—	—	25.70
Deduct for paper back lath	—	SY	—	—	-.16

Plumbing Fixtures and Equipment Costs are for good to better quality fixtures and trim installed with the accessories listed. All vitreous china, enameled steel cast iron, and fiberglass fixtures are white except as noted. Colored acrylic fiberglass fixtures will cost about 6% more. Colored enameled steel, cast iron, or vitreous fixtures will cost about 25% more. No rough plumbing included. Rough plumbing is the drain, waste and vent (DWV) piping and water supply piping that isn't exposed when construction is complete. See costs for rough plumbing at the end of this section. Minimum quality fixtures and trim will cost 10% to 20% less and deluxe fixtures will cost much more. Installation times assume standard quality work and uncomplicated layouts. For expanded coverage of plumbing cost estimates, see the *2006 National Plumbing & HVAC Estimator* at http://CraftsmanSiteLicense.com/

Bathtubs

Acrylic fiberglass one-piece bathtub and shower, 5' L, 32" W, 72" H. Includes polished chrome pop-up drain/overflow. Mixing valve, tub filler, and 2.5 GPM shower head

	Craft@Hrs	Unit	Material	Labor	Total
Acrylic tub, shower and trim	P1@4.50	Ea	388.00	135.00	523.00

Enameled cast iron bathtub, 5' L, 32" W, 15" H for recessed installation. Includes polished chrome pop-up drain/overflow, mixing valve, tub filler and 2.5 GPM shower head

Cast iron tub and trim	P1@3.50	Ea	461.00	105.00	566.00

Enameled steel bathtub, 5' L , 30" W, 15" H for recessed installations. With full wall flange, pop-up drain/overflow, chrome mixing valve, tub filler, shower flange and 2.5 GPM shower head

Enameled steel tub	P1@1.40	Ea	105.00	42.00	147.00
Single control tub and shower faucet set	P1@1.30	Ea	76.00	39.00	115.00

Enameled cast iron bathtub with chrome grab rails, 5' L, 32" W, 16" H. Includes polished chrome pop-up drain/overflow, mixing valve, tub filler, and 2.5 GPM shower head

Cast iron tub with handrail and trim	P1@5.25	Ea	1,030.00	157.00	1,187.00

	Craft@Hrs	Unit	Material	Labor	Total

Recessed whirlpool bathtub with built-in 2 HP 115-volt motor and pump, integral drain. With multidirectional and flow adjustable jets and air control. Acrylic with fiberglass reinforcement. White.

	Craft@Hrs	Unit	Material	Labor	Total
Whirlpool tub and trim	P1@5.50	Ea	735.00	165.00	900.00
Add for electrical connection	—	Ea	—	—	210.00

Indoor whirlpools Acrylic fiberglass drop-in whirlpool bath, includes chrome pop-up drain/overflow (at $139), mixing valve and tub filler (at $139).

	Craft@Hrs	Unit	Material	Labor	Total
Whirlpool acrylic tub, 6 jets, 60" x 36" x 20"	P1@4.0	Ea	1,670.00	120.00	1,790.00
Whirlpool acrylic tub, 8 jets, 72" x 42" x 22"	P1@5.0	Ea	2,090.00	150.00	2,240.00
Corner model, 6 jets, 60" x 60" x 20"	P1@5.0	Ea	1,770.00	150.00	1,920.00
Add for polished brass Roman tub filler	P1@1.5	Ea	263.00	45.00	308.00
Add for chrome Roman tub hand shower	P1@1.5	Ea	168.00	45.00	213.00
Add for electrical connection	—	Ea	—	—	210.00

Shower stalls

Acrylic fiberglass shower stall, 32" W, 32" D, 72" H, includes mixing valve, 2.5 GPM shower head and door

	Craft@Hrs	Unit	Material	Labor	Total
Fiberglass 32" shower stall and trim	P1@4.15	Ea	575.00	124.00	699.00

Acrylic fiberglass shower stall, 48" W, 35" D, 72" H with integral soap dishes, grab bar and drain (at $265), mixing valve, 2.5 GPM shower head (at $139) and door (at $300)

	Craft@Hrs	Unit	Material	Labor	Total
Fiberglass 48" shower stall and trim	P1@4.65	Ea	689.00	139.00	828.00

Acrylic fiberglass shower stall, 60" W, 35" D, 72" H with integral soap dishes, grab bar, drain and seat, mixing valve, 2.5 GPM shower head and door

	Craft@Hrs	Unit	Material	Labor	Total
Fiberglass 60" shower stall and trim	P1@4.80	Ea	844.00	144.00	988.00
Roll-in wheelchair stall and trim	P1@5.40	Ea	1,440.00	162.00	1,602.00
Add for shower cap (roof)	P1@.500	Ea	145.00	15.00	160.00

Handicapped tub and shower

Handicapped and elderly access fiberglass tub and shower, 60" x 30" H with integral soap dishes, seat and four grab bars, body washer shower, polished chrome pop-up drain/overflow, mixing valve and tub filler

	Craft@Hrs	Unit	Material	Labor	Total
Handicap shower-tub and trim	P1@4.75	Ea	3800.00	142.00	3,942.00
Add for shower curtain rod	P1@.572	Ea	29.80	17.10	46.90

Water closets

Floor-mounted, residential tank type 1.6 gal. vitreous china toilet, valved closet supply, and toilet seat with cover

	Craft@Hrs	Unit	Material	Labor	Total
Floor-mount tank type water closet and trim	P1@2.10	Ea	166.00	63.00	229.00

Floor-mounted, back outlet, residential tank type 1.6 gal. vitreous china toilet, valved closet supply, elongated bowl, and toilet seat with cover

	Craft@Hrs	Unit	Material	Labor	Total
Floor-mount, back outlet tank type WC and trim	P1@2.35	Ea	266.00	70.50	336.50

Wall-mounted, tank type 1.6 gal. vitreous china toilet, toilet carrier, valved closet supply, elongated bowl, and toilet seat with cover

	Craft@Hrs	Unit	Material	Labor	Total
Wall-mount, tank type WC, carrier and trim	P1@3.65	Ea	413.00	109.00	522.00

One piece, 1.6 gal. floor-mounted vitreous china toilet, including toilet seat and valved closet supply

	Craft@Hrs	Unit	Material	Labor	Total
One-piece water closet and trim	P1@2.35	Ea	244.00	70.50	314.50

Floor-mounted vitreous china 1.6 gal. pressure assist toilet, including toilet seat and valved closet supply

	Craft@Hrs	Unit	Material	Labor	Total
Pressure assist water closet and trim	P1@2.25	Ea	323.00	67.50	390.50
Add for elongated bowl models	—	Ea	60.00	—	60.00

Bidets

	Craft@Hrs	Unit	Material	Labor	Total

Bidets Standard vitreous china bidet with floor-mounted hardware, hot and cold valved supplies and polished chrome faucet

	Craft@Hrs	Unit	Material	Labor	Total
Vitreous china bidet, vertical spray, and trim	P1@2.50	Ea	440.00	75.00	515.00
Vitreous china bidet, over the rim, and trim	P1@2.50	Ea	398.00	75.00	473.00

Lavatories

Oval self-rimming enameled steel lavatory, hot and cold valved supplies, single lever faucet and pop-up drain fitting and trap assembly

Oval self-rim steel lavatory and trim	P1@2.50	Ea	222.00	75.00	297.00

Oval, self-rimming vitreous china lavatory, hot and cold valved supplies, single lever faucet and pop-up drain fitting and trap assembly

Oval self-rim china lavatory and trim	P1@2.00	Ea	265.00	60.00	325.00

Oval, pedestal-mounted vitreous china lavatory, hot and cold valved supplies, single lever faucet & pop-up drain fitting and trap

Oval pedestal-mount china lavatory and trim	P1@2.50	Ea	303.00	75.00	378.00

Rectangular, pedestal-mount vitreous china lavatory, hot and cold valved supplies, single lever faucet and pop-up drain fitting and trap

Rectangular pedestal china lavatory and trim	P1@2.50	Ea	352.00	75.00	427.00

Rectangular, self-rimming vitreous china lavatory, hot and cold valved supplies, single lever faucet & pop-up drain fitting and trap assembly

Counter type rectangular china lavatory & trim	P1@2.00	Ea	309.00	60.00	369.00

Rectangular, wall-hung vitreous china lavatory, hot and cold valved supplies, single lever faucet & pop-up drain fitting and trap assembly

Wall-hung china lavatory and trim	P1@2.50	Ea	288.00	75.00	363.00

Wheelchair access lavatory, wall-hung vitreous china, with carrier insulated hot and cold valved supplies, gooseneck faucet with wrist handles and pop-up drain and insulated offset trap

Wall-hung wheelchair lavatory and trim	P1@2.75	Ea	509.00	82.40	591.40

Bar sinks

Acrylic bar sink, 15" L, 15" W, 6" D, self-rimming, with chrome faucet, hot and cold water valved supplies and P trap assembly

Acrylic bar sink and trim	P1@2.00	Ea	239.00	60.00	299.00

Enameled cast iron bar sink, 16" L, 19" W, 6" D with chrome faucet, hot and cold water valved supplies and P trap assembly

Enameled cast iron bar sink and trim	P1@2.25	Ea	362.00	67.50	429.50

Stainless steel bar sink, 15" L, 15" W, 6" D , chrome faucet, (at $70)hot and cold valved supplies, and trap assembly

Stainless steel bar sink and trim	P1@2.00	Ea	298.00	60.00	358.00

Kitchen sinks

Single bowl, enameled cast iron, self-rimming, 25" L, 22" W, 8" D with chrome faucet and sprayer, P trap assembly and valved supplies

Single bowl cast iron sink and trim	P1@1.75	Ea	252.00	52.50	304.50

Single bowl, stainless steel, self-rimming sink, 20" L, 20" W, 8" D with chrome faucet and sprayer, strainer, P-trap assembly and valved supplies

Single bowl stainless sink and trim	P1@2.00	Ea	217.00	60.00	277.00

Double bowl, enameled cast iron, self-rimming sink, 33" L, 22" W, 8" D, chrome faucet and sprayer, two strainers, P-trap assembly and valved supplies

Double bowl cast iron sink and trim	P1@2.00	Ea	337.00	60.00	397.00

Double bowl, stainless steel, self-rimming sink, 32" L, 22" W, 8" D with chrome faucet and sprayer, two strainers, P-trap assembly and valved supplies

Double bowl stainless sink and trim	P1@2.25	Ea	258.00	67.50	325.50

	Craft@Hrs	Unit	Material	Labor	Total

Triple bowl, stainless steel, self-rimming sink, 43" L, 22" W with chrome faucet and sprayer, three strainers, P-trap assembly and valved supplies

	Craft@Hrs	Unit	Material	Labor	Total
Triple bowl stainless sink and trim	P1@2.75	Ea	488.00	82.40	570.40

Laundry sinks

Enameled cast iron laundry sink, 24" L, 20" D, 14" D, wall-mounted faucet with integral stops and vacuum breaker, strainer and P-trap assembly

Cast iron laundry sink and trim	P1@2.76	Ea	372.00	82.70	454.70

Acrylic laundry sink, 24" L, 20" D, 14" D, deck-mounted faucet valved supplies, and P-trap assembly

Acrylic laundry sink and trim	P1@2.00	Ea	114.00	60.00	174.00

Service sinks

Enameled cast iron service sink, 22" L, 18" D, wall mounted faucet with integral stops and vacuum breaker, strainer and cast iron P-trap

Cast iron service sink and trim	P1@3.00	Ea	746.00	89.90	835.90

Drinking fountains and electric water coolers

Wall-hung vitreous china fountain, 14" W, 13" H, complete with fittings and 1-1/4" trap

Wall hung vitreous fountain and trim	P1@2.00	Ea	355.00	60.00	415.00

Semi-recessed vitreous china, 15" W, 27" H, with fittings, wall cleanout and 1-1/4" trap

Semi-recessed vitreous fountain and trim	P1@2.50	Ea	625.00	75.00	700.00

Wall-hung electric water cooler, 13 gallon capacity, complete with trap and fittings (requires electrical outlet for plug-in connection, not included)

Electric water cooler and trim	P1@2.75	Ea	825.00	82.40	907.40

Wall-hung wheelchair access electric water cooler, front push bar, 7.5 gallon capacity, complete with trap and fittings (requires electrical outlet for plug-in connection, not included)

Wheelchair access water cooler and trim	P1@2.50	Ea	651.00	75.00	726.00

Fully recessed electric water cooler, 11.5 gallon capacity, complete with trap and fittings (requires electrical outlet for plug-in connection, not included)

Recessed electric water cooler and trim	P1@2.95	Ea	1,830.00	88.40	1,918.40

Garbage disposers

Small, 1/3 HP kitchen disposer, including fittings, no waste pipe

1/3 HP kitchen disposer	P1@2.00	Ea	84.80	60.00	144.80

Standard, 5/8 HP kitchen disposer, including fittings, no waste pipe

5/8 HP kitchen disposer	P1@2.00	Ea	121.00	60.00	181.00

Standard, 3/4 HP kitchen disposer, including fittings, no waste pipe

3/4 HP kitchen disposer	P1@2.00	Ea	165.00	60.00	225.00

Septic system, 3/4 HP kitchen disposer, including fittings, no waste pipe

3/4 HP kitchen disposer	P1@2.00	Ea	222.00	60.00	282.00

Large capacity, 1 HP kitchen disposer, including fittings, no waste pipe

1 HP kitchen disposer	P1@2.15	Ea	284.00	64.50	348.50
Add for electrical connection and switch	BE@1.00	Ea	26.80	33.90	60.70

Dishwashers Under counter installation, connected to existing drain.

Deluxe	P1@2.40	Ea	950.00	72.00	1,022.00
Standard	P1@2.40	Ea	650.00	72.00	722.00
Economy	P1@2.40	Ea	355.00	72.00	427.00
Add for electrical connection	BE@1.00	Ea	24.70	33.90	58.60

Water Heaters

	Craft@Hrs	Unit	Material	Labor	Total

Gas water heaters Includes labor to connect to existing water supply piping and gas line. 6 year warranty. Labor column includes cost of installing heater, making the gas and water connections but not the cost of those connections.

	Craft@Hrs	Unit	Material	Labor	Total
30 gallon tall, 53 GPH	P1@1.90	Ea	285.00	57.00	342.00
40 gallon short, 55 GPH	P1@2.00	Ea	325.00	60.00	385.00
40 gallon tall, 62 GPH	P1@2.00	Ea	350.00	60.00	410.00
50 gallon tall, 72 GPH	P1@2.15	Ea	400.00	64.50	464.50
75 gallon tall, 132 GPH	P1@3.50	Ea	650.00	105.00	755.00

Propane water heaters Includes labor to connect to existing water supply piping. 6 year warranty. Labor column includes cost of installing heater, making the water connections but not the cost of those connections.

	Craft@Hrs	Unit	Material	Labor	Total
40 gallon, 34 BTU, 62 GPH	P1@2.00	Ea	400.00	60.00	460.00

Electric water heaters Includes labor to connect to existing water supply piping. Add for the cost of these connection materials.

	Craft@Hrs	Unit	Material	Labor	Total
40 gallon 53 GPH 12 yr.	P1@2.15	Ea	290.00	64.50	354.50
50 gallon 63 GPH 12 yr.	P1@2.25	Ea	320.00	67.50	387.50
80 gallon 87 GPH 12 yr	P1@3.50	Ea	559.00	105.00	664.00
40 gallon 49 GPH 9 yr.	P1@2.00	Ea	282.00	60.00	342.00
50 gallon 55 GPH 9 yr.	P1@2.15	Ea	293.00	64.50	357.50
80 gallon tall, 80 GPH 9 yr.	P1@3.50	Ea	357.00	105.00	462.00
30 gallon 38 GPH 6 yr.	P1@1.90	Ea	199.00	57.00	256.00
40 gallon short, 42 GPH 6 yr.	P1@2.00	Ea	227.00	60.00	287.00

Solar water heating system Water storage tanks compatible with most closed-loop solar water heating systems, heat transfer by insulated all-copper heat exchanger, double wall construction and back-up electric heating element. Rheem

	Craft@Hrs	Unit	Material	Labor	Total
80 gallon, 59" high x 25" deep	P1@8.44	Ea	930.00	253.00	1,183.00

Water heaters, installed on a subcontract basis
Passive thermosiphon domestic hot water system with integral 80 gallon
water tank, retrofitted to an existing single story structure, with 5 year limited warranty

		Unit	Material	Labor	Total
Passive solar water heater system (retrofit)	—	Ea	—	—	3,610.00

Active solar hot water system for 100 or 120 gallon water supply tank, fitted with automatic drain retrofitted to existing single story, single family unit, 5 year warranty

		Unit	Material	Labor	Total
Active solar water heater system (retrofit)	—	Ea	—	—	4,130.00

Add for strengthening existing roof structure under solar collectors

		Unit	Material	Labor	Total
Per SF of roof strengthened	—	SF	—	—	1.18

Active solar hot water system for new one-story home

		Unit	Material	Labor	Total
Including design and engineering (new work)	—	Ea	—	—	3,550.00

For designs requiring special cosmetic treatment

		Unit	Material	Labor	Total
Add for concealment of pipe, collectors, etc.	—	%	—	—	15.50

Solar pool heater, active heater for pools with freeze tolerant collector, roof mounted

		Unit	Material	Labor	Total
Per SF of collector area	—	SF	—	—	5.77

Water softeners Installed on a subcontract basis

		Unit	Material	Labor	Total
9 gallons per minute, automatic	—	Ea	—	—	1,080.00
12 gallons per minute, automatic	—	Ea	—	—	1,130.00

Plumbing System Components

Vent through roof

Fixture vent

Fixture vent

Fixture vent

Hose bib

Horizontal branch supply

Soil stack

Second story supply extension

Individual fixture runouts

Remote fixture branch

Horizontal branch soil

Horizontal branch waste

Main house drain

Main hot water supply

Main cold water supply

To building sewer

Plumbing system components for a two story house

Whole House Plumbing Rough-In

Rough plumbing is the drain, waste and vent (DWV) piping and water supply piping that isn't visible when construction is complete. Rough plumbing costs vary with the number of fixtures and length of the pipe runs. The figures in this section show rough plumbing costs for a typical single family home using plastic drainage and copper water supply materials. No plumbing fixtures, equipment or their final connections are included in these costs. See Plumbing Fixtures and Equipment for fixture and fixture connection costs. These costs include the subcontractor's markup. Add the general contractor's markup when applicable. For more complete information on plumbing and piping costs, see *National Plumbing & HVAC Estimator* at http://CraftsmanSiteLicense.com/

	Craft@Hrs	Unit	Material	Labor	Total
Single story homes					
Hot and cold potable water rough-in	P1@15.0	Ea	492.00	450.00	942.00
Drainage and venting rough-in	P1@20.0	Ea	472.00	600.00	1,072.00
Single story, total plumbing rough-in	—@35.0	Ea	964.00	1,050.00	2,014.00
Two story homes					
Hot and cold potable water rough-in	P1@20.0	Ea	725.00	600.00	1,325.00
Drainage and venting rough-in	P1@25.0	Ea	834.00	750.00	1,584.00
Two story, total plumbing rough-in	—@45.0	Ea	1,559.00	1,350.00	2,909.00

Plumbing Fixture Rough-In Assemblies Includes labor and material costs for the installation of a typical water supply, drain and vent piping system on new residential construction projects. For renovation projects, add 20% to labor and 10% to material costs to allow for connection to existing service lines and working within existing spaces. Add the cost of any demolition or removal required before the new rough-in can be installed. No plumbing fixtures or fixture connections are included in these costs. See Plumbing Fixtures and Equipment on the following pages for fixture and fixture connection costs. These costs include the subcontractor's markup. Add the general contractor's mark-up.

3-piece washroom group rough-in assembly Includes the water supply, drain and vent piping to connect all three fixtures (tub, toilet and sink) in the group to one common waste, one vent, one hot and one cold supply line that will be connected to each respective main service drain, vent or supply pipe. These figures are based on the main service piping being within 5' of the fixture group. Add the cost of main water supply, drain and vent lines to which the fixture group will connect.

Hot and cold water supply: 26' of 1/2" type M hard copper pipe, (straight lengths) including associated soldered wrought copper fittings, solder, flux, and pipe clips.

Drains and vents: 5' of 3" drainage pipe on the water closet, 38' of 1-1/2" pipe serving all other drains and vents. 43' total plastic drain and vent pipe, with associated fittings, glue and strapping. Make additional allowances for final fixture connections. (See Plumbing Fixtures and Equipment for fixture and connection costs)

	Craft@Hrs	Unit	Material	Labor	Total
Hot and cold water supply	P1@2.50	Ea	34.00	75.00	109.00
Drains and vents	P1@3.35	Ea	66.70	100.00	166.70
Total washroom group rough-in assembly	—@5.85	Ea	100.70	175.00	275.70
Add for shower head rough-in	P1@.350	Ea	5.40	10.50	15.90

Kitchen sink rough-in assembly

Hot and cold water supply: 10' of 1/2" type M hard copper pipe, (straight lengths) including associated soldered wrought copper fittings, solder, flux, and pipe clips. Drains and vents: 15' of 1-1/2" plastic drain and vent pipe with associated fittings, glue and strapping. Make additional allowance for final fixture connection. (See Plumbing Fixtures and Equipment for fixture and connection costs)

	Craft@Hrs	Unit	Material	Labor	Total
Hot and cold water supply	P1@.750	Ea	12.60	22.50	35.10
Drains and vents	P1@1.00	Ea	20.50	30.00	50.50
Total kitchen sink rough-in assembly	—@1.75	Ea	33.10	52.50	85.60

Laundry tub and washing machine rough-in assembly

Hot and cold water supply: 16' of 1/2" type M hard copper pipe, (straight lengths) including associated fittings, solder, flux, and pipe clips. Drains and vents: 21' of 1-1/2" plastic drain and vent pipe with associated fittings, glue and strapping.

	Craft@Hrs	Unit	Material	Labor	Total
Hot and cold water supply	P1@1.00	Ea	20.10	30.00	50.10
Drains and vents	P1@1.35	Ea	28.60	40.50	69.10
Total tub & washer rough-in assembly	—@2.35	Ea	48.70	70.50	119.20

Hot water tank rough-in assembly

Hot and cold water supply: 10' of 3/4" type M hard copper pipe, (straight lengths) including associated soldered wrought copper fittings, solder, flux, and pipe clips

	Craft@Hrs	Unit	Material	Labor	Total
Hot and cold water supply	P1@.500	Ea	26.20	15.00	41.20
Add for 3/4" isolation valve	P1@.250	Ea	6.64	7.50	14.14

	Craft@Hrs	Unit	Material	Labor	Total

Water softener rough-in assembly

Hot and cold water supply: 10' of 3/4" type M hard copper pipe, (straight lengths) including associated soldered wrought copper fittings, solder, flux, and pipe clips

	Craft@Hrs	Unit	Material	Labor	Total
Cold water supply and return	P1@.500	Ea	26.20	15.00	41.20
Regeneration drain	P1@.500	Ea	13.10	15.00	28.10

For fixture rough-in assemblies not listed here, see plumbing assemblies, Mechanical Section 15

Natural gas appliance rough-in assembly

Includes labor and material to install and test 12' of 3/4" schedule 40 black steel piping, associated 150 lb malleable iron fittings and a plug valve to serve a typical residential gas fired appliance. Add for the natural gas main supply line running through the building, meter connection and final connection to the appliance. Assembly assumes the appliance is within 5' of the main supply line.

	Craft@Hrs	Unit	Material	Labor	Total
Gas fired domestic hot water heater	P1@.650	Ea	23.50	19.50	43.00
Gas fired hot water heating boiler	P1@.650	Ea	23.50	19.50	43.00
Gas fired forced air furnace	P1@.650	Ea	23.50	19.50	43.00
Gas fired stove and oven	P1@.650	Ea	23.50	19.50	43.00
Gas fired clothes dryer	P1@.650	Ea	23.50	19.50	43.00
Gas main, 1" dia. with fittings and hangers	P1@.120	LF	3.54	3.60	7.14
Gas meter connection (house side only)	P1@.750	Ea	19.00	22.50	41.50

Natural gas appliance connections

Includes labor and material to connect and test a typical residential natural gas fired appliance. Make additional allowances for the natural gas main supply line running through the building, the appliance gas supply rough-in and labor to set the appliance in place.

	Craft@Hrs	Unit	Material	Labor	Total
Gas fired domestic hot water heater	P1@.650	Ea	12.60	19.50	32.10
Gas fired hot water heating boiler	P1@.650	Ea	12.60	19.50	32.10
Gas fired forced air furnace	P1@.650	Ea	12.60	19.50	32.10
Gas fired grill, stove or oven	P1@.650	Ea	31.70	19.50	51.20
Gas fired clothes dryer	P1@.650	Ea	31.70	19.50	51.20

Hot Water Heating Systems
Whole house hot water heating system assembly

Includes 150' of 1-1/2" insulated supply and return mains, 150' of 3/4" insulated branch piping, 10 radiant baseboard convectors (50'), a 110 mbh high efficiency gas fired hot water boiler, a circulation pump, electric controls, installed and tested. (Based on 1,200 SF of heated space)

	Craft@Hrs	Unit	Material	Labor	Total
Distribution system	P1@46.0	Ea	672.00	1,380.00	2,052.00
Distribution system (second floor)	P1@16.0	Ea	290.00	480.00	770.00
Boiler, pump and controls	P1@16.0	Ea	2,720.00	480.00	3,200.00
Radiant baseboard and valves	P1@16.0	Ea	1,960.00	480.00	2,440.00
Radiant baseboard and valves (2nd floor)	P1@12.0	Ea	1,300.00	360.00	1,660.00
Total hot water heating system, bungalow	—@78.0	Ea	5,352.00	2,260.00	7,612.00
Total hot water heating system, 2 story	—@106	Ea	6,942.00	3,100.00	10,042.00

Gate Valves

1/2" bronze gate valves

	Craft@Hrs	Unit	Material	Labor	Total
125 lb, threaded	P1@.210	Ea	8.36	6.30	14.66
125 lb, soldered	P1@.200	Ea	8.01	6.00	14.01
150 lb, threaded	P1@.210	Ea	14.40	6.30	20.70
200 lb, threaded	P1@.210	Ea	18.60	6.30	24.90

Plumbing and Heating

	Craft@Hrs	Unit	Material	Labor	Total
3/4" bronze gate valves					
125 lb, threaded	P1@.250	Ea	9.94	7.50	17.44
125 lb, soldered	P1@.240	Ea	9.54	7.20	16.74
150 lb, threaded	P1@.250	Ea	19.50	7.50	27.00
200 lb, threaded	P1@.250	Ea	24.20	7.50	31.70
1" bronze gate valves					
125 lb, threaded	P1@.300	Ea	14.30	8.99	23.29
125 lb, soldered	P1@.290	Ea	13.90	8.69	22.59
150 lb, threaded	P1@.300	Ea	23.00	8.99	31.99
200 lb, threaded	P1@.300	Ea	29.30	8.99	38.29
1-1/2" bronze gate valves					
125 lb, threaded	P1@.450	Ea	23.50	13.50	37.00
125 lb, soldered	P1@.440	Ea	22.70	13.20	35.90
150 lb, threaded	P1@.450	Ea	41.50	13.50	55.00
200 lb, threaded	P1@.450	Ea	50.30	13.50	63.80
2" bronze gate valves					
125 lb, threaded	P1@.500	Ea	39.40	15.00	54.40
125 lb, soldered	P1@.480	Ea	38.00	14.40	52.40
150 lb, threaded	P1@.500	Ea	65.00	15.00	80.00
200 lb, threaded	P1@.500	Ea	73.90	15.00	88.90
Forged steel Class 800 bolted bonnet OS&Y gate valves, threaded ends					
1/2" valve	P1@.210	Ea	70.70	6.30	77.00
3/4" valve	P1@.250	Ea	96.50	7.50	104.00
1" valve	P1@.300	Ea	174.00	8.99	182.99
1-1/2" valve	P1@.450	Ea	219.00	13.50	232.50
2" valve	P1@.500	Ea	335.00	15.00	350.00

Globe Valves 125 lb and 150 lb ratings are saturated steam pressure. 200 lb rating is working steam pressure, union bonnet. 125 lb valves have bronze disks. 150 lb valves have composition disks. 300 lb valves have regrinding disks.

	Craft@Hrs	Unit	Material	Labor	Total
1/2" threaded bronze globe valves					
125 lb	P1@.210	Ea	20.70	6.30	27.00
150 lb	P1@.210	Ea	36.50	6.30	42.80
300 lb	P1@.210	Ea	55.00	6.30	61.30
3/4" threaded bronze globe valves					
125 lb	P1@.250	Ea	27.80	7.50	35.30
150 lb	P1@.250	Ea	49.70	7.50	57.20
300 lb	P1@.250	Ea	70.10	7.50	77.60
1" threaded bronze globe valves					
125 lb	P1@.300	Ea	39.90	8.99	48.89
150 lb	P1@.300	Ea	77.10	8.99	86.09
300 lb	P1@.300	Ea	99.90	8.99	108.89
1-1/2" threaded bronze globe valves					
125 lb	P1@.450	Ea	75.10	13.50	88.60
150 lb	P1@.450	Ea	146.00	13.50	159.50
300 lb	P1@.450	Ea	193.00	13.50	206.50
2" threaded bronze globe valves					
125 lb	P1@.500	Ea	122.00	15.00	137.00
150 lb	P1@.500	Ea	221.00	15.00	236.00
300 lb	P1@.500	Ea	268.00	15.00	283.00

	Craft@Hrs	Unit	Material	Labor	Total
Swing Check Valves					
1/2" bronze check valves					
125 lb, threaded, swing	P1@.210	Ea	10.70	6.30	17.00
125 lb, soldered, swing	P1@.200	Ea	10.40	6.00	16.40
125 lb, threaded, vertical lift check	P1@.210	Ea	53.40	6.30	59.70
150 lb, threaded, swing, bronze disk	P1@.210	Ea	35.90	6.30	42.20
200 lb, threaded, swing, bronze disk	P1@.210	Ea	70.60	6.30	76.90
3/4" bronze check valves					
125 lb, threaded, swing	P1@.250	Ea	15.40	7.50	22.90
125 lb, soldered, swing	P1@.240	Ea	14.80	7.20	22.00
125 lb, threaded, swing, vertical check	P1@.250	Ea	72.80	7.50	80.30
150 lb, threaded, swing, bronze disk	P1@.250	Ea	44.80	7.50	52.30
200 lb, threaded, swing, bronze disk	P1@.250	Ea	83.60	7.50	91.10
1" bronze check valves					
125 lb, threaded, swing	P1@.300	Ea	20.40	8.99	29.39
125 lb, soldered, swing	P1@.290	Ea	19.60	8.69	28.29
125 lb, threaded, swing, vertical check	P1@.300	Ea	91.60	8.99	100.59
150 lb, threaded, swing, bronze disk	P1@.300	Ea	60.60	8.99	69.59
200 lb, threaded, swing, bronze disk	P1@.300	Ea	128.00	8.99	136.99
1-1/2" bronze check valves					
125 lb, threaded, swing	P1@.450	Ea	40.10	13.50	53.60
125 lb, soldered, swing	P1@.440	Ea	39.20	13.20	52.40
125 lb, threaded, swing, vertical check	P1@.450	Ea	154.00	13.50	167.50
150 lb, threaded, swing, swing, bronze disk	P1@.450	Ea	104.00	13.50	117.50
200 lb, threaded, swing, bronze disk	P1@.450	Ea	209.00	13.50	222.50
2" bronze check valves					
125 lb, threaded, swing	P1@.500	Ea	67.20	15.00	82.20
125 lb, soldered, swing	P1@.480	Ea	65.30	14.40	79.70
125 lb, threaded, vertical check	P1@.500	Ea	229.00	15.00	244.00
150 lb, threaded, swing, bronze disk	P1@.500	Ea	153.00	15.00	168.00
200 lb, threaded, swing, bronze disk	P1@.500	Ea	294.00	15.00	309.00
Miscellaneous Valves and Regulators					
Angle valves, bronze, 150 lb, threaded, Teflon disk					
1/2" valve	P1@.210	Ea	47.60	6.30	53.90
3/4" valve	P1@.250	Ea	64.70	7.50	72.20
1" valve	P1@.300	Ea	93.50	8.99	102.49
1-1/4" valve	P1@.400	Ea	122.00	12.00	134.00
1-1/2" valve	P1@.450	Ea	197.00	13.50	210.50
Backflow preventers, reduced pressure					
3/4", threaded, bronze, with 2 gate valves	P1@1.00	Ea	200.00	30.00	230.00
1", threaded, bronze, with 2 gate valves	P1@1.25	Ea	243.00	37.50	280.50
2", threaded, with 2 gate valves	P1@2.13	Ea	358.00	63.90	421.90
Ball valves, bronze, 150 lb threaded, Teflon seat					
1/2" valve	P1@.210	Ea	6.11	6.30	12.41
3/4" valve	P1@.250	Ea	9.04	7.50	16.54
1" valve	P1@.300	Ea	12.80	8.99	21.79
1-1/4" valve	P1@.400	Ea	22.20	12.00	34.20
1-1/2" valve	P1@.450	Ea	27.10	13.50	40.60
2" valve	P1@.500	Ea	36.40	15.00	51.40

Plumbing Specialties

	Craft@Hrs	Unit	Material	Labor	Total
Lavatory supply valve with hand wheel shutoff, flexible compression connection tube with length as shown					
1/2" angle or straight stops					
12" tube	P1@.294	Ea	17.50	8.81	26.31
15" tube	P1@.294	Ea	18.70	8.81	27.51
20" tube	P1@.473	Ea	21.50	14.20	35.70
Gas stops, lever handle					
1/2" stop	P1@.210	Ea	8.43	6.30	14.73
3/4" stop	P1@.250	Ea	11.30	7.50	18.80
1" stop	P1@.300	Ea	17.40	8.99	26.39
1-1/4" stop	P1@.450	Ea	26.00	13.50	39.50
1-1/2" stop	P1@.500	Ea	45.20	15.00	60.20
2" stop	P1@.750	Ea	69.80	22.50	92.30
Hose bibbs, threaded					
3/4", brass	P1@.350	Ea	4.27	10.50	14.77
Pressure regulator valves, bronze, 300 lb threaded, 25 to 75 PSI, with "Y" strainer					
3/4" valve	P1@.250	Ea	108.00	7.50	115.50
1" valve	P1@.300	Ea	143.00	8.99	151.99
1-1/4" valve	P1@.400	Ea	223.00	12.00	235.00
1-1/2" valve	P1@.450	Ea	342.00	13.50	355.50
2" valve	P1@.500	Ea	432.00	15.00	447.00

Meters, Gauges and Indicators

	Craft@Hrs	Unit	Material	Labor	Total
Water meters					
Disc type, 1-1/2"	P1@1.00	Ea	150.00	30.00	180.00
Turbine type, 1-1/2"	P1@1.00	Ea	225.00	30.00	255.00
Displacement type, AWWA C7000, 1"	P1@.941	Ea	230.00	28.20	258.20
Displacement type, AWWA C7000, 1-1/2"	P1@1.23	Ea	794.00	36.90	830.90
Displacement type, AWWA C7000, 2"	P1@1.57	Ea	1,180.00	47.10	1,227.10
Curb stop and waste valve, 1"	P1@1.00	Ea	83.10	30.00	113.10
Thermoflow indicators, soldered					
1-1/4" indicator	P1@2.01	Ea	347.00	60.30	407.30
2" indicator	P1@2.01	Ea	396.00	60.30	456.30

Plumbing Specialties

	Craft@Hrs	Unit	Material	Labor	Total
Access doors for plumbing, painted steel, with cam lock, for masonry, drywall, tile, etc. Milcor					
8" x 8"	SW@.400	Ea	32.60	13.60	46.20
12" x 12"	SW@.500	Ea	37.80	17.00	54.80
18" x 18"	SW@.800	Ea	53.60	27.20	80.80
24" x 24"	SW@1.25	Ea	75.60	42.50	118.10
36" x 36"	SW@1.60	Ea	152.00	54.40	206.40
Add for fire rating	—	%	400.0	—	—
Add for stainless steel	—	%	300.0	—	—
Add for cylinder lock	—	Ea	15.00	—	15.00

Hangers and Supports

	Craft@Hrs	Unit	Material	Labor	Total
Angle bracket hangers, steel, by rod size					
3/8" angle bracket hanger	P1@.070	Ea	1.69	2.10	3.79
1/2" angle bracket hanger	P1@.070	Ea	1.89	2.10	3.99
5/8" angle bracket hanger	P1@.070	Ea	4.16	2.10	6.26
3/4" angle bracket hanger	P1@.070	Ea	6.39	2.10	8.49
7/8" angle bracket hanger	P1@.070	Ea	11.60	2.10	13.70

	Craft@Hrs	Unit	Material	Labor	Total
Top beam clamps, steel, by rod size					
3/8" top beam clamp	P1@.070	Ea	2.19	2.10	4.29
1/2" top beam clamp	P1@.070	Ea	3.56	2.10	5.66
5/8" top beam clamp	P1@.070	Ea	4.37	2.10	6.47
3/4" top beam clamp	P1@.070	Ea	6.38	2.10	8.48
7/8" top beam clamp	P1@.070	Ea	8.38	2.10	10.48
Clevis hangers, standard, by intended pipe size, no hanger rod or angle bracket included					
1/2" clevis hanger	P1@.156	Ea	1.72	4.68	6.40
3/4" clevis hanger	P1@.156	Ea	1.72	4.68	6.40
1" clevis hanger	P1@.156	Ea	1.82	4.68	6.50
1-1/4" clevis hanger	P1@.173	Ea	2.10	5.19	7.29
1-1/2" clevis hanger	P1@.173	Ea	2.28	5.19	7.47
2" clevis hanger	P1@.173	Ea	2.55	5.19	7.74
Copper tube riser clamps, by intended tube size					
1/2" riser clamp	P1@.100	Ea	5.61	3.00	8.61
3/4" riser clamp	P1@.100	Ea	5.87	3.00	8.87
1" riser clamp	P1@.100	Ea	7.55	3.00	10.55
1-1/4" riser clamp	P1@.105	Ea	7.84	3.15	10.99
1-1/2" riser clamp	P1@.110	Ea	9.51	3.30	12.81
2" riser clamp	P1@.115	Ea	10.10	3.45	13.55
Pipe clamps, medium duty, by intended pipe size					
1/2" or 3/4" pipe clamp	P1@.156	Ea	3.77	4.68	8.45
1" pipe clamp	P1@.156	Ea	3.97	4.68	8.65
1-1/4" or 1-1/2" pipe clamp	P1@.173	Ea	4.94	5.19	10.13
2" pipe clamp	P1@.173	Ea	6.10	5.19	11.29
Pipe straps, galvanized, two hole, lightweight					
1/2" pipe strap	P1@.072	Ea	.07	2.16	2.23
3/4" pipe strap	P1@.080	Ea	.13	2.40	2.53
1" pipe strap	P1@.088	Ea	.15	2.64	2.79
1-1/4" pipe strap	P1@.096	Ea	.23	2.88	3.11
1-1/2" pipe strap	P1@.104	Ea	.30	3.12	3.42
2" pipe strap	P1@.112	Ea	.36	3.36	3.72
Plumber's tape, galvanized					
26 gauge 3/4" x 10' roll	—	Ea	1.19	—	1.19
22 gauge 3/4" x 10' roll	—	Ea	1.60	—	1.60
Threaded hanger rod, by rod size, in 6' to 10' lengths, per LF					
3/8" rod	—	LF	1.20	—	1.20
1/2" rod	—	LF	1.85	—	1.85
5/8" rod	—	LF	2.60	—	2.60
3/4" rod	—	LF	4.45	—	4.45
U bolts, with hex nuts, light duty					
1/2" bolt	P1@.089	Ea	2.19	2.67	4.86
1" bolt	P1@.089	Ea	2.44	2.67	5.11
2" bolt	P1@.089	Ea	3.36	2.67	6.03

Closed Cell Elastomeric Pipe and Tubing Insulation Semi split, 1/2" wall thickness, no cover, by nominal pipe or tube diameter. R factor equals 3.58 at 220 degrees F. Manufactured by Rubatex. These costs do not include scaffolding. Also see fittings, flanges and valves at the end of this section.

	Craft@Hrs	Unit	Material	Labor	Total
1/4" pipe	P1@.039	LF	.34	1.17	1.51
3/8" pipe	P1@.039	LF	.36	1.17	1.53
1/2 " pipe	P1@.039	LF	.40	1.17	1.57
3/4" pipe	P1@.039	LF	.46	1.17	1.63

Plumbing Specialties

	Craft@Hrs	Unit	Material	Labor	Total
1" pipe	P1@.042	LF	.51	1.26	1.77
1-1/4" pipe	P1@.042	LF	.59	1.26	1.85
1-1/2" pipe	P1@.042	LF	.68	1.26	1.94
2" pipe	P1@.046	LF	.85	1.38	2.23

Pipe fittings and flanges:
   For each fitting or flange use the cost for 3 LF of pipe or tube of the same size.
Valves:
   For each valve body use the cost for 5 LF of pipe or tube of the same size.

Flexible PVC-Jacketed Copper Tubing For fuel oil, gas and propane transfer lines. Flame retardant and dielectric. Fungus, acid and moisture corrosion resistant. Acceptable for direct burial underground or under concrete. Installed in an open trench. By LDC Enviro Corp, UL listed, pat #5,868,191 and #6,135,159. www.hvacweb.com or email envirotube@aol.com

	Craft@Hrs	Unit	Material	Labor	Total
3/8" OD	PM@.010	LF	1.07	.35	1.42
1/2" OD	PM@.010	LF	1.16	.35	1.51
5/8" OD	PM@.010	LF	1.25	.35	1.60

Shrinkage kit for flexible PVC-jacketed copper tubing. Add for shrinkage tubing kit to allow for the coupling of sections with standard flare fittings while maintaining the integrity of the connection.

		Unit	Material	Labor	Total
3/8" kit	—	Ea	9.00	—	9.00
1/2" kit	—	Ea	9.90	—	9.90
5/8" kit	—	Ea	9.95	—	9.95

Fiberglass Pipe Insulation with AP-T Plus (All Purpose Self Sealing) Jacket By nominal pipe diameter. R factor equals 2.56 at 300 degrees F. Add the cost of scaffolding, if required. Also see fittings, flanges and valves at the end of this section. Available in 3 foot lengths, price is per linear foot.

	Craft@Hrs	Unit	Material	Labor	Total
1/2" diameter pipe					
1/2" thick insulation	P1@.038	LF	1.53	1.14	2.67
1" thick insulation	P1@.038	LF	1.82	1.14	2.96
1-1/2" thick insulation	P1@.040	LF	3.64	1.20	4.84
3/4" diameter pipe					
1/2" thick insulation	P1@.038	LF	1.69	1.14	2.83
1" thick insulation	P1@.038	LF	2.08	1.14	3.22
1-1/2" thick insulation	P1@.040	LF	3.84	1.20	5.04
1" diameter pipe					
1/2" thick insulation	P1@.040	LF	1.78	1.20	2.98
1" thick insulation	P1@.040	LF	1.82	1.20	3.02
1-1/2" thick insulation	P1@.042	LF	4.03	1.26	5.29
1-1/4" diameter pipe					
1/2" thick insulation	P1@.040	LF	2.41	1.20	3.61
1" thick insulation	P1@.040	LF	2.41	1.20	3.61
1-1/2" thick insulation	P1@.042	LF	4.33	1.26	5.59
1-1/2" diameter pipe					
1/2" thick insulation	P1@.042	LF	2.12	1.26	3.38
1" thick insulation	P1@.042	LF	2.73	1.26	3.99
1-1/2" thick insulation	P1@.044	LF	4.67	1.32	5.99
2" diameter pipe					
1/2" thick insulation	P1@.044	LF	2.26	1.32	3.58
1" thick insulation	P1@.044	LF	3.11	1.32	4.43
1-1/2" thick insulation	P1@.046	LF	5.01	1.38	6.39

	Craft@Hrs	Unit	Material	Labor	Total

Polyethylene Film, Clear or Black Material costs include 5% for waste and 10% for laps. Labor shown is for installation on grade.

4 mil (.004" thick)					
50 LF rolls, 3' to 20' wide	BL@.002	SF	.05	.05	.10
6 mil (.006" thick)					
50 LF rolls, 3' to 40' wide	BL@.002	SF	.06	.05	.11
Add for installation on walls, ceilings and roofs, tack stapled, typical	BL@.001	SF	—	.02	.02

Railings, Steel, Prefabricated See also Ornamental Iron. Porch or step rail, site fitted.

2'-10" high, 6" OC 1/2" solid twisted pickets, 4' or 6' sections

	Craft@Hrs	Unit	Material	Labor	Total
1" x 1/2" rails	Rl@.171	LF	3.14	5.53	8.67
1" x 1" rails	Rl@.180	LF	4.32	5.82	10.14
Add for newel post 3' H with hardware	—	Ea	11.90	—	12.50
Add lamb's tongue for end post	—	Ea	5.00	—	5.00

Range Hoods Material costs are for ducted or ductless range hood only, no electrical wiring included. Add for ducting or charcoal filters below. Labor costs are for installation of range hood only. No carpentry work included. All models are steel construction with baked enamel finish in various colors.

Vented economy range hood. Duct connector with built-in damper. Washable aluminum filter with light lens and replaceable charcoal filter pad. 190 CFM exhaust. 75 watt light.

	Craft@Hrs	Unit	Material	Labor	Total
30" wide, enamel finish ,	SW@1.42	Ea	76.00	48.30	124.30
30" wide, stainless steel finish	SW@1.42	Ea	98.30	48.30	146.60
36" wide, enamel finish	SW@1.42	Ea	68.00	48.30	116.30
36" wide, stainless steel	SW@1.42	Ea	89.00	48.30	137.30

Ductless economy range hood. Plug-in 2-speed motor. 8-1/4" x 11-1/4" grease-activated charcoal filter. 75 watt light.

	Craft@Hrs	Unit	Material	Labor	Total
24" wide, enamel finish	SW@1.42	Ea	47.30	48.30	95.60
30" wide, enamel finish	SW@1.42	Ea	35.70	48.30	84.00

Convertable range hood. Installs vertically or horizontally. 30" wide. Baked enamel finish. 3-1/4" x 10" duct, 7" round duct, or non-ducted. Aluminum filters for ducted operation. Charcoal replacement filters for non-ducted operation. 75 watt lamp.

	Craft@Hrs	Unit	Material	Labor	Total
160 CFM 3-speed exhaust	SW@1.42	Ea	76.70	48.30	125.00
200 CFM, 4.5 sones rating	SW@1.42	Ea	125.00	48.30	173.30

Quiet range hood. 170 to 220 CFM (depending on ducting). 3-1/4" x 10" duct, 7" round duct, or non-ducted. Dual 40 watt lamps. Aluminum filters for ducted operation. Charcoal replacement filters for non-ducted operation.

	Craft@Hrs	Unit	Material	Labor	Total
30" wide, enamel finish	SW@1.42	Ea	126.00	48.30	174.30
36" wide, enamel finish	SW@1.42	Ea	136.00	48.30	184.30
36" wide, stainless steel finish	SW@1.42	Ea	167.00	48.30	215.30
42" wide, enamel finish	SW@1.42	Ea	140.00	48.30	188.30

Custom range hood power package. Exhaust system components installed in a custom-built range hood 24" to 30" above the cooling surface. Stainless steel. With variable speed control, filters and halogen or incandescent lighting (bulbs not included). Add the cost of a decorative enclosure, ducting and electrical connection.

700 CFM, dual blowers, 7" duct, 1.5 to 4.6 sones

	Craft@Hrs	Unit	Material	Labor	Total
30"wide x 22"deep	SW@1.42	Ea	400.00	48.30	448.30
32"wide x 22"deep	SW@1.42	Ea	440.00	48.30	488.30
36"wide x 28"deep	SW@1.42	Ea	495.00	48.30	543.30
42"wide x 28"deep	SW@1.42	Ea	545.00	48.30	593.30
48"wide x 28"deep	SW@1.42	Ea	620.00	48.30	668.30

Ranges, Built-In

	Craft@Hrs	Unit	Material	Labor	Total
1300 CFM, dual blowers, twin 7" ducts, 1.5 to 4.6 sones					
30"wide x 22"deep	SW@1.42	Ea	650.00	48.30	698.30
32"wide x 22"deep	SW@1.42	Ea	715.00	48.30	763.30
36"wide x 28"deep	SW@1.42	Ea	805.00	48.30	853.30
42"wide x 28"deep	SW@1.42	Ea	885.00	48.30	933.30
48"wide x 28"deep	SW@1.42	Ea	1,010.00	48.30	1,058.30
Range splash guard. Mounts behind range with screws to protect adjacent wall.					
30" x 24", enamel finish	SW@.165	Ea	23.00	5.61	28.61
36" x 24", enamel finish	SW@.165	Ea	28.30	5.61	33.91
30" x 24", stainless steel finish	SW@.165	Ea	51.80	5.61	57.41
36" x 24", stainless steel finish	SW@.165	Ea	58.70	5.61	64.31
Add for ducting, including cap with back draft preventer at exterior wall or roof					
Back ducted straight through wall	SW@.526	Ea	15.80	17.90	33.70
Top or back ducted through wall with single bend, to 4'	SW@.785	Ea	53.00	26.70	79.70
Top or back ducted through roof straight through roof, to 4'	SW@.657	Ea	44.80	22.30	67.10
Top or back ducted through roof with single bend, to 4'	SW@.801	Ea	58.20	27.20	85.40
Add for each additional LF of straight duct over 4'	SW@.136	LF	6.60	4.62	11.22
Add for each additional bend	SW@.136	Ea	15.80	4.62	20.42
Add for charcoal filters, ductless installation	—	Ea	11.00	—	11.00

Ranges, Built-In Material costs are for good to better quality built-in appliances. Labor costs are for carpentry installation only. Add for gas and electric runs below. Note that all cooktops with grilles require exhaust venting. See Range Hoods above.

	Craft@Hrs	Unit	Material	Labor	Total
Counter-mounted cooktops					
Gas, pilot free ignition, 30" x 21" x 4-1/2"					
Baked enamel top	P1@1.92	Ea	230.00	57.60	287.60
Chrome plated top	P1@1.92	Ea	260.00	57.60	317.60
Gas, pilot free ignition, with downdraft, built-in downdraft fan requires outside vent but no hood					
Baked enamel top	P1@1.92	Ea	809.00	57.60	866.60
Electric					
Smooth ceramic cooktop with one 9" element, one 8" element and two 6" elements					
30" x 21" x 3-3/8"	BE@.903	Ea	396.00	30.60	426.60
Radiant heat type with smooth ceramic cooktop, one 9", two 8", and one 6" element					
29-3/8" x 20-1/2" x 3-3/8"	BE@.903	Ea	459.00	30.60	489.60
Radiant heat type with downdraft, built-in downdraft fan requires outside vent but no hood					
29-3/8" x 20-1/2" x 3-3/8"	BE@.903	Ea	538.00	30.60	568.60
Conventional coil type cooktop with two 8" and two 6" elements, 30-1/4" x 21-1/4" x 3-3/8"					
Baked enamel finish, various colors	BE@.903	Ea	215.00	30.60	245.60
Chrome plated finish	BE@.903	Ea	256.00	30.60	286.60
Double wall oven with adjustable cleaning levels. Precision electronic controls. Cook and hold setting.					
Delayed cook and clean options					
Black, 30" wide	BE@1.67	Ea	1,370.00	56.50	1,426.50
White, 27" wide clear glass	BE@1.67	Ea	1,710.00	56.50	1,766.50

	Craft@Hrs	Unit	Material	Labor	Total
Roof Coatings and Adhesives					
Roofing asphalt, 100 lb carton.					
145 to 165 degree (low melt)	—	Ea	18.40	—	18.40
185 degree, certified (high melt) - type IV	—	Ea	16.20	—	16.20
Asphalt emulsion coating, (35 SF per gallon)					
5 gallons	—	Ea	23.60	—	22.60
Asphalt primer, Henry #105, (150 to 300 SF per gallon)					
1 gallon	—	Ea	6.57	—	6.57
5 gallons	—	Ea	20.90	—	20.90
55 gallons	—	Ea	198.00	—	198.00
Wet patch roof cement, Western Colloid #101, (12.5 SF per gallon at 1/8" thick)					
1 gallon	—	Ea	11.20	—	11.20
5 gallons	—	Ea	31.90	—	31.90
Roof patch elastomeric, seals vertical and horizontal seams (67 SF per gallon)					
1 gallon	—	Ea	16.30	—	16.30
Elastomeric white roof coating (60 SF per gallon)					
1 gallon	—	Ea	16.70	—	16.70
5 gallons	—	Ea	52.80	—	52.80
Fibered aluminum roof coating, non-asbestos fiber, (50 SF per gallon on built-up or smooth surface, 65 SF per gallon on metal surface)					
1 gallon	—	Ea	5.55	—	5.55
5 gallons	—	Ea	17.80	—	17.80
Flashing compound, Western Colloid #106, (12.5 SF per gallon at 1/8" thick)					
1 gallon	—	Ea	6.93	—	6.93
3 gallons	—	Ea	18.40	—	18.40
5 gallons	—	Ea	25.50	—	25.50
Lap cement, Henry #108, (280 LF at 2" lap per gallon)					
1 gallon	—	Ea	6.78	—	6.78
5 gallons	—	Ea	20.10	—	20.10
Plastic roof cement (12-1/2 SF at 1/8" thick per gallon)					
1 gallon	—	Ea	5.13	—	5.13
5 gallons	—	Ea	17.20	—	17.20
Non-fibered roof coating, (80 SF per gallon on masonry surfaces or roofing, 100 SF per gallon on metal surfaces)					
1 gallon	—	Ea	6.80	—	6.80
5 gallons	—	Ea	16.80	—	16.80
Reflective fibered aluminum-flake roof coating (50 SF per gallon on built-up roof or smooth surface, 70 SF per gallon on metal surface)					
1 gallon	—	Ea	9.23	—	9.23
5 gallons	—	Ea	35.70	—	35.70
Wet-dry premium roof cement, (12-1/2 SF per gallon at 1/8" thick)					
11 oz cartridge for caulking gun	—	Ea	1.49	—	1.49
1 gallon	—	Ea	5.24	—	5.24
5 gallons	—	Ea	17.30	—	17.30
Cotton mat reinforcing for roof coatings					
6" x 150' roll	—	Ea	10.20	—	10.20
Polyester fabric reinforcing for roof coating, Henry					
3 oz per SY, 36" x 375'		Sq	13.10		13.10

Roofing Rule of Thumb

	Craft@Hrs	Unit	Material	Labor	Total

Roofing Rule of Thumb Typical subcontract prices per square (100 SF). Based on a 1,500 to 2,000 SF job not over two stories above ground level. Add the cost of flashing and deck preparation, if required. See detailed labor and material costs in the following sections. Many communities restrict use of wood roofing products.

	Craft@Hrs	Unit	Material	Labor	Total
Composition shingles (class A, minimum)	—	Sq	—	—	105.00
Composition shingles (class C, premium)	—	Sq	—	—	153.00
Built-up roof, 3 ply and gravel	—	Sq	—	—	151.00
Cedar shingles, #3, 16"	—	Sq	—	—	290.00
Cedar shingles, #1, 18", fire treated	—	Sq	—	—	395.00
Clay mission tile	—	Sq	—	—	333.00
Concrete tile	—	Sq	—	—	271.00
Metal tile	—	Sq	—	—	333.00
Perlite shakes	—	Sq	—	—	320.00

Obtaining Roof Area from Plan Area

Rise	Factor	Rise	Factor
3"	1.031	8"	1.202
3½"	1.042	8½"	1.225
4"	1.054	9"	1.250
4½"	1.068	9½"	1.275
5"	1.083	10"	1.302
5½"	1.100	10½"	1.329
6"	1.118	11"	1.357
6½"	1.137	11½"	1.385
7"	1.158	12"	1.414
7½"	1.179		

When a roof has to be figured from a plan only, and the roof pitch is known, the roof area may be fairly accurately computed from the table above. The horizontal or plan area (including overhangs) should be multiplied by the factor shown in the table opposite the rise, which is given in inches per horizontal foot. The result will be the roof area.

Built-Up Roofing Installed over existing suitable substrate. Add the cost of equipment rental (the kettle) and consumable supplies (mops, brooms, gloves). Typical cost for equipment and consumable supplies is $15 per 100 square feet applied.

Type 1: 2 ply felt and 1 ply 90-pound cap sheet, including 2 plies of hot mop asphalt

	Craft@Hrs	Unit	Material	Labor	Total
Base sheet, nailed down, per 100 SF	R1@.250	Sq	3.67	7.29	10.96
Hot mop asphalt, 30 lbs. per 100 SF	R1@.350	Sq	6.06	10.20	16.26
Asphalt felt, 15 lbs. per 100 SF	R1@.100	Sq	3.67	2.92	6.59
Hot mop asphalt, 30 lbs. per 100 SF	R1@.350	Sq	6.06	10.20	16.26
Mineral surface					
90 lb. cap sheet, per 100 SF	R1@.200	Sq	18.62	5.83	24.45
Type 1, total per 100 SF	R1@1.25	Sq	38.08	36.50	74.58

Type 2: 3 ply asphalt, 1 ply 30 lb felt, 2 plies 15 lb felt, 3 coats hot mop asphalt and gravel

	Craft@Hrs	Unit	Material	Labor	Total
Asphalt felt, 30 lbs. nailed down	R1@.300	Sq	7.36	8.75	16.11
Hot mop asphalt,					
30 lbs. per 100 SF	R1@.350	Sq	6.06	10.20	16.26
Asphalt felt, 15 lbs. per 100 SF	R1@.100	Sq	3.67	2.92	6.59
Hot mop asphalt,					
30 lbs. per 100 SF	R1@.350	Sq	6.06	10.20	16.26
Asphalt felt, 15 lbs. per 100 SF	R1@.100	Sq	3.67	2.92	6.59
Hot mop asphalt,					
30 lbs. per 100 SF	R1@.350	Sq	6.06	10.20	16.26
Gravel, 400 lbs. per 100 SF	R1@.600	Sq	25.23	17.50	42.73
Type 2, total per 100 SF	R1@2.15	Sq	58.11	62.70	120.81

Type 3: 4 ply asphalt, 4 plies 15 lb felt including 4 coats hot mop and gravel

	Craft@Hrs	Unit	Material	Labor	Total
Base sheet, nailed down, per 100 SF	R1@.250	Sq	3.67	7.29	10.96
Hot mop asphalt,					
30 lbs. per 100 SF	R1@.350	Sq	6.06	10.20	16.26
Asphalt felt, 15 pound, per 100 SF	R1@.100	Sq	3.67	2.92	6.59
Hot mop asphalt,					
30 lbs. per 100 SF	R1@.350	Sq	6.06	10.20	16.26
Asphalt felt, 15 pound, per 100 SF	R1@.100	Sq	3.67	2.92	6.59
Hot mop asphalt,					
30 lbs. per 100 SF	R1@.350	Sq	6.06	10.20	16.26
Asphalt felt, 15 lbs. per 100 SF	R1@.100	Sq	3.67	2.92	6.59
Hot mop asphalt,					
30 lbs. per 100 SF	R1@.350	Sq	6.06	10.20	16.26
Gravel, 400 lbs. per 100 SF	R1@.600	Sq	25.23	17.50	42.73
Type 3, total per 100 square feet	R1@2.55	Sq	64.15	74.40	138.55

	Craft@Hrs	Unit	Material	Labor	Total
Roofing asphalt					
Type III, 100-pound keg	—@.000	Ea	17.70	—	17.70
Type IV, 100-pound keg	—@.000	Ea	16.20	—	16.20
Mineral surface roll roofing Roll covers 1 square (100 square feet).					
90-pound fiberglass mat reinforced	R1@.200	Sq	16.60	5.83	22.43
GAF Mineral Guard	R1@.200	Sq	19.80	5.83	25.63
SBS-modified bitumen, Flintlastic	R1@.200	Sq	38.20	5.83	44.03
Add for mopping in 30-pound hot asphalt	R1@.350	Sq	4.95	10.20	15.15
Roofing felt Asphalt saturated. Per square. 15-pound roll covers 400 square feet. 30-pound roll covers 200 square feet. Nailed down.					
15-pound	R1@.140	Sq	3.61	4.08	7.69
15-pound, ASTM compliant	R1@.140	Sq	4.59	4.08	8.67
30-pound	R1@.150	Sq	7.22	4.38	11.60
30-pound, ASTM-D226	R1@.150	Sq	9.00	4.38	13.38
Add for mopping in 30-pound hot asphalt	R1@.350	Sq	4.95	10.20	15.15
Fiberglass ply IV roll roofing ASTM D-2178 Type IV. 500 square foot roll.					
Per 100 square feet	R1@.140	Sq	4.24	4.08	8.32
Add for mopping in 30-pound hot asphalt	R1@.350	Sq	4.95	10.20	15.15
Fiberglass base sheet ASTM D-4601. 300 square foot roll. Nailed down					
Per 100 square feet	R1@.300	Sq	5.60	8.75	14.35
Modified bitumen adhesive Rubberized. Use to adhere SBS modified bitumen membranes and rolled roofing. Blind nailing cement. Brush grade.					
5 gallons	—@.000	Ea	30.30	—	30.30
Cold process and lap cement Bonds plies of built-up roofing and gravel to bare spots on existing gravel roof. Restores severely worn roof. Gallon covers 50 square feet.					
1 gallon	—@.000	Ea	5.87	—	5.87
5 gallons	—@.000	Ea	21.00	—	21.00
Modified bitumen cold process adhesive For placing shingles and cementing asphalt-coated roofing sheets together. Use for bonding felts, smooth and granule-surfaced roll roofing. Quick-setting, waterproof adhesive with strong bonding action.					
5 gallons	—@.000	Ea	14.40	—	14.40
Cold-Ap® built-up roofing cement, Henry Bonds layers of conventional built-up roofing or SBS base and cap sheets. Replaces hot asphalt in built-up roofing. For embedding polyester fabric, gravel and granules. Not recommended for saturated felt. ASTM D3019 Type III. Two gallons cover 100 square feet.					
1 gallon	—@.000	Ea	9.25	—	9.25
5 gallons	—@.000	Ea	34.50	—	34.50
55 gallons	—@.000	Ea	245.00	—	245.00
3-ply SBS field adhesive, GAF Low-slope roofing. GAF Torch and Mop System.					
5 gallons	—@.000	Ea	27.90	—	27.90

Roofing, Adhesive

	Craft@Hrs	Unit	Material	Labor	Total

PBA™ Permanent Bond Adhesive, Henry
Cold-process waterproofing roofing adhesive. ASTM D3019 Type III. Two gallons cover 100 square feet.

	Craft@Hrs	Unit	Material	Labor	Total
5 gallons	—@.000	Ea	28.20	—	28.20

Roofing, Composition Shingle Material costs include 5% waste. Labor costs include roof loading and typical cutting and fitting for roofs of average complexity.

Architectural grade laminated shingles, Owens Corning Three bundles per square (100 square feet). Class A fire rating. 70 MPH wind resistance.

	Craft@Hrs	Unit	Material	Labor	Total
Oakridge, 30 year	R1@1.83	Sq	48.00	53.40	101.40
Oakridge, 30 year, algae-resistant	R1@1.83	Sq	52.20	53.40	105.60
Oakridge, 50 year, algae-resistant	R1@1.83	Sq	56.20	53.40	109.60

Hip and ridge shingles with sealant, Owens Corning Per linear foot of hip or ridge.

	Craft@Hrs	Unit	Material	Labor	Total
High Ridge	R1@.028	LF	1.62	.82	2.44
High Ridge, algae-resistant	R1@.028	LF	1.69	.82	2.51
High Style	R1@.028	LF	1.41	.82	2.23

3-tab asphalt fiberglass® shingles, Owens Corning Three bundles per square (100 square feet). Class A fire rating. 60 MPH wind resistance.

	Craft@Hrs	Unit	Material	Labor	Total
Supreme, 25 year	R1@1.83	Sq	38.40	53.40	91.80
Supreme, 25 year, algae-resistant	R1@1.83	Sq	41.60	53.40	95.00
Classic, 20 year	R1@1.83	Sq	34.90	53.40	88.30
Classic, 20 year, algae-resistant	R1@1.83	Sq	31.90	53.40	85.30

Hip and ridge shingles with sealant, Owens Corning Per linear foot of hip or ridge.

	Craft@Hrs	Unit	Material	Labor	Total
Asphalt-fiberglass	R1@.028	LF	.80	.82	1.62

Glaslock® interlocking shingles, Owens Corning Three bundles per square (100 square feet). Class A fire rating. 60 MPH wind resistance.

	Craft@Hrs	Unit	Material	Labor	Total
25 year rating	R1@1.83	Sq	43.90	53.40	97.30

Architectural grade laminated shingles, Timberline® Ultra®, GAF Class A fire rating. 110 MPH wind resistance. Four bundles per square (100 square feet).

	Craft@Hrs	Unit	Material	Labor	Total
Ultra, lifetime warranty	R1@1.83	Sq	77.10	53.40	130.50
Select, 40 year	R1@1.83	Sq	63.50	53.40	116.90
Select, 30 year	R1@1.83	Sq	73.10	53.40	126.50

Hip and ridge shingles, GAF Per linear foot of hip or ridge.

	Craft@Hrs	Unit	Material	Labor	Total
Timbertext(r)	R1@.028	LF	1.79	.82	2.61
Pacific Ridge	R1@.028	LF	1.47	.82	2.29
Timber Ridge	R1@.028	LF	1.36	.82	2.18

3-tab fiberglass asphalt shingles, GAF Class A fire rating. 60-mph wind warranty. 3 bundles cover 100 square feet at 5" exposure. Per square (100 square feet).

	Craft@Hrs	Unit	Material	Labor	Total
Royal Sovereign®, 25 year	R1@1.83	Sq	41.10	53.40	94.50
Royal Sovereign®, 25 year, algae-resistant	R1@1.83	Sq	41.60	53.40	95.00
Sentinel®, 20 year	R1@1.83	Sq	33.20	53.40	86.60
Sentinel®, 20 year, fungus- and algae-resistant	R1@1.83	Sq	39.80	53.40	93.20

	Craft@Hrs	Unit	Material	Labor	Total

Grand Sequoia®, GAF 3-tab laminate shingles Class A fire rating. 40-year limited warranty. 17" x 40". 4 bundles cover 100 square feet at 5" exposure. Hip and ridge bundle covers 20 linear feet. Price per bundle.

	Craft@Hrs	Unit	Material	Labor	Total
40 year shingles, per square	R1@1.83	Sq	82.00	53.40	135.40
Hip and ridge units, per linear foot	R1@.028	LF	1.49	.82	2.31

Asphalt shingles, CertainTeed Class A fire rated. Three bundles cover 100 square feet.

	Craft@Hrs	Unit	Material	Labor	Total
Sealdon, 25	R1@1.83	Sq	40.80	53.40	94.20

Shingle starter strip Peel 'n' stick SBS membrane. Rubberized to seal around nails and resist wind lift.

	Craft@Hrs	Unit	Material	Labor	Total
7" x 33' roll	R1@.460	Ea	9.25	13.40	22.65
Per linear foot	R1@.014	LF	.28	.41	.69

Allowance for felt, flashing, fasteners, and vents

	Craft@Hrs	Unit	Material	Labor	Total
Typical cost	—	Sq	—	—	6.50

Roofing Papers See also Building Paper. Labor laying paper on 3 in 12 pitch roof.

Asphalt roofing felt (115.5 SF covers 100 SF), labor to roll and mop

	Craft@Hrs	Unit	Material	Labor	Total
15 lb (432 SF roll)	R1@.360	Sq	3.67	10.50	14.17
30 lb (216 SF roll)	R1@.495	Sq	7.36	14.40	21.76

Glass base sheet (40" x 98' roll covers 300 SF)

	Craft@Hrs	Unit	Material	Labor	Total
Base for torch applied membranes	R1@.360	Sq	5.60	10.50	16.10
Roofing asphalt, 100 lbs/Sq	—	Sq	20.20	—	16.34

Mineral surfaced roll roofing, 36" x 36' roll covers 100 square feet (1 square)

	Craft@Hrs	Unit	Material	Labor	Total
90 lb, black or white	R1@1.03	Sq	18.60	30.00	48.60
90 lb, tan	R1@1.03	Sq	17.50	30.00	47.50

Mineral surfaced selvage edge roll roofing, 39.4" x 32.6' roll covers 100 square feet (1 square) with 2" selvage edge overlap

	Craft@Hrs	Unit	Material	Labor	Total
90 lb, white or colors	R1@1.03	Sq	20.50	30.00	50.50

Roofing Shakes and Shingles

Red cedar. No pressure treating. Labor includes flashing.

Taper split No. 1 medium cedar shakes, four bundles cover 100 square feet (1 square) at 10" exposure. Includes 10% for waste.

	Craft@Hrs	Unit	Material	Labor	Total
No. 1 medium shakes, $32.40 per bundle	R1@3.52	Sq	143.60	103.00	246.60
Add for pressure treated shakes	—	%	30.0	—	—
Deduct in the Pacific Northwest	—	%	-20.0	—	—
Shake felt, 30 lb	R1@.500	Sq	9.43	14.60	24.03

Hip and ridge shakes, medium, 20 per bundle, covers 16.7 LF at 10" and 20 LF at 12" exposure

	Craft@Hrs	Unit	Material	Labor	Total
Per bundle	R1@1.00	Ea	38.70	29.20	67.90
Per linear foot	R1@.060	LF	2.32	1.75	4.07

Sawn shakes, sawn 1 side, class "C" fire retardant

	Craft@Hrs	Unit	Material	Labor	Total
1/2" to 3/4" x 24" (4 bdle/sq at 10" exp.)	R1@3.52	Sq	194.00	103.00	297.00
3/4" to 5/4" x 24" (5 bdle/sq at 10" exp.)	R1@4.16	Sq	269.00	121.00	390.00

Cedar roofing shingles, No. 1 grade. Four bundles of Perfection grade 18" shingles cover 100 square feet (1 square) at 5-1/2" exposure. Five perfections are 2-1/4" thick. Four bundles of 5X Perfect grade shingles cover 100 square feet (1 square) at 5" exposure. Five perfects are 2" thick.

	Craft@Hrs	Unit	Material	Labor	Total
Perfections, $52.50 per bundle (Florida)	R1@3.52	Sq	231.00	103.00	334.00
Perfects, $45.90 per bundle	R1@3.52	Sq	202.00	103.00	305.00
No. 2 Perfections, $37.00per bundle	R1@3.52	Sq	163.00	103.00	266.00
Add for pitch over 6 in 12	—	%	—	40.0	—

Roofing, Sheets

	Craft@Hrs	Unit	Material	Labor	Total
No. 2 cedar ridge shingles, medium, bundle covers 16 linear feet					
Per bundle	R1@1.00	Ea	34.00	29.20	63.20
Per linear foot	R1@.062	Ea	2.12	1.81	3.93
Fire treated cedar shingles, 18", 5-1/2" exposure, 5 shingles are 2-1/4" thick, (5/2-1/4)					
No. 1 (houses)	R1@3.52	Sq	250.00	103.00	353.00
No. 2, red label (houses or garages)	R1@1.00	Sq	208.00	29.20	237.20
No. 3 (garages)	R1@1.00	Sq	179.00	29.20	208.20
Eastern white cedar shingles, smooth butt edge, four bundles cover 100 square feet (1 square) at 5" exposure					
White extras, $26.25 per bundle	R1@1.00	Sq	125.00	29.20	154.20
White clears, $42.00 per bundle	R1@1.00	Sq	99.60	29.20	128.80
White No. 2 clears, $14.25 per bundle	R1@1.00	Sq	67.60	29.20	96.80
Cedar shim shingles, tapered Western red cedar					
Builder's 12-pack, per pack	—	Ea	3.65	—	3.65

Roofing Sheets See also Fiberglass Panels.

	Craft@Hrs	Unit	Material	Labor	Total
Metal sheet roofing, utility gauge					
26" wide, 5-V crimp (includes 15% coverage loss)					
6' to 12' lengths	R1@.027	SF	.64	.79	1.43
26" wide, corrugated (includes 15% coverage loss)					
6' to 12' lengths	R1@.027	SF	.56	.79	1.35
27-1/2" wide, corrugated (includes 20% coverage loss)					
6' to 12' lengths	R1@.027	SF	.55	.79	1.34
Ridge roll, plain					
10" wide	R1@.030	LF	1.89	.88	2.77
Ridge cap, formed, plain, 12"					
12" x 10' roll	R1@.061	LF	.92	1.78	2.70
Sidewall flashing, plain					
3" x 4" x 10'	R1@.035	LF	.68	1.02	1.70
V-5 crimp closure strips					
24" long	R1@.035	Ea	1.02	1.02	2.04
Endwall flashing, 2-1/2" corrugated					
10" x 28"	R1@.035	LF	.96	1.02	1.98
Wood filler strip, corrugated					
7/8" x 7/8" x 6' (at $1.60 each)	R1@.035	LF	.27	1.02	1.29

Steel Panel Roofing Simulated tile, Met-Tile™. 15% waste included. Roofing panels, 2' to 20' long

	Craft@Hrs	Unit	Material	Labor	Total
26 gauge .019" thick	R1@.027	SF	1.48	.79	2.27
End cap					
4-3/4"	R1@.030	Ea	2.90	.88	3.78
5-3/4" round	R1@.030	Ea	2.90	.88	3.78
9-1/2" oversized	R1@.030	Ea	3.86	.88	4.74
Flashing, galvanized steel					
End wall	R1@.047	LF	1.48	1.37	2.85
Plain side wall	R1@.047	LF	1.48	1.37	2.85
Valley cover	R1@.047	LF	1.48	1.37	2.85
Valley pan	R1@.047	LF	1.97	1.37	3.34
Contoured end wall	R1@.047	LF	3.86	1.37	5.23
Ridge-hip-gables					
Spanish 18" x 14-1/2"	R1@.035	Ea	4.19	1.02	5.21
Standard 7-1/2" x 42"	R1@.030	Ea	8.38	.88	9.26

	Craft@Hrs	Unit	Material	Labor	Total
Trim					
Standard eave trim	R1@.035	LF	1.20	1.02	2.22
Flat edge trim	R1@.035	LF	1.20	1.02	2.22
Shake eave trim	R1@.035	LF	1.55	1.02	2.57
Hood vents					
18" x 22"	R1@.448	Ea	1.55	13.10	14.65

Roofing Slate Local delivery included. Costs will be higher where slate is not mined. Add freight cost at 800 to 1,000 pounds per 100 SF. Includes 20" long random width slate, 3/16" thick, 7-1/2" exposure. Meets Standard SS-S-451.

	Craft@Hrs	Unit	Material	Labor	Total
Semi-weathering green and gray	R1@11.3	Sq	367.00	330.00	697.00
Vermont black and gray black	R1@11.3	Sq	436.00	330.00	766.00
China black or gray	R1@11.3	Sq	374.00	330.00	704.00
Unfading and variegated purple	R1@11.3	Sq	495.00	330.00	825.00
Unfading mottled green and purple	R1@11.3	Sq	435.00	330.00	765.00
Unfading green	R1@11.3	Sq	405.00	330.00	735.00
Red slate	R1@13.6	Sq	1,270.00	397.00	1,667.00
Add for other specified widths and lengths	—	%	20.0	—	—

Fiber Cement Slate Roofing Eternit™, Stonit™, or Thrutone™, various colors. Non-asbestos formulation Laid over 1/2" sheathing. Includes 30 lb. felt and one copper storm anchor per slate. Add for sheathing, flashing, hip, ridge and valley units.

	Craft@Hrs	Unit	Material	Labor	Total
23-5/8" x 11-7/8", English (420 lb, 30 year)					
2" head lap (113 pieces per Sq)	R1@5.50	Sq	282.00	160.00	442.00
3" head lap (119 pieces per Sq)	R1@5.75	Sq	297.00	168.00	465.00
4" head lap (124 pieces per Sq)	R1@6.00	Sq	308.00	175.00	483.00
Add for hip and ridge units, in mastic	R1@0.10	LF	5.78	2.92	8.70
15-3/4" x 10-5/8", Continental (420 lb, 30 year)					
2" head lap (197 pieces per Sq)	R1@6.50	Sq	292.00	190.00	482.00
3" head lap (214 pieces per Sq)	R1@6.75	Sq	318.00	197.00	515.00
4" head lap (231 pieces per Sq)	R1@7.00	Sq	344.00	204.00	548.00
Add for hip and ridge units, in mastic	R1@0.10	LF	5.30	2.92	8.22
Copper storm anchors, box of 1,000 ($28.00)	—	Ea	.03	—	.03
Stainless steel slate hooks	—	Ea	.20	—	.20
Add for extra felt under 5 in 12 pitch	R1@0.20	Sq	12.50	5.83	18.33

Roofing Tile, Clay Material costs include felt and flashing. No freight or waste included. Costs will be higher where clay tile is not manufactured locally. U.S. Tile Co. Spanish tile, "S" shaped, 88 pieces per square,

	Craft@Hrs	Unit	Material	Labor	Total
(800 lbs per square at 11" centers and 15" exposure)					
Red clay tile	R1@3.46	Sq	105.00	101.00	206.00
Add for coloring	—	Sq	11.30	—	11.30
Red hip and ridge units	R1@.047	LF	.87	1.37	2.24
Color hip and ridge units	R1@.047	LF	1.25	1.37	2.62
Red rake units	R1@.047	LF	1.66	1.37	3.03
Color rake units	R1@.047	LF	1.82	1.37	3.19

Red clay mission tile 2-piece, 86 pans and 86 tops per square,

	Craft@Hrs	Unit	Material	Labor	Total
7-1/2" x 18" x 8-1/2" tiles at 11" centers and 15" exposure					
Red clay tile	R1@5.84	Sq	195.00	170.00	365.00
Add for coloring (costs vary widely)	—	Sq	43.80	—	43.80
Red hip and ridge units	R1@.047	LF	.92	1.37	2.29
Color hip and ridge units	R1@.047	LF	1.25	1.37	2.62

Roofing, Tile

	Craft@Hrs	Unit	Material	Labor	Total
Red rake units	R1@.047	LF	1.66	1.37	3.03
Color rake units	R1@.047	LF	1.82	1.37	3.19

Concrete Roof Tile Material includes felt, nails, and flashing. Approximately 90 pieces per square. Monier Lifetile

	Craft@Hrs	Unit	Material	Labor	Total
Shake, slurry color	R1@3.25	Sq	106.00	94.80	200.80
Slate, thru color	R1@3.25	Sq	128.00	94.80	222.80
Espana, slurry coated	R1@3.25	Sq	128.00	94.80	222.80
Monier 2000	R1@3.25	Sq	138.00	94.80	232.80
Vignette	R1@3.25	Sq	120.00	94.80	214.80
Collage	R1@3.25	Sq	120.00	94.80	214.80
Tapestry	R1@3.25	Sq	131.00	94.80	225.80
Split shake	R1@3.25	Sq	144.00	94.80	238.80
Villa, slurry coat	R1@3.25	Sq	119.00	94.80	213.80
Villa, Roma, thru color	R1@3.25	Sq	124.00	94.80	218.80
Trim tile					
Mansard "V" ridge or rake, slurry coated	—	Ea	1.69	—	1.69
Mansard, ridge or rake, thru color	—	Ea	1.77	—	1.77
Hipstarters, slurry coated	—	Ea	14.40	—	14.40
Hipstarters, thru color	—	Ea	18.40	—	18.40
Accessories					
Eave closure or birdstop	R1@.030	LF	1.05	.88	1.93
Hurricane or wind clips	R1@.030	Ea	.35	.88	1.23
Underlayment or felt	R1@.050	Sq	13.90	1.46	15.36
Metal flashing and nails	R1@.306	Sq	8.10	8.93	17.03
Pre-formed plastic flashings					
Hip (13" long)	R1@.020	Ea	27.70	.58	28.28
Ridge (39" long)	R1@.030	Ea	27.70	.88	28.58
Anti-ponding foam	R1@.030	LF	.23	.88	1.11
Batten extenders	R1@.030	Ea	.92	.88	1.80
Roof loading					
Add to load tile and accessories on roof	R1@.822	Sq	—	24.00	24.00

Wall Safes, Residential UL approved combination lock, flush combination dial, 1/2" thick door. Fits between 16" on center studs. Installed during new construction.

	Craft@Hrs	Unit	Material	Labor	Total
For 2 x 4 wall thickness, outside dimensions 13.5" x 14" x 3.75"					
Capacity 400 cu. inches	BC@.788	Ea	347.00	23.60	370.60
For 2 x 6 wall thickness, outside dimensions 13.5" x 14" x 5.75"					
Capacity 750 cu. inches	BC@.788	Ea	376.00	23.60	399.60
For 2 x 12 wall thickness, outside dimensions 13.5" x 14" x 10.25"					
Capacity 1350 cu. inches	BC@.950	Ea	406.00	28.50	434.50
For 2 x 12 wall thickness, outside dimensions 14" x 14" x 10.25"					
Capacity 1550 cu. inches	BC@.950	Ea	570.00	28.50	598.50
Optional equipment					
Electronic push button lock	—	Ea	165.00	—	165.00
Dual key lock - in place of combination lock	—	Ea	35.00	—	35.00
Installation kit	—	Ea	20.00	—	20.00
Combination change kit	—	Ea	10.00	—	10.00

	Craft@Hrs	Unit	Material	Labor	Total

Floor Safes, Residential 1/2" thick spring counter-balanced square door (13" x 13"), UL approved Group Two combination lock, metal cover plate - creates flush floor surface. Installed during new construction.

	Craft@Hrs	Unit	Material	Labor	Total
Outside dimensions 14.5" x 16" x 10.25"					
Capacity 1100 cu. inches	BC@1.10	Ea	384.00	33.00	417.00
Outside dimensions 14.5" x 16" x 16.25"					
Capacity 2325 cu. inches	BC@1.25	Ea	454.00	37.50	491.50
Outside dimensions 14.5" x 16" x 21.75"					
Capacity 3545 cu. inches	BC@1.35	Ea	528.00	40.50	568.50
Outside dimensions 14.5" x 25" x 16.25"					
Capacity 3970 cu. inches	BC@1.42	Ea	642.00	42.60	684.60
Outside dimensions 25" x 25" x 16.75"					
Capacity 7050 cu. inches	BC@1.55	Ea	748.00	46.50	794.50
Outside dimensions 14.5" x 48" x 16.25"					
Capacity 7775 cu. inches	BC@1.60	Ea	894.00	48.00	942.00

Sandblasting, Waterblasting, Subcontract The minimum charge for on-site (truck-mounted) sandblasting and waterblasting will usually be about $500.00. Add the cost of extra liability insurance and physical barriers to protect adjacent swimming pools, residences, process plants, etc. No scaffolding included. Based on sandblasting with 375 CFM, 125 PSI compressor. Portable sandblasting equipment will increase costs about 35%.

	Craft@Hrs	Unit	Material	Labor	Total
Sandblasting most surfaces					
Water soluble paints	—	SF	—	—	1.55
Oil paints	—	SF	—	—	1.80
Heavy mastic	—	SF	—	—	2.10
Sandblasting masonry					
Brick	—	SF	—	—	1.25
Block walls, most work	—	SF	—	—	1.25
Remove heavy surface grime	—	SF	—	—	1.70
Heavy, exposing aggregate	—	SF	—	—	1.90
Sandblasting concrete tilt-up panels					
Light blast	—	SF	—	—	1.20
Medium, exposing aggregate	—	SF	—	—	1.75
Heavy, exposing aggregate	—	SF	—	—	2.10
Sandblasting steel					
New, uncoated (commercial grade)	—	SF	—	—	1.70
New, uncoated (near white grade)	—	SF	—	—	2.35
Epoxy coated (near white grade)	—	SF	—	—	2.80
Sandblasting wood					
Medium blast, clean and texture	—	SF	—	—	1.55
Waterblast (hydroblast) with mild detergent					
To 5,000 PSI blast (4 hour minimum)	—	Hr	—	—	235.00
5,000 to 10,000 PSI blast (8 hour min.)	—	Hr	—	—	265.00
Over 10,000 PSI blast (8 hour minimum)	—	Hr	—	—	320.00
Wet sandblasting					
(4 hour minimum)	—	Hr	—	—	250.00

Handwritten note near "Heavy, exposing aggregate": $2.25 x 2.22 = /SF 1.02 X 1.15 X (1.90)

Sauna Rooms

	Craft@Hrs	Unit	Material	Labor	Total

Sauna Rooms Costs listed are for labor and materials to install sauna rooms as described. Units are shipped as a disassembled kit complete with kiln dried softwood paneling boards, sauna heater, temperature and humidity controls and monitors, pre-assembled benches, prehung door, water bucket and dipper, and interior light. No electrical work included.

Pre-cut western red cedar sauna room package with heater, controls, duckboard floor and accessories, ready for nailing to existing framed and insulated walls with exterior finish. http://www.amerec.com

	Craft@Hrs	Unit	Material	Labor	Total
4' x 4' room	B1@8.01	Ea	1,990.00	219.00	2,209.00
4' x 6' room	B1@8.01	Ea	2,420.00	219.00	2,639.00
6' x 6' room	B1@10.0	Ea	2,900.00	274.00	3,174.00
6' x 8' room	B1@12.0	Ea	3,460.00	328.00	3,788.00
8' x 8' room	B1@14.0	Ea	3,930.00	383.00	4,313.00
8' x 10' room	B1@15.0	Ea	4,410.00	410.00	4,820.00
8' x 12' room	B1@16.0	Ea	4,920.00	438.00	5,358.00
Add for electrical connection wiring	—	LS	—	—	293.00

Modular sauna room assembly. Includes the wired and insulated panels to construct a free-standing sauna on a waterproof floor. With heater, rocks, door, benches and accessories. http://www.amerec.com

	Craft@Hrs	Unit	Material	Labor	Total
4' x 4' room	B1@4.00	Ea	3,200.00	109.00	3,309.00
4' x 6' room	B1@4.00	Ea	3,900.00	109.00	4,009.00
6' x 6' room	B1@6.00	Ea	4,720.00	164.00	4,884.00
6' x 8' room	B1@8.01	Ea	5,640.00	219.00	5,859.00
6' x 10' room	B1@8.01	Ea	6,440.00	219.00	6,659.00
8' x 8' room	B1@10.0	Ea	6,500.00	274.00	6,774.00
8' x 10' room	B1@11.0	Ea	7,320.00	301.00	7,621.00
8' x 12' room	B1@12.0	Ea	8,200.00	328.00	8,528.00
Add for glass panel, 22-1/2" or 24"	—	Ea	70.00	—	70.00
Add for electrical connection wiring	—	LS	—	—	225.00

Pre-cut 100% vertical grain redwood or clear western red cedar sauna room package with heater, controls and accessories (including non-skid removable rubberized flooring). Ready for nailing to existing framed and insulated walls with exterior finish. Finlandia Sauna.

	Craft@Hrs	Unit	Material	Labor	Total
3' x 5' or 4' x 4' room	B1@6.68	Ea	1,650.00	183.00	1,833.00
4' x 6' room	B1@6.68	Ea	1,900.00	183.00	2,083.00
5' x 6' room	B1@6.68	Ea	2,100.00	183.00	2,283.00
6' x 6' room	B1@8.01	Ea	2,250.00	219.00	2,469.00
6' x 8' room	B1@10.7	Ea	2,750.00	293.00	3,043.00
8' x 8' room	B1@10.7	Ea	3,200.00	293.00	3,493.00
8' x 10' room	B1@10.7	Ea	3,700.00	293.00	3,993.00
8' x 12' room	B1@10.7	Ea	4,400.00	293.00	4,693.00
10' x 10' room	B1@10.7	Ea	4,300.00	293.00	4,593.00
10' x 12' room	B1@10.7	Ea	4,750.00	293.00	5,043.00
Add for electrical connection wiring	—	LS	—	—	293.00

Modular sauna room assembly. Completely pre-wired pre-insulated panels to construct a free-standing sauna on waterproof floor. Includes thermometer, sand timer, heater, rocks, door, benches and accessories. Finlandia Sauna http://www.finlandiasauna.com

	Craft@Hrs	Unit	Material	Labor	Total
4' x 4' room	B1@1.34	Ea	2,229.00	36.70	2,265.70
4' x 6' room	B1@1.34	Ea	2,607.00	36.70	2,643.70
5' x 6' room	B1@2.68	Ea	2,714.00	73.30	2,787.30
6' x 6' room	B1@2.68	Ea	3,074.00	73.30	3,147.30
6' x 8' room	B1@2.68	Ea	3,499.00	73.30	3,572.30
8' x 8' room	B1@4.02	Ea	4,327.00	110.00	4,437.00
8' x 10' room	B1@4.02	Ea	4,949.00	110.00	5,059.00
8' x 12' room	B1@4.02	Ea	5,553.00	110.00	5,663.00

	Craft@Hrs	Unit	Material	Labor	Total
10' x 10' room	B1@4.02	Ea	5,553.00	110.00	5,663.00
10' x 12' room	B1@4.02	Ea	6,085.00	110.00	6,195.00
Add for electrical connection wiring	—	LS	—	—	230.00
Add for delivery (typical)	—	%	—	—	5.0

	Alum mesh	Galv mesh	Bronze mesh	Fiberglass mesh
Screen Wire				
Cost per 100 LF roll				
18" wide roll	45.00	42.50	—	—
24" wide roll	62.50	60.00	172.00	35.00
30" wide roll	78.80	75.00	218.00	45.00
36" wide roll	95.00	83.50	260.00	50.00
42" wide roll	108.00	105.00	303.00	58.80
48" wide roll	126.00	118.00	347.00	67.50
54" wide roll	—	—	—	75.00
60" wide roll	—	—	—	86.30
72" wide roll	—	—	—	102.00
84" wide roll	—	—	—	116.00

	Craft@Hrs	Unit	Material	Labor	Total
Labor to install screen wire					
Includes measure, layout, cut and attach screen wire to an existing metal or wood frame.					
Per linear foot of perimeter of frame	BC@.017	LF	—	.51	.51

Security Alarms, Subcontract Costs are for securing building perimeter and all accessible openings with electronic alarm system, including all labor and materials. Costs assume that no interior wall, floor, or ceiling finishes are in place. All equipment is good to better quality.

		Unit			Total
Alarm control panel	—	Ea	—	—	421.00
Wiring, detectors per opening, with switch	—	Ea	—	—	73.00
Monthly monitoring charge	—	Mo	—	—	30.00
Alarm options (installed)					
Audio detectors	—	Ea	—	—	105.00
Communicator (central station service)	—	Ea	—	—	234.00
Digital touch pad control	—	Ea	—	—	128.00
Entry/exit delay	—	Ea	—	—	90.50
Fire/smoke detectors	—	Ea	—	—	156.00
Interior/exterior sirens	—	Ea	—	—	69.70
Motion detector	—	Ea	—	—	213.00
Panic button	—	Ea	—	—	52.00
Passive infrared detector	—	Ea	—	—	291.20
Pressure mat	—	LF	—	—	83.20
Security light control	—	Ea	—	—	512.00
Add for wiring detectors and controls in existing buildings					
With wall and ceiling finishes already in place	—	%	—	—	25.0

Security Guards, Subcontract Construction site security guards, unarmed, in uniform with two-way radio, backup patrol car, bond, and liability insurance. Per manhour.

		Unit			Total
Short term (1 night to 1 week)	—	Hr	—	—	34.50
Medium duration (1 week to 1 month)	—	Hr	—	—	34.50
Long term (1 to 6 months)	—	Hr	—	—	17.25
Add for licensed armed guard	—	Hr	—	—	5.15
Add for holidays	—	%	—	—	55.0

	Craft@Hrs	Unit	Material	Labor	Total

Construction site guard dog service for sites with totally enclosed perimeter, 24 hour per day, 7 day per week service. Local business codes may restrict or prohibit unattended guard dogs. Includes dog handling training session, liability insurance, on-site kennel. Cost per month.

Contractor feeding and tending dog	—	Mo	—	—	593.00

Daily delivery of guard dog at quitting time and pickup at starting time by guard dog service (special hours require advance notice)

Service feeding and tending dog	—	Mo	—	—	621.00

Construction site man and dog guard team, includes two-way radio communication with guard service office, backup patrol car, and liability insurance. Guard service may require that a telephone and guard shack be provided by the contractor. Cost for man and dog per hour.

Short term (1 week to 1 month)	—	Hr	—	—	22.00
Long term (1 to 6 months)	—	Hr	—	—	18.50

Septic Sewer Systems, Subcontract Soils testing (percolation test) by qualified engineer (required by some communities when applying for septic system permit). Does not include application fee.

Minimum cost	—	Ea	—	—	754.00

Residential septic sewer tanks. Costs include excavation for tank with good site conditions, placing of tank, inlet and outlet fittings, and backfill after hookup.

Steel reinforced concrete tanks					
1,250 gallons (3 or 4 bedroom house)	—	Ea	—	—	1,900.00
1,500 gallons (5 or 6 bedroom house)	—	Ea	—	—	2,010.00
Fiberglass tanks					
1,000 gallons (3 bedroom house)	—	Ea	—	—	1,900.00
1,250 gallons (4 bedroom house)	—	Ea	—	—	1,950.00
1,500 gallons (5 or 6 bedroom house)	—	Ea	—	—	2,080.00
Polyethylene tanks, 4" lines, 20" manhole diameter					
500 gallons, 42" H x 51" W x 100" L	—	Ea	—	—	922.00
750 gallons, 57" H x 52" W x 95" L	—	Ea	—	—	1,024.00
1,000 gallons, 66" H x 52" W x 100" L	—	Ea	—	—	1,250.00
1,250 gallons, 68" H x 56" W x 112" L	—	Ea	—	—	1,360.00
1,500 gallons, 65" H x 55" W x 143" L	—	Ea	—	—	1,700.00

Residential septic sewer drain fields (leach lines). Costs include labor, unsaturated paper, piping, gravel, excavation with good site conditions, backfill, and disposal of excess soil. 4" PVC pipe (ASTM D2729) laid in 3' deep by 1' wide trench.

With 12" gravel base	—	LF	—	—	7.10
With 24" gravel base	—	LF	—	—	8.34
With 36" gravel base	—	LF	—	—	9.44
Add for pipe laid 6' deep	—	LF	—	—	2.36

Add for piping from house to septic tank, and from septic tank to remote drain fields

4" PVC Schedule 40	—	LF	—	—	5.05

Low cost pumping systems for residential wastes, including fiberglass basin, pump, installation from septic tank or sewer line, 40' pipe run, and automatic float switch (no electric work or pipe included).

To 15' head	—	LS	—	—	2,450.00
To 25' head	—	LS	—	—	2,900.00
To 30' head	—	LS	—	—	3,750.00
Add for high water or pump failure alarm	—	LS	—	—	535.00

Better quality pump system with two alternating pumps, 700 to 800 gallon concrete or fiberglass basin, automatic float switch, indoor control panel, high water or pump failure alarm, explosion proof electrical system but no electric work or pipe.

Per pump system	—	LS	—	—	7,270.00

Pipe Locating Service

	Craft@Hrs	Unit	Material	Labor	Total

Pipe locating service to detect sewage and water leaks, breaks, and stoppages in pipes with diameter of 2" or more. Equipment can't be used where pipe has 90 degree turns. Maximum depth is 18" for cast iron and 30" for vitreous clay.

	Craft@Hrs	Unit	Material	Labor	Total
Residential 1/2 day rate	—	LS	—	—	364.00
Commercial 1/2 day rate	—	LS	—	—	489.00
Add for each additional pipe entered	—	LS	—	—	57.50
Plus per LF of pipe length	—	LF	—	—	2.10
Add for travel over 15 miles, per mile	—	Ea	—	—	.50

Sewer Connections, Subcontract Including typical excavation and backfill.

4" vitrified clay pipeline, house to property line

	Craft@Hrs	Unit	Material	Labor	Total
Long runs	—	LF	—	—	17.80
Short runs	—	LF	—	—	19.00

6" vitrified clay pipeline

	Craft@Hrs	Unit	Material	Labor	Total
Long runs	—	LF	—	—	20.10
Short runs	—	LF	—	—	23.70
Street work	—	LF	—	—	49.70

4" PVC sewer pipe, city installed, property line to main line

	Craft@Hrs	Unit	Material	Labor	Total
Up to 40 feet (connect in street)	—	LS	—	—	1,810.00
Up to 40 feet (connect in alley)	—	LS	—	—	1,300.00
Each foot over 40 feet	—	LF	—	—	38.90

6" PVC sewer pipe, city installed, property line to main line

	Craft@Hrs	Unit	Material	Labor	Total
Up to 40 feet (connect in street)	—	LS	—	—	2,900.00
Up to 40 feet (connect in alley)	—	LS	—	—	2,080.00
Each foot over 40 feet	—	LF	—	—	62.50

Sheet Metal Access Doors

Attic access doors

	Craft@Hrs	Unit	Material	Labor	Total
22" x 22"	SW@.363	Ea	55.80	12.30	68.10
22" x 30"	SW@.363	Ea	60.00	12.30	72.30
24" x 36"	SW@.363	Ea	60.00	12.30	72.30
30" x 30"	SW@.363	Ea	90.70	12.30	103.00

Wall access doors, nail-on type, galvanized or painted

	Craft@Hrs	Unit	Material	Labor	Total
24" x 18"	SW@.363	Ea	54.60	12.30	66.90
24" x 24"	SW@.363	Ea	64.40	12.30	76.70
24" x 18", screen only	—	Ea	16.00	—	16.00
24" x 24", screen only	—	Ea	18.30	—	18.30

Tub access doors

	Craft@Hrs	Unit	Material	Labor	Total
6" x 9", white ABS	SW@.363	Ea	11.80	12.30	24.10
14" x 14", white ABS	SW@.363	Ea	18.00	12.30	30.30
12" x 12", steel	SW@.363	Ea	14.80	12.30	27.10

Sheet Metal Area Walls Provides light well for basement windows, galvanized, corrugated.

	Craft@Hrs	Unit	Material	Labor	Total
12" deep, 6" projection, 37" wide	SW@.410	Ea	20.20	13.90	34.10
18" deep, 6" projection, 37" wide	SW@.410	Ea	26.90	13.90	40.80
24" deep, 6" projection, 37" wide	SW@.410	Ea	35.00	13.90	48.90
30" deep, 6" projection, 37" wide	SW@.410	Ea	41.70	13.90	55.60
Add for steel grille cover	—	Ea	47.30	—	47.30

Sheet Metal Flashing

Aluminum flashing. .0175" thickness is commercial grade. Mill finish except as noted.

	Craft@Hrs	Unit	Material	Labor	Total
Drip cap, 1-5/8" x 10' x .0175, white	SW@.350	Ea	4.20	11.90	16.10
Drip edge, 2-1/2" x 1" x .0145", 10' long	SW@.350	Ea	3.15	11.90	15.05

Copper Flashing

	Craft@Hrs	Unit	Material	Labor	Total
Drip edge, 6-1/2" x 1", 10' long	SW@.350	Ea	8.05	11.90	19.95
Flex corner, 2" x 2" x 3-1/2" radius	SW@.050	Ea	2.75	1.70	4.45
Gravel stop, 10' x 5-1/2", C-shape	SW@.350	Ea	4.20	11.90	16.10
Roll 6" x 10' x .011"	SW@.400	Ea	4.30	13.60	17.90
Roll 6" x 10' x .0175"	SW@.400	Ea	10.60	13.60	24.20
Roll 8" x 10' x .011"	SW@.400	Ea	5.40	13.60	19.00
Roll 8" x 10' x .0175"	SW@.400	Ea	14.70	13.60	28.30
Roll 10" x 10' x .011"	SW@.400	Ea	6.75	13.60	20.35
Roll 10" x 10' x .0175"	SW@.400	Ea	9.50	13.60	23.10
Roll 14" x 10' x .011"	SW@.400	Ea	8.55	13.60	22.15
Roll 14" x 10' x .0175"	SW@.400	Ea	10.95	13.60	24.55
Roll 20" x 10' x .011"	SW@.400	Ea	10.60	13.60	24.20
Roll 20" x 10' x .0175"	SW@.400	Ea	29.50	13.60	43.10
Roof edge, 2-1/2" x 1-1/2" x 10'	SW@.350	Ea	3.85	11.90	15.75
Roof edge, 4-1/2" x 1-1/2" x 10'	SW@.350	Ea	11.30	11.90	23.20
Sheet, 1' x 1', black	SW@.250	Ea	7.35	8.50	15.85
Sheet, 1' x 1', mill finish	SW@.250	Ea	6.30	8.50	14.80
Sheet, 1' x 2', black	SW@.250	Ea	14.40	8.50	22.90
Sheet, 1' x 2' x .020", lincaine perforated	SW@.250	Ea	14.50	8.50	23.00
Sheet, 2' x 3', mill finish	SW@.500	Ea	10.70	17.00	27.70
Sheet, 3' x 3', mill finish	SW@.500	Ea	16.20	17.00	33.20
Sheet, 3' x 3', lincaine perforated	SW@.500	Ea	21.90	17.00	38.90
Step flashing, bent, 2-1/2" x 2-1/2" x 7"	SW@.040	Ea	.34	1.36	1.70
Step flashing, bent, 4" x 4" x 8"	SW@.040	Ea	.40	1.36	1.76
Step flashing, bent, 5" x 7", pack of 10	SW@.400	Ea	2.00	13.60	15.60
Trim coil, 24" x 50', white and black	SW@2.50	Ea	60.90	85.00	145.90
Trim coil nails, box of 250	—@.000	Ea	3.70	—	3.70
Z-bar flashing, 2" x 5/8" x 3" x 10'	SW@.350	Ea	2.60	11.90	14.50

Copper flashing 16 ounce per square foot, .021 gauge, bright.

	Craft@Hrs	Unit	Material	Labor	Total
Roll 6" x 10'	SW@.500	Ea	33.60	17.00	50.60
Roll 8" x 10'	SW@.500	Ea	46.80	17.00	63.80
Roll 10" x 10'	SW@.500	Ea	55.50	17.00	72.50
Roll 12" x 100'	SW@5.00	Ea	395.00	170.00	565.00
Roll 14" x 10'	SW@.500	Ea	51.50	17.00	68.50
Roll 20" x 10'	SW@.500	Ea	64.60	17.00	81.60
Roll 24" x 50'	SW@2.50	Ea	415.00	85.00	500.00
Half round dormer, 12" x 24"	BC@.871	Ea	133.00	26.10	159.10
Roof edging, 1-1/2" x 2" x 10'	SW@.350	Ea	20.00	11.90	31.90
Roof edging, 2" x 2" x 10'	SW@.350	Ea	24.15	11.90	36.05
Roof-to-wall flashing, 4" x 6" x 10'	SW@.350	Ea	57.80	11.90	69.70
Shingle flashing, 5" x 7"	SW@.040	Ea	1.55	1.36	2.91
Shingle flashing, 8" x 12"	SW@.040	Ea	4.70	1.36	6.06
Step flashing, 4" x 6" x 14"	SW@.040	Ea	8.40	1.36	9.76
Step flashing, 5" x 7"	SW@.040	Ea	2.10	1.36	3.46
W-valley flashing, 18" x 10'	SW@.500	Ea	78.80	17.00	95.80
Copper nails, 1-1/4", box of 100	—@.000	Box	2.90	—	2.90

Paper-backed copper flashing 3 ounce per square foot. Compatible with ACQ treated lumber. Laminated to Kraft paper and reinforced with fiberglass scrim.

	Craft@Hrs	Unit	Material	Labor	Total
Roll 8" x 20'	SW@.500	Ea	14.40	17.00	31.40
Roll 10" x20'	SW@.500	Ea	18.60	17.00	35.60
Roll 12" x 20'	SW@.500	Ea	21.00	17.00	38.00

	Craft@Hrs	Unit	Material	Labor	Total

Galvanized steel flashing Galvanized mill (plain) finish except a noted. 26 gauge. Galvanized steel flashing has a zinc coating which resists corrosion. Bonderized steel flashing adds a colored phosphate coating which improves paint adhesion.

	Craft@Hrs	Unit	Material	Labor	Total
ACQ compatible flashing, 8" x 20'	SW@1.00	Ea	16.10	34.00	50.10
ACQ compatible flashing, 10" x 20'	SW@1.00	Ea	21.00	34.00	55.00
Angle flashing, 3" x 5" x 10', 90-degree	SW@.500	Ea	6.00	17.00	23.00
Angle flashing, 5" x 5" x 10', 90-degree	SW@.400	Ea	7.85	13.60	21.45
Angle hem flashing, 4" x 3-1/2" x 10'	SW@.400	Ea	5.15	13.60	18.75
Brick mold cover, 3/4" x 12', bonderized	SW@.500	Ea	8.30	17.00	25.30
Counter flash, 1/2" x 6" x 10', galvanized	SW@.400	Ea	6.50	13.60	20.10
Counter flash, 1/2" x 6" x 10', bonderized	SW@.400	Ea	6.80	13.60	20.40
Deck ledger, 2" x 2" x 10', ACQ lumber	SW@.400	Ea	5.75	13.60	19.35
Deck ledger, 2" x 2" x 8', galvanized	SW@.350	Ea	3.95	11.90	15.85
Dormer flashing, 7", black	SW@.250	Ea	7.80	8.50	16.30
Dormer flashing, 7", mill finish	SW@.250	Ea	5.20	8.50	13.70
Drip cap, 1-1/2" x 10'	SW@.400	Ea	2.40	13.60	16.00
Drip edge, 2-1/2" x 1" x 10', white	SW@.400	Ea	3.15	13.60	16.75
Drip edge, 5" x 2' x 10', mill finish	SW@.400	Ea	4.80	13.60	18.40
Drip edge, 6" x 10', Beige,	SW@.400	Ea	7.10	13.60	20.70
Endwall flashing, 3" x 4" x 10'	SW@.400	Ea	9.25	13.60	22.85
Fascia cap, 2" x 10', bonderized	SW@.400	Ea	5.80	13.60	19.40
Fungus preventor, 2-1/2" x 50' zinc strip	SW@1.25	Ea	18.30	42.50	60.80
Girder shield flashing, 2" x 8" x 2" x 10'	SW@.400	Ea	9.10	13.60	22.70
Gravel guard, 6" x 10', type "E"	SW@.400	Ea	3.85	13.60	17.45
Gravel stop, 2" x 4" x 10', bonderized	SW@.400	Ea	6.10	13.60	19.70
Gravel stop, 2" x 4" x 10', galvanized	SW@.400	Ea	5.35	13.60	18.95
Gutter apron, 2' x 3' x 10', galvanized	SW@.400	Ea	4.25	13.60	17.85
Gutter apron, 2" x 3" x 10', bonderized	SW@.400	ea	4.47	13.60	18.07
Half round square dormer, 12" x 12"	BC@.871	Ea	33.60	26.10	59.70
L-flashing, 2" x 2" x 10', bonderized	SW@.400	Ea	4.70	13.60	18.30
L-flashing, 2" x 2" x 10', galvanized	SW@.400	Ea	3.50	13.60	17.10
L-flashing, 4" x 6" x 10', bonderized	SW@.400	Ea	10.50	13.60	24.10
L-flashing, 4" x 6" x 10', galvanized	SW@.400	Ea	8.05	13.60	21.65
Moss preventor, 2-2/3" x 50', zinc	SW@1.25	Ea	26.00	42.50	68.50
Parapet coping, 2" x 8" x 2" x 10'	SW@.500	Ea	11.50	17.00	28.50
Vent pipe flash cone, 1/2" pipe	SW@.250	Ea	3.70	8.50	12.20
Vent pipe flash cone, 1" pipe	SW@.250	Ea	6.30	8.50	14.80
Vent pipe flash cone, 3" pipe	SW@.311	Ea	7.80	10.60	18.40
Vent pipe flash cone, 4" pipe	SW@.311	Ea	8.65	10.60	19.25
Vent pipe flash cone, 5" pipe	SW@.311	Ea	10.20	10.60	20.80
Rain diverter, 7-1/2" x 10'	SW@.400	Ea	5.40	13.60	19.00
Reglet, 3/4" x 10', snap top, bonderized	SW@.500	Ea	7.80	17.00	24.80
Reglet, 3/4" x 10', snap top, mill finish	SW@.500	Ea	5.20	17.00	22.20
Roll 6" x 10', galvanized	SW@.500	Ea	4.80	17.00	21.80
Roll 7" x 10', galvanized	SW@.500	Ea	6.42	17.00	23.42
Roll 8" x 10', bonderized	SW@.500	Ea	10.40	17.00	27.40
Roll 8" x 50', galvanized	SW@2.50	Ea	17.30	85.00	102.30
Roll 10" x 10', galvanized	SW@.500	Ea	8.55	17.00	25.55
Roll 14" x 10', galvanized	SW@.500	Ea	10.85	17.00	27.85
Roll 20" x 10', galvanized	SW@.500	Ea	15.20	17.00	32.20
Roll 20" x 50', galvanized	SW@.500	Ea	46.20	17.00	63.20
Roll valley, 20" x 50', bonderized	SW@2.50	ea	63.00	85.00	148.00
Roof apron, 5" x 10', bonderized	SW@.400	Ea	3.20	13.60	16.80

Lead Flashing

	Craft@Hrs	Unit	Material	Labor	Total
Roof edge, 1-1/2" x 2-" x 10', galvanized	SW@.400	Ea	3.45	13.60	17.05
Roof edge, 1-1/2" x 2 x 10, bonderized	SW@.500	Ea	3.75	17.00	20.75
Roof edge, 6" x 6" x 10', galvanized	SW@.400	Ea	10.65	13.60	24.25
Roof grip edge, 2" x 3" x 10', bonderized	SW@.400	Ea	5.36	13.60	18.96
Roof-to-wall flashing, 2" x 3" x 10'	SW@.400	Ea	4.50	13.60	18.10
Roof-to-wall flashing, 4" x 5" x 10'	SW@.400	Ea	9.90	13.60	23.50
Saddle stock, 4" x 14" x 10'	SW@.400	Ea	8.40	13.60	22.00
Sheet 3' x 4', 26-gauge	SW@.500	Ea	11.30	17.00	28.30
Shingle flashing, 5" x 7", box of 10	SW@.400	Box	8.55	13.60	22.15
Starter strip, 4-1/4" x 10', bonderized	SW@.500	Ea	3.45	17.00	20.45
Step flashing, bent 2" x 3" x 7", .031"	SW@.040	Ea	.25	1.36	1.61
Step flashing, bent, 4" x 12", bonderized	SW@.040	Ea	1.15	1.36	2.51
Termite shield, 8" x 2" x 10'	SW@.400	Ea	7.25	13.60	20.85
Tile roof pan, bent 3-1/2" x 6" x 1/2" x 10'	SW@.400	Ea	9.45	13.60	23.05
Tin repair shingles, 5" x 7", galvanized	SW@.040	Ea	.45	1.36	1.81
Tin repair singles, 8" x 12", bonderized	SW@.040	Ea	.97	1.36	2.33
Water table, 1-1/2" x 10', bonderized	SW@.400	Ea	5.20	13.60	18.80
Water table, 1-1/2" x 10', galvanized	SW@.400	Ea	4.90	13.60	18.50
W-valley, 18" x 10', 28 gauge	SW@.500	Ea	12.90	17.00	29.90
W-valley, 24" x 10', 28 gauge	SW@.500	Ea	15.10	17.00	32.10
W-valley, 24" x 10', bonderized	SW@.500	Ea	21.00	17.00	38.00
Z-bar flashing, 2" x 5/8" x 1" x 10'	SW@.400	Ea	2.55	13.60	16.15
Z-bar flashing, 2" x 1" x 3" x 10'	SW@.400	Ea	5.75	13.60	19.35

Lead flashing

	Craft@Hrs	Unit	Material	Labor	Total
Roll 8" wide, per linear foot	SW@.050	Ea	3.05	1.70	4.75
Roll 10" wide, per linear foot	SW@.066	Ea	3.65	2.24	5.89
Roll 12" wide, per linear foot	SW@.075	Ea	4.20	2.55	6.75
Vent pipe flash cone, 1-1/2" pipe	SW@.311	Ea	8.00	10.60	18.60
Vent pipe flash cone, 2" pipe	SW@.311	Ea	8.50	10.60	19.10
Vent pipe flash cone, 3" pipe	SW@.311	Ea	10.35	10.60	20.95
Vent pipe flash cone, 4" pipe	SW@.311	Ea	11.60	10.60	22.20

Vinyl flashing. For windows, fascia, soffits, siding and posts in contact with ACQ treated lumber.

	Craft@Hrs	Unit	Material	Labor	Total
Deck ledger, J-channel, 4" x 10' x .035"	SW@.350	Ea	10.50	11.90	22.40
Roll 20" x 50', tan	SW@3.50	Ea	42.00	119.00	161.00
Roof edge, 1-7/8" x 10'	SW@.400	Ea	4.80	13.60	18.40
Vent pipe flash cone, 1" to 3" pipe	SW@.250	Ea	4.65	8.50	13.15
Vent pipe flash cone, 3" to 4" pipe	SW@.311	Ea	5.75	10.60	16.35

Sheet Metal Hoods

Entrance hoods, galvanized

	Craft@Hrs	Unit	Material	Labor	Total
48" x 22" x 12", double mold	SW@1.00	Ea	108.00	34.00	142.00
60" x 24" x 12", double mold	SW@1.00	Ea	103.00	34.00	137.00
48" x 22" x 12", scalloped edge	SW@1.00	Ea	134.00	34.00	168.00
60" x 24" x 12", scalloped edge	SW@1.00	Ea	141.00	34.00	175.00

Side hoods, galvanized, square

	Craft@Hrs	Unit	Material	Labor	Total
4" x 4"	SW@.190	Ea	10.30	6.46	16.76
6" x 6"	SW@.216	Ea	17.40	7.34	24.74
7" x 7"	SW@.272	Ea	19.50	9.24	28.74

	Craft@Hrs	Unit	Material	Labor	Total

Shelving

Wood shelving, using 1" thick #3 & Btr paint grade board lumber. Costs shown include an allowance for ledger boards, nails and normal waste. Painting not included.

Closet shelves, with 1" diameter clothes pole and pole support brackets each 3' OC

	Craft@Hrs	Unit	Material	Labor	Total
10" wide shelf	BC@.080	LF	1.07	2.40	3.47
12" wide shelf	BC@.080	LF	1.10	2.40	3.50
18" wide shelf	BC@.100	LF	1.47	3.00	4.47

Linen cabinet shelves, cost of linen closet not included

	Craft@Hrs	Unit	Material	Labor	Total
18" wide shelf	BC@.125	LF	1.12	3.75	4.87
24" wide shelf	BC@.125	LF	1.50	3.75	5.25

Utility shelves, laundry room walls, garage walls, etc.

	Craft@Hrs	Unit	Material	Labor	Total
10" wide shelf	BC@.100	LF	.37	3.00	3.37
12" wide shelf	BC@.100	LF	.48	3.00	3.48
18" wide shelf	BC@.100	LF	.71	3.00	3.71
24" wide shelf	BC@.112	LF	.94	3.36	4.30

Wire shelving kits, vinyl coated steel wire 1" OC. Includes support braces and clamps.

Heavy-duty garage shelf

	Craft@Hrs	Unit	Material	Labor	Total
3' x 16"	BC@.800	Ea	34.00	24.00	58.00
6' x 16"	BC@.800	Ea	48.00	24.00	72.00

Linen shelf

	Craft@Hrs	Unit	Material	Labor	Total
4' x 12"	BC@.600	Ea	13.75	18.00	31.75

Closet organizer. Wire shelf above and clothes hanging rack below

	Craft@Hrs	Unit	Material	Labor	Total
5' wide shelf to 8' high	BC@1.00	Ea	45.20	30.00	75.20
5' wide, with free-slide clothes	BC@1.00	Ea	37.40	30.00	67.40

Glass shelving kits, includes wall support brackets

	Craft@Hrs	Unit	Material	Labor	Total
6" x 18", wall, brass support bracket	BC@.400	Ea	15.00	12.00	27.00
8" x 24", wall, brass support bracket	BC@.400	Ea	20.70	12.00	32.70
6" x 18", wall, white support bracket	BC@.400	Ea	10.70	12.00	22.70
8" x 24", wall, white support bracket	BC@.400	Ea	20.80	12.00	32.80
12" x 12", corner, brass support brackets	BC@.500	Ea	21.60	15.00	36.60
12" x 12", corner, white support brackets	BC@.500	Ea	22.50	15.00	37.50

Glass shelf, 3/8" glass. Add shelf supports below.

	Craft@Hrs	Unit	Material	Labor	Total
24" x 8", clear	BC@.100	Ea	21.00	3.00	24.00
24" x 12", clear	BC@.100	Ea	26.50	3.00	29.50
36" x 8", clear	BC@.100	Ea	27.90	3.00	30.90
36" x 12", clear	BC@.100	Ea	37.40	3.00	40.40
48" x 8", clear	BC@.100	Ea	31.60	3.00	34.60
48" x 12", clear	BC@.100	Ea	48.00	3.00	51.00

Brackets for glass shelf, per bracket

	Craft@Hrs	Unit	Material	Labor	Total
Clear, upright, 4"	BC@.250	Ea	1.70	7.50	9.20
Clear, upright, 24"	BC@.250	Ea	4.00	7.50	11.50
Silver, upright, 4"	BC@.250	Ea	1.70	7.50	9.20
Silver, upright, 24"	BC@.250	Ea	3.95	7.50	11.45

Shower and Tub Doors

Swinging shower doors, with hardware, anodized aluminum frame, tempered safety glass

Door 64" high, silver or gold tone

	Craft@Hrs	Unit	Material	Labor	Total
24" to 26" wide	BG@1.57	Ea	144.00	45.30	189.30
26" to 28" wide	BG@1.57	Ea	149.00	45.30	194.30

Door 66" high, silver or brass frame

	Craft@Hrs	Unit	Material	Labor	Total
27" to 31" wide	BG@1.57	Ea	180.00	45.30	225.30
31" to 36" wide	BG@1.57	Ea	197.00	45.30	242.30

Shower and Tub Enclosures, Subcontract

	Craft@Hrs	Unit	Material	Labor	Total
Door 73" high, silver or gold frame					
32" to 36" wide	BG@1.57	Ea	233.00	45.30	278.30
42" to 48" wide	BG@1.57	Ea	266.00	45.30	311.30
Bi-folding doors, 70" high,					
46" to 48" total width	BG@2.41	Ea	199.00	69.60	268.60
Sliding shower doors, 2 panels, 5/32" tempered glass, anodized aluminum frame, 70" high, outside towel bar					
Non-textured glass, 48" wide	BG@1.57	Ea	235.00	45.30	280.30
Non-textured glass, 56" wide	BG@1.57	Ea	246.00	45.30	291.30
Non-textured glass, 64" wide	BG@1.57	Ea	250.00	45.30	295.30
Texture, opaque, 48" wide	BG@1.57	Ea	270.00	45.30	315.30
Texture, opaque, 56" wide	BG@1.57	Ea	286.00	45.30	331.30
Texture, opaque, 64" wide	BG@1.57	Ea	295.00	45.30	340.30
One panel double mirrored, one bronze tinted	BG@1.57	Ea	305.00	45.30	350.30
Snap-on plastic trim kit, per set of doors	BG@.151	Ea	27.90	4.36	32.26
Sliding shower doors, 2 panels, 1/4" tempered glass, anodized aluminum frame, 70" high, outside towel bar					
Non-textured glass, 48" wide	BG@1.57	Ea	270.00	45.30	315.30
Non-textured glass, 56" wide	BG@1.57	Ea	286.00	45.30	331.30
Non-textured glass, 64" wide	BG@1.57	Ea	295.00	45.30	340.30
Frosted opaque design, 48" wide	BG@1.57	Ea	353.00	45.30	398.30
Frosted opaque design, 56" wide	BG@1.57	Ea	373.00	45.30	418.30
Frosted, opaque design, 64" wide	BG@1.57	Ea	394.00	45.30	439.30
Tub doors, sliding, tempered safety glass, anodized aluminum frame					
Three panels, 57"H x 60"W opening					
Mirrored center panel	BG@2.02	Ea	300.00	58.30	358.30
Reflective glass panels	BG@2.02	Ea	377.00	58.30	435.30
Two panels, 56"H x 60"W opening					
Shatter-resistant plastic panels	BG@1.77	Ea	83.40	51.10	134.50
Semi-clear, tempered pebble pattern	BG@1.77	Ea	122.00	51.10	173.10
Semi-clear, with design	BG@1.77	Ea	207.00	51.10	258.10
One semi-clear, one mirrored	BG@1.77	Ea	169.00	51.10	220.10
Reflective glass panels	BG@1.77	Ea	222.00	51.10	273.10
Accordion folding tub and shower doors, vinyl, chrome frame					
60" x 57" high, tub enclosure	BG@1.57	Ea	83.40	45.30	128.70
30" to 32", 69" high, shower enclosure	BG@1.57	Ea	77.70	45.30	123.00
33" to 36", 69" high, shower enclosure	BG@1.57	Ea	83.40	45.30	128.70
46" to 48", 69" high, shower enclosure	BG@1.57	Ea	88.80	45.30	134.10

Shower and Tub Enclosures, Subcontract No plumbing or plumbing fixtures included.
Shower enclosures, 48"W x 34"D x 66"H, tempered glass, installed, with composition receptor, door and towel bar

		Unit	Material	Labor	Total
Aluminum frame, opaque glass	—	Ea	—	—	521.00
Gold frame, clear glass	—	Ea	—	—	562.00
Bronze frame, bronze tint	—	Ea	—	—	610.00
Corner shower enclosures, 2-wall, aluminum frame, plastic panels, installed, with composition receptor, door and towel bar					
30" to 34"W, 72"H	—	Ea	—	—	603.00
35" to 45"W, 72"H	—	Ea	—	—	506.00
45" to 60"W, 72"H	—	Ea	—	—	550.00
Add for tempered glass panels	—	Ea	—	—	141.00

	Craft@Hrs	Unit	Material	Labor	Total
"Neo-Angle" corner shower cabinets, tempered glass					
3'W x 3'D x 72"H, hammered glass	—	Ea	—	—	574.00
3'4"W x 3'4"D x 72"H, bronze frame	—	Ea	—	—	756.00
Shower cabinet door, tempered opaque glass, installed					
Aluminum frame, 27"W x 66"H	—	Ea	—	—	262.00
Add for extra towel bars, installed	—	Ea	—	—	20.00
Add for clear or tinted glass	—	Ea	—	—	49.00
Tub enclosures, cultured marble					
60" high, 2 panels 30" wide and 1 panel 60" wide	—	Ea	—	—	1,040.00
Tub enclosures, fiberglass, installed, 70"H, 60"W Plexiglass sliding door					
Semi-clear	—	Ea	—	—	248.00
Colors, tints, designs	—	Ea	—	—	285.00
Tub enclosures, tempered reflective glass, installed, aluminum frame					
70" high, two panel frame	—	Ea	—	—	406.00

Fiberglass shower receptor One-piece molded fiberglass shower floor. Slip resistant with a starburst pattern. Drain molded-in with seal for 2" ABS, PVC, iron pipe and strainer. White. Single threshold 7" high.

	Craft@Hrs	Unit	Material	Labor	Total
36", neo-angle	PM@.600	Ea	94.00	21.20	115.20
38", neo-angle	PM@.800	Ea	100.50	28.20	128.70
32" x 32", square	PM@.600	Ea	75.49	21.20	96.69
36" x 36", square	PM@.600	Ea	86.13	21.20	107.33
32" x 48", rectangular	PM@.800	Ea	106.54	28.20	134.74
34" x 48", rectangular	PM@.800	Ea	98.00	28.20	126.20
34" x 60", rectangular	PM@.800	Ea	109.00	28.20	137.20
Cultured marble receptor, 36" x 48"	PM@2.00	Ea	417.00	70.50	487.50

Shutters, Louvered Co-polymer, Exterior Louvered, price per pair. If needed, add hinges and anchors from the installation section below.

	Craft@Hrs	Unit	Material	Labor	Total
15" x 39"	BC@.200	Pr	22.80	6.00	28.80
15" x 43"	BC@.200	Pr	23.70	6.00	29.70
15" x 48"	BC@.200	Pr	24.70	6.00	30.70
15" x 52"	BC@.320	Pr	27.80	9.60	37.40
15" x 55"	BC@.320	Pr	29.50	9.60	39.10
15" x 60"	BC@.320	Pr	30.50	9.60	40.10

Shutters, exterior, installation (Figures in the material column show the cost of anchors and hinges)

	Craft@Hrs	Unit	Material	Labor	Total
Installing shutters on hinges					
Wood frame construction, 1-3/8" throw					
25" to 51" high	BC@.400	Pr	26.00	12.00	38.00
52" to 80" high	BC@.634	Pr	39.10	19.00	58.10
Masonry construction, 4-1/4" throw					
25" to 51" high	BC@.598	Pr	41.70	17.90	59.60
52" to 80" high	BC@.827	Pr	55.60	24.80	80.40
Installing fixed shutters any size above	BC@.390	Pr	4.90	11.70	16.60

Shutters, Interior Movable 1-1/4" louver, 3/4" thick, faux wood, with knobs and hooks, panels attached with hanging strips hinged to panels, finished, priced per set of two shutters with two panels per shutter.

Width x Height	Craft@Hrs	Unit	Material	Labor	Total
23" x 24", per pair	BC@.500	Pr	34.70	15.00	49.70
27" x 20", per pair	BC@.500	Pr	43.90	15.00	58.90
27" x 24", per pair	BC@.500	Pr	50.00	15.00	65.00
27" x 28", per pair	BC@.500	Pr	56.90	15.00	71.90

Siding

	Craft@Hrs	Unit	Material	Labor	Total
27" x 36", per pair	BC@.750	Pr	67.10	22.50	89.60
29" x 20", per pair	BC@.500	Pr	45.00	15.00	60.00
29" x 24", per pair	BC@.500	Pr	51.10	15.00	66.10
29" x 28", per pair	BC@.500	Pr	57.80	15.00	72.80
31" x 20", per pair	BC@.500	Pr	46.10	15.00	61.10
31" x 24", per pair	BC@.750	Pr	48.40	22.50	70.90
31" x 28", per pair	BC@.750	Pr	57.60	22.50	80.10
31" x 36", per pair	BC@.750	Pr	70.00	22.50	92.50
35" x 20", per pair	BC@.750	Pr	48.20	22.50	70.70
35" x 24", per pair	BC@.750	Pr	53.30	22.50	75.80
35" x 28", per pair	BC@.750	Pr	59.80	22.50	82.30
35" x 36", per pair	BC@.750	Pr	73.20	22.50	95.70

Siding See also Siding in the Lumber Section, Plywood and Hardboard.

Fiber cement lap siding, Hardiplank® 6-1/4" width requires 20 planks per 100 square feet of wall. 7-1/4" width requires 17 planks per 100 square feet of wall. 8-1/4" width requires 15 planks per 100 square feet of wall. 9-1/4" width requires 13 planks per 100 square feet of wall. 12" width requires 10 planks per 100 square feet of wall.

	Craft@Hrs	Unit	Material	Labor	Total
6-1/4" x 12', cedarmill	B1@.048	SF	.94	1.31	2.25
6-1/4" x 12', smooth	B1@.048	SF	.97	1.31	2.28
7-1/4" x 12', cedarmill	B1@.048	SF	.88	1.31	2.19
8-1/4" x 12', colonial roughsawn	B1@.048	SF	.92	1.31	2.23
8-1/4" x 12', cedarmill	B1@.048	SF	.91	1.31	2.22
8-1/4" x 12', beaded cedarmill	B1@.048	SF	1.01	1.31	2.32
8-1/4" x 12', beaded	B1@.048	SF	.90	1.31	2.21
8-1/4" x 12', smooth	B1@.048	SF	1.00	1.31	2.31
9-1/4" x 12', cedarmill	B1@.046	SF	1.06	1.26	2.32
12" x 12', smooth	B1@.042	SF	.92	1.15	2.07
12" x 12', cedarmill	B1@.042	SF	.93	1.15	2.08

Fiber cement panel siding, primed, HardiPanel®, 4' x 8' panels. Includes 10% for waste and fasteners.

	Craft@Hrs	Unit	Material	Labor	Total
Sierra, Grooves 8" OC	BC@.042	SF	.83	1.26	2.09
Stucco	BC@.042	SF	.86	1.26	2.12
Primed cement fiber trim, HardiTrim®, 12' lengths, per linear foot					
Rustic, 3/4" x 4", 12'	BC@.032	LF	.78	.96	1.74
Rustic, 3/4" x 6", 12'	BC@.032	LF	1.22	.96	2.18
Primed cement fiber soffit, James Hardie					
Vented, 1/4" x 12" wide	BC@.031	LF	1.27	.93	2.20
Solid, 1/4" x 4' x 8' panel	BC@.031	SF	.85	.93	1.78

Hardboard panel siding Cost per square foot including 5% waste. 4' x 8' panels.

	Craft@Hrs	Unit	Material	Labor	Total
7/16" plain panel	B1@.022	SF	.62	.60	1.22
7/16" grooved 8" OC	B1@.022	SF	.78	.60	1.38
7/16" Sturdi-Panel grooved 8" OC	B1@.022	SF	.58	.60	1.18
7/16" textured board and batt, 12" OC	B1@.022	SF	.59	.60	1.19
7/16" cedar, grooved 8" OC	B1@.022	SF	.55	.60	1.15
7/16" cedar, plain, textured	B1@.022	SF	.63	.60	1.23
1/2" adobe panel	B1@.022	SF	1.03	.60	1.63
Add for 4' x 9' panels	—	SF	.13	—	.13

	Craft@Hrs	Unit	Material	Labor	Total

Hardboard lap siding Primed plank siding. With 8% waste, 6" x 16' covers 5.6 square feet, 8" x 16' covers 7.4 square feet, 9-1/2" x 16' covers 9.7 square feet.

	Craft@Hrs	Unit	Material	Labor	Total
7/16" x 6" x 16', smooth lap 7	B1@.033	SF	.88	.90	1.78
7/16" x 6" x 16', textured cedar	B1@.033	SF	.92	.90	1.82
7/16" x 8" x 16', textured cedar	B1@.031	SF	.88	.85	1.73
7/16" x 8" x 16', smooth	B1@.031	SF	.89	.85	1.74
1/2" x 8" x 16', self-aligning	B1@.031	SF	1.40	.85	2.25
1/2" x 8" x 16', old mill sure lock	B1@.031	SF	1.39	.85	2.24
1/2" x 9-1/2" x 16' designer craftsman	B1@.030	SF	1.34	.82	2.16
7/16" x 6" joint cover	—	Ea	0.51	.00	.51
7/16" x 6" corner	—	Ea	0.79	.00	.79
7/16" x 8" joint cover	—	Ea	0.93	.00	.93
7/16" x 8" corner	—	Ea	0.97	.00	.97
7/16" x 9-1/2" joint cover	—	Ea	0.93	.00	.93
7/16" x 9-1/2" corner	—	Ea	1.00	.00	1.00

OSB siding Oriented strand board. Includes 6% waste.

Lap siding, smooth finish

	Craft@Hrs	Unit	Material	Labor	Total
3/8" x 8" x 16' (190 LF per Sq) $8.40 ea	B1@.033	SF	.99	.90	1.89

Lap siding, rough sawn finish

	Craft@Hrs	Unit	Material	Labor	Total
7/16" x 6" x 16' (250 LF per Sq) $7.09 ea	B1@.033	SF	1.23	.90	2.13
7/16" x 8" x 16' (190 LF per Sq) $8.82 ea	B1@.031	SF	1.05	.85	1.90

Smart Panel II siding

	Craft@Hrs	Unit	Material	Labor	Total
3/8" x 4' x 8', smooth, $19.30 each	B1@.022	SF	.65	.60	1.25
3/8" x 4' x 9', 8" grooved, $28.30 each	B1@.022	SF	.94	.60	1.54
7/16" x 4' x 8', 4" grooved, $27.80 each	B1@.022	SF	.92	.60	1.52
7/16" x 4' x 8', 8" grooved, $28.30 each	B1@.022	SF	.94	.60	1.54
5/8" x 4' x 8', borate treated, $44.50 each	B1@.022	SF	1.47	.60	2.07

Primed Smart trim. Per linear foot of trim, including 10% waste.

	Craft@Hrs	Unit	Material	Labor	Total
4/4" x 4" x 16', $8.40 each	B1@.012	LF	.58	.33	.91
4/4" x 6" x 16'. $12.60 each	B1@.017	LF	.87	.47	1.34
4/4" x 8" x 16', $16.80 each	B1@.021	LF	1.16	.57	1.73

Pre-stained finish, Olympic machine coat, any of 85 colors, factory applied

	Craft@Hrs	Unit	Material	Labor	Total
One-coat application, 5-year warranty	—	SF	.05	—	.05
Two-coat application, 10-year warranty	—	SF	.14	—	.14

Oriented strand board soffit panels. Textured Smart Soffit. Including 10% waste.

	Craft@Hrs	Unit	Material	Labor	Total
3/8" x 4" x 8', $17.50 each	B1@.033	SF	.60	.90	1.50

OSB prefinished "splitless" and corrosion resistant nails

	Craft@Hrs	Unit	Material	Labor	Total
Siding nails (1.7 Lb per Sq)	—	SF	.03	—	.03
Trim nails (.85 Lb per LF)	—	LF	.02	—	.02

Geocel "stain match" acrylic latex caulking, 1/8" to 1/2" bead

	Craft@Hrs	Unit	Material	Labor	Total
Trim (1 tube per CLF)	B1@.001	LF	.03	.03	.06

Sheet metal flashing

	Craft@Hrs	Unit	Material	Labor	Total
7/8", 5/4", 1-5/8" x 10'	B1@.040	LF	.54	1.09	1.63
Brick mold, 10' lengths	B1@.040	LF	.92	1.09	2.01

Under eave soffit vents, screen type

	Craft@Hrs	Unit	Material	Labor	Total
4" or 8" x 16"	B1@.500	Ea	4.77	13.70	18.47

Continuous aluminum soffit vents

	Craft@Hrs	Unit	Material	Labor	Total
2" x 8'	B1@.050	LF	.67	1.37	2.04

Siding

	Craft@Hrs	Unit	Material	Labor	Total
Touch-up stain, Olympic, per coat, applied with brush or roller					
Prime coat (300 SF per gal)	PT@.006	SF	.05	.19	.24
Oil stain (400 SF per gal)	PT@.006	SF	.04	.19	.23
Penofin oil finish (400 SF and $19.00 per gal)	PT@.006	SF	.05	.19	.24
Hammerhead covers, plastic					
Per SF of siding (1 per Sq)	—	SF	.02	—	.02

Plywood siding, 4' x 8' panels. Includes caulking, nails and 6% waste. For T1-11 siding, see Plywood in the Lumber section.

	Craft@Hrs	Unit	Material	Labor	Total
11/32" satin bead, $18.40 each	B1@.025	SF	.62	.68	1.30
3/8" plybead, 16" grooved, $17.50 each	B1@.025	SF	.59	.68	1.27
3/8" southern pine, $20.00 each	B1@.025	SF	.68	.68	1.36
3/8" premium plain fir, $26.80 each	B1@.025	SF	.91	.68	1.59
3/8" rough sawn cedar, $33.60 each	B1@.025	SF	1.13	.68	1.81
5/8" rough sawn pine, $31.50 each	B1@.025	SF	1.06	.68	1.74
5/8" premium rough sawn, $37.80 each	B1@.025	SF	1.28	.68	1.96
5/8" premium fir, #303-6, $39.90 each	B1@.025	SF	1.35	.68	2.03

Yellow pine 4' x 8' plywood batten siding. Battens are 10" wide each 12" on center. Includes caulking, nails and 6% waste.

	Craft@Hrs	Unit	Material	Labor	Total
19/32" thick, $26.70 per panel	B1@.025	SF	.90	.68	1.58
5/8" ACQ treated, $43.20 per panel	B1@.025	SF	1.28	.68	1.96

Aluminum sheet siding

Aluminum corrugated 4-V x 2-1/2", plain or embossed finish. Includes 15% waste and coverage loss.

	Craft@Hrs	Unit	Material	Labor	Total
17 gauge, 26" x 6' to 24'	B1@.034	SF	2.23	.93	3.16
19 gauge, 26" x 6' to 24'	B1@.034	SF	2.30	.93	3.23
Rubber filler strip, 3/4" x 7/8" x 6'	—	LF	.34	—	.34

Flashing for corrugated aluminum siding, embossed

	Craft@Hrs	Unit	Material	Labor	Total
End wall, 10" x 52"	B1@.048	Ea	3.98	1.31	5.29
Side wall, 7-1/2" x 10'	B1@.015	LF	1.71	.41	2.12

Aluminum smooth 24 gauge, horizontal patterns, non-insulated

	Craft@Hrs	Unit	Material	Labor	Total
8" or double 4" widths, acrylic finish	B1@2.77	Sq	158.00	75.80	233.80
12" widths, bonded vinyl finish	B1@2.77	Sq	161.00	75.80	236.80
Add for foam backing	—	Sq	32.50	—	32.50
Starter strip	B1@.030	LF	.38	.82	1.20
Inside and outside corners	B1@.033	LF	1.14	.90	2.04
Casing and trim	B1@.033	LF	.38	.90	1.28
Drip cap	B1@.044	LF	.39	1.20	1.59

Galvanized steel siding

28 gauge, 27-1/2" wide (includes 20% coverage loss)

	Craft@Hrs	Unit	Material	Labor	Total
6' to 12' standard lengths	B1@.034	SF	1.06	.93	1.99

26 gauge, 26" wide (includes 15% coverage loss)

	Craft@Hrs	Unit	Material	Labor	Total
6' to 12' standard lengths	B1@.034	SF	1.30	.93	2.23

Shingle siding

Cedar rebutted sidewall shingles, Western red cedar
No. 1 western red cedar rebutted and rejointed 18" shingles, carton covers 50 square feet

	Craft@Hrs	Unit	Material	Labor	Total
at 7" exposure, $100 per carton	R1@3.52	Sq	202.00	103.00	305.00

	Craft@Hrs	Unit	Material	Labor	Total

Red cedar perfects 5X green, 16" long, 7-1/2" exposure, 5 shingles are 2" thick (5/2)
3 bundles cover 100 square feet (one square) including waste. Cost shown includes fasteners

	Craft@Hrs	Unit	Material	Labor	Total
#1	B1@3.86	Sq	202.00	106.00	308.00
#2 red label	B1@3.86	Sq	113.00	106.00	219.00
#3	B1@3.86	Sq	81.60	106.00	187.60

Panelized shingle siding, Shakertown, KD, regraded shingle on plywood backing, no waste included.

8' long x 7" wide single course

	Craft@Hrs	Unit	Material	Labor	Total
Colonial 1	B1@.033	SF	4.02	.90	4.92
Cascade	B1@.033	SF	3.89	.90	4.79

8' long x 14" wide, single course

	Craft@Hrs	Unit	Material	Labor	Total
Colonial 1	B1@.033	SF	4.02	.90	4.92

8' long x 14" wide, double course

	Craft@Hrs	Unit	Material	Labor	Total
Colonial 2	B1@.033	SF	4.40	.90	5.30
Cascade Classic	B1@.033	SF	3.89	.90	4.79

Decorative shingles, 5" wide x 18" long, Shakertown, fancy cuts, 9 hand-shaped patterns

	Craft@Hrs	Unit	Material	Labor	Total
7-1/2" exposure	B1@.035	SF	4.70	.96	5.66
Deduct for 10" exposure	—	SF	- .50	—	- .50

Vinyl siding Solid vinyl double 4" traditional lap profile, 12' 6" long panels, 8" exposure, embossed wood grain texture. Each panel covers 8.33 SF. Case of 24 panels covers 200 square feet. Cost per square (100 square feet).

	Craft@Hrs	Unit	Material	Labor	Total
White, .040"	B1@3.34	Sq	51.70	91.40	143.10
Colors, .040"	B1@3.34	Sq	53.00	91.40	144.40
White, .042"	B1@3.34	Sq	62.40	91.40	153.80
Colors, .042"	B1@3.34	Sq	61.80	91.40	153.20
Starter strip	B1@.030	LF	.34	.82	1.16
J channel trim at wall openings	B1@.033	LF	.35	.90	1.25
Outside or inside corner	B1@.033	LF	1.38	.90	2.28
Casing and trim	B1@.033	LF	.35	.90	1.25
Add for R-3 insulated vinyl siding	B1@.350	Sq	30.00	9.58	39.58

Thermoplastic resin siding Nailite™ 9/10" thick panels installed over sheathing. Add for trim board, corners, starter strips, j-channels and mortar fill. Based on five nails per piece. Includes 10% waste

	Craft@Hrs	Unit	Material	Labor	Total
Hand-split shake, 41-3/8" x 18-3/4"	B1@.043	SF	2.04	1.18	3.22
Hand-laid brick, 44-1/4" x 18-5/8"	B1@.043	SF	2.17	1.18	3.35
Hand-cut stone, 44-1/4" x 18-5/8"	B1@.043	SF	2.17	1.18	3.35
Rough sawn cedar, 59-1/4" x 15"	B1@.043	SF	2.26	1.18	3.44
Perfection plus cedar, 36-1/8" x 15"	B1@.043	SF	2.26	1.18	3.44

Ledge trim

	Craft@Hrs	Unit	Material	Labor	Total
Brick ledge trim	B1@.033	LF	4.71	.90	5.61
Stone ledge trim	B1@.033	LF	4.71	.90	5.61

Corners

	Craft@Hrs	Unit	Material	Labor	Total
Hand-split shake, 4" x 18"	B1@.030	LF	8.90	.82	9.72
Rough sawn cedar 3" x 26"	B1@.030	LF	9.43	.82	10.25
Hand-laid brick 4" x 18"	B1@.030	LF	9.43	.82	10.25
Hand-cut stone 4" x 18"	B1@.030	LF	9.43	.82	10.25
Perfection plus cedar 3" x 13"	B1@.030	LF	7.85	.82	8.67

Starter strips

	Craft@Hrs	Unit	Material	Labor	Total
12' universal	B1@.025	LF	.47	.68	1.15

Skylights

	Craft@Hrs	Unit	Material	Labor	Total
J-channels					
Hand-split shake 3/4" opening	B1@.025	LF	.72	.68	1.40
Hand-split shake 1-1/4" flexible opening	B1@.027	LF	1.42	.74	2.16
Rough sawn cedar 3/4" opening	B1@.025	LF	.53	.68	1.21
Rough sawn cedar 1-1/4" flexible opening	B1@.027	LF	1.42	.74	2.16
Hand-laid brick 3/4" opening	B1@.025	LF	.53	.68	1.21
Hand-cut stone 3/4" opening	B1@.025	LF	.53	.68	1.21
Perfection plus cedar 3/4" opening	B1@.025	LF	.53	.68	1.21
Thermoplastic siding accessories					
Mortar fill, 12 tubes and 200 LF per carton	—	Ea	46.60	—	46.60
Touch up paint, 6 oz. aerosol can	—	Ea	4.20		4.20

Skylights Polycarbonate dome, clear transparent or tinted, roof opening sizes, price each, Crestline Skylights, curb or flush mount, self-flashing. Comfort

	Craft@Hrs	Unit	Material	Labor	Total
Single dome, 2' x 2'	B1@2.58	Ea	51.50	70.60	122.10
Single dome, 2' x 4'	B1@2.58	Ea	73.20	70.60	143.80
Double dome, 4" x 4"	B1@2.84	Ea	146.00	77.70	223.70

Tempered over laminated low-E argon gas filled doubled insulated fixed glass skylights. Includes costs for flashing. Conforms to all building codes. For installation on roofs with 15 to 85 degree slope. Labor costs are for installation of skylight and flashing. Roll shade costs are for factory attached shades. No carpentry or roofing included.

Operable skylight, Velux®, VS® Comfort™ models. Prices include EDL® flashing	Craft@Hrs	Unit	Material	Labor	Total
21-1/2" x 38-1/2", #104	B1@3.42	Ea	712.00	93.60	805.60
21-1/2" x 46-3/8", #106	B1@3.42	Ea	759.00	93.60	852.60
21-1/2" x 55", #108	B1@3.42	Ea	813.00	93.60	906.60
30-5/8" x 38-1/2", #304	B1@3.42	Ea	781.00	93.60	874.60
30-5/8" x 46-3/8", #306	B1@3.42	Ea	813.00	93.60	906.60
30-5/8" x 55", #308	B1@3.52	Ea	854.00	96.30	950.30
44-3/4" x 46-1/2", #606	B1@3.52	Ea	921.00	96.30	1,017.30
Add for electric operation	—	%	—	—	200.0
Add for ComfortPlus™ Laminated Low E	—	%	10.0	—	—
Add for roll shade, natural					
21-1/2" x 27-1/2" to 55" width	—	Ea	58.00	—	58.00
30-5/8" x 38-1/2" to 55" width	—	Ea	66.70	—	66.70
Fixed skylight, Velux®, FS® Comfort™ models. Prices include EDL® flashing					
21-1/2" x 27-1/2", #101	B1@3.42	Ea	138.00	93.60	231.60
21-1/2" x 38-1/2", #104	B1@3.42	Ea	169.00	93.60	262.60
21-1/2" x 46-3/8", #106	B1@3.42	Ea	191.00	93.60	284.60
21-1/2" x 55", #108	B1@3.42	Ea	193.00	93.60	286.60
21-1/2" x 70-7/8", #112	B1@3.42	Ea	249.00	93.60	342.60
30-5/8" x 38-1/2", #304	B1@3.42	Ea	203.00	93.60	296.60
30-5/8" x 46-3/8", #306	B1@3.42	Ea	231.00	93.60	324.60
30-5/8" x 55", #308	B1@3.52	Ea	258.00	96.30	354.30
44-3/4" x 27-1/2", #601	B1@3.52	Ea	196.00	96.30	292.30
44-3/4" x 46-1/2", #606	B1@3.52	Ea	252.00	96.30	348.30
Add for roll shade, natural					
21-1/2" x 27-1/2" to 70-7/8" width	—	Ea	55.00	—	55.00
30-5/8" x 38-1/2" to 55" width	—	Ea	63.50	—	63.50

	Craft@Hrs	Unit	Material	Labor	Total

Fixed ventilation, with removable filter on ventilation flap, Velux® model FSF® ComfortPlus™. Prices include EDL® flashing

	Craft@Hrs	Unit	Material	Labor	Total
21-1/2" x 27-1/2", #101	B1@3.14	Ea	195.00	85.90	280.90
21-1/2" x 38-1/2", #104	B1@3.14	Ea	228.00	85.90	313.90
21-1/2" x 46-1/2", #106	B1@3.14	Ea	257.00	85.90	342.90
21-1/2" x 70-7/8", #108	B1@3.42	Ea	282.00	93.60	375.60
30-5/8" x 38-1/2", #304	B1@3.14	Ea	278.00	85.90	363.90
30-5/8" x 46-3/8", #306	B1@3.14	Ea	318.00	85.90	403.90
30-5/8" x 55", #308	B1@3.39	Ea	344.00	92.80	436.80
44-3/4" x 27-1/2", #601	B1@3.36	Ea	303.00	91.90	394.90
44-3/4" x 46-1/2", #606	B1@3.39	Ea	399.00	92.80	491.80

Add for roll shade, natural

		Unit	Material		Total
21-1/2" x 27-1/2" to 70-7/8" width	—	Ea	60.90	—	60.90
30-5/8" x 38-1/2" to 55" width	—	Ea	68.80	—	68.80

Curb mounted fixed skylight, Velux model FCM™

	Craft@Hrs	Unit	Material	Labor	Total
22-1/2" x 22-1/2", #2222	B1@3.14	Ea	110.00	85.90	195.90
22-1/2" x 34-1/2", #2234	B1@3.14	Ea	142.00	85.90	227.90
22-1/2" x 46-1/2", #2246	B1@3.14	Ea	154.00	85.90	239.90
30-1/2" x 30-1/2", #3030	B1@3.14	Ea	170.00	85.90	255.90
46-1/2" x 46-1/2", #4646	B1@3.14	Ea	261.00	85.90	346.90

Skywindows

Skywindow, E-Class. Ultraseal self-flashing units. Clear insulated glass, flexible flange, no mastic or step flashing required. Other glazings available. Integral skyshade optional. Wasco

	Craft@Hrs	Unit	Material	Labor	Total
22-1/2" x 22-1/2", fixed	B1@0.75	Ea	227.00	20.50	247.50
22-1/2" x 38-1/2", fixed	B1@0.75	Ea	260.00	20.50	280.50
22-1/2" x 46-1/2", fixed	B1@0.75	Ea	307.00	20.50	327.50
30-1/2" x 30-1/2", fixed	B1@0.75	Ea	307.00	20.50	327.50
30-1/2" x 46-1/2", fixed	B1@0.75	Ea	374.00	20.50	394.50
46-1/2" x 46-1/2", fixed	B1@0.75	Ea	420.00	20.50	440.50
22-1/2" x 22-1/2", vented	B1@0.75	Ea	429.00	20.50	449.50
22-1/2" x 38-1/2", vented	B1@0.75	Ea	465.00	20.50	485.50
22-1/2" x 46-1/2", vented	B1@0.75	Ea	526.00	20.50	546.50
30-1/2" x 30-1/2", vented	B1@0.75	Ea	498.00	20.50	518.50
30-1/2" x 46-1/2", vented	B1@0.75	Ea	571.00	20.50	591.50

Skywindow, Excel-10. Step flash/pan flashed units. Insulated clear tempered glass. Deck mounted step flash unit. Other glazings available. Wasco

	Craft@Hrs	Unit	Material	Labor	Total
22-1/2" x 22-1/2", fixed	B1@2.35	Ea	210.00	64.30	274.30
22-1/2" x 38-1/2", fixed	B1@2.35	Ea	221.00	64.30	285.30
22-1/2" x 46-1/2", fixed	B1@2.35	Ea	232.00	64.30	296.30
30-1/2" x 30-1/2", fixed	B1@2.35	Ea	249.00	64.30	313.30
30-1/2" x 46-1/2", fixed	B1@2.35	Ea	300.00	64.30	364.30
22-1/2" x 22-1/2", vented	B1@2.35	Ea	352.00	64.30	416.30
22-1/2" x 30-1/2", vented	B1@2.35	Ea	379.00	64.30	443.30
22-1/2" x 46-1/2", vented	B1@2.35	Ea	453.00	64.30	517.30
30-1/2" x 30-1/2", vented	B1@2.35	Ea	436.00	64.30	500.30
30-1/2" x 46-1/2", vented	B1@2.35	Ea	486.00	64.30	550.30
Add for flashing kit	—	Ea	71.40	—	71.40

Top Hinged Roof Windows

	Craft@Hrs	Unit	Material	Labor	Total

Skywindow, Genra-1. Low profile insulated glass skywindow.

Deck mount, clear tempered glass, self-flashing models. Other glazings available. Wasco

	Craft@Hrs	Unit	Material	Labor	Total
22-1/2" x 22-1/2", fixed	B1@2.35	Ea	187.00	64.30	251.30
22-1/2" x 38-1/2", fixed	B1@2.35	Ea	221.00	64.30	285.30
22-1/2" x 46-1/2", fixed	B1@2.35	Ea	271.00	64.30	335.30
30-1/2" x 30-1/2", fixed	B1@2.35	Ea	255.00	64.30	319.30
30-1/2" x 46-1/2", fixed	B1@2.35	Ea	311.00	64.30	375.30
46-1/2" x 46-1/2", fixed	B1@2.35	Ea	397.00	64.30	461.30
22-1/2" x 22-1/2", vented	B1@2.35	Ea	366.00	64.30	430.30

Top hinged roof windows Aluminum-clad wood frame, double insulated tempered over laminated low E gas filled double insulated glass, including prefabricated flashing, exterior awning, interior roller blind and insect screen. Sash rotates 180 degrees. For installation on roofs with 20 to 55 degree slope. Labor costs are for installation of roof window, flashing and sun screening accessories. No carpentry or roofing work other than curb included. Add for interior trim. Listed by actual unit dimensions, top hung, Velux® model GPL™ Comfort™ Roof Window

	Craft@Hrs	Unit	Material	Labor	Total
30-5/8" wide x 55" high, #308	B1@3.42	Ea	620.00	93.60	713.60
44-3/4" wide x 46-3/8" high, #606	B1@3.52	Ea	693.00	96.30	789.30

Soffit Systems

Baked enamel finish, 6" fascia, J-channel, 8" perforated or solid soffit, .019 gauge aluminum

	Craft@Hrs	Unit	Material	Labor	Total
12" soffit	B1@.047	LF	3.24	1.29	4.53
18" soffit	B1@.055	LF	3.75	1.50	5.25
24" soffit	B1@.060	LF	4.26	1.64	5.90
Vinyl soffit systems					
12" soffit	B1@.030	LF	2.78	.82	3.60
18" soffit	B1@.030	LF	3.17	.82	3.99
24" soffit	B1@.030	LF	3.59	.82	4.41
Aluminum fascia alone					
4" fascia	B1@.030	LF	.87	.82	1.69
6" fascia	B1@.030	LF	1.05	.82	1.87
8" fascia	B1@.030	LF	1.24	.82	2.06

Soil Testing

Field observation and testing

	Craft@Hrs	Unit	Material	Labor	Total
Soil technician (average)	—	Hr	—	—	60.00
Soil testing, per test					
Moisture	—	Ea	—	—	11.00
Bulk density (loose or core)	—	Ea	—	—	15.00
Bulk density (clod)	—	Ea	—	—	25.00
Porosity	—	Ea	—	—	15.00
Infiltration rate (in/hr)	—	Ea	—	—	25.00
Expansion index test	—	Ea	—	—	90.00
Report on compacted fill (costs vary widely)	—	Ea	—	—	1,500.00

Foundation investigation (bearing capacity, lateral loads, piles, seismic, stability, settlement for design purposes) and preliminary soils investigation (depth of fill, classification and profile of soil, expansion, shrinkage, grading, drainage recommendations). Including drilling and sampling costs

	Craft@Hrs	Unit	Material	Labor	Total
Drill rig (driller and helper)	—	Hr	—	—	140.00
Staff engineer	—	Hr	—	—	70.00
Project engineer, RCE	—	Hr	—	—	79.00
Principal engineer	—	Hr	—	—	132.00

	Craft@Hrs	Unit	Material	Labor	Total
Transportation	—	Mile	—	—	.75
Transportation	—	Hr	—	—	6.90
Laboratory testing (general)	—	Hr	—	—	60.00
Laboratory testing (varies on individual test basis)	—	Ea	—	—	100.00
Report preparation					
Analysis, conclusions, recommendations	—	Ea	—	—	2,000.00
Court preparation (Principal Engineer)	—	Hr	—	—	150.00
Expert witness court appearance	—	Hr	—	—	250.00
Add for hazardous studies	—	%	—	—	20.0

Soil Treatments, Subcontract Costs based on total square footage inside of, and including, foundation. Includes inside and outside perimeter of foundation, interior foundation walls, and underneath all slabs. Three-plus year re-application guarantee against existence of termites. Use $300 as a minimum job charge.
Dursban T.C, Primus, or Equity. Mix applied per manufacturers' and state and federal specifications

Complete application	—	SF	—	—	.35

Solar Photovoltaic Electrical Systems, Subcontract Costs are for solar electric generating systems for use where commercial power is not available. These estimates assume 8 clear sunlight hours per day. Costs will be higher where less sun time is available. Photocomm with Kyocera modules.
Small remote vacation home system (12 volt DC). 64 watt solar electric module (98 amp hours), digital voltage regulator, mounting rack and wiring

384 watt-hours per day	—	LS	—	—	1,150.00

Larger remote primary home system (120 volt AC). Includes eight L16 storage batteries (1,050 amp hours), voltage regulator, mounting rack, 2,400 watt DC to AC inverter

2,880 watt-hours per day	—	LS	—	—	6,890.00

Basic two room solar power package Simpler Solar Systems

Includes three 48 watt PV panels, one 12 amp voltage regulator, one 220 Ah battery with hydrocaps, one 600 watt inverter, mounting frame and hardware, and two receptacles

12 amp regulator, with 600 watt inverter	—	LS	—	—	2,780.00

Small sized home solar power package Simpler Solar Systems

90% solar powered system for 1,500 SF home includes six 60 watt PV panels, one 750 Ah battery, 20 amp voltage regulator and DC distribution box, and one 2.2 kW inverter. Costs also include one solar hot water system, one water source heat pump, one 19 cubic foot refrigerator freezer, one ringer type washing machine, six 12 volt lights and three 12 volt ceiling fans

20 amp regulator, 12 volt system	—	LS	—	—	14,500.00

Medium sized home solar power package Simpler Solar Systems. 80% solar powered system for 2,500 SF home includes ten 45 watt PV panels, one 750 Ah battery, one 30 amp voltage regulator, and one 2 kW inverter with charger. Costs also include one solar domestic hot water system and 22 assorted 12 volt light fixtures

30 amp regulator, 12 volt system	—	LS	—	—	8,780.00

Photovoltaic roofing systems Photovoltaic panels and shingles. Converts sunlight into electric energy and distributes it directly to the building. The solar electric roofing products are configured in series or parallel on the roof deck to form an array. The array is then integrated into the roof following standard roofing specifications. Installation time does not include electrical connections. United Solar Roofing Systems, http://www.unisolar.com

	Craft@Hrs	Unit	Material	Labor	Total
Architectural or structural standing photovoltaic seam panels, 9-1/2' x 16"					
60 watt panel	RR@.035	SF	72.40	1.18	73.58
64 watt panel	RR@.035	SF	77.40	1.18	78.58
Architectural or structural standing photovoltaic seam panels, 18'4" x 16"					
120 watt panel	RR@.033	SF	75.50	1.11	76.61
128 watt panel	RR@.033	SF	80.80	1.11	81.91

Solar Pool Heating System

	Craft@Hrs	Unit	Material	Labor	Total
Photovoltaic shingles, 12" x 86" panel					
17 watt shingle panel, 3 SF exposure	RR@.030	SF	31.10	1.01	32.11
Photovoltaic roofing system accessories					
Add for combiner box	—	Ea	577.00	—	577.00
Add for 2.5 kW inverter	—	Ea	3,150.00	—	3,150.00
Add for 4.0 kW inverter	—	Ea	3,910.00	—	3,910.00
Add for 6 volt DC 350 Ah storage batteries	—	Ea	155.00	—	155.00
Add for utility ground fault protection device	—	Ea	278.00	—	278.00
Solar pool heating system Simpler Solar Systems					
Pool system, eight panels, valves, and piping	—	LS	—	—	3,600.00
Spa system, one panel, valves and piping	—	LS	—	—	1,060.00
Spas, Fiberglass, Outdoor Includes pump and standard heater. Add the cost of a concrete slab base.					
http://www.holiday-pools.com					
5' x 7' spa, 18 jets, 210 Gal	P1@3.5	Ea	3,590.00	105.00	3,695.00
68" x 67" x 31" deep, 12 jets, seats 4, 200 Gal	P1@4.0	Ea	3,590.00	120.00	3,710.00
7' x 7' spa, 20 jets, 350 Gal	P1@4.0	Ea	3,830.00	120.00	3,950.00
44" x 90" x 30" deep, 24 jets, seats 2, 120 Gal	P1@4.0	Ea	4,190.00	120.00	4,310.00
89" x 120" x 36" deep, 41 jets, seats 8, 615 Gal	P1@4.0	Ea	9,940.00	120.00	10,060.00
Add for 50,000 Btu heater	P1@1.5	Ea	814.00	45.00	859.00
Add for electrical connection	—	Ea	—	—	200.00

Stairs

Factory cut and assembled straight closed box stairs, unfinished, 36" width

	Craft@Hrs	Unit	Material	Labor	Total
Oak treads with 7-1/2" risers, price per riser					
3'0" wide	B1@.324	Ea	105.00	8.86	113.86
3'6" wide	B1@.324	Ea	136.00	8.86	144.86
4'0" wide	B1@.357	Ea	146.00	9.77	155.77
Deduct for pine treads and plywood risers	—	%	-22.0	—	—
Add for prefinished assembled stair rail with					
balusters and newel, per riser	B1@.257	Ea	53.60	7.03	60.63
Add for prefinished handrail, with					
brackets and balusters	B1@.135	LF	15.70	3.69	19.39

Basement stairs, open riser, delivered to job site assembled, 12 treads, 2'11-3/4" wide.

	Craft@Hrs	Unit	Material	Labor	Total
13 risers at 7-13/16" riser height	B1@4.00	Ea	233.00	109.00	342.00
13 risers at 8" riser height	B1@4.00	Ea	239.00	109.00	348.00
13 risers at 8-3/16" riser height	B1@4.00	Ea	244.00	109.00	353.00
Add for handrail	B1@.161	LF	9.74	4.40	14.14

Curved stair, clear oak, self-supporting, 8'6" radii, factory cut, unfinished, unassembled

	Craft@Hrs	Unit	Material	Labor	Total
Open one side (1 side against wall)					
8'9" to 9'4" rise	B1@24.6	Ea	5,410.00	673.00	6,083.00
9'5" to 10'0" rise	B1@24.6	Ea	5,840.00	673.00	6,513.00
10'1" to 10'8" rise	B1@26.5	Ea	6,270.00	725.00	6,995.00
10'9" to 11'4" rise	B1@26.5	Ea	6,710.00	725.00	7,435.00
Open two sides					
8'9" to 9'4" rise	B1@30.0	Ea	9,740.00	821.00	10,561.00
9'5" to 10'0" rise	B1@30.0	Ea	10,500.00	821.00	11,321.00
10'1" to 10'8" rise	B1@31.9	Ea	11,100.00	873.00	11,973.00
10'9" to 11'4" rise	B1@31.9	Ea	11,800.00	873.00	12,673.00
Add for newel posts	B1@.334	Ea	67.60	9.14	76.74

	Craft@Hrs	Unit	Material	Labor	Total
Spiral stairs, aluminum, delivered to job site unassembled, 7-3/4" or 8-3/4" riser heights, 5' diameter					
85-1/4" to 96-1/4" with 10 treads	B1@9.98	Ea	1,680.00	273.00	1,953.00
93" to 105" with 11 treads	B1@10.7	Ea	1,840.00	293.00	2,133.00
100-3/4" to 113-3/4" with 12 treads	B1@11.3	Ea	1,990.00	309.00	2,299.00
108-1/2" to 122-1/2" with 13 treads	B1@12.0	Ea	2,140.00	328.00	2,468.00
116-1/4" to 131-1/4" with 14 risers	B1@12.6	Ea	2,280.00	345.00	2,625.00
124" to 140" with 15 risers	B1@12.8	Ea	2,460.00	350.00	2,810.00
131-3/4" to 148-3/4" with 16 risers	B1@13.5	Ea	2,590.00	369.00	2,959.00
138-1/2" to 157-1/2" with 17 risers	B1@14.2	Ea	2,740.00	389.00	3,129.00
147-1/4" to 166-1/4" with 18 risers	B1@14.8	Ea	2,910.00	405.00	3,315.00
Deduct for 4' diameter	—	%	-20.0	—	—
Deduct for 4'6" diameter	—	%	-10.0	—	—
Add for additional aluminum treads	—	Ea	155.00	—	155.00
Add for oak tread inserts	—	Ea	45.50	—	46.50
Spiral stairs, red oak, FOB factory, double handrails, 5' diameter					
74" to 82" with 9 risers	B1@16.0	Ea	3,180.00	438.00	3,618.00
82" to 90" with 10 risers	B1@16.0	Ea	3,420.00	438.00	3,858.00
90" to 98" with 11 risers	B1@16.0	Ea	3,480.00	438.00	3,918.00
98" to 106" with 12 risers	B1@16.0	Ea	3,540.00	438.00	3,978.00
106" to 114" with 13 risers	B1@20.0	Ea	3,630.00	547.00	4,177.00
114" to 122" with 14 risers	B1@20.0	Ea	3,750.00	547.00	4,297.00
122" to 130" with 15 risers	B1@20.0	Ea	3,950.00	547.00	4,497.00
Add for 6' diameter	—	%	20.0	—	—
Add for 8'6" diameter	—	%	25.0	—	—
Add for mahogany, any of above	—	%	10.0	—	—
Add for straight rail, unbored	B1@.060	LF	11.20	1.64	12.84
Add for straight nosing, unbored	B1@.060	LF	6.89	1.64	8.53
Add for balusters, 7/8" x 42"	B1@.334	Ea	4.49	9.14	13.63
Add for newel posts 1-7/8" x 44"	B1@.334	Ea	29.20	9.14	38.34
Spiral stairs, steel, tubular steel handrail, composition board treads, delivered unassembled, 5' diameter					
86" to 95" with 10 risers	B1@9.98	Ea	1,030.00	273.00	1,303.00
95" to 105" with 11 risers	B1@10.7	Ea	1,140.00	293.00	1,433.00
105" to 114" with 12 risers	B1@11.3	Ea	1,210.00	309.00	1,519.00
114" to 124" with 13 risers	B1@12.0	Ea	1,320.00	328.00	1,648.00
124" to 133" with 14 risers	B1@12.7	Ea	1,510.00	347.00	1,857.00
Deduct for 4' diameter	—	%	-20.0	—	—
Add for 6' diameter	—	%	24.0	—	—
Add for oak treads	—	%	25.0	—	—
Add for oak handrail	—	%	45.0	—	—
Straight stairs, red oak, factory cut, unassembled , 36" tread, includes two 1-1/2" x 3-1/2" handrails and 7/8" round balusters.					
70" to 77" with 10 risers	B1@12.0	Ea	1,440.00	328.00	1,768.00
77" to 85" with 11 risers	B1@12.0	Ea	1,640.00	328.00	1,968.00
85" to 93" with 12 risers	B1@12.0	Ea	1,770.00	328.00	2,098.00
93" to 100" with 13 risers	B1@14.0	Ea	1,920.00	383.00	2,303.00
100" to 108" with 14 risers	B1@14.0	Ea	2,120.00	383.00	2,503.00
108" to 116" with 15 risers	B1@14.0	Ea	2,210.00	383.00	2,593.00
116" to 124" with 16 risers	B1@16.0	Ea	2,400.00	438.00	2,838.00
124" to 131" with 17 risers	B1@16.0	Ea	2,490.00	438.00	2,928.00
131" to 139" with 18 risers	B1@16.0	Ea	2,790.00	438.00	3,228.00
Add for 44" tread, any of above	—	Ea	91.80	—	91.80
Add for straight rail, unbored	B1@.060	LF	10.80	1.64	12.44

Job-built Stairways

	Craft@Hrs	Unit	Material	Labor	Total
Add for straight nosing	B1@.060	LF	6.75	1.64	8.39
Add for round balusters, 7/8" x 42"	B1@.334	Ea	4.25	9.14	13.39
Add for square balusters,					
1-1/8" x 1-1/8" x 44"	B1@.334	Ea	8.75	9.14	17.89
Add for newel posts, 3" x 3" x 44"	B1@.334	Ea	38.50	9.14	47.64

Job-built stairways 3 stringers cut from 2" x 12" Douglas fir material, treads and risers from 3/4" CDX plywood. Cost per 7-1/2" rise, 36" wide

	Craft@Hrs	Unit	Material	Labor	Total
Using 2" Douglas fir, per MBF	—	MBF	728.00	—	728.00
Using 3/4" CDX plywood, per MSF	—	MSF	490.00	—	490.00
Using 1" x 6" white pine, per MBF	—	MBF	612.00	—	612.00
Straight run, 8'0" to 10'0" rise, per riser	B1@.530	Ea	9.49	14.50	23.99
"L" or "U" shape, per riser, add for landings	B1@.625	Ea	11.10	17.10	28.20
Semi-circular, repetitive tract work, per riser	B1@.795	Ea	11.10	21.80	32.90
Landings, from 2" x 6" Douglas fir and 3/4" CDX plywood surfaced,					
per SF of landing surface	B1@.270	SF	2.38	7.39	9.77

Stair treads and risers, unfinished. Including layout, cutting, fitting, and installation of treads, risers and typical tread nosing. Includes a miter return on one side of each tread. 9' rise is usually 14 treads and 15 risers

	Craft@Hrs	Unit	Material	Labor	Total
Yellow pine treads, 11-1/2" x 48"	—	Ea	11.40	—	11.40
Oak treads, 11-1/2" x 48"	—	Ea	18.00	—	18.00
Oak curved first tread, 10-3/8" x 48"	—	Ea	29.50	—	29.50
Oak risers, 8" x 48"	—	Ea	15.50	—	15.50
Oak nosing	—	LF	1.98	—	1.98
Oak return nosing, 13"	—	Ea	2.08	—	2.08
Oak treads, risers and nosing for 9' rise stairway	BC@2.50	Ea	550.00	75.00	625.00
Skirt board with dadoes for treads and risers	BC@3.75	Ea	23.00	113.00	136.00
Skirt board on open side, mitered risers	BC@5.50	Ea	23.00	165.00	188.00
Oak shoe rail with fillet, 1-1/4" plowed,					
2-1/2" x 3/4"	BC@.030	LF	2.51	.90	3.41
Oak landing tread, 1" x 5-1/2"	BC@.180	LF	5.16	5.40	10.56
False treads , 4-3/4" x 16-1/4" x 6"					
Right end	BC@.180	Ea	10.50	5.40	15.90
Left end	BC@.180	Ea	11.00	5.40	16.40
Wall tread, 4-3/4" x 13-1/4"	BC@.180	Ea	12.80	5.40	18.20

Stair rail, posts and balusters, unfinished. Including layout, cutting, fitting and installation
Newel posts, 3" x 3"x 48" high

	Craft@Hrs	Unit	Material	Labor	Total
Half newel, ball top, hemlock	BC@2.50	Ea	19.80	75.00	94.80
Half newel, ball top, oak	BC@2.50	Ea	25.80	75.00	100.80
Half newel, ball top, poplar	BC@2.50	Ea	25.80	75.00	100.80
Starting newel, ball top, hemlock	BC@2.50	Ea	28.80	75.00	103.80
Starting newel, ball top, oak	BC@2.50	Ea	39.00	75.00	114.00
Starting newel, peg top, oak	BC@2.50	Ea	39.30	75.00	114.30
Starting newel, peg top, poplar	BC@2.50	Ea	39.30	75.00	114.30
Landing newel, ball top, hemlock	BC@2.50	Ea	35.80	75.00	110.80
Landing newel, ball top, oak	BC@2.50	Ea	54.80	75.00	129.80
Starting newel, peg top, oak	BC@2.50	Ea	56.10	75.00	131.10
Starting newel, ball top, poplar	BC@2.50	Ea	32.10	75.00	107.10
Handrail, 2-3/8" D x 2-1/4" W, set between posts					
Oak, set between newel posts	BC@.150	LF	5.25	4.50	9.75
Oak, set over newel posts	BC@.250	LF	5.25	7.50	12.75
Oak with fillet, set over newel posts	BC@.250	LF	5.86	7.50	13.36

	Craft@Hrs	Unit	Material	Labor	Total
Hemlock, set between newel posts	BC@.150	LF	3.95	4.50	8.45
Poplar, set between newel posts	BC@.150	LF	3.90	4.50	8.40
Hemlock, set over newel posts	BC@.250	LF	3.90	7.50	11.40
Poplar, set over newel posts	BC@.250	LF	3.90	7.50	11.40
Flat bottom, set over newel posts	BC@.250	LF	2.28	7.50	9.78
Balusters, 1-1/4" D x 2-1/4" W, round top					
Oak, 34" high	BC@.330	Ea	5.02	9.90	14.92
Oak, 36" high	BC@.330	Ea	5.27	9.90	15.17
Poplar, 34" high	BC@.330	Ea	3.86	9.90	13.76
Poplar, 36" high	BC@.330	Ea	4.18	9.90	14.08
Hemlock, 31" high, square top	BC@.330	Ea	3.81	9.90	13.71
Hemlock, 34" high, tapered top	BC@.330	Ea	3.93	9.90	13.83
Hemlock, 36" high, tapered top	BC@.330	Ea	4.44	9.90	14.34
Hemlock, 36" high, square top	BC@.330	Ea	4.31	9.90	14.21
Hemlock, 41" high, tapered top	BC@.330	Ea	4.97	9.90	14.87
Hemlock, 41" high, square top	BC@.330	Ea	4.87	9.90	14.77
Handrail easings and transition fittings, oak					
Gooseneck riser	BC@1.75	LF	64.00	52.50	116.50
Upeasing	BC@1.75	Ea	20.10	52.50	72.60
Opening cap	BC@1.75	Ea	13.80	52.50	66.30
Quarter return	BC@1.75	Ea	13.80	52.50	66.30
Volute	BC@1.75	Ea	70.90	52.50	123.40
Turnout	BC@1.75	Ea	39.90	52.50	92.40
Starting easing	BC@1.75	Ea	27.90	52.50	80.40
Over easing	BC@1.75	Ea	20.60	52.50	73.10

Stairs, attic

Economy folding stairway, tri-fold design, sanded plywood door, full width main hinge, fir door panel, 3/16" handrail, preassembled, 1" x 4" x 13" treads, 1" x 5" stringer, 22" x 54" opening size

	Craft@Hrs	Unit	Material	Labor	Total
8'9" ceiling height, 64" landing	B1@5.00	Ea	69.10	137.00	206.10
10' ceiling height, 67" landing	B1@5.00	Ea	74.00	137.00	211.00

Economy all-aluminum tri-fold stairway, sanded plywood door, full width main hinge, double-riveted treads, 3/16" handrail, preassembled, 25" x 54" opening size

	Craft@Hrs	Unit	Material	Labor	Total
8'9" ceiling height, 64" landing	B1@5.00	Ea	129.00	137.00	266.00
10' ceiling height, 67" landing	B1@5.00	Ea	135.00	137.00	272.00

Good quality folding stairway, Bessler Stairways, full width main hinge, fir door panel, molded handrail, preassembled, 1" x 6" treads, 1" x 5" stringer and frame of select yellow pine, "space saver" for narrow openings, 22", 25-1/2", or 30" wide by 54" long

	Craft@Hrs	Unit	Material	Labor	Total
8'9" or 10' height	B1@5.00	Ea	330.00	137.00	467.00

Superior quality, Bessler Stairways, solid one-piece stringer sliding stairway, stringer and treads of clear yellow, pine, unassembled, full-width piano hinge at head of stairs, with springs and

Compact model 26/26T, for a 22-1/2" x 54" to 22-1/2" x 74" opening, 7'7" to 9'3" height

	Craft@Hrs	Unit	Material	Labor	Total
1" x 4" stringers and treads	B1@6.00	Ea	577.00	164.00	741.00

Standard unit, for openings from 25-1/2" x 5'6" to 2'6" x 6'0", 7'7" to 10'10" heights

	Craft@Hrs	Unit	Material	Labor	Total
1" x 6" stringers and treads	B1@6.00	Ea	577.00	164.00	741.00

Heavy-duty unit, for openings from 2'6" x 5'10" to 2'6" x 8'0", 7'7" to 12'10" heights

	Craft@Hrs	Unit	Material	Labor	Total
1" x 8" stringers and treads	B1@9.00	Ea	1,100.00	246.00	1,346.00

Steam Bath Generators Steam bath equipment, 240 Volts AC, single phase, UL approved. Costs include steam generator and controls. Labor costs include installation of steam unit and connections to enclosure, no rough plumbing or electrical included. Installed in existing, adequately sealed, tub enclosure. Maximum enclosure size as shown. http://www.amerec.com

	Craft@Hrs	Unit	Material	Labor	Total
100 cubic feet max	P1@10.0	Ea	808.00	300.00	1,108.00
160 cubic feet max	P1@10.0	Ea	878.00	300.00	1,178.00

Tarpaulins

	Craft@Hrs	Unit	Material	Labor	Total
220 cubic feet max	P1@10.0	Ea	885.00	300.00	1,185.00
300 cubic feet max	P1@10.0	Ea	983.00	300.00	1,283.00
400 cubic feet max	P1@10.0	Ea	1,010.00	300.00	1,310.00

Tarpaulins
Polyethylene, heavy duty. Water repellant, with rustproof grommets, cut sizes

	Craft@Hrs	Unit	Material	Labor	Total
6' x 8'	—	Ea	7.20	—	7.20
8' x 10'	—	Ea	5.00	—	5.00
9' x 12'	—	Ea	7.97	—	7.97
12" x 16'	—	Ea	12.00	—	12.00
16' x 24'	—	Ea	43.00	—	43.00
20' x 30'	—	Ea	70.00	—	70.00
40' x 60'	—	Ea	100.00	—	100.00

Canvas, 8 oz.

	Craft@Hrs	Unit	Material	Labor	Total
8' x 10'	—	Ea	26.50	—	26.50
10' x 12'	—	Ea	42.40	—	42.40
10' x 16'	—	Ea	58.20	—	58.20

Taxes, Payroll
Employer's cost for payroll taxes and insurance, expressed as a percent of total payroll. See also, Insurance for more detailed information.

Rule of thumb: When actual payroll taxes and insurance are not known, estimate that for each $1 paid in wages the employer must pay an additional $.30.

	Craft@Hrs	Unit	Material	Labor	Total
Payroll taxes and insurance	—	%	—	—	30.0

Specific taxes, percent of total payroll

	Craft@Hrs	Unit	Material	Labor	Total
Typical SUI, State unemployment insurance	—	%	—	—	3.70
FICA Social Security 6.20%, Medicare 1.45%	—	%	—	—	7.65
FUTA, Federal Unemployment Insurance	—	%	—	—	.80

Thresholds See the Mouldings section for oak threshold.
Aluminum with vinyl insert

1-1/8" high, low rug, 3-1/4" wide

	Craft@Hrs	Unit	Material	Labor	Total
36" long	BC@.515	Ea	9.30	15.50	24.80
48" long	BC@.515	Ea	10.20	15.50	25.70
60" long	BC@.730	Ea	12.50	21.90	34.40
72" long	BC@.730	Ea	13.00	21.90	34.90

Aluminum thresholds, 1-3/4" wide1-1/8" high, 1-3/4" wide

	Craft@Hrs	Unit	Material	Labor	Total
30" long	BC@.515	Ea	12.10	15.50	27.60
48" long	BC@.515	Ea	17.10	15.50	32.60
60" long	BC@.730	Ea	21.30	21.90	43.20
72" long	BC@.730	Ea	24.40	21.90	46.30

Combination threshold, vinyl door shoe and drip cap

Low rug

	Craft@Hrs	Unit	Material	Labor	Total
32" or 36" long	BC@.730	Ea	11.00	21.90	32.90
42" long	BC@.730	Ea	14.40	21.90	36.30
48" long	BC@.730	Ea	16.30	21.90	38.20
60" long	BC@.941	Ea	20.70	28.20	48.90
72" long	BC@.941	Ea	25.00	28.20	53.20

Combination threshold, vinyl door shoe and drip cap

High rug

	Craft@Hrs	Unit	Material	Labor	Total
32" long	BC@.730	Ea	13.70	21.90	35.60
36" long	BC@.730	Ea	15.20	21.90	37.10
42" long	BC@.730	Ea	18.30	21.90	40.20

	Craft@Hrs	Unit	Material	Labor	Total
48" long	BC@.730	Ea	21.10	21.90	43.00
60" long	BC@.941	Ea	26.50	28.20	54.70
72" long	BC@.941	Ea	31.50	28.20	59.70
Aluminum sill and interlocking threshold, low rug					
32" or 36"	BC@1.05	Ea	10.60	31.50	42.10
42"	BC@1.27	Ea	12.40	38.10	50.50
72"	BC@1.27	Ea	21.70	38.10	59.80
Door bottom caps, surface type, aluminum and vinyl					
32" or 36"	BC@.515	Ea	4.55	15.50	20.05
42"	BC@.515	Ea	6.85	15.50	22.35
48"	BC@.515	Ea	6.85	15.50	22.35
Rain drip					
32" or 36"	BC@.230	Ea	3.80	6.90	10.70
42"	BC@.230	Ea	5.22	6.90	12.12
48"	BC@.230	Ea	5.59	6.90	12.49

Tile Costs shown are typical costs for the tile only, including 10% allowance for breakage and cutting waste. Add costs for mortar or adhesive, grout and labor to install tile from the section that follows tile material costs.

Glazed ceramic wall tile (5/16" thickness)

	Craft@Hrs	Unit	Material	Labor	Total
2" x 2"					
Minimum quality	—	SF	2.05	—	2.05
Good quality	—	SF	4.51	—	4.51
Better quality	—	SF	7.90	—	7.90
3" x 3"					
Minimum quality	—	SF	2.02	—	2.02
Good quality	—	SF	3.94	—	3.94
Better quality	—	SF	7.04	—	7.04
4-1/4" x 4-1/4"					
Minimum quality	—	SF	2.04	—	2.04
Good quality	—	SF	3.93	—	3.93
Better quality	—	SF	5.75	—	5.75
6" x 6"					
Minimum quality	—	SF	1.96	—	1.96
Good quality	—	SF	3.83	—	3.83
Better quality	—	SF	6.43	—	6.43
8" x 8"					
Minimum quality	—	SF	2.31	—	2.31
Good quality	—	SF	4.31	—	4.31
Better quality	—	SF	7.02	—	7.02

Ceramic mosaic tile (1/4" thickness)

	Craft@Hrs	Unit	Material	Labor	Total
1" hexagonal					
Minimum quality	—	SF	2.02	—	2.02
Good quality	—	SF	2.74	—	2.74
Better quality	—	SF	4.69	—	4.69
2" hexagonal					
Minimum quality	—	SF	2.02	—	2.02
Good quality	—	SF	3.50	—	3.50
Better quality	—	SF	6.10	—	6.10

Tile

	Craft@Hrs	Unit	Material	Labor	Total
3/4" penny round					
Minimum quality	—	SF	2.53	—	2.53
Good quality	—	SF	2.61	—	2.61
Better quality	—	SF	4.08	—	4.08
7/8" x 7/8" square					
Minimum quality	—	SF	2.30	—	2.30
Good quality	—	SF	4.02	—	4.02
Better quality	—	SF	7.00	—	7.00
1-1/2" x 1-1/2" square					
Minimum quality	—	SF	2.03	—	2.03
Good quality	—	SF	3.39	—	3.39
Better quality	—	SF	5.94	—	5.94
2" x 2" square					
Minimum quality	—	SF	2.00	—	2.00
Good quality	—	SF	4.26	—	4.26
Better quality	—	SF	7.30	—	7.30
Ceramic fixtures					
Toothbrush/tumbler holder	TL@.250	SF	6.86	7.76	14.62
Soap dish	TL@.250	SF	6.86	7.76	14.62
Soap dish/washcloth holder	TL@.250	SF	10.70	7.76	18.46
Bath corner shelf, 7" x 7"	TL@.500	SF	11.20	15.50	26.70
Glazed ceramic floor tile					
8" x 8"					
Minimum quality	—	SF	.92	—	.92
Good quality	—	SF	2.70	—	2.70
Better quality	—	SF	5.40	—	5.40
12" x 12"					
Minimum quality	—	SF	1.40	—	1.40
Good quality	—	SF	3.10	—	3.10
Better quality	—	SF	5.40	—	5.40
Saltillo tile, 12" x 12"					
Minimum quality	—	SF	1.04	—	1.04
Good quality	—	SF	1.50	—	1.50
Quarry tile (1/2" or 3/4" thickness)					
6" x 6" or 8" x 8"					
Minimum quality	—	SF	2.12	—	2.12
Good quality	—	SF	4.14	—	4.14
Better quality	—	SF	8.69	—	8.69
Add for abrasive grain surfaces	—	SF	.66	—	.66
Paver tile (1" thickness)					
4" x 4" or 6" x 6"					
Minimum quality	—	SF	2.67	—	2.67
Good quality	—	SF	5.04	—	5.04
Better quality	—	SF	10.80	—	10.80
12" x 12"					
Minimum quality	—	SF	2.02	—	2.02
Good quality	—	SF	4.19	—	4.19
Better quality	—	SF	9.09	—	9.09
Mexican red pavers, 12" x 12"	—	SF	1.21	—	1.21
Saltillo paver, 12" x 12"	—	SF	1.86	—	1.86

	Craft@Hrs	Unit	Material	Labor	Total

Marble and granite tile, "Martile." Prices for marble and granite tile vary widely.

Marble, 3/8" x 12" x 12", polished one face

Low price	—	SF	5.50	—	5.50
Medium price	—	SF	9.35	—	9.35
High price	—	SF	18.20	—	18.20

Granite tile, natural, 3/8" x 12" x 12", polished on one face

Low price	—	SF	4.19	—	4.19
Medium price	—	SF	11.10	—	11.10
High price	—	SF	20.20	—	20.20
Granite, 3/4", cut to size, imported	—	SF	25.20	—	25.20
Marble, 3/4", cut to size, imported	—	SF	18.10	—	18.10
Travertine beige marble tile, 3/8" x 12" x 12", imported	—	SF	4.54	—	4.54
Carrera white marble tile, 3/8" x 12" x 12", imported	—	SF	4.99	—	4.99

Ceramic tile installation The costs below are for setting and grouting tile only. Add the cost of preparing the surface for installation (tile backer) and the tile. See 'Floor Leveling' for related costs to patch and level floor; See 'Membrane Ditra' for detachable membrane costs.

Ceramic tile adhesive. Premixed Type 1, Acryl-4000. 1 gallon covers 70 SF applied with #1 trowel (3/16" x 5/32" V-notch), 50 SF applied with #2 trowel (3-/16" x 1/4" V-notch) or 40 SF applied with #3 trowel (1/4" x 1/4" square-notch)

1/2 pint	—	Ea	2.80	—	2.80
1 gallon	—	Ea	11.50	—	11.50
3-1/2 gallons	—	Ea	32.10	—	32.10
Tile adhesive applied with #2 trowel	—	SF	.23	—	.23

Thin-set tile mortar. 50 pound bag covers 100 SF applied with #1 trowel (1/4" x 1/4" square-notch), 80 SF applied with #2 trowel (1/4" x 3/8" square-notch), 45 SF applied with #3 trowel (1/2" x 1/2" square notch). Cost per 50 pound bag

Standard, gray	—	Ea	8.42	—	8.42
Master-Blend, white	—	Ea	10.10	—	10.10
FlexiBond, gray	—	Ea	25.90	—	25.90
Marble and granite mix	—	Ea	21.50	—	21.50
Master-Blend applied with #3 trowel	—	SF	.22	—	.22

Tile grout. Polymer-modified dry grout for joints 1/16" to 1/2". Coverage varies with tile and joint size. Cost per 25 pound bag

Unsanded	—	Ea	15.80	—	15.80
Sanded, light colors	—	Ea	15.90	—	15.90
Sanded, dark colors	—	Ea	11.70	—	11.70

Installation of tile in adhesive Includes adhesive and grout. Costs are for adhesive and grout. Add the cost of preparing the surface for installation (tile backer) and the cost of tile.

Countertops

Ceramic mosaic field tile	TL@.203	SF	.27	6.30	6.57
4-1/4" x 4-1/4" to 6" x 6" glazed field tile	TL@.180	SF	.27	5.59	5.86
Countertop trim pieces and edge tile	TL@.180	LF	.05	5.59	5.64

Floors

Ceramic mosaic field tile	TL@.121	SF	.27	3.76	4.03
4-1/4" x 4-1/4" to 6" x 6" glazed field tile	TL@.110	SF	.27	3.42	3.69
Floor trim pieces and edge tile	TL@.110	LF	.05	3.42	3.47

Tile

	Craft@Hrs	Unit	Material	Labor	Total
Setting wall tile with adhesiveWalls					
Ceramic mosaic field tile	TL@.143	SF	.27	4.44	4.71
4-1/4" x 4-1/4" to 6" x 6" glazed field tile	TL@.131	SF	.27	4.07	4.34
Wall trim pieces and edge tile	TL@.131	LF	.05	4.07	4.12

Installation of tile in thin-set mortar Includes grout. Costs are for mortar and grout. Add the cost of surface preparation (backerboard) and tile.

	Craft@Hrs	Unit	Material	Labor	Total
Countertops					
Ceramic mosaic field tile	TL@.407	SF	.30	12.60	12.90
4-1/4" x 4-1/4" to 6" x 6" glazed field tile	TL@.352	SF	.30	10.90	11.20
Countertop trim pieces and edge tile	TL@.352	LF	.05	10.90	10.95
Floors					
Ceramic mosaic field tile	TL@.238	SF	.30	7.39	7.69
4-1/4" x 4-1/4" to 6" x 6" glazed field tile	TL@.210	SF	.30	6.52	6.82
8" x 8" to 12" x 12" glazed field tile	TL@.200	SF	.30	6.21	6.51
Quarry or paver	TL@.167	SF	.30	5.19	5.49
Marble or granite, 3/8" thick	TL@.354	SF	.30	11.00	11.30
Marble or granite, 3/4" thick	TL@.591	SF	.30	18.40	18.70
Floor trim pieces and edge tile	TL@.210	LF	.05	6.52	6.57
Walls					
Ceramic mosaic field tile	TL@.315	SF	.30	9.78	10.08
4-1/4" x 4-1/4" to 6" x 6" glazed field tile	TL@.270	SF	.30	8.38	8.68
Wall trim pieces and edge tile	TL@.270	LF	.05	8.38	8.43

Tile backer board (Durock™ or Wonderboard™). Water-resistant underlayment for ceramic tile on floors, walls, countertops and other interior wet areas. Material cost for 100 square feet (CSF) of board includes waterproofing membrane, the backer board (and 10% waste), 50 pounds of job mixed latex-fortified mortar for the joints and surface skim coat, 75 linear feet of fiberglass joint tape, 200 1-1/4" backerboard screws. For scheduling purposes, estimate that a crew of 2 can install, tape and apply the skim coat on the following quantity of backer board in an 8-hour day: countertops 180 SF, floors 525 SF and walls 350 SF. Use $250.00 as a minimum charge for this type work.

	Craft@Hrs	Unit	Material	Labor	Total
Using 15 Lb. felt waterproofing, per 400 SF roll	—	Roll	13.70	—	13.70
Using 1/4" or 1/2" backer board, per 100 SF	—	CSF	75.50	—	75.50
Using 50 Lb. of latex-fortified mortar each 100 SF, per 50 Lb. sack	—	Sack	10.10	—	10.10
Using 75 LF of backerboard tape each 100 SF per 50 LF roll	—	Roll	3.12	—	3.12
Using 200 backerboard screws for each 100 SF, per pack of 200 screws	—	Pack	6.57	—	6.57
Backer board with waterproofing, mortar, tape and screws					
Floors	T1@.030	SF	1.03	.84	1.87
Walls	T1@.045	SF	1.03	1.25	2.28
Countertops	T1@.090	SF	1.03	2.51	3.54

Vacuum Cleaning Systems, central, subcontract Costs listed are for better quality central vacuum power unit, piping to 4 remote outlets, hose and attachments (adequate for typical 2,000 SF home). Assumes that all vacuum tubes are installed before any interior finishes are applied.

	Craft@Hrs	Unit	Material	Labor	Total
Basic package with 23' hose	—	LS	—	—	1,380.00
Additional remote outlets	—	Ea	—	—	141.00
Add for electric or turbo power head	—	Ea	—	—	203.00

	Craft@Hrs	Unit	Material	Labor	Total

Vanities No countertops, lavatory bowls or plumbing included. Prefinished. See also Countertops and Marble Setting.

Good quality sink base, 3/4" furniture-grade sides with veneer interior. Overlay door hung on adjustable hinges

Vanity lavatory sink base, 21" deep x 31-1/2" high

1 door sink base, 21" deep x 31-1/2" high

	Craft@Hrs	Unit	Material	Labor	Total
24" wide, oak front, maple sides	BC@.400	Ea	104.00	12.00	116.00
24" side, white front and sides	BC@.400	Ea	104.00	12.00	116.00

2 drawer sink base, 21" deep x 31-1/2" high

30" wide, white front and sides	BC@.400	Ea	146.00	12.00	158.00
36" wide, white front and sides	BC@.400	Ea	167.00	12.00	179.00
30" wide, maple front and sides	BC@.400	Ea	167.00	12.00	179.00
36" wide, maple front and sides	BC@.400	Ea	188.00	12.00	200.00
30" wide, oak front, maple sides	BC@.400	Ea	145.00	12.00	157.00
36" wide, oak front, maple sides	BC@.400	Ea	167.00	12.00	179.00

2 door, 2 drawer sink base, 21" deep x 31-1/2" high

48" wide, white front and sides	BC@.500	Ea	219.00	15.00	234.00
48" wide, maple front and sides	BC@.500	Ea	251.00	15.00	266.00
48" wide, oak front, maple sides	BC@.500	Ea	219.00	15.00	234.00

3 drawer sink base, 21-1/2" deep x 33-1/2" high

12" wide, white front	BC@.350	Ea	135.00	10.50	145.50
12" wide, oak front	BC@.350	Ea	135.00	10.50	145.50

Over the toilet vanity cabinet, 8" deep x 24" high

24" wide, 2 doors, 1 shelf, oak	B1@1.37	Ea	89.00	37.50	126.50
24" wide, 2 doors, 1 shelf, matte white	B1@1.37	Ea	69.00	37.50	106.50

Standard quality vanity cabinets. Square corner raised-panel doors. 31-1/2" high. Frameless construction

18" wide x 16" deep, 1 door	BC@.400	Ea	64.90	12.00	76.90
24" wide x 18" deep, 2 doors	BC@.400	Ea	79.30	12.00	91.30
30" wide x 18" deep, 3 doors	BC@.400	Ea	93.80	12.00	105.80

Vents and Louvers

Attic and gable louver vents, UV stabilized copolymer, screened, rough opening sizes

12" x 12", 44 sq. inch vent area	SW@.448	Ea	17.80	15.20	33.00
12" x 18", 86 sq. inch vent area	SW@.448	Ea	20.40	15.20	35.60
12" x 24", 140 sq. inch vent area	SW@.871	Ea	28.80	29.60	58.40
21" x 27" oval	SW@.871	Ea	44.10	29.60	73.70
16" x 24", half round, 100 sq. inch vent area	SW@.871	Ea	43.60	29.60	73.20
16" round, 50 sq. inch vent area	SW@.448	Ea	38.60	15.20	53.80
12" x 18", round top 50 sq. inch vent area	SW@.448	Ea	32.20	15.20	47.40
16" octagon, 54 sq. inch vent area	SW@.871	Ea	37.60	29.60	67.20

Chimney caps

9" x 9" rectangular	SW@.324	Ea	24.70	11.00	35.70
9" x 13" rectangular	SW@.324	Ea	26.80	11.00	37.80
13" x 13" rectangular	SW@.324	Ea	28.30	11.00	39.30
13" x 13" rectangular, stainless	SW@.324	Ea	42.00	11.00	53.00
8" x 8" to 13" x 13" adjustable	SW@.324	Ea	28.30	11.00	39.30
13" x 18" fixed, black	SW@.324	Ea	31.40	11.00	42.40

Clothes dryer vent set, aluminum, with hood, inside plate, 8' of flexible duct pipe, clamps and collar

4" diameter vent kit	SW@.440	Ea	12.40	15.00	27.40
Flex hose, 4" diameter, 8' length	—	Ea	10.50	—	10.50
Flex hose clamps, 4" diameter	—	Ea	1.66	—	1.66

Vents and Louvers

	Craft@Hrs	Unit	Material	Labor	Total
Code caps					
4" diameter, top only	SW@.324	Ea	16.30	11.00	27.30
6" diameter, top only	SW@.324	Ea	20.50	11.00	31.50
8" diameter, top only	SW@.324	Ea	26.00	11.00	37.00
6" diameter shake base	SW@.324	Ea	11.00	11.00	22.00
8" diameter shake base	SW@.324	Ea	12.60	11.00	23.60
Dormer louver vents, galvanized, 1/4" mesh screen					
19" x 3", low profile, rectangular	BC@.871	Ea	21.80	26.10	47.90
24" x 12", half round	BC@.871	Ea	16.80	26.10	42.90
Add for 1/8" mesh	—	%	3.0	—	—
Under-eave louver vents, rectangular, with 1/4" mesh screen, 16" wide					
16" x 4", mill finish	SW@.255	Ea	1.27	8.66	9.93
16" x 8", mill finish	SW@.255	Ea	1.52	8.66	10.18
16" x 8", brown finish	SW@.255	Ea	1.59	8.66	10.25
16" x 4", white finish	SW@.255	Ea	1.33	8.66	9.99
16" x 8", white finish	SW@.255	Ea	1.65	8.66	10.31
Foundation screen vents, galvanized, with screen, no louvers					
14" x 6"	SW@.255	Ea	2.18	8.66	10.84
16" x 4"	SW@.255	Ea	2.17	8.66	10.83
16" x 6"	SW@.255	Ea	2.61	8.66	11.27
16" x 8"	SW@.255	Ea	2.82	8.66	11.48
Automatic foundation vent, fits 8" x 16" concrete block opening, aluminum mesh on front and molded plastic screen on back. Opens at 72 degrees, closes at 38 degrees. 1" frame each side					
Automatic open and close	SW@.255	Ea	8.27	8.66	16.93
Foundation vent with manual damper, sheet metal					
4" x 16"	SW@.255	Ea	5.54	8.66	14.20
6" x 16"	SW@.255	Ea	6.07	8.66	14.73
8" x 16"	SW@.255	Ea	6.27	8.66	14.93
Foundation access door					
24" x 24"	BC@.363	Ea	25.10	10.90	36.00
32" x 24"	BC@.363	Ea	31.40	10.90	42.30
Heater closet door vents, louver					
70 square inches, louvers only	SW@.220	Ea	4.61	7.48	12.09
70 square inches, louvers and screen	SW@.255	Ea	8.53	8.66	17.19
38 square inches, louvers only	SW@.190	Ea	4.47	6.46	10.93
38 square inches, louvers and screen	SW@.210	Ea	6.76	7.14	13.90
Circular louver vents, for moisture control in sidewalls, aluminum mill finish					
1" circular vent	SW@.255	Ea	4.56	8.66	13.22
2" circular vent	SW@.255	Ea	6.89	8.66	15.55
3" circular vent	SW@.255	Ea	9.06	8.66	17.72
Rafter vents, 1/8" mesh, galvanized, louver					
3" x 14"	SW@.220	Ea	2.72	7.48	10.20
3" x 22"	SW@.220	Ea	4.73	7.48	12.21
Plaster ground rafter vent four sides					
3" x 14"	SW@.220	Ea	2.72	7.48	10.20
3" x 22"	SW@.220	Ea	4.73	7.48	12.21
Flat rafter vents for sidingFlat type for siding					
3" x 14"	SW@.220	Ea	3.14	7.48	10.62
3" x 22"	SW@.220	Ea	4.73	7.48	12.21
Rafter inserts					
3" x 14"	SW@.220	Ea	2.72	7.48	10.20
3" x 22"	SW@.220	Ea	4.73	7.48	12.21

	Craft@Hrs	Unit	Material	Labor	Total
Rafter vent with ears					
3" x 14"	SW@.220	Ea	3.79	7.48	11.27
3" x 22"	SW@.220	Ea	4.50	7.48	11.98
6" x 22"	SW@.220	Ea	5.44	7.48	12.92
For FHA painted, add	—	%	5.0	—	—
For galvanized 1/4" mesh, deduct	—	%	-5.0	—	—
Rafter bay vent channel, keeps attic insulation away from vents, promotes circulation.					
Rafters 16" on center,					
14" wide, 15" net-free vent area	BC@.220	Ea	1.61	6.60	8.21
Rafters 24" on center,					
22" wide, 26" net-free vent area	BC@.220	Ea	2.05	6.60	8.65
Ridge ventilators, 1/8" louver opening, double baffle, aluminum					
10' long, black	BC@.650	Ea	13.00	19.50	32.50
10' long, brown	BC@.650	Ea	13.40	19.50	32.90
10' long, white	BC@.650	Ea	13.30	19.50	32.80
4' long, hinged for steep roof	BC@.255	Ea	9.50	7.65	17.15
Joint strap	BC@.010	Ea	.98	.30	1.28
End connector plug	BC@.010	Ea	1.89	.30	2.19
Roof vents, with mesh screen, for roof venting of kitchen and bath exhaust fans					
Square cap type, 26" L x 23" D x 5" H	SW@.448	Ea	51.20	15.20	66.40
Dormer type, 20" L x 10" D x 6" H	SW@.448	Ea	33.70	15.20	48.90
Rotary roof ventilators, with base, by turbine diameter					
12", galvanized	SW@.694	Ea	29.50	23.60	53.10
12", brown	SW@.694	Ea	31.80	23.60	55.40
14", galvanized	SW@.694	Ea	31.80	23.60	55.40
14", brown	SW@.694	Ea	34.80	23.60	58.40
16", galvanized	SW@.930	Ea	101.00	31.60	132.60
18", galvanized	SW@1.21	Ea	122.00	41.10	163.10
Round louver vents, 1/8" mesh, galvanized					
12" or 14" diameter	SW@.440	Ea	41.40	15.00	56.40
16" diameter	SW@.440	Ea	47.60	15.00	62.60
18" diameter	SW@.440	Ea	52.50	15.00	67.50
24" diameter	SW@.661	Ea	95.00	22.50	117.50
Soffit vents, aluminum louvers, 8' long, reversible					
2-5/8" wide, white finish	SW@.440	Ea	3.61	15.00	18.61
2-5/8" wide, mill finish	SW@.440	Ea	3.42	15.00	18.42
Wall louvers, reversible for flush or recessed mounting. White finish. Screened					
12" x 12", aluminum	SW@.440	Ea	7.81	15.00	22.81
12" x 18", aluminum	SW@.440	Ea	10.40	15.00	25.40
14" x 24", aluminum	SW@.440	Ea	13.10	15.00	28.10
18" x 24", aluminum	SW@.440	Ea	15.20	15.00	30.20
12" x 18", plastic	SW@.440	Ea	7.34	15.00	22.34
18" x 24", plastic	SW@.440	Ea	10.50	15.00	25.50

Wallboard Partitions, Subcontract Costs for interior partitions in one and two story buildings. 1/4" gypsum wallboard on 2 sides with "V-grooved" edges and plastic moulding at corners fastened to 25 gauge metal framing at 24" OC, including framing, per LF of wall.

		Unit	Material	Labor	Total
Vinyl covered 8' high	—	LF	—	—	72.80
Vinyl covered 10' high	—	LF	—	—	98.00
Paintable paper-covered 8' high	—	LF	—	—	53.00
Paintable paper-covered 10' high	—	LF	—	—	79.00
Add for glued vinyl base on two sides	—	LF	—	—	2.86

Wallcoverings

	Craft@Hrs	Unit	Material	Labor	Total

Wallcoverings Costs listed are for good to better quality materials installed on a clean, smooth surface and include adhesive, edge trimming, pattern matching, and normal cutting and fitting waste. Surface preparation such as patching and priming are extra. Typical roll is 8 yards long and 18" wide (approximately 36 square feet).

	Craft@Hrs	Unit	Material	Labor	Total
Scrape off old wallpaper, up to 2 layers	PP@.015	SF	—	.42	.42
Paper stripper (1,000 SF per gallon)	—	Gal	17.50	—	17.50
Patch holes, scrape marks	PP@.010	SF	.02	.28	.30
Blank stock (underliner)					
Bath, kitchen, and laundry room	PP@.668	Roll	9.73	18.90	28.63
Other rooms	PP@.558	Roll	9.73	15.80	25.53
Textured basket weave					
Bath, kitchen, and laundry rooms	PP@.995	Roll	18.00	28.20	46.20
Other rooms	PP@.830	Roll	18.00	23.50	41.50
Papers, vinyl-coated papers, trimmed, and pre-pasted					
Bath, kitchen, and laundry rooms	PP@.884	Roll	16.80	25.00	41.80
Other rooms	PP@.780	Roll	16.80	22.10	38.90
Floral patterns					
Bath, kitchen, and laundry rooms	PP@1.12	Roll	17.80	31.70	49.50
Other rooms	PP@.844	Roll	17.80	23.90	41.70
Designer paper and hand-painted prints					
Bath, kitchen, and laundry rooms	PP@1.68	Roll	75.70	47.50	123.20
Other rooms	PP@1.41	Roll	75.70	39.90	115.60
Vinyls, patterned					
Bath, kitchen, and laundry rooms	PP@.844	Roll	16.80	23.90	40.70
Other rooms	PP@.730	Roll	16.80	20.70	37.50
Vinyls, textured, no pattern					
Double cut seams, cloth backing	PP@.844	Roll	13.60	23.90	37.50
Coordinating borders, vinyl coated					
roll is 4" to 6" x 12' to 15'	PP@.017	LF	.69	.48	1.17
Wallpaper adhesive					
Clear, strippable (300 SF per gallon)	—	Gal	7.32	—	7.32
Clear, strippable, per SF	—	SF	.02	—	.02
Mildew-resistant (250 SF per gallon)	—	Gal	16.80	—	16.80
Adhesive, per SF	—	SF	.07	—	.07

Waterproofing

Below grade waterproofing, applied to concrete or masonry walls, no surface preparation, excavation, or backfill included, typical job

	Craft@Hrs	Unit	Material	Labor	Total
Bentonite waterproofing					
One layer, nailed in place	BL@.079	SF	.65	1.95	2.60
Cementitious waterproofing with protection board					
Spray or brush applied, 2 coats	BL@.077	SF	1.09	1.90	2.99
Crystalline waterproofing					
Spray or brush applied, 3 coats	BL@.110	SF	1.40	2.72	4.12
Elastomeric waterproofing 30 mm per coat					
Sprayed, troweled, or rolled, 2 coats	BL@.065	SF	1.09	1.61	2.70

Above grade waterproofing, applied to smooth concrete or plywood deck, no surface preparation included, typical job

	Craft@Hrs	Unit	Material	Labor	Total
Between slabs membrane					
Sprayed, troweled or rolled, 2 coats	PT@.047	SF	.78	1.50	2.28
Pedestrian walking surface, elastomeric membrane,					
Applied 4 coats thick with aggregate	PT@.086	SF	1.04	2.74	3.78

	Craft@Hrs	Unit	Material	Labor	Total
Chloroprene/Neoprene latex deck Surfacing	PT@.144	SF	2.15	4.59	6.74

Waterproofing system for foundations Wrapped over concrete foundation footing. System should be tied into a perimeter drain system at or around the footing. Cover top of system with at least 6 inches of soil and backfill immediately after installation. No surface preparation, excavation, or backfill included. MiraDRI™

	Craft@Hrs	Unit	Material	Labor	Total
MiraDRI™ 860/861 rubberized asphalt membrane	B1@.075	SF	.56	2.05	2.61
Latex-based primer (400 SF per gal)	B1@.075	SF	.08	2.05	2.13
LM-800 liquid membrane	B1@.034	SF	.02	.93	.95
MiraDRI™ M-800 mastic, 1/8" x 3/4", 130 LF per gallon	B1@.034	LF	.14	.93	1.07
Mira Drain™ 6200 (200 SF Roll)	B1@.075	SF	.75	2.05	2.80
Quick Drain™ (16" x 50' Roll)	B1@.075	SF	1.65	2.05	3.70

Weatherstripping Materials

	Craft@Hrs	Unit	Material	Labor	Total
Door bottom seals					
Spring brass in pinch grip carrier (KEL-EEZ®)					
32" x 1-3/4" brass	BC@.250	Ea	15.60	7.50	23.10
36" x 1-3/4"	BC@.250	Ea	17.50	7.50	25.00
Door frame seal sets					
Spring bronze, nailed every 1-3/8 inch					
36" x 6'8" door	BC@.500	Ea	19.10	15.00	34.10
36" x 7'0" door	BC@.500	Ea	20.40	15.00	35.40
Rigid PVC frame, with silicon bulb / screws	BC@.008	Ea	13.40	.24	13.64
Header and jamb side stripping, adhesive (Peel N Stick®)					
36" x 6'8" door, brass stripping	BC@.500	Ea	20.50	15.00	35.50
Window seal sets, spring brass, adhesive (Peel N Stick®), linear feet per side					
Double hung window, head and sill	BC@.008	LF	1.41	.24	1.65
Add to remove & replace window sash	BC@.250	Ea	—	7.50	7.50
Casement window, head and sill	BC@.008	LF	.66	.24	.90
Gray or black pile gasket	BC@.008	LF	.52	.24	.76
Gray of black pile gasket with adhesive	BC@.006	LF	.72	.18	.90
Garage door top and side seal					
Kry-O-Gem™ fiber gasket in aluminum or vinyl frame (KEL-EEZ®)					
9' x 7' door / screws	BC@.500	Ea	39.70	15.00	54.70
16' x 7' door / screws	BC@.600	Ea	51.50	18.00	69.50
Flexible PVC	BC@.050	LF	1.09	1.50	2.59
Garage door bottom seal. Set with adhesive					
9', neoprene or EPDM	BC@.383	Ea	16.50	11.50	28.00
16', neoprene or EPDM	BC@.500	Ea	28.90	15.00	43.90
Flexible PVC	BC@.025	LF	.93	.75	1.68
EPDM Rubber	BC@.025	LF	1.79	.75	2.54

Well Drilling, Subcontract Typical costs shown, based on well depth including welded steel casing (or continuous PVC casing). Actual costs will depend on specific conditions, i.e. hard-rock, alluvial river-bed or a combination of conditions. The amount of casing needed depends on soil conditions. Some types of rock require casings at the well surface only. Your local health department can probably provide information on the average well depth and subsurface conditions in the community. Note that unforeseen subsurface conditions can increase drilling costs substantially. Most subcontractors will estimate costs on an hourly basis.

	Craft@Hrs	Unit	Material	Labor	Total
Well hole with 4" ID F480 PVC casing	—	LF	—	—	27.00

Window and Door Mantels, Aluminum

	Craft@Hrs	Unit	Material	Labor	Total
Well hole with 6" ID steel casing	—	LF	—	—	46.40
Well hole with 8" ID steel casing	—	LF	—	—	64.00
6" well hole in sturdy rock (no casing)	—	LF	—	—	16.90
Add per well for 6" diameter bottom filter screen if needed due to site conditions					
Stainless steel	—	LF	—	—	134.00
Low carbon steel	—	LF	—	—	80.40
Add per well for drive shoe	—	LS	—	—	86.00
Add for 6" surface seal if needed due to local code or site conditions					
20' sanitary seal – for individual well	—	LS	—	—	730.00
50' sanitary seal – for community well	—	LS	—	—	2,400.00
Add per well for well drilling permit					
Individual water well	—	LS	—	—	440.00
Community water well with Source Assessment	—	LS	—	—	1,120.00
Add for automatic electric pumping system with pressure tank(s) domestic use, no electrical work incl.					
Individual (18 GPM @ 400 TDH), typical	—	LS	18,100.00	—	18,100.00
Community (60 GPM @ 400 TDH), typical	—	LS	18,100.00	—	18,100.00
Add for electrical service drop pole					
Typical cost 100 amp 230v 1 phase service	—	LS	—	—	1,240.00

Window and Door Mantels, Aluminum Includes hardware, aluminum crown mould and roof. Modern design, bright copper or primed aluminum finish.

	Craft@Hrs	Unit	Material	Labor	Total
Up to 24" width	SW@.500	Ea	104.00	17.00	121.00
25" to 36" width	SW@.500	Ea	138.00	17.00	155.00
37" to 48" width	SW@.500	Ea	174.00	17.00	191.00
49" to 60" width	SW@.600	Ea	208.00	20.40	228.40
61" to 72" width	SW@.600	Ea	244.00	20.40	264.40
73" to 84" width	SW@.600	Ea	279.00	20.40	299.40
85" to 96" width	SW@.700	Ea	315.00	23.80	338.80
97" to 108" width	SW@.700	Ea	350.00	23.80	373.80
109" to 120" width	SW@.700	Ea	386.00	23.80	409.80
Add for aged copper or brass aluminum finish	—	%	10.0	—	—
Add for units over 120"	—	LF	47.00	—	47.00

Window Sills, Marble Thassos white cultured marble, 1/2" radius front edge.

	Craft@Hrs	Unit	Material	Labor	Total
4" x 54" x 3/4"	B1@.687	Ea	19.10	18.80	37.90
4" x 74" x 3/4"	B1@.730	Ea	26.50	20.00	46.50
5" x 37" x 3/4"	B1@.687	Ea	15.60	18.80	34.40
5" x 54" x 3/4"	B1@.687	Ea	24.20	18.80	43.00
6" x 37" x 3/4"	B1@.687	Ea	18.80	18.80	37.60
6" x 54" x 3/4"	B1@.687	Ea	29.70	18.80	48.50
6" x 74" x 3/4"	B1@.730	Ea	39.00	20.00	59.00

Window Sills, Vinyl Solid vinyl. Set with latex or solvent base adhesive. Hard, scratch resistant surface. Ultraviolet protection to prevent yellowing. Gloss finish.

	Craft@Hrs	Unit	Material	Labor	Total
3-1/2" x 3'	BC@.129	Ea	11.90	3.87	15.77
3-1/2" x 4'	BC@.172	Ea	14.90	5.16	20.06
3-1/2" x 5'	BC@.215	Ea	17.60	6.45	24.05
3-1/2" x 6'	BC@.258	Ea	20.60	7.74	28.34
4-1/4" x 3'	BC@.129	Ea	12.80	3.87	16.67
4-1/4" x 4'	BC@.172	Ea	16.00	5.16	21.16
4-1/4" x 5'	BC@.215	Ea	20.40	6.45	26.85
4-1/4" x 6'	BC@.258	Ea	23.60	7.74	31.34

	Craft@Hrs	Unit	Material	Labor	Total
5-1/2" x 3'	BC@.129	Ea	15.80	3.87	19.67
5-1/2" x 4'	BC@.172	Ea	20.10	5.16	25.26
5-1/2" x 5'	BC@.215	Ea	24.00	6.45	30.45
5-1/2" x 6'	BC@.258	Ea	28.50	7.74	36.24
6-1/4" x 3'	BC@.129	Ea	16.70	3.87	20.57
6-1/4" x 4'	BC@.172	Ea	22.70	5.16	27.86
6-1/4" x 5'	BC@.215	Ea	27.10	6.45	33.55
6-1/4" x 6'	BC@.258	Ea	33.50	7.74	41.24

Window Treatments Blinds and shades. See also Draperies.
Horizontal blinds. Per square foot of blind, width x height. Includes head rail and wand tilter. Standard colors and custom sizes.

	Craft@Hrs	Unit	Material	Labor	Total
1" wide PVC	BC@.083	SF	1.24	2.49	3.73
1" wide aluminum	BC@.083	SF	1.90	2.49	4.39
1" wide basswood	BC@.083	SF	5.46	2.49	7.95
2" wide faux wood	BC@.083	SF	3.06	2.49	5.55
2" wide basswood	BC@.083	SF	4.57	2.49	7.06
2" wide American hardwood	BC@.083	SF	5.13	2.49	7.62
2-1/2" wide faux wood	BC@.083	SF	3.32	2.49	5.81
2-1/2" wide beveled basswood	BC@.083	SF	6.20	2.49	8.69

Roll-up blinds. Per square foot of blind, width x height. Includes aluminum head rail, valance and roller chain lift. Standard colors and custom sizes.

	Craft@Hrs	Unit	Material	Labor	Total
Woven wood (rattan and jute)	BC@.083	SF	5.15	2.49	7.64
Bamboo	BC@.083	SF	6.25	2.49	8.74
Vinyl-fiberglass, translucent	BC@.083	SF	2.75	2.49	5.24
Room darkening fabric	BC@.083	SF	6.00	2.49	8.49
Denim fabric	BC@.083	SF	2.75	2.49	5.24
Decorative woven cotton fabric	BC@.083	SF	4.40	2.49	6.89
Sun control vinyl fiberglass sheer weave	BC@.083	SF	4.50	2.49	6.99

Vertical blinds. 3-1/2" vanes. Per square foot of blind, width x height. Includes aluminum head rail, tension pulley and dust cover valance. Standard colors and custom sizes. Left or right pull.

	Craft@Hrs	Unit	Material	Labor	Total
Smooth surface PVC	BC@.083	SF	2.25	2.49	4.74
Textured surface PVC	BC@.083	SF	2.60	2.49	5.09
Premium textured surface PVC	BC@.083	SF	3.34	2.49	5.83
5 mm thick wood	BC@.083	SF	14.30	2.49	16.79
Add for center pull, per blind	—	Ea	20.00	—	20.00
Add for tile or molding cutout, per blind	—	Ea	20.00	—	—
Add for motorized remote control, per blind	—	Ea	100.00	—	—

Cellular shades, polyester. With head rail and mounting hardware.

	Craft@Hrs	Unit	Material	Labor	Total
3/8" double cell light filter	BC@.083	SF	3.85	2.49	6.34
3/8" single cell translucent	BC@.083	SF	4.80	2.49	7.29
3/8" double cell room darkening	BC@.083	SF	5.25	2.49	7.74
3/8" single cell blackout	BC@.083	SF	5.76	2.49	8.25
1/2" double cell light filter	BC@.083	SF	4.90	2.49	7.39
1/2" single cell light filter	BC@.083	SF	3.70	2.49	6.19
3/4" single cell translucent	BC@.083	SF	4.80	2.49	7.29
3/4" single cell blackout	BC@.083	SF	5.65	2.49	8.14
Add for width over 8'	—	%	20.0	—	—
Add for continuous loop lift cord	—	%	25.0	—	—
Add for no lift cord (child safety), per blind	—	Ea	40.00	—	—
Add for tile or molding cutout, per blind	—	Ea	20.00	—	—

Windows

	Craft@Hrs	Unit	Material	Labor	Total

Windows See also Skylights and Sky Windows.

Single-hung insulating glass vinyl windows. 5/8" insulating glass with grid between the lites. 4/4 grid means the upper lite is divided into four panes and the lower lite is divided into four panes. Tilt sash for easy cleaning. Includes half screen. Dimensions are rough opening sizes, width x height. White finish.

	Craft@Hrs	Unit	Material	Labor	Total
1'8" x 4'2", 4/4 grid	BC@.500	Ea	86.70	15.00	101.70
2'0" x 3'0", 4/4 grid	BC@.500	Ea	75.50	15.00	90.50
2'0" x 6'0", 6/4 grid	BC@.500	Ea	95.20	15.00	110.20
2'8" x 3'0", 6/6 grid	BC@.500	Ea	86.70	15.00	101.70
2'8" x 4'4", 6/6 grid	B1@1.00	Ea	94.40	27.40	121.80
2'8" x 5'0", 6/6 grid	B1@1.00	Ea	97.60	27.40	125.00
2'8" x 6'0", 6/6 grid	B1@1.00	Ea	107.00	27.40	134.40
3'0" x 3'0", 6/6 grid	BC@.500	Ea	90.00	15.00	105.00
3'0" x 4'0", 6/6 grid	B1@1.00	Ea	95.40	27.40	122.80
3'0" x 4'4", 6/6 grid	B1@1.00	Ea	97.60	27.40	125.00
3'0" x 5'0", 6/6 grid	B1@1.50	Ea	101.00	41.00	142.00
3'0" x 6'0", 9/6 grid	B1@1.50	Ea	108.00	41.00	149.00

Single-hung insulated low-E glass vinyl windows. Low-E (energy efficient) 5/8" insulating glass with grids between the glass. 4/4 grid means the upper light is divided into four panes and the lower light is divided into four panes. White finish. Tilt sash for easy cleaning. Includes half screen. Dimensions are rough opening sizes, width x height.

	Craft@Hrs	Unit	Material	Labor	Total
2'0" x 3'0", 4/4 grid	BC@.500	Ea	94.40	15.00	109.40
2'0" x 5'0", 4/4 grid	BC@.500	Ea	106.00	15.00	121.00
2'0" x 6'0", 6/4 grid	B1@1.00	Ea	120.00	27.40	147.40
2'8" x 3'0", 6/6 grid	BC@.500	Ea	108.00	15.00	123.00
2'8" x 4'4", 6/6 grid	B1@1.00	Ea	120.00	27.40	147.40
2'8" x 5'0", 6/6 grid	B1@1.00	Ea	124.00	27.40	151.40
2'8" x 6'0", 9/6 grid	B1@1.50	Ea	130.00	41.00	171.00
3'0" x 3'0", 6/6 grid	BC@.500	Ea	108.00	15.00	123.00
3'0" x 4'0", 6/6 grid	B1@1.00	Ea	120.00	27.40	147.40
3'0" x 4'4", 6/6 grid	B1@1.00	Ea	124.00	27.40	151.40
3'0" x 5'0", 6/6 grid	B1@1.50	Ea	130.00	41.00	171.00
3'0" x 6'0", 9/6 grid	B1@1.50	Ea	137.00	41.00	178.00

Double-hung insulating glass vinyl windows 5/8" insulating glass. Upper and lower lites are divided into panes by a vinyl grid between the sheets of glass. Sash tilts for easy cleaning. With full screen. 4-9/16" white jamb. Built-in J channel. Dimensions are rough opening sizes, width x height.

	Craft@Hrs	Unit	Material	Labor	Total
2'4" x 3'0"	BC@.500	Ea	126.00	15.00	141.00
2'4" x 3'1"	BC@.500	Ea	118.00	15.00	133.00
2'4" x 3'2"	BC@.500	Ea	119.00	15.00	134.00
2'4" x 4'6"	BC@.500	Ea	152.00	15.00	167.00
2'8" x 2'1"	BC@.500	Ea	110.00	15.00	125.00
2'8" x 3'2"	BC@.500	Ea	116.00	15.00	131.00
2'8" x 3'1"	BC@.500	Ea	129.00	15.00	144.00
2'8" x 3'2"	BC@.500	Ea	138.00	15.00	153.00
2'8" x 4'6"	B1@1.00	Ea	152.00	27.40	179.40
2'8" x 5'2"	B1@1.00	Ea	151.00	27.40	178.40
3'0" x 3'1"	BC@.500	Ea	139.00	15.00	154.00
3'0" x 3'2"	BC@.500	Ea	135.00	15.00	150.00
3'0" x 4'6"	B1@1.00	Ea	158.00	27.40	185.40
3'0" x 5'2"	B1@1.50	Ea	156.00	41.00	197.00

	Craft@Hrs	Unit	Material	Labor	Total

Double-hung insulated low-E glass vinyl windows. White, with full screen. Low-E (energy-efficient) insulating glass. By rough opening size, width x height. Actual size is 1/2" less. Upper and lower lites are divided into six panes by a vinyl grid between the sheets of glass. 4-9/16" wide jamb.

	Craft@Hrs	Unit	Material	Labor	Total
2'0" x 3'2"	BC@.500	Ea	120.00	15.00	135.00
2'4" x 2'10"	BC@.500	Ea	109.00	15.00	124.00
2'4" x 3'2"	BC@.500	Ea	117.00	15.00	132.00
2'4" x 3'2"	BC@.500	Ea	111.00	15.00	126.00
2'4" x 3'10"	BC@.500	Ea	123.00	15.00	138.00
2'4" x 4'6"	B1@1.00	Ea	134.00	27.40	161.40
2'8" x 2'10"	BC@.500	Ea	121.00	15.00	136.00
2'8" x 3'2"	BC@.500	Ea	119.00	15.00	134.00
2'8" x 3'10"	BC@.500	Ea	127.00	15.00	142.00
2'8" x 4'2"	B1@1.00	Ea	130.00	27.40	157.40
2'8" x 4'6"	B1@1.00	Ea	132.00	27.40	159.40
3'0" x 3'10"	B1@1.00	Ea	152.00	27.40	179.40
3'0" x 4'6"	B1@1.00	Ea	158.00	27.40	185.40

Horizontal sliding single-glazed vinyl windows. By nominal size, width x height. Actual size is 1/2" less. With screen.

	Craft@Hrs	Unit	Material	Labor	Total
2'0" x 2'0"	BC@.500	Ea	73.70	15.00	88.70
3'0" x 2'0"	BC@.500	Ea	81.10	15.00	96.10
3'0" x 3'0"	BC@.500	Ea	93.10	15.00	108.10
4'0" x 3'0"	B1@1.00	Ea	102.00	27.40	129.40
4'0" x 4'0"	B1@1.50	Ea	116.00	41.00	157.00
5'0" x 4'0"	B1@1.50	Ea	126.00	41.00	167.00

Horizontal sliding insulated low-E glass vinyl windows. Low-E (energy-efficient) insulating glass. Argon filled. For new construction. By rough opening size, width x height. With screen.

	Craft@Hrs	Unit	Material	Labor	Total
2'0" x 2'0"	BC@.500	Ea	86.50	15.00	101.50
2'0" x 3'0"	BC@.500	Ea	94.10	15.00	109.10
2'4" x 4'6"	BC@.500	Ea	117.00	15.00	132.00
2'8" x 4'6"	B1@1.00	Ea	125.00	27.40	152.40
3'0" x 2'0"	BC@.500	Ea	97.40	15.00	112.40
3'0" x 3'0"	BC@.500	Ea	108.00	15.00	123.00
3'0" x 4'0"	B1@1.00	Ea	125.00	27.40	152.40
3'0" x 5'2"	B1@1.50	Ea	138.00	41.00	179.00
3'0" x 6'0"	B1@1.50	Ea	140.00	41.00	181.00
4'0" x 2'0"	BC@.500	Ea	119.00	15.00	134.00
4'0" x 3'0"	B1@1.00	Ea	123.00	27.40	150.40
4'0" x 4'0"	B1@1.50	Ea	145.00	41.00	186.00
5'0" x 4'0"	B1@1.50	Ea	161.00	41.00	202.00

Casement insulated low-E glass vinyl windows. Low-E (energy-efficient) insulating glass. Argon filled. By window size, width x height. With screen. Sash opening direction when viewed from the exterior.

	Craft@Hrs	Unit	Material	Labor	Total
28-3/8" x 40-7/8", left opening	BC@.500	Ea	156.00	15.00	171.00
28-3/8" x 40-7/8", right opening	BC@.500	Ea	156.00	15.00	171.00
28-3/8" x 48", left opening	B1@1.00	Ea	174.00	27.40	201.40
28-3/8" x 48", right opening	B1@1.00	Ea	174.00	27.40	201.40
33-3/4" x 36", right opening	BC@.500	Ea	229.00	15.00	244.00
33-3/4" x 48", right opening	B1@1.00	Ea	291.00	27.40	318.40
40-3/4" x 40-7/8", right opening	B1@1.00	Ea	282.00	27.40	309.40
48" x 36", right opening	B1@1.00	Ea	269.00	27.40	296.40
48" x 40-7/8", right opening	B1@1.50	Ea	282.00	41.00	323.00
56-1/2" x 48", left opening	B1@1.50	Ea	309.00	41.00	350.00

Windows

	Craft@Hrs	Unit	Material	Labor	Total

Casement bay insulating glass vinyl windows. Fixed center lite and two casement flankers. White vinyl. Factory assembled. By nominal (opening) size, width x height. Actual size is 1/2" less in both width and height. Includes two screens.

	Craft@Hrs	Unit	Material	Labor	Total
69" x 50"	B1@4.00	Ea	906.00	109.00	1,015.00
74" x 50"	B1@4.00	Ea	914.00	109.00	1,023.00
93" x 50"	B1@4.50	Ea	1,040.00	123.00	1,163.00
98" x 50"	B1@4.50	Ea	1,100.00	123.00	1,223.00

Garden insulating glass vinyl windows. Single front awning vent. White vinyl construction with wood frame. Tempered glass shelf. 5/4" vinyl laminated head and seat board. With hardware and screen. By nominal (opening) size, width x height. Replacement window. Labor includes removing the old stops and sash and setting new J-trim as needed.

	Craft@Hrs	Unit	Material	Labor	Total
36" x 36"	B1@2.75	Ea	492.00	75.20	567.20
36" x 36", low-E glass	B1@2.75	Ea	532.00	75.20	607.20
36" x 48", low-E glass	B1@2.75	Ea	579.00	75.20	654.20

Fixed half circle (picture) insulating glass vinyl windows. White frame. Low-E (energy-efficient) glass. Stock units for new construction. By width x height.

	Craft@Hrs	Unit	Material	Labor	Total
20" x 10"	B1@2.75	Ea	429.00	75.20	504.20
30" x 16"	B1@2.75	Ea	429.00	75.20	504.20
40" x 20"	B1@2.75	Ea	571.00	75.20	646.20
50" x 26"	B1@2.75	Ea	676.00	75.20	751.20
60" x 30"	B1@2.75	Ea	768.00	75.20	843.20
80" x 40"	B1@2.75	Ea	1,110.00	75.20	1,185.20

Fixed (picture) low-E insulating glass vinyl windows. Argon filled. White frame. Dimensions are rough opening sizes, width x height. Clear glass.

	Craft@Hrs	Unit	Material	Labor	Total
24" x 36"	BC@.500	Ea	81.00	15.00	96.00
24" x 48"	BC@.500	Ea	95.00	15.00	110.00
24" x 60"	BC@.500	Ea	113.00	15.00	128.00
30" x 36"	BC@.500	Ea	91.60	15.00	106.60
30" x 48"	BC@.500	Ea	111.00	15.00	126.00
36" x 36"	BC@.500	Ea	100.00	15.00	115.00
36" x 48"	B1@1.00	Ea	125.00	27.40	152.40
36" x 48", sound rated	B1@1.50	Ea	265.00	41.00	306.00
36" x 60"	B1@1.00	Ea	145.00	27.40	172.40
48" x 48"	B1@1.50	Ea	149.00	41.00	190.00
48" x 48", sound rated	B1@2.00	Ea	332.00	54.70	386.70
48" x 60"	B1@1.50	Ea	174.00	41.00	215.00
60" x 48"	B1@1.50	Ea	180.00	41.00	221.00
60" x 60"	B1@1.50	Ea	207.00	41.00	248.00

Jalousie (louver) vinyl windows. Positive locking handle for security. By opening size, width x height.

	Craft@Hrs	Unit	Material	Labor	Total
24" x 35-1/2", clear	BC@.500	Ea	234.00	15.00	249.00
24" x 35-1/2", obscure glass	BC@.500	Ea	269.00	15.00	284.00
30" x 35-1/2", clear	BC@.500	Ea	249.00	15.00	264.00
30" x 35-1/2", obscure glass	BC@.500	Ea	293.00	15.00	308.00
36" x 35-1/2", clear	BC@.500	Ea	259.00	15.00	274.00
36" x 35-1/2", obscure glass	BC@.500	Ea	311.00	15.00	326.00

	Craft@Hrs	Unit	Material	Labor	Total

Hopper insulating glass vinyl windows. Low-E (energy-efficient) insulating glass. Argon filled. Frame depth 3-5/16". By opening size, width x height.

	Craft@Hrs	Unit	Material	Labor	Total
32" x 15"	BC@.500	Ea	75.60	15.00	90.60
32" x 17"	BC@.500	Ea	80.30	15.00	95.30
32" x 19"	BC@.500	Ea	83.00	15.00	98.00
32" x 23"	BC@.500	Ea	86.70	15.00	101.70

Vinyl-clad double-hung insulating glass wood windows. Prefinished pine interior. White vinyl exterior. Tilt sash. By nominal (opening) size, width x height. Actual width is 1-5/8" more. Actual height is 3-1/4" more. Includes factory-applied extension jamb for 4-9/16" wall. Add the cost of screens and colonial grilles.

	Craft@Hrs	Unit	Material	Labor	Total
2'0" x 3'0"	BC@.500	Ea	137.00	15.00	152.00
2'0" x 3'6"	BC@.500	Ea	149.00	15.00	164.00
2'4" x 3'0"	BC@.500	Ea	146.00	15.00	161.00
2'4" x 3'6"	BC@.500	Ea	158.00	15.00	173.00
2'4" x 4'0"	BC@.500	Ea	169.00	15.00	184.00
2'4" x 4'6"	B1@1.00	Ea	182.00	27.40	209.40
2'4" x 4'9"	B1@1.00	Ea	190.00	27.40	217.40
2'8" x 3'0"	B1@1.00	Ea	153.00	27.40	180.40
2'8" x 3'6"	B1@1.00	Ea	166.00	27.40	193.40
2'8" x 4'0'	B1@1.00	Ea	179.00	27.40	206.40
2'8" x 4'6"	B1@1.00	Ea	191.00	27.40	218.40
2'8" x 4'9"	B1@1.00	Ea	197.00	27.40	224.40
2'8" x 5'0"	B1@1.00	Ea	203.00	27.40	230.40
2'8" x 5'6"	B1@1.00	Ea	214.00	27.40	241.40
3'0" x 3'0"	B1@1.00	Ea	161.00	27.40	188.40
3'0" x 3'6"	B1@1.00	Ea	174.00	27.40	201.40
3'0" x 4'0"	B1@1.00	Ea	186.00	27.40	213.40
3'0" x 4'6"	B1@1.00	Ea	201.00	27.40	228.40
3'0" x 4'9"	B1@1.00	Ea	205.00	27.40	232.40
3'0" x 5'0"	B1@1.00	Ea	212.00	27.40	239.40
3'0" x 5'6"	B1@1.50	Ea	224.00	41.00	265.00
3'0" x 6'0"	B1@1.50	Ea	236.00	41.00	277.00
Add for low-E glazing, per SF of opening	—	SF	1.65	—	1.65

Screens for vinyl-clad wood windows. By window nominal size.

	Craft@Hrs	Unit	Material	Labor	Total
2'0" x 3'0"	BC@.050	Ea	15.40	1.50	16.90
2'0" x 3'6"	BC@.050	Ea	17.60	1.50	19.10
2'4" x 3'0"	BC@.050	Ea	16.60	1.50	18.10
2'4" x 3'6"	BC@.050	Ea	18.70	1.50	20.20
2'4" x 4'0"	BC@.050	Ea	19.80	1.50	21.30
2'4" x 4'6"	BC@.050	Ea	20.90	1.50	22.40
2'4" x 4'9"	BC@.050	Ea	22.10	1.50	23.60
2'8" x 3'0"	BC@.050	Ea	17.60	1.50	19.10
2'8" x 3'6"	BC@.050	Ea	19.80	1.50	21.30
2'8" x 4'0"	BC@.050	Ea	20.90	1.50	22.40
2'8" x 4'6"	BC@.050	Ea	22.10	1.50	23.60
2'8" x 4'9"	BC@.050	Ea	23.10	1.50	24.60
2'8" x 5'0"	BC@.050	Ea	24.30	1.50	25.80
2'8" x 5'6"	BC@.050	Ea	25.40	1.50	26.90
3'0" x 3'0"	BC@.050	Ea	18.70	1.50	20.20
3'0" x 3'6"	BC@.050	Ea	19.80	1.50	21.30
3'0" x 4'0"	BC@.050	Ea	22.10	1.50	23.60
3'0" x 4'6"	BC@.050	Ea	22.90	1.50	24.40
3'0" x 4'9"	BC@.050	Ea	24.30	1.50	25.80

Windows

	Craft@Hrs	Unit	Material	Labor	Total
3'0" x 5'0"	BC@.050	Ea	25.40	1.50	26.90
3'0" x 5'6"	BC@.050	Ea	23.20	1.50	24.70

Narrow Line® double-hung insulating glass vinyl-clad wood windows, Andersen. White vinyl Perma-Shield® exterior. Clear pine interior ready for finish. Weatherstripped. Stone locks and keepers. Integrated 360-degree nailing flange. Grilles and screens sold separately. By nominal size (and actual size), width x height.

	Craft@Hrs	Unit	Material	Labor	Total
2'0" x 3'2" (2'2-1/8" x 3'5-1/4")	BC@.500	Ea	162.00	15.00	177.00
2'4" x 2'10" (2'6-1/8" x 3'1-1/4")	BC@.500	Ea	168.00	15.00	183.00
2'4" x 3'2" (2'6-1/8" x 3'5-1/4")	BC@.500	Ea	173.00	15.00	188.00
2'4" x 3'10" (2'6-1/8" x 4'1-1/4")	B1@1.00	Ea	195.00	27.40	222.40
2'4" x 4'6" (2'6-1/8" x 4'9-1/4")	B1@1.00	Ea	182.00	27.40	209.40
2'8" x 2'10" (2'10-1/8" x 3'1-1/4")	BC@.500	Ea	175.00	15.00	190.00
2'8" x 3'2" (2'10-1/8" x 3'5-1/4")	BC@.500	Ea	172.00	15.00	187.00
2'8" x 3'10" (2'10-1/8" x 4'1-1/4")	B1@1.00	Ea	202.00	27.40	229.40
2'8" x 4'2" (2'10-1/8" x 4'5-1/4")	B1@1.00	Ea	207.00	27.40	234.40
2'8" x 4'6" (2'10-1/8" x 4'9-1/4")	B1@1.00	Ea	218.00	27.40	245.40
3'0" x 3'2" (3'2-1/8" x 3'5-1/4")	BC@.500	Ea	188.00	15.00	203.00
3'0" x 3'10" (3'2-1/8" x 4'1-1/4")	B1@1.00	Ea	207.00	27.40	234.40
3'0" x 4'2" (3'2-1/8" x 4'5-1/4")	B1@1.00	Ea	223.00	27.40	250.40
3'0" x 4'6" (3'2-1/8" x 4'9-1/4")	B1@1.00	Ea	225.00	27.40	252.40

Vinyl-clad insulating glass wood awning windows. Prefinished pine interior. White vinyl exterior. Add the cost of screens. By nominal (opening) size, width x height.

	Craft@Hrs	Unit	Material	Labor	Total
2'1" x 2'0"	BC@.500	Ea	168.00	15.00	183.00
3'1" x 2'0"	BC@.500	Ea	201.00	15.00	216.00
4'1" x 2'0"	BC@.500	Ea	226.00	15.00	241.00

White vinyl-clad insulating glass wood casement windows, Andersen. Prefinished pine interior. White vinyl exterior. Tilt sash. With nailing flange for new construction. By nominal (opening) size, width x height. Add the cost of screens. Hinge operation as viewed from the exterior.

	Craft@Hrs	Unit	Material	Labor	Total
2'1" x 3'0", right hinge	BC@.500	Ea	181.00	15.00	196.00
2'1" x 3'0", left hinge	BC@.500	Ea	181.00	15.00	196.00
2'1" x 3'6", right hinge	BC@.500	Ea	214.00	15.00	229.00
2'1" x 3'6", left hinge	BC@.500	Ea	214.00	15.00	229.00
2'1" x 4'0", right hinge	BC@.500	Ea	208.00	15.00	223.00
2'1" x 4'0", left hinge	BC@.500	Ea	208.00	15.00	223.00
2'1" x 5'0", right hinge	B1@1.00	Ea	238.00	27.40	265.40
2'1" x 5'0", left hinge	B1@1.00	Ea	238.00	27.40	265.40
4'1" x 3'0", left and right hinge	B1@1.00	Ea	272.00	27.40	299.40
4'1" x 3'5", left and right hinge	B1@1.00	Ea	378.00	27.40	405.40
4'1" x 4'0", left and right hinge	B1@1.00	Ea	418.00	27.40	445.40
2'1" x 3'6", right hinge, egress	BC@.500	Ea	214.00	15.00	229.00
2'1" x 3'6", left hinge, egress	BC@.500	Ea	214.00	15.00	229.00
2'1" x 4'0", right hinge, egress	BC@.500	Ea	239.00	15.00	254.00
3'5" x 3'5", left hinge, egress	B1@1.00	Ea	417.00	27.40	444.40

Terratone vinyl-clad insulating glass wood casement windows, Andersen. Tan vinyl Perma-Shield® exterior. Wood interior ready to finish. High-performance™ glass. With hardware. Add the cost of grilles and screens. By nominal size, width x height. Hinge operation as viewed from the exterior.

	Craft@Hrs	Unit	Material	Labor	Total
2'1" x 3'6", right hinge	BC@.500	Ea	200.00	15.00	215.00
2'1" x 3'6", left hinge	BC@.500	Ea	200.00	15.00	215.00
2'1" x 4'0", right hinge	BC@.500	Ea	227.00	15.00	242.00
2'1" x 4'0", left hinge	BC@.500	Ea	227.00	15.00	242.00

	Craft@Hrs	Unit	Material	Labor	Total
4'1" x 4'0", left and right hinge	B1@1.50	Ea	435.00	41.00	476.00
4'8" x 5'0", left and right hinge	B1@1.50	Ea	526.00	41.00	567.00
6'0" x 4'0", left and right hinge, center fixed	B1@1.50	Ea	516.00	41.00	557.00
6'0" x 5'0", left and right hinge, center fixed	B1@2.00	Ea	579.00	54.70	633.70

Vinyl-clad insulating glass wood basement windows, Andersen. Prefinished pine interior. White vinyl exterior. Tilt sash. With nailing flange for new construction. By nominal (opening) size, width x height. Add the cost of screens.

	Craft@Hrs	Unit	Material	Labor	Total
2'8" x 1'3"	BC@.500	Ea	80.90	15.00	95.90
2'8" x 1'7"	BC@.500	Ea	85.40	15.00	100.40
2'8" x 2'0"	BC@.500	Ea	91.80	15.00	106.80

Aluminum single-hung vertical sliding insulating glass windows. White or bronze frame. Tilt-in removable sash. Dual sash weatherstripping. Includes screen. By nominal (opening) size, width x height. With nailing flange for new construction.

	Craft@Hrs	Unit	Material	Labor	Total
2' x 3'	BC@.500	Ea	46.20	15.00	61.20
2' x 4'	BC@.500	Ea	66.40	15.00	81.40
2' x 5'	BC@.500	Ea	65.80	15.00	80.80
2'8" x 5'	B1@1.00	Ea	72.00	27.40	99.40
3' x 3'	BC@.500	Ea	57.60	15.00	72.60
3' x 4'	B1@1.00	Ea	67.80	27.40	95.20
3' x 5'	B1@1.00	Ea	77.90	27.40	105.30
3' x 6'	B1@1.50	Ea	87.00	41.00	128.00
4' x 5'	B1@1.50	Ea	103.00	41.00	144.00
4' x 3'	B1@1.00	Ea	78.80	27.40	106.20
4' x 4'	B1@1.50	Ea	97.20	41.00	138.20
5' x 3'	B1@1.00	Ea	93.70	27.40	121.10
5' x 4'	B1@1.50	Ea	111.00	41.00	152.00
6' x 4'	B1@1.50	Ea	131.00	41.00	172.00
8' x 4'	B1@2.00	Ea	169.00	54.70	223.70
Add for colonial grille between the lites	—	%	15.0	—	.00
Add for obscure glass in 2' windows	—	Ea	3.00	—	3.00

Aluminum single-hung single glazed vertical sliding windows. Mill finish. Includes screen. By nominal (opening) size, width x height. Clear glass except as noted. With nailing flange for new construction.

	Craft@Hrs	Unit	Material	Labor	Total
2' x 1', obscure glass	BC@.500	Ea	22.70	15.00	37.70
2' x 1'	BC@.500	Ea	19.70	15.00	34.70
2' x 2', obscure glass	BC@.500	Ea	30.80	15.00	45.80
2' x 2'	BC@.500	Ea	27.90	15.00	42.90
2' x 3', obscure glass	BC@.500	Ea	44.10	15.00	59.10
3' x 1', obscure glass	BC@.500	Ea	24.70	15.00	39.70
3' x 2'	BC@.500	Ea	33.10	15.00	48.10
3' x 3'	BC@.500	Ea	42.30	15.00	57.30
3' x 4'	B1@1.00	Ea	62.00	27.40	89.40
4' x 3'	B1@1.00	Ea	48.30	27.40	75.70
4' x 4'	B1@1.50	Ea	52.20	41.00	93.20
5' x 3'	B1@1.00	Ea	52.40	27.40	79.80
5' x 4'	B1@1.50	Ea	73.50	41.00	114.50
6' x 3'	B1@1.50	Ea	67.50	41.00	108.50
6' x 4'	B1@1.50	Ea	80.90	41.00	121.90

Windows

	Craft@Hrs	Unit	Material	Labor	Total

Aluminum horizontal sliding single glazed windows. Right sash is operable when viewed from the exterior. With screen. 7/8" nailing fin setback for 3-coat stucco or wood siding. By nominal (opening) size, width x height. Obscure glass is usually available in smaller (1' and 2') windows at no additional cost.

	Craft@Hrs	Unit	Material	Labor	Total
2' x 2', mill finish	BC@.500	Ea	24.80	15.00	39.80
2' x 2', white finish	BC@.500	Ea	34.10	15.00	49.10
2' x 3', mill finish	BC@.500	Ea	31.20	15.00	46.20
3' x 1', white finish	BC@.500	Ea	31.10	15.00	46.10
3' x 1' 6", white finish	BC@.500	Ea	35.20	15.00	50.20
3' x 2', mill finish	BC@.500	Ea	30.10	15.00	45.10
3' x 3', bronze finish	BC@.500	Ea	56.70	15.00	71.70
3' x 3', mill finish	BC@.500	Ea	36.20	15.00	51.20
3' x 3', white finish	BC@.500	Ea	52.00	15.00	67.00
3' x 4', bronze finish	B1@1.00	Ea	68.30	27.40	95.70
3' x 4', mill finish	B1@1.00	Ea	45.50	27.40	72.90
3' x 4', white finish	B1@1.00	Ea	54.80	27.40	82.20
4' x 1', white finish	BC@.500	Ea	39.90	15.00	54.90
4' x 3', bronze finish	B1@1.00	Ea	66.80	27.40	94.20
4' x 3', mill finish	B1@1.00	Ea	46.10	27.40	73.50
4' x 4', bronze finish	B1@1.50	Ea	75.60	41.00	116.60
4' x 4', mill finish	B1@1.50	Ea	51.80	41.00	92.80
4' x 4', white finish	B1@1.50	Ea	75.70	41.00	116.70
5' x 3', bronze finish	B1@1.00	Ea	75.60	27.40	103.00
5' x 3', mill finish	B1@1.00	Ea	58.10	27.40	85.50
5' x 3', white finish	B1@1.00	Ea	73.30	27.40	100.70
5' x 4', bronze finish	B1@1.50	Ea	86.70	41.00	127.70
5' x 4', mill finish	B1@1.50	Ea	59.20	41.00	100.20
5' x 4', white finish	B1@1.50	Ea	82.70	41.00	123.70
6' x 3', bronze finish	B1@1.50	Ea	83.90	41.00	124.90
6' x 3', mill finish	B1@1.50	Ea	54.00	41.00	95.00
6' x 3', white finish	B1@1.50	Ea	81.50	41.00	122.50
6' x 3' 6", bronze finish	B1@1.50	Ea	86.50	41.00	127.50
6' x 4', bronze finish	B1@1.50	Ea	113.00	41.00	154.00
6' x 4', mill finish	B1@1.50	Ea	52.80	41.00	93.80
6' x 4', white finish	B1@1.50	Ea	93.40	41.00	134.40

Aluminum horizontal sliding insulating glass windows. Right sash is operable when viewed from the exterior. With screen. Nailing fin for new construction. By nominal (opening) size, width x height. Obscure glass is usually available in smaller (1' and 2') windows at no additional cost. White or bronze finish.

	Craft@Hrs	Unit	Material	Labor	Total
2' x 2'	BC@.500	Ea	51.60	15.00	66.60
2' x 3'	BC@.500	Ea	57.20	15.00	72.20
3' x 2'	BC@.500	Ea	58.10	15.00	73.10
3' x 3'	BC@.500	Ea	72.70	15.00	87.70
3' x 4'	B1@1.00	Ea	88.30	27.40	115.70
4' x 3'	B1@1.00	Ea	82.10	27.40	109.50
4' x 4'	B1@1.50	Ea	100.00	41.00	141.00
5' x 3'	B1@1.00	Ea	97.20	27.40	124.60
5' x 4'	B1@1.50	Ea	113.00	41.00	154.00
6' x 3'	B1@1.50	Ea	108.00	41.00	149.00
8' x 4'	B1@2.00	Ea	181.00	54.70	235.70
Add for colonial grid between the lites	—	%	15.0	—	.00
Add for low-E (emissivity) glass	—	%	15.0	—	.00

	Craft@Hrs	Unit	Material	Labor	Total

Aluminum single glazed awning windows. With hardware and screen. Painted powder polyurethane finish. Solar bronze glass. Opening (nominal) sizes, width x height.

	Craft@Hrs	Unit	Material	Labor	Total
24" x 20-1/8", one lite high	BC@.500	Ea	37.90	15.00	52.90
24" x 28-3/4", two lites high	BC@.500	Ea	53.50	15.00	68.50
24" x 37-3/8", three lites high	B1@1.00	Ea	65.50	27.40	92.90
24" x 46", four lites high	B1@1.00	Ea	69.00	27.40	96.40
24" x 54-5/8", five lites high	B1@1.50	Ea	71.10	41.00	112.10
24" x 63-1/4", six lites high	B1@1.50	Ea	92.80	41.00	133.80
30" x 20-1/8", one lite high	BC@.500	Ea	47.60	15.00	62.60
30" x 28-3/4", two lites high	BC@.500	Ea	59.50	15.00	74.50
30" x 37-3/8", three lites high	B1@1.00	Ea	71.40	27.40	98.80
30" x 46", four lites high	B1@1.00	Ea	78.60	27.40	106.00
30" x 54-5/8", five lites high	B1@1.50	Ea	98.80	41.00	139.80
30" x 63-1/4", six lites high	B1@1.50	Ea	107.00	41.00	148.00
36" x 20-1/8", one lite high	BC@.500	Ea	53.50	15.00	68.50
36" x 28-3/4", two lites high	BC@.500	Ea	65.40	15.00	80.40
36" x 37-3/8", three lites high	B1@1.00	Ea	75.50	27.40	102.90
36" x 46", four lites high	B1@1.00	Ea	80.50	27.40	107.90
36" x 54-5/8", five lites high	B1@1.50	Ea	102.30	41.00	143.30
36" x 63-1/4", six lites high	B1@1.50	Ea	142.00	41.00	183.00

Double-hung insulating glass wood windows. Treated western pine. Natural wood interior. 4-9/16" jamb width. 5/8" insulating glass. Weatherstripped. Tilt-in, removable sash. With hardware. Nominal sizes, width x height. With screen.

	Craft@Hrs	Unit	Material	Labor	Total
2'0" x 3'2"	BC@.500	Ea	103.00	15.00	118.00
2'4" x 3'2"	BC@.500	Ea	111.00	15.00	126.00
2'4" x 4'6"	BC@.500	Ea	130.00	15.00	145.00
2'8" x 3'2"	BC@.500	Ea	114.00	15.00	129.00
2'8" x 3'10"	BC@.500	Ea	126.00	15.00	141.00
2'8" x 4'6"	B1@1.00	Ea	138.00	27.40	165.40
2'8" x 5'2"	B1@1.00	Ea	149.00	27.40	176.40
2'8" x 6'2"	B1@1.00	Ea	170.00	27.40	197.40
3'0" x 3'2"	BC@.500	Ea	129.00	15.00	144.00
3'0" x 3'10"	B1@1.00	Ea	135.00	27.40	162.40
3'0" x 4'6"	B1@1.00	Ea	147.00	27.40	174.40
3'0" x 5'2"	B1@1.00	Ea	158.00	27.40	185.40
3'0" x 6'2"	B1@1.00	Ea	184.00	27.40	211.40
2'8" x 4'6", twin	B1@1.00	Ea	264.00	27.40	291.40
3'0" x 4'6", twin	B1@1.00	Ea	324.00	27.40	351.40

Fixed octagon wood windows. Stock units for new construction.

	Craft@Hrs	Unit	Material	Labor	Total
24", 9-lite beveled glass	BC@1.50	Ea	111.00	45.00	156.00
24", insulating glass	BC@1.50	Ea	78.20	45.00	123.20
24", beveled glass	BC@1.50	Ea	63.00	45.00	108.00
24", single glazed	BC@1.50	Ea	55.60	45.00	100.60
24", venting insulating glass	BC@1.50	Ea	120.00	45.00	165.00

Angle bay windows. Fixed center lite and two double-hung flankers. 30 degree angle. Pine with aluminum cladding insulating glass. With screen. White or bronze finish. 4-9/16" wall thickness. Opening (nominal) sizes, width x height. Add the cost of interior trim and installing or framing the bay window roof. See the figures that follow.

	Craft@Hrs	Unit	Material	Labor	Total
6'4" x 4'8"	B1@4.00	Ea	1,310.00	109.00	1,419.00
7'2" x 4'8"	B1@4.00	Ea	1,320.00	109.00	1,429.00
7'10" x 4'8"	B1@4.50	Ea	1,440.00	123.00	1,563.00

Windows

	Craft@Hrs	Unit	Material	Labor	Total
8'6" x 4'8"	B1@4.50	Ea	1,650.00	123.00	1,773.00
8'10" x 4'8"	B1@4.50	Ea	1,680.00	123.00	1,803.00
9'2" x 4'8"	B1@5.00	Ea	1,700.00	137.00	1,837.00
9'10" x 4'8"	B1@5.00	Ea	1,790.00	137.00	1,927.00
Add for 45 degree angle windows	—	Ea	67.60	—	67.60
Add for roof framing kit	—	Ea	120.00	—	120.00

Bay window roof cover. Architectural copper roof system for standard angle bay windows. Includes hardware. Stock units for new construction.

	Craft@Hrs	Unit	Material	Labor	Total
Up to 48" width	SW@.500	Ea	403.00	17.00	420.00
49" to 60" width	SW@.500	Ea	458.00	17.00	475.00
61" to 72" width	SW@.600	Ea	513.00	20.40	533.40
73" to 84" width	SW@.600	Ea	569.00	20.40	589.40
85" to 96" width	SW@.600	Ea	625.00	20.40	645.40
97" to 108" width	SW@.700	Ea	679.00	23.80	702.80
109" to 120" width	SW@.700	Ea	735.00	23.80	758.80
121" to 132" width	SW@.800	Ea	846.00	27.20	873.20
133" to 144" width	SW@.800	Ea	901.00	27.20	928.20
145" to 156" width	SW@1.00	Ea	958.00	34.00	992.00
157" to 168" width	SW@1.00	Ea	1,010.00	34.00	1,044.00
Add for aged copper, brass or aluminum finish	—	%	20.0	—	—
Add for 24" to 26" projection	—	%	10.0	—	—
Add for 26" to 28" projection	—	%	15.0	—	—
Add for 28" to 30" projection	—	%	20.0	—	—
Add for brick soffit kit	—	%	45.0	—	—
Add for 90-degree bays	—	Ea	66.40	—	66.40

Casement bow windows. Insulating glass, assembled, with screens, weatherstripped, unfinished pine, 16" projection, 22" x 56" lites, 4-7/16" jambs. Add the cost of interior trim and installing or framing the bow window roof. Stock units for new construction. Generally a roof kit is required for projections of 12" or more. See the figures that follow.

	Craft@Hrs	Unit	Material	Labor	Total
8'1" wide x 4'8" (4 lites, 2 venting)	B1@4.75	Ea	1,400.00	130.00	1,530.00
8'1" wide x 5'4" (4 lites, 2 venting)	B1@4.75	Ea	1,510.00	130.00	1,640.00
8'1" wide x 6'0" (4 lites, 2 venting)	B1@4.75	Ea	1,620.00	130.00	1,750.00
10' wide x 4'8" (5 lites, 2 venting)	B1@4.75	Ea	1,700.00	130.00	1,830.00
10' wide x 5'4" (5 lites, 2 venting)	B1@4.75	Ea	1,840.00	130.00	1,970.00
10' wide x 6'0" (5 lites, 2 venting)	B1@4.75	Ea	2,020.00	130.00	2,150.00
Add for roof framing kit, 5 lites	—	Ea	110.00	—	110.00
Add for interior and exterior trim					
3, 4 or 5 lites	B1@2.00	Set	35.00	54.70	89.70

Bow or bay window roof framed on site. 1/2" plywood sheathing, fiberglass batt insulation, fascia board, cove moulding, roof flashing, 15-pound felt, 240-pound seal-tab asphalt shingles, and cleanup.

	Craft@Hrs	Unit	Material	Labor	Total
Up to 120" width	B1@3.50	Ea	66.20	95.80	162.00
Add per linear foot of width over 10'	B1@.350	LF	6.62	9.58	16.20

Basement storm windows. White aluminum frame with removable plexi-glass window. Clips on over aluminum or wood basement window. Screen included. Nominal window sizes, width x height.

	Craft@Hrs	Unit	Material	Labor	Total
32" x 14"	BC@.420	Ea	14.10	12.60	26.70
32" x 18"	BC@.420	Ea	16.00	12.60	28.60
32" x 22"	BC@.420	Ea	17.10	12.60	29.70

	Craft@Hrs	Unit	Material	Labor	Total

Aluminum storm windows. Removable glass inserts. Removable screen. Double track. Includes weatherstripping. Adaptable to aluminum single-hung windows.

	Craft@Hrs	Unit	Material	Labor	Total
24" x 39", mill finish	BC@.420	Ea	27.90	12.60	40.50
24" x 39", white finish	BC@.420	Ea	32.30	12.60	44.90
24" x 55", mill finish	BC@.420	Ea	32.50	12.60	45.10
24" x 55", white finish	BC@.420	Ea	39.50	12.60	52.10
28" x 39", mill finish	BC@.420	Ea	29.10	12.60	41.70
28" x 39", white finish	BC@.420	Ea	34.70	12.60	47.30
28" x 47", white finish	BC@.420	Ea	38.70	12.60	51.30
28" x 55", mill finish	BC@.420	Ea	32.50	12.60	45.10
28" x 55", white finish	BC@.420	Ea	39.10	12.60	51.70
32" x 39", mill finish	BC@.420	Ea	29.10	12.60	41.70
32" x 39", white finish	BC@.420	Ea	39.50	12.60	52.10
32" x 47", white finish	BC@.420	Ea	44.30	12.60	56.90
32" x 55", mill finish	BC@.420	Ea	32.50	12.60	45.10
32" x 55", white finish	BC@.420	Ea	44.80	12.60	57.40
32" x 63", mill finish	BC@.420	Ea	47.00	12.60	59.60
32" x 63", white finish	BC@.420	Ea	46.40	12.60	59.00
32" x 75", white finish	BC@.420	Ea	61.30	12.60	73.90
36" x 39", mill finish	BC@.420	Ea	29.10	12.60	41.70
36" x 39", white finish	BC@.420	Ea	39.30	12.60	51.90
36" x 55", mill finish	BC@.420	Ea	37.10	12.60	49.70
36" x 55", white finish	BC@.420	Ea	44.80	12.60	57.40
36" x 63", mill finish	BC@.420	Ea	48.00	12.60	60.60
36" x 63", white finish	BC@.420	Ea	58.20	12.60	70.80

Window wells ScapeWel egress window, meets code requirements for basement escape window. Snaps into place on site. For new construction and remodeling, attaches directly to foundation. Bilco.

	Craft@Hrs	Unit	Material	Labor	Total
Model 4048-42, 42 x 48" Two tier	B1@1.50	Ea	578.00	41.00	619.00
4042C Polycarbonate window well cover					
Fits Model 4048-42	—	Ea	310.00	—	310.000
Model 4048-54, 54 x 48" Two tier	B1@1.50	Ea	625.00	41.00	666.00
4054C Polycarbonate window well cover					
Fits Model 4048-54	—	Ea	340.00	—	340.000
Model 4048-66, 66 x 48" Two tier	B1@1.50	Ea	646.00	41.00	687.00
4066C Polycarbonate window well cover					
Fits Model 4048-66	—	Ea	368.00	—	368.000
Model 4862-42, 42 x 62" Three tier	B1@1.50	Ea	767.00	41.00	808.00
4842C Polycarbonate window well cover					
Fits Model 4862-42	—	Ea	331.00	—	331.000
Model 4862-54, 54 x 62" Three tier	B1@1.50	Ea	792.00	41.00	833.00
4854C Polycarbonate window well cover					
Fits Model 4862-54	—	Ea	341.00	—	341.000
Model 4862-66, 66 x 62" Three tier	B1@1.50	Ea	825.00	41.00	866.00
4866C Polycarbonate window well cover					
Fits Model 4862-66	—	Ea	378.00	—	378.00

Single Family Home Costs

Use the costs below to check your estimates and to verify bids submitted by subcontractors. These costs are based on a 1,600 square foot home (as described on page 321) built during the third quarter of 2005 at a cost of $87.18 per square foot. Figures in the column "% of Total" are the same as those listed in the Residential Rule of Thumb Construction Costs on page 321. These are averages and will vary widely with the type and quality of construction.

Item	Total manpower required	% of Total	Unit	$ Material	$ Labor	$ Equipment	Total $ per SF	Total $ 1600 SF
Excavation	3 men 2 days	1.2%	SF	—	$.79	$.26	$1.05	$1,680
Foundation, slab, piers	3 men 4 days	3.7%	SF	$1.29	1.62	.32	3.23	5,168
Flatwork, drive and walk	3 men 3 days	2.4%	SF	.84	1.05	.20	2.09	3,344
Brick hearth & veneer	2 men 1 day	0.7%	SF	.24	.31	.06	.61	976
Rough hardware		0.6%	SF	.21	.26	.05	.52	832
Finish hardware		0.2%	SF	.10	.07	—	.17	272
Rough lumber		8.4%	SF	7.32	—	—	7.32	11,712
Finish lumber		0.5%	SF	.44	—	—	.44	704
Rough carpentry labor	2 men 20 days	8.9%	SF	—	7.76	—	7.76	12,416
Finish carpentry labor	1 man 7 days	1.7%	SF	—	1.48	—	1.48	2,368
Countertops	1 man 2 days	1.5%	SF	.79	.52	—	1.31	2,096
Cabinets	2 men 2 days	3.7%	SF	2.58	.65	—	3.23	5,168
Insulation (R19 ceiling)	1 man 3 days	2.3%	SF	1.30	.70	—	2.00	3,200
Roofing	4 men 3 days	5.5%	SF	2.87	1.92	—	4.79	7,664
Painting	2 men 3 days	3.6%	SF	1.10	2.04	—	3.14	5,024
Shower and tub enclosure	1 man 1 day	0.5%	SF	.26	.18	—	.44	704
Prefabricated fireplace	1 man 1 day	0.9%	SF	.59	.16	.03	.78	1,248
Bath accessories	1 man 1 day	0.7%	SF	.41	.20	—	.61	976
Built-In appliances	1 man 1 day	1.6%	SF	1.25	.14	—	1.39	2,224
Heating and ducting	2 men 5 days	2.9%	SF	1.01	1.52	—	2.53	4,048
Plumbing & sewer, connection	3 men 5 days	7.3%	SF	2.54	3.18	.64	6.36	10,176
Doors	2 men 2 days	1.9%	SF	1.08	.58	—	1.66	2,656
Garage door	1 man 1 day	0.4%	SF	.26	.09	—	.35	560
Alum windows & sliding doors	1 man 2 days	1.2%	SF	.63	.42	—	1.05	1,680
Exterior stucco	3 men 3 days	6.4%	SF	3.63	1.67	.28	5.58	8,928
Gypsum wallboard	2 men 6 days	4.7%	SF	1.84	2.26	—	4.10	6,560
Resilient flooring	1 man 1 day	2.0%	SF	.82	.92	—	1.74	2,784
Carpeting	2 men 1 day	2.4%	SF	1.67	.42	—	2.09	3,344
Wiring (Romex)	1 man 10 days	3.2%	SF	1.12	1.67	—	2.79	4,464
Lighting fixtures		1.2%	SF	.84	.21	—	1.05	1,680
Subtotal, Direct Job Costs		**82.2%**		**$37.03**	**$32.79**	**$1.84**	**$71.66**	**$114,656**
Amount above, as a percent of total direct costs				51.6%	45.8%	2.6%	100.0%	

Indirect Costs	% of Total	Unit	$ Material	$ Labor	$ Equipment	Total $ per SF	Total $ 1600 SF
Insurance, payroll tax	2.8%	SF	$2.44	—	—	$2.44	$3,904
Plans & specs	0.4%	SF	.35	—	—	.35	560
Permits & utilities	1.7%	SF	1.48	—	—	1.48	2,368
Final cleanup	0.4%	SF	—	.35	—	.35	560
Subtotal, Indirect Job Costs	**5.3%**	**SF**	**$4.27**	**$.35**	**—**	**$4.62**	**$7,392**
Overhead & profit	**12.5%**	**SF**	**$10.90**	**—**	**—**	**$10.90**	**$17,434**
Total all costs, rounded (based on 1600 SF)	**100.0%**	**SF**	**$52.20**	**$33.14**	**$1.84**	**$87.18**	**$139,482**
Amounts above, as a percent of total costs			59.9%	38.0%	2.1%	100.0%	

The "Manpower" column assumes work is done in phases. Elapsed time, from breaking ground to final cleanup, will be about 10 weeks, plus delays due to weather, shortages and schedule conflicts.

Construction Economics Division
Construction Cost Index for New Single Family Homes

Period	% of 1967 Cost	$ per SF of Floor	Period	% of 1967 Cost	$ per SF of Floor	Period	% of 1967 Cost	$ per SF of Floor
1988			**1994**			**2000**		
1st quarter	398.6	55.78	1st quarter	459.5	64.33	1st quarter	516.8	72.30
2nd quarter	399.6	55.92	2nd quarter	460.7	64.50	2nd quarter	520.4	72.80
3rd quarter	402.8	56.37	3rd quarter	461.4	64.60	3rd quarter	524.2	73.33
4th quarter	407.1	56.97	4th quarter	461.6	64.62	4th quarter	525.7	73.55
1989			**1995**			**2001**		
1st quarter	407.5	57.02	1st quarter	462.1	64.68	1st quarter	527.4	73.78
2nd quarter	411.9	57.64	2nd quarter	463.6	64.90	2nd quarter	528.8	73.98
3rd quarter	422.6	59.13	3rd quarter	464.3	65.00	3rd quarter	530.3	74.19
4th quarter	429.0	60.02	4th quarter	464.8	65.07	4th quarter	533.1	74.58
1990			**1996**			**2002**		
1st quarter	432.4	60.49	1st quarter	469.4	65.72	1st quarter	530.11	74.16
2nd quarter	434.2	60.74	2nd quarter	474.1	66.37	2nd quarter	530.29	74.18
3rd quarter	435.3	60.89	3rd quarter	478.6	67.00	3rd quarter	532.80	74.52
4th quarter	435.2	60.74	4th quarter	480.6	672.5	4th quarter	535.31	74.89
1991			**1997**			**2003**		
1st quarter	433.0	60.62	1st quarter	482.0	67.53	1st quarter	539.22	75.45
2nd quarter	432.6	60.56	2nd quarter	487.4	68.20	2nd quarter	540.58	75.63
3rd quarter	433.2	60.65	3rd quarter	491.9	68.84	3rd quarter	541.11	75.71
4th quarter	433.0	60.62	4th quarter	493.5	69.09	4th quarter	554.32	77.56
1992			**1998**			**2004**		
1st quarter	432.6	60.56	1st quarter	495.2	69.30	1st quarter	557.54	78.01
2nd quarter	432.4	60.54	2nd quarter	498.4	69.76	2nd quarter	578.42	80.93
3rd quarter	432.3	60.52	3rd quarter	501.7	70.22	3rd quarter	587.05	82.10
4th quarter	436.9	61.17	4th quarter	503.5	70.45	4th quarter	611.14	85.51
1993			**1999**			**2005**		
1st quarter	443.9	62.15	1st quarter	504.3	70.56	1st quarter	608.70	85.17
2nd quarter	450.7	63.10	2nd quarter	505.2	70.69	2nd quarter	616.87	86.31
3rd quarter	456.4	63.90	3rd quarter	506.5	70.85	3rd quarter	623.00	87.18
4th quarter	458.0	64.12	4th quarter	511.1	71.50	4th quarter	—	—

The figures under the column "$ per SF of Floor" show construction costs for building a good quality home in a suburban area under competitive conditions in each calendar quarter since 1988. This home is described below under the section "Residential Rule of Thumb." These costs include the builder's overhead and profit and a 450 square foot garage but no basement. The cost of a finished basement per square foot will be approximately 40% of the square foot cost of living area. If the garage area is more than 450 square feet, use 50% of the living area cost to adjust for the larger or smaller garage. To find the total construction cost of the home, multiply the living area (excluding the garage) by the cost in the column "$ per SF of Floor."

Deduct for rural areas	5.0%
Add for 1,800 SF house (better quality)	4.0
Add for 2,000 SF house (better quality)	3.0
Deduct for over 2,400 SF house	3.0
Add for split level house	3.0
Add for 3-story house	10.0
Add for masonry construction	9.0

Construction costs are higher in some cities and lower in others. Square foot costs listed in the table above are national averages. To modify these costs to your job site, apply the appropriate area modification factor from pages 12 through 15 of this manual. But note that area modifications on pages 12 through 15 are based on recent construction and may not apply to work that was completed many years ago.

Residential Rule of Thumb

Construction Costs

The following figures are percentages of total construction cost for a good quality single family residence: 1,600 square foot single story non-tract 3-bedroom, 1¾ bath home plus attached two car (450 SF) garage. Costs assume a conventional floor, stucco exterior, wallboard interior, shake roof, single fireplace, forced air heat, copper water supply, ABS drain lines, range with double oven, disposer, dishwasher and 900 SF of concrete flatwork.

Item	Percent	Item	Percent	Item	Percent	Item	Percent	Item	Percent
Excavation	1.2	Finish lumber	.5	Painting	3.6	Doors	1.9	Wiring (Romex)	3.2
Flatwork (drive & walk)	2.4	Rough carpentry labor	8.9	Shower & tub enclosure	.5	Garage door	.4	Lighting fixtures	1.2
Foundation, slab, piers	3.7	Finish carpentry labor	1.7	Prefabricated fireplace	.9	Alum. windows, door	1.2	Insurance, tax	2.8
Brick hearth & veneer	.7	Countertops	1.5	Bath accessories	.7	Exterior stucco	6.4	Plans and specs	.4
Rough hardware	.6	Cabinets	3.7	Built-in appliances	1.6	Gypsum wallboard	4.7	Permits & utilities	1.7
Finish hardware	.2	Insulation (R19 ceiling)	2.3	Heating and ducting	2.9	Resilient flooring	2.0	Final cleanup	.4
Rough lumber	8.4	Roofing	5.5	Plumb. & sewer conn.	7.3	Carpeting	2.4	Overhead & profit	12.5

Industrial and Commercial Division Contents
A complete index begins on page 636

Industrial and Commercial Division
Hourly Labor Costs

The hourly labor costs shown in the column headed "Hourly Cost" have been used to compute the manhour costs for the crews on pages 8 and 9 and the costs in the "Labor" column on pages 320 to 635 of this book. All figures are in U.S. dollars per hour.

"Hourly Wage and Benefits" includes the wage, welfare, pension, vacation, apprentice and other mandatory contributions. The "Employer's Burden" is the cost to the contractor for Unemployment Insurance (FUTA), Social Security and Medicare (FICA), state unemployment insurance, workers' compensation insurance and liability insurance. Tax and insurance expense included in these labor-hour costs are itemized in the sections beginning on pages 180 and 298.

These hourly labor costs will apply within a few percent on many jobs. But wages may be much higher or lower on the job you are estimating. If the hourly cost on this page is not accurate, use the labor cost adjustment procedure on page 11.

If your hourly labor cost is not known and can't be estimated, use both labor and material figures in this book without adjustment. When all material and labor costs have been compiled, multiply the total by the appropriate figure in the area modification table on pages 12 through 15.

Craft	Hourly Wage and Benefits ($)	Typical Employer Burden (%)	Employer's Burden Per Hour ($)	Hourly Cost ($)
Air Tool Operator	27.95	28.02%	7.84	35.79
Asbestos Worker	38.56	40.95%	15.79	54.35
Boilermaker	39.78	36.90%	14.68	54.46
Bricklayer	36.01	27.58%	9.93	45.94
Bricklayer Tender	27.47	27.58%	7.58	35.05
Building Laborer	28.17	29.20%	8.22	36.39
Carpenter	35.56	28.10%	9.99	45.55
Cement Mason	33.81	28.88%	9.76	43.57
Crane Operator	36.44	35.56%	12.96	49.40
Drywall Installer	35.26	25.21%	8.89	44.15
Electrician	41.01	23.15%	9.50	50.51
Elevator Constructor	38.03	19.34%	7.35	45.38
Floor Layer	33.98	32.65%	11.10	45.08
Glazier	35.83	27.20%	9.75	45.58
Iron Worker (Structural)	38.57	40.86%	15.76	54.33
Lather	32.59	33.39%	10.88	43.47
Marble Setter	35.59	27.83%	9.91	45.50
Millwright	36.17	26.92%	9.73	45.90
Mosaic & Terrazzo Worker	33.47	27.83%	9.31	42.78
Painter	35.36	28.38%	10.03	45.39
Pile Driver	35.75	36.35%	13.00	48.75
Pipefitter	40.83	25.15%	10.27	51.10
Plasterer	36.61	30.44%	11.15	47.76
Plasterer Helper	30.21	30.44%	9.19	39.40
Plumber	41.51	25.70%	10.67	52.18
Reinforcing Ironworker	37.78	40.69%	15.37	53.15
Roofer	31.55	41.13%	12.98	44.53
Sheet Metal Worker	40.38	25.12%	10.14	50.52
Sprinkler Fitter	42.81	25.15%	10.77	53.58
Tractor Operator	36.31	36.56%	13.27	49.58
Truck Driver	29.43	36.56%	10.77	40.20

General Contractor's Markup Costs in this manual do not include the general contractor's markup. Costs in sections identified as *subcontract* are based on quotes from subcontractors and include the subcontractor's markup. See pages 3 through 5 for more on subcontracted work and markup. Typical markup for general contractors handling commercial and industrial projects is shown at the bottom of this page. The two sections that follow, Indirect Overhead and Direct Overhead, give a more detailed breakdown.

Indirect Overhead (home office overhead)
This cost, usually estimated as a percentage, is calculated by dividing average annual indirect overhead costs by average annual receipts. Indirect overhead usually doesn't increase or decrease as quickly as gross volume increases or decreases. Home office expense may be about the same even if annual sales double or drop by one-half or more.

The figures below show typical indirect overhead costs per $1,000 of estimated project costs for a general contractor handling $600,000 to $1,000,000 commercial and industrial projects. Indirect overhead costs vary widely but will be about 8% of gross for most profitable firms. Total indirect overhead cost ($20.00 per $1,000 for materials and $60.00 per $1,000 for labor) appears as 2.0% and 6.0% under Total General Contractor's Markup at the bottom of this page.

	Craft@Hrs	Unit	Material	Labor	Total
Rent, office supplies, utilities, equipment, advertising, etc.	—	M$	20.00	—	20.00
Office salaries and professional fees	—	M$	—	60.00	60.00
General Contractor's indirect overhead	**—**	**M$**	**20.00**	**60.00**	**80.00**

Direct Overhead (job site overhead)
The figures below show typical direct overhead costs per $1,000 of a project's total estimated costs for a general contractor handling a $600,000 to $1,000,000 job that requires 6 to 9 months for completion. Use these figures for preliminary estimates and to check final bids. Add the cost of mobilization, watchmen, fencing, hoisting, permits, bonds, scaffolding and testing. Total direct overhead cost ($14.30 per $1,000 for materials and $58.70 per $1,000 for labor) appears as 1.4% and 5.9% under Total General Contractor's Markup at the bottom of this page. These costs can be expected to vary and will usually be lower on larger projects.

	Craft@Hrs	Unit	Material	Labor	Total
Job superintendent ($4,400 per month)	—	M$	—	44.00	44.00
Pickup truck for superintendent ($400 per month)	—	M$	4.00	—	4.00
Temporary power, light and heat ($225 per month)	—	M$	2.50	—	2.50
Temporary water ($100 per month)	—	M$	1.00	—	1.00
Temporary phone ($100 per month)	—	M$	1.00	—	1.00
Job site toilets ($100 per month)	—	M$	1.00	—	1.00
Job site office trailer 8' x 30' ($150 per month)	—	M$	1.50	—	1.50
Storage bin and tool shed 8' x 24' ($80 per month)	—	M$	.80	—	.80
Job site cleanup & debris removal ($300 per month)	—	M$	—	3.00	3.00
Job signs & first aid equipment ($50 per month)	—	M$	.50	—	.50
Small tools, supplies ($200 per month)	—	M$	2.00	—	2.00
Taxes and insurance on wages (at 24.9%)	—	M$	—	11.70	11.70
Total General Contractor's direct overhead	**—**	**M$**	**14.30**	**58.70**	**73.00**

Total General Contractor's Markup Typical costs for commercial and industrial projects.

	Craft@Hrs	Unit	Material	Labor	Total
Indirect overhead (home office overhead)	—	%	2.0	6.0	8.0
Direct overhead (job site overhead)	—	%	1.4	5.9	7.3
Contingency (allowance for unknown conditions)	—	%	—	—	2.0
Profit (varies widely, mid-range shown)	—	%	—	—	7.5
Total General Contractor's Markup	**—**	**%**	**—**	**—**	**24.8**

General Requirements 1

	Craft@Hrs	Unit	Material	Labor	Total

Project Financing Construction loans are usually made for a term of up to 18 months. The maximum loan will be based on the value of the land and building when completed. Typically this maximum is 75% for commercial, industrial and apartment buildings and 77.5% for tract housing.

The initial loan disbursement will be about 50% of the land cost. Periodic disbursements are based on the percentage of completion as verified by voucher or inspection. The last 20% of loan proceeds is disbursed when an occupancy permit is issued. Fund control fees cover the cost of monitoring disbursements. Typical loan fees, loan rates and fund control fees are listed below.

Loan origination fee, based on amount of loan

	Craft@Hrs	Unit	Material	Labor	Total
Tracts	—	%	—	—	2.5
Apartments	—	%	—	—	2.0
Commercial, industrial buildings	—	%	—	—	3.0

Loan interest rate, prime rate plus

	Craft@Hrs	Unit	Material	Labor	Total
Tracts, commercial, industrial	—	%	—	—	2.0
Apartments	—	%	—	—	1.5

Fund control fees (costs per $1,000 disbursed)

	Craft@Hrs	Unit	Material	Labor	Total
To $200,000	—	LS	—	—	1,000.00
Over $200,000 to $500,000, $1,000 plus	—	M$	—	—	1.50
Over $500,000 to $1,000,000, $1,450 plus	—	M$	—	—	1.00
Over $1,000,000 to $3,000,000, $1,250 plus	—	M$	—	—	.85
Over $3,000,000 to $5,000,000, $1,150 plus	—	M$	—	—	.75
Over $5,000,000, $1,000 plus	—	M$	—	—	.55

Project Scheduling by CPM (Critical Path Method). Includes consultation, review of construction documents, development of construction logic, and graphic schedule.

Wood frame buildings, one or two stories

	Craft@Hrs	Unit	Material	Labor	Total
Simple schedule	—	LS	—	—	750.00
Complex schedule	—	LS	—	—	1,800.00

Tilt-up concrete buildings, 10,000 to 50,000 SF

	Craft@Hrs	Unit	Material	Labor	Total
Simple schedule	—	LS	—	—	1,020.00
Complex schedule	—	LS	—	—	1,530.00

Two to five story buildings (mid rise)

	Craft@Hrs	Unit	Material	Labor	Total
Simple schedule	—	LS	—	—	1,850.00
Complex schedule	—	LS	—	—	2,600.00

Five to ten story buildings (high rise)

	Craft@Hrs	Unit	Material	Labor	Total
Simple schedule	—	LS	—	—	3,000.00
Complex schedule	—	LS	—	—	7,500.00

Manufacturing plants and specialized use low rise buildings

	Craft@Hrs	Unit	Material	Labor	Total
Simple schedule	—	LS	—	—	1,330.00
Complex schedule	—	LS	—	—	1,840.00

Project scheduling software for in-house production

	Craft@Hrs	Unit	Material	Labor	Total
Complete package	—	LS	—	—	6,500.00

Comprehensive schedules to meet government or owner specifications may cost 20% to 40% more. Daily schedule updates can increase costs by 40% to 50%.

Sewer Connection Fees Check with the local sanitation district for actual charges. These costs are typical for work done by city or county crews and include excavation, re-compaction and repairs to the street. Note that these costs are set by local government and will vary widely. These costs do not include capacity fees or other charges which are often levied on connection to new construction. Cost per living unit may be $2,000 or more. For commercial buildings, figure 20 fixture units equal one living unit. A 6" sewer connection will be required for buildings that include more than 216 fixture units. (Bathtubs, showers and sinks are 2 fixture units, water closets are 4, clothes washers are 3 and lavatories are 1.) The costs shown assume that work will be done by sanitation district crews. Similar work done by a private contractor under a public improvement permit may cost 50% less.

	Craft@Hrs	Unit	Material	Labor	Total
Typical charges based on main up to 11' deep and lateral up to 5' deep					
4" connection and 40' run to main in street	—	LS	—	—	1,680.00
4" pipe runs over 40' to street	—	LF	—	—	38.00
6" connection and 40' run to main in street	—	LS	—	—	2,740.00
6" pipe runs over 40' to street	—	LF	—	—	65.00
4" connection and 15' run to main in alley	—	LS	—	—	1,220.00
6" connection and 15' run to main in alley	—	LS	—	—	1,980.00
Manhole cut-in	—	Ea	—	—	335.00
Add for main depths over 5'					
Over 5' to 8'	—	%	—	—	30.0
Over 8' to 11'	—	%	—	—	60.0
Over 11'	—	%	—	—	100.0

Water Meters Check with the local water district for actual charges. These costs are typical for work done by city or county crews and include excavation and pipe to 40' from the main, meter, vault, re-compaction and repairs to the street. These costs do not include capacity fees or other charges which are often levied on connection to new construction. The capacity fee per living unit is typically $5,000. For commercial buildings, figure 20 fixture units equal one living unit. (Bathtubs, showers and sinks are 2 fixture units, water closets are 4, clothes washers are 3 and lavatories are 1.) The cost for discontinuing service will usually be about the same as the cost for starting new service, but no capacity fee will be charged. Add the backflow device and capacity fee, if required.

	Craft@Hrs	Unit	Material	Labor	Total
1" service, 3/4" meter	—	LS	—	—	2,450.00
1" service, 1" meter	—	LS	—	—	2,540.00
Add for 1" service per LF over 40'	—	LF	—	—	59.00
2" service, 1-1/2" meter	—	LS	—	—	2,850.00
2" service, 2" meter	—	LS	—	—	2,950.00
Add for 2" service per LF over 40'	—	LF	—	—	63.20
Two 2" service lines and meter manifold	—	LS	—	—	4,671.00
Add for two 2" lines per LF over 40'	—	LF	—	—	86.70
Installation of meter only. Connected to existing water service if correct size line exists from the water main to the meter box location					
3/4" meter	—	LS	—	—	94.00
1" x 3/4" meter	—	LS	—	—	102.00
1" meter	—	LS	—	—	122.00
1-1/2" meter	—	LS	—	—	397.00
2" meter	—	LS	—	—	571.00
2" meter (double line)	—	LS	—	—	1,010.00
Backflow preventer with single check valve. Includes pressure vacuum breaker (PVB), two ball valves, one air inlet and one check valve					
1" pipe	—	LS	—	—	250.00
2" pipe	—	LS	—	—	495.00
Backflow preventer with double check valves. Includes pressure vacuum breaker (PVB), two ball or gate valves, one air inlet and two check valves					
3/4" pipe	—	LS	—	—	330.00
1" pipe	—	LS	—	—	330.00
1-1/2" pipe	—	LS	—	—	403.00
2" pipe	—	LS	—	—	540.00
Reduced pressure backflow preventer. Includes pressure reducing vacuum breaker (RPVB), two ball or gate valves, one air inlet and two check valves and a relief valve					
1" pipe	—	LS	—	—	872.00
2" pipe	—	LS	—	—	1,060.00
2-2" pipes (double installation)	—	LS	—	—	2,090.00

	Craft@Hrs	Unit	Material	Labor	Total
Surveying					
Surveying party, 2 or 3 technicians					
Typical cost	—	Day	—	—	1,150.00
Higher cost	—	Day	—	—	1,480.00
Data reduction and drafting	—	Hr	—	—	70.00

Surveys Including wood hubs or pipe markers as needed. These figures assume that recorded monuments are available adjacent to the site.

	Craft@Hrs	Unit	Material	Labor	Total
Residential lot, tract work, 4 corners	—	LS	—	—	700.00
Residential lot, individual, 4 corners	—	LS	—	—	1,190.00
Commercial lot, based on 30,000 SF lot	—	LS	—	—	1,290.00
Over 30,000 SF, add per acre	—	Acre	—	—	200.00

Lots without recorded markers cost more, depending on distance to the nearest recorded monument.

	Craft@Hrs	Unit	Material	Labor	Total
Add up to	—	%	—	—	100.0

Some states require that a corner record be prepared and filed with the county surveyor's office when new monuments are set. The cost of preparing and filing a corner record will be about $350.00 to $400.00.

Aerial Mapping Typical costs based on a scale of 1" to 40'.

	Craft@Hrs	Unit	Material	Labor	Total
Back up ground survey including flagging model		Ea	—	—	3,170.00

(Typical "model" is approximately 750' by 1,260' or 25 acres)

Aerial photo flight, plane and crew, per local flight. One flight can cover several adjacent models

	Craft@Hrs	Unit	Material	Labor	Total
Per local flight	—	Ea	—	—	395.00
Vertical photos (2 required per model)	—	Ea	—	—	42.00
Data reduction and drafting, per hour	—	Hr	—	—	58.00

Complete survey with finished drawing, per model, including both field work and flying

	Craft@Hrs	Unit	Material	Labor	Total
Typical cost per model	—	Ea	—	—	3,290.00
Deduct if ground control is marked on the ground by others					
Per model	—	LS	—	—	-1,030.00

Typical mapping cost per acre including aerial control, flight and compilation

	Craft@Hrs	Unit	Material	Labor	Total
Based on 25 acre coverage	—	Acre	—	—	152.00

Aerial photos, oblique (minimum 4 per job), one 8" x 10" print

	Craft@Hrs	Unit	Material	Labor	Total
Black and white	—	Ea	—	—	42.00
Color	—	Ea	—	—	90.00

Quantitative studies for earthmoving operations to evaluate soil quantities moved

	Craft@Hrs	Unit	Material	Labor	Total
Ground control and one flight	—	Ea	—	—	2,650.00
Each additional flight	—	Ea	—	—	1,800.50

Engineering Fees Typical billing rates.

Registered engineers (calculations, design and drafting)

	Craft@Hrs	Unit	Material	Labor	Total
Assistant engineer	—	Hr	—	—	77.00
Senior engineer	—	Hr	—	—	131.00

Principal engineers (cost control, project management, client contact

	Craft@Hrs	Unit	Material	Labor	Total
Assistant principal engineer	—	Hr	—	—	137.00
Senior principal engineer	—	Hr	—	—	185.00
Job site engineer (specification compliance)	—	Hr	—	—	77.30

Estimating Cost Compiling detailed labor and material cost estimate for commercial and industrial construction, in percent of job cost. Actual fees are based on time spent plus cost of support services.

Most types of buildings, typical fees

	Craft@Hrs	Unit	Material	Labor	Total
Under $100,000 (range of .75 to 1.5%)	—	%	—	—	1.1
Over $100,000 to $500,000 (range of .5 to 1.25%)	—	%	—	—	0.9
Over $500,000 to $1,000,000 (range of .5 to 1.0%)	—	%	—	—	0.8
Over $1,000,000 (range of .25 to .75%)	—	%	—	—	0.5

	Craft@Hrs	Unit	Material	Labor	Total

Complex, high-tech, research or manufacturing facilities or other buildings with many unique design features, typical fees

Under $100,000 (range of 2.3 to 3%)	—	%	—	—	2.7
Over 100,000 to $500,000 (range of 1.5 to 2.5%)	—	%	—	—	2.0
Over $500,000 to $1,000,000 (range of 1.0 to 2.0%)	—	%	—	—	1.5
Over $1,000,000 (range of .5 to 1.0%)	—	%	—	—	0.8
Typical estimator's fee per hour	—	Hr	—	—	50.00

Specifications Fee Preparing detailed construction specifications to meet design requirements for commercial and industrial buildings, in percent of job cost. Actual fees are based on time spent plus cost of support services, if any.

Typical fees for most jobs	—	%	—	—	0.5
Small or complex jobs					
Typical cost	—	%	—	—	1.0
High cost	—	%	—	—	1.5
Large jobs or repetitive work					
Typical cost	—	%	—	—	0.2
High cost	—	%	—	—	0.5

Job Layout Setting layout stakes for the foundation from monuments identified on the plans and existing on site. Per square foot of building first floor area. Typical costs. Actual costs will be based on time spent.

Smaller or more complex jobs	—	SF	—	—	.75
Large job, easy work	—	SF	—	—	.45

General Non-Distributable Supervision and Expense

Project manager, typical	—	Mo	—	—	5,520.00
Job superintendent	—	Mo	—	—	4,670.00
Factory representative	—	Day	—	—	660.00
Timekeeper	—	Mo	—	—	1,780.00
Job site engineer for layout	—	Mo	—	—	3,300.00
Office manager	—	Mo	—	—	2,000.00
Office clerk	—	Mo	—	—	1,470.00

Mobilization Typical costs for mobilization and demobilization within a 50 mile radius as a percent of total contract price. Includes cost of moving general contractor-owned equipment and job office to the site, setting up a fenced material yard, and removing equipment, office and fencing at job completion.

Allow, as a percentage of total contract price	—	%	—	—	0.5

Aggregate Testing Add $45.00 per hour for sample preparation.

Add for sample prep	—	Hr	—	—	45.00
Abrasion, L.A. Rattler, 100/500 cycles, ASTM C 131	—	Ea	—	—	110.00
Absorption, ASTM C 127, 128	—	Ea	—	—	55.00
Acid solubility	—	Ea	—	—	32.00
Aggregate test for mix design, including sieve analysis, specific gravity, number 200 wash, organic impurities	—	Ea	—	—	45.00
Clay lumps and friable particles, ASTM C 142	—	Ea	—	—	60.00
Coal and lignite, ASTM C 123	—	Ea	—	—	95.00
Material finer than #200 sieve, ASTM C 117	—	Ea	—	—	37.00
Organic impurities, ASTM C 40	—	Ea	—	—	47.00
Percent crushed particles, CAL 205	—	Ea	—	—	63.00
Percent flat or elongated particles, CRD C 119	—	Ea	—	—	100.00

	Craft@Hrs	Unit	Material	Labor	Total
Potential reactivity, chemical method, 3 determinations per series, ASTM C 289	—	Ea	—	—	315.00
Potential reactivity, mortar bar method, ASTM C 227	—	Ea	—	—	300.00
Sieve analysis, pit run aggregate	—	Ea	—	—	63.00
Lightweight aggregate ASTM C 123	—	Ea	—	—	70.00
Sand equivelant	—	Ea	—	—	80.00
Sieve analysis, processed (each size), ASTM C 136	—	Ea	—	—	73.00
Soft particles, ASTM C 235	—	Ea	—	—	63.00
Soundness, sodium or magnesium, 5 cycle, per series, ASTM C 88	—	Ea	—	—	135.00
Specific gravity, coarse, ASTM C 127	—	Ea	—	—	55.00
Specific gravity, fine, ASTM C 127	—	Ea	—	—	75.00
Unit weight per cubic foot, ASTM C 29	—	Ea	—	—	30.00

Asphaltic Concrete Testing Minimum fee is usually $40.00.

	Craft@Hrs	Unit	Material	Labor	Total
Asphaltic concrete core drilling, field or laboratory					
Per linear inch	—	Ea	—	—	5.00
Add for equipment and technician for core drilling	—	Hr	—	—	70.00
Asphaltic core density	—	Ea	—	—	25.00
Extraction, percent asphalt, (Method B) excluding ash correction, ASTM D 2172	—	Ea	—	—	60.00
Gradation on extracted sample (including wash), ASTM C 136	—	Ea	—	—	75.00
Maximum density, lab mixed, Marshall, 1559	—	Ea	—	—	175.00
Maximum density, premixed, Marshall, 1559	—	Ea	—	—	125.00
Maximum theoretical unit weight (Rice Gravity), ASTM 2041	—	Ea	—	—	75.00
Penetration, ASTM D 5	—	Ea	—	—	60.00
Stability, lab mixed, Marshall, ASTM D 1559	—	Ea	—	—	250.00
Stability, premixed, Marshall, ASTM D 1559	—	Ea	—	—	150.00

Concrete Testing Minimum fee is usually $40.00.

	Craft@Hrs	Unit	Material	Labor	Total
Aggregate tests for concrete mix designs only, including sieve analysis, specific gravity, number 200 wash, organic impurities, weight per cubic foot					
Per aggregate size	—	Ea	—	—	72.00
Amend or re-type existing mix designs					
Not involving calculations	—	Ea	—	—	42.00
Compression test, 2", 4", or 6" cores (excludes sample preparation) ASTM C 42	—	Ea	—	—	37.00
Compression test, 6" x 12" cylinders, including mold ASTM C 39	—	Ea	—	—	13.00
Core cutting in laboratory	—	Ea	—	—	42.00
Cylinder pickup within 40 miles of laboratory					
Cost per cylinder (minimum of 3)	—	Ea	—	—	9.00
Flexure test, 6" x 6" beams, ASTM C 78	—	Ea	—	—	53.00
Laboratory trial, concrete batch, ASTM C 192	—	Ea	—	—	345.00
Length change (3 bars, 4 readings, up to 90 days)					
ASTM C 157 modified	—	LS	—	—	240.00
Additional readings, per set of 3 bars	—	LS	—	—	37.00
Storage over 90 days, per set of 3 bars, per month	—	Mo	—	—	37.00
Modulus of elasticity test, static, ASTM C 469	—	Ea	—	—	73.00
Mix design, determination of proportions	—	Ea	—	—	79.00

	Craft@Hrs	Unit	Material	Labor	Total
Pick up aggregate sample					
Within 40 mile radius of laboratory, per trip	—	Ea	—	—	47.00
Pickup or delivery of shrinkage molds, per trip	—	Ea	—	—	32.00
Prepare special strength documentation for mix design	—	Ea	—	—	89.00
Proportional analysis, cement factor and percent of aggregate					
Per analysis	—	Ea	—	—	230.00
Review mix design prepared by others	—	Ea	—	—	68.00
Splitting tensile, 6" x 12" cylinder, ASTM C 496	—	Ea	—	—	32.00
Weight per cubic foot determination of lightweight concrete					
Cylinders	—	Ea	—	—	6.00

Masonry and Tile Testing

	Craft@Hrs	Unit	Material	Labor	Total
Brick, ASTM C 67					
Modulus of rupture (flexure) or compressive strength	—	Ea	—	—	27.00
Absorption initial rate, 5 hour or 24 hour	—	Ea	—	—	22.00
Boil, 1, 2, or 5 hours	—	Ea	—	—	22.00
Efflorescence	—	Ea	—	—	27.00
Dimensions, overall, coring, shell and web	—	Ea	—	—	22.00
Compression of core	—	Ea	—	—	37.00
Cores, shear, 6" and 8" diameter, 2 faces, per core	—	Ea	—	—	47.00
Coefficient of friction (slip test)	—	Ea	—	—	84.00
Concrete block, ASTM C 140					
Moisture content as received	—	Ea	—	—	22.00
Absorption	—	Ea	—	—	27.00
Compression	—	Ea	—	—	32.00
Shrinkage, modified British, ASTM C 426	—	Ea	—	—	68.00
Compression, 4", 6", 8" cores	—	Ea	—	—	37.00
Fireproofing, oven dry density	—	Ea	—	—	32.00
Gunite					
Compression, 2", 4", 6" cores, ASTM C 42	—	Ea	—	—	37.00
Pickup gunite field sample (40 mile trip maximum)	—	Ea	—	—	53.00
Building stone					
Compression	—	Ea	—	—	32.00
Flex strength	—	Ea	—	—	32.00
Modules/rupture	—	Ea	—	—	32.00
Specific gravity	—	Ea	—	—	32.00
Water absorption	—	Ea	—	—	32.00
Masonry prisms, ASTM E 447					
Compression test, grouted prisms	—	Ea	—	—	136.00
Pickup prisms, within 40 miles of lab	—	Ea	—	—	37.00
Mortar and grout, UBC Standard 24-22 & 24-28					
Compression test, 2" x 4" mortar cylinder	—	Ea	—	—	18.00
Compression test, 3" x 6" grout prisms	—	Ea	—	—	24.00
Compression test, 2" cubes, ASTM C 109	—	Ea	—	—	47.00
Roof fill, lightweight ASTM C 495					
Compression test	—	Ea	—	—	22.00
Density	—	Ea	—	—	16.00
Roofing tile					
Roofing tile breaking strength, UBC	—	Ea	—	—	22.00
Roofing tile absorption	—	Ea	—	—	22.00

	Craft@Hrs	Unit	Material	Labor	Total
Slate					
Modulus of rupture	—	Ea	—	—	58.00
Modulus of elasticity	—	Ea	—	—	37.00
Water absorption	—	Ea	—	—	32.00

Reinforcement Testing

	Craft@Hrs	Unit	Material	Labor	Total
Bend test, number 11 bar or smaller, ASTM 615	—	Ea	—	—	19.00
Modulus of elasticity for prestressing wire	—	Ea	—	—	105.00
Nick-break test, welded bar	—	Ea	—	—	58.00
Sampling at fabricator's plant (within 40 miles of lab), $50.00 minimum					
During normal business hours, per sample	—	Ea	—	—	18.00
Other than normal business hours, per sample	—	Ea	—	—	27.00
Tensile test, mechanically spliced bar	—	Ea	—	—	100.00
Tensile test, number 11 bar or smaller, ASTM 615	—	Ea	—	—	26.00
Tensile test, number 14 bar or smaller, ASTM 615	—	Ea	—	—	58.00
Tensile test, number 18 bar, ASTM 615	—	Ea	—	—	68.00
Tensile test, welded number 11 bar or smaller	—	Ea	—	—	37.00
Tensile test, welded number 14 bar	—	Ea	—	—	68.00
Tensile test, welded number 18 bar	—	Ea	—	—	84.00
Tensile and elongation test for prestress strands					
24" test, ASTM A 416	—	Ea	—	—	84.00
10" test, ASTM A 421	—	Ea	—	—	47.00

Soils and Aggregate Base Testing Minimum fee is usually $400.

	Craft@Hrs	Unit	Material	Labor	Total
Bearing ratio (excluding moisture-density curve)	—	Ea	—	—	110.00
Consolidation test (single point)	—	Ea	—	—	75.00
Consolidation test (without rate data)	—	Ea	—	—	95.00
Direct shear test					
Unconsolidated, undrained	—	Ea	—	—	65.00
Consolidated, undrained	—	Ea	—	—	100.00
Consolidated, drained	—	Ea	—	—	125.00
Durability index - coarse and fine	—	Ea	—	—	140.00
Expansion index test, UBC 29-2	—	Ea	—	—	110.00
Liquid limit or plastic limit, Atterberg limits	—	Ea	—	—	100.00
Mechanical analysis - ASTM D1140 (wash 200 sieve)	—	Ea	—	—	41.00
Mechanical analysis - sand and gravel (wash sieve)	—	Ea	—	—	116.00
Moisture content	—	Ea	—	—	15.00
Moisture-density curve for compacted fill, ASTM 1557					
4-inch mold	—	Ea	—	—	130.00
6-inch mold	—	Ea	—	—	140.00
Permeability (falling head)	—	Ea	—	—	225.00
Permeability (constant head)	—	Ea	—	—	175.00
Resistance value	—	Ea	—	—	220.00
Sand equivalent	—	Ea	—	—	65.00
Specific gravity - fine-grained soils	—	Ea	—	—	60.00
Unconfined compression test (undisturbed sample)	—	Ea	—	—	70.00
Unit dry weight and moisture content (undisturbed sample)	—	Ea	—	—	21.00

	Craft@Hrs	Unit	Material	Labor	Total
Steel Testing $100.00 minimum for machine time.					
Charpy impact test, reduced temperature	—	Ea	—	—	26.00
Charpy impact test, room temperature	—	Ea	—	—	13.00
Electric extensometer	—	Ea	—	—	63.00
Hardness tests, Brinell or Rockwell	—	Ea	—	—	22.00
Photo micrographs					
20x to 2000x	—	Ea	—	—	22.00
With negative	—	Ea	—	—	26.00
Tensile test, under 100,000 lbs	—	Ea	—	—	37.00
Welder Qualification Testing. Add $36 minimum per report					
Tensile test	—	Ea	—	—	37.00
Bend test	—	Ea	—	—	26.00
Macro etch	—	Ea	—	—	53.00
Fracture test	—	Ea	—	—	32.00
Machining for weld tests, 1/2" and less	—	Ea	—	—	37.00
Machining for weld tests, over 1/2"	—	Ea	—	—	53.00

Construction Photography Digital, professional quality work, hand printed, labeled, numbered and dated. Also high resolution photographs on CD-ROM or lower resolution for emailing purposes.

	Craft@Hrs	Unit	Material	Labor	Total
Job progress photos, includes shots of all building corners, two 8" x 10" color prints of each shot, per visit					
Typical cost	—	LS	—	—	540.00
High cost	—	LS	—	—	620.00
Periodic survey visits to job site to figure angles (for shots of all building corners)					
Minimum two photos, per visit	—	Ea	—	—	350.00
Exterior completion photographs, two 8" x 10" prints (two angles)					
Black and white	—	Ea	—	—	540.00
Color	—	Ea	—	—	540.00
Interior completion photographs, with lighting setup, two 8" x 10" prints of each shot (minimum three shots)					
Black and white	—	Ea	—	—	620.00
Color	—	Ea	—	—	620.00
Reprints or enlargements from existing digital file. Hand (custom) prints					
8" x 10", black and white	—	Ea	—	—	25.00
11" x 14", black and white	—	Ea	—	—	35.00
8" x 10", color	—	Ea	—	—	25.00
11" x 14", color	—	Ea	—	—	35.00

Retouching services (taking out phone lines, power poles, autos, signs, etc.) will run about $100.00 per hour of computer time.

Signs and Markers Pipe markers, self-stick vinyl, by overall outside diameter of pipe or covering, meets OSHA and ANSI requirements. Costs are based on standard wording and colors.

	Craft@Hrs	Unit	Material	Labor	Total
3/4" to 2-1/2" diameter, 3/4" letters, 7" long	PA@.077	Ea	1.75	3.50	5.25
2-1/2" to 8" diameter, 1-1/4" letters, 12" long	PA@.077	Ea	2.40	3.50	5.90
8" to 10" diameter, 2-1/2" letters, 24" long	PA@.077	Ea	4.00	3.50	7.50
10" and larger diameter, 3-1/2" letters, 32" long	PA@.095	Ea	4.50	4.31	8.81
Flow arrows, 2" wide tape x 108' roll, perforated every 8"					
(use $50.00 as minimum material cost)	PA@.095	LF	.52	4.31	4.83
Barrier tapes, 3" wide x 1,000' long. "Caution," "Danger," "Open Trench," and other stock wording (Use $35.00 as minimum material cost.) Labor shown is based on one man installing 500 LF in 1 hour					
2 mil thick, short-term or indoor use	PA@.002	LF	.02	.09	.11
4 mil thick, long-term or outdoor use	PA@.002	LF	.04	.09	.13

	Craft@Hrs	Unit	Material	Labor	Total
Traffic control signage, "No Parking Between Signs", "Customer Parking" and other stock traffic control signs, rectangular, 30 gauge, baked enamel					
6" x 12", non-reflective	PA@.301	Ea	6.30	13.70	20.00
6" x 12", reflective	PA@.301	Ea	14.50	13.70	28.20
12" x 18", non-reflective	PA@.455	Ea	17.00	20.70	37.70
12" x 18", reflective	PA@.455	Ea	25.00	20.70	45.70
18" x 24", non-reflective	PA@.455	Ea	32.40	20.70	53.10
18" x 24", reflective	PA@.455	Ea	24.20	20.70	44.90
Add for pole-mounted signs	CL@1.00	Ea	13.00	36.40	49.40

Temporary Utilities No permit fees included.

	Craft@Hrs	Unit	Material	Labor	Total
Drop pole for electric service					
Power pole installation, wiring, 6 months rent and pickup					
100 amp meter pole	—	Ea	—	—	440.00
200 amp meter pole	—	Ea	—	—	580.00
Electric drop line and run to contractor's panel (assuming proper phase, voltage and service are available within 100' of site)					
Overhead run to contractor's "T" pole (including removal)					
Single phase, 100 amps	—	LS	—	—	354.00
Single phase, over 100 to 200 amps	—	LS	—	—	478.00
Three phase, to 200 amps	—	LS	—	—	550.00
Underground run in contractor's trench and conduit (including removal)					
Single phase, 100 amps, "USA" wire	—	LS	—	—	510.00
Single phase, 101-200 amps, "USA" wire	—	LS	—	—	660.00
Three phase, to 200 amps, "USA" wire	—	LS	—	—	730.00
Typical power cost					
Per kilowatt hour ($5.00 minimum)	—	Ea	—	—	.12
Per 1000 SF of building per month of construction	—	MSF	—	—	2.00
Temporary water service, meter installed on fire hydrant. Add the cost of distribution hose or piping.					
Meter fee, 2" meter (excluding $400 deposit)	—	LS	—	—	73.00
Water cost, per 1,000 gallons	—	M	—	—	1.70
Job phone, typical costs including message unit charges					
cellular phone and one year service	—	LS	—	—	460.00
Portable job site toilets, including delivery, weekly service, and pickup, per 28 day month, six month contract					
Good quality (polyethylene)	—	Mo	—	—	57.00
Deluxe quality, with sink, soap dispenser, towel dispenser, mirror	—	Mo	—	—	88.00
Trailer-mounted male and female compartment double, 2 stalls in each, sink, soap dispenser, mirror and towel dispenser in each compartment					
Rental per week	—	Wk	—	—	260.00

Construction Elevators and Hoists

	Craft@Hrs	Unit	Material	Labor	Total
Hoisting towers to 200', monthly rent					
Material hoist	CO@176.	Mo	2,480.00	8,690.00	11,170.00
Personnel hoist	CO@176.	Mo	5,690.00	8,690.00	14,380.00
Tower cranes, self-climbing (electric), monthly rent					
5 ton lift at 180' radius, one operator	CO@176.	Mo	14,100.00	8,690.00	22,790.00
With two operators	CO@352.	Mo	14,100.00	17,400.00	31,500.00
7 ton lift at 170' radius, one operator	CO@176.	Mo	19,600.00	8,690.00	28,290.00
With two operators	CO@352.	Mo	19,600.00	17,400.00	37,000.00

	Craft@Hrs	Unit	Material	Labor	Total
Mobilize, erect and dismantle self-climbing tower cranes					
Good access	—	LS	—	—	38,200.00
Average access	—	LS	—	—	44,500.00
Poor access	—	LS	—	—	50,300.00

Hydraulic Truck Cranes Includes equipment rental and operator. Four hour minimum. Time charge begins at the rental yard. By rated crane capacity.

	Craft@Hrs	Unit	Material	Labor	Total
12-1/2 ton	—	Hr	—	—	114.00
15 ton	—	Hr	—	—	126.00
25 ton	—	Hr	—	—	162.00
50 ton	—	Hr	—	—	197.00
70 ton	—	Hr	—	—	246.00
80 ton	—	Hr	—	—	269.00
100 ton	—	Hr	—	—	305.00
125 ton	—	Hr	—	—	347.00
150 ton	—	Hr	—	—	449.00
Add for clamshell or breaker ball work	—	Hr	—	—	35.00

Conventional Cable Cranes Including equipment rental and operator. Four hour minimum. Time charge begins at the rental yard. By rated crane capacity with 80' boom.

	Craft@Hrs	Unit	Material	Labor	Total
50 tons	—	Hr	—	—	236.00
70 tons	—	Hr	—	—	267.00
90 tons	—	Hr	—	—	298.00
125 tons	—	Hr	—	—	312.00
140 tons	—	Hr	—	—	323.00

Equipment Rental Rates Costs are for equipment in good condition but exclude accessories. Includes the costs of damage waiver (about 10% of the rental cost) insurance, fuel, oil. Add the cost of an equipment operator, pickup, return to yard, and repairs if necessary. Delivery and pickup within 30 miles of the rental yard are usually offered at an additional cost of one day's rent. A rental "day" is assumed to be one 8-hour shift and begins when the equipment leaves the rental yard. A "week" is 40 hours in five consecutive days. A "month" is 176 hours in 30 consecutive days.

	Day	Week	Month
Air equipment rental			
Air compressors, wheel-mounted			
16 CFM, shop type, electric	42.00	126.00	378.00
30 CFM, shop type, electric	58.00	173.00	520.00
80 CFM, shop type, electric	68.00	205.00	614.00
100 CFM, gasoline unit	84.00	252.00	756.00
125 CFM, gasoline unit	105.00	315.00	945.00
Air compressors wheel-mounted diesel units			
to 159 CFM	79.00	236.00	709.00
160 - 249 CFM	105.00	315.00	945.00
250 - 449 CFM	158.00	473.00	1,420.00
450 - 749 CFM	210.00	630.00	1,890.00
750 - 1200 CFM	420.00	1,260.00	3,780.00
1400 CFM and over	525.00	1,575.00	5,360.00
Paving breakers (no bits included) hand-held, pneumatic			
To 40 lb	24.00	73.00	206.00
41 - 55 lb	23.00	67.00	196.00
56 - 70 lb	33.00	95.00	275.00
71 - 90 lb	34.00	100.00	285.00

	Day	Week	Month
Paving breakers jackhammer bits			
Moil points, 15" to 18"	6.00	13.00	25.00
Chisels, 3"	7.00	18.00	48.00
Clay spades, 5-1/2"	11.00	24.00	69.00
Asphalt cutters, 5"	8.00	25.00	52.00
Rock drills, jackhammers, hand-held, pneumatic			
to 29 lb	28.00	85.00	250.00
30 - 39 lb	31.00	100.00	280.00
40 - 49 lb	33.00	104.00	305.00
50 lb and over	45.00	135.00	380.00
Rock drill bits, hexagonal			
7/8", 2' long	8100	2500	45.00
1", 6' long	9.00	28.00	63.00
Pneumatic chippers, medium weight, 10 lb	2900	8900	2500
Pneumatic tampers, medium weight, 30 lb	35.00	115.00	318.00
Pneumatic grinders, hand-held, 7 lb (no stone)	20.00	700	1700
Air hose, 50 LF section			
5/8" air hose	6.00	19.00	55.00
3/4" air hose	6.00	20.00	45.00
1" air hose	8.00	22.00	55.00
1-1/2" air hose	17.00	52.00	130.00
Compaction equipment rental			
Vibro plate, 300 lb, 24" plate width, gas	67.00	210.00	605.00
Vibro plate, 600 lb, 32" plate width, gas	108.00	340.00	940.00
Rammer, 60 CPM, 200 lb, gas powered	60.00	190.00	535.00
Rollers, two axle steel drum, road type			
1 ton, gasoline	249.00	778.00	2,200.00
6 - 8 ton, gasoline	390.00	1,170.00	3,400.00
8 - 10 ton	458.00	1,420.00	4,240.00
Rollers, rubber tired, self propelled			
12 ton, 9 wheel, diesel	299.00	903.00	2,670.00
15 ton, 9 wheel, diesel	415.00	1,760.00	3,750.00
30 ton, 11 wheel, diesel	420.00	1,420.00	4,010.00
Rollers, towed, sheepsfoot type			
40" diameter, double drum, 28" wide	116.00	347.00	1,040.00
60" diameter, double drum, 54" wide	210.00	630.00	1,890.00
Rollers, self-propelled vibratory dual sheepsfoot drum, gas			
25 HP, 26" wide roll	310.00	1,050.00	3,170.00
100 HP, 36" wide roll	600.00	1,150.00	6,400.00
Rollers, vibrating steel drum, gas powered, walk behind			
7 HP, 1,000 lb, one drum, 2' wide	98.00	313.00	857.00
10 HP, 2,000 lb, two drum, 2'-6" wide	130.00	420.00	1,250.00
Rollers, self-propelled riding type, vibrating steel drum			
9,000 lb, gasoline, two drum	210.00	635.00	1,820.00
15,000 lb, diesel, two drum	384.00	1,200.00	3,550.00
25,000 lb, diesel, two drum	495.00	1,460.00	4,250.00
Concrete equipment rental			
Buggies, push type, 7 CF	44.00	150.00	400.00
Buggies, walking type, 12 CF	62.00	205.00	565.00
Buggies, riding type, 14 CF	90.00	275.00	730.00

	Day	Week	Month
Buggies, riding type, 18 CF	90.00	290.00	835.00
Vibrator, electric, 3 HP, flexible shaft	39.00	125.00	340.00
Vibrator, gasoline, 6 HP, flexible shaft	59.00	175.00	525.00
Troweling machine, 36", 4 paddle	50.00	160.00	450.00
Cement mixer, 6 CF	60.00	185.00	535.00
Towable mixer (10+ CF)	105.00	295.00	795.00
Power trowel 36", gas, with standard handle	50.00	160.00	450.00
Bull float	13.00	40.00	110.00
Concrete saws, gas powered, excluding blade cost			
10 HP, push type	54.00	188.00	515.00
20 HP, self propelled	95.00	307.00	875.00
37 HP, self propelled	140.00	455.00	1,320.00
Concrete bucket, bottom dump 1 CY	44.00	130.00	365.00
Plaster mixer, gas, portable, 5 CF	49.00	155.00	442.00
Concrete conveyor, belt type, portable, gas powered			
All sizes	170.00	625.00	1,680.00
Vibrating screed, 16', single beam	63.00	200.00	570.00
Column clamps, to 48", per set	—	—	5.45

Crane rental Add for clamshell, dragline, or pile driver attachments from below.

	Day	Week	Month
Cable controlled crawler-mounted lifting cranes, diesel			
45 ton	—	—	5,800.00
60 ton	—	2,500.00	6,840.00
100 ton (125 ton)	1,200.00	4,000.00	12,000.00
150 ton	2,000.00	5,900.00	20,000.00
Cable controlled, truck-mounted lifting cranes, diesel			
50 ton	675.00	1,990.00	6,800.00
75 ton	860.00	2,570.00	7,750.00
100 ton	1,250.00	3,740.00	9,740.00
Hydraulic - rough terrain			
10 ton, gasoline	420.00	1,210.00	3,500.00
17 ton, diesel	510.00	1,450.00	4,200.00
35 ton, diesel	700.00	2,020.00	5,420.00
50 ton, diesel	950.00	3,300.00	10,000.00

Clamshell, dragline, and pile driver attachments Add to crane rental costs shown above.

	Day	Week	Month
Clamshell buckets			
Rehandling type, 1 CY	—	320.00	855.00
General purpose, 2 CY	80.00	510.00	1,370.00
Dragline buckets			
1 CY bucket	44.00	110.00	310.00
2 CY bucket	83.00	210.00	630.00
Pile drivers, hammers			
25,000 - 49,999 ft-lb	—	2,090.00	4,650.00
75 - 99 tons vibratory	1,640.00	3,450.00	8,330.00
1,000 - 1,999 ft-lb extractor	—	710.00	1,840.00

Excavation equipment rental

	Day	Week	Month
Backhoe/loaders, crawler mounted, diesel			
3/8 CY backhoe	275.00	920.00	2,700.00
3/4 CY backhoe	330.00	1,040.00	3,030.00

	Day	Week	Month
1-1/2 CY backhoe	390.00	1,250.00	3,710.00
2 CY backhoe	575.00	1,870.00	5,600.00
2-1/2 CY backhoe	700.00	2,240.00	6,730.00
3 CY backhoe	965.00	2,940.00	9,400.00
Backhoe/loaders, wheel mounted, diesel or gasoline			
1/2 CY bucket capacity, 55 HP	185.00	585.00	1,660.00
1 CY bucket capacity, 65 HP	235.00	690.00	2,080.00
1-1/4 CY bucket capacity, 75 HP	180.00	610.00	1,810.00
1-1/2 CY bucket capacity, 100 HP	240.00	750.00	2,150.00
Mounted with 750-lb. breaker	420.00	1,470.00	4,410.00
Mounted with 2,000-lb. breaker	630.00	2,200.00	6,620.00
Crawler dozers, with angle dozer or bulldozer blade, diesel			
65 HP, D-3	310.00	950.00	2,800.00
90-105 HP, D-4 or D-5	405.00	1,260.00	3,710.00
140 HP, D-6	625.00	1,970.00	5,840.00
200 HP, D-7	950.00	2,900.00	8,700.00
335 HP, D-8	1,360.00	4,200.00	13,000.00
460 HP, D-9	1,800.00	6,000.00	17,500.00
Crawler loaders, diesel			
3/4 CY loader	330.00	1,040.00	3,030.00
1-1/2 CY loader	390.00	1,250.00	3,710.00
2 CY loader	575.00	1,870.00	5,600.00
Hydraulic excavators, crawler mounted with backhoe type arm, diesel			
.2 - .4 CY bucket	350.00	1,060.00	3,100.00
.4 - .6 CY bucket	420.00	1,200.00	3,550.00
.6 - .8 CY bucket	500.00	1,400.00	4,050.00
Gradall, truck mounted, diesel			
3/4 CY bucket	984.00	2,000.00	6,000.00
1 CY bucket	900.00	3,100.00	8,600.00
Front-end loader, diesel or gas (Skiploader)			
40-hp	110.00	415.00	1,200.00
60-hp	230.00	745.00	2,200.00
80-hp	205.00	650.00	1,900.00
Skid steer loaders			
Bobcat 643, 1,000-lb capacity	157.00	520.00	1,500.00
Bobcat 753, 1,350-lb capacity	160.00	530.00	1,550.00
Bobcat 763, 1,750-lb capacity	180.00	560.00	1,650.00
Bobcat 863, 1,900-lb capacity	205.00	640.00	1,850.00
Skid steer attachments			
Auger	61.00	198.00	585.00
Hydraulic breaker	150.00	485.00	1,400.00
Backhoe	104.00	340.00	980.00
Sweeper	100.00	314.00	930.00
Grapple bucket	53.00	169.00	500.00
Wheel loaders, front-end load and dump, diesel			
3/4 CY bucket, 4WD, articulated	235.00	680.00	2,020.00
1 CY bucket, 4WD, articulated	254.00	790.00	2,300.00
2 CY bucket, 4WD, articulated	310.00	975.00	2,840.00
3-1/4 CY bucket, 4WD, articulated	635.00	1,960.00	5,700.00
5 CY bucket, 4WD, articulated	900.00	2,800.00	8,350.00

	Day	Week	Month
Motor scraper-hauler, 2 wheel tractor			
Under 18 CY capacity, single engine	1,270.00	4,200.00	12,100.00
18 CY and over, single engine	1,820.00	6,100.00	18,100.00
Under 18 CY capacity, Two engine	1,840.00	5,820.00	17,100.00
18 CY and over, Two engine	2,430.00	7,400.00	23,000.00
Graders, diesel, pneumatic tired, articulated			
100 HP	390.00	1,250.00	3,650.00
150 HP	660.00	1,980.00	6,000.00
200 HP	910.00	2,600.00	7,750.00
Trenchers, inclined chain boom type, pneumatic tired			
15 HP, 12" wide, 48" max. depth, walking	165.00	540.00	1,620.00
20 HP, 12" wide, 60" max. depth, riding	250.00	820.00	2,400.00
55 HP, 18" wide, 96" max. depth, riding	475.00	1,450.00	4,220.00
100 HP, 24" x 96" deep, crawler mount	655.00	2,100.00	6,100.00
Trenching shields, steel, 8' high x 12' long			
Single wall	130.00	375.00	1,080.00
Hydraulic trench braces, 8' long, to 42" wide	19.00	56.00	171.00
Truck-mounted wagon drills, self propelled, crawler mounted			
4" drifter, straight boom	481.00	1,270.00	3,550.00
4-1/2" drifter, hydraulic swing boom	734.00	2,240.00	6,620.00
Truck rental			
1-1/2 ton truck	83.00	248.00	744.00
Flatbed with boom	448.00	1,370.00	3,950.00
Dump trucks, rental rate plus mileage			
3 CY	182.00	497.00	1,490.00
5 CY	198.00	761.00	2,430.00
10 CY	254.00	992.00	2,870.00
25 CY off highway	—	3,420.00	10,300.00
Forklift rental			
Rough terrain, towable, gas powered			
4,000 lb capacity	175.00	540.00	1,480.00
8,000 lb capacity	218.00	650.00	1,900.00
Extension boom, 2 wheel steering			
4,000 lb, 30' lift	225.00	700.00	2,010.00
6,000 lb, 35' lift	240.00	730.00	2,100.00
8,000 lb, 40' lift, 4 wheel steering	306.00	900.00	2,640.00
Aerial platform rental			
Rolling scissor lifts, 2' x 3' platform, 650 lb capacity, push around, electric powered			
30' high	70.00	195.00	523.00
40' high	135.00	375.00	975.00
50' high	197.00	548.00	1,500.00
Rolling scissor lifts, self-propelled, hydraulic, electric powered			
To 20'	90.00	250.00	700.00
21' - 30'	110.00	335.00	870.00
31' - 40'	200.00	650.00	2,000.00
Rolling scissor lifts, self-propelled, hydraulic, diesel powered			
To 20'	50.00	150.00	450.00
21' - 30'	160.00	450.00	1,250.00
31' - 40'	180.00	515.00	1,450.00

	Day	Week	Month
Boomlifts, telescoping and articulating booms, self-propelled, gas or diesel powered, 2-wheel drive			
21' - 30'	150.00	400.00	1,050.00
31' - 40'	185.00	555.00	1,600.00
41' - 50'	225.00	690.00	1,950.00
51' - 60'	330.00	927.00	2,670.00

Pump rental Hoses not included.

Submersible pump rental	Day	Week	Month
1-1/2" to 2", electric	34.00	105.00	298.00
3" to 4", electric	75.00	250.00	680.00
All diameters, hydraulic	245.00	765.00	2,300.00
Trash pumps, self-priming, gas			
1-1/2" connection	39.00	120.00	345.00
4" connection	75.00	225.00	580.00
6" connection	215.00	540.00	1,600.00
Diaphragm pumps			
2", gas, single action	40.00	130.00	355.00
4", gas, double action	83.00	265.00	640.00
Hose, coupled, 25 LF sections			
2" suction line	10.00	30.00	75.00
1-1/2" discharge line	7.00	20.00	55.00
3" discharge line	10.00	27.00	70.00

Sandblasting equipment rental

Compressor and hopper	Day	Week	Month
To 250 PSI	46.00	156.00	435.00
Over 250 to 300 PSI	78.00	240.00	690.00
Over 600 PSI to 1,000 PSI	150.00	500.00	1,420.00
Sandblasting accessories			
Hoses, 50', coupled			
3/4" sandblast hose or air hose	8.00	27.00	68.00
1" air hose	11.00	35.00	98.00
1-1/4" air hose or sandblast hose	18.00	50.00	130.00
Nozzles, all types	26.00	79.00	236.00
Valve, remote control (deadman valve), all sizes 137.00		37.00	46.00
Air-fed hood	21.00	63.00	189.00

Roadway equipment rental

	Day	Week	Month
Boom mounted pavement breaker, hydraulic 1,500 lb	355.00	1,150.00	3,200.00
Paving breaker, backhoe mount, pneumatic, 1,000 lb	350.00	1,060.00	2,900.00
Asphalt curbing machine	88.00	265.00	800.00
Paving machine, diesel, 10' width, self propelled	1,400.00	5,000.00	13,200.00
Distribution truck for asphalt prime coat, 3,000 gallon	918.00	2,740.00	8,100.00
Pavement striper, 1 line, walk behind	51.00	175.00	695.00
Water blast truck for 1,000 PSI pavement cleaning	1,320.00	3,790.00	11,800.00
Router-groover for joint or crack sealing in pavement	213.00	725.00	2,180.00

General Requirements 1

	Day	Week	Month
Sealant pot for joint or crack sealing	257.00	802.00	2,410.00
Self-propelled street sweeper, vacuum (7 CY max.)	740.00	2,650.00	7,550.00
Water trucks/tanks			
1,800-gallon water truck	185.00	550.00	1,780.00
3,500-gallon water truck	318.00	953.00	2,860.00
Water tower	100.00	400.00	1,200.00
Tow behind trailer, 500 gal.	59.00	252.00	750.00
Off-road water tanker, 10,000 gal.	1,030.00	3,750.00	9,900.00
Traffic control equipment rental			
Arrow board-25 lights	73.00	260.00	700.00
Barricades	4.00	8.00	18.00
Barricades with flashers	9.00	18.00	50.00
Delineators	4.00	7.00	12.00
Drill rental			
1/2" cordless drill, 18 volt	8.00	32.00	96.00
3/8" or 1/2", electric drill	8.00	32.00	96.00
3/4" electric drill	24.00	88.00	264.00
1/2" right angle electric drill	20.00	68.00	204.00
1/2" electric hammer drill	30.00	100.00	300.00
Electric generator rental			
2.5 kw, gasoline	37.00	140.00	420.00
5 kw, gasoline	58.00	200.00	600.00
15 kw, diesel	100.00	300.00	875.00
60 kw, diesel	230.00	830.00	1,850.00
Floor equipment rental			
Carpet blower	12.00	40.00	120.00
Carpet cleaner	22.00	80.00	240.00
Carpet kicker	14.00	44.00	132.00
Carpet power stretcher	25.00	88.00	264.00
Carpet stapler, electric	15.00	60.00	180.00
Carpet seamer, electric	14.00	44.00	132.00
Floor edger, 7", electric	36.00	120.00	360.00
Floor maintainer	30.00	100.00	300.00
Grinder 9"	24.00	80.00	240.00
Mini grinder, 4-1/2"	18.00	60.00	180.00
Pergo installation kit	30.00	100.00	300.00
Tile roller	18.00	60.00	180.00
Tile stripper, electric	52.00	180.00	540.00
Tile stripper, manual	8.00	32.00	96.00
Vacuum, 10 gallons, wet/dry	30.00	100.00	300.00
Hammer rental			
Electric brute breaker	52.00	180.00	540.00
Gas breaker	60.00	240.00	720.00
Demolition hammer, electric	45.00	160.00	480.00
Roto hammer, 7/8", electric	52.00	180.00	540.00
Roto hammer, 1-1/2", electric	40.00	140.00	420.00

	Day	Week	Month
Heater rental, fan forced			
Kerosene heater, to 200 MBtu	40.00	120.00	340.00
Kerosene heater, over 200 to 300 MBtu	37.00	130.00	350.00
Kerosene heater, over 500 to 1,000 MBtu	75.00	230.00	640.00
Salamanders, LP gas (no powered fan)	18.00	72.00	216.00
Ladder rental			
Step ladder, 6' or 8'	12.00	40.00	120.00
Step ladder, 10' or 14'	18.00	50.00	180.00
Extension ladder, 24'	44.00	160.00	480.00
Extension ladder, 28' or 32'	46.00	160.00	480.00
Extension ladder, 40'	44.00	160.00	480.00
Stairway step ladder	15.00	60.00	180.00
Landscape equipment rental			
Aerator, gas	52.00	180.00	540.00
Air broom backpack, gas	30.00	120.00	360.00
Auger, 2 man (gas)	58.00	200.00	600.00
3" chipper, gas	90.00	320.00	960.00
Brush cutter, gas	36.00	120.00	360.00
Brush hog, 24"	46.00	160.00	480.00
Clamshovel, manual	9.00	32.00	96.00
Digging bar, 5'	10.00	28.00	84.00
Edger, lawn, gas	25.00	100.00	300.00
Fertilizer spreader	8.00	32.00	96.00
Hand auger, 6"	10.00	32.00	96.00
Hedge trimmer, 24", electric	14.00	52.00	156.00
Hedge trimmer, 30", electric	24.00	80.00	240.00
Hedge trimmer, 30", gas	21.00	80.00	240.00
Lawn roller	12.00	40.00	120.00
Log splitter, gas	80.00	280.00	840.00
Overseeder, gas	52.00	180.00	540.00
Power rake, gas	52.00	180.00	540.00
Sod cutter	80.00	280.00	840.00
Stump grinder, 9 HP	98.00	340.00	1,020.00
Tiller, light duty, gas	52.00	180.00	540.00
Tiller, medium duty, gas	84.00	220.00	650.00
Tiller, heavy duty, gas	80.00	280.00	840.00
Tree pruning shears	12.00	40.00	120.00
Weed eater, gas	30.00	100.00	300.00
Wheelbarrow	12.00	40.00	120.00
Level rental			
Laser level, small	42.00	150.00	480.00
Level kit	58.00	200.00	600.00
Transit level kit	36.00	120.00	360.00
Light tower rental			
Floodlight, 1,000 watt, on stand	34.00	110.00	300.00
To 7,000 watt trailer mounted light set, gas	75.00	250.00	650.00
Over 7,000 watt trailer mounted light set, gas	105.00	320.00	900.00
Extension cord, 50'	8.00	20.00	60.00

General Requirements 1

	Day	Week	Month
Nailer and stapler rental			
Drywall screwdriver	14.00	48.00	144.00
Finish nailer, pneumatic or cordless	30.00	100.00	300.00
Floor stapler, pneumatic	36.00	120.00	360.00
Framing nailer, cordless	30.00	100.00	300.00
Framing nailer, pneumatic, 1-1/2" to 3-1/4"	30.00	100.00	300.00
Hardwood floor nailer, 3/4"	18.00	60.00	180.00
Low velocity nail gun, single	15.00	60.00	180.00
Low velocity gun, multi	30.00	100.00	300.00
Roofing nailer, pneumatic, 1/2" to 2"	38.00	120.00	360.00
Screw gun	35.00	140.00	420.00
Underlay spot nailer	15.00	60.00	180.00
Painting equipment rental			
Airless paint sprayer - large	74.00	260.00	780.00
Heat gun blower	9.00	32.00	96.00
Mud mixing paddle	8.00	24.00	72.00
Pressure washer - cold, 1,500 PSI	61.00	193.00	634.00
Pressure washer - cold, 2,000 PSI	64.00	220.00	660.00
Pressure washer - cold, 3,500 PSI	74.00	240.00	720.00
Texture sprayer with compressor	45.00	160.00	480.00
Wallpaper steamer, electric	30.00	100.00	300.00
Plumbing and piping equipment rental			
Basin wrench, 3/8" to 1-1/4"	9.00	32.00	96.00
Chain wrench	8.00	32.00	96.00
Conduit & pipe cutter	14.00	48.00	144.00
Drain cleaner, corded drill, 3/8"	36.00	120.00	360.00
Laser pipe level	80.00	245.00	720.00
Pipe reamer, 1/2" to 2"	14.00	48.00	144.00
Pipe stand	13.00	48.00	144.00
Pipe threader, 1/2" to 2"	14.00	48.00	144.00
Pipe wrench, 18" or 24"	14.00	48.00	144.00
Pipe wrench, 36"	18.00	60.00	180.00
Sewer snake, 100' x 5/8", electric	58.00	200.00	600.00
Sewer snake, 25' x 5/16", manual	20.00	80.00	240.00
Sewer snake, 50' x 1/2", electric	46.00	160.00	480.00
Sewer snake, 50' x 1/2", manual	30.00	100.00	300.00
Soil pipe cutter, ratchet, 6"	14.00	48.00	144.00
Toilet auger	9.00	32.00	96.00
Sanding equipment rental			
Belt sander, 3" x 21"	16.00	60.00	180.00
Belt sander, 4" x 24"	24.00	80.00	240.00
Drywall sander and vacuum	46.00	160.00	480.00
Floor sander, drum type	46.00	160.00	480.00
Floor sander, square	46.00	160.00	480.00
Orbit sander	12.00	40.00	120.00
Palm sander	12.00	40.00	120.00
Vibrator sander	18.00	60.00	180.00
Saw rental			
Band saw, electric	30.00	120.00	360.00
Bolt cutters, 24"	10.00	28.00	84.00

	Day	Week	Month
Bolt cutters, 36"	10.00	32.00	96.00
Chain saws, 18", gasoline	52.00	52.00	540.00
Chop saw, 14", electric	35.00	140.00	420.00
Circular saw, 8-1/4" (electric)	18.00	60.00	180.00
Compound miter saw	34.00	120.00	360.00
Jamb saw	25.00	88.00	264.00
Jig saw, electric	20.00	64.00	192.00
Masonry table saw, 2 HP	68.00	200.00	600.00
Reciprocating saw, electric	18.00	60.00	180.00
Router, 2 HP, fixed base	30.00	120.00	360.00
Sliding compound miter saw	40.00	140.00	420.00
Table saw, 10", electric	40.00	140.00	420.00
Tile cutter, manual	12.00	44.00	132.00
Tile saw and stand, wet cut	48.00	180.00	540.00

Scaffolding rental

	Day	Week	Month
5' high, with casters	30.00	30.00	90.00
10' high, with casters	44.00	44.00	132.00
15' high, with casters	62.00	62.00	186.00
5' high, with leg jacks	26.00	26.00	78.00
10' high with leg jacks	42.00	42.00	126.00
15' high, with leg jacks	58.00	58.00	174.00
14' wide with base plates	38.00	38.00	114.00
21' wide with base plates	54.00	54.00	162.00
14' wide with leg jacks	42.00	42.00	126.00
21' wide with leg jacks	58.00	58.00	174.00
14' wide with casters	44.00	44.00	132.00
21' wide with casters	62.00	62.00	186.00
Drywall panel lift, 5 piece	25.00	100.00	300.00
Drywall scaffold	28.00	100.00	300.00
Scissor lift	75.00	200.00	550.00

Welding equipment rental, with helmet

	Day	Week	Month
To 200 amp	60.00	180.00	510.00
201 to 300 amp	75.00	230.00	650.00

Miscellaneous equipment rental

	Day	Week	Month
Aluminum break, 10'	50.00	200.00	600.00
Appliance dolly	12.00	40.00	120.00
Chain hoist, 1 ton, 20' lift	20.00	80.00	240.00
Chain hoist, 5 ton, 20' lift	30.00	120.00	360.00
Come-a-long, 3000 pounds	15.00	60.00	180.00
Dehumidifier, electric	24.00	80.00	240.00
Dolly, 4 wheel	10.00	32.00	96.00
Fan, low profile	24.00	73.00	216.00
Fan, 30", with pedestal	35.00	94.00	270.00
Fishtape, 200', 1/8"	11.00	44.00	132.00
Hydraulic jack, 12 ton	12.00	48.00	144.00
Loading ramps	8.00	20.00	60.00
Metal detector	15.00	60.00	180.00
Pavement line marker	15.00	60.00	180.00
Screwjack, 10 ton	10.00	40.00	120.00

	Craft@Hrs	Unit	Material	Labor	Total

Concrete Pumping, Subcontract Small jobs, including trailer mounted pump and operator at 15 CY per hour. Based on 6 sack concrete mix designed for pumping. No concrete included. $125.00 minimum charge.

3/8" aggregate mix (pea gravel), using hose to 200'

	Craft@Hrs	Unit	Material	Labor	Total
Cost per cubic yard (pumping only)	—	CY	—	—	10.00
Add for hose over 200', per LF	—	LF	—	—	1.00

3/4" aggregate mix, using hose to 200'

	Craft@Hrs	Unit	Material	Labor	Total
Cost per cubic yard (pumping only)	—	CY	—	—	11.00
Add for hose over 200', per LF	—	LF	—	—	1.00

Concrete Pumping with a Boom Truck, Subcontract Includes truck rent, operator, local travel but no concrete. Add costs equal to 1 hour for equipment setup and 1 hour for cleanup. Use 4 hours as the minimum cost for 23, 28 and 32 meter boom trucks and 5 hours as the minimum cost for 36 through 52 meter boom trucks. Estimate the actual pour rate at 70% of the rated capacity on thicker slabs and 50% of the capacity on most other work.

Costs include operator and oiler where necessary. Costs shown include subcontractor's markup.

	Craft@Hrs	Unit	Material	Labor	Total
23 meter boom (75'), 70 CY per hour rating	—	Hr	—	—	100.00
Add per CY pumped with 23 meter boom	—	CY	—	—	2.00
28 meter boom (92'), 70 CY per hour rating	—	Hr	—	—	118.00
Add per CY pumped with 28 meter boom	—	CY	—	—	2.00
32 meter boom (105'), 90 CY per hour rating	—	Hr	—	—	136.00
Add per CY pumped with 32 meter boom	—	CY	—	—	2.00
36 meter boom (120'), 90 CY per hour rating	—	Hr	—	—	172.00
Add per CY pumped with 36 meter boom	—	CY	—	—	2.00
42 meter boom (138'), 100 CY per hour rating	—	Hr	—	—	227.00
Add per CY pumped with 42 meter boom	—	CY	—	—	2.50
52 meter boom (170'), 100 CY per hour rating	—	Hr	—	—	272.00
Add per CY pumped with 52 meter boom	—	CY	—	—	3.00

Temporary Enclosures Rented chain link fence and accessories. Costs are a one-time charge for up to six months usage on a rental basis. Costs include installation and one trip for removal and assume level site with truck access along pre-marked fence line. Add for gates and barbed wire as shown. Minimum charge is $350.

Chain link fence, 6' high

	Craft@Hrs	Unit	Material	Labor	Total
Less than 250 feet	—	LF	—	—	1.45
250 to 500 feet	—	LF	—	—	1.40
501 to 750 feet	—	LF	—	—	1.30
751 to 1,000 feet	—	LF	—	—	1.10
Over 1,000 feet	—	LF	—	—	98.00

Add for gates

	Craft@Hrs	Unit	Material	Labor	Total
6' x 10' single	—	Ea	—	—	90.70
6' x 12' single	—	Ea	—	—	115.00
6' x 15' single	—	Ea	—	—	150.00
6' x 20' double	—	Ea	—	—	173.00
6' x 24' double	—	Ea	—	—	231.00
6' x 30' double	—	Ea	—	—	288.00

Add for barbed wire

	Craft@Hrs	Unit	Material	Labor	Total
per strand, per linear foot	—	LF	—	—	.15

Contractor furnished items, installed and removed, based on single use and no salvage value

	Craft@Hrs	Unit	Material	Labor	Total
Railing on stairway, two sides, 2" x 4"	CL@.121	LF	.50	4.40	4.90

Guardrail at second and higher floors

	Craft@Hrs	Unit	Material	Labor	Total
toe rail, mid rail and top rail	CL@.124	LF	3.00	4.51	7.51

Plywood barricade fence

	Craft@Hrs	Unit	Material	Labor	Total
Bolted to pavement, 8' high	CL@.430	LF	12.00	15.60	27.60
Post in ground, 8' high	CL@.202	LF	10.00	7.35	17.35
8' high with 4' wide sidewalk cover	CL@.643	LF	32.00	23.40	55.40

	Craft@Hrs	Unit	Material	Labor	Total

Scaffolding Rental Exterior scaffolding, tubular steel, 60" wide, with 6'4" open end frames set at 7', with cross braces, base plates, mud sills, adjustable legs, post-mounted guardrail, climbing ladders and landings, brackets, clamps and building ties. Including two 2" x 10" scaffold planks on all side brackets and six 2" x 10" scaffold planks on scaffold where indicated. Add local delivery and pickup. Minimum rental is one month.

Costs are per square foot of building wall covered, per month

	Craft@Hrs	Unit	Material	Labor	Total
With plank on scaffold only	—	SF	.96	—	.96
With plank on side brackets and scaffold	—	SF	1.20	—	1.20

Add for erection and dismantling, level ground, truck accessible

	Craft@Hrs	Unit	Material	Labor	Total
With plank on scaffold only	CL@.008	SF	—	.29	.29
With planks on side brackets and scaffold	CL@.012	SF	—	.44	.44

Caster-mounted scaffolding, 30" wide by 7' or 10' long, rental. Minimum charge is one month

	Craft@Hrs	Unit	Material	Labor	Total
6' high	—	Mo	120.00	—	120.00
7' high	—	Mo	130.00	—	130.00
10' high	—	Mo	140.00	—	140.00
13' high	—	Mo	150.00	—	150.00
15' high	—	Mo	160.00	—	160.00
20' high	—	Mo	170.00	—	170.00
Hook-end scaffold plank, rental, each	—	Mo	10.00	—	10.00

Caster-mounted interior scaffold, purchase. Light duty

	Craft@Hrs	Unit	Material	Labor	Total
51" high x 4' long x 22" wide	—	Ea	184.00	—	184.00
72" high x 56" long x 25" wide	—	Ea	411.00	—	411.00

Plain-end micro-lam scaffold plank, rental each

	Craft@Hrs	Unit	Material	Labor	Total
9' length	—	Mo	3.00	—	3.00
12' length	—	Mo	4.50	—	4.50

Aluminum extension scaffold plank, purchase. 250-pound load capacity. 2"-deep box section extrusions. Plank length extends in 12" increments. OSHA compliant

	Craft@Hrs	Unit	Material	Labor	Total
8' to 13' long, 14" wide	—	Ea	193.00	—	193.00
Safety nets, nylon, 4" mesh, purchase	CL@.004	SF	1.60	.15	1.75
Safety nets, nylon, 1/2" mesh, purchase	CL@.004	SF	2.16	.15	2.31
Swinging stage, 10', complete, motor operated, purchase	CL@4.00	Ea	1,460.00	146.00	1,606.00

Heavy Duty Shoring Adjustable vertical tower type shoring. Rated capacity 5.5 tons per leg. Base frames are 4' wide by 6' high or 2' wide by 5' high. Extension frames are 4' wide by 5'4" high or 2' wide by 4'4" high. Frames are erected in pairs using cross-bracing to form a tower. Combinations of base frames and extension frames are used to reach the height required. Screw jacks with base plates are installed in the bottom of each leg. Similar screw jacks with "U" bracket heads are installed in the top of each leg. Material costs shown are rental rates based on a 1-month minimum. Add the cost of delivery and pickup. For scheduling purposes, estimate that a crew of 2 can unload, handle and erect 8 to 10 frames (including typical jacks and heads) per hour. Dismantling, handling and moving or loading will require nearly the same time.

Frames (any size), including braces and assembly hardware

	Craft@Hrs	Unit	Material	Labor	Total
Rental per frame per month	—	Mo	10.00	—	10.00

Screw jacks (four required per frame with base plate or "U" head)

	Craft@Hrs	Unit	Material	Labor	Total
Rental per jack per month	—	Mo	4.50	—	4.50
Add for erecting each frame	CL@.245	Ea	—	8.92	8.92
Add for dismantling each frame	CL@.197	Ea	—	7.17	7.17

	Craft@Hrs	Unit	Material	Labor	Total

Temporary Structures Portable job site office trailers with electric air conditioners and baseboard heat, built-in desks, plan tables, insulated and paneled walls, tiled floors, locking doors and windows with screens, closet, fluorescent lights, and electrical receptacles. Add $30 per month for units with toilet and lavatory. Monthly rental based on 6 month minimum. Dimensions are length, width and height overall and floor area (excluding hitch area).

	Craft@Hrs	Unit	Material	Labor	Total
16' x 8' x 7' high (108 SF)	—	Mo	—	—	155.00
20' x 8' x 7' high (160 SF)	—	Mo	—	—	165.00
24' x 8' x 7' high (192 SF)	—	Mo	—	—	175.00
28' x 8' x 7' high (240 SF)	—	Mo	—	—	185.00
28' x 10' x 10' high (360 SF)	—	Mo	—	—	226.00
40' x 10' x 10' high (440 SF)	—	Mo	—	—	257.00
48' x 10' x 10' high (500 SF)	—	Mo	—	—	288.00
40' x 12' x 10' high (552 SF)	—	Mo	—	—	288.00
60' x 12' x 10' high (720 SF)	—	Mo	—	—	370.00
60' x 14' x 10' high (840 SF)	—	Mo	—	—	380.00
Add for skirting, per LF of perimeter	—	LF	—	—	9.30
Deduct for unit without heating & cooling	—	Mo	—	—	-25.00
Add for delivery, setup, and dismantling of office trailers within 15 miles					
Delivery and setup	—	LS	—	—	180.00
Dismantle and pickup	—	LS	—	—	180.00
Add for delivery or pickup over 15 miles, per mile					
8' wide units	—	Mile	—	—	1.80
10' wide units	—	Mile	—	—	2.68
12' wide units	—	Mile	—	—	2.99
14' wide units	—	Mile	—	—	3.61

Portable steel storage containers (lockable), suitable for job site storage of tools and materials. Rental per month, based on 6 month rental

	Craft@Hrs	Unit	Material	Labor	Total
8' x 8' x 20'	—	Mo	—	—	69.00
8' x 8' x 24'	—	Mo	—	—	89.60
8' x 8' x 40'	—	Mo	—	—	139.60
Add for delivery and pickup, within 15 miles					
One time charge	—	LS	—	—	170.00

Portable job site shacks with lights, power receptacles, locking door and window. Rental per month, based on 6 month rental

	Craft@Hrs	Unit	Material	Labor	Total
12' x 8' x 8'	—	Mo	—	—	70.00
8' x 8' x 8'	—	Mo	—	—	60.00
Add for typical delivery and pickup, within 15 miles					
One time charge	—	LS	—	—	58.50

Cleanup

Progressive "broom clean" cleanup, as necessary, per 1,000 SF of floor per month

	Craft@Hrs	Unit	Material	Labor	Total
Typical cost	CL@.183	MSF	—	6.66	6.66
Final, total floor area (no glass cleaning)	CL@1.65	MSF	—	60.00	60.00

Glass cleaning, per 1,000 square feet of glass cleaned on one side. (Double these figures when both sides are cleaned.) Add the cost of staging or scaffolding when needed

	Craft@Hrs	Unit	Material	Labor	Total
Cleaning glass with sponge and squeegee	CL@1.34	MSF	14.00	48.80	62.80
Cleaning glass and window trim					
With solvent, towel, sponge and squeegee	CL@5.55	MSF	17.00	202.00	219.00
Removing paint and concrete splatter from windows and trim					
By scraping and solvent	CL@15.9	MSF	42.00	579.00	621.00
Mop resilient floor by hand	CL@1.20	MSF	1.60	43.70	45.30

General Requirements 1

	Craft@Hrs	Unit	Material	Labor	Total

Waste Disposal

Dump fees. Tippage charges for solid waste disposal at the dump vary from $30 to $120 per ton. For planning purposes, estimate waste disposal at $75 per ton plus the hauling cost. Call the dump or trash disposal company for actual charges. Typical costs are shown below.

	Craft@Hrs	Unit	Material	Labor	Total
Dumpster, 3 CY trash bin, emptied weekly	—	Mo	—	—	330.00
Dumpster, 40 CY solid waste bin (lumber, drywall, roofing)					
Hauling cost, per load	—	Ea	—	—	230.00
Add to per load charge, per ton	—	Ton	—	—	55.00
Low-boy, 14 CY solid waste container (asphalt, dirt, masonry, concrete)					
Hauling cost, per load (use 7 CY as maximum load)	—	Ea	—	—	230.00
Add to per load charge, per ton	—	Ton	—	—	55.00

Recycler fees. Recycling construction waste materials can substantially reduce disposal costs. Recycling charges vary from $95 to $120 per load, depending on the type material and the size of the load. Call the recycling company for actual charges. Typical recycler fees are shown below. Add the cost for hauling.

	Craft@Hrs	Unit	Material	Labor	Total
Greenwaste	—	Ton	—	—	30.00
Asphalt, per load (7 CY)	—	Ea	—	—	100.00
Concrete, masonry or rock, per load (7 CY)	—	Ea	—	—	100.00
Dirt, per load (7 CY)	—	Ea	—	—	100.00
Mixed loads, per load (7 CY)	—	Ea	—	—	100.00

Site Work 2

	Craft@Hrs	Unit	Material	Labor	Equipment	Total

Structure Moving Up to 5 mile haul, not including new foundation costs, utility hookup or finishing. These figures assume two stories maximum, no obstruction from trees or utility lines and that adequate right of way is available. Fees and charges imposed by government are not included. Costs shown are per SF of total floor area. Equipment cost includes one tow-truck equipped with dollies, jacks, support beams and hand tools. For estimating purpose, use 1,000 SF as a minimum job charge.

	Craft@Hrs	Unit	Material	Labor	Equipment	Total
Concrete or masonry structures, 12" maximum thick walls or floors						
1,000 to 2,000 SF	C1@.168	SF	—	6.24	7.47	13.71
2,000 to 4,000 SF	C1@.149	SF	—	5.54	6.62	12.16
Wood frame						
1,000 to 2,000 SF	C1@.142	SF	—	5.28	6.31	11.59
2,000 to 4,000 SF	C1@.131	SF	—	4.87	5.82	10.69
Steel frame						
1,000 to 2,000 SF	C1@.215	SF	—	7.99	9.56	17.55
2,000 to 4,000 SF	C1@.199	SF	—	7.39	8.85	16.24

Site Clearing Using a dozer as described below, a 3/4 CY wheel loader and two 10 CY dump trucks. Clearing costs will be higher on smaller jobs and where the terrain limits production. These costs include hauling 6 miles to a legal dump. Dump fees are not included. See Waste Disposal. Quantity shown in parentheses gives the approximate "loose" volume of the materials removed (volume being hauled to the dump). For estimating purposes, add the cost equal to 1 acre of the appropriate clearing operation to allow for moving the equipment on and off the job.

	Craft@Hrs	Unit	Material	Labor	Equipment	Total
Clear light brush and grub roots						
Using a 105 HP D-5 dozer (350 CY)	C2@34.8	Acre	—	1,500.00	1080.00	2,580.00
Clear medium brush and small trees, grub roots						
Using a 130 HP D-6 dozer (420 CY)	C2@41.8	Acre	—	1,810.00	1,400.00	3,210.00
Clear brush and trees to 6" trunk diameter						
Using a 335 HP D-8 dozer (455 CY)	C2@45.2	Acre	—	1,950.00	1,590.00	3,540.00

	Craft@Hrs	Unit	Material	Labor	Equipment	Total
Clear wooded area, pull stumps						
Using a 460 HP D-9 dozer (490 CY)	C2@48.7	Acre	—	2,100.00	2,790.00	4,890.00

Paving and Curb Demolition No salvage of materials. These costs include the cost of loading and hauling to a legal dump within 6 miles. Dump fees are not included. See Waste Disposal. Equipment cost includes one wheel-mounted air compressor, one paving breaker and jackhammer bits, one 55 HP wheel loader with integral backhoe and one 5 CY dump truck. The figures in parentheses give the approximate loose volume of the materials (volume after being demolished). Use $500.00 as a minimum charge.

	Craft@Hrs	Unit	Material	Labor	Equipment	Total
Bituminous paving, depths to 3" (85 SF per CY)						
Large area, with a wheel loader	C3@.028	SF	—	1.18	.72	1.90
Strips 24" wide for utility lines	C3@.029	SF	—	1.22	.74	1.96
Add for jobs under 500 SF	C3@.010	SF	—	.42	.26	.68
Bituminous curbs,						
to 12" width (65 LF per CY)	C3@.028	LF	—	1.18	.72	1.90
Concrete paving and slabs on grade						
4" concrete without rebars						
(1.25 CY waste per CY)	C3@.325	CY	—	13.70	8.31	22.01
4" concrete without rebars						
(1 CY waste per 65 SF)	C3@.004	SF	—	.17	.10	.27
6" concrete without rebars						
(1.25 CY waste per CY)	C3@.421	CY	—	17.70	10.80	28.50
6" concrete without rebars						
(1 CY waste per 45 SF)	C3@.007	SF	—	.29	.18	.47
8" concrete without rebars						
(1.25 CY waste per CY)	C3@.809	CY	—	34.00	20.70	54.70
8" concrete without rebars						
(1 CY waste per 32 SF)	C3@.020	SF	—	.84	.51	1.35
Over 8" to 12" thick concrete						
Per CY without rebars						
(1.25 CY waste per CY)	C3@1.37	CY	—	57.60	35.00	92.60
Per CY with rebars						
(1 CY waste per .68 SF)	C3@1.58	CY	—	66.50	40.40	106.90
Per SF without rebars						
(1 CY waste per 20 SF)	C3@.051	SF	—	2.15	1.30	3.45
Per SF with rebars						
(1 CY waste per 20 SF)	C3@.058	SF	—	2.44	1.48	3.92
Concrete curbs, demolition						
Curb and 24" monolithic gutter						
(7 LF per CY)	C3@.100	LF	—	4.21	2.56	6.77
Planter and batter type curbs, 6" wide						
(30 LF per CY)	C3@.043	LF	—	1.81	1.10	2.91
Removal of pavement markings by water blasting. Use $100.00 as a minimum charge						
Water blaster, per hour	—	Hour	—	—	10.60	10.10
4" wide strips	CL@.032	LF	—	1.16	.34	1.50
Per square foot	CL@.098	SF	—	3.57	.99	4.56

	Craft@Hrs	Unit	Material	Labor	Equipment	Total

Fencing demolition Equipment includes one 5 CY dump truck. No salvage except as noted. These costs include loading and hauling to a legal dump within 6 miles. Dump fees are not included. See Waste Disposal. Use $250.00 as a minimum charge.

	Craft@Hrs	Unit	Material	Labor	Equipment	Total
Remove and dispose chain link, 6' high	C4@.015	LF	—	.56	.15	.71
Remove and salvage chain link, 6' high	C4@.037	LF	—	1.39	.38	1.77
Remove and dispose board fence, 6' high	C4@.016	LF	—	.60	.16	.76

Highway type guardrail demolition These costs include loading and hauling to a legal dump within 6 miles. Dump fees are not included. See Waste Disposal. No salvage except as noted. Equipment includes one wheel-mounted air compressor, one paving breaker and jackhammer bits, one 55 HP wheel loader with integral backhoe and one 5 CY dump truck. Use $500.00 as a minimum charge

	Craft@Hrs	Unit	Material	Labor	Equipment	Total
Remove and dispose guardrail	C4@.043	LF	—	1.62	1.10	2.72
Remove guardrail in salvage condition	C4@.074	LF	—	2.79	1.89	4.68
Remove and dispose guardrail posts	C3@.245	Ea	—	10.30	6.26	16.56

Manhole, Piping and Underground Tank Demolition No salvage of materials except as noted. These costs include loading and hauling to a legal dump within 6 miles. Dump fees are not included. See Waste Disposal. Equipment includes one wheel-mounted air compressor, one paving breaker and jackhammer bits, one 55 HP wheel loader with integral backhoe and one 5 CY dump truck. Use $400.00 as a minimum charge.

Manholes and catch basins, demolition, to 10' deep. Break below collar and plug

	Craft@Hrs	Unit	Material	Labor	Equipment	Total
Brick	C3@5.16	Ea	—	217.00	132.00	349.00
Masonry	C3@5.33	Ea	—	224.00	136.00	360.00
Precast concrete	C3@6.64	Ea	—	279.00	170.00	449.00
Add for sand fill, any of above	C3@.125	CY	18.60	5.26	3.19	27.05

Frame and cover from manhole or catch basin

	Craft@Hrs	Unit	Material	Labor	Equipment	Total
Remove in salvage condition	C3@1.45	Ea	—	61.00	37.10	98.10
Remove and reset	C3@3.99	Ea	—	168.00	102.00	270.00

Fire hydrant demolition

	Craft@Hrs	Unit	Material	Labor	Equipment	Total
Remove and dispose	C3@5.57	Ea	—	234.00	142.00	376.00

Break out storm or sewer pipe, non-salvageable. Excavation or backfill not included

	Craft@Hrs	Unit	Material	Labor	Equipment	Total
Up to 12" diameter	C3@.114	LF	—	4.79	2.91	7.70
15" to 18"	C3@.134	LF	—	5.64	3.42	9.06
21" to 24"	C3@.154	LF	—	6.48	3.94	10.42
27" to 36"	C3@.202	LF	—	8.50	5.16	13.66

Remove welded steel pipe for salvage. Excavation or backfill not included

	Craft@Hrs	Unit	Material	Labor	Equipment	Total
4" diameter or smaller	C3@.111	LF	—	4.67	2.84	7.51
6" to 10" diameter	C3@.202	LF	—	8.50	5.16	13.66

Remove and haul away empty underground liquid storage tanks. (Draining and disposing of hazardous liquid in a tank may require special waste handling equipment.) Cost of draining not included. Includes excavation and backfill to 6' deep.

	Craft@Hrs	Unit	Material	Labor	Equipment	Total
50 to 250 gallon tank	C3@4.24	Ea	—	178.00	108.00	286.00
Over 250 to 600 gallon tank	C3@11.7	Ea	—	492.00	299.00	791.00
Over 600 to 1,000 gallon tank	C3@22.4	Ea	—	942.00	572.00	1,514.00
Add for sand fill, any of above	C3@.125	CY	18.60	5.26	3.19	27.05

	Craft@Hrs	Unit	Material	Labor	Equipment	Total

Miscellaneous Sitework Demolition No salvage of materials except as noted. These costs include loading and hauling to a legal dump within 6 miles. Dump fees are not included. See Waste Disposal.

Railroad demolition, siding quantities. Equipment includes one wheel-mounted air compressor, one paving breaker and jackhammer bits, one 55 HP wheel loader with integral backhoe and one 5 CY dump truck

	Craft@Hrs	Unit	Material	Labor	Equipment	Total
Remove track and ties for scrap	C5@.508	LF	—	20.60	9.74	30.34
Remove and dispose of ballast stone	C5@.111	CY	—	4.51	2.13	6.64
Remove wood ties alone	C5@.143	Ea	—	5.81	2.74	8.55

Remove light standards, flagpoles, playground poles, up to 30' high, including foundations. Equipment includes one wheel-mounted air compressor, one pneumatic breaker and jackhammer bits and one 5 CY dump truck. Use $400.00 as a minimum charge

	Craft@Hrs	Unit	Material	Labor	Equipment	Total
Remove item, in salvage condition	C4@6.24	Ea	—	235.00	95.80	330.80
Remove item, no salvage	C4@.746	Ea	—	28.10	11.50	39.60

Torch cutting steel plate. Equipment is an oxygen-acetylene manual welding and cutting torch with gases, regulator and goggles. Use $200.00 as a minimum charge

	Craft@Hrs	Unit	Material	Labor	Equipment	Total
To 3/8" thick	CL@.073	LF	—	2.66	.51	3.17

Building Demolition Costs for demolishing an entire building. Includes loading and hauling up to 6 miles but no dump fees. See Waste Disposal. Costs are by square foot of floor area based on 8' ceiling height. No salvage value assumed. Figures in parentheses give the approximate "loose" volume of the materials (volume after being demolished).

Light wood-frame structures Up to three stories in height. Based on 2,500 SF job. No basements included. Estimate each story separately. Equipment includes one 55 HP wheel loader with integral backhoe and one 5 CY dump truck. Use $3,200.00 as a minimum charge.

	Craft@Hrs	Unit	Material	Labor	Equipment	Total
First story (8 SF per CY)	C5@.038	SF	—	1.54	.58	2.12
Second story (8 SF per CY)	C5@.053	SF	—	2.15	.81	2.96
Third story (8 SF per CY)	C5@.070	SF	—	2.84	1.08	3.92

Building demolition with pneumatic tools Equipment includes one wheel-mounted air compressor, three breakers and jackhammer bits, two 55 HP wheel loaders with integral backhoes and two 5 CY dump trucks. Use $10,000.00 as a minimum charge.

	Craft@Hrs	Unit	Material	Labor	Equipment	Total
Concrete building (30 SF per CY)	C6@.089	SF	—	3.54	1.28	4.82
Reinforced concrete building						
(20 SF per CY)	C6@.101	SF	—	4.02	1.46	5.48
Masonry building (50 SF per CY)	C6@.074	SF	—	2.94	1.07	4.01

Building demolition with crane and headache ball Equipment includes one wheel-mounted air compressor, one pneumatic paving breaker and jackhammer bits, one 55 HP wheel loader with integral backhoe and backhoe mounted paving breaker, one 15-ton hydraulic crane with headache ball and three 5 CY dump trucks. Use $10,000.00 as a minimum charge.

	Craft@Hrs	Unit	Material	Labor	Equipment	Total
Concrete building (30 SF per CY)	C7@.031	SF	—	1.29	.96	2.25
Reinforced concrete building						
(20 SF per CY)	C7@.037	SF	—	1.55	1.14	2.69
Masonry building (50 SF per CY)	C7@.026	SF	—	1.09	.80	1.89

	Craft@Hrs	Unit	Material	Labor	Equipment	Total

Concrete foundation and footing demolition with D-6 crawler dozer and pneumatic tools. Equipment includes one wheel-mounted air compressor, one pneumatic paving breaker and jackhammer bits, one D-6 crawler dozer with attachments and one 5 CY dump truck. Use $5,000.00 as a minimum charge.

	Craft@Hrs	Unit	Material	Labor	Equipment	Total
Non-reinforced concrete						
(.75 CY per loose CY)	C3@1.10	CY	—	46.30	40.40	86.70
Reinforced concrete						
(.60 CY loose per CY)	C3@1.58	CY	—	66.50	58.00	124.50
Chip out concrete using paving breaker						
No dozer used, (.8 CF per loose CF)	CL@.203	CF	—	7.39	3.12	10.51

Gutting a building Interior finishes stripped back to the structural walls. Building structure to remain. No allowance for salvage value. These costs include loading and hauling up to 6 miles. Dump fees are not included. See Waste Disposal. Costs shown are per square foot of floor area based on 8' ceiling height. Costs will be about 50% less if a small tractor can be used and up to 25% higher if debris must be carried to ground level in an elevator. Equipment includes one air compressor, two pneumatic breakers with jackhammer bits and one 5 CY dump truck. Figures in parentheses give the approximate "loose" volume of the materials (volume after being demolished). Use $3,000.00 as a minimum charge.

	Craft@Hrs	Unit	Material	Labor	Equipment	Total
Residential buildings (125 SF per CY)	C4@.110	SF	—	4.14	4.07	8.21
Commercial buildings (140 SF per CY)	C4@.100	SF	—	3.77	3.73	7.50

Partition wall demolition Building structure to remain. No allowance for salvage value. Dump fees are not included. See Waste Disposal. Costs shown are per square foot of wall removed (as measured on one side). Knock down with pneumatic jackhammers and pile adjacent to site but no removal. Equipment includes one air compressor and two pneumatic breakers with jackhammer bits. Figures in parentheses give the approximate "loose" volume of the materials (volume after being demolished). Use $350.00 as a minimum charge.

	Craft@Hrs	Unit	Material	Labor	Equipment	Total
Brick or block partition demolition						
4" thick partition (60 SF per CY)	CL@.040	SF	—	1.46	.73	2.19
8" thick partition (30 SF per CY)	CL@.057	SF	—	2.07	1.04	3.11
12" thick partition (20 SF per CY)	CL@.073	SF	—	2.66	1.33	3.99
Concrete partition demolition						
Non-reinforced (18 CF per CY)	CL@.273	CF	—	9.93	4.99	14.92
Reinforced (20 CF per CY)	CL@.363	CF	—	13.20	6.63	19.83

Knock down with hand tools and pile adjacent to site but no removal.

	Craft@Hrs	Unit	Material	Labor	Equipment	Total
Stud partition demolition (typically 75 to 100 SF per CY)						
Gypsum or terra cotta on metal lath	CL@.027	SF	—	.98	—	.98
Drywall on metal or wood studs	CL@.028	SF	—	1.02	—	1.02
Plaster on metal studs	CL@.026	SF	—	.95	—	.95

Ceiling demolition Building structure to remain. No allowance for salvage value. Dump fees are not included. See Waste Disposal. Costs shown are per square foot of ceiling (as measured on one side). Figures in parentheses give the approximate "loose" volume of the materials (volume after being demolished). Use $300.00 as a minimum charge. Knock down with hand tools to remove ceiling finish on framing at heights to 9'.

	Craft@Hrs	Unit	Material	Labor	Equipment	Total
Plaster ceiling (typically 175 to 200 SF per CY)						
Including lath and furring	CL@.025	SF	—	.91	—	.91
Including suspended grid	CL@.020	SF	—	.73	—	.73
Acoustic tile ceiling (typically 200 to 250 SF per CY)						
Including suspended grid	CL@.010	SF	—	.36	—	.36
Including grid in salvage condition	CL@.019	SF	—	.69	—	.69
Including strip furring	CL@.014	SF	—	.51	—	.51
Tile glued to ceiling	CL@.015	SF	—	.55	—	.55
Drywall ceiling (typically 250 to 300 SF per CY)						
Nailed or attached with screws						
to joists	CL@.012	SF	—	.44	—	.44
Including strip furring	CL@.023	SF	—	.84	—	.84

	Craft@Hrs	Unit	Material	Labor	Equipment	Total

Roof demolition Building structure to remain. No allowance for salvage value. Dump fees are not included, refer to construction materials disposal section and add for same. Costs shown are per square foot of area removed (as measured on one side).

Roof surface removal, using hand tools. Figures in parentheses give the approximate "loose" volume of the materials (volume after being demolished). Use $900.00 as a minimum charge.

	Craft@Hrs	Unit	Material	Labor	Equipment	Total
Asphalt shingles (2.50 Sq per CY)	CL@.850	Sq	—	30.90	—	30.90
Built-up roofing, including sheathing and gravel (1.25 Sq per CY)	CL@2.91	Sq	—	106.00	—	106.00
Clay or concrete tile (.70 Sq per CY)	CL@1.03	Sq	—	37.50	—	37.50
Concrete plank, no covering (.80 Sq per CY)	CL@1.10	Sq	—	40.00	—	40.00
Gypsum plank, no covering (.70 Sq per CY)	CL@.820	Sq	—	29.80	—	29.80
Metal deck, no covering (.50 Sq per CY)	CL@2.27	Sq	—	82.60	—	82.60
Wood shingles or flashing (1.66 Sq per CY)	CL@.756	Sq	—	27.50	—	27.50
Remove gravel stop	CL@.070	LF		2.55	—	2.55

Floor slab demolition Using pneumatic jackhammers. Piled adjacent to site but no removal. Equipment includes one air compressor and two pneumatic breakers with jackhammer bits. Use $350.00 as a minimum charge.

	Craft@Hrs	Unit	Material	Labor	Equipment	Total
Slab on grade, 4" to 6" thick, reinforced						
With wire mesh (55 SF per CY)	CL@.057	SF	—	2.07	1.04	3.11
With number 4 bars (45 SF per CY)	CL@.063	SF	—	2.29	1.15	3.44
Slab, suspended, 6" to 8" thick, free fall (35 SF per CY)	CL@.092	SF	—	3.35	1.68	5.03
Slab fill, lightweight concrete fill						
On metal deck (45 SF per CY)	CL@.029	SF	—	1.06	.53	1.59
Topping, insulating (150 SF per CY)	CL@.023	SF	—	.84	.42	1.26

Floor covering demolition Using hand tools. Debris piled adjacent to site but no removal. Figures in parentheses give the approximate "loose" volume of the materials (volume after being demolished). Use $350.00 as a minimum charge.

	Craft@Hrs	Unit	Material	Labor	Equipment	Total
Ceramic or quarry tile, brick (200 SF per CY)	CL@.029	SF	—	1.06	—	1.06
Resilient materials only (270 SF per CY)	CL@.018	SF	—	.66	—	.66
Terrazzo tile (225 SF per CY)	CL@.032	SF	—	1.16	—	1.16
Wood block floor, laid in mastic (200 SF per CY)	CL@.051	SF	—	1.86	—	1.86
Wood, residential strip floor (225 SF per CY)	CL@.032	SF	—	1.16	—	1.16

Cutting openings in frame walls Using hand tools. Cost per square foot of wall measured on one face.

	Craft@Hrs	Unit	Material	Labor	Equipment	Total
Metal stud wall with stucco or plaster	CL@.094	SF	—	3.42	—	3.42
Wood stud wall with drywall	CL@.051	SF	—	1.86	—	1.86
Dust control partitions, 6 mil plastic	C8@.011	SF	.30	.45	—	.75

	Craft@Hrs	Unit	Material	Labor	Equipment	Total

Debris removal Break demolition debris into manageable size with hand tools, load into a 5 CF wheelbarrow, move to chute and dump. Costs shown are per cubic foot of material dumped.

	Craft@Hrs	Unit	Material	Labor	Equipment	Total
Wheelbarrow, 50' to trash chute and dump	CL@.018	CF	—	.66	—	.66
Wheelbarrow, 100' to trash chute and dump	CL@.022	CF	—	.80	—	.80
Wheelbarrow, 50' to elevator, descend 10 floors to trash chute and dump	CL@.024	CF	—	.87	—	.87

Trash chutes. Prefabricated steel chute installed and removed in a multi-story building. One-time charge for up to six-months usage. Use 50 LF as minimum charge

	Craft@Hrs	Unit	Material	Labor	Equipment	Total
18" diameter	—	LS	—	—	—	1,160.00
36" diameter	—	LS	—	—	—	1,735.00

Load demolition debris on a truck and haul 6 miles to a dump site. Equipment includes one 5 CY dump truck and one 1 CY articulating wheel loader. Includes truck cost but dump fees are not included. See Waste Disposal. Per cubic yard of debris.

	Craft@Hrs	Unit	Material	Labor	Equipment	Total
Truck loaded by hand	C4@.828	CY	—	31.20	8.52	39.72
Truck loaded using a wheel loader	C5@.367	CY	—	14.90	5.76	20.66

Removal of interior items Using hand tools. Items removed in salvageable condition do not include an allowance for salvage value. Figures in parentheses give the approximate "loose" volume of the materials (volume after being demolished.) Removal in non-salvageable condition (demolished) except as noted.

	Craft@Hrs	Unit	Material	Labor	Equipment	Total
Hollow metal door and frame in a masonry wall. (2 doors per CY)						
Single door to 4' x 7'	CL@1.00	Ea	—	36.40	—	36.40
Two doors, per opening to 8' x 7'	CL@1.50	Ea	—	54.60	—	54.60
Wood door and frame in a wood-frame wall (2 doors per CY)						
Single door to 4' x 7'	CL@.500	Ea	—	18.20	—	18.20
Two doors, per opening to 8' x 7'	CL@.750	Ea	—	27.30	—	27.30
Remove door and frame in masonry wall, salvage condition						
Hollow metal door to 4' x 7'	C8@2.00	Ea	—	81.90	—	81.90
Wood door to 4' x 7'	C8@1.00	Ea	—	41.00	—	41.00
Frame for resilient mat, metal						
Per SF of mat area	C8@.054	SF	—	2.21	—	2.21
Lockers, metal 12" W, 60" H, 15" D	C8@.500	Ea	—	20.50	—	20.50
Sink and soap dispenser, wall-hung	C8@.500	Ea	—	20.50	—	20.50
Toilet partitions, wood or metal						
Per partition	C8@.750	Ea	—	30.70	—	30.70
Urinal screens, wood or metal						
Per partition	C8@.500	Ea	—	20.50	—	20.50
Window and frame, wood or metal	C8@.076	SF	—	3.11	—	3.11
Remove and reset airlock doors	C8@5.56	Ea	—	228.00	—	228.00

Asbestos Hazard Surveys Building inspection, hazard identification, sampling and ranking of asbestos risk.

	Craft@Hrs	Unit	Material	Labor	Equipment	Total
Asbestos hazard survey and sample collection (10,000 SF per hour)						
4 hour minimum	—	Hour	—	—	—	80.00
Sample analysis (usually one sample per 1,000 SF)						
10 sample minimum	—	Ea	—	—	—	48.00
Report writing (per 1,000 SF of floor)						
$200 minimum	—	MSF	—	—	—	25.00

Site Work 2

	Craft@Hrs	Unit	Material	Labor	Equipment	Total

Asbestos Removal, Subcontract Typical costs including site preparation, monitoring, equipment, and removal of waste. Disposal of hazardous waste materials vary widely, consult your local waste disposal facility concerning prices and procedures.

Ceiling insulation in containment structure

	Craft@Hrs	Unit	Material	Labor	Equipment	Total
500 to 5,000 SF job	—	SF	—	—	—	37.00
5,000 to 20,000 SF job	—	SF	—	—	—	25.00

Pipe insulation in containment structure

100 to 1,000 LF of 6" pipe	—	LF	—	—	—	65.00
1,000 to 3,000 LF of 6" pipe	—	LF	—	—	—	42.00

Pipe insulation using glove bags

100 to 1,000 LF of 6" pipe	—	LF	—	—	—	55.00
1,000 to 3,000 LF of 6" pipe	—	LF	—	—	—	42.00

Wall Sawing, Subcontract Using an electric, hydraulic or air saw. Per LF of cut including overcuts at each corner equal to the depth of the cut. Costs shown assume electric power is available. Minimum cost will be $400.

	Unit	Brick or block	Concrete w/#4 bar	Concrete w/#7 bar
To 4" depth	LF	12.30	17.60	19.50
To 5" depth	LF	15.40	19.10	22.70
To 6" depth	LF	16.90	20.60	25.60
To 7" depth	LF	21.30	25.10	29.40
To 8" depth	LF	25.90	29.40	35.30
To 10" depth	LF	32.30	36.40	42.00
To 12" depth	LF	39.70	43.70	50.10
To 14" depth	LF	48.60	52.90	58.20
To 16" depth	LF	58.20	60.40	66.30
To 18" depth	LF	69.00	69.00	79.20

	Craft@Hrs	Unit	Material	Labor	Total

Concrete Core Drilling, Subcontract Using an electric, hydraulic or air drill. Costs shown are per LF of depth for diameter shown. Prices are based on one mat reinforcing with 3/8" to 5/8" bars and include cleanup. Difficult drill setups are higher. Prices assume availability of 110 volt electricity. Figure travel time at $95 per hour. Minimum cost will be $200.

	Craft@Hrs	Unit	Material	Labor	Total
1" or 1-1/2" diameter	—	LF	—	—	53.90
2" or 2-1/2" diameter	—	LF	—	—	55.70
3" or 3-1/2"diameter	—	LF	—	—	67.20
4" or 5" diameter	—	LF	—	—	81.60
6" or 8" diameter	—	LF	—	—	112.00
10" diameter	—	LF	—	—	154.00
12" diameter	—	LF	—	—	251.00
14" diameter	—	LF	—	—	288.00
16" diameter	—	LF	—	—	327.00
18" diameter	—	LF	—	—	365.00
20" diameter	—	LF	—	—	404.00
24" diameter	—	LF	—	—	481.00
30" diameter	—	LF	—	—	594.00
36" diameter	—	LF	—	—	710.00
Hourly rate, 2 hour minimum	—	Hr	—	—	121.00

Concrete Slab Sawing, Subcontract Using a gasoline powered saw. Costs per linear foot for cured concrete assuming a level surface with good access and saw cut lines laid out and pre-marked by others. Costs include local travel time. Minimum cost will be $250. Electric powered slab sawing will be approximately 40% higher for the same depth.

Depth	Unit	Under 200'	200'- 1000'	Over 1000'
1" deep	LF	.65	.57	.48
1-1/2" deep	LF	.96	.89	.74
2" deep	LF	1.28	1.12	.96
2-1/2" deep	LF	1.61	1.44	1.20
3" deep	LF	1.93	1.68	1.44
3-1/2" deep	LF	2.24	2.01	1.68
4" deep	LF	2.55	2.24	1.93
5" deep	LF	3.20	2.81	2.40
6" deep	LF	3.85	3.37	2.88
7" deep	LF	4.65	3.99	3.45
8" deep	LF	5.28	4.64	3.99
9" deep	LF	6.01	5.28	4.57
10" deep	LF	6.81	6.01	5.21
11" deep	LF	7.60	6.72	5.84
12" deep	LF	8.42	7.45	6.49

Green concrete (2 days old) sawing will usually cost 15 to 20% less. Work done on an hourly basis will cost $105 per hour for slabs up to 4" thick and $115 per hour for 5" or 6" thick slabs. A two hour minimum charge will apply on work done on an hourly basis.

Asphalt Sawing, Subcontract Using a gasoline powered saw. Minimum cost will be $170. Cost per linear foot of green or cured asphalt, assuming level surface with good access and saw cut lines laid out and pre-marked by others. Costs include local travel time. Work done on an hourly basis will cost $98 per hour for asphalt up to 4" thick and $108 per hour for 5" or 6" thick asphalt. A two hour minimum charge will apply on work done on an hourly basis.

Depth	Unit	Under 450'	450'- 1000'	Over 1000'
1" or 1-1/2" deep	LF	.41	.32	.25
2" deep	LF	.65	.48	.41
2-1/2" deep	LF	.79	.65	.48
3" deep	LF	.96	.74	.57
3-1/2" deep	LF	1.12	.89	.65
4" deep	LF	1.28	.96	.74
5" deep	LF	1.61	1.20	.89
6" deep	LF	1.93	1.44	1.05

	Craft@Hrs	Unit	Material	Labor	Equipment	Total

Rock Excavation Rock drilling costs. Based on using a pneumatic truck-mounted wagon drill. Equipment includes one wheel-mounted air compressor and one 4-1/2" pneumatic wagon drill with hydraulic swing boom and drill bits. Costs shown are per linear foot of hole drilled. Add blasting costs below. Use $2,500 as a minimum job charge. LF per hour shown is for a 2-man crew

	Craft@Hrs	Unit	Material	Labor	Equipment	Total
Easy work, 38 LF per hour	S1@.052	LF	—	2.24	2.45	4.69
Moderate work, 30 LF per hour	S1@.067	LF	—	2.88	2.15	5.03
Most work, 25 LF per hour	S1@.080	LF	—	3.44	3.76	7.20
Hard work, 20 LF per hour	S1@.100	LF	—	4.30	4.70	9.00
Dense rock, 15 LF per hour	S1@.132	LF	—	5.67	6.21	11.88
Drilling 2-1/2" hole for rock bolts, 24 LF/hour	S1@.086	LF	—	3.70	4.05	7.75

Site Work 2

	Craft@Hrs	Unit	Material	Labor	Equipment	Total

Blasting costs Based on two cycles per hour and 20 linear foot lifts. These costs assume 75% fill of holes with explosives. Equipment includes one flatbed truck. Costs shown are per cubic yard of area blasted. Loading or hauling of blasted material not included. Use $3,500 as a minimum job charge. Load explosives in holes and detonate, by pattern spacing. CY per hour shown is for a 5-man crew.

	Craft@Hrs	Unit	Material	Labor	Equipment	Total
Using explosives, per pound	—	Lb	3.30	—	—	3.30
6' x 6' pattern (26 CY per hour)	C1@.192	CY	9.90	7.13	.65	17.68
7' x 7' pattern (35 CY per hour)	C1@.144	CY	7.43	5.35	.48	13.26
7' x 8' pattern (47 CY per hour)	C1@.106	CY	5.78	3.94	.36	10.08
9' x 9' pattern (60 CY per hour)	C1@.084	CY	4.95	3.12	.28	8.35
10' x 10' pattern (72 CY per hour)	C1@.069	CY	4.13	2.56	.23	6.92
12' x 12' pattern (108 CY per hour)	C1@.046	CY	2.81	1.71	.15	4.67
14' x 14' pattern (147 CY per hour)	C1@.034	CY	1.98	1.26	.11	3.35

Ripping rock Based on using a tractor-mounted ripper. Equipment includes a D-8 dozer and a D-9 dozer. Costs shown are per CY of area ripped. Loading or hauling of material not included. Use $2,000 as a minimum job charge.

	Craft@Hrs	Unit	Material	Labor	Equipment	Total
Clay or glacial tills, D-9 cat tractor with 2-shank ripper (500 CY per hour)	T0@.002	CY	—	.10	.35	.45
Clay or glacial tills, D-8 cat tractor with 1-shank ripper (335 CY per hour)	T0@.003	CY	—	.15	.38	.53
Shale, sandstone, or limestone, D-9 cat tractor with 2-shank ripper (125 CY per hour)	T0@.008	CY	—	.40	1.40	1.80
Shale, sandstone, or limestone, D-8 cat tractor with 1-shank ripper (95 CY per hour)	T0@.011	CY	—	.55	1.38	1.93
Slate, metamorphic rock, D-9 cat tractor with 2-shank ripper (95 CY per hour)	T0@.011	CY	—	.55	1.93	2.48
Slate, metamorphic rock, D-8 cat tractor with 1-shank ripper (63 CY per hour)	T0@.016	CY	—	.79	2.01	2.80
Granite or basalt, D-9 cat tractor with 2-shank ripper (78 CY per hour)	T0@.013	CY	—	.64	2.28	2.92
Granite or basalt, D-8 cat tractor with 1-shank ripper (38 CY per hour)	T0@.027	CY	—	1.34	3.39	4.73

Rock loosening Based on using pneumatic jackhammers. Equipment includes one compressor and two jackhammers with chisel points. Costs shown are per CY of area loosened. Loading and hauling not included. CY per hour shown is for a 2-man crew. Use $800 as a minimum job charge.

	Craft@Hrs	Unit	Material	Labor	Equipment	Total
Igneous or dense rock (1.4 CY per hour)	CL@1.43	CY	—	52.00	25.50	77.50
Most weathered rock (2.4 CY per hour)	CL@.835	CY	—	30.40	14.90	45.30
Soft sedimentary rock (4 CY per hour)	CL@.500	CY	—	18.20	8.92	27.12

Site Grading Layout, staking, flagmen, lights, watering, ripping, rock breaking, loading or hauling not included. Site preparation grading based on using one crawler tractor. Use $3,000.00 as a minimum job charge. (One acre is 43,560 SF or 4,840 SY.)

	Craft@Hrs	Unit	Material	Labor	Equipment	Total
General area rough grading with 100 HP D-4 tractor (.5 acres per hour)	S1@4.00	Acre	—	172.00	88.90	260.90
General area rough grading with 335 HP D-8 tractor (1 acre per hour)	S1@2.00	Acre	—	86.00	125.50	211.50

Site Work 2

	Craft@Hrs	Unit	Material	Labor	Equipment	Total
Site grading based on using one 100 HP (10,000 lb) motor grader						
General area grading						
(.7 acres per hour)	S1@2.86	Acre	—	123.00	63.50	186.50
Fine grading subgrade to 1/10'						
(.45 acres per hour)	S1@4.44	Acre	—	191.00	98.50	289.50
Cut slope or shape embankment to 2' high						
(.25 acres per hour)	S1@8.00	Acre	—	344.00	178.00	522.00
Rough grade sub-base course on roadway						
(.2 acres per hour)	S1@10.0	Acre	—	430.00	222.00	652.00
Finish grade base or leveling course on roadway						
(.17 acres per hour)	S1@11.8	Acre	—	507.00	262.00	769.00
Finish grading for building slab, windrow excess						
(1,000 SF per hour)	S1@.002	SF	—	.09	.04	.13

Embankment Grading Earth embankment spreading, shaping, compacting and watering, and finish shaping. Use $15,000 as a minimum job charge, loading or hauling of earth fill not included.

Spreading and shaping Spread and shape earth from loose piles, based on using a D-8 tractor

6" to 10" lifts (164 CY per hour)	T0@.006	CY	—	.30	.75	1.05

Compacting and watering (Add for cost of water from below.) CY per hour shown is for a 3-man crew. Based on using a self-propelled 100 HP vibrating roller and a 3,500 gallon truck. Productivity assumes 3 passes at 7' wide

6" lifts (500 CY per hour)	C3@.006	CY	—	.25	.17	.42
8" lifts (750 CY per hour)	C3@.004	CY	—	.17	.11	.28
10" lifts (1,000 CY per hour)	C3@.003	CY	—	.13	.09	.22

Compacting with a sheepsfoot roller Towed behind a D-7 tractor and a 3500 gallon truck. Productivity assumes 3 passes at 5' wide

6" lifts (185 CY per hour)	C3@.016	CY	—	.67	.78	1.45
8" lifts (245 CY per hour)	C3@.012	CY	—	.50	.58	1.08

Cost of water for compacted earth embankments. Based on water at $2.50 per 1,000 gallons and 66 gallons per cubic yard of compacted material. Assumes optimum moisture at 10%, natural moisture of 2% and evaporation of 2%. Placed in conjunction with compaction shown above.

Cost per CY of compacted embankment	—	CY	.20	—	—	.20

Finish shaping Earth embankment slopes and swales up to 1 in 4 incline. Based on using a 150 HP grader and a 15 ton self-propelled rubber tired roller. SY per hour shown is for a 3-man crew

(200 SY per hour)	SS@.016	SY	—	.72	.66	1.38
Based on using a D-8 tractor						
(150 SY per hour)	T0@.007	SY	—	.35	.88	1.23
Finish shaping of embankment slopes and swales by hand. SY per hour shown is for 1 man						
Slopes up to 1 in 4 (16 SY per hour)	CL@.063	SY	—	2.29	—	2.29
Slopes over 1 in 4 (12.5 SY per hour)	CL@.080	SY	—	2.91	—	2.91

Site Work 2

	Craft@Hrs	Unit	Material	Labor	Equipment	Total

Trench Excavation and Backfill These costs and productivity are based on utility line trenches and continuous footings where the spoil is piled adjacent to the trench. Linear feet (LF), cubic yards (CY), square feet (SF), or square yards (SY) per hour shown are based on a 2-man crew. Increase costs by 10% to 25% when spoil is loaded in trucks. Hauling, shoring, dewatering or unusual conditions are not included.

12" wide bucket, for 12" wide trench. Depths 3' to 5'. Equipment is a wheel loader with 1 CY bucket and integral backhoe. Use $400 as a minimum job charge

	Craft@Hrs	Unit	Material	Labor	Equipment	Total
Light soil (60 LF per hour)	S1@.033	LF	—	1.42	.48	1.90
Medium soil (55 LF per hour)	S1@.036	LF	—	1.55	.52	2.07
Heavy or wet soil (35 LF per hour)	S1@.057	LF	—	2.45	.83	3.28

18" wide bucket, for 18" wide trench. Depths 3' to 5'. Equipment is a wheel loader with 1 CY bucket and integral backhoe. Use $400 as a minimum job charge

Light soil (55 LF per hour)	S1@.036	LF	—	1.55	.52	2.07
Medium soil (50 LF per hour)	S1@.040	LF	—	1.72	.58	2.30
Heavy or wet soil (30 LF per hour)	S1@.067	LF	—	2.88	.97	3.85

24" wide bucket, for 24" wide trench. Depths 3' to 5'. Equipment is a wheel loader with 1 CY bucket and integral backhoe. Use $400 as a minimum job charge

Light soil (50 LF per hour)	S1@.040	LF	—	1.72	.58	2.30
Medium soil (45 LF per hour)	S1@.044	LF	—	1.89	.64	2.53
Heavy or wet soil (25 LF per hour)	S1@.080	LF	—	3.44	1.16	4.60

Truck-mounted Gradall with 1 CY bucket. Use $700 as a minimum job charge

Light soil (40 CY per hour)	S1@.050	CY	—	2.15	2.49	4.64
Medium soil (34 CY per hour)	S1@.059	CY	—	2.54	2.94	5.48
Heavy or wet soil (27 CY per hour)	S1@.074	CY	—	3.18	3.68	6.86
Loose rock (23 CY per hour)	S1@.087	CY	—	3.74	4.33	8.07

Crawler-mounted hydraulic backhoe with 3/4 CY bucket. Use $500 as a minimum job charge

Light soil (33 CY per hour)	S1@.061	CY	—	2.62	1.48	4.10
Medium soil (27 CY per hour)	S1@.074	CY	—	3.18	1.79	4.97
Heavy or wet soil (22 CY per hour)	S1@.091	CY	—	3.91	2.21	6.12
Loose rock (18 CY per hour)	S1@.112	CY	—	4.81	2.72	7.53

Crawler-mounted hydraulic backhoe with 1 CY bucket. Use $500 as a minimum job charge

Light soil (65 CY per hour)	S1@.031	CY	—	1.33	.88	2.21
Medium soil (53 CY per hour)	S1@.038	CY	—	1.63	1.08	2.71
Heavy or wet soil (43 CY per hour)	S1@.047	CY	—	2.02	1.33	3.35
Loose rock (37 CY per hour)	S1@.055	CY	—	2.36	1.56	3.92
Blasted rock (34 CY per hour)	S1@.058	CY	—	2.49	1.65	4.14

Crawler-mounted hydraulic backhoe with 1-1/2 CY bucket . Use $500 as a minimum job charge

Light soil (83 CY per hour)	S1@.024	CY	—	1.03	.81	1.84
Medium soil (70 CY per hour)	S1@.029	CY	—	1.25	.98	2.23
Heavy or wet soil (57 CY per hour)	S1@.035	CY	—	1.50	1.19	2.69
Loose rock (48 CY per hour)	S1@.042	CY	—	1.81	1.42	3.23
Blasted rock (43 CY per hour)	S1@.047	CY	—	2.02	1.59	3.61

Crawler-mounted hydraulic backhoe with 2 CY bucket. Use $700 as a minimum job charge

Light soil (97 CY per hour)	S1@.021	CY	—	.90	.88	1.78
Medium soil (80 CY per hour)	S1@.025	CY	—	1.07	1.05	2.12
Heavy or wet soil (65 CY per hour)	S1@.031	CY	—	1.33	1.31	2.64
Loose rock (55 CY per hour)	S1@.036	CY	—	1.55	1.52	3.07
Blasted rock (50 CY per hour)	S1@.040	CY	—	1.72	1.69	3.41

Crawler-mounted hydraulic backhoe with 2-1/2 CY bucket. Use $750 as a minimum job charge

Light soil (122 CY per hour)	S1@.016	CY	—	.69	.98	1.67
Medium soil (100 CY per hour)	S1@.020	CY	—	.86	1.23	2.09
Heavy or wet soil (82 CY per hour)	S1@.024	CY	—	1.03	1.47	2.50
Loose rock (68 CY per hour)	S1@.029	CY	—	1.25	1.78	3.03
Blasted rock (62 CY per hour)	S1@.032	CY	—	1.38	1.96	3.34

	Craft@Hrs	Unit	Material	Labor	Equipment	Total
Chain-boom Ditch Witch digging trench to 12" wide and 5' deep. Use $400 as a minimum job charge						
Light soil (10 CY per hour)	S1@.200	CY	—	8.60	2.71	11.31
Most soils (8.5 CY per hour)	S1@.235	CY	—	10.10	3.18	13.28
Heavy soil (7 CY per hour)	S1@.286	CY	—	12.30	3.87	16.17
Trenchers, chain boom, 55 HP, digging trench to 18" wide and 8' deep. Use $500 as a minimum job charge						
Light soil (62 CY per hour)	S1@.032	CY	—	1.38	.71	2.09
Most soils (49 CY per hour)	S1@.041	CY	—	1.76	.91	2.67
Heavy soil (40 CY per hour)	S1@.050	CY	—	2.15	1.11	3.26
Trenchers, chain boom, 100 HP, digging trench to 24" wide and 8' deep. Use $500 as a minimum job charge						
Light soil (155 CY per hour)	S1@.013	CY	—	.56	.40	.96
Most soils (125 CY per hour)	S1@.016	CY	—	.69	.50	1.19
Heavy soil (100 CY per hour)	S1@.020	CY	—	.86	.62	1.48
Trim trench bottom to 1/10'						
By hand (200 SF per hour)	CL@.010	SF	—	.36	—	.36

Backfill trenches from loose material piled adjacent to a trench. No compaction included. Soil previously excavated.

	Craft@Hrs	Unit	Material	Labor	Equipment	Total
Backfill trenches by hand						
(8 CY per hour)	CL@.250	CY	—	9.10	—	9.10
Wheel loader 55 HP. Use $400 as a minimum job charge						
Typical soils (50 CY per hour)	S1@.040	CY	—	1.72	.61	2.33
D-3 crawler dozer. Use $475 as a minimum job charge						
Typical soils (25 CY per hour)	S1@.080	CY	—	3.44	1.44	4.88
3/4 CY crawler loader. Use $450 as a minimum job charge						
Typical soils (33 CY per hr)	S1@.061	CY	—	2.62	1.12	3.74
D-7 crawler dozer. Use $700 as a minimum job charge						
Typical soils (130 CY per hour)	S1@.015	CY	—	.64	.67	1.31
Wheel loader 55 HP. Sand or gravel bedding. Use $400 as a minimum job charge						
Typical soils (80 CY per hour)	S1@.025	CY	18.60	1.07	.38	20.05
Fine grade bedding by hand						
(22 SY per hr)	CL@.090	SY	—	3.28	—	3.28
Compaction of soil in trenches in 8" layers. Use $200 as a minimum job charge						
Pneumatic tampers (20 CY per hour)	CL@.050	CY	—	1.82	1.18	3.00
Vibrating rammers (10 CY per hour)	CL@.100	CY	—	3.64	.97	4.61

Dragline Excavation For mass excavation, footings or foundations. Costs shown include casting the excavated soil adjacent to the excavation or loading it into trucks. Hauling costs are not included. Cubic yards (CY) per hour are as measured in an undisturbed condition (bank measure) and are based on a 2-man crew. Use $3,000 as a minimum job charge.

	Craft@Hrs	Unit	Material	Labor	Equipment	Total
1-1/2 CY dragline with a 45 ton crawler crane						
Loam or light clay (73 CY per hour)	H2@.027	CY	—	1.21	.72	1.93
Sand or gravel (67 CY per hour)	H2@.030	CY	—	1.34	.80	2.14
Heavy clay (37 CY per hour)	H2@.054	CY	—	2.42	1.45	3.87
Unclassified soil (28 CY per hour)	H2@.071	CY	—	3.18	1.90	5.08
2 CY dragline with a 60 ton crawler crane						
Loam or light clay (88 CY per hour)	H2@.023	CY	—	1.03	.97	2.00
Sand or gravel (85 CY per hour)	H2@.024	CY	—	1.08	1.01	2.09
Heavy clay (48 CY per hour)	H2@.042	CY	—	1.88	1.77	3.65
Unclassified soil (36 CY per hour)	H2@.056	CY	—	2.51	2.36	4.87
2-1/2 CY dragline with a 100 ton crawler crane						
Loam or light clay (102 CY per hour)	H2@.019	CY	—	.85	1.23	2.08
Sand or gravel (98 CY per hour)	H2@.021	CY	—	.94	1.36	2.30
Heavy clay (58 CY per hour)	H2@.035	CY	—	1.57	2.27	3.84
Unclassified soil (44 CY per hour)	H2@.045	CY	—	2.02	2.92	4.94

	Craft@Hrs	Unit	Material	Labor	Equipment	Total
3 CY dragline with a 150 ton crawler crane						
Loam or light clay (116 CY per hour)	H2@.017	CY	—	.76	1.71	2.47
Sand or gravel (113 CY per hour)	H2@.018	CY	—	.81	1.81	2.62
Heavy clay (70 CY per hour)	H2@.029	CY	—	1.30	2.92	4.22
Unclassified soil (53 CY per hour)	H2@.038	CY	—	1.70	3.83	5.53

Backhoe Excavation For mass excavation, footings or foundations Costs shown include casting the excavated soil adjacent to the excavation or loading it into trucks. Hauling costs are not included. Cubic yards (CY) per hour are as measured in an undisturbed condition (bank measure) and are based on a 2-man crew. Equipment costs are based on using a crawler-mounted hydraulic backhoe. Use $2,500 as a minimum job charge

	Craft@Hrs	Unit	Material	Labor	Equipment	Total
1 CY backhoe						
Light soil (69 CY per hour)	S1@.029	CY	—	1.25	.50	1.75
Most soils (57 CY per hour)	S1@.035	CY	—	1.50	.60	2.10
Wet soil, loose rock (46 CY per hour)	S1@.043	CY	—	1.85	.74	2.59
1-1/2 CY backhoe						
Light soil (90 CY per hour)	S1@.022	CY	—	.95	.38	1.33
Most soils (77 CY per hour)	S1@.026	CY	—	1.12	.45	1.57
Wet soil, loose rock (60 CY per hour)	S1@.033	CY	—	1.42	.57	1.99
Blasted rock (54 CY per hour)	S1@.044	CY	—	1.89	.76	2.65
2 CY backhoe						
Light soil (100 CY per hour)	S1@.020	CY	—	.86	.69	1.55
Most soils (85 CY per hour)	S1@.023	CY	—	.99	.79	1.78
Wet soil, loose rock (70 CY per hour)	S1@.029	CY	—	1.25	1.00	2.25

Moving and Loading Excavated Materials Costs shown include moving the material 50' and dumping it into piles or loading it into trucks. Add 25% for each 50' of travel beyond the first 50'. Hauling costs are not included. Cubic yards (CY) per hour are for material in a loose condition (previously excavated) and are based on a 2-man crew performing the work. Equipment costs are based on using a wheel-mounted front end loader. Use $2,000 as a minimum job charge

	Craft@Hrs	Unit	Material	Labor	Equipment	Total
3/4 CY loader						
(32 CY per hour)	S1@.062	CY	—	2.67	.88	3.55
1 CY loader						
(55 CY per hour)	S1@.036	CY	—	1.55	.53	2.08
2 CY loader						
(90 CY per hour)	S1@.022	CY	—	.95	.38	1.33
3-1/4 CY loader						
(125 CY per hour)	S1@.016	CY	—	.69	.49	1.18
5 CY loader						
(218 CY per hour)	S1@.009	CY	—	.39	.40	.79

Hauling excavated material Using trucks. Costs shown include 4 minutes for loading, 3 minutes for dumping and travel time based on the one-way distance, speed and cycles per hour as noted. Costs for equipment to excavate and load the material are not included. Truck capacity shown is based on loose cubic yards of material. Allow the following percentage amounts when estimating quantities based on undisturbed (bank measure) materials for swell when exporting from the excavation site or shrinkage when importing from borrow site: clay (33%), common earth (25%), granite (65%), mud (21%), sand or gravel (12%).

	Craft@Hrs	Unit	Material	Labor	Equipment	Total
3 CY dump truck. Use $550 as a minimum job charge						
1 mile haul at 20 MPH						
(4.2 cycles and 12.6 CY per hour)	TD@.029	CY	—	1.17	.68	1.85
3 mile haul at 30 MPH						
(2.9 cycles and 8.7 CY per hour)	TD@.044	CY	—	1.77	1.04	2.81
6 mile haul at 40 MPH						
(2.1 cycles and 6.3 CY per hour)	TD@.058	CY	—	2.33	1.37	3.70

Site Work 2

	Craft@Hrs	Unit	Material	Labor	Equipment	Total
5 CY dump truck. Use $600 as a minimum job charge						
1 mile haul at 20 MPH						
(4.2 cycles and 21 CY per hour)	TD@.020	CY	—	.80	.62	1.42
3 mile haul at 30 MPH						
(2.9 cycles and 14.5 CY per hour)	TD@.028	CY	—	1.13	.86	1.99
6 mile haul at 40 MPH						
(2.1 cycles and 10.5 CY per hour)	TD@.041	CY	—	1.65	1.27	2.92
10 CY dump truck. Use $800 as a minimum job charge						
1 mile haul at 20 MPH						
(4.2 cycles and 42 CY per hour)	TD@.015	CY	—	.60	.56	1.16
3 mile haul at 30 MPH						
(2.9 cycles and 29 CY per hour)	TD@.022	CY	—	.88	.82	1.70
6 mile haul at 40 MPH						
(2.1 cycles and 21 CY per hour)	TD@.029	CY	—	1.17	1.08	2.25
25 CY off-highway dump truck. Use $2,000 as a minimum job charge						
1 mile haul at 20 MPH						
(4.2 cycles and 105 CY per hour)	TD@.009	CY	—	.36	.94	1.30
3 mile haul at 30 MPH						
(2.9 cycles and 72.5 CY per hour)	TD@.014	CY	—	.56	1.46	2.02
6 mile haul at 40 MPH						
(2.1 cycles and 52.5CY per hour)	TD@.019	CY	—	.76	1.98	2.74

Dozer Excavation Mass excavation using a crawler tractor with a dozing blade attached. Costs shown include excavation and pushing soil 150' to a stockpile. Based on good access and a maximum of 5% grade. Increase or reduce costs 20% for each 50' of push more or less than 150'. Cubic yards (CY) per hour shown are undisturbed bank measure and are based on a 1-man crew. Use the following as minimum job charges: $850 if using a D-4 or a D-6 dozer; $1,350 if using a D-7 or D-8 dozer; or $2,000 if using a D-9 dozer.

	Craft@Hrs	Unit	Material	Labor	Equipment	Total
Gravel or loose sand						
100 HP D-4 dozer with "S" blade						
(29 CY per hour)	TO@.034	CY	—	1.69	1.51	3.20
140 HP D-6 dozer with "S" blade						
(56 CY per hour)	TO@.018	CY	—	.89	1.15	2.04
200 HP D-7 dozer with "S" blade						
(78 CY per hour)	TO@.013	CY	—	.64	1.16	1.80
335 HP D-8 dozer with "U" blade						
(140 CY per hour)	TO@.007	CY	—	.35	.88	1.23
460 HP D-9 dozer with "U" blade						
(205 CY per hour)	TO@.005	CY	—	.25	.88	1.13
Loam or soft clay						
100 HP D-4 dozer with "S" blade						
(26 CY per hour)	TO@.038	CY	—	1.88	1.69	3.57
140 HP D-6 dozer with "S" blade						
(50 CY per hour)	TO@.020	CY	—	.99	1.28	2.27
200 HP D-7 dozer with "S" blade						
(70 CY per hour)	TO@.014	CY	—	.69	1.25	1.94
335 HP D-8 dozer with "U" blade						
(125 CY per hour)	TO@.008	CY	—	.40	1.00	1.40
460 HP D-9 dozer with "U" blade						
(185 CY per hour)	TO@.005	CY	—	.25	.88	1.13
Shale, sandstone or blasted rock						
100 HP D-4 dozer with "S" blade						
(20 CY per hour)	TO@.051	CY	—	2.53	2.27	4.80

	Craft@Hrs	Unit	Material	Labor	Equipment	Total
140 HP D-6 dozer with "S" blade						
(38 CY per hour)	TO@.026	CY	—	1.29	1.67	2.96
200 HP D-7 dozer with "S" blade						
(53 CY per hour)	TO@.019	CY	—	.94	1.70	2.64
335 HP D-8 dozer with "U" blade						
(95 CY per hour)	TO@.011	CY	—	.55	1.38	1.93
460 HP D-9 dozer with "U" blade						
(140 CY per hour)	TO@.007	CY	—	.35	1.23	1.58

Scraper-Hauler Excavation Mass excavation using a self-propelled scraper-hauler. Equipment costs include the scraper-hauler and a crawler tractor with a dozer blade attached pushing it 10 minutes each hour. Cubic yards (CY) per hour shown are undisturbed bank measure. Work done in clay, shale or soft rock will cost 10% to 25% more.

15 CY self-propelled 200 HP scraper-hauler and a D-8 335 HP tractor pushing. Use $6,000 as a minimum job charge

	Craft@Hrs	Unit	Material	Labor	Equipment	Total
1,000' haul						
(9 cycles and 135 CY per hr)	SS@.016	CY	—	.72	1.14	1.86
2,500' haul						
(6 cycles and 90 CY per hr)	SS@.024	CY	—	1.08	1.71	2.79
4,000' haul						
(4.5 cycles and 68 CY per hr)	SS@.032	CY	—	1.45	2.29	3.74

24 CY self-propelled 350 HP scraper-hauler and a D-9 460 HP tractor pushing. Use $8,000 as a minimum job charge

	Craft@Hrs	Unit	Material	Labor	Equipment	Total
1,000' haul						
(9 cycles and 225 CY per hr)	SS@.010	CY	—	.45	.96	1.41
2,500' haul						
(6 cycles and 150 CY per hr)	SS@.014	CY	—	.63	1.34	1.97
4,000' haul						
(4.5 cycles and 113 CY per hr)	SS@.019	CY	—	.86	1.82	2.68

32 CY self-propelled 550 HP scraper-hauler and a D-9 460 HP tractor pushing. Use $10,000 as a minimum job charge

	Craft@Hrs	Unit	Material	Labor	Equipment	Total
1,000' haul						
(9 cycles and 315 CY per hr)	SS@.007	CY	—	.32	.81	1.13
2,500' haul						
(6 cycles and 219 CY per hr)	SS@.010	CY	—	.45	1.16	1.61
4,000' haul						
(4.5 cycles and 158 CY per hr)	SS@.014	CY	—	.63	1.62	2.25

Roadway and Embankment Earthwork Cut & Fill Mass excavating, hauling, placing and compacting (cut & fill) using a 200 HP, 15 CY capacity self-propelled scraper-hauler with attachments. Productivity is based on a 1,500' haul with 6" lifts compacted to 95% per AASHO requirements. Cubic yards (CY) or square yards (SY) per hour shown are bank measurement. Use $3,000 as a minimum job charge.

	Craft@Hrs	Unit	Material	Labor	Equipment	Total
Cut, fill and compact, clearing or finishing not included						
Most soil types (45 CY per hour)	TO@.022	CY	—	1.09	2.76	3.85
Rock and earth mixed (40 CY per hour)	TO@.025	CY	—	1.24	3.14	4.38
Rippable rock (25 CY per hour)	TO@.040	CY	—	1.98	5.02	7.00
Earth banks and levees, clearing or finishing not included						
Cut, fill and compact (85 CY per hour)	TO@.012	CY	—	.59	1.51	2.10
Channelize compacted fill						
(55 CY per hour)	TO@.018	CY	—	.89	2.26	3.15
Roadway subgrade preparation						
Scarify and compact (91 CY per hour)	TO@.011	CY	—	.55	1.38	1.93
Roll and compact base course						
(83 CY per hour)	TO@.012	CY	—	.59	1.51	2.10
Excavation, open ditches						
(38 CY per hour)	TO@.026	CY	—	1.29	3.26	4.55

	Craft@Hrs	Unit	Material	Labor	Equipment	Total
Trimming and finishing						
Banks, swales or ditches						
(140 CY per hour)	T0@.007	CY	—	.35	.88	1.23
Scarify asphalt pavement						
(48 SY per hour)	T0@.021	SY	—	1.04	2.64	3.68

Slope Protection See also Soil Stabilization. Gabions as manufactured by Maccaferri Inc. Typical prices for gabions as described, placed by hand. Prices can be expected to vary based on quantity. Cost for stone is not included. Add for stone as required. Costs shown are per each gabion.

	Craft@Hrs	Unit	Material	Labor	Equipment	Total
Galvanized mesh						
6' x 3' x 3'	CL@2.70	Ea	67.70	98.30	—	166.00
9' x 3' x 3'	CL@2.90	Ea	97.83	106.00	—	203.83
12' x 3' x 3'	CL@3.10	Ea	122.00	113.00	—	235.00
Erosion and drainage control polypropylene geotextile fabric (Mirafi FW700)						
For use under riprap, hand placed	CL@.011	SY	2.15	.40	—	2.55

Riprap Dumped from trucks and placed using a 15 ton hydraulic crane and a 10 cubic yard dump truck. Riprap costs will vary widely. Add $.75 per CY per mile for trucking to the site. Cubic (CY) yards per hour shown in parentheses are based on a 7-man crew. Use $2,500 as a minimum job charge

	Craft@Hrs	Unit	Material	Labor	Equipment	Total
Typical riprap cost, per CY	—	CY	66.00	—	—	66.00
5 to 7 CF pieces (9.4 CY per hour)	S5@.745	CY	66.00	28.90	9.27	104.17
5 to 7 CF pieces, sacked and placed, excluding bag cost (3 CY per hour)	S5@2.33	CY	66.00	90.40	8.99	165.39
Loose small riprap stone, under 30 lbs each Hand placed (2 CY per hour)	CL@.500	CY	66.00	18.20	—	84.20

Ornamental large rock Rock prices include local delivery of 10 CY minimum. Cubic (CY) yards per hour shown in parentheses are based on a 7-man crew. Equipment is a 15 ton hydraulic crane and a 10 cubic yard dump truck. Use $2,500 as a minimum job charge. Expect wide local variations in costs.

	Craft@Hrs	Unit	Material	Labor	Equipment	Total
Volcanic cinder, 1,200 lbs per CY						
(8 CY per hour)	S5@.875	CY	300.00	33.90	10.89	344.79
Featherock, 1,000 lbs per CY						
(12 CY per hour)	S5@.583	CY	330.00	22.60	7.25	359.85

Rock fill Dumped from trucks and placed with a 100 HP D-4 tractor. Rock prices include local delivery of 10 CY minimum. Cubic (CY) yards per hour shown in parentheses are based on a 3-man crew. Use $1,350 as a minimum job charge

	Craft@Hrs	Unit	Material	Labor	Equipment	Total
Drain rock, 3/4" to 1-1/2"						
(12 CY per hour)	S6@.250	CY	25.00	10.20	3.70	38.90
Bank run gravel (12 CY per hour)	S6@.250	CY	22.00	10.20	3.70	35.90
Pea gravel (12 CY per hour)	S6@.250	CY	22.00	10.20	3.70	35.90
Straw bales secured to ground with reinforcing bars, bale is 3'6" long, 2' wide, 1'6" high.						
Per bale of straw, placed by hand	CL@.166	Ea	4.75	6.04	—	10.79
Sedimentation control fence, installed vertically at bottom of slope, without stakes attached						
36" high woven fabric (Mirafi 100X)	CL@.003	LF	.75	.11	—	.86
Add to fence above for #6 rebar stakes 6'0" long at 10'0" OC hand-driven 36" into the ground						
Per LF of fence	CL@.050	LF	.35	1.82	—	2.17

	Craft@Hrs	Unit	Material	Labor	Equipment	Total

Manual Excavation Based on one throw (except where noted) of shoveled earth from trench or pit. Shoring or dewatering not included

Trench or pit in light soil (silty sand or loess, etc.)

Up to 4' deep (.82 CY per hour)	CL@1.22	CY	—	44.40	—	44.40
Over 4' to 6' deep (.68 CY per hour)	CL@1.46	CY	—	53.10	—	53.10
Over 6' deep, two throws						
(.32 CY per hour)	CL@3.16	CY	—	115.00	—	115.00

Trench or pit in medium soil (sandy clay, for example)

Up to 4' deep (.68 CY per hour)	CL@1.46	CY	—	53.10	—	53.10
Over 4' to 6' deep (.54 CY per hour)	CL@1.86	CY	—	67.70	—	67.70
Over 6' deep, two throws						
(.27 CY per hour)	CL@3.66	CY	—	133.00	—	133.00

Trench or pit in heavy soil (clayey materials, shales, caliche, etc.)

Up to 4' deep (.54 CY per hour)	CL@1.86	CY	—	67.70	—	67.70
Over 4' to 6' deep (.46 CY per hour)	CL@2.16	CY	—	78.60	—	78.60
Over 6' deep, 2 throws						
(.25 CY per hour)	CL@3.90	CY	—	142.00	—	142.00

Wheelbarrow (5 CF) on firm ground, load by shovel from pile, wheel 300 feet and dump

Light soil (.74 CY per hour)	CL@1.35	CY	—	49.10	—	49.10
Medium soil (.65 CY per hour)	CL@1.54	CY	—	56.00	—	56.00
Heavy soil (.56 CY per hour)	CL@1.79	CY	—	65.10	—	65.10
Loose rock (.37 CY per hour)	CL@2.70	CY	—	98.30	—	98.30

Hand trim and shape

Around utility lines (15 CF hour)	CL@.065	CF	—	2.37	—	2.37
For slab on grade (63 SY per hour)	CL@.016	SY	—	.58	—	.58
Trench bottom (100 SF per hour)	CL@.010	SF	—	.36	—	.36

Soil Stabilization See also Slope Protection. Lime slurry injection treatment. Equipment is a 100 HP grader, a 1 ton self-propelled steel roller and a 10 ton 100 HP vibratory steel roller. Use $3,000 as a minimum job charge. Cost per CY of soil treated

(25 CY per hour)	S7@.240	CY	11.50	10.30	5.96	27.76

Vibroflotation treatment Equipment includes a 50 ton hydraulic crane, a 1 CY wheel loader for 15 minutes of each hour, and a vibroflotation machine with pumps. Use $3,500 as a minimum job charge. Typical cost per CY of soil treated

Low cost (92 CY per hour)	S7@.065	CY	—	2.79	2.52	5.31
High cost (46 CY per hour)	S7@.130	CY	—	5.59	5.05	10.64

Soil cement treatment Includes materials and placement for 7% cement mix. Equipment is a cross shaft mixer, a 2 ton 3 wheel steel roller and a vibratory roller. Costs shown are per CY of soil treated. Use $3,000 as a minimum job charge

On level ground

(5 CY per hour)	S7@1.20	CY	10.50	51.60	29.10	91.20

On slopes under 3 in 1

(4.3 CY per hour)	S7@1.40	CY	10.50	60.20	34.00	104.70

On slopes over 3 in 1

(4 CY per hour)	S7@1.50	CY	10.50	64.50	36.38	111.38

Construction fabrics (geotextiles) Mirafi products. Costs shown include 10% for lapover. Drainage fabric (Mirafi 140NC), used to line trenches, allows water passage but prevents soil migration into drains

Per SY of fabric placed by hand	CL@.016	SY	2.58	.58	—	3.16

Stabilization fabric (Mirafi 500X), used to prevent mixing of unstable soils with road base aggregate

Per SY of fabric placed by hand	CL@.008	SY	.88	.29	—	1.17

	Craft@Hrs	Unit	Material	Labor	Equipment	Total

Prefabricated drainage fabric (Miradrain), used to drain subsurface water from structural walls

	Craft@Hrs	Unit	Material	Labor	Equipment	Total
Per SY of fabric placed by hand	CL@.006	SY	.72	.22	—	.94

Erosion control mat (Miramat TM8), used to reduce surface soil erosion on slopes or in ditches while promoting seed growth

	Craft@Hrs	Unit	Material	Labor	Equipment	Total
Per SY of mat placed by hand	CL@.020	SY	4.14	.73	—	4.87

Pile Foundations These costs assume solid ground for support of the pile driving equipment with all site work performed by others at no cost prior to pile driving being performed. Costs shown per linear foot (LF) are per vertical foot of pile depth. Standby or idle time, access roads, rig mats, test piles or special engineering required by unusual conditions are not included.

Pile testing Testing costs will vary with the soil type, method of testing, and seismic requirements for any given job site. Use the following as guideline costs for estimating purposes. Typical costs including cost for moving test equipment on and off job site. Add cost below for various types of piling

	Craft@Hrs	Unit	Material	Labor	Equipment	Total
50 to 100 ton range	S8@224.	Ea	—	9,900.00	6,800.00	16,700.00
Over 100 ton to 200 ton range	S8@280.	Ea	—	12,400.00	8,500.00	20,900.00
Over 200 ton to 300 ton range	S8@336.	Ea	—	14,900.00	10,200.00	25,100.00

Prestressed concrete piles Equipment includes a 60 ton crawler mounted crane equipped with pile driving attachments, a 4-ton forklift, a truck with hand tools, a portable air compressor complete with hoses and pneumatic operated tools, a welding machine and an oxygen-acetylene cutting torch. Use $28,000 as minimum job charge, including 1,000 LF of piling and the cost for moving the pile driving equipment on and off the job site

	Craft@Hrs	Unit	Material	Labor	Equipment	Total
Minimum job charge	—	LS	—	—	—	28,000.00

For jobs requiring over 1,000 LF of piling, add to minimum charge as follows. (Figures in parentheses show approximate production per day based on a 7-man crew)

	Craft@Hrs	Unit	Material	Labor	Equipment	Total
12" square						
30' or 40' long piles (550 LF per day)	S8@.102	LF	13.30	4.51	3.52	21.33
50' or 60' long piles (690 LF per day)	S8@.082	LF	13.30	3.63	2.83	19.76
14" square						
30' or 40' long piles (500 LF per day)	S8@.113	LF	16.20	5.00	3.89	25.09
50' or 60' long piles (645 LF per day)	S8@.087	LF	16.20	3.85	3.00	23.05
16" square						
30' or 40' long piles (420 LF per day)	S8@.134	LF	23.10	5.92	4.62	33.64
50' or 60' long piles (550 LF per day)	S8@.102	LF	23.10	4.51	3.52	31.13
18" square						
30' or 40' long piles (400 LF per day)	S8@.143	LF	28.80	6.32	4.93	40.05
50' or 60' long piles (525 LF per day)	S8@.107	LF	28.80	4.73	3.69	37.22

Additional costs for prestressed concrete piles Pre-drilling holes for piles. Equipment includes a flatbed truck with a boom and a power operated auger. Use $2,200 as a minimum job charge, including drilling 100 LF of holes and the cost to move the drilling equipment on and off the job site

	Craft@Hrs	Unit	Material	Labor	Equipment	Total
Minimum job charge	—	LS	—	—	—	2,200.00

For jobs requiring more than 100 LF of holes drilled, add to minimum cost as shown below

	Craft@Hrs	Unit	Material	Labor	Equipment	Total
Pre-drilling first 10 LF, per hole	C9@1.50	LS	—	64.40	47.20	111.60
Add per LF of drilling, after first 10 LF	C9@.177	LF	—	7.59	5.57	13.16

Steel "HP" shape piles Equipment includes a 60 ton crawler mounted crane equipped with pile driving attachments, a 4-ton forklift, a truck with hand tools, a portable air compressor complete with hoses and pneumatic operated tools, a welding machine and an oxygen-acetylene cutting torch. Use $25,000 as minimum job charge, including 1,000 LF of piling and cost for moving pile driving equipment on and off the job site.

	Craft@Hrs	Unit	Material	Labor	Equipment	Total
Minimum job charge	—	LS	—	—	—	25,000.00

	Craft@Hrs	Unit	Material	Labor	Equipment	Total

For jobs requiring over 1,000 LF of piling add to minimum charge as follows. (Figures in parentheses show approximate production per day based on a 7-man crew)

HP8 8" x 8", 36 lbs per LF

	Craft@Hrs	Unit	Material	Labor	Equipment	Total
30' or 40' long piles (800 LF per day)	S8@.070	LF	11.80	3.09	2.41	17.30
50' or 60' long piles (880 LF per day)	S8@.064	LF	11.80	2.83	2.21	16.84

HP10 10" x 10", 42 lbs per LF

	Craft@Hrs	Unit	Material	Labor	Equipment	Total
30' or 40' long piles (760 LF per day)	S8@.074	LF	13.40	3.27	2.55	19.22
50' or 60' long piles (800 LF per day)	S8@.070	LF	13.40	3.09	2.41	18.90

HP10 10" x 10", 48 lbs per LF

	Craft@Hrs	Unit	Material	Labor	Equipment	Total
30' or 40' long piles (760 LF per day)	S8@.074	LF	15.40	3.27	2.55	21.22
50' or 60' long piles (800 LF per day)	S8@.070	LF	15.40	3.09	2.41	20.90

HP12 12" x 12", 53 lbs per LF

	Craft@Hrs	Unit	Material	Labor	Equipment	Total
30' or 40' long piles (730 LF per day)	S8@.077	LF	16.50	3.40	2.65	22.55
50' or 60' long piles (760 LF per day)	S8@.074	LF	16.50	3.27	2.55	22.32

HP12 12" x 12", 74 lbs per LF

	Craft@Hrs	Unit	Material	Labor	Equipment	Total
30' or 40' long piles (730 LF per day)	S8@.077	LF	23.00	3.40	2.65	29.05
50' or 60' long piles (760 LF per day)	S8@.074	LF	23.00	3.27	2.55	28.82

HP14 14" x 14", 73 lbs per LF

	Craft@Hrs	Unit	Material	Labor	Equipment	Total
30' or 40' long piles (690 LF per day)	S8@.081	LF	22.10	3.58	2.79	28.47
50' or 60' long piles (730 LF per day)	S8@.077	LF	22.10	3.40	2.65	28.15

HP14 14" x 14", 89 lbs per LF

	Craft@Hrs	Unit	Material	Labor	Equipment	Total
30' or 40' long piles (630 LF per day)	S8@.089	LF	26.90	3.93	3.07	33.90
50' or 60' long piles (690 LF per day)	S8@.081	LF	26.90	3.58	2.79	33.27

HP14 14" x 14", 102 lbs per LF

	Craft@Hrs	Unit	Material	Labor	Equipment	Total
30' or 40' long piles (630 LF per day)	S8@.089	LF	30.90	3.93	3.07	37.90
50' or 60' long piles (690 LF per day)	S8@.081	LF	30.90	3.58	2.79	37.27

HP14 14" x 14", 117 lbs per LF

	Craft@Hrs	Unit	Material	Labor	Equipment	Total
30' or 40' long piles (630 LF per day)	S8@.089	LF	35.40	3.93	3.07	42.40
50' or 60' long piles (690 LF per day)	S8@.081	LF	35.40	3.58	2.79	41.77

Additional costs related to any of above

Standard driving points or splices

	Craft@Hrs	Unit	Material	Labor	Equipment	Total
HP8	S8@1.50	Ea	37.80	66.30	20.40	124.50
HP10	S8@1.75	Ea	41.10	77.40	23.80	142.30
HP12	S8@2.00	Ea	51.60	88.40	27.20	167.20
HP14	S8@2.50	Ea	60.70	111.00	33.90	205.60

Cutting pile off to required elevation

	Craft@Hrs	Unit	Material	Labor	Equipment	Total
HP8 or HP10	S8@1.00	Ea	4.94	44.20	13.60	62.74
HP12 or HP14	S8@1.50	Ea	7.44	66.30	20.40	94.14

Steel pipe piles Equipment includes a 45 ton crawler mounted crane equipped with pile driving attachments, a 2-ton forklift, a truck with hand tools, a portable air compressor complete with hoses and pneumatic operated tools, a welding machine and an oxygen-acetylene cutting torch. Use $20,000 as minimum job charge, including 500 LF of piling and cost for moving pile driving equipment on and off the job site

	Craft@Hrs	Unit	Material	Labor	Equipment	Total
Minimum job charge	—	LS	—	—	—	20,000.00

For jobs requiring over 500 LF of piling add to minimum charge as follows. (Figures in parentheses show approximate production per day based on a 7-man crew)

Steel pipe piles, non-filled

	Craft@Hrs	Unit	Material	Labor	Equipment	Total
8" (660 LF per day)	S8@.085	LF	9.92	3.76	2.37	16.05
10" (635 LF per day)	S8@.088	LF	14.20	3.89	2.45	20.54
12" (580 LF per day)	S8@.096	LF	17.40	4.24	2.68	24.32
14" (500 LF per day)	S8@.112	LF	19.30	4.95	3.12	27.37
16" (465 LF per day)	S8@.120	LF	22.20	5.31	3.34	30.85
18" (440 LF per day)	S8@.128	LF	25.30	5.66	3.57	34.53

	Craft@Hrs	Unit	Material	Labor	Equipment	Total
Steel pipe piles, concrete filled						
8" (560 LF per day)	S8@.100	LF	11.70	4.42	2.79	18.91
10" (550 LF per day)	S8@.101	LF	14.70	4.47	2.81	21.98
12" (500 LF per day)	S8@.112	LF	19.40	4.95	3.12	27.47
14" (425 LF per day)	S8@.131	LF	20.90	5.79	3.65	30.34
16" (400 LF per day)	S8@.139	LF	24.80	6.15	3.87	34.82
18" (375 LF per day)	S8@.149	LF	28.60	6.59	4.15	39.34
Splices for steel pipe piles						
8" pile	S8@1.12	Ea	46.20	49.50	12.70	108.40
10" pile	S8@1.55	Ea	52.00	68.50	17.60	138.10
12" pile	S8@2.13	Ea	56.00	94.20	24.20	174.40
14" pile	S8@2.25	Ea	74.00	99.50	25.60	199.10
16" pile	S8@2.39	Ea	118.00	106.00	27.20	251.20
18" pile	S8@2.52	Ea	141.00	111.00	28.70	280.70
Standard points for steel pipe piles						
8" point	S8@1.49	Ea	71.00	65.90	17.00	153.90
10" point	S8@1.68	Ea	96.00	74.30	19.10	189.40
12" point	S8@1.84	Ea	129.00	81.30	20.90	231.20
14" point	S8@2.03	Ea	162.00	89.70	23.10	274.80
16" point	S8@2.23	Ea	238.00	98.60	25.40	362.00
18" point	S8@2.52	Ea	314.00	111.00	28.70	453.70

Wood piles End bearing Douglas fir, treated with 12 lb creosote. Piles up to 40' long are 8" minimum tip diameter. Over 45' long, 7" is the minimum tip diameter. Equipment includes a 45 ton crawler mounted crane equipped with pile driving attachments, a 2-ton forklift, a truck with hand tools, a portable air compressor complete with hoses and pneumatic operated tools. Use $20,000 as minimum job charge, including 1,000 LF of piling and cost for moving pile driving equipment on and off the job site

Minimum job charge	—	LS	—	—	—	20,000.00

For jobs requiring over 1,000 LF of piling, add to minimum charge as follows. (Figures in parentheses show approximate production per day based on a 7-man crew)

	Craft@Hrs	Unit	Material	Labor	Equipment	Total
To 30' long (725 LF per day)	S8@.077	LF	6.80	3.40	1.98	12.18
Over 30' to 40' (800 LF per day)	S8@.069	LF	6.91	3.05	1.77	11.73
Over 40' to 50' (840 LF per day)	S8@.067	LF	7.60	2.96	1.72	12.28
Over 50' to 60' (920 LF per day)	S8@.061	LF	8.13	2.70	1.56	12.39
Over 60' to 80' (950 LF per day)	S8@.059	LF	10.40	2.61	1.51	14.52
Add for drive shoe, per pile	—	Ea	31.50	—	—	31.50

Caisson Foundations Drilled in stable soil, no shoring required, filled with reinforced concrete. Add for belled bottoms, if required, from the following section. Equipment includes a truck-mounted A-frame hoist and a power operated auger and a concrete pump. Concrete is based on using 3,000 PSI pump mix delivered to the job site in ready-mix trucks and includes a 5% allowance for waste. Spiral caisson reinforcing is based on using 3/8" diameter hot rolled steel spirals with main vertical bars as shown, shop fabricated delivered ready to install, with typical engineering and drawings. Costs for waste disposal are not included. Refer to construction materials disposal section and add for same. Use $15,000 as a minimum charge, including 200 LF of reinforced concrete filled caissons and cost for moving the equipment on and off the job

Minimum job charge	—	LS	—	—	—	15,000.00

For jobs requiring over 200 LF of reinforced concrete-filled caissons add to minimum charge as follows. (Figures in parentheses show approximate production per day based on a 3-man crew)
16" diameter, drilled caisson filled with reinforced concrete

	Craft@Hrs	Unit	Material	Labor	Equipment	Total
Drilling						
(680 LF per day)	S6@.035	LF	—	1.43	2.31	3.74
Spiral reinforcing plus 6 #6 bars						
(590 LF per day)	S6@.041	LF	26.20	1.67	2.70	30.57

	Craft@Hrs	Unit	Material	Labor	Equipment	Total
Concrete filling						
(615 LF per day)	S6@.039	LF	5.68	1.59	2.57	9.84
Total for 16" caisson	S6@.115	LF	31.88	4.69	7.58	44.15

24" diameter, drilled caisson filled with reinforced concrete

	Craft@Hrs	Unit	Material	Labor	Equipment	Total
Drilling						
(570 LF per day)	S6@.042	LF	—	1.71	2.77	4.48
Spiral reinforcing plus 6 #6 bars						
(280 LF per day)	S6@.085	LF	44.70	3.47	5.60	53.77
Concrete filling						
(280 LF per day)	S6@.086	LF	12.70	3.51	5.67	21.88
Total for 24" caisson	S6@.213	LF	57.40	8.69	14.00	80.09

36" diameter, drilled caisson filled with reinforced concrete

	Craft@Hrs	Unit	Material	Labor	Equipment	Total
Drilling						
(440 LF per day)	S6@.055	LF	—	2.24	3.62	5.86
Spiral reinforcing plus 8 #10 bars						
(135 LF per day)	S6@.176	LF	81.10	7.18	11.60	99.88
Concrete filling						
(125 LF per day)	S6@.194	LF	28.60	7.91	12.78	49.29
Total for 36" caisson	S6@.425	LF	109.70	17.30	28.00	155.00

Bell-bottom footings, concrete filled, for pre-drilled caissons. Add to the cost of any of the caissons above. (Figures in parentheses show approximate production per day based on a 3-man crew)

	Craft@Hrs	Unit	Material	Labor	Equipment	Total
6' diameter bell bottom						
concrete filled (8 per day)	S6@3.00	Ea	343.00	122.00	198.00	663.00
8' diameter bell bottom						
concrete filled (4 per day	S6@6.00	Ea	813.00	245.00	395.00	1,453.00

Shoring, Bulkheads and Underpinning Steel sheet piling, driven and pulled, based on good driving conditions and using sheets 30' to 50' in length with 32 lbs of 38.5 KSI steel per SF. Material cost assumes the piling is salvaged Use $16,000 as a minimum job charge

	Craft@Hrs	Unit	Material	Labor	Equipment	Total
Using steel at, per ton	—	Ton	945.00	—	—	945.00
Assumed steel salvage value, per ton	—	Ton	711.00	—	—	711.00
Typical driving and						
pulling cost, per ton	—	Ton	—	—	180.00	180.00
Up to 20' deep	S8@.096	SF	3.83	4.24	3.62	11.69
Over 20' to 35' deep	S8@.088	SF	3.83	3.89	3.33	11.05
Over 35' to 50' deep	S8@.087	SF	3.83	3.85	3.28	10.96
Over 50' deep	S8@.085	SF	3.83	3.76	3.21	10.80

Extra costs for sheet steel piling
Add for:

	Craft@Hrs	Unit	Material	Labor	Equipment	Total
50' to 65' lengths	—	SF	.09	—	—	.09
65' to 100' lengths	—	SF	.23	—	—	.23
Epoxy coal tar 16 mil factory finish	—	SF	1.80	—	—	1.80
32 lb per SF piling material						
left in place	—	SF	4.10	—	—	4.10
38 lbs per SF (MZ 38) steel	—	SF	.86	—	—	.86
50 KSI high-strength steel	—	SF	.50	—	—	.50
Marine grade 50 KSI						
high strength steel	—	SF	.92	—	—	.92

Deductive costs for sheet steel piling
Deduct for:

	Craft@Hrs	Unit	Material	Labor	Equipment	Total
28 lbs per SF (PS 28) steel	—	SF	-.42	—	—	-.42
23 lbs per SF (PS 23) steel	—	SF	-.92	—	—	-.92
Deduct labor & equipment cost						
when piling is left in place	—	SF	—	-1.50	-.53	-2.03

	Craft@Hrs	Unit	Material	Labor	Equipment	Total

Timber trench sheeting and bracing per square foot of trench wall measured on one side of trench. Costs include pulling and salvage, depths to 12'. Equipment includes a 10 ton hydraulic crane and a 1 ton flat-bed truck. Use $4,000 as a minimum job charge

	Craft@Hrs	Unit	Material	Labor	Equipment	Total
Less than 6' wide, open bracing	S5@.053	SF	1.80	2.06	.45	4.31
6' to 10' wide, open bracing	S5@.064	SF	1.89	2.48	.55	4.92
11' to 16' wide, open bracing	S5@.079	SF	2.22	3.06	.68	5.96
Over 16' wide, open bracing	S5@.091	SF	2.44	3.53	.78	6.75
Less than 6' wide, closed sheeting	S5@.114	SF	2.06	4.42	.98	7.46
6' to 10' wide, closed sheeting	S5@.143	SF	1.83	5.55	1.22	8.60
11' to 16' wide, closed sheeting	S5@.163	SF	2.19	6.32	1.40	9.91
Over 16' wide, closed sheeting	S5@.199	SF	2.76	7.72	1.70	12.18

Additional costs for bracing on trenches over 12' deep

	Craft@Hrs	Unit	Material	Labor	Equipment	Total
Normal bracing, to 15' deep	S5@.050	SF	.37	1.94	.43	2.74
One line of bracing, over 15' to 22' deep	S5@.053	SF	.45	2.06	.45	2.96
Two lines of bracing, over 22' to 35' deep	S5@.061	SF	.75	2.37	.52	3.64
Three lines of bracing, over 35' to 45' deep	S5@.064	SF	.84	2.48	.55	3.87

Wellpoint Dewatering Based on header pipe connecting 2" diameter jetted wellpoints 5' on center. Includes header pipe, wellpoints, filter sand, pumps and swing joints. The header pipe length is usually equal to the perimeter of the area excavated.

Typical cost for installation and removal of wellpoint system with wellpoints 14' deep and placed 5' on center along 6" header pipe. Based on equipment rented for 1 month. Use 100 LF as a minimum job charge and add cost for operator as required. Also add (if required) for additional months of rental, fuel for the pumps, a standby pump, water truck for jetting, outflow pipe, permits and consultant costs.

6" header pipe and accessories, system per LF installed, rented 1 month and removed

	Craft@Hrs	Unit	Material	Labor	Equipment	Total
100 LF system	C5@.535	LF	1.35	21.70	44.40	67.45
200 LF system	C5@.535	LF	1.35	21.70	26.50	49.55
500 LF system	C5@.535	LF	1.35	21.70	23.00	46.05
1,000 LF system	C5@.535	LF	1.35	21.70	19.40	42.45
Add for 8" header pipe	C5@.010	LF	.06	.41	1.53	2.00
Add for 18' wellpoint depth	C5@.096	LF	.11	3.90	4.85	8.86
Add for second month	—	LF	—	—	14.30	15.00
Add for each additional month	—	LF	—	—	8.52	8.95
Add for operator, per hour	T0@1.00	Hr	—	49.60	—	49.60

Pipe Jacking Typical costs for jacking .50" thick wall pipe casing under an existing roadway. Costs include leaving casing in place. Add 15% when ground water is present. Add 100% for light rock conditions. Includes jacking pits on both sides. Equipment includes a 1 CY wheel mounted backhoe plus a 2-ton truck equipped for this type of work. Size shown is casing diameter. Use $2,200 as a minimum job charge.

	Craft@Hrs	Unit	Material	Labor	Equipment	Total
2" casing	C5@.289	LF	6.47	11.70	3.85	22.02
3" casing	C5@.351	LF	7.64	14.30	4.67	26.61
4" casing	C5@.477	LF	10.00	19.40	6.35	35.75
6" casing	C5@.623	LF	12.90	25.30	8.30	46.50
8" casing	C5@.847	LF	17.00	34.40	11.28	62.68
10" casing	C5@1.24	LF	25.30	50.40	16.50	92.20
12" casing	C5@1.49	LF	30.50	60.60	19.84	110.94
16" casing	C5@1.78	LF	36.50	72.30	23.70	132.50
17" casing	C5@2.09	LF	41.60	84.90	27.80	154.30
24" casing	C5@7.27	LF	147.00	295.00	96.80	538.80

	Craft@Hrs	Unit	Material	Labor	Equipment	Total
30" casing	C5@7.88	LF	160.00	320.00	105.00	585.00
36" casing	C5@8.61	LF	176.00	350.00	115.00	641.00
42" casing	C5@9.29	LF	187.00	378.00	124.00	689.00
48" casing	C5@10.2	LF	206.00	415.00	136.00	757.00

Gas Distribution Lines Installed in an open trench. Equipment includes a wheel mounted 1 CY backhoe for lifting and placing the pipe. No excavation, bedding material or backfill included. Use $2,000 as a minimum job charge.

Polyethylene pipe, 60 PSI

	Craft@Hrs	Unit	Material	Labor	Equipment	Total
1-1/4" diameter, coils	U1@.064	LF	.75	2.79	.55	4.09
1-1/2" diameter, coils	U1@.064	LF	1.06	2.79	.55	4.40
2" diameter, coils	U1@.071	LF	1.35	3.10	.61	5.06
3" diameter, coils	U1@.086	LF	2.69	3.75	.74	7.18
3" diameter, 38' long with couplings	U1@.133	LF	2.24	5.80	1.15	9.19
4" diameter, 38' long with couplings	U1@.168	LF	4.49	7.33	1.45	13.27
6" diameter, 38' long with couplings	U1@.182	LF	9.53	7.94	1.57	19.04
8" diameter, 38' long with couplings	U1@.218	LF	17.40	9.51	1.88	28.79

Meters and pressure regulators Natural gas diaphragm meters, direct digital reading, not including temperature and pressure compensation. For use with pressure regulators below

	Craft@Hrs	Unit	Material	Labor	Equipment	Total
425 CFH at 10 PSI	P6@4.38	Ea	257.00	194.00	0.00	451.00
425 CFH at 25 PSI	P6@4.38	Ea	339.00	194.00	0.00	533.00
Natural gas pressure regulators with screwed connections						
3/4" or 1" connection	P6@.772	Ea	26.30	34.20	0.00	60.50
1-1/4" or 1-1/2" connection	P6@1.04	Ea	138.00	46.10	0.00	184.10
2" connection	P6@1.55	Ea	1,130.00	68.60	0.00	1,198.60

Polyethylene Pipe Belled ends with rubber ring joints. Installed in an open trench. Equipment includes a wheel mounted 1/2 CY backhoe for lifting and placing the pipe. Excavation, bedding material, shoring, backfill or dewatering are not included. Use $2,000 as a minimum job charge.

160 PSI polyethylene pipe, Standard Dimension Ratio (SDR) 11 to 1

	Craft@Hrs	Unit	Material	Labor	Equipment	Total
3" pipe	U1@.108	LF	2.29	4.71	.84	7.84
4" pipe	U1@.108	LF	3.80	4.71	.84	9.35
5" pipe	U1@.108	LF	5.93	4.71	.84	11.48
6" pipe	U1@.108	LF	8.23	4.71	.84	13.78
8" pipe	U1@.145	LF	13.90	6.33	1.13	21.36
10" pipe	U1@.145	LF	21.90	6.33	1.13	29.36
12" pipe	U1@.145	LF	30.60	6.33	1.13	38.06

100 PSI polyethylene pipe, Standard Dimension Ratio (SDR) 17 to 1

	Craft@Hrs	Unit	Material	Labor	Equipment	Total
4" pipe	U1@.108	LF	2.58	4.71	.84	8.13
6" pipe	U1@.108	LF	5.54	4.71	.84	11.09
8" pipe	U1@.145	LF	9.36	6.33	1.13	16.82
10" pipe	U1@.145	LF	14.40	6.33	1.13	21.86
12" pipe	U1@.145	LF	20.40	6.33	1.13	27.86

Low pressure polyethylene pipe, Standard Dimension Ratio (SDR) 21 to 1

	Craft@Hrs	Unit	Material	Labor	Equipment	Total
4" pipe	U1@.108	LF	2.14	4.71	.84	7.69
5" pipe	U1@.108	LF	3.21	4.71	.84	8.76
6" pipe	U1@.108	LF	4.65	4.71	.84	10.20
7" pipe	U1@.145	LF	5.26	6.33	1.13	12.72
8" pipe	U1@.145	LF	7.76	6.33	1.13	15.22
10" pipe	U1@.145	LF	11.60	6.33	1.13	19.06
12" pipe	U1@.145	LF	16.80	6.33	1.13	24.26
14" pipe	U1@.169	LF	20.40	7.38	1.32	29.10

	Craft@Hrs	Unit	Material	Labor	Equipment	Total
16" pipe	U1@.169	LF	26.50	7.38	1.32	35.20
18" pipe	U1@.169	LF	33.10	7.38	1.32	41.80
20" pipe	U1@.212	LF	41.30	9.25	1.66	52.21
22" pipe	U1@.212	LF	50.00	9.25	1.66	60.91
24" pipe	U1@.212	LF	59.10	9.25	1.66	70.01
Fittings, molded, bell ends						
4" 90 degree ell, 160 PSI	U1@2.00	Ea	50.30	87.30	15.60	153.20
4" 45 degree ell, 160 PSI	U1@2.00	Ea	50.30	87.30	15.60	153.20
6" 90 degree ell, 160 PSI	U1@2.00	Ea	103.00	87.30	15.60	205.90
6" 45 degree ell, 160 PSI	U1@2.00	Ea	103.00	87.30	15.60	205.90
4" to 3" reducer, 160 PSI	U1@2.00	Ea	30.10	87.30	15.60	133.00
6" to 4" reducer, 160 PSI	U1@2.00	Ea	71.90	87.30	15.60	174.80
3" transition section, 160 PSI	U1@2.00	Ea	63.10	87.30	15.60	166.00
6" transition section, 160 PSI	U1@2.00	Ea	279.00	87.30	15.60	381.90
10" by 4" branch saddle, 160 PSI	U1@3.00	Ea	91.10	131.00	23.50	245.60
3" flanged adapter, low pressure	U1@1.00	Ea	47.50	43.60	7.82	98.92
4" flanged adapter, low pressure	U1@1.00	Ea	67.50	43.60	7.82	118.92
6" flanged adapter, low pressure	U1@1.00	Ea	88.70	43.60	7.82	140.12
8" flanged adapter, low pressure	U1@1.50	Ea	128.00	65.50	11.70	205.20
10" flanged adapter, low pressure	U1@1.50	Ea	181.00	65.50	11.70	258.20
4" flanged adapter, 160 PSI	U1@1.00	Ea	73.60	43.60	7.82	125.02
6" flanged adapter, 160 PSI	U1@1.00	Ea	104.00	43.60	7.82	155.42
Fittings, fabricated, belled ends						
10" 90 degree ell, 100 PSI	U1@3.00	Ea	520.00	131.00	23.50	674.50
10" 45 degree ell, 100 PSI	U1@3.00	Ea	289.00	131.00	23.50	443.50
4" 30 degree ell, 100 PSI	U1@2.00	Ea	84.00	87.30	15.60	186.90
8" to 6" reducer, 100 PSI	U1@3.00	Ea	154.00	131.00	23.50	308.50
10" to 8" reducer, 100 PSI	U1@3.00	Ea	181.00	131.00	23.50	335.50
4" 45 degree wye, 100 PSI	U1@2.00	Ea	224.00	87.30	15.60	326.90

Polyethylene Rigid Drain Pipe Standard 10' lengths, 1500 lb crush rating, plain or perforated and high impact ABS glued-on fittings. Installed in an open trench. No excavation, shoring, backfill or dewatering included.

	Craft@Hrs	Unit	Material	Labor	Equipment	Total
3" pipe	CL@.070	LF	.46	2.55	—	3.01
4" pipe	CL@.073	LF	.82	2.66	—	3.48
6" pipe	CL@.079	LF	1.55	2.87	—	4.42
Elbows fittings (1/4 bend)						
3" elbow	CL@.168	Ea	2.02	6.11	—	8.13
4" elbow	CL@.250	Ea	2.90	9.10	—	12.00
6" elbow	CL@.332	Ea	11.90	12.10	—	24.00
Tee fittings						
3" elbow	CL@.251	Ea	2.29	9.13	—	11.42
4" elbow	CL@.375	Ea	3.05	13.60	—	16.65
6" elbow	CL@.500	Ea	13.30	18.20	—	31.50

Polyethylene Flexible Drainage Tubing Corrugated drainage tubing, plain or perforated and snap-on ABS fittings. Installed in an open trench. Excavation, bedding material and backfill not included.

	Craft@Hrs	Unit	Material	Labor	Equipment	Total
3" (10' length)	CL@.009	LF	.43	.33	—	.76
3" (100' coil)	CL@.009	LF	.32	.33	—	.65
4" (10' length)	CL@.011	LF	.39	.40	—	.79
4" (100' coil)	CL@.011	LF	.33	.40	—	.73
6" (100' coil)	CL@.014	LF	.99	.51	—	1.50
8" (20' or 40' length)	CL@.017	LF	2.02	.62	—	2.64
10" (20' length)	CL@.021	LF	4.43	.76	—	5.19
12" (20' length)	CL@.022	LF	6.30	.80	—	7.10

	Craft@Hrs	Unit	Material	Labor	Equipment	Total
Polyethylene tube snap fittings, elbows or tees						
3" fitting	CL@.025	Ea	4.10	.91	—	5.01
4" fitting	CL@.025	Ea	4.90	.91	—	5.81
6" fitting	CL@.025	Ea	8.15	.91	—	9.06
Filter sock sieve. Keeps soil out of perforated pipe. Reduces need for aggregate filters. By pipe size. Per linear foot of sock						
3", 10' length	CL@.006	LF	.55	.22	—	.77
4", 10' length	CL@.007	LF	.63	.25	—	.88
4", 100' length	CL@.007	LF	.22	.25	—	.47

Hi-Q drain pipe Corrugated high-density polyethylene drainage tubing and culvert pipe installed in an open trench. Excavation, bedding material and backfill not included. 20' lengths

	Craft@Hrs	Unit	Material	Labor	Equipment	Total
4"	CL@.009	LF	.60	.33	—	.93
6"	CL@.014	LF	1.43	.51	—	1.94
12" culvert	CL@.022	LF	5.80	.80	—	6.60
15" culvert	CL@.026	LF	7.25	.95	—	8.20
18" culvert	CL@.045	LF	13.30	1.64	—	14.94
24" culvert	CL@.090	LF	25.10	3.28	—	28.38

Corrugated drain pipe fittings

	Craft@Hrs	Unit	Material	Labor	Equipment	Total
4" snap adapter	CL@.025	Ea	2.08	.91	—	2.99
6" snap adapter	CL@.025	Ea	4.18	.91	—	5.09
6" x 4" reducer	CL@.025	Ea	4.18	.91	—	5.09
6" snap coupling	CL@.025	Ea	3.62	.91	—	4.53
6" wye	CL@.037	Ea	10.50	1.35	—	11.85
6" tee	CL@.037	Ea	9.39	1.35	—	10.74
6" blind tee	CL@.037	Ea	7.64	1.35	—	8.99
6" split end cap	CL@.025	Ea	3.13	.91	—	4.04
12" split coupling	CL@.050	Ea	10.50	1.82	—	12.32
15" split coupling	CL@.090	Ea	13.60	3.28	—	16.88

Cast Iron Flanged Pipe AWWA C151 pipe with cast iron flanges on both ends. Pipe is laid and connected in an open trench. Equipment includes a wheel mounted 1 CY backhoe for lifting and placing the pipe and fittings. Trench excavation, dewatering, backfill and compaction are not included. Use $3,000 as a minimum job charge. See next section for fittings, valves and accessories.

	Craft@Hrs	Unit	Material	Labor	Equipment	Total
4" pipe, 260 pounds per 18' section	U1@.057	LF	10.90	2.49	.49	13.88
6" pipe, 405 pounds per 18' section	U1@.075	LF	12.80	3.27	.65	16.72
8" pipe, 570 pounds per 18' section	U1@.075	LF	18.20	3.27	.65	22.12
10" pipe, 740 pounds per 18' section	U1@.094	LF	24.80	4.10	.81	29.71
12" pipe, 930 pounds per 18' section	U1@.094	LF	33.60	4.10	.81	38.51

Cast Iron Fittings, Valves and Accessories Based on type ASA A21.10-A21-11-6 and 4C110-64, C111-64 fittings. Installed in an open trench and connected. Equipment includes a wheel mounted 1 CY backhoe for lifting and placing. Trench excavation, dewatering, backfill and compaction are not included. These items are also suitable for use with the Ductile Iron Pipe in the section that follows. Costs estimated using this section may be combined with the costs estimated from the Cast Iron Flanged Pipe section or the Ductile Iron Pipe section for arriving at the minimum job charge shown in those sections.

Cast iron mechanical joint 90 degree ells

	Craft@Hrs	Unit	Material	Labor	Equipment	Total
4" ell	U1@.808	Ea	81.60	35.30	6.96	123.86
6" ell	U1@1.06	Ea	124.00	46.30	9.13	179.43
8" ell	U1@1.06	Ea	179.00	46.30	9.13	234.43
10" ell	U1@1.32	Ea	270.00	57.60	11.40	339.00
12" ell	U1@1.32	Ea	337.00	57.60	11.40	406.00
14" ell	U1@1.32	Ea	773.00	57.60	11.40	842.00
16" ell	U1@1.65	Ea	884.00	72.00	14.20	970.20
Deduct for 1/8 or 1/16 bends	—	%	-20.0	—	—	—

	Craft@Hrs	Unit	Material	Labor	Equipment	Total
Cast iron mechanical joint tees						
4" x 4"	U1@1.22	Ea	127.00	53.20	10.50	190.70
6" x 4"	U1@1.59	Ea	161.00	69.40	13.70	244.10
6" x 6"	U1@1.59	Ea	176.00	69.40	13.70	259.10
8" x 8"	U1@1.59	Ea	253.00	69.40	13.70	336.10
10" x 10"	U1@1.59	Ea	367.00	69.40	13.70	450.10
12" x 12"	U1@2.48	Ea	477.00	108.00	21.40	606.40
Add for wyes	—	%	40.0	—	—	—
Add for crosses	—	%	30.0	33.0	33.0	—
Cast iron mechanical joint reducers						
6" x 4"	U1@1.06	Ea	93.40	46.30	9.13	148.83
8" x 6"	U1@1.06	Ea	137.00	46.30	9.13	192.43
10" x 8"	U1@1.32	Ea	180.00	57.60	11.40	249.00
12" x 10"	U1@1.32	Ea	243.00	57.60	11.40	312.00
AWWA mechanical joint gate valves						
3" valve	U1@3.10	Ea	310.00	135.00	26.70	471.70
4" valve	U1@4.04	Ea	438.00	176.00	34.80	648.80
6" valve	U1@4.80	Ea	697.00	209.00	41.30	947.30
8" valve	U1@5.14	Ea	1,190.00	224.00	44.30	1,458.30
10" valve	U1@5.67	Ea	1,860.00	247.00	48.80	2,155.80
12" valve	U1@6.19	Ea	3,160.00	270.00	53.30	3,483.30
Indicator post valve						
6" upright, adjustable	U1@6.50	Ea	581.00	284.00	56.00	921.00
Fire hydrant with spool						
6"	U1@5.00	Ea	983.00	218.00	43.10	1,244.10
Backflow preventers, installed at ground level						
3" backflow preventer	U1@3.50	Ea	1,610.00	153.00	30.10	1,793.10
4" backflow preventer	U1@4.00	Ea	2,180.00	175.00	34.50	2,389.50
6" backflow preventer	U1@5.00	Ea	2,480.00	218.00	43.10	2,741.10
8" backflow preventer	U1@6.00	Ea	5,500.00	262.00	51.70	5,813.70
Block structure for backflow preventer						
5' x 7' x 3'4"	M1@32.0	Ea	782.00	1,300.00	0.00	2,082.00
Shear gate						
8", with adjustable release handle	U1@2.44	Ea	1,370.00	106.00	21.00	1,497.00
Adjustable length valve boxes, for up to 20" valves						
5' depth	U1@1.75	Ea	83.50	76.40	15.10	175.00
Concrete thrust blocks, minimum formwork						
1/4 CY thrust block	C8@1.20	Ea	56.40	49.20	—	105.60
1/2 CY thrust block	C8@2.47	Ea	113.00	101.00	—	214.00
3/4 CY thrust block	C8@3.80	Ea	169.00	156.00	—	325.00
1 CY thrust block	C8@4.86	Ea	225.00	199.00	—	424.00
Tapping saddles, double strap, iron body, tap size to 2"						
Pipe to 4"	U1@2.38	Ea	31.20	104.00	20.50	155.70
8" pipe	U1@2.89	Ea	40.80	126.00	24.90	191.70
10" pipe	U1@3.15	Ea	41.60	137.00	27.10	205.70
12" pipe	U1@3.52	Ea	55.00	154.00	30.30	239.30
14" pipe, bronze body	U1@3.78	Ea	66.10	165.00	32.60	263.70

	Craft@Hrs	Unit	Material	Labor	Equipment	Total
Tapping valves, mechanical joint						
4" valve	U1@6.93	Ea	333.00	302.00	59.70	694.70
6" valve	U1@8.66	Ea	448.00	378.00	74.60	900.60
8" valve	U1@9.32	Ea	658.00	407.00	80.30	1,145.30
10" valve	U1@10.4	Ea	998.00	454.00	89.60	1,541.60
12" valve	U1@11.5	Ea	1,530.00	502.00	99.04	2,131.04
Labor and equipment tapping pipe, by hole size						
4" pipe	U1@5.90	Ea	—	257.00	50.80	307.80
6" pipe	U1@9.32	Ea	—	407.00	80.30	487.30
8" pipe	U1@11.7	Ea	—	511.00	101.00	612.00
10" pipe	U1@15.5	Ea	—	676.00	134.00	810.00
12" pipe	U1@23.4	Ea	—	1,020.00	203.00	1,223.00

Ductile Iron Pressure Pipe Mechanical joint 150 lb Class 52 ductile iron pipe. Installed in an open trench to 5' deep. Equipment includes a wheel mounted 1 CY backhoe for lifting and placing the pipe. Trench excavation, dewatering, backfill and compaction are not included. Use the Cast Iron Fittings and Valves from the preceding section as required. Use $4,000 as a minimum job charge.

	Craft@Hrs	Unit	Material	Labor	Equipment	Total
4" pipe	U1@.140	LF	11.40	6.11	1.21	18.72
6" pipe	U1@.180	LF	16.50	7.86	1.55	25.91
8" pipe	U1@.200	LF	22.90	8.73	1.72	33.35
10" pipe	U1@.230	LF	33.70	10.00	1.98	45.68
12" pipe	U1@.260	LF	46.20	11.30	2.24	59.74
14" pipe	U1@.300	LF	70.70	13.10	2.58	86.38
16" pipe	U1@.340	LF	90.20	14.80	2.93	107.93

Non-Reinforced Concrete Pipe Smooth wall, standard strength, 3' lengths with tongue and groove mortar joint ends. Installed in an open trench. Equipment includes a wheel mounted 1 CY backhoe backhoe for lifting and placing the pipe. Trench excavation, bedding material and backfill are not included. Use $3,500 as a minimum job charge.

	Craft@Hrs	Unit	Material	Labor	Equipment	Total
6" pipe	C5@.076	LF	2.89	3.09	.65	6.63
8" pipe	C5@.088	LF	4.46	3.58	.76	8.80
10" pipe	C5@.088	LF	5.31	3.58	.76	9.65
12" pipe	C5@.093	LF	7.40	3.78	.80	11.98
15" pipe	C5@.105	LF	9.22	4.27	.90	14.39
18" pipe	C5@.117	LF	11.60	4.75	1.01	17.36
21" pipe	C5@.117	LF	14.80	4.75	1.01	20.56
24" pipe	C5@.130	LF	21.70	5.28	1.12	28.10
30" pipe	C5@.143	LF	37.90	5.81	1.23	44.94
36" pipe	C5@.158	LF	41.10	6.42	1.36	48.88

Non-Reinforced Perforated Concrete Underdrain Pipe Perforated smooth wall, standard strength, 3' lengths with tongue and groove mortar joint ends. Installed in an open trench. Equipment includes a wheel mounted 1 CY backhoe for lifting and placing the pipe. Trench excavation, bedding material and backfill are not included. Use $2,200 as a minimum job charge.

	Craft@Hrs	Unit	Material	Labor	Equipment	Total
6" pipe	C5@.076	LF	3.42	3.09	.65	7.16
8" pipe	C5@.104	LF	5.18	4.23	.90	10.31

Reinforced Concrete Pipe Class III, 1350 "D" load, ASTM C-76, 8' lengths. Tongue and groove mortar joint ends. Installed in an open trench. Equipment cost shown is for a wheel mounted 1 CY backhoe for lifting and placing the pipe. Trench excavation, dewatering, backfill and compaction are not included. Use $3,000 as a minimum job charge.

	Craft@Hrs	Unit	Material	Labor	Equipment	Total
18" pipe	C5@.272	LF	14.800	11.10	2.34	28.24
24" pipe	C5@.365	LF	18.20	14.80	3.14	36.14
30" pipe	C5@.426	LF	25.10	17.30	3.67	46.07

	Craft@Hrs	Unit	Material	Labor	Equipment	Total
36" pipe	C5@.511	LF	35.20	20.80	4.40	60.40
42" pipe	C5@.567	LF	43.20	23.00	4.88	71.08
48" pipe	C5@.623	LF	52.30	25.30	5.37	82.97
54" pipe	C5@.707	LF	62.60	28.70	6.09	97.39
60" pipe	C5@.785	LF	76.10	31.90	6.76	114.76
66" pipe	C5@.914	LF	94.40	37.10	7.87	139.37
72" pipe	C5@.914	LF	109.00	37.10	7.87	153.97
78" pipe	C5@.914	LF	124.00	37.10	7.87	168.97
84" pipe	C5@1.04	LF	145.00	42.30	8.96	196.26
90" pipe	C5@1.04	LF	155.00	42.30	8.96	206.26
96" pipe	C5@1.18	LF	174.00	48.00	10.20	232.20

Reinforced Elliptical Concrete Pipe Class III, 1350 "D", C-502-72, 8' lengths, tongue and groove mortar joint ends. Installed in an open trench. Equipment includes a wheel mounted 1 CY backhoe for lifting and placing the pipe. Trench excavation, dewatering, backfill and compaction are not included. Use $3,500 as a minimum job charge.

	Craft@Hrs	Unit	Material	Labor	Equipment	Total
19" x 30" pipe, 24" pipe equivalent	C5@.356	LF	28.10	14.50	3.07	45.67
24" x 38" pipe, 30" pipe equivalent	C5@.424	LF	37.90	17.20	3.65	58.75
29" x 45" pipe, 36" pipe equivalent	C5@.516	LF	61.10	21.00	4.44	86.54
34" x 53" pipe, 42" pipe equivalent	C5@.581	LF	75.70	23.60	5.00	104.30
38" x 60" pipe, 48" pipe equivalent	C5@.637	LF	90.20	25.90	5.49	121.59
48" x 76" pipe, 60" pipe equivalent	C5@.797	LF	150.00	32.40	6.86	189.26
53" x 83" pipe, 66" pipe equivalent	C5@.923	LF	184.00	37.50	7.95	229.45
58" x 91" pipe, 72" pipe equivalent	C5@.923	LF	218.00	37.50	7.95	263.45

Flared Concrete End Sections For round concrete drain pipe, precast. Equipment includes a wheel mounted 1 CY backhoe for lifting and placing the items. Trench excavation, dewatering, backfill and compaction are not included. Use $2,500 as a minimum job charge.

	Craft@Hrs	Unit	Material	Labor	Equipment	Total
12" opening, 530 lbs	C5@2.41	Ea	274.00	97.90	20.80	392.70
15" opening, 740 lbs	C5@2.57	Ea	323.00	104.00	22.10	449.10
18" opening, 990 lbs	C5@3.45	Ea	360.00	140.00	29.70	529.70
24" opening, 1,520 lbs	C5@3.45	Ea	441.00	140.00	29.70	610.70
30" opening, 2,190 lbs	C5@3.96	Ea	604.00	161.00	34.10	799.10
36" opening, 4,100 lbs	C5@3.96	Ea	765.00	161.00	34.10	960.10
42" opening, 5,380 lbs	C5@4.80	Ea	1,080.00	195.00	41.30	1,316.30
48" opening, 6,550 lbs	C5@5.75	Ea	1,260.00	234.00	49.50	1,543.50
54" opening, 8,000 lbs	C5@7.10	Ea	1,620.00	289.00	61.10	1,970.10

Precast Reinforced Concrete Box Culvert Pipe ASTM C-850 with tongue and groove mortar joint ends. Installed in an open trench. Equipment includes a wheel mounted 1 CY backhoe for lifting and placing the pipe. Trench excavation, dewatering, backfill and compaction are not included. Laying length is 8' for smaller cross sections and 6' or 4' for larger cross sections. Use $8,500 as a minimum job charge.

	Craft@Hrs	Unit	Material	Labor	Equipment	Total
4' x 3'	C5@.273	LF	113.00	11.10	2.35	126.45
5' x 5'	C5@.336	LF	169.00	13.70	2.89	185.59
6' x 4'	C5@.365	LF	196.00	14.80	3.14	213.94
6' x 6'	C5@.425	LF	228.00	17.30	3.66	248.96
7' x 4'	C5@.442	LF	238.00	18.00	3.81	259.81
7' x 7'	C5@.547	LF	293.00	22.20	4.71	319.91
8' x 4'	C5@.579	LF	255.00	23.50	4.99	283.49
8' x 8'	C5@.613	LF	331.00	24.90	5.28	361.18
9' x 6'	C5@.658	LF	355.00	26.70	5.67	387.37
9' x 9'	C5@.773	LF	419.00	31.40	6.66	457.06
10' x 6'	C5@.787	LF	423.00	32.00	6.78	461.78
10' x 10'	C5@.961	LF	518.00	39.10	8.28	565.38

	Craft@Hrs	Unit	Material	Labor	Equipment	Total

Corrugated Metal Pipe, Galvanized 20' lengths, installed in an open trench. Costs include couplers. Equipment includes a wheel mounted 1 CY backhoe for lifting and placing the pipe. Trench excavation, dewatering, backfill and compaction are not included. Use $3,500 as a minimum job charge.

Round pipe

	Craft@Hrs	Unit	Material	Labor	Equipment	Total
8", 16 gauge (.064)	C5@.102	LF	9.22	4.15	.88	14.25
10", 16 gauge (.064)	C5@.102	LF	9.66	4.15	.88	14.69
12", 16 gauge (.064)	C5@.120	LF	10.00	4.88	1.03	15.91
15", 16 gauge (.064)	C5@.120	LF	12.30	4.88	1.03	18.21
18", 16 gauge (.064)	C5@.152	LF	15.00	6.18	1.31	22.49
24", 14 gauge (.079)	C5@.196	LF	23.20	7.97	1.69	32.86
30", 14 gauge (.079)	C5@.196	LF	28.60	7.97	1.69	38.26
36", 14 gauge (.079)	C5@.231	LF	34.80	9.39	1.99	46.18
42", 14 gauge (.079)	C5@.231	LF	40.40	9.39	1.99	51.78
48", 14 gauge (.079)	C5@.272	LF	45.90	11.10	2.34	59.34
60", 12 gauge (.109)	C5@.311	LF	78.30	12.60	2.68	93.58

Oval pipe

	Craft@Hrs	Unit	Material	Labor	Equipment	Total
18" x 11", 16 gauge (.064)	C5@.133	LF	14.70	5.41	1.15	21.26
22" x 13", 16 gauge (.064)	C5@.167	LF	17.50	6.79	1.44	25.73
29" x 18", 14 gauge (.079)	C5@.222	LF	27.40	9.02	1.91	38.33
36" x 22", 14 gauge (.079)	C5@.228	LF	34.00	9.27	1.96	45.23
43" x 27", 14 gauge (.079)	C5@.284	LF	36.60	11.50	2.45	50.55

Flared end sections for oval pipe

	Craft@Hrs	Unit	Material	Labor	Equipment	Total
18" x 11"	C5@1.59	Ea	79.50	64.60	13.70	157.80
22" x 13"	C5@1.75	Ea	94.70	71.10	15.00	180.80
29" x 18"	C5@1.85	Ea	143.00	75.20	16.00	234.20
36" x 22"	C5@2.38	Ea	230.00	96.70	20.50	347.20
43" x 27"	C5@2.85	Ea	380.00	116.00	24.50	520.50

Add for bituminous coating, with paved invert

	Craft@Hrs	Unit	Material	Labor	Equipment	Total
On round pipe	—	%	20.0	—	—	—
On oval pipe or flared ends	—	%	24.0	—	—	—

Deduct for aluminum pipe,

	Craft@Hrs	Unit	Material	Labor	Equipment	Total
round, oval or flared ends	—	%	-12.0	—	—	—

Corrugated Metal Nestable Pipe Split-in-half, type I, 16 gauge. Trench excavation, backfill, compaction and grading are not included. Equipment includes a wheel mounted 1 CY backhoe for lifting and placing the pipe. Use $5,500 as a minimum job charge.

	Craft@Hrs	Unit	Material	Labor	Equipment	Total
12" diameter, galvanized	C5@.056	LF	16.90	2.28	.48	19.66
18" diameter, galvanized	C5@.064	LF	21.80	2.60	.55	24.95
Add for bituminous coating, with paved invert	—	%	20.0	—	—	—

Drainage Tile Vitrified clay drainage tile pipe. Standard weight, 6' joint length with 1 coupler per joint. Installed in an open trench. Couplers consist of a rubber sleeve with stainless steel compression bands. Equipment includes a wheel mounted 1/2 CY backhoe for lifting and placing the pipe. Excavation, bedding material or backfill are not included. Use $2,000 as a minimum job charge.

	Craft@Hrs	Unit	Material	Labor	Equipment	Total
4" tile	S6@.095	LF	2.62	3.88	.99	7.49
6" tile	S6@.119	LF	5.64	4.85	1.24	11.73
8" tile	S6@.143	LF	8.05	5.83	1.49	15.37
10" tile	S6@.165	LF	12.90	6.73	1.72	21.35
Add for covering 4" to 10" tile, 30 lb felt building paper	CL@.048	LF	.10	1.75	—	1.85

	Craft@Hrs	Unit	Material	Labor	Equipment	Total
Elbows (1/4 bends), costs include one coupler						
4" elbow	S6@.286	Ea	17.30	11.70	2.98	31.98
6" elbow	S6@.360	Ea	30.50	14.70	3.75	48.95
8" elbow	S6@.430	Ea	51.80	17.50	4.48	73.78
10" elbow	S6@.494	Ea	89.30	20.20	5.15	114.65
Tees, costs include two couplers						
4" tee	S6@.430	Ea	28.90	17.50	4.48	50.88
6" tee	S6@.540	Ea	49.00	22.00	5.63	76.63
8" tee	S6@.643	Ea	72.10	26.20	6.71	105.01
10" tee	S6@.744	Ea	157.00	30.30	7.76	195.06

	Craft@Hrs	Unit	Material	Labor	Total
Fiberglass Trench Drain System System includes 16 sloped channels, each 6 feet long, 16 channel frames, 64 iron grates, each 18 inches long, 128 grate locking bolts, outlet and end caps. Neutral channels extend the system run beyond 96 feet. Corrosion resistant polyester or vinylester. Channel has 8" internal width and built-in slope of 1%. Installed in an open trench.					
Typical 96' fiberglass drainage system	P6@10.0	Ea	4,420.00	443.00	4,863.00
System components					
6' sloped channel	—	Ea	137.00	—	137.00
6' channel frame	—	Ea	55.00	—	55.00
18" cast iron grate	—	Ea	25.70	—	25.70
6" grate locking bolt	—	Ea	.70	—	.70
6" outlet cap	—	Ea	19.70	—	19.70
6" closed end cap	—	Ea	19.70	—	19.70
Add for 6' neutral channels	—	Ea	138.00	—	138.00

Precast Trench Drain System System includes 30 framed, interlocking tongue and groove connection channels made of precast polyester or vinylester, one meter long each, with integral cast-in metal rail edge, 30 snap-in galvanized steel slotted grates, 24" x 24" catch basin, 60 grate locking bolts, outlet and end caps. Channel has 4" internal width and built-in slope of .6%. All channels have preformed round and oval drillouts for vertical outlet connection with underground piping. Neutral channels extend the system run beyond 30 meters. Channel and grating combination certified to Load Class D (90,000 lbs). Installed in an open trench.

	Craft@Hrs	Unit	Material	Labor	Total
Typical 30 meter galvanized steel drainage system	P6@8.0	Ea	1,960.00	354.00	2,314.00
System components					
1 meter sloped channel with frame	—	Ea	46.80	—	46.80
1 meter galvanized steel grate	—	Ea	10.70	—	10.70
24" x 24" catch basin	—	Ea	75.90	—	75.90
Grate locking bolt	—	Ea	2.50	—	2.50
4" outlet cap	—	Ea	5.60	—	5.60
4" closed end cap	—	Ea	5.60	—	5.60
Add for 1 meter neutral channels	—	Ea	46.80	—	46.80

	Craft@Hrs	Unit	Material	Labor	Equipment	Total
Corrugated Polyethylene Culvert Pipe Heavy duty, 20' lengths. Installed in an open trench. Equipment includes a wheel mounted 1 CY backhoe for lifting and placing the pipe. Trench excavation, dewatering, backfill and compaction are not included. Based on 500' minimum job. Various sizes may be combined for total footage. Couplings are split type that snap onto the pipe. Use $2,500 as a minimum job charge.						
8" diameter pipe	C5@.035	LF	1.72	1.42	.30	3.44
10" diameter pipe	C5@.038	LF	3.81	1.54	.33	5.68
12" diameter pipe	C5@.053	LF	5.41	2.15	.46	8.02
18" diameter pipe	C5@.111	LF	10.90	4.51	.96	16.37
8" coupling	C5@.105	Ea	4.61	4.27	.90	9.78
10" coupling	C5@.114	Ea	4.47	4.63	.98	10.08
12" coupling	C5@.158	Ea	6.95	6.42	1.36	14.73
18" coupling	C5@.331	Ea	14.40	13.50	2.85	30.75

Site Work 2

	Craft@Hrs	Unit	Material	Labor	Equipment	Total

PVC Sewer Pipe 13' lengths, bell and spigot ends with rubber ring gasket in bell. Installed in open trench. Equipment includes a wheel mounted 1 CY backhoe for lifting and placing the pipe. Trench excavation, dewatering, backfill and compaction are not included. Use $2,500 as a minimum job charge.

	Craft@Hrs	Unit	Material	Labor	Equipment	Total
4" pipe	C5@.047	LF	.92	1.91	.40	3.23
6" pipe	C5@.056	LF	1.69	2.28	.48	4.45
8" pipe	C5@.061	LF	3.45	2.48	.53	6.46
10" pipe	C5@.079	LF	5.50	3.21	.68	9.39
12" pipe	C5@.114	LF	7.75	4.63	.98	13.36

PVC Water Pipe Installed in an open trench. Equipment includes a wheel mounted 1 CY backhoe for lifting and placing the pipe. Trench excavation, bedding material and backfill are not included. Use $2,500 as a minimum job charge.

Class 200 PVC cold water pressure pipe

	Craft@Hrs	Unit	Material	Labor	Equipment	Total
1-1/2" pipe	U1@.029	LF	.29	1.27	.25	1.81
2" pipe	U1@.036	LF	.44	1.57	.31	2.32
2-1/2" pipe	U1@.037	LF	.64	1.61	.32	2.57
3" pipe	U1@.065	LF	.93	2.84	.56	4.33
4" pipe	U1@.068	LF	1.55	2.97	.59	5.11
6" pipe	U1@.082	LF	3.38	3.58	.71	7.67

90 degree ells Schedule 40

	Craft@Hrs	Unit	Material	Labor	Equipment	Total
1-1/2" ells	P6@.385	Ea	.93	17.10	—	18.03
2" ells	P6@.489	Ea	1.51	21.70	—	23.21
2-1/2" ells	P6@.724	Ea	4.56	32.10	—	36.66
3" ells	P6@1.32	Ea	5.40	58.50	—	63.90
4" ells	P6@1.57	Ea	9.65	69.50	—	79.15
6" ells	P6@1.79	Ea	31.10	79.30	—	110.40
Deduct for 45 degree ells	—	%	-5.0	—	—	—

Tees

	Craft@Hrs	Unit	Material	Labor	Equipment	Total
1-1/2" tees	P6@.581	Ea	1.24	25.70	—	26.94
2" tees	P6@.724	Ea	1.87	32.10	—	33.97
2-1/2" tees	P6@.951	Ea	5.96	42.10	—	48.06
3" tees	P6@1.57	Ea	7.88	69.50	—	77.38
4" tees	P6@2.09	Ea	14.30	92.60	—	106.90
6" tees	P6@2.70	Ea	47.70	120.00	—	167.70

Vitrified Clay Pipe C-200, extra strength, installed in open trench. 4" through 12" are 6'0" joint length, over 12" are either 7' or 7'6" joint length. Costs shown include one coupler per joint. Couplers consist of a rubber sleeve with stainless steel compression bands. Equipment includes a wheel mounted 1 CY backhoe for lifting and placing the pipe. Trench excavation, dewatering, backfill and compaction are not included. Use $3,500 as a minimum job charge.

	Craft@Hrs	Unit	Material	Labor	Equipment	Total
4" pipe	C5@.081	LF	3.75	3.29	.70	7.74
6" pipe	C5@.175	LF	8.05	7.11	1.51	16.67
8" pipe	C5@.183	LF	11.60	7.44	1.58	20.62
10" pipe	C5@.203	LF	18.50	8.25	1.75	28.50
12" pipe	C5@.225	LF	26.20	9.14	1.94	37.28
15" pipe	C5@.233	LF	24.80	9.47	2.01	36.28
18" pipe	C5@.254	LF	35.80	10.30	2.19	48.29
21" pipe	C5@.283	LF	49.10	11.50	2.44	63.04
24" pipe	C5@.334	LF	65.10	13.60	2.88	81.58
27" pipe	C5@.396	LF	74.30	16.10	3.41	93.81
30" pipe	C5@.452	LF	90.40	18.40	3.89	112.69

	Craft@Hrs	Unit	Material	Labor	Equipment	Total
1/4 bends including couplers						
4" bend	C5@.494	Ea	24.80	20.10	4.25	49.15
6" bend	C5@.524	Ea	43.10	21.30	4.51	68.91
8" bend	C5@.549	Ea	57.70	22.30	4.73	84.73
10" bend	C5@.610	Ea	99.20	24.80	5.25	129.25
12" bend	C5@.677	Ea	127.00	27.50	5.83	160.33
15" bend	C5@.715	Ea	335.00	29.10	6.16	370.26
18" bend	C5@.844	Ea	459.00	34.30	7.27	500.57
21" bend	C5@.993	Ea	625.00	40.40	8.55	673.95
24" bend	C5@1.17	Ea	819.00	47.50	10.10	876.60
30" bend	C5@1.37	Ea	1,170.00	55.70	11.80	1,237.50
Wyes and tees including couplers						
4" wye or tee	C5@.744	Ea	32.00	30.20	6.41	68.61
6" wye or tee	C5@.787	Ea	54.60	32.00	6.78	93.38
8" wye or tee	C5@.823	Ea	80.70	33.40	7.09	121.19
10" wye or tee	C5@.915	Ea	174.00	37.20	7.88	219.08
12" wye or tee	C5@1.02	Ea	229.00	41.50	8.78	279.28
15" wye or tee	C5@1.10	Ea	375.00	44.70	9.47	429.17
18" wye or tee	C5@1.30	Ea	508.00	52.80	11.20	572.00
21" wye or tee	C5@1.53	Ea	666.00	62.20	13.18	741.38
24" wye or tee	C5@1.80	Ea	793.00	73.20	15.80	882.00
27" wye or tee	C5@2.13	Ea	964.00	86.60	18.34	1,068.94
30" wye or tee	C5@2.50	Ea	945.00	102.00	21.50	1,068.50

Sewer Main Cleaning Ball method, typical costs for cleaning new sewer mains prior to inspection, includes operator and equipment. Add $200 to total costs for setting up and removing equipment. Minimum cost will be $500. These costs include the subcontractor's overhead and profit. Typical costs per 1,000 LF of main diameter listed.

		Unit				Total
6" diameter	—	MLF	—	—	—	77.70
8" diameter	—	MLF	—	—	—	100.00
10" diameter	—	MLF	—	—	—	117.00
12" diameter	—	MLF	—	—	—	149.00
14" diameter	—	MLF	—	—	—	178.00
16" diameter	—	MLF	—	—	—	200.00
18" diameter	—	MLF	—	—	—	232.00
24" diameter	—	MLF	—	—	—	307.00
36" diameter	—	MLF	—	—	—	462.00
48" diameter	—	MLF	—	—	—	598.00

Water Main Sterilization Gas chlorination method, typical costs for sterilizing new water mains prior to inspection, includes operator and equipment. Add $200 to total costs for setting up and removing equipment. Minimum cost will be $500. These costs include the subcontractor's overhead and profit. Typical cost per 1,000 LF of main with diameter as shown.

		Unit				Total
3" or 4" diameter	—	MLF	—	—	—	44.30
6" diameter	—	MLF	—	—	—	77.70
8" diameter	—	MLF	—	—	—	105.00
10" diameter	—	MLF	—	—	—	120.00
12" diameter	—	MLF	—	—	—	149.00
14" diameter	—	MLF	—	—	—	177.00
16" diameter	—	MLF	—	—	—	188.00
18" diameter	—	MLF	—	—	—	232.00
24" diameter	—	MLF	—	—	—	307.00
36" diameter	—	MLF	—	—	—	440.00
48" diameter	—	MLF	—	—	—	597.00

	Craft@Hrs	Unit	Material	Labor	Equipment	Total
Soil Covers						
Using staked burlap and tar over straw	CL@.019	SY	.79	.69	—	1.48
Using copolymer-base sprayed-on liquid soil sealant						
On slopes for erosion control	CL@.012	SY	.41	.44	—	.85
On level ground for dust abatement	CL@.012	SY	.36	.44	—	.80

Catch Basins Catch basin including 6" concrete top and base. Equipment includes a wheel mounted 1 CY backhoe for lifting and placing the items. Trench excavation, bedding material and backfill are not included. Use $3,000 as a minimum job charge.

	Craft@Hrs	Unit	Material	Labor	Equipment	Total
4' diameter precast concrete basin						
4' deep	S6@19.0	Ea	1,340.00	775.00	218.00	2,333.00
6' deep	S6@23.7	Ea	1,680.00	967.00	272.00	2,919.00
8' deep	S6@28.9	Ea	2,060.00	1,180.00	332.00	3,572.00
Light duty grate, gray iron, asphalt coated						
Grate on pipe bell						
6" diameter, 13 lb	S6@.205	Ea	11.60	8.36	2.35	22.31
8" diameter, 25 lb	S6@.247	Ea	31.60	10.10	2.84	44.54
Frame and grate						
8" diameter, 55 lb	S6@.424	Ea	67.60	17.30	4.87	89.77
17" diameter, 135 lb	S6@.823	Ea	164.00	33.60	9.45	207.05
Medium duty, frame and grate						
11" diameter, 70 lb	S6@.441	Ea	81.60	18.00	5.06	104.66
15" diameter, 120 lb	S6@.770	Ea	146.00	31.40	8.84	186.24
Radial grate, 20" diameter, 140 lb	S6@1.68	Ea	169.00	68.50	19.30	256.80
Heavy duty frame and grate						
Flat grate						
11-1/2" diameter, 85 lb	S6@.537	Ea	107.00	21.90	6.17	135.07
20" diameter, 235 lb	S6@2.01	Ea	287.00	82.00	32.10	401.10
21" diameter, 315 lb	S6@2.51	Ea	350.00	102.00	28.80	480.80
24" diameter, 350 lb	S6@2.51	Ea	389.00	102.00	28.80	519.80
30" diameter, 555 lb	S6@4.02	Ea	570.00	164.00	46.20	780.20
Convex or concave grate						
20" diameter, 200 lb	S6@1.83	Ea	203.00	74.60	21.00	298.60
20" diameter, 325 lb	S6@2.51	Ea	332.00	102.00	28.80	462.80
Beehive grate and frame						
11" diameter, 80 lb	S6@.495	Ea	146.00	20.20	5.68	171.88
15" diameter, 120 lb	S6@1.83	Ea	140.00	74.60	21.00	235.60
21" diameter, 285 lb	S6@2.51	Ea	293.00	102.00	28.80	423.80
24" diameter, 375 lb	S6@2.87	Ea	406.00	117.00	33.00	556.00

Manholes Equipment includes a wheel mounted 1 CY backhoe for lifting and placing the items. Trench excavation, bedding material and backfill are not included. Use $3,000 as a minimum job charge.

Precast concrete manholes, including concrete top and base. No excavation or backfill included. Add for steps, frames and lids from the costs listed below

	Craft@Hrs	Unit	Material	Labor	Equipment	Total
3' x 6' to 8' deep	S6@17.5	Ea	1,590.00	714.00	201.00	2,505.00
3' x 9' to 12' deep	S6@19.4	Ea	1,840.00	791.00	223.00	2,854.00
3' x 13' to 16' deep	S6@23.1	Ea	2,380.00	942.00	265.00	3,587.00
3' diameter, 4' deep	S6@12.1	Ea	929.00	494.00	139.00	1,562.00
4' diameter, 5' deep	S6@22.8	Ea	1,120.00	930.00	262.00	2,312.00
4' diameter, 6' deep	S6@26.3	Ea	1,270.00	1,070.00	302.00	2,642.00
4' diameter, 7' deep	S6@30.3	Ea	1,390.00	1,240.00	348.00	2,978.00
4' diameter, 8' deep	S6@33.7	Ea	1,560.00	1,370.00	387.00	3,317.00
4' diameter, 9' deep	S6@38.2	Ea	1,640.00	1,560.00	439.00	3,639.00
4' diameter, 10' deep	S6@43.3	Ea	1,800.00	1,770.00	497.00	4,067.00

	Craft@Hrs	Unit	Material	Labor	Equipment	Total
Concrete block radial manholes, 4' inside diameter, no excavation or backfill included						
4' deep	M1@8.01	Ea	376.00	324.00	—	700.00
6' deep	M1@13.4	Ea	554.00	543.00	—	1,097.00
8' deep	M1@20.0	Ea	796.00	810.00	—	1,606.00
10' deep	M1@26.7	Ea	817.00	1,080.00	—	1,897.00
Depth over 10', add per LF	M1@2.29	LF	105.00	92.70	—	197.70
2' depth cone block for 30" grate	M1@5.34	LF	158.00	216.00	—	374.00
2'6" depth cone block for 24" grate	M1@6.43	LF	204.00	260.00	—	464.00
Manhole steps, cast iron, heavy type, asphalt coated						
10" x 14-1/2" in job-built manhole	M1@.254	Ea	17.00	10.30	—	27.30
12" x 10" in precast manhole	—	Ea	15.60	—	—	15.60
Gray iron manhole frames, asphalt coated, standard sizes						
Light duty frame and lid						
15" diameter, 65 lb	S6@.660	Ea	89.00	26.90	7.58	123.48
22" diameter, 140 lb	S6@1.42	Ea	195.00	57.90	16.30	269.20
Medium duty frame and lid						
11" diameter, 75 lb	S6@1.51	Ea	104.00	61.60	17.30	182.90
20" diameter, 185 lb	S6@3.01	Ea	239.00	123.00	34.60	396.60
Heavy duty frame and lid						
17" diameter, 135 lb	S6@2.51	Ea	182.00	102.00	28.80	312.80
21" diameter, 315 lb	S6@3.77	Ea	405.00	154.00	43.30	602.30
24" diameter, 375 lb	S6@4.30	Ea	468.00	175.00	49.40	692.40
Connect new drain line to existing manhole, no excavation or backfill included. Typical cost						
Per connection	S6@6.83	Ea	127.00	279.00	78.40	484.40
Connect existing drain line to new manhole, no excavation or backfill included. Typical cost						
Per connection	S6@3.54	Ea	77.10	144.00	40.70	261.80
Manhole repairs and alterations, typical costs						
Repair manhole leak with grout	S6@15.8	Ea	389.00	644.00	181.00	1,214.00
Repair inlet leak with grout	S6@15.8	Ea	459.00	644.00	181.00	1,284.00
Grout under manhole frame	S6@1.29	Ea	7.76	52.60	14.80	75.16
Drill and grout pump setup cost	S6@17.4	LS	459.00	710.00	200.00	1,369.00
Grout for pressure grouting	—	Gal	7.33	—	—	7.33
Replace brick in manhole wall	M1@8.79	SF	11.20	356.00	—	367.20
Replace brick under manhole frame	M1@8.79	LS	37.40	356.00	—	393.40
Raise existing frame and cover 2"	CL@4.76	LS	207.00	173.00	—	380.00
Raise existing frame and cover more than 2", per each 1" added	CL@.250	LS	71.50	9.10	—	80.60
TV inspection of pipe interior, subcontract	—	LF	—	—	—	3.50
Grouting concrete pipe joints						
6" to 30" diameter	S6@.315	Ea	21.20	12.80	3.62	37.62
33" to 60" diameter	S6@1.12	Ea	41.80	45.70	12.90	100.40
66" to 72" diameter	S6@2.05	Ea	84.40	83.60	23.50	191.50

Accessories for Site Utilities Equipment includes a wheel mounted 1 CY backhoe for lifting and placing the items. Trench excavation, bedding material and backfill are not included. Costs estimated using this section may be combined with costs estimated from other site work utility sections for arriving at a minimum job charge.

	Craft@Hrs	Unit	Material	Labor	Equipment	Total
Curb inlets, gray iron, asphalt coated, heavy duty frame, grate and curb box						
20" x 11", 260 lb	S6@2.75	Ea	323.00	112.00	31.60	466.60
20" x 16.5", 300 lb	S6@3.04	Ea	352.00	124.00	34.90	510.90
20" x 17", 400 lb	S6@4.30	Ea	445.00	175.00	49.40	669.40
19" x 18", 500 lb	S6@5.06	Ea	533.00	206.00	58.10	797.10
30" x 17", 600 lb	S6@6.04	Ea	576.00	246.00	69.40	891.40

	Craft@Hrs	Unit	Material	Labor	Equipment	Total
Gutter inlets, gray iron, asphalt coated, heavy duty frame and grate						
8" x 11.5", 85 lb	S6@.444	Ea	109.00	18.10	5.10	132.20
22" x 17", 260 lb, concave	S6@1.84	Ea	309.00	75.10	21.10	405.20
22.3" x 22.3", 475 lb	S6@2.87	Ea	334.00	117.00	33.00	484.00
29.8" x 17.8", 750 lb	S6@3.37	Ea	513.00	137.00	38.70	688.70
Trench inlets, ductile iron, light duty frame and grate for pedestrian traffic						
53.5" x 8.3", 100 lb	S6@.495	Ea	108.00	20.20	5.68	133.88
Trench inlets, gray iron, asphalt coated frame and grate or solid cover						
Light duty, 1-1/4"						
8" wide grate	S6@.247	LF	44.50	10.10	2.84	57.44
12" wide grate	S6@.275	LF	64.30	11.20	3.16	78.66
8" wide solid cover	S6@.247	LF	49.40	10.10	2.85	62.35
12" wide solid cover	S6@.275	LF	70.50	11.20	3.16	84.86
Heavy duty, 1-3/4"						
8" wide grate	S6@.275	LF	48.30	11.20	3.16	62.66
12" wide grate	S6@.309	LF	71.80	12.60	3.55	87.95
8" wide solid cover	S6@.275	LF	55.70	11.20	3.16	70.06
12" wide solid cover	S6@.309	LF	76.70	12.60	3.55	92.85

Asphalt Paving To arrive at the total cost for new asphalt paving jobs, add the cost of the subbase preparation and aggregate to the costs for the asphalt binder course, the rubberized interliner and the wear course. Cost for site preparation is not included. These costs are intended to be used for on-site roadways and parking lots. Use $7,500 as a minimum charge.

Subbase and aggregate Equipment includes a pneumatic tired articulated 10,000 pound motor grader, a 4 ton two axle steel drum roller and a 3,500 gallon water truck.

	Craft@Hrs	Unit	Material	Labor	Equipment	Total
Fine grading and compacting existing subbase						
To within plus or minus 1/10'	P5@.066	SY	.01	2.63	2.12	4.76
Stone aggregate base, Class 2. Includes grading and compacting of base material						
Class 2 stone aggregate base, per CY	—	CY	25.10	—	—	25.10
Class 2 stone aggregate base, per Ton	—	Ton	17.90	—	—	17.90
1" thick base (36 SY per CY)	P5@.008	SY	.70	.32	.26	1.28
4" thick base (9 SY per CY)	P5@.020	SY	2.78	.80	.64	4.22
6" thick base (6 SY per CY)	P5@.033	SY	4.18	1.31	1.06	6.55
8" thick base (4.5 SY per CY)	P5@.039	SY	5.57	1.55	1.25	8.37
10" thick base (3.6 SY per CY)	P5@.042	SY	6.96	1.67	1.35	9.98
12" thick base (3 SY per CY)	P5@.048	SY	8.35	1.91	1.54	11.80

Asphalt materials Equipment includes a 10' wide self-propelled paving machine and a 4 ton two axle steel drum roller.

	Craft@Hrs	Unit	Material	Labor	Equipment	Total
Binder course, spread and rolled						
Binder course, per ton	—	Ton	57.80	—	—	57.80
1" thick (18 SY per ton)	P5@.028	SY	3.19	1.11	.95	5.25
1-1/2" thick (12 SY per ton)	P5@.036	SY	5.32	1.43	1.37	8.12
2" thick (9 SY per ton)	P5@.039	SY	7.12	1.55	1.48	10.15
3" thick (6 SY per ton)	P5@.050	SY	10.70	1.99	1.90	14.59
Rubberized asphalt interlayer with tack coat. Based on using 10' wide rolls 3/16" thick x 500' long and one coat of mopped-on bituminous tack coat, MC70 (0.15 gallon per SY) before allowance for waste						
Rubberized asphalt inner layer, per 5,000 SF roll	—	Roll	178.00	—	—	178.00
Bituminous tack coat, per gallon	—	Gal	2.53	—	—	2.53
Laid, cut, mopped and rolled	P5@.005	SY	.70	.20	.19	1.09

	Craft@Hrs	Unit	Material	Labor	Equipment	Total
Wear course						
Bituminous wear course, per ton	—	Ton	61.50	—	—	61.50
1" thick (18 SY per ton)	P5@.028	SY	3.39	1.11	1.06	5.56
1-1/2" thick (12 SY per ton)	P5@.036	SY	5.09	1.43	1.37	7.89
2" thick (9 SY per ton)	P5@.039	SY	6.82	1.55	1.48	9.85
3" thick (6 SY per ton)	P5@.050	SY	10.20	1.99	1.90	14.09
Add to total installed cost of paving as described, if required						
Bituminous prime coat	P5@.001	SY	.63	.04	.04	.71
Tar emulsion protective seal coat	P5@.034	SY	2.17	1.35	1.29	4.81
Asphalt slurry seal	P5@.043	SY	2.75	1.71	1.63	6.09
Asphaltic concrete curb, straight, placed during paving. Equipment includes an asphalt curbing machine						
8" x 6", placed by hand	P5@.141	LF	1.42	5.61	—	7.03
8" x 6", machine extruded	P5@.067	LF	1.30	2.67	.49	4.46
8" x 8", placed by hand	P5@.141	LF	2.27	5.61	—	7.88
8" x 8", machine extruded	P5@.068	LF	1.73	2.71	.24	4.68
Add for curved sections	P5@.034	LF	—	1.35	.12	1.47
Asphaltic concrete speed bumps, 3" high, 24" wide (135 LF and 2 tons per CY), placed on parking lot surface and rolled						
100 LF job	P5@.198	LF	.92	7.88	1.91	10.71
Painting speed bumps, 24" wide, white	P5@.022	LF	.34	.88	—	1.22

See also the section on Speed Bumps and Parking Blocks following Pavement Striping and Marking.

Asphalt Paving for Roadways, Assembly Costs On-site roadway, with 4" thick aggregate base, 2" thick asphalt binder course, rubberized asphalt interliner, and 2" thick wear course. This is often called 4" paving over 4" aggregate base.

	Craft@Hrs	Unit	Material	Labor	Equipment	Total
Fine grading and compact existing subbase	P5@.066	SY	.01	2.63	2.50	5.14
Stone aggregate base, Class 2, 4" thick	P5@.020	SY	2.78	.80	.76	4.34
Binder course 2" thick (9 SY per ton)	P5@.039	SY	7.12	1.55	1.48	10.15
Rubberized asphalt interlayer	P5@.005	SY	.38	.20	.19	.77
Wear course 2" thick (9 SY per ton)	P5@.039	SY	7.12	1.55	1.48	10.15
Tar emulsion protective seal coat	P5@.034	SY	2.17	1.35	1.29	4.81
Total estimated cost for typical 10' wide roadway, 4" paving over 4" aggregate base						
Cost per SY of roadway	P5@.203	SY	19.58	8.08	7.70	35.36
Cost per SF roadway	P5@.023	SF	2.18	.92	.87	3.97
Cost per LF, 10' wide roadway	P5@.226	LF	21.70	8.99	7.58	38.27

Asphalt Paving for Parking Areas, Assembly Costs Equipment includes a 10' wide self-propelled paving machine and a 4 ton two axle steel drum roller. On-site parking lot with 6" thick aggregate base, 3" thick asphalt binder course, rubberized asphalt interliner, and 3" thick wear course. This is often called 6" paving over 6" aggregate base.

	Craft@Hrs	Unit	Material	Labor	Equipment	Total
Fine grading and compact existing subbase	P5@.066	SY	.01	2.63	2.50	5.14
Stone aggregate base, Class 2, 6" thick	P5@.033	SY	4.18	1.31	1.25	6.74
Binder course 3" thick	P5@.050	SY	10.70	1.99	1.90	14.59
Rubberized asphalt interlayer	P5@.005	SY	.38	.20	.19	.77
Wear course 3" thick	P5@.050	SY	10.70	1.99	1.90	14.59
Tar emulsion protective seal coat	P5@.034	SY	2.17	1.35	1.29	4.81
Total estimated cost of typical 1,000 SY parking lot, 6" paving over 6" aggregate base						
Cost per SY of parking lot	P5@.238	SY	28.14	9.47	9.03	46.64
Cost per SF of parking lot	P5@.026	SF	3.13	1.03	.99	5.15

	Craft@Hrs	Unit	Material	Labor	Equipment	Total

Cold Milling of Asphalt Surface to 1" Depth Equipment includes a motorized street broom, a 20" pavement miller, two 5 CY dump trucks and a pickup truck. Hauling and disposal not included. Use $3,000 as a minimum charge.

	Craft@Hrs	Unit	Material	Labor	Equipment	Total
Milling asphalt to 1" depth	C6@.017	SY	—	.68	.60	1.28

Asphalt Pavement Repairs to conform to existing street, strips of 100 to 500 SF. Equipment includes one wheel-mounted 150 CFM air compressor, paving breaker, jackhammer bits, one 55 HP wheel loader with integral backhoe, one medium weight pneumatic tamper, one 1 ton two axle steel roller and one 5 CY dump truck. These costs include the cost of loading and hauling to a legal dump within 6 miles. Dump fees are not included. See Waste Disposal.

Asphalt pavement repairs, assembly costs Use $3,600 as a minimum job charge.

	Craft@Hrs	Unit	Material	Labor	Equipment	Total
Cut and remove 2" asphalt and 36" subbase	P5@.250	SY	—	9.95	5.68	15.63
Place 30" fill compacted to 90%, 6" aggregate base, 2" thick asphalt and tack coat						
30" fill (30 SY per CY at $16.90 CY)	P5@.099	SY	.59	3.94	2.25	6.78
6" thick base (6 SY per CY)	P5@.033	SY	4.18	1.31	.75	6.24
2" thick asphalt (9 SY per ton)	P5@.039	SY	7.12	1.55	.89	9.56
Tar emulsion protective seal coat	P5@.034	SY	2.17	1.35	.77	4.29
Total per SY	**--@.455**	**SY**	**14.06**	**20.60**	**10.34**	**45.00**

Asphalt overlay on existing asphalt pavement. Equipment includes a 10' wide self-propelled paving machine and a 4 ton two axle steel drum roller. Costs shown assume unobstructed access to area receiving overlay. Use $6,000 as a minimum job charge.

Rubberized asphalt interlayer with tack coat. Based on using 10' wide rolls 3/16" thick x 500' long and one coat of mopped-on bituminous tack coat, MC70 (with .15 gallon per SY) before allowance for waste

	Craft@Hrs	Unit	Material	Labor	Equipment	Total
Rubberized asphalt inner layer, per 5,000 SF roll	—	Roll	178.00	—	—	178.00
Bituminous tack coat, per gallon	—	Gal	2.53	—	—	2.53
Laid, cut, mopped and rolled	P5@.005	SY	.75	.20	.19	1.14
Wear course						
Bituminous wear course, per ton	—	Ton	64.20	—	—	64.20
1" thick (18 SY per ton)	P5@.028	SY	3.54	1.11	1.06	5.71
1-1/2" thick (12 SY per ton)	P5@.036	SY	5.32	1.43	1.37	8.12
2" thick (9 SY per ton)	P5@.039	SY	7.12	1.55	1.48	10.15
Add to overlaid paving as described above, if required						
Bituminous prime coat, MC70	P5@.001	SY	.60	.04	.04	.68
Tar emulsion protective seal coat	P5@.034	SY	2.21	1.35	1.29	4.85
Asphalt slurry seal	P5@.043	SY	2.78	1.71	1.63	6.12

Concrete Paving No grading or base preparation included.

	Craft@Hrs	Unit	Material	Labor	Equipment	Total
Steel edge forms to 15", rented	C8@.028	LF	—	1.15	.69	1.84

2,000 PSI concrete placed directly from the chute of a ready-mix truck, vibrated and broom finished. Costs include 4% allowance for waste but no forms, reinforcing, or joints.

	Craft@Hrs	Unit	Material	Labor	Equipment	Total
4" thick (80 SF per CY)	P8@.013	SF	1.19	.52	.02	1.73
6" thick (54 SF per CY)	P8@.016	SF	1.79	.64	.02	2.45
8" thick (40 SF per CY)	P8@.019	SF	2.39	.76	.02	3.17
10" thick (32 SF per CY)	P8@.021	SF	2.99	.84	.03	3.86
12" thick (27 SF per CY)	P8@.024	SF	3.59	.96	.03	4.58
15" thick (21 SF per CY)	P8@.028	SF	4.56	1.12	.03	5.71

	Craft@Hrs	Unit	Material	Labor	Equipment	Total

Clean and Sweep Pavement Equipment cost is $53.00 per hour for a self-propelled vacuum street sweeper, 7 CY maximum capacity. Use $300 as a minimum charge for sweeping using this equipment,

	Craft@Hrs	Unit	Material	Labor	Equipment	Total
Pavement sweeping, using equipment	TO@.200	MSF	—	9.92	16.58	26.50
Clean and sweep pavement, by hand	CL@1.44	MSF	—	52.40	—	52.40

Joint Treatment for Concrete Pavement Seal joint in concrete pavement with ASTM D-3569 PVC coal tar including 5% waste). Cost per linear foot (LF), by joint size

	Craft@Hrs	Unit	Material	Labor	Equipment	Total
PVC coal tar, per gallon	—	Gal	20.00	—	—	20.00
3/8" x 1/2" (128 LF per gallon)	PA@.002	LF	.17	.09	—	.26
1/2" x 1/2" (96 LF per gallon)	PA@.002	LF	.22	.09	—	.31
1/2" x 3/4" (68 LF per gallon)	PA@.003	LF	.30	.14	—	.44
1/2" x 1" (50 LF per gallon)	PA@.003	LF	.42	.14	—	.56
1/2" x 1-1/4" (40 LF per gallon)	PA@.004	LF	.53	.18	—	.71
1/2" x 1-1/2" (33 LF per gallon)	PA@.006	LF	.64	.27	—	.91
3/4" x 3/4" (46 LF per gallon)	PA@.004	LF	.45	.18	—	.63
3/4" x 1-1/4" (33 LF per gallon)	PA@.006	LF	.64	.27	—	.91
3/4" x 1-1/2" (27 LF per gallon)	PA@.007	LF	.78	.32	—	1.10
3/4" x 2" (22 LF per gallon)	PA@.008	LF	.96	.36	—	1.32
1" x 1" (25 LF per gallon)	PA@.008	LF	.84	.36	—	1.20

Backer rod for joints in concrete pavement

	Craft@Hrs	Unit	Material	Labor	Equipment	Total
1/2" backer rod	PA@.002	LF	.03	.09	—	.12
3/4" or 1" backer rod	PA@.002	LF	.05	.09	—	.14

Clean out old joint sealer in concrete pavement by sandblasting and reseal with PVC coal tar sealer, including new backer rod. Use $400 as a minimum job charge

	Craft@Hrs	Unit	Material	Labor	Equipment	Total
1/2" x 1/2"	PA@.009	LF	.21	.41	.23	.85
1/2" x 3/4"	PA@.013	LF	.30	.59	.33	1.22
1/2" x 1"	PA@.013	LF	.40	.59	.33	1.32
1/2" x 1-1/4"	PA@.018	LF	.49	.82	.46	1.77
1/2" x 1-1/2"	PA@.022	LF	.59	1.00	.56	2.15

Pavement Striping and Marking Use $150 as a minimum job charge. Pavement line markings

	Craft@Hrs	Unit	Material	Labor	Equipment	Total
Single line striping, 4" wide solid	PA@.005	LF	.07	.23	.05	.35
Single line striping, 4" wide skip	PA@.005	LF	.04	.23	.05	.32
Red curb painting	PA@.006	LF	.18	.27	.06	.51

Parking lot spaces. For estimating purposes figure one space per 300 SF of parking pavement area

	Craft@Hrs	Unit	Material	Labor	Equipment	Total
Single line striping, approx 25 LF per space	PA@.115	Ea	1.76	5.22	1.07	8.05
Dual line striping, approx 45 LF per space	PA@.175	Ea	3.16	7.94	1.63	12.73

Traffic symbols

	Craft@Hrs	Unit	Material	Labor	Equipment	Total
Arrows	PA@.250	Ea	6.55	11.30	2.33	20.18
Lettering, 2' x 8' template	PA@.500	Ea	42.00	22.70	4.66	69.36
Handicapped symbol						
One color	PA@.250	Ea	5.32	11.30	2.33	18.95
Two color	PA@.500	Ea	6.84	22.70	4.66	34.20

	Craft@Hrs	Unit	Material	Labor	Total

Speed Bumps and Parking Blocks Prefabricated solid plastic, pre-drilled. Based on Super Saver Bumps or Super Saver Blocks as manufactured by The Parking Block Store (http://www.parkingblock.com) Costs shown include installation hardware. Labor includes drilling holes in substrate for anchors. Available in yellow, gray, blue, white.

	Craft@Hrs	Unit	Material	Labor	Total
Speed bumps, based on quantity of 16 bumps					
6' x 2" x 10" (25 lbs each)	CL@.750	Ea	84.00	27.30	111.30
8' x 2" x 12" (45 lbs each)	CL@1.00	Ea	100.00	36.40	136.40

	Craft@Hrs	Unit	Material	Labor	Total
Parking blocks, based on quantity of 25 blocks					
6' x 4" x 6" (16 lbs each)	CL@.750	Ea	46.00	27.30	73.30

Traffic Signs Reflectorized steel, with a high strength U-channel galvanized steel pipe post 10' long set 2' into the ground. Includes digging of hole with a manual auger and backfill.

	Craft@Hrs	Unit	Material	Labor	Total
Stop, 24" x 24"	CL@1.25	Ea	130.00	45.50	175.50
Speed limit, Exit, etc, 18" x 24"	CL@1.25	Ea	115.00	45.50	160.50
Warning, 24" x 24"	CL@1.25	Ea	118.00	45.50	163.50
Yield, 30" triangle	CL@1.25	Ea	140.00	45.50	185.50
Add for breakaway galvanized square sign post	—	Ea	35.00	—	35.00

Brick Paving Extruded hard red brick. These costs do not include site preparation. Material costs shown include 2% for waste. (Figures in parentheses show quantity of brick before allowance for waste.) For expanded coverage of brick paving, see the *National Concrete & Masonry Estimator*, http://www.CraftsmanSiteLicense.com

	Craft@Hrs	Unit	Material	Labor	Total
Brick laid flat without mortar					
Using brick, per thousand	—	M	357.00	—	357.00
Running bond					
4" x 8" x 2-1/4" (4.5 per SF)	M1@.114	SF	1.61	4.62	6.23
4" x 8" x 1-5/8" (4.5 per SF)	M1@.112	SF	1.58	4.54	6.12
3-5/8" x 7-5/8" x 2-1/4" (5.2 per SF)	M1@.130	SF	1.83	5.27	7.10
3-5/8" x 7-5/8" x 1-5/8" (5.2 per SF)	M1@.127	SF	1.83	5.14	6.97
Add for herringbone pattern	M1@.033	SF	—	1.34	1.34
Basketweave					
4" x 8" x 2-1/4" (4.5 per SF)	M1@.133	SF	1.64	5.39	7.03
4" x 8" x 1-5/8" (4.5 per SF)	M1@.130	SF	1.58	5.27	6.85
Brick laid on edge without mortar					
Running bond					
4" x 8" x 2-1/4" (8 per SF)	M1@.251	SF	2.91	10.20	13.11
3-5/8" x 7-5/8" x 2-1/4" (8.4 per SF)	M1@.265	SF	2.96	10.70	13.66
Brick laid flat with 3/8" mortar joints and mortar setting bed. Mortar included in material cost.					
Running bond					
4" x 8" x 2-1/4" (3.9 per SF)	M1@.127	SF	1.97	5.14	7.11
4" x 8" x 1-5/8" (3.9 per SF)	M1@.124	SF	1.92	5.02	6.94
3-5/8" x 7-5/8" x 2-1/4" (4.5 per SF)	M1@.144	SF	2.19	5.83	8.02
3-5/8" x 7-5/8" x 1-5/8" (4.5 per SF)	M1@.140	SF	2.13	5.67	7.80
Add for herringbone pattern	M1@.033	SF	—	1.34	1.34
Basketweave					
4" x 8" x 2-1/4" (3.9 per SF)	M1@.147	SF	1.42	5.95	7.37
4" x 8" x 1-5/8" (3.9 per SF)	M1@.144	SF	1.37	5.83	7.20
3-5/8" x 7-5/8" x 2-1/4" (4.5 per SF)	M1@.147	SF	1.64	5.95	7.59
3-5/8" x 7-5/8" x 1-5/8" (4.5 per SF)	M1@.144	SF	1.58	5.83	7.41
Brick laid on edge with 3/8" mortar joints and setting bed, running bond, including mortar					
4" x 8" x 2-1/4" (6.5 per SF)	M1@.280	SF	2.96	11.30	14.26
3-5/8" x 7-5/8" x 2-1/4" (6.9 per SF)	M1@.314	SF	3.11	12.70	15.81
Mortar color for brick laid in mortar, add per SF of paved area					
Red or yellow	—	SF	.09	—	.09
Black or brown	—	SF	.14	—	.14
Green	—	SF	.37	—	.37
Base underlayment for paving brick					
Add for 2" sand base (124 SF per ton)	M1@.009	SF	.09	.36	.45
Add for 15 lb felt underlayment	M1@.002	SF	.05	.08	.13
Add for asphalt tack coat (80 SF per gallon)	M1@.005	SF	.05	.20	.25
Add for 2% neoprene tack coat (40 SF per gal.)	M1@.005	SF	.15	.20	.35

	Craft@Hrs	Unit	Material	Labor	Total
Brick edging, 8" deep, set in concrete with dry joints					
Using brick, per thousand	—	M	356.00	—	356.00
Headers 4" x 8" x 2-1/4" (3 per LF)	M1@.147	LF	1.07	5.95	7.02
Headers 3-5/8" x 7-5/8" x 2-1/4" (3.3 per LF)	M1@.159	LF	1.18	6.44	7.62
Rowlocks 3-5/8" x 7-5/8" x 2-1/4" (5.3 per LF)	M1@.251	LF	1.89	10.20	12.09
Rowlocks 4" x 8" x 2-1/4" (5.3 per LF)	M1@.265	LF	1.89	10.70	12.59
Brick edging, 8" deep, set in concrete with 3/8" mortar joints					
Using brick, per thousand	—	M	356.00	—	356.00
Headers 4" x 8" x 2-1/4" (2.75 per LF)	M1@.159	LF	.98	6.44	7.42
Headers 3-5/8" x 7-5/8" x 2-1/4" (3 per LF)	M1@.173	LF	1.07	7.01	8.08
Rowlocks 3-5/8" x 7-5/8" x 2-1/4" (4.6 per LF)	M1@.298	LF	1.64	12.10	13.74
Rowlocks 4" x 8" x 2-1/4" (4.6 per LF)	M1@.314	LF	1.64	12.70	14.34

Masonry, Stone and Slate Paving These costs do not include site preparation. Costs for stone and slate vary widely. (Quantities shown in parentheses are before waste allowance.) Use $500 as a minimum job charge.

	Craft@Hrs	Unit	Material	Labor	Total
Interlocking 9" x 4-1/2" concrete pavers, dry joints, natural gray					
2-3/8" thick (3.5 per SF)	M1@.058	SF	2.32	2.35	4.67
3-1/8" thick (3.5 per SF)	M1@.063	SF	2.59	2.55	5.14
Add for standard colors (3.5 per SF)	—	SF	.18	—	.18
Add for custom colors (3.5 per SF)	—	SF	.37	—	.37
Patio blocks, 8" x 16" x 2", dry joints					
Natural gray (1.125 per SF)	M1@.052	SF	1.22	2.11	3.33
Standard colors (1.125 per SF)	M1@.052	SF	1.35	2.11	3.46
Flagstone pavers in sand bed with dry joints					
Random ashlar, 1-1/4"	M1@.193	SF	2.56	7.82	10.38
Random ashlar, 2-1/2"	M1@.228	SF	3.89	9.23	13.12
Irregular fitted, 1-1/4"	M1@.208	SF	2.15	8.42	10.57
Irregular fitted, 2-1/2"	M1@.251	SF	3.30	10.20	13.50
Flagstone pavers in mortar bed with mortar joints. Add the cost of mortar below					
Random ashlar, 1-1/4"	M1@.208	SF	2.67	8.42	11.09
Random ashlar, 2-1/2"	M1@.251	SF	4.05	10.20	14.25
Irregular fitted, 1-1/4"	M1@.228	SF	2.26	9.23	11.49
Irregular fitted, 2-1/2"	M1@.277	SF	3.42	11.20	14.62
Granite pavers, gray, sawn, thermal finish, 3/4" joint. Add for bedding below					
4" x 4" x 2" thick	M1@.184	SF	20.40	7.45	27.85
4" x 4" x 3" thick	M1@.184	SF	24.50	7.45	31.95
4" x 4" x 4" thick	M1@.184	SF	27.70	7.45	35.15
4" x 8" x 2" thick	M1@.171	SF	17.60	6.93	24.53
4" x 8" x 3" thick	M1@.184	SF	20.00	7.45	27.45
4" x 8" x 4" thick	M1@.184	SF	22.20	7.45	29.65
8" x 8" x 2" thick	M1@.171	SF	14.50	6.93	21.43
8" x 8" x 3" thick	M1@.171	SF	15.60	6.93	22.53
8" x 8" x 4" thick	M1@.184	SF	16.70	7.45	24.15
12" x 12" x 2" thick	M1@.171	SF	7.13	6.93	14.06
12" x 12" x 3" thick	M1@.171	SF	8.47	6.93	15.40
12" x 12" x 4" thick	M1@.171	SF	9.03	6.93	15.96
Mortar bed, 1/2" thick, 24 SF per CF of mortar					
Per SF of bed	M1@.039	SF	.21	1.58	1.79
Sand bed, 2" thick, 150 SF per CY of sand					
Per SF of bed	M1@.024	SF	.09	.97	1.06

	Craft@Hrs	Unit	Material	Labor	Total
Limestone flag paving					
Sand bed, dry joints					
Random ashlar, 1-1/4"	M1@.193	SF	2.67	7.82	10.49
Random ashlar, 2-1/2"	M1@.228	SF	5.09	9.23	14.32
Irregular fitted, 1-1/2"	M1@.208	SF	2.32	8.42	10.74
Irregular fitted, 2-1/2"	M1@.251	SF	3.94	10.20	14.14
Mortar bed and joints					
Random ashlar, 1"	M1@.208	SF	2.37	8.42	10.79
Random ashlar, 2-1/2"	M1@.251	SF	5.22	10.20	15.42
Irregular fitted, 1"	M1@.228	SF	2.15	9.23	11.38
Irregular fitted, 2-1/2"	M1@.277	SF	3.99	11.20	15.19
Slate flag paving, natural cleft					
Sand bed, dry joints					
Random ashlar, 1-1/4"	M1@.173	SF	8.52	7.01	15.53
Irregular fitted, 1-1/4"	M1@.193	SF	7.88	7.82	15.70
Mortar bed and joints					
Random ashlar, 3/4"	M1@.173	SF	6.71	7.01	13.72
Random ashlar, 1"	M1@.208	SF	7.88	8.42	16.30
Irregular fitted, 3/4"	M1@.193	SF	6.25	7.82	14.07
Irregular fitted, 1"	M1@.228	SF	7.30	9.23	16.53
Add for sand rubbed slate	—	SF	1.60	—	1.60

Curbs, Gutters and Driveway Aprons Costs assume 3 uses of the forms and 2,000 PSI concrete placed directly from the chute of a ready-mix truck. Concrete cast-in-place curb. Excavation or backfill not included. Use $750 as a minimum job charge.

	Craft@Hrs	Unit	Material	Labor	Equipment	Total
Curbs						
Using concrete, per cubic yard	—	CY	89.60	—	—	89.60
Vertical curb						
6" x 12" straight curb	P9@.109	LF	3.28	4.56	1.36	9.20
6" x 12" curved curb	P9@.150	LF	3.34	6.28	1.88	11.50
6" x 18" straight curb	P9@.119	LF	4.86	4.98	1.49	11.33
6" x 18" curved curb	P9@.166	LF	4.95	6.95	1.08	12.98
6" x 24" straight curb	P9@.123	LF	6.46	5.15	1.54	13.15
6" x 24" curved curb	P9@.179	LF	6.53	7.49	2.24	16.26
Rolled curb and gutter, 6" roll						
6" x 18" base, straight curb	P9@.143	LF	6.46	5.98	1.79	14.23
6" x 18" base, curved curb	P9@.199	LF	6.53	8.33	2.49	17.35
6" x 24" base, straight curb	P9@.168	LF	8.13	7.03	2.10	17.26
6" x 24" base, curved curb	P9@.237	LF	8.22	9.92	2.96	21.10

Driveway Aprons Concrete cast-in-place driveway aprons, including 4% waste, wire mesh, forms and finishing. Per square foot of apron, no site preparation included. Use $200 as a minimum job charge.

	Craft@Hrs	Unit	Material	Labor	Equipment	Total
4" thick (80 SF per CY)	P9@.024	SF	2.05	1.00	.18	3.23
6" thick (54 SF per CY)	P9@.032	SF	2.53	1.34	.24	4.11

Walkways Concrete, 2,000 PSI placed directly from the chute of a ready mix truck, vibrated, plain scored, and broom finished. Costs include 2% allowance for waste but no forms and reinforcing.

	Craft@Hrs	Unit	Material	Labor	Equipment	Total
4" thick (80 SF per CY)	P8@.013	SF	1.12	.52	.02	1.66
Add for integral colors, most pastels	—	SF	1.12	—	—	1.10
Add for 1/2" color top course	P8@.012	SF	1.25	.48	.02	1.75
Add for steel trowel finish	CM@.011	SF	—	.48	.03	.51
Add for seeded aggregate finish	CM@.011	SF	.11	.48	.03	.62
Add for exposed aggregate wash process	CM@.006	SF	.05	.26	.02	.33

Asphalt walkway, temporary, including installation and breakout, but no hauling. Use $500 as a minimum job charge.

	Craft@Hrs	Unit	Material	Labor	Equipment	Total
2" to 2-1/2" thick	P5@.015	SF	.95	.60	.49	2.04

	Craft@Hrs	Unit	Material	Labor	Equipment	Total
Headers and Dividers						
2" x 4", treated pine	C8@.027	LF	1.05	1.11	—	2.16
2" x 6", treated pine	C8@.028	LF	1.71	1.15	—	2.86
2" x 6", redwood, B grade	C8@.028	LF	1.81	1.15	—	2.96
2" x 4", patio type dividers, untreated	C8@.030	LF	.63	1.23	—	1.86

Porous Paving Systems Thin-walled HDPE plastic rings connected by an interlocking geogrid structure, installed on a porous base course for commercial applications. Rings transfer loads from the surface to the grid structure to an engineered course base. Equipment includes a 1/2 CY backhoe. Installed on an existing prepared base. http://www.invisiblestructures.com

Porous paving system, grass. Rolled out over base course and seeded with grass mixture. Includes fertilizer and soil polymer mix for spreading over base

	Craft@Hrs	Unit	Material	Labor	Equipment	Total
6.6' wide by 164' long rolls	S6@.100	CSF	130.00	4.08	1.43	135.51
Add for 1" sand fill	S6@.030	CSF	4.76	1.22	—	5.98

Porous paving system, gravel. Rolled out over base course and covered with decorative fill gravel

	Craft@Hrs	Unit	Material	Labor	Equipment	Total
6.6' wide by 164' long rolls	S6@.110	CSF	130.00	4.49	1.58	136.07
Add for 1" gravel fil	S6@.030	CSF	4.40	1.22	—	5.62

Post Holes Drilling 2' deep by 10" to 12" diameter fence post holes. Equipment is a gasoline powered 12" auger. Costs will be higher on slopes or where access is limited. These costs do not include layout, setting of fence posts or disposal of excavated material.

Light to medium soil, laborer working with a hand auger

	Craft@Hrs	Unit	Material	Labor	Equipment	Total
4 holes per hour	CL@.250	Ea	—	9.10	2.81	11.91

Medium to heavy soil, laborer working with a breaking bar and hand auger

	Craft@Hrs	Unit	Material	Labor	Equipment	Total
2 holes per hour	CL@.500	Ea	—	18.20	5.62	23.82

Medium to heavy soil, 2 laborers working with a gas powered auger

	Craft@Hrs	Unit	Material	Labor	Equipment	Total
15 holes per hour	CL@.133	Ea	—	4.84	1.49	6.33

Broken rock, laborer and equipment operator working with a compressor and jackhammer

	Craft@Hrs	Unit	Material	Labor	Equipment	Total
2 holes per hour	S1@1.00	Ea	—	43.00	7.45	50.45

Rock, laborer and an equipment operator working with a truck-mounted drill

	Craft@Hrs	Unit	Material	Labor	Equipment	Total
12 holes per hour	S1@.166	Ea	—	7.14	3.73	10.87

Fencing, Chain Link Fence, industrial grade 9 gauge 2" x 2" galvanized steel chain link fabric and framework, including 2" or 2-3/8" line posts as indicated set at 10' intervals, set in concrete in augured post holes and 1-5/8" top rail, tension panels at corners and abrupt grade changes. Add for gates, gate posts and corner posts as required. Equipment is a gasoline powered 12" auger. Use $600 as a minimum job charge. For scheduling purposes, estimate that a 2-man crew will install 125 to 130 LF of fence per 8-hour day

	Craft@Hrs	Unit	Material	Labor	Equipment	Total
4' high galvanized fence, 2" line posts	C4@.073	LF	11.70	2.75	.27	14.72
5' high galvanized fence, 2" line posts	C4@.091	LF	13.10	3.43	.34	16.87
6' high galvanized fence, 2" line posts	C4@.106	LF	14.30	3.99	.40	18.69
6' high galvanized fence, 2-3/8" line posts	C4@.109	LF	16.00	4.10	.41	20.51
7' high galvanized fence, 2-3/8" line posts	C4@.127	LF	16.20	4.78	.48	21.46
8' high galvanized fence, 2-3/8" line posts	C4@.146	LF	19.00	5.50	.55	25.05
10' high galvanized fence, 2-3/8" line posts	C4@.182	LF	22.00	6.85	.68	29.53
12' high galvanized fence, 2-3/8" line posts	C4@.218	LF	25.60	8.21	.82	34.63
Add for vertical aluminum privacy slats	C4@.016	SF	1.33	.60	—	1.93
Add for vinyl coated fabric and posts	—	%	50.0	—	—	—
Add for under 200 LF quantities	C4@.006	LF	.60	.23	.02	.85

	Craft@Hrs	Unit	Material	Labor	Equipment	Total
Deduct for highway quantities	—	LF	-1.19	—	—	-1.19
Deduct for 11 gauge chain link fabric	—	SF	-1.19	—	—	-1.19
Deduct for "C" section line posts	—	LF	-1.70	—	—	-1.70

Gates Driveway or walkway. 9 gauge 2" x 2" galvanized steel chain link with frame and tension bars. Costs shown are per square foot of frame area. Add for hardware and gate posts, from below.

	Craft@Hrs	Unit	Material	Labor	Equipment	Total
With 1-3/8" diameter pipe frame	C4@.025	SF	6.44	.94	—	7.38
With 1-5/8" diameter pipe frame	C4@.025	SF	7.44	.94	—	8.38
Gate hardware, per gate						
Walkway gate	C4@.250	Ea	29.90	9.42	—	39.32
Driveway gate	C4@.250	Ea	73.08	9.42	—	82.50
Add for vinyl coated fabric and frame	—	SF	.62	—	—	.62

Corner, end or gate posts, complete, heavyweight, 2-1/2" outside diameter galvanized, with fittings and brace. Costs include 2/3 CF sack of concrete per post

	Craft@Hrs	Unit	Material	Labor	Equipment	Total
4' high	C4@.612	Ea	39.60	23.00	2.29	64.89
5' high	C4@.668	Ea	44.90	25.20	2.50	72.60
6' high	C4@.735	Ea	44.20	27.70	2.75	74.65
7' high	C4@.800	Ea	45.60	30.10	3.00	78.70
8' high	C4@.882	Ea	60.10	33.20	3.30	96.60
10' high	C4@.962	Ea	68.00	36.20	3.60	107.80
12' high, with middle rail	C4@1.00	Ea	74.80	37.70	3.75	116.25
Add for vinyl coated posts						
Per vertical foot	—	VLF	.60	—	—	.60
Deduct for 2" outside diameter	—	VLF	-.84	—	—	-.84

Barbed wire topper for galvanized chain link fence

	Craft@Hrs	Unit	Material	Labor	Equipment	Total
Single strand barbed wire	C4@.004	LF	.22	.15	—	.37
3 strands on one side, with support arm	C4@.008	LF	1.37	.30	—	1.67
3 strand on each side, with support arm	C4@.013	LF	2.80	.49	—	3.29
Double coil of 24" to 30" barbed tape	C4@.038	LF	10.90	1.43	—	12.33

Guardrails and Bumpers Equipment is a 4000 lb forklift. Use $4,000 as a minimum job charge.

	Craft@Hrs	Unit	Material	Labor	Equipment	Total
Steel guardrail, standard beam	H1@.250	LF	19.00	11.60	1.55	32.15
Posts, wood, treated, 4" x 4" x 36" high	H1@1.00	Ea	14.00	46.50	6.21	66.71
Guardrail two sides, channel rail & side blocks	H1@1.00	LF	29.10	46.50	6.21	81.81
Sight screen	H1@.025	LF	1.50	1.16	.16	2.82
Highway median barricade, precast concrete						
2'8" high x 10' long	H1@.500	Ea	299.00	23.20	3.11	325.31
Add for reflectors, 2 per 10' section	—	LS	12.00	—	—	12.00
Parking bumper, precast concrete, including dowels						
3' wide	CL@.381	Ea	24.00	13.90	4.73	42.63
6' wide	CL@.454	Ea	42.20	16.50	5.64	64.34

	Craft@Hrs	Unit	Material	Labor	Equipment	Total

Bollards Posts mounted in walkways to limit vehicle entry. Equipment is a 4000 lb capacity forklift. Add for concrete footings, if required, from costs shown at the end of this section. Use $1,000 as a minimum job charge.

Cast iron bollards, ornamental, surface mounted

	Craft@Hrs	Unit	Material	Labor	Equipment	Total
12" diameter base, 42" high	S1@1.00	Ea	801.00	43.00	12.40	856.40
17" diameter base, 42" high	S1@1.00	Ea	1,340.00	43.00	12.40	1,395.40
Lighted, electrical work not included,						
17" diameter base, 42" high	S1@1.00	Ea	1,510.00	43.00	12.40	1,565.40

Concrete bollards, precast, exposed aggregate, surface mounted

Square, 12" x 12" x 30" high	S1@1.00	Ea	317.00	43.00	12.40	372.40
Round, 12" diameter x 30" high	S1@1.00	Ea	324.00	43.00	12.40	379.40

Granite bollards, doweled to concrete foundation or slab (foundation not included.) Expect these costs to vary widely.

Square, smooth matte finish, flat top						
16" x 16" x 30", 756 lbs	S1@1.00	Ea	875.00	43.00	12.40	930.40
16" x 16" x 54", 1,320 lbs	S1@1.50	Ea	1,110.00	64.50	18.60	1,193.10
Square, rough finish, pyramid top						
12" x 12" x 24", 330 lbs	S1@1.00	Ea	391.00	43.00	12.40	446.40
Round, smooth matte finish, flat top						
12" diameter, 18" high	S1@1.00	Ea	846.00	43.00	12.40	901.40
Octagonal, smooth matte finish, flat top,						
24" x 24" x 20", 740 lbs	S1@1.00	Ea	1,270.00	43.00	12.40	1,325.40

Pipe bollards, concrete filled steel pipe, painted yellow, 8' long, set 4' in concrete, includes digging hole

6" diameter pipe	S1@.750	Ea	96.40	32.20	9.32	137.92
8" diameter pipe	S1@.750	Ea	159.00	32.20	9.32	200.52
12" diameter pipe	S1@.750	Ea	258.00	32.20	9.32	299.52

Wood bollards, pressure treated timber, includes hand-digging and backfill of hole but no disposal of excess soil. Length shown is portion exposed, costs include 3' bury depth

6" x 6" x 36" high	S1@.750	Ea	33.00	32.20	9.32	74.52
8" x 8" x 30" high	S1@.750	Ea	54.00	32.20	9.32	95.52
8" x 8" x 36" high	S1@.750	Ea	58.90	32.20	9.32	100.42
8" x 8" x 42" high	S1@.750	Ea	63.80	32.20	9.32	105.32
12" x 12" x 24" high	S1@.750	Ea	110.00	32.20	9.32	151.52
12" x 12" x 36" high	S1@.750	Ea	132.00	32.20	9.32	173.52
12" x 12" x 42" high	S1@.750	Ea	144.00	32.20	9.32	185.52

Footings for bollards, add if required

Permanent concrete footing	C8@.500	Ea	29.00	20.50	—	49.50
Footing for removable bollards	C8@.500	Ea	130.00	20.50	—	150.50

Baseball Backstops Galvanized 9 gauge mesh chain link backstop with 2-5/8" OD galvanized vertical end posts, 2-3/8" OD galvanized vertical back posts and 1-5/8" OD galvanized framing. Includes 4' deep overhang, 20 LF behind home plate, 16 LF on both first and third base lines, rear and wing planks 3' high and two 12' long benches. Equipment is a 4,000 lb forklift.

Traditional backstop, 20' H x 58' W x 14' D	S6@24.0	Ea	5,050.00	979.00	199.00	6,228.00
Traditional backstop, 30' H x 68' W x 14' D	S6@24.0	Ea	7,360.00	979.00	199.00	8,538.00
Arched backstop, 20' H x 62' W x 20' D	S6@44.0	Ea	5,700.00	1,790.00	199.00	7,689.00
Arched backstop, 18' H x 62' W x 20' D	S6@44.0	Ea	5,300.00	1,790.00	199.00	7,289.00

Add for concrete footings, 12" diameter x 24" deep, includes digging by hand and spreading excavated material adjacent to hole, using fence post mix bagged concrete

Per footing	S6@4.00	Ea	5.50	163.00	—	168.50

	Craft@Hrs	Unit	Material	Labor	Equipment	Total

Playing Fields Typical subcontract costs. Use 3,000 SF as a minimum job charge.

Playing fields, synthetic surface

	Craft@Hrs	Unit	Material	Labor	Equipment	Total
Turf, including base and turf	—	SF	—	—	—	11.00
Turf, 3 layer, base foam, turf	—	SF	—	—	—	12.00
Uniturf, embossed running surface, 3/8	—	SF	—	—	—	8.10

Running track

	Craft@Hrs	Unit	Material	Labor	Equipment	Total
Volcanic cinder, 7"	—	SF	—	—	—	2.15
Bituminous and cork, 2"	—	SF	—	—	—	3.50

	Craft@Hrs	Unit	Material	Labor	Total

Artificial grass turf, glue-down carpet installation includes cleaning application surface, adhesive, seam seal and rolling. Based on 100 SY job. Challenger Industries, http://www.challengerind.com.

	Craft@Hrs	Unit	Material	Labor	Total
Economy grade non-directional turf	F9@.055	SY	5.25	2.24	7.49
Mid grade non-directional weather resistant turf	F9@.055	SY	6.30	2.24	8.54
High performance non-directional UV stabilized turf	F9@.055	SY	7.17	2.24	9.41
Miniature golf style weather resistant turf	F9@.055	SY	7.35	2.24	9.59
Premium spike proof non-directional UV stabilized turf	F9@.065	SY	18.00	2.65	20.65
Padded spike proof non-directional UV stabilized turf	F9@.065	SY	20.20	2.65	22.85
Carpet adhesive (25 SY per gallon)	—	Gal	12.70	—	12.70
Seam sealer (800 LF per gallon)	—	Gal	43.70	—	43.70
Spike proof indoor carpeting					
Economy grade patterned carpet	F9@.060	SY	14.40	2.44	16.84
Mid grade patterned carpet	F9@.060	SY	15.80	2.44	18.24
Mid grade insignia carpet	F9@.060	SY	21.50	2.44	23.94
Premium patterned carpet	F9@.060	SY	25.00	2.44	27.44

	Craft@Hrs	Unit	Material	Labor	Equipment	Total

Playground Equipment Equipment is a 4,000 lb capacity forklift. Use $2,500 as a minimum job charge.

Balance beams. All are supported 12" above the ground by 2-3/8" OD galvanized pipe uprights

	Craft@Hrs	Unit	Material	Labor	Equipment	Total
10' long Z-shaped beam	S6@6.00	Ea	313.00	245.00	49.70	607.70
10' long curved beam	S6@6.00	Ea	407.00	245.00	49.70	701.70
10' long straight beam	S6@6.00	Ea	308.00	245.00	49.70	602.70
15' long straight beam	S6@6.00	Ea	247.00	245.00	49.70	541.70
30' long Z-shaped beam	S6@9.00	Ea	480.00	367.00	74.60	921.60

Geodesic dome beehive climber, 1" galvanized pipe rungs

	Craft@Hrs	Unit	Material	Labor	Equipment	Total
8' diameter, 4' high	S6@24.0	Ea	426.00	979.00	199.00	1,604.00
13' diameter, 5' high	S6@36.0	Ea	1,290.00	1,470.00	298.00	3,058.00
18' diameter, 7' high	S6@48.0	Ea	1,600.00	1,960.00	399.00	3,959.00

Horizontal ladders, 20" wide with 1-1/16" OD galvanized pipe rungs 12" OC and 2-3/8" OD galvanized pipe headers and top rails

	Craft@Hrs	Unit	Material	Labor	Equipment	Total
12' long, 6' high	S6@6.00	Ea	543.00	245.00	49.70	837.70
18' long, 7' high	S6@9.00	Ea	667.00	367.00	74.60	1,108.60

Rope and pole climbs. Horizontal vertical end posts are 3-1/2" OD galvanized steel tubing

	Craft@Hrs	Unit	Material	Labor	Equipment	Total
10' H x 10' W with three 1-5/8" diameter poles	S6@24.0	Ea	800.00	979.00	199.00	1,978.00
10' H x 10' W with three 1" diameter ropes	S6@24.0	Ea	866.00	979.00	199.00	2,044.00

	Craft@Hrs	Unit	Material	Labor	Equipment	Total
16' H x 12' W						
with three 1-5/8" diameter poles	S6@24.0	Ea	1,060.00	979.00	199.00	2,238.00
16' H x 12' W						
with three 1-1/2" diameter ropes	S6@24.0	Ea	1,220.00	979.00	199.00	2,398.00
Sandboxes, prefabricated. Painted galvanized steel, 12" high, pine seats, sand not included						
6' x 6'	S6@9.00	Ea	337.00	367.00	74.60	778.60
12' x 12'	S6@12.0	Ea	478.00	489.00	99.40	1,066.40
Masonry sand	—	CY	25.00	—	—	25.00
Washed plaster sand	—	CY	32.00	—	—	32.00
Concrete sand	—	CY	41.50	—	—	41.50
White silica sand	—	CY	55.00	—	—	55.00
Sand scoopers, crane simulator. Heavy duty metal construction. Two handles control crane arm height and sand scooper, set in concrete						
3' high miniature scooper with seat	CL@2.15	Ea	370.00	78.20	53.40	501.60
4' high rotating stand up scooper	CL@2.75	Ea	382.00	100.00	68.30	550.30
Masonry sand	—	CY	25.00	—	—	25.00
Seesaws, galvanized pipe beam with polyethylene saddle or animal seats. Seesaw beam pivots between two coil springs. Double units have seesaw beams spaced 90 degrees apart						
Single unit, 7' metal base						
set in concrete	S6@5.00	Ea	324.00	204.00	41.40	569.40
Double unit, 4 saddle seats	S6@9.00	Ea	1,070.00	367.00	74.60	1,511.60
Double unit, 2 animal						
& 2 saddle seats	S6@9.00	Ea	1,550.00	367.00	74.60	1,991.60
Double unit, 4 animal seats	S6@9.00	Ea	2,020.00	367.00	74.60	2,461.60
Spring saws, steel frame with aluminum seats, set in concrete						
2' x 12' with two saddle seats	S6@6.00	Ea	753.00	245.00	49.70	1,047.70
12' diameter with two animal seats	S6@6.00	Ea	512.00	245.00	49.70	806.70
12' diameter with four saddle seats	S6@8.50	Ea	1,530.00	347.00	70.40	1,947.40
12' diameter with four animal seats	S6@8.50	Ea	655.00	347.00	70.40	1,072.40
12' diameter with six saddle seats	S6@9.00	Ea	2,270.00	367.00	74.60	2,711.60
12' diameter with six animal seats	S6@9.00	Ea	895.00	367.00	74.60	1,336.60
Slides						
Traditional slide with 18" wide 16 gauge stainless steel slide and galvanized steel stairway						
Chute 5' H x 10' L,						
overall length is 13'	S6@12.0	Ea	979.00	489.00	99.40	1,567.40
Chute 6' H x 12' L,						
overall length is 15'	S6@12.0	Ea	1,060.00	489.00	99.40	1,648.40
Chute 8' H x 16' L,						
overall length is 20'	S6@12.0	Ea	1,440.00	489.00	99.40	2,028.40
Swings, belt seats, galvanized pipe frame						
8' high two-leg ends						
2 seats wood framed	S6@8.00	Ea	1,090.00	326.00	66.30	1,482.30
2 seats	S6@12.0	Ea	471.00	489.00	99.40	1,059.40
4 seats	S6@18.0	Ea	808.00	734.00	149.00	1,691.00
6 seats	S6@24.0	Ea	1,120.00	979.00	199.00	2,298.00
10' high two-leg ends						
2 seats	S6@18.0	Ea	513.00	734.00	149.00	1,396.00
4 seats	S6@24.0	Ea	878.00	979.00	199.00	2,056.00
6 seats	S6@30.0	Ea	1,240.00	1,220.00	249.00	2,709.00
10' high three-leg ends						
4 seats	S6@24.0	Ea	1,090.00	979.00	199.00	2,268.00
6 seats	S6@30.0	Ea	1,350.00	1,220.00	250.00	2,820.00
8 seats	S6@36.0	Ea	1,930.00	1,470.00	298.00	3,698.00

	Craft@Hrs	Unit	Material	Labor	Equipment	Total
12' high three-leg ends						
3 seats	S6@30.0	Ea	921.00	1,220.00	249.00	2,390.00
6 seats	S6@36.0	Ea	1,380.00	1,470.00	299.00	3,149.00
8 seats	S6@42.0	Ea	1,910.00	1,710.00	348.00	3,968.00
Add for platform for wheelchair,						
any of above	S6@12.0	Ea	534.00	489.00	68.00	1,091.00
Add for seat replacements						
Seat replacement	—	Ea	32.80	—	—	32.80
Child's rubber bucket seat	—	Ea	75.20	—	—	75.20
Tetherball sets, galvanized pipe post, ground sleeve and chain,						
complete with nylon rope and ball	S6@3.00	Ea	136.00	122.00	24.90	282.90
Sand for pit (masonry sand)	—	CY	25.00	—	—	25.00

Tower climber. Complete with 8' high fireman's slide pole, "loft" platform, cargo net, horizontal ladders and vertical ladders. Frame is 1-1/4" square steel tubing with painted finish, bracing, ladders and slide pole are 1" OD steel pipe with painted finish. The platform consists of two 30" x 96" fiberglass planks with integral color located 5'0" above the ground. Available in mixed or matched colors.

	Craft@Hrs	Unit	Material	Labor	Equipment	Total
8' L x 6' W x 8' H	S6@36.0	Ea	2,590.00	1,470.00	298.00	4,358.00
Arch climbers, 1.9" OD galvanized steel pipe, set in concrete						
2'2" x 10'7"	S6@6.75	Ea	480.00	275.00	22.80	777.80
2'2" x 13'11"	S6@7.00	Ea	531.00	286.00	58.00	875.00
2'2" x 15'1"	S6@7.00	Ea	610.00	286.00	58.00	954.00
15' x 20' four way climber	S6@10.0	Ea	815.00	408.00	82.80	1,305.80
Funnel ball, 3-1/2" OD galvanized steel post with yellow polyethylene hopper						
10' high, set in concrete	S6@4.00	Ea	618.00	163.00	33.10	814.10
Chin up bars, 1-1/2" OD galvanized steel post with 54" and 66" and 84" bar height levels						
12' width, set in concrete	S6@4.75	Ea	359.00	194.00	39.40	592.40
Parallel bars, 1-7/8" OD galvanized steel post, 2'9" high.						
2' x 9'2" set in concrete	S6@6.00	Ea	220.00	245.00	49.70	514.70
Add for motion beam attachment	S6@.250	Ea	149.00	10.20	2.07	161.27

	Craft@Hrs	Unit	Material	Labor	Total
Park Grills Steel grill, set in concrete. Add for cost of concrete.					
Rotating adjustable grill, 40" high with					
300 sq. inch cooking area	CM@2.50	Ea	167.00	109.00	276.00
Large grill with two adjustable grates,					
40" high with 1008 sq. inch cooking area	CM@2.75	Ea	330.00	120.00	450.00
Hooded rotating grill, 40" tall with hood vents,					
40" high with 500 sq. inch cooking area	CM@3.00	Ea	431.00	131.00	562.00
Shelterhouse grill with shelves and two grates,					
40" high with 1368 sq. inch cooking area	CM@3.50	Ea	612.00	152.00	764.00
Campfire rings with 300 sq. inch cooking area and security brackets set in concrete					
Hinged grate	CM@2.25	Ea	144.00	98.00	242.00
Adjustable grate	CM@2.25	Ea	167.00	98.00	265.00

	Craft@Hrs	Unit	Material	Labor	Equipment	Total
Athletic Equipment Installed with a 4,000 lb. capacity forklift.						
Football goals, costs shown are for two goals for one field						
Regulation goal, galvanized steel pipe						
4-1/2" center support post	S6@22.0	Pr	2,310.00	897.00	182.00	3,389.00
6-5/8" center support post	S6@24.0	Pr	4,170.00	979.00	199.00	5,348.00
Combination football/soccer goal, regulation size, galvanized						
steel pipe, dual uprights	S6@24.0	Pr	2,070.00	979.00	199.00	3,248.00
Soccer goals, regulation size, costs shown are for two goals for one field						
Aluminum frame with net	S6@24.0	Pr	2,340.00	979.00	199.00	3,518.00

	Craft@Hrs	Unit	Material	Labor	Equipment	Total
Basketball goals, single support, adjustable, 6' backboard extension aluminum fan backboard, hoop and net						
Single goal	S6@12.0	Ea	800.00	489.00	99.40	1,388.40
Double goal, back to back	S6@15.0	Ea	1,120.00	612.00	124.00	1,856.00
Volleyball nets, with galvanized posts net, and ground sleeves						
Per net	S6@18.0	Ea	853.00	734.00	149.00	1,736.00

Tennis Courts Typical costs for championship size single playing court installation including normal fine grading. These costs do not include site preparation, excavation, drainage, or retaining walls. Equipment, where shown, is a 4,000 lb capacity forklift. Use $22,000 as a minimum job charge. Court is 60' x 120' with asphaltic concrete or Portland cement slab (4" thickness built up to 6" thickness at perimeter). Includes compacted base material and wire mesh reinforcing in concrete

	Craft@Hrs	Unit	Material	Labor	Equipment	Total
Typical cost per court, slab-on-grade	—	LS	—	—	—	17,000.00
Court surface seal coat (7,200 SF)						
One color plus 2" wide boundary lines	—	LS	—	—	—	2,500.00
Fence, 10' high galvanized chain link (360 LF) at court perimeter						
Fencing including gates	—	LS	—	—	—	6,500.00
Lighting						
Typical installation	—	LS	—	—	—	8,000.00
Net posts, with tension reel						
Galvanized steel, 3-1/2" diameter	S6@6.00	Pr	313.00	245.00	49.70	607.70
Galvanized steel, 4-1/2" diameter	S6@6.00	Pr	353.00	245.00	49.70	647.70
Tennis nets						
Nylon	S6@1.50	Ea	241.00	61.20	12.40	314.60
Steel	S6@1.50	Ea	542.00	61.20	12.40	615.60
Ball wall, fiberglass, 8' x 5'						
Attached to existing wall	S6@1.50	Ea	656.00	61.20	12.40	729.60
Windbreak, 9' high, closed mesh polypropylene						
Lashed to existing chain link fence	S6@.003	SF	.18	.12	.02	.32

	Craft@Hrs	Unit	Material	Labor	Total
Athletic Benches Backed athletic benches, galvanized pipe frame, stationary					
Aluminum seat and back					
7' long	C8@.930	Ea	257.00	38.10	295.10
8' long	C8@1.16	Ea	297.00	47.50	344.50
15' long	C8@1.31	Ea	474.00	53.70	527.70
Fiberglass seat and back					
10' long	C8@.930	Ea	369.00	38.10	407.10
12' long	C8@1.16	Ea	424.00	47.50	471.50
16' long	C8@1.31	Ea	568.00	53.70	621.70
Galvanized steel seat and back					
15' long	C8@1.31	Ea	339.00	53.70	392.70
Backless athletic bench, galvanized pipe frame, stationary, 12" wide x 10' long					
Wood seat	C8@.776	Ea	180.00	31.80	211.80
Steel seat	C8@.776	Ea	199.00	31.80	230.80
Aluminum seat	C8@.776	Ea	228.00	31.80	259.80
Surface Preparation and Striping for Athletic Courts Use $150 as a minimum job charge.					
Surface preparation					
Acid etch concrete surface	PA@.013	SY	.19	.59	.78
Primer on asphalt	PA@.006	SY	.24	.27	.51
Primer on concrete	PA@.006	SY	.36	.27	.63
Asphalt emulsion filler, 1 coat	PA@.008	SY	.45	.36	.81
Acrylic filler, 1 coat	PA@.008	SY	.69	.36	1.05
Rubber/acrylic compound, per coat	PA@.012	SY	1.79	.54	2.33

	Craft@Hrs	Unit	Material	Labor	Total
Acrylic emulsion texture, per coat					
Sand filled	PA@.006	SY	1.16	.27	1.43
Rubber filled	PA@.006	SY	1.22	.27	1.49
Acrylic emulsion color coat, per coat					
Tan, brown, or red	PA@.006	SY	.38	.27	.65
Greens	PA@.006	SY	.43	.27	.70
Blue	PA@.006	SY	.48	.27	.75
Rubber cushioned concrete pavers					
18" x 18", dry joints	FL@.039	SF	14.70	1.76	16.46
Striping for athletic courts					
Marked with white or yellow striping paint					
Lines 3" wide	PA@.005	LF	.05	.23	.28
Lines 4" wide	PA@.005	LF	.07	.23	.30
Lines 4" wide, reflectorized	PA@.002	LF	.13	.09	.22
Lines 8" wide, reflectorized	PA@.004	LF	.28	.18	.46
Marked with reflectorized thermoplastic tape					
Lines 4" wide	PA@.004	LF	1.13	.18	1.31
Lines 6" wide	PA@.006	LF	1.71	.27	1.98
Symbols, lettering					
Within 2'-0" x 8'-0" template	PA@2.50	Ea	39.90	113.00	152.90

Bicycle, Moped and Motorcycle Security Racks Installation kit includes tamper proof bolts and quick setting cement. Locks supplied by users. Labor is to install units on concrete surface. Cost per bicycle or motorcycle of capacity. Add cost for drilling holes in concrete as shown. These costs are based on a minimum purchase of six racks.

	Craft@Hrs	Unit	Material	Labor	Total
Basic bicycle rack					
5' long, looped steel, holds 8 bikes	C8@1.00	Ea	234.00	41.00	275.00
7' long, looped steel, holds 12 bikes	C8@1.25	Ea	287.00	51.20	338.20
10' long, single sided, holds 9 bikes	C8@.862	Ea	253.00	35.30	288.30
20' long, single sided, holds 18 bikes	C8@1.50	Ea	451.00	61.50	512.50
10' long, double sided, holds 18 bikes	C8@1.50	Ea	325.00	61.50	386.50
20' long, double sided, holds 36 bikes	C8@1.00	Ea	572.00	41.00	613.00
Moped or motorcycle rack					
With 24" security chain (RR-700)	C8@.575	Ea	116.00	23.60	139.60

Add for mounting track for use on asphalt or non-concrete surfaces. No foundation work included. Each 10' track section holds 6 bicycle racks or 4 moped or motorcycle racks

	Craft@Hrs	Unit	Material	Labor	Total
Per 10' track section, including drilling holes	C8@2.50	Ea	121.00	102.00	223.00
Add for purchasing less than six mounting racks	—	Ea	14.20	—	14.20
Add for drilling mounting holes in concrete,					
2 required per rack, cost per rack	C8@.125	Ea	—	5.12	5.12

Bicycle racks, galvanized pipe rack, embedded or surface mounted to concrete, cost per rack

	Craft@Hrs	Unit	Material	Labor	Total
Double entry					
5' long, 7 bikes	C8@3.31	Ea	264.00	136.00	400.00
10' long, 14 bikes	C8@3.31	Ea	380.00	136.00	516.00
20' long, 28 bikes	C8@4.88	Ea	693.00	200.00	893.00
30' long, 42 bikes	C8@6.46	Ea	1,010.00	265.00	1,275.00
Single side entry					
5' long, 4 bikes	C8@2.41	Ea	253.00	98.70	351.70
10' long, 8 bikes	C8@2.41	Ea	286.00	98.70	384.70
20' long, 16 bikes	C8@3.55	Ea	517.00	145.00	662.00

	Craft@Hrs	Unit	Material	Labor	Total
Wood and metal rack, embedded or surface mounted					
Double entry					
10' long, 14 bikes	C8@3.31	Ea	414.00	136.00	550.00
20' long, 28 bikes	C8@4.88	Ea	776.00	200.00	976.00
Single entry					
10' long, 8 bikes	C8@2.41	Ea	345.00	98.70	443.70
20' long, 10 bikes	C8@3.55	Ea	620.00	145.00	765.00
Bike post bollard,					
dimensional wood, concrete footing	C8@.223	Ea	188.00	9.14	197.14
Single tube galvanized steel, looped wave pattern, 30" high, surface mounted					
3' long, 5 bikes	C8@3.45	Ea	616.00	141.00	757.00
5' long, 7 bikes	C8@3.45	Ea	705.00	141.00	846.00
7' long, 9 bikes	C8@4.61	Ea	920.00	189.00	1,109.00
9' long, 11 bikes	C8@4.61	Ea	1,090.00	189.00	1,279.00
Bonded color, add	—	Ea	94.00	—	94.00
Add for embedded mounting	C8@1.77	Ea	23.40	72.50	95.90

	Craft@Hrs	Unit	Material	Labor	Equipment	Total
Sprinkler Irrigation Systems Typical complete system costs, including PVC pipe, heads, valves, fittings, trenching and backfill. Per SF of area watered. Add 10% for irrigation systems installed in areas subject to freezing hazard. Equipment is a 55 HP pneumatic tired riding type trencher.						
Large areas using pop-up impact heads, 200 to 300 SF per head, manual system						
To 5,000 SF job	S4@.005	SF	.17	.18	.08	.43
Over 5,000 SF to 10,000 SF	S4@.004	SF	.14	.15	.07	.36
Over 10,000 SF	S4@.003	SF	.12	.11	.05	.28
Add for automatic system	S4@.001	SF	.06	.04	.02	.12
Small areas, 5,000 SF and under, spray heads						
Strip, automatic, shrub type	S4@.008	SF	.27	.29	.13	.69
Commercial type, manual	S4@.007	SF	.25	.25	.12	.62
Residential type, manual	S4@.007	SF	.24	.25	.12	.61
Add for automatic system	S4@.001	SF	.06	.04	.02	.12
Trenching for sprinkler systems, in normal non-compacted topsoil						
Main lines with a 55 HP riding type trencher, 18" wide						
12" deep, 143 LF per hour	S1@.007	LF	—	.30	.17	.47
18" deep, 125 LF per hour	S1@.008	LF	—	.34	.20	.54
24" deep, 100 LF per hour	S1@.010	LF	—	.43	.25	.68
Lateral lines with a 20 HP riding type trencher, 12" wide						
8" deep, 200 LF per hour	S1@.005	LF	—	.21	.08	.29
12" deep, 125 LF per hour	S1@.008	LF	—	.34	.13	.47
Add for hard soil						
(shelf, rock field, hardpan)	S1@.006	LF	—	.26	.10	.36
Boring under walkways and pavement						
To 8' wide, by hand, 8 LF per hour	CL@.125	LF	—	4.55	—	4.55
Backfill and compact trench, by hand						
Mains to 24" deep, 125 LF per hour	CL@.008	LF	—	.29	—	.29
Laterals to 12" deep, 190 LF per hour	CL@.005	LF	—	.18	—	.18

	Craft@Hrs	Unit	Material	Labor	Total
Restore turf above pipe after sprinkler installation					
Replace and roll sod	CL@.019	SY	4.00	.69	4.69
Reseed by hand, cover, water and mulch	CL@.010	SY	.18	.36	.54

	Craft@Hrs	Unit	Material	Labor	Total
Connection to existing water line					
Residential or small commercial tap	CL@.939	Ea	20.00	34.20	54.20
Residential or small commercial stub	CL@.683	Ea	15.50	24.90	40.40
Medium commercial stub	CL@1.25	Ea	42.10	45.50	87.60
Atmospheric vacuum breaker, brass					
3/4"	CL@.470	Ea	11.80	17.10	28.90
1"	CL@.500	Ea	15.80	18.20	34.00
1-1/2"	CL@.575	Ea	40.10	20.90	61.00
2"	CL@.625	Ea	69.80	22.70	92.50
Pressure type vacuum breaker, with threaded ball valves					
3/4"	CL@.700	Ea	104.00	25.50	129.50
1"	CL@.750	Ea	117.00	27.30	144.30
1-1/2"	CL@.850	Ea	284.00	30.90	314.90
2"	CL@.939	Ea	324.00	34.20	358.20
Double check valve assembly with two ball valves					
1-1/2" valve	CL@.750	Ea	312.00	27.30	339.30
2" valve	CL@.939	Ea	372.00	34.20	406.20
3" valve	CL@1.25	Ea	1,290.00	45.50	1,335.50
4" valve	CL@1.50	Ea	2,000.00	54.60	2,054.60
Reduced pressure backflow preventer with gate valves					
1" backflow preventer	CL@.750	Ea	184.00	27.30	211.30
1-1/2" backflow preventer	CL@.939	Ea	345.00	34.20	379.20
2" backflow preventer	CL@.939	Ea	387.00	34.20	421.20
3" backflow preventer	CL@1.25	Ea	1,430.00	45.50	1,475.50
4" backflow preventer	CL@1.50	Ea	2,220.00	54.60	2,274.60
6" backflow preventer	CL@1.88	Ea	2,870.00	68.40	2,938.40
Valve boxes, plastic box with plastic lid					
7" round	CL@.314	Ea	5.33	11.40	16.73
10" round	CL@.375	Ea	11.60	13.60	25.20
Sprinkler controllers, solid state electronics, 115-volt/24-volt electric timer, add electrical connection and feeder wire to valves					
Indoor mount, 4 station	CL@.960	Ea	31.30	34.90	66.20
Indoor mount, 6 station	CL@.960	Ea	38.90	34.90	73.80
Indoor mount, 9 station	CL@1.10	Ea	50.00	40.00	90.00
Indoor mount, 12 station	CL@1.10	Ea	59.80	40.00	99.80
Outdoor mount, 8 station	CL@1.10	Ea	105.00	40.00	145.00
Outdoor mount, 12 station	CL@1.10	Ea	125.00	40.00	165.00
Pedestal for outdoor units, set in concrete	CL@1.00	Ea	50.90	36.40	87.30
Underground control wire for automatic valves, includes waterproof connectors, wire laid in an open trench, two conductors per valve, costs are per pair of conductors					
4 conductors	CL@.001	LF	.14	.04	.18
5 conductors	CL@.001	LF	.16	.04	.20
7 conductors	CL@.002	LF	.22	.07	.29
Anti-siphon solenoid flow control valves, with atmospheric vacuum breaker. UV-resistant PVC, 24 volt. Add feeder wire as required					
3/4", residential	CL@.341	Ea	26.80	12.40	39.20
1" residential	CL@.341	Ea	27.90	12.40	40.30
3/4" commercial	CL@.418	Ea	26.80	15.20	42.00
1" commercial	CL@.418	Ea	15.30	15.20	30.50
1-1/2" commercial	CL@.796	Ea	33.60	29.00	62.60
2" commercial	CL@.866	Ea	47.10	31.50	78.60

	Craft@Hrs	Unit	Material	Labor	Total
Manual sprinkler control valves, anti-siphon, with union					
3/4" valve	CL@.518	Ea	29.20	18.90	48.10
1" valve	CL@.610	Ea	37.40	22.20	59.60
1-1/4" valve	CL@.677	Ea	59.90	24.60	84.50
1-1/2" valve	CL@.796	Ea	65.80	29.00	94.80
2" valve	CL@.866	Ea	99.80	31.50	131.30
Angle valves, with union					
3/4" valve	CL@.518	Ea	22.10	18.90	41.00
1" valve	CL@.610	Ea	29.10	22.20	51.30
1-1/4" valve	CL@.677	Ea	68.50	24.60	93.10
1-1/2" valve	CL@.796	Ea	82.70	29.00	111.70
2" valve	CL@.866	Ea	105.00	31.50	136.50

PVC Schedule 40 pipe and fittings, except as noted. Solvent welded. Plain end. Schedule 40 denotes wall thickness. Class 315, 200 and 160 identify the normal working pressure, 315, 200 or 160 PSI. Add the cost of excavation and backfill

	Craft@Hrs	Unit	Material	Labor	Total
1/2" pipe	CL@.005	LF	.14	.18	.32
1/2" pipe, Class 315	CL@.005	LF	.10	.18	.28
1/2" ell	CL@.120	Ea	.20	4.37	4.57
1/2" tee	CL@.140	Ea	.21	5.09	5.30
3/4" pipe	CL@.005	LF	.18	.18	.36
3/4" pipe, Class 200	CL@.005	LF	.11	.18	.29
3/4" ell	CL@.160	Ea	.21	5.82	6.03
3/4" tee	CL@.180	Ea	.26	6.55	6.81
1" pipe	CL@.006	LF	.27	.22	.49
1" pipe, Class 200	CL@.006	LF	.14	.22	.36
1" ell	CL@.180	Ea	.41	6.55	6.96
1" tee	CL@.220	Ea	.51	8.01	8.52
1-1/4" pipe	CL@.006	LF	.37	.22	.59
1-1/4" pipe, Class 160	CL@.006	LF	.20	.22	.42
1-1/4" ell	CL@.180	Ea	.81	6.55	7.36
1-1/4" tee	CL@.250	Ea	.96	9.10	10.06
1-1/2" pipe	CL@.007	LF	.44	.25	.69
1-1/2" pipe, Class 160	CL@.007	LF	.26	.25	.51
1-1/2" ell	CL@.180	Ea	.93	6.55	7.48
1-1/2" tee	CL@.250	Ea	1.17	9.10	10.27
2" pipe	CL@.008	LF	.56	.29	.85
2" ell	CL@.200	Ea	1.35	7.28	8.63
2" tee	CL@.300	Ea	1.66	10.90	12.56
3" pipe	CL@.008	LF	1.06	.29	1.35
3" ell	CL@.240	Ea	5.33	8.73	14.06
3" tee	CL@.360	Ea	6.29	13.10	19.39
Polyethylene pipe, 80 PSI					
3/4" diameter	CL@.006	LF	.11	.22	.33
1" diameter	CL@.006	LF	.13	.22	.35
1-1/4" diameter	CL@.007	LF	.25	.25	.50
1-1/2" diameter	CL@.008	LF	.28	.29	.57
2" diameter	CL@.009	LF	.35	.33	.68
Polyethylene pipe, 100 PSI					
1-1/2" diameter	CL@.008	LF	.36	.29	.65
2" diameter	CL@.009	LF	.59	.33	.92
Add for 125 PSI polyethylene pipe	—	%	30.0	—	—

	Craft@Hrs	Unit	Material	Labor	Total
Shrub Sprinklers and Bubblers These costs include a plastic pipe riser as noted at each head.					
Bubblers (5' spacing, 12" riser), plastic	CL@.032	Ea	2.03	1.16	3.19
Flex riser	CL@.035	Ea	1.66	1.27	2.93
Double swing joint	CL@.035	Ea	2.01	1.27	3.28
Triple swing joint	CL@.035	Ea	3.01	1.27	4.28
Add for brass head and metal riser	—	Ea	3.76	—	3.76
Shrub spray (8' to 18' spacing, 12" riser)	CL@.038	Ea	1.43	1.38	2.81
Flex riser	CL@.038	Ea	1.75	1.38	3.13
Double swing joint	CL@.038	Ea	2.01	1.38	3.39
Triple swing joint	CL@.041	Ea	3.18	1.49	4.67
Add for brass head and metal riser	—	Ea	3.76	—	3.76
Stream spray (12' to 22' spacing, 12" riser)					
Plastic riser	CL@.041	Ea	2.36	1.49	3.85
Flex riser	CL@.041	Ea	2.66	1.49	4.15
Double swing joint	CL@.045	Ea	2.98	1.64	4.62
Triple swing joint	CL@.045	Ea	3.38	1.64	5.02
Add for brass head and metal riser	—	Ea	3.76	—	3.76
Rotary head, plastic (20' to 40' spacing, 12" riser)					
Plastic riser	CL@.063	Ea	8.40	2.29	10.69
Flex riser	CL@.063	Ea	8.96	2.29	11.25
Double swing joint	CL@.067	Ea	9.20	2.44	11.64
Triple swing joint	CL@.067	Ea	9.68	2.44	12.12
Rotary head, metal (20' to 40' spacing, 12" riser)					
Plastic riser	CL@.063	Ea	12.60	2.29	14.89
Flex riser	CL@.063	Ea	13.00	2.29	15.29
Double swing joint	CL@.067	Ea	13.20	2.44	15.64
Triple swing joint	CL@.067	Ea	13.60	2.44	16.04
Hi-pop spray, 6" pop-up (6" riser)					
Plastic riser	CL@.041	Ea	5.22	1.49	6.71
Flex riser	CL@.043	Ea	5.59	1.56	7.15
Double swing joint	CL@.045	Ea	5.89	1.64	7.53
Triple swing joint	CL@.045	Ea	7.89	1.64	9.53
Rotary, hi-pop, 10" to 12" pop-up					
Plastic riser	CL@.076	Ea	22.20	2.77	24.97
Flex riser	CL@.076	Ea	22.60	2.77	25.37
Double swing joint	CL@.079	Ea	23.50	2.87	26.37
Triple swing joint	CL@.079	Ea	23.70	2.87	26.57
Add for shrub sprinklers and bubblers					
Installed in existing shrub and flower beds	—	%	—	25.0	—
Lawn Sprinklers These costs include a plastic pipe riser as noted at each head.					
1" pop-up spray, plastic head (10' to 12' spacing)					
Single riser	CL@.045	Ea	2.62	1.64	4.26
Double swing joint	CL@.048	Ea	3.38	1.75	5.13
Triple swing joint	CL@.048	Ea	3.78	1.75	5.53
2" pop-up spray, plastic head (10' to 12' spacing)					
Single riser	CL@.045	Ea	3.10	1.64	4.74
Double swing joint	CL@.048	Ea	4.07	1.75	5.82
Triple swing joint	CL@.048	Ea	4.74	1.75	6.49
Add for brass head and metal riser	—	Ea	7.75	—	7.75

Site Work 2

	Craft@Hrs	Unit	Material	Labor	Total
Rotary pop-up spray (20' to 30' spacing)					
Regular	CL@.094	Ea	12.60	3.42	16.02
Double swing joint	CL@.098	Ea	14.20	3.57	17.77
Triple swing joint	CL@.102	Ea	15.10	3.71	18.81
Rotary pop-up spray (30' to 50' spacing)					
Regular	CL@.124	Ea	15.00	4.51	19.51
Double swing joint	CL@.132	Ea	16.40	4.80	21.20
Triple swing joint	CL@.133	Ea	17.50	4.84	22.34
Rotary head (50' to 70' spacing), full or part circle					
Without valve	CL@.130	Ea	58.60	4.73	63.33
With integral check valve	CL@.130	Ea	75.60	4.73	80.33
With integral automatic electric valve	CL@.137	Ea	79.00	4.99	83.99
Rotary head (70' to 80' spacing), full or part circle					
Without valve	CL@.149	Ea	75.60	5.42	81.02
With integral check valve	CL@.149	Ea	91.20	5.42	96.62
With integral automatic electric valve	CL@.159	Ea	110.00	5.79	115.79
With metal cover	—	Ea	4.94	—	4.94
Quick coupling valves, add swing joint costs below					
3/4" regular	CL@.089	Ea	20.30	3.24	23.54
3/4" 2-piece	CL@.089	Ea	23.80	3.24	27.04
1" regular	CL@.094	Ea	28.90	3.42	32.32
1" 2-piece	CL@.094	Ea	34.50	3.42	37.92
1-1/4" regular	CL@.108	Ea	36.30	3.93	40.23
1-1/2" regular	CL@.124	Ea	43.70	4.51	48.21
Add for double swing joint	CL@.132	Ea	3.78	4.80	8.58
Add for triple swing joint	CL@.133	Ea	6.26	4.84	11.10
Add for locking vinyl cover	—	Ea	9.38	—	9.38
Hose bibb, 3/4" on 12" galvanized riser	CL@.041	Ea	9.12	1.49	10.61

	Craft@Hrs	Unit	Material	Labor	Equipment	Total

Planting Bed Preparation

Spreading topsoil from piles on site, based on topsoil delivered to the site. Topsoil prices can be expected to vary widely.

	Craft@Hrs	Unit	Material	Labor	Equipment	Total
Hand work, 10 CY job, level site, 10' throw	CL@.500	CY	30.00	18.20	—	48.20
Move 25' in wheelbarrow	CL@1.00	CY	30.00	36.40	—	66.40
With 3/4 CY loader						
100 CY job, level site	TO@.080	CY	30.00	3.97	2.34	36.31

Spreading granular or powdered soil conditioners, based on fertilizer at $.20 per pound. Spread 20 pounds per 1,000 square feet (MSF). Add cost for mixing soil and fertilizer from below.

	Craft@Hrs	Unit	Material	Labor	Equipment	Total
By hand, 2,250 SF per hour	CL@.445	MSF	4.00	16.20	—	20.20
Hand broadcast spreader, 9,000 SF per hour	CL@.111	MSF	4.00	4.04	.25	8.29
Push gravity spreader, 3,000 SF per hour	CL@.333	MSF	4.00	12.10	1.21	17.31
Push broadcast spreader, 20,000 SF per hour	CL@.050	MSF	4.00	1.82	.18	6.00
Tractor drawn broadcast spreader, 87 MSF per hour	CL@.011	MSF	4.00	.40	.35	4.75

Spreading organic soil conditioners, based on peat humus at $.15 per pound. Spread 200 pounds per 1,000 square feet (MSF). Add mixing cost from below

	Craft@Hrs	Unit	Material	Labor	Equipment	Total
By hand, 1,000 SF per hour	CL@1.00	MSF	40.00	36.40	—	76.40

	Craft@Hrs	Unit	Material	Labor	Equipment	Total
Push gravity spreader,						
2,750 SF per hour	CL@.363	MSF	40.00	13.20	1.32	54.52
Manure spreader,						
34,000 SF per hour	CL@.029	MSF	40.00	1.06	.39	41.45
Fertilizer and soil conditioners						
Most nitrate fertilizers,						
ammonium sulfate	—	Lb	.17	—	—	.17
Lawn and garden fertilizer,						
calcium nitrate	—	Lb	.18	—	—	.18
Hydrated lime sodium nitrate	—	Lb	.15	—	—	.15
Ground limestone	—	Ton	145.00	—	—	145.00
Ground dolomitic limestone	—	Ton	150.00	—	—	150.00
Composted manure	—	Lb	.03	—	—	.03
Peat humus	—	Lb	.03	—	—	.03
Vermiculite, perlite	—	CF	2.65	—	—	2.65

Spreading lime at 70 pounds per 1,000 SF. Based on an 8 acre job using a tractor-drawn spreader at $25.00 per hour and 1 acre per hour. Add mixing cost from below. Note: 1 acre equals 43,560 square feet.

	Craft@Hrs	Unit	Material	Labor	Equipment	Total
Lime	TO@1.00	Acre	250.00	49.60	32.00	331.60
Lime	TO@.656	Ton	142.00	32.50	21.00	195.50

Mixing soil and fertilizer in place. Medium soil. Light soil (sand or soft loam) will decrease costs shown by 20% to 30%. Heavy soil (clay, wet or rocky soil) will increase costs shown by 20% to 40%. Add soil conditioner cost from sections preceding this section

	Craft@Hrs	Unit	Material	Labor	Equipment	Total
Mixing by hand						
2" deep, 9 SY per hour	CL@.111	SY	—	4.04	—	4.04
4" deep, 7.5 SY per hour	CL@.133	SY	—	4.84	—	4.84
6" deep, 5.5 SY per hour	CL@.181	SY	—	6.59	—	6.59
Mixing with a 11 HP rear-tined hydraulic tiller						
2" deep, 110 SY per hour	CL@.009	SY	—	.33	.16	.49
4" deep, 90 SY per hour	CL@.011	SY	—	.40	.19	.59
6" deep, 65 SY per hour	CL@.015	SY	—	.55	.26	.81
8" deep, 45 SY per hour	CL@.022	SY	—	.80	.39	1.19
Mixing with a 3' wide tractor driven tiller 20 HP tractor						
2" deep, 50,000 SF per hour	TO@.020	MSF	—	.99	.49	1.48
4" deep, 40,000 SF per hour	TO@.025	MSF	—	1.24	.62	1.86
6" deep, 30,000 SF per hour	TO@.033	MSF	—	1.64	.82	2.46
8" deep, 20,000 SF per hour	TO@.050	MSF	—	2.48	1.24	3.72
Mixing with a 6' wide tractor driven tiller and 40 HP tractor						
2" deep, 90,000 SF per hour	TO@.011	MSF	—	.55	.31	.86
4" deep, 70,000 SF per hour	TO@.014	MSF	—	.69	.39	1.08
6" deep, 53,000 SF per hour	TO@.019	MSF	—	.94	.53	1.47
8" deep, 35,000 SF per hour	TO@.028	MSF	—	1.39	.78	2.17
Leveling the surface for lawn or planting bed						
By hand, 110 SY per hour	CL@.009	SY	—	.33	—	.33
With 6' drag harrow,						
at 60,000 SF per hour	TO@.017	MSF	—	.84	.43	1.27
With 12' drag harrow,						
at 115,000 SF per hour	TO@.009	MSF	—	.45	.37	.82
Plant bed preparation						
With a 3/4 CY wheel loader,						
at 12.5 CY per hour	TO@.080	CY	—	3.97	2.95	6.92
By hand to 18" deep, 8.5 SY/hr	CL@.118	SY	—	4.29	—	4.29
Mixing planting soil						
By hand, .9 CY per hour	CL@1.11	CY	—	40.40	—	40.40

	Craft@Hrs	Unit	Material	Labor	Equipment	Total
With 3/4 CY wheel loader,						
7 CY per hour	T0@.143	CY	—	7.09	5.26	12.35
With a soil shredder,						
11 CY per hour	T0@.090	CY	—	4.46	3.31	7.77

Seeding and Planting

Seeding with a hand broadcast spreader, 10 pounds per 1,000 SF. The cost of grass seed will vary greatly depending on variety – see below

	Craft@Hrs	Unit	Material	Labor	Equipment	Total
4,000 SF per hour,						
Hand seeding	CL@.250	MSF	14.00	9.10	.56	23.66

Seeding with push broadcast spreader, 5 pounds per 1,000 SF

	Craft@Hrs	Unit	Material	Labor	Equipment	Total
10,000 SF per hour,						
Push spreader seeding	CL@.100	MSF	7.00	3.64	.36	11.00

Seeding with a mechanical seeder at $28 per hour and 175 pounds per acre, 5 acre job

	Craft@Hrs	Unit	Material	Labor	Equipment	Total
Mechanical seeding	T0@13.1	Acre	578.00	649.00	413.00	1,640.00
Mechanical seeding	T0@.300	MSF	16.50	14.90	9.45	40.85

Hydroseeding (spray application of seed, binder and fertilizer slurry), 3 pounds of Alta Fescue seed applied in a 12 pound mix per 1,000 SF. Equipment is a 1-1/2 ton truck and a hydroseeding unit . Use $1,500 as a minimum job charge.

	Craft@Hrs	Unit	Material	Labor	Equipment	Total
10,000 SF job	C4@1.12	MSF	22.00	42.20	9.36	73.56
25,000 SF job	C4@.645	MSF	22.00	24.30	5.39	51.69
50,000 SF job	C4@.399	MSF	22.00	15.00	2.83	39.83
Add for work in congested areas	—	%	—	—	—	50.0

Sealing soil, application by a hydroseeding unit. Use $750 as a minimum job charge.

	Craft@Hrs	Unit	Material	Labor	Equipment	Total
Copolymer based liquid						
for erosion control	C4@.020	SY	.25	.75	.17	1.17

Grass seed

	Craft@Hrs	Unit	Material	Labor	Equipment	Total
Kentucky bluegrass	—	Lb	2.70	—	—	2.70
Bermuda	—	Lb	3.00	—	—	3.00
Most fescue	—	Lb	1.35	—	—	1.35
Most ryegrass, timothy, orchard grass	—	Lb	1.90	—	—	1.90
Annual ryegrass	—	Lb	.84	—	—	.84
Heat and drought resistant						
premium blend	—	Lb	4.20	—	—	4.20
Argentine bahia	—	Lb	4.00	—	—	4.00

Sodding, placed on level ground at 25 SY per hour. Add delivery cost below. Use 50 SY as a minimum job charge.

	Craft@Hrs	Unit	Material	Labor	Equipment	Total
Northeast, Midwest, bluegrass blend	CL@.040	SY	1.60	1.46	—	3.06
Great Plains, South, bluegrass	CL@.040	SY	1.90	1.46	—	3.36
West, bluegrass, tall fescue	CL@.040	SY	2.90	1.46	—	4.36
South, Bermuda grass,						
centipede grass	CL@.040	SY	3.25	1.46	—	4.71
South, Zoysia grass, St. Augustine	CL@.040	SY	5.00	1.46	—	6.46
Typical delivery cost, to 90 miles	—	SY	.50	—	—	.50
Add for staking sod on slopes	CL@.010	SY	.10	.36	—	.46

Work grass seed into soil, no seed included

	Craft@Hrs	Unit	Material	Labor	Equipment	Total
By hand, 220 SY per hour	CL@.005	SY	—	.18	—	.18
6' harrow and 40 HP tractor						
60 MSF per hour	T0@.017	MSF	—	.84	.60	1.44
12' harrow and 60 HP tractor						
100 MSF per hour	T0@.010	MSF	—	.50	.38	.88

Apply top dressing over seed or stolons

	Craft@Hrs	Unit	Material	Labor	Equipment	Total
300 pounds per 1,000 SF (2.7 pounds per SY)						
By hand, 65 SY per hour	CL@.015	SY	.43	.55	—	.98
With manure spreader,						
10,000 SF per hour	CL@.125	MSF	48.00	4.55	—	52.55

Site Work 2

	Craft@Hrs	Unit	Material	Labor	Equipment	Total
Roll sod or soil surface with a roller						
With a push roller, 400 SY per hour	CL@.002	SY	—	.07	—	.07
With a 20 HP tractor, 25,000 SF per hour	TO@.040	MSF	—	1.98	.90	2.88
Sprigging, by hand						
6" spacing, 50 SY per hour	CL@.020	SY	.21	.73	—	.94
9" spacing, 100 SY per hour	CL@.010	SY	.14	.36	—	.50
12" spacing, 150 SY per hour	CL@.007	SY	.11	.25	—	.36
Hybrid Bermuda grass stolons, per bushel	—	Ea	3.65	—	—	3.65
Ground cover, large areas, no surface preparation						
Ice plant (80 SF per flat)	CL@.008	SF	.19	.29	—	.48
Strawberry (75 SF per flat)	CL@.008	SF	.22	.29	—	.51
Ivy and similar (70 SF per flat)	CL@.008	SF	.23	.29	—	.52
Gravel bed, pea gravel (1.4 tons per CY)						
Spread by hand	CL@.555	CY	31.50	20.20	—	51.70
Trees and shrubs. See also Landscaping in the Residential Division						
Shrubs, most varieties						
1 gallon, 100 units	CL@.085	Ea	8.00	3.09	—	11.09
1 gallon, over 100 units	CL@.076	Ea	5.15	2.77	—	7.92
5 gallons, 100 units	CL@.443	Ea	18.90	16.10	—	35.00
Trees, most varieties, complete, staked, typical costs						
5 gallon, 4' to 6' high	CL@.716	Ea	24.00	26.10	—	50.10
15 gallon, 8' to 10' high	CL@1.37	Ea	78.80	49.90	—	128.70
20" x 24" box, 10' to 12' high	CL@4.54	Ea	300.00	165.00	—	465.00
Specimen size 36" box, 14' to 20' high	CL@11.6	Ea	775.00	422.00	—	1,197.00
Guaranteed establishment of plants						
Add, as a % of total contract price	—	%	—	—	—	12.0
Edging						
Redwood benderboard, 6" x 3/8", staked	CL@.013	LF	.85	.47	—	1.32
Redwood benderboard, 4" x 5/16", staked	CL@.013	LF	.55	.47	—	1.02
Redwood headerboard, 2" x 4", staked Specification grade	CL@.127	LF	5.41	4.62	—	10.03
Decomposed granite edging, Saturated and rolled, 3" deep	CL@.011	SF	.37	.40	—	.77
Landscape stepping stones (concrete pavers)						
Laid on level ground, 14" x 14", 30 lb. each	CL@.024	SF	2.00	.87	—	2.87
Pressure treated piles in 2' to 5' sections,						
9" to 12" diameter	CL@.045	LF	3.29	1.64	—	4.93

	Craft@Hrs	Unit	Material	Labor	Total
Tree Grates Two piece concrete grates surrounding a tree trunk, set in concrete. Add for delivery.					
36" x 36" x 2-3/4" square (270 lbs.)	C5@4.00	Ea	225.00	163.00	388.00
48" x 48" x 2-3/4" square (512 lbs.)	C5@5.00	Ea	250.00	203.00	453.00
60" x 60" x 2-3/4" square (750 lbs.)	C5@6.50	Ea	280.00	264.00	544.00
36" diameter x 2-3/4" round (230 lbs.)	C5@4.00	Ea	230.00	163.00	393.00
48" diameter x 2-3/4" round (480 lbs.)	C5@5.00	Ea	265.00	203.00	468.00
Add for cast steel frame	—	%	45.0	20.0	—

	Craft@Hrs	Unit	Material	Labor	Equipment	Total

Excavation for Concrete Work These costs do not include shoring or disposal. Typical soil conditions.

Batterboards, lay out for footings,

	Craft@Hrs	Unit	Material	Labor	Equipment	Total
per corner	C8@1.24	Ea	7.00	50.80	—	57.80

Trenching with a 1/2 CY utility backhoe/loader, small jobs, good soil conditions, no backfill included

	Craft@Hrs	Unit	Material	Labor	Equipment	Total
18" x 24" depth,						
.111 CY per LF (135 LF/Hr)	S1@.015	LF	—	.64	.23	.87
24" x 36" depth,						
.222 CY per LF (68 LF/Hr)	S1@.029	LF	—	1.25	.45	1.70
36" x 48" depth,						
.444 CY per LF (34 LF/Hr)	S1@.059	LF	—	2.54	.92	3.46
48" x 60" depth,						
.741 CY per LF (20 LF/Hr)	S1@.099	LF	—	4.26	1.55	5.81
Per cubic yard,						
15 CY per hour	S1@.133	CY	—	5.72	2.08	7.80

Hand labor

	Craft@Hrs	Unit	Material	Labor	Equipment	Total
Clearing trench shoulders						
of obstructions	CL@.017	LF	—	.62	—	.62
Hard pan or rock outcropping						
Breaking and stacking	CL@3.23	CY	—	118.00	—	118.00
Column pads and small piers,						
average soil	CL@1.32	CY	—	48.00	—	48.00
Backfilling against foundations, no import or export of materials						
Hand, including hand tamp	CL@.860	CY	—	31.30	—	31.30

Backfill with a 1/2 CY utility backhoe/loader

	Craft@Hrs	Unit	Material	Labor	Equipment	Total
Minimum compaction, wheel-rolled	S1@.055	CY	—	2.36	.86	3.22

Backfill with 1/2 CY utility backhoe/loader, 150CFM compressor & pneumatic tamper

	Craft@Hrs	Unit	Material	Labor	Equipment	Total
Spec grade compaction	S1@.141	CY	—	6.06	3.76	9.82

Disposal of excess backfill material with a 1/2 CY backhoe, no compaction

	Craft@Hrs	Unit	Material	Labor	Equipment	Total
Spot spread, 100' haul	S1@.034	CY	—	1.46	.53	1.99
Area spread, 4" deep						
(covers 81 SF per CY)	S1@.054	CY	—	2.32	.84	3.16

Grading for slabs

	Craft@Hrs	Unit	Material	Labor	Equipment	Total
Using a D-4 crawler tractor with dozer blade						
(1000 SF per hour)	S1@.001	SF	—	.04	.02	.06
Fine grading, by hand, light to medium soil						
(165 to 170 SF per hour)	CL@.006	SF	—	.22	—	.22

Capillary fill, one CY covers 81 SF at 4" deep

	Craft@Hrs	Unit	Material	Labor	Equipment	Total
4", hand graded and rolled,						
small job	CL@.014	SF	.28	.51	—	.79
4", machine graded and rolled,						
larger job	S1@.006	SF	.28	.26	.12	.66
Add for each extra 1" hand graded	CL@.004	SF	.08	.15	—	.23
Add for each extra 1" machine graded	S1@.001	SF	.08	.04	.02	.14

Sand fill, 1 CY covers 162 SF at 2" deep

	Craft@Hrs	Unit	Material	Labor	Equipment	Total
2" sand cushion, hand spread, 1 throw	CL@.003	SF	.12	.11	—	.23
Add for each additional 1"	CL@.002	SF	.06	.07	—	.13

Waterproof membrane, polyethylene, over sand bed, including 20% lap and waste

	Craft@Hrs	Unit	Material	Labor	Equipment	Total
.004" (4 mil) clear or black	CL@.001	SF	.04	.04	—	.08
.006" (6 mil) clear or black	CL@.001	SF	.05	.04	—	.09

Formwork for Concrete Labor costs include the time needed to prepare formwork sketches at the job site, measure for the forms, fabricate, erect, align, brace, strip, clean and stack the forms. Costs for reinforcing and concrete are not included. Multiple use of forms shows the cost per use when a form is used several times on the same job without being totally disassembled or discarded. Normal cleaning and repairs are included in the labor cost on lines that show multiple use of forms. Cost will be higher if plans are not detailed enough to identify the work to be performed. No salvage value is assumed except as noted. Costs listed are per square foot of form in contact with the concrete (SFCA). These costs assume standard and better lumber price of $510 per MBF and $848 per MSF for 3/4" plyform before waste allowance. For more detailed coverage of concrete formwork, see National Concrete & Masonry Estimator, http://CraftsmanSiteLicense.com

	Craft@Hrs	Unit	Material	Labor	Total

Wall footing, grade beam or tie beam forms These figures assume nails, stakes and form oil costing $.25 per square foot and 2.5 board feet of lumber are used for each square foot of form in contact with the concrete (SFCA). To calculate the quantity of formwork required, multiply the depth of footing in feet by the length of footing in feet. Then double the results if two sides will be formed. For scheduling purposes, estimate that a crew of 5 can lay out, fabricate and erect 600 to 700 SF of footing, grade beam or tie beam forms in an 8-hour day.

	Craft@Hrs	Unit	Material	Labor	Total
1 use	F5@.070	SFCA	1.80	2.93	4.73
3 uses	F5@.050	SFCA	1.03	2.09	3.12
5 uses	F5@.040	SFCA	.87	1.68	2.55
Add for stepped footings	F5@.028	SFCA	.51	1.17	1.68
Add for keyed joint, 1 use	F5@.020	LF	.74	.84	1.58

Reinforcing bar supports for footing, grade beam and tie beam forms. Bars suspended from 2" x 4" lumber, including stakes. Based on .86 BF of lumber per LF

	Craft@Hrs	Unit	Material	Labor	Total
1 use	F5@.060	LF	.78	2.51	3.29
3 uses	F5@.050	LF	.52	2.09	2.61
5 uses	F5@.040	LF	.46	1.68	2.14

Integral starter wall forms (stem walls) formed monolithic with footings, heights up to 4'0", with 3 BF of lumber per SFCA plus an allowance of $0.25 for nails, stakes and form oil

	Craft@Hrs	Unit	Material	Labor	Total
1 use	F5@.100	SFCA	2.11	4.19	6.30
3 uses	F5@.075	SFCA	1.18	3.14	4.32
5 uses	F5@.067	SFCA	.99	2.81	3.80

Bulkheads or pour-stops. When integral starter walls are formed, forms will usually be required at the ends of each wall or footing. These are usually called bulkheads or pour-stops. These figures assume the use of 1.1 SF of 3/4" plyform per SFCA plus an allowance of $0.25 per SFCA for nails, stakes and bracing

	Craft@Hrs	Unit	Material	Labor	Total
1 use	F5@.150	SFCA	1.75	6.28	8.03
3 uses	F5@.120	SFCA	1.00	5.03	6.03
5 uses	F5@.100	SFCA	.85	4.19	5.04

Column footing or pile cap forms These figures include nails, stakes and form oil and 3.5 board feet of lumber for each square foot of form in contact with the concrete. For scheduling purposes, estimate that a crew of 5 can lay out, fabricate and erect 500 to 600 SF of square or rectangular footing forms in an 8-hour day or 350 to 450 SF of octagonal, hexagonal or triangular footing forms in an 8-hour day.

Square or rectangular column forms

	Craft@Hrs	Unit	Material	Labor	Total
1 use	F5@.090	SFCA	2.42	3.77	6.19
3 uses	F5@.070	SFCA	1.34	2.93	4.27
5 uses	F5@.060	SFCA	1.12	2.51	3.63

Octagonal, hexagonal or triangular forms

	Craft@Hrs	Unit	Material	Labor	Total
1 use	F5@.117	SFCA	2.42	4.90	7.32
3 uses	F5@.091	SFCA	1.34	3.81	5.15
5 uses	F5@.078	SFCA	1.12	3.27	4.39

Reinforcing bar supports for column forms. Bars suspended from 2" x 4" lumber, including stakes. Based on .86 BF of lumber per LF

	Craft@Hrs	Unit	Material	Labor	Total
1 use	F5@.060	LF	.53	2.51	3.04

Anchor bolt templates or dowel supports for column forms. Includes cost of nails, stakes, bracing, and form oil and 1.1 SF of 3/4" plyform for each SF of column base or dowel support. No anchor bolts included

	Craft@Hrs	Unit	Material	Labor	Total
1 use	F5@.150	SF	1.75	6.28	8.03
3 uses	F5@.100	SF	1.00	4.19	5.19
5 uses	F5@.090	SF	.85	3.77	4.62

	Craft@Hrs	Unit	Material	Labor	Total

Slab-on-grade forms These costs assume nails, stakes, form oil, and accessories are used for each board foot of lumber. Note that costs listed below are per linear foot (LF) of form. Figures in parentheses show the average square feet of contact area (SFCA) per linear foot of form. Use this figure to convert costs and man hours per linear foot to costs and manhours per square foot. For scheduling purposes, estimate that a crew of 5 will lay out, fabricate and erect the following quantities of slab-on-grade edge forms in an 8-hour day:

Slabs to 6" high at 615 to 720 LF. Slabs over 6" to 12" high at 450 to 500 LF

Slabs over 12" to 24" high at 320 to 350 LF. Slabs over 24" to 36" high, 240 to 260 LF

Edge forms to 6" high, 1.5 BF per LF of form (.5 SFCA per LF)

	Craft@Hrs	Unit	Material	Labor	Total
1 use	F5@.065	LF	1.18	2.72	3.90
3 uses	F5@.061	LF	.72	2.56	3.28
5 uses	F5@.055	LF	.62	2.30	2.92

Edge forms over 6" high to 12" high, 2.25 BF per LF of form (average .75 SFCA per LF)

1 use	F5@.090	LF	1.65	3.77	5.42
3 uses	F5@.086	LF	.95	3.60	4.55
5 uses	F5@.080	LF	.81	3.35	4.16

Edge forms over 12" high to 24" high, 4.5 BF per LF of form (average 1.5 SFCA per LF)

1 use	F5@.124	LF	3.04	5.19	8.23
3 uses	F5@.119	LF	1.65	4.98	6.63
5 uses	F5@.115	LF	1.37	4.82	6.19

Edge forms over 24" high to 36" high, 7.5 BF per LF of forms (average 2.5 SFCA per LF)

1 use	F5@.166	LF	4.90	6.95	11.85
3 uses	F5@.160	LF	2.58	6.70	9.28
5 uses	F5@.155	LF	2.11	6.49	8.60

Add for 2" x 2" tapered keyed joint, one-piece, .38 BF per LF

1 use	F5@.044	LF	.24	1.84	2.08
3 uses	F5@.040	LF	.12	1.68	1.80
5 uses	F5@.035	LF	.09	1.47	1.56

Blockout and slab depression forms. Figure the linear feet required. Use the linear foot costs for slab edge forms for the appropriate height and then add per linear foot

Blockouts	F5@.030	LF	.63	1.26	1.89

Wall forms Formwork over 6' high includes an allowance for a work platform and handrail built on one side of the form for use by the concrete placing crew.

Heights to 4', includes 1.1 SF of plyform, 1.5 BF of lumber and allowance for nails, ties and oil per SFCA

1 use	F5@.119	SFCA	2.68	4.98	7.66
3 uses	F5@.080	SFCA	1.46	3.35	4.81
5 uses	F5@.069	SFCA	1.22	2.89	4.11
Add for 1 side battered	F5@.016	SFCA	.27	.67	.94
Add for 2 sides battered	F5@.024	SFCA	.54	1.01	1.55

Heights over 4' to 6', includes 1.1 SF of plyform, 2.0 BF of lumber and allowance for nails, ties and oil

1 use	F5@.140	SFCA	2.99	5.86	8.85
3 uses	F5@.100	SFCA	1.62	4.19	5.81
5 uses	F5@.080	SFCA	1.34	3.35	4.69
Add for 1 side battered	F5@.016	SFCA	.30	.67	.97
Add for 2 sides battered	F5@.024	SFCA	.60	1.01	1.61

Heights over 6' to 12', includes 1.2 SF of plyform, 2.5 BF of lumber and allowance for nails, ties and oil

1 use	F5@.160	SFCA	3.43	6.70	10.13
3 uses	F5@.110	SFCA	1.84	4.61	6.45
5 uses	F5@.100	SFCA	1.52	4.19	5.71
Add for 1 side battered	F5@.016	SFCA	.34	.67	1.01
Add for 2 sides battered	F5@.024	SFCA	.69	1.01	1.70

Heights over 12' to 16', includes 1.2 SF of plyform, 3.0 BF of lumber and allowance for nails, ties and oil

1 use	F5@.180	SFCA	3.74	7.54	11.28
3 uses	F5@.130	SFCA	2.00	5.45	7.45
5 uses	F5@.110	SFCA	1.65	4.61	6.26
Add for 1 side battered	F5@.016	SFCA	.37	.67	1.04
Add for 2 sides battered	F5@.024	SFCA	.75	1.01	1.76

	Craft@Hrs	Unit	Material	Labor	Total
Heights over 16', includes 1.3 SF of plyform, 3.5 BF of lumber and allowance for nails, ties and oil					
1 use	F5@.199	SFCA	4.19	8.34	12.53
3 uses	F5@.140	SFCA	2.22	5.86	8.08
5 uses	F5@.119	SFCA	1.83	4.98	6.81
Add for 1 side battered	F5@.016	SFCA	.42	.67	1.09
Add for 2 sides battered	F5@.024	SFCA	.84	1.01	1.85
Architectural form liner for wall forms					
Low cost liners	F5@.020	SFCA	1.25	.84	2.09
Average cost liners	F5@.100	SFCA	2.00	4.19	6.19
Reveal strips 1" deep by 2" wide, one-piece, wood					
1 use	F5@.069	LF	.12	2.89	3.01
3 uses	F5@.050	LF	.07	2.09	2.16
5 uses	F5@.020	LF	.06	.84	.90
Blockouts for openings. Form area is the opening perimeter times the depth. These figures assume that nails and form oil costing $.25 per square foot, 1.2 SF of plyform and .5 board feet of lumber are used per SF of form. Blockouts usually can be used only once.					
1 use	F5@.250	SF	2.19	10.50	12.69
Bulkheads or pour-stops. These costs assume that nails, form oil and accessories, 1.1 SF of plyform and 1 board foot of lumber are used per SF of form					
1 use	F5@.220	SF	2.37	9.22	11.59
3 uses	F5@.149	SF	1.31	6.24	7.55
5 uses	F5@.130	SF	1.10	5.45	6.55
Add for keyed wall joint, two-piece, tapered, 1.2 BF per SF					
1 use	F5@.100	SF	.99	4.19	5.18
3 uses	F5@.069	SF	.62	2.89	3.51
5 uses	F5@.061	SF	.55	2.56	3.11
Curved wall forms					
Smooth radius, add to straight wall cost	F5@.032	SFCA	.86	1.34	2.20
8' chord sections, add to straight wall cost	F5@.032	SFCA	.52	1.34	1.86
Haunches or ledges. Area is the width of the ledge times the length. These costs assume nails, form oil and accessories, 2 SF of plyform, and .5 board foot of lumber are used per SF of ledge					
1 use	F5@.300	SF	3.28	12.60	15.88
3 uses	F5@.210	SF	1.77	8.80	10.57
5 uses	F5@.180	SF	1.46	7.54	9.00

Steel framed plywood forms Rented steel framed plywood forms can reduce forming costs on many jobs. Where rented forms are used 3 times a month, use the wall forming costs shown for 3 uses at the appropriate height but deduct 25% from the material cost and 50% from the labor manhours and labor costs. Savings will be smaller where layouts change from one use to the next, when form penetrations must be made and repaired before returning a form, where form delivery costs are high, and where non-standard form sizes are needed. For more detailed coverage of steel framed plywood forms, see National Concrete & Masonry Estimator, http://CraftsmanSiteLicense.com

Column forms for square or rectangular columns Quantities shown in parentheses give cross section area of a square column in square inches. When estimating a rectangular column, use the costs for the square column with closest cross section area. For scheduling purposes estimate that a crew of 5 can lay out, fabricate and erect about 550 to 600 SFCA of square or rectangular column forms in an 8-hour day.

Up to 12" x 12" (144 square inches), using nails, snap ties, oil and column clamps with 1.15 SF of plyform and 2 BF of lumber per SFCA					
1 use	F5@.130	SFCA	3.06	5.45	8.51
3 uses	F5@.090	SFCA	1.65	3.77	5.42
5 uses	F5@.080	SFCA	1.37	3.35	4.72
Over 12" x 12" to 16" x 16" (256 square inches), using nails, snap ties, oil and column clamps, with 1.125 SF of plyform and 2.1 BF of lumber per SFCA					
1 use	F5@.100	SFCA	3.08	4.19	7.27
3 uses	F5@.069	SFCA	1.67	2.89	4.56
5 uses	F5@.061	SFCA	1.38	2.56	3.94

	Craft@Hrs	Unit	Material	Labor	Total

Over 16" x 16" to 20" x 20" (400 square inches), using nails, snap ties, oil and column clamps, with 1.1 SF of plyform and 2.2 BF of lumber per SFCA

	Craft@Hrs	Unit	Material	Labor	Total
1 use	F5@.080	SFCA	3.11	3.35	6.46
3 uses	F5@.061	SFCA	1.68	2.56	4.24
5 uses	F5@.050	SFCA	1.39	2.09	3.48

Over 20" x 20" to 24" x 24" (576 square inches), using nails, snap ties, oil and column clamps, with 1.1 SF of plyform and 2.4 BF of lumber per SFCA

	Craft@Hrs	Unit	Material	Labor	Total
1 use	F5@.069	SFCA	3.24	2.89	6.13
3 uses	F5@.050	SFCA	1.74	2.09	3.83
5 uses	F5@.040	SFCA	1.44	1.68	3.12

Over 24" x 24" to 30" x 30" (900 square inches), using nails, snap ties, oil and column clamps, with 1.1 SF of plyform and 2.4 BF of lumber per SFCA

	Craft@Hrs	Unit	Material	Labor	Total
1 use	F5@.061	SFCA	3.24	2.56	5.80
3 uses	F5@.040	SFCA	1.74	1.68	3.42
5 uses	F5@.031	SFCA	1.44	1.30	2.74

Over 30" x 30" to 36" x 36" (1,296 square inches), using nails, snap ties, oil and column clamps, with 1.1 SF of plyform and 2.4 BF of lumber per SFCA

	Craft@Hrs	Unit	Material	Labor	Total
1 use	F5@.050	SFCA	3.24	2.09	5.33
3 uses	F5@.040	SFCA	1.74	1.68	3.42
5 uses	F5@.031	SFCA	1.44	1.30	2.74

Over 36" x 36" to 48" x 48" (2,304 square inches), using nails, snap ties, oil and column clamps, with 1.1 SF of plyform and 2.5 BF of lumber per SFCA

	Craft@Hrs	Unit	Material	Labor	Total
1 use	F5@.040	SFCA	3.30	1.68	4.98
3 uses	F5@.031	SFCA	1.77	1.30	3.07
5 uses	F5@.020	SFCA	1.54	.84	2.38

Column capitals for square columns Capital forms for square columns usually have four symmetrical sides. Length and width at the top of the capital are usually twice the length and width at the bottom of the capital. Height is usually the same as the width at the capital base. These costs assume capitals installed not over 12' above floor level, use of nails, form oil, shores and accessories with 1.5 SF of plyform and 2 BF of lumber per SF of contact area (SFCA). Complexity of these forms usually makes more than 3 uses impractical.

For scheduling purposes estimate that a crew of 5 can lay out, fabricate and erect the following quantities of capital formwork in an 8-hour day:

70 to 80 SF for 12" to 16" columns   90 to 100 SF for 20" to 24" columns
110 to 120 SF for 30" to 36" columns   160 to 200 SF for 48" columns

Up to 12" x 12" column, 6.0 SFCA

	Craft@Hrs	Unit	Material	Labor	Total
1 use of forms	F5@4.01	Ea	21.20	168.00	189.20
3 uses of forms	F5@3.00	Ea	10.60	126.00	136.60

Over 12" x 12" to 16" x 16" column, 10.7 SFCA

	Craft@Hrs	Unit	Material	Labor	Total
1 use of forms	F5@6.00	Ea	37.80	251.00	288.80
3 uses of forms	F5@4.01	Ea	18.90	168.00	186.90

Over 16" x 16" to 20" x 20" column, 16.6 SFCA

	Craft@Hrs	Unit	Material	Labor	Total
1 use of forms	F5@8.31	Ea	58.60	348.00	406.60
3 uses of forms	F5@6.00	Ea	29.30	251.00	280.30

Over 20" x 20" to 24" x 24" column, 24.0 SFCA

	Craft@Hrs	Unit	Material	Labor	Total
1 use of forms	F5@12.0	Ea	84.80	503.00	587.80
3 uses of forms	F5@9.00	Ea	42.40	377.00	419.40

Over 24" x 24" to 30" x 30" column, 37.5 SFCA

	Craft@Hrs	Unit	Material	Labor	Total
1 use of forms	F5@16.0	Ea	134.00	670.00	804.00
3 uses of forms	F5@12.0	Ea	66.60	503.00	569.60

Over 30" x 30" to 36" x 36" column, 54.0 SFCA

	Craft@Hrs	Unit	Material	Labor	Total
1 use of forms	F5@20.0	Ea	191.00	838.00	1,029.00
3 uses of forms	F5@16.0	Ea	95.40	670.00	765.40

Over 36" x 36" to 48" x 48", 96.0 SFCA

	Craft@Hrs	Unit	Material	Labor	Total
1 use of forms	F5@24.0	Ea	339.00	1,010.00	1,349.00
3 uses of forms	F5@18.0	Ea	170.00	754.00	924.00

Concrete 3

	Craft@Hrs	Unit	Material	Labor	Total

Column forms for round columns Use the costs below to estimate round fiber tube forms (Sonotube is one manufacturer). Add the cost of column footings or foundations. These forms are peeled off when the concrete has cured. Costs shown include setting, aligning, bracing and stripping and assume that bracing and collars at top and bottom can be used 3 times. Column forms over 12'0" long will cost more per linear foot. Costs are per linear foot for standard weight column forms. Plastic lined forms have one vertical seam. For scheduling purposes, estimate that a crew of 5 can lay out, erect and brace the following quantities of 10' to 12' high round fiber tube forms in an 8-hour day: twenty 8" to 12" diameter columns; sixteen 14" to 20" columns and twelve 24" to 48" columns. For more detailed coverage of round column forms, see National Concrete & Masonry Estimator, http://CraftsmanSiteLicense.com.

	Craft@Hrs	Unit	Material	Labor	Total
Bracing per LF (add to each cost below)	—	LF	1.64	—	1.64
8" spiral type	F5@.166	LF	1.44	6.95	8.39
10" spiral type	F5@.182	LF	1.79	7.62	9.41
12" spiral type	F5@.207	LF	2.17	8.67	10.84
14" spiral type	F5@.224	LF	2.59	9.38	11.97
16" spiral type	F5@.232	LF	2.93	9.72	12.65
20" spiral type	F5@.250	LF	4.48	10.50	14.98
24" spiral type	F5@.275	LF	5.66	11.50	17.16
30" spiral type	F5@.289	LF	8.76	12.10	20.86
36" spiral type	F5@.308	LF	10.50	12.90	23.40
48" spiral type	F5@.334	LF	32.50	14.00	46.50
8" plastic lined	F5@.166	LF	5.82	6.95	12.77
10" plastic lined	F5@.177	LF	6.12	7.41	13.53
12" plastic lined	F5@.207	LF	8.13	8.67	16.80
14" plastic lined	F5@.224	LF	9.00	9.38	18.38
16" plastic lined	F5@.232	LF	10.60	9.72	20.32
20" plastic lined	F5@.250	LF	13.90	10.50	24.40
24" plastic lined	F5@.275	LF	16.70	11.50	28.20
30" plastic lined	F5@.289	LF	19.80	12.10	31.90
36" plastic lined	F5@.308	LF	21.50	12.90	34.40
48" plastic lined	F5@.334	LF	32.50	14.00	46.50

Column capitals for round columns

Use these figures to estimate costs for rented prefabricated steel conical shaped capital forms. Costs assume a minimum rental period of 1 month and include an allowance for nails and form oil. Labor shown assumes the capital will be supported from the deck formwork above the capital and includes laying out and cutting the hole in the deck to receive the capital. Capital bottom diameter is sized to fit into the column tube form. The column form size (diameter) does not significantly affect the capital form rental cost. Dimensions shown below are the top diameter. Form weights are shown in parentheses. Only 3 uses of these forms are usually possible in a 30-day period. Setting forms and installing reinforcing steel takes one day, concrete placing takes another day, curing requires 7 to 10 days, removing and cleaning forms takes another day. For scheduling purposes estimate that a crew of 5 can lay out, cut openings in floor deck formwork and install an average of 8 or 9 round column capital forms in an 8-hour day.

	Craft@Hrs	Unit	Material	Labor	Total
3'6" (100 pounds) 1 use	F5@4.01	Ea	193.00	168.00	361.00
3'6" (100 pounds) 3 uses	F5@4.01	Ea	64.60	168.00	232.60
4'0" (125 pounds) 1 use	F5@4.01	Ea	208.00	168.00	376.00
4'0" (125 pounds) 3 uses	F5@4.01	Ea	69.50	168.00	237.50
4'6" (150 pounds) 1 use	F5@4.30	Ea	216.00	180.00	396.00
4'6" (150 pounds) 3 uses	F5@4.30	Ea	72.10	180.00	252.10
5'0" (175 pounds) 1 use	F5@4.51	Ea	227.00	189.00	416.00
5'0" (175 pounds) 3 uses	F5@4.51	Ea	81.00	189.00	270.00
5'6" (200 pounds) 1 use	F5@4.51	Ea	261.00	189.00	450.00
5'6" (200 pounds) 3 uses	F5@4.51	Ea	87.40	189.00	276.40
6'0" (225 pounds) 1 use	F5@4.70	Ea	273.00	197.00	470.00
6'0" (225 pounds) 3 uses	F5@4.70	Ea	86.30	197.00	283.30

Concrete 3

	Craft@Hrs	Unit	Material	Labor	Total

Beam and girder forms Using nails, snap ties and oil, 1.3 SF of 3/4" plyform and 2.1 BF of lumber per SFCA. For scheduling purposes, estimate that a crew of 5 can lay out, fabricate and erect 250 to 300 SF of beam and girder forms in an 8-hour day. For more detailed coverage of beam and girder forms, see National Concrete & Masonry Estimator, http://CraftsmanSiteLicense.com.

	Craft@Hrs	Unit	Material	Labor	Total
1 use	F5@.149	SFCA	3.32	6.24	9.56
3 uses	F5@.141	SFCA	1.79	5.91	7.70
5 uses	F5@.120	SFCA	1.48	5.03	6.51

Shores for beams and girders Using 4" x 4" wooden posts. These figures assume an average shore height of 12'0" with a 2'0" long horizontal 4" x 4" head and 2" x 6" diagonal brace (30 BF of lumber per shore) and include nails, adjustable post clamps and accessories. For scheduling purposes estimate that a crew of 5 can lay out, fabricate and install 30 to 35 adjustable wood shores in an 8-hour day when shores have an average height of 12'0". Quantity and spacing of shores is dictated by the size of the beams and girders. Generally, allow one shore for each 6 LF of beam or girder. Costs for heavy duty shoring are listed under General Requirements, Heavy Duty Shoring.

	Craft@Hrs	Unit	Material	Labor	Total
1 use	F5@1.50	Ea	28.40	62.80	91.20
3 uses	F5@1.00	Ea	15.70	41.90	57.60
5 uses	F5@.748	Ea	13.20	31.30	44.50

Elevated slab forms These are costs for forms per square foot of finished slab area and include shoring and stripping at heights to 12'. Bays are assumed to be 20'0" x 20'0" bays.

Waffle slab-joist pans, monolithic Using rented forms (30-day rental period minimum). For scheduling purposes, estimate that a crew of 5 can lay out, fabricate and install 500 to 600 SF of waffle slab-joist pan forms and shoring in an 8-hour day. Two uses of the forms in a 30-day period is usually the maximum.

Metal pans, 10,000 to 20,000 SF

	Craft@Hrs	Unit	Material	Labor	Total
1 use	F5@.100	SF	4.66	4.19	8.85
2 uses	F5@.066	SF	2.61	2.76	5.37

Fiberglass pans, 10,000 to 20,000 SF

	Craft@Hrs	Unit	Material	Labor	Total
1 use	F5@.077	SF	4.66	3.23	7.89
2 uses	F5@.061	SF	2.67	2.56	5.23

Flat slab with beams, monolithic Including nails, braces, stiffeners and oil 1.2 SF of plyform and 3.2 BF of lumber per SFCA for one-way beam systems. Two-way beam systems require approximately 40% more forming materials than one-way beam systems. For scheduling purposes, estimate that a crew of 5 can lay out, fabricate and erect 350 to 400 SF of one-way beam systems and 250 to 300 SF of two-way beam systems in an 8-hour day, including shoring.

	Craft@Hrs	Unit	Material	Labor	Total
One-way beam 1 use	F5@.110	SFCA	3.87	4.61	8.48
One-way beam 2 uses	F5@.080	SFCA	2.97	3.35	6.32
Two-way beam 1 use	F5@.152	SFCA	5.32	6.37	11.69
Two-way beam 2 uses	F5@.100	SFCA	3.13	4.19	7.32

Structural flat slab Using nails, braces, stiffeners and oil costing $.25, 1.2 SF of plyform and 2.25 BF of lumber per SFCA. For scheduling purposes, estimate that a crew of 5 can lay out, fabricate and erect 500 to 600 SFCA of structural flat slab forms in an 8-hour day, including shoring.

	Craft@Hrs	Unit	Material	Labor	Total
Edge forms	F5@.072	LF	1.65	3.02	4.67
Slab forms 1 use	F5@.069	SFCA	3.28	2.89	6.17
Slab forms 3 uses	F5@.048	SFCA	1.76	2.01	3.77
Slab forms 5 uses	F5@.043	SFCA	1.46	1.80	3.26

Additional forming costs for elevated slabs, if required Based on single use of plyform and lumber. For scheduling purposes estimate that a 5 man crew can install 500 SFCA per 8-hour day.

	Craft@Hrs	Unit	Material	Labor	Total
Control joints	F5@.031	LF	1.76	1.30	3.06
Curbs & pads	F5@.083	SFCA	3.28	3.48	6.76
Depressions	F5@.083	SFCA	.98	3.48	4.46
Keyed joints	F5@.031	LF	.74	1.30	2.04

Concrete 3

	Craft@Hrs	Unit	Material	Labor	Total

Miscellaneous formwork Based on single use of plyform and lumber. For scheduling purposes, estimate that a crew of 5 can lay out, fabricate and erect 225 to 250 SFCA of these types of forms in an 8-hour day.

	Craft@Hrs	Unit	Material	Labor	Total
Flat soffits & landings	F5@.166	SFCA	5.08	6.95	12.03
Sloping soffits	F5@.166	SFCA	5.25	6.95	12.20
Stair risers (steps)	F5@.120	SFCA	1.97	5.03	7.00

Grade 60 ASTM A615 Rebars, 20' lengths, material costs only. Add the cost of delivery, shop fabrication, cutting, bending, setting, lap and tying bars.

	Craft@Hrs	Unit	Material	Labor	Total
3/8" x 20'	--@.000	Ea	3.25	—	3.25
1/2" x 20'	--@.000	Ea	5.30	—	5.30
5/8" x 20'	--@.000	Ea	8.20	—	8.20
3/4" x 20'	--@.000	Ea	9.75	—	9.75
1" x 20'	--@.000	Ea	22.90	—	22.90

Reinforcing for Cast-in-Place Concrete Steel reinforcing bars (rebar), ASTM A615 Grade 60. Material costs are for deformed steel reinforcing rebar, including 10% lap allowance, cutting and bending. These costs also include detailed shop drawings and delivery to jobsite with identity tags per shop drawings. Add for epoxy or galvanized coating of rebar, chairs, splicing, spiral caissons and round column reinforcing, if required, from the sections following the data below. Costs per pound (Lb) and per linear foot (LF) including tie wire and tying.

Reinforcing steel placed and tied in footings, foundations and grade beams

	Craft@Hrs	Unit	Material	Labor	Total
1/4" diameter, #2 rebar	RB@.015	Lb	.78	.80	1.58
1/4" diameter, #2 rebar (.17 lb per LF)	RB@.003	LF	.14	.16	.30
3/8" diameter, #3 rebar	RB@.011	Lb	.47	.58	1.05
3/8" diameter, #3 rebar (.38 lb per LF)	RB@.004	LF	.18	.21	.39
1/2" diameter, #4 rebar	RB@.010	Lb	.45	.53	.98
1/2" diameter, #4 rebar (.67 lb per LF)	RB@.007	LF	.30	.37	.67
5/8" diameter, #5 rebar	RB@.009	Lb	.40	.48	.88
5/8" diameter, #5 rebar (1.04 lb per LF)	RB@.009	LF	.42	.48	.90
3/4" diameter, #6 rebar	RB@.008	Lb	.37	.43	.80
3/4" diameter, #6 rebar (1.50 lb per LF)	RB@.012	LF	.56	.64	1.20
7/8" diameter, #7 rebar	RB@.008	Lb	.45	.43	.88
7/8" diameter, #7 rebar (2.04 lb per LF)	RB@.016	LF	.92	.85	1.77
1" diameter, #8 rebar	RB@.008	Lb	.43	.43	.86
1" diameter, #8 rebar (2.67 lb per LF)	RB@.021	LF	1.15	1.12	2.27
1-1/8" diameter, #9 rebar	RB@.008	Lb	.48	.43	.91
1-1/8" diameter, #9 rebar (3.40 lb per LF)	RB@.027	LF	1.63	1.44	3.07
1-1/4" diameter, #10 rebar	RB@.007	Lb	.48	.37	.85
1-1/4" diameter, #10 rebar (4.30 lb per LF)	RB@.030	LF	2.06	1.59	3.65
1-3/8" diameter, #11 rebar	RB@.007	Lb	.48	.37	.85
1-3/8" diameter, #11 rebar (5.31 lb per LF)	RB@.037	LF	2.55	1.97	4.52

Reinforcing steel placed and tied in structural slabs

	Craft@Hrs	Unit	Material	Labor	Total
1/4" diameter, #2 rebar	RB@.014	Lb	.78	.74	1.52
1/4" diameter, #2 rebar (.17 lb per LF)	RB@.002	LF	.14	.11	.25
3/8" diameter, #3 rebar	RB@.010	Lb	.47	.53	1.00
3/8" diameter, #3 rebar (.38 lb per LF)	RB@.004	LF	.18	.21	.39
1/2" diameter, #4 rebar	RB@.009	Lb	.45	.48	.93
1/2" diameter, #4 rebar (.67 lb per LF)	RB@.006	LF	.30	.32	.62
5/8" diameter, #5 rebar	RB@.008	Lb	.40	.43	.83
5/8" diameter, #5 rebar (1.04 lb per LF)	RB@.008	LF	.42	.43	.85
3/4" diameter, #6 rebar	RB@.007	Lb	.37	.37	.74
3/4" diameter, #6 rebar (1.50 lb per LF)	RB@.011	LF	.56	.58	1.14
7/8" diameter, #7 rebar	RB@.007	Lb	.45	.37	.82
7/8" diameter, #7 rebar (2.04 lb per LF)	RB@.014	LF	.92	.74	1.66

Concrete 3

	Craft@Hrs	Unit	Material	Labor	Total
1" diameter, #8 rebar	RB@.007	Lb	.43	.37	.80
1" diameter, #8 rebar (2.67 lb per LF)	RB@.019	LF	1.15	1.01	2.16
1-1/8" diameter, #9 rebar	RB@.007	Lb	.48	.37	.85
1-1/8" diameter, #9 rebar (3.40 lb per LF)	RB@.024	LF	1.63	1.28	2.91
1-1/4" diameter, #10 rebar	RB@.006	Lb	.48	.32	.80
1-1/4" diameter, #10 rebar (4.30 lb per LF)	RB@.026	LF	2.06	1.38	3.44
1-3/8" diameter, #11 rebar	RB@.006	Lb	.48	.32	.80
1-3/8" diameter, #11 rebar (5.31 lb per LF)	RB@.032	LF	2.55	1.70	4.25
Reinforcing steel placed and tied in columns, stairs and walls					
1/4" diameter, #2 rebar	RB@.017	Lb	.78	.90	1.68
1/4" diameter, #2 rebar (.17 lb per LF)	RB@.003	LF	.14	.16	.30
3/8" diameter, #3 rebar	RB@.012	Lb	.47	.64	1.11
3/8" diameter, #3 rebar (.38 lb per LF)	RB@.005	LF	.18	.27	.45
1/2" diameter, #4 rebar	RB@.011	Lb	.45	.58	1.03
1/2" diameter, #4 rebar (.67 lb per LF)	RB@.007	LF	.30	.37	.67
5/8" diameter, #5 rebar	RB@.010	Lb	.40	.53	.93
5/8" diameter, #5 rebar (1.04 lb per LF)	RB@.010	LF	.42	.53	.95
3/4" diameter, #6 rebar	RB@.009	Lb	.37	.48	.85
3/4" diameter, #6 rebar (1.50 lb per LF)	RB@.014	LF	.56	.74	1.30
7/8" diameter, #7 rebar	RB@.009	Lb	.45	.48	.93
7/8" diameter, #7 rebar (2.04 lb per LF)	RB@.018	LF	.92	.96	1.88
1" diameter, #8 rebar	RB@.009	Lb	.43	.48	.91
1" diameter, #8 rebar (2.67 lb per LF)	RB@.024	LF	1.15	1.28	2.43
1-1/8" diameter, #9 rebar	RB@.009	Lb	.48	.48	.96
1-1/8" diameter, #9 rebar (3.40 lb per LF)	RB@.031	LF	1.63	1.65	3.28
1-1/4" diameter, #10 rebar	RB@.008	Lb	.48	.43	.91
1-1/4" diameter, #10 rebar (4.30 lb per LF)	RB@.034	LF	2.06	1.81	3.87
1-3/8" diameter, #11 rebar	RB@.008	Lb	.48	.43	.91
1-3/8" diameter, #11 rebar (5.31 lb per LF)	RB@.037	LF	2.55	1.97	4.52
Any size rebar, add	—	Lb	.47	—	.47
Galvanized coating					
Any #2 rebar, #3 rebar or #4 rebar, add	—	Lb	.36	—	.36
Any #5 rebar or #6 rebar, add	—	Lb	.34	—	.34
Any #7 rebar or #8 rebar, add	—	Lb	.32	—	.32
Any #9 rebar, #10 rebar or #11 rebar, add	—	Lb	.30	—	.30
Pre-bent rebar, 90 degree except as noted					
3/8" diameter, #3 rebar, 18" x 18"	RB@.014	Ea	.75	.74	1.49
3/8" diameter, #3 rebar, 24" x 24"	RB@.014	Ea	.95	.74	1.69
1/2" diameter, #4 rebar, 10" x 3'	RB@.021	Ea	2.25	1.12	3.37
1/2" diameter, #4 rebar, 2' x 3'	RB@.021	Ea	2.85	1.12	3.97
5/8" diameter, #5 rebar, 18" x 18"	RB@.014	Ea	1.70	.74	2.44
5/8" diameter, #5 rebar, 24" x 24"	RB@.014	Ea	2.10	.74	2.84
5/8" diameter, #5 rebar, 1' x 4'	RB@.035	Ea	3.50	1.86	5.36
5/8" diameter, #5 rebar, 3' x 3'	RB@.043	Ea	5.25	2.29	7.54
5/8" diameter, #5 rebar, 30" x 9" J-bar	RB@.028	Ea	1.80	1.49	3.29
Rebar stirrups					
3/8" #3 bar, 6" x 12"	RB@.014	Ea	1.05	.74	1.79
3/8" #3 bar, 6" x 14"	RB@.014	Ea	1.20	.74	1.94
3/8" #3 bar, 6" x 18"	RB@.014	Ea	1.40	.74	2.14
3/8" #3 bar, 6" x 24"	RB@.014	Ea	1.45	.74	2.19
5/8" #8 bar, 6" x 48"	RB@.014	Ea	2.75	.74	3.49

	Craft@Hrs	Unit	Material	Labor	Total
Rebar rings					
Rectangular, 6" x 3-1/2"	RB@.014	Ea	1.00	.74	1.74
Rectangular, 9" x 5"	RB@.014	Ea	1.20	.74	1.94
Rectangular, 11" x 5"	RB@.014	Ea	1.20	.74	1.94
Rectangular, 12" x 3-1/2"	RB@.014	Ea	1.20	.74	1.94
Rectangular, 14" x 4"	RB@.014	Ea	1.20	.74	1.94
Rectangular, 14" x 5"	RB@.014	Ea	1.20	.74	1.94
Round, 6" diameter	RB@.014	Ea	.95	.74	1.69
Round, 8" diameter	RB@.014	Ea	1.00	.74	1.74
Round, 10" diameter	RB@.014	Ea	1.20	.74	1.94
Square, 8"	RB@.014	Ea	1.05	.74	1.79

Add for reinforcing steel chairs. Labor for placing rebar shown above includes time for setting chairs.

	Craft@Hrs	Unit	Material	Labor	Total
Chairs, individual type, cost each					
3" high					
Steel	—	Ea	.47	—	.47
Plastic	—	Ea	.21	—	.21
Galvanized steel	—	Ea	.72	—	.72
5" high					
Steel	—	Ea	.72	—	.72
Plastic	—	Ea	.56	—	.56
Galvanized steel	—	Ea	.93	—	.93
8" high					
Steel	—	Ea	1.52	—	1.52
Plastic	—	Ea	1.14	—	1.14
Galvanized steel	—	Ea	2.03	—	2.03
12" high					
Steel	—	Ea	2.96	—	2.96
Plastic	—	Ea	1.96	—	1.96
Galvanized steel	—	Ea	3.64	—	3.64
Chairs, continuous type, with legs 8" OC, cost per linear foot					
3" high					
Steel	—	LF	.65	—	.65
Plastic	—	LF	.46	—	.46
Galvanized steel	—	LF	.84	—	.84
6" high					
Steel	—	LF	.89	—	.89
Plastic	—	LF	.65	—	.65
Galvanized steel	—	LF	1.17	—	1.17
8" high					
Steel	—	LF	2.99	—	2.99
Plastic	—	LF	1.19	—	1.19
Galvanized steel	—	LF	1.65	—	1.65
12" high					
Steel	—	LF	2.99	—	2.99
Plastic	—	LF	1.93	—	1.93
Galvanized steel	—	LF	3.41	—	3.41

Reinforcing bar weld splicing. Cost per splice

	Craft@Hrs	Unit	Material	Labor	Total
#8 bars, #9 bars or #10 bars	RB@.527	Ea	3.62	28.00	31.62

Reinforcing bar clip splicing, sleeve and wedge, with hand-held hydraulic ram. Cost per splice

	Craft@Hrs	Unit	Material	Labor	Total
Number 4 bars, 1/2"	RB@.212	Ea	6.50	11.30	17.80
Number 5 bars, 5/8"	RB@.236	Ea	15.30	12.50	27.80
Number 6 bars, 3/4"	RB@.258	Ea	28.50	13.70	42.20

Waterstop, 3/8"

	Craft@Hrs	Unit	Material	Labor	Total
Rubber, 6"	B2@.059	LF	5.15	1.67	6.82
Rubber, 9"	B2@.059	LF	7.70	1.67	9.37
PVC, 9"	B2@.059	LF	4.60	1.67	6.27

Concrete 3

	Craft@Hrs	Unit	Material	Labor	Total

Spiral caisson and round column reinforcing, 3/8" diameter hot rolled steel spirals with main vertical bars as shown. Costs include typical engineering and shop drawings. Shop fabricated spirals delivered, tagged, ready to install.
Cost per vertical linear foot (VLF)

	Craft@Hrs	Unit	Material	Labor	Total
16" diameter with 6 #6 bars, 16.8 lbs per VLF	RB@.041	VLF	26.20	2.18	28.38
24" diameter with 6 #6 bars, 28.5 lbs per VLF	RB@.085	VLF	44.70	4.52	49.22
36" diameter with 8 #10 bars, 51.2 lbs per VLF	RB@.216	VLF	81.10	11.50	92.60

Welded wire mesh steel, electric weld, including 15% waste and overlap

	Craft@Hrs	Unit	Material	Labor	Total
2" x 2" W.9 x W.9 (#12 x #12), slabs	RB@.004	SF	.85	.21	1.06
2" x 2" W.9 x W.9 (#12 x #12), beams and columns	RB@.020	SF	.82	1.06	1.88
4" x 4" W1.4 x W1.4 (#10 x #10), slabs	RB@.003	SF	.39	.16	.55
4" x 4" W2.0 x W2.0 (#8 x #8), slabs	RB@.004	SF	.46	.21	.67
4" x 4" W2.9 x W2.9 (#6 x #6), slabs	RB@.005	SF	.51	.27	.78
4" x 4" W4.0 x W4.0 (#4 x #4), slabs	RB@.006	SF	.67	.32	.99
6" x 6" W1.4 x W1.4 (#10 x #10), slabs	RB@.003	SF	.18	.16	.34
6" x 6" W2.0 x W2.0 (#8 x #8), slabs	RB@.004	SF	.30	.21	.51
6" x 6" W2.9 x W2.9 (#6 x #6), slabs	RB@.004	SF	.34	.21	.55
6" x 6" W4.0 x W4.0 (#4 x #4), slabs	RB@.005	SF	.48	.27	.75
Add for lengthwise cut, LF of cut	RB@.002	LF	—	.11	.11

Welded wire mesh, galvanized steel, electric weld, including 15% waste and overlap

	Craft@Hrs	Unit	Material	Labor	Total
2" x 2" W.9 x W.9 (#12 x #12), slabs	RB@.004	SF	1.65	.21	1.86
2" x 2" W.9 x W.9 (#12 x #12), beams and columns	RB@.020	SF	1.65	1.06	2.71
4" x 4" W1.4 x W1.4 (#10 x #10), slabs	RB@.003	SF	.78	.16	.94
4" x 4" W2.0 x W2.0 (#8 x #8), slabs	RB@.004	SF	.91	.21	1.12
4" x 4" W2.9 x W2.9 (#6 x #6), slabs	RB@.005	SF	1.03	.27	1.30
4" x 4" W4.0 x W4.0 (#4 x #4), slabs	RB@.006	SF	1.33	.32	1.65
6" x 6" W1.4 x W1.4 (#10 x #10), slabs	RB@.003	SF	.42	.16	.58
6" x 6" W2.0 x W2.0 (#8 x #8), slabs	RB@.004	SF	.55	.21	.76
6" x 6" W2.9 x W2.9 (#6 x #6), slabs	RB@.004	SF	.67	.21	.88
6" x 6" W4.0 x W4.0 (#4 x #4), slabs	RB@.005	SF	.97	.27	1.24
Add for lengthwise cut, LF of cut	RB@.002	LF	—	.11	.11

Welded wire mesh, epoxy coated steel, electric weld, including 15% waste and overlap

	Craft@Hrs	Unit	Material	Labor	Total
2" x 2" W.9 x W.9 (#12 x #12), slabs	RB@.004	SF	3.38	.21	3.59
2" x 2" W.9 x W.9 (#12 x #12), beams and columns	RB@.020	SF	3.38	1.06	4.44
4" x 4" W1.4 x W1.4 (#10 x #10), slabs	RB@.003	SF	2.93	.16	3.09
4" x 4" W2.0 x W2.0 (#8 x #8), slabs	RB@.004	SF	2.99	.21	3.20
4" x 4" W2.9 x W2.9 (#6 x #6), slabs	RB@.005	SF	3.04	.27	3.31
4" x 4" W4.0 x W4.0 (#4 x #4), slabs	RB@.006	SF	3.20	.32	3.52
6" x 6" W1.4 x W1.4 (#10 x #10), slabs	RB@.003	SF	2.74	.16	2.90
6" x 6" W2.0 x W2.0 (#8 x #8), slabs	RB@.004	SF	2.81	.21	3.02
6" x 6" W2.9 x W2.9 (#6 x #6), slabs	RB@.004	SF	2.86	.21	3.07
6" x 6" W4.0 x W4.0 (#4 x #4), slabs	RB@.005	SF	3.02	.27	3.29
Add for lengthwise cut, LF of cut	RB@.002	LF	—	.11	.11

Reinforcing mesh mini-mats, electric weld, including 15% waste and overlap

	Craft@Hrs	Unit	Material	Labor	Total
6" x 6" W1.4 x W1.4 (#10 x #10), 4' x 8'	RB@.003	SF	.19	.16	.35

	Craft@Hrs	Unit	Material	Labor	Total

Ready-Mix Concrete Ready-mix delivered by truck. Typical prices for most cities. Includes delivery up to 20 miles for 10 CY or more, 3" to 4" slump. Material cost only, no placing or pumping included. All concrete material costs in this manual are based on these figures.

	Craft@Hrs	Unit	Material	Labor	Total
5.0 sack mix, 2,000 PSI	—	CY	89.60	—	89.60
6.0 sack mix, 3,000 PSI	—	CY	95.80	—	95.80
6.6 sack mix, 3,500 PSI	—	CY	97.40	—	97.40
7.1 sack mix, 4,000 PSI	—	CY	100.00	—	100.00
8.5 sack mix, 5,000 PSI	—	CY	106.00	—	106.00
Extra costs for ready-mix concrete					
Add for less than 10 CY per load	—	CY	20.00	—	20.00
Add for delivery over 20 miles	—	Mile	3.50	—	3.50
Add for standby charge in excess of 5 minutes per CY delivered, per minute of extra time	—	Ea	1.75	—	1.75
Add for super plasticized mix, 7"-8" slump	—	%	8.0	—	—
Add for high early strength concrete					
5 sack mix	—	CY	10.90	—	10.90
6 sack mix	—	CY	14.10	—	14.10
Add for lightweight aggregate, typical	—	CY	39.00	—	39.00
Add for pump mix (pea-gravel aggregate)	—	CY	8.00	—	8.00
Add for granite aggregate, typical	—	CY	5.75	—	5.75
Add for white cement (architectural)	—	CY	49.20	—	49.20
Add for 1% calcium chloride	—	CY	1.50	—	1.50
Add for chemical compensated shrinkage	—	CY	15.00	—	15.00

Add for colored concrete, typical prices. The ready-mix supplier charges for cleanup of the ready-mix truck used for delivery of colored concrete. The usual practice is to provide one truck per day for delivery of all colored concrete for a particular job. Add the cost below to the cost per cubic yard for the design mix required. Also add the cost for truck cleanup.

	Craft@Hrs	Unit	Material	Labor	Total
Colored concrete, truck cleanup, per day	—	LS	60.00	—	60.00
Adobe	—	CY	19.00	—	19.00
Black	—	CY	32.80	—	32.80
Blended red	—	CY	18.00	—	18.00
Brown	—	CY	25.00	—	25.00
Green	—	CY	27.50	—	27.50
Yellow	—	CY	23.50	—	23.50

Ready-Mix Concrete and Placing No forms, finishing or reinforcing included. Material cost is based on 3,000 PSI concrete. Use $900 minimum charge for boom truck and $125 minimum for trailer mounted pump.

	Craft@Hrs	Unit	Material	Labor	Equipment	Total
Columns						
By crane (60 ton crawler crane)	H3@.873	CY	95.80	35.50	11.50	142.80
By pump (28 meter boom truck and operator)	M2@.062	CY	105.00	2.45	5.10	112.55
Slabs-on-grade						
Direct from chute	CL@.430	CY	95.80	15.60	—	111.40
By crane (60 ton crawler crane)	H3@.594	CY	95.80	24.20	7.80	127.80
By pump (trailer mounted. 20 CY/ hour)	M2@.188	CY	105.00	7.41	5.58	117.99
With buggy	CL@.543	CY	83.30	19.80	6.77	109.87
Elevated slabs						
By crane (60 ton crawler crane)	H3@.868	CY	95.80	35.30	11.40	142.50
By pump (28 meter boom truck and operator)	M2@.054	CY	105.00	2.13	4.70	111.83
Footings, pile caps, foundations						
Direct from chute	CL@.564	CY	95.80	20.50	—	116.30
By pump (trailer mounted. 30 CY/ hour)	M2@.125	CY	105.00	4.93	1.91	111.84
With buggy	CL@.701	CY	95.80	25.50	8.74	130.04

	Craft@Hrs	Unit	Material	Labor	Equipment	Total
Beams and girders						
By pump (28 meter boom truck and operator)	M2@.062	CY	105.00	2.45	5.10	112.55
By crane (60 ton crawler crane)	H3@.807	CY	95.80	32.90	10.60	139.30
Stairs						
Direct from chute	CL@.814	CY	95.80	29.60	—	125.40
By crane (60 ton crawler crane)	H3@1.05	CY	95.80	42.80	13.80	152.40
By pump (28 meter boom truck and operator)	M2@.108	CY	105.00	4.26	7.40	116.66
With buggy	CL@.948	CY	95.80	34.50	11.80	142.10
Walls and grade beams to 4' high						
Direct from chute	CL@.479	CY	95.80	17.40	—	113.20
By pump (trailer mounted. 20 CY/ hour)	M2@.188	CY	105.00	7.41	1.90	114.31
With buggy	CL@.582	CY	95.80	21.20	7.26	124.26
Walls over 4' to 8' high						
By crane (60 ton crawler crane)	H3@.821	CY	95.80	33.40	10.80	140.00
By pump (28 meter boom truck and operator)	M2@.062	CY	105.00	2.45	3.10	110.55
Walls over 8' to 16' high						
By crane (60 ton crawler crane)	H3@.895	CY	95.80	36.40	11.80	144.00
By pump (28 meter boom truck and operator)	M2@.078	CY	105.00	3.08	5.90	113.98
Walls over 16' high						
By crane (60 ton crawler crane)	H3@.986	CY	95.80	40.10	12.90	148.80
By pump (28 meter boom truck and operator)	M2@.097	CY	105.00	3.83	6.85	115.68

Concrete Slab-on-Grade Assemblies Typical costs including fine grading, edge forms, 3,000 PSI concrete, wire mesh, finishing and curing.

	Craft@Hrs	Unit	Material	Labor	Equipment	Total
4" thick 4' x 6' equipment pad						
Hand trimming and shaping, 24 SF	CL@.512	LS	—	18.60	—	18.60
Lay out, set and strip edge forms, 20 LF, 5 uses	C8@1.30	LS	12.40	53.30	—	65.70
Place W2.9 x W2.9 x 6" x 6" mesh, 24 SF	RB@.097	LS	12.20	5.16	—	17.36
Place ready-mix concrete, from chute .3 CY	CL@.127	LS	28.70	4.62	—	33.32
Finish concrete, broom finish	CM@.591	LS	—	25.70	—	25.70
Cure concrete, curing paper	CL@.124	LS	1.00	4.51	—	5.51
Total job cost for 4" thick 24 SF pad	**--@2.75**	**LS**	**54.30**	**111.89**	**—**	**166.19**
Cost per SF for 4" thick 24 SF pad	--@.115	SF	2.26	4.66	—	6.92
Cost per CY of 4" thick concrete, .3 CY job	--@9.17	CY	181.00	373.00	—	554.00
8" thick 10' x 10' equipment pad						
Hand trimming and shaping, 100 SF	CL@2.79	LS	—	102.00	—	102.00
Lay out, set and strip edge forms, 40 LF, 5 uses	C8@3.28	LS	32.40	134.00	—	166.40
Place W2.9 x W2.9 x 6" x 6" mesh, 100 SF	RB@.400	LS	51.00	21.30	—	72.30

	Craft@Hrs	Unit	Material	Labor	Equipment	Total
Place ready-mix concrete, from chute						
2.5 CY	CL@1.08	LS	240.00	39.30	—	279.30
Finish concrete, broom finish	CM@2.11	LS	—	91.90	—	91.90
Cure concrete, curing paper	CL@.515	LS	4.00	18.70	—	22.70
Total job cost for 8" thick 100 SF pad	**--@10.2**	**LS**	**327.40**	**407.20**	**—**	**734.60**
Cost per SF for 8" thick 100 SF job	--@.102	SF	3.27	4.63	—	7.91
Cost per CY of 8" thick concrete,						
2.5 CY job	--@4.08	CY	130.96	162.88	—	293.84

Note: Ready-mix concrete used in both examples assumes a minimum of 10 cubic yards will be delivered (per load) with the excess used elsewhere on the same job.

Add for each CY						
less than 10 delivered	—	CY	20.00	—	—	20.00

Floor Slab Assemblies Typical reinforced concrete slab-on-grade including excavation, gravel fill, forms, vapor barrier, wire mesh, 3,000 PSI concrete, finishing and curing.
Based on 100' x 75' slab (7,500 SF)

	Craft@Hrs	Unit	Material	Labor	Equipment	Total
3,000 PSI concrete	—	CY	95.80	—	—	95.80
4" thick slab	--@.024	SF	2.17	1.09	.11	3.37
5" thick slab	--@.025	SF	2.71	1.13	.11	3.95
6" thick slab	--@.026	SF	3.25	1.18	.12	4.55

Detailed cost breakdown slab as described above:

4" thick 100' x 75' slab

	Craft@Hrs	Unit	Material	Labor	Equipment	Total
Grade sandy loam site using a D-4 tractor,						
140 CY, balanced job,						
(no import or export)	S1@2.66	LS	—	114.00	59.10	173.10
Buy and spread 6" crushed rock base using a D-4 tractor,						
140 CY	S1@15.8	LS	2,506.00	679.00	351.00	3,536.00
Lay out, set and strip edge forms,						
350 LF, 5 uses	C8@19.6	LS	217.00	803.00	—	1,020.00
Place .006" polyethylene vapor barrier						
7,500 SF	CL@8.84	LS	418.00	322.00	—	740.00
Place W2.9 x W2.9 x 6" x 6" mesh,						
7,500 SF	RB@30.0	LS	3,825.00	1,590.00	—	5,415.00
Place and remove 2" x 4" keyway,						
200 LF, 1 use	C8@8.80	LS	48.00	361.00	—	409.00
Place 4" concrete, 93 CY, from chute	CL@39.9	LS	8,909.00	1,450.00	—	10,359.00
Float finish 7,500 SF	CM@52.4	LS	—	2,280.00	416.00	2,696.00
Cure with slab curing paper	CM@5.58	LS	375.00	243.00	—	618.00
Total job cost 4" thick						
100' x 75' floor slab	**--@180**	**LS**	**16,298.00**	**7,842.00**	**826.10**	**24,966.10**
Cost per SF for 4" thick 7,500 SF job	--@.024	SF	2.17	1.05	.11	3.33
Cost per CY for 4" thick 93 CY job	--@1.94	CY	175.24	84.32	8.88	268.44
Cost per each						
additional 1" of concrete	--@.001	SF	.54	.05	.01	.60

	Craft@Hrs	Unit	Material	Labor	Equipment	Total

Cast-in-Place Concrete, Subcontract Typical costs including forms, concrete, reinforcing, finishing and curing. These costs include the subcontractor's overhead and profit but no excavation, shoring or backfill. Use these figures only for preliminary estimates and on jobs that require a minimum of 1,000 CY of concrete.

	Craft@Hrs	Unit	Material	Labor	Equipment	Total
Foundations						
Institutional or office buildings	—	CY	—	—	—	493.00
Heavy engineered structures	—	CY	—	—	—	535.00
Structural concrete walls						
Single story, 8" wall	—	CY	—	—	—	664.00
Single story, 10" wall	—	CY	—	—	—	733.00
Multi-story, 8" wall	—	CY	—	—	—	855.00
Multi-story, 10"	—	CY	—	—	—	896.00
Slip formed 8" wall	—	CY	—	—	—	753.00
Structural slabs, including shoring,						
6", 1 way beams,						
100 pounds reinforcing	—	CY	—	—	—	783.00
6", 2 way beams,						
225 pounds reinforcing	—	CY	—	—	—	722.00
8", flat, with double mat reinforcing over steel frame						
by others, 100 pounds reinforcing	—	CY	—	—	—	654.00
12" flat, 250 pounds						
reinforcing bars per CY	—	CY	—	—	—	591.00
7", post-tensioned,						
to 1 pound tempered bars	—	CY	—	—	—	734.00
6" to 8" including beam jacketing	—	CY	—	—	—	710.00
6" to 8" on permanent metal form	—	CY	—	—	—	643.00
8" concurrent						
with slipform construction	—	CY	—	—	—	683.00
Lift slab construction 6" to 8"	—	SF	—	—	—	15.00
Pan and joist slabs 3" thick, 30" pan, including shoring						
6" x 16" joist	—	CY	—	—	—	755.00
6" x 12" joist	—	CY	—	—	—	668.00
6" x 10" joist	—	CY	—	—	—	637.00
Pan and joist slabs 4-1/2" thick, 20" pan, including shoring						
6" x 20" joist	—	CY	—	—	—	637.00
6" x 16" joist	—	CY	—	—	—	623.00
6" x 12" joist	—	CY	—	—	—	618.00
Parapet and fascia, 6" thick	—	CY	—	—	—	960.00
Loading docks,						
based on 8" walls and 8" slab	—	CY	—	—	—	441.00
Beams and girders, not slab integrated						
12" x 24", typical	—	CY	—	—	—	1,290.00
18" x 24", typical	—	CY	—	—	—	988.00
Columns, average reinforcing						
Square, wood formed, with chamfer						
12" column	—	CY	—	—	—	1,600.00
16" column	—	CY	—	—	—	1,470.00
18" column	—	CY	—	—	—	1,340.00
20" column	—	CY	—	—	—	1,230.00
24" column	—	CY	—	—	—	1,070.00
Round, fiber tube formed, 12" to 18"	—	CY	—	—	—	1,060.00

Concrete 3

Cast-in-Place Concrete Steps on Grade Assemblies Typical in-place costs, including layout, fabrication and placing forms, setting and tying steel reinforcing, installation of steel nosing, placing 2,000 PSI concrete directly from the chute of a ready-mix truck, finishing, stripping forms and curing. Costs assume excavation, back filling and compaction have been completed by others. Crew is a carpenter, laborer and finisher. Formwork is based on using standard & better lumber and concrete and includes a 5% waste allowance. For scheduling purposes, estimate that a crew of 3 can set forms and place steel for and pour 4 to 5 CY of concrete cast-in-place steps in an 8-hour day. Costs shown are for concrete steps with a 6" riser height and 12" tread depth supported by a 6" thick monolithic slab. The "tricks of the trade" below explain how to use these unit prices to estimate stairways of various widths and heights. Note: Total cost per CY has been rounded. For more detailed coverage of formed concrete stairs, see the National Concrete & Masonry Estimator, http://CraftsmanSiteLicense.com.

	Craft@Hrs	Unit	Material	Labor	Total
Fabricate and erect forms, 2 uses of forms, includes stakes, braces, form oil and nails at $.13 per LF of forms					
Side forms, 1.7 BF per LF of tread	P9@.080	LF	.70	3.35	4.05
Riser forms, 1.5 BF per LF of tread	P9@.040	LF	.61	1.67	2.28
Rebar, #3, Grade 60, 1 pound per LF of tread	P9@.009	LF	.47	.38	.85
Embedded steel step nosing, 2-1/2" x 2-1/2" x 1/4" black mild steel					
Angle iron, 4.5 pounds per LF of tread.	P9@.004	LF	3.55	.17	3.72
Concrete, .03 CY per LF of tread	P9@.030	LF	2.87	1.26	4.13
Total cost per LF of risers	**--@.163**	**LF**	**8.20**	**6.83**	**15.03**
Total cost per CY of concrete	**--@5.43**	**CY**	**273.33**	**248.66**	**521.99**

Tricks of the trade:

To quickly estimate the cost of cast-in-place concrete steps on grade, do this:
Multiply the width of the steps (in feet) times their height in feet and multiply the result times 2.
This will give the total linear feet (LF) of risers in the steps. Multiply the total LF by the costs per LF.

Sample estimates for cast-in-place concrete steps on grade using the procedure and costs from above:
2'6" wide by 3' high (6 steps) would be: 2.5' x 3' x 2 = 15 LF x $15.03 = $225.45
3' wide by 3' high (6 steps) would be: 3' x 3' x 2 = 18 LF x $15.03 = $270.54
4' wide by 4' high (8 steps) would be: 4' x 4' x 2 = 32 LF x $15.03 = $480.96
7' wide by 4'6" high (9 steps) would be: 7' x 4.5' x 2 = 63 LF x $15.03 = $946.89

Tip for using these estimates with The National Estimator computer estimating program: Calculate the linear feet of stair, copy and paste the line "Total cost per LF of risers" and type the quantity.

Concrete Accessories

Bentonite					
50 pound bag, granules	—	Ea	30.00	—	30.00
3 gallon Volclay Bentoseal	—	Ea	118.00	—	118.00
4' x 15' Bentonite geotextile roll	—	Ea	64.30	—	64.30
Bond breaker and form release, Thompson's CBA, spray on					
Lumber & plywood form release, 200 SF/gal	CL@.006	SF	.04	.22	.26
Metal form release, 600 SF per gallon	CL@.003	SF	.02	.11	.13
Casting bed form breaker, 300 SF per gallon	CL@.005	SF	.03	.18	.21
Cure coat, 300 SF per gallon	CL@.005	SF	.03	.18	.21
Chamfer strip					
3/4" x 3/4" x 1", 10' lengths	CC@.015	LF	.16	.68	.84
Chamfer corner forms					
3/4" corner forms	CC@.015	LF	8.60	.68	9.28

Concrete 3

	Craft@Hrs	Unit	Material	Labor	Total
Column clamps for plywood forms, per set, by column size, Econ-o-clamp					
8" x 8" to 24" x 24", purchase	—	Set	111.00	—	111.00
24" x 24" to 48" x 48", purchase	—	Set	145.00	—	145.00
Monthly rental cost as % of purchase price	—	%	5.0	—	—
Column clamps for lumber forms, per set, by column size					
10" x 10" to 28" x 28", purchase	—	Set	134.00	—	134.00
16" x 16" to 40" x 40", purchase	—	Set	168.00	—	168.00
21" x 21" to 50" x 50", purchase	—	Set	281.00	—	281.00
10" x 10" to 28" x 28", rent per month, per set	—	Mo	4.63	—	4.63
16" x 16" to 40" x 40", rent per month, per set	—	Mo	5.12	—	5.12
21" x 21" to 50" x 50", rent per month, per set	—	Mo	8.51	—	8.51
Concrete sleeves, circular polyethylene liner, any slab thickness					
1-1/2" to 3" diameter	—	Ea	1.89	—	1.89
4" to 6" diameter	—	Ea	3.62	—	3.62
Embedded iron, installation only	C8@.045	Lb	—	1.84	1.84
Expansion joints, cane fiber, asphalt impregnated, pre-molded					
1/2" x 4", slabs	C8@.028	LF	.14	1.15	1.29
1/2" x 6", in walls	C8@.035	LF	.19	1.43	1.62
1/2" x 8", in walls	C8@.052	LF	.26	2.13	2.39
Concrete footing form wedge ties					
6" wall thickness, pack of 50	—	Pack	19.60	—	19.60
8" wall thickness, pack of 50	—	Pack	21.50	—	21.50
Tie wedge, pack of 100	—	Pack	14.20	—	14.20
Form spreaders, 4" to 10"	—	Ea	.31	—	.31
Form oil, 800 to 1,200 SF per gallon	—	Gal	4.50	—	4.50
Grout for machine bases mixed and placed (add for edge forms if required, from below)					
Five Star epoxy					
per SF for each 1" of thickness	C8@.170	SF	12.10	6.96	19.06
Five Star NBEC non-shrink, non-metallic					
per SF for each 1" of thickness	C8@.170	SF	2.65	6.96	9.61
Embeco 636, metallic aggregate					
per SF for each 1" of thickness	C8@.163	SF	6.28	6.68	12.96
Forms for grout, based on using 2" x 4" lumber,					
1 use	C8@.195	LF	1.69	7.99	9.68
Inserts, Unistrut, average	C8@.068	Ea	7.79	2.79	10.58
She bolt form clamps, with cathead and bolt, purchase					
17" rod	—	Ea	4.31	—	4.31
21" rod	—	Ea	7.17	—	7.17
24" rod	—	Ea	7.72	—	7.72
Monthly rental as % of purchase price	—	%	15.0	—	—
Snap ties, average, 4,000 pound, 6" to 10"	—	Ea	.43	—	.41
Water stop, 3/8"					
Rubber, 6"	C8@.059	LF	2.90	2.42	5.32
Rubber, 9"	C8@.059	LF	3.90	2.42	6.32
PVC, 9"	C8@.059	LF	1.55	2.42	3.97
Concrete waler brackets					
2" x 4"	—	Ea	3.24	—	2.94
2" x 6"	—	Ea	4.16	—	3.94
Waler jacks (scaffold support on form), purchase	—	Ea	48.00	—	48.00
Rental per month, each	—	Mo	5.25	—	5.25

	Craft@Hrs	Unit	Material	Labor	Total
Concrete Slab Finishes					
Float finish	CM@.009	SF	—	.39	.39
Trowel finishing					
Steel, machine work	CM@.014	SF	—	.61	.61
Steel, hand work	CM@.017	SF	—	.74	.74
Broom finish	CM@.012	SF	—	.52	.52
Scoring concrete surface, hand work	CM@.005	LF	—	.22	.22
Exposed aggregate (washed, including finishing),					
no disposal of slurry	CM@.015	SF	.29	.65	.94
Color hardener, nonmetallic, trowel application					
Standard colors	—	Lb	.73	—	.73
Non standard colors	—	Lb	.99	—	.99
Light duty floors (40 pounds per 100 SF)	CM@.008	SF	—	.35	.35
Pastel shades (60 pounds per 100 SF)	CM@.012	SF	—	.52	.52
Heavy duty (90 pounds per 100 SF)	CM@.015	SF	—	.65	.65
Add for colored wax, 800 SF per gal	CM@.003	SF	.03	.13	.16
Sidewalk grits, abrasive, 25 pounds per 100 SF					
Aluminum oxide 100 pound bag	CM@.008	SF	.22	.35	.57
Silicon carbide, 100 pound bag	CM@.008	SF	.39	.35	.74
Metallic surface hardener, dry shake, Masterplate 200					
Moderate duty, 1 pound per SF	CM@.008	SF	.75	.35	1.10
Heavy duty, 1.5 pounds per SF	CM@.009	SF	1.14	.39	1.53
Concentrated traffic, 2 pounds per SF	CM@.010	SF	1.52	.44	1.96
Liquid curing and sealing compound, sprayed-on,					
Mastercure, 400 SF	CL@.004	SF	.07	.15	.22
Sweep, scrub and wash down	CL@.006	SF	.02	.22	.24
Finish treads and risers					
No abrasives, no plastering, per SF of tread	CM@.038	SF	—	1.66	1.66
With abrasives, plastered, per SF of tread	CM@.058	SF	.75	2.53	3.28
Wax coating	CL@.002	SF	.05	.07	.12
Concrete Wall Finishes					
Cut back ties and patch	CM@.010	SF	.12	.44	.56
Remove fins	CM@.007	LF	.05	.30	.35
Grind smooth	CM@.020	SF	.09	.87	.96
Sack, simple	CM@.012	SF	.06	.52	.58
Bush hammer light, green concrete	CM@.020	SF	.13	.87	1.00
Bush hammer standard, cured concrete	CM@.029	SF	.13	1.26	1.39
Bush hammer, heavy, cured concrete	CM@.077	SF	.13	3.35	3.48
Needle gun treatment, large areas	CM@.058	SF	.13	2.53	2.66
Wire brush, green	CM@.014	SF	.05	.61	.66
Wash with acid and rinse	CM@.004	SF	.28	.17	.45
Break fins, patch voids, Carborundum rub	CM@.032	SF	.07	1.39	1.46
Break fins, patch voids, burlap grout rub	CM@.024	SF	.07	1.05	1.12
Sandblasting, Subcontract. No scaffolding included, flat vertical surfaces					
Light sandblast	—	SF	1.25	—	1.25
Medium, to expose aggregate	—	SF	1.50	—	1.50
Heavy, with dust control requirements	—	SF	2.30	—	2.30
Note: Minimum charge for sandblasting	—	LS	—	—	850.00
Miscellaneous Concrete Finishes					
Monolithic natural aggregate topping					
1/16" topping	P9@.006	SF	.17	.25	.42
3/16" topping	P9@.018	SF	.39	.75	1.14
1/2" topping	P9@.021	SF	1.03	.88	1.91

	Craft@Hrs	Unit	Material	Labor	Total
Integral colors,					
Typically 8 pounds per sack of cement	—	Lb	2.07	—	2.07
Mono rock, 3/8" wear course only, typical cost	—	SF	1.21	—	1.21
Kalman 3/4", wear course only, typical cost	—	SF	1.84	—	1.84
Acid etching, 5% muriatic acid, typical cost	P9@.005	SF	.35	.21	.56
Felton sand, for use with white cement	—	CY	38.00	—	38.00

Specially Placed Concrete Typical subcontract prices.
Gunite, no forms and no reinforcing included

	Craft@Hrs	Unit	Material	Labor	Total
Flat plane, vertical areas, 1" thick	—	SF	—	—	3.81
Curved arch, barrel, per 1" thickness	—	SF	—	—	4.75
Pool or retaining walls, per 1" thickness	—	SF	—	—	3.66
Add for architectural finish	—	SF	—	—	1.45
Wire mesh for gunite 2" x 2", W.9 x W.9	—	SF	—	—	2.25
Pressure grouting, large quantity, 50/50 mix	—	CY	—	—	553.00

Precast Concrete Panel costs are based on 3,000 PSI natural gray concrete with a smooth or board finish on one side. Placing (lifting) costs include handling, hoisting, alignment, bracing and permanent connections. Erection costs vary with the type of crane. Heavier lifts over a longer radius (reach) require larger cranes. Equipment costs include the crane and a 300 amp gas powered welding machine. Add the cost of grouting or caulking joints. Costs will be higher on small jobs (less than 35,000 SF).

Precast wall panels Costs include delivery to 40 miles, typical reinforcing steel and embedded items but no lifting. Use 2,500 SF as a minimum job charge. See placing costs below.

	Craft@Hrs	Unit	Material	Labor	Equipment	Total
4" thick wall panels (52 lbs per SF)	—	SF	10.20	—	—	10.20
5" thick wall panels (65 lbs per SF)	—	SF	11.10	—	—	11.10
6" thick wall panels (78 lbs per SF)	—	SF	12.20	—	—	12.20
8" thick wall panels (105 lbs per SF)	—	SF	12.80	—	—	12.80
Add for insulated sandwich wall panels						
2-1/2" of cladding and 2" of fiberboard	—	SF	4.00	—	—	4.00
Add for high strength concrete						
3,500 lb concrete mix	—	%	2.0	—	—	—
4,000 lb concrete mix	—	%	3.0	—	—	—
4,500 lb concrete mix	—	%	4.0	—	—	—
5,000 lb concrete mix	—	%	6.0	—	—	—
Add for white facing or integral white mix	—	%	20.0	—	—	—
Add for broken rib						
or fluted surface design	—	%	20.0	—	—	—
Add for exposed aggregate finish	—	%	17.0	—	—	—
Add for sandblasted surface finish	—	%	12.0	—	—	—

Placing precast structural wall panels Figures in parentheses show the panels placed per day with a 100 ton crawler-mounted crane and welding machine. Use $9,000 as a minimum job charge.

	Craft@Hrs	Unit	Material	Labor	Equipment	Total
1 ton panels to 180' reach (23 per day)	H4@2.79	Ea	—	145.00	66.20	211.20
1 to 3 tons, to 121' reach (21 per day)	H4@3.05	Ea	—	158.00	72.40	230.40
3 to 5 tons, to 89' reach (18 per day)	H4@3.54	Ea	—	184.00	84.00	268.00
5 to 7 tons, to 75' reach (17 per day)	H4@3.76	Ea	—	195.00	89.20	284.20
7 to 9 tons, to 65' reach (16 per day)	H4@4.00	Ea	—	208.00	94.90	302.90
9 to 11 tons, to 55' reach (15 per day)	H4@4.28	Ea	—	222.00	102.00	324.00
11 to 13 tons, to 50' reach (14 per day)	H4@4.56	Ea	—	237.00	108.00	345.00

	Craft@Hrs	Unit	Material	Labor	Equipment	Total
13 to 15 tons, to 43' reach (13 per day)	H4@4.93	Ea	—	256.00	117.00	373.00
15 to 20 tons, to 36' reach (11 per day)	H4@5.81	Ea	—	302.00	138.00	440.00
20 to 25 tons to 31' reach (10 per day)	H4@6.40	Ea	—	332.00	152.00	484.00
For placing panels with cladding, add	—	%	—	5.00	5.00	—
For placing sandwich wall panels, add	—	%	—	10.00	10.00	—
If a 45 ton crane can be used, deduct	—	%	—	—	-22.0	—

Placing precast partition wall panels Figures in parentheses show the panels placed per day with a 45 ton truck-mounted crane and welding machine. Use $6,000 as a minimum job charge.

	Craft@Hrs	Unit	Material	Labor	Equipment	Total
1 ton panels to 100' reach (23 per day)	H4@2.79	Ea	—	145.00	32.50	177.50
1 to 2 tons, to 82' reach (21 per day)	H4@3.05	Ea	—	158.00	35.60	193.60
2 to 3 tons, to 65' reach (18 per day)	H4@3.54	Ea	—	184.00	41.30	225.30
3 to 4 tons, to 57' reach (16 per day)	H4@4.00	Ea	—	208.00	46.60	254.60
4 to 5 tons, to 50' reach (15 per day)	H4@4.28	Ea	—	222.00	49.90	271.90
5 to 6 tons, to 44' reach (14 per day)	H4@4.56	Ea	—	237.00	53.20	290.20
6 to 7 tons, to 40' reach (13 per day)	H4@4.93	Ea	—	256.00	57.50	313.50
7 to 8 tons, to 36' reach (11 per day)	H4@5.81	Ea	—	302.00	67.80	369.80
8 to 10 tons, to 31' reach (10 per day)	H4@6.40	Ea	—	332.00	74.60	406.60
10 to 12 tons to 26' reach (9 per day)	H4@7.11	Ea	—	369.00	82.90	451.90
12 to 14 tons, to 25' reach (8 per day)	H4@8.02	Ea	—	417.00	93.50	510.50
14 to 16 tons, to 21' reach (7 per day)	H4@9.14	Ea	—	475.00	107.00	582.00
If a 35 ton crane can be used, deduct	—	%	—	—	-5.0	—

Precast flat floor slab or floor plank Costs include delivery to 40 miles, 3,000 PSI concrete, typical reinforcing steel and embedded items. Use $2,000 as a minimum job charge. See placing costs below.

	Craft@Hrs	Unit	Material	Labor	Equipment	Total
4" thick floor panels (52 lbs per SF)	—	SF	7.88	—	—	7.88
5" thick floor panels (65 lbs per SF)	—	SF	9.36	—	—	9.36
6" thick floor panels (78 lbs per SF)	—	SF	12.10	—	—	12.10
8" thick floor panels (105 lbs per SF)	—	SF	12.50	—	—	12.50

Placing panels with a 100 ton crawler-mounted crane and welding machine at $775 per day. Figures in parentheses show the number of panels placed per day. Use $8,500 as a minimum job charge

	Craft@Hrs	Unit	Material	Labor	Equipment	Total
2 ton panels to 150' reach (23 per day)	H4@2.79	Ea	—	145.00	45.60	190.60
2 to 3 tons, to 121' reach (21 per day)	H4@3.05	Ea	—	158.00	49.90	207.90
3 to 4 tons, to 105' reach (19 per day)	H4@3.37	Ea	—	175.00	55.10	230.10
4 to 5 tons, to 88' reach (18 per day)	H4@3.56	Ea	—	185.00	58.20	243.20
5 to 6 tons, to 82' reach (17 per day)	H4@3.76	Ea	—	195.00	61.50	256.50
6 to 7 tons, to 75' reach (16 per day)	H4@4.00	Ea	—	208.00	65.40	273.40
7 to 8 tons, to 70' reach (15 per day)	H4@4.28	Ea	—	222.00	70.00	292.00
8 to 9 tons, to 65' reach (14 per day)	H4@4.56	Ea	—	237.00	74.60	311.60
9 to 10 tons, to 60' reach (11 per day)	H4@5.82	Ea	—	302.00	95.20	397.20
20 to 25 tons to 31' reach (10 per day)	H4@6.40	Ea	—	332.00	105.00	437.00

Precast combination beam and slab units Beam and slab units with beam on edge of slab. Weight of beams and slabs will be about 4,120 lbs per CY. Use 40 CY as a minimum job charge. See placing costs below.

	Craft@Hrs	Unit	Material	Labor	Equipment	Total
Slabs with no intermediate beam	—	CY	627.00	—	—	627.00

	Craft@Hrs	Unit	Material	Labor	Equipment	Total
Slabs with one intermediate beam	—	CY	662.00	—	—	662.00
Slabs with two intermediate beams	—	CY	722.00	—	—	722.00

Placing deck units with a 100 ton crawler-mounted crane and a welding machine. Figures in parentheses show the number of units placed per day. Use $8,500 as a minimum job charge.

	Craft@Hrs	Unit	Material	Labor	Equipment	Total
5 ton units to 88' reach (21 per day)	H4@3.05	Ea	—	158.00	50.00	208.00
5 to 10 tons, to 60' reach (19 per day)	H4@3.37	Ea	—	175.00	55.10	230.10
10 to 15 tons, to 43' reach (17 per day)	H4@3.76	Ea	—	195.00	61.50	256.50
15 to 20 tons, to 36' reach (15 per day)	H4@4.28	Ea	—	222.00	70.00	292.00
20 to 25 tons, to 31' reach (14 per day)	H4@4.57	Ea	—	237.00	74.70	311.70
If a 150 ton crane is required, add	—	%	—	—	16.0	—

Precast beams, girders and joists Beams, girders and joists, costs per linear foot of span, including 3,000 PSI concrete, reinforcing steel, embedded items and delivery to 40 miles. Use 400 LF as a minimum job charge. See placing costs below.

	Craft@Hrs	Unit	Material	Labor	Equipment	Total
1,000 lb load per foot, 15' span	—	LF	75.60	—	—	75.60
1,000 lb load per foot, 20' span	—	LF	83.70	—	—	83.70
1,000 lb load per foot, 30' span	—	LF	96.50	—	—	96.50
3,000 lb load per foot, 10' span	—	LF	75.60	—	—	75.60
3,000 lb load per foot, 20' span	—	LF	96.50	—	—	96.50
3,000 lb load per foot, 30' span	—	LF	116.00	—	—	116.00
5,000 lb load per foot, 10' span	—	LF	75.60	—	—	75.60
5,000 lb load per foot, 20' span	—	LF	106.00	—	—	106.00
5,000 lb load per foot, 30' span	—	LF	124.00	—	—	124.00

Placing beams, girders and joists with a 100 ton crawler-mounted crane and a welding machine. Figures in parentheses show the number of units placed per day. Use $8,500 as a minimum job charge.

	Craft@Hrs	Unit	Material	Labor	Equipment	Total
5 ton units to 88' reach (18 per day)	H4@3.56	Ea	—	185.00	58.20	243.20
5 to 10 tons, to 60' reach (17 per day)	H4@3.76	Ea	—	195.00	61.50	256.50
10 to 15 tons, to 43' reach (16 per day)	H4@4.00	Ea	—	208.00	65.40	273.40
15 to 20 tons, to 36' reach (15 per day)	H4@4.27	Ea	—	222.00	69.80	291.80
If a 150 ton crane is required, add	—	%	—	—	16.0	—

Precast concrete columns Precast concrete columns, 12" x 12" to 36" x 36" including concrete, reinforcing steel, embedded items and delivery to 40 miles. Use 40 CY as a minimum job charge. See placing costs below.

	Craft@Hrs	Unit	Material	Labor	Equipment	Total
Columns with 500 lbs of reinforcement per CY	—	CY	651.00	—	—	651.00
Columns with 350 lbs of reinforcement per CY	—	CY	639.00	—	—	639.00
Columns with 200 lbs of reinforcement per CY	—	CY	976.00	—	—	976.00

Placing concrete columns with a 100 ton crawler-mounted crane and a welding machine. Figures in parentheses show the number of units placed per day. Use $8,500 as a minimum job charge.

	Craft@Hrs	Unit	Material	Labor	Equipment	Total
3 ton columns to 121' reach (29 per day)	H4@2.21	Ea	—	115.00	36.10	151.10
3 to 5 tons, to 89' reach (27 per day)	H4@2.37	Ea	—	123.00	38.80	161.80
5 to 7 tons, to 75' reach (25 per day)	H4@2.55	Ea	—	132.00	41.70	173.70
7 to 9 tons, to 65' reach (23 per day)	H4@2.79	Ea	—	145.00	45.60	190.60
9 to 11 tons, to 55' reach (22 per day)	H4@2.92	Ea	—	152.00	47.80	199.80
11 to 13 tons, to 50' reach (20 per day)	H4@3.20	Ea	—	166.00	52.30	218.30
13 to 15 tons, to 43' reach (18 per day)	H4@3.56	Ea	—	185.00	58.20	243.20

Concrete 3

	Craft@Hrs	Unit	Material	Labor	Equipment	Total

Precast concrete stairs Costs include concrete, reinforcing, embedded steel, nosing, and delivery to 40 miles. Use $500 as a minimum job charge. See placing costs below.

Stairs, 44" to 48" wide, 10' to 12' rise, "U" or "L" shape, including typical landings,

	Craft@Hrs	Unit	Material	Labor	Equipment	Total
Cost per each step (7" rise)	—	Ea	37.30	—	—	37.30

Placing precast concrete stairs with a 45 ton truck-mounted crane and a welding machine. Figures in parentheses show the number of stair flights placed per day. Use $6,000 as a minimum job charge.

	Craft@Hrs	Unit	Material	Labor	Equipment	Total
2 ton stairs to 82' reach (21 per day)	H4@3.05	Ea	—	158.00	35.60	193.60
2 to 3 tons, to 65' reach (19 per day)	H4@3.37	Ea	—	175.00	39.30	214.30
3 to 4 tons, to 57' reach (18 per day)	H4@3.54	Ea	—	184.00	41.30	225.30
4 to 5 tons, to 50' reach (17 per day)	H4@3.76	Ea	—	195.00	43.90	238.90
5 to 6 tons, to 44' reach (16 per day)	H4@4.00	Ea	—	208.00	46.60	254.60
6 to 7 tons, to 40' reach (15 per day)	H4@4.28	Ea	—	222.00	49.90	271.90
7 to 8 tons, to 36' reach (14 per day)	H4@4.56	Ea	—	237.00	53.20	290.20
8 to 9 tons, to 33' reach (10 per day)	H4@6.40	Ea	—	332.00	74.60	406.60
If a 35 ton crane can be used, deduct	—	%	—	—	-10.0	—

Precast double tees 8' wide, 24" deep. Design load is shown in pounds per square foot (PSF). Costs include lifting and placing but no slab topping. Equipment cost is for a 45 ton truck crane and a welding machine. Use $6,000 as a minimum job charge.

	Craft@Hrs	Unit	Material	Labor	Equipment	Total
30' to 35' span, 115 PSF	H4@.019	SF	8.56	.99	.22	9.77
30' to 35' span, 141 PSF	H4@.019	SF	9.36	.99	.22	10.57
35' to 40' span, 78 PSF	H4@.018	SF	8.56	.94	.21	9.71
35' to 40' span, 98 to 143 PSF	H4@.018	SF	9.36	.94	.21	10.51
40' to 45' span, 53 PSF	H4@.017	SF	8.56	.88	.20	9.64
40' to 45' span, 69 to 104 PSF	H4@.017	SF	9.36	.88	.20	10.44
40' to 45' span, 134 PSF	H4@.017	SF	9.72	.88	.20	10.80
45' to 50' span, 77 PSF	H4@.016	SF	9.36	.83	.19	10.38
45' to 50' span, 101 PSF	H4@.016	SF	9.72	.83	.19	10.74

Concrete topping, based on 350 CY job 3,000 PSI design mix with lightweight aggregate, no forms or finishing included.

	Craft@Hrs	Unit	Material	Labor	Equipment	Total
By crane	H3@.868	CY	95.80	35.30	11.40	142.50
By pump (28 meter boom truck and operator)	M2@.080	CY	105.00	3.16	6.00	114.16

Hollow core roof planks, not including topping, 35 PSF live load, 10' to 15' span

	Craft@Hrs	Unit	Material	Labor	Equipment	Total
4" thick	H4@.018	SF	5.14	.94	.18	6.26
6" thick	H4@.018	SF	6.29	.94	.18	7.41
8" thick	H4@.020	SF	7.36	1.04	.20	8.60
10" thick	H4@.026	SF	7.67	1.35	.26	9.28

Prestressed concrete support poles Costs shown are per linear foot for sizes and lengths shown. Equipment cost is for a 45 ton truck crane. These costs do not include base or excavation. Lengths other than as shown will cost more per foot. Use $6,000 as a minimum job charge.

10" x 10" poles with four 7/16" 270 KSI ultimate tendon strand strength

	Craft@Hrs	Unit	Material	Labor	Equipment	Total
10', 14', 20', 25' or 30' long	H4@.028	LF	11.90	1.45	.29	13.64

12" x 12" poles with six 7/16" 270 KSI ultimate tendon strand strength

	Craft@Hrs	Unit	Material	Labor	Equipment	Total
10', 14' or 20' long	H4@.028	LF	14.40	1.45	.29	16.14
25' or 30' long	H4@.028	LF	15.90	1.45	.29	17.64

Concrete 3

	Craft@Hrs	Unit	Material	Labor	Equipment	Total

Tilt-Up Concrete Construction Costs will be higher if access is not available from all sides or obstructions delay the work. Except as noted, no grading, compacting, engineering, design, permit or inspection fees are included.

Foundations and footings for tilt-up Costs assume normal soil conditions with no forming or shoring required and are based on concrete poured directly from the chute of a ready-mix truck. Soil excavated and not needed for backfill can be stored on site and used by others. For scheduling purposes, estimate that a crew of 5 can set forms for and pour 40 to 50 CY of concrete footings and foundations in an 8-hour day.

Lay out foundation from known point on site, set stakes and mark for excavation.

	Craft@Hrs	Unit	Material	Labor	Equipment	Total
Per SF of floor slab	C8@.001	SF	—	.04	—	.04

Excavation with a 1/2 CY utility backhoe

	Craft@Hrs	Unit	Material	Labor	Equipment	Total
27 CY per hour	S1@.075	CY	—	3.22	1.01	4.23

Set and tie grade A60 reinforcing bars, typical foundation has 50 to 60 pounds of

	Craft@Hrs	Unit	Material	Labor	Equipment	Total
reinforcing bar per CY of concrete	RB@.008	Lb	.60	.43	—	1.03

Place stringer supports for reinforcing bars (at 1.1 LF of support per LF of rebar), using 2" x 4" or 2" x 6" lumber. Includes typical stakes and hardware

	Craft@Hrs	Unit	Material	Labor	Equipment	Total
Based on 3 uses of supports	C8@.017	LF	.46	.70	—	1.16

(Inspection of reinforcing may be required before concrete is placed.)

Make and place steel column anchor bolt templates (using 1.1 SF of plywood per SF of column base). Based on 2 uses of 3/4" plyform. If templates are furnished by the steel column fabricator, include only the labor cost to place the templates as shown

	Craft@Hrs	Unit	Material	Labor	Equipment	Total
Make and place templates	C8@.160	SF	.55	6.56	—	7.11
Place templates furnished by others	C8@.004	SF	—	.16	—	.16

Blockout form at foundation, step footings, columns, etc., per SF of contact area, using 2" x 6" lumber. Includes allowance for waste, typical stakes and hardware.

	Craft@Hrs	Unit	Material	Labor	Equipment	Total
Based on 3 uses of forms	C8@.018	SFCA	1.33	.74	—	2.07

Column anchor bolts, usually 4 per column. J-hook bolt includes flat washer and two nuts

	Craft@Hrs	Unit	Material	Labor	Equipment	Total
1/2" x 6"	C8@.037	Ea	.63	1.52	—	2.15
3/4" x 12"	C8@.037	Ea	2.25	1.52	—	3.77

Concrete, 2,000 PSI design mix, including 3% allowance for waste,

	Craft@Hrs	Unit	Material	Labor	Equipment	Total
Placed directly from chute	CL@.552	CY	89.60	20.10	—	109.70

Remove, clean and stack

	Craft@Hrs	Unit	Material	Labor	Equipment	Total
reinforcing supports	CL@.020	LF	—	.73	—	.73

Dry pack non-shrink grout under column bases. Typical grout area is

	Craft@Hrs	Unit	Material	Labor	Equipment	Total
2' x 2' x 4" (1.33 CF of grout)	CM@.625	Ea	28.30	27.20	—	55.50

Panel support pads. (Supports panel when grout is placed.) One needed per panel. Based on

	Craft@Hrs	Unit	Material	Labor	Equipment	Total
3 CF of 2,000 PSI concrete per pad	C8@.141	Ea	8.31	5.78	—	14.09

Backfill at foundation with 1/2 CY utility backhoe.

	Craft@Hrs	Unit	Material	Labor	Equipment	Total
90 CY per hour	S1@.022	CY	—	.95	.34	1.29

Dispose of excess soil on site,

	Craft@Hrs	Unit	Material	Labor	Equipment	Total
using backhoe	S1@.011	CY	—	.47	.17	.64

Concrete test cylinders, typically 5 required

	Craft@Hrs	Unit	Material	Labor	Equipment	Total
per 100 CY of concrete	—	CY	.75	—	—	.75

Floor slabs for tilt-up These costs assume that the site has been graded and compacted by others before slab work begins. The costs below include fine grading the site, placing a 2" layer of sand, a vapor barrier and a second 2" layer of sand over the barrier. Each layer of sand should be screeded level. Reinforcing may be either wire mesh or bars. When sand and vapor barrier are used, concrete may have to be pumped in place. If inspection of the reinforcing is required, the cost will usually be included in the building permit fee. Floor slabs are usually poured 6" deep in 14' to 18' wide strips with edge forms on both sides and ends. Edge forms are used as supports for the mechanical screed which consolidates the concrete with pneumatic vibrators as the surface is leveled. The screed is pulled forward by a winch and cable. Edge forms should have an offset edge to provide a groove which is filled by the adjacent slab. The slab perimeter should end 2' to 3' inside the finished building walls. This "pour strip" is filled in with concrete when the walls are up and braces have been removed. About 2 days after pouring, the slab should be saw-cut 1-1/2" to 2" deep and 1/4" wide each 20' both ways. This controls cracking as the slab cures. For scheduling purposes, estimate that a finishing crew of 10 workers can place and finish 10,000 to 12,000 square feet of slab in 8 hours.

	Craft@Hrs	Unit	Material	Labor	Equipment	Total
Fine grade the building pad with a utility tractor with blade and bucket						
2,500 SF per hour	S6@.001	SF	—	.04	.01	.05
Spread sand 4" deep in 2" layers, (1.35 tons is 1 cubic yard and covers 81 SF at 4" depth), using a utility tractor						
Cost of sand	—	Ton	13.80	—	—	13.80
60 SF per hour	S6@.006	SF	.23	.24	.06	.53
Vapor barrier, 6 mil polyethylene	CL@.001	SF	.06	.04	—	.10
Welded wire mesh, 6" x 6", W1.4 x W1.4						
Including 10% for overlap and waste	RB@.004	SF	.18	.21	—	.39
Reinforcing bars, grade 60, typically #4 bars on 2' centers each way						
With 1.05 LF bars per SF	RB@.004	SF	.30	.21	—	.51
Dowel out for #4 bars						
at construction joints	RB@.004	LF	.18	.21	—	.39
Edge forms with keyed joint, rented	F5@.013	LF	—	.54	.09	.63
Ready-mix concrete for the slab (with 3% waste allowance)						
Placed from the chute						
of the ready-mix truck	CL@.421	CY	98.70	15.30	—	114.00
Pump mix, including pumping cost	M2@.188	CY	106.00	7.41	9.40	122.81
Finish the floor slab, using power trowels	CM@.003	SF	—	.13	.03	.16
Vibrating screed and accessories, (includes screeds, fuel, air compressor and hoses) rented						
Per SF of slab	—	SF	—	—	.13	.13
Curing compound, spray applied,						
250 SF per gallon	CL@.001	SF	.03	.04	—	.07
Saw cut green concrete,						
1-1/2" to 2" deep	C8@.010	LF	—	.41	.08	.49
Strip, clean and stack edge forms	CL@.011	LF	—	.40	—	.40
Ready-mix concrete for the pour strip including 3% waste						
Placed from the chute of the						
ready-mix truck	CL@.421	CY	95.80	15.30	—	111.10
Pump mix, including pumping cost	M2@.188	CY	105.00	7.41	9.40	121.81
Finish the pour strip	CM@.004	SF	—	.17	.04	.21
Special inspection of the concrete pour (if required)						
Based on 10,000 SF in 8 hours	—	SF	—	—	—	.04
Clean and caulk the saw joints, based on 50 LF per hour and						
caulk at 1,500 LF per gallon	CL@.020	LF	.03	.73	—	.76
Clean slab (broom clean)	CL@.001	SF	—	.04	—	.04

Concrete 3

Tilt-up wall panels Wall thickness is usually a nominal 6" (5-1/2" actual) or a nominal 8" (7-1/2"). Panel heights over 40' are uncommon. Typical panel width is 20'. When the floor area isn't large enough to form all wall panels on the floor slab, stacking will be required. Construction period will be longer and the costs higher than the figures shown below when panels must be poured and cured in stacked multiple layers. When calculating the volume of concrete required, the usual practice is to multiply overall panel dimensions (width, length and thickness) and deduct only for panel openings that exceed 25 square feet. When pilasters (thickened wall sections) are formed as part of the panel, add the cost of extra formwork and the additional concrete. Use the production rates that follow for scheduling a job.

Production rates for tilt-up

Lay out and form panels, place reinforcing bars and install embedded lifting devices:
2,000 to 3,000 SF of panel face (measured one side) per 8-hour day for a crew of 5. Note that reinforcing may have to be inspected before concrete is poured.

Place and finish concrete: 4,000 to 5,000 SF of panel per 8-hour day for a crew of 5.
Continuous inspection of the concrete pour by a licensed inspector may be required.
The minimum charge will usually be 4 hours at $50 per hour plus travel cost.

Install ledgers with embedded anchor bolts on panels before the concrete hardens:
150 to 200 LF of ledger per 8-hour day for a crew of 2.

Sack and patch exposed panel face before panel is erected:
2,000 to 3,000 SF of panel face per 8-hour day for a crew of 2.
Install panel braces before lifting panels:
30 to 40 braces per 8-hour day for a crew of 2. Panels usually need 3 or 4 braces. Braces are usually rented for 30 days.

Tilt up and set panels 20 to 24 panels per 8-hour day for a crew of 11 (6 workers, the crane operator, oiler and 3 riggers). If the crew sets 22 panels averaging 400 SF each in an 8-hour day, the crew is setting 1,100 SF per hour. These figures include time for setting and aligning panels, drilling the floor slab for brace bolts, and securing the panel braces. Panel tilt-up will go slower if the crane can't work from the floor slab when lifting panels. Good planning of the panel lifting sequence and easy exit of the crane from the building interior will increase productivity. Size and weight of the panel has little influence on the setting time if the crane is capable of making the lift and is equipped with the right rigging equipment, spreader bars and strongbacks. Panels will usually have to cure for 14 days before lifting unless high early strength concrete is used.

Weld or bolt steel embedded in panels: 16 to 18 connections per 8-hour day for a crew of 2 working from ladders. Work more than 30 feet above the floor may take more time.

Sack and patch panel faces after erection: 400 SF of face per 8-hour day for a crew of 2.

Tilt-up panel costs	Craft@Hrs	Unit	Material	Labor	Equipment	Total
Engineering fee, per panel, typical	—	Ea	—	—	—	36.00
Fabricate and lay out forms for panel edges and blockouts, assumes 2 uses and includes						
typical hardware and chamfer						
2" x 6" per MBF	—	MBF	627.00	—	—	627.00
2" x 8" per MBF	—	MBF	639.00	—	—	639.00
2" x 6" per LF	F5@.040	LF	.63	1.68	—	2.31
2" x 8" per LF	F5@.040	LF	.85	1.68	—	2.53
Reveals (1" x 4" accent strips on exterior panel face)						
based on 1 use of lumber,						
nailed to slab	C8@.018	LF	.22	.74	—	.96
Clean and prepare the slab prior to placing the concrete for the panels,						
1,250 SF per hour	CL@.001	SF	—	.04	—	.04
Releasing agent (bond breaker), sprayed on,						
250 SF and $7.50 per gallon	CL@.001	SF	.04	.04	—	.08
Place and tie grade 60 reinforcing bars and supports. Based on using #3 and #4 bars, including waste and laps. A						
typical panel has 80 to 100 pounds of reinforcing steel per CY of concrete in the panel						
Reinforcing per CY of concrete	RB@.500	CY	41.80	26.60	—	68.40
Reinforcing per SF of 6" thick panel	RB@.009	SF	.77	.48	—	1.25
Reinforcing per SF of 8" thick panel	RB@.012	SF	1.04	.64	—	1.68

	Craft@Hrs	Unit	Material	Labor	Equipment	Total

Ledger boards with anchor bolts. Based on 4" x 12" treated lumber and 3/4" x 12" bolts placed each 2 feet, installed prior to lifting panel

	Craft@Hrs	Unit	Material	Labor	Equipment	Total
Using 4" x 12" treated lumber	—	MBF	1,020.00	—	—	1,020.00
Typical ledgers and anchor bolts	C8@.353	LF	1.82	14.50	—	16.32

Embedded steel for pickup and brace point hardware. Usually 10 to 12 points per panel, quantity and type depend on panel size

	Craft@Hrs	Unit	Material	Labor	Equipment	Total
Embedded pickup & brace hardware, typical	C8@.001	SF	.54	.04	—	.58

Note: Inspection of reinforcing may be required before concrete is placed.

Concrete, 3,000 PSI, placed direct from chute

	Craft@Hrs	Unit	Material	Labor	Equipment	Total
Concrete only (before waste allowance)	—	CY	95.80	—	—	95.80
6" thick panels (54 SF per CY)	CL@.008	SF	1.82	.29	.08	2.19
8" thick panels (40.5 SF per CY)	CL@.010	SF	2.08	.36	.10	2.54

Concrete, 3,000 PSI, pumped in place

	Craft@Hrs	Unit	Material	Labor	Equipment	Total
Concrete only (before waste allowance)	—	CY	91.60	—	—	91.60
6" thick panels (54 SF per CY)	M2@.008	SF	1.70	.32	.03	2.05
8" thick panels (40.5 SF per CY)	M2@.011	SF	2.43	.43	.04	2.90
Finish concrete panels, one face before lifting	P8@.004	SF	—	.16	.02	.18

Continuous inspection of pour, if required, based on 5,000 to 6,000 SF being placed per 8-hour day

	Craft@Hrs	Unit	Material	Labor	Equipment	Total
Typical cost for continuous inspection of pour	—	SF	—	—	—	.07

Curing wall panels with spray-on curing compound, per SF of face,

	Craft@Hrs	Unit	Material	Labor	Equipment	Total
based on 250 SF per gallon	CL@.001	SF	.03	.04	—	.07
Strip, clean and stack 2" edge forms	CL@.030	LF	—	1.09	—	1.09

Braces for panels

Install rented panel braces, based on an average of three braces per panel for panels averaging 400 to 500 SF each. See also, Remove panel braces

	Craft@Hrs	Unit	Material	Labor	Equipment	Total
Typical cost for braces, per SF of panel	F5@.005	SF	—	.21	.10	.31

Miscellaneous equipment rental, typical costs per SF of wall panel

	Craft@Hrs	Unit	Material	Labor	Equipment	Total
Rent power trowels and vibrators	—	SF	—	—	.05	.04
Rent compressors, hoses, rotohammers, fuel	—	SF	—	—	.06	.05

Lifting panels into place, with crane and riggers, using a 140 ton truck crane

	Craft@Hrs	Unit	Material	Labor	Equipment	Total
Move crane on and off site, typical	—	LS	—	—	2,420.00	2,420.00
Hourly cost (4 hr minimum applies), per hour	T2@11.0	Hr	—	498.00	323.00	821.00
Set, brace and align panels, per SF of panel measured on one face, Based on 1,100 SF per hour	T2@.010	SF	—	.45	.33	.78

Allowance for shims, bolts for panel braces and miscellaneous hardware, labor cost is included with associated items

	Craft@Hrs	Unit	Material	Labor	Equipment	Total
Typical cost per SF of panel	—	SF	.25	—	—	.25

Weld or bolt panels together, typically 2 connections are required per panel, equipment is a welding machine.

	Craft@Hrs	Unit	Material	Labor	Equipment	Total
Typical cost per panel connection	T3@.667	Ea	70.00	29.90	3.61	103.51

Continuous inspection of welding, if required.

	Craft@Hrs	Unit	Material	Labor	Equipment	Total
Inspection of welds, per connection (3 per hour)	—	Ea	—	—	—	25.00

	Craft@Hrs	Unit	Material	Labor	Equipment	Total

Set grout forms at exterior wall perimeter. Based on using 2" x 12" boards. Costs shown include stakes and assume 3 uses of the forms

Using 2" x 12" boards (per MBF)	—	MBF	713.00	—	—	713.00
Cost of forms per linear foot (LF)	F5@.019	LF	.93	.80	—	1.73

Pour grout under panel bottoms. Place 2,000 PSI concrete (grout) directly from the chute of ready-mix truck.

Typically 1.85 CY per 100 LF of wall	M2@.005	LF	1.86	.20	—	2.06

Sack and patch wall, per SF of area (measure area to be sacked and patched, both inside and outside),

per SF of wall face(s)	P8@.007	SF	.07	.28	—	.35

Caulking and backing strip for panel joints. Cost per LF for caulking both inside & outside face of joint,

per linear foot (LF) of joint	PA@.055	LF	.20	2.50	—	2.70

Remove panel braces,

stack & load onto trucks	CL@.168	Ea	—	6.11	—	6.11

Final cleanup & patch floor,

per SF of floor	P8@.003	SF	.01	.12	—	.13

Additional costs for tilt-up panels, if required

Architectural form liner, minimum cost	F5@.001	SF	1.40	.04	—	1.44
Architectural form liner, typical cost	F5@.003	SF	1.80	.13	—	1.93
Sandblast, light and medium, subcontract	—	SF	—	—	—	1.50
Sandblast, heavy, subcontract	—	SF	—	—	—	2.25
Truck dock bumpers, installation only	C8@2.00	Ea	—	81.90	—	81.90
Dock levelers, installation only	C8@7.46	Ea	—	306.00	—	306.00
Angle iron guard around door opening, 2" x 2" x 1/4", mild steel (4' lengths)	C8@.120	LF	2.10	4.92	—	7.02
Pipe guard at door opening, 6" black steel, concrete filled	C8@.281	LF	25.00	11.50	—	36.50
Embedded reglet at parapet wall, aluminum	F5@.019	LF	1.00	.80	—	1.80
Integral color in concrete, light colors, add to concrete. Price per CY of concrete to be colored	—	CY	22.00	—	—	22.00
Backfill outside perimeter wall with a utility tractor. 90 CY per hour	S1@.022	CY	—	.95	.24	1.19

Cast-in-place columns for tilt-up Concrete columns are often designed as part of the entryway of a tilt-up building or to support lintel panels over a glass storefront. Both round and square freestanding columns are used. They must be formed and poured before the lintel panels are erected. Round columns are formed with plastic-treated fiber tubes (Sonotube is one manufacturer) that are peeled off when the concrete has cured. Square columns are usually formed with plywood and 2" x 4" wood braces or adjustable steel column clamps.

Round plastic-lined fiber tube forms, set and aligned, based on 12' long tubes. Longer tubes will cost more per LF

12" diameter tubes	M2@.207	LF	8.13	8.16	—	16.29
20" diameter tubes	M2@.250	LF	13.90	9.86	—	23.76
24" diameter tubes	M2@.275	LF	16.70	10.80	—	27.50

Rectangular plywood forms, using 1.1 SF of 3/4" plyform at $21.70 per 4' x 8' sheet ($0.678 per SF), and 2" x 4" (at $621.00 per MBF) bracing 3' OC

Per SF of contact area based on 3 uses	F5@.083	SF	.72	3.48	—	4.20

Column braces, 2" x 4" at $621.00 per MBF, with 5 BF per LF of column height

Per LF of column height based on 3 uses	F5@.005	LF	1.50	.20	—	1.73

	Craft@Hrs	Unit	Material	Labor	Equipment	Total
Column clamps rented at $4.50 for 30 days, 1 per LF of column height, cost per LF of column height						
Per LF of column						
based on 2 uses a month	F5@.040	LF	—	1.68	2.73	4.41
Reinforcing bars, grade 60, placed and tied, typically 150 to 200 pounds per CY of concrete						
Cages, round or rectangular	RB@1.50	CY	83.50	79.70	—	163.20
Vertical bars						
attached to cages	RB@1.50	CY	83.50	79.70	—	163.20
Ready-mix concrete, 3,000 PSI pump mix						
Placed by pump (includes 3% waste)	M2@.094	CY	105.00	3.71	6.70	115.41
Strip off fiber forms and dismantle bracing, per LF of height						
12" or 16" diameter	CL@.075	LF	—	2.73	—	2.73
24" diameter	CL@.095	LF	—	2.78	—	2.78
Strip wood forms and dismantle bracing						
Per SF of contact area, no salvage	CL@.011	SF	—	.40	—	.40
Clean and stack column clamps	CL@.098	Ea	—	3.57	—	3.57
Sack and patch concrete column face	P8@.011	SF	.12	.44	—	.56
Allowance for miscellaneous equipment rental (vibrators, compressors, hoses, etc.)						
Per CY of concrete in column	—	CY	—	—	2.50	2.50

Trash enclosures for tilt-up These usually consist of a 6" concrete slab measuring 5' x 10' and precast concrete tilt-up panels 6' high. A 6" ledger usually runs along the wall interior. Overall wall length is usually 20 feet. Cost for the trash enclosure foundation, floor slab and wall will be approximately the same as for the tilt-up foundation, floor slab and wall. The installed cost of a two-leaf steel gate measuring 6' high and 8' long overall will be about $400 including hardware.

Equipment Foundations Except as noted, no grading, compacting, engineering, design, permit or inspection fees are included. Costs assume normal soil conditions with no shoring or dewatering required and are based on concrete poured directly from the chute of a ready-mix truck. Costs will be higher if access is not available from all sides or if obstructions delay the work.

Labor costs include the time needed to prepare formwork sketches at the job site, measure for the forms, fabricate, erect, align and brace the forms, place and tie the reinforcing steel and embedded steel items, pour and finish the concrete and strip, clean and stack the forms.

Reinforcing steel costs assume an average of #3 and #4 bars at 12" on center each way. Weights shown include an allowance for waste and laps.

Form costs assume 3 uses of the forms and show the cost per use when a form is used on the same job without being totally disassembled. Normal cleaning and repairs are included. No salvage value is assumed. Costs listed are per square foot of form in contact with the concrete, assume using #2 & Btr lumber

For scheduling purposes, estimate that a crew of 5 can set the forms and place and tie the reinforcing steel for and pour 8 to 10 CY of concrete equipment foundations in an 8-hour day.

	Craft@Hrs	Unit	Material	Labor	Equipment	Total

Rule-of-thumb estimates The costs below are based on the volume of concrete in the foundation.
If plans for the foundation have not been prepared, estimate the quantity of concrete required from the weight of the equipment using the rule-of-thumb estimating data below. Design considerations may alter the actual quantities required. Based on the type foundation required, use the cost per CY given in the detailed cost breakdown section which follows the table on the next page.

Estimating concrete quantity from the weight of equipment:

1. Rectangular pad type foundations —
   Boilers and similar equipment mounted at grade:
      Allow 1 CY per ton

   Centrifugal pumps and compressors, including weight of motor:
      Up to 2 HP, allow .5 CY per ton
      Over 2 to 10 HP, allow 1 CY per ton
      Over 10 to 30 HP, allow 2 CY per ton
      Over 30 HP, allow 3 CY per ton

   Reciprocating pumps and compressors, including weight of motor:
      Up to 2 HP, allow .75 CY per ton
      Over 2 to 10 HP, allow 1.5 CY per ton
      Over 10 to 30 HP, allow 2.5 CY per ton
      Over 30 HP, allow 4.5 CY per ton

2. Rectangular pad and stem wall type foundations —
   Horizontal tanks or shell and tube heat exchangers, allow 6 CY per ton

3. Square pad and pedestal type foundations — Vertical tanks not over 3' in diameter, allow 6 CY per ton

4. Octagonal pad and pedestal type foundations — Vertical tanks over 3' in diameter, allow 4 CY per ton of empty weight

	Craft@Hrs	Unit	Material	Labor	Equipment	Total

Rectangular pad type foundations For pumps, compressors, boilers and similar equipment mounted at grade. Form costs assume nails, stakes and form oil costing $.50 per square foot and 1.1 SF of 3/4" plyform plus 2.2 BF of lumber per square foot of contact surface. Costs shown are per cubic yard (CY) of concrete within the foundation.

	Craft@Hrs	Unit	Material	Labor	Equipment	Total
Lay out foundation from known point on site, set stakes and mark for excavation, with 1 CY per CY of concrete	C8@.147	CY	.05	6.02	—	6.07
Excavation with utility backhoe, at 28 CY per hour, with 1.5 CY per CY of concrete	S1@.108	CY	—	4.64	3.38	8.02
Formwork, material (3 uses) and labor at .150 manhours per SF with 15 SF per CY of concrete	F5@2.25	CY	23.40	94.30	—	117.70
Reinforcing bars, Grade 60, set and tied, material and labor .006 manhours per lb with 25 lb per CY of concrete	RB@.150	CY	12.50	7.97	—	20.47
Anchor bolts, sleeves, embedded steel, material and labor at .040 manhours per lb with 5 lb per CY of concrete	C8@.200	CY	6.25	8.19	—	14.44
Ready-mix concrete, 3,000 PSI mix after allowance for 10% waste with 1 CY per CY of concrete	C8@.385	CY	105.00	15.80	—	120.80
Concrete test cylinders including test reports, with 5 per 100 CY of concrete	—	CY	.50	—	—	.50
Finish, sack and patch concrete, material and labor at .010 manhours per SF with 15 SF per CY of concrete	P8@.150	CY	.75	6.00	—	6.75
Backfill around foundation with utility backhoe at 40 CY per hour, with .5 CY per CY of concrete	S1@.025	CY	—	1.07	.39	1.46
Dispose of excess soil on site with utility backhoe at 80 CY per hour with 1 CY per CY of concrete	S1@.025	CY	—	1.07	.39	1.46
Rectangular pad type foundation						
Total cost per CY of concrete	**--@3.44**	**CY**	**148.45**	**145.06**	**4.16**	**297.67**

Concrete 3

	Craft@Hrs	Unit	Material	Labor	Equipment	Total

Rectangular pad and stem wall type foundations For horizontal cylindrical or shell and tube heat exchangers mounted above grade. Form costs assume nails, stakes and form oil costing $.60 per square foot and 1.1 SF of 3/4" plyform plus 4.5 BF of lumber per square foot of contact surface. Costs shown are per cubic yard (CY) of concrete within the foundations.

	Craft@Hrs	Unit	Material	Labor	Equipment	Total
Lay out foundation from known point on site, set stakes and mark for excavation, with 1 CY per CY of concrete	C8@.147	CY	.04	6.02	—	6.06
Excavation with utility backhoe, at $28 per hour, with 1 CY per CY of concrete	S1@.072	CY	—	3.10	1.13	4.23
Formwork, material (3 uses) per SF and labor at .200 manhours per SF with 60 SF per CY of concrete	F5@12.0	CY	165.00	503.00	—	668.00
Reinforcing bars, Grade 60, set and tied, material per lb and labor at .006 manhours per lb with 45 lb per CY of concrete	RB@.270	CY	22.50	14.40	—	36.90
Anchor bolts, sleeves, embedded steel, material and labor at .040 manhours per lb with 5 lb per CY of concrete	C8@.200	CY	6.25	8.19	—	14.44
Ready-mix concrete, 3,000 PSI mix, after 10% waste allowance with 1 CY per CY	C8@.385	CY	105.00	15.80	—	120.80
Concrete test cylinders including test reports, with 5 per 100 CY	—	CY	.50	—	—	.50
Finish, sack and patch concrete, material per SF and labor at .010 manhours per SF with 60 SF per CY of concrete	P8@.600	CY	3.00	24.00	—	27.00
Backfill around foundation with utility backhoe, at 40 CY per hour, with .5 CY per CY of concrete	S1@.025	CY	—	1.07	.39	1.46
Dispose of excess soil on site with utility backhoe, at 80 CY per hour, with .5 CY per CY of concrete	S1@.025	CY	—	1.07	.39	1.46
Rectangular pad and stem wall type foundation **Total cost per CY of concrete**	**--@13.7**	**CY**	**302.29**	**576.65**	**1.91**	**880.85**

Square pad and pedestal type foundations For vertical cylindrical tanks not over 3' in diameter. Form costs assume nails, stakes and form oil costing $.50 per square foot and 1.1 SF of 3/4" plyform plus 3 BF of lumber per square foot of contact surface. Costs shown are per cubic yard (CY) of concrete within the foundations.

	Craft@Hrs	Unit	Material	Labor	Equipment	Total
Lay out foundation from known point on site, set stakes and mark for excavation, with 1 CY per CY of concrete	C8@.147	CY	.05	6.02	—	6.07
Excavation with utility backhoe, at 28 CY per hour, with 1 CY per CY of concrete	S1@.072	CY	—	3.10	1.13	4.23
Formwork, material and labor at .192 manhours per SF with 13 SF per CY of concrete	F5@2.50	CY	28.30	105.00	—	133.30
Reinforcing bars, Grade 60, set and tied, material and labor at .006 manhours per lb with 20 lb per CY of concrete	RB@.120	CY	10.00	6.38	—	16.38
Anchor bolts, sleeves, embedded steel, material and labor at .040 manhours per lb with 2 lb per CY of concrete	C8@.080	CY	2.50	3.28	—	5.78
Ready-mix concrete, 3,000 PSI mix, after 10% waste allowance with 1 CY per CY	C8@.385	CY	105.00	15.80	—	120.80
Concrete test cylinders including test reports, with 5 per 100 CY	—	CY	.50	—	—	.50
Finish, sack and patch concrete, material per SF and labor at .010 manhours per SF with 60 SF per CY of concrete	P8@.600	CY	3.00	24.00	—	27.00
Backfill around foundation with utility backhoe, at 40 CY per hour, with .5 CY per CY of concrete	S1@.025	CY	—	1.07	.39	1.46
Dispose of excess soil on site with utility backhoe, at 80 CY per hour, with .5 CY per CY of concrete	S1@.025	CY	—	1.07	.39	1.46
Square pad and pedestal type foundation **Total cost per CY of concrete**	**--@3.95**	**CY**	**149.35**	**165.72**	**1.91**	**316.98**

Concrete 3

	Craft@Hrs	Unit	Material	Labor	Equipment	Total

Octagonal pad and pedestal type foundations For vertical cylindrical tanks over 3' in diameter. Form costs assume nails, stakes and form oil costing $.50 per square foot and 1.1 SF of 3/4" plyform plus 4.5 BF of lumber per square foot of contact surface. Costs shown are per cubic yard (CY) of concrete within the foundation.

Lay out foundation from known point on site, set stakes and mark for excavation,
| with 1 CY per CY of concrete | C8@.147 | CY | .05 | 6.02 | — | 6.07 |

Excavation with utility backhoe, at $28 CY per hour,
| with 1 CY per CY of concrete | S1@.072 | CY | — | 3.10 | 1.13 | 4.23 |

Formwork, material SF and labor at .250 manhours per SF
| with 10 SF per CY of concrete | F5@2.50 | CY | 26.40 | 105.00 | — | 131.40 |

Reinforcing bars, Grade 60, set and tied, material and labor at .006 manhours per lb
| with 20 lb per CY of concrete | RB@.120 | CY | 10.00 | 6.38 | — | 16.38 |

Anchor bolts, sleeves, embedded steel, material and labor at .040 manhours per lb
| with 2 lb per CY of concrete | C8@.080 | CY | 2.50 | 3.28 | — | 5.78 |

Ready-mix concrete, 3,000 PSI mix, after 10% waste allowance
| with 1 CY per CY of concrete | C8@.385 | CY | 105.00 | 15.80 | — | 120.80 |

Concrete test cylinders including test reports,
| with 5 per 100 CY of concrete | — | CY | .50 | — | — | .50 |

Finish, sack and patch concrete, material and labor at .010 manhours per SF
| with 60 SF per CY of concrete | P8@.600 | CY | 3.00 | 24.00 | — | 27.00 |

Backfill around foundation with utility backhoe, at 40 CY per hour,
| with .5 CY per CY of concrete | S1@.025 | CY | — | 1.07 | .39 | 1.46 |

Dispose of excess soil on site with utility backhoe, at 80 CY per hour,
| with .5 CY per CY of concrete | S1@.025 | CY | — | 1.07 | .39 | 1.46 |

Octagonal pad and pedestal type foundation
| **Total cost per CY of concrete** | --@3.95 | CY | 147.45 | 165.72 | 1.91 | 315.08 |

Cementitious Decks, Subcontract Typical costs including the subcontractor's overhead and profit.

Insulating concrete deck, poured on existing substrate, 1-1/2" thick
Under 3,000 SF	—	SF	—	—	—	1.94
Over 3,000 SF to 10,000 SF	—	SF	—	—	—	1.67
Over 10,000 SF	—	SF	—	—	—	1.41

Poured on existing metal deck, 2-5/8" thick, lightweight
| Concrete fill, trowel finish | — | SF | — | — | — | 2.28 |

Gypsum decking,
| poured over form board, screed and float | — | SF | — | — | — | 2.06 |

Vermiculite,
| screed and float, field mixed | — | SF | — | — | — | 1.68 |

Exterior concrete deck for walking service, troweled
Lightweight, 2-5/8", 110 lbs per CF	—	SF	—	—	—	2.94
Add for 4,000 PSI mix	—	SF	—	—	—	.22
Add for 5,000 PSI mix	—	SF	—	—	—	.35

Pool deck at pool side,
| 1/8" to 3/16" thick, colors, finished | — | SF | — | — | — | 2.65 |

Fiber deck, tongue and groove, cementitious planks, complete
2" thick	—	SF	—	—	—	2.24
2-1/2" thick	—	SF	—	—	—	2.72
3" thick	—	SF	—	—	—	2.97

Concrete 3

	Craft@Hrs	Unit	Material	Labor	Total

Concrete Surfacing, Subcontract Load-bearing, architectural, three-part, acrylic polymer cementitious resurfacing system. 6,000 PSI compressive strength. Applied over existing concrete, aggregate, masonry, steel, asphalt, foam or wood surfaces. Installed on surfaces that do not require crack repair, brick or tile patterns or the removal of heavy grease contaminants. Includes stain sealer and colored texture coat. Based on a typical 1,000 square foot project. www.permacrete.com

	Craft@Hrs	Unit	Material	Labor	Total
Applied over new surface	—	SF	—	—	.79
Applied over old concrete, includes cleaning	—	SF	—	—	1.43
Applied over pool interior, includes cleaning	—	SF	—	—	1.69
Applied over vertical surface	—	SF	—	—	1.95

Surface Imprinted Concrete Paving, Subcontract Patterned and colored to simulate slate, granite, limestone, sandstone, wood or cobblestone. 3 to 4 inch depth. Costs include imprinting and texturing, concrete placement and finish, color hardener, release agent and sealer application. Add the cost of grading, forms and form stripping. www.bomanite.com

	Craft@Hrs	Unit	Material	Labor	Total
Regular grade	—	SF	—	—	15.80
Heavy grade (for heavy road traffic and freeze areas)	—	SF	—	—	19.50

Troweled-on Cementitious Coating, Subcontract For direct use on interior flooring, accent areas or walls. Bonds to concrete, wood, metal, plastic or asphalt.

	Craft@Hrs	Unit	Material	Labor	Total
Cementitious coating	—	SF	—	—	11.80

Masonry 4

	Craft@Hrs	Unit	Material	Labor	Total

Brick Wall Assemblies Typical costs for smooth red clay brick walls, laid in running bond with 3/8" concave joints. These costs include the bricks, mortar for bricks and cavities, typical ladder type reinforcing, wall ties and normal waste. Foundations are not included. Wall thickness shown is the nominal size based on using the type bricks described. "Wythe" means the quantity of bricks in the thickness of the wall. Costs shown are per square foot (SF) of wall measured on one face. Deduct for openings over 10 SF in size. The names and dimensions of bricks can be expected to vary, depending on the manufacturer. For more detailed coverage of brick wall assemblies, see National Concrete & Masonry Estimator, http://CraftsmanSiteLicense.com.

	Craft@Hrs	Unit	Material	Labor	Total
Standard bricks, 3-3/4" wide x 2-1/4" high x 8" long					
4" thick wall, single wythe, veneer facing	M1@.211	SF	3.64	8.55	12.19
8" thick wall, double wythe, cavity filled	M1@.464	SF	7.14	18.80	25.94
12" thick wall, triple wythe, cavity filled	M1@.696	SF	10.60	28.20	38.80
Modular bricks, 3" wide x 3-1/2" x 11-1/2" long					
3-1/2" thick wall, single wythe, veneer facing	M1@.156	SF	4.18	6.32	10.50
7-1/2" thick wall, double wythe, cavity filled	M1@.343	SF	8.23	13.90	22.13
11-1/2" thick wall, triple wythe, cavity filled	M1@.515	SF	12.30	20.90	33.20
Colonial bricks, 3" wide x 3-1/2" x 10" long (at $719 per M)					
3-1/2" thick wall, single wythe, veneer facing	M1@.177	SF	3.93	7.17	11.10
7-1/2" thick wall, double wythe, cavity filled	M1@.389	SF	7.73	15.80	23.53
11-1/2" thick wall, triple wythe, cavity filled	M1@.584	SF	11.50	23.70	35.20
Add to any brick wall assembly above for:					
American bond or stacked bond	—	%	5.0	10.0	—
Basketweave bond or soldier bond	—	%	10.0	10.0	—
Flemish bond	—	%	10.0	15.0	—
Herringbone pattern	—	%	15.0	15.0	—
Institutional inspection	—	%	—	20.0	—
Short runs or cut-up work	—	%	5.0	15.0	—

Masonry 4

	Craft@Hrs	Unit	Material	Labor	Total

Chemical Resistant Brick or Tile, Subcontract (also known as acid-proof brick). Typical prices for chemical resistant brick or tile meeting ASTM C-279, types I, II, and III requirements. Installed with 1/8" thick bed and mortar joints over a membrane of either elastomeric bonded to the floors or walls, sheet membrane with a special seam treatment, squeegee hot-applied asphalt, cold-applied adhesive troweled on, or special spray-applied material. The setting bed and joint materials can be either 100% carbon-filled furan or phenolic, sodium or potassium silicates, silica-filled epoxy, polyester, vinyl ester or other special mortars. Expansion joints are required in floor construction, particularly where construction joints exist in the concrete. Additional expansion joints may be required, depending on the service conditions to which the floor will be subjected. Brick or tile will be either 3-3/4" x 8" face, 1-1/8" thick (splits); 3-3/4" x 8" face, 2-1/4" thick (singles); or 4" x 8" face, 1-3/8" thick (pavers). Costs are based on a 2,000 SF job. For smaller projects, increase the cost 5% for each 500 SF or portion of 500 SF less than 2,000 SF.
For scheduling purposes estimate that a crew of 4 can install 75 to 100 SF of membrane and brick in an 8-hour day. These costs assume that the membrane is laid on a suitable surface prepared by others and include the subcontractor's overhead and profit.

Typical installed costs for chemical resistant brick or tile (acid-proof brick), installed

Less complex jobs	—	SF	—	—	26.20
More complex jobs	—	SF	—	—	31.70
Minimum job cost	—	LS	—	—	17,600.00

Quarry Tile Unglazed natural red tile set in a portland cement bed with mortar joints.

Quarry floor tile

4" x 4" x 1/2", 1/8" straight joints	M1@.112	SF	3.30	4.54	7.84
6" x 6" x 1/2", 1/4" straight joints	M1@.101	SF	3.09	4.09	7.18
6" x 6" x 3/4", 1/4" straight joints	M1@.105	SF	4.04	4.25	8.29
8" x 8" x 3/4", 3/8" straight joints	M1@.095	SF	3.82	3.85	7.67
8" x 8" x 1/2", 1/4" hexagon joints	M1@.112	SF	4.35	4.54	8.89

Quarry wall tile

4" x 4" x 1/2", 1/8" straight joints	M1@.125	SF	3.28	5.06	8.34
6" x 6" x 3/4", 1/4" straight joints	M1@.122	SF	4.04	4.94	8.98

Quarry tile stair treads

6" x 6" x 3/4", 12" wide tread	M1@.122	LF	4.45	4.94	9.39

Quarry tile window sill

6" x 6" x 3/4" tile on 6" wide sill	M1@.078	LF	4.27	3.16	7.43

Quarry tile trim or cove base

5" x 6" x 1/2" straight top	M1@.083	LF	3.28	3.36	6.64
6" x 6" x 3/4" round top	M1@.087	LF	3.82	3.52	7.34
Deduct for tile set in epoxy bed without grout joints	—	SF	—	-1.91	-1.91
Add for tile set in epoxy bed with grout joints	—	SF	1.22	—	1.22

Pavers and Floor Tile

Brick, excluding platform cost

Plate, glazed	M1@.114	SF	2.66	4.62	7.28
Laid in sand	M1@.089	SF	2.03	3.60	5.63
Pavers, on concrete, grouted	M1@.107	SF	3.58	4.33	7.91
Add for special patterns	—	%	25.0	35.0	—
Slate	M1@.099	SF	3.13	4.01	7.14

Terrazzo tiles, 1/4" thick

Standard	M1@.078	SF	5.05	3.16	8.21
Granite	M1@.085	SF	8.23	3.44	11.67
Brick steps	M1@.186	SF	3.19	7.53	10.72

	Craft@Hrs	Unit	Material	Labor	Total
Interlocking Paving Stones These costs include fine sand to fill joints and machine vibration, but no excavation.					
Interlocking pavers, rectangular, 60mm thick, Orion	M1@.083	SF	1.46	3.36	4.82
Interlocking pavers, hexagonal, 80mm thick, Omni	M1@.085	SF	1.79	3.44	5.23
Interlocking pavers, multi-angle, 80mm thick, Venus	M1@.081	SF	1.53	3.28	4.81
Concrete masonry grid pavers (erosion control)	M1@.061	SF	2.57	2.47	5.04
Add for a 1" sand cushion	M1@.001	SF	.05	.04	.09
Add for a 2" sand cushion	M1@.002	SF	.09	.08	.17
Deduct for over 5,000 SF	—	%	-3.0	-7.0	—

Concrete Block Wall Assemblies Typical costs for standard natural gray medium weight masonry block walls including blocks, mortar, typical reinforcing and normal waste. Foundations are not included. For more detailed coverage of concrete block masonry, see National Concrete & Masonry Estimator, http://CraftsmanSiteLicense.com.

Walls constructed with 8" x 16" blocks laid in running bond

	Craft@Hrs	Unit	Material	Labor	Total
4" thick wall	M1@.090	SF	1.65	3.65	5.30
6" thick wall	M1@.100	SF	2.01	4.05	6.06
8" thick wall	M1@.120	SF	2.41	4.86	7.27
12" thick wall	M1@.150	SF	3.55	6.08	9.63

Additional costs for concrete block wall assemblies

Add for grouting concrete block cores at 36" intervals, poured by hand

	Craft@Hrs	Unit	Material	Labor	Total
4" thick wall	M1@.007	SF	.43	.28	.71
6" thick wall	M1@.007	SF	.65	.28	.93
8" thick wall	M1@.007	SF	.87	.28	1.15
12" thick wall	M1@.010	SF	1.29	.41	1.70

Add for 2" thick caps, natural gray concrete

	Craft@Hrs	Unit	Material	Labor	Total
4" thick wall	M1@.027	LF	.69	1.09	1.78
6" thick wall	M1@.038	LF	.80	1.54	2.34
8" thick wall	M1@.046	LF	.82	1.86	2.68
12" thick wall	M1@.065	LF	1.35	2.63	3.98

Add for detailed block, 3/8" score

	Craft@Hrs	Unit	Material	Labor	Total
Single score, one side	—	SF	.42	—	.42
Single score, two sides	—	SF	.84	—	.84
Multi-scores, one side	—	SF	.42	—	.42
Multi-scores, two sides	—	SF	.84	—	.84

Add for color block, any of above

	Craft@Hrs	Unit	Material	Labor	Total
Light colors	—	%	12.0	—	—
Medium colors	—	%	18.0	—	—
Dark colors	—	%	25.0	—	—
Add for other than running bond	—	%	—	20.0	—

Glazed Concrete Block Prices are per square foot of wall and include allowance for mortar and waste but no foundation or excavation. 8" high x 16" long, 3/8" joints, no reinforcing or grout included.
Based on 1.05 blocks per square foot (SF).
Glazed 1 side

	Craft@Hrs	Unit	Material	Labor	Total
2" wide	M1@.102	SF	10.20	4.13	14.33
4" wide	M1@.107	SF	11.80	4.33	16.13
6" wide	M1@.111	SF	11.70	4.50	16.20
8" wide	M1@.118	SF	12.40	4.78	17.18
Glazed 2 sides					
4" wide	M1@.122	SF	17.70	4.94	22.64
6" wide	M1@.127	SF	18.70	5.14	23.84
8" wide	M1@.136	SF	19.00	5.51	24.51
Glazed cove base, glazed 1 side, 8" high					
4" length	M1@.127	LF	34.70	5.14	39.84
6" length	M1@.137	LF	26.20	5.55	31.75
8" length	M1@.147	LF	20.10	5.95	26.05

Masonry 4

	Craft@Hrs	Unit	Material	Labor	Total

Decorative Concrete Block Prices are per square foot of wall and include allowance for mortar and waste but no foundation, reinforcing, grout or excavation.

Regular weight units, 1.05 blocks per SF.
Split face hollow block, 8" high x 16" long, 3/8" joints

	Craft@Hrs	Unit	Material	Labor	Total
4" wide	M1@.110	SF	1.43	4.46	5.89
6" wide	M1@.121	SF	1.51	4.90	6.41
8" wide	M1@.133	SF	1.73	5.39	7.12
10" wide	M1@.140	SF	2.48	5.67	8.15
12" wide	M1@.153	SF	2.44	6.20	8.64

Split rib (combed) hollow block, 8" high x 16" long, 3/8" joints

	Craft@Hrs	Unit	Material	Labor	Total
4" wide	M1@.107	SF	2.11	4.33	6.44
6" wide	M1@.115	SF	2.19	4.66	6.85
8" wide	M1@.121	SF	2.42	4.90	7.32
10" wide	M1@.130	SF	3.17	5.27	8.44
12" wide	M1@.146	SF	3.12	5.91	9.03

Screen block, 3/8" joints

	Craft@Hrs	Unit	Material	Labor	Total
6" x 6" x 4" wide	M1@.210	SF	2.81	8.51	11.32
8" x 8" x 4" wide (2.25 per SF)	M1@.140	SF	4.25	5.67	9.92
12" x 12" x 4" wide (1 per SF)	M1@.127	SF	2.15	5.14	7.29
8" x 16" x 8" wide (1.11 per SF)	M1@.115	SF	2.24	4.66	6.90

Acoustical slotted block, by rated Noise Reduction Coefficient (NRC)

	Craft@Hrs	Unit	Material	Labor	Total
4" wide, with metal baffle, NRC, .45 to .55	M1@.121	SF	3.50	4.90	8.40
6" wide, with metal baffle, NRC, .45 to .55	M1@.130	SF	4.16	5.27	9.43
8" wide, with metal baffle, NRC, .45 to .55	M1@.121	SF	5.16	4.90	10.06
8" wide, with fiber filler, NRC, .65 to .75	M1@.143	SF	6.67	5.79	12.46

Concrete Block Bond Beams For door and window openings, no grout or reinforcing included.
Cost per linear foot of bond beam block

	Craft@Hrs	Unit	Material	Labor	Total
6" wide, 8" high, 16" long	M1@.068	LF	1.72	2.75	4.47
8" wide, 8" high, 16" long	M1@.072	LF	1.89	2.92	4.81
10" wide, 8" high, 16" long	M1@.079	LF	2.40	3.20	5.60
12" wide, 8" high, 16" long	M1@.089	LF	2.40	3.60	6.00
12" wide, 8" high double bond beam, 16" long	M1@.102	LF	2.40	4.13	6.53
8" wide, 16" high, 16" long	M1@.127	LF	3.69	5.14	8.83

Grout for Concrete Block Grout for block cores, type "S". Costs shown are based on 2 cores per 16" block and include a 3% allowance for waste.
Grout per SF of wall, all cells filled, pumped in place and rodded

	Craft@Hrs	Unit	Material	Labor	Total
4" wide block, .11 CF per SF	M1@.018	SF	.71	.73	1.44
6" wide block, .22 CF per SF	M1@.025	SF	1.42	1.01	2.43
8" wide block, .32 CF per SF	M1@.035	SF	2.07	1.42	3.49
10" wide block, .45 CF per SF	M1@.035	SF	2.91	1.42	4.33
12" wide block, .59 CF per SF	M1@.039	SF	3.81	1.58	5.39

Grout per LF of cell filled measured vertically

	Craft@Hrs	Unit	Material	Labor	Total
4" wide block, .06 CF per LF	M1@.010	LF	.39	.41	.80
6" wide block, .11 CF per LF	M1@.015	LF	.71	.61	1.32
8" wide block, .16 CF per LF	M1@.021	LF	1.03	.85	1.88
10" wide block, .23 CF per LF	M1@.022	LF	1.49	.89	2.38
12" wide block, .30 CF per LF	M1@.024	LF	1.94	.97	2.91
Add for type M mortar (2,500 PSI)	—	%	6.0	—	—

	Craft@Hrs	Unit	Material	Labor	Total

Grout for bond beams, 3,000 PSI concrete, by block size and linear foot of beam, no block or reinforcing included, based on block cavity at 50% of block volume

	Craft@Hrs	Unit	Material	Labor	Total
8" high x 4" wide, .10 CF per LF	M1@.005	LF	.65	.20	.85
8" high x 6" wide, .15 CF per LF	M1@.010	LF	.97	.41	1.38
8" high x 8" wide, .22 CF per LF	M1@.017	LF	1.42	.69	2.11
8" high x 10" wide, .30 CF per LF	M1@.023	LF	1.94	.93	2.87
8" high x 12" wide, .40 CF per LF	M1@.032	LF	2.58	1.30	3.88
16" high x 8" wide, .44 CF per LF	M1@.061	LF	2.84	2.47	5.31
16" high x 10" wide, .60 CF per LF	M1@.058	LF	3.88	2.35	6.23
16" high x 12" wide, .80 CF per LF	M1@.079	LF	5.17	3.20	8.37

Grout hollow metal door frame in a masonry wall

	Craft@Hrs	Unit	Material	Labor	Total
Single door	M1@.668	Ea	8.73	27.10	35.83
Double door	M1@.851	Ea	11.60	34.50	46.10

Additional Costs for Concrete Block

	Craft@Hrs	Unit	Material	Labor	Total
Sill blocks	M1@.087	LF	1.50	3.52	5.02
Cutting blocks	M1@.137	LF	—	5.55	5.55
Stack bond	—	%	5.0	10.0	—
Add for cut-up jobs	—	%	5.0	5.0	—
Add for institutional grade work	—	%	5.0	15.0	—
Add for colored block	—	%	15.0	—	—

Masonry Reinforcing and Flashing Reinforcing bars for concrete block, ASTM A615 grade 60 bars, cost per linear foot including 10% overlap and waste.

	Craft@Hrs	Unit	Material	Labor	Total
#3 bars (3/8", 5,319 LF per ton), horizontal	M1@.005	LF	.21	.20	.41
#3 bars placed vertically	M1@.006	LF	.21	.24	.45
#3 galvanized bars, horizontal	M1@.005	LF	.36	.20	.56
#3 galvanized bars placed vertically	M1@.006	LF	.36	.24	.60
#4 bars (1/2", 2,994 LF per ton), horizontal	M1@.008	LF	.35	.32	.67
#4 bars placed vertically	M1@.010	LF	.35	.41	.76
#4 galvanized bars, horizontal	M1@.008	LF	.61	.32	.93
#4 galvanized bars placed vertically	M1@.010	LF	.61	.41	1.02
#5 bars (5/8", 1,918 LF per ton), horizontal	M1@.009	LF	.46	.36	.82
#5 bars placed vertically	M1@.012	LF	.46	.49	.95
#5 galvanized bars, horizontal	M1@.009	LF	.83	.36	1.19
#5 galvanized bars placed vertically	M1@.012	LF	.83	.49	1.32
#6 bars (3/4", 1,332 LF per ton), horizontal	M1@.010	LF	.57	.41	.98
#6 bars placed vertically	M1@.013	LF	.57	.53	1.10
#6 galvanized bars, horizontal	M1@.014	LF	1.12	.57	1.69
#6 galvanized bars placed vertically	M1@.014	LF	1.12	.57	1.69

Wall ties

	Craft@Hrs	Unit	Material	Labor	Total
Rectangular, galvanized	M1@.005	Ea	.33	.20	.53
Rectangular, copper coated	M1@.005	Ea	.39	.20	.59
Cavity "Z" type, galvanized	M1@.005	Ea	.31	.20	.51
Cavity "Z" type, copper coated	M1@.005	Ea	.35	.20	.55

Masonry anchors, steel, including bolts

	Craft@Hrs	Unit	Material	Labor	Total
3/16" x 18" long	M1@.173	Ea	3.80	7.01	10.81
3/16" x 24" long	M1@.192	Ea	4.77	7.78	12.55
1/4" x 18" long	M1@.216	Ea	4.91	8.75	13.66
1/4" x 24" long	M1@.249	Ea	5.97	10.10	16.07

	Craft@Hrs	Unit	Material	Labor	Total
Wall reinforcing					
Truss type, plain					
4" wall	M1@.002	LF	.18	.08	.26
6" wall	M1@.002	LF	.22	.08	.30
8" wall	M1@.003	LF	.24	.12	.36
10" wall	M1@.003	LF	.26	.12	.38
12" wall	M1@.004	LF	.28	.16	.44
Truss type, galvanized					
4" wall	M1@.002	LF	.23	.08	.31
6" wall	M1@.002	LF	.25	.08	.33
8" wall	M1@.003	LF	.31	.12	.43
10" wall	M1@.003	LF	.33	.12	.45
12" wall	M1@.004	LF	.37	.16	.53
Ladder type, galvanized, Class 3					
2" wide, for 4" wall	M1@.003	LF	.12	.12	.24
4" wide	M1@.003	LF	.25	.12	.37
6" wide	M1@.003	LF	.26	.12	.38
8" wide	M1@.003	LF	.28	.12	.40
Control joints					
Cross-shaped PVC	M1@.024	LF	2.20	.97	3.17
8" wide PVC	M1@.047	LF	2.78	1.90	4.68
Closed cell 1/2" joint filler	M1@.012	LF	.28	.49	.77
Closed cell 3/4" joint filler	M1@.012	LF	.56	.49	1.05
Through the wall flashing					
5 ounce copper	M1@.023	SF	2.18	.93	3.11
.030" elastomeric sheeting	M1@.038	SF	.48	1.54	2.02

Expansion Shields Masonry anchors set in block or concrete including drilling. With nut and washer. Length indicates drilled depth.

	Craft@Hrs	Unit	Material	Labor	Total
One wedge anchors, 1/4" diameter					
1-3/4" long	CL@.079	Ea	.63	2.87	3.50
2-1/4" long	CL@.114	Ea	.74	4.15	4.89
3-1/4" long	CL@.171	Ea	.99	6.22	7.21
One wedge anchors, 1/2" diameter					
4-1/4" long	CL@.227	Ea	1.58	8.26	9.84
5-1/2" long	CL@.295	Ea	2.09	10.70	12.79
7" long	CL@.409	Ea	2.56	14.90	17.46
One wedge anchors, 3/4" diameter					
5-1/2" long	CL@.286	Ea	4.29	10.40	14.69
8-1/2" long	CL@.497	Ea	6.13	18.10	24.23
10" long	CL@.692	Ea	7.24	25.20	32.44
One wedge anchors, 1" diameter					
6" long	CL@.295	Ea	10.50	10.70	21.20
9" long	CL@.515	Ea	13.00	18.70	31.70
12" long	CL@.689	Ea	15.00	25.10	40.10
Flush self-drilling anchors					
1/4" x 1-1/4"	CL@.094	Ea	.63	3.42	4.05
5/16" x 1-1/4"	CL@.108	Ea	.74	3.93	4.67
3/8" x 1-1/2"	CL@.121	Ea	.99	4.40	5.39
1/2" x 1-1/2"	CL@.159	Ea	1.47	5.79	7.26
5/8" x 2-3/8"	CL@.205	Ea	2.33	7.46	9.79
3/4" x 2-3/8"	CL@.257	Ea	3.80	9.35	13.15

Clay Backing Tile Load bearing, 12" x 12"

	Craft@Hrs	Unit	Material	Labor	Total
4" thick	M1@.066	SF	3.63	2.67	6.30
6" thick	M1@.079	SF	4.10	3.20	7.30
8" thick	M1@.091	SF	4.80	3.69	8.49
Deduct for non-load bearing tile	—	%	-8.0	-10.0	—

	Craft@Hrs	Unit	Material	Labor	Total

Structural Glazed Tile Heavy-duty, fire-safe, structural glazed facing tile for interior surfaces. No foundations or reinforcing included.

Glazed one side, all colors, 5-1/3" x 12"

	Craft@Hrs	Unit	Material	Labor	Total
2" thick	M1@.125	SF	4.44	5.06	9.50
4" thick	M1@.136	SF	5.35	5.51	10.86
6" thick	M1@.141	SF	7.92	5.71	13.63
8" thick	M1@.172	SF	10.10	6.97	17.07

Glazed one side, all colors, 8" x 16" block

	Craft@Hrs	Unit	Material	Labor	Total
2" thick	M1@.072	SF	4.99	2.92	7.91
4" thick	M1@.072	SF	5.35	2.92	8.27
6" thick	M1@.079	SF	7.37	3.20	10.57
8" thick	M1@.085	SF	8.64	3.44	12.08

Acoustical glazed facing tile, with small holes on a smooth ceramic glazed facing, and one mineral fiberglass sound batt located in the hollow core behind perforated face

	Craft@Hrs	Unit	Material	Labor	Total
8" x 8" face size unit, 4" thick	M1@.136	SF	8.76	5.51	14.27
8" x 16" face size unit, 4" thick	M1@.141	SF	7.79	5.71	13.50

Reinforced glazed security tile, with vertical cores and knockouts for horizontal reinforcing, 8" x 16" block

	Craft@Hrs	Unit	Material	Labor	Total
4" thick	M1@.092	SF	6.82	3.73	10.55
6" thick	M1@.095	SF	8.83	3.85	12.68
8" thick	M1@.101	SF	10.60	4.09	14.69

Add to 8" security tile for horizontal and vertical reinforcing and grout cores

	Craft@Hrs	Unit	Material	Labor	Total
to meet high security standards	M1@.047	SF	3.15	1.58	4.28
Add for glazed 2 sides	—	%	55.0	10.0	—
Base, glazed 1 side	M1@.190	LF	6.62	7.70	14.32
Cap, glazed 2 sides	M1@.193	LF	10.60	7.82	18.42
Deduct for large areas	—	%	-5.0	-5.0	—
Add for institutional inspection	—	%	—	10.0	—

Glass Block Clear 3-7/8" thick block installed in walls to 8' high. Costs shown include 1/4" mortar joints, caulking, wall ties and 6% waste. Heights over 8' to 10' require additional scaffolding and handling and may increase labor costs up to 40%. Mortar is 1 part portland cement, 1/2 part lime, and 4 parts sand, plus integral waterproofer. Costs for continuous panel reinforcing in horizontal joints are listed below.

Flat glass block, under 1,000 SF job

	Craft@Hrs	Unit	Material	Labor	Total
6" x 6", 4 per SF	M1@.323	SF	24.90	13.10	38.00
8" x 8", 2.25 per SF	M1@.245	SF	14.40	9.92	24.32
12" x 12", 1 per SF	M1@.202	SF	19.00	8.18	27.18
Deduct for jobs over 1,000 SF to 5,000 SF	—	%	-4.0	-4.0	—
Deduct for jobs over 5,000 SF	—	%	-10.0	-10.0	—
Deduct for Thinline interior glass block	—	%	-10.0	—	—
Add for thermal control fiberglass inserts	—	%	5.0	5.0	—

Replacing 8" x 8" x 4" clear glass block, 10 unit minimum

	Craft@Hrs	Unit	Material	Labor	Total
Per block replaced	M1@.851	Ea	9.08	34.50	43.58

Additional costs for glass block

	Craft@Hrs	Unit	Material	Labor	Total
Fiberglass or polymeric expansion joints	—	LF	.37	—	.37
Reinforcing steel mesh	—	LF	.46	—	.46

Wall anchors, 2' long

	Craft@Hrs	Unit	Material	Labor	Total
Used where blocks aren't set into a wall	—	Ea	1.80	—	1.80
Asphalt emulsion, 600 LF per gallon	—	Gal	6.00	—	6.00
Waterproof sealant, 180 LF per gallon	—	Gal	8.00	—	8.00

Cleaning block after installation, using sponge and water only

	Craft@Hrs	Unit	Material	Labor	Total
Per SF cleaned	CL@.023	SF	—	.84	.84

Masonry 4

	Craft@Hrs	Unit	Material	Labor	Total

Stone Work The delivered price of stone can be expected to vary widely. These costs include mortar, based on 1/2" joints, but no scaffolding. Figures in parentheses show typical quantities installed per 8-hour day by a crew of two.

Rough stone veneer, 4", placed over stud wall

	Craft@Hrs	Unit	Material	Labor	Total
Lava stone (79 SF)	M4@.202	SF	5.93	8.27	14.20
Rubble stone (50 SF)	M4@.400	SF	8.40	16.40	24.80
Most common veneer (60 SF)	M4@.267	SF	10.50	10.90	21.40

Cut stone

	Craft@Hrs	Unit	Material	Labor	Total
Thinset granite tile, 1/2" thick (120 SF)	M4@.133	SF	12.40	5.45	17.85
Granite, 7/8" thick, interior (70 SF)	M4@.229	SF	20.30	9.38	29.68
Granite, 1-1/4" thick, exterior (60 SF)	M4@.267	SF	15.10	10.90	26.00
Limestone, 2" thick (70 SF)	M4@.229	SF	18.60	9.38	27.98
Limestone, 3" thick (60 SF)	M4@.267	SF	22.80	10.90	33.70
Marble, 7/8" thick, interior (70 SF)	M4@.229	SF	23.90	9.38	33.28
Marble, 1-1/4" thick, interior (70 SF)	M4@.229	SF	29.40	9.38	38.78
Sandstone, 2" thick (70 SF)	M4@.229	SF	16.80	9.38	26.18
Sandstone, 3" thick (60 SF)	M4@.267	SF	21.90	10.90	32.80
Add for honed finish	—	SF	2.00	—	2.00
Add for polished finish	—	SF	3.50	—	3.50

Marble specialties, typical prices

	Craft@Hrs	Unit	Material	Labor	Total
Floors, 7/8" thick, exterior (70 SF)	M1@.229	SF	17.90	9.27	27.17
Thresholds, 1-1/4" thick (70 LF)	M1@.229	LF	9.64	9.27	18.91
Bases, 7/8" x 6" (51 LF)	M1@.314	LF	8.57	12.70	21.27
Columns, plain (15 CF)	M1@1.06	CF	99.60	42.90	142.50
Columns, fluted (13 CF)	M1@1.23	CF	140.00	49.80	189.80
Window stools, 5" x 1" (60 LF)	M1@.267	LF	10.50	10.80	21.30
Toilet partitions (1.5 Ea)	M1@10.6	Ea	557.00	429.00	986.00
Stair treads, 1-1/4" x 11" (50 LF)	M1@.320	LF	23.50	13.00	36.50
Limestone, rough cut large blocks (39 CF)	M1@.410	CF	28.90	16.60	45.50

Manufactured stone veneer, including mortar, Cultured Stone®

	Craft@Hrs	Unit	Material	Labor	Total
Prefitted types (160 SF)	M1@.100	SF	6.01	4.05	10.06
Brick veneer type (120 SF)	M1@.133	SF	4.99	5.39	10.38
Random cast type (120 SF)	M1@.133	SF	5.50	5.39	10.89

Masonry Lintels and Coping

Reinforced precast concrete lintels

	Craft@Hrs	Unit	Material	Labor	Total
4" wide x 8" high, to 6'6" long	M1@.073	LF	5.88	2.96	8.84
6" wide x 8" high, to 6'6" long	M1@.078	LF	10.50	3.16	13.66
8" wide x 8" high, to 6'6" long	M1@.081	LF	10.90	3.28	14.18
10" wide x 8" high, to 6'6" long	M1@.091	LF	14.10	3.69	17.79
4" wide x 8" high, over 6'6" to 10' long	M1@.073	LF	11.00	2.96	13.96
6" wide x 8" high, over 6'6" to 10' long	M1@.077	LF	12.90	3.12	16.02
8" wide x 8" high, over 6'6" to 10' long	M1@.081	LF	13.50	3.28	16.78
10" wide x 8" high, over 6'6" to 10' long	M1@.091	LF	18.70	3.69	22.39

Precast concrete coping, 5" average thickness, 4' to 8' long

	Craft@Hrs	Unit	Material	Labor	Total
10" wide, gray concrete	M1@.190	LF	9.81	7.70	17.51
12" wide, gray concrete	M1@.200	LF	11.00	8.10	19.10
12" wide, white concrete	M1@.226	LF	27.20	9.15	36.35

	Craft@Hrs	Unit	Material	Labor	Total

Cleaning and Pointing Masonry These costs assume the masonry surface is in fair to good condition with no unusual damage. Add the cost of protecting adjacent surfaces such as trim or the base of the wall and the cost of scaffolding, if required. Labor required to presoak or saturate the area cleaned is included in the labor cost. Work more than 12' above floor level will increase costs.

Brushing (hand cleaning) masonry, includes the cost of detergent or chemical solution

	Craft@Hrs	Unit	Material	Labor	Total
Light cleanup (100 SF per manhour)	M1@.010	SF	.02	.41	.43
Medium scrub (75 SF per manhour)	M1@.013	SF	.03	.53	.56
Heavy (50 SF per manhour)	M1@.020	SF	.04	.81	.85

Water blasting masonry, using rented 400 to 700 PSI power washer with 3 to 8 gallon per minute flow rate, includes blaster rental at $50 per day

	Craft@Hrs	Unit	Material	Labor	Total
Smooth face (250 SF per manhour)	M1@.004	SF	.05	.16	.21
Rough face (200 SF per manhour)	M1@.005	SF	.07	.20	.27

Sandblasting masonry, using 150 PSI compressor (with accessories and sand) at $150.00 per day

	Craft@Hrs	Unit	Material	Labor	Total
Smooth face (50 SF per manhour)	M1@.020	SF	.25	.81	1.06
Rough face (40 SF per manhour)	M1@.025	SF	.28	1.01	1.29

Steam cleaning masonry, using rented steam cleaning rig (with accessories) at $52.50 per day

	Craft@Hrs	Unit	Material	Labor	Total
Smooth face (75 SF per manhour)	M1@.013	SF	.04	.53	.57
Rough face (55 SF per manhour)	M1@.018	SF	.06	.73	.79
Add for masking adjacent surfaces	—	%	—	5.0	—
Add for difficult stain removal	—	%	50.0	50.0	—
Add for working from scaffold	—	%	—	20.0	—

Repointing brick, cut out joint, mask (blend-in), and regrout (tuck pointing)

	Craft@Hrs	Unit	Material	Labor	Total
30 SF per manhour	M1@.033	SF	.05	1.34	1.39

Masonry Cleaning Costs to clean masonry surfaces using commercial cleaning agents, add for pressure washing and scaffolding equipment.

Typical cleaning of surfaces

	Craft@Hrs	Unit	Material	Labor	Total
Granite, sandstone, terra cotta, brick	CL@.015	SF	.28	.55	.83

Cleaning surfaces of heavily carbonated

	Craft@Hrs	Unit	Material	Labor	Total
Limestone or cast stone	CL@.045	SF	.38	1.64	2.02
Typical wash with acid and rinse	CL@.004	SF	.44	.15	.59

Fireplaces Common brick, including the chimney foundation, damper, flue lining and stack to 15'

	Craft@Hrs	Unit	Material	Labor	Total
30" box, brick to 5', large quantity	M1@30.3	Ea	819.00	1,230.00	2,049.00
36" box, brick to 5', custom	M1@33.7	Ea	989.00	1,360.00	2,349.00
42" box, brick to 8', custom	M1@48.1	Ea	1,150.00	1,950.00	3,100.00
48" box, brick to 8', custom	M1@53.1	Ea	1,330.00	2,150.00	3,480.00

Flue lining, 1' lengths

	Craft@Hrs	Unit	Material	Labor	Total
8-1/2" x 17"	M1@.148	LF	5.61	5.99	11.60
10" x 14"	M1@.206	LF	5.61	8.34	13.95
13" x 13"	M1@.193	LF	6.55	7.82	14.37

Prefabricated cast concrete

	Craft@Hrs	Unit	Material	Labor	Total
30" box and 15' chimney	M1@5.43	Ea	1,160.00	220.00	1,380.00
Face brick, mantle to ceiling	M1@.451	SF	7.25	18.30	25.55

Pargeting, Waterproofing and Dampproofing

	Craft@Hrs	Unit	Material	Labor	Total
Pargeting, 2 coats, 1/2" thick	M3@.070	SF	.16	3.05	3.21
Pargeting, 2 coats, waterproof, 3/4" thick	M3@.085	SF	.17	3.70	3.87

Dampproofing, asphalt primer at 1 gallon per 100 SF,

	Craft@Hrs	Unit	Material	Labor	Total
Dampproof coat at 30 pounds per 100 SF	CL@.029	SF	.04	1.06	1.10

	Craft@Hrs	Unit	Material	Labor	Equipment	Total

Pre-Engineered Steel Buildings 26 gauge colored galvanized steel roof and siding with 4 in 12 (20 pound live load) roof. Cost per SF of floor area. These costs do not include foundation or floor slab. Add delivery cost to the site. Equipment is a 15 ton hydraulic crane and a 2-ton truck equipped for work of this type.

	Craft@Hrs	Unit	Material	Labor	Equipment	Total
40' x 100' (4,000 SF)						
14' eave height	H5@.074	SF	3.83	3.42	1.02	8.27
16' eave height	H5@.081	SF	4.31	3.74	1.11	9.16
20' eave height	H5@.093	SF	4.87	4.29	1.28	10.44
60' x 100' (6,000 SF)						
14' eave height	H5@.069	SF	3.72	3.19	.95	7.86
16' eave height	H5@.071	SF	3.79	3.28	.98	8.05
20' eave height	H5@.081	SF	4.39	3.74	1.11	9.24
80' x 100' (8,000 SF)						
14' eave height	H5@.069	SF	3.64	3.19	.95	7.78
16' eave height	H5@.071	SF	3.78	3.28	.98	8.04
20' eave height	H5@.080	SF	4.39	3.69	1.10	9.18
100' x 100' (10,000 SF)						
14' eave height	H5@.067	SF	3.61	3.09	.92	7.62
16' eave height	H5@.071	SF	3.78	3.28	.98	8.04
20' eave height	H5@.080	SF	4.24	3.69	1.10	9.03
100' x 150' (15,000 SF)						
14' eave height	H5@.065	SF	3.10	3.00	.89	6.99
16' eave height	H5@.069	SF	3.28	3.19	.95	7.42
20' eave height	H5@.074	SF	3.65	3.42	1.02	8.09
100' x 200' (20,000 SF)						
14' eave height	H5@.063	SF	3.02	2.91	.87	6.80
16' eave height	H5@.066	SF	3.16	3.05	.91	7.12
20' eave height	H5@.071	SF	3.40	3.28	.98	7.66
140' x 150' (21,000 SF)						
14' eave height	H5@.058	SF	2.80	2.68	.80	6.28
16' eave height	H5@.059	SF	2.89	2.72	.81	6.42
20' eave height	H5@.066	SF	3.23	3.05	.91	7.19
140' x 175' (24,500 SF)						
14' eave height	H5@.057	SF	2.70	2.63	.78	6.11
16' eave height	H5@.058	SF	2.80	2.68	.80	6.28
20' eave height	H5@.065	SF	3.09	3.00	.89	6.98
160' x 200' (32,000 SF)						
14' eave height	H5@.055	SF	2.62	2.54	.76	5.92
16' eave height	H5@.056	SF	2.70	2.59	.77	6.06
20' eave height	H5@.061	SF	2.97	2.82	.84	6.63
200' x 200' (40,000 SF)						
14' eave height	H5@.052	SF	2.53	2.40	.71	5.64
16' eave height	H5@.053	SF	2.60	2.45	.73	5.78
20' eave height	H5@.058	SF	2.80	2.68	.80	6.28
Additive costs:						
Personnel door, 3'0" x 6'8"	H5@2.83	Ea	481.00	131.00	38.90	650.90
Overhead door, 8'0" x 10'0"	H5@8.02	Ea	1,180.00	370.00	110.00	1,660.00
Interior liner panel, 26 gauge						
Per SF of wall	H5@.012	SF	1.58	.55	.16	2.29

	Craft@Hrs	Unit	Material	Labor	Equipment	Total
Foam insulated sandwich panel roofing and siding						
Per SF of roof and sidewall	H5@.005	SF	4.86	.23	.07	5.16
Plastic skylight roof panel						
Per SF of panel	H5@.013	SF	4.20	.60	.18	4.98
R-11 blanket insulation,						
3-1/2", wall or floor	H5@.005	SF	.33	.23	.07	.63
Ridge ventilator, 9" throat	H5@.193	LF	40.10	8.91	2.65	51.66
24 gauge colored eave gutter	H5@.028	LF	6.43	1.29	.38	8.10
4" x 4" colored downspouts	H5@.014	LF	5.74	.65	.19	6.58
30 pound live load roof design	—	%	12.0	—	—	—

Railings, Aluminum, Brass and Wrought Iron See also Steel Railings.

Aluminum railings. 3'6" high with 1-1/2" square horizontal rails top and bottom, 5/8" square vertical bars at 4" OC, and 1-1/2" square posts spaced 4' OC. Welded construction with standard mill finish.

	Craft@Hrs	Unit	Material	Labor	Equipment	Total
Installed on balconies	H6@.092	LF	33.30	4.17	—	37.47
Installed on stairways	H6@.115	LF	37.60	5.22	—	42.82
Add for anodized finish,						
bronze or black	—	LF	12.00	—	—	12.00

Brass railings. Round tubular type, bright finish, standard 12' sections, .050" wall commercial quality with supports at 3' OC attached to existing structure using brass screws. Labor includes layout, measure, cut and end preparation

	Craft@Hrs	Unit	Material	Labor	Equipment	Total
1-1/2" tubing	H6@.058	LF	4.65	2.63	—	7.28
2" tubing	H6@.058	LF	5.86	2.63	—	8.49
Fittings, slip-on type, no soldering or brazing required						
1-1/2" elbows	H6@.092	Ea	13.40	4.17	—	17.57
1-1/2" tees	H6@.137	Ea	14.30	6.21	—	20.51
1-1/2" crosses	H6@.277	Ea	17.50	12.60	—	30.10
2" elbows	H6@.115	Ea	14.70	5.22	—	19.92
2" tees	H6@.171	Ea	19.90	7.76	—	27.66
2" crosses	H6@.346	Ea	22.20	15.70	—	37.90

Wrought iron railings. Standard 3'6" high, welded steel construction, shop prime painted, attached to existing structure

	Craft@Hrs	Unit	Material	Labor	Equipment	Total
Stock patterns	H6@.171	LF	14.10	7.76	—	21.86
Custom patterns	H6@.287	LF	21.10	13.00	—	34.10
Stairway handrails	H6@.343	LF	23.20	15.60	—	38.80

Space Frame System 10,000 SF or more. Equipment cost shown is for a 15 ton truck crane and a 2 ton truck equipped for work of this type.

	Craft@Hrs	Unit	Material	Labor	Equipment	Total
5' module 4.5 lb live load	H5@.146	SF	18.10	6.74	2.01	26.85
4' module, add	H5@.025	SF	2.60	1.15	.34	4.09

Steel Floor and Roof Decking These costs assume a 40,000 SF job not over six stories high. Weight and cost include 10% for waste and connections. Equipment cost shown is for a 15 ton truck crane and a 2 ton truck equipped for work of this type.

Floor and roof decking, 1-1/2" standard rib, simple non-composite section, non-cellular, shop prime painted finish

	Craft@Hrs	Unit	Material	Labor	Equipment	Total
22 gauge, 2.1 lbs per SF	H5@.013	SF	1.54	.60	.18	2.32
20 gauge, 2.5 lbs per SF	H5@.013	SF	1.62	.60	.18	2.40
18 gauge, 3.3 lbs per SF	H5@.013	SF	2.03	.60	.18	2.81
16 gauge, 4.1 lbs per SF	H5@.013	SF	2.25	.60	.18	3.03
Add for galvanized finish	—	SF	.26	—	—	.26

	Craft@Hrs	Unit	Material	Labor	Equipment	Total

Steel floor deck system. Consists of 18 gauge, 1-1/2" standard rib, shop prime painted steel decking, installed over a structural steel beam and girder support system (cost of support structure not included) covered with 3" thick 6.0 sack mix, 3,000 PSI lightweight ready-mix concrete, placed by pump, reinforced with 6" x 6", W1.4 x W1.4. welded wire mesh. Costs shown assume shoring is not required and include curing and finishing the concrete.

	Craft@Hrs	Unit	Material	Labor	Equipment	Total
18 gauge decking	H5@.013	SF	1.62	.60	.18	2.40
Welded wire mesh	RB@.004	SF	.15	.21	—	.36
Lightweight concrete	M2@.006	SF	1.45	.24	.08	1.77
Cure and finish concrete	P8@.011	SF	.15	.44	—	.59
Total floor deck system	**--@.034**	**SF**	**3.37**	**1.49**	**.26**	**5.12**
Add for 16 gauge decking		SF	.30	—	—	.30

Steel roof deck system. Consists of 18 gauge, 1-1/2" standard rib, shop prime painted steel decking, installed over open web steel joists 4' OC (costs of joists not included) covered with 1-1/4" R-8.3 urethane insulation board overlaid with .045" loose laid membrane EPDM (ethylene propylene diene monomer) elastomeric roofing. Joist pricing assumes H or J series at 9.45 lbs per LF spanning 20'0", spaced 4' OC.

	Craft@Hrs	Unit	Material	Labor	Equipment	Total
18 gauge decking	H5@.013	SF	2.03	.60	.18	2.81
Insulation board	A1@.011	SF	1.48	.50	—	1.98
EPDM roofing	R3@.006	SF	1.18	.25	—	1.43
Total roof deck system	**--@.030**	**SF**	**4.69**	**1.35**	**.18**	**6.22**
Add for 16 gauge decking	—	SF	.30	—	—	.30
Add open web steel joists	H5@.005	SF	.83	.23	.07	1.13

Steel Handrails, Kickplates, Ladders, Landings, Railings, and Stairways Material costs assume shop fabrication and assembly, prime painted and erection bolts. Equipment cost includes a 10 ton hydraulic truck-mounted crane, an electric powered welding machine, an oxygen/acetylene welding rig, a diesel powered 100 CFM air compressor and accessories.

Safety cage   Ladder   Stairway with handrail   Landing

Handrails. Wall mounted, with brackets 5' OC, based on 12' lengths, welded steel, shop prime painted

	Craft@Hrs	Unit	Material	Labor	Equipment	Total
1-1/4" diameter rail	H7@.092	LF	6.65	4.85	2.33	13.83
1-1/2" diameter rail	H7@.102	LF	7.78	5.37	2.58	15.73
Add for galvanized finish	—	LF	1.42	—	—	1.42

Kickplates, 1/4" thick flat steel plate, shop prime painted, attached to steel railings, stairways or landings

	Craft@Hrs	Unit	Material	Labor	Equipment	Total
4" high	H7@.058	LF	5.34	3.06	1.47	9.87
6" high	H7@.069	LF	5.80	3.64	1.75	11.19
Add for galvanized finish	—	LF	1.05	—	—	1.05

Metals 5

	Craft@Hrs	Unit	Material	Labor	Equipment	Total
Ladders, vertical type. 1'6" wide, attached to existing structure						
Without safety cage	H7@.100	LF	36.10	5.27	2.53	43.90
With steel safety cage	H7@.150	LF	42.80	7.90	3.80	54.50
Add for galvanized finish	—	LF	4.21	—	—	4.21
Landings. Costs assume standard C shape structural steel channels for joists and frame, 10 gauge steel pan deck areas, shop prime painted. Add for support legs from Structural Steel Section						
Cost per square foot	H7@.250	SF	130.00	13.20	6.34	149.54
Add for galvanized finish	—	SF	14.80	—	—	14.80
Pipe and chain railings. Welded steel pipe posts with U-brackets attached to accept 1/4" chain, shop prime painted. Installed in existing embedded sleeves. Posts are 5'0" long (1'6" in the sleeve and 3'6" above grade). Usual spacing is 4' OC. Add chain per lineal foot						
1-1/4" diameter post	H7@.287	Ea	6.60	15.10	—	21.70
1-1/2" diameter post	H7@.287	Ea	8.05	15.10	—	23.15
2" diameter post	H7@.343	Ea	10.80	18.10	—	28.90
3" diameter post	H7@.343	Ea	21.50	18.10	—	39.60
Add for galvanized finish	—	Ea	2.80	—	—	2.80
Chain, 1/4" diameter galvanized	H7@.058	LF	2.18	3.06	—	5.24
Railings. Floor mounted, 3'6" high with posts 5' OC, based on 12' length, welded steel, shop prime painted						
1-1/4" diameter rails and posts						
2-rail type	H7@.115	LF	27.50	6.06	2.91	36.47
3-rail type	H7@.125	LF	35.50	6.59	3.17	45.26
1-1/2" diameter rails and posts						
2-rail type	H7@.125	LF	32.90	6.59	3.17	42.66
3-rail type	H7@.137	LF	43.50	7.22	3.47	54.19
Add for galvanized finish	—	LF	7.00	—	—	7.00

Spiral stairways Cost per LF of vertical rise, floor to floor. Costs assume a 6" diameter round tubular steel center column, 12 gauge steel pan type 10" treads with 7-1/2" open risers, and 2" x 2" tubular steel handrail with 1" x 1" tubular steel posts at each riser, shop prime painted. Add for railings at top of stairways as required.

	Craft@Hrs	Unit	Material	Labor	Equipment	Total
4'0" diameter	H7@.151	LF	274.00	7.96	3.83	285.79
6'0" diameter	H7@.151	LF	374.00	7.96	3.83	385.79
8'0" diameter	H7@.151	LF	471.00	7.96	3.83	482.79

Straight stairways Cost per LF based on sloping length. Rule of thumb: the sloping length of stairways is the vertical rise times 1.6. Costs assume standard C shape structural steel channels for stringers, 12 gauge steel pan 10" treads with 7-1/2" closed risers. Posts are 2" diameter tubular steel, 5' OC. Handrails are 3'6" high, 1-1/2" diameter 2-rail type, on both sides of the stairs. Installation assumes stairs and landings are attached to an existing structure on two sides.

	Craft@Hrs	Unit	Material	Labor	Equipment	Total
3'0" wide	H7@.469	LF	194.00	24.70	11.90	230.60
4'0" wide	H7@.630	LF	257.00	33.20	16.00	306.20
5'0" wide	H7@.781	LF	322.00	41.20	19.80	383.00

Structural Steel Costs shown below are for 60 to 100 ton jobs up to six stories high where the fabricated structural steel is purchased from a steel fabricator located within 50 miles of the job site. Material costs are for ASTM A36 steel, fabricated according to plans, specifications and AISC recommendations and delivered to the job site with connections attached, piece marked and ready for erection. Labor includes time to unload and handle the steel, erect and plumb, install permanent bolts, weld and all other work usually associated with erecting the steel including field touch-up painting. Equipment includes a 60 ton crawler mounted crane, plus an electric powered welding machine, an oxygen/acetylene welding and cutting torch, a diesel powered 100 CFM air compressor with pneumatic grinders, ratchet wrenches, hoses, and other tools usually associated with work of this type. Structural steel is usually estimated by the tons of steel in the project. Labor for erection will vary with the weight of steel per linear foot. For example a member weighing 25 pounds per foot will require more manhours per ton than one weighing 60 pounds per foot.

	Craft@Hrs	Unit	Material	Labor	Equipment	Total

Open web steel joists Estimate the total weight of "bare" joists and then add 10% to allow for connections and accessories.

"H" and "J" series, 8H3 thru 26H10 or 8J3 thru 30J8

	Craft@Hrs	Unit	Material	Labor	Equipment	Total
Under 10 lbs per LF	H8@10.0	Ton	1,400.00	512.00	186.00	2,098.00
Over 10 to 20 lbs per LF	H8@8.00	Ton	1,390.00	409.00	149.00	1,948.00

"LH" and "LJ" series, 18LH02 thru 48LH10 or 18LJ02 thru 48LJ11

	Craft@Hrs	Unit	Material	Labor	Equipment	Total
Under 20 lbs per LF	H8@8.00	Ton	1,290.00	409.00	149.00	1,848.00
Over 20 to 50 lbs per LF	H8@7.00	Ton	1,250.00	358.00	130.00	1,738.00
Over 50 lbs per LF	H8@6.00	Ton	1,200.00	307.00	112.00	1,619.00

"DLH" or "DLJ" series, 52DLH10 thru 72DLH18 or 52DLJ12 thru 52DLJ20

	Craft@Hrs	Unit	Material	Labor	Equipment	Total
20 to 50 lbs per LF	H8@8.00	Ton	1,390.00	409.00	149.00	1,948.00
Over 50 lbs per LF	H8@6.00	Ton	1,310.00	307.00	112.00	1,729.00

Structural shapes Estimate the total weight of "bare" fabricated structural shapes and then add 15% to allow for the weight of connections.

Beams, purlins, and girts

	Craft@Hrs	Unit	Material	Labor	Equipment	Total
Under 20 pounds per LF	H8@14.0	Ton	2,610.00	716.00	261.00	3,587.00
From 20 to 50 lbs per LF	H8@11.0	Ton	1,680.00	563.00	205.00	2,448.00
Over 50 to 75 lbs per LF	H8@9.40	Ton	1,580.00	481.00	175.00	2,236.00
Over 75 to 100 lbs per LF	H8@8.00	Ton	1,550.00	409.00	149.00	2,108.00

Columns

	Craft@Hrs	Unit	Material	Labor	Equipment	Total
Under 20 pounds per LF	H8@12.0	Ton	2,450.00	614.00	223.00	3,287.00
From 20 to 50 lbs per LF	H8@10.0	Ton	1,580.00	512.00	186.00	2,278.00
Over 50 to 75 lbs per LF	H8@9.00	Ton	1,520.00	460.00	167.00	2,147.00
Over 75 to 100 lbs per LF	H8@7.00	Ton	1,520.00	358.00	130.00	2,008.00

Sag rods and X type bracing

	Craft@Hrs	Unit	Material	Labor	Equipment	Total
Under 1 pound per LF	H8@82.1	Ton	3,960.00	4,200.00	1,530.00	9,690.00
From 1 to 3 lbs per LF	H8@48.0	Ton	3,670.00	2,460.00	893.00	7,023.00
Over 3 to 6 lbs per LF	H8@34.8	Ton	3,420.00	1,780.00	648.00	5,848.00
Over 6 to 10 lbs per LF	H8@28.0	Ton	3,130.00	1,430.00	521.00	5,081.00

Trusses and girders

	Craft@Hrs	Unit	Material	Labor	Equipment	Total
Under 20 pounds per LF	H8@10.0	Ton	2,420.00	512.00	186.00	3,118.00
From 20 to 50 lbs per LF	H8@7.50	Ton	1,580.00	384.00	140.00	2,104.00
Over 50 to 75 lbs per LF	H8@6.00	Ton	1,520.00	307.00	112.00	1,939.00
Over 75 to 100 lbs per LF	H8@5.00	Ton	1,470.00	256.00	93.00	1,819.00

	Craft@Hrs	Unit	Material	Labor	Equipment	Total

Structural Steel, Subcontract, by Square Foot of Floor Use these figures only for preliminary estimates on buildings with 20,000 to 80,000 SF per story, 100 PSF live load and 14' to 16' story height. A "bay" is the center to center distance between columns. Costs assume steel at $2,300 to $2,500 per ton including installation labor and material and the subcontractor's overhead and profit. Include all area under the roof when calculating the total square footage.

Single story

	Craft@Hrs	Unit	Material	Labor	Equipment	Total
20' x 20' bays (8 to 9 lbs per SF)	—	SF	—	—	—	9.45
30' x 30' bays (10 to 11 lbs per SF)	—	SF	—	—	—	11.70
40' x 40' bays (12 to 13 lbs per SF)	—	SF	—	—	—	14.00

Two story to six story

	Craft@Hrs	Unit	Material	Labor	Equipment	Total
20' x 20' bays (9 to 10 lbs per SF)	—	SF	—	—	—	10.50
30' x 30' bays (11 to 12 lbs per SF)	—	SF	—	—	—	12.70
40' x 40' bays (13 to 14 lbs per SF)	—	SF	—	—	—	14.90

Tubular Columns Rectangular, round, and square structural columns. When estimating total weight of columns, figure the column weight and add 30% to allow for a shop attached U-bracket on the top, a square base plate on the bottom and four anchor bolts. This should be sufficient for columns 15' to 20' in length. Equipment cost includes a 10 ton hydraulic truck-mounted crane plus an electric powered welding machine, an oxygen/acetylene welding and cutting torch, a diesel powered 100 CFM air compressor with pneumatic grinders, ratchet wrenches, hoses, and other tools usually associated with work of this type.

Rectangular tube columns, in pounds per LF

	Craft@Hrs	Unit	Material	Labor	Equipment	Total
3" x 2" to 12" x 4", to 20 lbs per LF	H7@9.00	Ton	2,410.00	474.00	228.00	3,112.00
6" x 4" to 16" x 8", 21 to 50 lbs per LF	H7@8.00	Ton	1,930.00	422.00	203.00	2,555.00
12 " x 6" to 20" x 12", 51 to 75 lbs per LF	H7@6.00	Ton	1,880.00	316.00	152.00	2,348.00
14" x 10" to 20" x 12", 76 to 100 lbs/LF	H7@5.00	Ton	1,840.00	263.00	127.00	2,230.00

Round columns, in pounds per LF

	Craft@Hrs	Unit	Material	Labor	Equipment	Total
3" to 6" diameter pipe, to 20 lbs per LF	H7@9.00	Ton	2,550.00	474.00	228.00	3,252.00
4" to 12" pipe, 20 to 50 lbs per LF	H7@8.00	Ton	2,050.00	422.00	203.00	2,675.00
6" to 12" pipe, 51 to 75 lbs per LF	H7@6.00	Ton	1,980.00	316.00	152.00	2,448.00

Square tube columns, in pounds per LF

	Craft@Hrs	Unit	Material	Labor	Equipment	Total
2" x 2" to 8" x 8", to 20 lbs per LF	H7@9.00	Ton	2,300.00	474.00	228.00	3,002.00
4" x 4" to 12" x 12", 21 to 50 lbs per LF	H7@8.00	Ton	1,880.00	422.00	203.00	2,505.00
8" x 8" to 16" x 16", 51 to 75 lbs per LF	H7@6.00	Ton	1,850.00	316.00	152.00	2,318.00
10" x 10" to 16" x 16", 76 to 100 lbs/LF	H7@5.00	Ton	1,810.00	263.00	127.00	2,200.00

Add for concrete fill in columns, 3,000 PSI design pump mix,

	Craft@Hrs	Unit	Material	Labor	Equipment	Total
Placed by pump (including 10% waste)	M2@.738	CY	114.00	29.10	7.30	150.40

	Craft@Hrs	Unit	Material	Labor	Total

Fabricated metals

Catch basin grating and frame

	Craft@Hrs	Unit	Material	Labor	Total
24" x 24", standard duty	CL@8.68	Ea	272.00	316.00	588.00

Manhole rings with cover

	Craft@Hrs	Unit	Material	Labor	Total
24", 330 lbs	CL@6.80	Ea	262.00	247.00	509.00
30", 400 lbs	CL@6.80	Ea	272.00	247.00	519.00
38", 730 lbs	CL@6.80	Ea	748.00	247.00	995.00

	Craft@Hrs	Unit	Material	Labor	Total
Welded steel grating Weight shown is approximate. Costs shown are per square foot of grating, shop prime painted or galvanized after fabrication as shown					
Grating bars as shown, 1/2" cross bars at 4" OC					
1-1/4" x 3/16" at 1-1/4" OC, 9 lbs per SF, painted	IW@.019	SF	6.70	1.03	7.73
1-1/4" x 3/16" at 1-1/4" OC, 9 lbs per SF, galvanized	IW@.019	SF	8.27	1.03	9.30
Add for banding on edges	—	LF	3.51	—	3.51
2" x 1/4" at 1-1/4" OC, 18.7 lbs per SF, painted	IW@.026	SF	15.80	1.41	17.21
2" x 1/4" at 1-1/4" OC, 18.7 lbs per SF, galvanized	IW@.026	SF	21.20	1.41	22.61
Add for banding on edges	—	LF	5.23	—	5.23
Add for toe-plates, steel, shop prime painted					
1/4" thick x 4" high attached after grating is in place	IW@.048	LF	6.88	2.61	9.49
Checkered-steel floor plate Shop prime painted or galvanized after fabrication as shown					
1/4", 11-1/4 lbs per SF, painted	IW@.025	SF	12.60	1.36	13.96
3/8", 16-1/2 lbs per SF, painted	IW@.025	SF	18.10	1.36	19.46
1/4", 11-1/4 lbs per SF, galvanized	IW@.025	SF	13.70	1.36	15.06
3/8", 16-1/2 lbs per SF, galvanized	IW@.025	SF	20.10	1.36	21.46
Checkered-steel trench cover plate Shop prime painted or galvanized after fabrication as shown					
1/4", 11-1/4 lbs per SF, primed	IW@.045	SF	16.50	2.44	18.94
1/4", 11-1/4 lbs per SF, galvanized	IW@.045	SF	19.10	2.44	21.54
3/8", 16-1/2 lbs per SF, primed	IW@.045	SF	23.80	2.44	26.24
3/8", 16-1/2 lbs per SF, galvanized	IW@.045	SF	28.90	2.44	31.34
Steel fabrications					
Canopy framing, steel	IW@.009	Lb	1.04	.49	1.53
Miscellaneous supports, steel	IW@.019	Lb	1.69	1.03	2.72
Miscellaneous supports, aluminum	IW@.061	Lb	10.90	3.31	14.21
Embedded steel (embedded in concrete)					
Light, to 20 lbs per LF	T3@.042	Lb	2.06	1.88	3.94
Medium, over 20 to 50 lbs per LF	T3@.035	Lb	1.86	1.57	3.43
Heavy, over 50 lbs per LF	T3@.025	Lb	1.57	1.12	2.69
Channel sill, 1/8" steel, 1-1/2" x 8" wide	T3@.131	LF	16.10	5.86	21.96
Angle sill, 1/8" steel, 1-1/2" x 1-1/2"	T3@.095	LF	5.08	4.25	9.33
Fire escapes, ladder and balcony, per floor	IW@16.5	Ea	1,900.00	896.00	2,796.00
Grey iron foundry items	T3@.014	Lb	.93	.63	1.56
Stair nosing, 2-1/2" x 2-1/2", set in concrete	T3@.139	LF	7.97	6.22	14.19
Cast iron					
Lightweight sections	T3@.012	Lb	.98	.54	1.52
Heavy section	T3@.010	Lb	.80	.45	1.25
Cast column bases					
Iron, 16" x 16", custom	T3@.759	Ea	83.40	34.00	117.40
Iron, 32" x 32", custom	T3@1.54	Ea	347.00	68.90	415.90
Aluminum, 8" column, stock	T3@.245	Ea	18.90	11.00	29.90
Aluminum, 10" column, stock	T3@.245	Ea	23.80	11.00	34.80
Wheel type corner guards, per LF of height	T3@.502	LF	60.30	22.50	82.80
Angle corner guards, with anchors	T3@.018	Lb	1.81	.81	2.62
Channel door frames, with anchors	T3@.012	Lb	1.67	.54	2.21
Steel lintels, painted	T3@.013	Lb	1.00	.58	1.58
Steel lintels, galvanized	T3@.013	Lb	1.32	.58	1.90

	Craft@Hrs	Unit	Material	Labor	Total
Anodizing Add to cost of fabrication					
Aluminum	—	%	15.0	—	—
Bronze	—	%	30.0	—	—
Black	—	%	50.0	—	—
Galvanizing, add to cost of fabrication	—	Lb	.43	—	.43

Ornamental Metal

	Craft@Hrs	Unit	Material	Labor	Total
Ornamental screens, aluminum	H6@.135	SF	29.20	6.12	35.32
Ornamental screens extruded metal	H6@.151	SF	24.40	6.85	31.25
Gates, wrought iron, 6' x 7', typical installed cost	—	Ea	—	—	1,250.00
Sun screens, aluminum stock design, manual	H6@.075	SF	31.40	3.40	34.80
Sun screens, aluminum stock design, motorized	H6@.102	SF	57.10	4.63	61.73
Aluminum work, general fabrication	H6@.160	Lb	15.50	7.26	22.76
Brass and bronze work, general fabrication	H6@.079	Lb	22.60	3.58	26.18
Stainless steel, general fabrication	H6@.152	Lb	18.40	6.89	25.29

Security Gratings

	Craft@Hrs	Unit	Material	Labor	Total
Door grating, 1" tool steel vertical bars at 4" OC, 3/8" x 2-1/4" steel frame and horizontal bars at 18" OC,					
3' x 7' overall	H6@10.5	Ea	1,560.00	476.00	2,036.00
Add for butt hinges and locks, per door	H6@2.67	Ea	474.00	121.00	595.00
Add for field painting, per door	H6@1.15	Ea	8.13	52.20	60.33
Window grating, tool steel vertical bars at 4" OC, 3/8" x 2-1/4" steel frame, horizontal bars 6" OC,					
20" x 30" overall set with 1/2" bolts at 6" OC	H6@3.21	Ea	208.00	146.00	354.00
Louver vent grating, 1" x 3/16", "I" bearing bars 1-3/16" OC with 3/16" spaced bars 4" OC and banded ends,					
set with 1/2" bolts at 1" OC, prime painted	H6@.114	SF	25.50	5.17	30.67
Steel plate window covers, hinged, 3/8" plate steel, 32" W x 48" H overall, 10 lbs per SF,					
including hardware	H6@.150	SF	11.60	6.80	18.40

Steel fabrication on site These costs are for work at the job site and assume engineering and design, including detail drawings sufficient to do the work, have been prepared by others at no cost to the installing contractor. Costs are based on ASTM A36 all-purpose carbon steel, which is widely used in building construction. Equipment cost includes a 1-1/4-ton flatbed truck equipped with a diesel powered welding machine, an oxygen/acetylene welding and cutting torch, a diesel powered 100 CFM air compressor with pneumatic grinders, ratchet wrenches, hoses, and other tools usually associated with work of this type. Material cost for the steel is not included.

	Craft@Hrs	Unit	Material	Labor	Equipment	Total
Flat plate, cutting and drilling holes						
Flame cutting. Labor shown includes layout, marking, cutting and grinding of finished cut						
Square cut, thickness of plate as shown						
1/4", 5/16" or 3/8"	IW@.020	LF	—	1.09	.90	1.99
1/2", 5/8" or 3/4"	IW@.025	LF	—	1.36	1.13	2.49
7/8", 1" or 1-1/4"	IW@.030	LF	—	1.63	1.36	2.99
Bevel cut 30 to 45 degrees, thickness of plate as shown						
1/4", 5/16" or 3/8"	IW@.030	LF	—	1.63	1.36	2.99
1/2", 5/8" or 3/4"	IW@.035	LF	—	1.90	1.58	3.48
7/8", 1" or 1-1/4"	IW@.043	LF	—	2.34	1.94	4.28
Drill holes. Using a hand-held drill motor. Labor shown includes layout, drill and ream holes with plate thickness and hole size as shown (note: hole size is usually 1/16" greater than bolt diameter)						
1/4", 5/16" or 3/8" plate thickness,						
5/16", 3/8" or 1/2" hole	IW@.084	Ea	—	4.56	3.80	8.36

	Craft@Hrs	Unit	Material	Labor	Equipment	Total
1/2", 5/8" or 3/4" plate thickness,						
9/16", 3/4" or 13/16" hole	IW@.134	Ea	—	7.28	6.06	13.34
7/8", 1" or 1-1/4" plate thickness,						
1", 1-1/16" or 1-5/16" hole	IW@.200	Ea	—	10.90	9.04	19.94

Welding Using manual inert gas shielded arc electric welding process utilizing coated electrodes (rod) of appropriate size and type with welding machine amperage suitable for the welding rod being used. Labor includes layout, fitup and weld.

Fillet welds for attaching base plates, clip angles, gussets. Welding in flat position

	Craft@Hrs	Unit	Material	Labor	Equipment	Total
Size of continuous fillet weld						
3/16"	IW@.300	LF	—	16.30	13.60	29.90
1/4"	IW@.400	LF	—	21.70	18.10	39.80
3/8"	IW@.585	LF	—	31.80	26.50	58.30
1/2"	IW@.810	LF	—	44.00	36.60	80.60
5/8"	IW@1.00	LF	—	54.30	45.20	99.50
3/4"	IW@1.20	LF	—	65.20	54.30	119.50

Buttweld joints, 100% penetration, flat plates

	Craft@Hrs	Unit	Material	Labor	Equipment	Total
Thickness at joint						
3/16"	IW@.200	LF	—	10.90	9.04	19.94
1/4"	IW@.250	LF	—	13.60	11.30	24.90
3/8"	IW@.440	LF	—	23.90	19.90	43.80
1/2"	IW@.660	LF	—	35.90	29.80	65.70
5/8"	IW@.825	LF	—	44.80	37.30	82.10
3/4"	IW@1.05	LF	—	57.00	47.50	104.50

Wood and Plastic 6

	Craft@Hrs	Unit	Material	Labor	Total

Rough Carpentry Industrial and commercial work. Costs include 10% waste. Standard and Better dimension lumber based on the following prices:

	Craft@Hrs	Unit	Material	Labor	Total
2" x 4"	—	MBF	621.00	—	621.00
2" x 6"	—	MBF	620.00	—	620.00
2" x 8"	—	MBF	640.00	—	640.00
2" x 10"	—	MBF	657.00	—	657.00
2" x 12"	—	MBF	738.00	—	738.00
4" x 4"	—	MBF	960.00	—	960.00
4" x 6"	—	MBF	974.00	—	974.00
4" x 8"	—	MBF	819.00	—	819.00
4" x 10"	—	MBF	812.00	—	812.00
4" x 12"	—	MBF	828.00	—	828.00
6" x 6"	—	MBF	1,280.00	—	1,280.00
6" x 8"	—	MBF	1,280.00	—	1,280.00
6" x 10"	—	MBF	1,280.00	—	1,280.00
6" x 12"	—	MBF	1,280.00	—	1,280.00
ACQ treatment	—	MBF	220.00	—	220.00

Stud walls, including plates, blocking, headers and diagonal bracing, per SF of wall, 16" OC

	Craft@Hrs	Unit	Material	Labor	Total
2" x 4", 8' high, 1.1 BF per SF	C8@.021	SF	.75	.86	1.61
2" x 4", 10' high, 1.0 BF per SF	C8@.021	SF	.68	.86	1.54
2" x 6", 8' high, 1.6 BF per SF	C8@.031	SF	1.09	1.27	2.36
2" x 6", 10' high, 1.5 BF per SF	C8@.030	SF	1.02	1.23	2.25

	Craft@Hrs	Unit	Material	Labor	Total
12" OC, 2" x 6", 8' high, 2.0 BF per SF	C8@.037	SF	1.36	1.52	2.88
8" OC, 2" x 4", staggered on 2" x 6" plates, 1.8 BF per SF	C8@.034	SF	1.23	1.39	2.62
Pier caps, .60 ACQ treated					
2" x 6"	C8@.024	LF	.84	.98	1.82
Mud sills, treated material, add bolts or fastener					
Bolted, 2" x 4"	C8@.023	LF	.84	.94	1.78
Bolted, 2" x 6"	C8@.024	LF	.84	.98	1.82
Shot, 2" x 4"	C8@.018	LF	1.18	.74	1.92
Shot, 2" x 6"	C8@.020	LF	.84	.82	1.66
Plates (add for .60 ACQ treatment if required)					
2" x 4"	C8@.013	LF	.45	.53	.98
2" x 6"	C8@.016	LF	.68	.66	1.34
Posts, #1 structural					
4" x 4"	C8@.110	LF	1.41	4.51	5.92
6" x 6"	C8@.145	LF	4.22	5.94	10.16
Headers, 4" x 12"	C8@.067	LF	3.64	2.74	6.38
Diagonal bracing, 1" x 4", let in	C8@.021	LF	.31	.86	1.17
Blocking and nailers, 2" x 6"	C8@.034	LF	.68	1.39	2.07
Blocking, solid diagonal bracing, 2" x 4"	C8@.023	LF	.45	.94	1.39
Beams and girders, #2 & Btr					
4" x 8" to 4" x 12"	C8@.094	LF	3.57	3.85	7.42
6" x 8" to 6" x 12"	C8@.115	LF	8.45	4.71	13.16
Ceiling joists and blocking, including rim joists, Std & Btr, per LF of joist					
2" x 6"	C8@.028	LF	.68	1.15	1.83
2" x 8"	C8@.037	LF	.94	1.52	2.46
2" x 10"	C8@.046	LF	1.20	1.88	3.08
2" x 12"	C8@.057	LF	1.62	2.34	3.96
Floor joists and blocking including rim joists and 6% waste, per LF of joist					
2" x 6", 1.06 BF per LF	C8@.017	LF	.68	.70	1.38
2" x 8", 1.41 BF per LF	C8@.018	LF	.94	.74	1.68
2" x 10", 1.77 BF per LF	C8@.019	LF	1.20	.78	1.98
2" x 12", 2.12 BF per LF	C8@.021	LF	1.62	.86	2.48
Purlins, struts, ridge boards, Std & Btr					
4" x 10"	C8@.050	LF	2.98	2.05	5.03
4" x 12"	C8@.055	LF	3.64	2.25	5.89
6" x 10"	C8@.066	LF	7.04	2.70	9.74
6" x 12"	C8@.066	LF	8.45	2.70	11.15
Rafters and outriggers, to 4 in 12 pitch, Std & Btr					
2" x 4" to 2" x 8"	C8@.033	LF	.94	1.35	2.29
2" x 10" to 2" x 12"	C8@.036	LF	1.62	1.47	3.09
Add for over 4 in 12 pitch roof	—	%	—	30.0	—
Rafters, including frieze blocks, ribbon, strongbacks, 24" OC, per SF of roof plan area, Std & Btr					
2" x 4", .44 BF per SF	C8@.015	SF	.27	.61	.88
2" x 6", .65 BF per SF	C8@.021	SF	.40	.86	1.26
2" x 8", .86 BF per SF	C8@.027	SF	.55	1.11	1.66
2" x 10", 1.10 BF per SF	C8@.031	SF	.72	1.27	1.99
2" x 12", 1.40 BF per SF	C8@.041	SF	1.03	1.68	2.71

Sills and Ledgers Bolted to Concrete or Masonry Walls Costs are per linear foot of standard and better pressure treated sill or ledger with anchor bolt holes drilled each 2'6" center. Labor costs include handling, measuring, cutting to length and drilling holes at floor level and installation working on a scaffold 10' to 20' above floor level. Figures in parentheses show the board feet per linear foot of sill or ledger. Costs shown include 5% waste.

For scheduling purposes, estimate that a crew of 2 can install 150 to 200 LF of sills with a single row of anchor bolts in an 8-hour day when working from scaffolds. Estimate that a crew of 2 can install 150 to 175 LF of ledgers with a single row of anchor bolt holes or 50 to 60 LF of ledgers with a double row of anchor bolt holes in an 8-hour day working from scaffolds. Add the cost of anchor bolts and scaffold, if required.

	Craft@Hrs	Unit	Material	Labor	Total
Sills with a single row of anchor bolt holes					
2" x 4" (0.70 BF per LF)	C8@.040	LF	.66	1.64	2.30
2" x 6" (1.05 BF per LF)	C8@.040	LF	.91	1.64	2.55
2" x 8" (1.40 BF per LF)	C8@.040	LF	1.28	1.64	2.92
2" x 10" (1.75 BF per LF)	C8@.040	LF	1.72	1.64	3.36
2" x 12" (2.10 BF per LF)	C8@.040	LF	2.11	1.64	3.75
4" x 6" (2.10 BF per LF)	C8@.051	LF	2.16	2.09	4.25
4" x 8" (2.80 BF per LF)	C8@.051	LF	3.18	2.09	5.27
4" x 10" (3.33 BF per LF)	C8@.059	LF	3.87	2.42	6.29
4" x 12" (4.20 BF per LF)	C8@.059	LF	5.19	2.42	7.61
Ledgers with a single row of anchor bolt holes					
2" x 4" (0.70 BF per LF)	C8@.059	LF	.66	2.42	3.08
4" x 6" (2.10 BF per LF)	C8@.079	LF	2.16	3.24	5.40
4" x 8" (2.80 BF per LF)	C8@.079	LF	3.18	3.24	6.42
Ledgers with a double row of anchor bolt holes					
4" x 4" (1.40 BF per LF)	C8@.120	LF	1.39	4.92	6.31
4" x 6" (2.10 BF per LF)	C8@.160	LF	2.16	6.56	8.72
4" x 8" (2.80 BF per LF)	C8@.160	LF	3.18	6.56	9.74
4" x 12" (4.20 BF per LF)	C8@.250	LF	5.19	10.20	15.39

Sheathing Includes nails and normal waste.

	Craft@Hrs	Unit	Material	Labor	Total
Floors, hand nailed, underlayment C-D grade plywood					
1/2"	C8@.012	SF	.52	.49	1.01
5/8"	C8@.012	SF	.61	.49	1.10
3/4"	C8@.013	SF	.79	.53	1.32
Floors, machine nailed OSB					
7/16" x 4' x 8'	C8@.009	SF	.34	.37	.71
7/16" x 4' x 9'	C8@.009	SF	.40	.37	.77
7/16" x 4' x 10'	C8@.009	SF	.39	.37	.76
15/32" 4' x 8'	C8@.009	SF	.39	.37	.76
1/2" x 4' x 8'	C8@.009	SF	.38	.37	.75
5/8" x 4' x 8'	C8@.010	SF	.51	.41	.92
5/8" x 4' x 8', tongue and groove	C8@.010	SF	.69	.41	1.10
23/32" 4' x 8'	C8@.010	SF	.69	.41	1.10
23/32" 4' x 8' subfloor	C8@.010	SF	.82	.41	1.23
3/4" x 4' x 8', tongue and groove	C8@.011	SF	.61	.45	1.06
1-1/8" x 4' x 8', tongue and groove	C8@.015	SF	1.24	.61	1.85
15/32" x 4' x 8', hibor treated	C8@.011	SF	1.19	.45	1.64
Walls, B-D, sanded, standard, interior grade plywood					
1/4"	C8@.015	SF	.46	.61	1.07
3/8"	C8@.015	SF	.53	.61	1.14
1/2"	C8@.016	SF	.65	.66	1.31
5/8"	C8@.018	SF	.81	.74	1.55
3/4"	C8@.018	SF	.96	.74	1.70

Wood and Plastic 6

	Craft@Hrs	Unit	Material	Labor	Total
Walls, exterior grade CD plywood					
1/2"	C8@.016	SF	.52	.66	1.18
5/8"	C8@.018	SF	.61	.74	1.35
3/4"	C8@.020	SF	.79	.82	1.61
3/8" rough sawn T-1-11	C8@.018	SF	.62	.74	1.36
5/8" rough sawn T-1-11	C8@.018	SF	.56	.74	1.30
5/8" rough sawn RBB	C8@.018	SF	.80	.74	1.54
Walls, OSB					
11/32" or 3/8" x 4' x 8'	C8@.013	SF	.34	.53	.87
7/16" x 4' x 8'	C8@.015	SF	.34	.61	.95
7/16" x 4' x 9'	C8@.015	SF	.40	.61	1.01
7/16" x 4' x 10'	C8@.015	SF	.39	.61	1.00
15/32" 4' x 8'	C8@.015	SF	.39	.61	1.00
1/2" x 4' x 8'	C8@.018	SF	.38	.74	1.12
5/8" x 4' x 8'	C8@.018	SF	.51	.74	1.25
5/8" x 4' x 8', tongue and groove	C8@.018	SF	.69	.74	1.43
23/32" x 4' x 8'	C8@.020	SF	.69	.82	1.51
15/32" x 4' x 10', Structural 1	C8@.020	SF	.52	.82	1.34
Add for nailing shear walls	—	%	—	20.0	—
Roofs, CD exterior plywood, machine nailed					
3/8"	C8@.009	SF	.40	.37	.77
1/2"	C8@.009	SF	.52	.37	.89
5/8"	C8@.009	SF	.61	.37	.98
3/4"	C8@.011	SF	.79	.45	1.24
Roofs, OSB, machine nailed					
11/32" or 3/8" x 4' x 8'	C8@.009	SF	.34	.37	.71
7/16" x 4' x 8'	C8@.009	SF	.34	.37	.71
7/16" x 4' x 9'	C8@.009	SF	.40	.37	.77
7/16" x 4' x 10'	C8@.009	SF	.39	.37	.76
15/32" 4' x 8'	C8@.009	SF	.39	.37	.76
1/2" x 4' x 8'	C8@.009	SF	.38	.37	.75
5/8" x 4' x 8'	C8@.010	SF	.51	.41	.92
5/8" x 4' x 8', tongue and groove	C8@.010	SF	.69	.41	1.10
23/32" x 4' x 8'	C8@.010	SF	.69	.41	1.10
3/4" x 4' x 8', tongue and groove	C8@.011	SF	.61	.45	1.06
Blocking for panel joints,					
2" x 4"	C8@.010	LF	.45	.41	.86
Lumber floor sheathing					
Diagonal, 1" x 6" Standard and Better					
1.14 BF per SF	C8@.018	SF	1.01	.74	1.75
Straight 2" x 6" commercial grade T&G fir					
2.28 BF per SF	C8@.032	SF	1.41	1.31	2.72
Straight 2" x 6" commercial grade T&G cedar					
2.28 BF per SF	C8@.032	SF	6.51	1.31	7.82
Timber deck, Southern Pine, 5/4" x 6", premium grade					
1.25 BF per SF	C8@.012	SF	2.87	.49	3.36
Lumber wall sheathing, diagonal, 1" x 6"					
Standard and Better, 1.14 BF per SF	C8@.024	SF	1.53	.98	2.51
Lumber roof sheathing, diagonal, 1" x 6"					
Standard and Better, 1.14 BF per SF	C8@.018	SF	1.53	.74	2.27

Panelized Wood Roof Systems The costs shown will apply on 20,000 to 30,000 SF jobs where roof panels are fabricated near the point of installation and moved into place with a forklift. Costs will be higher on smaller jobs and jobs with poor access. 4' x 8' and 4' x 10' roof panels are made of plywood or OSB nailed to 2" x 4" or 2" x 6" stiffeners spaced 24" on centers and running the long dimension of the panel. Based on fabricated panels lifted by forklift (with metal hangers attached) and nailed to 4" x 12" or 4" x 14" roof purlins spaced 8' on center. Equipment cost is for a 4-ton forklift (40' lift), a truck with hand tools, a gas powered air compressor with air hoses and two pneumatic-operated nailing guns. Costs are per 1,000 square feet (MSF) of roof surface and do not include roof supports except as noted.

	Craft@Hrs	Unit	Material	Labor	Equipment	Total

For scheduling purposes, estimate that a crew of 5 can fabricate 6,000 to 6,400 SF of 4' x 8' or 4' x 10' panels in an 8-hour day. Estimate that the same crew of 5 can install and nail down 7,600 to 8,000 SF of panels in an 8-hour day. Based on the following material costs:

	Craft@Hrs	Unit	Material	Labor	Equipment	Total
2" x 4" stiffeners	—	MBF	621.00	—	—	621.00
2" x 6" stiffeners	—	MBF	620.00	—	—	620.00
Purlins	—	MBF	652.00	—	—	652.00
1/2" CDX plywood	—	MSF	519.00	—	—	519.00
1/2" OSB sheathing	—	MSF	422.00	—	—	422.00

4' x 8' roof panels
With 1/2" CDX structural #1 plywood attached to 2" x 4" construction grade fir stiffeners

	Craft@Hrs	Unit	Material	Labor	Equipment	Total
Panels fabricated, ready to install	F6@6.60	MSF	778.00	282.00	76.10	1,136.10
Equipment & labor to install and nail panels	F6@5.21	MSF	—	222.00	60.10	282.10
Total installed cost, 4' x 8' panels	**--@11.8**	**MSF**	**778.00**	**504.00**	**136.20**	**1,418.20**

With 1/2" OSB structural sheathing attached to 2" x 4" construction grade fir stiffeners

	Craft@Hrs	Unit	Material	Labor	Equipment	Total
Panels fabricated, ready to install	F6@6.60	MSF	682.00	282.00	76.10	1,040.10
Equipment & labor to install and nail panels	F6@5.21	MSF	—	222.00	60.10	282.10
Total installed cost, 4' x 8' panels	**--@11.8**	**MSF**	**682.00**	**504.00**	**136.20**	**1,322.20**

4' x 10' roof panels
With 1/2" CDX structural #1 plywood attached to 2" x 6" construction grade fir stiffeners

	Craft@Hrs	Unit	Material	Labor	Equipment	Total
Panels fabricated, ready to install	F6@6.20	MSF	904.00	265.00	71.70	1,240.70
Equipment & labor to install and nail panels	F6@4.90	MSF	—	209.00	56.60	265.60
Total installed cost, 4' x 10' panels	**--@11.1**	**MSF**	**904.00**	**474.00**	**128.30**	**1,506.30**

With 1/2" OSB structural sheathing attached to 2" x 6" construction grade fir stiffeners

	Craft@Hrs	Unit	Material	Labor	Equipment	Total
Panels fabricated, ready to install	F6@6.20	MSF	807.00	265.00	71.70	1,143.70
Equipment & labor to install and nail panels	F6@4.90	MSF	—	209.00	56.60	265.60
Total installed cost, 4' x 10' panels	**--@11.1**	**MSF**	**807.00**	**474.00**	**128.30**	**1,409.30**

8' x 20' and 8' x 32' roof panels
These panels can be fabricated on site using the same crew and equipment as shown in the preceding section. However, the stiffeners run across the short dimension of the panel and one purlin is attached to the outside edge of the long dimension of each panel before lifting into place. Figures below include the cost of purlins but not supporting beams. For scheduling purposes, estimate that a crew of 5 can fabricate 6,400 to 7,000 SF of 8' x 20' or 8' x 32' panels in an 8-hour day. Estimate that the same crew of 5 can install and nail down 10,000 to 11,000 SF of panels in an 8-hour day.

8' x 20' roof panels with purlins attached
4' x 10' x 1/2" CDX structural #1 plywood attached to 2" x 4" const. grade fir stiffeners and 4" x 12" #1 grade fir purlins

	Craft@Hrs	Unit	Material	Labor	Equipment	Total
Panels fabricated, ready to install	F6@6.23	MSF	1,550.00	266.00	71.70	1,887.70
Equipment & labor to install and nail panels	F6@3.96	MSF	—	169.00	45.70	214.70
Total installed cost, 8' x 20' panels	**--@10.2**	**MSF**	**1,550.00**	**435.00**	**117.40**	**2,102.40**

	Craft@Hrs	Unit	Material	Labor	Equipment	Total
8' x 20' x 1/2" OSB structural sheathing attached to 2" x 4" const. grade fir stiffeners and 4" x 12" #1 grade fir purlins						
Panels fabricated, ready to install	F6@6.23	MSF	1,450.00	266.00	71.70	1,787.70
Equipment & labor						
to install and nail panels	F6@3.96	MSF	—	169.00	45.70	214.70
Total installed cost, 8' x 20' panels	**--@10.2**	**MSF 1,450.00**		**435.00**	**117.40**	**2,002.40**

8' x 32' panel with purlins attached

	Craft@Hrs	Unit	Material	Labor	Equipment	Total
4' x 8' x 1/2" CDX structural #1 plywood attached to 2" x 4" const. grade fir stiffeners and 4" x 12" #1 grade fir purlins						
Panels fabricated, ready to install	F6@5.82	MSF	1,620.00	248.00	67.00	1,935.00
Equipment & labor						
to install and nail panels	F6@3.54	MSF	—	151.00	41.00	192.00
Total installed cost, 8' x 32' panels	**--@9.36**	**MSF 1,620.00**		**399.00**	**108.00**	**2,127.00**
4' x 8' x 1/2" OSB structural sheathing attached to 2" x 4" const. grade fir stiffeners and 4" x 12" #1 grade fir purlins						
Panels fabricated, ready to install	F6@5.82	MSF	1,520.00	248.00	67.00	1,835.00
Equipment & labor						
to install and nail panels	F6@3.54	MSF	—	151.00	41.00	192.00
Total installed cost, 8' x 32' panels	**--@9.36**	**MSF 1,520.00**		**399.00**	**108.00**	**2,027.00**
Adjust to any of the panel costs for 5/8" CDX #1 structural plywood, instead of 1/2" attached before						
panels are erected	—	MSF	84.00	—	—	84.00
Adjust to any of the panel costs for 15/32" OSB Structural I sheathing, instead of 1/2" attached before						
panels are erected	—	MSF	132.00	—	—	132.00
Add for roof hatch opening, includes reinforced opening and curb. Does not include roof hatch, see below						
2" x 4" roof system	F6@.055	SF	2.84	2.35	—	5.19
2" x 6" roof system	F6@.060	SF	3.87	2.56	—	6.43

Panelized Wood Wall Panels These costs will apply on jobs with 20,000 to 30,000 SF of wall panels. Panel material costs are for prefabricated panels purchased from a modular plant located within 50 miles of the job site. Costs will be higher on smaller jobs and on jobs with poor access. Costs assume wall panels are fabricated according to job plans and specifications, delivered piece-marked, ready to install and finish. Labor costs for installation of panels are listed at the end of this section and include unloading and handling, sorting, moving the panels into position using a forklift, installing and nailing. Maximum panel height is usually 12'. Equipment cost includes a forklift, hand tools, a portable air compressor with air hoses and two pneumatic-operated nailing guns. Costs are per square foot of wall, measured on one face. For scheduling purposes, estimate that a crew of 4 can place and fasten 3,500 to 4,500 SF of panels in an 8-hour day.

	Craft@Hrs	Unit	Material	Labor	Equipment	Total
2" x 4" studs 16" OC with gypsum wallboard on both sides						
1/2" wallboard	—	SF	1.29	—	—	1.29
5/8" wallboard	—	SF	1.43	—	—	1.43
2" x 4" studs 16" OC with gypsum wallboard inside and plywood or OSB outside						
1/2" wallboard and 1/2" CDX plywood	—	SF	1.52	—	—	1.52
1/2" wallboard and 1/2" OSB sheathing	—	SF	1.42	—	—	1.42
1/2" wallboard and 5/8" CDX plywood	—	SF	1.63	—	—	1.63
1/2" wallboard and 5/8" OSB sheathing	—	SF	1.58	—	—	1.58
5/8" wallboard and 1/2" CDX plywood	—	SF	1.59	—	—	1.59
5/8" wallboard and 1/2" OSB sheathing	—	SF	1.49	—	—	1.49
5/8" wallboard and 5/8" CDX plywood	—	SF	1.70	—	—	1.70
5/8" wallboard and 5/8" OSB sheathing	—	SF	1.65	—	—	1.65
2" x 6" studs 12" OC with gypsum wallboard on both sides						
1/2" wallboard	—	SF	1.63	—	—	1.63
5/8" wallboard	—	SF	1.70	—	—	1.70

	Craft@Hrs	Unit	Material	Labor	Equipment	Total
2" x 6" studs 12" OC with gypsum wallboard inside and plywood or OSB outside						
1/2" wallboard and 1/2" CDX plywood	—	SF	2.13	—	—	2.13
1/2" wallboard and 1/2" OSB sheathing	—	SF	2.04	—	—	2.04
1/2" wallboard and 5/8" CDX plywood	—	SF	2.24	—	—	2.24
1/2" wallboard and 5/8" OSB sheathing	—	SF	2.19	—	—	2.19
5/8" wallboard and 1/2" CDX plywood	—	SF	2.20	—	—	2.20
5/8" wallboard and 1/2" OSB sheathing	—	SF	2.11	—	—	2.11
5/8" wallboard and 5/8" CDX plywood	—	SF	2.31	—	—	2.31
5/8" wallboard and 5/8" OSB sheathing	—	SF	2.26	—	—	2.26
2" x 6" studs 16" OC with gypsum wallboard on both sides						
1/2" wallboard	—	SF	1.36	—	—	1.36
5/8" wallboard	—	SF	1.43	—	—	1.43
2" x 6" studs 16" OC with gypsum wallboard inside and plywood or OSB outside						
1/2" wallboard and 1/2" CDX plywood	—	SF	1.85	—	—	1.85
1/2" wallboard and 1/2" OSB sheathing	—	SF	1.76	—	—	1.76
1/2" wallboard and 5/8" CDX plywood	—	SF	1.97	—	—	1.97
1/2" wallboard and 5/8" OSB sheathing	—	SF	1.92	—	—	1.92
5/8" wallboard and 1/2" CDX plywood	—	SF	1.92	—	—	1.92
5/8" wallboard and 1/2" OSB sheathing	—	SF	1.83	—	—	1.83
5/8" wallboard and 5/8" CDX plywood	—	SF	2.04	—	—	2.04
5/8" wallboard and 5/8" OSB sheathing	—	SF	1.99	—	—	1.99
Add to any of the above for the following, installed by panel fabricator:						
R-19 insulation	—	SF	.77	—	—	.77
R-11 insulation	—	SF	.51	—	—	.51
Vapor barrier	—	SF	.42	—	—	.42
Door or window openings to 16 SF each	—	SF	1.87	—	—	1.87
Door or window openings over 16 SF each	—	SF	1.73	—	—	1.73
Electrical receptacles	—	Ea	36.20	—	—	36.20
Electrical switches	—	Ea	41.40	—	—	41.40
Labor and equipment cost to install panels						
Up to 8' high	F7@.003	SF	—	.13	.03	.16
Over 8' to 12' high	F7@.002	SF	—	.09	.02	.11

	Craft@Hrs	Unit	Material	Labor	Total
Miscellaneous Rough Carpentry					
Furring, Utility grade, unit costs incl. 5% allowance for waste and nails					
2" x 4" on masonry walls	C8@.024	LF	.43	.98	1.41
1" x 4" machine nailed to ceiling	C8@.017	LF	.30	.70	1.00
1" x 4" nailed on concrete	C8@.016	LF	.30	.66	.96
Skip sheathing, on roof rafters, machine nailed, #2 and Better, unit costs incl. 7% allowance for waste & nails					
1" x 6" at 9" centers (.67 BF per SF)					
Per SF of roof area	C8@.020	SF	.63	.82	1.45
1" x 4" at 7" centers (.50 BF per SF)					
Per SF of roof area	C8@.030	SF	.46	1.23	1.69
Hardboard, 4' x 8', used as underlayment					
1/4" x 4' x 8' sheets	C8@.011	SF	.36	.45	.81
1/2" x 4' x 8' sheets	C8@.011	SF	.42	.45	.87
5/8" x 4' x 8' sheets	C8@.011	SF	.48	.45	.93
3/4" x 4' x 8' sheets	C8@.011	SF	.52	.45	.97
Add for attaching to metal frame	C8@.005	SF	—	.20	.20

	Craft@Hrs	Unit	Material	Labor	Total
15 pound felt on walls	C8@.003	SF	.04	.12	.16
Kraft flashing with foil 2 sides, 30 pound	C8@.004	SF	.10	.16	.26
Insulation board on walls, 4' x 8' Foamular® 150 polystyrene, including 5% waste					
1/2" thick (R-2.5), $9.04 per 4' x 8' panel	C8@.010	SF	.28	.41	.69
3/4" thick (R-3.8), $11.10 per 4' x 8' panel	C8@.011	SF	.35	.45	.80
1" thick (R-5.0), $15.60 per 4 x 8 panel	C8@.011	SF	.49	.45	.94
1-1/2" thick (R-7.5), $16.70 per 4' x 8' panel	C8@.015	SF	.52	.61	1.13
2" thick (R-10), $22.50 per 4' x 8' panel	C8@.015	SF	.70	.61	1.31

	Craft@Hrs	Unit	Material	Labor	Equipment	Total

Wood Trusses Material costs are for prefabricated wood trusses shipped to job site fully assembled ready for installation. Equipment is a 6000 lb capacity fork lift. Labor includes unloading and handling and installing the trusses not over 20' above grade.

Truss joists, TJI type at 50 PSF load design, at 16" OC

	Craft@Hrs	Unit	Material	Labor	Equipment	Total
9-1/2" TJI/15	C8@.017	SF	1.93	.70	.26	2.89
11-7/8" TJI/15	C8@.017	SF	2.12	.70	.26	3.08
14" TJI/35	C8@.018	SF	2.99	.74	.27	4.00
16" TJI/35	C8@.018	SF	3.26	.74	.27	4.27
Add for job under 3,000 SF	C8@.004	SF	.52	.16	.06	.74

Roof trusses, 24" OC, any slope from 3 in 12 to 12 in 12, total height not to exceed 12' high from bottom chord to highest point on truss. Prices for trusses over 12' high will be up to 100% higher. Square foot (SF) costs, where shown, are per square foot of roof area to be covered.

Scissor truss, 2" x 4" top and bottom chords

	Craft@Hrs	Unit	Material	Labor	Equipment	Total
Up to 38' span	F5@.022	SF	2.38	.92	.13	3.43
40' to 50' span	F5@.028	SF	2.98	1.17	.17	4.32

Fink truss "W" (conventional roof truss), 2" x 4" top and bottom chords

	Craft@Hrs	Unit	Material	Labor	Equipment	Total
Up to 38' span	F5@.017	SF	2.06	.71	.10	2.87
40' to 50' span	F5@.022	SF	2.43	.92	.13	3.48

Fink truss with 2" x 6" top and bottom chords

	Craft@Hrs	Unit	Material	Labor	Equipment	Total
Up to 38' span	F5@.020	SF	2.48	.84	.12	3.44
40' to 50' span	F5@.026	SF	2.99	1.09	.16	4.24

Fink truss with gable fill at 16" OC

	Craft@Hrs	Unit	Material	Labor	Equipment	Total
28' span, 5 in 12 slope	F5@.958	Ea	168.00	40.10	5.76	213.86
32' span, 5 in 12 slope	F5@1.26	Ea	208.00	52.80	7.58	268.38
40' span, 5 in 12 slope	F5@1.73	Ea	298.00	72.50	10.40	380.90

Laminated Glued Beams (Glu-lam beams) Typical costs for horizontal glu-lam beams. Figures in parentheses show the approximate weight per linear foot of beam. Equipment cost is based on using a 4,000 lb capacity forklift with 30' lift.

Steel connectors and shoes for glu-lam beams, including bolts,

	Craft@Hrs	Unit	Material	Labor	Equipment	Total
Typical per end or splice connection	—	Ea	86.30	—	—	86.30

Drill holes in glu-lam beams and attach connectors at grade before installing beam, based on number of holes in the connector, with work performed at job site

	Craft@Hrs	Unit	Material	Labor	Equipment	Total
Labor per each hole	F6@.236	Ea	—	10.10	—	10.10
Install beams with connections attached	F6@.133	LF	—	5.68	.70	6.38

	Craft@Hrs	Unit	Material	Labor	Equipment	Total

Stock glu-lam beams, typical costs for West Coast Douglas fir, 24F-V4, H=240 PSI, exterior glue, camber is 3500' R for beams. Industrial grade, without wrap. For other grades see the adjustments that follow.

	Craft@Hrs	Unit	Material	Labor	Equipment	Total
3-1/8" x 6" (4.4 pounds per LF)	—	LF	5.52	—	—	5.52
3-1/8" x 9" (6.6 pounds per LF)	—	LF	8.17	—	—	8.17
3-1/8" x 10-1/2" (7.5 pounds per LF)	—	LF	9.55	—	—	9.55
3-1/8" x 12" (8.0 pounds per LF)	—	LF	10.80	—	—	10.80
3-1/8" x 15" (10.0 pounds per LF)	—	LF	13.50	—	—	13.50
3-1/8" x 18" (13.2 pounds per LF)	—	LF	16.10	—	—	16.10
5-1/8" x 6" (7.3 pounds per LF)	—	LF	7.94	—	—	7.94
5-1/8" x 9" (10.7 pounds per LF)	—	LF	12.00	—	—	12.00
5-1/8" x 12" (14.5 pounds per LF)	—	LF	16.00	—	—	16.00
5-1/8" x 15" (18.1 pounds per LF)	—	LF	20.00	—	—	20.00
5-1/8" x 18" (21.7 pounds per LF)	—	LF	23.90	—	—	23.90
5-1/8" x 21" (25.4 pounds per LF)	—	LF	27.80	—	—	27.80
5-1/8" x 24" (29.0 pounds per LF)	—	LF	31.70	—	—	31.70
6-3/4" x 12" (19.1 pounds per LF)	—	LF	21.20	—	—	21.20
6-3/4" x 15" (23.8 pounds per LF)	—	LF	26.50	—	—	26.50
6-3/4" x 18" (28.6 pounds per LF)	—	LF	31.70	—	—	31.70
6-3/4" x 21" (33.3 pounds per LF)	—	LF	37.00	—	—	37.00
6-3/4" x 24" (38.1 pounds per LF)	—	LF	42.30	—	—	42.30
Add to any of the beam costs shown in this section for:						
Industrial grade, clean, wrapped, add	—	%	12.5	—	—	—
Architectural grade, wrapped, add	—	%	20.0	—	—	—
One-coat rez-latex stain, add	—	%	10.0	—	—	—
Transportation to job site, per delivery, maximum 50 miles one way, add	—	LS	275.00	—	—	275.00

	Craft@Hrs	Unit	Material	Labor	Total

Stairs Job-built stairways, 3 stringers cut from 2" x 12" material, treads and risers from 3/4" CDX plywood. Material costs based on:

	Craft@Hrs	Unit	Material	Labor	Total
2" x 12"	—	MBF	752.00	—	752.00
1" x 6"	—	MSF	1,510.00	—	1,510.00
3/4" CDX plywood	—	MSF	675.00	—	675.00
Cost per 7-1/2" rise, 36" wide					
Straight run, 8'0" to 10'0" rise, per riser	C8@.530	Ea	12.90	21.70	34.60
"L" or "U" shape, per riser, add for landings	C8@.625	Ea	14.50	25.60	40.10
Landings, framed and 3/4" CDX plywood surfaced					
Per SF of landing surface	C8@.270	SF	2.92	11.10	14.02
Factory cut and assembled straight closed box stairs, unfinished, 36" width					
Oak treads with 7-1/2" risers, price per riser					
3'0" to 3'6" wide, pre riser	C8@.324	Ea	106.00	13.30	119.30
4'0" wide, per riser	C8@.357	Ea	128.00	14.60	142.60
Add for prefinished assembled stair rail with balusters and newel, per riser	C8@.257	Ea	46.70	10.50	57.20
Add for prefinished handrail, brackets and balusters	C8@.135	LF	13.80	5.53	19.33
Spiral stairs, 5' diameter. Steel, tubular handrail, composition board treads, prefabricated, shipped unassembled					
86" to 95" with 10 risers	C8@9.98	Ea	1,020.00	409.00	1,429.00
95" to 105" with 11 risers	C8@10.7	Ea	1,120.00	438.00	1,558.00
105" to 114" with 12 risers	C8@11.3	Ea	1,170.00	463.00	1,633.00
114" to 124" with 13 risers	C8@12.0	Ea	1,290.00	492.00	1,782.00
124" to 133" with 14 risers	C8@12.7	Ea	1,480.00	520.00	2,000.00
Add for 6' diameter	—	%	24.0	—	—
Add for oak treads	—	%	25.0	—	—
Add for oak handrail	—	%	45.0	—	—

	Craft@Hrs	Unit	Material	Labor	Total
Dampproofing					
Asphalt wall primer, per coat, gallon					
covers 250 SF	CL@.010	SF	.08	.36	.44
Asphalt emulsion, wall, per coat, gallon					
covers 33 SF, brush on	CL@.013	SF	.14	.47	.61
Hot mop concrete wall, 2 coats and glass fabric	CL@.033	SF	.30	1.20	1.50
Hot mop deck, 2 ply, felt, typical subcontract price	—	Sq	—	—	93.20
Hot mop deck, 3 ply, felt, typical subcontract price	—	Sq	—	—	118.00
Add for 1/2" asphalt fiberboard, subcontract price	—	Sq	—	—	62.50
Bituthene waterproofing membrane,					
60 mil plain surface	CL@.025	SF	.64	.91	1.55
Waterproof baths in high rise	CL@.025	SF	1.30	.91	2.21
Iron compound, internal coating, 2 coats	CL@.012	SF	1.20	.44	1.64
Roof vapor barrier, .004" polyethylene	CL@.002	SF	.05	.07	.12

Exterior Wall Insulation and Finish System (EIFS) Polymer based exterior non-structural wall finish applied to concrete, masonry, stucco or exterior grade gypsum sheathing (such as Dryvit). System consists of a plaster base coat approximately 1/4" thick, and a glass fiber mesh that is embedded in the adhesive base coat. The base coat is applied over polystyrene insulation boards that are bonded to an existing flat vertical wall surface with adhesive. Costs shown are per SF of finished wall surface. These costs do not include surface preparation or wall costs. Cost for scaffolding is not included: Add 50% to manhours and labor costs if scaffolding is required.

For scheduling purposes, estimate that a crew of three can field apply 400 SF of exterior wall insulation and finish systems in an 8-hour day.

	Craft@Hrs	Unit	Material	Labor	Total
Adhesive mixture, 2 coats					
2.1 SF per SF of wall area including waste	F8@.013	SF	.59	.58	1.17
Glass fiber mesh					
1.1 SF per SF of wall area including waste	F8@.006	SF	.33	.27	.60
Insulation board, 1" thick					
1.05 SF per SF of wall area including waste	F8@.013	SF	.21	.58	.79
Textured finish coat, with integral color					
1.05 SF per SF of wall area including waste	F8@.024	SF	.58	1.08	1.66
Total with 1" thick insulation	F8@.056	SF	1.67	2.52	4.19
Add for 2" thick insulation board	—	SF	.23	—	.23
Add for 3" thick insulation board	—	SF	.44	—	.44
Add for 4" thick insulation board	—	SF	.65	—	.65

Insulation Roof insulation. Local regulation can affect the cost and availability of urethanes, isocyanurates, phenolics and polystyrene roof insulation board.

	Craft@Hrs	Unit	Material	Labor	Total
Fiberglass board roof insulation					
3/4", R-2.80, C-0.36	R3@.570	Sq	45.00	23.80	68.80
1", R-4.20, C-0.24	R3@.636	Sq	54.00	26.60	80.60
1-3/8", R-5.30, C-0.19	R3@.684	Sq	69.40	28.60	98.00
1-5/8", R-6.70, C-0.15	R3@.706	Sq	78.50	29.50	108.00
2-1/4", R-8.30, C-0.12	R3@.797	Sq	95.40	33.30	128.70
Polystyrene rigid foam roof insulation. R-19 rating when used with batt insulation					
1/2" 4' x 8' panels	R3@1.10	Sq	21.00	46.00	67.00
3/4" 4' x 8' panels	R3@1.10	Sq	24.50	46.00	70.50
1" 4' x 8' panels	R3@1.10	Sq	37.20	46.00	83.20
1-1/2" 4' x 8' panels	R3@1.10	Sq	43.20	46.00	89.20
2" 4' x 8' panels	R3@1.10	Sq	57.90	46.00	103.90

	Craft@Hrs	Unit	Material	Labor	Total
Phenolic board roof insulation					
1-1/4", R-10.0, C-0.10	R3@.500	Sq	59.60	20.90	80.50
1-1/2", R-12.5, C-0.08	R3@.592	Sq	66.90	24.80	91.70
1-3/4", R-14.6, C-0.07	R3@.592	Sq	72.80	24.80	97.60
2", R-16.7, C-0.06	R3@.684	Sq	83.90	28.60	112.50
2-1/2", R-20.0, C-0.05	R3@.728	Sq	91.50	30.40	121.90
3", R-25.0, C-0.04	R3@.889	Sq	122.00	37.20	159.20
3-1/4", R-28.0, C-0.035	R3@1.05	Sq	157.00	43.90	200.90
Expanded polystyrene board roof insulation					
3/4", R-5.30, C-0.19	R3@.478	Sq	24.60	20.00	44.60
1", R-6.70, C-0.15	R3@.478	Sq	29.50	20.00	49.50
1-1/2", R-11.1, C-0.09	R3@.570	Sq	43.20	23.80	67.00
2", R-14.3, C-0.07	R3@.684	Sq	71.70	28.60	100.30
2-1/4", R-16.7, C-0.06	R3@.684	Sq	79.20	28.60	107.80
Perlite board roof insulation					
1", R-2.80, C-0.36	R3@.639	Sq	49.90	26.70	76.60
1-1/2", R-4.20, C-0.24	R3@.728	Sq	73.20	30.40	103.60
2", R-5.30, C-0.19	R3@.820	Sq	89.80	34.30	124.10
2-1/2", R-6.70, C-0.15	R3@.956	Sq	112.00	40.00	152.00
3", R-8.30, C-0.12	R3@1.03	Sq	136.00	43.10	179.10
4", R-10.0, C-0.10	R3@1.23	Sq	180.00	51.40	231.40
5-1/4", R-14.3, C-0.07	R3@1.57	Sq	239.00	65.70	304.70
Cellular Foamglass (isocyanurates) board insulation, on roofs					
1-1/2"	R3@.775	Sq	180.00	32.40	212.40
Tapered 3/16" per foot	R3@.775	Sq	164.00	32.40	196.40
Polystyrene board roof insulation					
1", R-4.20, C-0.24	R3@.545	Sq	20.80	22.80	43.60
1-1/2", R-5.9, C-0.16	R3@.639	Sq	25.40	26.70	52.10
2", R-8.30, C-0.12	R3@.797	Sq	34.30	33.30	67.60
Fasteners for board type insulation on roofs					
Typical 2-1/2" to 3-1/2" screws with 3" metal disks					
I-60	R3@.053	Sq	2.59	2.22	4.81
I-90	R3@.080	Sq	3.92	3.35	7.27
Add for plastic disks	—	Sq	.65	—	.65
Fiberglass batts, placed between or over framing members					
3-1/2" kraft faced, R-11, between ceiling joists	A1@.006	SF	.33	.27	.60
3-1/2" kraft faced, R-11, on suspended ceiling, working off ladders from below	A1@.018	SF	.33	.82	1.15
3-1/2" kraft faced, R-11, in crawl space	A1@.020	SF	.33	.91	1.24
3-1/2" kraft faced, R-13, between ceiling joists	A1@.006	SF	.36	.27	.63
3-1/2" kraft faced, R-13, on suspended ceiling, working off ladders from below	A1@.018	SF	.36	.82	1.18
3-1/2" kraft faced, R-13, in crawl space	A1@.020	SF	.36	.91	1.27
3-1/2" kraft faced, R-15, between ceiling joists	A1@.006	SF	.66	.27	.93
3-1/2" kraft faced, R-15, on suspended ceiling, working off ladders from below	A1@.018	SF	.66	.82	1.48
3-1/2" kraft faced, R-15, in crawl space	A1@.020	SF	.66	.91	1.57
6-1/4" kraft faced, R-19, between ceiling joists	A1@.006	SF	.44	.27	.71
6-1/4" kraft faced, R-19, on suspended ceiling, working off ladders from below	A1@.018	SF	.44	.82	1.26
6-1/4" kraft faced, R-19, in crawl space	A1@.020	SF	.44	.91	1.35
6-1/4" kraft faced, R-19, between studs	A1@.006	SF	.44	.27	.71

	Craft@Hrs	Unit	Material	Labor	Total
5-1/2" kraft faced, R-21, between ceiling joists	A1@.006	SF	.59	.27	.86
5-1/2" kraft faced, R-21, on suspended ceiling, working off ladders from below	A1@.018	SF	.59	.82	1.41
5-1/2" kraft faced, R-21, in crawl space	A1@.020	SF	.59	.91	1.50
5-1/2" kraft faced, R-21, between studs	A1@.006	SF	.59	.27	.86
8-1/4" kraft faced, R-30, between ceiling joists	A1@.006	SF	.80	.27	1.07
8-1/4" kraft faced, R-30, on suspended ceiling, working off ladders from below	A1@.018	SF	.80	.82	1.62
8-1/4" kraft faced, R-30, in crawl space	A1@.020	SF	.80	.91	1.71
8-1/4" kraft faced, R-30, between studs	A1@.006	SF	.80	.27	1.07
10-1/4" kraft faced, R-38, between ceiling joists	A1@.006	SF	.89	.27	1.16
10-1/4" kraft faced, R-38, on suspended ceiling, working off ladders from below	A1@.018	SF	.89	.82	1.71
10-1/4" kraft faced, R-38, in crawl space	A1@.020	SF	.89	.91	1.80
10-1/4" kraft faced, R-38, between studs	A1@.006	SF	.89	.27	1.16
Add for supporting batts on wire rods, 1 per SF	A1@.001	SF	.03	.05	.08
Add for foil one side	—	SF	.03	—	.03
Add for foil two sides	—	SF	.06	—	.06
Deduct for unfaced one side	—	SF	-.02	—	-.02
Add for Kraft paper two sides	—	SF	.02	—	.02
Fasteners for batts placed under a ceiling deck, including flat washers					
2-1/2" long	A1@.025	Ea	.16	1.13	1.29
4-1/2" long	A1@.025	Ea	.19	1.13	1.32
6-1/2" long	A1@.025	Ea	.26	1.13	1.39
Domed self-locking washers	A1@.005	Ea	.14	.23	.37
Blown fiberglass or mineral wool, over ceiling joists. Add blowing equipment cost at $300 per day					
R-11, 5"	A1@.008	SF	.19	.36	.55
R-13, 6"	A1@.010	SF	.24	.45	.69
R-19, 9"	A1@.014	SF	.32	.64	.96
Expanded polystyrene board installed on perimeter walls					
1", R-4.20, C-0.24	A1@.010	SF	.23	.45	.68
1", R-6.70, C-0.15	A1@.010	SF	.25	.45	.70
1-1/2", R-6.30, C-0.16	A1@.010	SF	.27	.45	.72
1-1/2", R-11.0, C-0.09	A1@.010	SF	.43	.45	.88
2", R-8.30, C-0.12	A1@.011	SF	.39	.50	.89
2", R-14.3, C-0.07	A1@.011	SF	.72	.50	1.22
Perlite or vermiculite, poured in concrete block cores					
Perlite only	—	CF	2.61	—	2.61
4" wall, 8.1 SF per CF	M1@.004	SF	.34	.16	.50
6" wall, 5.4 SF per CF	M1@.007	SF	.51	.28	.79
8" wall, 3.6 SF per CF	M1@.009	SF	.76	.36	1.12
10" wall, 3.0 SF per CF	M1@.011	SF	.91	.45	1.36
12" wall, 2.1 SF per CF	M1@.016	SF	1.31	.65	1.96
Poured perlite or vermiculite in wall cavities	M1@.032	CF	2.75	1.30	4.05
Sound board, 1/2", on walls	C8@.010	SF	.25	.41	.66
Sound board, 1/2", on floors	C8@.009	SF	.26	.37	.63
Cold box insulation					
2" polystyrene	A1@.012	SF	.33	.54	.87
1" cork	A1@.013	SF	.99	.59	1.58
Add for overhead work or poor access areas	—	SF	—	.07	.07
Add for enclosed areas	—	SF	—	.07	.07
Add for 1 hour fire rating	—	SF	.20	—	.20

	Craft@Hrs	Unit	Material	Labor	Total

Aluminum Wall Cladding System, Exterior These costs assume installation by a manufacturer-trained contractor, on a framed and insulated exterior building surface, including track system, panels and trim. Alply, Inc.

Techwall solid aluminum panel, 1/8" thick, joints sealed with silicon sealer

	Craft@Hrs	Unit	Material	Labor	Total
Kynar 500 paint finish	SM@.158	SF	21.30	7.98	29.28
Anodized aluminum finish	SM@.158	SF	21.00	7.98	28.98
Add for radius panels	SM@.032	SF	4.38	1.62	6.00

Cladding, Preformed Roofing and Siding Applied on metal framing.
Architectural insulated metal panels with smooth .040 aluminum exterior face with Kynar 500 finish, 2" thick isocyanurate insulation, .040 mill finish aluminum interior sheet, shop applied extrusion and dry seal gasketed joint system, shop fabricated. Alply, Inc.

	Craft@Hrs	Unit	Material	Labor	Total
Kynar 500 paint finish	SM@.136	SF	15.40	6.87	22.27
Add for radius panels	SM@.020	SF	2.19	1.01	3.20

Corrugated or ribbed roofing, colored galvanized steel, 9/16" deep, uninsulated

	Craft@Hrs	Unit	Material	Labor	Total
18 gauge	SM@.026	SF	2.64	1.31	3.95
20 gauge	SM@.026	SF	2.31	1.31	3.62
22 gauge	SM@.026	SF	2.07	1.31	3.38
24 gauge	SM@.026	SF	1.76	1.31	3.07

Corrugated siding, colored galvanized steel, 9/16" deep, uninsulated

	Craft@Hrs	Unit	Material	Labor	Total
18 gauge	SM@.034	SF	2.64	1.72	4.36
20 gauge	SM@.034	SF	2.31	1.72	4.03
22 gauge	SM@.034	SF	2.07	1.72	3.79

Ribbed siding, colored galvanized steel, 1-3/4" deep, box rib

	Craft@Hrs	Unit	Material	Labor	Total
18 gauge	SM@.034	SF	2.79	1.72	4.51
20 gauge	SM@.034	SF	2.44	1.72	4.16
22 gauge	SM@.034	SF	2.15	1.72	3.87

Corrugated aluminum roofing, natural finish

	Craft@Hrs	Unit	Material	Labor	Total
.020" aluminum	SM@.026	SF	1.71	1.31	3.02
.032" aluminum	SM@.026	SF	2.10	1.31	3.41
Add for factory painted finish	—	SF	.23	—	.23

Corrugated aluminum siding, natural finish

	Craft@Hrs	Unit	Material	Labor	Total
.020" aluminum	SM@.034	SF	1.71	1.72	3.43
.032" aluminum	SM@.034	SF	2.10	1.72	3.82
Add for factory painted finish	—	SF	.23	—	.23

Vinyl siding

	Craft@Hrs	Unit	Material	Labor	Total
Solid vinyl, .042", horizontal siding	SM@.026	SF	.63	1.31	1.94
Add for insulated siding backer panel	SM@.001	SF	.30	.05	.35
Vinyl window and door trim	SM@.040	LF	.35	2.02	2.37
Vinyl siding and fascia system	SM@.032	SF	.85	1.62	2.47

Laminated sandwich panel siding, enameled 22 gauge aluminum face one side, polystyrene core

	Craft@Hrs	Unit	Material	Labor	Total
1" core	SM@.069	SF	4.42	3.49	7.91
1-1/2" core	SM@.069	SF	4.92	3.49	8.41

Membrane Roofing (One square "Sq" = 100 square feet)

Built-up roofing, asphalt felt

	Craft@Hrs	Unit	Material	Labor	Total
3 ply, smooth surface top sheet	R3@1.61	Sq	39.80	67.30	107.10
4 ply, base sheet, 3-ply felt, smooth surface top sheet	R3@1.85	Sq	65.70	77.40	143.10
5 ply, 20 year	R3@2.09	Sq	71.40	87.40	158.80

	Craft@Hrs	Unit	Material	Labor	Total
Built-up fiberglass felt roof					
Base and 3-ply felt	R3@1.93	Sq	84.00	80.70	164.70
Add for each additional ply	R3@.188	Sq	4.04	7.86	11.90
Add for light rock dress off (250 lbs per Sq)	R3@.448	Sq	15.80	18.70	34.50
Add for heavy rock dress off (400 lbs per Sq)	R3@.718	Sq	25.23	30.00	55.23
Remove and replace rock (400 lbs per Sq)					
Including new flood coat	R3@1.32	Sq	29.70	55.20	84.90
Aluminized coating for built-up roofing	R3@.599	Sq	18.70	25.10	43.80
Fire rated type	R3@.609	Sq	23.80	25.50	49.30
Modified APP, SBS flashing membrane	R3@.033	SF	1.12	1.38	2.50
Cap sheet	R3@.277	Sq	30.50	11.60	42.10
Modified asphalt, APP,					
SBS base sheet and membrane	R3@1.97	Sq	113.00	82.40	195.40
Roll roofing					
90 lb mineral surface	R3@.399	Sq	16.70	16.70	33.40
19" wide selvage edge mineral surface (110 lb)	R3@.622	Sq	66.50	26.00	92.50
Double coverage selvage edge roll (140 lb)	R3@.703	Sq	71.20	29.40	100.60
Asphalt impregnated walkway for built-up roofing					
1/2"	R3@.016	SF	2.14	.67	2.81
3/4"	R3@.017	SF	2.71	.71	3.42
1"	R3@.020	SF	3.27	.84	4.11
Strip off existing 4-ply roof, no disposal included	R3@1.32	Sq	—	55.20	55.20
Elastomeric Roofing					
Butyl					
1/16"	R3@1.93	Sq	141.00	80.70	221.70
1/32"	R3@1.93	Sq	122.00	80.70	202.70
Neoprene, 1/16"	R3@2.23	Sq	211.00	93.30	304.30
Acrylic-urethane foam roof system					
Remove gravel and prepare existing built-up roof	R3@1.26	Sq	5.42	52.70	58.12
Apply base and top coats of acrylic (.030")	R3@.750	Sq	117.00	31.40	148.40
Urethane foam 1" thick (R-7.1)	R3@.280	Sq	104.00	11.70	115.70
Urethane foam 2" thick (R-14.3)	R3@1.10	Sq	175.00	46.00	221.00
Spray-on mineral granules, 40 lbs per CSF	R3@.600	Sq	16.90	25.10	42.00
EPDM roofing system (Ethylene Propylene Diene Monomer)					
.045" loose laid membrane	R3@.743	Sq	96.70	31.10	127.80
.060" adhered membrane (excluding adhesive)	R3@1.05	Sq	110.00	43.90	153.90
Bonding adhesive	R3@.305	Sq	44.90	12.80	57.70
Lap splice cement, per 100 LF	R3@.223	CLF	26.50	9.33	35.83
Ballast rock, 3/4" to 1-1/2" gravel	R3@.520	Sq	18.20	21.70	39.90
Neoprene sheet flashing, .060"	R3@.021	SF	2.40	.88	3.28
CSPE (Chloro-sulfonated Polyethylene)					
.045" single ply membrane	R3@.743	Sq	153.00	31.10	184.10
Add for mechanical fasteners and washers					
6" long, 12" OC at laps	—	Ea	.69	—	.69
12" long, 12" OC at laps	—	Ea	1.41	—	1.41

	Craft@Hrs	Unit	Material	Labor	Total

Cementitious Roofing System Tectum™ roof deck systems for use on open beam ceilings with exposed joists. Provides a nailable top surface, a core of Styrofoam brand insulation and an exposed interior Tectum™ ceiling plank substrate. Tectum™ roof deck is highly moisture resistant and possesses high compressive strength, excellent sound absorption, insulative properties and a finished interior. Costs shown include nails, screws, and adhesive. Includes normal cutting waste.

Tectum I with T&G edges and square ends. Standard roof deck with foam insulation. Available in lengths up to 14' and widths to 3'

	Craft@Hrs	Unit	Material	Labor	Total
1-1/2" thick	C8@.015	SF	1.66	.61	2.27
2" thick	C8@.017	SF	1.94	.70	2.64
2-1/2" thick	C8@.019	SF	2.38	.78	3.16
3" thick	C8@.021	SF	2.88	.86	3.74

Tectum III with T&G edges and square ends. Composite roof deck used where greater insulative properties are required. Tectum III provides a finished interior surface, and foam insulation. Width is 47" and lengths are available from 4" to 16"

	Craft@Hrs	Unit	Material	Labor	Total
3-1/2" thick	C8@.015	SF	3.66	.61	4.27
4" thick	C8@.021	SF	4.14	.86	5.00
5" thick	C8@.027	SF	4.50	1.11	5.61
6" thick	C8@.033	SF	5.18	1.35	6.53
7" thick	C8@.039	SF	5.94	1.60	7.54
8" thick	C8@.045	SF	6.63	1.84	8.47
9" thick	C8@.051	SF	7.31	2.09	9.40
10" thick	C8@.057	SF	8.87	2.34	11.21
11" thick	C8@.063	SF	9.69	2.58	12.27

Cant Strips

	Craft@Hrs	Unit	Material	Labor	Total
3" fiber	R3@.017	LF	.14	.71	.85
6" fiber	R3@.017	LF	.28	.71	.99
4" wood	R3@.017	LF	.70	.71	1.41
6" wood	R3@.017	LF	.70	.71	1.41

Flashing and Sheet Metal

Prefinished sheet metal fascia and mansards, standing beam and batten

	Craft@Hrs	Unit	Material	Labor	Total
Straight or simple	SM@.054	SF	3.77	2.73	6.50
Curved or complex	SM@.082	SF	5.69	4.14	9.83
Coping, gravel stop and flashing	SM@.037	SF	2.75	1.87	4.62
Sheet metal wainscot, galvanized	SM@.017	SF	.76	.86	1.62

Aluminum flashing .032",

	Craft@Hrs	Unit	Material	Labor	Total
Coping and wall cap, 16" girth	SM@.069	LF	3.13	3.49	6.62
Counter flash, 8"	SM@.054	LF	2.75	2.73	5.48
Reglet flashing, 8"	SM@.058	LF	1.13	2.93	4.06
Neoprene gasket for flashing	SM@.015	LF	.86	.76	1.62
Gravel stop and fascia, 10"	SM@.084	LF	3.28	4.24	7.52
Side wall flashing, 9"	SM@.027	LF	1.71	1.36	3.07

Gravel stop

	Craft@Hrs	Unit	Material	Labor	Total
4"	SM@.058	LF	3.72	2.93	6.65
6"	SM@.063	LF	4.21	3.18	7.39
8"	SM@.068	LF	5.10	3.44	8.54
12"	SM@.075	LF	8.47	3.79	12.26
Valley, 24"	SM@.070	LF	2.19	3.54	5.73

Copper flashing 16 oz,

	Craft@Hrs	Unit	Material	Labor	Total
Coping and wall cap, 16" girth	SM@.074	LF	8.19	3.74	11.93
Counter flash, 8"	SM@.057	LF	5.09	2.88	7.97

	Craft@Hrs	Unit	Material	Labor	Total
Gravel stop					
4"	SM@.058	LF	6.35	2.93	9.28
6"	SM@.063	LF	7.38	3.18	10.56
8"	SM@.068	LF	8.64	3.44	12.08
10"	SM@.075	LF	10.60	3.79	14.39
Reglet flashing, 8"	SM@.054	LF	3.07	2.73	5.80
Neoprene gasket for flashing	SM@.015	LF	.94	.76	1.70
Side wall flashing, 6"	SM@.027	LF	2.79	1.36	4.15
Base, 20 oz	SM@.072	LF	5.76	3.64	9.40
Valley, 24"	SM@.069	LF	5.15	3.49	8.64
Sheet lead flashing					
Plumbing vents, soil stacks, 4 lbs per square foot					
4" pipe size	P6@.709	Ea	20.60	31.40	52.00
6" pipe size	P6@.709	Ea	23.90	31.40	55.30
8" pipe size	P6@.709	Ea	29.40	31.40	60.80
Roof drains, flat pan type, 36" x 36"					
2.5 lbs/SF material	P6@1.49	Ea	25.20	66.00	91.20
4 lbs/SF material	P6@1.49	Ea	42.40	66.00	108.40
6 lbs/SF material	P6@1.49	Ea	60.40	66.00	126.40
Stainless steel flashing					
Fascia roof edge, .018"	SM@.054	LF	4.32	2.73	7.05
Base, .018"	SM@.054	LF	5.18	2.73	7.91
Counter flash, .015"	SM@.054	LF	4.08	2.73	6.81
Valley, .015", 24"	SM@.069	LF	4.06	3.49	7.55
Reglets, .020", 8"	SM@.035	LF	3.53	1.77	5.30
Galvanized sheet metal flashing					
Cap and counter flash, 8"	SM@.046	LF	1.25	2.32	3.57
Coping and wall cap, 16" girth	SM@.069	LF	1.62	3.49	5.11
Gravel stop, 6"	SM@.060	LF	.49	3.03	3.52
Neoprene gasket	SM@.005	LF	.71	.25	.96
Pitch pockets, 24 gauge, filled					
4" x 4"	SM@.584	Ea	18.60	29.50	48.10
6" x 6"	SM@.584	Ea	23.50	29.50	53.00
8" x 8"	SM@.584	Ea	27.70	29.50	57.20
8" x 10"	SM@.777	Ea	43.80	39.30	83.10
8" x 12"	SM@.777	Ea	38.60	39.30	77.90
Reglet flashing, 8"	SM@.054	LF	1.20	2.73	3.93
Shingles (chimney flash)	RF@.064	LF	.95	2.85	3.80
Side wall flashing, 9"	SM@.027	LF	.92	1.36	2.28
"W" valley, 24"	SM@.069	LF	1.84	3.49	5.33
Plumbers counter flash cone, 4" diameter	P8@.204	Ea	5.84	8.16	14.00
Roof safes and caps					
4" galvanized sheet metal	SM@.277	Ea	11.80	14.00	25.80
4" aluminum	SM@.277	Ea	13.50	14.00	27.50
Pitch pockets, 16 oz copper, filled					
4" x 4"	SM@.547	Ea	39.90	27.60	67.50
6" x 6"	SM@.547	Ea	50.80	27.60	78.40
8" x 8"	SM@.547	Ea	59.50	27.60	87.10
8" x 10"	SM@.679	Ea	69.90	34.30	104.20
8" x 12"	SM@.679	Ea	82.20	34.30	116.50
Galvanized corner guards, 4" x 4"	SM@.039	LF	12.00	1.97	13.97

	Craft@Hrs	Unit	Material	Labor	Total
Copper sheet metal roofing, 16 oz					
Batten seam	SM@.065	SF	4.30	3.28	7.58
Standing seam	SM@.056	SF	4.99	2.83	7.82
Gutters and Downspouts					
Aluminum, .032", including hangers					
Fascia gutter, 5"	SM@.050	LF	1.18	2.53	3.71
Box gutter, 4"	SM@.050	LF	1.09	2.53	3.62
Dropouts, elbows, for either of above	SM@.101	Ea	3.92	5.10	9.02
Downspouts, to 24' height					
2" x 3"	SM@.037	LF	.98	1.87	2.85
3" x 4"	SM@.047	LF	1.18	2.37	3.55
4" diameter, round, 12' to 24' high	SM@.047	LF	1.51	2.37	3.88
Add for height over 24'	SM@.017	LF	—	.86	.86
Copper, 16 oz, including hangers					
Box gutter, 4"	SM@.054	LF	5.10	2.73	7.83
Half round gutter					
4" gutter	SM@.039	LF	4.41	1.97	6.38
5" gutter	SM@.051	LF	5.53	2.58	8.11
6" gutter	SM@.056	LF	6.84	2.83	9.67
Dropouts, elbows,					
for either of above	SM@.154	Ea	11.10	7.78	18.88
Downspouts, to 24' height					
2" x 3"	SM@.036	LF	4.86	1.82	6.68
3" x 4"	SM@.047	LF	6.59	2.37	8.96
3" diameter	SM@.037	LF	3.25	1.87	5.12
4" diameter	SM@.047	LF	4.95	2.37	7.32
5" diameter	SM@.049	LF	5.99	2.48	8.47
Add for heights over 24'	SM@.017	LF	—	.86	.86
Scuppers, copper, 16 oz,					
8" x 8"	SM@.972	Ea	126.00	49.10	175.10
10" x 10"	SM@.972	Ea	149.00	49.10	198.10
Galvanized steel, 26 gauge, including hangers					
Box gutter 4"	SM@.054	LF	1.35	2.73	4.08
Dropouts, elbows	SM@.082	Ea	2.07	4.14	6.21
Fascia gutter					
5" face	SM@.054	LF	1.60	2.73	4.33
7" face	SM@.062	LF	2.14	3.13	5.27
Dropouts, elbows, for either of above	SM@.088	Ea	3.15	4.45	7.60
Downspouts, with all fittings, height to 24'					
2" x 3"	SM@.036	LF	1.14	1.82	2.96
3" x 4"	SM@.047	LF	1.73	2.37	4.10
3" round	SM@.037	LF	1.73	1.87	3.60
6" round	SM@.055	LF	1.62	2.78	4.40
Add for heights over 24'	SM@.017	LF	—	.86	.86
Roof sump, 18" x 18" x 5"	SM@.779	Ea	8.79	39.40	48.19
Scupper, 6" x 6" x 8"	SM@.591	Ea	26.20	29.90	56.10
Stainless steel, .015" thick, including hangers					
Box gutter, 5" wide, 4-1/2" high	SM@.054	LF	8.35	2.73	11.08
Dropouts, elbows	SM@.076	Ea	12.30	3.84	16.14
Downspout, 4" x 5", to 24' high	SM@.051	LF	7.48	2.58	10.06
Add for heights over 24'	SM@.017	LF	—	.86	.86

	Craft@Hrs	Unit	Material	Labor	Total
Vents, Louvers and Screens					
Fixed louvers, with screen, typical	SM@.402	SF	11.30	20.30	31.60
Door louvers, typical	SM@.875	Ea	24.40	44.20	68.60
Manual operating louvers, typical	SM@.187	SF	18.00	9.45	27.45
Cooling tower screens	SM@.168	SF	10.00	8.49	18.49
Bird screens with frame	SM@.028	SF	1.51	1.41	2.92
Concrete block vents, aluminum	M1@.370	SF	13.40	15.00	28.40
Frieze vents, with screen, 14" x 4"	C8@.198	Ea	2.77	8.11	10.88
Foundation vents, 6" x 14", ornamental	C8@.317	Ea	7.49	13.00	20.49
Attic vents, with louvers, 14" x 24"	C8@.377	Ea	21.80	15.40	37.20
Architectural facade screen, aluminum	SM@.187	SF	20.90	9.45	30.35
Add for enamel or light anodized	—	SF	2.23	—	2.23
Add for porcelain or heavy anodized	—	SF	5.40	—	5.40

Rain Dispersing Gutters Rustproof aluminum, installed on mounting brackets. Disperses rain into a 4 foot area. No downspouts needed. Resist clogging by leaves and debris. Rainhandler

	Craft@Hrs	Unit	Material	Labor	Total
Rain disperser	RF@.100	LF	4.64	4.45	9.09
Mounting brackets (use one each 18")	—	Ea	1.04	—	1.04
Horizontal adapter for roof edges that extend outward more than 2-1/2" from building (use one each 18")	RF@.020	Ea	2.07	.89	2.96
Vertical adapter for vertical fascias, under 3" from building (use one each 5')	RF@.020	Ea	2.07	.89	2.96
"Doorbrella" for installation above doors and windows	RF@.250	Ea	15.90	11.10	27.00
Drip edge extension for roof edges that extend less than 1-1/2" from vertical fascia (use one each 18")	RF@.020	Ea	4.19	.89	5.08
Roof valley rain diverter	RF@.050	Ea	6.31	2.23	8.54
Fascia adapter for angled fascia boards	RF@.050	Ea	2.07	2.23	4.30

	Craft@Hrs	Unit	Material	Labor	Total
Roof Accessories					
Roof hatches, steel, not including ladder					
30" x 30"	C8@3.09	Ea	429.00	127.00	556.00
30" x 72"	C8@4.69	Ea	1,030.00	192.00	1,222.00
36" x 36"	C8@3.09	Ea	490.00	127.00	617.00
36" x 72"	C8@4.69	Ea	1,090.00	192.00	1,282.00
Ceiling access hatches					
30" x 30"	C8@1.59	Ea	259.00	65.10	324.10
42" x 42"	C8@1.89	Ea	309.00	77.40	386.40
Smoke vents					
Aluminum, 48" x 48"	C8@3.09	Ea	1,230.00	127.00	1,357.00
Galvanized, 48" x 48"	C8@3.09	Ea	1,070.00	127.00	1,197.00
Fusible "shrink-out" heat and smoke vent (PVC dome in aluminum frame)					
4' x 8'	C8@5.15	Ea	973.00	211.00	1,184.00
Roof scuttle, aluminum, 2'6" x 3'0"	C8@3.50	Ea	483.00	143.00	626.00

	Craft@Hrs	Unit	Material	Labor	Total
Ventilators					
Rotary, wind driven					
6" diameter	SM@.913	Ea	63.20	46.10	109.30
12" diameter	SM@.913	Ea	77.10	46.10	123.20
24" diameter	SM@1.36	Ea	219.00	68.70	287.70

	Craft@Hrs	Unit	Material	Labor	Total
Ventilators, mushroom type, motorized, single speed, including damper and bird screen but no electrical work					
8", 180 CFM	SM@2.93	Ea	307.00	148.00	455.00
12", 1,360 CFM	SM@2.93	Ea	509.00	148.00	657.00
18", 2,000 CFM	SM@4.00	Ea	731.00	202.00	933.00
24", 4,000 CFM	SM@4.86	Ea	1,120.00	246.00	1,366.00
Add for two-speed motor	—	%	35.0	—	—
Add for explosive proof units	—	Ea	421.00	—	421.00

Sealants and Caulking By joint size. Costs assume 10% waste. Twelve 10.6 ounce cartridges equal one gallon and yield 231 cubic inches of caulk or sealant. Productivity shown assumes one roofer using a bulk dispenser. Add the cost of joint cleaning, if required.

	Craft@Hrs	Unit	Material	Labor	Total
Silicone-latex caulk					
1/4" x 1/4", 351 LF/gallon, 63 LF per hour	RF@.016	LF	.10	.71	.81
1/4" x 3/8", 234 LF/gallon, 60 LF per hour	RF@.017	LF	.15	.76	.91
1/4" x 1/2", 176 LF/gallon, 58 LF per hour	RF@.017	LF	.20	.76	.96
3/8" x 3/8", 156 LF/gallon, 58 LF per hour	RF@.017	LF	.23	.76	.99
3/8" x 1/2", 117 LF/gallon, 57 LF per hour	RF@.017	LF	.30	.76	1.06
3/8" x 5/8", 94 LF/gallon, 56 LF per hour	RF@.018	LF	.37	.80	1.17
3/8" x 3/4", 76 LF/gallon, 52 LF per hour	RF@.019	LF	.49	.85	1.34
1/2" x 1/2", 87 LF/gallon, 56 LF per hour	RF@.018	LF	.40	.80	1.20
1/2" x 5/8", 70 LF/gallon, 50 LF per hour	RF@.020	LF	.50	.89	1.39
1/2" x 3/4", 58 LF/gallon, 48 LF per hour	RF@.021	LF	.61	.94	1.55
1/2" x 7/8", 51 LF/gallon, 46 LF per hour	RF@.022	LF	.69	.98	1.67
1/2" x 1", 44 LF/gallon, 45 LF per hour	RF@.022	LF	.80	.98	1.78
3/4" x 3/4", 39 LF/gallon, 44 LF per hour	RF@.023	LF	.90	1.02	1.92
1" x 1", 22 LF/gallon, 30 LF per hour	RF@.033	LF	1.60	1.47	3.07
Acrylic latex plus silicone caulk					
1/4" x 1/4", 351 LF/gallon, 63 LF per hour	RF@.016	LF	.07	.71	.78
1/4" x 3/8", 234 LF/gallon, 60 LF per hour	RF@.017	LF	.11	.76	.87
1/4" x 1/2", 176 LF/gallon, 58 LF per hour	RF@.017	LF	.15	.76	.91
3/8" x 3/8", 156 LF/gallon, 58 LF per hour	RF@.017	LF	.17	.76	.93
3/8" x 1/2", 117 LF/gallon, 57 LF per hour	RF@.017	LF	.22	.76	.98
3/8" x 5/8", 94 LF/gallon, 56 LF per hour	RF@.018	LF	.27	.80	1.07
3/8" x 3/4", 76 LF/gallon, 52 LF per hour	RF@.019	LF	.34	.85	1.19
1/2" x 1/2", 87 LF/gallon, 56 LF per hour	RF@.018	LF	.30	.80	1.10
1/2" x 5/8", 70 LF/gallon, 50 LF per hour	RF@.020	LF	.37	.89	1.26
1/2" x 3/4", 58 LF/gallon, 48 LF per hour	RF@.021	LF	.44	.94	1.38
1/2" x 7/8", 51 LF/gallon, 46 LF per hour	RF@.022	LF	.51	.98	1.49
1/2" x 1", 44 LF/gallon, 45 LF per hour	RF@.022	LF	.59	.98	1.57
3/4" x 3/4", 39 LF/gallon, 44 LF per hour	RF@.023	LF	.66	1.02	1.68
1" x 1", 22 LF/gallon, 30 LF per hour	RF@.033	LF	1.17	1.47	2.64
100 percent silicone caulk					
1/4" x 1/4", 351 LF/gallon, 63 LF per hour	RF@.016	LF	.15	.71	.86
1/4" x 3/8", 234 LF/gallon, 60 LF per hour	RF@.017	LF	.22	.76	.98
1/4" x 1/2", 176 LF/gallon, 58 LF per hour	RF@.017	LF	.29	.76	1.05
3/8" x 3/8", 156 LF/gallon, 58 LF per hour	RF@.017	LF	.33	.76	1.09
3/8" x 1/2", 117 LF/gallon, 57 LF per hour	RF@.017	LF	.44	.76	1.20
3/8" x 5/8", 94 LF/gallon, 56 LF per hour	RF@.018	LF	.55	.80	1.35
3/8" x 3/4", 76 LF/gallon, 52 LF per hour	RF@.019	LF	.67	.85	1.52
1/2" x 1/2", 87 LF/gallon, 56 LF per hour	RF@.018	LF	.59	.80	1.39
1/2" x 5/8", 70 LF/gallon, 50 LF per hour	RF@.020	LF	.73	.89	1.62
1/2" x 3/4", 58 LF/gallon, 48 LF per hour	RF@.021	LF	.88	.94	1.82

	Craft@Hrs	Unit	Material	Labor	Total
1/2" x 7/8", 51 LF/gallon, 46 LF per hour	RF@.022	LF	1.01	.98	1.99
1/2" x 1", 44 LF/gallon, 45 LF per hour	RF@.022	LF	1.16	.98	2.14
3/4" x 3/4", 39 LF/gallon, 44 LF per hour	RF@.023	LF	1.31	1.02	2.33
1" x 1", 22 LF/gallon, 30 LF per hour	RF@.033	LF	2.33	1.47	3.80
Polyurethane sealant					
1/4" x 1/4", 351 LF/gallon, 63 LF per hour	RF@.016	LF	.16	.71	.87
1/4" x 3/8", 234 LF/gallon, 60 LF per hour	RF@.017	LF	.25	.76	1.01
1/4" x 1/2", 176 LF/gallon, 58 LF per hour	RF@.017	LF	.33	.76	1.09
3/8" x 3/8", 156 LF/gallon, 58 LF per hour	RF@.017	LF	.37	.76	1.13
3/8" x 1/2", 117 LF/gallon, 57 LF per hour	RF@.017	LF	.49	.76	1.25
3/8" x 5/8", 94 LF/gallon, 56 LF per hour	RF@.018	LF	.61	.80	1.41
3/8" x 3/4", 76 LF/gallon, 52 LF per hour	RF@.019	LF	.76	.85	1.61
1/2" x 1/2", 87 LF/gallon, 56 LF per hour	RF@.018	LF	.66	.80	1.46
1/2" x 5/8", 70 LF/gallon, 50 LF per hour	RF@.020	LF	.82	.89	1.71
1/2" x 3/4", 58 LF/gallon, 48 LF per hour	RF@.021	LF	.99	.94	1.93
1/2" x 7/8", 51 LF/gallon, 46 LF per hour	RF@.022	LF	1.13	.98	2.11
1/2" x 1", 44 LF/gallon, 45 LF per hour	RF@.022	LF	1.31	.98	2.29
3/4" x 3/4", 39 LF/gallon, 44 LF per hour	RF@.023	LF	1.48	1.02	2.50
1" x 1", 22 LF/gallon, 30 LF per hour	RF@.033	LF	2.62	1.47	4.09
Backing rods for sealant					
Closed cell, 1/4" rod	RF@.010	LF	.07	.45	.52
Closed cell, 3/8" rod	RF@.011	LF	.15	.49	.64
Closed cell, 1/2" rod	RF@.011	LF	.15	.49	.64
Closed cell, 5/8" rod	RF@.012	LF	.23	.53	.76
Closed cell, 7/8" rod	RF@.012	LF	.36	.53	.89
Closed cell, 1" rod	RF@.014	LF	.41	.62	1.03
Closed cell, 1-1/4" rod	RF@.014	LF	.74	.62	1.36
Closed cell, 1-1/2" rod	RF@.015	LF	1.10	.67	1.77
Closed cell, 2" rod	RF@.015	LF	1.67	.67	2.34

Expansion Joints, Preformed

Wall and ceiling, aluminum cover					
Drywall or panel type	CC@.067	LF	14.40	3.05	17.45
Plaster type	CC@.067	LF	14.40	3.05	17.45
Floor to wall, 3" x 3"	CC@.038	LF	7.90	1.73	9.63
Floor, 3"	CC@.038	LF	9.86	1.73	11.59
Neoprene joint with aluminum cover	CC@.079	LF	21.00	3.60	24.60
Wall to roof joint	CC@.094	LF	8.70	4.28	12.98
Bellows type expansion joints, butyl with neoprene backer					
16 oz copper, 4" wide	CC@.046	LF	13.30	2.10	15.40
28 gauge stainless steel, 4" wide	CC@.046	LF	13.10	2.10	15.20
26 gauge galvanized sheet metal, 4" wide	CC@.046	LF	7.75	2.10	9.85
16 oz copper, 6" wide	CC@.046	LF	15.30	2.10	17.40
28 gauge stainless steel, 6" wide	CC@.046	LF	15.10	2.10	17.20
26 gauge galvanized sheet metal, 6" wide	CC@.046	LF	9.36	2.10	11.46
Neoprene gaskets, closed cell					
1/8" x 2"	CC@.019	LF	.68	.87	1.55
1/8" x 6"	CC@.023	LF	1.59	1.05	2.64
1/4" x 2"	CC@.022	LF	.78	1.00	1.78
1/4" x 6"	CC@.024	LF	1.68	1.09	2.77
1/2" x 6"	CC@.025	LF	2.51	1.14	3.65
1/2" x 8"	CC@.028	LF	3.68	1.28	4.96

	Craft@Hrs	Unit	Material	Labor	Total
Acoustical caulking, drywall or plaster type	CC@.038	LF	.25	1.73	1.98
Polyisobutylene tapes, non-drying					
Polybutene	CC@.028	LF	.24	1.28	1.52
Polyisobutylene/butyl, preformed	CC@.020	LF	.12	.91	1.03

Expansion Joint Covers
Surface type, aluminum

	Craft@Hrs	Unit	Material	Labor	Total
Floor, 1-1/2"	CC@.194	LF	23.50	8.84	32.34
Wall and ceiling, 1-1/2"	CC@.185	LF	13.30	8.43	21.73
Gymnasium base	CC@.099	LF	10.70	4.51	15.21
Roof, typical	CC@.145	LF	22.50	6.60	29.10

Cast-in-Place Concrete Fireproofing. Costs per linear foot (LF) of steel beam, column or girder. Use these figures to estimate the cubic feet (CF) of concrete per linear foot and square feet (SF) of form required per linear foot for each type of beam, column or girder. Quantities include concrete required to fill the void between the web and flange and provide 2" protection at the flanges. A small change in concrete thickness at flanges has very little effect on cost. Use $7,500.00 as a minimum job cost for work of this type. Concrete thickness for fireproofing will usually be:

   Members at least 6" x 6" but less than 8" x 8", 3" for a 4 hour rating, 2" for a 3 hour rating, 1-1/2" for a 2 hour rating, and 1" for a 1 hour rating.

   Members at least 8" x 8" but less than 12" x 12", 2-1/2" for a 4 hour rating, 2" for a 3 hour rating, 1" for a 1 or 2 hour rating.

   Members 12" x 12" or greater, 2" for a 4 hour rating and 1" for a 1, 2, or 3 hour rating.

Labor costs include the time needed to prepare formwork sketches at the job site, measure for the forms, fabricate, erect, align and brace the forms, cut, bend, place and tie the reinforcing steel, install embedded steel items, place and finish the concrete and strip, clean and stack the forms. Labor costs assume a concrete forming, placing and finishing crew of 1 carpenter, 1 laborer and 1 cement finisher. Concrete placing equipment assumes the use of a concrete pump.

For scheduling purposes, estimate that a crew of 3 will fireproof the following quantities of steel in an 8-hour day:
   15 LF of W21 through W36 members,
   25 LF of W8 through W18 members, 22 LF of S12 through S24 members, and 42 LF of M4 through W6 or S6 through S8 members.

Material costs assume reinforcing steel is #3 and #4 bars at 12" on center each way and includes waste and laps. Form cost assumes 3 uses without completely disassembling the form. Normal cleaning and repairs are included. No salvage value is assumed. Cost for forms, before allowance for waste, is based on Standard & Better lumber and 3/4" plyform.

	Craft@Hrs	Unit	Material	Labor	Equipment	Total
Concrete and accessories for fireproofing						
Concrete, 3,000 PSI pump mix with 5% allowance for waste						
including pump cost	P9@.600	CY	116.00	25.10	9.15	150.25
Grade A60 reinforcing bars, set and tied,						
with 50 lbs per CY of concrete	P9@.790	CY	18.50	33.10	—	51.60
Hangers, snap-ties, misc. embedded steel						
with 10 lb per CY of concrete	P9@.660	CY	42.50	27.60	—	70.10
Concrete test cylinders including test reports,						
with 5 per 100 CY	—	CY	.85	—	—	.85
Total for concrete and accessories	P9@2.05	CY	177.85	85.80	9.15	272.80

	Craft@Hrs	Unit	Material	Labor	Equipment	Total

Forms for fireproofing. These costs assume the following materials are used per square foot of contact area (SFCA): Nails, clamps and form oil costing $.12 per board foot of lumber, 1.2 SF of 3/4" plyform and 3.7 BF of lumber per SFCA.

Make, erect, align and strip forms,

	Craft@Hrs	Unit	Material	Labor	Equipment	Total
3 uses	P9@.150	SF	.75	6.28	—	7.03
Sack and patch concrete SF	P9@.010	SF	.12	.42	—	.54
Total for forms and finishing, per use	**P9@.160**	**SF**	**.87**	**6.70**	**—**	**7.57**

Sample fireproofing estimate for ten W24 x 55 beams 20' long. Note from the table below that this beam requires 2.7 cubic feet of concrete per linear foot and 6.9 square feet of form per linear foot. Using labor and material prices from above to estimate 200 linear feet of beam. (Total costs have been rounded)

	Craft@Hrs	Unit	Material	Labor	Equipment	Total
Concrete (200 x 2.7 = 540 CF or 20 CY)	P9@41.0	LS	3,613.00	1,720.00	305.00	5,638.00
Form and finish (200 x 6.9 = 1,380 SF)	P9@221	LS	966.00	9,250.00	—	10,216.00
Total job cost shown above	**--@262**	**LS**	**4,579.00**	**10,970.00**	**305.00**	**15,854.00**
Cost per linear foot of beam	**--@1.31**	**LF**	**22.90**	**54.85**	**1.53**	**79.28**

Cast-in-Place Concrete Fireproofing Cost per LF, including concrete and forming.

	Craft@Hrs	Unit	Material	Labor	Equipment	Total
W36 x 135 to W36 x 300						
5.13 CF and 10.07 SF per LF	P9@2.00	LF	41.40	83.70	1.74	126.84
W33 x 118 to W30 x 241						
4.59 CF and 9.35 SF per LF	P9@1.84	LF	37.30	77.00	1.56	115.86
W30 x 99 to W30 x 211						
3.78 CF and 8.4 SF per LF	P9@1.63	LF	31.20	68.20	1.28	100.68
W27 x 84 to W27 x 178						
3.51 CF and 8.19 SF per LF	P9@1.57	LF	29.20	65.70	1.19	96.09
W24 x 55 to W24 x 162						
2.7 CF and 6.9 SF per LF	P9@1.31	LF	22.90	54.80	.92	78.62
W21 x 44 to W21 x 147						
2.43 CF and 6.75 SF per LF	P9@1.26	LF	21.00	52.70	.82	74.52
W18 x 37 to W18 x 71						
1.62 CF and 5.34 SF per LF	P9@.974	LF	14.60	40.80	.55	55.95
W16 x 26 to W16 x 100						
1.62 CF and 5.4 SF per LF	P9@.987	LF	14.60	41.30	.55	56.45
W14 x 90 to W14 x 730						
2.97 CF and 7.15 SF per LF	P9@1.37	LF	24.90	57.30	1.01	83.21
W14 x 22 to W14 x 82						
1.35 CF and 4.7 SF per LF	P9@.854	LF	12.30	35.70	.46	48.46
W12 x 65 to W12 x 336						
2.16 CF and 6.08 SF per LF	P9@1.13	LF	18.70	47.30	.73	66.73
W12 x 40 to W12 x 58						
1.62 CF and 5.4 SF per LF	P9@.987	LF	14.60	41.30	.55	56.45
W12 x 14 to W12 x 35						
1.08 CF and 4.44 SF per LF	P9@.790	LF	10.30	33.10	.37	43.77
W10 x 33 to W10 x 112						
1.35 CF and 4.75 SF per LF	P9@.862	LF	12.40	36.10	.46	48.96
W10 x 12 to W10 x 30						
1.08 CF and 4.56 SF per LF	P9@.808	LF	10.40	33.80	.37	44.57
W8 x 24 to W8 x 67						
1.08 CF and 4.4 SF per LF	P9@.784	LF	10.30	32.80	.37	43.47
W8 x 10 to W8 x 21						
0.81 CF and 3.87 SF per LF	P9@.678	LF	8.13	28.40	.27	36.80

	Craft@Hrs	Unit	Material	Labor	Equipment	Total
W6 x 15 to W6 x 25						
0.81 CF and 3.84 SF per LF	P9@.672	LF	8.11	28.10	.27	36.48
W6 x 9 to W6 x 16						
0.54 CF and 2.92 SF per LF	P9@.507	LF	5.66	21.20	.18	27.04
W5 x 16 to W5 x 19						
0.54 CF and 2.86 SF per LF	P9@.499	LF	5.62	20.90	.18	26.70
W4 x 13						
0.54 CF and 3.2 SF per LF	P9@.552	LF	5.85	23.10	.18	29.13
M4 x 13 to M14 x 18						
0.54 CF and 2.9 SF per LF	P9@.504	LF	5.64	21.10	.18	26.92
S24 x 80 to S24 x 121						
2.43 CF and 7.11 SF per LF	P9@1.32	LF	21.20	55.20	.82	77.22
S20 x 66 to S20 x 96						
1.89 CF and 6.09 SF per LF	P9@1.12	LF	16.90	46.90	.64	64.44
S15 x 43 to S18 x 70						
1.62 CF and 5.82 SF per LF	P9@1.05	LF	14.90	43.90	.55	59.35
S12 x 35 to S12 x 50						
1.08 CF and 4.44 SF per LF	P9@.790	LF	10.30	33.10	.37	43.77
S8 x 23 to S10 x 35						
0.81 CF and 3.81 SF per LF	P9@.670	LF	8.09	28.00	.27	36.36
S6 x 13 to S7 x 20						
0.54 CF and 2.91 SF per LF	P9@.504	LF	5.65	21.10	.18	26.93

Spray-Applied Fireproofing, Subcontract Fire endurance coating made from inorganic vermiculite and portland cement. Costs assume a 10,000 board foot job. (One BF is one square foot covered 1" thick.) For smaller jobs, increase the cost by 5% for each 500 BF less than 10,000 BF. Use $2,000 as a minimum subcontract price. For thickness other than 1", adjust these costs proportionately. For scheduling purposes, estimate that a crew of 2 plasterers and 1 helper can apply 200 to 250 board feet per hour.

	Craft@Hrs	Unit	Material	Labor	Total
Structural steel columns	—	BF	—	—	2.43
Structural steel beams	—	BF	—	—	2.17
Purlins, girts, and miscellaneous members	—	BF	—	—	2.00
Decks, ceilings or walls	—	BF	—	—	1.84
Add for 18 gauge 2" hex mesh reinforcing	—	BF	—	—	.51
Add for key coat bonded on primed surfaces	—	SF	—	—	.49

Rule-of-thumb method for estimating spray-applied fireproofing on bare structural steel by member size:

Beams and Columns	Beams	Cost LF	Columns	Cost LF
W36 x 135 to 300	10 BF/LF	21.70	11 BF/LF	26.70
W33 x 118 to 241	9 BF/LF	19.50	10 BF/LF	24.30
W30 x 99 to 211	8 BF/LF	17.40	9 BF/LF	21.90
W27 x 84 to 178	7 BF/LF	15.20	8 BF/LF	19.40
W24 x 55 to 182	6.5 BF/LF	14.10	7.5 BF/LF	18.20
W21 x 44 to 147	6 BF/LF	13.00	7 BF/LF	17.00
W18 x 35 to 118	5 BF/LF	10.90	6 BF/LF	14.60
W16 x 28 to 57	3.5 BF/LF	7.60	4.5 BF/LF	10.90
W14 x 61 to 132	5.5 BF/LF	11.90	6.5 BF/LF	15.80
W14 x 22 to 63	4 BF/LF	8.68	5 BF/LF	12.20
W12 x 65 to 190	5 BF/LF	10.90	6 BF/LF	14.60
W12 x 40 to 58	4 BF/LF	8.68	5 BF/LF	12.20

Beams and Columns

Beams and Columns	Beams	Cost LF	Columns	Cost LF
W12 x 14 to 35	3 BF/LF	6.51	4 BF/LF	9.72
W10 x 49 to 112	4 BF/LF	8.68	5 BF/LF	12.20
W10 x 22 to 45	3 BF/LF	6.51	4 BF/LF	9.72
W10 x 12 to 19	2 BF/LF	4.34	3 BF/LF	7.29
W8 x 24 to 67	3 BF/LF	6.51	4 BF/LF	9.72
W6 x 9 to 25	2 BF/LF	4.34	3 BF/LF	7.29
W5 x 16 to 19	2 BF/LF	4.34	2.5 BF/LF	6.08
W4 x 13	1 BF/LF	2.17	2 BF/LF	4.86

Purlins and Girts

Purlins and Girts	Purlins or Girts	Cost LF
MC 18 members	4 BF/LF	8.00
MC 10 to MC 13 members	3 BF/LF	6.00
MC 8 members	2 BF/LF	4.00
C 10 to C 15 members	3 BF/LF	6.00
C 7 to C 9 members	2 BF/LF	4.00

Flexible Strip Doors Suitable for interior or exterior use. High strength USDA-approved clear transparent PVC plastic strips, with aluminum mounting hardware. Typical uses include truck dock door openings, railway car entrances, freezer doors and as positive draft protection to check the flow of unwanted air at +140 degrees to –40 degrees F. Costs shown per SF are per square foot of flexible strip area. To calculate the flexible strip area, add 2 linear feet to the opening width and height to allow for coverage at the sides and top. Costs shown per LF are per linear foot of mounting hardware. To calculate the length of the mounting hardware, add 2 LF to the opening width. Labor costs are based on work performed using ladders. Add for scaffolding costs if required. Based on Thermoscreen™ by Environmental Products.

	Craft@Hrs	Unit	Material	Labor	Total
Openings to 8' in height, 8" wide strips x .080" thick					
With 1/2 strip overlap	C8@.058	SF	3.03	2.38	5.41
With 3/4 strip overlap	C8@.076	SF	3.36	3.11	6.47
With full strip overlap	C8@.096	SF	4.69	3.93	8.62
Openings over 8' to 14' in height, 12" wide strips x .120" thick					
With 2/3 strip overlap	C8@.100	SF	5.37	4.10	9.47
With full strip overlap	C8@.115	SF	6.44	4.71	11.15
Openings over 14' to 20' in height, 16" wide strips x .120" thick					
With 1/2 strip overlap	C8@.075	SF	5.74	3.07	8.81
With 3/4 strip overlap	C8@.100	SF	7.52	4.10	11.62
With full strip overlap	C8@.125	SF	8.59	5.12	13.71
Add for mounting hardware, aluminum, pre-drilled, complete with nuts, bolts, washers and flex strip mounting pins					
Ceiling or header mounted (1.5 lbs per LF)	C8@.400	LF	7.72	16.40	24.12
Wall mounted (2.5 lbs per LF)	C8@.500	LF	28.40	20.50	48.90

	Craft@Hrs	Unit	Material	Labor	Total

Complete Hollow Metal Door Assembly These figures show the costs normally associated with installing an exterior hollow core steel door.

Hollow metal exterior 3' x 7' flush door, with frame, hardware and trim, complete

	Craft@Hrs	Unit	Material	Labor	Total
Total cost door, frame and trim as described below	C8@7.62	Ea	892.95	312.30	1,205.25
Stock hollow metal flush door, 16 gauge, 3' x 7' x 1-3/4", non-rated	C8@.721	Ea	346.00	29.50	375.50
Closer plate reinforcing on door	—	Ea	13.60	—	13.60
Stock 18 gauge frame, 6" jamb, non-rated	C8@.944	Ea	102.00	38.70	140.70
Closer plate on frame	—	Ea	5.35	—	5.35
Three hinges, 4-1/2" x 4-1/2"	C8@.578	LS	56.00	23.70	79.70
Lockset, mortise type	C8@.949	Ea	213.00	38.90	251.90
Saddle type threshold, aluminum, 3'	C8@.276	Ea	16.70	11.30	28.00
Standard duty closer	C8@.787	Ea	87.10	32.20	119.30
Weatherstripping, bronze and neoprene	C8@2.58	LS	42.80	106.00	148.80
Paint with primer and 2 coats enamel	C8@.781	LS	10.40	32.00	42.40

Right Hand Doors — Hinges on right opens inward

Right Hand Doors Reverse — Hinges on right opens outward

Left Hand Doors — Hinges on left opens inward

Left Hand Doors Reverse — Hinges on left opens outward

Hollow Metal Exterior Doors Commercial quality doors. These costs do not include the frame, hinges, lockset, trim or finishing.

	Craft@Hrs	Unit	Material	Labor	Total
6'8" high hollow flush doors, 1-3/8" thick, non-rated					
2'6" wide, 20 gauge	C8@.681	Ea	196.00	27.90	223.90
2'8" wide, 20 gauge	C8@.683	Ea	202.00	28.00	230.00
3'0" wide, 20 gauge	C8@.721	Ea	216.00	29.50	245.50
6'8" high hollow flush doors, 1-3/4" thick, non-rated					
2'6" wide, 20 gauge	C8@.681	Ea	237.00	27.90	264.90
2'8" wide, 20 gauge	C8@.697	Ea	244.00	28.60	272.60
2'8" wide, 18 gauge	C8@.697	Ea	282.00	28.60	310.60
2'8" wide, 16 gauge	C8@.697	Ea	330.00	28.60	358.60
3'0" wide, 20 gauge	C8@.721	Ea	256.00	29.50	285.50
3'0" wide, 18 gauge	C8@.721	Ea	296.00	29.50	325.50
3'0" wide, 16 gauge	C8@.721	Ea	346.00	29.50	375.50
7'0" high hollow flush doors, 1-3/4" thick, non-rated					
2'6" or 2'8" wide, 20 gauge	C8@.757	Ea	250.00	31.00	281.00
2'8" or 3'0" wide, 18 gauge	C8@.757	Ea	301.00	31.00	332.00
2'8" or 3'0" wide, 16 gauge	C8@.757	Ea	352.00	31.00	383.00
3'0" wide, 20 gauge	C8@.816	Ea	268.00	33.40	301.40
Additional costs for hollow metal doors, cost per door					
Add for factory applied steel astragal	—	Ea	35.30	—	35.30
Add for steel astragal set on site	C8@.496	Ea	35.30	20.30	55.60
Add for 90 minute "B" fire label rating	—	Ea	39.50	—	39.50
Add for R-7 polyurethane foam core	—	Ea	95.00	—	95.00

	Craft@Hrs	Unit	Material	Labor	Total
Add for 10" x 10" wired glass panel	—	Ea	68.80	—	68.80
Add for closer reinforcing plate	—	Ea	13.60	—	13.60
Add for chain or bolt reinforcing plate	—	Ea	102.00	—	102.00
Add for rim exit latch reinforcing	—	Ea	29.10	—	29.10
Add for vertical exit latch reinforcing	—	Ea	52.80	—	52.80
Add for pull plate reinforcing	—	Ea	21.30	—	21.30
Add for galvanizing 1-3/8" doors	—	Ea	26.80	—	26.80
Add for galvanizing 1-3/4" door	—	Ea	32.80	—	32.80
Add for cutouts to 4 SF	—	Ea	55.00	—	55.00
Add for cutouts over 4 SF	—	SF	11.60	—	11.60
Add for stainless steel doors	—	Ea	924.00	—	924.00
Add for baked enamel finish	—	Ea	69.80	—	69.80
Add for porcelain enamel finish	—	Ea	166.00	—	166.00
Add for larger sizes	—	SF	3.80	—	3.80
Add for special dapping	—	Ea	29.90	—	29.90

Hollow Metal Door Frames 18 gauge prefinished hollow metal door frames, non-rated, stock sizes, for 1-3/4" or 1-3/8" doors. These costs do not include the door, hinges, lockset, trim or finishing.

6'8" high	Craft@Hrs	Unit	Material	Labor	Total
To 3'6" wide, to 4-1/2" jamb	C8@.944	Ea	91.20	38.70	129.90
To 3'6" wide, 4-3/4" to 6" jamb	C8@.944	Ea	96.00	38.70	134.70
To 3'6" wide, over 6" jamb	C8@.944	Ea	129.00	38.70	167.70
Over 4' wide, 4-1/2" jamb	C8@.944	Ea	110.00	38.70	148.70
Over 4' wide, 4-3/4" to 6" jamb	C8@.944	Ea	116.00	38.70	154.70
Over 4' wide, over 6" jamb	C8@.944	Ea	149.00	38.70	187.70
7' high					
To 3'6" wide, to 4-1/2" jamb	C8@.944	Ea	99.90	38.70	138.60
To 3'6" wide, 4-3/4" to 6" jamb	C8@.944	Ea	105.00	38.70	143.70
To 3'6" wide, over 6" jamb	C8@.944	Ea	137.00	38.70	175.70
Over 4' wide, 4-1/2" jamb	C8@.944	Ea	118.00	38.70	156.70
Over 4' wide, 4-3/4" to 6" jamb	C8@.944	Ea	122.00	38.70	160.70
Over 4' wide, over 6" jamb	C8@.944	Ea	156.00	38.70	194.70
8' high					
To 3'6" wide, to 4-1/2" jamb	C8@1.05	Ea	114.00	43.00	157.00
To 3'6" wide, 4-3/4" to 6" jamb	C8@1.05	Ea	120.00	43.00	163.00
To 3'6" wide, over 6" jamb	C8@1.05	Ea	155.00	43.00	198.00
Over 4' wide, 4-1/2" jamb	C8@1.05	Ea	133.00	43.00	176.00
Over 4' wide, 4-3/4" to 6" jamb	C8@1.05	Ea	136.00	43.00	179.00
Over 4' wide, over 6" jamb	C8@1.05	Ea	175.00	43.00	218.00

Cost additions or deductions for hollow metal door frames, per frame

	Craft@Hrs	Unit	Material	Labor	Total
Add for 90 minute UL label	—	Ea	27.60	—	27.60
Add for aluminum casing	—	Ea	22.90	—	22.90
Add for communicator frame (back-to-back doors)	—	Ea	16.80	—	16.80
Add for galvanized finish	—	%	12.0	—	—
Add for lengthening, 7' to 8'10"	—	Ea	14.70	—	14.70
Add for extra hinge reinforcing	—	Ea	5.01	—	5.01
Add for stainless steel frames	—	Ea	482.00	—	482.00
Add for porcelain enamel finish	—	Ea	156.00	—	156.00
Add for reinforcing for chain and bolt	—	Ea	6.43	—	6.43
Add for closer plate reinforcing	—	Ea	5.67	—	5.67
Add for exit latch reinforcing	—	Ea	10.50	—	10.50
Add for concrete filled frames	C8@.382	Ea	7.19	15.70	22.89
Add for borrowed lite	C8@.086	SF	6.87	3.52	10.39

Doors and Windows 8

	Craft@Hrs	Unit	Material	Labor	Total
Add for fixed transom lite	C8@.099	SF	9.05	4.06	13.11
Add for movable transom lite	C8@.107	SF	10.40	4.38	14.78
Add for window frame sections	C8@.071	LF	5.56	2.91	8.47
Add for wall frame sections	C8@.071	LF	6.32	2.91	9.23
Deduct for 22 gauge frames	—	Ea	-12.20	—	-12.20

Prehung Steel Doors 18 gauge primed insulated 1-3/4" thick entry doors with sweep and 18 gauge steel frame. Costs shown Include three 4" x 4" x 1/4" hinges but no lockset.

Flush doors

	Craft@Hrs	Unit	Material	Labor	Total
Jamb to 5-1/2" wide, 2'8" x 6'8", or 3'0" x 6'8"	C8@1.05	Ea	398.00	43.00	441.00
Jamb over 5-1/2" wide, 2'8" x 6'8", or 3'0" x 6'8"	C8@1.05	Ea	407.00	43.00	450.00

6 or 8 panel doors

Jamb to 5-1/2" wide, 2'8" x 6'8", or 3'0" x 6'8"	C8@1.05	Ea	411.00	43.00	454.00
Jamb over 5-1/2" wide, 2'8" x 6'8", or 3'0" x 6'8"	C8@1.05	Ea	417.00	43.00	460.00

Additional costs for prehung steel doors

90 minute Factory Mutual fire rating	—	Ea	20.40	—	20.40
60 minute Factory Mutual fire rating	—	Ea	15.70	—	15.70
Add for 16 gauge door and frame	—	Ea	12.40	—	12.40
Deduct for 22 gauge and door	—	Ea	-3.38	—	-3.38
Add for galvanized steel finish	—	Ea	10.80	—	10.80
Deduct for unfinished door	—	Ea	-12.60	—	-12.60
Add for installed weatherstripping	—	Ea	31.40	—	31.40
Add for aluminum threshold	—	Ea	31.10	—	31.10

Doors and Frames Commercial and institutional quality. Unfinished, no hardware included.

Wood hollow core 1-3/8" thick prehung flush interior doors

2'6" x 6'8", hardboard face	C8@.936	Ea	58.10	38.30	96.40
2'8" x 6'8", hardboard face	C8@.936	Ea	60.20	38.30	98.50
3'0" x 6'8", hardboard face	C8@.982	Ea	62.30	40.20	102.50
2'6" x 6'8", hardwood face	C8@.936	Ea	69.80	38.30	108.10
2'8" x 6'8", hardwood face	C8@.936	Ea	71.60	38.30	109.90
3'0" x 6'8", hardwood face	C8@.982	Ea	73.90	40.20	114.10

Wood solid core 1-3/4" thick flush prehung exterior doors

2'8" x 6'8", hardboard face	C8@1.10	Ea	98.00	45.10	143.10
3'0" x 6'8", hardboard face	C8@1.10	Ea	101.00	45.10	146.10
2'8" x 6'8", hardwood face	C8@1.10	Ea	102.00	45.10	147.10
3'0" x 6'8", hardwood face	C8@1.10	Ea	104.00	45.10	149.10
2'6" x 6'8", hardboard face, B fire label	C8@1.10	Ea	134.00	45.10	179.10
2'8" x 6'8", hardboard face, B fire label	C8@1.10	Ea	140.00	45.10	185.10
3'0" x 6'8", hardboard face, B fire label	C8@1.10	Ea	140.00	45.10	185.10

Steel foam core 1-3/4" thick prehung exterior doors

2'6" x 6'8", flush face	C8@.982	Ea	146.00	40.20	186.20
2'8" x 6'8", flush face	C8@.982	Ea	140.00	40.20	180.20
3'0" x 6'8", flush face	C8@1.01	Ea	146.00	41.40	187.40
2'6" x 6'8", colonial panel face	C8@.982	Ea	146.00	40.20	186.20
2'8" x 6'8", colonial panel face	C8@.982	Ea	146.00	40.20	186.20
3'0" x 6'8", colonial panel face	C8@1.01	Ea	146.00	41.40	187.40

Fiberglass foam core 1-3/4" thick prehung exterior doors

2'8" x 6'8", colonial panel face	C8@.982	Ea	156.00	40.20	196.20
3'0" x 6'8", colonial panel face	C8@1.01	Ea	156.00	41.40	197.40
2'8" x 6'8", French 15-lite	C8@.982	Ea	240.00	40.20	280.20
3'0" x 6'8", French 15-lite	C8@1.01	Ea	240.00	41.40	281.40

Doors and Windows 8

	Craft@Hrs	Unit	Material	Labor	Equipment	Total

Special Doors These costs do not include electrical work. Equipment is a 4000 lb. forklift.

Flush panel doors for curtain walls

	Craft@Hrs	Unit	Material	Labor	Equipment	Total
3' x 7'	H5@7.14	Ea	1,570.00	330.00	43.30	1,943.30
3'6" x 7'	H5@7.14	Ea	1,630.00	330.00	43.30	2,003.30
Pair 2'6" x 7', per opening	H5@10.5	Ea	2,300.00	485.00	63.70	2,848.70
Pair 3' x 7', per opening	H5@10.5	Ea	2,620.00	485.00	63.70	3,168.70
Add for bronze anodized aluminum	—	%	15.0	—	—	—

Sliding metal clad fire doors including electric motor

	Craft@Hrs	Unit	Material	Labor	Equipment	Total
Light duty, minimum	H5@.249	SF	32.10	11.50	1.51	45.11
Light duty, maximum	H5@.405	SF	55.90	18.70	2.46	77.06
Heavy duty, minimum	H5@.600	SF	64.10	27.70	3.64	95.44
Heavy duty, maximum	H5@.750	SF	109.00	34.60	4.55	148.15

Fire doors, overhead roll-up type, sectional steel, fusible link, motor operated, UL label

	Craft@Hrs	Unit	Material	Labor	Equipment	Total
6' x 7', manual operation	F7@13.4	Ea	1,860.00	593.00	102.00	2,555.00
5' x 8', manual operation	F7@13.4	Ea	1,260.00	593.00	102.00	1,955.00
14' x 14', chain hoist operated	F7@26.2	Ea	3,640.00	1,160.00	199.00	4,999.00
18' x 14', chain hoist operated	F7@32.9	Ea	5,040.00	1,460.00	250.00	6,750.00

Sliding metal fire doors. Motor operated, 3 hour rating, with fusible link. Add the cost of electrical hookup

	Craft@Hrs	Unit	Material	Labor	Equipment	Total
3' x 6'8"	H5@16.0	Ea	2,980.00	739.00	97.10	3,816.10
3'8" x 6'8"	H5@16.0	Ea	3,170.00	739.00	97.10	4,006.10
4' x 8'	H5@16.0	Ea	3,280.00	739.00	97.10	4,116.10
5' x 8'	H5@16.0	Ea	3,410.00	739.00	97.10	4,246.10
6' x 8'	H5@16.0	Ea	4,070.00	739.00	97.10	4,906.10

Hangar doors, 25' panels including hardware and motorized opener. Costs based on 150' wide doors, using a 25 ton hydraulic truck crane

	Craft@Hrs	Unit	Material	Labor	Equipment	Total
To 20' high	H5@.213	SF	28.40	9.83	6.52	44.75
20' to 40' high	H5@.142	SF	37.00	6.56	4.35	47.91
40' to 60' high	H5@.085	SF	42.60	3.92	2.60	49.12
60' to 80' high	H5@.061	SF	51.20	2.82	1.87	55.89
More than 80' high	H5@.047	SF	85.30	2.17	1.44	88.91

Refrigerator doors, manually operated, with hardware and standard frame

	Craft@Hrs	Unit	Material	Labor	Equipment	Total
Galvanized, cooler, 3' x 7', hinged	F7@6.60	Ea	1,570.00	292.00	50.10	1,912.10
Stainless, cooler, 3' x 7', hinged	F7@6.60	Ea	1,810.00	292.00	50.10	2,152.10
Stainless, freezer, 3' x 7', hinged	F7@6.60	Ea	2,340.00	292.00	50.10	2,682.10
Stainless, cooler, 4' x 7', hinged	F7@6.60	Ea	2,040.00	292.00	50.10	2,382.10
Galvanized, cooler, 5' x 7', sliding	F7@7.93	Ea	2,290.00	351.00	60.20	2,701.20
Galvanized, freezer, 5' x 7' sliding	F7@7.93	Ea	2,680.00	351.00	60.20	3,091.20

Revolving doors, 7' diameter, complete. Typical prices

	Craft@Hrs	Unit	Material	Labor	Equipment	Total
Aluminum	F7@50.3	Ea	28,100.00	2,230.00	382.00	30,712.00
Stainless steel	F7@50.3	Ea	32,100.00	2,230.00	382.00	34,712.00
Bronze	F7@69.0	Ea	41,100.00	3,050.00	524.00	44,674.00

Security grilles, aluminum, overhead roll-up type, horizontal rods at 2" OC, chain hoist operated

	Craft@Hrs	Unit	Material	Labor	Equipment	Total
8' x 8', clear anodized aluminum	F7@18.5	Ea	2,140.00	819.00	140.00	3,099.00
8' x 8', medium bronze aluminum	F7@18.5	Ea	3,030.00	819.00	140.00	3,989.00
10' x 10', anodized aluminum	F7@21.8	Ea	2,860.00	965.00	165.00	3,990.00
18' x 8', anodized aluminum	F7@26.3	Ea	4,410.00	1,160.00	200.00	5,770.00
18' x 8', anodized, motor operated	F7@28.7	Ea	5,160.00	1,270.00	218.00	6,648.00

Service doors, aluminum, overhead roll-up type

	Craft@Hrs	Unit	Material	Labor	Equipment	Total
4' x 4', manual counter shutter	F7@13.6	Ea	854.00	602.00	103.00	1,559.00
8' x 4', manual operation	F7@12.7	Ea	1,240.00	562.00	96.40	1,898.40
10' x 10', chain hoist operated	F7@12.7	Ea	1,900.00	562.00	96.40	2,558.40

	Craft@Hrs	Unit	Material	Labor	Equipment	Total
Service doors, steel, overhead roll-up, chain hoist operated						
8 'x 8'	F7@11.7	Ea	1,250.00	518.00	88.80	1,856.80
10' x 10'	F7@11.7	Ea	1,530.00	518.00	88.80	2,136.80
12' x 12'	F7@14.7	Ea	2,050.00	651.00	112.00	2,813.00
14' x 14'	F7@18.2	Ea	2,250.00	806.00	138.00	3,194.00
18' x 18'	F7@22.0	Ea	3,030.00	974.00	167.00	4,171.00
18' x 14', 240 volt motor operated	F7@25.0	Ea	3,960.00	1,110.00	190.00	5,260.00
Service doors, weather-stripped, steel, overhead roll-up type, chain hoist operated, bottom and side astragal, with hood baffles but no insulation						
10' x 10'	F7@12.8	Ea	1,480.00	567.00	97.10	2,144.10
14' x 14'	F7@26.9	Ea	2,050.00	1,190.00	204.00	3,444.00
Add for insulated weather-stripped doors	—	%	150.0	—	—	—
Telescoping doors with enamel finish, electric operated						
10' high x 10' wide	H5@15.0	Ea	8,170.00	693.00	91.10	8,954.10
20' high x 10' wide	H5@19.3	Ea	13,100.00	891.00	117.00	14,108.00
20' high x 12' wide	H5@24.0	Ea	14,200.00	1,110.00	146.00	15,456.00
20' high x 16' wide	H5@26.4	Ea	17,100.00	1,220.00	160.00	18,480.00
Accordion folding doors, 2 layers including track and fittings						
Vinyl covered	C8@.505	SF	10.40	20.70	—	31.10
Woven mahogany color vinyl	C8@.505	SF	13.50	20.70	—	34.20
Economy vinyl	C8@.505	SF	8.88	20.70	—	29.58
Rigid PVC	C8@.505	SF	14.90	20.70	—	35.60
Add for custom sizes	—	%	26.0	—	—	—
Add for fire retardant accordion folding doors	—	%	15.0	10.0	—	—
Sliding glass doors with 1/4" tempered float glass including frame, trim, and hardware						
6-0' wide economy grade	C8@2.75	Ea	800.00	113.00	—	913.00
6-0' wide premium grade	C8@3.00	Ea	919.00	123.00	—	1,042.00
12-0' wide economy grade	C8@4.80	Ea	1,271.00	197.00	—	1,468.00
12-0' wide premium grade	C8@5.00	Ea	1,859.00	205.00	—	2,064.00
Sliding glass doors with 5/8" insulating glass including frame, trim, and hardware						
6-0' wide economy grade	C8@2.75	Ea	1,124.00	113.00	—	1,237.00
6-0' wide premium grade	C8@3.00	Ea	1,365.00	123.00	—	1,488.00
12-0' wide economy grade	C8@4.80	Ea	1,523.00	197.00	—	1,720.00
12-0' wide premium grade	C8@5.00	Ea	1,859.00	205.00	—	2,064.00
Air curtains. Add the cost of motor starters, transformers, door switches and temperature controls. Unheated, with minimal wind stoppage						
8-0' high, multiples of 3-0' wide	C8@4.40	Ea	220.00	180.00	—	400.00
8-0' high, multiples of 5-0' wide	C8@4.40	Ea	195.00	180.00	—	375.00
10-0' high, multiples of 4-0' wide	C8@4.40	Ea	203.00	180.00	—	383.00
12-0' high, multiples of 3'-6" wide	C8@4.50	Ea	237.00	184.00	—	421.00
12-0' high, 12-0' wide	C8@4.80	Ea	262.00	197.00	—	459.00
16-0' high, multiples of 3'-6" wide	C8@5.00	Ea	285.00	205.00	—	490.00
16-0' high, 12-0' wide	C8@5.00	Ea	253.00	205.00	—	458.00

	Craft@Hrs	Unit	Material	Labor	Total
Steel access doors, with casing and ground					
12" x 12"	CC@.150	Ea	17.20	6.83	24.03
18" x 18"	CC@.155	Ea	24.00	7.06	31.06
24" x 24"	CC@.160	Ea	36.50	7.29	43.79

	Craft@Hrs	Unit	Material	Labor	Equipment	Total

Stock room doors, double-action .063" aluminum, with hardware and bumper strips both sides, 7' high doors, plastic laminate finish, 12" high baseplate in each door

	Craft@Hrs	Unit	Material	Labor	Equipment	Total
2' wide opening, 1 door	F7@1.70	Ea	203.00	75.30	12.90	291.20
2'6" wide opening, 1 door	F7@1.70	Ea	227.00	75.30	12.90	315.20
3' wide opening, 1 door	F7@1.80	Ea	250.00	79.70	13.70	343.40
3'6" wide opening, 1 door	F7@1.90	Ea	279.00	84.10	14.40	377.50
4' wide opening, 2 doors, per pair	F7@3.36	Pr	406.00	149.00	25.50	580.50
5' wide opening, 2 doors, per pair	F7@3.36	Pr	455.00	149.00	25.50	629.50
6' wide opening, 2 doors, per pair	F7@3.98	Pr	500.00	176.00	30.20	706.20
7' wide opening, 2 doors, per pair	F7@3.98	Pr	559.00	176.00	30.20	765.20

Vault doors, minimum security

	Craft@Hrs	Unit	Material	Labor	Equipment	Total
3' x 7', 2 hour fire rating	F7@22.0	Ea	2,300.00	974.00	167.00	3,441.00
4' x 7', 2 hour fire rating	F7@23.4	Ea	3,290.00	1,040.00	178.00	4,508.00
3' x 7', 4 hour fire rating	F7@27.2	Ea	2,510.00	1,200.00	206.00	3,916.00
4' x 7', 4 hour fire rating	F7@28.7	Ea	3,430.00	1,270.00	218.00	4,918.00

	Craft@Hrs	Unit	Material	Labor	Total

Commercial and Industrial Grade Steel Windows, glazed.

	Craft@Hrs	Unit	Material	Labor	Total
Industrial grade, fixed 100%	G1@.086	SF	15.80	3.53	19.33
Industrial grade, vented 50%	G1@.086	SF	20.40	3.53	23.93
Projected, vented 50%	G1@.086	SF	22.80	3.53	26.33
Add for screen, SF of screen	—	SF	2.68	—	2.68

Hardware Commercial and industrial quality. Installation costs include drilling and routing of doors where required.

Rule of thumb. Use these typical finish hardware costs for preliminary estimates only.

Hardware for commercial buildings, per square foot of floor area

	Craft@Hrs	Unit	Material	Labor	Total
Economy grade, for doors, cabinets, and toilet rooms	CC@.002	SF	.17	.09	.26
Standard grade, for doors, cabinets, and toilet rooms	CC@.004	SF	.37	.18	.55
Hospital, not including panic hardware, per door	CC@2.46	Ea	457.00	112.00	569.00
Office, not including panic hardware, per door	CC@1.57	Ea	219.00	71.50	290.50
School, not including panic hardware, per door	CC@1.94	Ea	305.00	88.40	393.40

Detail cost for individual hardware components

Cabinet hardware, per LF of face, typical prices

	Craft@Hrs	Unit	Material	Labor	Total
Commercial grade	CC@.145	LF	10.40	6.60	17.00
Institutional grade	CC@.181	LF	13.00	8.24	21.24

Closer, surface mounted. Based on Schlage Lock

	Craft@Hrs	Unit	Material	Labor	Total
Interior doors	CC@.616	Ea	77.40	28.10	105.50
Exterior doors	CC@.709	Ea	96.80	32.30	129.10
Exterior, heavy duty	CC@.894	Ea	112.00	40.70	152.70
Floor mounted	CC@1.16	Ea	194.00	52.80	246.80

Deadbolts and deadlatches, chrome finish

	Craft@Hrs	Unit	Material	Labor	Total
Standard deadbolt, single cylinder	CC@.745	Ea	28.80	33.90	62.70
Standard deadbolt, double cylinder	CC@.745	Ea	39.60	33.90	73.50
Standard turnbolt, no cylinder	CC@.745	Ea	22.10	33.90	56.00
Nightlatch, single cylinder	CC@.745	Ea	69.70	33.90	103.60
Nightlatch, double cylinder	CC@.745	Ea	77.00	33.90	110.90
Nightlatch, no cylinder, turnbolt	CC@.745	Ea	60.10	33.90	94.00
Heavy duty deadbolt, single cylinder	CC@.745	Ea	54.00	33.90	87.90
Heavy duty deadbolt, double cylinder	CC@.745	Ea	67.30	33.90	101.20

	Craft@Hrs	Unit	Material	Labor	Total
Heavy duty turnbolt, no cylinder	CC@.745	Ea	54.00	33.90	87.90
Extra heavy deadbolt, single cylinder	CC@.745	Ea	85.50	33.90	119.40
Extra heavy deadbolt, double cylinder	CC@.745	Ea	101.00	33.90	134.90
Add for brass or bronze finish	—	Ea	2.45	—	2.45
Door stops					
Rubber tip, screw base	CC@.060	Ea	3.93	2.73	6.66
Floor mounted, with holder	CC@.279	Ea	16.30	12.70	29.00
Wall mounted, with holder	CC@.279	Ea	16.00	12.70	28.70
Overhead mounted	CC@.528	Ea	93.60	24.10	117.70
Hinges, butt type					
3-1/2" x 3-1/2"	CC@.357	Pr	8.97	16.30	25.27
5" x 5", hospital swing clear	CC@.547	Pr	118.00	24.90	142.90
4", spring, single acting	CC@.243	Pr	29.20	11.10	40.30
4-1/2", heavy duty	CC@.347	Pr	54.80	15.80	70.60
7", spring, double acting	CC@.491	Pr	71.90	22.40	94.30
Floor-mounted, interior, per door	CC@.665	Ea	240.00	30.30	270.30
Floor-mounted, exterior, per 3' door	CC@.665	Ea	259.00	30.30	289.30
Floor-mounted, exterior, over 3' wide door	CC@.799	Ea	547.00	36.40	583.40
Invisible, blind doors, interior, per door	CC@.574	Ea	28.40	26.10	54.50
Invisible, for blind doors, exterior	CC@.574	Pr	36.30	26.10	62.40
Kick plates (see also Plates below)					
10" x 34" 16 gauge, bronze	CC@.452	Ea	39.30	20.60	59.90
10" x 34" 18 gauge, stainless	CC@.452	Ea	32.70	20.60	53.30
Letter drop, for wood doors					
brass plated	CC@.369	Ea	27.40	16.80	44.20
Locksets					
Heavy duty residential or light duty commercial locksets, chrome finish					
Classroom or storeroom, key lock	CC@.745	Ea	85.60	33.90	119.50
Dummy knob	CC@.247	Ea	16.10	11.30	27.40
Entrance lockset, key lock	CC@.745	Ea	78.40	33.90	112.30
Passage latch, no lock	CC@.689	Ea	37.20	31.40	68.60
Privacy latch, button lock	CC@.689	Ea	44.60	31.40	76.00
Add for lever handle (handicapped)	—	Ea	30.50	—	30.50
Add for brass or bronze finish	—	Ea	4.52	—	4.52
Heavy duty commercial, chrome finish					
Bored entrance lockset	CC@.745	Ea	197.00	33.90	230.90
Grip handle, with trim	CC@.855	Ea	217.00	38.90	255.90
Single deadbolt-lockset	CC@.745	Ea	95.60	33.90	129.50
Double deadbolt-lockset	CC@.745	Ea	125.00	33.90	158.90
Mortise deadbolt-lockset	CC@.855	Ea	217.00	38.90	255.90
Add for lever handle (handicapped)	—	Ea	20.40	—	20.40
Add for brass or bronze finish	—	Ea	7.50	—	7.50
Panic type exit door hardware, with trim					
Mortise type, for aluminum, steel or wood doors					
Satin aluminum finish	CC@2.41	Ea	396.00	110.00	506.00
Dark bronze finish	CC@2.41	Ea	408.00	110.00	518.00
Polished chrome finish	CC@2.41	Ea	483.00	110.00	593.00
Add for external lockset	CC@.401	Ea	54.00	18.30	72.30
Rim lock type, for aluminum, steel or wood doors					
Satin aluminum finish	CC@1.50	Ea	290.00	68.30	358.30
Dark bronze finish	CC@1.50	Ea	304.00	68.30	372.30
Polished chrome finish	CC@1.50	Ea	375.00	68.30	443.30
Add for external lockset	CC@.401	Ea	40.80	18.30	59.10

	Craft@Hrs	Unit	Material	Labor	Total
Vertical rod type, satin aluminum finish, with trim					
Aluminum doors, concealed rod	CC@2.14	Ea	385.00	97.50	482.50
Metal or wood doors, external rod	CC@1.79	Ea	402.00	81.50	483.50
Wood doors, concealed rod	CC@2.14	Ea	473.00	97.50	570.50
Add for external lockset	CC@.401	Ea	52.40	18.30	70.70
Add for bronze or black finish	—	%	37.5	—	—
Add for chrome finish	—	%	70.0	—	—
Plates (see also Kick Plates above)					
Pull plates, bronze, 4" x 16"	CC@.190	Ea	32.60	8.65	41.25
Push plates bronze, 4" x 16"	CC@.190	Ea	25.10	8.65	33.75
Surface bolts					
4" bolt	CC@.308	Ea	5.03	14.00	19.03
6" bolt	CC@.310	Ea	6.84	14.10	20.94
Threshold					
Aluminum, 36"	CC@.241	Ea	16.20	11.00	27.20
Bronze, 36"	CC@.241	Ea	66.10	11.00	77.10

Weatherstripping

	Craft@Hrs	Unit	Material	Labor	Total
Bronze and neoprene, 3' x 7' door					
Wood door	CC@1.53	Ea	18.40	69.70	88.10
Steel door, adjustable	CC@2.32	Ea	44.40	106.00	150.40
Astragal, mortise mounted, bronze, adjustable	CC@.064	LF	11.40	2.92	14.32
Glue-back foam for 3' x 7' door	CC@.006	LF	.32	.27	.59

Glazing Based on Atascadero Glass, http://atascaderoglass.com

	Craft@Hrs	Unit	Material	Labor	Total
Sheet (window) glass					
Single strength "B"					
To 60" width plus length	G1@.057	SF	1.98	2.34	4.32
Over 60" to 70" width plus length	G1@.057	SF	2.03	2.34	4.37
Over 70" to 80" width plus length	G1@.057	SF	2.21	2.34	4.55
Over 80" width plus length	G1@.057	SF	2.25	2.34	4.59
Double strength window glass					
Double strength "B"					
To 60" width plus length	G1@.057	SF	2.57	2.34	4.91
Over 60" to 70" width plus length	G1@.057	SF	2.67	2.34	5.01
Over 70" to 80" width plus length	G1@.057	SF	2.76	2.34	5.10
Over 80" width plus length	G1@.057	SF	2.85	2.34	5.19
Tempered "B"	G1@.057	SF	4.19	2.34	6.53
Obscure	G1@.057	SF	3.96	2.34	6.30
3/16" glass					
Crystal (clear)	G1@.068	SF	3.43	2.79	6.22
Tempered	G1@.068	SF	5.66	2.79	8.45
1/4" glass					
Float, clear, quality 3	G1@.102	SF	3.27	4.18	7.45
Float, bronze or gray, quality 3	G1@.102	SF	3.44	4.18	7.62
Float, obscure	G1@.102	SF	4.97	4.18	9.15
Float, heat absorbing	G1@.102	SF	5.43	4.18	9.61
Float, safety, laminated, clear	G1@.102	SF	5.99	4.18	10.17
Float, tempered	G1@.102	SF	5.43	4.18	9.61
Spandrel, plain	G1@.102	SF	7.31	4.18	11.49
Spandrel, tinted gray or bronze	G1@.102	SF	8.04	4.18	12.22
Tempered, reflective	G1@.102	SF	8.36	4.18	12.54

	Craft@Hrs	Unit	Material	Labor	Total
3/8" glass					
Float, clear	G1@.128	SF	5.57	5.25	10.82
Float, tinted	G1@.128	SF	7.82	5.25	13.07
Obscure, tempered	G1@.128	SF	8.36	5.25	13.61
Corrugated glass	G1@.128	SF	7.17	5.25	12.42
1/2" float, tempered	G1@.157	SF	20.50	6.44	26.94
1" bullet resistant, 12" x 12" panels	G1@1.32	SF	79.80	54.10	133.90
2" bullet resistant, 15 to 20 SF	G1@1.62	SF	47.70	66.40	114.10
Insulating glass, 2 layers of 1/8", 1/2" overall, 10 to 15 square foot lites					
"B" quality sheet glass	G1@.131	SF	6.34	5.37	11.71
Float, polished bronze or gray	G1@.131	SF	8.83	5.37	14.20
Insulating glass, 2 layers of 1/4" float, 1" overall, 30 to 40 square foot lites					
Clear float glass	G1@.136	SF	9.73	5.57	15.30
Acrylic plastic sheet (Plexiglass), clear except as noted					
8" x 10" .093" thick	—	Ea	2.07	—	2.07
11"x14" .093" thick	—	Ea	2.39	—	2.39
12" x 24" .093" thick	—	Ea	3.97	—	3.97
18" x 24" .093" thick	—	Ea	6.59	—	6.59
24" x 28" .093" thick	—	Ea	8.89	—	8.89
24" x 48" .093" thick	—	Ea	18.80	—	18.80
30" x 36" .093" thick	—	Ea	14.10	—	14.10
36" x 48" .093" thick	—	Ea	22.00	—	22.00
36" x 72" .093" thick	—	Ea	33.50	—	33.50
18" x 24" .236" thick	—	Ea	14.10	—	14.10
24" x 48" .236" thick	—	Ea	41.90	—	41.90
30" x 36" .236" thick	—	Ea	33.30	—	33.30
36" x 72" .236" thick	—	Ea	75.30	—	75.30
30" x 36" .093" thick, bronze	—	Ea	12.70	—	12.70
32" x 44" .093" thick, bronze	—	Ea	17.30	—	17.30
Wired glass, 1/4", type 3					
Clear	G1@.103	SF	6.90	4.22	11.12
Hammered	G1@.103	SF	6.45	4.22	10.67
Obscure	G1@.103	SF	6.20	4.22	10.42
Mirrors, unframed					
Sheet glass, 3/16"	G1@.103	SF	5.01	4.22	9.23
Float glass, 1/4"	G1@.106	SF	7.76	4.34	12.10
Reflective, 1 way, in wood stops	G1@.118	SF	14.30	4.84	19.14
Small lites, adhesive mount	G1@.057	SF	3.27	2.34	5.61
Polycarbonate sheet glazing. 4' x 8' panels, Cyro					
Impact resistant, weather resistant sheets					
Clear 1/8"	G1@.027	SF	4.11	1.11	5.22
Clear 3/16"	G1@.027	SF	4.30	1.11	5.41
Clear 1/4"	G1@.027	SF	5.56	1.11	6.67
Clear 1/2"	G1@.035	SF	10.10	1.43	11.53
Add for tints		%	17.0	—	—
Chemical and abrasion resistant sheets					
Clear 1/8"	G1@.027	SF	6.27	1.11	7.38
Clear 3/16"	G1@.027	SF	7.14	1.11	8.25
Clear 1/4"	G1@.027	SF	10.40	1.11	11.51
Clear 1/2"	G1@.035	SF	16.00	1.43	17.43

	Craft@Hrs	Unit	Material	Labor	Total
Add for tints		%	17.0	—	—

Shock absorbent, high light transmission sheets (greenhouse applications)

	Craft@Hrs	Unit	Material	Labor	Total
Clear 5/16"	G1@.027	SF	6.27	1.11	7.38
Clear 5/8"	G1@.035	SF	7.14	1.43	8.57

Curtain Walls Typical costs.

Structural sealant curtain wall systems with aluminum or steel frame

	Craft@Hrs	Unit	Material	Labor	Total
Regular weight, float glass	—	SF	—	—	32.80
Heavy duty, float glass	—	SF	—	—	35.90
Corning glasswall glass	—	SF	—	—	37.90
Add for spandrel glass	—	SF	—	—	4.80
Add for heat absorbing glass	—	SF	—	—	8.60
Add for low transmission glass	—	SF	—	—	5.60
Add for 1" insulating glass	—	SF	—	—	8.10
Add for bronze anodizing	—	%	—	—	15.0
Add for black anodizing	—	%	—	—	18.0

Furring and Lathing For commercial applications. See plaster costs in the following section.

Furring, no plaster included

Walls, steel hat channel at 16" OC galvanized

	Craft@Hrs	Unit	Material	Labor	Total
3/4" x 5/8"	C8@.243	SY	2.71	9.96	12.67
3/4" x 3/4"	C8@.243	SY	2.42	9.96	12.38
1-1/2" x 1-1/2"	C8@.269	SY	3.29	11.00	14.29

Beams and columns, steel hat channel at 16" OC galvanized

	Craft@Hrs	Unit	Material	Labor	Total
3/4" x 3/4" at 12" OC	C8@.423	SY	3.04	17.30	20.34
3/4" x 3/4" at 16" OC	C8@.382	SY	2.39	15.70	18.09

Ceilings, main runners at 4' OC, furring channels at 16" OC, 8 gauge wire at 4' OC

	Craft@Hrs	Unit	Material	Labor	Total
1-1/2" x 3/4", standard centering	C8@.344	SY	3.51	14.10	17.61
1-1/2" x 7/8", hat channel, standard centering	C8@.363	SY	3.38	14.90	18.28
3-1/4" x 1-1/2" x 3/4", triple hung	C8@.683	SY	6.43	28.00	34.43
1-1/2" x 3/4", coffered	C8@.887	SY	6.09	36.30	42.39
Add for resilient spring system	—	SY	11.40	—	11.40

Lathing, walls and ceilings, nailed or wired in place

	Craft@Hrs	Unit	Material	Labor	Total
2.5 lb, standard diamond galvanized	F8@.099	SY	1.72	4.45	6.17
2.5 lb, painted	F8@.099	SY	1.53	4.45	5.98
3.4 lb, standard diamond galvanized	F8@.099	SY	1.88	4.45	6.33
3.4 lb, painted copper alloy	F8@.099	SY	1.81	4.45	6.26
1/8" flat rib, 3.4 lb, galvanized	F8@.099	SY	2.08	4.45	6.53
1/8" flat rib, 2.75 lb, painted	F8@.099	SY	2.01	4.45	6.46
3/8" rib lath, galvanized, 3.4 lb	F8@.099	SY	2.42	4.45	6.87
3/8" painted copper alloy, 3.4 lb	F8@.099	SY	2.04	4.45	6.49

Wire mesh, 15 lb felt and line wire

	Craft@Hrs	Unit	Material	Labor	Total
1" mesh, 18 gauge	F8@.079	SY	1.43	3.55	4.98
1-1/2" mesh, 17 gauge	F8@.079	SY	1.50	3.55	5.05
Paper back lath	F8@.063	SY	1.94	2.83	4.77
Stucco rite 2" corner bead, 16 gauge	F8@.028	LF	.14	1.26	1.40
Aqua lath, 2" x 16 gauge	F8@.076	SY	2.01	3.42	5.43

Plastering These costs do not include the furring or lath

Gypsum interior plaster, lime putty trowel finish

	Craft@Hrs	Unit	Material	Labor	Total
One coat application, on walls, extra smooth	F8@.330	SY	6.17	14.80	20.97
One coat application, on ceilings, extra smooth	F8@.381	SY	6.17	17.10	23.27

	Craft@Hrs	Unit	Material	Labor	Total
Two coat application, on ceilings, regular	F8@.381	SY	6.17	17.10	23.27
Two coat application, on walls regular	F8@.330	SY	6.17	14.80	20.97
Three coat application, on ceilings regular	F8@.448	SY	8.11	20.10	28.21
Three coat application, on walls regular	F8@.405	SY	8.11	18.20	26.31
Gypsum interior vermiculite plaster, trowel finish					
Two coat application, on ceilings	F8@.381	SY	6.48	17.10	23.58
Two coat application, on walls	F8@.330	SY	6.48	14.80	21.28
Three coat application, on ceilings	F8@.448	SY	8.44	20.10	28.54
Three coat application, on walls	F8@.405	SY	8.44	18.20	26.64
Keene's cement plaster, troweled lime putty medium hard finish					
Two coat application, on ceilings	F8@.435	SY	5.54	19.60	25.14
Two coat application, on walls	F8@.381	SY	5.54	17.10	22.64
Three coat application, on ceilings	F8@.521	SY	6.82	23.40	30.22
Three coat application, on walls	F8@.459	SY	6.82	20.60	27.42
Portland cement stucco on exterior walls, 3 coats totaling 1" thick					
Natural gray, sand float finish	F8@.567	SY	5.24	25.50	30.74
Natural gray, trowel finish	F8@.640	SY	5.24	28.80	34.04
White cement, sand float finish	F8@.675	SY	6.01	30.40	36.41
White cement, trowel finish	F8@.735	SY	6.19	33.10	39.29
Portland cement stucco on soffits, 3 coats totaling 1" thick					
Natural gray, sand float finish	F8@.678	SY	5.49	30.50	35.99
Natural gray, trowel finish	F8@.829	SY	5.49	37.30	42.79
White cement, sand float finish	F8@.878	SY	6.44	39.50	45.94
White cement, trowel finish	F8@1.17	SY	6.62	52.60	59.22
Brown and scratch coat base for tile	F8@.308	SY	4.32	13.90	18.22
Scratch coat only for tile	F8@.152	SY	2.07	6.84	8.91
Thin coat plaster, 2 coats, with rock lath but not including studs					
3/8" rock lath (gypsum lath)	F8@.265	SY	9.56	11.90	21.46
Simulated acoustic texture, sometimes called "popcorn" or "cottage cheese"					
On ceilings	F8@.052	SY	1.20	2.34	3.54
Texture coat on exterior walls and soffits	F8@.045	SY	1.20	2.02	3.22
Patching gypsum plaster, including lath repair but no studding					
To 5 SF repairs	P5@.282	SF	1.40	11.20	12.60
Over 5 SF repairs	P5@.229	SF	1.40	9.11	10.51
Repair cracks only ($150 per job minimum typical)	P5@.055	LF	.25	2.19	2.44
Plastering bead and expansion joint, galvanized					
Stop and casing, square nose	F8@.028	LF	.40	1.26	1.66
Corner bead, 3/4" radius	F8@.023	LF	.44	1.03	1.47
Corner bead, expanded, zinc nose	F8@.024	LF	.85	1.08	1.93
Expansion joint, 3/4", 26 gauge, 1 piece	F8@.025	LF	.85	1.12	1.97
Expansion joint, 1-1/2", 2 piece	F8@.028	LF	1.47	1.26	2.73
Plastering accessories					
Base screed, 1/2", 26 gauge	F8@.032	LF	.53	1.44	1.97
Vents, 1-1/2", galvanized	F8@.042	LF	1.47	1.89	3.36
Vents, 4", galvanized	F8@.061	LF	1.74	2.74	4.48
Archbead, plastic nose	F8@.028	LF	.89	1.26	2.15
Plaster moldings, ornate designs					
2"	CC@.090	LF	3.77	4.10	7.87
4"	CC@.115	LF	7.69	5.24	12.93
6"	CC@.132	LF	11.22	6.01	17.23

	Craft@Hrs	Unit	Material	Labor	Total

Steel Studding (Hollow Metal Studs) Framing Systems

Hollow Metal Stud Partitions for plaster or gypsum wallboard. Primed metal studs, including studs, top and bottom track, screws and normal waste. Do not subtract for openings less than 16' wide. Add door and window opening framing, backing, let-in bracing, and sheathing for shear walls. These costs do not include the lath, plaster or gypsum wallboard. Costs per square foot of wall area, measured on one side. The figures in parentheses show production per day for a 2-man crew. Prices based on a 1,000 SF job.

Non-load bearing partitions, 25 gauge, 16" on center spacing

	Craft@Hrs	Unit	Material	Labor	Total
1-5/8" wide studs (534 SF per day)	C8@.015	SF	.24	.61	.85
2-1/2" wide studs (500 SF per day)	C8@.016	SF	.31	.66	.97
3-5/8" wide studs (445 SF per day)	C8@.018	SF	.40	.74	1.14
4" wide studs (445 SF per day)	C8@.018	SF	.44	.74	1.18
6" wide studs (400 SF per day)	C8@.020	SF	.55	.82	1.37

Non-load bearing partitions, 25 gauge, 24" on center spacing

	Craft@Hrs	Unit	Material	Labor	Total
1-5/8" wide studs (571 SF per day)	C8@.014	SF	.24	.57	.81
2-1/2" wide studs (534 SF per day)	C8@.015	SF	.27	.61	.88
3-5/8" wide studs (500 SF per day)	C8@.016	SF	.34	.66	1.00
4" wide studs (500 SF per day)	C8@.016	SF	.40	.66	1.06
6" wide studs (445 SF per day)	C8@.018	SF	.46	.74	1.20

Non-load bearing partitions, 20 gauge, 16" on center spacing

	Craft@Hrs	Unit	Material	Labor	Total
2-1/2" wide studs (470 SF per day)	C8@.017	SF	.80	.70	1.50
3-5/8" wide studs (400 SF per day)	C8@.020	SF	.86	.82	1.68
4" wide studs (385 SF per day)	C8@.021	SF	1.01	.86	1.87
6" wide studs (334 SF per day)	C8@.024	SF	1.16	.98	2.14

Non-load bearing partitions, 20 gauge, 24" on center spacing

	Craft@Hrs	Unit	Material	Labor	Total
2-1/2" wide studs (534 SF per day)	C8@.015	SF	.65	.61	1.26
3-5/8" wide studs (470 SF per day)	C8@.017	SF	.70	.70	1.40
4" wide studs (445 SF per day)	C8@.018	SF	.86	.74	1.60
6" wide studs (400 SF per day)	C8@.020	SF	.99	.82	1.81

Load bearing partitions, 18 gauge "C" section, 16" on center spacing

	Craft@Hrs	Unit	Material	Labor	Total
4" wide studs (267 SF per day)	C8@.030	SF	1.14	1.23	2.37
6" wide studs (250 SF per day)	C8@.032	SF	1.31	1.31	2.62

Load bearing partitions. 18 gauge "C" section, 24" on center spacing

	Craft@Hrs	Unit	Material	Labor	Total
4" wide studs (308 SF per day)	C8@.026	SF	.86	1.07	1.93
6" wide studs (267 SF per day)	C8@.030	SF	1.03	1.23	2.26

Load bearing partitions, 16 gauge "C" section, 16" on center spacing

	Craft@Hrs	Unit	Material	Labor	Total
2-1/2" wide studs (300 SF per day)	C8@.027	SF	1.21	1.11	2.32
3-5/8" wide studs (267 SF per day)	C8@.030	SF	1.37	1.23	2.60
4" wide studs (258 SF per day)	C8@.031	SF	1.50	1.27	2.77
6" wide studs (235 SF per day)	C8@.034	SF	1.76	1.39	3.15

Load bearing partitions, 16 gauge "C" section, 24" on center spacing

	Craft@Hrs	Unit	Material	Labor	Total
3-5/8" wide studs (300 SF per day)	C8@.027	SF	1.12	1.11	2.23
6" wide studs (243 SF per day)	C8@.033	SF	1.39	1.35	2.74

Stud adapters (internal slotted standards for supporting shelf brackets) Assembly includes two load-bearing 3-5/8" wide 16 gauge galvanized "C" section hollow metal studs spaced 1" apart, a steel or aluminum stud adapter, screws and normal waste. These costs assume that studs and adapter are installed at the same time. Capitol Hardware A-Line 1700. Costs per vertical linear foot of assembly height

	Craft@Hrs	Unit	Material	Labor	Total
.080" steel adapter, galvanized, 1/2" slots 1" OC	C8@.042	VLF	15.20	1.72	16.92
.080" aluminum adapter, mill finish, 1/2" slots 1" OC	C8@.042	VLF	14.10	1.72	15.82
Add for caps and foot pieces to meet fire code requirements, press-fit type					
Per set of top and bottom caps for each adapter	C8@.021	Set	1.78	.86	2.64

	Craft@Hrs	Unit	Material	Labor	Total

Door opening steel framing Using 16 gauge load bearing steel studs (hollow metal studs), based on walls 8' high. Figures in parentheses indicate typical built-up header size. Costs shown are per door opening and include header (built up using steel studs), double vertical steel studs each side of the opening less than 8' wide (triple vertical steel studs each side of openings 8' wide or wider), with steel stud cripples and blocking installed using screws. These costs include the studs, screws and normal waste. Width shown is size of finished opening.

3-5/8" wall studs, opening size as shown

	Craft@Hrs	Unit	Material	Labor	Total
Up to 3'0" wide (4" x 4" header)	C8@1.50	Ea	21.60	61.50	83.10
Up to 4'0" wide (4" x 6" header)	C8@1.60	Ea	30.50	65.60	96.10
Up to 5'0" wide (4" x 6" header)	C8@1.80	Ea	34.50	73.70	108.20
Up to 6'0" wide (4" x 8" header)	C8@2.00	Ea	37.00	81.90	118.90
Up to 8'0" wide (4" x 10" header)	C8@2.30	Ea	57.90	94.20	152.10
Up to 10'0" wide (4" x 12" header)	C8@2.40	Ea	70.30	98.30	168.60
Up to 12'0" wide (4" x 14" header)	C8@2.50	Ea	98.60	102.00	200.60
Add per foot of vertical height, walls over 8' high	C8@.250	VLF	3.11	10.20	13.31
Deduct for 2-1/2" wall studs	—	%	-15.0	-10.0	—
Add for 6" wall studs	—	%	30.0	10.0	—

Window opening steel framing Using 16 gauge load bearing steel studs (hollow metal studs), based on walls 8' high. Figures in parentheses indicate typical built-up header size. Costs shown are per window opening and include header (built up using steel studs), sub-sill plate (double sub-sill if opening is 8' wide or wider), double vertical steel studs each side of openings less than 8' wide (triple vertical steel studs each side of openings 8' wide or wider), with steel stud top and bottom cripples and blocking installed using screws. These costs include the studs, screws and normal waste.

3-5/8" wall studs, opening size shown is width of finished opening

	Craft@Hrs	Unit	Material	Labor	Total
Up to 2'0" wide (4" x 4" header)	C8@1.90	Ea	29.70	77.80	107.50
Up to 3'0" wide (4" x 4" header)	C8@2.20	Ea	33.00	90.10	123.10
Up to 4'0" wide (4" x 6" header)	C8@2.50	Ea	50.01	102.00	152.01
Up to 5'0" wide (4" x 6" header)	C8@2.70	Ea	55.70	111.00	166.70
Up to 6'0" wide (4" x 8" header)	C8@3.00	Ea	59.60	123.00	182.60
Up to 7'0" wide (4" x 8" header)	C8@3.30	Ea	76.10	135.00	211.10
Up to 8'0" wide (4" x 10" header)	C8@3.50	Ea	85.00	143.00	228.00
Up to 10'0" wide (4" x 12" header)	C8@3.80	Ea	102.00	156.00	258.00
Up to 12'0" wide (4" x 14" header)	C8@4.00	Ea	133.00	164.00	297.00
Add per foot of vertical height, walls over 8' high	C8@.400	VLF	5.07	16.40	21.47
Deduct for 2-1/2" wall studs	—	%	-15.0	-10.0	—
Add for 6" wall studs	—	%	30.0	10.0	—

Bracing for steel framing See also sheathing for plywood bracing and shear panels

Let-in wall bracing, using 16 gauge load bearing steel studs installed using screws
These costs include screws and normal waste

	Craft@Hrs	Unit	Material	Labor	Total
2-1/2" brace	C8@.027	LF	.53	1.11	1.64
3-5/8" brace	C8@.035	LF	.59	1.43	2.02
6" brace	C8@.035	LF	.78	1.43	2.21

Steel strap bracing, 1-1/4" wide,

	Craft@Hrs	Unit	Material	Labor	Total
9'6" or 11'6" lengths	C8@.010	LF	.45	.41	.86
Steel "V" bracing, 3/4" x 3/4"	C8@.010	LF	.72	.41	1.13

Temporary steel stud frame wall bracing, assumes salvage at 50% and 3 uses

	Craft@Hrs	Unit	Material	Labor	Total
2-1/2", 16 gauge load bearing steel studs	C8@.010	LF	.06	.41	.47
3-5/8", 16 gauge load bearing steel studs	C8@.012	LF	.10	.49	.59
6", 16 gauge load bearing steel studs	C8@.018	LF	.13	.74	.87

	Craft@Hrs	Unit	Material	Labor	Total

Gypsum Wallboard Commercial grade work. Material costs include 6% for waste.
Costs per square foot of area covered.

Gypsum wallboard nailed or screwed to wood framing or wood furring, no taping or finishing included

	Craft@Hrs	Unit	Material	Labor	Total
3/8" on walls	CD@.007	SF	.23	.31	.54
3/8" on ceilings	CD@.009	SF	.23	.40	.63
3/8" on furred columns and beams	CD@.013	SF	.23	.57	.80
1/2" on walls	CD@.007	SF	.23	.31	.54
1/2" on ceilings	CD@.009	SF	.23	.40	.63
1/2" on furred columns and beams	CD@.015	SF	.23	.66	.89
5/8" on walls	CD@.007	SF	.30	.31	.61
5/8" on ceilings	CD@.010	SF	.30	.44	.74
5/8" on furred columns and beams	CD@.015	SF	.30	.66	.96
Add for taping and finishing wall joints	CD@.007	SF	.04	.31	.35
Add for taping and finishing ceiling joints	CD@.009	SF	.04	.40	.44

Gypsum wallboard clipped to metal furring, no taping or finishing included

	Craft@Hrs	Unit	Material	Labor	Total
3/8" on wall furring	CD@.008	SF	.23	.35	.58
3/8" on ceiling furring	CD@.011	SF	.23	.49	.72
1/2" on wall furring	CD@.009	SF	.23	.40	.63
1/2" on ceiling furring	CD@.011	SF	.23	.49	.72
1/2" on column or beam furring	CD@.018	SF	.23	.79	1.02
5/8" on wall furring	CD@.009	SF	.30	.40	.70
5/8" on ceiling furring	CD@.011	SF	.30	.49	.79
5/8" on column or beam furring	CD@.018	SF	.30	.79	1.09
1/2", two layers on ceiling furring	CD@.022	SF	.46	.97	1.43
1/2", two layers on furring	CD@.017	SF	.46	.75	1.21
1/4" sound board	CD@.008	SF	.25	.35	.60
Add for taping and finishing wall joints	CD@.007	SF	.04	.31	.35
Add for taping and finishing ceiling joints	CD@.009	SF	.04	.40	.44
Add for taping only, no joint finishing	CD@.003	SF	.02	.13	.15

Additional costs for gypsum wallboard

	Craft@Hrs	Unit	Material	Labor	Total
Add for foil-backed board	—	SF	.05	—	.05
Add for fire resistant board	—	SF	.03	—	.03
Add for water resistant board	—	SF	.13	—	.13
Add for 10' or 12' wall heights	CD@.001	SF	.02	.04	.06
Add for school jobs	CD@.002	SF	—	.09	.09
Add for trowel textured finish	CD@.009	SF	.40	.40	.80
Add for adhesive application, 1/4" bead					
Studs 16" on center	CD@.002	SF	.02	.09	.11
Joists 16" on center	CD@.002	SF	.02	.09	.11

Bead, casing and channels

	Craft@Hrs	Unit	Material	Labor	Total
Corner bead, 1-1/4" x 1-1/4"	CD@.017	LF	.17	.75	.92
Stop or casing	CD@.021	LF	.25	.93	1.18
Jamb casing	CD@.023	LF	.25	1.02	1.27
RC-1 channel or 7/8" hat channel	CD@.011	LF	.17	.49	.66

Vinyl clad gypsum board, adhesive or clip application on walls

	Craft@Hrs	Unit	Material	Labor	Total
1/2", no mouldings included	CD@.010	SF	.55	.44	.99
5/8", no mouldings included	CD@.011	SF	.58	.49	1.07

	Craft@Hrs	Unit	Material	Labor	Total

Metal Framed Shaft Walls Shaft walls are used to enclose vertical shafts that surround pipe chases, electrical conduit, elevators or stairwells. A 2-hour wall can be made from 1" gypsum wallboard screwed to a 2-1/2" metal stud partition on the shaft side with two layers of 5/8" gypsum wallboard screwed to the partition exterior face. The wall cavity is filled with fiberglass batt insulation and the drywall is taped and finished on both sides. This wall will be 4-3/4" thick. Labor includes installing metal studs, installing the insulation, hanging, taping and finishing the drywall, and cleanup. Metal studs and insulation include a 10% allowance for waste. Wallboard costs include corner and edge trim and a 15% allowance for waste. Costs shown are per square foot of wall measured on one side. For scheduling purposes, estimate that a crew of 4 can install metal studs and insulation, hang, tape and finish 320 SF of shaft wall per 8-hour day.

Metal studs, "C" section, 2-1/2" wide, 20 gauge, 24" on center

	Craft@Hrs	Unit	Material	Labor	Total
Complete with top runner and bottom plate	C8@.036	SF	.74	1.47	2.21
1" type X gypsum shaftboard	CD@.029	SF	.55	1.28	1.83
2 layers 5/8" type X gypsum wallboard	CD@.027	SF	.66	1.19	1.85
Fiberglass insulation, foil faced,	A1@.006	SF	.30	.27	.57
Total for metal framed shaft wall as described	**--@.098**	**SF**	**2.25**	**4.21**	**6.46**

Ceramic Tile Set in adhesive and grouted. Add the cost of adhesive below. No scratch or brown coat included. Based on standard U.S. grades and stock colors. Custom colors, designs and imported tile will cost more.

	Craft@Hrs	Unit	Material	Labor	Total
4-1/4" x 4-1/4" glazed wall tile					
Smooth gloss glaze, minimum quality	T4@.131	SF	2.05	5.19	7.24
Smooth gloss glaze, standard quality	T4@.131	SF	4.51	5.19	9.70
Matte finish, better quality	T4@.131	SF	5.93	5.19	11.12
High gloss finish	T4@.131	SF	7.90	5.19	13.09
Commercial quality					
Group 1 (matte glaze)	T4@.131	SF	3.61	5.19	8.80
Group 2 (bright glaze)	T4@.131	SF	4.42	5.19	9.61
Group 3 (crystal glaze)	T4@.131	SF	5.08	5.19	10.27
6" wall tile, smooth glaze					
6" x 6", gray or brown	T4@.131	SF	2.42	5.19	7.61
6" x 8", quilted look	T4@.131	SF	4.91	5.19	10.10
6" x 8", fume look	T4@.131	SF	4.33	5.19	9.52
Trim pieces for glazed wall tile					
Surface bullnose	T4@.077	LF	1.65	3.05	4.70
Surface bullnose corner	T4@.039	Ea	1.07	1.54	2.61
Sink rail or cap	T4@.077	LF	4.78	3.05	7.83
Radius bullnose	T4@.077	LF	2.18	3.05	5.23
Quarter round or bead	T4@.077	LF	2.58	3.05	5.63
Outside corner or bead	T4@.039	Ea	.85	1.54	2.39
Base	T4@.077	LF	5.85	3.05	8.90
Soap holder, soap dish	T4@.265	Ea	11.60	10.50	22.10
Tissue holder, towel bar	T4@.268	Ea	17.00	10.60	27.60
Glazed floor tile, 1/4" thick					
Minimum quality, standard colors	T4@.210	SF	2.33	8.31	10.64
Better quality, patterns	T4@.210	SF	2.78	8.31	11.09
Commercial quality, 6" x 6" to 12" x 12"					
Standard colors and reds and browns	T4@.210	SF	2.68	8.31	10.99
Blues and greens	T4@.210	SF	3.27	8.31	11.58
Add for abrasive surface	—	SF	.28	—	.28
1" x 1" mosaic tile, back-mounted, natural clays					
Group I colors (browns, reds)	T4@.143	SF	2.32	5.66	7.98
Group II colors (tans, grays)	T4@.143	SF	2.54	5.66	8.20
Group III colors (charcoal, blues)	T4@.143	SF	2.77	5.66	8.43

	Craft@Hrs	Unit	Material	Labor	Total
Ceramic tile adhesive, premixed Type 1, Acryl-4000. 1 gallon covers 70 SF applied with #1 trowel (3/16" x 5/32" V-notch), 50 SF applied with #2 trowel (3-/16" x 1/4" v-notch) or 40 SF applied with #3 trowel (1/4" x 1/4" square-notch)					
1/2 pint	—	Ea	2.80	—	2.80
1 gallon	—	Ea	11.50	—	11.50
3-1/2 gallons	—	Ea	32.10	—	32.10
Tile adhesive applied with #2 trowel	—	SF	.23	—	.23
Thin-set tile mortar. 50 pound bag covers 100 SF applied with #1 trowel (1/4" x 1/4" square-notch), 80 SF applied with #2 trowel (1/4" x 3/8" square-notch), 45 SF applied with #3 trowel (1/2" x 1/2" square notch). Cost per 50 pound bag.					
Standard, gray	—	Ea	8.49	—	8.49
Master-Blend, white	—	Ea	10.00	—	10.00
FlexiBond, gray	—	Ea	26.00	—	26.00
Marble and granite mix	—	Ea	21.60	—	21.60
Master-Blend applied with #3 trowel	—	SF	.22	—	.22
Tile grout, polymer-modified dry grout for joints 1/16" to 1/2" Coverage varies with tile and joint size, cost per 25 pound bag					
Unsanded	—	Ea	15.80	—	15.80
Sanded, light colors	—	Ea	15.90	—	15.90
Sanded, dark colors	—	Ea	11.70	—	11.70
Decorator tile panels, institutional quality, 6" x 9" x 3/4", typical costs					
Unglazed domestic	T4@.102	SF	5.53	4.04	9.57
Wash-glazed domestic	T4@.102	SF	9.20	4.04	13.24
Full glazed domestic	T4@.102	SF	9.73	4.04	13.77
Glazed imported, decorative	T4@.102	SF	23.10	4.04	27.14
Deduct for imported tile	—	%	-20.0	—	—
Plastic tile					
4-1/4" x 4-1/4" x .110"	T4@.052	SF	1.75	2.06	3.81
4-1/4" x 4-1/4" x .050"	T4@.052	SF	1.35	2.06	3.41
Aluminum tile, 4-1/4"	T4@.073	SF	2.86	2.89	5.75
Copper on aluminum, 4-1/4" x 4-1/4"	T4@.076	SF	3.14	3.01	6.15
Stainless steel tile, 4-1/4" x 4-1/4"	T4@.076	SF	5.36	3.01	8.37
Quarry tile set in Portland cement					
5" high base tile	T4@.109	LF	2.74	4.32	7.06
4" x 4" x 1/2" floor tile	T4@.096	SF	3.23	3.80	7.03
6" x 6" x 1/2" floor tile	T4@.080	SF	2.65	3.17	5.82
Quarry tile set in furan resin					
5" high base tile	T4@.096	LF	1.90	3.80	5.70
6" x 6" x 3/4" floor tile	T4@.093	SF	4.55	3.68	8.23
Add for abrasive finish	—	SF	.42	—	.42

Terrazzo

	Craft@Hrs	Unit	Material	Labor	Total
Floors, 1/2" terrazzo topping on an underbed bonded to an existing concrete slab, #1 and #2 chips in gray Portland cement. Add divider strips below					
2" epoxy or polyester	T4@.128	SF	2.86	5.07	7.93
2" conductive	T4@.134	SF	3.71	5.31	9.02
2" with #3 and larger chips	T4@.134	SF	4.92	5.31	10.23
3" with 15 lb felt and sand cushion	T4@.137	SF	3.71	5.42	9.13
2-3/4" with mesh, felt and sand cushion	T4@.137	SF	3.61	5.42	9.03
Additional costs for terrazzo floors. Add for:					
White Portland cement	—	SF	.47	—	.47
Non-slip abrasive, light	—	SF	1.03	—	1.03
Non-slip abrasive, heavy	—	SF	1.55	—	1.55
Countertops, mud set	T4@.250	SF	9.84	9.90	19.74

	Craft@Hrs	Unit	Material	Labor	Total
Wainscot, precast, mud set	T4@.198	SF	8.63	7.84	16.47
Cove base, mud set	T4@.198	LF	2.09	7.84	9.93
Add for 2,000 SF job	—	%	10.0	15.0	—
Add for 1,000 SF job	—	%	40.0	60.0	—
Add for waterproof membrane	—	SF	1.20	—	1.20
Divider strips					
Brass, 12 gauge	T4@.009	LF	2.43	.36	2.79
White metal, 12 gauge	T4@.009	LF	1.12	.36	1.48
Brass 4' OC each way	T4@.004	SF	1.27	.16	1.43
Brass 2' OC each way	T4@.007	SF	2.47	.28	2.75

Acoustical Treatment

Tile board and panels, no suspension grid included, applied to ceilings with staples, not including furring
   Decorative (non-acoustic) Armstrong ceiling tile, 12" x 12" x 1/2", T&G

	Craft@Hrs	Unit	Material	Labor	Total
Washable white	C8@.017	SF	.57	.70	1.27
Grenoble	C8@.017	SF	.62	.70	1.32
Tin design	C8@.017	SF	1.33	.70	2.03
Glenwood	C8@.017	SF	.77	.70	1.47
Windstone	C8@.017	SF	.75	.70	1.45

Wood fiber ceiling tile, Advantage Series, USG. Class C panels for non-acoustical and non fire-rated applications. Tongue-and-groove edging hides staples for a smooth, clean look. 12" x 12" x 1/2" with staple flange. Stapled or set with adhesive.

	Craft@Hrs	Unit	Material	Labor	Total
Lace	C8@.017	SF	0.59	.70	1.29
Tivoli	C8@.017	SF	0.75	.70	1.45
White wood fiber	C8@.017	SF	0.78	.70	1.48

Acoustical ceiling tile adhesive, 237, Henry. For use on 12" x 12" or 12" x 24" tile. Bonds to concrete, concrete block, drywall, plaster and brick. 30-minute working time. One gallon covers 60 square feet

	Craft@Hrs	Unit	Material	Labor	Total
Per square foot	--@.000	SF	.23	—	.23

Sound-absorbing fabric covered wall panels with splines for attaching to structural walls,
   Armstrong Soundsoak, 9' or 10' high
     Embossed 60, with internal splines, 30" wide,

	Craft@Hrs	Unit	Material	Labor	Total
3/4" thick	C8@.038	SF	6.09	1.56	7.65
Soundsoak 80, with internal splines, 24" wide,					
1" thick	C8@.038	SF	8.35	1.56	9.91
Vinyl Soundsoak, 24" wide, 3/4" thick	C8@.038	SF	5.33	1.56	6.89
Encore, 30" wide, 3/4" thick	C8@.038	SF	8.43	1.56	9.99
Aluminum starter spline for panels	C8@.020	LF	1.04	.82	1.86
Aluminum internal spline for panels	C8@.021	LF	1.08	.86	1.94

Suspended Ceiling Grid Systems 5,000 SF job. Add for ceiling tile below.

"T" bar suspension system, no tile included, suspended from ceiling joists with wires

	Craft@Hrs	Unit	Material	Labor	Total
2' x 2' grid	C8@.011	SF	.65	.45	1.10
2' x 4' grid	C8@.010	SF	.48	.41	.89
4' x 4' grid	C8@.009	SF	.32	.37	.69
Add for jobs under 5,000 SF					
Less than 2,500 SF	C8@.007	SF	.40	.29	.69
2,500 SF to 4,500 SF	C8@.005	SF	.30	.20	.50

Ceiling Tile Tile laid in suspended ceiling grid. No suspended ceiling grid system included. Labor column shows the cost of laying tile in a suspended ceiling grid.

Acoustical rated suspended ceiling panels

	Craft@Hrs	Unit	Material	Labor	Total
2' x 2', fissured, "A" fire rated	C8@.012	SF	.37	.49	.86
2' x 2', random texture, "A" fire rated	C8@.012	SF	.58	.49	1.07
2' x 2' "A" fire rated, with Bioguard™	C8@.012	SF	.89	.49	1.38

	Craft@Hrs	Unit	Material	Labor	Total
2' x 2', "A" fire rated, Armstrong "Sahara"	C8@.012	SF	1.00	.49	1.49
2' x 2', Armstrong "Classic"	C8@.012	SF	1.35	.49	1.84
2' x 2' x 3/4", foil back, USG "Cheyenne"	C8@.012	SF	1.40	.49	1.89
2' x 2' x 3/4", USG "Luna", fine textured	C8@.012	SF	1.41	.49	1.90
2' x 4' x 5/8", USG "5th Avenue"	C8@.007	SF	.38	.29	.67
2' x 4' x 5/8", fissured fire code	C8@.007	SF	.44	.29	.73
2' x 4', fiberglass class "A" fire rated	C8@.007	SF	.48	.29	.77
2' x 4' x 5/8", "A" fire rated, USG "5th Ave"	C8@.007	SF	.53	.29	.82
Non-acoustical suspended ceiling panels					
2' x 2' x 5/8", USG "ClimaPlus"	C8@.012	SF	.33	.49	.82
2' x 2' x 5/8", USG "Ceramic ClimaPlus"	C8@.012	SF	1.13	.49	1.52
2' x 2' x 3/4", USG "Saville Row"	C8@.012	SF	1.27	.49	1.76
2' x 2', grid pattern, Armstrong "Prestige"	C8@.012	SF	1.50	.49	1.99
2' x 4', mineral fiber, USG "Plateau"	C8@.007	SF	.27	.29	.56
2' x 4' x 9/16", mineral, USG "Stonehurst"	C8@.007	SF	.43	.29	.72
2' x 4' random texture, "A" fire rated	C8@.007	SF	.96	.29	1.25
2' x 4' clean room, non-perforated	C8@.007	SF	1.70	.29	1.99
Translucent panels, 2' x 4'					
Clear or frosted prismatic	C8@.004	SF	.48	.16	.64
Arctic opal or cracked ice	C8@.004	SF	.48	.16	.64
Clear acrylic	C8@.004	SF	1.05	.16	1.21
White acrylic	C8@.004	SF	1.05	.16	1.21

Wood Flooring

	Craft@Hrs	Unit	Material	Labor	Total
Oak, 3/4" x 2-1/4", unfinished					
Select grade, white	F9@.062	SF	6.36	2.53	8.89
Select, grade, red	F9@.062	SF	7.72	2.53	10.25
Select, grade, ash	F9@.062	SF	7.72	2.53	10.25
Parquet, prefinished, 5/16"					
Ash or quartered oak, select	F9@.053	SF	16.50	2.16	18.66
Ash or quartered oak, natural	F9@.053	SF	12.00	2.16	14.16
Cherry or walnut select grade	F9@.053	SF	18.90	2.16	21.06
Cherry or walnut natural grade	F9@.053	SF	14.30	2.16	16.46
Oak, select plain	F9@.053	SF	13.60	2.16	15.76
Oak, natural plain	F9@.053	SF	11.80	2.16	13.96
Concrete moisture vapor barriers					
Koester VAP primer (3 coat system)	FL@.026	SF	2.75	1.17	3.92
Koester VAP primer (4 coat system	FL@.033	SF	4.25	1.49	5.74
Gym floors, 3/4", maple, including finishing					
On sleepers and membrane	F9@.086	SF	6.23	3.50	9.73
On steel spring system	F9@.114	SF	7.40	4.64	12.04
Connected with steel channels	F9@.097	SF	6.49	3.95	10.44
Softwood floors, fir, unfinished					
1" x 2"	F9@.031	SF	4.01	1.26	5.27
1" x 3"	F9@.031	SF	4.01	1.26	5.27
Finishing wood flooring					
Fill and stain	F9@.010	SF	.09	.41	.50
Two coats shellac	F9@.047	SF	.25	1.91	2.16
Wax coating	F9@.001	SF	.03	.04	.07
Sand, scrape and edge hardwood floor					
Large room or hall, using 12" drum sander and 7" disc sander					
Three cuts	F9@.009	SF	.05	.37	.42
Four cuts	F9@.011	SF	.06	.45	.51

	Craft@Hrs	Unit	Material	Labor	Total
Closet or 4' x 10' area, using 7" disc sander					
Three cuts	F9@.010	SF	.04	.41	.45
Four cuts	F9@.012	SF	.04	.49	.53

Resilient Flooring

Asphalt tile					
9" x 9" x 1/8"	F9@.018	SF	1.17	.73	1.90
Vinyl composition and vinyl tile, 12" x 12"					
No wax, self-stick, Metro, 3 year warranty	F9@.016	SF	.64	.65	1.29
No wax, self-stick, Themes, 5 year warranty	F9@.016	SF	1.05	.65	1.70
No wax, self-stick, urethane, 25 year warranty	F9@.016	SF	1.55	.65	2.20

Sheet vinyl flooring, Armstrong. Includes 10% waste and adhesive

Medley series .08" thick. Urethane no-wax wear layer. Includes 10% waste and adhesive					
Chalk White	BF@.300	SY	16.10	9.08	25.18
Sage Stone	BF@.300	SY	16.50	9.08	25.58
White, Edinburgh	BF@.300	SY	16.30	9.08	25.38
Caspian series .07" thick. Urethane no-wax wear layer. Includes 10% waste and adhesive					
Black dot	BF@.300	SY	13.90	9.08	22.98
Green walkway	BF@.300	SY	13.90	9.08	22.98
Honey Spice, Decker Station	BF@.300	SY	13.90	9.08	22.98
Hunter, corner dot	BF@.300	SY	14.10	9.08	23.18
Navy, corner dot	BF@.300	SY	14.10	9.08	23.18
Park Moss, Mt. Bethel	BF@.300	SY	14.00	9.08	23.08
Spice, Shavano Park	BF@.300	SY	13.80	9.08	22.88
Taupe, corner dot	BF@.300	SY	14.20	9.08	23.28
Taupe, walkway	BF@.300	SY	14.10	9.08	23.18
Warm Beige, walkway	BF@.300	SY	14.00	9.08	23.08
Themes series .08" thick. 2.9 Performance Appearance Rating					
Beige Cameo	BF@.300	SY	11.60	9.08	20.68
Everest White	BF@.300	SY	11.50	9.08	20.58
Green and Natural, Alexandria V	BF@.300	SY	11.70	9.08	20.78
Green, Florantino	BF@.300	SY	11.50	9.08	20.58
Light Shale, Leed's Landing	BF@.300	SY	11.50	9.08	20.58
Natural White, Prescott	BF@.300	SY	11.70	9.08	20.78
Sandstone	BF@.300	SY	11.60	9.08	20.68
White Crystal, Marble Wisp	BF@.300	SY	11.70	9.08	20.78
Sundial series .098 thick. No adhesive needed					
Black/White, Florence 2	BF@.300	SY	12.75	9.08	21.83
Emerald and White, Florence 2	BF@.300	SY	12.90	9.08	21.98
Light Neutral, Antigua 2	BF@.300	SY	12.50	9.08	21.58
Light Oak, Myrtlewood	BF@.300	SY	12.80	9.08	21.88
Light Wood, Parquet de Luxe	BF@.300	SY	12.80	9.08	21.88
Limestone, Quinault Light	BF@.300	SY	12.80	9.08	21.88
Mushroom, Bologna 2	BF@.300	SY	12.90	9.08	21.98
Rust, Chestnut Corner	BF@.300	SY	12.80	9.08	21.88
Stone/White, Romanesque 2	BF@.300	SY	12.70	9.08	21.78
Metro series .045" thick. Vinyl no-wax wear layer. 2.3 Performance Appearance Rating					
Adobe Tan	BF@.300	SY	6.87	9.08	15.95
Blue Green, colorweave	BF@.300	SY	9.30	9.08	18.38
Bluelake	BF@.300	SY	6.89	9.08	15.97
Cocoa	BF@.300	SY	6.85	9.08	15.93
Jewel Flower, colorweave	BF@.300	SY	9.12	9.08	18.20

	Craft@Hrs	Unit	Material	Labor	Total
Media Heights, Sandstone	BF@.300	SY	7.34	9.08	16.42
Navy	BF@.300	SY	9.04	9.08	18.12
Sand	BF@.300	SY	6.84	9.08	15.92
Royelle series .05" thick. 1.3 Performance Appearance Rating					
Black and White, Sheffley	BF@.300	SY	4.54	9.08	13.62
Dark Oak, Augusta	BF@.300	SY	4.63	9.08	13.71
Seashell, Kaley's Korner	BF@.300	SY	4.62	9.08	13.70

Plastic Interlocking Safety Floor Tiles Based on Duragrid, for a job with 1,000 to 5,000 SF of floor tiles.

	Craft@Hrs	Unit	Material	Labor	Total
Firm, smooth tiles, for athletic courts, gym floors, industrial areas					
12" x 12" x 1/2"	F9@.003	SF	3.73	.12	3.85
2" x 12" x 1/2" line strips	F9@.003	LF	1.81	.12	1.93
Firm, non-slip (nippled) tile for tennis courts, wet-weather surface,					
12" x 12" x 1/2"	F9@.003	SF	4.77	.12	4.89
2" x 12" x 1/2" line strips	F9@.003	LF	1.81	.12	1.93
Soft tiles, for anti-fatigue matting, household or deck areas					
12" x 12" x 1/2"	F9@.003	SF	5.19	.12	5.31
2" x 12" x 1/2" line strips	F9@.003	LF	1.81	.12	1.93
Mitered edging strip, 2" x 12" x 1/2"	F9@.003	LF	1.21	.12	1.33
Monogramming/graphics pegs	F9@.085	C	3.59	3.46	7.05
Add for jobs under 1,000 SF	—	%	10.0	20.0	—
Deduct for jobs over 5,000 SF	—	%	-5.0	-5.0	—

Rubber Flooring and Accessories

	Craft@Hrs	Unit	Material	Labor	Total
Rubber tile flooring					
17-13/16" x 17-13/16" x 1/8" circular pattern	F9@.021	SF	8.11	.86	8.97
Base, top set rubber, 1/8" thick					
2-1/2" high	F9@.018	LF	.82	.73	1.55
4" high	F9@.018	LF	1.03	.73	1.76
6" high	F9@.018	LF	1.34	.73	2.07
Base corners, top set rubber, 1/8" thick					
2-1/2" or 4" high	F9@.028	Ea	1.34	1.14	2.48
6" high	F9@.028	Ea	1.69	1.14	2.83
Stair treads, molded rubber, 12-1/2" tread width, per LF of tread length					
1/4" circle design, colors	F9@.067	LF	12.40	2.73	15.13
Rib design, light duty	F9@.067	LF	13.50	2.73	16.23
Tread with abrasive strip, light duty	F9@.071	LF	15.00	2.89	17.89
Tread with abrasive strip, heavy duty	F9@.071	LF	19.40	2.89	22.29
Stair risers, molded rubber, 7" x 1/8"	F9@.038	LF	3.93	1.55	5.48
Stringer cover, .100"					
10" high	F9@.106	LF	3.93	4.32	8.25
Add for project under 2,000 SF	F9@.010	LF	.22	.41	.63
Underlayment, plywood, 1/2"	C8@.009	SF	.37	.37	.74

Shock-absorbing resilient tile, rubber Installed over asphalt, concrete or compacted gravel base. Indoor or outdoor use, for playgrounds, factory floors, aerobic centers, recreational decks and docks. Unit costs below include single component polyurethane adhesive (also listed separately below). CSSI. http://www.carlsurf.com

	Craft@Hrs	Unit	Material	Labor	Total
2-1/4" thickness, meets CPSC guidelines for 6' drop heights					
Full tile, 24" x 24"	F9@.014	SF	8.52	.57	9.09
Diagonal tile	F9@.014	Ea	20.20	.57	20.77
Transition ramp, 48" x 8"	F9@.056	Ea	24.30	2.28	26.58
Accessible ramp, 24" x 24"	F9@.056	Ea	37.80	2.28	40.08
90-degree molded outside corner 16" x 8"	F9@.112	Ea	18.50	4.56	23.06
90-degree molded inside corner 12" x 8"	F9@.110	Ea	18.10	4.48	22.58

	Craft@Hrs	Unit	Material	Labor	Total
45-degree molded outside corner 8" x 8"	F9@.110	Ea	17.90	4.48	22.38
45-degree molded inside corner 8" x 8"	F9@.109	Ea	16.60	4.44	21.04
3-3/4" thickness, meets CPSC guidelines for 11' drop heights					
Full tile, 24" x 24"	F9@.017	SF	14.60	.69	15.29
Ramp, 48" x 12"	F9@.056	Ea	57.30	2.28	59.58
Add for one part polyurethane adhesive (30 SF per gal)	—	Gal	39.80	—	39.80

Composition Flooring 1/4" thick

	Craft@Hrs	Unit	Material	Labor	Total
Acrylic	F9@.075	SF	3.32	3.06	6.38
Epoxy terrazzo	F9@.105	SF	3.78	4.28	8.06
Conductive epoxy terrazzo	F9@.132	SF	5.20	5.38	10.58
Troweled neoprene	F9@.082	SF	4.26	3.34	7.60
Polyester	F9@.065	SF	2.84	2.65	5.49

Carpeting Glue-down carpet installation includes sweeping the floor and applying adhesive. Carpet over pad installation includes laying the pad and stapling or spot adhesive.

	Craft@Hrs	Unit	Material	Labor	Total
Nylon, cut pile, with 3/8" nova pad					
30 oz	F9@.114	SY	14.20	4.64	18.84
36 oz	F9@.114	SY	16.60	4.64	21.24
40 oz	F9@.114	SY	18.90	4.64	23.54
Nylon, level loop, with 3/8" nova pad					
20 oz	F9@.114	SY	14.20	4.64	18.84
24 oz	F9@.114	SY	16.60	4.64	21.24
Base grade, level loop, with 50 oz pad					
18 oz	F9@.114	SY	7.38	4.64	12.02
20 oz	F9@.114	SY	8.27	4.64	12.91
26 oz	F9@.114	SY	8.27	4.64	12.91
Antron nylon, 20 oz, anti-static, no pad	F9@.068	SY	10.00	2.77	12.77
Lee's "Faculty IV" with unibond back, direct glue	F9@.114	SY	19.40	4.64	24.04

Carpet Tile
Modular carpet tiles

	Craft@Hrs	Unit	Material	Labor	Total
"Faculty 11" tiles, 26 oz face weight	F9@.170	SY	30.30	6.93	37.23
"Surfaces" tiles, 38 oz face weight	F9@.170	SY	33.50	6.93	40.43
ESD 26 oz loop pile for computer floors	F9@.170	SY	35.70	6.93	42.63

Velvet Tile Recycled automotive tire material for entrances, stairs, wet areas, etc. Installation includes adhesive. Add for border strips below. El-Do Velvet Tile

	Craft@Hrs	Unit	Material	Labor	Total
12" x 12" tiles	F9@.044	SF	6.49	1.79	8.28
Aluminum border strips for velvet tile					
Add to any tile above	F9@.031	LF	3.24	1.26	4.50

Decontamination and Surface Preparation Dustless removal of undesirable surface contamination and paint from steel and concrete substrates. Applications include PCBs, radioactivity, toxic chemicals and lead-based paints. The removal system is a mechanical process and no water, chemicals or abrasive grit are used. Containment or special ventilation is not required. Based on Pentek Inc. For estimating purposes figure one 55-gallon drum of contaminated waste per 2,500 square feet of coatings removed from surfaces with an average coating thickness of 12 mils. Waste disposal costs are not included. Unit prices shown are based on a three-man crew. Use $4,500 as a minimum charge for work of this type.

Steel and concrete surfaces
Equipment is based on using three pneumatic needle scaler units attached to one air-powered vacuum-waste packing unit. Equipment cost includes one vacuum-waste packing unit, three needle scalers, and one 150 CFM gas or diesel powered air compressor including all hoses and connectors. Add the one-time charge for move-on, move-off and refurbishing of equipment as shown.

	Craft@Hrs	Unit	Material	Labor	Equipment	Total
Steel surfaces Steel Structures Painting Council Surface Preparation (SSPC-SP) specification codes are shown						
Steel surfaces, SSPC-SP-11 (coating mechanically removed, bare metal exposed)						
Overhead surfaces, box girders, obstructed areas						
30 SF of surface area per hour	PA@.100	SF	—	4.54	2.99	7.53
Large unobstructed structural steel members						
40 SF of surface area per hour	PA@.075	SF	—	3.40	2.24	5.64
Above ground storage tanks						
50 SF of surface area per hour	PA@.060	SF	—	2.72	1.80	4.52
Steel surfaces, SSPC-SP-3 (coating mechanically removed, mill scale left in place)						
Overhead surfaces, box girders, obstructed areas						
50 SF of surface area per hour	PA@.060	SF	—	2.72	1.80	4.52
Large unobstructed structural steel members						
70 SF of surface area per hour	PA@.042	SF	—	1.91	1.26	3.17
Above ground storage tanks						
90 SF of surface area per hour	PA@.033	SF	—	1.50	.99	2.49
Door or window frames						
To 3'0" x 6'8"	PA@1.50	Ea	—	68.10	44.90	113.00
Over 3'0" x 6'8" to 6'0" x 8'0"	PA@2.00	Ea	—	90.80	59.90	150.70
Over 6'0" x 8'0" to 12'0 x 20'0"	PA@3.00	Ea	—	136.00	89.80	225.80
Concrete Surfaces						
Concrete lintels, per SF of area	PA@.033	SF	—	1.50	.99	2.49
Concrete walls, measured on one face of wall						
80 SF of surface area per hour	PA@.038	SF	—	1.72	1.14	2.86
Concrete block walls, measured on one face of wall						
70 SF of surface area per hour	PA@.042	SF	—	1.91	1.26	3.17
Add to total estimated costs arrived at using above, one-time cost						
Move-on, move-off						
and refurbish equipment	—	LS	—	—	—	3,400.00

Concrete floor and slab surfaces

Equipment is based on using three pneumatic manually-operated wheel-mounted scabblers attached to one air-powered vacuum-waste packing unit to scarify concrete floors and slabs. Equipment cost includes one vacuum-waste packing unit, three scabblers, and one 150 CFM gas or diesel powered air compressor including all hoses and connectors. Add the one-time charge for move-on, move-off and refurbishing of equipment as shown.

	Craft@Hrs	Unit	Material	Labor	Equipment	Total
Large unobstructed areas; warehouses and aircraft hanger floor slabs						
90 SF of surface area per hour	PA@.033	SF	—	1.50	1.41	2.91
Obstructed areas; narrow aisles, areas in pits						
60 SF of surface area per hour	PA@.050	SF	—	2.27	2.13	4.40
Small areas; around equipment, piping and conduit						
40 SF of surface area per hour	PA@.075	SF	—	3.40	3.20	6.60
Add to total estimated costs arrived at using above, one-time cost						
Move-on, move-off						
and refurbish equipment	—	LS	—	—	—	4,450.00

	Craft@Hrs	Unit	Material	Labor	Total
Painting These costs do not include surface preparation, scaffolding or masking unless noted.					
Rule of thumb, in place costs, by SF of floor area, typical costs					
Commercial or industrial, 2 coats flat and enamel	—	SF	—	—	1.15
Institutional, 2 coats flat, 3 coats enamel	—	SF	—	—	1.40
Hospitals, public buildings	—	SF	—	—	1.50
Deduct for roller application	—	%	—	-25.0	—

	Craft@Hrs	Unit	Material	Labor	Total
Deduct for spray application	—	%	—	-65.0	—
Add for spray application	—	%	25.0	—	—
Add for panel or carved doors	—	%	10.0	35.0	—
Masonry walls, 1 coat primer, 2 coats acrylic latex					
Brush	PA@.019	SF	.20	.86	1.06
Roller	PA@.010	SF	.22	.45	.67
Spray	PA@.003	SF	.22	.14	.36
Masonry walls, 2 coats acrylic latex					
Brush	PA@.014	SF	.16	.64	.80
Roller	PA@.006	SF	.17	.27	.44
Spray	PA@.002	SF	.18	.09	.27
Metal ladders and cages, per linear foot of height, widths to 4', brush work					
Wire brush ladder	PA@.027	LF	—	1.23	1.23
Wire brush ladder and cage	PA@.054	LF	—	2.45	2.45
Primer, 2 coats enamel, ladder or cage	PA@.019	LF	.65	.86	1.51
Metal trim with brush, figure trim less than 6" wide as being 1 square foot for each linear foot					
1 coat primer, 1 coat enamel	PA@.012	SF	.18	.54	.72
Metal walls, 1 coat primer, 2 coats enamel					
Brush	PA@.013	SF	.15	.59	.74
Roller	PA@.008	SF	.16	.36	.52
Spray	PA@.003	SF	.19	.14	.33
Metal walls, 2 coats enamel					
Brush	PA@.009	SF	.11	.41	.52
Roller	PA@.006	SF	.12	.27	.39
Spray	PA@.001	SF	.12	.05	.17
Most metals, primer and 2 coats enamel					
Brush	PA@.023	SF	.16	1.04	1.20
Roller	PA@.010	SF	.17	.45	.62
Spray	PA@.004	SF	.19	.18	.37
Miscellaneous metals, 2 coats enamel					
Brush	PA@.015	SF	.12	.68	.80
Roller	PA@.007	SF	.12	.32	.44
Spray	PA@.003	SF	.14	.14	.28
Pipe, figure 1 SF per LF for pipe 4" diameter or less					
Per coat, brush or glove	PA@.008	SF	.08	.36	.44
Per coat, spray	PA@.004	SF	.12	.18	.30
Pipe insulation, glue size, primer, enamel	PA@.013	SF	.14	.59	.73
Plaster or wallboard ceilings, 2 coats acrylic latex (water-based)					
Brush	PA@.011	SF	.11	.50	.61
Roller	PA@.007	SF	.13	.32	.45
Spray	PA@.003	SF	.18	.14	.32
Plaster or wallboard walls, enamel coats, per coat (oil-based)					
Brush	PA@.005	SF	.06	.23	.29
Roller	PA@.003	SF	.07	.14	.21
Spray	PA@.001	SF	.08	.05	.13
Plaster or wallboard walls, primer coat (water-based)					
Brush	PA@.005	SF	.06	.23	.29
Roller	PA@.003	SF	.06	.14	.20
Spray	PA@.001	SF	.07	.05	.12

Finishes 9

	Craft@Hrs	Unit	Material	Labor	Total
Plaster or wallboard walls, primer coat (oil-based)					
Brush	PA@.005	SF	.07	.23	.30
Roller	PA@.003	SF	.07	.14	.21
Spray	PA@.001	SF	.08	.05	.13
Roofing, 2 coats aluminum paint					
Roller	PA@.005	SF	.13	.23	.36
Spray	PA@.001	SF	.18	.05	.23
Steel tank exteriors, 2 coats red lead and 1 coat enamel					
Brush	PA@.013	SF	.34	.59	.93
Roller	PA@.008	SF	.42	.36	.78
Spray	PA@.002	SF	.43	.09	.52
Structural steel, brush, medium or heavy members					
Primer and 2 coats enamel	PA@.012	SF	.17	.54	.71
2 coats enamel	PA@.008	SF	.12	.36	.48
2 coats primer	PA@.008	SF	.12	.36	.48
2 coats enamel, spray	PA@.004	SF	.18	.18	.36
Epoxy touch-up and 1 coat enamel	PA@.008	SF	.34	.36	.70
Structural steel, brush, light members					
Primer and 2 coats enamel	PA@.024	SF	.17	1.09	1.26
2 coats enamel	PA@.018	SF	.12	.82	.94
2 coats primer	PA@.018	SF	.12	.82	.94
Stucco walls, 2 coats acrylic latex					
Brush	PA@.010	SF	.12	.45	.57
Roller	PA@.006	SF	.15	.27	.42
Spray	PA@.003	SF	.16	.14	.30
Window sash and trim, per linear foot of sash or trim painted each side, brush					
Wood, 1 coat primer, 2 coats oil base	PA@.022	LF	.17	1.00	1.17
Wood, 2 coats oil base	PA@.010	LF	.12	.45	.57
Metal, 1 coat primer, 2 coats enamel	PA@.018	LF	.16	.82	.98
Metal, 2 coats enamel	PA@.012	LF	.11	.54	.65
Wood or plywood ceilings, brush application					
2 coats sealer, 2 coats enamel	PA@.023	SF	.26	1.04	1.30
2 coats enamel	PA@.010	SF	.13	.45	.58
Wood or plywood walls, 2 coats primer, 1 coat alkyd					
Brush	PA@.013	SF	.11	.59	.70
Roller	PA@.007	SF	.13	.32	.45
Spray	PA@.005	SF	.16	.23	.39
Wood or plywood walls, 2 coats primer					
Brush	PA@.009	SF	.11	.41	.52
Roller	PA@.005	SF	.11	.23	.34
Spray	PA@.003	SF	.15	.14	.29
Wood or plywood walls, brush application					
2 coats sealer and varnish, 2 coats enamel	PA@.019	SF	.25	.86	1.11
2 coats enamel	PA@.008	SF	.11	.36	.47
Sealer and varnish, 3 coats	PA@.010	SF	.23	.45	.68
Wood trim, brush, figure trim less than 6" wide at 1 SF per LF					
1 coat primer, 1 coat enamel	PA@.012	SF	.16	.54	.70
Sealer and varnish, 3 coats	PA@.013	SF	.21	.59	.80

	Craft@Hrs	Unit	Material	Labor	Total

Wallcoverings Based on 2,000 SF job. These costs include 10% waste allowance. For estimating purposes figure an average roll of wallpaper will cover 40 SF.

	Craft@Hrs	Unit	Material	Labor	Total
Paperhanging, average quality, labor only	PA@.011	SF	—	.50	.50
Paperhanging, good quality, labor only	PA@.013	SF	—	.59	.59
Vinyl wallcovering, typical costs					
7 oz, light grade	PA@.012	SF	.46	.54	1.00
14 oz, medium grade	PA@.013	SF	.82	.59	1.41
22 oz, heavy grade, premium quality	PA@.017	SF	1.08	.77	1.85
Vinyl, 14 oz, on aluminum backer	PA@.035	SF	.86	1.59	2.45
Flexwood, many veneers, typical job	PA@.053	SF	2.59	2.41	5.00
Cloth wallcovering					
Linen, acrylic backer, scotch guarded	PA@.021	SF	1.08	.95	2.03
Grasscloth	PA@.021	SF	1.70	.95	2.65
Felt	PA@.027	SF	2.15	1.23	3.38
Cork sheathing, 1/8", typical	PA@.027	SF	1.32	1.23	2.55
Blank stock (underliner)	PA@.011	SF	.31	.50	.81
Paper backed vinyl, prepasted, typical	PA@.012	SF	.75	.54	1.29
Untrimmed paper, hand prints, foils, custom	PA@.029	SF	3.05	1.32	4.37
Gypsum impregnated jute fabric Flexi-wall™, with #500 adhesive. Use 360 SF as minimum job size					
Medium weight plaster wall liner #605	PA@.010	SF	.58	.45	1.03
Heavy duty plaster wall liner #609	PA@.010	SF	.66	.45	1.11
Classics line	PA@.012	SF	.68	.54	1.22
Images line	PA@.012	SF	.85	.54	1.39
Trimmed paper, hand prints, foils	PA@.012	SF	1.05	.54	1.59
Add for job less than 200 SF (5 rolls)	—	%	—	20.0	—
Add for small rooms (kitchen, bath)	—	%	10.0	20.0	—
Add for patterns	—	%	—	15.0	—
Deduct for job over 2,000 SF (50 rolls)	—	%	—	-5.0	—

Vinyl Wallcovering and Borders Solid vinyl, prepasted, standard patterns and colors.

	Craft@Hrs	Unit	Material	Labor	Total
Gold floral on green	PA@.028	SF	.52	1.27	1.79
Red, green and gold	PA@.028	SF	.50	1.27	1.77
Ivy and berries	PA@.035	SF	.50	1.59	2.09
Pastel shells on sand	PA@.017	SF	.41	.77	1.18
Coordinated border	PA@.028	SF	.32	1.27	1.59

Hardboard Wallcovering Vinyl clad, printed.

	Craft@Hrs	Unit	Material	Labor	Total
3/16" pegboard	C8@.025	SF	.46	1.02	1.48
1/4" perforated white garage liner	C8@.025	SF	1.16	1.02	2.18
Add for adhesive, 1/4" bead, 16" OC	—	SF	.11	—	.11
Add for metal trim	—	LF	.40	—	.40

Specialties 10

Chalkboards, Whiteboards, and Tackboards

School and institutional grade chalkboards

Chalkboard with 1/4" hardboard backing

	Craft@Hrs	Unit	Material	Labor	Total
4' x 6'	CC@.034	SF	4.70	1.55	6.25
4' x 8'	CC@.034	SF	4.60	1.55	6.15
4' x 10'	CC@.034	SF	4.80	1.55	6.35
4' x 12'	CC@.034	SF	4.70	1.55	6.25

	Craft@Hrs	Unit	Material	Labor	Total
Chalkboard with 1/2" particleboard backing					
4' x 6'	CC@.034	SF	7.70	1.55	9.25
4' x 8'	CC@.034	SF	7.40	1.55	8.95
4' x 10'	CC@.034	SF	8.00	1.55	9.55
4' x 12'	CC@.034	SF	7.80	1.55	9.35
4' x 16'	CC@.034	SF	7.60	1.55	9.15
Melamine whiteboards					
Aluminum frame					
3' x 4'	CC@.034	SF	3.60	1.55	5.15
3' x 5'	CC@.034	SF	3.90	1.55	5.45
4' x 5'	CC@.034	SF	4.10	1.55	5.65
4' x 6'	CC@.034	SF	4.00	1.55	5.55
4' x 9'	CC@.034	SF	3.40	1.55	4.95
4' x 10'	CC@.034	SF	4.90	1.55	6.45
4' x 12'	CC@.034	SF	4.40	1.55	5.95
Wood frame					
3' x 4'	CC@.034	SF	3.60	1.55	5.15
3' x 5'	CC@.034	SF	4.95	1.55	6.50
4' x 5'	CC@.034	SF	4.45	1.55	6.00
4' x 6'	CC@.034	SF	4.10	1.55	5.65
4' x 8'	CC@.034	SF	3.90	1.55	5.45
Corkboards					
Natural cork – aluminum frame					
3' x 5'	CC@.032	SF	3.80	1.46	5.26
4' x 5'	CC@.032	SF	3.40	1.46	4.86
4' x 6'	CC@.032	SF	3.10	1.46	4.56
4' x 8'	CC@.032	SF	3.10	1.46	4.56
4' x 10'	CC@.032	SF	3.00	1.46	4.46
4' x 12'	CC@.032	SF	2.95	1.46	4.41
Natural cork – wood frame					
3' x 5'	CC@.032	SF	4.60	1.46	6.06
4' x 5'	CC@.032	SF	4.10	1.46	5.56
4' x 6'	CC@.032	SF	3.70	1.46	5.16
4' x 8'	CC@.032	SF	3.75	1.46	5.21
4' x 10'	CC@.032	SF	4.00	1.46	5.46
4' x 12'	CC@.032	SF	3.75	1.46	5.21
Map and display rail, aluminum					
1" x 6'	CC@.032	LF	2.32	1.46	3.78
1" x 8'	CC@.032	LF	2.35	1.46	3.81
1" x 10'	CC@.032	LF	2.37	1.46	3.83
1" x 12'	CC@.032	LF	2.32	1.46	3.78
2" x 6'	CC@.032	LF	3.67	1.46	5.13
2" x 8'	CC@.032	LF	3.67	1.46	5.13
2" x 10'	CC@.032	LF	3.59	1.46	5.05
2" x 12'	CC@.032	LF	2.32	1.46	3.78
3" x 6'	CC@.032	LF	4.62	1.46	6.08
3" x 8'	CC@.032	LF	4.49	1.46	5.95
3" x 10'	CC@.032	LF	4.65	1.46	6.11
3" x 12'	CC@.032	LF	4.56	1.46	6.02
Map and display rail, oak					
2" x 6'	CC@.032	LF	4.08	1.46	5.54

	Craft@Hrs	Unit	Material	Labor	Total

Cigarette Disposal Units

20 gauge stainless steel with grate. For cigarettes and cigars

	Craft@Hrs	Unit	Material	Labor	Total
5-1/4" x 5-3/16" x 8-3/8" wall mounted	CM@.500	Ea	133.00	21.80	154.80
5-1/4" x 5-3/16" x 8-3/8" pole mounted	CM@.750	Ea	143.00	32.70	175.70
13" x 16" x 43" wall mounted	CM@.600	Ea	163.00	26.10	189.10
13" x 16" x 43" pole mounted	CM@.900	Ea	184.00	39.20	223.20
15" x 16" x 43" wall mounted with trash can	CM@.750	Ea	383.00	32.70	415.70
15" x 16" x 43" pole mounted with trash can	CM@1.25	Ea	400.00	54.50	454.50

Compartments and Cubicles

Toilet partitions, floor mounted overhead braced. standard sizes, includes 1 panel 1 door and all pilasters and hardware

	Craft@Hrs	Unit	Material	Labor	Total
Baked enamel	T5@2.66	Ea	530.00	116.00	646.00
Solid plastic (polymer)	T5@4.00	Ea	690.00	174.00	864.00
Laminated plastic	T5@2.66	Ea	545.00	116.00	661.00
Stainless steel	T5@2.66	Ea	753.00	116.00	869.00
Add for wheel chair units	—	Ea	48.40	—	48.40

Urinal screens, wall mounted with brackets, 18" wide

	Craft@Hrs	Unit	Material	Labor	Total
Standard quality, baked enamel	T5@1.95	Ea	115.00	84.70	199.70
Solid plastic (polymer)	T5@1.50	Ea	245.00	65.20	310.20
Laminated plastic	T5@1.03	Ea	165.00	44.80	209.80
Stainless steel	T5@1.03	Ea	386.00	44.80	430.80
Add for floor mounted, 24" wide	T5@.476	Ea	31.50	20.70	52.20

Sight screens, 58" x 58"

	Craft@Hrs	Unit	Material	Labor	Total
Baked enamel, floor mounted	T5@1.28	Ea	82.00	55.60	137.60
Solid plastic (polymer), floor mounted	T5@2.00	Ea	170.00	86.90	256.90
Laminated plastic, floor mounted	T5@1.28	Ea	90.00	55.60	145.60
Stainless steel, floor mounted	T5@1.28	Ea	250.00	55.60	305.60
Add for ceiling-mounted units	—	Ea	42.60	—	42.60

Accessories mounted on partitions

	Craft@Hrs	Unit	Material	Labor	Total
Coat hooks and door stop, chrome	T5@.166	Ea	9.10	7.21	16.31
Purse shelf, chrome, 5" x 14"	T5@.166	Ea	20.40	7.21	27.61

Dressing cubicles, 80" high, with curtain, floor mounted, overhead braced

	Craft@Hrs	Unit	Material	Labor	Total
Baked enamel	T5@3.41	Ea	328.00	148.00	476.00
Solid plastic (polymer)	T5@4.52	Ea	510.00	196.00	706.00
Laminated plastic	T5@3.41	Ea	400.00	148.00	548.00
Stainless steel	T5@3.41	Ea	668.00	148.00	816.00

Shower compartments, industrial type, with receptor but no door or plumbing

	Craft@Hrs	Unit	Material	Labor	Total
Baked enamel	T5@2.68	Ea	884.00	116.00	1,000.00
Solid plastic (polymer)	T5@4.04	Ea	1,020.00	176.00	1,196.00
Laminated plastic	T5@2.68	Ea	995.00	116.00	1,111.00
Stainless steel	T5@2.68	Ea	1,670.00	116.00	1,786.00
Add for soap dish	T5@.166	Ea	10.50	7.21	17.71
Add for curtain rod	T5@.166	Ea	7.27	7.21	14.48

Doors for industrial shower compartments

	Craft@Hrs	Unit	Material	Labor	Total
Wire or tempered glass, 24" x 72"	G1@1.03	Ea	110.00	42.20	152.20
Plastic, 24" x 72"	G1@1.03	Ea	74.70	42.20	116.90
Plastic, 1 panel door and 1 side panel	G1@1.03	Ea	155.00	42.20	197.20

Specialties 10

	Craft@Hrs	Unit	Material	Labor	Total
Shower stalls, with receptor, 32" x 32", with door but no plumbing					
Fiberglass	P6@2.65	Ea	290.00	117.00	407.00
Painted metal	P6@2.65	Ea	304.00	117.00	421.00
Hospital cubicle curtain track, ceiling mount	MW@.060	LF	5.19	2.75	7.94

Baby Changing Stations Polyethylene with steel on steel hinges. Holds up to 350 lbs. Changing stations require 22" horizontal wall space. Costs include sanitary bed liner dispenser, safety straps, necessary hardware and metal door plaques.

	Craft@Hrs	Unit	Material	Labor	Total
Horizontal baby changing station, 20" H x 35" W	BL@.750	Ea	225.00	18.50	243.50
Add for sanitary bed liners, pack of 500	—	Ea	50.00	—	50.00
Add for changing stations replacement strap	—	Ea	18.00	—	18.00

Child Protection Seat Polyethylene with steel on steel hinges. Holds up to 150 lbs. Requires 19 square inches wall space. Costs include safety straps and necessary hardware.

	Craft@Hrs	Unit	Material	Labor	Total
Child protection seat, 19" H x 12" W	BL@.550	Ea	75.00	13.60	88.60
Add for replacement safety strap	—	Ea	8.75	—	8.75

Wall Protection Systems These costs include a .063" extruded aluminum retainer, with a .100" extruded plastic (Acrovyn) covering, installed with adhesives or mechanical fasteners over existing drywall, doors or door frames. No substrate costs are included.

	Craft@Hrs	Unit	Material	Labor	Total
Flush-mounted corner guards, 3" x 3" wide legs	CC@.194	LF	10.10	8.84	18.94
Surface-mounted corner guards, 3" x 3" wide legs	CC@.069	LF	5.67	3.14	8.81
Handrail	CC@.111	LF	11.20	5.06	16.26
Bumper guards, bed bumpers	CC@.138	LF	7.58	6.29	13.87
Door frame protectors	CC@.056	LF	11.40	2.55	13.95
Door protectors	CC@.083	SF	3.64	3.78	7.42
Protective wallcovering (vinyl only)	PA@.087	SF	2.87	3.95	6.82

Access Flooring Standard 24" x 24" steel flooring panels with 1/8" high pressure laminate surface. Costs shown include the floor panels and the supporting understructures listed. Add for accessories below. Raised Floor Installation, Inc.

	Craft@Hrs	Unit	Material	Labor	Total
Stringerless system, up to 10" finish floor height					
Under 1,000 SF job	D3@.058	SF	12.60	2.64	15.24
1,000 to 5,000 SF job	D3@.046	SF	10.80	2.09	12.89
Over 5,000 SF job	D3@.039	SF	9.50	1.78	11.28
Snap-on stringer system, up to 10" finish floor height					
Under 1,000 SF job	D3@.058	SF	13.80	2.64	16.44
1,000 to 5,000 SF job	D3@.046	SF	11.70	2.09	13.79
Over 5,000 SF job	D3@.039	SF	10.50	1.78	12.28
Rigid bolted stringer system, up to 18" finish floor height					
Under 1,000 SF job	D3@.058	SF	14.60	2.64	17.24
1,000 to 5,000 SF job	D3@.046	SF	12.30	2.09	14.39
Over 5,000 SF job	D3@.039	SF	11.10	1.78	12.88
Corner lock system, under 2,500 SF office areas, 6" to 10" finish floor height, blank panels	D3@.036	SF	6.67	1.64	8.31
Used flooring, most floor system types					
Unrefurbished, typical	D3@.058	SF	9.10	2.64	11.74
Refurbished, new panel surface and trim	D3@.058	SF	12.50	2.64	15.14
Extras for all floor systems, add to the costs above					
Carpeted panels	—	SF	1.25	—	1.25
Concrete-filled steel panels	—	SF	.73	—	.73
Ramps, per SF of actual ramp area	D3@.104	SF	7.99	4.74	12.73

	Craft@Hrs	Unit	Material	Labor	Total
Non-skid tape on ramp surface	D3@.033	SF	4.90	1.50	6.40
Portable ramp for temporary use	—	SF	54.10	—	54.10
Steps, per SF of actual tread area	D3@.208	SF	13.20	9.47	22.67
Fascia at steps and ramps, to 18" high	D3@.117	LF	9.24	5.33	14.57
Cove base, 4" high, black or brown	D3@.026	LF	1.51	1.18	2.69
Guardrail	D3@.573	LF	54.60	26.10	80.70
Handrail	D3@.573	LF	25.10	26.10	51.20
Cable cutouts					
Standard "L" type trim	D3@.407	Ea	8.68	18.50	27.18
Type "F" trim, under floor plenum systems	D3@.407	Ea	10.00	18.50	28.50
Perforated air distribution panels	—	SF	3.49	—	3.49
Air distribution grilles, 6" x 18"					
With adjustable damper, including cutout	D3@.673	Ea	65.00	30.60	95.60
Fire extinguishing systems, smoke detectors, halon type, subcontract					
With halon gas suppression	—	SF	—	—	5.20
Automatic fire alarm	—	SF	—	—	.81

Flagpoles Including erection cost but no foundation. See foundation costs below. By pole height above ground, except as noted. Buried dimension is usually 10% of pole height.

Fiberglass, semi-gloss white, tapered, external halyard, commercial grade

	Craft@Hrs	Unit	Material	Labor	Total
25' with 5.90" butt diameter	D4@8.19	Ea	1,000.00	337.00	1,337.00
30' with 6.50" butt diameter	D4@9.21	Ea	1,200.00	379.00	1,579.00
35' with 7.00" butt diameter	D4@9.96	Ea	1,500.00	410.00	1,910.00
39' with 7.00" butt diameter	D4@10.2	Ea	1,760.00	420.00	2,180.00
50' with 10.00" butt diameter	D4@15.1	Ea	4,000.00	621.00	4,621.00
60' with 10.50" butt diameter	D4@18.5	Ea	5,790.00	761.00	6,551.00
70' with 11.40" butt diameter	D4@21.7	Ea	8,000.00	893.00	8,893.00
80' with 16.00" butt diameter	D4@22.8	Ea	9,180.00	938.00	10,118.00

As above but with internal halyard, jam cleat type

	Craft@Hrs	Unit	Material	Labor	Total
25' with 6.00" butt diameter	D4@8.19	Ea	1,290.00	337.00	1,627.00
30' with 6.50" butt diameter	D4@9.21	Ea	1,440.00	379.00	1,819.00
35' with 7.00" butt diameter	D4@9.96	Ea	1,790.00	410.00	2,200.00
39' with 7.00" butt diameter	D4@10.2	Ea	1,990.00	420.00	2,410.00
50' with 10.00" butt diameter	D4@15.1	Ea	5,390.00	621.00	6,011.00
60' with 10.50" butt diameter	D4@18.5	Ea	6,430.00	761.00	7,191.00
70' with 11.40" butt diameter	D4@21.7	Ea	8,250.00	893.00	9,143.00
80' with 16.00" butt diameter	D4@22.8	Ea	10,400.00	938.00	11,338.00
Add for nautical yardarm mast, based on height of flagpole					
Per foot of pole height	—	LF	28.00	—	28.00

Aluminum, satin finish, tapered, external halyard, heavy duty type

	Craft@Hrs	Unit	Material	Labor	Total
20' x .188" wall, with 5" butt diam.	D4@8.19	Ea	868.00	337.00	1,205.00
25' x .188" wall, with 5" butt diam.	D4@8.19	Ea	1,010.00	337.00	1,347.00
30' x .188" wall, with 6" butt diam.	D4@9.21	Ea	1,200.00	379.00	1,579.00
35' x .188" wall, with 7" butt diam.	D4@9.96	Ea	1,560.00	410.00	1,970.00
40' x .188" wall, with 8" butt diam.	D4@10.2	Ea	2,060.00	420.00	2,480.00
45' x .188" wall, with 8" butt diam.	D4@10.5	Ea	2,300.00	432.00	2,732.00
50' x .188" wall, with 10" butt diam.	D4@15.1	Ea	3,150.00	621.00	3,771.00
60' x .250" wall, with 12" butt diam.	D4@18.5	Ea	6,070.00	761.00	6,831.00
70' x .250" wall, with 12" butt diam.	D4@21.7	Ea	6,760.00	893.00	7,653.00
80' x .375" wall, with 12" butt diam.	D4@25.2	Ea	8,710.00	1,040.00	9,750.00

Specialties 10

	Craft@Hrs	Unit	Material	Labor	Total
Wall-mounted aluminum poles, including wall mount on concrete or masonry structure					
8' high	D4@7.95	Ea	875.00	327.00	1,202.00
10' high	D4@9.69	Ea	890.00	399.00	1,289.00
15' with wall support	D4@11.1	Ea	1,190.00	457.00	1,647.00
20' with wall support	D4@14.4	Ea	1,560.00	593.00	2,153.00
Steel poles, white enamel finish, lightweight					
12' high	D4@6.41	Ea	55.20	264.00	319.20
18' high	D4@6.41	Ea	73.60	264.00	337.60
22' high	D4@7.13	Ea	152.00	293.00	445.00
30' high	D4@8.97	Ea	317.00	369.00	686.00
Concrete poles, off-white or brown finish, with internal halyard, overall heights including buried portion					
20' high	D4@6.41	Ea	1,240.00	264.00	1,504.00
30' high	D4@8.83	Ea	1,660.00	363.00	2,023.00
40' high	D4@10.4	Ea	2,110.00	428.00	2,538.00
50' high	D4@14.4	Ea	4,790.00	593.00	5,383.00
60' high	D4@16.0	Ea	6,670.00	658.00	7,328.00
70' high	D4@19.2	Ea	8,550.00	790.00	9,340.00
80' high	D4@22.3	Ea	10,500.00	918.00	11,418.00
90' high	D4@27.2	Ea	12,700.00	1,120.00	13,820.00
100' high	D4@31.8	Ea	14,100.00	1,310.00	15,410.00
Add for yardarm nautical mast					
11' aluminum yardarm	—	Ea	928.00	—	928.00
13'6" aluminum yardarm	—	Ea	982.00	—	982.00
12' fiberglass yardarm	—	Ea	803.00	—	803.00
Add for locking cleat box	D4@2.42	Ea	107.00	99.60	206.60
Add for eagle mounted on spindle rod					
12" bronze	—	Ea	59.90	—	59.90
12" natural	—	Ea	74.00	—	74.00
16" bronze	—	Ea	101.00	—	101.00
16" natural	—	Ea	125.00	—	125.00

		Clear	Bronze	Black
Add for anodized finish, aluminum or steel pole				
Per foot of pole height, heights to 50' per LF	LF	10.00	13.00	15.00
Per foot of pole height, heights over 50' per LF	LF	15.00	20.00	25.00

	Craft@Hrs	Unit	Material	Labor	Total
Poured concrete flagpole foundations, including excavation, metal collar, steel casing, base plate, sand fill and ground spike, by pole height above ground, typical prices					
30' high or less	P9@10.5	Ea	64.40	439.00	503.40
35' high	P9@10.8	Ea	70.10	452.00	522.10
40' high	P9@12.5	Ea	75.90	523.00	598.90
45' high	P9@13.2	Ea	81.90	552.00	633.90
50' high	P9@15.8	Ea	99.30	661.00	760.30
60' high	P9@19.2	Ea	117.00	803.00	920.00
70' high	P9@22.5	Ea	151.00	941.00	1,092.00
80' high	P9@26.0	Ea	188.00	1,090.00	1,278.00
90' to 100' high	P9@30.1	Ea	234.00	1,260.00	1,494.00
Direct embedded pole base, augered in place, no concrete required, by pole height					
20' to 25' pole	D4@1.88	Ea	319.00	77.40	396.40
25' to 40' pole	D4@2.71	Ea	456.00	112.00	568.00
45' to 55' pole	D4@2.99	Ea	513.00	123.00	636.00
60' to 70' pole	D4@3.42	Ea	570.00	141.00	711.00

	Craft@Hrs	Unit	Material	Labor	Total

Identifying Devices

Architectural signage, engraved, self-adhesive plastic signs with high contrast white lettering

	Craft@Hrs	Unit	Material	Labor	Total
4" x 12", stock wording	PA@.191	Ea	20.50	8.67	29.17
4" x 12", custom wording	PA@.191	Ea	27.00	8.67	35.67
3" x 8", stock wording	PA@.191	Ea	17.90	8.67	26.57
3" x 8", custom wording	PA@.191	Ea	24.20	8.67	32.87
Add for aluminum mounting frame					
4" x 12"	PA@.096	Ea	17.10	4.36	21.46
3" x 8"	PA@.096	Ea	10.30	4.36	14.66

Directory boards, felt-covered changeable letter boards, indoor mount

	Craft@Hrs	Unit	Material	Labor	Total
Open faced, aluminum frame					
18" x 2', 1 door	C8@.635	Ea	115.00	26.00	141.00
2' x 3', 1 door	C8@1.31	Ea	164.00	53.70	217.70
4' x 3', 2 doors	C8@1.91	Ea	257.00	78.30	335.30
Open faced, wood frame					
18" x 2', 1 door	C8@.635	Ea	98.50	26.00	124.50
2' x 3', 1 door	C8@1.31	Ea	136.00	53.70	189.70
4' x 3', 2 doors	C8@1.91	Ea	202.00	78.30	280.30
Acrylic door, aluminum frame					
18" x 2', 1 door	C8@.635	Ea	257.00	26.00	283.00
2' x 3', 1 door	C8@1.31	Ea	362.00	53.70	415.70
4' x 3', 2 doors	C8@1.91	Ea	674.00	78.30	752.30

Hand-painted lettering, typical subcontract prices

	Craft@Hrs	Unit	Material	Labor	Total
Standard, per square foot	—	SF	—	—	15.00
Gold leaf, per square foot	—	SF	—	—	29.90

Illuminated letters, porcelain enamel finish, stock letters, aluminum with Plexiglass face, add electrical work

	Craft@Hrs	Unit	Material	Labor	Total
12" high x 2-1/2" wide, 1 tube per letter	C8@.849	Ea	139.00	34.80	173.80
24" high x 5" wide, 2 tubes per letter	C8@1.61	Ea	283.00	66.00	349.00
36" high x 6-1/2" wide, 2 tubes per letter	C8@3.28	Ea	439.00	134.00	573.00

Fabricated back-lighted letters, aluminum with neon illumination, baked enamel finish, script, add electrical work

	Craft@Hrs	Unit	Material	Labor	Total
12" high x 3" deep, 1 tube	G1@.834	Ea	75.30	34.20	109.50
24" high x 6" deep, 2 tubes	G1@1.41	Ea	166.00	57.80	223.80
36" high x 8" deep, 2 tubes	G1@2.53	Ea	301.00	104.00	405.00

Cast aluminum letters, various styles, For use on building interior or exterior

	Craft@Hrs	Unit	Material	Labor	Total
2" high	C8@.215	Ea	14.50	8.81	23.31
6" high	C8@.320	Ea	61.00	13.10	74.10
12" high	C8@.320	Ea	127.00	13.10	140.10
14" high	C8@.456	Ea	193.00	18.70	211.70
Add for baked enamel finish	—	%	2.0	—	—
Add for satin bronze finish	—	%	75.0	—	—

Braille signage, meets ADA and CABO/ANSI A117.1 requirements, various colors, interior (See ADA section)

	Craft@Hrs	Unit	Material	Labor	Total
8" x 8"	PA@.250	Ea	37.50	11.30	48.80
2" x 8"	PA@.250	Ea	22.70	11.30	34.00

Plaques, including standard lettering, expansion or toggle bolt applied

	Craft@Hrs	Unit	Material	Labor	Total
Aluminum, 12" x 4"	G1@.900	Ea	173.00	36.90	209.90
Aluminum, 24" x 18"	G1@2.70	Ea	878.00	111.00	989.00
Aluminum, 24" x 36"	G1@2.91	Ea	1,600.00	119.00	1,719.00
Bronze, 12" x 4"	G1@.900	Ea	232.00	36.90	268.90
Bronze, 24" x 18"	G1@2.70	Ea	1,330.00	111.00	1,441.00
Bronze, 24" x 36"	G1@2.91	Ea	2,470.00	119.00	2,589.00

	Craft@Hrs	Unit	Material	Labor	Total
Safety and warning signs, stock signs ("High Voltage," "Fire Exit," etc.). Indoor, high performance plastic					
10" x 7"	PA@.099	Ea	13.40	4.49	17.89
14" x 10"	PA@.099	Ea	19.30	4.49	23.79
20" x 14"	PA@.099	Ea	30.90	4.49	35.39
Add for aluminum	—	%	40.0		—
Vandal resistant 14" x 10"	PA@.099	Ea	41.00	4.49	45.49
Indoor, pressure sensitive vinyl signs					
10" x 7"	PA@.099	Ea	10.60	4.49	15.09
14" x 10"	PA@.099	Ea	12.70	4.49	17.19

Turnstiles

	Craft@Hrs	Unit	Material	Labor	Total
Non-register type, manual	D4@2.78	Ea	640.00	114.00	754.00
Register type, manual	D4@2.78	Ea	910.00	114.00	1,024.00
Attached ticket collection box, register type, add	—	Ea	230.00	—	225.00
Register type, portable	—	Ea	1,400.00	—	1,400.00

Lockers, Metal Baked enamel finish, key lock 12" wide x 60" high x 15" deep overall dimensions

	Craft@Hrs	Unit	Material	Labor	Total
Single tier, cost per locker	D4@.571	Ea	130.00	23.50	153.50
Double tier, cost per 2 lockers	D4@.571	Ea	150.00	23.50	173.50
Five tier, cost per 5 lockers	D4@.571	Ea	170.00	23.50	193.50
Six tier, cost per 6 lockers	D4@.571	Ea	200.00	23.50	223.50
Add for 72" height	—	%	10.0	—	—
Add for 15" width	—	%	25.0	—	—

Benches, Locker Room Wooden, floor-mounted.

	Craft@Hrs	Unit	Material	Labor	Total
Maple, 6' long	D4@.753	Ea	205.00	31.00	236.00

Benches, Park For public and commercial use. Cast aluminum with treated 1-1/2" by 2-1/4" pine wood slats. Central Park or Victorian style.

	Craft@Hrs	Unit	Material	Labor	Total
4' long	D4@.750	Ea	229.00	30.90	259.90
5' long	D4@.750	Ea	259.00	30.90	289.90
6' long	D4@.750	Ea	289.00	30.90	319.90

Postal Specialties

Letter boxes, cost per each, recessed mounted

	Craft@Hrs	Unit	Material	Labor	Total
Aluminum					
15" x 19" private access	D4@1.10	Ea	135.00	45.30	180.30
15" x 19" USPS access	D4@1.10	Ea	125.00	45.30	170.30
Brass					
15" x 19" private access	D4@1.10	Ea	135.00	45.30	180.30
15" x 19" USPS access	D4@1.10	Ea	125.00	45.30	170.30

Brass indoor mailboxes for private use. Solid brass die-cast doors with lock, 1/4" glass window and name plate

	Craft@Hrs	Unit	Material	Labor	Total
Rear loading 30 door unit, 3-1/2" W x 5" H door	C8@1.10	Ea	580.00	45.10	625.10
Rear loading 14 door unit, 3-1/2" W x 5" H door	C8@1.00	Ea	540.00	41.00	581.00
Add for front loading unit	—	%	20.0	—	—

Aluminum rack ladder system, private use, 1/4" extruded aluminum doors. 20 gauge cold rolled steel compartments, open back. Unit dimensions 23-1/4" W x 12-1/8" H x 15-1/2" D

	Craft@Hrs	Unit	Material	Labor	Total
Rear loading 12 door unit, 3-1/2" W x 5-1/2" H door	C8@1.50	Ea	160.00	61.50	221.50
Rear loading 8 door unit, 5-1/2" W x 5-1/2" H door	C8@1.50	Ea	150.00	61.50	211.50
Rear loading 4 door unit, 10-3/4" W x 5-1/2" H door	C8@1.50	Ea	140.00	61.50	201.50
Rack ladder, 61-7/8" H x 15-1/2" D	C8@.500	Ea	70.00	20.50	90.50
Steel rotary mail center, 30" W x 74" H x 24" D	C8@1.00	Ea	900.00	41.00	941.00

	Craft@Hrs	Unit	Material	Labor	Total

Vertical mailboxes, aluminum, USPS approved, surface mounted or recessed. Door size 5-1/2" W x 15-3/4" H, front loading

	Craft@Hrs	Unit	Material	Labor	Total
7 door unit, 40-3/4" W x 19-1/2" H x 7-1/4" D	C8@1.00	Ea	175.00	41.00	216.00
5 door unit, 29-3/4" W x 19-1/2" H x 7-1/4" D	C8@1.00	Ea	125.00	41.00	166.00
3 door unit, 18-3/4" W x 19-1/2" H x 7-1/4" D	C8@1.00	Ea	75.00	41.00	116.00

Pedestal mailboxes, aluminum neighborhood delivery and collection box units with weather protection hood. USPS approved, rear access. Doors are 6-3/8" W x 5-1/4" H

	Craft@Hrs	Unit	Material	Labor	Total
16 door unit, 26-1/2" W x 22-1/2" H x 22" D	C8@1.10	Ea	1,175.00	45.10	1,220.10
12 door unit, 20" W x 22-1/2" H x 22" D	C8@1.10	Ea	1,125.00	45.10	1,170.10
32" replacement pedestal	C8@1.00	Ea	75.00	41.00	116.00

Movable Office Partitions Prefabricated units. Cost per SF of partition wall measured one side. Use 100 SF as minimum job size.

Gypsum board on 2-1/2" metal studs (includes finish on both sides)

	Craft@Hrs	Unit	Material	Labor	Total
1/2" board, unpainted, STC 38	C8@.028	SF	3.94	1.15	5.09
5/8" board, unpainted, STC 40	C8@.028	SF	4.02	1.15	5.17
1/2" or 5/8", unpainted, STC 45	C8@.028	SF	4.44	1.15	5.59
1/2" vinyl-covered board, STC 38	C8@.028	SF	4.75	1.15	5.90
5/8" vinyl-covered board, STC 40	C8@.028	SF	4.99	1.15	6.14
Metal-covered, baked enamel finish	C8@.028	SF	10.50	1.15	11.65
Add for factory vinyl wrap, 15 to 22 oz	—	SF	.40	—	.40
Windows, including frame and glass, add to above, factory installed					
3'6" x 2'0"	—	Ea	93.60	—	93.60
3'6" x 4'0"	—	Ea	134.00	—	134.00
Wood doors, hollow core, prefinished, add to above, factory installed					
1-3/4", 3'0" x 7'0"	—	Ea	265.00	—	265.00
Metal door jambs, 3'6" x 7'0"	—	Ea	151.00	—	151.00
Metal door jambs, 6'0" x 7'0"	—	Ea	259.00	—	259.00
Passage hardware set	—	Ea	60.60	—	60.60
Lockset hardware	—	LS	81.40	—	81.40
Corner for gypsum partitions	—	Ea	100.00	—	100.00
Corner for metal-covered partitions	—	Ea	197.00	—	197.00
Starter, gypsum partitions	—	Ea	33.40	—	33.40
Starter, metal-covered partitions	—	Ea	71.30	—	71.30
Metal base	—	LF	2.02	—	2.02
Cubicles and welded booths, 5' x 5' x 5'	C8@.028	SF	11.40	1.15	12.55

Banker type divider partition, subcontract

	Craft@Hrs	Unit	Material	Labor	Total
5'6" high including 18" glass top, 2" thick, vinyl finish, not including door or gates	—	SF	—	—	10.30

Demountable Partitions Prefinished units. Cost per SF of partition measured one side. Use 100 SF as minimum job size. Labor shown applies to installing or removing.

Modular panels, spring pressure, mounted to floor and ceiling with removable tracks

	Craft@Hrs	Unit	Material	Labor	Total
Standard panel, non-rated	C8@.124	SF	13.50	5.08	18.58

Accordion Folding Partitions

Vinyl-covered particleboard core, top hung, including structural supports and track

	Craft@Hrs	Unit	Material	Labor	Total
Economy, to 8' high, STC 36	—	SF	10.10	—	10.10
Economy, to 30' wide x 17' high, STC 41	—	SF	13.70	—	13.70
Good quality, to 30' W x 17' H, STC 43	—	SF	16.10	—	16.10

	Craft@Hrs	Unit	Material	Labor	Total
Better quality, to 30' W x 17' H, STC 44	—	SF	16.60	—	16.60
Better quality, large openings, STC 45	—	SF	15.40	—	15.40
Better quality, extra large openings, STC 47	—	SF	16.60	—	16.60
Installation of folding partitions	C8@.647	LF	—	26.50	26.50
Add for prefinished birch or ash wood slats accordion folding partitions					
With vinyl hinges, 15' W x 8' H maximum	—	SF	6.60	—	6.60

Folding fire partitions, accordion, with automatic closing system, baked enamel finish, with standard hardware and motor, cost per 5' x 7' module

	Craft@Hrs	Unit	Material	Labor	Total
20 minute rating	C8@22.6	Ea	4,260.00	926.00	5,186.00
60 minute rating	C8@22.6	Ea	5,410.00	926.00	6,336.00
90 minute rating	C8@22.6	Ea	5,820.00	926.00	6,746.00
2 hour rating	C8@22.6	Ea	6,840.00	926.00	7,766.00

Folding leaf partitions, typical subcontract prices. Not including header or floor track recess, vinyl covered

	Craft@Hrs	Unit	Material	Labor	Total
Metal panels, 7.5 lbs per SF, top hung, hinged,					
(STC 52), 16' high x 60' wide maximum	—	SF	—	—	49.90
Metal panels, 7.5 lbs per SF, top hung, single panel pivot,					
(STC 48), 18' high x 60' wide maximum	—	SF	—	—	51.00
Wood panels, hinged, floor or ceiling hung, 6 lbs per SF,					
(STC 40), 12' high x 36' wide maximum	—	SF	—	—	41.20
Add for laminated plastic	—	SF	—	—	2.30
Add for unfinished wood veneer	—	SF	—	—	2.80
Add for chalkboard, large area	—	SF	—	—	2.25

Woven Wire Partitions 10 gauge painted, 1-1/2" steel mesh.

Wall panels, 8' high, installed on a concrete slab

	Craft@Hrs	Unit	Material	Labor	Total
1' wide	H6@.437	Ea	87.90	19.80	107.70
2' wide	H6@.491	Ea	103.00	22.30	125.30
3' wide	H6@.545	Ea	120.00	24.70	144.70
4' wide	H6@.545	Ea	130.00	24.70	154.70
5' wide	H6@.598	Ea	138.00	27.10	165.10
Add for galvanized mesh	—	SF	.53	—	.53
Corner posts for 8' high walls	H6@.270	Ea	23.70	12.20	35.90
Stiffener channel posts for 8' walls	H6@.277	Ea	54.20	12.60	66.80
Service window panel, 8' H x 5' W panel	H6@.836	Ea	310.00	37.90	347.90

Sliding doors, 8' high

	Craft@Hrs	Unit	Material	Labor	Total
3' wide	H6@2.94	Ea	446.00	133.00	579.00
4' wide	H6@3.29	Ea	475.00	149.00	624.00
5' wide	H6@3.29	Ea	506.00	149.00	655.00
6' wide	H6@3.58	Ea	596.00	162.00	758.00

Hinged doors, 8' high

	Craft@Hrs	Unit	Material	Labor	Total
3' wide	H6@2.85	Ea	360.00	129.00	489.00
4' wide	H6@3.14	Ea	398.00	142.00	540.00

Dutch doors, 8' high

	Craft@Hrs	Unit	Material	Labor	Total
3' wide	H6@2.85	Ea	545.00	129.00	674.00
4' wide	H6@3.14	Ea	637.00	142.00	779.00

Complete wire partition installation including posts and typical doors, per SF of wall and ceiling panel

	Craft@Hrs	Unit	Material	Labor	Total
Painted wall panels	H6@.029	SF	4.07	1.32	5.39
Galvanized wall panels	H6@.029	SF	4.66	1.32	5.98
Painted ceiling panels	H6@.059	SF	3.72	2.68	6.40
Galvanized ceiling panels	H6@.059	SF	4.23	2.68	6.91

	Craft@Hrs	Unit	Material	Labor	Total
Truck Scales With dial and readout. Steel platform. These costs do not include concrete work or excavation.					
5 ton, 6' x 8' platform	D4@45.9	Ea	7,400.00	1,890.00	9,290.00
20 ton, 10' x 24' platform	D4@70.2	Ea	10,800.00	2,890.00	13,690.00
50 ton, 10' x 60' platform	D4@112.	Ea	18,900.00	4,610.00	23,510.00
80 ton, 10' x 60' platform	D4@112.	Ea	21,900.00	4,610.00	26,510.00

Recessed Wall and Floor Safes Safes with 1" thick steel door, 1/4" hardplate protecting locking mechanism, welded deadbolts, and key or combination lock, internal relock device.

In-floor safes	Craft@Hrs	Unit	Material	Labor	Total
15" L x 14" W x 15" D, 12" x 12" door	D4@2.33	Ea	555.00	95.90	650.90
15" L x 14" W x 20" D, 12" x 12" door	D4@2.33	Ea	568.00	95.90	663.90
24" L x 24" W x 17" D, 12" x 12" door	D4@2.33	Ea	790.00	95.90	885.90
Add for installation in existing building	D4@2.33	Ea	—	95.90	95.90

Storage Shelving

Retail store gondolas (supermarket shelving units), material prices vary widely, check local source

	Craft@Hrs	Unit	Material	Labor	Total
Install only	D4@.333	LF	—	13.70	13.70

Steel industrial shelving, enamel finish, 2 sides & back, 12" deep x 30" wide x 75" high

	Craft@Hrs	Unit	Material	Labor	Total
5 shelves high	D4@.758	Ea	95.20	31.20	126.40
8 shelves high	D4@1.21	Ea	131.00	49.80	180.80

Steel industrial shelving, enamel finish, open on sides, 12" deep x 36" wide x 75" high

	Craft@Hrs	Unit	Material	Labor	Total
5 shelves high	D4@.686	Ea	55.40	28.20	83.60
8 shelves high	D4@1.06	Ea	84.60	43.60	128.20

Stainless steel modular shelving, 20" deep x 36" wide x 63" high

	Craft@Hrs	Unit	Material	Labor	Total
3 shelves, 16 gauge	D4@1.19	Ea	932.00	49.00	981.00
Stainless steel wall shelf, 12" deep, bracket mount	D4@.675	LF	42.70	27.80	70.50

Stainless steel work counters, with under-counter storage shelf, 16 gauge

	Craft@Hrs	Unit	Material	Labor	Total
4' wide x 2' deep x 3' high	D4@1.36	Ea	1000.00	56.00	1,056.00
6' wide x 2' deep x 3' high	D4@2.01	Ea	1,520.00	82.70	1,602.70

Library type metal shelving, 90" high, 8" deep, 5 shelves

	Craft@Hrs	Unit	Material	Labor	Total
30" long units	D4@1.17	Ea	165.00	48.10	213.10
36" long units	D4@1.30	Ea	179.00	53.50	232.50

Telephone Enclosures Typical subcontract prices

	Craft@Hrs	Unit	Material	Labor	Total
Shelf type, wall mounted, 15" D x 28" W x 30" H	—	Ea	—	—	700.00

Wall mounted stainless steel directory shelf. Typical subcontract price

	Craft@Hrs	Unit	Material	Labor	Total
3 directory	—	Ea	—	—	660.00

Toilet and Bath Accessories Material cost only. See labor following this section

Combination towel/waste units	Unit	Commercial Industrial Grade	Institutional Grade
Recessed paper towel dispenser	Ea	151.00	187.00
Recessed roll towel dispenser	Ea	265.00	409.00
Surface paper towel dispenser	Ea	43.00	59.00
Surface mounted roll towel dispenser	Ea	163.00	246.00
Recessed roll towel dispenser with waste bin	Ea	429.00	499.00
Recessed paper towel dispenser with waste bin	Ea	119.00	354.00

		Commercial Industrial Grade	Institutional Grade
Waste receptacles			
Recessed, 12 gallon	Ea	154.00	187.00
Surface mounted	Ea	60.00	147.00
Floor standing	Ea	124.00	200.00
Countertop	Ea	—	275.00
Wall urns			
Recessed	Ea	121.00	141.00
Surface mounted	Ea	81.00	111.00
Floor standing with waste	Ea	—	132.00
Sanitary napkin dispenser, recessed	Ea	260.00	409.00
Soap dispensers			
Liquid, 20 oz, lavatory mounted	Ea	31.00	31.00
Liquid, 34 oz, lavatory mounted	Ea	36.00	36.00
Powder	Ea	33.00	46.00
Under-counter reservoir system, with 3 stations	Ea	—	901.00
Add for additional stations	Ea	—	77.00
Liquid, surface mounted	Ea	14.50	31.00
Powdered, surface mounted	Ea	11.50	72.00
Toilet paper dispensers			
Multi roll, recessed	Ea	62.00	94.00
Multi roll, surface mounted	Ea	48.00	73.00
Toilet seat cover dispensers			
Recessed	Ea	54.00	96.00
Surfaced mounted	Ea	28.00	40.00
Soap dishes			
With grab bar, recessed	Ea	18.50	23.00
With grab bar, surface mounted	Ea	30.00	37.00
Grab bars, stainless steel, wall mounted			
1-1/2" x 12"	Ea	—	21.00
1-1/2" x 18"	Ea	—	20.00
1-1/2" x 24"	Ea	—	23.00
1-1/2" x 30"	Ea	—	24.00
1-1/2" x 36"	Ea	—	25.00
1-1/2" x 48"	Ea	—	27.00
1-1/4" x 18"	Ea	—	48.00
1-1/4" x 24"	Ea	—	52.00
1-1/4" x 36"	Ea	—	64.00
1", 90 degree angle, 30"	Ea	—	75.00
1-1/4", 90 degree angle, 30"	Ea	—	92.00
Towel bars			
18" chrome	Ea	—	33.00
24" chrome	Ea	—	34.00
18" stainless steel	Ea	—	35.00
24" stainless steel	Ea	—	37.00
Medicine cabinets, with mirror			
Swing door, 16" x 22", wall hung	Ea	64.00	244.50
Swing door, 16" x 22", recessed	Ea	75.00	208.00
Mirrors, unframed, polished stainless steel surface			
16" x 20"	SF	43.00	—
16" x 24"	SF	49.00	—
18" x 24"	SF	52.00	—
18" x 30"	SF	62.00	—

		Commercial Industrial Grade	Institutional Grade
24" x 30"	SF	80.00	—
24" x 36"	SF	101.00	—
Mirrors, stainless steel frame, 1/4" glass surface			
16" x 24"	Ea	—	38.10
18" x 24"	Ea	—	40.40
18" x 30"	Ea	—	48.50
18" x 36"	Ea	—	63.60
24" x 30"	Ea	—	67.10
24" x 36"	Ea	—	78.50
24" x 48"	Ea	—	97.60
24" x 60"	Ea	—	137.70
72" x 36"	Ea	—	232.60
Toilet partitions, based on recessed layout with 3 compartments up to 36" wide x 60" deep.			
Floor anchored or ceiling hung	Ea	468.00	667.00
Overhead braced	Ea	489.00	689.00
Floor to ceiling anchored	Ea	725.00	875.00
Urinal partitions			
Floor anchored			
Up to 24"	Ea	247.00	284.00
25" to 36"	Ea	268.00	325.00
37" to 48"	Ea	284.00	405.00
49" to 60"	Ea	305.00	436.00
Post to ceiling			
Up to 24"	Ea	185.00	247.00
25" to 36"	Ea	302.00	281.00
37" to 48"	Ea	218.00	300.00
49" to 60"	Ea	240.00	320.00
Ceiling hung			
Up to 24"	Ea	247.00	357.00
25" to 36"	Ea	268.00	389.00
37" to 48"	Ea	284.00	403.00
49" to 60"	Ea	305.00	430.00
Wall hung			
Up to 18"	Ea	113.00	168.00
19" to 24"	Ea	117.00	170.00
Shelves, stainless steel, 6" deep			
18" long	Ea	44.00	51.00
24" long	Ea	46.00	53.00
36" long	Ea	59.00	68.00
72" long	Ea	97.00	127.00
Shower rod, end flanges, vinyl curtain			
6' long, straight	Ea	46.30	57.80
Robe hooks			
Single	Ea	6.70	9.90
Double	Ea	8.20	10.80
Single with door stop	Ea	7.60	12.40
Straddle bar, stainless steel			
24" x 24" x 20"	Ea	98.20	114.00
Urinal bar, 24" long, stainless steel	Ea	79.90	95.90
Hot air hand dryer, 115 volt			
add electric connection	Ea	510.00	614.00

Specialties 10

	Craft@Hrs	Unit	Material	Labor	Total
Labor to install toilet and bath accessories					
Toilet paper dispensers					
Surface mounted	CC@.182	Ea	—	8.29	8.29
Recessed	CC@.310	Ea	—	14.10	14.10
Toilet seat cover dispensers					
Surface mounted	CC@.413	Ea	—	18.80	18.80
Recessed	CC@.662	Ea	—	30.20	30.20
Feminine napkin dispenser, surface mounted	CC@.760	Ea	—	34.60	34.60
Feminine napkin disposer, recessed	CC@.882	Ea	—	40.20	40.20
Soap dishes	CC@.230	Ea	—	10.50	10.50
Soap dispensers, surface mounted	CC@.254	Ea	—	11.60	11.60
Soap dispensers, recessed	CC@.628	Ea	—	28.60	28.60
Soap dispenser system	CC@2.91	Ea	—	133.00	133.00
Paper towel dispensers and waste receptacles					
Semi-recessed	CC@1.96	Ea	—	89.30	89.30
Recessed	CC@2.57	Ea	—	117.00	117.00
Medicine cabinets with mirror, swing door					
Wall hung	CC@.943	Ea	—	43.00	43.00
Mirrors					
To 5 SF	CG@.566	Ea	—	25.80	25.80
5 to 10 SF	CG@.999	Ea	—	45.50	45.50
Over 10 SF	CG@.125	SF	—	5.70	5.70
Towel bars and grab bars	CC@.286	Ea	—	13.00	13.00
Shelves, stainless steel, to 72"	CC@.569	Ea	—	25.90	25.90
Robe hooks	CC@.182	Ea	—	8.29	8.29
Straddle and urinal bars	CC@.484	Ea	—	22.00	22.00
Hot air hand dryer, add for electric circuit wiring	CE@.908	Ea	—	45.90	45.90

Coat and Hat Racks Assembly and installation of prefabricated coat and hat racks. Bevco.
Wall-mounted coat and hat rack with pilfer-resistant hangers, chrome (#200-SHB series)

	Craft@Hrs	Unit	Material	Labor	Total
24" length, 9 hangers	CC@.665	Ea	73.60	30.30	103.90
36" length, 13 hangers	CC@.748	Ea	85.30	34.10	119.40
48" length, 17 hangers	CC@.833	Ea	98.70	37.90	136.60
60" length, 21 hangers	CC@.916	Ea	115.00	41.70	156.70
72" length, 25 hangers	CC@.997	Ea	133.00	45.40	178.40

Church Glass Material cost includes fabrication of the lite. Labor cost includes setting in a prepared opening only.
Minimum labor cost is usually $150 per job. Add the cost of scaffold or a hoist, if needed.

	Craft@Hrs	Unit	Material	Labor	Total
Stained glass					
Simple artwork	G1@.184	SF	69.10	7.54	76.64
Moderate artwork	G1@.318	SF	175.00	13.00	188.00
Elaborate artwork	G1@.633	SF	556.00	25.90	581.90
Faceted glass					
Simple artwork	G1@.184	SF	62.50	7.54	70.04
Moderate artwork	G1@.318	SF	123.00	13.00	136.00
Elaborate artwork	G1@.633	SF	170.00	25.90	195.90
Colored glass, no artwork					
Single pane	G1@.056	SF	6.83	2.30	9.13
Patterned	G1@.184	SF	14.40	7.54	21.94
Small pieces	G1@.257	SF	36.90	10.50	47.40

	Craft@Hrs	Unit	Material	Labor	Total
Bank Equipment					
Vault door and frame, 78" x 32", 10", single					
2 hour, 10 bolt	D4@11.9	Ea	3,350.00	490.00	3,840.00
4 hour, 10 bolt	D4@11.9	Ea	3,770.00	490.00	4,260.00
6 hour, 10 bolt	D4@11.9	Ea	4,420.00	490.00	4,910.00
Vault door and frame, 78" x 40", 10", double					
4 hour, 18 bolt	D4@20.8	Ea	4,930.00	856.00	5,786.00
6 hour, 18 bolt	D4@20.8	Ea	5,290.00	856.00	6,146.00
Teller windows					
Drive-up, motorized drawer, projected	D4@26.8	Ea	11,100.00	1,100.00	12,200.00
Walk-up, one teller	D4@17.6	Ea	5,660.00	724.00	6,384.00
Walk-up, two teller	D4@17.6	Ea	6,770.00	724.00	7,494.00
After hours depository doors					
Envelope and bag, whole chest, complete	D4@37.1	Ea	9,570.00	1,530.00	11,100.00
Envelope only	D4@4.72	Ea	1,140.00	194.00	1,334.00
Flush mounted, bag only	D4@5.44	Ea	1,700.00	224.00	1,924.00
Teller counter, modular components	C8@.532	LF	106.00	21.80	127.80
Square check desks, 48" x 48", 4 person	C8@2.00	Ea	1,260.00	81.90	1,341.90
Round check desks, 48" diameter, 4 person	C8@2.00	Ea	1,390.00	81.90	1,471.90
Rectangular check desks, 72" x 24", 4 person	C8@2.00	Ea	1,180.00	81.90	1,261.90
Rectangular check desks, 36" x 72", 8 person	C8@2.00	Ea	2,310.00	81.90	2,391.90
Safe deposit boxes, modular units					
18 openings, 2" x 5"	D4@3.42	Ea	1,250.00	141.00	1,391.00
30 openings, 2" x 5"	D4@3.42	Ea	1,580.00	141.00	1,721.00
42 openings, 2" x 5"	D4@3.42	Ea	2,030.00	141.00	2,171.00
Base, 32" x 24" x 3"	D4@1.11	Ea	120.00	45.70	165.70
Canopy top	D4@.857	Ea	68.60	35.30	103.90
Cash dispensing automatic teller, with phone	CE@84.3	Ea	45,600.00	4,260.00	49,860.00
Video camera surveillance system, 3 cameras	CE@15.6	Ea	2,950.00	788.00	3,738.00
Perimeter alarm system, typical price	CE@57.7	Ea	7,400.00	2,910.00	10,310.00

Ecclesiastical Equipment Labor includes unpacking, layout and complete installation of prefinished furniture and equipment.

	Craft@Hrs	Unit	Material	Labor	Total
Lecterns, 16" x 24"					
Plain	C8@4.48	Ea	576.00	184.00	760.00
Deluxe	C8@4.48	Ea	1,070.00	184.00	1,254.00
Premium	C8@4.48	Ea	2,010.00	184.00	2,194.00
Pulpits					
Plain	C8@4.77	Ea	653.00	195.00	848.00
Deluxe	C8@4.77	Ea	1,030.00	195.00	1,225.00
Premium	C8@4.77	Ea	2,530.00	195.00	2,725.00
Arks, with curtain					
Plain	C8@5.97	Ea	677.00	245.00	922.00
Deluxe	C8@5.97	Ea	885.00	245.00	1,130.00
Premium	C8@5.97	Ea	1,470.00	245.00	1,715.00
Arks, with door					
Plain	C8@8.95	Ea	766.00	367.00	1,133.00
Deluxe	C8@8.95	Ea	1,300.00	367.00	1,667.00
Premium	C8@8.95	Ea	2,010.00	367.00	2,377.00
Pews, bench type					
Plain	C8@.382	LF	42.40	15.70	58.10
Deluxe	C8@.382	LF	46.20	15.70	61.90
Premium	C8@.382	LF	52.60	15.70	68.30

	Craft@Hrs	Unit	Material	Labor	Total
Pews, seat type					
Plain	C8@.464	LF	53.20	19.00	72.20
Deluxe	C8@.491	LF	69.30	20.10	89.40
Premium	C8@.523	LF	82.40	21.40	103.80
Kneelers					
Plain	C8@.185	LF	10.50	7.58	18.08
Deluxe	C8@.185	LF	13.40	7.58	20.98
Premium	C8@.185	LF	20.70	7.58	28.28
Cathedral chairs, shaped wood					
Without accessories	C8@.235	Ea	136.00	9.63	145.63
With book rack	C8@.235	Ea	147.00	9.63	156.63
With book rack and kneeler	C8@.235	Ea	163.00	9.63	172.63
Cathedral chairs, upholstered					
Without accessories	C8@.285	Ea	142.00	11.70	153.70
With book rack	C8@.285	Ea	152.00	11.70	163.70
With book rack and kneeler	C8@.285	Ea	184.00	11.70	195.70
Adaptable fabric upholstered chair seating (Sauder)					
Hardwood laminated frame, interlocking legs, book rack, stackable, "Modlok"					
Standard	C8@.066	Ea	115.00	2.70	117.70
Deluxe	C8@.066	Ea	121.00	2.70	123.70
Solid oak frame, stackable, "Oaklok"	C8@.066	Ea	94.60	2.70	97.30
Laminated beech frame, stackable, "Plylok"	C8@.066	Ea	103.00	2.70	105.70
Laminated beech frame, stacks & folds, "Plyfold"	C8@.066	Ea	89.40	2.70	92.10
Confessionals, single, with curtain					
Plain	C8@9.11	Ea	2,440.00	373.00	2,813.00
Deluxe	C8@9.11	Ea	2,930.00	373.00	3,303.00
Premium	C8@9.11	Ea	3,350.00	373.00	3,723.00
Add for door in place of curtain	—	Ea	390.00	—	390.00
Confessionals, double, with curtain					
Plain	C8@13.7	Ea	3,870.00	561.00	4,431.00
Deluxe	C8@13.7	Ea	4,480.00	561.00	5,041.00
Premium	C8@13.7	Ea	5,510.00	561.00	6,071.00
Communion rail, hardwood, with standards					
Plain	C8@.315	LF	44.70	12.90	57.60
Deluxe	C8@.315	LF	56.20	12.90	69.10
Premium	C8@.315	LF	67.00	12.90	79.90
Communion rail, carved oak, with standards					
Embellished	C8@.315	LF	72.80	12.90	85.70
Ornate	C8@.315	LF	97.30	12.90	110.20
Premium	C8@.315	LF	160.00	12.90	172.90
Communion rail, bronze or stainless steel,					
With standards	H6@.285	LF	105.00	12.90	117.90
Altars, hardwood, sawn					
Plain	C8@5.97	Ea	786.00	245.00	1,031.00
Deluxe	C8@5.97	Ea	927.00	245.00	1,172.00
Premium	C8@5.97	Ea	1,260.00	245.00	1,505.00
Altars, carved hardwood					
Plain	C8@5.97	Ea	2,350.00	245.00	2,595.00
Deluxe	C8@6.73	Ea	5,420.00	276.00	5,696.00
Premium	C8@7.30	Ea	6,160.00	299.00	6,459.00
Altars, with Nimmerillium fonts, marble	D4@27.8	Ea	4,010.00	1,140.00	5,150.00

	Craft@Hrs	Unit	Material	Labor	Total
Altars, marble or granite					
Plain	D4@29.1	Ea	4,170.00	1,200.00	5,370.00
Deluxe	D4@29.1	Ea	6,060.00	1,200.00	7,260.00
Premium	D4@29.1	Ea	8,980.00	1,200.00	10,180.00
Altars, marble legs and base					
Deluxe	D4@29.1	Ea	2,500.00	1,200.00	3,700.00
Premium	D4@29.1	Ea	3,580.00	1,200.00	4,780.00

Theater & Stage Curtains

Straight track for manual curtains, see lifting and drawing equipment on next page

	Craft@Hrs	Unit	Material	Labor	Total
Standard duty					
24' wide stage	D4@.486	LF	24.70	20.00	44.70
40' wide stage	D4@.315	LF	24.70	13.00	37.70
60' wide stage	D4@.233	LF	24.70	9.59	34.29
Medium duty					
24' wide stage	D4@.486	LF	33.60	20.00	53.60
40' wide stage	D4@.315	LF	33.60	13.00	46.60
60' wide stage	D4@.233	LF	33.60	9.59	43.19
Heavy duty					
24' wide stage	D4@.486	LF	34.30	20.00	54.30
40' wide stage	D4@.315	LF	34.30	13.00	47.30
60' wide stage	D4@.233	LF	34.30	9.59	43.89

Curved track, manual operation, see lifting and drawing equipment following

	Craft@Hrs	Unit	Material	Labor	Total
Standard duty					
24' wide stage	D4@.486	LF	45.80	20.00	65.80
40' wide stage	D4@.371	LF	45.80	15.30	61.10
60' wide stage	D4@.315	LF	45.80	13.00	58.80
Heavy duty					
24' wide stage	D4@.584	LF	140.00	24.00	164.00
40' wide stage	D4@.427	LF	119.00	17.60	136.60
60' wide stage	D4@.342	LF	119.00	14.10	133.10

Cyclorama track, manual, see lifting and drawing equipment following

	Craft@Hrs	Unit	Material	Labor	Total
24' wide stage	D4@.486	LF	29.50	20.00	49.50
40' wide stage	D4@.315	LF	29.50	13.00	42.50
60' wide stage	D4@.233	LF	29.50	9.59	39.09

Stage curtains, Subcontract Complete costs for custom fabricated, flameproof stage curtains including labor to hang. No track or curtain lifting or drawing equipment included. By percent of added fullness

	Craft@Hrs	Unit	Material	Labor	Total
Heavyweight velour					
0% fullness	—	SF	—	—	4.30
50% fullness	—	SF	—	—	5.50
75% fullness	—	SF	—	—	5.60
100% fullness	—	SF	—	—	6.10
Lightweight velour					
0% fullness	—	SF	—	—	3.50
50% fullness	—	SF	—	—	3.90
75% fullness	—	SF	—	—	4.30
100% fullness	—	SF	—	—	5.25
Muslin, 0% fullness	—	SF	—	—	2.50
Repp cloth					
0% fullness	—	SF	—	—	2.70
50% fullness	—	SF	—	—	3.80
75% fullness	—	SF	—	—	4.10
100% fullness	—	SF	—	—	4.55

	Craft@Hrs	Unit	Material	Labor	Total
Stage curtain lifting and drawing equipment Labor is equipment installation, see electrical connection below					
Lift curtain equipment					
1/3 HP	D4@7.50	Ea	2,370.00	309.00	2,679.00
1/2 HP	D4@7.50	Ea	4,690.00	309.00	4,999.00
3/4 HP	D4@7.50	Ea	4,970.00	309.00	5,279.00
Draw curtain equipment					
1/3 HP	D4@7.50	Ea	3,210.00	309.00	3,519.00
1/2 HP	D4@7.50	Ea	3,500.00	309.00	3,809.00
3/4 HP	D4@7.50	Ea	3,780.00	309.00	4,089.00
Add for electrical connection. Typical subcontract price for 40' wire run with conduit, circuit breaker, junction box and starter switch					
Per electrical connection	—	Ea	—	—	1,350.00
Educational Equipment Labor includes unpacking, layout and complete installation of prefinished furniture and equipment.					
Wardrobes, 40" x 78" x 26"					
Teacher type	C8@2.00	Ea	1,170.00	81.90	1,251.90
Student type	C8@2.00	Ea	587.00	81.90	668.90
Seating, per seat					
Lecture room, pedestal, with folding arm	C8@2.00	Ea	153.00	81.90	234.90
Horizontal sections	C8@2.00	Ea	73.40	81.90	155.30
Tables, fixed pedestal					
48" x 16"	C8@2.00	Ea	332.00	81.90	413.90
With chairs, 48" x 16"	C8@2.00	Ea	476.00	81.90	557.90
Projection screen					
Ceiling mounted, pull down	D4@.019	SF	3.70	.78	4.48
Videotape recorder	CE@.500	Ea	153.00	25.30	178.30
T.V. camera, color	CE@.800	Ea	612.00	40.40	652.40
T.V. monitor, wall mounted, color	CE@2.00	Ea	357.00	101.00	458.00
Language laboratory study carrels, plastic laminated wood					
Individual, 48" x 30" x 54"	C8@1.43	Ea	209.00	58.60	267.60
Two position, 73" x 30" x 47"	C8@1.93	Ea	559.00	79.10	638.10
Island, 4 position, 66" x 66" x 47"	C8@3.17	Ea	631.00	130.00	761.00
Checkroom Equipment					
Automatic electric checkroom conveyor, 350 coat capacity					
Conveyor and controls	D4@15.2	Ea	3,070.00	625.00	3,695.00
Add for electrical connection for checkroom conveyor, typical subcontract price					
40' wire run with conduit, circuit breaker, and junction box	—	LS	—	—	388.00
Food Service Equipment These costs do not include electrical work or plumbing. Add the cost of hookup when needed. Based on Superior Products Company. (Figure in parentheses is equipment weight.)					
Beverage dispenser, non-carbonated, refrigerated, countertop model					
5 gallon, 1 bowl	P6@2.00	Ea	700.00	88.60	788.60
5 gallon, 2 bowls	P6@2.00	Ea	980.00	88.60	1,068.60
2.4 gallon, 2 mini bowls	P6@2.00	Ea	845.00	88.60	933.60
2.4 gallon, 4 mini bowls	P6@2.00	Ea	1,370.00	88.60	1,458.60
Charbroiler, lava rock, countertop model, burner and control every 12", natural or LP gas					
40,000 Btu, 15" W (121 lbs)	P6@2.00	Ea	605.00	88.60	693.60
80,000 Btu, 24" W (186 lbs)	P6@2.00	Ea	720.00	88.60	808.60
120,000 Btu, 36" W (230 lbs)	P6@2.00	Ea	975.00	88.60	1,063.60

	Craft@Hrs	Unit	Material	Labor	Total
Charbroiler, radiant, countertop model, burner and control every 12", natural or LP gas					
40,000 Btu, 15" W (103 lbs)	P6@2.00	Ea	785.00	88.60	873.60
80,000 Btu, 24" W (155 lbs)	P6@2.00	Ea	970.00	88.60	1,058.60
120,000 Btu, 36" W (211 lbs)	P6@2.00	Ea	1,290.00	88.60	1,378.60
Deep fat fryer, free standing floor model, natural or LP gas, 36" height, stainless steel front and top.					
35 lbs capacity (200 lbs)	P6@2.00	Ea	760.00	88.60	848.60
45 lbs capacity (200 lbs)	P6@2.00	Ea	1,170.00	88.60	1,258.60
65 lbs capacity (225 lbs)	P6@2.00	Ea	1,460.00	88.60	1,548.60
Add for filter ready	—	Ea	275.00	—	275.00
Dishwashers					
Under counter, free standing, by rack capacity per hour					
18 racks per hour (245 lbs)	P6@4.00	Ea	3,815.00	177.00	3,992.00
30 racks per hour (245 lbs)	P6@4.00	Ea	3,820.00	177.00	3,997.00
40 racks per hour (255 lbs)	P6@4.00	Ea	4,530.00	177.00	4,707.00
Add for detergent and rinse pump	—	Ea	425.00		425.00
Upright, heavy duty, semi-automatic, 53 racks per hour					
Straight-thru model (450 lbs)	P6@6.00	Ea	6,195.00	266.00	6,461.00
Corner model (450 lbs)	P6@6.00	Ea	6,295.00	266.00	6,561.00
Add for booster heater – 70 degrees	—	Ea	1,400.00	—	1,400.00
Exhaust hoods, 16 gauge type 304 stainless steel with #3 polished finish. Conforms to NFPA code #96 and the National Sanitation Foundation requirements. Add for exhaust fan from next section					
48" deep angled front hood, 24" high with length as shown					
6' long (205 lbs)	T5@2.00	Ea	1,020.00	86.90	1,106.90
8' long (250 lbs)	T5@3.00	Ea	1,320.00	130.00	1,450.00
10' long (285 lbs)	T5@4.00	Ea	1,630.00	174.00	1,804.00
Additives to any hood above					
Installation kit	—	Ea	175.00	—	175.00
Grease filters, aluminum, cleanable, 2' deep	T5@.125	LF	38.50	5.43	43.93
Exhaust fans, 2,800 CFM, 1,265 RPM. Wall or roof mounted, for use with exhaust hoods. Heavy duty aluminum with 115 volt AC motor enclosed in weathertight enclosure					
These costs are based on installing the fan on an existing opening, electrical hookup not included					
Fan for 6' long hood, 18-1/2" square curb size					
1/2 HP (75 lbs)	T5@2.00	Ea	897.00	86.90	983.90
Fan for 8' or 10' long hood, 23-1/2" square curb size					
1/2 HP (100 lbs)	T5@2.00	Ea	957.00	86.90	1,043.90
Additives to either fan above					
Roof curb	T5@1.00	Ea	270.00	43.50	313.50
Variable speed motor controller, 115 volt AC	—	Ea	121.00	—	121.00
Lights, incandescent, pre-wired, 115 volt AC	—	Ea	69.00	—	69.00
Ductwork, aluminized steel	T5@.100	LF	33.00	4.35	37.35
90 degree elbow for ductwork, aluminized steel	T5@.500	Ea	220.00	21.70	241.70
Food mixing machines, floor mounted					
20 quart, bench mount (210 lbs)	D4@2.00	Ea	3,960.00	82.30	4,042.30
30 quart (350 lbs)	D4@3.00	Ea	6,440.00	123.00	6,563.00
Food stations, vinyl clad steel, with stainless steel work area and clear acrylic breath/sneeze guard, 24" D x 35" H, mounted on 4" casters					
Hot food station, 60" long (182 lbs)	D4@1.00	Ea	1,710.00	41.20	1,751.20
Hot food station, 46" long (156 lbs)	D4@1.00	Ea	1,540.00	41.20	1,581.20
Cold food station, 60" long (166 lbs)	D4@1.00	Ea	1,470.00	41.20	1,511.20
Cold food station, 46" long (147 lbs)	D4@1.00	Ea	1,290.00	41.20	1,331.20
Plate rest with mounting kit (10 lbs)	D4@.250	Ea	290.00	10.30	300.30
Incandescent light assembly (20 lbs)	D4@.250	Ea	531.00	10.30	541.30

Equipment 11

	Craft@Hrs	Unit	Material	Labor	Total
Food warmers, stainless steel top, hardwood cutting board, 12" x 20" openings, waterless					
58-1/2", natural gas, 4 openings (135 lbs)	P6@1.00	Ea	935.00	44.30	979.30
58-1/2", LP gas, 4 openings (135 lbs)	P6@1.00	Ea	935.00	44.30	979.30
58-1/2", electric, 3000 watt, 4 openings, (135 lbs)	E4@1.00	Ea	950.00	43.50	993.50
Freezers, under counter type, stainless steel front and top, CFC free refrigeration system					
6.2 cu. ft. (162 lbs)	P6@.500	Ea	1,180.00	22.10	1,202.10
11.8 cu. ft. (239 lbs)	P6@1.00	Ea	1,610.00	44.30	1,654.30
Garbage disposers, includes control switch, solenoid valve, siphon breaker and flow control valve					
1/2 HP (56 lbs)	P6@1.00	Ea	935.00	44.30	979.30
3/4 HP (57 lbs)	P6@1.00	Ea	1,150.00	44.30	1,194.30
1 HP (60 lbs)	P6@1.00	Ea	1,250.00	44.30	1,294.30
1-1/2 HP (76 lbs)	P6@2.00	Ea	1,580.00	88.60	1,668.60
2 HP (81 lbs)	P6@2.00	Ea	1,740.00	88.60	1,828.60
3 HP (129 lbs)	P6@3.00	Ea	2,060.00	133.00	2,193.00
Gas ranges, free standing, floor mounted, 29-3/4" D x 59" H x 62-3/4" W, with twin ovens and 24" raised griddle/broiler					
6 burners, 20,000 Btu each (505 lbs)	P6@4.00	Ea	1,130.00	177.00	1,307.00
Add for stainless steel shelf above range	—	Ea	100.00	—	100.00
Griddles, gas, 27,500 Btu, 1" thick x 24" deep griddle plate, countertop mounted					
24" wide (295 lbs)	P6@2.00	Ea	655.00	88.60	743.60
36" wide (385 lbs)	P6@2.00	Ea	825.00	88.60	913.60
48" wide (520 lbs)	P6@3.00	Ea	1,100.00	133.00	1,233.00
60" wide (650 lbs)	P6@3.00	Ea	2,830.00	133.00	2,963.00
Hot chocolate dispenser, continuous flow, 3 cups per minute, 8" W x 18" D x 31" H, countertop model					
For powder (33 lbs)	D4@1.00	Ea	628.00	41.20	669.20
For liquid (33 lbs)	D4@1.00	Ea	644.00	41.20	685.20
Ice-maker, free standing, floor mounted model					
270 lbs per day capacity (140 lbs)	P6@2.00	Ea	1,790.00	88.60	1,878.60
690 lbs per day capacity (174 lbs)	P6@2.00	Ea	2,530.00	88.60	2,618.60
810 lbs per day capacity (209 lbs)	P6@2.00	Ea	3,390.00	88.60	3,478.60
1370 lbs per day capacity (307 lbs)	P6@2.00	Ea	4,420.00	88.60	4,508.60
Ice storage unit, self service for bagged ice, 40 cu. ft. capacity, floor mounted. Includes refrigeration unit					
Indoor type	E4@3.00	Ea	1,850.00	130.00	1,980.00
Outdoor type, automatic defrost	E4@3.00	Ea	1,400.00	130.00	1,530.00
Meat and deli case, self-contained with gravity cooling, stainless steel top, 54" H x 34" D					
45" long (482 lbs)	D4@3.00	Ea	2,760.00	123.00	2,883.00
69" long (610 lbs)	D4@4.00	Ea	3,200.00	165.00	3,365.00
93" long (805 lbs)	D4@4.00	Ea	3,500.00	165.00	3,665.00
117" long (960 lbs)	D4@4.00	Ea	3,720.00	165.00	3,885.00
Ovens, free standing, floor mounted					
Convection oven, natural gas, stainless steel, porcelain interior, 38" W x 36" D x 54-1/4" H					
54,000 Btu (550 lbs)	D4@4.00	Ea	2,940.00	165.00	3,105.00
Conveyer oven, electric, stainless steel body, 18" H x 50" W x 31-3/8" D, 208 or 240 V AC					
16" W conveyor, 6 KW, UL listed (190 lbs)	D4@4.00	Ea	4,140.00	165.00	4,305.00
Entry & exit shelves	—	Ea	185.00	—	165.00
Pot and pan racks, galvanized steel, pot hooks placed 4 per linear foot					
Ceiling mounted, 5'	C8@2.00	Ea	280.00	81.90	361.90
Wall mounted, 5'	C8@2.00	Ea	195.00	81.90	276.90
Refrigerators, CFC free refrigeration system 115 V AC					
Under counter, stainless steel front and top					
6.2 cu. ft. (157 lbs)	P6@3.00	Ea	970.00	133.00	1,103.00
11.8 cu. ft. (226 lbs)	P6@4.00	Ea	1,410.00	177.00	1,587.00

	Craft@Hrs	Unit	Material	Labor	Total
Reach-in, bottom mounted, stainless steel doors, anodized aluminum sides					
Single door, 23 cu. ft. (254 lbs)	P6@4.00	Ea	1,630.00	177.00	1,807.00
Double door, 49 cu. ft. (410 lbs)	P6@4.00	Ea	2,160.00	177.00	2,337.00
Triple door, 72 cu. ft. (576 lbs)	P6@4.00	Ea	2,900.00	177.00	3,077.00
Sandwich preparation units free standing, stainless steel top, sides and back, 115 V AC, refrigerated 30" D x 43" H, full length 12" deep cutting board					
6.5 cu. ft. 27-1/2" long	D4@2.00	Ea	1,170.00	82.30	1,252.30
12 cu. ft. 48" long	D4@2.00	Ea	1,750.00	82.30	1,832.30
Pizza preparation units free standing, stainless steel top, sides and back, 115 V AC, refrigerated 35-1/2" D x 41-1/2" H, full length 19-1/2" deep cutting board					
11.4 cu. ft. 44-1/2" long, one door	D4@2.00	Ea	1,930.00	82.30	2,012.30
20.6 cu. ft. 67" long, two doors	D4@2.00	Ea	2,750.00	82.30	2,832.30
30.9 cu. ft. 93-1/4" long, three doors	D4@2.00	Ea	3,620.00	82.30	3,702.30
Pan kit for prep tables	—	Ea	58.00	—	58.00
Shelving, aluminum, free standing, 2000 lb capacity lower shelf, 400 lb capacity uppers					
18" x 36" x 68" H, 3 tier	C8@2.00	Ea	160.00	81.90	241.90
18" x 48" x 68" H, 3 tier	C8@2.00	Ea	182.00	81.90	263.90
18" x 36" x 76" H, 4 tier	C8@2.00	Ea	193.00	81.90	274.90
18" x 48" x 76" H, 4 tier	C8@2.00	Ea	220.00	81.90	301.90
Sinks, stainless steel, free standing, floor mounted, with 7" high backsplash and faucet holes drilled on 8" centers, no trim included					
Scullery sink, 16" x 20" tubs, 12" deep					
40" long, 1 tub, 1 drainboard (65 lbs)	P6@2.00	Ea	810.00	88.60	898.60
58" long, 2 tubs, 1 drainboard (80 lbs)	P6@2.00	Ea	1,040.00	88.60	1,128.60
103" long, 3 tubs, 2 drainboards (110 lbs)	P6@2.00	Ea	1,200.00	88.60	1,288.60
122" long, 4 tubs, 2 drainboards (150 lbs)	P6@2.00	Ea	1,640.00	88.60	1,728.60
Mop sink, faucet not included					
21" x 21" x 41", single compartment	P6@2.00	Ea	230.00	88.60	318.60
Toaster, conveyor type, stainless steel, fused quartz sheathing, countertop model					
350 slices per hour, 115 V AC, 1,400 watts	D4@3.00	Ea	605.00	123.00	728.00
Work tables, 16 gauge type 430 stainless steel, 30" high. Includes two 12" deep over-shelves, full depth under-shelf and pot and ladle rack					
48" long (118 lbs)	D4@1.00	Ea	695.00	41.20	736.20
60" long (135 lbs)	D4@1.00	Ea	755.00	41.20	796.20
72" long (154 lbs)	D4@1.00	Ea	825.00	41.20	866.20
Add to any work table above					
20" x 20" x 5" pull flange drawer (13 lbs)	D4@.250	Ea	77.00	10.30	87.30
Stainless steel pot hooks	D4@.250	Ea	7.60	10.30	17.90

Gymnasium and Athletic Equipment

	Craft@Hrs	Unit	Material	Labor	Total
Basketball backstops					
Fixed, wall mounted	D4@8.00	Ea	339.00	329.00	668.00
Swing-up, wall mounted	D4@10.0	Ea	894.00	412.00	1,306.00
Ceiling suspended, to 20', swing-up manual	D4@16.0	Ea	1,970.00	658.00	2,628.00
Add for glass backstop, fan shape	—	Ea	—	—	615.00
Add for glass backstop, rectangular	—	Ea	—	—	690.00
Add for power operation	—	Ea	—	—	610.00
Gym floors, typical subcontract prices, not including subfloor or base					
Synthetic gym floor, 3/16"	—	SF	—	—	5.71
Synthetic gym floor, 3/8"	—	SF	—	—	7.50
Built-up wood "spring system" maple floor	—	SF	—	—	10.70
Rubber cushioned maple floor	—	SF	—	—	9.49

	Craft@Hrs	Unit	Material	Labor	Total
Maple floor on sleepers and membrane					
Unfinished	—	SF	—	—	7.40
Apparatus inserts	—	Ea	—	—	39.80
Bleachers, telescoping, manual, per seat	D4@.187	Ea	43.90	7.70	51.60
Bleachers, hydraulic operated, per seat	D4@.235	Ea	59.90	9.67	69.57
Scoreboards, basketball, single face					
Economy	CE@7.01	Ea	1,290.00	354.00	1,644.00
Good	CE@13.0	Ea	1,780.00	657.00	2,437.00
Premium	CE@21.6	Ea	2,810.00	1,090.00	3,900.00

Rolling Ladders Rolling ladders for mercantile, library or industrial use. Putnam Rolling Ladder.
Oak rolling ladder with top and bottom rolling fixtures, non-slip treads, pre-finished. By floor to track height. Painted aluminum hardware. Add cost of track from below

	Craft@Hrs	Unit	Material	Labor	Total
7'6" high	D4@.819	Ea	394.00	33.70	427.70
9'0" high	D4@.819	Ea	360.00	33.70	393.70
11'8" high	D4@1.09	Ea	456.00	44.90	500.90
12'0" high	D4@1.09	Ea	478.00	44.90	522.90
13'9" high	D4@1.09	Ea	499.00	44.90	543.90
14'0" high	D4@1.09	Ea	522.00	44.90	566.90
Add for bend at ladder top or bottom	—	Ea	73.00	—	73.00
Add for brass or chrome hardware	—	Ea	93.50	—	93.50
Add for ladder work shelf	—	Ea	54.90	—	54.90
Add for handrail	—	Ea	41.20	—	41.20
Rolling ladder track, 7/8" slotted steel, per foot of length, 9' to 14' high					
Painted aluminum finish	D4@.089	LF	4.29	3.66	7.95
Painted black finish	D4@.089	LF	5.38	3.66	9.04
Brass or chrome plated finish	D4@.089	LF	12.40	3.66	16.06
Corner bend track, 90 degree, 30" radius					
Painted finish	D4@.363	Ea	42.90	14.90	57.80
Chrome or brass finish	D4@.363	Ea	69.50	14.90	84.40

Service Station Equipment These costs do not include excavation, concrete work, piping or electrical work

	Craft@Hrs	Unit	Material	Labor	Total
Air compressor, 1-1/2 HP with receiver	PF@5.54	Ea	1,700.00	283.00	1,983.00
Air compressor, 3 HP with receiver	PF@5.54	Ea	2,030.00	283.00	2,313.00
Hose reels with hose					
Air hose, 50', heavy duty	PF@8.99	Ea	854.00	459.00	1,313.00
Water hose, 50' and valve	PF@8.01	Ea	870.00	409.00	1,279.00
Lube rack, 3 hose, remote pump	PF@48.4	Ea	4,930.00	2,470.00	7,400.00
Lube rack, 5 hose, remote pump	PF@48.4	Ea	9,030.00	2,470.00	11,500.00
Air-operated pumps for hose reels, fit 55 gallon drum					
Motor oil or gear oil	PF@1.00	Ea	857.00	51.10	908.10
Lube oil	PF@1.00	Ea	1,830.00	51.10	1,881.10
Gasoline pumps, with cabinet, industrial type,					
15 gallons per minute	PF@5.93	Ea	1,730.00	303.00	2,033.00
Add for totalizer	—	Ea	304.00	—	304.00
Submerged turbine, 1/3 HP, serves 4 dispensers	PF@22.5	Ea	845.00	1,150.00	1,995.00
Submerged turbine, 3/4 HP, serves 8 dispensers	PF@22.5	Ea	1,170.00	1,150.00	2,320.00
Gasoline dispenser, computing, single hose, single product					
not including rough-in of pit box	PF@9.94	Ea	5,950.00	508.00	6,458.00
Gasoline dispenser, computing, dual hose, dual product,					
not including rough-in of pit box	PF@9.94	Ea	8,080.00	508.00	8,588.00

	Craft@Hrs	Unit	Material	Labor	Total
Pit box for submerged pump, 22" x 22",					
installed in existing pit	PF@1.26	Ea	135.00	64.40	199.40
Fill boxes, 12" diameter, cast iron	PF@.949	Ea	271.00	48.50	319.50
Air and water service, post type with auto-inflator and disappearing hoses,					
hoses included	PF@9.58	Ea	1,510.00	490.00	2,000.00
Add for electric thermal unit, factory installed	—	Ea	88.70	—	88.70
Auto hoist. Prices vary with the rating and capacity					
Hoist, single post frame contact					
8,000 lb semi hydraulic	D4@71.3	Ea	3,630.00	2,930.00	6,560.00
Hoist, single post frame contact					
8,000 lb fully hydraulic	D4@71.3	Ea	3,880.00	2,930.00	6,810.00
Hoist, two post, pneumatic, 11,000 lb	D4@114.	Ea	7,070.00	4,690.00	11,760.00
Hoist, two post, pneumatic, 24,000 lb	D4@188.	Ea	10,000.00	7,740.00	17,740.00
Two-post fully hydraulic hoists					
11,000 lb capacity	D4@188.	Ea	5,550.00	7,740.00	13,290.00
13,000 lb capacity	D4@188.	Ea	5,670.00	7,740.00	13,410.00
18,500 lb capacity	D4@188.	Ea	7,590.00	7,740.00	15,330.00
24,000 lb capacity	D4@188.	Ea	10,200.00	7,740.00	17,940.00
26,000 lb capacity	D4@188.	Ea	11,700.00	7,740.00	19,440.00
Service station equipment.					
Cash box and pedestal stand	D4@2.55	Ea	271.00	105.00	376.00
Tire changer, 120 PSI pneumatic, auto tire	D4@10.6	Ea	2,560.00	436.00	2,996.00
Tire changer, hydraulic, truck tire	D4@21.4	Ea	7,940.00	881.00	8,821.00
Exhaust fume system, underground, complete					
Per station	D5@9.52	Ea	560.00	432.00	992.00

Parking Control Equipment Costs are for parking equipment only, no concrete work, paving, or electrical work included. Labor costs include installation and testing.

	Craft@Hrs	Unit	Material	Labor	Total
Parking gates, controllable by attendant, vehicle detector, ticket dispenser, card reader, or coin/token machine					
Gate with 10' arm	D4@9.13	Ea	3,060.00	376.00	3,436.00
Folding arm gate (low headroom), 10'	D4@9.13	Ea	3,400.00	376.00	3,776.00
Vehicle detector, for installation in approach lane					
including cutting and grouting	D4@4.96	Ea	384.00	204.00	588.00
Ticket dispenser (spitter), with date/time imprinter					
actuated by vehicle detector or hand button	D4@9.13	Ea	5,790.00	376.00	6,166.00
Card readers. Non-programmable, parking areas with long term parking privileges					
Per station	D4@3.39	Ea	602.00	139.00	741.00
Programmable, parking areas for users paying a fee to renew parking privileges					
Per reader	D4@3.39	Ea	4,670.00	139.00	4,809.00
Lane spikes, dragon teeth, installation includes cutting and grouting					
6' wide	D4@9.69	Ea	1,220.00	399.00	1,619.00
Lighted warning sign for lane spikes	D4@5.71	Ea	562.00	235.00	797.00

	Craft@Hrs	Unit	Material	Labor	Equipment	Total
Loading Dock Equipment Costs shown assume that equipment is set in prepared locations. No excavation, concrete work or electrical work included. Equipment cost shown is based on using a 4000 lb forklift						
Move equipment on or off job						
Allow, per move	—	LS	—	—	100.00	75.00
Dock levelers, Systems, Inc.						
Edge-of-dock leveler, manual	D6@2.25	Ea	574.00	106.00	18.60	698.60

	Craft@Hrs	Unit	Material	Labor	Equipment	Total
Mechanical platform dock leveler, recessed, 20,000 lb capacity						
6' wide x 6' long	D6@3.00	Ea	2,320.00	141.00	24.90	2,485.90
6' wide x 8' long	D6@3.00	Ea	2,760.00	141.00	24.90	2,925.90
7' wide x 6' long	D6@3.00	Ea	2,720.00	141.00	24.90	2,885.90
7' wide x 8' long	D6@3.00	Ea	2,720.00	141.00	24.90	2,885.90
Hydraulic, pit type						
6' wide x 6' long	D6@4.50	Ea	3,680.00	212.00	37.30	3,929.30
6' wide x 8' long	D6@4.50	Ea	3,910.00	212.00	37.30	4,159.30
Dock seals, foam pad type, 12" projection, with anchor brackets (bolts not included), door opening size as shown, Systems, Inc.						
8' wide x 8' high	D6@3.75	Ea	584.00	177.00	31.10	792.10
8' wide x 10' high	D6@3.75	Ea	674.00	177.00	31.10	882.10
Dock lifts, portable, electro-hydraulic scissor lifts, Systems, Inc.						
4,000 lb capacity, 6' x 6' dock	D6@2.25	Ea	6,790.00	106.00	18.60	6,914.60
Dock lifts, pit recessed, electro-hydraulic scissor lifts, Advance Lifts						
5,000 lb capacity, 6' x 8' platform	D6@12.0	Ea	5,540.00	566.00	99.40	6,205.40
6,000 lb capacity, 6' x 8' platform	D6@12.0	Ea	5,670.00	566.00	99.40	6,335.40
8,000 lb capacity						
6' x 8' platform	D6@12.0	Ea	8,210.00	566.00	99.40	8,875.40
8' x 10' platform	D6@12.0	Ea	9,250.00	566.00	99.40	9,915.40
10,000 lb capacity						
6' x 8' platform	D6@12.0	Ea	9,340.00	566.00	99.40	10,005.40
8' x 10' platform	D6@12.0	Ea	10,300.00	566.00	99.40	10,965.40
12,000 lb capacity						
6' x 10' platform	D6@15.0	Ea	12,100.00	707.00	124.00	12,931.00
8' x 12' platform	D6@15.0	Ea	12,800.00	707.00	124.00	13,631.00
15,000 lb capacity						
6' x 10' platform	D6@15.0	Ea	13,400.00	707.00	124.00	14,231.00
8' x 12' platform	D6@15.0	Ea	14,600.00	707.00	124.00	15,431.00
18,000 lb capacity						
6' x 10' platform	D6@15.0	Ea	14,100.00	707.00	124.00	14,931.00
8' x 12' platform	D6@15.0	Ea	15,000.00	707.00	124.00	15,831.00
20,000 lb capacity						
6' x 12' platform	D6@15.0	Ea	15,700.00	707.00	124.00	16,531.00
8' x 12' platform	D6@15.0	Ea	17,100.00	707.00	124.00	17,931.00
Combination dock lift and dock leveler, electro-hydraulic scissor lifts, pit recessed, 5,000 lb capacity, Advance Lifts						
6' x 7'2" platform	D6@12.0	Ea	6,370.00	566.00	99.40	7,035.40
6' x 8' platform	D6@12.0	Ea	5,170.00	566.00	99.40	5,835.40
Dock bumpers, laminated rubber, not including masonry anchors						
10" high x 14" wide x 4-1/2" thick	D4@.248	Ea	55.10	10.20	—	65.30
10" high x 14" wide x 6" thick	D4@.299	Ea	74.30	12.30	—	86.60
10" high x 24" wide x 4-1/2" thick	D4@.269	Ea	63.40	11.10	—	74.50
10" high x 24" wide x 6" thick	D4@.358	Ea	89.40	14.70	—	104.10
10" high x 36" wide x 4-1/2" thick	D4@.416	Ea	75.00	17.10	—	92.10
10" high x 36" wide x 6" thick	D4@.499	Ea	108.00	20.50	—	128.50
12" high x 14" wide x 4-1/2" thick	D4@.259	Ea	64.90	10.70	—	75.60
12" high x 24" wide x 4-1/2" thick	D4@.328	Ea	75.30	13.50	—	88.80
12" high x 36" wide x 4-1/2" thick	D4@.432	Ea	93.00	17.80	—	110.80
22" high x 22" wide x 3" thick	D4@.251	Ea	81.70	10.30	—	92.00
Add for 4 masonry anchors	D4@1.19	LS	2.44	49.00	—	51.44

Equipment 11

	Craft@Hrs	Unit	Material	Labor	Equipment	Total
Barrier posts, 8' long concrete-filled steel posts set 4' underground in concrete						
3" diameter	P8@.498	Ea	76.70	19.90	—	96.60
4" diameter	P8@.498	Ea	109.00	19.90	—	128.90
6" diameter	P8@.752	Ea	136.00	30.10	—	166.10
8" diameter	P8@1.00	Ea	165.00	40.00	—	205.00
Loading dock shelter, fabric covered, steel frame,						
24" extension for 10' x 10' door	D4@9.91	Ea	919.00	408.00	—	1,327.00
Perimeter door seal,						
12" x 12", vinyl cover	D4@.400	LF	37.40	16.50	—	53.90

	Craft@Hrs	Unit	Material	Labor	Total
Medical Refrigeration Equipment Labor costs include unpacking, assembly and installation. Jewett Refrigerator. Blood bank refrigerators, upright enameled steel cabinets with locking glass doors, 7 day recording thermometer, LED temperature display, failure alarm, interior lights, and removable drawers. 2 to 4 degree C operating range					
5 drawer, 240 bag, 16.9 CF, 29" wide x 30" deep x 74" high, #BBR17	D4@2.25	Ea	6,130.00	92.60	6,222.60
6 drawer, 360 bag, 24.8 CF, 29" wide x 36" deep x 83" high, #BBR25	D4@2.25	Ea	6,400.00	92.60	6,492.60
10 drawer, 480 bag, 37.4 CF, 59" wide x 30" deep x 74" high, #BBR37	D4@3.42	Ea	8,870.00	141.00	9,011.00
12 drawer, 720 bag, 55.0 CF, 59" wide x 36" deep x 83" high, #BBR55	D4@3.42	Ea	9,110.00	141.00	9,251.00
Blood plasma storage freezers, upright enameled steel cabinets with locking steel doors, 7 day recording thermometer, LED temperature display, failure alarm, and removable drawers. -30 degree C operating temperature					
6 drawer, 265 pack, 13.2 CF, 34" wide x 30" deep x 74" high, #BPL13	D4@2.25	Ea	10,400.00	92.60	10,492.60
8 drawer, 560 pack, 24.8 CF, 29" wide x 36" deep x 83" high, #BPL25	D4@2.25	Ea	9,110.00	92.60	9,202.60
16 drawer, 1,120 pack, 55 CF, 59" wide x 36" deep x 83" high, #BPL55	D4@3.42	Ea	14,400.00	141.00	14,541.00
Laboratory refrigerators, upright enameled steel cabinets with locking steel doors, enameled steel interior, and adjustable stainless steel shelves. 2 to 4 degree C operating temperature					
4 shelf, 16.9 CF, 29" wide x 30" deep x 74" high #LR17B	D4@2.25	Ea	4,180.00	92.60	4,272.60
8 shelf, 37.4 CF, 59" wide x 30" deep x 74" high #LR37B	D4@3.42	Ea	6,400.00	141.00	6,541.00
Pharmacy refrigerators, upright enameled steel cabinets with locking steel doors, enameled steel interior, and adjustable stainless steel drawers and shelves. -2 degree C operating temperature					
6 drawer, 1 shelf, 24.8 CF, 29" wide x 36" deep x 83" high, #PR25B	D4@2.25	Ea	4,680.00	92.60	4,772.60
9 drawer, 5 shelf, 55.0 CF, 59" wide x 36" deep x 83" high, #PR55B	D4@3.42	Ea	7,440.00	141.00	7,581.00
Hospital/Lab/Pharmacy undercounter refrigerators, stainless steel interior and exterior, locking door. 24" wide x 24" deep x 34-1/2" high. 2 degree C operating temperature					
5.4 CF, with blower coil cooling system, stainless steel racks, and auto-defrost, #UC5B	D4@1.67	Ea	2,150.00	68.70	2,218.70
5.4 CF with cold wall cooling system, stainless steel racks, and auto-defrost, #UC5C	D4@1.69	Ea	2,740.00	69.50	2,809.50

Equipment 11

	Craft@Hrs	Unit	Material	Labor	Total
Hospital/Lab/Pharmacy undercounter freezers, stainless steel interior and exterior, locking door. 24" wide x 24" deep x 34-1/2" high. -20 degree C operating temperature					
4.6 CF, stainless steel racks, Auto-defrost system, #UCF5B	D4@1.69	Ea	3,030.00	69.50	3,099.50
4.6 CF, stainless steel racks, manual hot gas defrost system, #UCF5C	D4@1.69	Ea	3,580.00	69.50	3,649.50
Morgue refrigerators					
1 or 2 body roll-in type, 39" wide x 96" deep x 76" high, with 1/2 HP, 4380 Btu/hour condenser, #1SPEC	D4@15.8	Ea	9,650.00	650.00	10,300.00
2 body sliding tray type, 39" wide x 96" deep x 76" high, with 1/2 HP, 4380 Btu/hour condenser, #2EC	D4@18.0	Ea ·	9,530.00	741.00	10,271.00
2 body side opening type, 96" wide x 39" deep x 76" high, with 1/2 HP, 4380 Btu/hour condenser, #2SC	D4@18.0	Ea	10,500.00	741.00	11,241.00
4 or 6 body with roll-in type lower compartments, sliding tray type upper compartments, 73" wide x 96" deep x 102" high, 6060 Btu/hr, 3/4 HP condenser, #4SPEC2W	D4@22.5	Ea	16,500.00	926.00	17,426.00

Furnishings 12

Cabinets Cost per linear foot of face or back dimension. Base cabinets are 34" high x 24" deep, wall cabinets are 42" high x 12" deep and full height cabinets are 94" high x 24" deep. Labor shown is for installing shop fabricated units. Add for countertops at the end of this section.

	Craft@Hrs	Unit	Material	Labor	Total
Metal cabinets, shop and commercial type, with hardware but no countertop					
Base cabinet, drawer, door and shelf	D4@.379	LF	93.40	15.60	109.00
Wall cabinet with door and 2 shelves	D4@.462	LF	85.50	19.00	104.50
Full height cabinet, 5 shelves	D4@.595	LF	175.00	24.50	199.50
Library shelving, 48" high x 8" deep modules	D4@.336	LF	61.00	13.80	74.80
Wardrobe units, 72" high x 42" deep	D4@.398	LF	151.00	16.40	167.40
Classroom type wood cabinets, laminated plastic face, with hardware but no countertop					
Base cabinet, drawer, door and shelf	C8@.391	LF	173.00	16.00	189.00
Wall cabinet with door and 2 shelves	C8@.475	LF	151.00	19.50	170.50
Full height cabinet with doors	C8@.608	LF	244.00	24.90	268.90
Hospital type wood cabinets, laminated plastic face, with hardware but no countertop					
Base cabinet, drawer, door and shelf	C8@.982	LF	171.00	40.20	211.20
Wall, with 2 shelves and door	C8@1.15	LF	157.00	47.10	204.10
Full height with doors	C8@1.35	LF	233.00	55.30	288.30
Laboratory type metal cabinets, with hardware but no countertops					
Base cabinets, doors	D4@.785	LF	88.40	32.30	120.70
Base cabinets, drawers	D4@.785	LF	164.00	32.30	196.30
Base cabinets, island base	D4@.785	LF	164.00	32.30	196.30
Wall cabinets, doors	D4@.785	LF	143.00	32.30	175.30
Wardrobe or storage cabinets, 80" high	D4@.939	LF	162.00	38.60	200.60
Fume hoods, steel, without duct, typical price	T5@3.88	Ea	661.00	169.00	830.00
Add for duct, without electric work, typical price	T5@16.6	LF	999.00	721.00	1,720.00
Custom made wood cabinets, no hardware or countertops included, unfinished, 36" wide modules					
Standard grade, mahogany or laminated plastic face					
Base	C8@.256	LF	73.40	10.50	83.90
Wall	C8@.380	LF	55.90	15.60	71.50
Full height	C8@.426	LF	126.00	17.50	143.50

	Craft@Hrs	Unit	Material	Labor	Total
Custom grade, birch					
Base	C8@.361	LF	112.00	14.80	126.80
Wall	C8@.529	LF	95.40	21.70	117.10
Full height	C8@.629	LF	176.00	25.80	201.80
Sink fronts	C8@.361	LF	93.00	14.80	107.80
Drawer units, 4 drawers	C8@.361	LF	185.00	14.80	199.80
Apron (knee space)	C8@.148	LF	23.80	6.06	29.86
Additional costs for custom cabinets					
Ash face	—	LF	3.98	—	3.98
Walnut face	—	LF	60.00	—	60.00
Edge banding	—	LF	6.50	—	6.50
Prefinished exterior	—	LF	5.90	—	5.90
Prefinished interior	—	LF	9.95	—	9.95
Standard hardware, installed	—	LF	4.70	—	4.70
Institutional grade hardware, installed	—	LF	6.85	—	6.85
Drawer roller guides, per drawer	—	Ea	8.00	—	8.00
Drawer roller guides, suspension	—	Ea	25.00	—	25.00
Extra drawers					
12" wide	—	Ea	50.80	—	50.80
18" wide	—	Ea	59.40	—	59.40
24" wide	—	Ea	65.00	—	65.00

Countertops, shop fabricated, delivered ready for installation
Laminated plastic tops 2'0" wide over plywood

	Craft@Hrs	Unit	Material	Labor	Total
Custom work, square edge front, 4" splash	C8@.187	LF	26.70	7.66	34.36
Add for square edge 9" backsplash	—	LF	8.66	—	8.66
Add for solid colors	—	LF	2.58	—	2.58
Add for acid proof tops	—	LF	19.80	—	19.80
Stainless steel, 2'0" wide with backsplash	C8@.366	LF	70.30	15.00	85.30

Window Treatment
Lead mesh fabrics, measure glass size only. Includes allowance for overlap, pleating and all hardware.

	Craft@Hrs	Unit	Material	Labor	Total
Lead mesh (soundproofing)	—	SY	—	—	22.00
Lead mesh (x-ray) 2.5 lbs per SF	—	SY	—	—	25.00

Draperies, 8' height, fabric quality will cause major variation in prices, per LF of opening, typical installed prices

	Craft@Hrs	Unit	Material	Labor	Total
Standard quality, window or sliding door	—	LF	—	—	19.32
Better quality, window	—	LF	—	—	64.00
Better quality, sliding door	—	LF	—	—	60.98
Drapery lining	—	LF	—	—	7.25
Fiberglass fabric	—	LF	—	—	62.19
Filter light control type	—	LF	—	—	56.15
Flameproofed	—	LF	—	—	60.38
Velour, grand	—	LF	—	—	79.70

Recessed Open Link Entry Mats These costs assume that mats are mounted in an aluminum frame recessed into a concrete walkway and do not include concrete work, special designs or lettering

	Craft@Hrs	Unit	Material	Labor	Total
Aluminum link, 3/8" thick	D4@.560	SF	14.10	23.00	37.10
Steel link, 3/8" thick	D4@.560	SF	6.39	23.00	29.39
Aluminum hinged strips, carpet inserts, 1/2"	D4@.560	SF	30.80	23.00	53.80
Black Koroseal, 1/2" thick, open link	D4@.560	SF	12.20	23.00	35.20
Aluminum frame for entry mats	D4@.086	LF	5.39	3.54	8.93

	Craft@Hrs	Unit	Material	Labor	Total

Entrance Mats and Recessed Foot Grilles These costs assume installation of an extruded aluminum tread foot grille on a level surface or over a catch basin with or without a drain

Pedigrid/Pedimat extruded aluminum entrance mat

	Craft@Hrs	Unit	Material	Labor	Total
Grid mat with carpet treads	D4@.150	SF	32.00	6.17	38.17
Grid mat with vinyl treads	D4@.150	SF	34.90	6.17	41.07
Grid mat with pool or shower vinyl treads	D4@.150	SF	38.90	6.17	45.07
Grid mat with abrasive serrated aluminum treads	D4@.150	SF	39.00	6.17	45.17
Recessed catch basin with 2" drain	D4@.302	SF	45.30	12.40	57.70
Recessed catch basin without drain	D4@.242	SF	40.60	9.96	50.56
Level base recessed mount	D4@.150	SF	38.80	6.17	44.97
Surface-mount frame, vinyl	—	LF	3.88	—	3.80
Recessed-mount frame, bronze aluminum	D4@.092	LF	8.87	3.79	12.66

Seating These costs do not include concrete work

	Craft@Hrs	Unit	Material	Labor	Total
Theater style, economy	D4@.424	Ea	111.00	17.40	128.40
Theater style, lodge, rocking type	D4@.470	Ea	222.00	19.30	241.30

Bleachers, closed riser, extruded aluminum, cost per 18" wide seat

	Craft@Hrs	Unit	Material	Labor	Total
Portable bleachers	D4@.090	Ea	48.60	3.70	52.30
Portable bleachers with guardrails	D4@.136	Ea	54.00	5.60	59.60
Team bleachers, stationary	D4@.090	Ea	41.70	3.70	45.40

Stadium seating, on concrete foundation by others, extruded aluminum, cost per 18" seat width

	Craft@Hrs	Unit	Material	Labor	Total
10" flat bench seating, seatboard only	D4@.090	Ea	14.00	3.70	17.70
12" contour bench seating, seatboard only	D4@.090	Ea	15.00	3.70	18.70
Chair seating, molded plastic	D4@.310	Ea	81.00	12.80	93.80

Replacement bench seating, stadium or bleacher, extruded aluminum

Cost per 18" seat width

	Craft@Hrs	Unit	Material	Labor	Total
10" wide flat seatboard only	D4@.069	Ea	8.89	2.84	11.73
10" wide flat seatboard and footboard	D4@.136	Ea	19.00	5.60	24.60
12" contour seatboard only	D4@.069	Ea	10.60	2.84	13.44

Special Construction 13

Access for the Disabled

Use the following cost estimates when altering an existing building to meet requirements of the Americans with Disabilities Act (ADA). The ADA requires that existing public and private buildings be altered to accommodate the physically handicapped when that can be done without unreasonable difficulty or expense.

Ramps, flared aprons and landings Reinforced concrete, 2,000 PSI, placed directly from the chute of a ready-mix truck, vibrated and finished. Costs include typical wire mesh reinforcing and an allowance for waste but no excavation, backfill or foundations.

	Craft@Hrs	Unit	Material	Labor	Equipment	Total

Access ramps, based on ramps with 1" rise per 1' of run, 4" thick walls and 4" thick slab. Includes pipe sleeves for railings, sand fill in cavity and non-slip surface finish

	Craft@Hrs	Unit	Material	Labor	Equipment	Total
24" vertical rise (4' wide x 25' long)	C8@40.0	Ea	969.00	1,640.00	—	2,609.00
36" vertical rise (4' wide x 38' long)	C8@60.0	Ea	1,870.00	2,460.00	—	4,330.00
48" vertical rise (4' wide x 50' long)	C8@96.0	Ea	2,750.00	3,930.00	—	6,680.00

Access aprons at curbs, flared and inclined, 2' long x 8' wide, inclined

	Craft@Hrs	Unit	Material	Labor	Equipment	Total
4" thick (80 SF per CY)	P9@.024	SF	2.39	1.00	—	3.39
6" thick (54 SF per CY)	P9@.032	SF	3.59	1.29	—	4.41

	Craft@Hrs	Unit	Material	Labor	Equipment	Total

Landings, 4" thick walls on three sides and 4" thick top slab cast against an existing structure on one side. Costs include pipe sleeves for railings, sand fill in cavity and an allowance for waste

	Craft@Hrs	Unit	Material	Labor	Equipment	Total
3'0" high with 6' x 6' square slab	C8@32.0	Ea	800.00	1,310.00	—	2,110.00
Add per vertical foot of height over 3'0"	—	VLF	151.00	—	—	151.00

Snow melting cable for ramps and landings. Includes 208-277 volt AC self-regulating heater cable designed to be encased in concrete. Based on cable connected to an existing power source adjacent to the installation

	Craft@Hrs	Unit	Material	Labor	Equipment	Total
Heater cable (1.1 LF per SF of pavement)	CE@.015	SF	8.02	.76	—	8.78
Power connection kit	CE@.500	Ea	29.10	25.30	—	54.40
Thermostat	CE@.500	Ea	245.00	25.30	—	270.30

Railing for ramps and landings Welded steel. Equipment cost shown is for a flatbed truck and a welding machine. Wall-mounted handrails with brackets 5' OC, based on 12' lengths

	Craft@Hrs	Unit	Material	Labor	Equipment	Total
1-1/4" diameter rail, shop prime painted	H6@.092	LF	6.17	4.17	.50	10.84
1-1/2" diameter rail, shop prime painted	H6@.102	LF	7.22	4.63	.55	12.40
Add for galvanized finish	—	%	20.0	—	—	—

Floor mounted railings 3'6" high with posts 5' OC, based on 12' length

	Craft@Hrs	Unit	Material	Labor	Equipment	Total
1-1/4" diameter rails and posts, shop prime painted						
2-rail type	H6@.115	LF	25.30	5.22	.62	31.14
3-rail type	H6@.125	LF	32.80	5.67	.68	39.15
1-1/2" diameter rails and posts, shop prime painted						
2-rail type	H6@.125	LF	30.50	5.67	.68	36.85
3-rail type	H6@.137	LF	40.20	6.21	.74	47.15
Add for galvanized finish	—	%	20.0	—	—	—

Drinking fountains and water coolers for the disabled

Remove an existing drinking fountain or water cooler and disconnect waste and water piping at the wall adjacent to the fixture for use with the new fixture. No salvage value assumed

	Craft@Hrs	Unit	Material	Labor	Equipment	Total
Floor or wall mounted fixture	P6@1.00	Ea	—	44.30	—	44.30

Install a wall-hung white vitreous china fountain with chrome plated spout, 14" wide, 13" high, non-electric, self-closing dispensing valve and automatic volume control

	Craft@Hrs	Unit	Material	Labor	Equipment	Total
Drinking fountain, trim and valves	P6@2.80	Ea	692.00	124.00	—	816.00

Wheelchair wall hung water cooler, vitreous china, refrigerated, (no electrical work included)

	Craft@Hrs	Unit	Material	Labor	Equipment	Total
Wall hung water cooler, and trim (add rough-in)	P6@2.50	Ea	767.00	111.00	—	878.00

Install floor or wall-mounted electric water cooler, 8 gallons per hour, cold water only, connected to existing electrical outlet adjacent to the cooler (no electrical work included)

	Craft@Hrs	Unit	Material	Labor	Equipment	Total
Water cooler, trim and valves	P6@2.80	Ea	1,350.00	124.00	—	1,474.00

Parking stall marking for the disabled

Remove existing pavement marking by water blasting

	Craft@Hrs	Unit	Material	Labor	Equipment	Total
4" wide stripes (allow 40 LF per stall)	CL@.032	LF	—	1.16	.52	1.68

Markings, per square foot,

	Craft@Hrs	Unit	Material	Labor	Equipment	Total
(minimum 10 SF)	CL@.098	SF	—	3.57	1.60	5.17

Mark parking stall with handicapped symbol painted on, including layout

Reflectorized stripes and symbol, one color. Equipment to include a compressor, hose and spray gun, minimum daily rental $75.00.

	Craft@Hrs	Unit	Material	Labor	Equipment	Total
Per stall	PA@.425	Ea	6.85	19.30	—	26.15

	Craft@Hrs	Unit	Material	Labor	Equipment	Total

Parking lot handicapped sign, 18" x 24" 10 gauge metal, reflective lettering and handicapped symbol

Sign on 2" galvanized steel pipe post 10' long, set 2' into the ground. Includes digging of hole

	Craft@Hrs	Unit	Material	Labor	Equipment	Total
Using a manual auger and backfill	CL@1.25	Ea	82.00	45.50	—	127.50
On walls with mechanical fasteners	PA@.455	Ea	24.20	20.70	—	44.90

Paving and curb removal These costs include breaking out the paving or curb, loading and hauling debris to a legal dump within 6 miles but no dump fees. Equipment cost per hour is for one 90 CFM air compressor, one paving breaker and jackhammer bits, one 55 HP wheel loader with integral backhoe and one 5 CY dump truck. The figures in parentheses show the approximate "loose" volume of the materials (volume after being demolished). Use $250 as a minimum charge for this type work and add dump charges as a separate item

	Craft@Hrs	Unit	Material	Labor	Equipment	Total
Asphalt paving, depths to 3"						
(27 SF per CY)	C3@.040	SF	—	1.68	.99	2.67
Asphalt curbs, to 12" width						
(18 LF per CY)	C3@.028	LF	—	1.18	.70	1.88
Concrete paving and slabs on grade						
4" concrete without rebars						
(54 SF per CY)	C3@.060	SF	—	2.52	1.49	4.01
6" concrete without rebars						
(45 SF per CY)	C3@.110	SF	—	4.63	2.74	7.37
Add for removal of bar reinforced paving	C3@.200	SF	—	8.41	4.97	13.38

Curb, gutter and paving for the disabled These costs assume 3 uses of forms and 2,000 PSI concrete placed directly from the chute of a ready-mix truck. Use $250 as a minimum charge for this type work. See also Snow melting cable in the Electrical section.

Concrete cast-in-place vertical curb, including 2% waste, forms and finishing. These costs include typical excavation and backfill with excess excavation spread on site

	Craft@Hrs	Unit	Material	Labor	Equipment	Total
6" x 12" straight curb	P9@.136	LF	3.99	5.69	—	9.68
6" x 18" straight curb	P9@.149	LF	5.92	6.23	—	12.15
6" x 24" straight curb	P9@.154	LF	7.85	6.44	—	14.29
Add for curved sections	—	%	—	40.0		—
Curb and 24" monolithic gutter						
(7 LF per CY)	C3@.100	LF	7.07	4.21	—	11.28
Concrete paving and walkway						
4" thick	P8@.018	SF	1.90	.72	—	2.62
Add for integral colors, most pastels	—	SF	1.00	—	—	1.00

Plumbing fixtures for the disabled

Remove existing plumbing fixtures. Disconnect from existing waste and water piping at the wall adjacent to the fixture. Waste and water piping to remain in place for connection to a new fixture. No salvage value assumed

	Craft@Hrs	Unit	Material	Labor	Equipment	Total
Lavatories	P6@1.00	Ea	—	44.30	—	44.30
Urinals	P6@1.00	Ea	—	44.30	—	44.30
Water closets, floor or wall mounted	P6@1.00	Ea	—	44.30	—	44.30

Install new fixtures. White vitreous china fixtures with typical fittings, hangers, trim and valves. Connected to waste and supply piping from lines adjacent to the fixture

	Craft@Hrs	Unit	Material	Labor	Equipment	Total
Lavatories, wall hung	P6@1.86	Ea	517.00	82.40	—	599.40
Water closets, tank type, with elongated bowl, toilet seat, trim and valves						
Floor mounted, with bowl ring	P6@1.86	Ea	288.00	82.40	—	370.40
Wall hung, with carrier	P6@2.80	Ea	608.00	124.00	—	732.00

Plumbing partitions and grab bars for the disabled

Remove a standard toilet partition, no salvage value included. Floor or ceiling mounted

	Craft@Hrs	Unit	Material	Labor	Equipment	Total
Remove one wall and one door	T5@1.00	Ea	—	43.50	—	43.50

	Craft@Hrs	Unit	Material	Labor	Equipment	Total

Install toilet partition designed for wheel chair access. Floor or ceiling mounted. These costs include one wall and one door, mechanical fasteners and drilling of holes

	Craft@Hrs	Unit	Material	Labor	Equipment	Total
Baked enamel	T5@2.66	Ea	361.00	116.00	—	477.00
Solid plastic	T5@2.66	Ea	558.00	116.00	—	674.00
Laminated plastic	T5@2.66	Ea	464.00	116.00	—	580.00
Stainless steel	T5@2.66	Ea	802.00	116.00	—	918.00

Grab bars, stainless steel, wall mounted, commercial grade. Set of two required for each stall. Including mechanical fasteners and drilling of holes

	Craft@Hrs	Unit	Material	Labor	Equipment	Total
1-1/4" x 24" urinal bars	CC@1.00	Set	83.60	45.60	—	129.20
1-1/4" x 36" toilet partition bars	CC@1.00	Set	102.00	45.60	—	147.60

	Craft@Hrs	Unit	Material	Labor	Total

Signage in braille Standard signs (Rest Room, Elevator, Telephone, Stairway, etc.) mounted with self-stick foam, velcro or in frame. Add the cost of frames, if required, from below

5/8" lettering raised 1/32". Grade 2 braille raised 1/32"

	Craft@Hrs	Unit	Material	Labor	Total
2" x 8" sign, raised letters and tactile braille	PA@0.25	Ea	21.70	11.30	33.00
8" x 8" sign with words, graphic symbols and tactile braille standard graphic symbols	PA@0.25	Ea	35.90	11.30	47.20

Sign frames. Signs snap into frames mounted with mechanical fasteners

	Craft@Hrs	Unit	Material	Labor	Total
2" x 8" frames, black, almond or grey	PA@0.25	Ea	16.10	11.30	27.40
8" x 8" frames, black, almond or grey	PA@0.25	Ea	29.50	11.30	40.80

Elevator plates. Sign with graphic symbols and tactile braille, standard colors

	Craft@Hrs	Unit	Material	Labor	Total
Door jamb plates, 3" x 4", two per opening	PA@0.75	Set	44.80	34.00	78.80

Operation plates, Open Door, Close Door, Emergency Stop, Alarm,

	Craft@Hrs	Unit	Material	Labor	Total
1-1/4" x 1-1/4", set of four required per car	PA@1.00	Set	60.00	45.40	105.40

Operation plates, Floor Number, 1-1/4" x 1-1/4"

	Craft@Hrs	Unit	Material	Labor	Total
One required for each floor per car	PA@0.25	Ea	15.00	11.30	26.30

Bus Stop Shelters Prefabricated structure systems, shipped knocked-down. Complete with all hardware and foundation anchors. Labor cost is for assembly on site. These costs do not include the foundation or floor slab. Dimensions are approximate outside measurements. Based on Columbia Equipment Company.

Three-sided open front type. Clear satin silver anodized aluminum structure and fascia, 1/4" clear acrylic sheet glazing in panels and white baked enamel finish aluminum V-beam roof

9'2" L x 5'3" W x 7'5" H

	Craft@Hrs	Unit	Material	Labor	Total
Two rear panels, single panel each side	C8@8.00	Ea	3,050.00	328.00	3,378.00
Four rear panels, two panels each side	C8@12.0	Ea	3,200.00	492.00	3,692.00

11'3" L x 5'3" W x 7'5" H

	Craft@Hrs	Unit	Material	Labor	Total
Four rear panels, two panels each side	C8@12.0	Ea	3,730.00	492.00	4,222.00

Add to three-sided open front shelters for the following:

Wind break front entrance panel

	Craft@Hrs	Unit	Material	Labor	Total
5'3" W x 7'5" H, with two glazed panels	C8@2.00	Ea	368.00	81.90	449.90

Bench, all aluminum vandal-resistant seat & backrest

	Craft@Hrs	Unit	Material	Labor	Total
8'0" long	C8@1.00	Ea	305.00	41.00	346.00
10'0" long	C8@1.50	Ea	389.00	61.50	450.50

Light fixture with unbreakable Lexan diffuser, wiring not included

	Craft@Hrs	Unit	Material	Labor	Total
100 watt incandescent, photoelectric cell control	C8@2.00	Ea	166.00	81.90	247.90

Radiant heater, with vandal-resistant electric heating element in heavy duty enclosure and metal mesh guard over opening, wiring not included

	Craft@Hrs	Unit	Material	Labor	Total
2500 watt unit (no controls)	C8@1.50	Ea	660.00	61.50	721.50
5000 watt unit (no controls)	C8@1.50	Ea	775.00	61.50	836.50
Electronic timer control, add to above	C8@1.00	Ea	333.00	41.00	374.00

	Craft@Hrs	Unit	Material	Labor	Total

Integrated map/schedule display panel, anodized aluminum frame, tamper-proof fasteners and "spanner-head" tool for opening
Full panel width x 30" high

	Craft@Hrs	Unit	Material	Labor	Total
3/16" clear Plexiglass	C8@1.00	Ea	110.00	41.00	151.00
3/16" polycarbonate with mar-resistant coating	C8@1.00	Ea	155.00	41.00	196.00
Graphics, words "Bus Stop" in white letters, on fascia					
First side	—	LS	91.00	—	91.00
Each additional side	—	Ea	31.50	—	31.50

Metal shelter structures Prefabricated structure systems, shipped knocked-down, complete with all necessary hardware. Labor cost is for assembly on site. Typical use is for picnic pavilions. These costs do not include foundation or floor slab. Dimensions are approximate outside measurements. Based on Americana Building Products. http://www.americana.com
Four-sided hip roof shelter. Steel structural rafters and columns, aluminum roof decking with stucco embossed, painted finish, 8' height at perimeter

	Craft@Hrs	Unit	Material	Labor	Total
12' L x 12' W	S4@32.0	Ea	3,870.00	1,160.00	5,030.00
16' L x 16' W	S4@32.0	Ea	4,740.00	1,160.00	5,900.00
16' L x 32' W	S4@36.0	Ea	6,400.00	1,310.00	7,710.00
20' L x 20' W	S4@36.0	Ea	5,950.00	1,310.00	7,260.00
20' L x 30' W	S4@36.0	Ea	7,610.00	1,310.00	8,920.00
20' L x 40' W	S4@36.0	Ea	9,380.00	1,310.00	10,690.00
24' L x 24' W	S4@36.0	Ea	7,610.00	1,310.00	8,920.00
24' L x 48' W	S4@40.0	Ea	11,000.00	1,460.00	12,460.00
Add for 48" x 48" cupola with tubular metal frame	S4@1.50	Ea	772.00	54.60	826.60
Add for 6' x 12' cupola with tubular metal frame	S4@3.50	Ea	3,200.00	127.00	3,327.00

Gable roof shelter. Steel structural rafters and columns, aluminum roof and trim. 8' height, 3 in 12 roof pitch

	Craft@Hrs	Unit	Material	Labor	Total
20' L x 20' W	S4@36.0	Ea	3,530.00	1,310.00	4,840.00
20' L x 24' W	S4@36.0	Ea	5,080.00	1,310.00	6,390.00
20' L x 30' W	S4@36.0	Ea	5,410.00	1,310.00	6,720.00
20' L x 50' W	S4@40.0	Ea	9,150.00	1,460.00	10,610.00
22' L x 20' W	S4@36.0	Ea	3,740.00	1,310.00	5,050.00
22' L x 30' W	S4@36.0	Ea	5,730.00	1,310.00	7,040.00
22' L x 40' W	S4@36.0	Ea	7,610.00	1,310.00	8,920.00
22' L x 50' W	S4@40.0	Ea	9,490.00	1,460.00	10,950.00

Modular Building Construction (Manufactured housing)
Relocatable structures and institutional type housing for remote areas. Costs are per square foot covered by the roof. Based on 2" x 4" wood studs 16" OC, 1/2" gypsum wallboard inside and 1/2" CDX plywood outside, with minimum plumbing, heating and electrical systems for the intended use. Costs include factory assembly, delivery by truck to 50 miles and setup on site. Add 1.5% for each additional 50 miles of delivery. No site preparation, earthwork, foundations, or furnishings included. Cost for smaller, more complex structures will be higher.
Portable structures (temporary offices, school classrooms, etc.)

	Craft@Hrs	Unit	Material	Labor	Total
Medium quality	—	SF	—	—	46.20
Better quality	—	SF	—	—	57.20
Institutional housing, four-plex modules					
Medium quality	—	SF	—	—	59.70
Better quality	—	SF	—	—	60.30
Construction or mining camp barracks, mess halls, kitchens, etc.					
Single story structures	—	SF	—	—	52.50
Add for 2nd story (per SF of floor)	—	SF	—	—	5.30

	Craft@Hrs	Unit	Material	Labor	Total

Air-supported Structures

Air-supported pool or tennis court enclosures, polyester reinforced vinyl dome, with zippered entry air lock doors, cable tiedowns and blower, typical cost per square foot of area covered.

	Craft@Hrs	Unit	Material	Labor	Total
Under 12,000 SF	D4@.030	SF	10.80	1.23	12.03
12,000 to 24,000 SF	D4@.030	SF	10.00	1.23	11.23
24,000 to 30,000 SF	D4@.030	SF	9.29	1.23	10.52
30,000 to 36,000 SF	D4@.030	SF	8.77	1.23	10.00
36,000 to 45,000 SF	D4@.030	SF	8.66	1.23	9.89
45,000 to 80,000 SF	D4@.030	SF	8.55	1.23	9.78
80,000 to 110,000 SF	D4@.030	SF	8.30	1.23	9.53

X-Ray Viewing Panels, Clear Lead-Plastic

Based on CLEAR-Pb, by Cardinal Health. Based on SF of panel (rounded up to the next higher whole square foot). For panels 12 square feet or larger, add a crating charge. Panels larger than 72" x 96" are special order items. Weights shown are approximate

	Craft@Hrs	Unit	Material	Labor	Total
Crating charge	—	Ea	85.00	—	85.00
7mm thick, 0.3mm lead equivalence, 2.3 lbs/SF	G1@.115	SF	127.00	4.71	131.71
12mm thick, 0.5mm lead equivalence, 3.9 lbs/SF	G1@.195	SF	170.00	7.99	177.99
18mm thick, 0.8mm lead equivalence, 5.9 lbs/SF	G1@.294	SF	185.00	12.10	197.10
22mm thick, 1.0mm lead equivalence, 7.2 lbs/SF	G1@.361	SF	196.00	14.80	210.80
35mm thick, 1.5mm lead equivalence, 11.5 lbs/SF	G1@.574	SF	213.00	23.50	236.50
46mm thick, 2.0mm lead equivalence, 15.0 lbs/SF	G1@.751	SF	289.00	30.80	319.80
70mm thick, 3.0mm lead equivalence, 18.0 lbs/SF	G1@1.00	SF	408.00	41.00	449.00

Mobile X-Ray Barriers

Clear lead-plastic window panels on the upper portion and opaque panels on the lower portion, mounted within a framework with casters on the bottom. Based on Cardinal Health. Labor shown is to uncrate factory assembled barrier and attach casters.

30" W x 75" H overall

	Craft@Hrs	Unit	Material	Labor	Total
0.5mm lead equiv. window panel 30" x 24" and 0.8mm lead equiv. opaque panel 30" x 48"	MW@.501	Ea	1,930.00	23.00	1,953.00

48" W x 75" H overall

	Craft@Hrs	Unit	Material	Labor	Total
0.5mm lead equiv. window panel 48" x 36" and 0.8mm lead equiv. opaque panel 48" x 36"	MW@.501	Ea	3,560.00	23.00	3,583.00
1.0mm lead equiv. window panel 48" x 36" and 1.5mm lead equiv. opaque panel 48" x 36"	MW@.501	Ea	4,210.00	23.00	4,233.00

72" W x 75" H, overall

	Craft@Hrs	Unit	Material	Labor	Total
0.5mm lead equiv. window panel 72" x 36" and 0.8mm lead equiv. opaque panel 72" x 36"	MW@.751	Ea	4,200.00	34.50	4,234.50
1.0mm lead equiv. window panel 72" x 36" and 1.5mm lead equiv. opaque panel 72" x 36"	MW@.751	Ea	5,250.00	34.50	5,284.50

Modular X-Ray Barriers

Panels are mounted within a framework for attaching to floor, wall or ceiling. Shipped unassembled. Based on Cardinal Health. Structural supports not included. Costs shown are based on typical 36" wide x 84" high panel sections. Clear lead-plastic window panels 48" high are the upper portion and opaque leaded panels 36" high are the bottom portion.

1-section barrier, 36" W x 84" H overall

	Craft@Hrs	Unit	Material	Labor	Total
0.5mm lead equiv. panels	G1@2.78	Ea	2,920.00	114.00	3,034.00
0.8mm lead equiv. panels	G1@2.78	Ea	3,090.00	114.00	3,204.00
1.0mm lead equiv. panels	G1@3.34	Ea	3,160.00	137.00	3,297.00
1.5mm lead equiv. panels	G1@4.46	Ea	3,420.00	183.00	3,603.00

	Craft@Hrs	Unit	Material	Labor	Total
2-section barrier, 72" W x 84" H overall					
0.5mm lead equiv. panels	G1@5.56	Ea	6,210.00	228.00	6,438.00
0.8mm lead equiv. panels	G1@5.56	Ea	6,540.00	228.00	6,768.00
1.0mm lead equiv. panels	G1@6.68	Ea	6,690.00	274.00	6,964.00
1.5mm lead equiv. panels	G1@8.90	Ea	7,210.00	365.00	7,575.00
3-section barrier, 108" W x 84" H overall					
0.5mm lead equiv. panels	G1@8.34	Ea	9,570.00	342.00	9,912.00
0.8mm lead equiv. panels	G1@8.34	Ea	10,100.00	342.00	10,442.00
1.0mm lead equiv. panels	G1@10.0	Ea	10,300.00	410.00	10,710.00
1.5mm lead equiv. panels	G1@13.4	Ea	11,100.00	549.00	11,649.00
Larger than 3-section barriers, add to the cost of 1-section barriers for each section over 3					
Add per 18" W x 84" H section	—	%	52.0	—	—
Add per 36" W x 84" H section	—	%	104.0	—	—

Swimming Pools Complete gunite pools including filter, chlorinator, ladder, diving board, basic electrical system, but excluding deck beyond edge of coping. Typical subcontract prices per SF of water surface. Use the prices for preliminary estimates.

	Craft@Hrs	Unit	Material	Labor	Total
Apartment building	—	SF	—	—	48.60
Community	—	SF	—	—	62.60
Hotel or resort	—	SF	—	—	58.90
School, training (42' x 75')	—	SF	—	—	65.90
School, Olympic (42' x 165')	—	SF	—	—	100.00

Water Tanks, Elevated Typical subcontract prices for 100' high steel tanks, including design, fabrication and erection. Complete tank includes 36" pipe riser with access manhole, 8" overflow to ground, ladder with protective cage, vent, balcony catwalk, concrete foundation, painting and test. No well, pump, fencing, drainage or water distribution piping included. Costs are per tank with gallon capacity as shown. Tanks built in areas without earthquake or high wind risk may cost 15% to 25% less.

	Craft@Hrs	Unit	Material	Labor	Total
75,000 gallon	—	Ea	—	—	299,000.00
100,000 gallon	—	Ea	—	—	320,000.00
150,000 gallon	—	Ea	—	—	381,000.00
200,000 gallon	—	Ea	—	—	499,000.00
300,000 gallon	—	Ea	—	—	618,000.00
400,000 gallon	—	Ea	—	—	715,000.00
Add or subtract for each foot of height more or less than 100, per 100 gallons of capacity	—	Ea	—	—	.82
Cathodic protection system, per tank	—	LS	—	—	17,300.00

Electric service, including float switches, obstruction marker lighting system, and hookup of cathodic protection system

	Craft@Hrs	Unit	Material	Labor	Total
Per tank	—	LS	—	—	13,100.00

Solar Water Heating Systems

	Craft@Hrs	Unit	Material	Labor	Total
Solar collector panels, no piping included					
Typical 24 SF panel	P6@.747	Ea	269.00	33.10	302.10
Supports for solar panels, per panel	P6@2.25	Ea	28.80	99.70	128.50
Control panel for solar heating system	P6@1.56	Ea	99.20	69.10	168.30
Solar collector system pumps, no wiring or piping included					
5.4 GPM, 7.9' head	P6@1.89	Ea	198.00	83.70	281.70
10.5 GPM, 11.6' head or 21.6 GPM, 8.5' head	P6@2.25	Ea	358.00	99.70	457.70
12.6 GPM, 20.5' head or 22.5 GPM, 13.3' head	P6@3.26	Ea	550.00	144.00	694.00
46.5 GPM, 13.7' head	P6@4.59	Ea	823.00	203.00	1,026.00
60.9 GPM, 21.0' head	P6@5.25	Ea	941.00	233.00	1,174.00

	Craft@Hrs	Unit	Material	Labor	Total
Solar hot water storage tanks, no piping and no pad included					
40 gallon, horizontal	P6@1.12	Ea	210.00	49.60	259.60
110 gallon, horizontal	P6@3.54	Ea	860.00	157.00	1,017.00
180 gallon, vertical	P6@5.43	Ea	1,400.00	240.00	1,640.00
650 gallon, horizontal	P6@5.84	Ea	2,060.00	259.00	2,319.00
1,250 gallon, horizontal	P6@5.84	Ea	3,030.00	259.00	3,289.00
1,500 gallon, vertical	P6@11.4	Ea	4,790.00	505.00	5,295.00
2,200 gallon, horizontal	P6@17.1	Ea	4,360.00	757.00	5,117.00
2,700 gallon, vertical	P6@14.9	Ea	4,960.00	660.00	5,620.00

Dumbwaiters These costs do not include allowance for constructing the hoistway walls or electrical work.

Manual, 2 stop, 25 feet per minute, no doors, by rated capacity, 24" x 24" x 36" high car

	Craft@Hrs	Unit	Material	Labor	Total
25 pounds to 50 pounds	CV@39.8	Ea	1,760.00	1,810.00	3,570.00
75 pounds to 200 pounds	CV@42.7	Ea	2,390.00	1,940.00	4,330.00
Add for each additional stop	—	Ea	—	—	1,330.00

Electric, with machinery mounted above, floor loading, no security gates included

	Craft@Hrs	Unit	Material	Labor	Total
50 lbs, 25 FPM, 2 stop, no doors	CV@42.2	Ea	2,100.00	1,920.00	4,020.00
50 lbs, 25 FPM, 2 stop, manual doors	CV@106.	Ea	5,230.00	4,810.00	10,040.00
50 lbs, 50 FPM, 2 stop, manual doors	CV@107.	Ea	5,230.00	4,860.00	10,090.00
75 lbs, 25 FPM, 2 stop, no doors	CV@42.2	Ea	2,100.00	1,920.00	4,020.00
75 lbs, 25 FPM, 2 stop, manual doors	CV@107.	Ea	6,300.00	4,860.00	11,160.00
75 lbs, 50 FPM, 2 stop, manual doors	CV@111.	Ea	6,300.00	5,040.00	11,340.00
100 lbs, 25 FPM, 2 stop, no doors	CV@50.6	Ea	3,060.00	2,300.00	5,360.00
100 lbs, 25 FPM, 2 stop, manual doors	CV@107.	Ea	6,480.00	4,860.00	11,340.00
100 lbs, 50 FPM, 2 stop, manual doors	CV@111.	Ea	6,620.00	5,040.00	11,660.00
100 lbs, 100 FPM, 5 stop, manual doors	CV@162.	Ea	6,940.00	7,350.00	14,290.00
200 lbs, 25 FPM, 2 stop, no doors	CV@52.3	Ea	3,140.00	2,370.00	5,510.00
200 lbs, 25 FPM, 2 stop, manual doors	CV@111.	Ea	6,620.00	5,040.00	11,660.00
200 lbs, 100 FPM, 5 stop, manual doors	CV@162.	Ea	8,100.00	7,350.00	15,450.00
300 lbs, 50 FPM, 2 stop, manual doors	CV@113.	Ea	13,700.00	5,130.00	18,830.00
300 lbs, 100 FPM, 5 stop, manual doors	CV@171.	Ea	18,300.00	7,760.00	26,060.00
400 lbs, 50 FPM, 2 stop, manual doors	CV@114.	Ea	13,900.00	5,170.00	19,070.00
400 lbs, 100 FPM, 5 stop, manual doors	CV@174.	Ea	20,300.00	7,900.00	28,200.00
500 lbs, 50 FPM, 2 stop, manual doors	CV@117.	Ea	14,500.00	5,310.00	19,810.00
500 lbs, 100 FPM, 5 stop, manual doors	CV@178.	Ea	21,400.00	8,080.00	29,480.00

Elevators, Passenger Typical subcontract costs excluding the shaftwall.

Hydraulic, office type

100 FPM, 10 to 13 passenger, automatic exit door, 7' x 5' cab, 2,500 lb capacity

	Craft@Hrs	Unit	Material	Labor	Total
3 stop	—	LS	—	—	53,500.00
4 stop	—	LS	—	—	57,900.00
Each additional stop over 4	—	Ea	—	—	5,180.00

150 FPM, 10 to 13 passenger, center opening automatic door, 7' x 5' cab, 2,500 lb capacity

	Craft@Hrs	Unit	Material	Labor	Total
3 stop	—	LS	—	—	56,900.00
4 stop	—	LS	—	—	60,200.00
Each additional stop over 4	—	Ea	—	—	7,390.00

Geared, 350 FPM, 3,500 lb automatic, selective collective, 5' x 8' cab

	Craft@Hrs	Unit	Material	Labor	Total
5 stop	—	LS	—	—	97,100.00
6 stop	—	LS	—	—	103,000.00
10 stop	—	LS	—	—	124,000.00
15 stop	—	LS	—	—	166,000.00
Each additional stop	—	Ea	—	—	6,320.00
Add for 5,000 lb capacity	—	LS	—	—	3,330.00

	Craft@Hrs	Unit	Material	Labor	Total
Gearless, 500 FPM, 3,500 lb, 6' x 9' cab					
10 stop	—	LS	—	—	197,000.00
15 stop	—	LS	—	—	230,000.00
Add for each additional stop	—	Ea	—	—	5,570.00
Gearless, 700 FPM, 4,500 lb, 6' x 9' cab, passenger					
10 stop	—	LS	—	—	214,000.00
15 stop	—	LS	—	—	241,000.00
20 stop	—	LS	—	—	268,000.00
Add for each additional stop	—	Ea	—	—	6,660.00
Gearless, 1,000 FPM, 4,500 lb, 6' x 9' cab, passenger					
10 stop	—	LS	—	—	228,000.00
Add for each additional stop	—	Ea	—	—	8,770.00
Sidewalk elevators, 2 stops, 2,500 pounds	—	LS	—	—	40,100.00

Elevators, Freight Typical subcontract costs excluding the shaftwall.

	Craft@Hrs	Unit	Material	Labor	Total
Hydraulic, 2,500 pound capacity, 100 FPM, single entry cab					
Manual vertical doors, 2 stops	—	LS	—	—	43,600.00
Hydraulic, 6,000 pound capacity, 100 FPM, single entry cab					
Manual vertical doors, 5 stops	—	LS	—	—	64,100.00
Hydraulic, 6,000 pound capacity, 100 FPM, single entry cab					
Powered vertical doors, 5 stops	—	LS	—	—	75,500.00
Hydraulic, 10,000 pound capacity, 100 FPM, single entry cab					
Manual vertical doors, 5 stops	—	LS	—	—	71,800.00
Powered vertical doors, 5 stops	—	LS	—	—	81,100.00
Geared, 2,500 pound capacity, 100 FPM, single entry cab					
Manual doors, 2 stops	—	LS	—	—	69,500.00

Hoists and Cranes These costs do not include electrical work.

	Craft@Hrs	Unit	Material	Labor	Total
Electric hoists, manual trolley, 20 FPM lift. Add cost of monorail below					
1/2 ton, swivel mount, 25' lift	D4@9.34	Ea	4,140.00	384.00	4,524.00
1/2 ton, swivel mount, 45' lift	D4@13.7	Ea	4,400.00	564.00	4,964.00
1/2 ton, swivel mount, 75' lift	D4@18.0	Ea	4,890.00	741.00	5,631.00
1 ton, geared trolley, 20' lift	D4@11.9	Ea	4,940.00	490.00	5,430.00
1 ton, geared trolley, 30' lift	D4@15.8	Ea	5,230.00	650.00	5,880.00
1 ton, geared trolley, 55' lift	D4@19.2	Ea	5,750.00	790.00	6,540.00
2 ton, geared trolley, 15' lift	D4@15.8	Ea	6,060.00	650.00	6,710.00
2 ton, geared trolley, 35' lift	D4@15.8	Ea	6,440.00	650.00	7,090.00
2 ton, geared trolley, 55' lift	D4@15.8	Ea	7,020.00	650.00	7,670.00
2 ton, geared trolley, 70' lift	D4@15.8	Ea	7,360.00	650.00	8,010.00
Electric hoists, power trolley, 20 FPM trolley speed. Add cost of monorail from below					
1/2 ton, 50 FPM lift to 25'	D4@8.94	Ea	4,890.00	368.00	5,258.00
1/2 ton, 100 FPM lift to 25'	D4@15.8	Ea	7,190.00	650.00	7,840.00
1/2 ton, 50 FPM lift to 45'	D4@8.94	Ea	5,170.00	368.00	5,538.00
1/2 ton, 100 FPM lift to 45'	D4@15.8	Ea	7,530.00	650.00	8,180.00
1/2 ton, 50 FPM lift to 75'	D4@11.9	Ea	5,640.00	490.00	6,130.00
1/2 ton, 100 FPM lift to 75'	D4@18.0	Ea	7,910.00	741.00	8,651.00
1 ton, 50 FPM lift to 20'	D4@15.8	Ea	4,980.00	650.00	5,630.00
1 ton, 50 FPM lift to 30'	D4@15.8	Ea	5,290.00	650.00	5,940.00
1 ton, 50 FPM lift to 55'	D4@18.0	Ea	5,780.00	741.00	6,521.00
2 ton, 50 FPM lift to 15'	D4@17.9	Ea	6,000.00	737.00	6,737.00
2 ton, 50 FPM lift to 35'	D4@19.9	Ea	6,590.00	819.00	7,409.00
2 ton, 50 FPM lift to 55'	D4@19.9	Ea	7,060.00	819.00	7,879.00
2 ton, 50 FPM lift to 70'	D4@19.9	Ea	7,150.00	819.00	7,969.00

Conveying Systems 14

	Craft@Hrs	Unit	Material	Labor	Total
Monorail for electric hoists, channel type					
100 pounds per LF	D4@.443	LF	9.70	18.20	27.90
200 pounds per LF	D4@.555	LF	15.60	22.80	38.40
300 pounds per LF	D4@.761	LF	23.30	31.30	54.60
Jib cranes, self-supporting, swinging 8' boom, 220 degree rotation					
1,000 pounds	D4@7.90	Ea	1,540.00	325.00	1,865.00
2,000 pounds	D4@11.9	Ea	1,640.00	490.00	2,130.00
3,000 pounds	D4@13.5	Ea	1,850.00	556.00	2,406.00
4,000 pounds	D4@13.5	Ea	2,170.00	556.00	2,726.00
6,000 pounds	D4@15.8	Ea	2,330.00	650.00	2,980.00
10,000 pounds	D4@19.7	Ea	3,150.00	811.00	3,961.00
Jib cranes, wall mounted, swinging 8' boom, 180 degree rotation					
1,000 pounds	D4@8.19	Ea	828.00	337.00	1,165.00
2,000 pounds	D4@13.5	Ea	888.00	556.00	1,444.00
4,000 pounds	D4@13.5	Ea	1,360.00	556.00	1,916.00
6,000 pounds	D4@15.8	Ea	1,540.00	650.00	2,190.00
10,000 pounds	D4@19.7	Ea	2,800.00	811.00	3,611.00

Material Handling Systems

Conveyors, typical subcontract price. Foundations, support structures or electrical work not included

	Craft@Hrs	Unit	Material	Labor	Total
Belt type, 24" width	—	LF	—	—	191.00
Mail conveyors, automatic, electronic					
Horizontal	—	LF	—	—	1,650.00
Vertical, per 12' floor	—	Ea	—	—	21,000.00

Chutes, linen or solid waste handling, prefabricated unit including roof vent, 1-1/2 hour "B" rated doors, discharge and sprinkler system, gravity feed. Costs shown are for each floor based on 8' to 10' floor to floor height

	Craft@Hrs	Unit	Material	Labor	Total
Light duty, aluminum, 20" diameter	D4@5.85	Ea	781.00	241.00	1,022.00
Standard, 18 gauge steel, 24" diameter	D4@5.85	Ea	869.00	241.00	1,110.00
Standard, 18 gauge steel, 30" diameter	D4@5.85	Ea	1,050.00	241.00	1,291.00
Heavy duty, 18 gauge stainless steel					
24" diameter	D4@5.85	Ea	1,310.00	241.00	1,551.00
30" diameter	D4@5.85	Ea	1,390.00	241.00	1,631.00
Manual door with stainless steel rim	D4@2.17	Ea	374.00	89.30	463.30
Disinfecting and sanitizing unit	D4@2.17	Ea	176.00	89.30	265.30
Discharge storage unit					
Aluminum	D4@1.00	Ea	656.00	41.20	697.20
Stainless steel	D4@1.00	Ea	1,200.00	41.20	1,241.20

Turntables These costs do not include electrical work.

Baggage handling carousels

	Craft@Hrs	Unit	Material	Labor	Total
Round, 20' diameter	D4@208.	Ea	31,200.00	8,560.00	39,760.00
Round, 25' diameter	D4@271.	Ea	42,000.00	11,200.00	53,200.00

Moving Stairs and Walks Typical subcontract prices.

Escalators, 90 FPM, steel trim, 32" step width, opaque balustrade

	Craft@Hrs	Unit	Material	Labor	Total
13' rise	—	Ea	—	—	100,000.00
15' rise	—	Ea	—	—	105,000.00
17' rise	—	Ea	—	—	111,000.00
19' rise	—	Ea	—	—	115,000.00
21' rise	—	Ea	—	—	134,000.00
Add for 48" width	—	%	—	—	12.60

	Craft@Hrs	Unit	Material	Labor	Total
Pneumatic Tube Systems Typical subcontract prices.					
3" diameter, two station, single pipe 100' between stations	—	LS	—	—	5,090.00
3" diameter, two station, twin pipe, 100' between stations	—	LS	—	—	6,510.00

Mechanical 15

Black Steel Pipe and Fittings Pipe is ERW – A53 (electric resistance welded) with a factory-welded seam. Installation heights indicate the working height above the building floor. Based on orders aggregating 1,000 foot quantities. Assembly costs include hangers as required for the pipe diameter, a reducing tee and an elbow each 16 feet for 1/2" pipe to each 50 feet for 12" pipe. Expect to pay as much as double these prices for smaller quantities purchased at retail. For more detailed coverage, see National Plumbing & HVAC Estimator, http://CraftsmanSiteLicense.Com.

1/2" black steel (A53) threaded pipe and fittings

	Craft@Hrs	Unit	Material	Labor	Total
Pipe, no fittings or supports included, 1/2", threaded ends					
Schedule 40 to 10' high	M5@.060	LF	.48	2.63	3.11
Schedule 40 over 10' to 20'	M5@.070	LF	.48	3.06	3.54
Schedule 80 to 10' high	M5@.070	LF	.59	3.06	3.65
Schedule 80 over 10' to 20'	M5@.080	LF	.59	3.50	4.09
Pipe assembly, Sch. 40, fittings, hangers and supports	M5@.130	LF	1.36	5.69	7.05
Add for seamless pipe (A106)	—	%	50.0	—	—
90 degree ells, 1/2", threaded					
150 lb, malleable iron	M5@.120	Ea	.46	5.25	5.71
300 lb, malleable iron	M5@.132	Ea	3.08	5.78	8.86
45 degree ells, 1/2", threaded					
150 lb, malleable iron	M5@.120	Ea	.78	5.25	6.03
300 lb, malleable iron	M5@.132	Ea	4.66	5.78	10.44
Tees, 1/2", threaded					
150 lb, malleable iron	M5@.180	Ea	.62	7.88	8.50
300 lb, malleable iron	M5@.198	Ea	5.20	8.66	13.86
150 lb, reducing, malleable iron	M5@.170	Ea	1.02	7.44	8.46
300 lb, reducing, malleable iron	M5@.180	Ea	7.75	7.88	15.63
Caps, 1/2" threaded					
150 lb, malleable iron	M5@.090	Ea	.52	3.94	4.46
300 lb, malleable iron	M5@.100	Ea	3.27	4.38	7.65
Couplings, 1/2" threaded					
150 lb, malleable iron	M5@.120	Ea	.59	5.25	5.84
300 lb, malleable iron	M5@.132	Ea	2.48	5.78	8.26
Flanges, forged steel, 1/2"					
150 lb, threaded flange	M5@.489	Ea	10.10	21.40	31.50
300 lb, threaded flange	M5@.781	Ea	13.00	34.20	47.20
600 lb, threaded flange	M5@.859	Ea	45.40	37.60	83.00
Unions, 1/2"					
150 lb, malleable iron	M5@.140	Ea	2.16	6.13	8.29
300 lb, malleable iron	M5@.150	Ea	5.11	6.56	11.67
Add for galvanized 1/2" pipe	—	%	32.0	—	—
Add for galvanized 1/2" 150 lb. malleable fittings	—	%	15.0	—	—
Add for galvanized 1/2" 300 lb. malleable fittings	—	%	55.0	—	—
Add for galvanized 1/2" pipe assembly, Sch. 40	—	%	25.0	—	—

	Craft@Hrs	Unit	Material	Labor	Total
3/4" black steel (A53) threaded pipe and fittings					
Pipe, no fittings or supports included, 3/4", threaded					
Schedule 40 to 10' high	M5@.070	LF	.53	3.06	3.59
Schedule 40 over 10' to 20'	M5@.080	LF	.53	3.50	4.03
Schedule 80 to 10' high	M5@.080	LF	.71	3.50	4.21
Schedule 80 over 10' to 20'	M5@.095	LF	.71	4.16	4.87
Pipe assembly, Sch. 40, fittings,					
hangers and supports	M5@.140	LF	1.64	6.13	7.77
Add for seamless pipe (A106)	—	%	55.0	—	—
90 degree ells, 3/4" threaded					
150 lb, malleable iron	M5@.130	Ea	.48	5.69	6.17
300 lb, malleable iron	M5@.143	Ea	4.69	6.26	10.95
45 degree ells, 3/4" threaded					
150 lb, malleable iron	M5@.130	Ea	1.29	5.69	6.98
300 lb, malleable iron	M5@.143	Ea	5.32	6.26	11.58
Tees, 3/4" threaded					
150 lb, malleable iron	M5@.190	Ea	.90	8.31	9.21
300 lb, malleable iron	M5@.209	Ea	5.95	9.14	15.09
150 lb, reducing, malleable iron	M5@.180	Ea	1.31	7.88	9.19
300 lb, reducing, malleable iron	M5@.205	Ea	8.42	8.97	17.39
Caps, 3/4" threaded					
150 lb, malleable iron	M5@.100	Ea	.95	4.38	5.33
300 lb, malleable iron	M5@.110	Ea	4.05	4.81	8.86
Couplings, 3/4" threaded					
150 lb, malleable iron	M5@.130	Ea	.66	5.69	6.35
300 lb, malleable iron	M5@.143	Ea	3.90	6.26	10.16
Unions, 3/4"					
150 lb, malleable iron	M5@.150	Ea	2.36	6.56	8.92
300 lb, malleable iron	M5@.165	Ea	6.03	7.22	13.25
Flanges, forged steel, 3/4"					
150 lb, threaded flange	M5@.585	Ea	10.70	25.60	36.30
300 lb, threaded flange	M5@.781	Ea	14.10	34.20	48.30
600 lb, threaded flange	M5@.859	Ea	45.40	37.60	83.00
Add for galvanized 3/4" pipe	—	%	32.0	—	—
Add for galvanized 3/4" 150 lb. malleable fittings	—	%	25.0	—	—
Add for galvanized 3/4" 300 lb. malleable fittings	—	%	55.0	—	—
Add for galvanized 3/4" pipe assembly, Sch. 40	—	%	30.0	—	—
1" black steel (A53) threaded pipe and fittings					
Pipe, no fittings or supports included, 1", threaded					
Schedule 40 to 10' high	M5@.090	LF	.76	3.94	4.70
Schedule 40 over 10' to 20'	M5@.110	LF	.76	4.81	5.57
Schedule 80 to 10' high	M5@.100	LF	.99	4.38	5.37
Schedule 80 over 10' to 20'	M5@.120	LF	.99	5.25	6.24
Pipe assembly, Sch. 40, fittings,					
hangers and supports	M5@.160	LF	2.29	7.00	9.29
Add for seamless pipe (A106)	—	%	50.0	—	—
90 degree ells, 1" threaded					
150 lb, malleable iron	M5@.180	Ea	1.04	7.88	8.92
300 lb, malleable iron	M5@.198	Ea	5.89	8.66	14.55

	Craft@Hrs	Unit	Material	Labor	Total
45 degree ells, 1" threaded					
150 lb, malleable iron	M5@.180	Ea	1.60	7.88	9.48
300 lb, malleable iron	M5@.198	Ea	6.03	8.66	14.69
Tees, 1" threaded					
150 lb, malleable iron	M5@.230	Ea	1.56	10.10	11.66
300 lb, malleable iron	M5@.253	Ea	7.22	11.10	18.32
150 lb, reducing, malleable iron	M5@.205	Ea	1.75	8.97	10.72
300 lb, reducing, malleable iron	M5@.220	Ea	10.50	9.63	20.13
Caps, 1" threaded					
150 lb, malleable iron	M5@.140	Ea	.86	6.13	6.99
300 lb, malleable iron	M5@.155	Ea	5.09	6.78	11.87
Couplings, 1" threaded					
150 lb, malleable iron	M5@.180	Ea	1.10	7.88	8.98
300 lb, malleable iron	M5@.198	Ea	4.04	8.66	12.70
Flanges, forged steel, 1"					
150 lb, threaded flange	M5@.731	Ea	9.35	32.00	41.35
300 lb, threaded flange	M5@1.05	Ea	15.60	45.90	61.50
600 lb, threaded flange	M5@1.15	Ea	45.40	50.30	95.70
Unions, 1"					
150 lb, malleable iron	M5@.210	Ea	3.19	9.19	12.38
300 lb, malleable iron	M5@.230	Ea	7.92	10.10	18.02
Add for galvanized 1" pipe	—	%	32.0	—	—
Add for galvanized 1" 150 lb. malleable fittings	—	%	20.0	—	—
Add for galvanized 1" 300 lb. malleable fittings	—	%	60.0	—	—
Add for galvanized 1" pipe assembly, Sch. 40	—	%	30.0	—	—

1-1/4" black steel (A53) threaded pipe and fittings

	Craft@Hrs	Unit	Material	Labor	Total
Pipe, no fittings or supports included, 1-1/4", threaded					
Schedule 40 to 10' high	M5@.100	LF	.99	4.38	5.37
Schedule 40 over 10' to 20'	M5@.120	LF	.99	5.25	6.24
Schedule 80 to 10' high	M5@.110	LF	1.33	4.81	6.14
Schedule 80 over 10' to 20'	M5@.130	LF	1.33	5.69	7.02
Pipe assembly, Sch. 40, fittings, hangers and supports	M5@.200	LF	2.81	8.75	11.56
Add for seamless pipe (A106)	—	%	45.0	—	—
90 degree ells, 1-1/4" threaded					
150 lb, malleable iron	M5@.240	Ea	1.71	10.50	12.21
300 lb, malleable iron	M5@.264	Ea	7.91	11.60	19.51
45 degree ells, 1-1/4" threaded					
150 lb, malleable iron	M5@.240	Ea	2.96	10.50	13.46
300 lb, malleable iron	M5@.264	Ea	8.26	11.60	19.86
Tees, 1-1/4" threaded					
150 lb, malleable iron	M5@.310	Ea	2.54	13.60	16.14
300 lb, malleable iron	M5@.341	Ea	9.64	14.90	24.54
150 lb, reducing, malleable iron	M5@.290	Ea	2.96	12.70	15.66
300 lb, reducing, malleable iron	M5@.310	Ea	13.30	13.60	26.90
Caps, 1-1/4" threaded					
150 lb, malleable iron	M5@.180	Ea	1.25	7.88	9.13
300 lb, malleable iron	M5@.200	Ea	6.34	8.75	15.09
Couplings, 1-1/4" threaded					
150 lb, malleable iron	M5@.240	Ea	1.48	10.50	11.98
300 lb, malleable iron	M5@.264	Ea	5.05	11.60	16.65

Mechanical 15

	Craft@Hrs	Unit	Material	Labor	Total
Unions, 1-1/4"					
150 lb, malleable iron	M5@.280	Ea	4.45	12.30	16.75
300 lb, malleable iron	M5@.310	Ea	11.70	13.60	25.30
Flanges, forged steel, 1-1/4"					
150 lb, threaded flange	M5@.788	Ea	14.70	34.50	49.20
300 lb, threaded flange	M5@1.04	Ea	21.60	45.50	67.10
600 lb, threaded flange	M5@1.14	Ea	47.70	49.90	97.60
Add for galvanized 1-1/4" pipe	—	%	30.0	—	—
Add for galvanized 1-1/4" 150 lb. malleable fittings	—	%	20.0	—	—
Add for galvanized 1-1/4" 300 lb. malleable fittings	—	%	65.0	—	—
Add for galvanized 1-1/4" pipe assembly, Sch. 40	—	%	30.0	—	—

1-1/2" black steel (A53) threaded pipe and fittings

	Craft@Hrs	Unit	Material	Labor	Total
Pipe, no fittings or supports included, 1-1/2", threaded					
Schedule 40 to 10' high	M5@.110	LF	1.21	4.81	6.02
Schedule 40 over 10' to 20'	M5@.130	LF	1.21	5.69	6.90
Schedule 80 to 10' high	M5@.120	LF	1.62	5.25	6.87
Schedule 80 over 10' to 20'	M5@.140	LF	1.62	6.13	7.75
Pipe assembly, Sch. 40, fittings, hangers and supports	M5@.230	LF	3.28	10.10	13.38
Add for seamless pipe (A106)	—	%	45.0	—	—
90 degree ells, 1-1/2" threaded					
150 lb, malleable iron	M5@.300	Ea	2.27	13.10	15.37
300 lb, malleable iron	M5@.333	Ea	10.00	14.60	24.60
45 degree ells, 1-1/2" threaded					
150 lb, malleable iron	M5@.390	Ea	3.24	17.10	20.34
300 lb, malleable iron	M5@.429	Ea	10.70	18.80	29.50
Tees, 1-1/2" threaded					
150 lb, malleable iron	M5@.390	Ea	3.25	17.10	20.35
300 lb, malleable iron	M5@.429	Ea	11.30	18.80	30.10
150 lb, reducing, malleable iron	M5@.360	Ea	3.61	15.80	19.41
300 lb, reducing, malleable iron	M5@.390	Ea	15.60	17.10	32.70
Caps, 1-1/2" threaded					
150 lb, malleable iron	M5@.230	Ea	1.60	10.10	11.70
300 lb, malleable iron	M5@.250	Ea	8.44	10.90	19.34
Couplings, 1-1/2" threaded					
150 lb, malleable iron	M5@.300	Ea	1.90	13.10	15.00
300 lb, malleable iron	M5@.330	Ea	6.87	14.40	21.27
Unions, 1-1/2"					
150 lb, malleable iron	M5@.360	Ea	5.65	15.80	21.45
300 lb, malleable iron	M5@.400	Ea	12.20	17.50	29.70
Flanges, forged steel, 1-1/2"					
150 lb, threaded flange	M5@.840	Ea	14.70	36.80	51.50
300 lb, threaded flange	M5@1.15	Ea	22.10	50.30	72.40
600 lb, threaded flange	M5@1.26	Ea	47.70	55.10	102.80
Add for galvanized 1-1/2" pipe	—	%	30.0	—	—
Add for galvanized 1-1/2" 150 lb. malleable fittings	—	%	20.0	—	—
Add for galvanized 1-1/2" 300 lb. malleable fittings	—	%	55.0	—	—
Add for galvanized 1-1/2" pipe assembly, Sch. 40	—	%	30.0	—	—

	Craft@Hrs	Unit	Material	Labor	Total
2" black steel (A53) threaded pipe and fittings					
Pipe, no fittings or supports included, 2", threaded					
Schedule 40 to 10' high	M5@.120	LF	1.61	5.25	6.86
Schedule 40 over 10' to 20'	M5@.140	LF	1.61	6.13	7.74
Schedule 80 to 10' high	M5@.130	LF	2.23	5.69	7.92
Schedule 80 over 10' to 20'	M5@.150	LF	2.23	6.56	8.79
Pipe assembly, Sch. 40, fittings, hangers and supports	M5@.250	LF	4.24	10.90	15.14
Add for seamless pipe (A106)	—	%	40.0	—	—
90 degree ells, 2" threaded					
150 lb, malleable iron	M5@.380	Ea	3.36	16.60	19.96
300 lb, malleable iron	M5@.418	Ea	13.80	18.30	32.10
45 degree ells, 2" threaded					
150 lb, malleable iron	M5@.380	Ea	5.07	16.60	21.67
300 lb, malleable iron	M5@.418	Ea	15.30	18.30	33.60
Tees, 2" threaded					
150 lb, malleable iron	M5@.490	Ea	4.82	21.40	26.22
300 lb, malleable iron	M5@.539	Ea	17.50	23.60	41.10
150 lb, reducing, malleable iron	M5@.460	Ea	5.18	20.10	25.28
300 lb, reducing, malleable iron	M5@.500	Ea	23.50	21.90	45.40
Caps, 2" threaded					
150 lb, malleable iron	M5@.290	Ea	2.25	12.70	14.95
300 lb, malleable iron	M5@.320	Ea	9.75	14.00	23.75
Couplings, 2" threaded					
150 lb, malleable iron	M5@.380	Ea	2.76	16.60	19.36
300 lb, malleable iron	M5@.418	Ea	10.90	18.30	29.20
Flanges, forged steel, 2"					
150 lb, threaded flange	M5@.290	Ea	15.90	12.70	28.60
300 lb, threaded flange	M5@.310	Ea	25.30	13.60	38.90
600 lb, threaded flange	M5@.320	Ea	65.20	14.00	79.20
Unions, 2"					
150 lb, malleable iron	M5@.450	Ea	5.72	19.70	25.42
300 lb, malleable iron	M5@.500	Ea	15.90	21.90	37.80
Add for galvanized 2" pipe, add	—	%	30.0	—	—
Add for galvanized 2" 50 lb. malleable fittings	—	%	20.0	—	—
Add for galvanized 2" 300 lb. malleable fittings	—	%	55.0	—	—
Add for galvanized 2" pipe assembly, Sch. 40	—	%	30.0	—	—
3" black steel pipe (A53) and welded fittings					
Pipe, no fittings or supports included, 3", plain end					
Schedule 40 to 10' high	MI@.190	LF	3.78	8.78	12.56
Schedule 40 over 10' to 20' high	MI@.230	LF	3.78	10.60	14.38
Schedule 80 to 10' high	MI@.280	LF	4.97	12.90	17.87
Schedule 80 over 10' to 20' high	MI@.335	LF	4.97	15.50	20.47
Pipe assembly, Sch. 40, fittings, hangers and supports	MI@.380	LF	6.29	17.60	23.89
Add for seamless pipe (A106)	—	%	40.0	—	—
Add for galvanized pipe assembly, Sch. 40	—	%	30.0	—	—

	Craft@Hrs	Unit	Material	Labor	Total
Butt welded joints, 3" pipe					
Schedule 40 pipe	Ml@.930	Ea	—	43.00	43.00
Schedule 80 pipe	Ml@1.11	Ea	—	51.30	51.30
90 degree ells, 3"					
Schedule 40 carbon steel, butt weld, long turn	Ml@1.33	Ea	7.53	61.40	68.93
Schedule 80 carbon steel, butt weld, long turn	Ml@1.77	Ea	14.30	81.80	96.10
45 degree ells, 3"					
Schedule 40, carbon steel, butt weld	Ml@1.33	Ea	5.16	61.40	66.56
Schedule 80, carbon steel, butt weld	Ml@1.77	Ea	11.00	81.80	92.80
Tees, 3"					
Schedule 40, carbon steel, butt weld	Ml@2.00	Ea	16.30	92.40	108.70
Schedule 80, carbon steel, butt weld	Ml@2.66	Ea	29.20	123.00	152.20
Reducing, Schedule 40, carbon steel, butt weld	Ml@1.89	Ea	27.40	87.30	114.70
Caps, 3"					
Schedule 40, carbon steel, butt weld	Ml@.930	Ea	5.65	43.00	48.65
Schedule 80, carbon steel, butt weld	Ml@1.11	Ea	6.83	51.30	58.13
Reducing couplings					
Schedule 40, carbon steel, butt weld	Ml@1.20	Ea	10.20	55.40	65.60
Weld-o-lets, 3"					
Schedule 40, carbon steel	Ml@2.00	Ea	14.60	92.40	107.00
Welded pipe flanges, forged steel, 3" butt weld					
150 lb, slip-on flange	Ml@.730	Ea	15.20	33.70	48.90
300 lb, slip-on flange	Ml@.980	Ea	20.90	45.30	66.20
600 lb, slip-on flange	Ml@1.03	Ea	41.70	47.60	89.30
150 lb, weld neck flange	Ml@.730	Ea	19.00	33.70	52.70
300 lb, weld neck flange	Ml@.980	Ea	23.60	45.30	68.90
600 lb, weld neck flange	Ml@1.03	Ea	40.40	47.60	88.00
150 lb, threaded flange	Ml@.460	Ea	27.60	21.30	48.90
300 lb, threaded flange	Ml@.500	Ea	33.70	23.10	56.80
600 lb, threaded flange	Ml@.510	Ea	55.20	23.60	78.80
Flange bolt & gasket kits, 3"					
Full face, red rubber gasket, 150 lb	Ml@.750	Ea	4.28	34.70	38.98
Full face, red rubber gasket, 300 lb	Ml@1.00	Ea	8.26	46.20	54.46
4" black steel pipe (A53) and welded fittings					
Pipe, no fittings or supports included, 4", plain end					
Schedule 40 to 10' high	Ml@.260	LF	5.19	12.00	17.19
Schedule 40 over 10' to 20' high	Ml@.360	LF	5.19	16.60	21.79
Schedule 80 to 10' high	Ml@.340	LF	7.49	15.70	23.19
Schedule 80 over 10' to 20' high	Ml@.400	LF	7.49	18.50	25.99
Pipe assembly, Sch. 40, fittings, hangers and supports	Ml@.430	LF	9.18	19.90	29.08
Add for seamless pipe (A106)	—	%	35.0	—	—
Add for galvanized pipe assembly, Sch. 40	—	%	30.0	—	—
Butt welded joints, 4" pipe					
Schedule 40 pipe	Ml@1.25	Ea	—	57.80	57.80
Schedule 80 pipe	Ml@1.50	Ea	—	69.30	69.30

	Craft@Hrs	Unit	Material	Labor	Total
90 degree ells, 4"					
Schedule 40 carbon steel, butt weld, long turn	Ml@1.33	Ea	12.10	61.40	73.50
Schedule 80 carbon steel, butt weld, long turn	Ml@2.37	Ea	23.90	109.00	132.90
45 degree ells, 4"					
Schedule 40, carbon steel, butt weld	Ml@1.33	Ea	9.46	61.40	70.86
Schedule 80, carbon steel, butt weld	Ml@2.37	Ea	16.70	109.00	125.70
Tees, 4"					
Schedule 40, carbon steel, butt weld	Ml@2.67	Ea	24.40	123.00	147.40
Schedule 80, carbon steel, butt weld	Ml@3.55	Ea	47.40	164.00	211.40
Reducing, Schedule 40 carbon steel, butt weld	Ml@2.44	Ea	34.70	113.00	147.70
Caps, 4"					
Schedule 40, carbon steel, butt weld	Ml@1.25	Ea	7.43	57.80	65.23
Schedule 80, carbon steel, butt weld	Ml@1.50	Ea	9.87	69.30	79.17
Reducing couplings, 4"					
Schedule 40, carbon steel, butt weld	Ml@1.60	Ea	11.50	73.90	85.40
Weld-o-lets, 4"					
Schedule 40, carbon steel	Ml@2.67	Ea	20.50	123.00	143.50
Welded flanges, forged steel, 4"					
150 lb, slip-on flange	Ml@.980	Ea	20.50	45.30	65.80
300 lb, slip-on flange	Ml@1.30	Ea	33.00	60.10	93.10
600 lb, slip-on flange	Ml@1.80	Ea	80.70	83.20	163.90
150 lb, weld neck flange	Ml@.980	Ea	25.40	45.30	70.70
300 lb, weld neck flange	Ml@1.30	Ea	36.50	60.10	96.60
600 lb, weld neck flange	Ml@1.80	Ea	72.20	83.20	155.40
150 lb, threaded flange	Ml@.600	Ea	29.30	27.70	57.00
300 lb, threaded flange	Ml@.640	Ea	37.30	29.60	66.90
600 lb, threaded flange	Ml@.650	Ea	94.00	30.00	124.00
Flange bolt & gasket kits, 4"					
Full face, red rubber gasket, 150 lb	Ml@1.00	Ea	7.69	46.20	53.89
Full face, red rubber gasket, 300 lb	Ml@1.20	Ea	11.30	55.40	66.70
6" black steel pipe (A53) and welded fittings					
Pipe, no fittings or supports included, 6", plain end					
Schedule 40 to 10' high	M8@.420	LF	9.25	19.90	29.15
Schedule 40 over 10' to 20'	M8@.500	LF	9.25	23.70	32.95
Schedule 80 to 10' high	M8@.460	LF	15.40	21.80	37.20
Schedule 80 over 10' to 20'	M8@.550	LF	15.40	26.10	41.50
Pipe assembly, Sch. 40, fittings, hangers and supports	M8@.800	LF	23.40	37.90	61.30
Add for seamless pipe (A106)	—	%	40.0	—	—
Add for galvanized pipe assembly, Sch. 40	—	%	30.0	—	—
Butt welded joints, 6" pipe					
Schedule 40 pipe	M8@1.87	Ea	—	88.70	88.70
Schedule 80 pipe	M8@2.24	Ea	—	106.00	106.00
90 degree ells, 6"					
Schedule 40 carbon steel, butt weld, long turn	M8@2.67	Ea	35.50	127.00	162.50
Schedule 80 carbon steel, butt weld, long turn	M8@3.55	Ea	62.00	168.00	230.00
45 degree ells, 6"					
Schedule 40, carbon steel, butt weld	M8@2.67	Ea	22.10	127.00	149.10
Schedule 80, carbon steel, butt weld	M8@3.55	Ea	47.00	168.00	215.00

	Craft@Hrs	Unit	Material	Labor	Total
Tees, 6"					
Schedule 40, carbon steel, butt weld	M8@4.00	Ea	50.90	190.00	240.90
Schedule 80, carbon steel, butt weld	M8@5.32	Ea	81.60	252.00	333.60
Reducing, Schedule 40 carbon steel, butt weld	M8@3.78	Ea	77.80	179.00	256.80
Caps, 6"					
Schedule 40, carbon steel, butt weld	M8@1.87	Ea	14.90	88.70	103.60
Schedule 80, carbon steel, butt weld	M8@2.24	Ea	21.40	106.00	127.40
Reducing couplings, 6"					
Schedule 40, carbon steel, butt weld	M8@2.30	Ea	27.10	109.00	136.10
Weld-o-lets, 6"					
Schedule 40, carbon steel	M8@3.65	Ea	79.90	173.00	252.90
Welded flanges, forged steel, 6"					
150 lb, slip-on flange	M8@1.47	Ea	32.50	69.70	102.20
300 lb, slip-on flange	M8@1.95	Ea	61.60	92.50	154.10
150 lb, weld neck flange	M8@1.47	Ea	42.60	69.70	112.30
300 lb, weld neck flange	M8@1.95	Ea	65.80	92.50	158.30
Flange bolt & gasket kits, 6"					
Full face, red rubber gasket, 150 lb	M8@1.20	Ea	13.00	56.90	69.90
Full face, red rubber gasket, 300 lb	M8@1.65	Ea	20.90	78.20	99.10

8" black steel pipe (A53) and welded fittings

	Craft@Hrs	Unit	Material	Labor	Total
Pipe, no fittings or supports included, 8" plain end					
Schedule 40 to 10' high	M8@.510	LF	14.00	24.20	38.20
Schedule 40 over 10' to 20'	M8@.600	LF	14.00	28.50	42.50
Schedule 80 to 10' high	M8@.560	LF	23.60	26.60	50.20
Schedule 80 over 10' to 20' high	M8@.670	LF	23.60	31.80	55.40
Add for seamless pipe (A106)	—	%	40.0	—	—
Butt welded joints, 8" pipe					
Schedule 40 pipe	M8@2.24	Ea	—	106.00	106.00
Schedule 80 pipe	M8@2.68	Ea	—	127.00	127.00
90 degree ells, 8"					
Schedule 40 carbon steel, butt weld, long turn	M8@3.20	Ea	69.00	152.00	221.00
Schedule 80 carbon steel, butt weld, long turn	M8@4.26	Ea	117.00	202.00	319.00
45 degree ells, 8"					
Schedule 40, carbon steel, butt weld	M8@3.20	Ea	43.00	152.00	195.00
Schedule 80, carbon steel, butt weld	M8@4.26	Ea	78.10	202.00	280.10
Tees, 8"					
Schedule 40, carbon steel, butt weld	M8@4.80	Ea	101.00	228.00	329.00
Schedule 80, carbon steel, butt weld	M8@6.38	Ea	168.00	303.00	471.00
Reducing, Schedule 40, carbon steel, butt weld	M8@4.40	Ea	146.00	209.00	355.00
Caps, 8"					
Schedule 40, carbon steel, butt weld	M8@2.24	Ea	23.10	106.00	129.10
Schedule 80, carbon steel, butt weld	M8@2.68	Ea	36.80	127.00	163.80
Reducing couplings, 8"					
Schedule 40, carbon steel, butt weld	M8@2.88	Ea	38.30	137.00	175.30
Weld-o-lets, 8"					
Schedule 40, carbon steel	M8@4.76	Ea	170.00	226.00	396.00
Welded flanges, forged steel, 8"					
150 lb, slip-on flange	M8@1.77	Ea	50.80	83.90	134.70
300 lb, slip-on flange	M8@2.35	Ea	96.80	111.00	207.80
150 lb, weld neck flange	M8@1.77	Ea	70.80	83.90	154.70
300 lb, weld neck flange	M8@2.35	Ea	103.00	111.00	214.00

	Craft@Hrs	Unit	Material	Labor	Total
Flange bolt & gasket kits, 8"					
Full face, red rubber gasket, 150 lb	M8@1.40	Ea	14.30	66.40	80.70
Full face, red rubber gasket, 300 lb	M8@1.85	Ea	22.50	87.70	110.20

10" black steel pipe (A53) and welded fittings

	Craft@Hrs	Unit	Material	Labor	Total
Pipe, no fittings or supports included, 10", plain end					
Schedule 40 to 20' high	M8@.720	LF	20.90	34.10	55.00
Schedule 80 to 20' high	M8@.830	LF	39.80	39.40	79.20
Add for seamless pipe (A106)	—	%	40.0	—	—
Butt welded joints, 10" pipe					
Schedule 40 pipe	M8@2.70	Ea	—	128.00	128.00
Schedule 80 pipe	M8@3.22	Ea	—	153.00	153.00
90 degree ells, 10"					
Schedule 40 carbon steel, butt weld, long turn	M8@4.00	Ea	128.00	190.00	318.00
Schedule 80 carbon steel, butt weld, long turn	M8@5.75	Ea	216.00	273.00	489.00
45 degree ells, 10"					
Schedule 40, carbon steel, butt weld	M8@4.00	Ea	83.80	190.00	273.80
Schedule 80, carbon steel, butt weld	M8@5.75	Ea	124.00	273.00	397.00
Tees, 10"					
Schedule 40, carbon steel, butt weld	M8@6.00	Ea	178.00	285.00	463.00
Schedule 80, carbon steel, butt weld	M8@8.61	Ea	269.00	408.00	677.00
Reducing, Schedule 40, carbon steel, butt weld	M8@5.60	Ea	239.00	266.00	505.00
Caps, 10"					
Schedule 40, carbon steel, butt weld	M8@2.80	Ea	40.40	133.00	173.40
Schedule 80, carbon steel, butt weld	M8@3.62	Ea	59.00	172.00	231.00
Reducing couplings, 10"					
Schedule 40, carbon steel, butt weld	M8@3.60	Ea	52.10	171.00	223.10
Weld-o-lets, 10"					
Schedule 40, carbon steel	M8@6.35	Ea	248.00	301.00	549.00
Welded pipe flanges, forged steel, 10"					
150 lb, slip-on flange	M8@2.20	Ea	83.80	104.00	187.80
150 lb, weld neck flange	M8@2.20	Ea	115.00	104.00	219.00
300 lb, weld neck flange	M8@2.90	Ea	233.00	138.00	371.00
Flange bolt & gasket kits, 10"					
Full face, red rubber gasket, 150 lb	M8@1.80	Ea	28.30	85.40	113.70
Full face, red rubber gasket, 300 lb	M8@2.20	Ea	35.90	104.00	139.90

12" black steel pipe (A53) and welded fittings

	Craft@Hrs	Unit	Material	Labor	Total
Pipe, no fittings or supports included, 12", plain end					
Schedule 40 to 20' high	M8@.910	LF	34.80	43.20	78.00
Schedule 80 to 20' high	M8@1.05	LF	57.30	49.80	107.10
Add for seamless pipe (A106)	—	%	40.0	—	—
Butt welded joints, 12" pipe					
Schedule 40 pipe	M8@3.25	Ea	—	154.00	154.00
Schedule 80 pipe	M8@3.88	Ea	—	184.00	184.00
90 degree ells, 12"					
Schedule 40 carbon steel, butt weld, long turn	M8@4.80	Ea	249.00	228.00	477.00
Schedule 80 carbon steel, butt weld, long turn	M8@7.76	Ea	389.00	368.00	757.00
45 degree ells, 12"					
Schedule 40, carbon steel, butt weld	M8@4.80	Ea	191.00	228.00	419.00
Schedule 80, carbon steel, butt weld	M8@7.76	Ea	246.00	368.00	614.00

	Craft@Hrs	Unit	Material	Labor	Total
Tees, 12"					
Schedule 40, carbon steel, butt weld	M8@7.20	Ea	365.00	341.00	706.00
Schedule 80, carbon steel, butt weld	M8@11.6	Ea	515.00	550.00	1,065.00
Reducing, Schedule 40, carbon steel, butt weld	M8@6.80	Ea	477.00	322.00	799.00
Caps, 12"					
Schedule 40, carbon steel, butt weld	M8@3.36	Ea	66.50	159.00	225.50
Schedule 80, carbon steel, butt weld	M8@4.88	Ea	111.00	231.00	342.00
Reducing couplings, 12"					
Schedule 40, carbon steel, butt weld	M8@4.30	Ea	108.00	204.00	312.00
Weld-o-lets, 12"					
Schedule 40, carbon steel	M8@8.48	Ea	471.00	402.00	873.00
Welded flanges, forged steel, 12"					
150 lb, slip-on flange	M8@2.64	Ea	170.00	125.00	295.00
300 lb, slip-on flange	M8@3.50	Ea	287.00	166.00	453.00
150 lb, weld neck flange	M8@2.64	Ea	218.00	125.00	343.00
300 lb, weld neck flange	M8@3.50	Ea	381.00	166.00	547.00
Flange bolt & gasket kits, 12"					
Full face, red rubber gasket, 150 lb	M8@2.20	Ea	37.20	104.00	141.20
Full face, red rubber gasket, 300 lb	M8@2.60	Ea	48.20	123.00	171.20

Copper Pressure Pipe and Soldered Copper Fittings Note that copper pipe and fitting prices can change very quickly as the price of copper changes. No hangers included. See pipe hangers in the sections that follow. Labor cost assumes pipe is installed either horizontally or vertically in a building and up to 10' above floor level. Add 25% to the labor cost for heights over 10' above floor level. For more detailed coverage of copper pressure pipe, see National Plumbing & HVAC Estimator, http://www.CraftsmanSiteLicense.Com

Type M hard copper pipe. Used for water supply pressure piping above ground inside residential buildings. Smaller quantities purchased from building material retailers may cost more. Larger quantities purchased from specialty suppliers will cost less. Add the cost of soldered wrought copper fittings and supports in the sections that follow

	Craft@Hrs	Unit	Material	Labor	Total
1/2" pipe	P6@.032	LF	.58	1.42	2.00
3/4" pipe	P6@.035	LF	.93	1.55	2.48
1" pipe	P6@.038	LF	1.22	1.68	2.90
1-1/4" pipe	P6@.042	LF	1.89	1.86	3.75
1-1/2" pipe	P6@.046	LF	2.59	2.04	4.63
2" pipe	P6@.053	LF	4.12	2.35	6.47
2-1/2" pipe	P6@.060	LF	5.81	2.66	8.47
3" pipe	P6@.066	LF	7.40	2.92	10.32
4" pipe	P6@.080	LF	12.50	3.54	16.04

Type L hard copper pipe. Type L is commonly used for water supply pressure piping in commercial buildings, either above or below ground. Larger quantities will cost less. Smaller quantities purchased from building material retailers will cost more. Add the cost of soldered wrought copper fittings and supports in the sections that follow

	Craft@Hrs	Unit	Material	Labor	Total
1/2" pipe	P6@.032	LF	.78	1.42	2.20
3/4" pipe	P6@.035	LF	1.22	1.55	2.77
1" pipe	P6@.038	LF	1.82	1.68	3.50
1-1/4" pipe	P6@.042	LF	2.37	1.86	4.23
1-1/2" pipe	P6@.046	LF	3.27	2.04	5.31
2" pipe	P6@.053	LF	5.30	2.35	7.65
2-1/2" pipe	P6@.060	LF	7.67	2.66	10.33
3" pipe	P6@.066	LF	10.40	2.92	13.32
4" pipe	P6@.080	LF	16.50	3.54	20.04

Mechanical 15

	Craft@Hrs	Unit	Material	Labor	Total

Type L soft copper tube is commonly used for water supply pressure piping in commercial buildings, either above or below ground. Commercial quantities will cost less. Smaller quantities purchased from building material retailers will cost more. Add the cost of copper fittings and supports in the sections that follow

	Craft@Hrs	Unit	Material	Labor	Total
1/4", 60' coil	P6@.020	LF	.52	.89	1.41
3/8", per foot	P6@.020	LF	1.51	.89	2.40
3/8", 60' coil	P6@.020	LF	.87	.89	1.76
1/2", per foot	P6@.025	LF	1.51	1.11	2.62
1/2", 10' coil	P6@.025	LF	1.54	1.11	2.65
1/2", 20' coil	P6@.025	LF	1.13	1.11	2.24
1/2", 60' coil	P6@.025	LF	1.14	1.11	2.25
1/2", 100' coil	P6@.025	LF	.93	1.11	2.04
5/8", 20' coil	P6@.030	LF	1.31	1.33	2.64
3/4", 60' coil	P6@.030	LF	1.65	1.33	2.98
3/4", 100' coil	P6@.030	LF	1.46	1.33	2.79
1", per foot	P6@.035	LF	2.58	1.55	4.13
1", 100' coil	P6@.035	LF	2.04	1.55	3.59
1-1/4", 60' coil	P6@.042	LF	3.13	1.86	4.99
1-1/2", 60' coil	P6@.046	LF	3.83	2.04	5.87

Type K hard copper pipe. Installed in an open trench to 5' deep. No excavation or backfill included. Type K is acceptable for applications above and below ground, inside and outside of buildings. Use corporation (compression) fittings rather than soldered joints in underground applications. Based on commercial quantity purchases. Smaller quantities will cost up to 35% more. Add the cost of fittings and hangers when required

	Craft@Hrs	Unit	Material	Labor	Total
1/2" pipe	P6@.032	LF	1.07	1.42	2.49
3/4" pipe	P6@.035	LF	2.15	1.55	3.70

Type K soft copper tube. Type K is acceptable for applications above and below ground, inside and outside of buildings. Use corporation (compression) fittings rather than soldered joints in underground applications. Commercial quantities will cost less. Smaller quantities purchased from building material retailers will cost more. Add the cost of copper fittings and supports in the sections that follow

	Craft@Hrs	Unit	Material	Labor	Total
1/2", 60' coil	P6@.025	LF	1.14	1.11	2.25
3/4", 60' coil	P6@.030	LF	1.95	1.33	3.28
3/4", 100' coil	P6@.030	LF	1.53	1.33	2.86
1", 100' coil	P6@.035	LF	1.91	1.55	3.46

Copper pressure fittings Soldered wrought copper fittings. Solder is 95% tin and 5% antimony (lead free). Installed either horizontally or vertically in a building up to 10' above floor level

1/2" copper pressure fittings

	Craft@Hrs	Unit	Material	Labor	Total
1/2" 90 degree ell	P6@.107	Ea	.30	4.74	5.04
1/2" 45 degree ell	P6@.107	Ea	.54	4.74	5.28
1/2" tee	P6@.129	Ea	.66	5.71	6.37
1/2" x 1/2" x 3/8" reducing tee	P6@.121	Ea	3.65	5.36	9.01
1/2" cap	P6@.069	Ea	.24	3.06	3.30
1/2" coupling	P6@.107	Ea	.22	4.74	4.96
1/2" copper x male pipe thread adapter	P6@.075	Ea	.79	3.32	4.11

3/4" copper pressure fittings

	Craft@Hrs	Unit	Material	Labor	Total
3/4" 90 degree ell	P6@.150	Ea	.67	6.64	7.31
3/4" 45 degree ell	P6@.150	Ea	.89	6.64	7.53
3/4" tee	P6@.181	Ea	1.24	8.02	9.26
3/4" x 3/4" x 1/2" reducing tee	P6@.170	Ea	1.23	7.53	8.76

	Craft@Hrs	Unit	Material	Labor	Total
3/4" cap	P6@.096	Ea	.37	4.25	4.62
3/4" coupling	P6@.150	Ea	.38	6.64	7.02
3/4" copper x male pipe thread adapter	P6@.105	Ea	.86	4.65	5.51
1" copper pressure fittings					
1" 90 degree ell	P6@.193	Ea	2.04	8.55	10.59
1" 45 degree ell	P6@.193	Ea	2.95	8.55	11.50
1" tee	P6@.233	Ea	4.78	10.30	15.08
1" x 1" x 3/4" reducing tee	P6@.219	Ea	5.12	9.70	14.82
1" cap	P6@.124	Ea	1.29	5.49	6.78
1" coupling	P6@.193	Ea	1.24	8.55	9.79
1" copper x male pipe thread adapter	P6@.134	Ea	3.83	5.93	9.76
1-1/4" copper pressure fittings					
1-1/4" 90 degree ell	P6@.236	Ea	3.29	10.50	13.79
1-1/4" 45 degree ell	P6@.236	Ea	4.47	10.50	14.97
1-1/4" tee	P6@.285	Ea	7.62	12.60	20.22
1-1/4" x 1-1/4" x 1" reducing tee	P6@.268	Ea	8.66	11.90	20.56
1-1/4" cap	P6@.151	Ea	1.86	6.69	8.55
1-1/4" coupling	P6@.236	Ea	2.14	10.50	12.64
1-1/4" copper x male pipe thread adapter	P6@.163	Ea	7.15	7.22	14.37
1-1/2" copper pressure fittings					
1-1/2" 90 degree ell	P6@.278	Ea	5.01	12.30	17.31
1-1/2" 45 degree ell	P6@.278	Ea	5.36	12.30	17.66
1-1/2" tee	P6@.337	Ea	11.00	14.90	25.90
1-1/2" x 1-1/2" x 1-1/4" reducing tee	P6@.302	Ea	10.40	13.40	23.80
1-1/2" cap	P6@.178	Ea	2.71	7.88	10.59
1-1/2" coupling	P6@.278	Ea	3.13	12.30	15.43
1-1/2" copper x male pipe thread adapter	P6@.194	Ea	7.35	8.59	15.94
2" copper pressure fittings					
2" 90 degree ell	P6@.371	Ea	8.90	16.40	25.30
2" 45 degree ell	P6@.371	Ea	8.73	16.40	25.13
2" tee	P6@.449	Ea	17.60	19.90	37.50
2" x 2" x 1-1/2" reducing tee	P6@.422	Ea	14.00	18.70	32.70
2" cap	P6@.238	Ea	4.94	10.50	15.44
2" coupling	P6@.371	Ea	4.78	16.40	21.18
2" copper x male pipe thread adapter	P6@.259	Ea	10.50	11.50	22.00
2-1/2" copper pressure fittings					
2-1/2" 90 degree ell	P6@.457	Ea	18.60	20.20	38.80
2-1/2" 45 degree ell	P6@.457	Ea	22.60	20.20	42.80
2-1/2" tee	P6@.552	Ea	36.00	24.40	60.40
2-1/2" x 2 1/2" x 2" reducing tee	P6@.519	Ea	42.40	23.00	65.40
2-1/2" cap	P6@.292	Ea	16.00	12.90	28.90
2-1/2" coupling	P6@.457	Ea	11.00	20.20	31.20
2-1/2" copper x male pipe thread adapter	P6@.319	Ea	33.70	14.10	47.80
3" copper pressure fittings					
3" 90 degree ell	P6@.543	Ea	23.40	24.00	47.40
3" 45 degree ell	P6@.543	Ea	29.50	24.00	53.50
3" tee	P6@.656	Ea	55.20	29.10	84.30
3" x 3" x 2" reducing tee	P6@.585	Ea	48.30	25.90	74.20
3" cap	P6@.347	Ea	19.40	15.40	34.80
3" coupling	P6@.543	Ea	16.80	24.00	40.80
3" copper x male pipe thread adapter	P6@.375	Ea	56.70	16.60	73.30

	Craft@Hrs	Unit	Material	Labor	Total
4" copper pressure fittings					
4" 90 degree ell	P6@.714	Ea	56.70	31.60	88.30
4" 45 degree ell	P6@.714	Ea	61.80	31.60	93.40
4" tee	P6@.863	Ea	120.00	38.20	158.20
4" x 4" x 3" reducing tee	P6@.811	Ea	90.40	35.90	126.30
4" cap	P6@.457	Ea	40.80	20.20	61.00
4" coupling	P6@.714	Ea	39.50	31.60	71.10
4" copper x male pipe thread adapter	P6@.485	Ea	121.00	21.50	142.50

Schedule 40 PVC Pressure Pipe and Socket-weld Fittings Installed in a building vertically or horizontally up to 10' above floor level.

	Craft@Hrs	Unit	Material	Labor	Total
1/2" Schedule 40 PVC pressure pipe and fittings					
1/2" pipe	P6@.020	LF	.14	.89	1.03
1/2" 90 degree ell	P6@.100	Ea	.20	4.43	4.63
1/2" tee	P6@.130	Ea	.21	5.76	5.97
3/4" Schedule 40 PVC pressure pipe and fittings					
3/4" pipe	P6@.025	LF	.18	1.11	1.29
3/4" 90 degree ell	P6@.115	Ea	.21	5.09	5.30
3/4" tee	P6@.140	Ea	.26	6.20	6.46
1" Schedule 40 PVC pressure pipe and fittings					
1" pipe	P6@.030	LF	.27	1.33	1.60
1" 90 degree ell	P6@.120	Ea	.41	5.31	5.72
1" tee	P6@.170	Ea	.51	7.53	8.04
1-1/4" Schedule 40 PVC pressure pipe and fittings					
1-1/4" pipe	P6@.035	LF	.37	1.55	1.92
1-1/4" 90 degree ell	P6@.160	Ea	.81	7.09	7.90
1-1/4" tee	P6@.220	Ea	.96	9.74	10.70
1-1/2" Schedule 40 PVC pressure pipe and fittings					
1-1/2" pipe	P6@.040	LF	.44	1.77	2.21
1-1/2" 90 degree ell	P6@.180	Ea	.93	7.97	8.90
1-1/2" tee	P6@.250	Ea	1.17	11.10	12.27
2" Schedule 40 PVC pressure pipe and fittings					
2" pipe	P6@.046	LF	.56	2.04	2.60
2" 90 degree ell	P6@.200	Ea	1.35	8.86	10.21
2" tee	P6@.280	Ea	1.66	12.40	14.06
2-1/2" Schedule 40 PVC pressure pipe and fittings					
2-1/2" pipe	P6@.053	LF	.69	2.35	3.04
2-1/2" 90 degree ell	P6@.250	Ea	4.09	11.10	15.19
2-1/2" tee	P6@.350	Ea	5.47	15.50	20.97
3" Schedule 40 PVC pressure pipe and fittings					
3" pipe	P6@.055	LF	1.66	2.44	4.10
3" 90 degree ell	P6@.300	Ea	5.33	13.30	18.63
3" tee	P6@.420	Ea	6.29	18.60	24.89
4" Schedule 40 PVC pressure pipe and fittings					
4" pipe	P6@.070	LF	1.30	3.10	4.40
4" 90 degree ell	P6@.500	Ea	8.76	22.10	30.86
4" tee	P6@.560	Ea	13.00	24.80	37.80

	Craft@Hrs	Unit	Material	Labor	Total
Schedule 80 PVC Pressure Pipe and Socket-weld Fittings Installed in a building vertically or horizontally up to 10' above floor level.					
1/2" Schedule 80 PVC pressure pipe and fittings					
1/2" pipe	P6@.021	LF	.26	.93	1.19
1/2" 90 degree ell	P6@.105	Ea	.65	4.65	5.30
1/2" tee	P6@.135	Ea	1.83	5.98	7.81
3/4" Schedule 80 PVC pressure pipe and fittings					
3/4" pipe	P6@.026	LF	.29	1.15	1.44
3/4" 90 degree ell	P6@.120	Ea	.84	5.31	6.15
3/4" tee	P6@.145	Ea	1.90	6.42	8.32
1" Schedule 80 PVC pressure pipe and fittings					
1" pipe	P6@.032	LF	.46	1.42	1.88
1" 90 degree ell	P6@.125	Ea	1.33	5.54	6.87
1" tee	P6@.180	Ea	2.38	7.97	10.35
1-1/4" Schedule 80 PVC pressure pipe and fittings					
1-1/4" pipe	P6@.037	LF	.53	1.64	2.17
1-1/4" 90 degree ell	P6@.170	Ea	1.77	7.53	9.30
1-1/4" tee	P6@.230	Ea	6.56	10.20	16.76
1-1/2" Schedule 80 PVC pressure pipe and fittings					
1-1/2" pipe	P6@.042	LF	.72	1.86	2.58
1-1/2" 90 degree ell	P6@.190	Ea	1.90	8.42	10.32
1-1/2" tee	P6@.265	Ea	6.56	11.70	18.26
2" Schedule 80 PVC pressure pipe and fittings					
2" pipe	P6@.047	LF	.97	2.08	3.05
2" 90 degree ell	P6@.210	Ea	2.31	9.30	11.61
2" tee	P6@.295	Ea	8.20	13.10	21.30
2-1/2" Schedule 80 PVC pressure pipe and fittings					
2-1/2" pipe	P6@.052	LF	1.56	2.30	3.86
2-1/2" 90 degree ell	P6@.262	Ea	5.39	11.60	16.99
2-1/2" tee	P6@.368	Ea	8.93	16.30	25.23
3" Schedule 80 PVC pressure pipe and fittings					
3" pipe	P6@.057	LF	2.09	2.52	4.61
3" 90 degree ell	P6@.315	Ea	6.07	14.00	20.07
3" tee	P6@.440	Ea	11.10	19.50	30.60
4" Schedule 80 PVC pressure pipe and fittings					
4" pipe	P6@.074	LF	3.08	3.28	6.36
4" 90 degree ell	P6@.520	Ea	9.21	23.00	32.21
4" tee	P6@.585	Ea	12.90	25.90	38.80

CPVC Chlorinated Pressure Pipe and Socket-Weld Fittings Installed in a building vertically or horizontally up to 10' above floor level. CPVC pipe is the plastic alternative to traditional copper pipe. For hot and cold water distribution to 210 degrees F. Tan in color. Complies with ASTM D 2846 and has the same outside diameter as copper tubing. SDR 11 pressure pipe has the same pressure rating for all pipe diameters. FlowGuard Gold® pipe has a gold stripe.

	Craft@Hrs	Unit	Material	Labor	Total
1/2" CPVC chlorinated pressure pipe and fittings					
#4120 SDR 11 CPVC pipe	P6@.020	LF	.25	.89	1.14
FlowGuard Gold pipe	P6@.020	LF	.32	.89	1.21
90 degree elbow	P6@.100	Ea	.21	4.43	4.64
90 degree drop ear elbow	P6@.100	Ea	.84	4.43	5.27
FlowGuard drop ear reducing elbow	P1@.100	Ea	.50	3.00	3.50
45 degree elbow	P6@.100	Ea	.30	4.43	4.73
Coupling	P6@.100	Ea	.21	4.43	4.64

	Craft@Hrs	Unit	Material	Labor	Total
90 degree street elbow	P6@.100	Ea	.25	4.43	4.68
Tee, 1/2" x 1/2" x 1/2"	P6@.130	Ea	.24	5.76	6.00
Tee, 1/2" x 3/4" x 3/4"	P6@.150	Ea	.64	6.64	7.28
Union	P6@.200	Ea	2.40	8.86	11.26
Compression coupling	P6@.200	Ea	2.60	8.86	11.46
Bushing, 1/2" x 1/8" bushing	P6@.080	Ea	.53	3.54	4.07
Cap	P6@.060	Ea	.23	2.66	2.89

3/4" CPVC chlorinated pressure pipe and fittings

	Craft@Hrs	Unit	Material	Labor	Total
#4120 SDR 11 CPVC pipe	P6@.025	LF	.40	1.11	1.51
FlowGuard Gold pipe	P6@.025	LF	.52	1.11	1.63
90 degree elbow	P6@.115	Ea	.36	5.09	5.45
45 degree elbow	P6@.115	Ea	.38	5.09	5.47
Coupling	P6@.115	Ea	.23	5.09	5.32
90 degree street elbow	P6@.115	Ea	.41	5.09	5.50
90 degree reducing elbow, 3/4" x 1/2"	P6@.115	Ea	.41	5.09	5.50
Tee, 3/4" x 3/4" x 3/4"	P6@.140	Ea	.51	6.20	6.71
Tee, 3/4" x 1/2" x 1/2"	P6@.140	Ea	.61	6.20	6.81
Tee, 3/4" x 1/2" x 3/4"	P6@.150	Ea	.53	6.64	7.17
Tee, 3/4" x 3/4" x 1/2"	P6@.150	Ea	.60	6.64	7.24
Tee, 3/4" x 1/2" x 1/2"	P6@.150	Ea	.61	6.64	7.25
Cap	P6@.070	Ea	.27	3.10	3.37
Bushing, 1" x 1/2"	P6@.080	Ea	.62	3.54	4.16
Bushing, 3/4" x 1/2"	P6@.080	Ea	.30	3.54	3.84
Union	P6@.250	Ea	3.54	11.10	14.64
Reducing coupling, 3/4" x 1/2"	P6@.115	Ea	.27	5.09	5.36
Compression coupling	P6@.200	Ea	3.07	8.86	11.93

1" CPVC chlorinated pressure pipe and fittings

	Craft@Hrs	Unit	Material	Labor	Total
FlowGuard Gold pipe	P6@.030	Ea	11.40	1.33	12.73
45 degree elbow	P6@.120	Ea	1.00	5.31	6.31
Coupling	P6@.120	Ea	1.02	5.31	6.33
Tee, 1" x 1" x 1"	P6@.170	Ea	2.56	7.53	10.09
Tee, 1" x 1" x 3/4"	P6@.170	Ea	2.59	7.53	10.12
Cap	P6@.070	Ea	0.71	3.10	3.81
Bushing, 1" x 3/4"	P6@.080	Ea	1.02	3.54	4.56

CPVC and brass transition fittings. FIPT means female iron pipe thread. MIPT means male iron pipe thread.

	Craft@Hrs	Unit	Material	Labor	Total
1/2" union, slip x FIPT	P6@.150	Ea	3.64	6.64	10.28
1/2" drop ear 90 degree elbow, slip x FIPT	P6@.090	Ea	3.92	3.99	7.91
1/2" union, slip x MIPT	P6@.250	Ea	3.13	11.10	14.23
1/2" x 3/4" union, slip x MIPT	P6@.250	Ea	4.71	11.10	15.81
3/4" union, slip x FIPT	P6@.170	Ea	4.42	7.53	11.95
3/4" union, slip x MIPT	P6@.250	Ea	5.21	11.10	16.31

CPVC to copper pipe transition fittings. One slip joint and one threaded joint. FIPT means female iron pipe thread. MIPT means male iron pipe thread.

	Craft@Hrs	Unit	Material	Labor	Total
1/2" slip x FIPT	P6@.080	Ea	3.13	3.54	6.67
3/4" slip x FIPT	P6@.085	Ea	4.11	3.76	7.87
1/2" slip x MIPT	P6@.080	Ea	3.13	3.54	6.67
3/4" slip x MIPT	P6@.085	Ea	4.03	3.76	7.79

CPVC adapter fittings. One slip joint and one threaded joint. Joins CVPV pipe to threaded fittings or valves. FIPT means female iron pipe thread. MIPT means male iron pipe thread.

	Craft@Hrs	Unit	Material	Labor	Total
1/2" slip x FIPT	P6@.090	Ea	.78	3.99	4.77
1/2" slip x MIPT	P6@.090	Ea	.24	3.99	4.23
1/2" SPIG x compression	P6@.090	Ea	.56	3.99	4.55

	Craft@Hrs	Unit	Material	Labor	Total
3/4" slip x MIPT	P6@.100	Ea	.36	4.43	4.79
3/4" slip x FIPT	P6@.100	Ea	.96	4.43	5.39
1" slip x FIPT	P6@.120	Ea	1.22	5.31	6.53
1" slip x MIPT	P6@.120	Ea	1.74	5.31	7.05

CPVC snap-on tubing strap. Secures CPVC pipe in a horizontal or vertical position. Polypropylene.

	Craft@Hrs	Unit	Material	Labor	Total
1/2" pipe	P6@.110	Ea	.08	4.87	4.95
3/4" pipe	P6@.115	Ea	.10	5.09	5.19

Schedule 80 CPVC Chlorinated Pressure Pipe and Socket-weld Fittings Installed in a building vertically or horizontally up to 10' above floor level. Typical use is in fire protection systems.

3/4" Schedule 80 CPVC chlorinated pressure pipe and fittings

	Craft@Hrs	Unit	Material	Labor	Total
3/4" pipe	SP@.055	LF	1.18	2.95	4.13
3/4" 90 degree ell	SP@.100	Ea	1.46	5.36	6.82
3/4" tee	SP@.150	Ea	2.73	8.04	10.77

1" Schedule 80 CPVC chlorinated pressure pipe and fittings

	Craft@Hrs	Unit	Material	Labor	Total
1" pipe	SP@.055	LF	1.56	2.95	4.51
1" 90 degree ell	SP@.120	Ea	2.31	6.43	8.74
1" tee	SP@.170	Ea	3.34	9.11	12.45

1-1/4" Schedule 80 CPVC chlorinated pressure pipe and fittings

	Craft@Hrs	Unit	Material	Labor	Total
1-1/4" pipe	SP@.065	LF	2.46	3.48	5.94
1-1/4" 90 degree ell	SP@.170	Ea	5.08	9.11	14.19
1-1/4" tee	SP@.230	Ea	7.86	12.30	20.16

1-1/2" Schedule 80 CPVC chlorinated pressure pipe and fittings

	Craft@Hrs	Unit	Material	Labor	Total
1-1/2" pipe	SP@.075	LF	2.58	4.02	6.60
1-1/2" 90 degree ell	SP@.180	Ea	5.59	9.64	15.23
1-1/2" tee	SP@.250	Ea	8.07	13.40	21.47

2" Schedule 80 CPVC chlorinated pressure pipe and fittings

	Craft@Hrs	Unit	Material	Labor	Total
2" pipe	SP@.085	LF	3.56	4.55	8.11
2" 90 degree ell	SP@.200	Ea	6.75	10.70	17.45
2" tee	SP@.280	Ea	8.99	15.00	23.99

Copper DWV Pipe and Fittings Straight rigid lengths. For commercial drain, waste and vent (DWV) systems. Labor assumes soldered joints with pipe installed vertically in a building up to 10' above floor level. Add the cost of supports.

1-1/4" copper DWV pipe and fittings

	Craft@Hrs	Unit	Material	Labor	Total
Cut pipe, per foot	P6@.042	LF	2.04	1.86	3.90
10' lengths	P6@.042	LF	1.89	1.86	3.75
Coupling	P6@.200	Ea	1.46	8.86	10.32
90 degree elbow	P6@.200	Ea	2.04	8.86	10.90
45 degree elbow	P6@.200	Ea	2.40	8.86	11.26
Tee	P6@.240	Ea	4.68	10.60	15.28
Trap adapter	P6@.145	Ea	4.94	6.42	11.36

1-1/2" copper DWV pipe and fittings

	Craft@Hrs	Unit	Material	Labor	Total
Cut pipe, per foot	P6@.045	LF	3.11	1.99	5.10
10' lengths	P6@.045	LF	2.53	1.99	4.52
Coupling	P6@.200	Ea	1.74	8.86	10.60
90 degree elbow	P6@.350	Ea	4.11	15.50	19.61
45 degree elbow	P6@.200	Ea	3.06	8.86	11.92
Wye, cast brass	P6@.240	Ea	13.60	10.60	24.20
Tee	P6@.240	Ea	6.27	10.60	16.87
Test cap	P6@.075	Ea	0.51	3.32	3.83

	Craft@Hrs	Unit	Material	Labor	Total
Male adapter, C x MIP	P6@.150	Ea	4.15	6.64	10.79
Female adapter, FTG x C	P6@.200	Ea	3.54	8.86	12.40
Trap adapter	P6@.155	Ea	5.84	6.86	12.70
Trap adapter, with cleanout	P6@.235	Ea	20.30	10.40	30.70
2" copper DWV pipe and fittings					
Cut pipe, per foot	P6@.056	LF	3.31	2.48	5.79
10' lengths	P6@.056	LF	2.07	2.48	4.55
Coupling	P6@.270	Ea	2.05	12.00	14.05
90 degree elbow	P6@.270	Ea	6.80	12.00	18.80
45 degree elbow	P6@.270	Ea	5.13	12.00	17.13
Reducing bushing, 2" to 1-1/2"	P6@.140	Ea	5.26	6.20	11.46
Wye, cast brass	P6@.320	Ea	15.80	14.20	30.00
Tee, 2" x 2" x 2"	P6@.320	Ea	7.25	14.20	21.45
Tee, 2" x 2" x 1-1/2"	P6@.240	Ea	5.75	10.60	16.35
Male adapter C x MIP	P6@.185	Ea	5.08	8.19	13.27
Female adapter C x FIP	P6@.259	Ea	6.28	11.50	17.78
Trap adapter, with cleanout	P6@.325	Ea	26.70	14.40	41.10

PVC SDR 35 DWV Drainage Pipe Solvent weld, installed in an open trench inside a building, (main building drain) including typical fittings (one wye and 45 degree elbow every 20'). SDR 35 is thin wall pipe suitable for below grade applications only. Add for excavation and backfill. Use these figures for preliminary estimates.

	Craft@Hrs	Unit	Material	Labor	Total
3" pipe assembly below grade	P6@.160	LF	2.68	7.09	9.77
4" pipe assembly below grade	P6@.175	LF	4.07	7.75	11.82
6" pipe assembly below grade	P6@.220	LF	8.52	9.74	18.26
Add for interior excavation and backfill (shallow)	—	LF	—	6.56	6.56

ABS DWV Pipe and Fittings Acrylonitrile Butadiene Styrene (ASTM D 2661-73) pipe is made from rubber-based resins and used primarily for drain, waste and vent systems. The Uniform Plumbing Code limits ABS to residential buildings and below grade applications in commercial and public buildings. Solvent-weld ABS is easier to install than cast iron or copper drainage pipe. ABS does not rot, rust, corrode or collect waste and resists mechanical damage, even at low temperatures.

	Craft@Hrs	Unit	Material	Labor	Total
1-1/2" pipe, per linear foot	P6@.020	LF	.55	.89	1.44
1-1/2" pipe, 10' lengths	P6@.020	LF	.43	.89	1.32
1-1/2" pipe, 20' lengths	P6@.020	LF	.32	.89	1.21
2" pipe, per linear foot	P6@.025	LF	.71	1.11	1.82
2" pipe, 10' lengths	P6@.025	LF	.55	1.11	1.66
2" pipe, 20' lengths	P6@.025	LF	.42	1.11	1.53
3" pipe, per linear foot	P6@.030	LF	1.47	1.33	2.80
3" pipe, 10' lengths	P6@.030	LF	1.10	1.33	2.43
3" pipe, 20' lengths	P6@.030	LF	.97	1.33	2.30
4" pipe, per linear foot	P6@.040	LF	2.12	1.77	3.89
4" pipe, 10' lengths	P6@.040	LF	1.54	1.77	3.31
4" pipe, 20' lengths	P6@.040	LF	1.41	1.77	3.18
Deduct for below grade applications	—	%	—	-15.0	—

ABS 90 degree elbow (1/4 bend), hub and hub except as noted. Street elbows are hub and spigot

	Craft@Hrs	Unit	Material	Labor	Total
1-1/2" ell	P6@.060	Ea	.72	2.66	3.38
1-1/2" long sweep	P6@.200	Ea	1.41	8.86	10.27
1-1/2" street ell	P6@.120	Ea	1.13	5.31	6.44
1-1/2" venting	P6@.150	Ea	1.03	6.64	7.67
2" ell	P6@.125	Ea	.97	5.54	6.51
2", long sweep	P6@.250	Ea	1.77	11.10	12.87
2", street ell	P6@.125	Ea	1.60	5.54	7.14
2", venting	P6@.150	Ea	1.85	6.64	8.49

	Craft@Hrs	Unit	Material	Labor	Total
3" ell	P6@.175	Ea	2.59	7.75	10.34
3", long sweep	P6@.500	Ea	3.79	22.10	25.89
3", street ell	P6@.150	Ea	3.05	6.64	9.69
3", venting	P6@.150	Ea	4.57	6.64	11.21
3" x 3" x 1-1/2", hub x hub x hub, side inlet	P6@.500	Ea	7.50	22.10	29.60
3" x 3" x 2", hub x hub x hub, side inlet	P6@.500	Ea	6.97	22.10	29.07
3" x 3" x 2" low heel elbow	P6@.500	Ea	4.56	22.10	26.66
3" x 3" x 1-1/2" low heel elbow	P6@.175	Ea	6.17	7.75	13.92
4" ell	P6@.175	Ea	4.89	7.75	12.64
4", long sweep	P6@.500	Ea	8.03	22.10	30.13
4", street ell	P6@.175	Ea	7.09	7.75	14.84
4" x 3" closet elbow, spigot x hub	P6@.150	Ea	6.79	6.64	13.43
4" x 3" closet elbow, hub x hub	P6@.260	Ea	3.76	11.50	15.26

ABS 60 degree elbow (1/6 bend), hub and hub

	Craft@Hrs	Unit	Material	Labor	Total
1-1/2" bend	P6@.200	Ea	1.32	8.86	10.18
2" bend	P6@.250	Ea	1.80	11.10	12.90
3" bend	P6@.500	Ea	6.43	22.10	28.53
4" bend	P6@.500	Ea	9.38	22.10	31.48

ABS 45 degree elbow (1/8 bend). Regular bends are hub x hub. Street bends are hub and spigot

	Craft@Hrs	Unit	Material	Labor	Total
1-1/2" bend	P6@.120	Ea	.91	5.31	6.22
1-1/2" street bend	P6@.120	Ea	.87	5.31	6.18
2" bend	P6@.125	Ea	1.17	5.54	6.71
2" street bend	P6@.125	Ea	1.20	5.54	6.74
3" bend	P6@.175	Ea	2.47	7.75	10.22
3" street bend	P6@.150	Ea	2.54	6.64	9.18
4" bend	P6@.175	Ea	4.06	7.75	11.81
4" street bend	P6@.175	Ea	5.54	7.75	13.29

ABS 22-1/2 degree elbow (1/16 bend). Regular bends are hub x hub. Street bends are hub and spigot

	Craft@Hrs	Unit	Material	Labor	Total
1-1/2" bend	P6@.120	Ea	1.07	5.31	6.38
1-1/2", street bend	P6@.120	Ea	2.85	5.31	8.16
2" bend	P6@.125	Ea	1.56	5.54	7.10
2", street bend	P6@.125	Ea	2.73	5.54	8.27
3" bend	P6@.170	Ea	4.88	7.53	12.41
3" street bend	P6@.150	Ea	5.42	6.64	12.06
4" bend	P6@.175	Ea	5.03	7.75	12.78
4" street bend	P6@.175	Ea	7.38	7.75	15.13

ABS coupling, hub x hub.

	Craft@Hrs	Unit	Material	Labor	Total
1-1/2"	P6@.120	Ea	.37	5.31	5.68
2"	P6@.125	Ea	.55	5.54	6.09
2" x 1-1/2", reducing	P6@.125	Ea	1.34	5.54	6.88
3"	P6@.175	Ea	1.18	7.75	8.93
3" x 2", reducing	P6@.175	Ea	3.22	7.75	10.97
4"	P6@.250	Ea	2.11	11.10	13.21
4" x 3", reducing	P6@.250	Ea	4.74	11.10	15.84

ABS wye, connects two branch lines to a main horizontal drain line. Hub x hub x hub

	Craft@Hrs	Unit	Material	Labor	Total
1-1/2"	P6@.150	Ea	1.71	6.64	8.35
2"	P6@.190	Ea	2.19	8.42	10.61
2" x 1-1/2"	P6@.190	Ea	2.81	8.42	11.23
3"	P6@.260	Ea	4.20	11.50	15.70
3" x 1-1/2"	P6@.240	Ea	4.02	10.60	14.62
3" x 2"	P6@.260	Ea	3.52	11.50	15.02
4"	P6@.350	Ea	7.58	15.50	23.08
4" x 2"	P6@.350	Ea	6.39	15.50	21.89
4" x 3"	P6@.360	Ea	7.40	15.90	23.30

	Craft@Hrs	Unit	Material	Labor	Total
ABS double wye, joins three ABS branch lines to a main horizontal drain line. Hub x hub x hub x hub					
1-1/2"	P6@.260	Ea	3.14	11.50	14.64
2"	P6@.290	Ea	3.90	12.80	16.70
3"	P6@.325	Ea	9.82	14.40	24.22
4" x 3"	P6@.350	Ea	13.50	15.50	29.00
ABS sanitary tee (TY), connects a branch line to a vertical ABS drain line. Regular sanitary tees are hub x hub x hub. Street sanitary tees are spigot x hub x hub					
1-1/2"	P6@.150	Ea	1.00	6.64	7.64
1-1/2", long sweep	P6@.150	Ea	3.19	6.64	9.83
1-1/2", street tee	P6@.150	Ea	3.09	6.64	9.73
2" x 1-1/2"	P6@.160	Ea	1.16	7.09	8.25
2" x 1-1/2" x 1-1/2"	P6@.160	Ea	1.48	7.09	8.57
2" x 1-1/2" x 2"	P6@.160	Ea	1.59	7.09	8.68
2"	P6@.190	Ea	1.54	8.42	9.96
2", long sweep	P6@.190	Ea	3.87	8.42	12.29
2" x 1-1/2", long sweep	P6@.190	Ea	3.57	8.42	11.99
2" street tee	P6@.190	Ea	3.61	8.42	12.03
2" x 1-1/2" x 1-1/2" street tee	P6@.190	Ea	4.08	8.42	12.50
3"	P6@.260	Ea	3.35	11.50	14.85
3", long sweep	P6@.260	Ea	6.32	11.50	17.82
3" x 1-1/2"	P6@.210	Ea	3.31	9.30	12.61
3" x 1-1/2", long sweep	P6@.200	Ea	5.46	8.86	14.32
3" x 2"	P6@.235	Ea	3.27	10.40	13.67
3" x 2", long sweep	P6@.260	Ea	3.87	11.50	15.37
4"	P6@.350	Ea	6.43	15.50	21.93
4", long sweep	P6@.350	Ea	9.57	15.50	25.07
4" x 2"	P6@.275	Ea	7.63	12.20	19.83
4" x 2", long sweep	P6@.350	Ea	10.10	15.50	25.60
4" x 3"	P6@.325	Ea	7.98	14.40	22.38
4" x 3", long sweep	P6@.360	Ea	11.30	15.90	27.20
ABS bungalow fitting (Wisconsin tee). Connects vent (3" top inlet) and horizontal water closet discharge (3" branch) and bathtub discharge (1-1/2" branch) to a vertical soil stack. One fitting connects discharge and venting for a set of washroom fixtures					
3", 1-1/2" left inlet	P6@.325	Ea	8.83	14.40	23.23
3", 2" right inlet	P6@.325	Ea	7.77	14.40	22.17
ABS double sanitary tee (double TY/cross), brings two ABS-DWV branch lines into a vertical waste or soil stack. All hub except as noted					
1-1/2"	P6@.250	Ea	2.37	11.10	13.47
2" x 1-1/2"	P6@.260	Ea	3.16	11.50	14.66
2"	P6@.290	Ea	4.22	12.80	17.02
2" double fixture tee	P6@.290	Ea	5.76	12.80	18.56
2" x 1-1/2" x 1-1/2" x 1-1/2"	P6@.260	Ea	5.54	11.50	17.04
3"	P6@.325	Ea	9.02	14.40	23.42
3" double fixture tee, spigot x hub	P6@.450	Ea	11.90	19.90	31.80
3" x 1-1/2"	P6@.310	Ea	4.74	13.70	18.44
3" x 2"	P6@.335	Ea	6.39	14.80	21.19
4"	P6@.350	Ea	17.20	15.50	32.70
ABS cleanout tee, two-way					
3", hub x hub x hub	P6@.175	Ea	8.37	7.75	16.12
4", hub x hub x hub	P6@.250	Ea	19.20	11.10	30.30

	Craft@Hrs	Unit	Material	Labor	Total

ABS test tee, provides a cleanout opening flush with a finished wall. Includes a cleanout plug. FIPT means female iron pipe thread

	Craft@Hrs	Unit	Material	Labor	Total
1-1/2", hub x hub x cleanout	P6@.130	Ea	3.19	5.76	8.95
2", hub x hub x cleanout	P6@.170	Ea	5.24	7.53	12.77
3", hub x hub x cleanout	P6@.220	Ea	9.48	9.74	19.22
4", hub x hub x cleanout	P6@.240	Ea	15.30	10.60	25.90
1-1/2", hub x hub x FIPT	P6@.150	Ea	2.68	6.64	9.32
2", hub x hub x FIPT	P6@.190	Ea	3.15	8.42	11.57
3", hub x hub x FIPT	P6@.260	Ea	6.05	11.50	17.55
4", hub x hub x FIPT	P6@.350	Ea	9.41	15.50	24.91

ABS test cap

	Craft@Hrs	Unit	Material	Labor	Total
1-1/2"	P6@.100	Ea	.43	4.43	4.86
2"	P6@.100	Ea	.53	4.43	4.96
3"	P6@.100	Ea	.62	4.43	5.05
4"	P6@.100	Ea	.70	4.43	5.13

ABS test plug with galvanized wing nut, galvanized steel flanges, molded rubber gasket.

	Craft@Hrs	Unit	Material	Labor	Total
1-1/2"	P6@.030	Ea	2.39	1.33	3.72
2"	P6@.030	Ea	2.79	1.33	4.12
3"	P6@.030	Ea	3.80	1.33	5.13
4"	P6@.030	Ea	4.02	1.33	5.35

ABS closet flange. Closet flanges are used for mounting water closets.

	Craft@Hrs	Unit	Material	Labor	Total
4" x 3", with knock out	P6@.225	Ea	3.16	9.97	13.13
4" x 3", flush fit	P6@.225	Ea	4.99	9.97	14.96
4" x 3", flange x hub	P6@.225	Ea	5.99	9.97	15.96
4" x 3", flange x spigot	P6@.225	Ea	8.45	9.97	18.42
4" x 3", offset, adjustable	P6@.225	Ea	10.30	9.97	20.27
4" closet flange riser	P6@.180	Ea	3.12	7.97	11.09

ABS female adapter, joins ABS pipe to standard male threaded pipe. FIPT means female iron pipe thread

	Craft@Hrs	Unit	Material	Labor	Total
1-1/2", hub x FIPT	P6@.080	Ea	.89	3.54	4.43
1-1/2", spigot x FIPT	P6@.080	Ea	.87	3.54	4.41
2", hub x FIPT	P6@.085	Ea	1.50	3.76	5.26
2", spigot x FIPT	P6@.085	Ea	1.16	3.76	4.92
3", hub x FIPT	P6@.155	Ea	2.66	6.86	9.52
3", spigot x FIPT	P6@.155	Ea	2.65	6.86	9.51
4", hub x FIPT	P6@.220	Ea	4.57	9.74	14.31
4", spigot x FIPT	P6@.220	Ea	4.66	9.74	14.40

ABS male adapter, joins ABS pipe to standard female threaded pipe. MIPT means male iron pipe thread

	Craft@Hrs	Unit	Material	Labor	Total
1-1/2", hub x MIPT	P6@.080	Ea	.60	3.54	4.14
2", hub x MIPT	P6@.085	Ea	1.12	3.76	4.88
3", hub x MIPT	P6@.155	Ea	2.58	6.86	9.44
4", hub x MIPT	P6@.220	Ea	6.05	9.74	15.79

ABS trap adapter, joins ABS pipe to sink or bath waste outlets

	Craft@Hrs	Unit	Material	Labor	Total
1-1/2", hub x slip jam nut	P6@.200	Ea	1.00	8.86	9.86
1-1/2", spigot x slip jam nut	P6@.200	Ea	1.07	8.86	9.93
1-1/2" x 1-1/4", spigot x slip jam nut	P6@.200	Ea	.98	8.86	9.84
1-1/2" x 1-1/4", hub x slip jam nut	P6@.200	Ea	1.13	8.86	9.99

ABS cleanout adapter, fits inside pipe, flush finish

	Craft@Hrs	Unit	Material	Labor	Total
2", insert	P6@.100	Ea	3.13	4.43	7.56
3", insert	P6@.150	Ea	4.18	6.64	10.82

	Craft@Hrs	Unit	Material	Labor	Total
ABS-DWV cleanout plug, male pipe threads					
1-1/2"	P6@.090	Ea	.60	3.99	4.59
2"	P6@.120	Ea	.72	5.31	6.03
3"	P6@.240	Ea	1.17	10.60	11.77
4"	P6@.320	Ea	2.24	14.20	16.44
4", Tom Kap	P6@.320	Ea	5.26	14.20	19.46
2", countersunk	P6@.120	Ea	1.62	5.31	6.93
3", countersunk	P6@.240	Ea	2.03	10.60	12.63
ABS soil pipe adapter, joins ABS DWV to cast iron pipe soil pipe					
3" ABS hub x cast iron spigot	P6@.155	Ea	4.27	6.86	11.13
4" ABS hub to cast iron spigot	P6@.250	Ea	5.55	11.10	16.65
4" cast iron spigot to 3" DWV hub	P6@.220	Ea	5.31	9.74	15.05
ABS P-trap, used at fixture waste outlets to prevent sewer gas from entering living space via the fixture drain					
1-1/2", all hub	P6@.220	Ea	2.18	9.74	11.92
1-1/2", hub x hub, with union	P6@.280	Ea	2.73	12.40	15.13
1-1/2", hub, with cleanout	P6@.155	Ea	5.48	6.86	12.34
1-1/2", hub x spigot, with union	P6@.220	Ea	4.06	9.74	13.80
2", all hub	P6@.280	Ea	3.55	12.40	15.95
2", hub x hub, with union	P6@.280	Ea	5.62	12.40	18.02
2", hub, with cleanout	P6@.195	Ea	8.06	8.64	16.70
ABS return bend, bottom portion of P-trap, no cleanout					
1-1/2", all hub	P6@.120	Ea	2.84	5.31	8.15
2", all hub	P6@.125	Ea	4.19	5.54	9.73
ABS cap, permanently caps ABS-DWV piping					
1-1/2"	P6@.030	Ea	1.58	1.33	2.91
2"	P6@.040	Ea	2.51	1.77	4.28
3"	P6@.050	Ea	4.13	2.21	6.34
4"	P6@.090	Ea	5.45	3.99	9.44
ABS flush bushing, inserts into ABS fitting hub to reduce to a smaller pipe size					
2" x 1-1/2"	P6@.062	Ea	.54	2.75	3.29
3" x 1-1/2"	P6@.088	Ea	2.16	3.90	6.06
3" x 2"	P6@.088	Ea	1.65	3.90	5.55
4" x 2"	P6@.125	Ea	4.84	5.54	10.38
4" x 3"	P6@.125	Ea	4.19	5.54	9.73
ABS heavy duty floor drain					
3", less sediment bucket	P6@.280	Ea	31.20	12.40	43.60
4", less sediment bucket	P6@.300	Ea	33.20	13.30	46.50
Hangers and Supports					
Angle bracket hangers, steel, by rod size					
3/8" angle bracket hanger	M5@.070	Ea	1.62	3.06	4.68
1/2" angle bracket hanger	M5@.070	Ea	1.81	3.06	4.87
5/8" angle bracket hanger	M5@.070	Ea	3.99	3.06	7.05
3/4" angle bracket hanger	M5@.070	Ea	6.14	3.06	9.20
7/8" angle bracket hanger	M5@.070	Ea	11.00	3.06	14.06
Angle supports, wall mounted, welded steel					
12" x 18", medium weight	M5@.950	Ea	74.60	41.60	116.20
18" x 24", medium weight	M5@1.10	Ea	93.20	48.10	141.30
24" x 30", medium weight	M5@1.60	Ea	118.00	70.00	188.00
12" x 30", heavy weight	M5@1.10	Ea	106.00	48.10	154.10
18" x 24", heavy weight	M5@1.10	Ea	149.00	48.10	197.10
24" x 30", heavy weight	M5@1.60	Ea	168.00	70.00	238.00

	Craft@Hrs	Unit	Material	Labor	Total
C-clamps with lock nut, steel, by rod size					
3/8" C-clamp	M5@.070	Ea	1.79	3.06	4.85
1/2" C-clamp	M5@.070	Ea	2.17	3.06	5.23
5/8" C-clamp	M5@.070	Ea	4.40	3.06	7.46
3/4" C-clamp	M5@.070	Ea	6.35	3.06	9.41
7/8" C-clamp	M5@.070	Ea	17.10	3.06	20.16
Top beam clamps, steel, by rod size					
3/8" top beam clamp	M5@.070	Ea	2.11	3.06	5.17
1/2" top beam clamp	M5@.070	Ea	3.42	3.06	6.48
5/8" top beam clamp	M5@.070	Ea	4.18	3.06	7.24
3/4" top beam clamp	M5@.070	Ea	6.13	3.06	9.19
7/8" top beam clamp	M5@.070	Ea	8.02	3.06	11.08
Clevis hangers, standard duty, by intended pipe size, no threaded rod or beam clamp included					
1/2" clevis hanger	M5@.156	Ea	1.66	6.83	8.49
3/4" clevis hanger	M5@.156	Ea	1.66	6.83	8.49
1" clevis hanger	M5@.156	Ea	1.75	6.83	8.58
1-1/4" clevis hanger	M5@.173	Ea	2.00	7.57	9.57
1-1/2" clevis hanger	M5@.173	Ea	2.19	7.57	9.76
2" clevis hanger	M5@.173	Ea	2.44	7.57	10.01
2-1/2" clevis hanger	M5@.189	Ea	2.73	8.27	11.00
3" clevis hanger	M5@.189	Ea	3.42	8.27	11.69
4" clevis hanger	M5@.189	Ea	4.22	8.27	12.49
6" clevis hanger	M5@.202	Ea	7.09	8.84	15.93
8" clevis hanger	M5@.202	Ea	10.40	8.84	19.24
Add for epoxy coated	—	%	25.0	—	—
Add for copper plated	—	%	35.0	—	—
Deduct for light duty	—	%	-20.0	—	—
Swivel ring hangers, light duty, by intended pipe size, band hanger, galvanized steel, no hanger rod, beam clamp or insert included					
1/2" swivel ring hanger	M5@.125	Ea	.41	5.47	5.88
3/4" swivel ring hanger	M5@.125	Ea	.41	5.47	5.88
1" swivel ring hanger	M5@.125	Ea	.41	5.47	5.88
1-1/4" swivel ring hanger	M5@.125	Ea	.48	5.47	5.95
1-1/2" swivel ring hanger	M5@.125	Ea	.56	5.47	6.03
2" swivel ring hanger	M5@.125	Ea	.67	5.47	6.14
2-1/2" swivel ring hanger	M5@.135	Ea	.90	5.91	6.81
3" swivel ring hanger	M5@.135	Ea	1.09	5.91	7.00
4" swivel ring hanger	M5@.135	Ea	1.50	5.91	7.41
6" swivel ring hanger	M5@.145	Ea	2.81	6.34	9.15
8" swivel ring hanger	M5@.145	Ea	3.38	6.34	9.72
Add for epoxy coating	—	%	45.0	—	—
Add for copper plated	—	%	55.0	—	—
Riser clamps, black steel, by intended pipe size					
1/2" riser clamp	M5@.100	Ea	1.41	4.38	5.79
3/4" riser clamp	M5@.100	Ea	2.08	4.38	6.46
1" riser clamp	M5@.100	Ea	2.14	4.38	6.52
1-1/4" riser clamp	M5@.115	Ea	2.54	5.03	7.57
1-1/2" riser clamp	M5@.115	Ea	2.70	5.03	7.73
2" riser clamp	M5@.125	Ea	2.81	5.47	8.28
2-1/2" riser clamp	M5@.135	Ea	2.98	5.91	8.89
3" riser clamp	M5@.140	Ea	3.27	6.13	9.40
4" riser clamp	M5@.145	Ea	4.11	6.34	10.45
6" riser clamp	M5@.155	Ea	7.14	6.78	13.92

	Craft@Hrs	Unit	Material	Labor	Total
8" riser clamp	M5@.165	Ea	11.60	7.22	18.82
Add for epoxy coating	—	%	25.0	—	—
Add for copper plated	—	%	35.0	—	—
Pipe clamps, medium duty, by intended pipe size					
3" pipe clamp	M5@.189	Ea	4.28	8.27	12.55
4" pipe clamp	M5@.189	Ea	6.08	8.27	14.35
6" pipe clamp	M5@.202	Ea	14.90	8.84	23.74
Pipe rolls					
3" pipe roll, complete	M5@.403	Ea	13.30	17.60	30.90
4" pipe roll, complete	M5@.518	Ea	13.90	22.70	36.60
6" pipe roll, complete	M5@.518	Ea	13.90	22.70	36.60
8" pipe roll, complete	M5@.578	Ea	24.70	25.30	50.00
3" pipe roll stand, complete	M5@.323	Ea	35.50	14.10	49.60
4" pipe roll stand, complete	M5@.604	Ea	53.20	26.40	79.60
6" pipe roll stand, complete	M5@.775	Ea	53.20	33.90	87.10
8" pipe roll stand, complete	M5@.775	Ea	92.90	33.90	126.80
Pipe straps, galvanized, two hole, lightweight					
2-1/2" pipe strap	M5@.119	Ea	1.70	5.21	6.91
3" pipe strap	M5@.128	Ea	2.19	5.60	7.79
3-1/2" pipe strap	M5@.135	Ea	2.78	5.91	8.69
4" pipe strap	M5@.144	Ea	3.10	6.30	9.40
Plumber's tape, galvanized, 3/4" x 10' roll					
26 gauge	—	LF	1.50	—	1.50
22 gauge	—	LF	2.02	—	2.02
Sliding pipe guides, welded					
3" pipe guide	M5@.354	Ea	15.30	15.50	30.80
4" pipe guide	M5@.354	Ea	20.60	15.50	36.10
6" pipe guide	M5@.354	Ea	29.80	15.50	45.30
8" pipe guide	M5@.471	Ea	44.60	20.60	65.20
Threaded hanger rod, by rod size, in 6' to 10' lengths, per LF					
3/8" rod	—	LF	.37	—	.37
1/2" rod	—	LF	.57	—	.57
5/8" rod	—	LF	.79	—	.79
3/4" rod	—	LF	1.38	—	1.38
U bolts, with hex nuts, light duty					
4" U-bolt	M5@.089	Ea	2.85	3.89	6.74
6" U-bolt	M5@.135	Ea	5.24	5.91	11.15
Wall sleeves, cast iron					
4" sleeve	M5@.241	Ea	6.60	10.50	17.10
6" sleeve	M5@.284	Ea	8.20	12.40	20.60
8" sleeve	M5@.377	Ea	10.80	16.50	27.30

Gate Valves
Bronze trim, non-rising stem, Red & White

	Craft@Hrs	Unit	Material	Labor	Total
2-1/2" 125 lb, cast iron, flanged	MI@.600	Ea	153.00	27.70	180.70
2-1/2" 250 lb, cast iron, flanged	MI@.600	Ea	302.00	27.70	329.70
3" 125 lb, cast iron, flanged	MI@.750	Ea	166.00	34.70	200.70
4" 125 lb, cast iron, flanged	MI@1.35	Ea	244.00	62.40	306.40
4" 250 lb, cast iron, flanged	MI@1.35	Ea	628.00	62.40	690.40

	Craft@Hrs	Unit	Material	Labor	Total
Steel Gate, Globe and Check Valves					
Cast steel Class 150 gate valves, OS&Y, flanged ends					
2-1/2" valve	Ml@.600	Ea	624.00	27.70	651.70
3" valve	Ml@.750	Ea	624.00	34.70	658.70
4" valve	Ml@1.35	Ea	760.00	62.40	822.40
Cast steel Class 300 gate valves, OS&Y, flanged ends					
2-1/2" valve	Ml@.600	Ea	814.00	27.70	841.70
3" valve	Ml@.750	Ea	814.00	34.70	848.70
4" valve	Ml@1.35	Ea	1,130.00	62.40	1,192.40
Outside Stem and Yoke (OS&Y) All Iron Gate Valves					
2-1/2" flanged iron OS&Y gate valves					
125 lb	Ml@.600	Ea	216.00	27.70	243.70
250 lb	Ml@.600	Ea	486.00	27.70	513.70
3" flanged iron OS&Y gate valves					
125 lb	Ml@.750	Ea	229.00	34.70	263.70
250 lb	Ml@.750	Ea	559.00	34.70	593.70
4" flanged iron OS&Y gate valves					
125 lb	Ml@1.35	Ea	320.00	62.40	382.40
250 lb	Ml@1.35	Ea	825.00	62.40	887.40
6" flanged iron OS&Y gate valves					
125 lb	Ml@2.50	Ea	542.00	116.00	658.00
250 lb	Ml@2.50	Ea	1,350.00	116.00	1,466.00

Butterfly Valves, Ductile Iron
Wafer body, flanged, with lever lock, 150 lb valves have aluminum bronze disk, stainless steel stem and EPDM seals. 200 lb nickel plated ductile iron disk and Buna-N seals.

	Craft@Hrs	Unit	Material	Labor	Total
2-1/2" butterfly valve					
150 lb	Ml@.300	Ea	76.30	13.90	90.20
200 lb	Ml@.400	Ea	108.00	18.50	126.50
3" butterfly valve					
150 lb	Ml@.400	Ea	83.80	18.50	102.30
200 lb	Ml@.450	Ea	114.00	20.80	134.80
4" butterfly valve					
150 lb	Ml@.500	Ea	105.00	23.10	128.10
200 lb	Ml@.550	Ea	142.00	25.40	167.40
6" butterfly valve					
150 lb	Ml@.750	Ea	163.00	34.70	197.70
200 lb	Ml@.750	Ea	229.00	34.70	263.70
8" butterfly valve, with gear operator					
150 lb	Ml@.950	Ea	230.00	43.90	273.90
200 lb	Ml@.950	Ea	423.00	43.90	466.90
10" butterfly valve, with gear operator					
150 lb	Ml@1.40	Ea	324.00	64.70	388.70
200 lb	Ml@1.40	Ea	540.00	64.70	604.70

Globe Valves Flanged, iron body, OS&Y (outside stem & yoke) 125 psi steam pressure and 200 psi cold water pressure rated.

	Craft@Hrs	Unit	Material	Labor	Total
3" flanged OS&Y, globe valve	Ml@.750	Ea	309.00	34.70	343.70
3" 250 lb	Ml@.750	Ea	597.00	34.70	631.70

	Craft@Hrs	Unit	Material	Labor	Total
4" flanged OS&Y, globe valve	Ml@1.35	Ea	438.00	62.40	500.40
4" 250 lb	Ml@1.35	Ea	566.00	62.40	628.40
6" flanged OS&Y, globe valve	Ml@2.50	Ea	814.00	116.00	930.00
6" 250 lb	Ml@2.50	Ea	1,490.00	116.00	1,606.00
8" flanged OS&Y, globe valve	Ml@3.00	Ea	1,430.00	139.00	1,569.00
8" 250 lb	Ml@3.00	Ea	2,530.00	139.00	2,669.00

Swing Check Valves

Flanged, iron body, bolted cap, 125 psi steam/200 psi cold water pressure rated or 250 psi steam/500 psi cold water pressure rated.

	Craft@Hrs	Unit	Material	Labor	Total
2-1/2" flanged iron body check valve					
Swing check, bronze trim, 125 psi steam	Ml@.600	Ea	129.00	27.70	156.70
Swing check, iron trim, 125 psi steam	Ml@.600	Ea	151.00	27.70	178.70
Swing check, bronze trim, 250 psi steam	Ml@.750	Ea	394.00	34.70	428.70
3" flanged iron body check valve 3" flanged iron body check valve					
Swing check, bronze trim, 125 psi steam	Ml@.750	Ea	156.00	34.70	190.70
Swing check, iron trim, 125 psi steam	Ml@.750	Ea	208.00	34.70	242.70
Swing check, bronze trim, 250 psi steam	Ml@1.25	Ea	489.00	57.80	546.80
4" flanged iron body check valve					
Swing check, bronze trim, 125 psi steam	Ml@1.35	Ea	253.00	62.40	315.40
Swing check, iron trim, 125 psi steam	Ml@1.35	Ea	342.00	62.40	404.40
Swing check, bronze trim, 250 psi steam	Ml@1.85	Ea	623.00	85.50	708.50
6" flanged iron body check valve					
Swing check, bronze trim, 125 psi steam	Ml@2.50	Ea	410.00	116.00	526.00
Swing check, iron trim, 125 psi steam	Ml@2.50	Ea	545.00	116.00	661.00
Swing check, bronze trim, 250 psi steam	Ml@3.00	Ea	1,160.00	139.00	1,299.00
8" flanged iron body check valve 8" flanged iron body check valve					
Swing check, bronze trim, 125 psi steam	Ml@3.00	Ea	758.00	139.00	897.00
Swing check, iron trim, 125 psi steam	Ml@3.00	Ea	1,010.00	139.00	1,149.00
10" flanged iron body check valve 10" flanged iron body check valve					
Swing check, bronze trim, 125 psi steam	Ml@4.00	Ea	1,540.00	185.00	1,725.00
Swing check, iron trim, 125 psi steam	Ml@4.00	Ea	2,060.00	185.00	2,245.00
12" flanged iron body check valve 12" flanged iron body check valve					
Swing check, bronze trim, 125 psi steam	Ml@4.50	Ea	2,030.00	208.00	2,238.00
Swing check, iron trim, 125 psi steam	Ml@4.50	Ea	2,680.00	208.00	2,888.00

Miscellaneous Valves and Regulators

	Craft@Hrs	Unit	Material	Labor	Total
Backflow preventers reduced pressure					
2-1/2", with 2 gate valves, flanged, iron body	Ml@2.00	Ea	1,420.00	92.40	1,512.40
3", with 2 gate valves, flanged, iron body	Ml@3.00	Ea	1,800.00	139.00	1,939.00
4", with 2 gate valves, flanged, iron body	Ml@4.50	Ea	2,130.00	208.00	2,338.00
6", without gate valves, flanged, iron body	Ml@8.00	Ea	3,520.00	370.00	3,890.00
6", with 2 gate valves, flanged, iron body	Ml@9.00	Ea	4,060.00	416.00	4,476.00
8", without gate valves, flanged, iron body	Ml@11.5	Ea	6,150.00	531.00	6,681.00
8", with 2 gate valves, flanged, iron body	Ml@12.5	Ea	8,860.00	578.00	9,438.00
Ball valves, bronze, 150 lb steam working pressure, threaded, Teflon seat					
2-1/2" valve	Ml@.750	Ea	150.00	34.70	184.70
3" valve	Ml@.950	Ea	173.00	43.90	216.90
Hose gate valves, with cap, 200 lb					
2-1/2", non-rising stem	P6@1.00	Ea	65.40	44.30	109.70

	Craft@Hrs	Unit	Material	Labor	Total
Pressure regulator valves, bronze, threaded, Class 300 PSI					
2-1/2" valve	P6@.750	Ea	784.00	33.20	817.20
3" valve	P6@.950	Ea	911.00	42.10	953.10
Water control valves, threaded, 3-way with actuator					
2-1/2", CV 54	MI@.750	Ea	1,350.00	34.70	1,384.70
3", CV 80	MI@.950	Ea	1,640.00	43.90	1,683.90
4", CV 157	MI@1.30	Ea	3,120.00	60.10	3,180.10

Piping Specialties

	Craft@Hrs	Unit	Material	Labor	Total
Stainless steel bellows type expansion joints, 150 lb. Flanged units include bolt and gasket sets					
3" pipe, 4" travel, welded	MI@1.33	Ea	836.00	61.40	897.40
3" pipe, 6" travel, welded	MI@1.33	Ea	913.00	61.40	974.40
4" pipe, 3" travel, welded	MI@1.78	Ea	873.00	82.20	955.20
4" pipe, 6" travel, welded	MI@1.78	Ea	1,090.00	82.20	1,172.20
5" pipe, 4" travel, flanged	MI@2.20	Ea	2,310.00	102.00	2,412.00
5" pipe, 7" travel, flanged	MI@2.20	Ea	2,390.00	102.00	2,492.00
6" pipe, 4" travel, flanged	MI@2.67	Ea	2,460.00	123.00	2,583.00
8" pipe, 8" travel, flanged	MI@3.20	Ea	3,650.00	148.00	3,798.00
Ball joints, welded ends, 150 lb.					
3" ball joint	MI@2.40	Ea	370.00	111.00	481.00
4" ball joint	MI@3.35	Ea	543.00	155.00	698.00
5" ball joint	MI@4.09	Ea	920.00	189.00	1,109.00
6" ball joint	MI@4.24	Ea	985.00	196.00	1,181.00
8" ball joint	MI@4.79	Ea	1,730.00	221.00	1,951.00
10" ball joint	MI@5.81	Ea	2,120.00	268.00	2,388.00
14" ball joint	MI@7.74	Ea	2,970.00	358.00	3,328.00

In-Line Circulating Pumps Flanged, for hot or cold water circulation. By flange size and pump horsepower rating. See flange costs below. No electrical work included.

	Craft@Hrs	Unit	Material	Labor	Total
Horizontal drive, iron body					
3/4" to 1-1/2", 1/12 HP	MI@1.25	Ea	168.00	57.80	225.80
2", 1/4 HP	MI@1.40	Ea	291.00	64.70	355.70
2-1/2" or 3", 1/4 HP	MI@1.40	Ea	417.00	64.70	481.70
3/4" to 1-1/2", 1/6 HP	MI@1.25	Ea	247.00	57.80	304.80
3/4" to 1-1/2", 1/4 HP	MI@1.40	Ea	330.00	64.70	394.70
3", 1/3 HP	MI@1.50	Ea	564.00	69.30	633.30
3", 1/2 HP	MI@1.70	Ea	593.00	78.50	671.50
3", 3/4 HP	MI@1.80	Ea	683.00	83.20	766.20
Horizontal drive, all bronze					
3/4" to 1-1/2", 1/12 HP	MI@1.25	Ea	256.00	57.80	313.80
3/4" to 1-1/2", 1/6 HP	MI@1.25	Ea	384.00	57.80	441.80
3/4" to 2", 1/4 HP	MI@1.40	Ea	456.00	64.70	520.70
2-1/2" or 3", 1/4 HP	MI@1.50	Ea	755.00	69.30	824.30
3", 1/3 HP	MI@1.50	Ea	1,000.00	69.30	1,069.30
3", 1/2 HP	MI@1.70	Ea	1,030.00	78.50	1,108.50
3", 3/4 HP	MI@1.80	Ea	1,070.00	83.20	1,153.20
Vertical drive, iron body					
3/4" to 1/2", 1/6 HP	MI@1.25	Ea	339.00	57.80	396.80
2", 1/4 HP	MI@1.40	Ea	402.00	64.70	466.70
2-1/2", 1/4 HP	MI@1.40	Ea	475.00	64.70	539.70
3", 1/4 HP	MI@1.40	Ea	475.00	64.70	539.70

Pump connectors, vibration isolators In line, stainless steel braided connections, threaded

1/2" x 9"	M5@.210	Ea	5.12	9.19	14.31
1/2" x 12"	M5@.300	Ea	5.44	13.10	18.54
1/2" x 16"	M5@.400	Ea	5.88	17.50	23.38
1/2" x 20"	M5@.450	Ea	6.01	19.70	25.71
1/2" x 30"	MI@.500	Ea	7.25	23.10	30.35
1/2" x 36"	MI@.750	Ea	8.91	34.70	43.61

Flanges for circulating pumps Per set of two, including bolts and gaskets

3/4" to 1-1/2", iron	M5@.290	Ea	7.47	12.70	20.17
2", iron	M5@.290	Ea	10.60	12.70	23.30
2-1/2" to 3", iron	M5@.420	Ea	33.80	18.40	52.20
3/4" to 1", bronze	M5@.290	Ea	16.60	12.70	29.30
2", bronze	M5@.290	Ea	66.60	12.70	79.30
2-1/2" to 3", iron	M5@.420	Ea	34.50	18.40	52.90
3/4" copper sweat flanges	M5@.290	Ea	10.40	12.70	23.10
1" copper sweat flanges	M5@.290	Ea	15.40	12.70	28.10
1-1/2" copper sweat flanges	M5@.290	Ea	17.70	12.70	30.40

Base Mounted Centrifugal Pumps Single stage cast iron pumps with flanges but no electrical connection.

50 GPM at 100' head	MI@3.15	Ea	2,140.00	146.00	2,286.00
50 GPM at 200' head	MI@4.15	Ea	2,430.00	192.00	2,622.00
50 GPM at 300' head	MI@5.83	Ea	2,720.00	269.00	2,989.00
200 GPM at 100' head	MI@6.70	Ea	2,760.00	310.00	3,070.00
200 GPM at 200' head	MI@7.15	Ea	2,840.00	330.00	3,170.00
200 GPM at 300' head	MI@10.2	Ea	3,080.00	471.00	3,551.00
240 GPM, 7-1/2 HP	MI@6.20	Ea	1,800.00	286.00	2,086.00
300 GPM, 5 HP	MI@3.91	Ea	2,050.00	181.00	2,231.00
340 GPM, 7-1/2 HP	MI@6.72	Ea	2,350.00	310.00	2,660.00
370 GPM, 5 HP	MI@4.02	Ea	2,100.00	186.00	2,286.00
375 GPM, 7-1/2 HP	MI@7.14	Ea	2,340.00	330.00	2,670.00
465 GPM, 10 HP	MI@7.63	Ea	2,440.00	353.00	2,793.00

Vertical Turbine Pumps 3,550 RPM These costs do not include electrical work.

Cast iron single stage vertical turbine pumps

50 GPM at 50' head	MI@9.9	Ea	3,730.00	457.00	4,187.00
50 GPM at 100' head	MI@9.9	Ea	4,310.00	457.00	4,767.00
100 GPM at 300' head	MI@9.9	Ea	5,550.00	457.00	6,007.00
200 GPM at 50' head	MI@9.9	Ea	4,670.00	457.00	5,127.00

Cast iron multi-stage vertical turbine pumps

50 GPM at 100' head	D9@15.6	Ea	4,240.00	667.00	4,907.00
50 GPM at 200' head	D9@15.6	Ea	4,880.00	667.00	5,547.00
50 GPM at 300' head	D9@15.6	Ea	5,540.00	667.00	6,207.00
100 GPM at 100' head	D9@15.6	Ea	4,580.00	667.00	5,247.00
100 GPM at 200' head	D9@15.6	Ea	5,130.00	667.00	5,797.00
100 GPM at 300' head	D9@15.6	Ea	5,920.00	667.00	6,587.00
200 GPM at 100' head	D9@15.6	Ea	5,000.00	667.00	5,667.00
200 GPM at 200' head	D9@20.5	Ea	5,920.00	876.00	6,796.00
200 GPM at 300' head	D9@20.5	Ea	6,310.00	876.00	7,186.00

Bronze single stage vertical turbine pumps

50 GPM at 50' head	D9@15.6	Ea	3,710.00	667.00	4,377.00
100 GPM at 50' head	D9@15.6	Ea	3,980.00	667.00	4,647.00
200 GPM at 50' head	D9@15.6	Ea	4,310.00	667.00	4,977.00

	Craft@Hrs	Unit	Material	Labor	Total
Bronze multi-stage vertical turbine pumps					
50 GPM at 50' head	D9@15.6	Ea	4,240.00	667.00	4,907.00
50 GPM at 100' head	D9@15.6	Ea	4,580.00	667.00	5,247.00
50 GPM at 150' head	D9@15.6	Ea	4,680.00	667.00	5,347.00
100 GPM at 50' head	D9@15.6	Ea	4,240.00	667.00	4,907.00
100 GPM at 100' head	D9@15.6	Ea	4,580.00	667.00	5,247.00
100 GPM at 150' head	D9@15.6	Ea	4,800.00	667.00	5,467.00
200 GPM at 50' head	D9@15.6	Ea	4,580.00	667.00	5,247.00
200 GPM at 100' head	D9@15.6	Ea	5,000.00	667.00	5,667.00
200 GPM at 150' head	D9@15.6	Ea	5,500.00	667.00	6,167.00

Sump Pumps Bronze 1,750 RPM sump pumps. These costs do not include electrical work.

	Craft@Hrs	Unit	Material	Labor	Total
25 GPM at 25' head	MI@17.1	Ea	2,990.00	790.00	3,780.00
25 GPM at 50' head	MI@17.1	Ea	3,120.00	790.00	3,910.00
25 GPM at 100' head	MI@17.1	Ea	3,280.00	790.00	4,070.00
25 GPM at 150' head	MI@17.1	Ea	4,350.00	790.00	5,140.00
50 GPM at 25' head	MI@20.1	Ea	2,890.00	929.00	3,819.00
50 GPM at 50' head	MI@20.1	Ea	3,020.00	929.00	3,949.00
50 GPM at 100' head	MI@20.1	Ea	3,110.00	929.00	4,039.00
50 GPM at 150' head	MI@20.1	Ea	4,460.00	929.00	5,389.00
100 GPM at 25' head	MI@22.8	Ea	3,070.00	1,050.00	4,120.00
100 GPM at 50' head	MI@22.8	Ea	3,030.00	1,050.00	4,080.00
100 GPM at 100' head	MI@22.8	Ea	3,900.00	1,050.00	4,950.00
100 GPM at 150' head	MI@22.8	Ea	4,440.00	1,050.00	5,490.00

Miscellaneous Pumps These costs do not include electrical work.

	Craft@Hrs	Unit	Material	Labor	Total
Salt water pump, 135 GPM at 300' head	D9@19.9	Ea	12,300.00	851.00	13,151.00
Sewage pumps, vertical mount, centrifugal					
1/2 HP, 25 GPM at 25' head	P6@8.00	Ea	953.00	354.00	1,307.00
1-1/2 HP, 100 GPM at 70' head	P6@14.6	Ea	4,160.00	647.00	4,807.00
10 HP, 300 GPM at 70' head	P6@18.3	Ea	6,800.00	811.00	7,611.00
Medical vacuum pump, duplex, tank mount with 30 gallon tank,					
1/2 HP, 20" vacuum	D4@20.6	Ea	7,140.00	848.00	7,988.00
Oil pump, base mounted, 1-1/2 GPM at 15' head,					
1/3 HP, 120 volt	D4@1.50	Ea	207.00	61.70	268.70

Closed Cell Elastomeric Pipe Insulation Semi split, thickness as shown, by nominal pipe diameter. R factor equals 3.58 at 220 degrees F. Make additional allowances for scaffolding if required. Also see fittings, flanges and valves at the end of this section. For more detailed coverage of pipe insulation, see National Plumbing & HVAC Estimator, http://www.CraftsmanSiteLicense.Com

	Craft@Hrs	Unit	Material	Labor	Total
2-1/2", 3/8" thick	P6@.056	LF	1.33	2.48	3.81
2-1/2", 1/2" thick	P6@.056	LF	1.78	2.48	4.26
2-1/2", 3/4" thick	P6@.056	LF	2.31	2.48	4.79
3", 3/8" thick	P6@.062	LF	1.73	2.75	4.48
3", 1/2" thick	P6@.062	LF	2.02	2.75	4.77
3", 3/4" thick	P6@.062	LF	2.61	2.75	5.36
4", 3/8" thick	P6@.068	LF	1.78	3.01	4.79
4", 1/2" thick	P6@.068	LF	2.37	3.01	5.38
4", 3/4" thick	P6@.068	LF	3.07	3.01	6.08

Additional cost for insulating pipe fittings and flanges

For each fitting or flange, add the cost of insulating 3 LF of pipe of the same size.

Additional cost for insulating valves

Valve body only. Add the cost of insulating 5 LF of pipe of the same size.

Body and bonnet or yoke valves. Add the cost of insulating 10 LF of pipe of the same size.

	Craft@Hrs	Unit	Material	Labor	Total
Meters, Gauges and Indicators					
Water meters					
Disc type, 1-1/2"	P6@1.00	Ea	155.00	44.30	199.30
Turbine type, 1-1/2"	P6@1.00	Ea	234.00	44.30	278.30
Displacement type, AWWA C7000, 1"	P6@.941	Ea	239.00	41.70	280.70
Displacement type, AWWA C7000, 1-1/2"	P6@1.23	Ea	824.00	54.50	878.50
Displacement type, AWWA C7000, 2"	P6@1.57	Ea	1,240.00	69.50	1,309.50
Curb stop and waste valve, 1"	P6@1.00	Ea	86.80	44.30	131.10
Thermoflow indicators, soldered					
1-1/4" indicator	M5@2.01	Ea	339.00	87.90	426.90
2" indicator	M5@2.01	Ea	385.00	87.90	472.90
2-1/2" indicator	MI@1.91	Ea	565.00	88.20	653.20
Cast steel steam meters, in-line, flanged, 300 PSI					
1" pipe, threaded	M5@1.20	Ea	3,230.00	52.50	3,282.50
2" pipe	MI@1.14	Ea	3,660.00	52.70	3,712.70
3" pipe	MI@2.31	Ea	4,080.00	107.00	4,187.00
4" pipe	MI@2.71	Ea	4,520.00	125.00	4,645.00
Cast steel by-pass steam meters, 2", by line pipe size					
6" line	MI@16.7	Ea	7,840.00	772.00	8,612.00
8" line	MI@17.7	Ea	8,060.00	818.00	8,878.00
10" line	MI@18.5	Ea	8,390.00	855.00	9,245.00
12" line	MI@19.7	Ea	8,710.00	910.00	9,620.00
14" line	MI@21.4	Ea	9,220.00	989.00	10,209.00
16" line	MI@22.2	Ea	9,500.00	1,030.00	10,530.00
Add for steam meter pressure-compensated counter	—	Ea	1,330.00	—	1,330.00
Add for contactor used with dial counter	—	Ea	619.00	—	619.00
Add for wall or panel-mounted remote totalizer with contactor	—	Ea	1,020.00	—	1,020.00
Add for direct reading pressure gauge	M5@.250	Ea	76.20	10.90	87.10
Add for direct reading thermometer, stainless steel case, with trim, 2% accuracy					
3" dial	M5@.250	Ea	57.60	10.90	68.50
Plumbing Specialties					
Access doors, primed steel, cam-lock. Non-fire rated					
8" x 8"	SM@.400	Ea	33.10	20.20	53.30
12" x 12"	SM@.500	Ea	38.10	25.30	63.40
16" x 16"	SM@.600	Ea	46.90	30.30	77.20
18" x 18"	SM@.800	Ea	53.90	40.40	94.30
24" x 24"	SM@1.25	Ea	79.70	63.20	142.90
36" x 36"	SM@1.60	Ea	147.00	80.80	227.80
Add for Allen key lock	—	Ea	4.73	—	4.73
Add for cylinder lock	—	Ea	11.70	—	11.70
Add for stainless steel	—	%	300.0	—	—
Access doors – 2 hour fire rating, painted steel, with screwdriver catch					
8" x 8"	SM@.400	Ea	79.00	20.20	99.20
12" x 12"	SM@.500	Ea	84.30	25.30	109.60
18" x 18"	SM@.800	Ea	116.00	40.40	156.40
24" x 24"	SM@1.25	Ea	135.00	63.20	198.20
36" x 36"	SM@1.60	Ea	271.00	80.80	351.80

	Craft@Hrs	Unit	Material	Labor	Total

Cleanouts For drainage piping. A cleanout may be required: at the end of each waste line, or where a main building drain exits the building, or at regular 100' intervals for straight runs longer than 100', or at the base of soil or waste stacks, or where a 90 degree change in direction of a main building drain takes place. Consult your plumbing code for the regulations that apply to your area.

Finished grade end cleanouts (Malcolm) with Neo-Loc & Dura-Coated cast iron body

	Craft@Hrs	Unit	Material	Labor	Total
2" pipe, 4-1/8" round top	P6@.500	Ea	68.60	22.10	90.70
3" pipe, 5-1/8" round top	P6@.500	Ea	91.40	22.10	113.50
4" pipe, 6-1/8" round top	P6@.500	Ea	91.40	22.10	113.50

Cast iron in-line cleanouts (Barrett). Make additional allowances for access doors

2" pipe	P6@.350	Ea	25.50	15.50	41.00
4" pipe	P6@.400	Ea	47.60	17.70	65.30
6" pipe	P6@.400	Ea	169.00	17.70	186.70

Cleanout access cover accessories

Round polished bronze cover	—	Ea	15.20	—	15.20
Square chrome plated cover	—	Ea	139.00	—	139.00
Square bronze hinged cover box	—	Ea	99.20	—	99.20

Floor drains, body and grate only, add rough-in

3" x 2" plastic grate and body	P6@.25	Ea	12.60	11.10	23.70
5" x 3" plastic grate and body	P6@.25	Ea	17.60	11.10	28.70
3" x 2" cast iron body, nickel-bronze grate	P6@.50	Ea	32.20	22.10	54.30
5" x 3" cast iron body, nickel-bronze grate	P6@.50	Ea	37.00	22.10	59.10
6" x 4" cast iron body, nickel-bronze grate	P6@.50	Ea	43.10	22.10	65.20
Add for a sediment bucket	—	Ea	7.55	—	7.55
Add for plastic backwater valve	—	Ea	26.40	—	26.40
Add for cast iron backwater valve	—	Ea	221.00	—	221.00
Add for trap seal primer fitting	—	Ea	5.88	—	5.88

Drain, general use, medium duty, PVC plastic, 3" or 4" pipe

3" x 4", stainless steel strainer	P6@1.16	Ea	8.16	51.40	59.56
3" x 4", brass strainer	P6@1.32	Ea	20.20	58.50	78.70
4" with 5" brass strainer & spigot outlet	P6@1.55	Ea	37.50	68.60	106.10

Roof drains, cast iron mushroom strainer, bottom outlet

2" to 4" drain	P6@1.75	Ea	151.00	77.50	228.50
5" or 6" drain	P6@1.98	Ea	214.00	87.70	301.70
8" drain	P6@2.36	Ea	278.00	105.00	383.00

Roof (jack) flashing, round

4" jack	SM@.311	Ea	6.92	15.70	22.62
6" jack	SM@.416	Ea	7.52	21.00	28.52
8" jack	SM@.516	Ea	10.20	26.10	36.30

Plumbing Equipment Valves, supports, vents, gas or electric hookup and related equipment are not included except as noted.

Garbage disposers, commercial, stainless steel, with water flow interlock switch. Insinkerator

1/2 HP, 1-1/2" drain connection	PL@1.47	Ea	71.40	76.70	148.10
3/4 HP, 1-1/2" drain connection	PL@2.09	Ea	177.00	109.00	286.00
1 HP, 1-1/2" drain connection	PL@3.92	Ea	209.00	205.00	414.00
1-1/2 HP, 2" drain connection	PL@5.24	Ea	314.00	273.00	587.00

Water heaters, commercial, gas fired 3 year warranty, Glasslined. Set in place only. Make additional allowances for pipe, recirculating pump, gas and flue connections

Standard efficiency

50 gallons, 98 MBtu, 95 GPH recovery	P6@2.00	Ea	508.00	88.60	596.60
60 gallons, 55 MBtu, 50 GPH recovery	P6@2.00	Ea	907.00	88.60	995.60
67 gallons, 114 MBtu, 108 GPH recovery	P6@2.00	Ea	1,070.00	88.60	1,158.60

	Craft@Hrs	Unit	Material	Labor	Total
76 gallons, 180 MBtu, 175 GPH recovery	P6@2.25	Ea	1,450.00	99.70	1,549.70
82 gallons, 156 MBtu, 151 GPH recovery	P6@2.25	Ea	1,380.00	99.70	1,479.70
91 gallons, 300 MBtu, 291 GPH recovery	P6@2.50	Ea	1,710.00	111.00	1,821.00
Energy Miser. Meets ASHRAE 90.1b-1992 standards					
50 gallons, 98 MBtu, 95 GPH recovery	P6@2.00	Ea	1,320.00	88.60	1,408.60
65 gallons, 360 MBtu, 350 GPH recovery	P6@2.00	Ea	1,930.00	88.60	2,018.60
67 gallons, 114 MBtu, 108 GPH recovery	P6@2.00	Ea	1,350.00	88.60	1,438.60
72 gallons, 250 MBtu, 242 GPH recovery	P6@2.25	Ea	1,800.00	99.70	1,899.70
76 gallons, 180 MBtu, 175 GPH recovery	P6@2.25	Ea	1,670.00	99.70	1,769.70
82 gallons, 156 MBtu, 151 GPH recovery	P6@2.25	Ea	1,640.00	99.70	1,739.70
85 gallons, 400 MBtu, 388 GPH recovery	P6@2.25	Ea	2,020.00	99.70	2,119.70
91 gallons, 200 MBtu, 194 GPH recovery	P6@2.50	Ea	1,800.00	111.00	1,911.00
91 gallons, 300 MBtu, 291 GPH recovery	P6@2.50	Ea	1,830.00	111.00	1,941.00
Power vent kit	P6@2.00	Ea	426.00	88.60	514.60
Equal flow manifold kit, duplex	—	Ea	293.00	—	293.00
Equal flow manifold kit, triplex	—	Ea	584.00	—	584.00
Add for propane fired units, any of above	—	Ea	75.10	—	75.10

Water heaters, commercial, electric 208/240 volts, with surface mounted thermostat, Rheem Glasslined Energy Miser. Set in place only. Make additional allowances for pipe, recirculating pump, and electrical connections.

	Craft@Hrs	Unit	Material	Labor	Total
50 gallons, 9 kW	P6@1.75	Ea	1,170.00	77.50	1,247.50
50 gallons, 27 kW	P6@1.75	Ea	1,450.00	77.50	1,527.50
85 gallons, 18 kW	P6@2.00	Ea	1,390.00	88.60	1,478.60
120 gallons, 45 kW	P6@3.00	Ea	2,100.00	133.00	2,233.00

Instantaneous water heaters, electric Maximum operating pressure of 150 PSI. Includes plumbing hookup only. Add for electrical work. 1/2" compression fittings used except where noted. http://www.eemaxinc.com
120/208/240 volt single point for point of use handwashing. 33-degree F. temperature rise at 1/2 GPM. Includes 3/8" compression fittings

	Craft@Hrs	Unit	Material	Labor	Total
2.4/3.0 kW unit	P6@1.00	Ea	307.00	44.30	351.30

208/240 volt flow controlled for kitchen, bar sink and dual lavatory sinks. 38-degree F temperature rise at 1.0 GPM

8.3 kW unit	P6@1.00	Ea	213.00	44.30	257.30

240/277 volt thermostatic for applications where accurate temperature control is required. May be used in combination with other water heating units. 41-degree F temperature rise at 1.0 GPM

6.0 kW unit	P6@1.00	Ea	273.00	44.30	317.30

240 volt twin heating module for residential showers, mop sinks, industrial processes, campgrounds and resort cabins. May be used in combination with other water heating units. 51-degree F temperature rise at 2.0 GPM

15.0 kW unit	P6@1.00	Ea	396.00	44.30	440.30

277 volt high capacity instantaneous water heater. May be used in combination with other water heating units. 35-degree F temperature rise at 3.5 GPM

18.0 kW unit	P6@1.00	Ea	882.00	44.30	926.30

240 volt high purity for microchip manufacturing, pharmaceutical production, ultrasonic cleaning and batch chemical mixing. Maintains water purity up to 18 MEG OHM quality. 38-degree F temperature rise at 1.0 GPM

5.5 kW unit	P6@1.00	Ea	710.00	44.30	754.30

Water heater stands Installed at same time as water heater. Raises flame elements 18" off floor. Secures to wall with seismic clips.

	Craft@Hrs	Unit	Material	Labor	Total
80 gallon capacity	—	Ea	37.00	—	37.00
100 gallon capacity	—	Ea	42.00	—	42.00
Fully enclosed, 40 gallon capacity	—	Ea	31.20	—	31.20

	Craft@Hrs	Unit	Material	Labor	Total

Water heater safety pans

1" side outlet drain, 16 gauge steel, installed at same time as water heater

	Craft@Hrs	Unit	Material	Labor	Total
14" diameter	—	Ea	13.70	—	13.70
16" diameter	—	Ea	15.30	—	15.30
19" diameter	—	Ea	18.70	—	18.70
22" diameter	—	Ea	20.90	—	20.90
24" diameter	—	Ea	23.10	—	23.10
26-1/2" diameter	—	Ea	25.50	—	25.50

1" bottom outlet drain, 16 gauge steel, installed at same time as water heater

	Craft@Hrs	Unit	Material	Labor	Total
16" diameter	—	Ea	13.30	—	13.30
19" diameter	—	Ea	17.30	—	17.30
22" diameter	—	Ea	20.30	—	20.30
24" diameter	—	Ea	22.60	—	22.60
26-1/2" diameter	—	Ea	24.90	—	24.90

Water heater earthquake restraints State approved, installed at the same time as water heater

	Craft@Hrs	Unit	Material	Labor	Total
Up to 52 gal heater, double strap	—	Ea	18.40	—	18.40
Up to 100 gal heater, double strap	—	Ea	21.60	—	21.60
Free standing restraint system – 75 gal capacity	—	Ea	152.00	—	152.00
Free standing restraint system – 100 gal capacity	—	Ea	164.00	—	164.00
Wall mount platform with restraints – 75 gal capacity	—	Ea	145.00	—	145.00
Wall mount platform with restraints – 100 gal capacity	—	Ea	161.00	—	161.00

Water heater connection assembly Includes typical pipe, fittings, valves, gages, and hangers

	Craft@Hrs	Unit	Material	Labor	Total
Copper pipe connection, 3/4" supply	P6@2.25	Ea	226.00	99.70	325.70
Copper pipe connection, 1" supply	P6@2.75	Ea	299.00	122.00	421.00
Copper pipe connection, 1-1/4" supply	P6@3.50	Ea	367.00	155.00	522.00
Copper pipe connection, 1-1/2" supply	P6@3.75	Ea	379.00	166.00	545.00
Copper pipe connection, 2" supply	P6@4.50	Ea	407.00	199.00	606.00
Copper pipe connection, 2-1/2" supply	P6@5.75	Ea	848.00	255.00	1,103.00
Copper pipe connection, 3" supply	P6@6.50	Ea	1,300.00	288.00	1,588.00
Black steel gas pipe connection, 3/4"	P6@1.40	Ea	108.00	62.00	170.00
Black steel gas pipe connection, 1"	P6@1.50	Ea	141.00	66.40	207.40
Black steel gas pipe connection, 1-1/2"	P6@2.00	Ea	208.00	88.60	296.60
Double wall vent pipe connection, 4"	P6@1.40	Ea	128.00	62.00	190.00

Grease interceptors

	Craft@Hrs	Unit	Material	Labor	Total
2" cast iron, 4 GPM, 8 lb capacity	P6@4.00	Ea	391.00	177.00	568.00
2" cast iron, 10 GPM, 20 lb capacity	P6@5.00	Ea	635.00	221.00	856.00
2" steel, 15 GPM, 30 lb capacity	P6@7.00	Ea	942.00	310.00	1,252.00
3" steel, 20 GPM, 40 lb capacity	P6@8.00	Ea	1,160.00	354.00	1,514.00

Hair interceptors 1-1/2" cast iron

	Craft@Hrs	Unit	Material	Labor	Total
1-1/2" cast iron	P6@.650	Ea	143.00	28.80	171.80
1-1/2" nickel bronze	P6@.750	Ea	201.00	33.20	234.20

Sewage ejector packaged systems with vertical pumps, float switch controls, poly tank, cover and fittings, no electrical work or pipe connections included

Simplex sewage ejector systems (single pump)

	Craft@Hrs	Unit	Material	Labor	Total
1/2 HP, 2" line, 20" x 30" tank	P6@4.75	Ea	426.00	210.00	636.00
3/4 HP, 2" line, 24" x 36" tank	P6@4.75	Ea	1,070.00	210.00	1,280.00
1 HP, 3" line, 24" x 36" tank	P6@6.00	Ea	1,250.00	266.00	1,516.00

	Craft@Hrs	Unit	Material	Labor	Total
Duplex sewage ejector systems (two pumps), alternator not included					
1/2 HP, 2" line, 30" x 36" tank	P6@6.50	Ea	1,650.00	288.00	1,938.00
3/4 HP, 2" line, 30" x 36" tank	P6@6.50	Ea	2,140.00	288.00	2,428.00
1 HP, 3" line, 30" x 36" tank	P6@8.00	Ea	2,410.00	354.00	2,764.00
Duplex pump controls (alternator)	—	Ea	902.00	—	902.00
Sump pumps With float switch, no electrical work included					
Submersible type					
1/4 HP, 1-1/2" outlet, ABS plastic	P6@2.00	Ea	103.00	88.60	191.60
1/3 HP, 1-1/4" outlet, cast iron	P6@2.00	Ea	150.00	88.60	238.60
1/3 HP, 1-1/4" outlet, ABS plastic	P6@2.00	Ea	123.00	88.60	211.60
Upright type					
1/3 HP, 1-1/4" outlet, cast iron	P6@2.00	Ea	119.00	88.60	207.60
1/3 HP, 1-1/2" outlet, ABS plastic	P6@2.00	Ea	88.20	88.60	176.80
Tanks See also ASME tanks					
Hot water storage tanks Insulated, floor mounted, includes pipe connection					
80 gallons	P6@4.00	Ea	509.00	177.00	686.00
120 gallons	P6@4.50	Ea	781.00	199.00	980.00
200 gallons	P6@5.25	Ea	2,100.00	233.00	2,333.00
980 gallons	P6@20.0	Ea	7,210.00	886.00	8,096.00
1,230 gallons	P6@22.0	Ea	8,580.00	974.00	9,554.00
1,615 gallons	P6@26.0	Ea	11,900.00	1,150.00	13,050.00
Septic tanks Includes typical excavation and piping					
500 gallons, polyethylene	U2@15.2	Ea	512.00	633.00	1,145.00
1,000 gallons, fiberglass	U2@16.1	Ea	729.00	671.00	1,400.00
2,000 gallons, concrete	U2@17.0	Ea	1,100.00	708.00	1,808.00
5,000 gallons, concrete	U2@22.3	Ea	4,100.00	929.00	5,029.00
Water softeners Automatic two tank unit, 3/4" valve					
1.0 CF, 30,000 grain capacity	P6@3.59	Ea	540.00	159.00	699.00
2.0 CF, 60,000 grain capacity	P6@3.59	Ea	879.00	159.00	1,038.00
Water filters In-line, filters for taste, odor, chlorine and sediment to 20 microns, with by-pass assembly. Includes pipe connection					
Housing, by-pass and carbon filter, complete	P6@2.50	Ea	128.00	111.00	239.00
Water filter housing only	P6@1.00	Ea	32.00	44.30	76.30
By-pass assembly only, 3/4" copper	P6@1.25	Ea	69.40	55.40	124.80
Carbon core cartridge, 20 microns	P6@.250	Ea	16.00	11.10	27.10
Ceramic disinfection and carbon cartridge	P6@.250	Ea	42.60	11.10	53.70
Lead, chlorine, taste and odor cartridge	P6@.250	Ea	32.00	11.10	43.10
Water sterilizers Ultraviolet, point of use water. Add the cost of electrical and pipe connections. In gallons per minute (GPM) of rated capacity. 2 GPM and larger units include light-emitting diode (LED)					
1/2 GPM point of use filter, 1/2" connection	P6@2.00	Ea	214.00	88.60	302.60
3/4 GPM point of use filter, 1/2" connection	P6@2.00	Ea	227.00	88.60	315.60
2 GPM, 1/2" connection	P6@2.50	Ea	252.00	111.00	363.00
2 GPM, with alarm, 1/2" connection	P6@2.50	Ea	328.00	111.00	439.00
2 GPM, with alarm and monitor, 1/2" connection	P6@3.00	Ea	486.00	133.00	619.00
5 GPM, 3/4" connection	P6@2.50	Ea	462.00	111.00	573.00

	Craft@Hrs	Unit	Material	Labor	Total
5 GPM, with alarm, 3/4" connection	P6@2.50	Ea	525.00	111.00	636.00
5 GPM, with alarm and monitor, 3/4" connection	P6@3.00	Ea	599.00	133.00	732.00
8 GPM, with alarm, 3/4" connection	P6@2.50	Ea	489.00	111.00	600.00
8 GPM, with alarm and monitor, 3/4" connection	P6@3.00	Ea	651.00	133.00	784.00
12 GPM, with alarm, 3/4" connection	P6@2.50	Ea	620.00	111.00	731.00
12 GPM, with alarm and monitor, 3/4" connection	P6@3.00	Ea	825.00	133.00	958.00

Reverse osmosis water filters Economy system. Add the cost of electrical and pipe connections. In gallons per day of rated capacity

	Craft@Hrs	Unit	Material	Labor	Total
12 gallons per day	P6@3.50	Ea	236.00	155.00	391.00
16 gallons per day	P6@3.50	Ea	252.00	155.00	407.00
30 gallons per day	P6@3.50	Ea	355.00	155.00	510.00
60 gallons per day	P6@3.50	Ea	394.00	155.00	549.00
100 gallons per day	P6@3.50	Ea	599.00	155.00	754.00

Plumbing Assemblies

Rough-in Plumbing Assemblies: Includes rough-ins for commercial industrial and residential buildings. This section shows costs for typical rough-ins of drain, waste and vent (DWV) and water supply piping using either plastic or metal pipe and fittings. Plastic ABS or PVC DWV pipe can be combined with plastic "poly b" supply pipe for a complete residential rough-in assembly. Cast iron or copper drainage DWV pipe can be combined with copper supply pipe for a complete commercial rough-in assembly. (Check local plumbing and building codes to determine acceptable materials for above and below grade installations in your area.) Detailed estimates of the piping requirements for each type plumbing fixture rough-in were made using the Uniform Plumbing Code as a guideline. For scheduling purposes, estimate that a plumber and an apprentice can install rough-in piping for 3 to 4 plumbing fixtures in an 8-hour day. The same plumber and apprentice will install, connect and test 3 to 4 fixtures a day. When roughing-in less than 4 fixtures in a building, increase the rough-in cost 25% for each fixture less than 4. Add the cost of the fixture and final connection assembly (as shown in Fixtures and Final Connection Assemblies) to the rough-in assembly cost.

Rule of thumb: Use plastic drainage and supply pipe in economy residential applications. Use plastic drainage and copper supply in quality residential applications. Use cast iron or copper DWV drainage and copper supply pipe in quality commercial applications.

Tricks of the trade: In some jurisdictions, plastic drainage pipe is allowed below grade for commercial applications only. In residential applications, plastic drainage pipe may be allowed in above and below grade applications. The difference is the liability of the owner to the patrons of a public building vs. the liability of the owner of a private residence in case of fire. ABS pipe is toxic when it burns.

Water closets

Floor mounted, tank type, rough-in includes 7' of 3", 8' of 1-1/2" DWV pipe and 8' of 1/2" poly b supply pipe with typical fittings, hangers, and strapping

	Craft@Hrs	Unit	Material	Labor	Total
Plastic DWV and poly b supply rough-in	P6@1.90	Ea	34.50	84.20	118.70
Plastic DWV and copper supply rough-in	P6@2.00	Ea	37.20	88.60	125.80
Cast iron, copper DWV and copper supply rough-in	P6@2.25	Ea	150.00	99.70	249.70

Floor mounted flush valve type, rough-in includes 7' of 3", 8' of 1-1/2" DWV pipe and 8' of 1" supply pipe with typical fittings, hangers, and strapping

	Craft@Hrs	Unit	Material	Labor	Total
Plastic DWV and PVC schedule 40 supply rough-in	P6@1.90	Ea	34.50	84.20	118.70
Plastic DWV and copper supply rough-in	P6@2.00	Ea	37.20	88.60	125.80
Cast iron, copper DWV and copper supply rough-in	P6@2.25	Ea	150.00	99.70	249.70

Mechanical 15

	Craft@Hrs	Unit	Material	Labor	Total
Wall hung, tank type, rough-in includes 5' of 4", 8' of 1-1/2" DWV pipe and 8' of 1/2" supply pipe with typical fittings, hangers, and strapping					
Plastic DWV and poly b supply rough-in	P6@1.80	Ea	24.40	79.70	104.10
Plastic DWV and copper supply rough-in	P6@1.90	Ea	30.80	84.20	115.00
Cast iron, copper DWV and copper supply rough-in	P6@2.35	Ea	173.00	104.00	277.00
Add for fixture carrier	P6@.750	Ea	153.00	33.20	186.20
Wall hung flush valve type, rough-in includes 5' of 4", 6' of 1-1/2" DWV pipe and 8' of 1" supply pipe with typical fittings, hangers, and strapping					
Plastic DWV and PVC schedule 40 supply rough-in	P6@1.70	Ea	33.50	75.30	108.80
Plastic DWV and copper supply rough-in	P6@1.00	Ea	35.90	44.30	80.20
Cast iron, copper DWV and copper supply rough-in	P6@1.95	Ea	88.30	86.40	174.70
Add for fixture carrier	P6@.750	Ea	153.00	33.20	186.20

Urinals

	Craft@Hrs	Unit	Material	Labor	Total
Wall hung urinal, rough-in includes 15' of DWV pipe and 8' of supply pipe with typical fittings, hangers and strapping					
Plastic DWV and supply rough-in	P6@2.00	Ea	64.20	88.60	152.80
Plastic DWV and copper supply rough-in	P6@2.10	Ea	70.50	93.00	163.50
Copper DWV and copper supply rough-in	P6@2.35	Ea	150.00	104.00	254.00
Add for fixture carrier	P6@.850	Ea	116.00	37.60	153.60
Floor mounted urinal, rough-in includes 15' of DWV pipe and 8' of supply pipe with typical fittings, hangers and strapping					
Plastic DWV and supply rough-in	P6@2.15	Ea	83.30	95.20	178.50
Plastic DWV and copper supply rough-in	P6@2.25	Ea	89.90	99.70	189.60
Copper DWV and copper supply rough-in	P6@2.50	Ea	186.00	111.00	297.00
Lavatory rough-in includes 15' of 1-1/2" DWV pipe and 15' of 1/2" supply pipe with typical fittings, hangers and strapping					
Plastic DWV and poly b supply rough-in	P6@1.70	Ea	25.60	75.30	100.90
Plastic DWV and copper supply rough-in	P6@1.75	Ea	28.20	77.50	105.70
Copper DWV and copper supply rough-in	P6@1.90	Ea	105.00	84.20	189.20
Add for carrier	P6@.750	Ea	116.00	33.20	149.20

Bathtub and shower
Rough-in includes 15' of 1-1/2" DWV pipe and 15' of 1/2" supply pipe with typical fittings, hangers, and strapping

	Craft@Hrs	Unit	Material	Labor	Total
Plastic DWV and poly b supply rough-in	P6@1.95	Ea	25.60	86.40	112.00
Plastic DWV and copper supply rough-in	P6@2.10	Ea	28.20	93.00	121.20
Copper DWV and copper supply rough-in	P6@2.35	Ea	161.00	104.00	265.00

Kitchen sink
Rough-in includes 15' of 1-1/2" DWV pipe and 15' of 1/2" supply pipe with typical, fittings, hangers and strapping

	Craft@Hrs	Unit	Material	Labor	Total
Plastic DWV and poly b supply rough-in	P6@1.70	Ea	28.20	75.30	103.50
Plastic DWV and copper supply rough-in	P6@1.75	Ea	30.80	77.50	108.30
Copper DWV and copper supply rough-in	P6@1.90	Ea	112.00	84.20	196.20

Drinking fountain and water cooler
Rough-in includes 15' of 1-1/2" DWV pipe and 10' of 1/2" supply pipe with typical, fittings, strapping and hangers

	Craft@Hrs	Unit	Material	Labor	Total
Plastic DWV and poly b supply rough-in	P6@1.80	Ea	23.00	79.70	102.70
Plastic DWV and copper supply rough-in	P6@2.00	Ea	25.60	88.60	114.20
Copper DWV and copper supply rough-in	P6@2.20	Ea	83.30	97.40	180.70

Wash fountain
Rough-in includes 20' of 1-1/2" DWV pipe and 20' of 1/2" supply pipe with typical fittings, hangers and strapping

	Craft@Hrs	Unit	Material	Labor	Total
Plastic DWV and supply rough-in	P6@1.90	Ea	43.60	84.20	127.80
Plastic DWV and copper supply rough-in	P6@2.00	Ea	48.80	88.60	137.40
Copper DWV and copper supply rough-in	P6@2.10	Ea	123.00	93.00	216.00

	Craft@Hrs	Unit	Material	Labor	Total

Service sink Rough-in includes 15' of DWV pipe and 15' of supply pipe with typical fittings, hangers, and strapping
Wall hung slop sink

	Craft@Hrs	Unit	Material	Labor	Total
Plastic DWV and supply rough-in	P6@2.25	Ea	53.90	99.70	153.60
Plastic DWV and copper supply rough-in	P6@2.35	Ea	57.60	104.00	161.60
Cast iron DWV and copper supply rough-in	P6@2.45	Ea	109.00	109.00	218.00
Floor mounted mop sink					
Plastic DWV and supply rough-in	P6@2.35	Ea	70.50	104.00	174.50
Plastic DWV and copper supply rough-in	P6@2.45	Ea	77.00	109.00	186.00
Cast iron DWV and copper supply rough-in	P6@2.60	Ea	135.00	115.00	250.00

Emergency drench shower and eye wash station Rough-in includes a valved water supply with 15' of supply pipe, typical fittings and hangers. Add the cost of a floor drain if required

	Craft@Hrs	Unit	Material	Labor	Total
Plastic poly b pipe	P6@1.25	Ea	19.20	55.40	74.60
Copper pipe	P6@1.50	Ea	25.60	66.40	92.00
Add for floor drain	P6@3.25	Ea	122.00	144.00	266.00

Surgeons' scrub sink Rough-in includes 7' of 2" and 8' of 1-1/2" DWV pipe and 15' of 3/4" supply pipe with typical fittings and hangers

	Craft@Hrs	Unit	Material	Labor	Total
Plastic DWV and supply rough-in	P6@2.25	Ea	70.50	99.70	170.20
Plastic DWV and copper supply rough-in	P6@2.35	Ea	83.30	104.00	187.30
Cast iron DWV and copper supply rough-in	P6@2.75	Ea	153.00	122.00	275.00

Floor sink, floor drain Rough-in includes 15' of drain and 15' of vent pipe, P trap, typical fittings and hangers

	Craft@Hrs	Unit	Material	Labor	Total
2" plastic DWV rough-in	P6@1.50	Ea	44.80	66.40	111.20
2" cast iron DWV rough-in	P6@2.00	Ea	96.20	88.60	184.80
3" plastic DWV rough-in	P6@1.75	Ea	64.20	77.50	141.70
3" cast iron DWV rough-in	P6@2.50	Ea	116.00	111.00	227.00
4" cast iron DWV rough-in	P6@2.50	Ea	153.00	111.00	264.00

Fixtures and final connection assemblies

Fixture costs are based on good to better quality white fixtures of standard dimensions and include all trim and valves. Add 30% for colored fixtures. Detailed estimates of the piping requirements for each type of plumbing fixture were made using the Uniform Plumbing Code as a guideline.

Water closets

Floor mounted, tank type, elongated vitreous china bowl

	Craft@Hrs	Unit	Material	Labor	Total
Water closet, trim and valves (add rough-in)	P6@2.10	Ea	196.00	93.00	289.00
Deduct for round bowl	—	Ea	-33.62	—	-33.62

Floor mounted flush valve type, elongated vitreous china bowl

	Craft@Hrs	Unit	Material	Labor	Total
Water closet, trim and valves (add rough-in)	P6@2.60	Ea	253.00	115.00	368.00

Floor mounted flush valve type, elongated vitreous china bowl, meets ADA requirements (18" high)

	Craft@Hrs	Unit	Material	Labor	Total
Water closet, trim and valves (add rough-in)	P6@2.60	Ea	300.00	115.00	415.00

Wall hung, tank type, elongated vitreous china bowl, cast iron carrier

	Craft@Hrs	Unit	Material	Labor	Total
Water closet, trim and valves (add rough-in)	P6@3.00	Ea	572.00	133.00	705.00
Deduct for round bowl	—	Ea	-66.17	—	-66.17

Wall hung flush valve type elongated vitreous china bowl, cast iron carrier

	Craft@Hrs	Unit	Material	Labor	Total
Water closet, trim and valves (add rough-in)	P6@3.55	Ea	404.00	157.00	561.00
Add for bed pan cleanser (flush valve and rim lugs)	P6@1.50	Ea	176.00	66.40	242.40

	Craft@Hrs	Unit	Material	Labor	Total

Urinals

Wall hung urinal, vitreous china, washout flush action, 3/4" top spud with hand operated flush valve/vacuum breaker

	Craft@Hrs	Unit	Material	Labor	Total
Urinal, trim and valves (add rough-in)	P6@2.35	Ea	329.00	104.00	433.00
Add for fixture carrier	P6@1.25	Ea	95.30	55.40	150.70

Floor mounted urinal, 18" wide, sloping front vitreous china, washout flush action, 3/4" top spud with hand operated flush valve/vacuum breaker

	Craft@Hrs	Unit	Material	Labor	Total
Urinal, trim and valves (add rough-in)	P6@5.25	Ea	794.00	233.00	1,027.00

Water conserving urinals, faucets and flush valves

Wall mounted urinal, ABS-structural plastic. Fit between studs framed at 16" on center.
Add the cost of wall finishing and rough-in plumbing

	Craft@Hrs	Unit	Material	Labor	Total
10 oz water per flush	P6@1.50	Ea	244.00	66.40	310.40

Battery powered sensor faucet, 6 volt DC. Continuous discharge up to 10 seconds. Less than one quart per cycle. www.totousa.com

	Craft@Hrs	Unit	Material	Labor	Total
Standard spout	P6@1.50	Ea	324.00	66.40	390.40
Gooseneck spout	P6@1.50	Ea	357.00	66.40	423.40

AC powered sensor faucet, 12 volt. Continuous discharge up to 10 seconds. Less than one quart per cycle

	Craft@Hrs	Unit	Material	Labor	Total
Standard spout	P6@1.50	Ea	324.00	66.40	390.40
Gooseneck spout	P6@1.50	Ea	357.00	66.40	423.40

Battery powered sensor flush valves, 6 volt DC, exposed. Light flushing after brief use and full flushing after extended use. Includes rough fittings

	Craft@Hrs	Unit	Material	Labor	Total
Urinal flush valve	P6@2.0	Ea	292.00	88.60	380.60
Toilet flush valve, 1.6 or 3.5 gallons	P6@2.0	Ea	319.00	88.60	407.60

AC powered sensor flush valves, 24 volt. Light flushing after brief use and full flushing after extended use. Includes rough fittings

	Craft@Hrs	Unit	Material	Labor	Total
Urinal flush valve, exposed	P6@2.0	Ea	314.00	88.60	402.60
Urinal flush valve, concealed	P6@2.0	Ea	321.00	88.60	409.60
Toilet flush valve	P6@2.0	Ea	340.00	88.60	428.60

Lavatories

Wall hung vitreous china lavatory with single handle faucet and pop-up waste

	Craft@Hrs	Unit	Material	Labor	Total
Lavatory, trim and valves (add rough-in)	P6@2.50	Ea	335.00	111.00	446.00
Add for wheelchair type (ADA standards)	—	Ea	155.00	—	155.00

Countertop mounted vitreous china lavatory with single handle water and pop-up waste

	Craft@Hrs	Unit	Material	Labor	Total
Lavatory, trim and valves (add rough-in)	P6@2.00	Ea	249.00	88.60	337.60

Countertop mounted enameled steel lavatory with single handle water faucet and pop-up waste

	Craft@Hrs	Unit	Material	Labor	Total
Lavatory, trim and valves (add rough-in)	P6@2.00	Ea	210.00	88.60	298.60

Bathtubs and Showers

Recessed enameled steel tub with shower

	Craft@Hrs	Unit	Material	Labor	Total
Tub, trim and valves (add rough-in)	P6@2.50	Ea	352.00	111.00	463.00

Recessed enameled cast iron tub with shower

	Craft@Hrs	Unit	Material	Labor	Total
Tub, trim and valves (add rough-in)	P6@4.50	Ea	688.00	199.00	887.00

Acrylic one-piece tub and shower enclosure

	Craft@Hrs	Unit	Material	Labor	Total
Enclosure, trim and valves (add rough-in)	P6@4.75	Ea	646.00	210.00	856.00

Acrylic three-piece tub and shower enclosure (renovation)

	Craft@Hrs	Unit	Material	Labor	Total
Enclosure, trim and valves (add rough-in)	P6@5.75	Ea	868.00	255.00	1,123.00

Fiberglass one-piece tub and shower enclosure

	Craft@Hrs	Unit	Material	Labor	Total
Enclosure, trim and valves (add rough-in)	P6@4.75	Ea	477.00	210.00	687.00

Fiberglass three-piece tub and shower enclosure

	Craft@Hrs	Unit	Material	Labor	Total
Enclosure, trim and valves (add rough-in)	P6@5.75	Ea	719.00	255.00	974.00

Mechanical 15

	Craft@Hrs	Unit	Material	Labor	Total
Acrylic one-piece shower stall					
Shower stall, trim and valves (add rough-in)	P6@3.50	Ea	614.00	155.00	769.00
Add for larger stall	—	Ea	106.00	—	106.00
Acrylic three-piece shower stall (renovation)					
Shower stall, trim and valves (add rough-in)	P6@4.25	Ea	804.00	188.00	992.00
Add for larger stall	—	Ea	126.00	—	126.00
Fiberglass one-piece shower stall					
Shower stall, trim and valves (add rough-in)	P6@3.50	Ea	414.00	155.00	569.00
Add for larger stall	—	Ea	84.70	—	84.70
Add for corner entry	—	Ea	68.90	—	68.90
Fiberglass three-piece shower stall (renovation)					
Shower stall, trim and valves (add rough-in)	P6@4.25	Ea	530.00	188.00	718.00
Add for larger stall	—	Ea	117.00	—	117.00
Add for corner entry	—	Ea	95.30	—	95.30
Acrylic shower base					
Base, trim and valves (add rough-in)	P6@2.50	Ea	259.00	111.00	370.00
Add for larger base	—	Ea	10.60	—	10.60
Add for corner entry	—	Ea	10.60	—	10.60
Fiberglass shower base					
Base, trim and valves (add rough-in)	P6@2.50	Ea	292.00	111.00	403.00
Add for larger base	—	Ea	10.60	—	10.60
Add for corner entry	—	Ea	10.60	—	10.60
Tub and shower doors					
Three panel sliding tub doors	P6@.500	Ea	265.00	22.10	287.10
Three panel sliding shower doors	P6@.500	Ea	232.00	22.10	254.10
Two panel sliding shower doors	P6@.500	Ea	196.00	22.10	218.10
One panel swinging shower door	P6@.500	Ea	95.30	22.10	117.40

Kitchen sinks

Single compartment kitchen sink, 20" long, 21" wide, 8" deep, self rimming, countertop mounted, stainless steel with single handle faucet and deck spray

	Craft@Hrs	Unit	Material	Labor	Total
Sink, trim and valves (add rough-in)	P6@2.00	Ea	209.00	88.60	297.60

Double compartment kitchen sink, 32" long, 21" wide, 8" deep, self rimming, countertop mounted, stainless steel with single handle faucet and deck spray

	Craft@Hrs	Unit	Material	Labor	Total
Sink, trim and valves (add rough-in)	P6@2.15	Ea	248.00	95.20	343.20

Drinking fountain and water cooler

Wall hung drinking fountains, 10" wide, 10" high, non-electric, white vitreous china

	Craft@Hrs	Unit	Material	Labor	Total
Wall hung fountain, and trim (add rough-in)	P6@2.00	Ea	343.00	88.60	431.60

Recessed drinking fountains, 14" wide, 26" high, non-electric, white vitreous china

	Craft@Hrs	Unit	Material	Labor	Total
Recessed fountain, and trim (add rough-in)	P6@2.50	Ea	607.00	111.00	718.00

Semi-recessed water cooler, vitreous china, refrigerated (no electrical work included)

	Craft@Hrs	Unit	Material	Labor	Total
Semi-recessed cooler, and trim (add rough-in)	P6@2.50	Ea	105.00	111.00	216.00
Add for fully recessed	P6@.750	Ea	788.00	33.20	821.20

Free standing water cooler, refrigerated (no electrical work included)

	Craft@Hrs	Unit	Material	Labor	Total
Wall hung water cooler, and trim (add rough-in)	P6@2.50	Ea	695.00	111.00	806.00

Wash fountains

54" circular wash fountain, precast stone (terrazzo), with foot operated water spray control

	Craft@Hrs	Unit	Material	Labor	Total
Wash fountain, trim and valves (add rough-in)	P6@3.65	Ea	2,280.00	162.00	2,442.00
Add for stainless steel in lieu of precast stone	—	Ea	525.00	—	525.00

	Craft@Hrs	Unit	Material	Labor	Total
54" semi-circular wash fountain, precast stone (terrazzo), with foot operated water spray control					
Wash fountain, trim and valves (add rough-in)	P6@3.65	Ea	2,050.00	162.00	2,212.00
Add for stainless steel in lieu of precast stone	—	Ea	368.00	—	368.00
36" circular wash fountain, precast stone (terrazzo), with foot operated water spray control					
Wash fountain, trim and valves (add rough-in)	P6@3.50	Ea	1,830.00	155.00	1,985.00
Add for stainless steel in lieu of precast stone	—	Ea	315.00	—	315.00
36" semi-circular wash fountain, precast stone, (terrazzo), with foot operated water spray control					
Wash fountain, trim and valves (add rough-in)	P6@3.50	Ea	1,670.00	155.00	1,825.00
Add for stainless steel in lieu of precast stone	—	Ea	208.00	—	208.00

Service sinks

Wall hung enameled cast iron slop sink, with wall-mounted faucet, hose and P trap standard

	Craft@Hrs	Unit	Material	Labor	Total
Slop sink, trim and valves (add rough-in)	P6@3.30	Ea	644.00	146.00	790.00

Floor mounted mop sink, precast molded stone (terrazzo), with wall mounted faucet and hose

	Craft@Hrs	Unit	Material	Labor	Total
Mop sink, 24" x 24", 6" curbs (add rough-in)	P6@2.60	Ea	359.00	115.00	474.00
Mop sink, 32" x 32", 6" curbs (add rough-in)	P6@2.60	Ea	386.00	115.00	501.00
Mop sink, 36" x 36", 6" curbs (add rough-in)	P6@2.60	Ea	421.00	115.00	536.00
Add for 12" curbs, 6" drop front & stainless steel caps	—	Ea	204.00	—	204.00

Drench shower and emergency eye wash stations

	Craft@Hrs	Unit	Material	Labor	Total
Drench shower, plastic pipe and head, wall mounted	P6@2.00	Ea	146.00	88.60	234.60
Drench shower galvanized pipe, free standing	P6@2.50	Ea	314.00	111.00	425.00
Drench shower, emergency eyewash combination	P6@2.75	Ea	849.00	122.00	971.00
Emergency eyewash station	P6@2.00	Ea	340.00	88.60	428.60
Walk-thru station, 18 heads	P6@12.5	Ea	2,190.00	554.00	2,744.00

Surgeons' scrub sink

Wall hung wash sink, white enameled cast iron, 48" long, 18" deep, trough-type with 2 twin-handled faucets and two soap dishes

	Craft@Hrs	Unit	Material	Labor	Total
Wash sink, trim and valves (add rough-in)	P6@2.50	Ea	923.00	111.00	1,034.00
Add for foot or knee pedal operation	P6@.500	Ea	106.00	22.10	128.10
Add for electronic infrared hands free operation	P6@.500	Ea	132.00	22.10	154.10
Add for stainless steel in lieu of enameled cast iron	—	Ea	160.00	—	160.00

Floor sinks

	Craft@Hrs	Unit	Material	Labor	Total
12" square sink, cast iron top and body (add rough-in)	P6@1.00	Ea	163.00	44.30	207.30
Add for an acid-resistant finish to the sink	—	Ea	27.10	—	27.10

Fire Protection

Sprinkler systems Typical subcontract prices including the subcontractor's overhead and profit. These prices include design drawings, valves and connection to piping to within 5'0" outside the building. Costs will be higher where room sizes are smaller or where coverage per head averages less than 110 SF. Make additional allowances if a booster pump is required

	Craft@Hrs	Unit	Material	Labor	Total
Exposed systems, wet, complete, cost per SF of floor protected					
5,000 SF	—	SF	—	—	2.67
Over 5,000 to 15,000 SF	—	SF	—	—	2.26
15,000 SF or more	—	SF	—	—	2.00
Concealed systems, wet, complete, cost per SF of floor protected					
5,000 SF	—	SF	—	—	2.96
Over 5,000 to 15,000 SF	—	SF	—	—	2.38
15,000 SF or more	—	SF	—	—	2.26
Add for dry systems	—	%	—	—	10.5

	Craft@Hrs	Unit	Material	Labor	Total
Fire sprinkler system components Make additional allowances to shut down and drain the system when required					
Sprinkler heads only (remove and replace)					
Brass pendent or upright head, 155-200 degree	SP@.350	Ea	7.64	18.80	26.44
Brass pendent or upright head, 286 degree	SP@.350	Ea	8.18	18.80	26.98
Brass pendent or upright head, 360 degree	SP@.350	Ea	19.50	18.80	38.30
Brass pendent or upright head, 400-500 degree	SP@.350	Ea	23.50	18.80	42.30
Chrome pendent or upright head, 155-200 degree	SP@.350	Ea	8.76	18.80	27.56
Chrome pendent or upright head, 286 degree	SP@.350	Ea	9.07	18.80	27.87
Dry pendent or upright head	SP@.350	Ea	69.80	18.80	88.60
Relocate wet sprinkler head and branch drop	SP@1.50	Ea	33.30	80.40	113.70
Add for heads more than 14' above floor	—	%	—	20.0	—
Wet system components (where freezing is not a hazard)					
Zone valves					
2" OS & Y gate valve, flanged or grooved	SP@1.75	Ea	228.00	93.80	321.80
3" OS & Y gate valve, flanged or grooved	SP@2.00	Ea	276.00	107.00	383.00
4" OS & Y gate valve, flanged or grooved	SP@2.25	Ea	382.00	121.00	503.00
6" OS & Y gate valve, flanged or grooved	SP@3.00	Ea	602.00	161.00	763.00
8" OS & Y gate valve, flanged or grooved	SP@3.50	Ea	1,240.00	188.00	1,428.00
4" alarm valve, flanged or grooved	SP@2.65	Ea	519.00	142.00	661.00
6" alarm valve, flanged or grooved	SP@3.50	Ea	647.00	188.00	835.00
8" alarm valve, flanged or grooved	SP@4.25	Ea	931.00	228.00	1,159.00
Alarm valve trim only (retard chamber and gauges)	SP@1.25	Ea	291.00	67.00	358.00
4" alarm valve package, complete	SP@4.00	Ea	2,340.00	214.00	2,554.00
6" alarm valve package, complete	SP@4.50	Ea	2,560.00	241.00	2,801.00
8" alarm valve package, complete	SP@5.25	Ea	3,350.00	281.00	3,631.00
Retard pressure switch	SP@1.75	Ea	276.00	93.80	369.80
Check valves					
3" swing check valve	SP@2.00	Ea	207.00	107.00	314.00
4" swing check valve	SP@2.25	Ea	221.00	121.00	342.00
6" swing check valve	SP@3.00	Ea	419.00	161.00	580.00
3" wafer check valve	SP@2.00	Ea	173.00	107.00	280.00
4" wafer check valve	SP@2.25	Ea	191.00	121.00	312.00
6" wafer check valve	SP@3.00	Ea	304.00	161.00	465.00
8" wafer check valve	SP@3.50	Ea	403.00	188.00	591.00
10" wafer check valve	SP@4.25	Ea	774.00	228.00	1,002.00
3" double check detector assembly	SP@3.75	Ea	1,790.00	201.00	1,991.00
4" double check detector assembly	SP@4.50	Ea	2,430.00	241.00	2,671.00
6" double check detector assembly	SP@5.75	Ea	3,730.00	308.00	4,038.00
8" double check detector assembly	SP@6.25	Ea	6,920.00	335.00	7,255.00
Dry system components (distribution piping holds no water)					
Dry valves (deluge)					
3" dry pipe valve	SP@2.25	Ea	988.00	121.00	1,109.00
4" dry pipe valve	SP@2.65	Ea	1,190.00	142.00	1,332.00
6" dry pipe valve	SP@3.50	Ea	1,600.00	188.00	1,788.00
Dry valve trim and gauges	SP@1.00	Ea	394.00	53.60	447.60
Wall hydrants for sprinkler systems, brass					
Single outlet, 2-1/2" x 2-1/2"	SP@2.00	Ea	324.00	107.00	431.00
2-way outlet, 2-1/2" x 2-1/2" x 4"	SP@2.50	Ea	521.00	134.00	655.00
3-way outlet, 2-1/2" x 2-1/2" x 3-1/3" x 4"	SP@2.75	Ea	948.00	147.00	1,095.00
Fire dept. connection, 4" (Siamese)	SP@2.50	Ea	521.00	134.00	655.00
Pumper connection, 6"	SP@3.00	Ea	1,030.00	161.00	1,191.00
Roof manifold, with valves	SP@3.90	Ea	608.00	209.00	817.00

	Craft@Hrs	Unit	Material	Labor	Total
Fire hose cabinets, recessed, 24" x 30" x 5-1/2" with rack, full glass door, 1-1/2" hose, valve and nozzle. Standard					
Painted steel, 75' hose, 24" x 30"	SP@1.95	Ea	400.00	104.00	504.00
Painted steel, 100' hose, 24" x 36"	SP@1.95	Ea	447.00	104.00	551.00
Aluminum, 75' hose	SP@1.95	Ea	619.00	104.00	723.00
Stainless steel, 75' hose	SP@1.95	Ea	798.00	104.00	902.00
Fire extinguishers Factory charged, complete with hose and horn and wall mounting bracket					
5 lb. carbon dioxide	C8@.488	Ea	254.00	20.00	274.00
10 lb. carbon dioxide	C8@.488	Ea	367.00	20.00	387.00
20 lb. carbon dioxide	C8@.488	Ea	528.00	20.00	548.00
5 lb. dry chemical	C8@.488	Ea	80.70	20.00	100.70
10 lb. dry chemical	C8@.488	Ea	174.00	20.00	194.00
20 lb. dry chemical	C8@.488	Ea	199.00	20.00	219.00
Extinguisher cabinets No extinguishers included, painted steel, recessed					
9" x 24" x 5" full glass door	C8@1.67	Ea	108.00	68.40	176.40
9" x 24" x 5" break glass door	C8@1.67	Ea	129.00	68.40	197.40
12" x 27" x 8" full glass door	C8@1.67	Ea	125.00	68.40	193.40
12" x 27" x 8" break glass door	C8@1.67	Ea	147.00	68.40	215.40
Add for semi-recessed or surface mount	—	%	10.0	—	—
Add for chrome cabinets	—	%	15.0	—	—
Add for stainless steel cabinets	—	Ea	168.00		168.00
Miscellaneous sprinkler system components					
Air maintenance device	SP@.781	Ea	189.00	41.80	230.80
Low air supervisory unit	SP@1.11	Ea	584.00	59.50	643.50
Low air pressure switch	SP@.784	Ea	189.00	42.00	231.00
Pressure switch (double circuit, open and close)	SP@1.92	Ea	183.00	103.00	286.00
1/2" ball drip at check valve and fire dept. connection	SP@.193	Ea	21.20	10.30	31.50
Water powered gong (local alarm)	SP@1.94	Ea	228.00	104.00	332.00
Cabinet with 6 spare heads and wrench	SP@.291	LS	87.60	15.60	103.20
Inspector's test connection	SP@.973	Ea	73.00	52.10	125.10
Escutcheon for sprinkler heads, chrome	SP@.071	Ea	1.24	3.80	5.04
Field testing and flushing, subcontract	—	LS	—	—	165.00
Disinfection of distribution system, subcontract	—	LS	—	—	165.00
Hydraulic design fee, subcontract					
Sprinklers in a high density or high risk area, typical	—	LS	—	—	2,540.00
Underground valve vault (if required), subcontract	—	LS	—	—	3,050.00
Tanks Includes fittings but no excavation, dewatering, shoring, supports, piping, backfill, paving, or concrete					
Underground steel oil storage tanks					
550 gallon	MI@4.86	Ea	1,300.00	225.00	1,525.00
2,500 gallon	M8@18.5	Ea	3,030.00	877.00	3,907.00
5,000 gallon	M8@24.0	Ea	5,410.00	1,140.00	6,550.00
10,000 gallon	M8@37.3	Ea	9,080.00	1,770.00	10,850.00
Above ground steel oil storage tanks					
275 gallon	MI@2.98	Ea	452.00	138.00	590.00
500 gallon	MI@3.80	Ea	963.00	176.00	1,139.00
1,000 gallon	M8@7.64	Ea	1,620.00	362.00	1,982.00
1,500 gallon	M8@8.29	Ea	1,790.00	393.00	2,183.00
2,000 gallon	M8@10.5	Ea	2,300.00	498.00	2,798.00
5,000 gallon	M8@29.6	Ea	6,360.00	1,400.00	7,760.00
Add for vent caps	M8@.155	Ea	1.45	7.35	8.80
Add for filler cap	M8@.155	Ea	32.20	7.35	39.55

	Craft@Hrs	Unit	Material	Labor	Total
Underground fiberglass oil storage tanks					
550 gallon	MI@8.36	Ea	3,060.00	386.00	3,446.00
1,000 gallon	M8@16.8	Ea	5,530.00	797.00	6,327.00
2,000 gallon	M8@16.8	Ea	6,250.00	797.00	7,047.00
4,000 gallon	M8@25.9	Ea	7,600.00	1,230.00	8,830.00
6,000 gallon	M8@33.5	Ea	8,470.00	1,590.00	10,060.00
8,000 gallon	M8@33.5	Ea	9,540.00	1,590.00	11,130.00
10,000 gallon	M8@33.5	Ea	10,400.00	1,590.00	11,990.00
12,000 gallon	M8@84.4	Ea	12,900.00	4,000.00	16,900.00
15,000 gallon	M8@89.0	Ea	15,100.00	4,220.00	19,320.00
20,000 gallon	M8@89.0	Ea	18,700.00	4,220.00	22,920.00
Propane tanks with valves, underground					
1,000 gallon	M8@10.0	Ea	2,260.00	474.00	2,734.00

Heat Generation

Gas fired heating equipment No gas pipe or electric runs included

Unit heaters, gas fired, ceiling suspended, with flue and gas valve, propeller fan with aluminized heat exchanger, Modine PAE series

	Craft@Hrs	Unit	Material	Labor	Total
50 MBtu input	M5@4.75	Ea	646.00	208.00	854.00
75 MBtu input	M5@4.75	Ea	705.00	208.00	913.00
125 MBtu input	M5@5.25	Ea	886.00	230.00	1,116.00
175 MBtu input	M5@5.50	Ea	992.00	241.00	1,233.00
225 MBtu input	MI@6.75	Ea	1,120.00	312.00	1,432.00
300 MBtu input	MI@7.50	Ea	1,410.00	347.00	1,757.00
400 MBtu input	MI@8.75	Ea	1,850.00	404.00	2,254.00
Add for single stage intermittent ignition pilot	—	LS	146.00	—	146.00
Add for high efficiency units (PV series)	—	%	35.0	—	—
Add for stainless steel burner	—	%	20.0	—	—
Add for air blower style fan	—	%	10.0	—	—
Add for switch and thermostat	CE@1.00	Ea	39.70	50.50	90.20
Add for gas supply pipe connection	M5@1.75	Ea	64.60	76.60	141.20
Add for typical electric run, BX	E4@2.50	Ea	80.30	109.00	189.30
Furnaces, wall type, upflow, gas fired with gas valve, flue vent and electronic ignition					
35 MBtu input	M5@4.25	Ea	676.00	186.00	862.00
50 MBtu input	M5@4.25	Ea	763.00	186.00	949.00
65 MBtu input	M5@4.25	Ea	770.00	186.00	956.00
Add for gas supply pipe connection	M5@1.75	Ea	64.40	76.60	141.00
Add for power vent kit (direct sidewall vent)	M5@2.50	Ea	87.90	109.00	196.90
Add for thermostat	CE@1.00	Ea	35.20	50.50	85.70
Furnaces, non-condensing gas fired, multi-positional, mid-efficiency with gas valve, flue vent and electronic ignition					
40 MBtu input, 80% AFUE	M5@4.25	Ea	776.00	186.00	962.00
60 MBtu input, 80% AFUE	M5@4.25	Ea	851.00	186.00	1,037.00
80 MBtu input, 80% AFUE	M5@4.25	Ea	879.00	186.00	1,065.00
100 MBtu input, 80% AFUE	M5@4.50	Ea	908.00	197.00	1,105.00
120 MBtu input, 80% AFUE	M5@4.50	Ea	967.00	197.00	1,164.00
140 MBtu input, 80% AFUE	M5@4.50	Ea	1,080.00	197.00	1,277.00
Add for gas supply pipe connection	M5@1.75	Ea	64.40	76.60	141.00
Add for power vent kit (direct sidewall vent)	M5@2.50	Ea	83.70	109.00	192.70
Add for thermostat	CE@1.00	Ea	35.20	50.50	85.70

	Craft@Hrs	Unit	Material	Labor		Total

Furnaces, condensing gas fired, upflow, high-efficiency with gas valve, flue vent and electronic ignition

	Craft@Hrs	Unit	Material	Labor	Total
50 MBtu input, 90% AFUE	M5@4.25	Ea	1,380.00	186.00	1,566.00
70 MBtu input, 90% AFUE	M5@4.25	Ea	1,550.00	186.00	1,736.00
90 MBtu input, 90% AFUE	M5@4.25	Ea	1,560.00	186.00	1,746.00
110 MBtu input, 90% AFUE	M5@4.50	Ea	1,680.00	197.00	1,877.00
130 MBtu input, 90% AFUE	M5@4.50	Ea	1,680.00	197.00	1,877.00
Add for gas supply pipe connection	M5@1.75	Ea	64.40	76.60	141.00
Add for power vent kit (direct sidewall vent)	M5@2.50	Ea	87.90	109.00	196.90

Air curtain, gas fired, ceiling suspended, includes flue, vent, and electrical ignition

	Craft@Hrs	Unit	Material	Labor	Total
256 MBtu, 2,840 CFM	MI@7.50	Ea	1,540.00	347.00	1,887.00
290 MBtu, 4,740 CFM	MI@8.25	Ea	1,680.00	381.00	2,061.00
400 MBtu, 5,660 CFM	MI@9.75	Ea	1,850.00	450.00	2,300.00
525 MBtu, 7,580 CFM	MI@11.0	Ea	1,950.00	508.00	2,458.00
630 MBtu, 9,480 CFM	MI@12.5	Ea	2,160.00	578.00	2,738.00
850 MBtu, 11,400 CFM	MI@16.0	Ea	3,080.00	739.00	3,819.00
Add for gas supply pipe connection	M5@1.75	Ea	64.40	76.60	141.00

Electric Heat

Air curtain electric heaters, wall mounted, includes electrical connection only

	Craft@Hrs	Unit	Material	Labor	Total
37" long, 9.5 kW	H9@6.33	Ea	642.00	320.00	962.00
49" long, 12.5 kW	H9@8.05	Ea	907.00	407.00	1,314.00
61" long, 16.0 kW	H9@11.2	Ea	1,090.00	566.00	1,656.00

Baseboard electric heaters, with remote thermostat

	Craft@Hrs	Unit	Material	Labor	Total
185 watts per LF	CE@.175	LF	21.60	8.84	30.44

	Craft@Hrs	Unit	Material	Labor	Equipment	Total

Hydronic hot water generators 30 PSI, with cast iron burners, factory installed boiler trim and circulating pump. For commercial and light industrial applications. Equipment cost is for a 2 ton capacity forklift

	Craft@Hrs	Unit	Material	Labor	Equipment	Total
4 BHP, 133 MBtu/Hr	M5@4.00	Ea	1,800.00	175.00	49.70	2,024.70
8 BHP, 263 MBtu/Hr	M5@4.00	Ea	2,390.00	175.00	49.70	2,614.70
12 BHP, 404 MBtu/Hr	M5@4.00	Ea	3,100.00	175.00	49.70	3,324.70
15 BHP, 514 MBtu/Hr	M5@5.00	Ea	4,680.00	219.00	62.10	4,961.10
19 BHP, 624 MBtu/Hr	M5@5.00	Ea	6,210.00	219.00	62.10	6,491.10
22 BHP, 724 MBtu/Hr	M5@6.00	Ea	6,460.00	263.00	74.60	6,797.60
29 BHP, 962 MBtu/Hr	M5@6.00	Ea	7,700.00	263.00	74.60	8,037.60
34 BHP, 1,125 MBtu/Hr	MI@8.00	Ea	8,590.00	370.00	66.30	9,026.30
40 BHP, 1,336 MBtu/Hr	MI@8.00	Ea	9,910.00	370.00	66.30	10,346.30
47 BHP, 1,571 MBtu/Hr	MI@9.00	Ea	10,300.00	416.00	74.60	10,790.60
53 BHP, 1,758 MBtu/Hr	MI@10.0	Ea	11,100.00	462.00	82.80	11,644.80
63 BHP, 2,100 MBtu/Hr	MI@12.0	Ea	11,600.00	554.00	99.40	12,253.40
89 BHP, 3,001 MBtu/Hr	MI@18.0	Ea	14,900.00	832.00	149.00	15,881.00
118 BHP, 4,001 MBtu/Hr	MI@22.0	Ea	15,300.00	1,020.00	182.00	16,502.00

	Craft@Hrs	Unit	Material	Labor	Total

Hydronic Heating

Hot water baseboard fin tube radiation, copper tube and aluminum fin element, commercial quality, 18 gauge enclosure. Make additional allowances for control wiring

	Craft@Hrs	Unit	Material	Labor	Total
9-7/8" enclosure and 1-1/4" element	P6@.280	LF	19.40	12.40	31.80
13-7/8" enclosure and 1-1/4" element, 2 rows	P6@.320	LF	33.50	14.20	47.70
9-7/8" enclosure only	P6@.130	LF	6.08	5.76	11.84
13-7/8" enclosure only	P6@.150	LF	9.35	6.64	15.99
1-1/4" element only	P6@.150	LF	13.80	6.64	20.44

	Craft@Hrs	Unit	Material	Labor	Total
1" element only	P6@.150	LF	6.92	6.64	13.56
3/4" element only	P6@.140	LF	5.55	6.20	11.75
Add for pipe connection and control valve	P6@2.25	Ea	127.00	99.70	226.70
Add for corners, fillers and caps, average per foot	—	%	10.0	—	—
End cap, 4"	P6@.090	Ea	4.19	3.99	8.18
End cap, 2"	P6@.090	Ea	2.10	3.99	6.09
Valve access door	P6@.250	Ea	10.50	11.10	21.60
Inside corner	P6@.120	Ea	4.17	5.31	9.48
Outside corner	P6@.120	Ea	5.92	5.31	11.23
Filler sleeve	P6@.090	Ea	2.61	3.99	6.60

Hot water baseboard fin tube radiation, steel element, commercial quality, 16 gauge, sloping top cover. Make additional allowances for control wiring

	Craft@Hrs	Unit	Material	Labor	Total
9-7/8" enclosure and 1-1/4" steel element	P6@.330	LF	22.40	14.60	37.00
9-7/8" enclosure only	P6@.130	LF	6.74	5.76	12.50
1-1/4" steel element only	P6@.160	LF	15.60	7.09	22.69
Add for pipe connection and control valve	P6@2.25	Ea	127.00	99.70	226.70
Add for corners, fillers and caps, average per foot	—	%	10.0	—	—
End cap, 4"	P6@.090	Ea	8.04	3.99	12.03
End cap, 2"	P6@.090	Ea	3.84	3.99	7.83
Valve access door	P6@.250	Ea	10.50	11.10	21.60
Inside corner	P6@.120	Ea	7.04	5.31	12.35
Outside corner	P6@.120	Ea	9.48	5.31	14.79
Filler sleeve	P6@.090	Ea	5.11	3.99	9.10

Unit heaters, hot water, horizontal, fan driven

	Craft@Hrs	Unit	Material	Labor	Total
12.5 MBtu, 200 CFM	M5@2.00	Ea	281.00	87.50	368.50
17 MBtu, 300 CFM	M5@2.00	Ea	312.00	87.50	399.50
25 MBtu, 500 CFM	M5@2.50	Ea	354.00	109.00	463.00
30 MBtu, 700 CFM	M5@3.00	Ea	373.00	131.00	504.00
50 MBtu, 1,000 CFM	M5@3.50	Ea	433.00	153.00	586.00
60 MBtu, 1,300 CFM	M5@4.00	Ea	515.00	175.00	690.00

Unit heaters, hot water, vertical, fan driven

	Craft@Hrs	Unit	Material	Labor	Total
12.5 MBtu, 200 CFM	M5@2.00	Ea	315.00	87.50	402.50
17 MBtu, 300 CFM	M5@2.00	Ea	321.00	87.50	408.50
25 MBtu, 500 CFM	M5@2.50	Ea	416.00	109.00	525.00
30 MBtu, 700 CFM	M5@3.00	Ea	434.00	131.00	565.00
50 MBtu, 1,000 CFM	M5@3.50	Ea	523.00	153.00	676.00
60 MBtu, 1,300 CFM	M5@4.00	Ea	575.00	175.00	750.00

Steam Heating

Steam baseboard radiation, per linear foot, wall mounted

	Craft@Hrs	Unit	Material	Labor	Total
1/2" tube and cover	P6@.456	LF	7.47	20.20	27.67
3/4" tube and cover	P6@.493	LF	8.02	21.80	29.82
3/4" tube and cover, high capacity	P6@.493	LF	9.49	21.80	31.29
1/2" tube alone	P6@.350	LF	4.06	15.50	19.56
3/4" tube alone	P6@.377	LF	4.62	16.70	21.32
Cover only	P6@.137	LF	4.86	6.07	10.93

Unit heater, steam, horizontal, fan driven

	Craft@Hrs	Unit	Material	Labor	Total
18 MBtu, 300 CFM	M5@2.00	Ea	257.00	87.50	344.50
45 MBtu, 500 CFM	M5@2.50	Ea	426.00	109.00	535.00
60 MBtu, 700 CFM	M5@3.00	Ea	481.00	131.00	612.00
85 MBtu, 1,000 CFM	M5@3.50	Ea	507.00	153.00	660.00

	Craft@Hrs	Unit	Material	Labor	Total
Unit heater, steam, vertical, fan driven					
12.5 MBtu, 200 CFM	M5@2.00	Ea	234.00	87.50	321.50
17 MBtu, 300 CFM	M5@2.00	Ea	237.00	87.50	324.50
40 MBtu, 500 CFM	M5@2.50	Ea	461.00	109.00	570.00
60 MBtu, 700 CFM	M5@3.00	Ea	534.00	131.00	665.00
70 MBtu, 1,000 CFM	M5@3.50	Ea	674.00	153.00	827.00

	Craft@Hrs	Unit	Material	Labor	Equipment	Total
ASME tanks Add the cost of a concrete pad and an insulation jacket						
Expansion tanks, galvanized steel, ASME code, 125 PSI						
15 gallons	P6@1.75	Ea	465.00	77.50	—	542.50
30 gallons	P6@2.00	Ea	504.00	88.60	—	592.60
60 gallons	P6@2.70	Ea	618.00	120.00	—	738.00
80 gallons	P6@2.95	Ea	664.00	131.00	—	795.00

Vertical cylindrical, epoxy-lined, with pipe legs, two 1/2" openings, a 1" temperature and pressure relief valve opening and a 1" drain opening. Equipment cost is for a 2 ton forklift. Make additional allowances for pipe connections.

	Craft@Hrs	Unit	Material	Labor	Equipment	Total
115 gallons, 24" dia	MI@4.00	Ea	1,810.00	185.00	49.70	2,044.70
261 gallons, 30" dia	MI@4.00	Ea	2,480.00	185.00	49.70	2,714.70
327 gallons, 36" dia	MI@4.00	Ea	2,900.00	185.00	49.70	3,134.70
462 gallons, 42" dia	MI@4.00	Ea	4,380.00	185.00	49.70	4,614.70
534 gallons, 42" dia	MI@4.00	Ea	4,600.00	185.00	49.70	4,834.70
606 gallons, 42" dia	MI@5.00	Ea	5,010.00	231.00	62.10	5,303.10
712 gallons, 48" dia	MI@5.00	Ea	5,950.00	231.00	62.10	6,243.10
803 gallons, 54" dia	MI@5.00	Ea	7,110.00	231.00	62.10	7,403.10
922 gallons, 54" dia	MI@6.00	Ea	7,460.00	277.00	74.60	7,811.60
1,017 gallons, 60" dia	MI@6.00	Ea	8,380.00	277.00	74.60	8,731.60
1,164 gallons, 60" dia	MI@6.00	Ea	8,780.00	277.00	74.60	9,131.60
1,311 gallons, 60" dia	MI@8.00	Ea	9,220.00	370.00	99.40	9,689.40
1,486 gallons, 60" dia	MI@8.00	Ea	9,600.00	370.00	99.40	10,069.40
1,616 gallons, 66" dia	MI@8.00	Ea	11,000.00	370.00	99.40	11,469.40
1,953 gallons, 72" dia	MI@8.00	Ea	12,600.00	370.00	99.40	13,069.40
2,378 gallons, 72" dia	MI@8.00	Ea	14,700.00	370.00	99.40	15,169.40
2,590 gallons, 72" dia	MI@8.00	Ea	15,200.00	370.00	99.40	15,669.40
3,042 gallons, 84" dia	MI@8.00	Ea	18,000.00	370.00	99.40	18,469.40
4,500 gallons, 96" dia	MI@8.00	Ea	24,000.00	370.00	99.40	24,469.40
5,252 gallons, 96" dia	MI@8.00	Ea	31,800.00	370.00	99.40	32,269.40

Compressors Costs shown include the concrete mounting pad, vibration isolation and anchor bolting, steel stand and structural supports, rigging, bolts and gasket sets for flanges, integrated intercooling and lubrication systems, typical vendor engineering and calibration, and inlet air cleaners. Add the cost of electrical connections, control valves, service bypass/isolation piping and the cost of piping to load and control valves. Equipment cost is for a 2-ton forklift.

	Craft@Hrs	Unit	Material	Labor	Equipment	Total
16.5 CFM at 175 PSI, 5 HP, 80 gallon tank	MI@3.50	Ea	2,130.00	162.00	43.50	2,335.50
34.4 CFM at 175 PSI, 10 HP, 120 gallon tank	MI@5.00	Ea	3,650.00	231.00	62.10	3,943.10
53.7 CFM at 175 PSI, 15 HP, 120 gallon tank	MI@8.00	Ea	5,170.00	370.00	99.40	5,639.40
91.0 CFM at 175 PSI, 25 HP, 120 gallon tank	MI@10.0	Ea	6,430.00	462.00	124.00	7,016.00

	Craft@Hrs	Unit	Material	Labor	Total
Hydronic Heating Specialties					
Air eliminator-purger, screwed connections					
3/4", cast iron, steam & water, 150 PSI	PF@.250	Ea	331.00	12.80	343.80
3/4", cast iron, 150 PSI water	PF@.250	Ea	107.00	12.80	119.80
1", cast iron, 150 PSI water	PF@.250	Ea	121.00	12.80	133.80
3/8", brass, 125 PSI steam	PF@.250	Ea	87.00	12.80	99.80
1/2", cast iron, 250 PSI steam	PF@.250	Ea	180.00	12.80	192.80
3/4", cast iron, 250 PSI steam	PF@.250	Ea	226.00	12.80	238.80
Airtrol fitting, 3/4"	PF@.300	Ea	38.30	15.30	53.60
Air eliminator vents, 1/4"	P1@.150	Ea	12.90	4.50	17.40
Atmospheric vacuum breakers, screwed					
1/2" vacuum breaker	PF@.210	Ea	18.50	10.70	29.20
3/4" vacuum breaker	PF@.250	Ea	19.80	12.80	32.60
1" vacuum breaker	PF@.300	Ea	31.40	15.30	46.70
1-1/4" vacuum breaker	PF@.400	Ea	52.40	20.40	72.80
1-1/2" vacuum breaker	PF@.450	Ea	61.60	23.00	84.60
2" vacuum breaker	PF@.500	Ea	91.80	25.60	117.40
Circuit balancing valves, brass, threaded					
1/2" circuit balancing valve	PF@.210	Ea	22.60	10.70	33.30
3/4" circuit balancing valve	PF@.250	Ea	29.00	12.80	41.80
1" circuit balancing valve	PF@.300	Ea	37.10	15.30	52.40
1-1/2" circuit balancing valve	PF@.450	Ea	48.70	23.00	71.70
Flow check valves, brass, threaded, horizontal type					
3/4" check valve	PF@.250	Ea	18.50	12.80	31.30
1" check valve	PF@.300	Ea	26.80	15.30	42.10
1-1/2" valve	PF@.350	Ea	59.00	17.90	76.90
2" valve	PF@.400	Ea	59.20	20.40	79.60
Pressure reducing valves, bronze body					
1/2", 25 – 75 PSI	PF@.210	Ea	79.00	10.70	89.70
1/2", 10 – 35 PSI	PF@.210	Ea	91.80	10.70	102.50
3/4", 25 – 75 PSI	PF@.250	Ea	94.70	12.80	107.50
3/4", 10 – 35 PSI	PF@.250	Ea	107.00	12.80	119.80
1", 25 – 75 PSI	PF@.300	Ea	149.00	15.30	164.30
1", 10 – 35 PSI	PF@.300	Ea	163.00	15.30	178.30
1-1/4", 25 – 75 PSI	PF@.350	Ea	299.00	17.90	316.90
1-1/4", 10 – 35 PSI	PF@.350	Ea	326.00	17.90	343.90
1-1/2", 25 – 75 PSI	PF@.250	Ea	322.00	12.80	334.80
1-1/2", 10 – 35 PSI	PF@.250	Ea	368.00	12.80	380.80
2", 25 – 75 PSI	PF@.250	Ea	481.00	12.80	493.80
2", 10 – 35 PSI	PF@.250	Ea	507.00	12.80	519.80
Pressure regulators, feed water 25 – 75 PSI, cast iron, screwed connections					
3/4" regulator	PF@.250	Ea	57.80	12.80	70.60
1" regulator	PF@.300	Ea	89.40	15.30	104.70
1-1/4" regulator	PF@.400	Ea	121.00	20.40	141.40
1-1/2" regulator	PF@.450	Ea	174.00	23.00	197.00
2" regulator	PF@.600	Ea	279.00	30.70	309.70
2-1/2" regulator	PF@.700	Ea	419.00	35.80	454.80
Combination pressure relief and reducing valves					
1/2", brass body	PF@.210	Ea	71.30	10.70	82.00
3/4", cast iron body	PF@.250	Ea	111.00	12.80	123.80

	Craft@Hrs	Unit	Material	Labor	Total
Wye pattern strainers, screwed connections, 250 PSI, cast iron body					
3/4" strainer	PF@.260	Ea	8.55	13.30	21.85
1" strainer	PF@.330	Ea	10.80	16.90	27.70
1-1/4" strainer	PF@.440	Ea	15.60	22.50	38.10
1-1/2" strainer	PF@.495	Ea	18.20	25.30	43.50
2" strainer	PF@.550	Ea	28.50	28.10	56.60
Wye pattern strainers, screwed connections, 250 PSI, cast bronze body					
1/2" strainer	PF@.210	Ea	19.20	10.70	29.90
3/4" strainer	PF@.250	Ea	26.00	12.80	38.80
1" strainer	PF@.300	Ea	33.10	15.30	48.40
1-1/4" strainer	PF@.400	Ea	46.80	20.40	67.20
1-1/2" strainer	PF@.450	Ea	60.70	23.00	83.70
2" strainer	PF@.500	Ea	104.00	25.60	129.60
Wye pattern strainers, screwed connections, 600 PSI, stainless steel body					
3/4" strainer	PF@.260	Ea	205.00	13.30	218.30
1" strainer	PF@.330	Ea	281.00	16.90	297.90
1-1/4" strainer	PF@.440	Ea	341.00	22.50	363.50
1-1/2" strainer	PF@.495	Ea	469.00	25.30	494.30
2" strainer	PF@.550	Ea	651.00	28.10	679.10
Wye pattern strainers, flanged connections, 125 PSI, cast iron body					
2" strainer	PF@.500	Ea	71.30	25.60	96.90
2-1/2" strainer	PF@.600	Ea	79.40	30.70	110.10
3" strainer	PF@.750	Ea	92.00	38.30	130.30
4" strainer	PF@1.35	Ea	158.00	69.00	227.00
6" strainer	PF@2.50	Ea	319.00	128.00	447.00
8" strainer	PF@3.00	Ea	541.00	153.00	694.00
Tempering valves, screwed connections, high temperature					
3/4", screwed	PF@.250	Ea	317.00	12.80	329.80
1", screwed	PF@.300	Ea	453.00	15.30	468.30
1-1/4", screwed	PF@.400	Ea	656.00	20.40	676.40
1-1/2", screwed	PF@.450	Ea	783.00	23.00	806.00
2", screwed	PF@.500	Ea	987.00	25.60	1,012.60
Thermostatic mixing valves, 110 to 150 degrees F					
1/2", soldered	PF@.240	Ea	40.40	12.30	52.70
3/4", threaded	PF@.250	Ea	52.10	12.80	64.90
3/4", soldered	PF@.300	Ea	44.40	15.30	59.70
1", threaded	PF@.300	Ea	203.00	15.30	218.30
Float and thermostatic traps, cast iron body, parallel connection					
3/4" (15 to 75 PSI)	PF@.250	Ea	108.00	12.80	120.80
1" (15 to 30 PSI)	PF@.300	Ea	128.00	15.30	143.30
3/4" – 1" (125 PSI)	PF@.250	Ea	166.00	12.80	178.80
3/4" – 1" (175 PSI)	PF@.300	Ea	364.00	15.30	379.30
1-1/4" (15 to 30 PSI)	PF@.450	Ea	181.00	23.00	204.00
1-1/4" – 1-1/2" (75 PSI)	PF@.450	Ea	305.00	23.00	328.00
1-1/4" – 1-1/2" (125 PSI)	PF@.500	Ea	305.00	25.60	330.60
1-1/4" – 1-1/2" (175 PSI)	PF@.500	Ea	441.00	25.60	466.60
1-1/2" (15 to 30 PSI)	PF@.450	Ea	270.00	23.00	293.00
2" (15 to 30 PSI)	PF@.450	Ea	922.00	23.00	945.00
2" (75 PSI)	PF@.550	Ea	921.00	28.10	949.10
2" (125 PSI)	PF@.600	Ea	921.00	30.70	951.70
2-1/2" (15 to 250 PSI)	PF@.800	Ea	2,370.00	40.90	2,410.90

Mechanical 15

	Craft@Hrs	Unit	Material	Labor	Total
Liquid level gauges					
3/4", aluminum	PF@.250	Ea	185.00	12.80	197.80
3/4", 125 PSI PVC	PF@.250	Ea	194.00	12.80	206.80
1/2", 175 PSI bronze	PF@.210	Ea	36.40	10.70	47.10
3/4", 150 PSI stainless steel	PF@.250	Ea	179.00	12.80	191.80
1", 150 PSI stainless steel	PF@.300	Ea	201.00	15.30	216.30

Electrical 16

	Craft@Hrs	Unit	Material	Labor	Total

Typical In-Place Costs Preliminary estimates per square foot of floor area for all electrical work, including service entrance, distribution and lighting fixtures, by building type. These costs include the subcontractor's overhead and profit.

	Craft@Hrs	Unit	Material	Labor	Total
Commercial stores	—	SF	—	—	7.86
Market buildings	—	SF	—	—	8.79
Recreation facilities	—	SF	—	—	9.47
Schools	—	SF	—	—	11.00
Colleges	—	SF	—	—	15.00
Complete electrical systems for buildings					
Hospitals	—	SF	—	—	25.40
Office buildings	—	SF	—	—	10.00
Warehouses	—	SF	—	—	4.57
Parking lots	—	SF	—	—	1.72
Garages	—	SF	—	—	7.44

Underground Power and Communication Duct Type BE LTMC8 (encased burial) PVC duct in 20' lengths. Add for excavation, shoring, dewatering, spacers, concrete envelope around duct, backfill, compaction, wire, pavement removal and re-paving as required. Direct burial (DB) power and communication duct will cost about 25% more.

	Craft@Hrs	Unit	Material	Labor	Total
Single duct run					
2-1/2" duct	E1@.053	LF	.40	2.37	2.77
3" duct	E1@.055	LF	.95	2.46	3.41
4" duct	E1@.070	LF	1.43	3.13	4.56

Concrete Envelope for Duct No wire, duct, excavation, pavement removal or replacement included. Based on ready-mix concrete. Includes allowance for waste

	Craft@Hrs	Unit	Material	Labor	Total
9" x 9" (.56 CF per linear foot)	E2@.020	LF	1.72	.87	2.59
12" x 12" (1.00 CF per linear foot)	E2@.036	LF	3.09	1.56	4.65
12" x 24" (2.00 CF per linear foot)	E2@.072	LF	6.17	3.13	9.30
Per CY	E2@1.21	CY	83.30	52.60	135.90

Precast Concrete Pull and Junction Box Reinforced concrete with appropriate inserts, cast iron frame, cover and pulling hooks. Overall dimensions as shown.

	Craft@Hrs	Unit	Material	Labor	Total
Precast handholes, 4' deep					
2' wide, 3' long	E1@4.24	Ea	837.00	189.00	1,026.00
3' wide, 3' long	E1@5.17	Ea	1,250.00	231.00	1,481.00
4' wide, 4' long	E1@13.2	Ea	1,750.00	590.00	2,340.00
Precast power manholes, 7' deep					
4' wide, 6' long	E1@24.6	Ea	2,580.00	1,100.00	3,680.00
6' wide, 8' long	E1@26.6	Ea	2,850.00	1,190.00	4,040.00
8' wide, 10' long	E1@26.6	Ea	3,010.00	1,190.00	4,200.00

	Craft@Hrs	Unit	Material	Labor	Total

Light and Power Circuits Costs per circuit including stranded THHN copper wire (except where noted), conduit, handy box, cover, straps, fasteners and connectors, but without fixture, receptacle or switch, except as noted.

Circuits for an exit light, fire alarm bell or suspended ceiling lighting system, 20 amp, with 4" junction box and 30' of 1/2" EMT conduit and copper THHN wire

3 #12 solid wire	E4@2.68	Ea	15.70	116.00	131.70
4 #12 solid wire	E4@2.90	Ea	17.20	126.00	143.20
5 #12 solid wire	E4@3.14	Ea	18.70	136.00	154.70

Under slab circuit, with 4" square junction box, 10' of 1/2" RSC and 20' of 1/2" PVC conduit

3 #12 wire, 20 amp	E4@1.75	Ea	23.80	76.00	99.80
4 #12 wire, 20 amp	E4@1.97	Ea	25.30	85.60	110.90
5 #12 wire, 20 amp	E4@2.21	Ea	26.80	96.00	122.80
3 #10 wire, 30 amp	E4@2.00	Ea	27.40	86.90	114.30
4 #10 wire, 30 amp	E4@2.31	Ea	30.10	100.00	130.10

Under slab circuit, with 4" square junction box, 10' of 1/2" RSC conduit, 20' of 1/2" PVC conduit and 6' of liquid tight flexible conduit

3 #12 wire, 20 amp	E4@2.41	Ea	29.000	105.00	134.00
4 #12 wire, 20 amp	E4@2.69	Ea	30.50	117.00	147.50
3 #10 wire, 30 amp	E4@2.75	Ea	32.60	119.00	151.60
4 #10 wire, 30 amp	E4@3.81	Ea	35.30	166.00	201.30

20 amp switch circuit, 277 volt, with 30' of 1/2" EMT conduit, switch, cover and 2 indenter connectors
One gang box

2 #12 wire, Single pole switch	E4@2.82	Ea	17.50	123.00	140.50
3 #12 wire, Three-way switch	E4@2.93	Ea	21.30	127.00	148.30
3 #12 wire, Four-way switch	E4@3.14	Ea	30.20	136.00	166.20

Two gang box

2 #12 wire, Single pole switch	E4@2.95	Ea	32.90	128.00	160.90
3 #12 wire, Three-way switch	E4@3.08	Ea	39.90	134.00	173.90
3 #12 wire, Four-way switch	E4@3.24	Ea	56.60	141.00	197.60

15 amp switch circuit, 125 volt, with 30' of #14 two conductor Romex non-metallic cable (NM) with ground, switch, cover and switch connection
One gang box

Single pole switch	E4@1.05	Ea	13.90	45.60	59.50
Three-way switch	E4@1.18	Ea	16.20	51.30	67.50

Two gang box

Single pole switch	E4@1.17	Ea	22.60	50.80	73.40
Single pole switch & receptacle	E4@1.36	Ea	28.90	59.10	88.00
Single pole & 3-way switch	E4@1.43	Ea	27.00	62.10	89.10
One two pole & 3-way switch	E4@1.66	Ea	30.30	72.10	102.40

20 amp 2 pole duplex receptacle outlet circuit, 125 volt, 30' of 1/2" EMT conduit, wire and specification grade receptacle, connectors and box

3 #12 wire	E4@2.80	Ea	21.30	122.00	143.30
4 #12 wire	E4@3.06	Ea	22.40	133.00	155.40
5 #12 wire	E4@3.26	Ea	24.30	142.00	166.30

Duplex receptacle outlet circuit with 30' of Romex non-metallic cable (NM) with ground, receptacle, cover and connection

15 amp, 125 volt, #14/2 wire	E4@1.11	Ea	21.50	48.20	69.70
15 amp, 125 volt, #14/2 wire with ground fault circuit interrupter (GFCI)	E4@1.30	Ea	41.80	56.50	98.30
20 amp, 125 volt, #12/2 wire	E4@1.36	Ea	23.90	59.10	83.00
30 amp, 250 volt, #10/3 wire, dryer circuit	E4@1.77	Ea	38.30	76.90	115.20
50 amp, 250 volt, #8/3 wire, range circuit	E4@2.06	Ea	41.00	89.50	130.50

	Craft@Hrs	Unit	Material	Labor	Total

Rigid Galvanized Steel Conduit, standard wall. Installed exposed in a building either vertically or horizontally up to 10' above floor level. No wire, fittings or supports included. Each 10' length comes with one coupling. Add the cost of fittings, boxes, supports and wire.

	Craft@Hrs	Unit	Material	Labor	Total
1/2" conduit	E4@.043	LF	1.49	1.87	3.36
3/4" conduit	E4@.054	LF	1.67	2.35	4.02
1" conduit	E4@.066	LF	2.63	2.87	5.50
1-1/4" conduit	E4@.077	LF	3.71	3.35	7.06
1-1/2" conduit	E4@.100	LF	4.23	4.35	8.58
2" conduit	E4@.116	LF	5.30	5.04	10.34
2-1/2" conduit	E4@.143	LF	4.15	6.21	10.36
3" conduit	E4@.164	LF	12.70	7.13	19.83
4" conduit	E4@.221	LF	16.70	9.60	26.30
Add for red plastic coating	—	%	80.0	—	—
Deduct for rigid steel conduit installed in a concrete slab or open trench	—	%	—	-40.0	—

90 or 45 degree ells, standard weight rigid galvanized steel, threaded both ends

	Craft@Hrs	Unit	Material	Labor	Total
1/2" elbow	E4@.170	Ea	2.42	7.39	9.81
3/4" elbow	E4@.190	Ea	2.53	8.26	10.79
1" elbow	E4@.210	Ea	2.88	9.12	12.00
1-1/4" elbow	E4@.311	Ea	3.75	13.50	17.25
1-1/2" elbow	E4@.378	Ea	4.32	16.40	20.72
2" elbow	E4@.448	Ea	7.21	19.50	26.71
2-1/2" elbow	E4@.756	Ea	13.00	32.80	45.80
3" elbow	E4@1.10	Ea	18.10	47.80	65.90
4" elbow	E4@1.73	Ea	29.50	75.20	104.70
Add for PVC coated ells	—	%	40.0	—	—

Rigid steel threaded couplings

	Craft@Hrs	Unit	Material	Labor	Total
1/2"	E4@.062	Ea	1.18	2.69	3.87
3/4"	E4@.080	Ea	1.36	3.48	4.84
1"	E4@.092	Ea	2.04	4.00	6.04
1-1/4"	E4@.102	Ea	2.32	4.43	6.75
1-1/2"	E4@.140	Ea	2.78	6.08	8.86
2"	E4@.170	Ea	3.12	7.39	10.51
2-1/2"	E4@.202	Ea	6.41	8.78	15.19
3"	E4@.230	Ea	10.10	9.99	20.09
4"	E4@.425	Ea	16.00	18.50	34.50

Three-piece "Erickson" unions, by pipe size

	Craft@Hrs	Unit	Material	Labor	Total
1"	E4@.394	Ea	4.76	17.10	21.86

Insulated grounding bushing

	Craft@Hrs	Unit	Material	Labor	Total
1/2"	E4@.062	Ea	1.78	2.69	4.47
3/4"	E4@.080	Ea	2.09	3.48	5.57
1"	E4@.092	Ea	2.87	4.00	6.87
1-1/4"	E4@.102	Ea	3.15	4.43	7.58
1-1/2"	E4@.140	Ea	3.73	6.08	9.81
2"	E4@.170	Ea	4.24	7.39	11.63
3"	E4@.230	Ea	8.64	9.99	18.63
4"	E4@.425	Ea	12.70	18.50	31.20

Insulated thermoplastic bushing

	Craft@Hrs	Unit	Material	Labor	Total
1/2"	E4@.020	Ea	.14	.87	1.01
3/4"	E4@.025	Ea	.15	1.09	1.24
1"	E4@.030	Ea	.30	1.30	1.60
1-1/4"	E4@.040	Ea	.41	1.74	2.15
1-1/2"	E4@.045	Ea	.53	1.96	2.49
2"	E4@.050	Ea	.67	2.17	2.84
2-1/2"	E4@.050	Ea	.74	2.17	2.91

	Craft@Hrs	Unit	Material	Labor	Total
Type A, C, LB, E, LL or LR threaded conduit body with cover and gasket					
1/2"	E4@.191	Ea	5.04	8.30	13.34
3/4"	E4@.238	Ea	5.85	10.30	16.15
1"	E4@.285	Ea	8.61	12.40	21.01
1-1/4"	E4@.358	Ea	12.80	15.60	28.40
1-1/2"	E4@.456	Ea	13.80	19.80	33.60
2"	E4@.598	Ea	22.40	26.00	48.40
2-1/2"	E4@1.43	Ea	52.50	62.10	114.60
3"	E4@1.77	Ea	58.30	76.90	135.20
4"	E4@2.80	Ea	120.00	122.00	242.00
Type T threaded conduit body, with cover and gasket					
1/2"	E4@.285	Ea	6.25	12.40	18.65
3/4"	E4@.358	Ea	7.59	15.60	23.19
1"	E4@.433	Ea	9.59	18.80	28.39
1-1/4"	E4@.534	Ea	13.20	23.20	36.40
1-1/2"	E4@.679	Ea	17.30	29.50	46.80
2"	E4@.889	Ea	21.10	38.60	59.70
2-1/2"	E4@2.15	Ea	56.60	93.40	150.00
3"	E4@2.72	Ea	81.70	118.00	199.70
4"	E4@4.22	Ea	151.00	183.00	334.00
Service entrance cap, threaded					
1/2"	E4@.319	Ea	3.11	13.90	17.01
3/4"	E4@.373	Ea	3.64	16.20	19.84
1"	E4@.376	Ea	3.68	16.30	19.98
1-1/4"	E4@.482	Ea	4.43	20.90	25.33
1-1/2"	E4@.536	Ea	6.77	23.30	30.07
2"	E4@.591	Ea	13.50	25.70	39.20
2-1/2"	E4@.671	Ea	24.30	29.20	53.50
3"	E4@.749	Ea	46.90	32.50	79.40
4"	E4@.912	Ea	63.60	39.60	103.20
Add for heights over 10' to 20'	—	%	—	20.0	—

Electric Metallic Tubing (EMT) Installed exposed in a building either vertically or horizontally up to 10' above floor level. No wire, fittings or supports included except as noted.

	Craft@Hrs	Unit	Material	Labor	Total
1/2" conduit	E4@.036	LF	.27	1.56	1.83
3/4" conduit	E4@.043	LF	.55	1.87	2.42
1" conduit	E4@.055	LF	1.00	2.39	3.39
1-1/4" conduit	E4@.067	LF	1.51	2.91	4.42
1-1/2" conduit	E4@.070	LF	1.84	3.04	4.88
2" conduit	E4@.086	LF	2.21	3.74	5.95
2-1/2" conduit	E4@.106	LF	3.30	4.61	7.91
3" conduit	E4@.127	LF	4.12	5.52	9.64
4" conduit	E4@.170	LF	4.26	7.39	11.65
EMT 90 degree elbow					
1/2" pull elbow	E4@.143	Ea	2.43	6.21	8.64
3/4" pull elbow	E4@.170	Ea	3.57	7.39	10.96
1" combination elbow	E4@.190	Ea	6.33	8.26	14.59
1-1/4" combination elbow	E4@.216	Ea	8.42	9.39	17.81
1-1/2" pull elbow	E4@.288	Ea	3.71	12.50	16.21
2" service entrance elbow	E4@.358	Ea	22.90	15.60	38.50
2-1/2" pull elbow	E4@.648	Ea	14.60	28.20	42.80
3" pull elbow	E4@.858	Ea	23.60	37.30	60.90
4" pull elbow	E4@1.33	Ea	37.90	57.80	95.70

	Craft@Hrs	Unit	Material	Labor	Total
EMT set screw connector					
1/2" EMT connector	E4@.085	Ea	.43	3.69	4.12
3/4" EMT connector	E4@.096	Ea	.55	4.17	4.72
1" EMT connector	E4@.124	Ea	.95	5.39	6.34
1-1/4" EMT connector	E4@.143	Ea	1.34	6.21	7.55
1-1/2" EMT connector	E4@.162	Ea	1.79	7.04	8.83
2" EMT connector	E4@.237	Ea	2.64	10.30	12.94
2-1/2" EMT connector	E4@.363	Ea	6.99	15.80	22.79
3" EMT connector	E4@.479	Ea	8.43	20.80	29.23
4" EMT connector	E4@.655	Ea	12.70	28.50	41.20
Add for insulated set-screw connector	—	%	25.0	—	—
EMT compression connector					
1/2" EMT connector	E4@.085	Ea	.53	3.69	4.22
3/4" EMT connector	E4@.096	Ea	.86	4.17	5.03
1" EMT connector	E4@.124	Ea	1.26	5.39	6.65
1-1/4" EMT connector	E4@.143	Ea	2.22	6.21	8.43
1-1/2" EMT connector	E4@.162	Ea	2.75	7.04	9.79
2" EMT connector	E4@.237	Ea	3.91	10.30	14.21
2-1/2" EMT connector	E4@.363	Ea	12.70	15.80	28.50
3" EMT connector	E4@.479	Ea	16.90	20.80	37.70
4" EMT connector	E4@.655	Ea	26.40	28.50	54.90
Add for insulated compression connector	—	%	25.0	—	—
EMT set screw coupling					
1/2" EMT coupling	E4@.085	Ea	.43	3.69	4.12
3/4" EMT coupling	E4@.096	Ea	.63	4.17	4.80
1" EMT coupling	E4@.124	Ea	1.05	5.39	6.44
1-1/4" EMT coupling	E4@.143	Ea	1.47	6.21	7.68
1-1/2" EMT coupling	E4@.162	Ea	2.18	7.04	9.22
2" EMT coupling	E4@.237	Ea	2.85	10.30	13.15
2-1/2" EMT coupling	E4@.363	Ea	5.50	15.80	21.30
3" EMT coupling	E4@.479	Ea	6.56	20.80	27.36
4" EMT coupling	E4@.655	Ea	10.60	28.50	39.10
EMT compression coupling					
1/2" EMT coupling	E4@.085	Ea	.67	3.69	4.36
3/4" EMT coupling	E4@.096	Ea	.89	4.17	5.06
1" EMT coupling	E4@.124	Ea	1.31	5.39	6.70
1-1/4" EMT coupling	E4@.143	Ea	2.43	6.21	8.64
1-1/2" EMT coupling	E4@.162	Ea	3.49	7.04	10.53
2" EMT coupling	E4@.237	Ea	4.19	10.30	14.49
2-1/2" EMT coupling	E4@.363	Ea	15.80	15.80	31.60
3" EMT coupling	E4@.479	Ea	20.10	20.80	40.90
4" EMT coupling	E4@.655	Ea	31.70	28.50	60.20
Add for heights over 10' to 20'	—	%	—	20.0	—

EMT Conduit with Wire THHN wire pulled in EMT conduit bent each 10', with one tap-on coupling each 10', one-set screw connector each 50' and a strap and lag screw each 6'. Installed exposed on walls or ceilings. Add 15% to the labor cost for heights over 10' to 20'. Use these figures for preliminary estimates.
1-1/2" EMT conduit and wire as shown below

	Craft@Hrs	Unit	Material	Labor	Total
2 #1 aluminum, 1 #8 copper	E4@.140	LF	3.35	6.08	9.43
3 #1 aluminum, 1 #8 copper	E4@.156	LF	3.93	6.78	10.71
4 #1 aluminum, 1 #8 copper	E4@.173	LF	4.51	7.52	12.03
4 #4 copper, 1 #8 copper	E4@.162	LF	3.68	7.04	10.72
2 #3 copper, 1 #8 copper	E4@.140	LF	3.12	6.08	9.20
3 #3 copper, 1 #8 copper	E4@.155	LF	3.58	6.73	10.31

	Craft@Hrs	Unit	Material	Labor	Total
4 #3 copper, 1 #8 copper	E4@.170	LF	4.04	7.39	11.43
2 #1 copper, 1 #8 copper	E4@.146	LF	3.68	6.34	10.02
3 #1 copper, 1 #8 copper	E4@.164	LF	4.42	7.13	11.55
4 #1 copper, 1 #8 copper	E4@.183	LF	5.16	7.95	13.11
2" EMT conduit and wire as shown below					
2 #2/0 aluminum, 1 #6 copper	E4@.170	LF	4.37	7.39	11.76
3 #2/0 aluminum, 1 #6 copper	E4@.190	LF	5.19	8.26	13.45
4 #2/0 aluminum, 1 #6 copper	E4@.210	LF	6.01	9.12	15.13
2 #3/0 aluminum, 1 #6 copper	E4@.174	LF	4.77	7.56	12.33
3 #3/0 aluminum, 1 #6 copper	E4@.195	LF	5.79	8.47	14.26
4 #3/0 aluminum, 1 #6 copper	E4@.218	LF	6.80	9.47	16.27
2 #1/0 copper, 1 #6 copper	E4@.171	LF	4.47	7.43	11.90
3 #1/0 copper, 1 #6 copper	E4@.193	LF	5.34	8.39	13.73
4 #1/0 copper, 1 #6 copper	E4@.213	LF	6.21	9.25	15.46
2 #3/0 copper, 1 #4 copper	E4@.175	LF	5.50	7.60	13.10
3 #3/0 copper, 1 #4 copper	E4@.197	LF	6.83	8.56	15.39
2-1/2" EMT conduit and wire as shown below					
2 250 MCM alum, 1 #4 copper	E4@.210	LF	7.13	9.12	16.25
3 250 MCM alum, 1 #4 copper	E4@.237	LF	8.51	10.30	18.81
4 #3 copper, 1 #4 copper	E4@.216	LF	6.21	9.39	15.60
2 #4 copper, 1 #2 copper	E4@.186	LF	5.32	8.08	13.40
3 #4 copper, 1 #2 copper	E4@.198	LF	5.69	8.60	14.29
4 #4 copper, 1 #2 copper	E4@.210	LF	6.06	9.12	15.18
3 250 MCM copper, 1 #2 copper	E4@.253	LF	10.60	11.00	21.60
3" EMT conduit and wire as shown below					
4 250 MCM alum, 1 #4 copper	E4@.288	LF	10.90	12.50	23.40
4 250 MCM copper, 1 #2 copper	E4@.290	LF	13.60	12.60	26.20
5 250 MCM copper, 1 #2 copper	E4@.319	LF	15.50	13.90	29.40

EMT Conduit Circuits Cost per circuit based on 30' run from the panel. Includes THHN copper wire pulled in conduit, 2 compression connectors, 2 couplings, conduit bending, straps, bolts and washers. Three-wire circuits are 120, 240 or 277 volt 1 phase, with 2 conductors and ground. Four-wire circuits are 120/208 or 480 volt 3 phase, with 3 conductors and ground. Five-wire circuits are 120/208 or 480 volt 3 phase, with 3 conductors, neutral and ground. Installed exposed in vertical or horizontal runs to 10' above floor level in a building. Add 15% to the manhours and labor costs for heights over 10' to 20'. Use these figures for preliminary estimates. Based on commercial quantity discounts.

	Craft@Hrs	Unit	Material	Labor	Total
15 amp circuits					
3 #14 wire, 1/2" conduit	E4@2.39	Ea	13.90	104.00	117.90
4 #14 wire, 1/2" conduit	E4@2.58	Ea	15.40	112.00	127.40
20 amp circuits					
3 #12 wire, 1/2" conduit	E4@2.49	Ea	14.80	108.00	122.80
4 #12 wire, 1/2" conduit	E4@2.72	Ea	16.60	118.00	134.60
5 #12 wire, 1/2" conduit	E4@2.95	Ea	18.40	128.00	146.40
30 amp circuits					
3 #10 wire, 1/2" conduit	E4@2.75	Ea	18.40	119.00	137.40
4 #10 wire, 1/2" conduit	E4@3.08	Ea	21.40	134.00	155.40
5 #10 wire, 3/4" conduit	E4@3.58	Ea	33.20	156.00	189.20
50 amp circuits					
3 #8 wire, 3/4" conduit	E4@2.98	Ea	34.40	129.00	163.40
4 #8 wire, 3/4" conduit	E4@3.32	Ea	39.80	144.00	183.80
5 #8 wire, 1" conduit	E4@4.02	Ea	59.90	175.00	234.90

	Craft@Hrs	Unit	Material	Labor	Total
65 amp circuits					
2 #6, 1 #8 wire, 1" conduit	E4@3.42	Ea	53.90	149.00	202.90
3 #6, 1 #8 wire, 1" conduit	E4@3.76	Ea	61.70	163.00	224.70
4 #6, 1 #8 wire, 1-1/4" conduit	E4@4.48	Ea	85.90	195.00	280.90
85 amp circuits in 1-1/4" conduit					
2 #4, 1 #8 wire	E4@3.89	Ea	76.90	169.00	245.90
3 #4, 1 #8 wire	E4@4.25	Ea	88.00	185.00	273.00
4 #4, 1 #8 wire	E4@4.48	Ea	99.10	195.00	294.10
100 amp circuits in 1-1/2" conduit					
2 #3, 1 #8 wire	E4@4.20	Ea	93.60	182.00	275.60
3 #3, 1 #8 wire	E4@4.66	Ea	107.00	202.00	309.00
4 #3, 1 #8 wire	E4@5.10	Ea	121.00	222.00	343.00
125 amp circuits in 1-1/2" conduit					
2 #1, 1 #8 wire	E4@4.40	Ea	110.00	191.00	301.00
3 #1, 1 #8 wire	E4@4.92	Ea	133.00	214.00	347.00
4 #1, 1 #8 wire	E4@5.47	Ea	155.00	238.00	393.00
150 amp circuits in 2" conduit					
2 #1/0, 1 #6 wire	E4@5.16	Ea	134.00	224.00	358.00
3 #1/0, 1 #6 wire	E4@5.78	Ea	160.00	251.00	411.00
4 #1/0, 1 #6 wire	E4@6.43	Ea	186.00	279.00	465.00
200 amp circuits in 2" conduit					
2 #3/0, 1 #4 wire	E4@5.49	Ea	165.00	239.00	404.00
3 #3/0, 1 #4 wire	E4@6.27	Ea	205.00	272.00	477.00
4 #3/0, 1 #4 wire	E4@7.80	Ea	245.00	339.00	584.00
225 amp circuits in 2-1/2" conduit					
2 #4/0, 1 #2 wire	E4@6.48	Ea	236.00	282.00	518.00
3 #4/0, 1 #2 wire	E4@7.33	Ea	286.00	318.00	604.00
4 #4/0, 1 #2 wire	E4@8.19	Ea	335.00	356.00	691.00
250 amp circuits in 2-1/2" conduit					
2 250 MCM, 1 #2 wire	E4@6.63	Ea	257.00	288.00	545.00
3 250 MCM, 1 #2 wire	E4@8.24	Ea	317.00	358.00	675.00
4 250 MCM, 1 #2 wire	E4@9.56	Ea	377.00	415.00	792.00

Intermediate Metal Conduit (IMC) Installed exposed in a building either vertically or horizontally up to 10' above floor level. No wire, fittings or supports included.

	Craft@Hrs	Unit	Material	Labor	Total
1/2" conduit	E4@.038	LF	1.11	1.65	2.76
3/4" conduit	E4@.048	LF	1.32	2.09	3.41
1" conduit	E4@.056	LF	2.01	2.43	4.44
1-1/4" conduit	E4@.073	LF	2.56	3.17	5.73
1-1/2" conduit	E4@.086	LF	3.02	3.74	6.76
2" conduit	E4@.101	LF	3.92	4.39	8.31
2-1/2" conduit	E4@.127	LF	10.80	5.52	16.32
3" conduit	E4@.146	LF	11.60	6.34	17.94

PVC Schedule 40 Conduit Installed in or under a building slab. No wire, fittings or supports included.

	Craft@Hrs	Unit	Material	Labor	Total
1/2"	E4@.030	LF	.13	1.30	1.43
3/4"	E4@.036	LF	.17	1.56	1.73
1"	E4@.046	LF	.26	2.00	2.26
1-1/4"	E4@.056	LF	.35	2.43	2.78
1-1/2"	E4@.061	LF	.42	2.65	3.07
2"	E4@.072	LF	.57	3.13	3.70
2-1/2"	E4@.089	LF	.87	3.87	4.74
3"	E4@.105	LF	1.14	4.56	5.70
4"	E4@.151	LF	.84	6.56	7.40
5"	E4@.176	LF	1.80	7.65	9.45
6"	E4@.208	LF	2.33	9.04	11.37

	Craft@Hrs	Unit	Material	Labor	Total
PVC Schedule 80 Conduit Installed in or under a building slab. No wire, fittings or supports included.					
1/2"	E4@.030	LF	.24	1.30	1.54
3/4"	E4@.036	LF	.32	1.56	1.88
1"	E4@.046	LF	.39	2.00	2.39
1-1/4"	E4@.056	LF	.56	2.43	2.99
2-1/2"	E4@.089	LF	1.32	3.87	5.19
3"	E4@.105	LF	1.54	4.56	6.10
Non-metallic conduit coupling					
1/2" coupling	E4@.020	Ea	.21	.87	1.08
3/4" coupling	E4@.051	Ea	.25	2.22	2.47
1" coupling	E4@.062	Ea	.36	2.69	3.05
1-1/4" coupling	E4@.078	Ea	.57	3.39	3.96
2" coupling	E4@.102	Ea	.86	4.43	5.29
2-1/2" coupling	E4@.113	Ea	1.35	4.91	6.26
3" coupling	E4@.129	Ea	1.75	5.61	7.36
4" coupling	E4@.156	Ea	3.15	6.78	9.93
2-1/2" long line coupling	E4@.113	Ea	3.17	4.91	8.08
4" long line coupling	E4@.156	Ea	5.29	6.78	12.07
90 degree Schedule 40 non-metallic elbow					
1/2" ell	E4@.082	Ea	.63	3.56	4.19
3/4" ell	E4@.102	Ea	.71	4.43	5.14
1" ell	E4@.124	Ea	1.07	5.39	6.46
1-1/4" ell	E4@.154	Ea	1.49	6.69	8.18
1-1/2" ell	E4@.195	Ea	1.97	8.47	10.44
2" ell	E4@.205	Ea	3.02	8.91	11.93
2-1/2" ell	E4@.226	Ea	4.84	9.82	14.66
3" ell	E4@.259	Ea	8.08	11.30	19.38
4" ell	E4@.313	Ea	14.80	13.60	28.40
1/2" access pull ell	E4@.082	Ea	2.54	3.56	6.10
3/4" access pull ell	E4@.102	Ea	2.76	4.43	7.19
1-1/4" ell, 24" radius	E4@.154	Ea	6.31	6.69	13.00
1-1/2" ell, 24" radius	E4@.195	Ea	7.58	8.47	16.05
2" ell, 24" radius	E4@.205	Ea	8.51	8.91	17.42
2-1/2" ell, 24" radius	E4@.226	Ea	10.60	9.82	20.42
3" ell, 24" radius	E4@.259	Ea	11.50	11.30	22.80
2" ell, 36" radius	E4@.205	Ea	12.60	8.91	21.51
2-1/2" ell, 36" radius	E4@.226	Ea	16.10	9.82	25.92
Deduct for 45 or 30 degree ells	—	%	-10.0	—	—
90 degree Schedule 80 non-metallic elbow.					
1" ell	E4@.124	Ea	2.14	5.39	7.53
1-1/4" ell	E4@.154	Ea	2.98	6.69	9.67
1-1/2" ell	E4@.195	Ea	4.05	8.47	12.52
2" ell	E4@.205	Ea	5.64	8.91	14.55
2-1/2" ell	E4@.226	Ea	11.10	9.82	20.92
1" ell, 24" radius	E4@.124	Ea	4.77	5.39	10.16
2" ell, 24" radius	E4@.205	Ea	9.87	8.91	18.78
2-1/2" ell, 24" radius	E4@.226	Ea	15.50	9.82	25.32
2" ell, 36" radius	E4@.205	Ea	21.10	8.91	30.01
PVC conduit end bell					
1" bell end	E4@.105	Ea	1.41	4.56	5.97
1-1/4" bell end	E4@.115	Ea	1.63	5.00	6.63
1-1/2" bell end	E4@.124	Ea	1.80	5.39	7.19
2" bell end	E4@.143	Ea	2.20	6.21	8.41
2-1/2" bell end	E4@.221	Ea	2.90	9.60	12.50
3" bell end	E4@.285	Ea	3.38	12.40	15.78

	Craft@Hrs	Unit	Material	Labor	Total
PVC bell end reducer					
3" to 2-1/2"	E4@.200	Ea	7.85	8.69	16.54
4" to 3"	E4@.250	Ea	10.50	10.90	21.40
Thermoplastic insulating bushing					
1" bushing	E4@.030	Ea	.28	1.30	1.58
1-1/4" bushing	E4@.040	Ea	.38	1.74	2.12
1-1/2" bushing	E4@.040	Ea	.39	1.74	2.13
2" bushing	E4@.050	Ea	.92	2.17	3.09
2-1/2" bushing	E4@.050	Ea	1.04	2.17	3.21
3" bushing	E4@.070	Ea	1.20	3.04	4.24
PVC reducer bushing, Schedule 40, socket ends					
3/4" x 1/2" bushing	E4@.030	Ea	.76	1.30	2.06
1" x 1/2" bushing	E4@.030	Ea	1.44	1.30	2.74
1" x 3/4" bushing	E4@.040	Ea	1.64	1.74	3.38
1-1/4" x 1" bushing	E4@.050	Ea	1.87	2.17	4.04
1-1/2" x 1" bushing	E4@.050	Ea	1.75	2.17	3.92
1-1/2" x 1-1/4" bushing	E4@.050	Ea	2.39	2.17	4.56
2" x 1-1/4" bushing	E4@.050	Ea	2.51	2.17	4.68
2" x 1-1/2" bushing	E4@.070	Ea	2.65	3.04	5.69
3" x 2" bushing	E4@.100	Ea	5.91	4.35	10.26
PVC conduit box adapter					
1/2" adapter	E4@.050	Ea	.43	2.17	2.60
3/4" adapter	E4@.060	Ea	.48	2.61	3.09
1" adapter	E4@.080	Ea	.52	3.48	4.00
1-1/4" adapter	E4@.100	Ea	.67	4.35	5.02
1-1/2" adapter	E4@.100	Ea	.75	4.35	5.10
2" adapter	E4@.150	Ea	.95	6.52	7.47
PVC female conduit adapter, for joining conduit to threaded fittings					
1/2" adapter	E4@.050	Ea	.43	2.17	2.60
3/4" adapter	E4@.060	Ea	.48	2.61	3.09
1" adapter	E4@.080	Ea	.52	3.48	4.00
1-1/4" adapter	E4@.100	Ea	.67	4.35	5.02
1-1/2" adapter	E4@.100	Ea	.75	4.35	5.10
2" adapter	E4@.150	Ea	.95	6.52	7.47
2-1/2" adapter	E4@.178	Ea	1.31	7.73	9.04
4" adapter	E4@.244	Ea	3.29	10.60	13.89
PVC male conduit adapter, for joining conduit to threaded fittings					
1/2" adapter	E4@.050	Ea	.23	2.17	2.40
3/4" adapter	E4@.060	Ea	.38	2.61	2.99
1" adapter	E4@.080	Ea	.54	3.48	4.02
1-1/4" adapter	E4@.100	Ea	.68	4.35	5.03
1-1/2" adapter	E4@.100	Ea	.71	4.35	5.06
2" adapter	E4@.150	Ea	.99	6.52	7.51
2-1/2" adapter	E4@.178	Ea	1.48	7.73	9.21
3" adapter	E4@.244	Ea	2.34	10.60	12.94
Type T PVC conduit body (access fitting)					
1/2" conduit body	E4@.150	Ea	2.49	6.52	9.01
3/4" conduit body	E4@.184	Ea	2.87	7.99	10.86
1" conduit body	E4@.225	Ea	3.40	9.78	13.18
1-1/4" conduit body	E4@.300	Ea	5.28	13.00	18.28
1-1/2" conduit body	E4@.350	Ea	6.15	15.20	21.35
2" conduit body	E4@.400	Ea	8.87	17.40	26.27

	Craft@Hrs	Unit	Material	Labor	Total
Type C PVC conduit body (access fitting)					
1/2" conduit body	E4@.100	Ea	2.16	4.35	6.51
3/4" conduit body	E4@.132	Ea	3.11	5.74	8.85
1" conduit body	E4@.164	Ea	4.75	7.13	11.88
1-1/4" conduit body	E4@.197	Ea	5.27	8.56	13.83
1-1/2" conduit body	E4@.234	Ea	6.95	10.20	17.15
2" conduit body	E4@.263	Ea	11.60	11.40	23.00
Type LB PVC conduit body (access fitting)					
1/2" conduit body	E4@.100	Ea	2.16	4.35	6.51
3/4" conduit body	E4@.132	Ea	2.54	5.74	8.28
1" conduit body	E4@.164	Ea	3.08	7.13	10.21
1-1/4" conduit body	E4@.197	Ea	4.46	8.56	13.02
1-1/2" conduit body	E4@.234	Ea	5.66	10.20	15.86
2" conduit body	E4@.263	Ea	8.05	11.40	19.45
3" conduit body	E4@.300	Ea	25.00	13.00	38.00
4" conduit body	E4@.350	Ea	37.00	15.20	52.20
Type LL PVC conduit body (access fitting)					
1/2" conduit body	E4@.100	Ea	2.07	4.35	6.42
3/4" conduit body	E4@.132	Ea	2.63	5.74	8.37
1" conduit body	E4@.164	Ea	2.62	7.13	9.75
1-1/4" conduit body	E4@.197	Ea	4.38	8.56	12.94
1-1/2" conduit body	E4@.234	Ea	4.70	10.20	14.90
2" conduit body	E4@.263	Ea	9.68	11.40	21.08
Type LR PVC conduit body (access fitting)					
1/2" conduit body	E4@.100	Ea	1.97	4.35	6.32
3/4" conduit body	E4@.132	Ea	2.70	5.74	8.44
1" conduit body	E4@.164	Ea	2.73	7.13	9.86
1-1/4" conduit body	E4@.197	Ea	4.41	8.56	12.97
1-1/2" conduit body	E4@.234	Ea	5.52	10.20	15.72
2" conduit body	E4@.263	Ea	10.00	11.40	21.40
Two-hole PVC clamp					
1/2" clamp	E4@.100	Ea	.16	4.35	4.51
3/4" clamp	E4@.100	Ea	.21	4.35	4.56
1" clamp	E4@.125	Ea	.25	5.43	5.68
1-1/4" clamp	E4@.150	Ea	.34	6.52	6.86
1-1/2" clamp	E4@.150	Ea	.36	6.52	6.88
2" clamp	E4@.175	Ea	.40	7.60	8.00
3" clamp	E4@.175	Ea	.60	7.60	8.20

PVC Conduit with Wire Includes THHN wire pulled in Type 40 rigid PVC heavy wall conduit placed in a slab. Use these figures for preliminary estimates. One rigid coupling every 10'. Add for fittings as required. Based on commercial quantity discounts.

	Craft@Hrs	Unit	Material	Labor	Total
1/2" PVC conduit with wire					
3 #12 solid copper	E4@.054	LF	.28	2.35	2.63
4 #12 solid copper	E4@.061	LF	.33	2.65	2.98
5 #12 solid copper	E4@.069	LF	.38	3.00	3.38
3 #10 solid copper	E4@.062	LF	.40	2.69	3.09
4 #10 solid copper	E4@.073	LF	.49	3.17	3.66
3/4" PVC conduit with wire					
5 #10 solid copper	E4@.089	LF	.62	3.87	4.49
3 #8 copper	E4@.069	LF	.71	3.00	3.71
4 #8 copper	E4@.080	LF	.89	3.48	4.37

	Craft@Hrs	Unit	Material	Labor	Total
1" PVC conduit with wire					
2 #6 copper, 1 #8 copper	E4@.081	LF	.96	3.52	4.48
3 #6 copper, 1 #8 copper	E4@.092	LF	1.22	4.00	5.22
5 #8 copper	E4@.100	LF	1.16	4.35	5.51
1-1/2" PVC conduit with wire					
4 #6 copper, 1 #8 copper	E4@.117	LF	1.64	5.08	6.72
2 #4 copper, 1 #8 copper	E4@.097	LF	1.34	4.21	5.55
2 #1 alum, 1 #8 copper	E4@.102	LF	1.76	4.43	6.19
3 #1 alum, 1 #8 copper	E4@.119	LF	2.33	5.17	7.50
4 #4 copper, 1 #8 copper	E4@.124	LF	2.08	5.39	7.47
3 #3 copper, 1 #8 copper	E4@.116	LF	1.98	5.04	7.02
4 #3 copper, 1 #8 copper	E4@.132	LF	2.44	5.74	8.18
2 #1 copper, 1 #8 copper	E4@.108	LF	2.08	4.69	6.77
3 #1 copper, 1 #8 copper	E4@.125	LF	2.82	5.43	8.25
4 #1 copper, 1 #8 copper	E4@.143	LF	3.56	6.21	9.77
2" PVC conduit with wire					
4 #1 alum, 1 #8 copper	E4@.154	LF	3.06	6.69	9.75
2 2/0 alum, 1 #6 copper	E4@.123	LF	2.47	5.34	7.81
3 2/0 alum, 1 #6 copper	E4@.143	LF	3.29	6.21	9.50
4 2/0 alum, 1 #6 copper	E4@.163	LF	4.11	7.08	11.19
2 3/0 alum, 1 #6 copper	E4@.127	LF	2.87	5.52	8.39
3 3/0 alum, 1 #6 copper	E4@.148	LF	3.88	6.43	10.31
2" PVC conduit with wire					
4 3/0 alum, 1 #6 copper	E4@.171	LF	4.90	7.43	12.33
2 1/0 & 1 #6 copper	E4@.124	LF	2.22	5.39	7.61
3 1/0 & 1 #6 copper	E4@.146	LF	2.91	6.34	9.25
4 1/0 & 1 #6 copper	E4@.167	LF	3.60	7.26	10.86
2 3/0 & 1 #4 copper	E4@.136	LF	2.98	5.91	8.89
3 3/0 & 1 #4 copper	E4@.162	LF	3.99	7.04	11.03
2-1/2" PVC conduit with wire					
2 250 MCM alum, 1 #4 copper	E4@.156	LF	4.00	6.78	10.78
3 250 MCM alum, 1 #4 copper	E4@.183	LF	5.38	7.95	13.33
4 3/0 alum, 1 #4 copper	E4@.206	LF	4.29	8.95	13.24
2 4/0 alum, 1 #2 copper	E4@.164	LF	3.72	7.13	10.85
3 4/0 alum, 1 #2 copper	E4@.193	LF	4.85	8.39	13.24
4 4/0 alum, 1 #2 copper	E4@.221	LF	5.98	9.60	15.58
2 250 MCM copper, 1 #2 copper	E4@.170	LF	5.45	7.39	12.84
3" PVC conduit with wire					
4 250 MCM alum, 1 #4 copper	E4@.226	LF	7.03	9.82	16.85
2 350 MCM alum, 1 #2 copper	E4@.186	LF	5.60	8.08	13.68
3 350 MCM alum, 1 #2 copper	E4@.217	LF	7.54	9.43	16.97
3 250 MCM alum, 1 #2 copper	E4@.217	LF	5.86	9.43	15.29
4 250 MCM alum, 1 #2 copper	E4@.248	LF	7.24	10.80	18.04
3-1/2" PVC conduit with wire					
4 350 MCM alum, 1 #2 copper	E4@.264	LF	9.63	11.50	21.13
4" PVC conduit with wire					
6 350 MCM alum, 1 #2 copper	E4@.293	LF	13.10	12.70	25.80
7 350 MCM alum, 1 #2 copper	E4@.391	LF	15.00	17.00	32.00
3 500 MCM alum, 1 1/0 copper	E4@.311	LF	9.21	13.50	22.71
4 500 MCM alum, 1 1/0 copper	E4@.355	LF	11.70	15.40	27.10

	Craft@Hrs	Unit	Material	Labor	Total

Flexible Aluminum Conduit Type RWA installed in a building to 10' above floor level. No wire, fittings or supports included.

	Craft@Hrs	Unit	Material	Labor	Total
3/8" conduit, 25' coil at $9.20	E4@.025	LF	.37	1.09	1.46
1/2" conduit, cut length	E4@.028	LF	.56	1.22	1.78
1/2" conduit, 50' coil at $19.80	E4@.028	LF	.40	1.22	1.62
1/2" conduit, 500' reel at $101.00	E4@.028	LF	.20	1.22	1.42
3/4" conduit, cut length	E4@.030	LF	.87	1.30	2.17
3/4" conduit, 25' coil at $14.10	E4@.030	LF	.56	1.30	1.86
1" conduit, cut length	E4@.033	LF	.92	1.43	2.35
1" conduit, 50' coil at $43.70	E4@.033	LF	.87	1.43	2.30
1-1/4" conduit, cut length	E4@.035	LF	1.25	1.52	2.77
1-1/4" conduit, 50' coil at $82.90	E4@.035	LF	1.66	1.52	3.18
1-1/2" conduit, 25' coil at $82.90	E4@.038	LF	3.32	1.65	4.97
2" conduit, cut length	E4@.040	LF	3.25	1.74	4.99
2" conduit, 25' coil at $74.80	E4@.040	LF	3.00	1.74	4.74

Flexible Steel Conduit installed in a building to 10' above floor level. No wire, fittings or supports included.

	Craft@Hrs	Unit	Material	Labor	Total
3/8" conduit, cut length	E4@.025	LF	.45	1.09	1.54
3/8" conduit, 25' coil at $8.34	E4@.025	LF	.34	1.09	1.43
3/8" conduit, 100' coil at $26.60	E4@.025	LF	.27	1.09	1.36
3/8" conduit, 250' coil at $67.20	E4@.025	LF	.27	1.09	1.36
1/2" conduit, cut length	E4@.028	LF	.57	1.22	1.79
1/2" conduit, 25' coil at $9.25	E4@.028	LF	.37	1.22	1.59
1/2" conduit, 100' coil at $29.20	E4@.028	LF	.30	1.22	1.52
3/4" conduit, cut length	E4@.030	LF	.63	1.30	1.93
3/4" conduit, 25' coil at $11.90	E4@.030	LF	.38	1.30	1.68
3/4" conduit, 100' coil at $42.80	E4@.030	LF	.48	1.30	1.78
3/4" conduit, 500' reel at $213.00	E4@.030	LF	.43	1.30	1.73
1" conduit, cut length	E4@.033	LF	1.09	1.43	2.52
1" conduit, 50' coil at $41.90	E4@.033	LF	.84	1.43	2.27
1-1/4" conduit, cut length	E4@.035	LF	1.39	1.52	2.91
1-1/4" conduit, 50' coil at $61.90	E4@.035	LF	1.24	1.52	2.76
1-1/2" conduit, cut length	E4@.038	LF	2.34	1.65	3.99
1-1/2" conduit, 25' coil at $53.80	E4@.038	LF	2.15	1.65	3.80
2" conduit, cut length	E4@.040	LF	3.15	1.74	4.89
2" conduit, 25' coil at $64.70	E4@.040	LF	2.59	1.74	4.33
90 degree flex screw-in connector					
3/8" connector	E4@.070	Ea	.71	3.04	3.75
1/2" connector	E4@.086	Ea	.71	3.74	4.45
3/4" connector	E4@.100	Ea	1.62	4.35	5.97
1" connector	E4@.113	Ea	4.76	4.91	9.67
1-1/4" connector	E4@.148	Ea	4.70	6.43	11.13
1-1/2" connector	E4@.191	Ea	9.61	8.30	17.91
2" connector	E4@.260	Ea	10.50	11.30	21.80
Flex screw-in coupling					
1/2" coupling	E4@.086	Ea	.45	3.74	4.19
3/4" coupling	E4@.100	Ea	1.05	4.35	5.40
1" coupling	E4@.113	Ea	1.28	4.91	6.19
1-1/4" coupling	E4@.148	Ea	2.54	6.43	8.97
1-1/2" coupling	E4@.191	Ea	3.19	8.30	11.49
2" coupling	E4@.260	Ea	5.85	11.30	17.15
Add for height over 10' to 20'	—	%	—	20.0	—

	Craft@Hrs	Unit	Material	Labor	Total

Flexible Non-metallic Conduit, (ENT) blue conduit installed in a building to 10' above floor level. NER-290 adopted by BOCA, ICBO and SBCC codes. No wire, fittings or supports included.

	Craft@Hrs	Unit	Material	Labor	Total
1/2" conduit, cut length	E4@.021	LF	.29	.91	1.20
1/2" conduit, 10' coil at $2.31	E4@.021	LF	.24	.91	1.15
1/2" conduit, 200' coil at $39.40	E4@.021	LF	.20	.91	1.11
3/4" conduit, cut length	E4@.022	LF	.45	.96	1.41
3/4" conduit, 10' coil at $3.32	E4@.022	LF	.33	.96	1.29
3/4" conduit, 100' coil at $30.90	E4@.022	LF	.31	.96	1.27
1" conduit, 10' coil at $2.90	E4@.030	LF	.29	1.30	1.59

Liquid-Tight Flexible Metal Conduit installed exposed in a building either vertically or horizontally up to 10' above floor level. No wire, fittings or supports included.

	Craft@Hrs	Unit	Material	Labor	Total
1/2" conduit, cut length	E4@.045	LF	1.02	1.96	2.98
3/4" conduit, 25' coil at $26.10	E4@.047	LF	1.00	2.04	3.04
1" conduit, cut length	E4@.050	LF	2.06	2.17	4.23
1-1/4" conduit, cut length	E4@.055	LF	2.51	2.39	4.90
1-1/2" conduit, cut length	E4@.065	LF	3.15	2.82	5.97
2" conduit, cut length	E4@.070	LF	3.56	3.04	6.60
Liquid-tight straight flex connector					
1/2" connector	E4@.129	Ea	2.10	5.61	7.71
3/4" connector	E4@.140	Ea	3.03	6.08	9.11
1" connector	E4@.221	Ea	4.53	9.60	14.13
1-1/4" connector	E4@.244	Ea	7.00	10.60	17.60
1-1/2" connector	E4@.293	Ea	10.30	12.70	23.00
2" connector	E4@.492	Ea	17.30	21.40	38.70
90 degree liquid-tight flex connector					
1/2" connector	E4@.129	Ea	3.24	5.61	8.85
3/4" connector	E4@.140	Ea	4.91	6.08	10.99
1" connector	E4@.221	Ea	8.33	9.60	17.93
1-1/4" connector	E4@.244	Ea	11.90	10.60	22.50
1-1/2" connector	E4@.293	Ea	14.00	12.70	26.70
2" connector	E4@.492	Ea	21.60	21.40	43.00

Non-metallic Liquid-Tight Flexible Conduit installed exposed in a building either vertically or horizontally up to 10' above floor level. No wire, fittings or supports included.

	Craft@Hrs	Unit	Material	Labor	Total
1/2" conduit, cut length	E4@.051	LF	.61	2.22	2.83
1/2" conduit, 25' coil at $11.40	E4@.051	LF	.46	2.22	2.68
1/2" conduit, 100' coil at $42.00	E4@.051	LF	.42	2.22	2.64
1/2" conduit, 500' reel at $172.00	E4@.051	LF	.34	2.22	2.56
3/4" conduit, cut length	E4@.067	LF	.87	2.91	3.78
3/4" conduit, 100' coil at $52.40	E4@.067	LF	.53	2.91	3.44
1" conduit, cut length	E4@.067	LF	1.05	2.91	3.96
1" conduit, 50' coil at $39.60	E4@.067	LF	.80	2.91	3.71
1" conduit, 100' coil at $79.80	E4@.067	LF	.80	2.91	3.71
Non-metallic liquid-tight flex connector					
1/2" zero to 90 degree connector	E4@.060	Ea	2.11	2.61	4.72
3/4" zero to 90 degree connector	E4@.070	Ea	3.32	3.04	6.36
1" zero to 90 degree connector	E4@.080	Ea	2.64	3.48	6.12
1-1/4" zero to 90 degree connector	E4@.100	Ea	5.27	4.35	9.62
1-1/2" zero to 90 degree connector	E4@.100	Ea	6.99	4.35	11.34
2" zero to 90 degree connector	E4@.140	Ea	11.90	6.08	17.98
1/2" 90 degree connector	E4@.060	Ea	2.11	2.61	4.72
3/4" 90 degree connector	E4@.070	Ea	3.53	3.04	6.57

	Craft@Hrs	Unit	Material	Labor	Total
1" 90 degree connector	E4@.080	Ea	4.24	3.48	7.72
1-1/4" 90 degree connector	E4@.100	Ea	9.40	4.35	13.75
1-1/2" 90 degree connector	E4@.100	Ea	11.90	4.35	16.25
2" 90 degree connector	E4@.140	Ea	17.20	6.08	23.28
Add for height over 10' to 20'	—	%	—	20.0	—

Conduit Hangers and Supports By conduit size. Installed exposed in a building to 10' above floor level. Hangers and supports installed at heights over 10' will add about 20% to the labor cost.

Right angle conduit supports

	Craft@Hrs	Unit	Material	Labor	Total
3/8" support	E4@.030	Ea	.35	1.30	1.65
1/2" support or 3/4" support	E4@.040	Ea	.35	1.74	2.09
1" support	E4@.050	Ea	.53	2.17	2.70
1-1/4" support	E4@.060	Ea	.59	2.61	3.20
1-1/2" support	E4@.060	Ea	.64	2.61	3.25
2" support	E4@.100	Ea	.85	4.35	5.20
2-1/2" support	E4@.100	Ea	1.01	4.35	5.36
3" support	E4@.150	Ea	1.26	6.52	7.78
3-1/2" support	E4@.150	Ea	1.68	6.52	8.20
4" support	E4@.150	Ea	3.50	6.52	10.02
Add for parallel conduit supports	—	%	50.0	—	—
Add for edge conduit supports	—	%	120.0	—	—

Conduit hangers, rigid with bolt, labor includes the cost of cutting threaded rod to length and attaching the rod. Add the material cost of the threaded rod from below.

	Craft@Hrs	Unit	Material	Labor	Total
1/2" hanger	E4@.097	Ea	.60	4.21	4.81
3/4" hanger	E4@.097	Ea	.62	4.21	4.83
1" hanger	E4@.108	Ea	1.02	4.69	5.71
1-1/4" hanger	E4@.108	Ea	1.43	4.69	6.12
1-1/2" hanger	E4@.108	Ea	1.73	4.69	6.42
2" hanger	E4@.119	Ea	1.88	5.17	7.05
2-1/2" hanger	E4@.119	Ea	2.43	5.17	7.60
3" hanger	E4@.119	Ea	2.90	5.17	8.07
3-1/2" hanger	E4@.129	Ea	4.14	5.61	9.75
4" hanger	E4@.129	Ea	9.74	5.61	15.35

All threaded rod, plated steel, per linear foot of rod

	Craft@Hrs	Unit	Material	Labor	Total
1/4", 20 thread	—	LF	.34	—	.34
5/16", 18 thread	—	LF	.42	—	.42
3/8", 16 thread	—	LF	.58	—	.58
1/2", 13 thread	—	LF	1.02	—	1.02
5/8", 11 thread	—	LF	1.65	—	1.65
Deduct for plain steel rod	—	%	-10.0	—	—

Rod beam clamps, for 1/4" or 3/8" rod

	Craft@Hrs	Unit	Material	Labor	Total
1" flange	E4@.234	Ea	1.56	10.20	11.76
1-1/2" flange	E4@.234	Ea	4.08	10.20	14.28
2" flange	E4@.234	Ea	5.74	10.20	15.94
2-1/2" flange	E4@.234	Ea	7.93	10.20	18.13

Conduit clamps for EMT conduit

	Craft@Hrs	Unit	Material	Labor	Total
1/2" or 3/4" clamp	E4@.062	Ea	.19	2.69	2.88
1" clamp	E4@.069	Ea	.26	3.00	3.26
1-1/4" clamp	E4@.069	Ea	.39	3.00	3.39
1-1/2" clamp	E4@.069	Ea	.41	3.00	3.41
2" clamp	E4@.075	Ea	.46	3.26	3.72

	Craft@Hrs	Unit	Material	Labor	Total
One hole heavy duty stamped steel conduit straps					
1/2" strap	E4@.039	Ea	.14	1.69	1.83
3/4" strap	E4@.039	Ea	.17	1.69	1.86
1" strap	E4@.039	Ea	.30	1.69	1.99
1-1/4" strap	E4@.039	Ea	.46	1.69	2.15
1-1/2" strap	E4@.039	Ea	.60	1.69	2.29
2" strap	E4@.039	Ea	.78	1.69	2.47
2-1/2" strap	E4@.053	Ea	2.71	2.30	5.01
3" strap	E4@.053	Ea	3.78	2.30	6.08
3-1/2" strap	E4@.059	Ea	5.67	2.56	8.23
4" strap	E4@.059	Ea	5.87	2.56	8.43
Add for malleable iron one hole straps	—	%	30.0	—	—
Deduct for nail drive straps	—	%	-20.0	-30.0	—
Deduct for two hole stamped steel EMT straps	—	%	-70.0	—	—
Hanger channel, 1-5/8" x 1-5/8", based on 12" length					
No holes, solid back,					
12 gauge steel	E4@.119	Ea	3.29	5.17	8.46
Holes 1-1/2" on center,					
12 gauge aluminum	E4@.119	Ea	3.78	5.17	8.95
Channel strap for rigid steel conduit or EMT conduit, by conduit size, with bolts					
1/2" strap	E4@.020	Ea	.95	.87	1.82
3/4" strap	E4@.020	Ea	1.06	.87	1.93
1" strap	E4@.023	Ea	1.14	1.00	2.14
1-1/4" strap	E4@.023	Ea	1.32	1.00	2.32
1-1/2" strap	E4@.023	Ea	1.50	1.00	2.50
2" strap	E4@.023	Ea	1.70	1.00	2.70
2-1/2" strap	E4@.039	Ea	1.85	1.69	3.54
3" strap	E4@.039	Ea	2.04	1.69	3.73
3-1/2" strap	E4@.039	Ea	2.49	1.69	4.18
4" strap	E4@.062	Ea	2.77	2.69	5.46
5" strap	E4@.062	Ea	3.42	2.69	6.11
6" strap	E4@.062	Ea	4.58	2.69	7.27
Lag screws, flattened end with stove bolt					
1/4" x 3"	E4@.092	Ea	.75	4.00	4.75
1/4" x 4"	E4@.092	Ea	.80	4.00	4.80
5/16" x 3-1/2"	E4@.092	Ea	1.11	4.00	5.11
3/8" x 4"	E4@.119	Ea	1.02	5.17	6.19
1/2" x 5"	E4@.129	Ea	1.16	5.61	6.77
Lag screw short expansion shields, without screws. Labor is for drilling hole					
1/4" or 5/16" shield	E4@.175	Ea	.20	7.60	7.80
3/8" shield	E4@.272	Ea	.34	11.80	12.14
1/2" shield	E4@.321	Ea	.54	13.90	14.44
5/8" shield	E4@.321	Ea	.91	13.90	14.81
3/4" shield	E4@.345	Ea	1.27	15.00	16.27
Self drilling masonry anchors					
1/4" anchor	E4@.127	Ea	.60	5.52	6.12
5/16" anchor	E4@.127	Ea	.80	5.52	6.32
3/8" anchor	E4@.191	Ea	.93	8.30	9.23
1/2" anchor	E4@.191	Ea	1.45	8.30	9.75
5/8" anchor	E4@.259	Ea	2.42	11.30	13.72
3/4" anchor	E4@.259	Ea	4.01	11.30	15.31

	Craft@Hrs	Unit	Material	Labor	Total
Wiremold Raceway Steel, one-piece, installed on a finished wall in a building.					
Raceway including one coupling each 10 feet					
V500 2 wire surface mounted raceway	E4@.059	LF	.84	2.56	3.40
V700 2 wire surface mounted raceway	E4@.067	LF	1.08	2.91	3.99
Wiremold fittings					
V700 90 degree flat elbow	E4@.050	Ea	4.21	2.17	6.38
V700 90 degree twist elbow	E4@.050	Ea	7.25	2.17	9.42
V700 90 degree inside elbow	E4@.050	Ea	4.21	2.17	6.38
V700 90 degree outside elbow	E4@.050	Ea	4.19	2.17	6.36
V700 90 degree internal pull elbow	E4@.050	Ea	11.40	2.17	13.57
V700 90 degree internal twist elbow	E4@.150	Ea	3.09	6.52	9.61
V500 45 degree flat elbow	E4@.150	Ea	4.17	6.52	10.69
V500 90 degree flat elbow	E4@.150	Ea	1.74	6.52	8.26
V500/700 adjustable offset connector	E4@.150	Ea	8.35	6.52	14.87
V500 conduit connector	E4@.150	Ea	3.01	6.52	9.53
V700 connection cover	E4@.200	Ea	.42	8.69	9.11
V700 3-way branch T-fitting	E4@.100	Ea	5.30	4.35	9.65
V700 to 1/2" conduit transition fitting	E4@.050	Ea	5.19	2.17	7.36
V500/700 flexible section, 18"	E4@.350	Ea	16.30	15.20	31.50
V500/700 galvanized box connector	E4@.200	Ea	4.19	8.69	12.88
V500/700 coupling	E4@.200	Ea	.28	8.69	8.97
V500/700 support clip	—@.000	Ea	.40	—	.40
Wiremold device box, V500/700 surface raceway					
1 gang, 1-3/8" diameter, 1 KO	E4@.250	Ea	6.29	10.90	17.19
1 gang, 2-1/4" diameter, 1 KO	E4@.250	Ea	13.60	10.90	24.50
1 gang, 4-5/8" x 2-7/8"	E4@.200	Ea	5.79	8.69	14.48
2 gang, 4-3/4" x 4-3/4"	E4@.300	Ea	10.50	13.00	23.50
2 gang, 15/16" diameter, 1 KO	E4@.300	Ea	4.77	13.00	17.77
2 gang, 1-3/4" diameter, 2 KO	E4@.300	Ea	11.10	13.00	24.10
2 gang, 2-1/4" diameter, 1 KO	E4@.300	Ea	15.70	13.00	28.70
Corner box, 2-1/2" diameter	E4@.250	Ea	12.60	10.90	23.50
Duplex receptacle box, 4-1/8" x 2" x 1-3/8"	E4@.360	Ea	11.70	15.60	27.30
Fixture box, 4-3/4" diameter	E4@.250	Ea	9.41	10.90	20.31
Fixture box, 6-3/8" diameter	E4@.250	Ea	11.00	10.90	21.90
Fixture box, 5-1/2" diameter	E4@.250	Ea	9.45	10.90	20.35
Junction and pull box, 1" x 5-1/2" diameter	E4@.250	Ea	8.38	10.90	19.28
Extension box, 4-3/4" diameter	E4@.250	Ea	9.42	10.90	20.32
Wiremold outlet box, V700 surface raceway					
Regular outlet box	E4@.250	Ea	6.04	10.90	16.94
Deep outlet box	E4@.250	Ea	7.30	10.90	18.20
Extra deep outlet box	E4@.250	Ea	6.43	10.90	17.33
Fan box	E4@.250	Ea	10.60	10.90	21.50
Starter box	E4@.250	Ea	6.17	10.90	17.07
Utility box, 2" x 4-1/8" x 1-3/8"	E4@.360	Ea	7.31	15.60	22.91
Plugmould® V2000 multi-outlet raceway					
Entrance end fitting	E4@.100	Ea	5.21	4.35	9.56
6 outlets, 36" long	E4@.450	Ea	31.60	19.60	51.20
6 outlets, 40" long	E4@.450	Ea	36.90	19.60	56.50
8 outlets, 52" long	E4@.550	Ea	47.00	23.90	70.90
10 outlets, 60" long	E4@.650	Ea	36.80	28.20	65.00

	Craft@Hrs	Unit	Material	Labor	Total
Cable Tray and Duct					
Cable tray, steel, ladder type					
4" wide	E4@.144	LF	15.30	6.26	21.56
6" wide	E4@.166	LF	29.50	7.21	36.71
8" wide	E4@.187	LF	46.30	8.13	54.43
10" wide	E4@.211	LF	85.30	9.17	94.47
12" wide	E4@.238	LF	65.10	10.30	75.40
Steel underfloor duct, including typical supports, fittings and accessories					
3-1/4" wide, 1 cell	E4@.097	LF	6.79	4.21	11.00
3-1/4" wide, 2 cell	E4@.104	LF	13.80	4.52	18.32
7-1/4" wide, 1 cell	E4@.148	LF	16.00	6.43	22.43
7-1/4" wide, 2 cell	E4@.148	LF	24.40	6.43	30.83

Wire and Cable

Copper wire and cable prices can change very quickly, sometimes as much as 10% in a week. The prices that follow were representative at the time of publication. Your material supplier will be able to quote more current wire and cable costs.

Service Entrance Cable 600 volt stranded service entrance wire pulled in conduit except as noted. No excavation or conduit included. By American Wire Gauge.

SEU service entrance cable. Type XHHW wire wound in a concentrically applied neutral and jacketed with gray sunlight-resistant PVC.

	Craft@Hrs	Unit	Material	Labor	Total
8 gauge, 3 conductor, copper	E4@.012	LF	.55	.52	1.07
6 gauge, 3 conductor, copper	E4@.012	LF	.98	.52	1.50
6, 6, and 8 gauge conductors, copper	E4@.012	LF	.99	.52	1.51
6 gauge, 3 conductor, aluminum	E4@.012	LF	.78	.52	1.30
4, 4, 4, and 6 gauge conductor, copper	E4@.014	LF	2.27	.61	2.88
4-4 gauge, 4 conductor, copper	E4@.014	LF	1.76	.61	2.37
4-4 gauge, 4 conductor, aluminum	E4@.014	LF	.83	.61	1.44
3-3-3 gauge, 5 conductor, copper	E4@.015	LF	3.25	.65	3.90
2-2-2 gauge, 4 conductor, copper	E4@.016	LF	3.53	.70	4.23
2-2-2 gauge, 4 conductor, aluminum	E4@.016	LF	1.43	.70	2.13
2-2 gauge, 2 conductor, copper	E4@.016	LF	2.63	.70	3.33
1-1-1 gauge, 3 conductor, aluminum	E4@.018	LF	1.65	.78	2.43
4/0, 4/0, 4/0, 2/0 conductor, aluminum	E4@.026	LF	3.15	1.13	4.28

USE underground service entrance wire. Stranded conductor insulated with abrasion, heat, moisture, and sunlight-resistant cross-linked polyethylene (XLP).

	Craft@Hrs	Unit	Material	Labor	Total
8 gauge, copper	E4@.009	LF	.36	.39	.75
6 gauge, copper	E4@.010	LF	.47	.43	.90
2 gauge, copper	E4@.016	LF	.77	.70	1.47
1/0 gauge, copper	E4@.017	LF	1.47	.74	2.21
2/0 gauge, copper	E4@.018	LF	1.79	.78	2.57
4/0 gauge, copper	E4@.018	LF	2.11	.78	2.89
250 MCM, copper	E4@.020	LF	1.89	.87	2.76
500 MCM, copper	E4@.022	LF	3.69	.96	4.65
Deduct for aluminum USE service entrance wire	—	%	-40.0	—	—

URD triplex direct-burial aluminum service entrance cable

	Craft@Hrs	Unit	Material	Labor	Total
6 gauge, Erskine	E4@.011	LF	.52	.48	1.00
4 gauge, Vassar	E4@.012	LF	.63	.52	1.15
2 gauge, Ramapo	E4@.013	LF	.74	.56	1.30
2/0 gauge, Converse	E4@.017	LF	1.04	.74	1.78

	Craft@Hrs	Unit	Material	Labor	Total

Bare Copper Groundwire Pulled in conduit with conductor wires. No excavation or duct included. By American Wire Gauge. Cost per linear foot based on typical discounts for the quantities listed.

	Craft@Hrs	Unit	Material	Labor	Total
12 gauge, solid, cut length	E4@.004	LF	.11	.17	.28
12 gauge, solid, 1250' reel at $73.30	E4@.004	LF	.06	.17	.23
10 gauge, solid, cut length	E4@.005	LF	.15	.22	.37
10 gauge, solid, 800' reel at $73.60	E4@.005	LF	.09	.22	.31
8 gauge, cut length	E4@.005	LF	.23	.22	.45
8 gauge, solid, 500' coil at $73.10	E4@.005	LF	.15	.22	.37
6 gauge, solid, cut length	E4@.005	LF	.36	.22	.58
6 gauge, solid, 315' coil at $71.10	E4@.005	LF	.23	.22	.45
4 gauge, solid, cut length	E4@.006	LF	.63	.26	.89
4 gauge, solid, 198' coil at $67.80	E4@.006	LF	.34	.26	.60
4 gauge, stranded, cut length	E4@.006	LF	.63	.26	.89
4 gauge, stranded, 198' coil at $73.10	E4@.006	LF	.37	.26	.63
2 gauge, cut length	E4@.008	LF	.69	.35	1.04
2 gauge, stranded, cut length	E4@.008	LF	.76	.35	1.11
2 gauge, solid, 125' reel at $68.90	E4@.008	LF	.55	.35	.90
2 gauge, stranded, 125' reel at $71.20	E4@.008	LF	.57	.35	.92

THHN/THWN-2 Copper Building Wire Pulled in conduit for feeder and branch circuits. No conduit included. THHN has a heat-resistant thermoplastic cover approved for working temperatures up to 90 degrees Celsius in dry locations. THWN resists moisture, oil and gasoline and is approved for temperatures to 75 degrees Celsius in dry, wet or oil locations. THHN/THWN is made from soft annealed copper and insulated with heat- and moisture-resistant PVC and covered with nylon (polyamide). Available in colors to simplify identification of conductors after wire is pulled in conduit. By American Wire Gauge (AWG). Stranded wire has 19 strands. Cost per linear foot based on typical discounts for the quantities listed. Copper wire prices are highly volatile.

	Craft@Hrs	Unit	Material	Labor	Total
14 gauge, solid, cut length	E4@.006	LF	.10	.26	.36
14 gauge, stranded, cut length	E4@.006	LF	.12	.26	.38
14 gauge, solid, 50' coil at $7.44	E4@.006	LF	.15	.26	.41
14 gauge, solid, 100' coil at $10.60	E4@.006	LF	.11	.26	.37
14 gauge, solid, 500' coil at $16.80	E4@.006	LF	.04	.26	.30
14 gauge, stranded, 500' coil at $20.60	E4@.006	LF	.04	.26	.30
14 gauge, stranded, 2500' reel at $116	E4@.006	LF	.05	.26	.31
12 gauge, solid, cut length	E4@.007	LF	.15	.30	.45
12 gauge, stranded, cut length	E4@.007	LF	.17	.30	.47
12 gauge, solid, 50' coil at $9.00	E4@.007	LF	.18	.30	.48
12 gauge, stranded, 50' coil at $11.20	E4@.007	LF	.22	.30	.52
12 gauge, solid, 100' coil at $15.90	E4@.007	LF	.16	.30	.46
12 gauge, stranded, 100' coil at $21.20	E4@.007	LF	.21	.30	.51
12 gauge, solid, 500' coil at $26.00	E4@.007	LF	.05	.30	.35
12 gauge, stranded, 500' coil at $31.90	E4@.007	LF	.06	.30	.36
12 gauge, solid, 2500' reel at $148.00	E4@.007	LF	.06	.30	.36
12 gauge, stranded, 2500' reel at $172.00	E4@.007	LF	.07	.30	.37
10 gauge, solid, cut length	E4@.008	LF	.17	.35	.52
10 gauge, stranded, cut length	E4@.008	LF	.22	.35	.57
10 gauge, stranded, 50' coil at $16.90	E4@.008	LF	.34	.35	.69
10 gauge, stranded, 100' coil at $29.70	E4@.008	LF	.30	.35	.65
10 gauge, solid, 500' coil at $46.60	E4@.008	LF	.09	.35	.44
10 gauge, stranded, 500' coil at $51.60	E4@.008	LF	.10	.35	.45
10 gauge, stranded, 2500' reel at $255	E4@.008	LF	.10	.35	.45
8 gauge, solid, cut length	E4@.011	LF	.29	.48	.77
8 gauge, stranded, cut length	E4@.009	LF	.27	.39	.66
8 gauge, stranded, 500' coil at $89.40	E4@.009	LF	.18	.39	.57
8 gauge, stranded, 1000' coil at $180	E4@.009	LF	.18	.39	.57

	Craft@Hrs	Unit	Material	Labor	Total
6 gauge, stranded, cut length	E4@.010	LF	.46	.43	.89
6 gauge, stranded, 500' coil at $130	E4@.010	LF	.26	.43	.69
6 gauge, stranded, 1000' reel at $261	E4@.010	LF	.26	.43	.69
4 gauge, stranded, cut length	E4@.012	LF	.52	.52	1.04
4 gauge, stranded, 500' reel at $186	E4@.012	LF	.37	.52	.89
4 gauge, stranded, 1000' reel at $371	E4@.012	LF	.37	.52	.89
4 gauge, stranded, 2500' reel at $928	E4@.012	LF	.37	.52	.89
3 gauge, stranded, cut length	E4@.013	LF	.58	.56	1.14
3 gauge, stranded, 500' reel at $230	E4@.013	LF	.46	.56	1.02
2 gauge, stranded, cut length	E4@.013	LF	.63	.56	1.19
2 gauge, stranded, 500' reel at $291	E4@.013	LF	.58	.56	1.14
2 gauge, stranded, 1000' reel at $582	E4@.013	LF	.58	.56	1.14
2 gauge, stranded, 2500' reel at $1,450	E4@.013	LF	.58	.56	1.14
1 gauge, stranded, cut length	E4@.014	LF	.83	.61	1.44
1 gauge, stranded, 500' reel at $371	E4@.014	LF	.74	.61	1.35
1/0 gauge, stranded, cut length	E4@.015	LF	1.03	.65	1.68
1/0 gauge, stranded, 500' reel at $437	E4@.015	LF	.87	.65	1.52
1/0 gauge, stranded, 1000' reel at $875	E4@.015	LF	.87	.65	1.52
2/0 gauge, stranded, cut length	E4@.016	LF	1.22	.70	1.92
2/0 gauge, stranded, 500' reel at $540	E4@.016	LF	1.08	.70	1.78
2/0 gauge, stranded, 1000' reel at $1,080	E4@.016	LF	1.08	.70	1.78
3/0 gauge, stranded, cut length	E4@.017	LF	1.54	.74	2.28
3/0 gauge, stranded, 500' reel at $667	E4@.017	LF	1.33	.74	2.07
3/0 gauge, stranded, 1000' reel at $1,340	E4@.017	LF	1.34	.74	2.08
4/0 gauge, stranded, cut length	E4@.018	LF	1.77	.78	2.55
4/0 gauge, stranded, 500' reel at $827	E4@.018	LF	1.65	.78	2.43
4/0 gauge, stranded, 1000' reel at $1,650	E4@.018	LF	1.65	.78	2.43
250 MCM, stranded, cut length	E4@.020	LF	2.18	.87	3.05
250 MCM, stranded, 500' reel at $1,050	E4@.020	LF	2.11	.87	2.98
250 MCM, stranded, 1000' reel at $2,000	E4@.020	LF	2.00	.87	2.87
350 MCM, stranded, cut length	E4@.038	LF	3.02	1.65	4.67
350 MCM, stranded, 1000' reel at $2,650	E4@.038	LF	2.65	1.65	4.30
500 MCM, stranded, cut length	E4@.046	LF	4.24	2.00	6.24
500 MCM, stranded, 1000' reel $3,870	E4@.046	LF	3.87	2.00	5.87

XHHW Copper Building Wire Pulled in conduit for feeder and branch circuits. No conduit included. Abrasion-, moisture- and heat-resistant thermoplastic intended for harsh environments. Maximum operating temperature is 90 degrees Celsius in dry locations and 75 degrees Celsius in wet locations. Flame-retardant cross-linked synthetic polymer insulation. XHHW has no outer jacket or covering. Available in colors to simplify identification of conductors after wire is pulled in conduit. By American Wire Gauge (AWG). Stranded wire has 19 strands. Cost per linear foot based on typical discounts for the quantities listed. Copper wire prices are highly volatile.

	Craft@Hrs	Unit	Material	Labor	Total
14 gauge stranded, cut length	E4@.006	LF	.15	.26	.41
14 gauge stranded, 500' coil at $46.60	E4@.006	LF	.09	.26	.35
12 gauge solid, cut length	E4@.007	LF	.16	.30	.46
12 gauge stranded, cut length	E4@.007	LF	.19	.30	.49
12 gauge, solid, 500' coil at $48.80	E4@.007	LF	.10	.30	.40
12 gauge, stranded, 500' coil at $60.40	E4@.007	LF	.12	.30	.42
12 gauge, solid, 2500' reel at $244	E4@.007	LF	.10	.30	.40
10 gauge, solid, cut length	E4@.008	LF	.17	.35	.52
10 gauge, solid, 500' coil at $58.30	E4@.008	LF	.12	.35	.47
4 gauge, stranded, cut length	E4@.012	LF	.40	.52	.92
4 gauge, stranded, 500' reel at $190	E4@.012	LF	.38	.52	.90

	Craft@Hrs	Unit	Material	Labor	Total
2 gauge, stranded, cut length	E4@.014	LF	.61	.61	1.22
2 gauge, stranded, 500' reel at $290	E4@.014	LF	.58	.61	1.19
1/0 gauge, stranded, cut length	E4@.015	LF	1.05	.65	1.70
1/0 gauge, stranded, 500' reel at $444	E4@.015	LF	.89	.65	1.54
2/0 gauge, stranded, cut length	E4@.016	LF	1.37	.70	2.07
2/0 gauge, stranded, 500' reel at $541	E4@.016	LF	1.08	.70	1.78
3/0 gauge, stranded, cut length	E4@.017	LF	1.65	.74	2.39
4/0 gauge, stranded, cut length	E4@.018	LF	2.11	.78	2.89
4/0 gauge, stranded, 500' reel at $858	E4@.018	LF	1.72	.78	2.50

Type UF-B Direct Burial Sheathed Cable installed in an open trench. No excavation or backfill included. UF-B is designed for burial in a trench but may also be used to wire interior branch circuits in residential and agricultural buildings. The cable jacket is sunlight-, fungus- and moisture-resistant gray PVC. Full-size ground wire. Copper wire prices are highly volatile.

	Craft@Hrs	Unit	Material	Labor	Total
14 gauge, 2 conductor, cut length	E4@.005	LF	.36	.22	.58
14 gauge, 2 conductor, 25' coil at $8.65	E4@.005	LF	.35	.22	.57
14 gauge, 2 conductor, 50' coil at $15.50	E4@.005	LF	.31	.22	.53
14 gauge, 2 conductor, 100' coil at $27.50	E4@.005	LF	.28	.22	.50
14 gauge, 2 conductor, 250' coil at $39.30	E4@.005	LF	.16	.22	.38
14 gauge, 2 conductor, 1000' coil at $158	E4@.005	LF	.16	.22	.38
14 gauge, 3 conductor, cut length	E4@.005	LF	.46	.22	.68
14 gauge, 3 conductor, 250' coil at $60.80	E4@.005	LF	.24	.22	.46
14 gauge, 3 conductor, 1000' reel at $234	E4@.005	LF	.23	.22	.45
12 gauge, 2 conductor, cut length	E4@.005	LF	.44	.22	.66
12 gauge, 2 conductor, 25' coil at $10.60	E4@.005	LF	.43	.22	.65
12 gauge, 2 conductor, 50' coil at $18.80	E4@.005	LF	.38	.22	.60
12 gauge, 2 conductor, 100' coil at $34.30	E4@.005	LF	.35	.22	.57
12 gauge, 2 conductor, 250' coil at $51.80	E4@.005	LF	.21	.22	.43
12 gauge, 2 conductor, 1000' reel at $208	E4@.005	LF	.21	.22	.43
12 gauge, 3 conductor, cut length	E4@.005	LF	.54	.22	.76
12 gauge, 3 conductor, 100' coil at $49.40	E4@.005	LF	.50	.22	.72
12 gauge, 3 conductor, 250' coil at $80.90	E4@.005	LF	.33	.22	.55
12 gauge, 3 conductor, 1000' coil at $310	E4@.005	LF	.31	.22	.53
10 gauge, 2 conductor, cut length	E4@.005	LF	.80	.22	1.02
10 gauge, 2 conductor, 100' coil at $57.00	E4@.005	LF	.57	.22	.79
10 gauge, 2 conductor, 250' coil at $103.00	E4@.005	LF	.41	.22	.63
10 gauge, 2 conductor, 1000' reel at $406	E4@.005	LF	.41	.22	.63
10 gauge, 3 conductor, cut length	E4@.006	LF	.96	.26	1.22
10 gauge, 3 conductor, 100' coil at $72.90	E4@.006	LF	.73	.26	.99
10 gauge, 3 conductor, 250' coil at $129	E4@.006	LF	.52	.26	.78
10 gauge, 3 conductor, 1000' reel at $490	E4@.006	LF	.49	.26	.75
8 gauge, 2 conductor, cut length	E4@.007	LF	1.42	.30	1.72
8 gauge, 2 conductor, 500' reel at $420	E4@.007	LF	.84	.30	1.14
8 gauge, 3 conductor, cut length	E4@.007	LF	1.83	.30	2.13
8 gauge, 3 conductor, 125' reel at $175	E4@.007	LF	1.40	.30	1.70
8 gauge, 3 conductor, 500' reel at $581	E4@.007	LF	1.16	.30	1.46
6 gauge, 2 conductor, cut length	E4@.008	LF	1.75	.35	2.10
6 gauge, 2 conductor, 500' reel at $572	E4@.008	LF	1.14	.35	1.49
6 gauge, 3 conductor, 125' reel at $223	E4@.008	LF	1.78	.35	2.13
6 gauge, 3 conductor, 500' reel at $777	E4@.008	LF	1.56	.35	1.91

	Craft@Hrs	Unit	Material	Labor	Total

Type NM-B (Romex) Sheathed Cable installed in wood frame walls for outlets and switched fixtures, including boring out and pulling cable. NM-B cable can also be run through voids in masonry block or tile walls if little moisture is expected. Conductor insulation is color-coded PVC. 14 AWG (American Wire Gauge) is white, 12 AWG is yellow, 10 AWG is orange, and 8 and 6 AWG are black. Rated at 15 amps. With full-size ground wire. Cost per linear foot based on typical discounts for the quantities listed.

	Craft@Hrs	Unit	Material	Labor	Total
14 gauge, 2 conductor, cut length	E4@.027	LF	.30	1.17	1.47
14 gauge, 2 conductor, 25' coil at $6.60	E4@.027	LF	.27	1.17	1.44
14 gauge, 2 conductor, 50' coil at $12.40	E4@.027	LF	.25	1.17	1.42
14 gauge, 2 conductor, 100' coil at $20.60	E4@.027	LF	.21	1.17	1.38
14 gauge, 2 conductor, 1000' reel at $114	E4@.027	LF	.12	1.17	1.29
14 gauge, 3 conductor, cut length	E4@.030	LF	.46	1.30	1.76
14 gauge, 3 conductor, 25' coil at $10.90	E4@.030	LF	.44	1.30	1.74
14 gauge, 3 conductor, 50' coil at $19.60	E4@.030	LF	.39	1.30	1.69
14 gauge, 3 conductor, 100' coil at $36.10	E4@.030	LF	.36	1.30	1.66
14 gauge, 3 conductor, 1000' coil at $181	E4@.030	LF	.18	1.30	1.48
12 gauge, 2 conductor, cut length	E4@.030	LF	.37	1.30	1.67
12 gauge, 2 conductor, 25' coil at $9.00	E4@.030	LF	.36	1.30	1.66
12 gauge, 2 conductor, 50' coil at $17.30	E4@.030	LF	.35	1.30	1.65
12 gauge, 2 conductor, 100' coil at $29.90	E4@.030	LF	.30	1.30	1.60
12 gauge, 3 conductor, cut length	E4@.031	LF	.61	1.35	1.96
12 gauge, 3 conductor, 25' coil at $14.50	E4@.031	LF	.58	1.35	1.93
12 gauge, 3 conductor, 50' coil at $26.50	E4@.031	LF	.53	1.35	1.88
12 gauge, 3 conductor, 100' coil at $46.70	E4@.031	LF	.47	1.35	1.82
12 gauge, 3 conductor, 250' coil at $66.00	E4@.031	LF	.27	1.35	1.62
12 gauge, 3 conductor, 1000' coil at $264	E4@.031	LF	.27	1.35	1.62
10 gauge, 2 conductor, cut length	E4@.031	LF	.68	1.35	2.03
10 gauge, 2 conductor, 25' coil at $17.10	E4@.031	LF	.69	1.35	2.04
10 gauge, 2 conductor, 50' coil at $29.80	E4@.031	LF	.60	1.35	1.95
10 gauge, 2 conductor, 100' coil at $54.90	E4@.031	LF	.55	1.35	1.90
10 gauge, 2 conductor, 250' coil at $81.30	E4@.031	LF	.33	1.35	1.68
10 gauge, 2 conductor, 1000' coil at $325	E4@.031	LF	.33	1.35	1.68
10 gauge, 3 conductor, cut length	E4@.035	LF	.85	1.52	2.37
10 gauge, 3 conductor, 25' coil at $21.60	E4@.035	LF	.87	1.52	2.39
10 gauge, 3 conductor, 50' coil at $39.30	E4@.035	LF	.80	1.52	2.32
10 gauge, 3 conductor, 100' coil at $71.00	E4@.035	LF	.71	1.52	2.23
10 gauge, 3 conductor, 1000' reel at $433	E4@.035	LF	.44	1.52	1.96
8 gauge, 2 conductor, cut length	E4@.040	LF	.88	1.74	2.62
8 gauge, 2 conductor, 125' coil at $81.10	E4@.040	LF	.65	1.74	2.39
8 gauge, 2 conductor, 500' reel at $271	E4@.040	LF	.55	1.74	2.29
8 gauge, 3 conductor, 125' coil at $109	E4@.043	LF	.87	1.87	2.74
8 gauge, 3 conductor, 500' reel at $483	E4@.043	LF	.97	1.87	2.84
6 gauge, 2 conductor, cut length	E4@.043	LF	1.30	1.87	3.17
6 gauge, 2 conductor, 125' coil at $109	E4@.043	LF	.87	1.87	2.74
6 gauge, 2 conductor, 500' reel at $392	E4@.043	LF	.79	1.87	2.66
6 gauge, 3 conductor, cut length	E4@.047	LF	1.64	2.04	3.68
6 gauge, 3 conductor, 125' reel at $149	E4@.047	LF	1.19	2.04	3.23
6 gauge, 3 conductor, 500' reel at $578	E4@.047	LF	1.16	2.04	3.20
4 gauge, 3 conductor, cut length	E4@.051	LF	2.38	2.22	4.60
2 gauge, 3 conductor, cut length	E4@.056	LF	3.15	2.43	5.58
2 gauge, 3 conductor, 500' reel at $1,330	E4@.056	LF	2.66	2.43	5.09

	Craft@Hrs	Unit	Material	Labor	Total
Deduct for no ground wire					
#14 and #12 cable	—	%	-5.0	-5.0	—
#10, #8 and #6 wire	—	%	-15.0	-5.0	—

MC Aluminum-Clad Cable For use in dry sites. Exposed, concealed, or in approved raceway. Open or messenger-supported aerial runs. For powering lighting, control and signal circuits. Includes neutral, ground and one or two phase conductors. Installed with clamps or staples in a frame building or embedded in masonry, including approved bushings at terminations. Note that wire prices can change very quickly.

	Craft@Hrs	Unit	Material	Labor	Total
14 gauge, 2 conductor, cut length	E4@.009	LF	.55	.39	.94
14 gauge, 2 conductor, 25' coil at $12.50	E4@.009	LF	.50	.39	.89
14 gauge, 2 conductor, 50' coil at $19.80	E4@.009	LF	.40	.39	.79
14 gauge, 2 conductor, 100' coil at $33.30	E4@.009	LF	.34	.39	.73
14 gauge, 2 conductor, 250' coil at $63.40	E4@.009	LF	.25	.39	.64
14 gauge, 2 conductor, 500' reel at $134	E4@.009	LF	.27	.39	.66
14 gauge, 3 conductor, cut length	E4@.010	LF	.81	.43	1.24
14 gauge, 3 conductor, 25' coil at $14.70	E4@.010	LF	.59	.43	1.02
14 gauge, 3 conductor, 100' coil at $51.00	E4@.010	LF	.51	.43	.94
14 gauge, 3 conductor, 250' coil at $97.80	E4@.010	LF	.39	.43	.82
14 gauge, 3 conductor, 500' reel at $202	E4@.010	LF	.40	.43	.83
12 gauge, 2 conductor, cut length	E4@.010	LF	.66	.43	1.09
12 gauge, 2 conductor, 25' coil at $13.60	E4@.010	LF	.54	.43	.97
12 gauge, 2 conductor, 50' coil at $23.20	E4@.010	LF	.47	.43	.90
12 gauge, 2 conductor, 100' coil at $39.40	E4@.010	LF	.40	.43	.83
12 gauge, 2 conductor, 250' coil at $68.30	E4@.010	LF	.28	.43	.71
12 gauge, 2 conductor, 500' reel at $138	E4@.010	LF	.28	.43	.71
12 gauge, 2 conductor, 1000' reel at $285	E4@.010	LF	.29	.43	.72
12 gauge, 3 conductor, cut length	E4@.010	LF	.91	.43	1.34
12 gauge, 3 conductor, 25' coil at $17.30	E4@.010	LF	.69	.43	1.12
12 gauge, 3 conductor, 100' coil at $57.80	E4@.010	LF	.58	.43	1.01
12 gauge, 3 conductor, 250' coil at $112	E4@.010	LF	.45	.43	.88
12 gauge, 3 conductor, 500' reel at $237	E4@.010	LF	.47	.43	.90
10 gauge, 2 conductor, cut length	E4@.011	LF	.89	.48	1.37
10 gauge, 2 conductor, 250' reel at $137	E4@.011	LF	.59	.48	1.07
10 gauge, 3 conductor, cut length	E4@.011	LF	1.42	.48	1.90
10 gauge, 3 conductor 250' reel at $222	E4@.011	LF	.98	.48	1.46

Flexible Armored AC Cable, glavanized steel, 16 gauge integral bond wire to armor grounding path. Paper wrap conductor insulation covering. By American Wire Gauge and number of conductors. Installed exposed in a building either vertically or horizontally up to 10' above floor level. Sometimes referred to as "BX" cable.

	Craft@Hrs	Unit	Material	Labor	Total
#14 wire, 2 conductor	E4@.030	LF	.37	1.30	1.67
#14 wire, 3 conductor	E4@.030	LF	.67	1.30	1.97
#12 wire, 2 conductor	E4@.030	LF	.63	1.30	1.93
#12 wire, 3 conductor	E4@.035	LF	.98	1.52	2.50
#12 wire, 4 conductor	E4@.038	LF	.71	1.65	2.36
#10 wire, 2 conductor	E4@.031	LF	1.04	1.35	2.39
#10 wire, 3 conductor	E4@.039	LF	1.30	1.69	2.99
# 6 wire, 3 conductor	E4@.047	LF	1.70	2.04	3.74
# 4 wire, 3 conductor	E4@.051	LF	2.87	2.22	5.09
# 2 wire, 3 conductor	E4@.056	LF	3.01	2.43	5.44

	Craft@Hrs	Unit	Material	Labor	Total

Power Cable Single conductor medium voltage ozone resistant stranded copper cable pulled in conduit. No conduit included. Labor cost assumes three bundled conductors are pulled at one time on runs up to 100 feet. Labor cost will be higher on longer cable pulls and about 50% lower when cable is laid in an open trench. No splicing included. Listed by American Wire Gauge. Note that wire prices can change very quickly. Check with supplier.

5,000 volt, tape shielded, crosslinked polyethylene (XLP) insulated, with PVC jacket

	Craft@Hrs	Unit	Material	Labor	Total
#6 gauge	E4@.014	LF	1.99	.61	2.60
#4 gauge	E4@.016	LF	2.30	.70	3.00
#2 gauge	E4@.021	LF	2.81	.91	3.72
#1/0 gauge	E4@.026	LF	3.59	1.13	4.72
#2/0 gauge	E4@.028	LF	4.38	1.22	5.60
#4/0 gauge	E4@.038	LF	5.86	1.65	7.51
350 MCM	E4@.047	LF	8.76	2.04	10.80
500 MCM	E4@.053	LF	11.20	2.30	13.50

15,000 volt, tape shielded, ethylene propylene rubber (EPR) insulated, with PVC jacket

	Craft@Hrs	Unit	Material	Labor	Total
#2 gauge	E4@.024	LF	3.71	1.04	4.75
#1/0	E4@.030	LF	4.61	1.30	5.91
#2/0	E4@.032	LF	5.20	1.39	6.59
350 MCM	E4@.055	LF	9.72	2.39	12.11
500 MCM	E4@.062	LF	12.00	2.69	14.69

Snow Melting Cable For concrete pavement. Self-regulating heater cable 208-277 volt AC, encased in concrete pavement for walkways, steps, loading ramps or parking garages. Based on Raychem ElectroMelt™ System for concrete pavement. Cable is cut to length at the site and laid in a serpentine pattern using 12-inch center-to-center spacing. Use nylon cable ties to fasten cable to the top of reinforcing before concrete is placed. For scheduling purposes estimate that a crew of one can lay out, cut to length, install and tie 600 LF of cable in an 8-hour day. Based on 1.1 LF of cable per SF of pavement including ties, "return bends" and waste, this yields 545 SF of pavement. These costs do not include reinforcing, concrete or concrete placing. A rule of thumb for sizing circuit breakers: For startup at 0 degrees F with 220 volts AC, allow 0.20 amps per LF of cable.

	Craft@Hrs	Unit	Material	Labor	Total
Heater cable (1.1 LF per SF of pavement)	CE@.015	SF	8.02	.76	8.78
Power connection kit, including end seals	CE@.500	Ea	29.10	25.30	54.40
Cable splice kit	CE@.250	Ea	26.20	12.60	38.80
Cable expansion joint kit	CE@.250	Ea	20.90	12.60	33.50
ElectroMelt™ junction box	CE@.500	Ea	113.00	25.30	138.30
System controller, automatic	CE@1.50	Ea	1,900.00	75.80	1,975.80
Thermostat, line sensing	CE@.500	Ea	245.00	25.30	270.30
Ground fault protection device	CE@.969	Ea	362.00	48.90	410.90

Ice Melting Cable For roofs, gutters and downspouts. Self-regulating heater cable 120 or 208-277 volts AC, run exposed on the surface of the roof or within gutters. Based on Raychem IceStop™ System for roofs and gutters. Cable is cut to length on site and laid in a serpentine pattern using 24-inch center-to-center spacing and 32-inch loop height. Fasten to the roof with Raychem clips. For scheduling purposes estimate that a crew of one can cut to length and install 400 LF of cable in an 8-hour day. Based on 1.2 LF of cable per SF of protected roof area including clips, "return bends" and waste, this yields 330 SF of protected area. Figure each LF of gutter as equal to one SF of roof. Figure each downspout as equal to 15 SF of roof. These costs do not include the roofing, gutters, or downspouts. A rule of thumb for sizing circuit breakers: For startup at 0 degrees F with 120 volts AC, allow 0.17 amps per LF of cable. For startup at 0 degrees F with 208-277 volts AC, allow 0.11 amps per LF of cable.

	Craft@Hrs	Unit	Material	Labor	Total
Heater cable (1.2 LF per SF of protected area)	CE@.033	LF	6.80	1.67	8.47
Power connection kit	CE@.500	Ea	28.40	25.30	53.70
Cable splice kit	CE@.374	Ea	26.30	18.90	45.20
End seal kit	CE@.374	Ea	10.80	18.90	29.70
Downspout hanger	CE@.374	Ea	17.20	18.90	36.10
Metal roof mounting kit	CE@.500	Ea	136.00	25.30	161.30

	Craft@Hrs	Unit	Material	Labor	Total

Electrical Outlet Boxes Steel boxes installed on an exposed wall or ceiling. The material cost for phenolic, PVC and fiberglass boxes will be 50% to 70% less. These costs include fasteners but no switch or receptacle.

	Craft@Hrs	Unit	Material	Labor	Total
Square outlet boxes					
4" x 4" x 1-1/2" deep with					
1/2" and 3/4" knockouts	E4@.187	Ea	.99	8.13	9.12
Side mounting bracket	E4@.187	Ea	1.99	8.13	10.12
4" x 4" x 2-1/8" deep with					
1/2" and 3/4" knockouts	E4@.245	Ea	2.01	10.60	12.61
Side mounting bracket	E4@.245	Ea	1.98	10.60	12.58
Add for					
1-1/2" deep extension rings	E4@.120	Ea	1.79	5.21	7.00
Romex or BX clamps	—	Ea	.38	—	.38
Steel flush cover blanks	E4@.039	Ea	.90	1.69	2.59
Plaster rings to 3/4" deep	E4@.080	Ea	1.62	3.48	5.10
Octagon outlet boxes					
4" x 1-1/2" deep with					
1/2" and 3/4" knockouts	E4@.187	Ea	1.33	8.13	9.46
Romex clamps	E4@.187	Ea	.38	8.13	8.51
Mounting bracket, horizontal	E4@.187	Ea	2.62	8.13	10.75
4" x 2-1/8" deep with					
1/2" and 3/4" knockouts	E4@.245	Ea	4.20	10.60	14.80
Romex clamps	E4@.245	Ea	.38	10.60	10.98
Add for					
1-1/2" deep extension rings	E4@.120	Ea	1.75	5.21	6.96
1-1/2" with 18.5" adjustable bar set	E4@.282	Ea	2.05	12.30	14.35
3" deep concrete ring	E4@.516	Ea	2.10	22.40	24.50
4" steel flush cover blanks	E4@.039	Ea	.68	1.69	2.37
Plaster rings to 3/4" deep	E4@.080	Ea	1.63	3.48	5.11
Handy boxes, 4" x 2-1/8"					
1-1/2" deep, 1/2" knockouts	E4@.187	Ea	1.55	8.13	9.68
1-7/8" deep, 3/4" knockouts	E4@.187	Ea	.83	8.13	8.96
2-1/2" deep, 3/4" knockouts	E4@.245	Ea	2.10	10.60	12.70
2-1/8" deep, with side mounting bracket	E4@.245	Ea	1.05	10.60	11.65
1-1/2" deep extension rings	E4@.120	Ea	2.05	5.21	7.26
Blank or switch cover	E4@.039	Ea	.61	1.69	2.30
Weatherproof box and cover	E4@.059	Ea	3.50	2.56	6.06
Switch boxes, 3" x 2", square corner gangable boxes with mounting ears					
2" deep, 1/2" knockouts	E4@.245	Ea	1.62	10.60	12.22
2-1/2" deep, 1/2" knockouts	E4@.245	Ea	3.80	10.60	14.40
3-1/2" deep, 3/4" knockouts	E4@.324	Ea	3.42	14.10	17.52
Gang switch boxes, 2" x 3" x 1-5/8"					
2 gang	E4@.242	Ea	9.42	10.50	19.92
3 gang	E4@.342	Ea	10.90	14.90	25.80
4 gang	E4@.342	Ea	15.30	14.90	30.20
5 gang	E4@.482	Ea	21.20	20.90	42.10
6 gang	E4@.482	Ea	38.80	20.90	59.70
Floor boxes, watertight, cast iron, round					
4-3/16" x 3-3/4" deep, non-adjustable	E4@1.54	Ea	42.50	66.90	109.40
3-3/4" x 2" deep, semi-adjustable	E4@1.54	Ea	51.90	66.90	118.80
4-3/16" x 3-3/4" deep, adjustable	E4@1.54	Ea	61.80	66.90	128.70
2-1/8" single, round floor cover plates	E4@.156	Ea	29.30	6.78	36.08

	Craft@Hrs	Unit	Material	Labor	Total
Galvanized or gray enamel NEMA class 1 (indoor) pull boxes, with screw cover, 4" deep					
4" x 4"	E4@.373	Ea	7.14	16.20	23.34
4" x 6"	E4@.373	Ea	8.53	16.20	24.73
6" x 6"	E4@.373	Ea	10.10	16.20	26.30
6" x 8"	E4@.391	Ea	11.80	17.00	28.80
8" x 8"	E4@.391	Ea	14.00	17.00	31.00
8" x 10"	E4@.444	Ea	15.70	19.30	35.00
8" x 12"	E4@.483	Ea	17.80	21.00	38.80
10" x 10"	E4@.492	Ea	23.40	21.40	44.80
10" x 12"	E4@.539	Ea	26.90	23.40	50.30
12" x 12"	E4@.593	Ea	30.40	25.80	56.20
12" x 16"	E4@.663	Ea	42.50	28.80	71.30
12" x 18"	E4@.785	Ea	44.30	34.10	78.40
Galvanized or gray enamel NEMA class 1 (indoor) pull boxes, with screw cover, 6" deep					
6" x 6"	E4@.373	Ea	11.10	16.20	27.30
8" x 8"	E4@.391	Ea	15.40	17.00	32.40
8" x 10"	E4@.444	Ea	17.20	19.30	36.50
10" x 10"	E4@.492	Ea	20.40	21.40	41.80
10" x 12"	E4@.539	Ea	22.40	23.40	45.80
12" x 12"	E4@.593	Ea	25.40	25.80	51.20
12" x 16"	E4@.663	Ea	34.60	28.80	63.40
12" x 18"	E4@.785	Ea	37.30	34.10	71.40
16" x 16"	E4@.929	Ea	42.20	40.40	82.60
18" x 18"	E4@1.10	Ea	45.80	47.80	93.60
18" x 24"	E4@1.30	Ea	83.70	56.50	140.20
24" x 24"	E4@1.54	Ea	105.00	66.90	171.90
Cast aluminum NEMA class 3R (weatherproof) screw cover junction boxes, recessed, flush mounted, with cover					
6" x 6" x 4" deep	E4@.930	Ea	166.00	40.40	206.40
6" x 8" x 4" deep	E4@.930	Ea	204.00	40.40	244.40
6" x 6" x 6" deep	E4@.930	Ea	204.00	40.40	244.40
6" x 12" x 6" deep	E4@1.42	Ea	302.00	61.70	363.70
8" x 8" x 4" deep	E4@.982	Ea	232.00	42.70	274.70
8" x 8" x 6" deep	E4@1.14	Ea	249.00	49.50	298.50
8" x 12" x 6" deep	E4@1.30	Ea	378.00	56.50	434.50
10" x 10" x 6" deep	E4@1.33	Ea	214.00	57.80	271.80
12" x 12" x 6" deep	E4@1.35	Ea	448.00	58.70	506.70
12" x 12" x 8" deep	E4@1.73	Ea	526.00	75.20	601.20
12" x 24" x 6" deep	E4@1.97	Ea	1,000.00	85.60	1,085.60
18" x 36" x 8" deep	E4@2.56	Ea	2,310.00	111.00	2,421.00
Hinged cover panel enclosures, NEMA 1, enamel, with cover					
16" x 12" x 7" deep	E4@1.30	Ea	90.30	56.50	146.80
20" x 20" x 7" deep	E4@1.30	Ea	123.00	56.50	179.50
30" x 20" x 7" deep	E4@1.77	Ea	114.00	76.90	190.90
24" x 20" x 9" deep	E4@1.30	Ea	149.00	56.50	205.50
30" x 24" x 9" deep	E4@1.77	Ea	169.00	76.90	245.90
36" x 30" x 9" deep	E4@1.77	Ea	216.00	76.90	292.90

	Craft@Hrs	Unit	Material	Labor	Total

Electrical Receptacles

Standard commercial grade except where noted. Ivory or brown receptacles with cover and screws. White or gray receptacles will cost about 10% more. No outlet boxes included. Labor includes connecting wire, securing device in the box, and attaching cover.

15 amp, 125 volt self-grounding, 2 pole, 3 wire duplex receptacles

	Craft@Hrs	Unit	Material	Labor	Total
Residential quality grounded duplex	CE@.162	Ea	.78	8.18	8.96
Screwless, not self-grounding	CE@.162	Ea	6.61	8.18	14.79
Side terminals	CE@.162	Ea	6.96	8.18	15.14
Back and side terminals	CE@.162	Ea	10.50	8.18	18.68
Feed thru wiring, back and side terminals	CE@.162	Ea	17.20	8.18	25.38
Safety ground, side terminals	CE@.162	Ea	25.10	8.18	33.28
Ground fault circuit interrupter receptacle	CE@.324	Ea	38.70	16.40	55.10
Add for NEMA 5 single receptacles	—	%	15.0	—	—

20 amp, 125 volt self-grounding, 2 pole, 3 wire duplex receptacles

	Craft@Hrs	Unit	Material	Labor	Total
Side terminals	CE@.162	Ea	10.50	8.18	18.68
Back and side terminals	CE@.162	Ea	14.80	8.18	22.98
Feed thru wiring, back and side terminals	CE@.162	Ea	20.70	8.18	28.88
Hospital grade	CE@.162	Ea	24.90	8.18	33.08
Ground fault circuit interrupter receptacle	CE@.324	Ea	41.00	16.40	57.40
Add for NEMA 5 single receptacles	—	%	15.0	—	—

250 volt receptacles, self-grounding, 2 pole, 3 wire, back & side terminals

	Craft@Hrs	Unit	Material	Labor	Total
15 amp, duplex	CE@.187	Ea	12.50	9.44	21.94
15 amp, duplex	CE@.187	Ea	21.10	9.44	30.54
15 amp, single	CE@.187	Ea	14.60	9.44	24.04
20 amp, duplex	CE@.187	Ea	16.40	9.44	25.84
20 amp, duplex, feed thru wiring	CE@.187	Ea	25.80	9.44	35.24
20 amp, single	CE@.187	Ea	17.50	9.44	26.94
Clock receptacle, 2 pole, 15 amp, 125 volt	CE@.329	Ea	21.20	16.60	37.80
120/208 volt 20 amp 4 pole, duplex receptacle	CE@.795	Ea	30.30	40.10	70.40

125/250 volt 3 pole receptacles, flush mount

	Craft@Hrs	Unit	Material	Labor	Total
15 amp/10 amp	CE@.714	Ea	17.10	36.10	53.20
20 amp	CE@.820	Ea	26.80	41.40	68.20
277 volt 30 amp 2 pole receptacle	CE@.268	Ea	88.70	13.50	102.20
Dryer receptacles, 250 volt, 30/50 amp, 3 wire	CE@.536	Ea	44.50	27.10	71.60

Accessories for 50 amp dryer receptacles

	Craft@Hrs	Unit	Material	Labor	Total
Flush mounted plug	CE@.288	Ea	81.50	14.50	96.00
Surface mounted plug	CE@.288	Ea	55.80	14.50	70.30

Cord sets for dryer receptacles

	Craft@Hrs	Unit	Material	Labor	Total
36 inch, three #10 wires	CE@.288	Ea	15.20	14.50	29.70
48 inch, three #10 wires	CE@.288	Ea	17.20	14.50	31.70
60 inch, three #10 wires	CE@.288	Ea	19.90	14.50	34.40

Electrical Switches Commercial grade, 120 to 277 volt rating, except where noted. Ivory or brown. Add 10% for white or gray. Includes cover plate but no fixture boxes except as noted. Labor includes connecting wire to switch, securing switch in box and attaching cover plate.

15 amp switches, back and side wired

	Craft@Hrs	Unit	Material	Labor	Total
One pole quiet switch, residential (side wired only)	CE@.112	Ea	.84	5.66	6.50
One pole switch, commercial, minimum quality	CE@.112	Ea	2.41	5.66	8.07
Two pole switch	CE@.309	Ea	7.30	15.60	22.90
Three-way switch	CE@.227	Ea	4.49	11.50	15.99
Four-way switch	CE@.309	Ea	10.80	15.60	26.40

	Craft@Hrs	Unit	Material	Labor	Total
20 amp switches, back and side wired					
One pole switch	CE@.187	Ea	4.48	9.44	13.92
Two pole switch	CE@.433	Ea	16.30	21.90	38.20
Three-way switch	CE@.291	Ea	5.55	14.70	20.25
Four-way switch	CE@.435	Ea	30.20	22.00	52.20
30 amp switches, side wired					
One pole switch	CE@.246	Ea	15.60	12.40	28.00
Two pole switch	CE@.475	Ea	23.10	24.00	47.10
Three-way switch	CE@.372	Ea	22.50	18.80	41.30
Four-way switch	CE@.475	Ea	33.60	24.00	57.60
20 amp weatherproof switches, lever handle, with cover					
One pole switch	CE@.187	Ea	10.30	9.44	19.74
Two pole switch	CE@.433	Ea	19.80	21.90	41.70
Three-way switch	CE@.291	Ea	13.60	14.70	28.30
Single pole, two gang, 10 amp	CE@.358	Ea	23.20	18.10	41.30
Dimmer switches, push for off					
600 watt, one pole	CE@.417	Ea	6.46	21.10	27.56
600 watt, three way	CE@.626	Ea	11.00	31.60	42.60
Fluorescent dimmer,					
10 lamp load	CE@.626	Ea	55.40	31.60	87.00
Astro dial time switch,					
40 amp, with box	CE@3.00	Ea	154.00	152.00	306.00
15 minute timer switch,					
wall box mounted	CE@.426	Ea	18.80	21.50	40.30
Single pole, 1 throw time switch, 277 volt	CE@.890	Ea	62.90	44.90	107.80
Float switches for sump pumps, automatic					
10A, 125/250/480 VAC	CE@1.03	Ea	30.50	52.00	82.50
One way 15 amp toggle switch					
with neon pilot	CE@.327	Ea	9.72	16.50	26.22
Three way 15 amp toggle switch					
with neon pilot	CE@.372	Ea	16.20	18.80	35.00
Photoelectric switches, flush, with wall plate					
120 volt, 1,000 watt	CE@.426	Ea	8.47	21.50	29.97
208 volt, 1,800 watt	CE@.626	Ea	10.20	31.60	41.80
480 volt, 3,000 watt	CE@.795	Ea	15.90	40.10	56.00
Button control stations, surface mounted, NEMA class 1, standard duty, 120/240 volt, with enclosure					
Start-stop switch, 2 button	CE@.624	Ea	22.50	31.50	54.00
Start-stop switch with lockout, 2 button	CE@.624	Ea	33.60	31.50	65.10
Forward, reverse and stop buttons, 3 button	CE@.624	Ea	44.90	31.50	76.40
Forward, reverse, stop and lockout, 3 button	CE@.624	Ea	44.90	31.50	76.40
Manual toggle starter switches, surface mounted, NEMA class 1, 120/240 volt, non-reversing, with enclosure					
Size 0 motors, 2 pole	CE@.613	Ea	108.00	31.00	139.00
Size 1 motors, 3 pole	CE@.698	Ea	140.00	35.20	175.20
Size 1-1/2 motors, 2 pole	CE@.901	Ea	161.00	45.50	206.50
Manual button starter switches, surface mounted, NEMA class 1, 110 to 240 volt, 2 pole, 1 phase, with enclosure, start, stop, reset, with relay					
Size 00 motors	CE@.712	Ea	145.00	36.00	181.00
Size 0 motors	CE@.712	Ea	157.00	36.00	193.00
Size 1 motors	CE@.820	Ea	174.00	41.40	215.40
Size 1-1/2 motors	CE@.975	Ea	210.00	49.20	259.20

	Craft@Hrs	Unit	Material	Labor	Total
Grounding Devices No excavation or concrete included.					
Copperweld grounding rods, driven					
5/8" x 8'	E4@.365	Ea	13.30	15.90	29.20
5/8" x 10'	E4@.420	Ea	26.70	18.20	44.90
3/4" x 8'	E4@.365	Ea	21.70	15.90	37.60
3/4" x 10'	E4@.420	Ea	26.70	18.20	44.90
Grounding locknuts					
1/2" locknuts	E4@.050	Ea	1.00	2.17	3.17
3/4" locknuts	E4@.060	Ea	1.26	2.61	3.87
1" locknuts	E4@.080	Ea	1.71	3.48	5.19
1-1/4" locknuts	E4@.100	Ea	2.29	4.35	6.64
1-1/2" locknuts	E4@.100	Ea	3.09	4.35	7.44
2" locknuts .	E4@.150	Ea	4.20	6.52	10.72
2-1/2" locknuts	E4@.200	Ea	8.46	8.69	17.15
3" locknuts	E4@.200	Ea	10.70	8.69	19.39
4" locknuts	E4@.300	Ea	22.40	13.00	35.40
Ground rod clamps					
5/8" clamp	E4@.249	Ea	1.53	10.80	12.33
3/4" clamp	E4@.249	Ea	1.93	10.80	12.73
Coupling for threaded ground rod					
5/8" clamp	E4@.249	Ea	8.18	10.80	18.98
3/4" clamp	E4@.249	Ea	12.20	10.80	23.00
Copper bonding connector straps, 3/4" x .050"					
1" strap	E4@.286	Ea	5.47	12.40	17.87
2" strap	E4@.300	Ea	7.03	13.00	20.03
3" strap	E4@.315	Ea	9.43	13.70	23.13
4" strap	E4@.330	Ea	11.20	14.30	25.50
Brazed connections for wire					
#6 wire	E4@.194	Ea	1.06	8.43	9.49
#2 wire	E4@.194	Ea	1.32	8.43	9.75
#1/0 wire	E4@.295	Ea	3.34	12.80	16.14
#4/0 wire	E4@.391	Ea	8.24	17.00	25.24
Grounding using a 5/8" x 8' copper clad ground rod, includes 10' grounding conductor, clamps, connectors, locknut, and rod.					
100 amp service, #8 copper wire	E4@1.13	Ea	23.80	49.10	72.90
150 amp service, #6 copper wire	E4@1.16	Ea	25.00	50.40	75.40
Electric Motors General purpose, 3 phase, open drip-proof, industrial duty, 1,700 - 1,780 RPM, 208-230/460 volt, rigid welded or solid base, furnished and placed, no hookup included. Based on Dayton motors.					
1/2 HP	E4@1.99	Ea	253.00	86.50	339.50
3/4 HP	E4@1.99	Ea	312.00	86.50	398.50
1 HP	E4@1.99	Ea	334.00	86.50	420.50
1-1/2 HP	E4@1.99	Ea	416.00	86.50	502.50
2 HP	E4@1.99	Ea	461.00	86.50	547.50
3 HP	E4@1.99	Ea	506.00	86.50	592.50
5 HP	E4@2.22	Ea	588.00	96.50	684.50
7-1/2 HP	E4@2.40	Ea	904.00	104.00	1,008.00
10 HP	E4@2.51	Ea	1,120.00	109.00	1,229.00

	Craft@Hrs	Unit	Material	Labor	Total

Electric Motor Connection Includes a NEMA-1 fusible heavy duty safety switch, connectors, fittings, ells, adapters, supports, anchors and conduit with wire appropriate for the load connected but no starter or motor.
Single phase, using 10' RSC, 20' PVC and 6' flex conduit, NEMA class 1 (indoor) general duty switch, 240 V

	Craft@Hrs	Unit	Material	Labor	Total
For motor to 2 HP, 20 amps	E4@6.35	Ea	123.00	276.00	399.00
For 2.5 to 3 HP motor, 30 amps	E4@6.63	Ea	130.00	288.00	418.00
For 3 HP motor, 45 amps	E4@6.68	Ea	289.00	290.00	579.00
For 8 to 15 HP motor, 90 amps	E4@13.4	Ea	385.00	582.00	967.00

Three phase wiring using 10' RSC, 20' PVC and 6' flex conduit, NEMA class 1 (indoor) switch, 600 V

For motor to 2 HP, 20 amps	E4@7.18	Ea	266.00	312.00	578.00
For 3 to 5 HP motor, 30 amps	E4@7.90	Ea	272.00	343.00	615.00
For 7 to 10 HP motor, 45 amps	E4@8.30	Ea	332.00	361.00	693.00
For 25 to 50 HP motor, 90 amps	E4@9.30	Ea	670.00	404.00	1,074.00
For 50 to 100 HP motor, 135 amps	E4@14.1	Ea	975.00	613.00	1,588.00
For 100 to 200 HP motor, 270 amps	E4@21.9	Ea	2,500.00	952.00	3,452.00

Single phase, using 10' RSC, 20' PVC and 6' flex conduit, NEMA class 3R (weatherproof) switch, 240 V

For motor to 2 HP, 20 amps	E4@6.63	Ea	161.00	288.00	449.00
For 2 to 5 HP motor, 30 amps	E4@6.97	Ea	167.00	303.00	470.00
For 5 to 7.5 HP motor, 45 amps	E4@6.94	Ea	241.00	302.00	543.00

Three phase using 10' RSC, 20' PVC and 6' flex conduit, NEMA class 3R (weatherproof) switch, 600 V

For motor to 5 HP motor, 20 amps	E4@7.10	Ea	420.00	308.00	728.00
For 1 to 10 HP motor, 30 amps	E4@7.25	Ea	426.00	315.00	741.00
For 10 to 25 HP motor, 45 amps	E4@9.51	Ea	507.00	413.00	920.00
For 25 to 50 HP motor, 90 amps	E4@9.51	Ea	850.00	413.00	1,263.00
For 60 to 100 HP motor, 135 amps	E4@12.1	Ea	1,200.00	526.00	1,726.00
For 125 to 200 HP motor, 270 amps	E4@21.6	Ea	3,020.00	939.00	3,959.00

Electric Motor Connection Complete installation, including magnetic motor starter, overload relay, all wiring, a heavy duty fusible safety switch, junction box and cover, connectors, fittings, adapters, supports, anchors and conduit appropriate for the load connected. No motor included.
Single phase, using 10' RSC, 20' PVC and 6' flex conduit, NEMA class 1 (indoor) starter, 230 V

For motor to 2 HP, 20 amps	E4@8.11	Ea	259.00	352.00	611.00
For 2 to 3 HP motor, 30 amps	E4@8.32	Ea	296.00	362.00	658.00
For 3 to 8 HP motor, 45 amps	E4@8.45	Ea	433.00	367.00	800.00
For 8 to 15 HP motor, 90 amps	E4@15.5	Ea	715.00	673.00	1,388.00

Three phase wiring using 10' RSC, 20' PVC and 6' flex conduit, NEMA class 1 (indoor) starter, 480 V

For motor to 2 HP, 20 amps	E4@9.90	Ea	717.00	430.00	1,147.00
For 2 to 5 HP motor, 30 amps	E4@10.6	Ea	760.00	461.00	1,221.00
For 7 to 25 HP motor, 45 amps	E4@20.5	Ea	1,180.00	891.00	2,071.00
For 25 to 50 HP motor, 90 amps	E4@23.1	Ea	2,010.00	1,000.00	3,010.00

Single phase, using 10' RSC, 20' PVC and 6' flex conduit, NEMA Type 3R (weatherproof) starter, 230 V

For motor to 2 HP, 20 amps	E4@8.91	Ea	362.00	387.00	749.00
For 2 to 5 HP motor, 30 amps	E4@9.15	Ea	416.00	398.00	814.00
For 5 to 7.5 HP motor, 45 amps	E4@9.25	Ea	597.00	402.00	999.00

Three phase using 10' RSC, 20' PVC and 6' flex conduit, NEMA Type 3R (weatherproof) starter, 480 V

For motor to 5 HP, 20 amps	E4@10.9	Ea	1,050.00	474.00	1,524.00
For 5 to 10 HP motor, 30 amps	E4@11.7	Ea	1,100.00	508.00	1,608.00
For 10 to 25 HP motor, 45 amps	E4@22.5	Ea	1,720.00	978.00	2,698.00
For 25 to 50 HP motor, 90 amps	E4@25.4	Ea	2,900.00	1,100.00	4,000.00

Motor Starters Magnetic operated full voltage non-reversing motor controllers with thermal overload relays and enclosure.
Two pole, 1 phase contactors, NEMA class 1 (indoor), electrically held with holding interlock, one reset

Size 00, 1/3 HP, 9 amp	E4@1.47	Ea	217.00	63.90	280.90
Size 0, 1 HP, 18 amp	E4@1.62	Ea	251.00	70.40	321.40

	Craft@Hrs	Unit	Material	Labor	Total
Size 1, 2 HP, 27 amp	E4@1.73	Ea	285.00	75.20	360.20
Size 2, 3 HP, 45 amp	E4@1.73	Ea	365.00	75.20	440.20
Size 3, 7 1/2 HP, 90 amp	E4@2.02	Ea	561.00	87.80	648.80

Three pole polyphase NEMA class 1 (indoor) AC magnetic combination starters with fusible disconnect and overload relays but no heaters, 208 to 240 volts

	Craft@Hrs	Unit	Material	Labor	Total
Size 0, 3 HP, 30 amps	E4@2.75	Ea	685.00	119.00	804.00
Size 1, 7.5 HP, 30 amps	E4@3.89	Ea	721.00	169.00	890.00
Size 1, 10 HP, 60 amps	E4@3.89	Ea	721.00	169.00	890.00
Size 2, 10 HP, 60 amps	E4@5.34	Ea	1,140.00	232.00	1,372.00
Size 2, 25 HP, 100 amps	E4@5.34	Ea	1,140.00	232.00	1,372.00
Size 3, 25 HP, 100 amps	E4@6.76	Ea	1,890.00	294.00	2,184.00
Size 3, 50 HP, 200 amps	E4@6.76	Ea	1,890.00	294.00	2,184.00
Size 4, 40 HP, 200 amps	E4@12.0	Ea	3,650.00	521.00	4,171.00
Size 4, 75 HP, 400 amps	E4@12.0	Ea	3,650.00	521.00	4,171.00

On and off switches for starter enclosures, any motor starter above

	Craft@Hrs	Unit	Material	Labor	Total
Switch kit without pilot light	E4@.744	Ea	35.30	32.30	67.60
Switch kit with pilot light	E4@.744	Ea	53.00	32.30	85.30

Accessories for any motor starter

	Craft@Hrs	Unit	Material	Labor	Total
Add for 440 to 600 volt starters	—	%	2.0	—	—
Add for NEMA Type 3R (rainproof) enclosure	—	%	40.0	10.0	—
Add for NEMA Type 4 (waterproof) enclosure	—	%	40.0	20.0	—
Add for NEMA Type 12 (dust-tight) enclosure	—	%	15.0	15.0	—
Add for starters in an oversize enclosure	—	%	21.0	—	—
Deduct for starters with circuit breakers	—	%	-4.0	—	—

Safety Switches Wall mounted switches with enclosures as noted. No fuses or hubs included.

Heavy duty (NEMA-1) 600 volt 2, 3 or 4 pole fusible safety switches

	Craft@Hrs	Unit	Material	Labor	Total
30 amp	E4@3.06	Ea	211.00	133.00	344.00
60 amp	E4@3.89	Ea	255.00	169.00	424.00
100 amp	E4@4.20	Ea	476.00	182.00	658.00
200 amp	E4@6.92	Ea	687.00	301.00	988.00
400 amp	E4@11.4	Ea	1,690.00	495.00	2,185.00
600 amp	E4@14.3	Ea	2,510.00	621.00	3,131.00
800 amp	E4@19.6	Ea	4,140.00	852.00	4,992.00
1200 amp	E4@22.7	Ea	5,820.00	986.00	6,806.00

Heavy duty rainproof (NEMA-3R) 600 volt 2, 3 or 4 pole fusible safety switches

	Craft@Hrs	Unit	Material	Labor	Total
30 amp	E4@3.32	Ea	357.00	144.00	501.00
60 amp	E4@4.61	Ea	422.00	200.00	622.00
100 amp	E4@4.77	Ea	657.00	207.00	864.00
200 amp	E4@7.51	Ea	905.00	326.00	1,231.00
400 amp	E4@12.4	Ea	2,180.00	539.00	2,719.00
600 amp	E4@15.6	Ea	4,370.00	678.00	5,048.00
800 amp	E4@21.4	Ea	6,560.00	930.00	7,490.00
1200 amp	E4@25.8	Ea	7,200.00	1,120.00	8,320.00

Heavy duty watertight (NEMA-4) 600 volt 3 pole, 4 wire fusible safety switches

	Craft@Hrs	Unit	Material	Labor	Total
30 amp	E4@3.76	Ea	961.00	163.00	1,124.00
60 amp	E4@5.21	Ea	1,330.00	226.00	1,556.00
100 amp	E4@5.34	Ea	2,080.00	232.00	2,312.00
200 amp	E4@8.52	Ea	2,930.00	370.00	3,300.00
400 amp	E4@13.7	Ea	5,800.00	595.00	6,395.00
600 amp	E4@17.0	Ea	8,330.00	739.00	9,069.00

	Craft@Hrs	Unit	Material	Labor	Total
Heavy duty dust-tight (NEMA-12) 600 volt 3 pole or 4 pole solid neutral fusible safety switches					
30 amp	E4@3.32	Ea	374.00	144.00	518.00
60 amp	E4@4.20	Ea	375.00	182.00	557.00
100 amp	E4@4.48	Ea	585.00	195.00	780.00
200 amp	E4@7.51	Ea	914.00	326.00	1,240.00
400 amp	E4@12.7	Ea	2,080.00	552.00	2,632.00
600 amp	E4@15.9	Ea	3,520.00	691.00	4,211.00
General duty (NEMA-1) 240 volt 3 pole, 4 wire non-fusible safety switches					
30 amp	E4@2.58	Ea	71.60	112.00	183.60
100 amp	E4@3.60	Ea	215.00	156.00	371.00
200 amp	E4@6.50	Ea	456.00	282.00	738.00
400 amp	E4@10.8	Ea	1,190.00	469.00	1,659.00
600 amp	E4@12.7	Ea	2,220.00	552.00	2,772.00
General duty rainproof (NEMA-3R) 240 volt 3 pole, 4 wire non-fusible safety switches					
30 amp	E4@2.88	Ea	111.00	125.00	236.00
60 amp	E4@3.76	Ea	169.00	163.00	332.00
100 amp	E4@3.89	Ea	312.00	169.00	481.00
200 amp	E4@7.23	Ea	557.00	314.00	871.00
400 amp	E4@11.8	Ea	1,400.00	513.00	1,913.00
600 amp	E4@14.0	Ea	3,010.00	608.00	3,618.00
Heavy duty watertight (NEMA-4) 240 volt 3 pole, 4 wire fusible safety switches					
30 amp	E4@3.19	Ea	808.00	139.00	947.00
60 amp	E4@3.76	Ea	961.00	163.00	1,124.00
100 amp	E4@4.33	Ea	1,950.00	188.00	2,138.00
200 amp	E4@7.80	Ea	2,640.00	339.00	2,979.00
400 amp	E4@13.0	Ea	5,820.00	565.00	6,385.00
600 amp	E4@15.4	Ea	8,230.00	669.00	8,899.00
Heavy duty dust-tight (NEMA-12) 240 volt 3 pole, 4 wire fusible safety switches					
30 amp	E4@2.88	Ea	242.00	125.00	367.00
60 amp	E4@3.76	Ea	312.00	163.00	475.00
100 amp	E4@4.04	Ea	447.00	176.00	623.00
200 amp	E4@7.23	Ea	599.00	314.00	913.00
400 amp	E4@11.8	Ea	1,500.00	513.00	2,013.00
600 amp	E4@14.0	Ea	2,510.00	608.00	3,118.00
Add for conduit hubs					
3/4" to 1-1/2", to 100 amp	—	Ea	7.88	—	7.88
2", 200 amp	—	Ea	13.80	—	13.80
2-1/2", 200 amp	—	Ea	23.20	—	23.20

Service Entrance and Distribution Switchboards MA Class 1 indoor, 600 volt, 3 phase, 4 wire, for 240/480 volt insulated case main breakers. Basic structure is 90" high by 21" deep. Width varies with equipment capacity. Labor cost includes setting and leveling on a prepared pad but excludes the pad cost. Add breaker, instrumentation and accessory costs for a complete installation. Based on Westinghouse Pow-R-Gear. These switchboards are custom designed for each installation. Costs can vary widely. Multiple units ordered at the same time may reduce costs per unit by 25% or more.

	Craft@Hrs	Unit	Material	Labor	Total
600 amp bus	E4@22.9	Ea	3,330.00	995.00	4,325.00
1,000 amp bus	E4@22.9	Ea	3,440.00	995.00	4,435.00
1,200 amp bus	E4@22.9	Ea	4,170.00	995.00	5,165.00
1,600 amp bus	E4@22.9	Ea	4,820.00	995.00	5,815.00
2,000 amp bus	E4@22.9	Ea	5,280.00	995.00	6,275.00
2,500 amp bus	E4@22.9	Ea	6,430.00	995.00	7,425.00
3,000 amp bus	E4@22.9	Ea	7,370.00	995.00	8,365.00
4,000 amp bus	E4@22.9	Ea	9,490.00	995.00	10,485.00

	Craft@Hrs	Unit	Material	Labor	Total
240/480 volt draw-out main breakers, 100K amp interrupt capacity. Includes connecting and testing breakers only. Based on Pow-R-Trip					
100 to 800 amp, manual operation	E4@10.2	Ea	9,820.00	443.00	10,263.00
1,000 to 1,600 amp, manual operation	E4@17.5	Ea	20,600.00	760.00	21,360.00
2,000 amp, manual operation	E4@23.2	Ea	25,200.00	1,010.00	26,210.00
2,500 amp, manual operation	E4@23.2	Ea	42,900.00	1,010.00	43,910.00
3,000 amp, manual operation	E4@23.2	Ea	47,900.00	1,010.00	48,910.00
4,000 amp, manual operation	E4@23.2	Ea	57,700.00	1,010.00	58,710.00
100 to 800 amp electric operation	E4@10.2	Ea	13,300.00	443.00	13,743.00
1,000 to 1,600 amp electric operation	E4@17.5	Ea	22,600.00	760.00	23,360.00
2,000 amp electric operation	E4@23.2	Ea	29,900.00	1,010.00	30,910.00
2,500 amp electric operation	E4@23.2	Ea	48,900.00	1,010.00	49,910.00
3,000 amp electric operation	E4@23.2	Ea	55,800.00	1,010.00	56,810.00
Additional space for future draw-out breakers					
100 to 800 amp	—	Ea	2,620.00	—	2,620.00
1,000 to 1,600 amp	—	Ea	3,830.00	—	3,830.00
2,000 amp	—	Ea	4,310.00	—	4,310.00
2,500 amp	—	Ea	4,600.00	—	4,600.00
3,000 amp	—	Ea	8,380.00	—	8,380.00

Service Entrance and Distribution Switchboards, Weatherproof

600 volt, 3 phase, 4 wire, for 480 volt insulated case main breakers. Basic structure is 90" high by 21" deep. Width varies with equipment capacity. Labor cost includes setting and leveling on a prepared pad but excludes the pad cost. Add breaker, instrumentation and accessory costs from below for a complete installation. These switchboards are custom designed for each installation. Costs can vary widely. Multiple units ordered at the same time can reduce costs per unit by 25% or more

	Craft@Hrs	Unit	Material	Labor	Total
600 amp bus	E4@23.2	Ea	3,930.00	1,010.00	4,940.00
1,000 amp bus	E4@23.2	Ea	4,610.00	1,010.00	5,620.00
1,200 amp bus	E4@23.2	Ea	4,920.00	1,010.00	5,930.00
1,600 amp bus	E4@23.2	Ea	5,680.00	1,010.00	6,690.00
2,000 amp bus	E4@23.2	Ea	6,280.00	1,010.00	7,290.00
2,500 amp bus	E4@23.2	Ea	7,560.00	1,010.00	8,570.00
3,000 amp bus	E4@23.2	Ea	7,690.00	1,010.00	8,700.00
4,000 amp bus	E4@23.2	Ea	11,200.00	1,010.00	12,210.00
480 volt manual operated draw-out breakers. By amp interrupt capacity (AIC) as noted. Includes connecting and testing factory installed breakers only					
50 to 800 amp, 30K AIC	E4@10.2	Ea	13,300.00	443.00	13,743.00
50 to 1,600 amp, 50K AIC	E4@17.5	Ea	28,100.00	760.00	28,860.00
2,000 amp	E4@23.2	Ea	35,500.00	1,010.00	36,510.00
1,200 to 3,200 amp	E4@23.2	Ea	62,500.00	1,010.00	63,510.00
4,000 amp	E4@23.2	Ea	99,500.00	1,010.00	100,510.00
480 volt manual operated draw-out breakers. 200K amp interrupt capacity (AIC) Includes connecting and testing factory installed breakers only					
50 to 800 amp	E4@10.2	Ea	20,400.00	443.00	20,843.00
50 to 1,600 amp	E4@17.5	Ea	36,400.00	760.00	37,160.00
1,200 to 3,200 amp	E4@23.2	Ea	92,300.00	1,010.00	93,310.00
4,000 amp	E4@23.2	Ea	150,000.00	1,010.00	151,010.00
480 volt electrically operated draw-out breakers. By amp interrupt capacity (AIC) as noted. Includes connecting and testing factory installed breakers only					
50 to 800 amp, 30K AIC	E4@10.2	Ea	18,100.00	443.00	18,543.00
50 to 1,600 amp, 50K AIC	E4@17.5	Ea	37,500.00	760.00	38,260.00
2,000 amp	E4@23.2	Ea	47,800.00	1,010.00	48,810.00
1,200 to 3,200 amp	E4@23.2	Ea	71,800.00	1,010.00	72,810.00
4,000 amp	E4@23.2	Ea	111,000.00	1,010.00	112,010.00

	Craft@Hrs	Unit	Material	Labor	Total

480 volt electrically operated draw-out breakers. 200K amp interrupt capacity (AIC)
Includes connecting and testing factory installed breakers only

	Craft@Hrs	Unit	Material	Labor	Total
50 to 800 amp	E4@10.2	Ea	24,800.00	443.00	25,243.00
50 to 1,600 amp	E4@17.5	Ea	50,100.00	760.00	50,860.00
1,200 to 3,200 amp	E4@23.2	Ea	103,000.00	1,010.00	104,010.00
4,000 amp	E4@23.2	Ea	150,000.00	1,010.00	151,010.00

Additional space for future draw-out breakers

	Craft@Hrs	Unit	Material	Labor	Total
50 to 800 amp	—	Ea	3,080.00	—	3,080.00
50 to 1,600 amp	—	Ea	5,000.00	—	5,000.00
2,000 amp	—	Ea	6,840.00	—	6,840.00
1,200 to 3,200 amp	—	Ea	10,600.00	—	10,600.00
4,000 amp	—	Ea	19,500.00	—	19,500.00

Switchboard Instrumentation and Accessories Add these costs to the cost of the basic enclosure and breakers to find the complete switchboard cost. These figures are for factory-installed instrumentation and accessories and include the cost of connecting and testing only.

Bus duct connection, 3 phase, 4 wire

	Craft@Hrs	Unit	Material	Labor	Total
225 amp	CE@.516	Ea	595.00	26.10	621.10
400 amp	CE@.516	Ea	693.00	26.10	719.10
600 amp	CE@.516	Ea	839.00	26.10	865.10
800 amp	CE@.516	Ea	1,330.00	26.10	1,356.10
1,000 amp	CE@.516	Ea	1,530.00	26.10	1,556.10
1,350 amp	CE@1.01	Ea	1,980.00	51.00	2,031.00
1,600 amp	CE@1.01	Ea	2,300.00	51.00	2,351.00
2,000 amp	CE@1.01	Ea	2,810.00	51.00	2,861.00
2,500 amp	CE@1.01	Ea	3,350.00	51.00	3,401.00
3,000 amp	CE@1.01	Ea	3,820.00	51.00	3,871.00
4,000 amp	CE@1.01	Ea	4,230.00	51.00	4,281.00

Recording meters

	Craft@Hrs	Unit	Material	Labor	Total
Voltmeter	CE@1.01	Ea	8,460.00	51.00	8,511.00
Ammeter	CE@1.01	Ea	8,620.00	51.00	8,671.00
Wattmeter	CE@1.01	Ea	10,800.00	51.00	10,851.00
Varmeter	CE@1.01	Ea	11,500.00	51.00	11,551.00
Power factor meter	CE@1.01	Ea	11,300.00	51.00	11,351.00
Frequency meter	CE@1.01	Ea	11,300.00	51.00	11,351.00
Watt-hour meter, 3 element	CE@1.01	Ea	14,600.00	51.00	14,651.00

Non-recording meters

	Craft@Hrs	Unit	Material	Labor	Total
Voltmeter	CE@.516	Ea	1,430.00	26.10	1,456.10
Ammeter for incoming line	CE@.516	Ea	1,460.00	26.10	1,486.10
Ammeter for feeder circuits	CE@.516	Ea	608.00	26.10	634.10
Wattmeter	CE@1.01	Ea	4,730.00	51.00	4,781.00
Varmeter	CE@1.01	Ea	4,280.00	51.00	4,331.00
Power factor meter	CE@1.01	Ea	3,460.00	51.00	3,511.00
Frequency meter	CE@1.01	Ea	4,080.00	51.00	4,131.00
Watt-hour meter, 3 element	CE@1.01	Ea	4,280.00	51.00	4,331.00
Instrument phase select switch	CE@.516	Ea	737.00	26.10	763.10
Adjustable short time pickup and delay	CE@.516	Ea	1,110.00	26.10	1,136.10
Ground fault trip pickup and delay, 4 wire	CE@.525	Ea	1,180.00	26.50	1,206.50
Ground fault test panel	CE@.516	Ea	1,280.00	26.10	1,306.10
Shunt trip for breakers	CE@.516	Ea	852.00	26.10	878.10
Key interlock for breakers	CE@.516	Ea	852.00	26.10	878.10
Breaker lifting and transport truck	—	Ea	5,620.00	—	5,620.00
Breaker lifting device	—	Ea	4,890.00	—	4,890.00

	Craft@Hrs	Unit	Material	Labor	Total
Current transformers, by primary capacity					
800 amps and less	CE@1.01	Ea	1,030.00	51.00	1,081.00
1,000 to 1,500 amps	CE@1.01	Ea	1,510.00	51.00	1,561.00
2,000 to 6,000 amps	CE@1.01	Ea	1,790.00	51.00	1,841.00
Current transformer mount	—	Ea	486.00	—	486.00
Potential transformer	CE@1.01	Ea	1,330.00	51.00	1,381.00
Potential transformer mount	—	Ea	306.00	—	306.00

Feeder Section Breaker Panels

Flush or surface mounted 277/480 volt, 3 phase, 4 wire, FA frame panels.

	Craft@Hrs	Unit	Material	Labor	Total
225 amp, with 22K amp interrupt capacity breaker	E4@6.76	Ea	1,410.00	294.00	1,704.00
225 amp, with 25K amp interrupt capacity breaker	E4@6.76	Ea	2,620.00	294.00	2,914.00
400 amp, with 30K amp interrupt capacity breaker	E4@10.8	Ea	3,150.00	469.00	3,619.00
400 amp, with 35K amp interrupt capacity breaker	E4@10.8	Ea	3,760.00	469.00	4,229.00
225 amp, with main lugs only	E4@6.76	Ea	476.00	294.00	770.00
400 amp, with main lugs only	E4@10.8	Ea	670.00	469.00	1,139.00
600 amp, with main lugs only	E4@19.3	Ea	893.00	839.00	1,732.00

Transformers Indoor dry type transformer, floor mounted, including connections.

Single phase light and power circuit transformer, 240/480 volt primary, 120/240 secondary, no taps for voltage change

	Craft@Hrs	Unit	Material	Labor	Total
1 KVA	E4@3.76	Ea	259.00	163.00	422.00
2 KVA	E4@4.20	Ea	348.00	182.00	530.00
3 KVA	E4@4.33	Ea	375.00	188.00	563.00
5 KVA	E4@5.47	Ea	513.00	238.00	751.00
7.5 KVA	E4@8.34	Ea	660.00	362.00	1,022.00
10 KVA	E4@9.38	Ea	803.00	408.00	1,211.00
15 KVA	E4@12.6	Ea	1,000.00	547.00	1,547.00
25 KVA	E4@12.6	Ea	1,580.00	547.00	2,127.00

Three phase light and power circuit transformers, 480 volt primary, 120/208 volt secondary, with 6 taps for voltage change

	Craft@Hrs	Unit	Material	Labor	Total
30 KVA	E4@10.5	Ea	751.00	456.00	1,207.00
45 KVA	E4@12.6	Ea	941.00	547.00	1,488.00
50 KVA	E4@12.6	Ea	1,300.00	547.00	1,847.00
75 KVA	E4@14.8	Ea	1,330.00	643.00	1,973.00
112.5 KVA	E4@15.7	Ea	2,060.00	682.00	2,742.00
150 KVA	E4@16.7	Ea	2,590.00	726.00	3,316.00
225 KVA	E4@18.9	Ea	3,700.00	821.00	4,521.00
300 KVA	E4@21.2	Ea	4,720.00	921.00	5,641.00

Three phase 5000 volt load center transformers, 4160 volt primary, 480/277 volt secondary, with voltage change taps, insulation class H, 80 degree system

	Craft@Hrs	Unit	Material	Labor	Total
225 KVA	E4@29.5	Ea	12,300.00	1,280.00	13,580.00
300 KVA	E4@33.7	Ea	14,600.00	1,460.00	16,060.00
500 KVA	E4@42.5	Ea	22,200.00	1,850.00	24,050.00
750 KVA	E4@53.1	Ea	28,900.00	2,310.00	31,210.00
1,000 KVA	E4@63.2	Ea	37,000.00	2,750.00	39,750.00

	Craft@Hrs	Unit	Material	Labor	Total

Panelboards Surface mounted lighting and appliance distribution panelboards for bolt-on circuit breakers. No breakers included, add for same as required.

3 wire 120/240 volt, 250 amp, single phase AC panelboards with main lugs only for bolt-on breakers

	Craft@Hrs	Unit	Material	Labor	Total
18 circuits	E4@10.7	Ea	220.00	465.00	685.00
30 circuits	E4@15.3	Ea	224.00	665.00	889.00
42 circuits	E4@20.9	Ea	295.00	908.00	1,203.00
Deduct for 3 wire 120/240 panelboards					
With main lugs only	—	%	-10.0	-15.0	—

4 wire 120/208 volt, 250 amp, 3 phase AC panelboards for bolt-on breakers, with main circuit breaker only

	Craft@Hrs	Unit	Material	Labor	Total
18 circuits	E4@12.9	Ea	178.00	561.00	739.00
30 circuits	E4@21.5	Ea	253.00	934.00	1,187.00
42 circuits	E4@30.1	Ea	329.00	1,310.00	1,639.00
Deduct for 3 wire panelboards					
With circuit breaker mains only	—	%	-10.0	-15.0	—

Loadcenters Surface mounted 240 volt NEMA 1 (indoor) distribution panels for plug-in branch breakers. Add the cost of distribution breakers and conduit hubs.

Single phase 3 wire 120/240 volt loadcenters with main circuit breaker only

	Craft@Hrs	Unit	Material	Labor	Total
100 amps, 14 circuits	E4@1.20	Ea	183.00	52.10	235.10
100 amps, 18 circuits	E4@1.20	Ea	197.00	52.10	249.10
125 amps, 22 circuits	E4@1.52	Ea	368.00	66.00	434.00
150 amps, 24 circuits	E4@2.30	Ea	379.00	99.90	478.90
150 amps, 32 circuits	E4@2.30	Ea	405.00	99.90	504.90
200 amps, 24 circuits	E4@2.60	Ea	379.00	113.00	492.00
200 amps, 32 circuits	E4@2.60	Ea	417.00	113.00	530.00
200 amps, 42 circuits	E4@2.93	Ea	478.00	127.00	605.00
225 amps, 32 circuits	E4@2.93	Ea	549.00	127.00	676.00
225 amps, 42 circuits	E4@2.93	Ea	572.00	127.00	699.00
Add for NEMA-4 raintight loadcenters	—	%	25.0	—	—

Three phase 4 wire 120/208 volt loadcenters with main circuit breaker only

	Craft@Hrs	Unit	Material	Labor	Total
100 amps, 12 circuits	E4@1.21	Ea	301.00	52.60	353.60
100 amps, 18 circuits	E4@1.60	Ea	405.00	69.50	474.50
125 amps, 30 circuits	E4@1.91	Ea	802.00	83.00	885.00
150 amps, 24 circuits	E4@2.18	Ea	775.00	94.70	869.70
150 amps, 42 circuits	E4@2.82	Ea	921.00	123.00	1,044.00
200 amps, 42 circuits	E4@3.63	Ea	921.00	158.00	1,079.00
225 amps, 42 circuits	E4@3.78	Ea	968.00	164.00	1,132.00
400 amps, 24 circuits	E4@3.91	Ea	1,950.00	170.00	2,120.00
400 amps, 42 circuits	E4@4.51	Ea	2,260.00	196.00	2,456.00
Add for NEMA-4 raintight loadcenters	—	%	25.0	—	—

Single phase 3 wire, 120/240 volt loadcenters with main lugs only

	Craft@Hrs	Unit	Material	Labor	Total
150 amps, 12 circuits	E4@1.20	Ea	105.00	52.10	157.10
150 amps, 16 circuits	E4@1.20	Ea	133.00	52.10	185.10
150 amps, 20 circuits	E4@1.52	Ea	139.00	66.00	205.00
200 amps, 30 circuits	E4@1.70	Ea	192.00	73.90	265.90
200 amps, 40 circuits	E4@2.03	Ea	237.00	88.20	325.20
225 amps, 24 circuits	E4@2.50	Ea	310.00	109.00	419.00
225 amps, 42 circuits	E4@2.77	Ea	382.00	120.00	502.00

Three phase 4 wire 120/208 volt loadcenters with main lugs only

	Craft@Hrs	Unit	Material	Labor	Total
125 amps, 12 circuits	E4@1.19	Ea	134.00	51.70	185.70
150 amps, 18 circuits	E4@1.47	Ea	201.00	63.90	264.90
200 amps, 30 circuits	E4@1.90	Ea	272.00	82.60	354.60
200 amps, 42 circuits	E4@2.18	Ea	362.00	94.70	456.70

	Craft@Hrs	Unit	Material	Labor	Total
225 amps, 42 circuits	E4@2.75	Ea	382.00	119.00	501.00
400 amps, 24 circuits	E4@3.63	Ea	697.00	158.00	855.00
400 amps, 42 circuits	E4@3.78	Ea	775.00	164.00	939.00
600 amps, 24 circuits	E4@3.91	Ea	848.00	170.00	1,018.00
600 amps, 42 circuits	E4@6.53	Ea	1,040.00	284.00	1,324.00

Circuit Breakers No enclosures included. 10,000 amp interrupt capacity except as noted.

Plug-in molded case 120 volt, 100 amp frame circuit breakers, 1" or 1/2" module

	Craft@Hrs	Unit	Material	Labor	Total
15 to 50 amps, single pole	CE@.270	Ea	8.69	13.60	22.29
15 to 60 amps, two pole	CE@.374	Ea	20.10	18.90	39.00
70 amps, two pole	CE@.563	Ea	38.60	28.40	67.00
80 to 100 amps, two pole	CE@.680	Ea	54.20	34.30	88.50

Plug-in molded case 240 volt, 100 amp frame circuit breakers

	Craft@Hrs	Unit	Material	Labor	Total
15 thru 60 amps, single pole	CE@.417	Ea	11.50	21.10	32.60
15 thru 60 amps, two pole	CE@.577	Ea	75.80	29.10	104.90
15 thru 60 amps, three pole	CE@.863	Ea	94.00	43.60	137.60
60 thru 100 amps, two pole	CE@.577	Ea	125.00	29.10	154.10
60 thru 100 amps, three pole	CE@.897	Ea	140.00	45.30	185.30

Plug-in molded case 480 volt, 100 amp frame circuit breakers

	Craft@Hrs	Unit	Material	Labor	Total
70 thru 100 amps, two pole	CE@1.22	Ea	287.00	61.60	348.60
70 thru 100 amps, three pole	CE@1.71	Ea	335.00	86.40	421.40

Plug-in molded case 600 volt, 100 amp frame circuit breakers

	Craft@Hrs	Unit	Material	Labor	Total
15 thru 60 amps, two pole	CE@.577	Ea	240.00	29.10	269.10
15 thru 60 amps, three pole	CE@.897	Ea	308.00	45.30	353.30
70 thru 100 amps, two pole	CE@1.22	Ea	321.00	61.60	382.60
70 thru 100 amps, three pole	CE@1.71	Ea	404.00	86.40	490.40

Plug-in molded case 250/600 volt, 225 amp frame circuit breakers

	Craft@Hrs	Unit	Material	Labor	Total
125 thru 225 amps, two pole	CE@2.06	Ea	749.00	104.00	853.00
125 thru 225 amps, three pole	CE@2.57	Ea	940.00	130.00	1,070.00

Plug-in molded case 250/600 volt, 400 amp frame circuit breakers

	Craft@Hrs	Unit	Material	Labor	Total
250 thru 400 amps, two pole	CE@2.88	Ea	1,180.00	145.00	1,325.00
250 thru 400 amps, three pole	CE@4.33	Ea	1,410.00	219.00	1,629.00

Plug-in molded case 250/600 volt, 800 amp frame circuit breakers

	Craft@Hrs	Unit	Material	Labor	Total
400 thru 600 amps, three pole	CE@6.04	Ea	2,590.00	305.00	2,895.00
700 thru 800 amps, two pole	CE@4.73	Ea	2,620.00	239.00	2,859.00
700 thru 800 amps, three pole	CE@6.71	Ea	3,370.00	339.00	3,709.00

Plug-in molded case 600 volt, 1,200 amp frame circuit breakers

	Craft@Hrs	Unit	Material	Labor	Total
600 thru 1,200 amps, two pole	CE@9.60	Ea	2,450.00	485.00	2,935.00
600 thru 1,200 amps, three pole	CE@11.9	Ea	3,130.00	601.00	3,731.00

Space only for factory assembled panels

	Craft@Hrs	Unit	Material	Labor	Total
Single pole space	—	Ea	19.50	—	19.50
Two pole space	—	Ea	17.90	—	17.90
Three pole space	—	Ea	24.10	—	24.10

Bolt-on molded case 120/240 volt circuit breakers

	Craft@Hrs	Unit	Material	Labor	Total
15 thru 60 amps, one pole	CE@.320	Ea	15.10	16.20	31.30
15 thru 60 amps, two pole	CE@.484	Ea	33.20	24.40	57.60
15 thru 60 amps, three pole	CE@.638	Ea	103.00	32.20	135.20
70 thru 100 amps, two pole	CE@.712	Ea	125.00	36.00	161.00
70 thru 100 amps, three pole	CE@.946	Ea	148.00	47.80	195.80

Bolt-on molded case 480 volt circuit breakers, 65,000 amp interrupt capacity

	Craft@Hrs	Unit	Material	Labor	Total
15 thru 60 amps, two pole	CE@.484	Ea	124.00	24.40	148.40
15 thru 60 amps, three pole	CE@.712	Ea	217.00	36.00	253.00
70 thru 100 amps, three pole	CE@.804	Ea	330.00	40.60	370.60

Electrical 16

	Craft@Hrs	Unit	Material	Labor	Total
Circuit Breaker Panels 4 wire 3 phase 240 volts, flush-mounted, with main breaker, assembled					
100 amp service, 12 circuits	E4@4.60	Ea	448.00	200.00	648.00
100 amp service, 18 circuits	E4@6.90	Ea	475.00	300.00	775.00
100 amp service, 24 circuits	E4@8.10	Ea	514.00	352.00	866.00
100 amp service, 30 circuits	E4@8.20	Ea	554.00	356.00	910.00
100 amp service, 36 circuits	E4@8.30	Ea	591.00	361.00	952.00
100 amp service, 42 circuits	E4@8.40	Ea	631.00	365.00	996.00
225 amp service, 12 circuits	E4@8.50	Ea	978.00	369.00	1,347.00
225 amp service, 18 circuits	E4@8.60	Ea	1,000.00	374.00	1,374.00
225 amp service, 24 circuits	E4@8.70	Ea	1,040.00	378.00	1,418.00
225 amp service, 30 circuits	E4@10.0	Ea	1,080.00	435.00	1,515.00
225 amp service, 36 circuits	E4@12.8	Ea	112.00	556.00	668.00
225 amp service, 42 circuits	E4@15.1	Ea	1,150.00	656.00	1,806.00
Fuses No enclosures included.					
250 volt cartridge fuses, one time non-renewable					
To 30 amps	CE@.077	Ea	1.17	3.89	5.06
35 to 60 amps	CE@.077	Ea	1.75	3.89	5.64
70 to 100 amps	CE@.077	Ea	7.91	3.89	11.80
110 to 200 amps	CE@.077	Ea	20.80	3.89	24.69
225 to 400 amps	CE@.077	Ea	34.30	3.89	38.19
450 to 600 amps	CE@.077	Ea	54.80	3.89	58.69
600 volt cartridge fuses, one time non-renewable					
To 30 amps	CE@.077	Ea	6.89	3.89	10.78
35 to 60 amps	CE@.077	Ea	10.30	3.89	14.19
70 to 100 amps	CE@.077	Ea	21.60	3.89	25.49
110 to 200 amps	CE@.077	Ea	43.20	3.89	47.09
225 to 400 amps	CE@.087	Ea	86.00	4.39	90.39
450 to 600 amps	CE@.124	Ea	130.00	6.26	136.26
250 volt class H renewable fuses					
To 10 amps	CE@.077	Ea	9.04	3.89	12.93
12 to 30 amps	CE@.077	Ea	7.13	3.89	11.02
35 to 60 amps	CE@.077	Ea	16.90	3.89	20.79
70 to 100 amps	CE@.077	Ea	38.00	3.89	41.89
110 to 200 amps	CE@.077	Ea	87.70	3.89	91.59
225 to 400 amps	CE@.080	Ea	159.00	4.04	163.04
450 to 600 amps	CE@.122	Ea	256.00	6.16	262.16
600 volt class H renewable fuses					
To 10 amps	CE@.077	Ea	25.40	3.89	29.29
12 to 30 amps	CE@.077	Ea	20.30	3.89	24.19
35 to 60 amps	CE@.077	Ea	33.70	3.89	37.59
70 to 100 amps	CE@.077	Ea	75.90	3.89	79.79
110 to 200 amps	CE@.077	Ea	149.00	3.89	152.89
225 to 400 amps	CE@.153	Ea	314.00	7.73	321.73
450 to 600 amps	CE@.223	Ea	413.00	11.30	424.30
250 volt class RK5 non-renewable cartridge fuses					
10 to 30 amps	CE@.077	Ea	7.10	3.89	10.99
35 to 60 amps	CE@.077	Ea	7.44	3.89	11.33
70 to 100 amps	CE@.077	Ea	18.90	3.89	22.79
110 to 200 amps	CE@.077	Ea	43.80	3.89	47.69
225 to 400 amps	CE@.230	Ea	85.90	11.60	97.50
450 to 600 amps	CE@.297	Ea	134.00	15.00	149.00
650 to 1,200 amps	CE@.372	Ea	166.00	18.80	184.80
1,400 to 1,600 amps	CE@.804	Ea	225.00	40.60	265.60
1,800 to 2,000 amps	CE@1.00	Ea	287.00	50.50	337.50

	Craft@Hrs	Unit	Material	Labor	Total
600 volt class RK5 non-renewable cartridge fuses					
10 to 30 amps	CE@.077	Ea	9.81	3.89	13.70
35 to 60 amps	CE@.077	Ea	16.80	3.89	20.69
70 to 100 amps	CE@.077	Ea	34.70	3.89	38.59
110 to 200 amps	CE@.129	Ea	69.10	6.51	75.61
225 to 400 amps	CE@.257	Ea	138.00	13.00	151.00
450 to 600 amps	CE@.354	Ea	198.00	17.90	215.90
650 to 1,200 amps	CE@.372	Ea	315.00	18.80	333.80
1,400 to 1,600 amps	CE@.797	Ea	440.00	40.20	480.20
1,800 to 2,000 amps	CE@1.00	Ea	520.00	50.50	570.50
Add for type K5 600 volt fuses	—	%	22.4	—	—
Deduct for type J 600 volt fuses	—	%	-50.0	—	—
600 volt class L fuses, bolt-on					
600 to 1,200 amps	CE@.333	Ea	487.00	16.80	503.80
1,500 to 1,600 amps	CE@.615	Ea	656.00	31.10	687.10
1,400 to 2,000 amps	CE@.809	Ea	868.00	40.90	908.90
2,500 amps	CE@1.10	Ea	868.00	55.60	923.60
3,000 amps	CE@1.25	Ea	1,000.00	63.10	1,063.10
3,500 to 4,000 amps	CE@1.74	Ea	1,660.00	87.90	1,747.90
5,000 amps	CE@2.33	Ea	2,130.00	118.00	2,248.00

	Craft@Hrs	Unit	Material	Labor	Equipment	Total
Overhead Electrical Distribution						

Poles, pressure treated wood, class 4, type C, set with crane or boom, including augured holes. Equipment is a flatbed truck with a boom and an auger.

	Craft@Hrs	Unit	Material	Labor	Equipment	Total
25' high	E1@7.62	Ea	286.00	340.00	89.60	715.60
30' high	E1@7.62	Ea	369.00	340.00	89.60	798.60
35' high	E1@9.72	Ea	478.00	434.00	114.00	1,026.00
40' high	E1@11.5	Ea	568.00	514.00	135.00	1,217.00
Cross arms, wood, with typical hardware						
4' high	E1@2.86	Ea	68.00	128.00	33.60	229.60
5' high	E1@2.86	Ea	85.20	128.00	33.60	246.80
6' high	E1@4.19	Ea	123.00	187.00	49.80	359.80
8' high	E1@4.91	Ea	163.00	219.00	57.70	439.70

Transformers, pole mounted, single phase 7620/13200 Y primary, 120/240 secondary, with two 2-1/2% taps above and below

	Craft@Hrs	Unit	Material	Labor	Equipment	Total
10 KVA	E1@4.14	Ea	867.00	185.00	48.70	1,100.70
15 KVA	E1@4.14	Ea	997.00	185.00	48.70	1,230.70
25 KVA	E1@4.14	Ea	1,220.00	185.00	48.70	1,453.70
36.5 KVA	E1@4.14	Ea	1,630.00	185.00	48.70	1,863.70
50 KVA	E1@4.14	Ea	1,860.00	185.00	48.70	2,093.70
75 KVA	E1@6.16	Ea	2,760.00	275.00	72.40	3,107.40
100 KVA	E1@6.16	Ea	3,140.00	275.00	72.40	3,487.40
167 KVA	E1@6.16	Ea	4,830.00	275.00	72.40	5,177.40
250 KVA	E1@6.16	Ea	4,940.00	275.00	72.40	5,287.40
333 KVA	E1@6.16	Ea	5,340.00	275.00	72.40	5,687.40
500 KVA	E1@6.16	Ea	7,440.00	275.00	72.40	7,787.40
Fused cutouts, pole mounted, 5 KV						
50 amp	E1@2.56	Ea	98.40	114.00	30.10	242.50
100 amp	E1@2.56	Ea	99.70	114.00	30.10	243.80
250 amp	E1@2.56	Ea	113.00	114.00	30.10	257.10

	Craft@Hrs	Unit	Material	Labor	Equipment	Total

Switches, disconnect, pole mounted, pole arm throw, 5 KV, set of 3 with throw and lock

	Craft@Hrs	Unit	Material	Labor	Equipment	Total
400 amp	E1@45.8	Ea	1,860.00	2,050.00	538.00	4,448.00
600 amp	E1@45.8	Ea	3,520.00	2,050.00	538.00	6,108.00
1,200 amp	E1@51.4	Ea	6,840.00	2,300.00	604.00	9,744.00

	Craft@Hrs	Unit	Material	Labor	Total

Fluorescent Lighting Commercial, industrial and architectural grade. Rapid start. With ballasts but no lamps, wire or conduit. Includes setting and connecting prewired fixtures only.

Surface mounted fluorescent fixtures, better quality, baked enamel finish, with hinged wraparound acrylic prismatic lens suitable for office or classroom use, 4' long, Lithonia LB series

	Craft@Hrs	Unit	Material	Labor	Total
Two 40 watt lamp fixture (#240-A)	CE@.773	Ea	59.60	39.00	98.60
Four 40 watt lamp fixture (#440-A)	CE@.773	Ea	113.00	39.00	152.00
Eight 40 watt lamp fixture, 8' long (#8T440-A)	CE@.960	Ea	172.00	48.50	220.50

Surface mounted fluorescent fixtures, baked enamel finish, with wraparound acrylic prismatic lens suitable for industrial use, 4' long, Lithonia series

	Craft@Hrs	Unit	Material	Labor	Total
One 40 watt lamp fixture (CB140A)	CE@.806	Ea	53.40	40.70	94.10
Two 40 watt lamp fixture (LB240A)	CE@.858	Ea	45.80	43.30	89.10
Four 40 watt lamp fixture (LB440A)	CE@1.07	Ea	78.00	54.00	132.00
Add for 40 watt lamps	—	Ea	3.05	—	3.05

Surface mounted steel sided fluorescent fixtures with acrylic prismatic diffuser, 7-1/16" x 4-1/2"

	Craft@Hrs	Unit	Material	Labor	Total
Two 40 watt lamps	CE@.955	Ea	65.40	48.20	113.60
Four 40 watt lamps	CE@1.05	Ea	91.20	53.00	144.20
Add for 40 watt straight lamps	—	Ea	3.05	—	3.05

Wall surface mounted fluorescent fixtures, baked enamel finish, 2 lamps, 8" wide 5" high, hinged acrylic prismatic lens, Lightolier

	Craft@Hrs	Unit	Material	Labor	Total
26" long fixture	CE@.737	Ea	136.00	37.20	173.20
38" long fixture	CE@.795	Ea	158.00	40.10	198.10
50" long fixture	CE@.858	Ea	154.00	43.30	197.30
Add for 24" 20 watt lamps	—	Ea	2.79	—	2.79
Add for 36" 30 watt lamps	—	Ea	4.33	—	4.33
Add for 48" 40 watt lamps	—	Ea	1.98	—	1.98

Industrial pendant or surface mounted open striplights, baked enamel finish, Lithonia C series

	Craft@Hrs	Unit	Material	Labor	Total
One or two 40 watt lamps, 4' long (C240)	CE@.865	Ea	31.00	43.70	74.70
One or two 75 watt lamps, 8' long (C296)	CE@1.02	Ea	45.80	51.50	97.30
One or two 55 watt lamps, 6' long (C272)	CE@1.02	Ea	45.80	51.50	97.30
Add for 40 watt lamps	—	Ea	2.13	—	2.13
Add for 75 or 55 watt lamps	—	Ea	5.05	—	5.05
Add for 24" stem and canopy set, per fixture	—	Ea	5.05	—	5.05

Industrial pendant or surface mounted fluorescent fixtures, 12" wide baked enamel, steel reflector-shield, Lithonia L/LA series

	Craft@Hrs	Unit	Material	Labor	Total
Two 40 watt lamps, 4' long (L-240-120)	CE@.903	Ea	33.30	45.60	78.90
Two 75 watt lamps, 8' long (L-296-120)	CE@1.21	Ea	65.40	61.10	126.50
Add for 40 watt lamps	—	Ea	2.13	—	2.13
Add for 75 watt lamps	—	Ea	5.09	—	5.09

Ceiling grid mounted fluorescent troffer fixtures, baked enamel finish, with hinged acrylic prismatic lens with energy saving ballast, Day-Brite

	Craft@Hrs	Unit	Material	Labor	Total
Two 40 watt "U" lamp fixture, 2' x 2'	CE@.908	Ea	48.30	45.90	94.20
Two 40 watt lamp fixture, 2' x 4'	CE@.998	Ea	54.70	50.40	105.10
Four 40 watt lamp fixture, 2' x 4'	CE@1.05	Ea	59.20	53.00	112.20
Add for plaster frames, per fixture	—	Ea	7.06	—	7.06

Air-handling grid mounted floating door, heat transfer troffer fixtures with acrylic lens, no ductwork included

	Craft@Hrs	Unit	Material	Labor	Total
1' x 4', two 40 watt lamps	CE@.908	Ea	92.00	45.90	137.90
2' x 2', one U-shape 40 watt lamp	CE@.809	Ea	96.60	40.90	137.50
2' x 4', two 40 watt lamps	CE@1.00	Ea	122.00	50.50	172.50

	Craft@Hrs	Unit	Material	Labor	Total
2' x 4', three 40 watt lamps	CE@1.05	Ea	151.00	53.00	204.00
2' x 4', four 40 watt lamps	CE@1.05	Ea	151.00	53.00	204.00
Add for 40 watt 4' lamps	—	Ea	2.13	—	2.13
Add for 40 watt "U" lamps	—	Ea	8.88	—	8.88

Surface mounted vandal resistant luminaire with tamper-resistant screw-type latching, Sylvania

	Craft@Hrs	Unit	Material	Labor	Total
1 lamp	CE@.761	Ea	100.00	38.40	138.40
2 lamps	CE@.761	Ea	102.00	38.40	140.40
4 lamps	CE@.761	Ea	174.00	38.40	212.40

Parabolic fluorescent luminaire, 6" x 24" x 24" with semi-specular anodized aluminum reflector and parabolic louvers, I.C.E. AR series

	Craft@Hrs	Unit	Material	Labor	Total
Two 40 watt lamp, bracket or pendant mount	CE@.955	Ea	63.50	48.20	111.70
Four 40 watt lamps, bracket or pendant mount	CE@.955	Ea	101.00	48.20	149.20

Round surface mounted fluorescent fixtures, aluminum housing, matte black finish, polycarbonate or acrylic opal globe, Circleline

	Craft@Hrs	Unit	Material	Labor	Total
22 to 32 watt, 14" diameter	CE@.478	Ea	40.00	24.10	64.10
32 to 40 watt, 18" diameter	CE@.478	Ea	55.20	24.10	79.30
Add for circline tubes	—	Ea	6.32	—	6.32

Outdoor sign fluorescent fixtures, anodized aluminum housing with parabolic specular reflector and acrylic lens, with high output lamps and remote ballast

	Craft@Hrs	Unit	Material	Labor	Total
One 60 watt lamp. 48" long fixture	CE@2.39	Ea	192.00	121.00	313.00
One 85 watt lamp, 72" long fixture	CE@2.39	Ea	203.00	121.00	324.00
One 110 watt lamp, 96" long fixture	CE@2.39	Ea	224.00	121.00	345.00
Two 85 watt lamps, 144" long fixture	CE@2.39	Ea	303.00	121.00	424.00
Add for high output T12 800 MA lamps, sign white					
60 watt	—	Ea	5.37	—	5.37
85 watt	—	Ea	5.37	—	5.37
110 watt	—	Ea	6.08	—	6.08
Cost per linear foot of fixture	CE@.179	LF	51.80	9.04	60.84
Add for 90-degree elbows	CE@.475	Ea	133.00	24.00	157.00
Add for in-line connectors	CE@.475	Ea	67.00	24.00	91.00

Wet or damp location surface mounted fluorescent fixtures, hinge mounted acrylic diffuse lens, fiberglass or plastic housing with gaskets, Hubbell

	Craft@Hrs	Unit	Material	Labor	Total
Two 40 watt lamps, 4' long fixture	CE@1.21	Ea	115.00	61.10	176.10
Two 40 watt lamps, 8' long fixture	CE@1.86	Ea	249.00	93.90	342.90
Add for 40 watt lamps	—	Ea	2.13	—	2.13

High Intensity Discharge Lighting

Commercial and architectural grade fixtures. With ballasts but no lamps, wire, conduit or concrete except as noted. See lamp costs below. Labor includes setting and connecting prewired fixtures only

Recess mounted commercial luminaire, 2' x 2' x 13", aluminum reflector glass lens, Wide-Lite series

	Craft@Hrs	Unit	Material	Labor	Total
175 watt, metal halide	CE@1.15	Ea	538.00	58.10	596.10
250 watt, high pressure sodium	CE@1.15	Ea	618.00	58.10	676.10
400 watt, metal halide	CE@1.15	Ea	543.00	58.10	601.10
400 watt, high pressure sodium	CE@1.15	Ea	677.00	58.10	735.10

Surface mounted commercial luminaire, 2' x 2' x 15", baked enamel finish with tempered glass lens, Holophane series

	Craft@Hrs	Unit	Material	Labor	Total
70 watt, high pressure sodium	CE@1.42	Ea	325.00	71.70	396.70
100 watt, high pressure sodium	CE@1.42	Ea	348.00	71.70	419.70
150 watt, high pressure sodium	CE@1.42	Ea	357.00	71.70	428.70
175 watt, metal halide	CE@1.42	Ea	342.00	71.70	413.70
400 watt, metal halide	CE@1.42	Ea	368.00	71.70	439.70

Recess mounted indirect metal halide luminaire, aluminum reflector with glass lens

	Craft@Hrs	Unit	Material	Labor	Total
175 watt, square	CE@1.71	Ea	538.00	86.40	624.40
250 watt, square	CE@1.71	Ea	538.00	86.40	624.40

	Craft@Hrs	Unit	Material	Labor	Total
400 watt, square	CE@1.71	Ea	543.00	86.40	629.40
175 watt, round	CE@1.71	Ea	517.00	86.40	603.40
250 watt, round	CE@1.71	Ea	517.00	86.40	603.40
400 watt, round	CE@1.71	Ea	522.00	86.40	608.40
Surface mounted indirect high pressure sodium luminaire, aluminum reflector with glass lens					
250 watt, square	CE@1.71	Ea	594.00	86.40	680.40
400 watt, square	CE@1.71	Ea	651.00	86.40	737.40
250 watt, round	CE@1.71	Ea	572.00	86.40	658.40
400 watt, round	CE@1.71	Ea	629.00	86.40	715.40
Outdoor wall pack, aluminum housing with hinged polycarbonate access door and lens, Sylvania					
100 watt, high pressure sodium	CE@.955	Ea	192.00	48.20	240.20
150 watt, high pressure sodium	CE@.955	Ea	170.00	48.20	218.20
175 watt, mercury vapor	CE@.955	Ea	165.00	48.20	213.20
175 watt, metal halide	CE@.955	Ea	211.00	48.20	259.20
Utility wall pack, aluminum housing with hinged polycarbonate or tempered glass lens, Holophane Wallpack 2					
70 watt, high pressure sodium	CE@.955	Ea	325.00	48.20	373.20
100 watt, high pressure sodium	CE@.955	Ea	348.00	48.20	396.20
150 watt, high pressure sodium	CE@.955	Ea	357.00	48.20	405.20
175 watt, metal halide	CE@.955	Ea	342.00	48.20	390.20
250 watt, mercury vapor	CE@.955	Ea	315.00	48.20	363.20
Add for photo cell automatic control, factory installed	—	Ea	20.30	—	20.30
High bay metal halide industrial fixture, cast aluminum housing, spun aluminum reflector, lens is tempered glass					
400 watt	CE@2.61	Ea	193.00	132.00	325.00
1,000 watt	CE@2.61	Ea	329.00	132.00	461.00
Add for power hook and cord	CE@.250	Ea	60.20	12.60	72.80
Add for thru wire power hook receptacle	CE@.199	Ea	51.20	10.00	61.20
Add for wire guard on lens	CE@.199	Ea	60.20	10.00	70.20
Low bay high pressure sodium industrial fixture, cast aluminum with epoxy finish, acrylic lens					
100 watt fixture	CE@1.42	Ea	284.00	71.70	355.70
150 watt fixture	CE@1.39	Ea	295.00	70.20	365.20
250 watt fixture	CE@1.39	Ea	421.00	70.20	491.20
Add for power hook and cord	CE@.250	Ea	55.50	12.60	68.10
Add for thru wire power hook receptacle	CE@.199	Ea	48.70	10.00	58.70
Add for wire guard on lens	CE@.199	Ea	66.10	10.00	76.10
Low bay metal halide industrial downlight, cast aluminum with epoxy finish, shock resistant glass lens					
100 watt fixture	CE@1.07	Ea	354.00	54.00	408.00
400 watt fixture	CE@1.39	Ea	444.00	70.20	514.20
Add for power hook and cord	CE@.250	Ea	55.50	12.60	68.10
Add for thru wire power hook receptacle	CE@.199	Ea	48.70	10.00	58.70
Add for wire guard on lens	CE@.199	Ea	67.50	10.00	77.50
Lamps for high intensity discharge luminaires					
400 watt, mercury vapor	—	Ea	54.70	—	54.70
1,000 watt, mercury vapor	—	Ea	99.80	—	99.80
35 watt, high pressure sodium	—	Ea	36.40	—	36.40
70 watt, high pressure sodium	—	Ea	36.50	—	36.50
100 watt, high pressure sodium	—	Ea	41.10	—	41.10
150 watt, high pressure sodium	—	Ea	35.60	—	35.60
250 watt, high pressure sodium	—	Ea	36.30	—	36.30
400 watt, high pressure sodium	—	Ea	31.00	—	31.00
1,000 watt, high pressure sodium	—	Ea	112.00	—	112.00

	Craft@Hrs	Unit	Material	Labor	Total
175 watt, metal halide	—	Ea	67.20	—	67.20
250 watt, metal halide	—	Ea	39.90	—	39.90
400 watt, metal halide	—	Ea	37.10	—	37.10
1,000 watt, metal halide	—	Ea	96.70	—	96.70
Add for diffuse coated lamps	—	Ea	5.57	—	5.57

Incandescent Lighting Commercial and architectural grade. No wire, conduit or concrete included. Includes setting and connecting prewired fixtures only.

Step illumination light fixtures, 16 gauge steel housing with white enamel finish, 5" high, 4" deep, 11" long

	Craft@Hrs	Unit	Material	Labor	Total
Polycarbonate or tempered glass lens fixture	CE@1.07	Ea	152.00	54.00	206.00
Louvered sheet metal front fixture	CE@1.07	Ea	130.00	54.00	184.00
Add for brushed aluminum trim, 2 lamps	—	Ea	36.40	—	36.40

Adjustable recessed spotlight, 20 gauge steel housing with matt finish, 30 degree adjustable with 350 degree rotation

	Craft@Hrs	Unit	Material	Labor	Total
150 watt fixture	CE@1.07	Ea	125.00	54.00	179.00

Recessed round eyelid wall wash, 6" milligroove baffle, Prescolite

	Craft@Hrs	Unit	Material	Labor	Total
150 watt fixture	CE@1.07	Ea	128.00	54.00	182.00

Exterior wall fixtures, cast anodized aluminum with white polycarbonate globe, gaskets between globe and housing and between housing and base, Prescolite

	Craft@Hrs	Unit	Material	Labor	Total
Cylinder globe fixture	CE@.320	Ea	21.60	16.20	37.80
Round globe fixture	CE@.430	Ea	25.70	21.70	47.40
Add for 100 watt lamp	—	Ea	1.68	—	1.68
Add for 150 watt lamp	—	Ea	2.46	—	2.46

Obstruction light, meets Federal Aviation Administration Specification L-810, cast aluminum housing, two one-piece red, heat-resistant fresnel globes, with photoelectric control, mounted on 1" rigid steel conduit, including junction box and mounting plate

	Craft@Hrs	Unit	Material	Labor	Total
Two 100 watt lamp fixture	CE@4.91	Ea	107.00	248.00	355.00

Vandal Resistant Lighting Commercial and architectural grade. Includes setting and connecting prewired fixtures but no wire, conduit or lamps.

Ceiling mounted 12" x 12" x 6" fixture, polycarbonate prismatic lens

	Craft@Hrs	Unit	Material	Labor	Total
Two 100 watt incandescent lamps	CE@.764	Ea	74.20	38.60	112.80
35 watt, high pressure sodium lamp	CE@.764	Ea	216.00	38.60	254.60
50 watt, high pressure sodium lamp	CE@.764	Ea	221.00	38.60	259.60
70 watt, high pressure sodium lamp	CE@.764	Ea	232.00	38.60	270.60
22 watt fluorescent circline lamp	CE@.955	Ea	52.10	48.20	100.30

Wall mounted 6" wide, 9" high, 7" deep fixture, polycarbonate prismatic diffuser

	Craft@Hrs	Unit	Material	Labor	Total
One 100 watt incandescent lamp	CE@.572	Ea	52.40	28.90	81.30
13 watt fluorescent lamp	CE@.764	Ea	82.30	38.60	120.90
35 watt, high pressure sodium lamp	CE@.475	Ea	222.00	24.00	246.00
50 watt, high pressure sodium lamp	CE@.475	Ea	222.00	24.00	246.00
70 watt, high pressure sodium lamp	CE@.475	Ea	212.00	24.00	236.00

Lamps for vandal resistant fixtures

	Craft@Hrs	Unit	Material	Labor	Total
100 watt incandescent lamp	—	Ea	1.67	—	1.67
13 watt fluorescent lamp	—	Ea	8.06	—	8.06
35 to 70 watt high pressure sodium lamp	—	Ea	57.60	—	57.60

Lighted Exit Signs Includes setting and connecting prewired fixtures but no wire or conduit.

Universal mount (wall or ceiling), Thinline series, 6" letters on one side, stencil face

	Craft@Hrs	Unit	Material	Labor	Total
Single	CE@.764	Ea	60.30	38.60	98.90
Twin	CE@.764	Ea	60.40	38.60	99.00

	Craft@Hrs	Unit	Material	Labor	Total
Universal mount, 6" letters on two sides of polycarbonate housing, two 15 watt lamps	CE@.764	Ea	67.70	38.60	106.30
Add for two circuit supply	—	Ea	8.37	—	8.37
Add for 24" stem hanger	—	Ea	11.40	—	11.40
Add for battery backup for exit signs	CE@.381	Ea	125.00	19.20	144.20

Explosion Proof Lighting

No wire, conduit or concrete included. Includes setting and connecting prewired fixtures only.

Explosion proof fixtures, cast aluminum housing, heat-resistant prestressed globe, with fiberglass reinforced polyester reflector

	Craft@Hrs	Unit	Material	Labor	Total
60-200 watt incandescent fixture	CE@1.74	Ea	67.50	87.90	155.40
70 watt HID fixture	CE@1.74	Ea	262.00	87.90	349.90
150 watt HID fixture	CE@1.74	Ea	370.00	87.90	457.90

Explosion proof (Class 1 Division 1) fixtures, cast aluminum housing, heat-resistant prestressed globe, with fiberglass reinforced polyester reflector

	Craft@Hrs	Unit	Material	Labor	Total
250 watt, high pressure sodium fixture	CE@2.57	Ea	726.00	130.00	856.00
175 watt, metal halide fixture	CE@2.57	Ea	558.00	130.00	688.00
400 watt, metal halide fixture	CE@2.57	Ea	698.00	130.00	828.00

Explosion proof (Class 1 Division 2) fixtures, cast aluminum housing, heat-resistant prestressed globe, with fiberglass reinforced polyester reflector

	Craft@Hrs	Unit	Material	Labor	Total
70 watt, high pressure sodium fixture	CE@2.57	Ea	375.00	130.00	505.00
250 watt, high pressure sodium fixture	CE@2.57	Ea	514.00	130.00	644.00
175 watt, metal halide fixture	CE@2.57	Ea	348.00	130.00	478.00
400 watt, metal halide fixture	CE@2.57	Ea	493.00	130.00	623.00

Emergency Lighting Includes setting and connecting prewired fixtures but no wire or conduit.

Shelf or wall mounted solid state battery pack with charger, lead-acid battery, baked enamel finish, with 2 halogen lamps, 90 minute operation, indoor use only

	Craft@Hrs	Unit	Material	Labor	Total
6 volt, 15 watt	CE@1.74	Ea	88.50	87.90	176.40
6 volt, 30 watt	CE@1.74	Ea	135.00	87.90	222.90
6 volt, 50 watt	CE@1.74	Ea	160.00	87.90	247.90
6 volt, 100 watt	CE@1.74	Ea	216.00	87.90	303.90
12 volt, 100 watt	CE@1.74	Ea	216.00	87.90	303.90

Self diagnostic emergency lighting fixtures, microprocessor based, continuously monitors battery/lamps

	Craft@Hrs	Unit	Material	Labor	Total
Emergency fixture, two lamps, wall mount	CE@.825	Ea	165.00	41.70	206.70
LED exit sign with emergency fixture, wall mount	CE@.825	Ea	289.00	41.70	330.70
LED exit sign only, wall mount	CE@.825	Ea	246.00	41.70	287.70

Yard and Street Lighting Includes setting and connecting prewired fixtures with ballast but no pole, concrete work, excavation, wire, lamps or conduit. See pole costs below. Equipment is a flatbed truck with a boom & auger.

Mast arm mounted rectangular high pressure sodium flood, aluminum housing with baked enamel finish, anodized aluminum reflector, tempered glass lens, slip-fitting for mast mount, photoelectric cell

	Craft@Hrs	Unit	Material	Labor	Equipment	Total
70 watt	E1@1.55	Ea	375.00	69.20	18.20	462.40
100 watt	E1@1.55	Ea	388.00	69.20	18.20	475.40
150 watt	E1@1.55	Ea	388.00	69.20	18.20	475.40
250 watt	E1@1.55	Ea	415.00	69.20	18.20	502.40
400 watt	E1@1.55	Ea	495.00	69.20	18.20	582.40
Add for 6' x 2" mounting arm	E1@.547	Ea	57.90	24.40	6.43	88.73

Yoke mounted high intensity discharge flood, die-cast anodized aluminum housing with tempered glass lens

	Craft@Hrs	Unit	Material	Labor	Equipment	Total
250 watt, high pressure sodium	E1@2.71	Ea	326.00	121.00	31.90	478.90
250 watt, metal halide	E1@2.71	Ea	286.00	121.00	31.90	438.90
400 watt, high pressure sodium	E1@2.71	Ea	336.00	121.00	31.90	488.90

	Craft@Hrs	Unit	Material	Labor	Equipment	Total
400 watt, metal halide	E1@2.71	Ea	291.00	121.00	31.90	443.90
1,000 watt, high pressure sodium	E1@2.71	Ea	548.00	121.00	31.90	700.90
1,000 watt, metal halide	E1@2.71	Ea	439.00	121.00	31.90	591.90
1,000 watt, mercury vapor	E1@2.71	Ea	71.70	121.00	31.90	224.60

Round flood, aluminum reflector, impact-resistant glass lens

	Craft@Hrs	Unit	Material	Labor	Equipment	Total
400 watt, high pressure sodium	E1@2.71	Ea	380.00	121.00	31.90	532.90
1,000 watt, high pressure sodium	E1@2.71	Ea	608.00	121.00	31.90	760.90
400 watt, metal halide	E1@2.71	Ea	349.00	121.00	31.90	501.90
1,000 watt, metal halide	E1@2.71	Ea	477.00	121.00	31.90	629.90

Cobra head high intensity flood, anodized aluminum reflector, glass lens, die cast aluminum housing, with photoelectric control but no bulb or mounting arm

	Craft@Hrs	Unit	Material	Labor	Equipment	Total
1,000 watt, high pressure sodium	E1@2.71	Ea	554.00	121.00	31.90	706.90
1,000 watt, mercury vapor	E1@2.71	Ea	406.00	121.00	31.90	558.90
1,000 watt, metal halide	E1@2.71	Ea	438.00	121.00	31.90	590.90

Square metal halide flood, anodized aluminum housing, tempered glass door, trunnion or yoke mount

	Craft@Hrs	Unit	Material	Labor	Equipment	Total
400 watt	E1@2.71	Ea	471.00	121.00	31.90	623.90
1,000 watt	E1@2.71	Ea	557.00	121.00	31.90	709.90

Yard and Street Lighting Poles

Hole for a pole foundation dug with a truck-mounted auger, no soil disposal included

	Craft@Hrs	Unit	Material	Labor	Equipment	Total
Per CF of undisturbed soil	E1@.064	CF	—	2.86	.75	3.61

Concrete pole foundations, formed, poured and finished, with anchor bolts. Material costs shown include concrete, forms and anchor bolts

	Craft@Hrs	Unit	Material	Labor	Equipment	Total
12" diameter, 36" deep, for 12' pole	E1@.746	Ea	26.90	33.30	8.77	68.97
24" diameter, 72" deep, for 30' pole	E1@1.14	Ea	108.00	50.90	13.40	172.30
30" diameter, 76" deep, for 40' pole	E1@1.54	Ea	141.00	68.80	18.10	227.90

Yard and street light poles, including base plate but no top brackets. See labor costs below. Crouse-Hinds

		Unit	Round steel	Square steel	Aluminum
12' high		Ea	540.00	350.00	241.00
16' high		Ea	630.00	279.00	298.00
20' high, 6" wide		Ea	860.00	516.00	424.00
25' high, 6" wide		Ea	1,030.00	630.00	940.00
30' high, 8" wide		Ea	1,180.00	957.00	1,060.00
35' high, 8" wide		Ea	1,520.00	1,190.00	1,640.00
40' high, 9" wide		Ea	1,790.00	1,350.00	1,860.00
50' high, 10" wide		Ea	1,980.00	1,940.00	—

	Craft@Hrs	Unit	Material	Labor	Equipment	Total
Labor setting metal light poles Equipment is a flatbed truck with a boom & auger						
To 20' high	E1@2.89	Ea	—	129.00	34.00	163.00
25' to 30' high	E1@4.35	Ea	—	194.00	51.10	245.10
35' to 40' high	E1@5.73	Ea	—	256.00	67.40	323.40
50' high	E1@8.57	Ea	—	383.00	101.00	484.00
Deduct for aluminum poles	—	%	—	-20.0	-20.0	—
Top brackets for steel and aluminum light poles, installed prior to setting pole						
Single arm	CE@.500	Ea	82.20	25.30	5.88	113.38
Two arm	CE@.500	Ea	96.70	25.30	5.88	127.88
Three arm	CE@.991	Ea	125.00	50.00	11.70	186.70
Four arm	CE@.991	Ea	156.00	50.00	11.70	217.70

Electrical 16

	Craft@Hrs	Unit	Material	Labor	Equipment	Total

Area Lighting, Installed on Poles

These costs include excavation for pole, pole foundation (2' x 2' x 3' deep), 150' of 2" PVC conduit run with 40' of RSC, 600' of #8 copper wire and terminations, 35' x 8" square steel pole and one arm and 480 volt mercury vapor 1,000 watt luminaire with lamp for each fixture. Equipment is a flatbed truck with a boom & auger.

Cost per pole including fixtures as shown

	Craft@Hrs	Unit	Material	Labor	Equipment	Total
1 fixture per pole	E1@30.4	Ea	3,110.00	1,360.00	357.00	4,827.00
2 fixtures per pole	E1@32.5	Ea	3,340.00	1,450.00	382.00	5,172.00
3 fixtures per pole	E1@34.8	Ea	3,990.00	1,550.00	409.00	5,949.00
4 fixtures per pole	E1@36.8	Ea	4,640.00	1,640.00	432.00	6,712.00

Standby Power Generators

Typical subcontract prices including the subcontractor's overhead and profit. Use these prices for preliminary estimates only.

Diesel generators including exhaust system, fuel tank and pump, automatic transfer switch, equipment pad, vibration isolators, associated piping, conduit, and lifting into position

	Craft@Hrs	Unit	Material	Labor	Equipment	Total
10 kW	—	Ea	—	—	—	12,500.00
30 kW	—	Ea	—	—	—	22,200.00
45 kW	—	Ea	—	—	—	26,800.00
50 kW	—	Ea	—	—	—	28,000.00
60 kW	—	Ea	—	—	—	33,600.00
75 kW	—	Ea	—	—	—	35,000.00
90 kW	—	Ea	—	—	—	40,800.00
100 kW	—	Ea	—	—	—	45,200.00
125 kW	—	Ea	—	—	—	46,100.00
150 kW	—	Ea	—	—	—	56,700.00
175 kW	—	Ea	—	—	—	61,100.00
200 kW	—	Ea	—	—	—	61,600.00
300 kW	—	Ea	—	—	—	89,700.00
350 kW	—	Ea	—	—	—	106,000.00
400 kW	—	Ea	—	—	—	107,000.00
450 kW	—	Ea	—	—	—	150,000.00
500 kW	—	Ea	—	—	—	154,000.00
600 kW	—	Ea	—	—	—	188,000.00
750 kW	—	Ea	—	—	—	258,000.00
900 kW	—	Ea	—	—	—	313,000.00
1,000 kW	—	Ea	—	—	—	318,000.00
1,100 kW	—	Ea	—	—	—	387,000.00

Automatic Transfer Switches for standby power systems.

Single phase, 3 wire, solid neutral, 50 to 60 cycle. NEMA Class 1 (indoor)

	Craft@Hrs	Unit	Material	Labor	Equipment	Total
100 amp, 250 volt	E4@4.97	Ea	592.00	216.00	—	808.00
150 amp, 250 volt	E4@4.97	Ea	3,070.00	216.00	—	3,286.00
200 amp, 250 volt	E4@4.97	Ea	2,300.00	216.00	—	2,516.00
100 amp, 600 volt	E4@4.97	Ea	3,250.00	216.00	—	3,466.00
200 amp, 600 volt	E4@8.76	Ea	4,440.00	381.00	—	4,821.00
400 amp, 600 volt	E4@15.1	Ea	6,160.00	656.00	—	6,816.00

	Craft@Hrs	Unit	Material	Labor	Total

Security and Alarm Systems

Sensor costs include connection but no wire runs.

Monitor panels, including accessory section and connection to signal and power supply wiring, cost per panel.

	Craft@Hrs	Unit	Material	Labor	Total
1 zone wall-mounted cabinet with 1 monitor panel, tone, standard line supervision, and 115 volt power supply	E4@2.02	Ea	902.00	87.80	989.80
5 zone wall-mounted cabinet with 5 monitor panels, 10 zone monitor rack, 5 monitor panel blanks, tone, standard line supervision and 115 volt power supply	E4@8.21	Ea	5,110.00	357.00	5,467.00

	Craft@Hrs	Unit	Material	Labor	Total
10 zone wall-mounted cabinet with 10 monitor panels, 10 zone monitor rack, tone, standard line supervision and 115 volt power supply	E4@12.1	Ea	5,880.00	526.00	6,406.00
10 zone monitor rack and 10 monitor panels with tone and standard line supervision but no power supply and no cabinet	E4@11.1	Ea	4,860.00	482.00	5,342.00
1 zone wall-mounted cabinet with 1 monitor panel, tone, high security line supervision and 115 volt power supply	E4@2.02	Ea	1,200.00	87.80	1,287.80
5 zone wall-mounted cabinet with 10 zone monitor rack, 5 monitor panel blanks, tone, high security line supervision but no power supply	E4@7.82	Ea	6,410.00	340.00	6,750.00
10 zone wall-mounted cabinet with 10 monitor panels, 10 zone monitor rack, tone, high security line supervision and 115 volt power supply	E4@12.0	Ea	7,510.00	521.00	8,031.00
Emergency power indicator for monitor panel	E4@1.14	Ea	240.00	49.50	289.50
Monitor panels					
Panel with accessory section, tone and standard line supervision	E4@.990	Ea	396.00	43.00	439.00
Panel with accessory section, tone and high security line supervision	E4@.904	Ea	596.00	39.30	635.30
Monitor racks					
1 zone with 115 volt power supply	E4@.904	Ea	326.00	39.30	365.30
10 zone with 115 volt power supply	E4@.904	Ea	1,680.00	39.30	1,719.30
10 zone with no power supply	E4@.904	Ea	1,290.00	39.30	1,329.30
Monitor cabinets					
1 zone monitor for wall mounted cabinet	E4@.990	Ea	511.00	43.00	554.00
5 zone monitor for wall mounted cabinet	E4@.990	Ea	656.00	43.00	699.00
10 zone monitor for wall mounted cabinet	E4@.990	Ea	800.00	43.00	843.00
20 zone monitor for wall mounted cabinet	E4@.990	Ea	1,080.00	43.00	1,123.00
50 zone monitor for wall mounted cabinet	E4@.990	Ea	2,420.00	43.00	2,463.00
50 zone monitor for floor mounted cabinet	E4@1.04	Ea	2,640.00	45.20	2,685.20
Balanced magnetic door switches					
Surface mounted	E4@2.02	Ea	100.00	87.80	187.80
Surface mounted with remote test	E4@2.02	Ea	140.00	87.80	227.80
Flush mounted	E4@2.02	Ea	96.60	87.80	184.40
Mounted bracket or spacer	E4@.249	Ea	5.79	10.80	16.59
Photoelectric sensors					
500' range, 12 volt DC	E4@2.02	Ea	274.00	87.80	361.80
800' range, 12 volt DC	E4@2.02	Ea	308.00	87.80	395.80
Fence type, 6 beam, without wire or trench	E4@3.89	Ea	9,960.00	169.00	10,129.00
Fence type, 9 beam, without wire or trench	E4@3.89	Ea	12,500.00	169.00	12,669.00
Capacitance wire grid systems					
Surface type	E4@.990	Ea	70.20	43.00	113.20
Duct type	E4@.990	Ea	57.10	43.00	100.10
Tube grid kit	E4@.990	Ea	104.00	43.00	147.00
Vibration sensors, to 30 per zone	E4@2.02	Ea	123.00	87.80	210.80
Audio sensors, to 30 per zone	E4@2.02	Ea	145.00	87.80	232.80
Inertia sensors, outdoor, without trenching	E4@2.02	Ea	96.40	87.80	184.20
Inertia sensors, indoor	E4@2.02	Ea	64.40	87.80	152.20
Electric sensor cable, 300 meters, no trenching	E4@3.89	Ea	1,460.00	169.00	1,629.00
Ultrasonic transmitters, to 20 per zone					
Omni-directional	E4@2.02	Ea	62.70	87.80	150.50
Directional	E4@2.02	Ea	70.20	87.80	158.00

	Craft@Hrs	Unit	Material	Labor	Total
Ultrasonic transceivers, to 20 per zone					
Omni-directional	E4@2.02	Ea	70.20	87.80	158.00
Directional	E4@2.02	Ea	75.10	87.80	162.90
High security	E4@2.02	Ea	255.00	87.80	342.80
Standard security	E4@2.02	Ea	104.00	87.80	191.80
Microwave perimeter sensor, to 4 per zone	E4@3.89	Ea	4,100.00	169.00	4,269.00
Passive interior infrared sensors, to 20 per zone					
Wide pattern	E4@2.02	Ea	568.00	87.80	655.80
Narrow pattern	E4@2.02	Ea	539.00	87.80	626.80
Access and secure control units					
For balanced magnetic door switches	E4@3.06	Ea	310.00	133.00	443.00
For photoelectric sensors	E4@3.06	Ea	568.00	133.00	701.00
For capacitance sensors	E4@3.06	Ea	576.00	133.00	709.00
For audio and vibration sensors	E4@3.06	Ea	665.00	133.00	798.00
For inertia sensors	E4@3.06	Ea	596.00	133.00	729.00
For nimmer (m.b.) detector	E4@3.06	Ea	496.00	133.00	629.00
For electric cable sensors	E4@3.06	Ea	2,990.00	133.00	3,123.00
For ultrasonic sensors	E4@3.06	Ea	912.00	133.00	1,045.00
For microwave sensors	E4@3.06	Ea	561.00	133.00	694.00
Accessories					
Tamper assembly for monitor cabinet	E4@.990	Ea	62.70	43.00	105.70
Monitor panel blank	E4@.249	Ea	7.00	10.80	17.80
Audible alarm	E4@.990	Ea	73.20	43.00	116.20
Audible alarm control	E4@.990	Ea	290.00	43.00	333.00
Termination screw terminal cabinet, 25 pair	E4@.990	Ea	210.00	43.00	253.00
Termination screw terminal cabinet, 50 pair	E4@1.36	Ea	342.00	59.10	401.10
Termination screw terminal cabinet, 150 pair	E4@3.45	Ea	564.00	150.00	714.00
Universal termination with remote test for cabinets and control panels	E4@1.47	Ea	51.50	63.90	115.40
Universal termination without remote test	E4@.990	Ea	30.10	43.00	73.10
High security line supervision termination					
With remote test and tone	E4@1.47	Ea	261.00	63.90	324.90
With tone only	E4@1.47	Ea	240.00	63.90	303.90
12" door cord for capacitance sensor	E4@.249	Ea	8.03	10.80	18.83
Insulation block kit for capacitance sensor	E4@.990	Ea	40.00	43.00	83.00
Termination block for capacitance sensor	E4@.249	Ea	8.50	10.80	19.30
200 zone graphic display	E4@3.06	Ea	4,000.00	133.00	4,133.00
210 zone multiplexer event recorder unit	E4@3.89	Ea	3,810.00	169.00	3,979.00
Guard alert display	E4@3.06	Ea	897.00	133.00	1,030.00
Uninterruptable power supply (battery)	E4@2.02	Ea	1,840.00	87.80	1,927.80
12 volt, 40VA plug-in transformer	E4@.495	Ea	34.50	21.50	56.00
18 volt, 40VA plug-in transformer	E4@.495	Ea	23.60	21.50	45.10
24 volt, 40VA plug-in transformer	E4@.495	Ea	17.10	21.50	38.60
Test relay	E4@.332	Ea	53.10	14.40	67.50
Sensor test control (average for sensors)	E4@.990	Ea	69.60	43.00	112.60
Cable, 2 pair 22 gauge, twisted, 1 pair shielded, run in exposed walls or per LF per pull, pulled in raceway and connected	E4@.008	LF	.52	.35	.87

	Craft@Hrs	Unit	Material	Labor	Total
Fire Alarm and Detection Systems No wiring included except as noted.					
Pedestal mounted master fire alarm box	E4@3.89	Ea	1,250.00	169.00	1,419.00
Automatic coded signal transmitter	E4@3.89	Ea	658.00	169.00	827.00
Fire alarm control panel, simplex	E4@3.89	Ea	616.00	169.00	785.00
Fire detection annunciator panel					
8 zone drop	E4@3.89	Ea	408.00	169.00	577.00
12 zone drop	E4@5.03	Ea	580.00	219.00	799.00
16 zone drop	E4@5.93	Ea	833.00	258.00	1,091.00
8 zone lamp panel only	E4@4.33	Ea	135.00	188.00	323.00
12 zone lamp panel only	E4@5.93	Ea	199.00	258.00	457.00
16 zone lamp panel only	E4@7.23	Ea	264.00	314.00	578.00
Battery charger and cabinet, simplex	E4@6.19	Ea	355.00	269.00	624.00
Add for nickel cadmium batteries	E4@3.89	Ea	385.00	169.00	554.00
Fire alarm pull stations, manual operation	E4@.497	Ea	19.90	21.60	41.50
Fire alarm 10" bell with outlet box	E4@.497	Ea	64.70	21.60	86.30
Magnetic door holder	E4@1.33	Ea	60.70	57.80	118.50
Combination door holder and closer	E4@2.48	Ea	245.00	108.00	353.00
Detex door lock	E4@2.72	Ea	29.10	118.00	147.10
Thermodetector	E4@.497	Ea	8.78	21.60	30.38
Ionization AC smoke detector, with wiring	E4@.710	Ea	150.00	30.80	180.80
Fixed temp. and rate of rise smoke detector	E4@.746	Ea	13.50	32.40	45.90
Carbon dioxide pressure switch	E4@.746	Ea	4.63	32.40	37.03
Telephone Wiring					
Plywood backboard mount for circuit wiring, set on masonry wall					
4' x 4' x 3/4"	E4@.938	Ea	15.00	40.80	55.80
4' x 8' x 3/4"	E4@1.10	Ea	30.00	47.80	77.80
Cable tap in a manhole or junction box					
25 to 50 pair cable	E4@7.80	Ea	101.00	339.00	440.00
100 to 200 pair cable	E4@10.1	Ea	115.00	439.00	554.00
300 pair cable	E4@12.0	Ea	117.00	521.00	638.00
400 pair cable	E4@13.0	Ea	127.00	565.00	692.00
Cable terminations in a manhole or junction box					
25 pair cable	E4@1.36	Ea	9.15	59.10	68.25
50 pair cable	E4@1.36	Ea	10.40	59.10	69.50
100 pair cable	E4@2.72	Ea	20.80	118.00	138.80
150 pair cable	E4@3.45	Ea	31.40	150.00	181.40
200 pair cable	E4@4.61	Ea	47.20	200.00	247.20
300 pair cable	E4@5.03	Ea	62.60	219.00	281.60
400 pair cable	E4@5.34	Ea	83.20	232.00	315.20
Communications cable, pulled in conduit or walls and connected					
2 pair cable	E4@.008	LF	.05	.35	.40
25 pair cable	E4@.030	LF	.59	1.30	1.89
50 pair cable	E4@.043	LF	1.25	1.87	3.12
75 pair cable	E4@.043	LF	1.56	1.87	3.43
100 pair cable	E4@.043	LF	1.82	1.87	3.69
150 pair cable	E4@.045	LF	2.53	1.96	4.49
200 pair cable	E4@.051	LF	3.30	2.22	5.52
300 pair cable	E4@.051	LF	4.87	2.22	7.09
Telephone outlets (no junction box included)					
Wall outlet	E4@.396	Ea	5.68	17.20	22.88
Floor outlet, flush mounted	E4@.495	Ea	7.06	21.50	28.56

	Craft@Hrs	Unit	Material	Labor	Total
Closed Circuit Television Systems					
Outdoor television camera in weatherproof housing, cable preinstalled.					
Wall or roof mount professional	E4@.500	Ea	1,620.00	21.70	1,641.70
Wall or roof mount economy	E4@.500	Ea	225.00	21.70	246.70
Self-terminating outlet box, cover plate	E4@.245	Ea	5.06	10.60	15.66
Coaxial cable (RG59/J), 75 ohms, no raceway included	E4@.006	LF	.16	.26	.42
In-line cable taps (PTU) for 36" TV system	E4@.689	Ea	8.41	29.90	38.31
Cable blocks for in-line taps	—	Ea	4.37	—	4.37
Desk type black and white TV monitor, 12" diagonal screen, 75 ohms, labor includes hookup only					
4 camera inputs	E4@.500	Ea	400.00	21.70	421.70
Public Address Paging Systems					
Signal cable, #22 solid copper polyethylene insulated, PVC jacketed wire					
Pulled in conduit, no conduit included	E4@.008	LF	.08	.35	.43
Microphone, desk type, push to talk, with					
7' of three conductor cable	E4@.250	Ea	106.00	10.90	116.90
Wall mounted enameled steel amplifier cabinet (18" high, 24" wide, 6" deep) with electrical receptacle					
and lockable door	E4@1.67	Ea	300.00	72.60	372.60
Amplifier, 100 watts, 4 inputs, installation includes hookup only, no wire pull					
Complete unit	E4@.500	Ea	313.00	21.70	334.70
Wall mounted paging horns, transformer type, 250 to 13,000 Hertz, 8 ohms, 30 Watts, 10" diameter, 12" deep, gray baked enamel, labor includes hookup only					
Interior/exterior 15/30 Watts	E4@.250	Ea	79.00	10.90	89.90
Ceiling mounted 8" speaker with 12" grill, labor includes installation and hookup only, no wire pull					
8" ceiling speaker	E4@.330	Ea	25.00	14.30	39.30
8" ceiling speaker with volume	E4@.330	Ea	35.00	14.30	49.30
Outdoor weather resistant speakers					
Landscape weather resistant	E4@.330	Ea	150.00	14.30	164.30
Rock landscape (high end)	E4@.330	Ea	405.00	14.30	419.30

Computer Network Cable Pulled in conduit or walls. Add for connectors and cable testing. Prices are for cable purchased in bulk on 500' spools.

	Craft@Hrs	Unit	Material	Labor	Total
Coaxial cable					
Arcnet, 75 ohm (RG62). Rarely used in modern networks					
PVC cable	E4@.008	LF	.21	.35	.56
Plenum cable	E4@.008	LF	.55	.35	.90
Thin Ethernet, 50 ohm, 10BASE2 (RG58). Rarely used in modern networks					
PVC cable	E4@.008	LF	.30	.35	.65
Plenum cable	E4@.008	LF	.59	.35	.94
Fiber optic cable					
Singlemode fiber optic cable, duplex is 4 strand, simplex is 1 strand					
Singlemode duplex fiber optic cables (8.3/125)	E4@.012	LF	.17	.52	.69
Singlemode simplex fiber optic cables (8.3/125)	E4@.012	LF	.11	.52	.63
Multimode fiber optic cable, duplex is a zip type					
Multimode duplex fiber optic cables (50/125)	E4@.012	LF	.21	.52	.73
Multimode duplex fiber optic cables (62.5/125)	E4@.012	LF	.26	.52	.78
Multimode simplex fiber optic cables (50/125)	E4@.012	LF	.19	.52	.71
Multimode simplex fiber optic cables (62.5/125)	E4@.012	LF	.28	.52	.80

	Craft@Hrs	Unit	Material	Labor	Total

Twisted pair Category 5, 100 Mbps, material price based on purchase of 1000' rolls, use solid for horizontal runs. STP is shielded

	Craft@Hrs	Unit	Material	Labor	Total
4 pair UTP PVC stranded cable	E4@.008	LF	.13	.35	.48
4 pair UTP PVC solid cable	E4@.008	LF	.09	.35	.44
4 pair UTP plenum solid cable	E4@.008	LF	.24	.35	.59
4 pair STP PVC stranded cable	E4@.008	LF	.22	.35	.57
4 pair STP PVC solid cable	E4@.008	LF	.22	.35	.57
4 pair STP plenum solid cable	E4@.008	LF	.48	.35	.83

Twisted pair Category 5e, 350 Mbps, material price based on purchase of 1000' rolls, use solid for horizontal runs. STP is shielded

	Craft@Hrs	Unit	Material	Labor	Total
4 pair UTP PVC solid	E4@.008	LF	.10	.35	.45
4 pair UTP PVC stranded	E4@.008	LF	.13	.35	.48
4 pair UTP plenum solid	E4@.008	LF	.26	.35	.61
4 pair STP PVC stranded	E4@.008	LF	.23	.35	.58
4 pair STP PVC solid	E4@.008	LF	.23	.35	.58
4 pair STP plenum solid	E4@.008	LF	.50	.35	.85

Twisted pair Category 6, 1 Gbps, material price based on purchase of 1000' rolls, use solid for horizontal runs

	Craft@Hrs	Unit	Material	Labor	Total
4 pair UTP PVC stranded	E4@.008	LF	.21	.35	.56
4 pair UTP PVC solid	E4@.008	LF	.15	.35	.50
4 pair UTP plenum solid	E4@.008	LF	.33	.35	.68

Add for multiple cable pulls of thin Ethernet or unshielded twisted pair cables when cables are pulled simultaneously in plenum or PVC, based on quantity of cables pulled

	Craft@Hrs	Unit	Material	Labor	Total
Add for each cable pulled simultaneously, over one	E4@.250	Ea	—	10.90	10.90

Punch-down blocks

Plywood backboard mount for patch panel or punch-down blocks, set on masonry wall

	Craft@Hrs	Unit	Material	Labor	Total
4' x 4' x 3/4"	E4@.938	Ea	12.10	40.80	52.90
4' x 8' x 3/4"	E4@1.10	Ea	24.20	47.80	72.00

Distribution frame for punch-down blocks

	Craft@Hrs	Unit	Material	Labor	Total
Holds 5 punch-down blocks	E4@.250	Ea	32.40	10.90	43.30
Holds 10 punch-down blocks	E4@.250	Ea	38.60	10.90	49.50

Punch-down blocks, labor is to install block in distribution frame only, add labor to punch-down wire pairs as shown below

	Craft@Hrs	Unit	Material	Labor	Total
25 pair	E4@.250	Ea	10.10	10.90	21.00
50 pair	E4@.250	Ea	10.30	10.90	21.20
Punch-down wire pairs, labor only, per pair	E4@.020	Ea	—	.87	.87

Computer network cable connectors

Coax

Thin Ethernet or Arcnet

	Craft@Hrs	Unit	Material	Labor	Total
BNC type, crimp on, female	E4@.100	Ea	3.60	4.35	7.95

Fiber optic cable, terminated, 10' length

	Craft@Hrs	Unit	Material	Labor	Total
ST-ST singlemode duplex riser	E4@.100	Ea	63.00	4.35	67.35
ST-SC singlemode duplex riser	E4@.100	Ea	63.00	4.35	67.35
SC-SC singlemode duplex riser	E4@.100	Ea	63.00	4.35	67.35
FC-SC singlemode duplex riser	E4@.100	Ea	63.00	4.35	67.35
ST-ST singlemode simplex riser	E4@.100	Ea	32.00	4.35	36.35
ST-SC singlemode simplex riser	E4@.100	Ea	32.00	4.35	36.35
SC-SC singlemode simplex riser	E4@.100	Ea	32.00	4.35	36.35
FC-SC singlemode simplex riser	E4@.100	Ea	32.00	4.35	36.35
ST-ST multimode duplex riser	E4@.100	Ea	37.00	4.35	41.35
ST-SC multimode duplex riser	E4@.100	Ea	41.00	4.35	45.35
SC-SC multimode duplex riser	E4@.100	Ea	44.00	4.35	48.35

	Craft@Hrs	Unit	Material	Labor	Total
FC-SC multimode duplex riser	E4@.100	Ea	50.00	4.35	54.35
FDDI-FDDI multimode duplex riser	E4@.100	Ea	62.00	4.35	66.35
FDDI OFNP plenum	E4@.100	Ea	100.00	4.35	104.35
Fiber optic cable connectors, special tools required					
SC multimode connector	E4@.150	Ea	2.75	6.52	9.27
SC singlemode connector	E4@.150	Ea	5.20	6.52	11.72
FC multimode connector	E4@.150	Ea	5.15	6.52	11.67
FC singlemode connector	E4@.150	Ea	5.72	6.52	12.24
ST multimode connector	E4@.150	Ea	2.31	6.52	8.83
ST singlemode connector	E4@.150	Ea	5.02	6.52	11.54
FDDI for round cable	E4@.150	Ea	9.25	6.52	15.77
FDDI for Zip cable	E4@.150	Ea	9.25	6.52	15.77
Shielded twisted pair (STP)					
IBM type 2, 6 pair data connector	E4@.100	Ea	6.72	4.35	11.07
IBM type 6, 2 pair data connector	E4@.150	Ea	17.50	6.52	24.02
IBM type 9, 2 pair data connector	E4@.150	Ea	17.40	6.52	23.92
Twisted pair connectors (UTP) and (STP), modular hand tool required, material prices based on package of 50					
RJ11	E4@.100	Ea	.39	4.35	4.74
Cat5 rated RJ45 connectors for solid wire	E4@.100	Ea	.68	4.35	5.03
Cat5 rated RJ45 connectors for stranded wire	E4@.100	Ea	.68	4.35	5.03
Cat6 rated RJ45 connectors for stranded wire	E4@.100	Ea	.68	4.35	5.03
Shielded RJ45 connectors for solid or stranded wire	E4@.100	Ea	1.90	4.35	6.25
Cable testing Testing, with dedicated cable tester					
Coax cable, per conductor	E4@.125	Ea	—	5.43	5.43
Fiber optic cable, per segment	E4@.160	Ea	—	6.95	6.95
Shielded twisted pair (STP), per pair	E4@.125	Ea	—	5.43	5.43
Unshielded twisted pair (UTP), per pair	E4@.125	Ea	—	5.43	5.43
Wall plates					
Coax wall plates with jacks (no junction box included)					
One or two jacks, including termination	E4@.396	Ea	7.95	17.20	25.15
Modular wall plates with jacks for UTP (no junction box included), including termination					
One 4 or 6 wire jack (RJ11)	E4@.396	Ea	4.95	17.20	22.15
One 8 wire jack (RJ45)	E4@.396	Ea	4.95	17.20	22.15
Two 8 wire jacks (RJ45)	E4@.450	Ea	11.00	19.60	30.60

Index

642

649

Practical References for Builders

Construction Forms & Contracts

125 forms you can copy and use — or load into your computer (from the FREE disk enclosed). Then you can customize the forms to fit your company, fill them out, and print. Loads into *Word* for *Windows*™, *Lotus 1-2-3*, *WordPerfect*, *Works*, or *Excel* programs. You'll find forms covering accounting, estimating, fieldwork, contracts, and general office. Each form comes with complete instructions on when to use it and how to fill it out. These forms were designed, tested and used by contractors, and will help keep your business organized, profitable and out of legal, accounting and collection troubles. Includes a CD-ROM for *Windows*™ and Mac. **400 pages, 8½ x 11, $41.75**

National Painting Cost Estimator

A complete guide to estimating painting costs for just about any type of residential, commercial, or industrial painting, whether by brush, spray, or roller. Shows typical costs and bid prices for fast, medium, and slow work, including material costs per gallon; square feet covered per gallon; square feet covered per manhour; labor, material, overhead, and taxes per 100 square feet; and how much to add for profit.

Includes a CD-ROM with an electronic version of the book with *National Estimator*, a stand-alone *Windows*™ estimating program, plus an interactive multimedia video that shows how to use the disk to compile construction cost estimates. **440 pages, 8½ x 11, $53.00. Revised annually**

2005 National Electrical Code

This new electrical code incorporates sweeping improvements to make the code more functional and user-friendly. Here you'll find the essential foundation for electrical code requirements for the 21st century. With hundreds of significant and widespread changes, this 2005 *NEC* contains all the latest electrical technologies, recently developed techniques, and enhanced safety standards for electrical work. This is the standard all electricians are required to know, even if it hasn't yet been adopted by their local or state jurisdictions. **784 pages, 8½ x 11, $65.00**

Contractor's Guide to QuickBooks Pro 2005

This user-friendly manual walks you through QuickBooks Pro's detailed setup procedure and explains step-by-step how to create a first-rate accounting system. You'll learn in days, rather than weeks, how to use *QuickBooks Pro* to get your contracting business organized, with simple, fast accounting procedures. On the CD included with the book you'll find a *QuickBooks Pro* file preconfigured for a construction company (you drag it over onto your computer and plug in your own company's data). You'll also get a complete estimating program, including a database, and a job costing program that lets you export your estimates to *QuickBooks Pro*. It even includes many useful construction forms to use in your business. **344 pages, 8½ x 11, $49.75**

Also available: **Contractor's Guide to QuickBooks Pro 2001, $45.25**
Contractor's Guide to QuickBooks Pro 2003, $47.75
Contractor's Guide to QuickBooks Pro 2004, $48.50

CD Estimator

If your computer has *Windows*™ and a CD-ROM drive, CD Estimator puts at your fingertips over 100,000 construction costs for new construction, remodeling, renovation & insurance repair, home improvement, electrical, concrete & masonry, painting, and plumbing & HVAC. Monthly cost updates are available at no charge on the Internet. You'll also have the *National Estimator* program — a stand-alone estimating program for *Windows*™ that Remodeling magazine called a "computer wiz," and Job Cost Wizard, a program that lets you export your estimates to QuickBooks Pro for actual job costing. A 60-minute interactive video teaches you how to use this CD-ROM to estimate construction costs. And to top it off, to help you create professional-looking estimates, the disk includes over 40 construction estimating and bidding forms in a format that's perfect for nearly any *Windows*™ word processing or spreadsheet program. **CD Estimator is $78.50**

Contractor's Guide to the Building Code Revised

This new edition was written in collaboration with the International Conference of Building Officials, writers of the code. It explains in plain English exactly what the latest edition of the *Uniform Building Code* requires. Based on the 1997 code, it explains the changes and what they mean for the builder. Also covers the *Uniform Mechanical Code* and the *Uniform Plumbing Code*. Shows how to design and construct residential and light commercial buildings that'll pass inspection the first time. Suggests how to work with an inspector to minimize construction costs, what common building shortcuts are likely to be cited, and where exceptions may be granted. **320 pages, 8½ x 11, $39.00**

Basic Engineering for Builders

If you've ever been stumped by an engineering problem on the job, yet wanted to avoid the expense of hiring a qualified engineer, you should have this book. Here you'll find engineering principles explained in non-technical language and practical methods for applying them on the job. With the help of this book you'll be able to understand engineering functions in the plans and how to meet the requirements, how to get permits issued without the help of an engineer, and anticipate requirements for concrete, steel, wood and masonry. See why you sometimes have to hire an engineer and what you can undertake yourself: surveying, concrete, lumber loads and stresses, steel, masonry, plumbing, and HVAC systems. This book is designed to help the builder save money by understanding engineering principles that you can incorporate into the jobs you bid. **400 pages, 8½ x 11, $36.50**

Steel-Frame House Construction

Framing with steel has obvious advantages over wood, yet building with steel requires new skills that can present challenges to the wood builder. This new book explains the secrets of steel framing techniques for building homes, whether pre-engineered or built stick by stick. It shows you the techniques, the tools, the materials, and how you can make it happen. Includes hundreds of photos and illustrations, plus a CD-ROM with steel framing details, a database of steel materials and manhours, with an estimating program. **320 pages, 8½ x 11, $39.75**

Construction Estimating Reference Data

Provides the 300 most useful manhour tables for practically every item of construction. Labor requirements are listed for sitework, concrete work, masonry, steel, carpentry, thermal and moisture protection, doors and windows, finishes, mechanical and electrical. Each section details the work being estimated and gives appropriate crew size and equipment needed. Includes a CD-ROM with an electronic version of the book with *National Estimator*, a stand-alone *Windows*™ estimating program, plus an interactive multimedia video that shows how to use the disk to compile construction cost estimates. **432 pages, 11 x 8½, $39.50**

Basic Lumber Engineering for Builders

Beam and lumber requirements for many jobs aren't always clear, especially with changing building codes and lumber products. Most of the time you rely on your own "rules of thumb" when figuring spans or lumber engineering. This book can help you fill the gap between what you can find in the building code span tables and what you need to pay a certified engineer to do. With its large, clear illustrations and

examples, this book shows you how to figure stresses for pre-engineered wood or wood structural members, how to calculate loads, and how to design your own girders, joists and beams. Included FREE with the book — an easy-to-use limited version of NorthBridge Software's *Wood Beam Sizing* program. **272 pages, 8½ x 11, $38.00**

Roof Framing

Shows how to frame any type of roof in common use today, even if you've never framed a roof before. Includes using a pocket calculator to figure any common, hip, valley, or jack rafter length in seconds. Over 400 illustrations cover every measurement and every cut on each type of roof: gable, hip, Dutch, Tudor, gambrel, shed, gazebo, and more. **480 pages, 5½ x 8½, $24.50**

National Repair & Remodeling Estimator

The complete pricing guide for dwelling reconstruction costs. Reliable, specific data you can apply on every repair and remodeling job. Up-to-date material costs and labor figures based on thousands of jobs across the country. Provides recommended crew sizes; average production rates; exact material, equipment, and labor costs; a total unit cost and a total price including overhead and profit. Separate listings for high- and low-volume builders, so prices shown are specific for any size business. Estimating tips specific to repair and remodeling work to make your bids complete, realistic, and profitable. Includes a CD-ROM with an electronic version of the book with *National Estimator*, a stand-alone *Windows*™ estimating program, plus an interactive multimedia video that shows how to use the disk to compile construction cost estimates. **296 pages, 8½ x 11, $53.50. Revised annually**

Roofing Construction & Estimating

Installation, repair and estimating for nearly every type of roof covering available today in residential and commercial structures: asphalt shingles, roll roofing, wood shingles and shakes, clay tile, slate, metal, built-up, and elastomeric. Covers sheathing and underlayment techniques, as well as secrets for installing leakproof valleys. Many estimating tips help you minimize waste, as well as insure a profit on every job. Troubleshooting techniques help you identify the true source of most leaks. Over 300 large, clear illustrations help you find the answer to just about all your roofing questions. **432 pages, 8½ x 11, $38.00**

Moving to Commercial Construction

In commercial work, a single job can keep you and your crews busy for a year or more. The profit percentages are higher, but so is the risk involved. This book takes you step-by-step through the process of setting up a successful commercial business; finding work, estimating and bidding, value engineering, getting through the submittal and shop drawing process, keeping a stable work force, controlling costs, and promoting your business. Explains the design/build and partnering business concepts and their advantage over the competitive bid process. Includes sample letters, contracts, checklists and forms that you can use in your business, plus a CD-ROM with blank copies in several word-processing formats for both Mac and PC computers. **256 pages, 8½ x 11, $42.00**

Craftsman's Construction Installation Encyclopedia

Step-by-step installation instructions for just about any residential construction, remodeling or repair task, arranged alphabetically, from Acoustic tile to Wood flooring. Includes hundreds of illustrations that show how to build, install, or remodel each part of the job, as well as manhour tables for each work item so you can estimate and bid with confidence. Also includes a CD-ROM with all the material in the book, handy look-up features, and the ability to capture and print out for your crew the instructions and diagrams for any job. **792 pages, 8½ x 11, $65.00**

CD Estimator — Heavy

CD Estimator — Heavy has a complete 780-page heavy construction cost estimating volume for each of the 50 states. Select the cost database for the state where the work will be done. Includes thousands of cost estimates you won't find anywhere else, and in-depth coverage of demolition, hazardous materials remediation, tunneling, site utilities, precast concrete, structural framing, heavy timber construction, membrane waterproofing, industrial windows and doors, specialty finishes, built-in commercial and industrial equipment, and HVAC and electrical systems for commercial and industrial buildings. **CD Estimator — Heavy is $74.00**

Finish Carpentry: Efficient Techniques for Custom Interiors

Professional finish carpentry demands expert skills, precise tools, and a solid understanding of how to do the work. This new book explains how to install moldings, paneled walls and ceilings, and just about every aspect of interior trim - including doors and windows. Covers built-in bookshelves, coffered ceilings, and skylight wells and soffits, including paneled ceilings with decorative beams.
288 pages, 8½ x 11, $34.95

How to Succeed With Your Own Construction Business

Everything you need to start your own construction business: setting up the paperwork, finding the work, advertising, using contracts, dealing with lenders, estimating, scheduling, finding and keeping good employees, keeping the books, and coping with success. If you're considering starting your own construction business, all the knowledge, tips, and blank forms you need are here. **336 pages, 8½ x 11, $28.50**

Building Contractor's Exam Preparation Guide

Passing today's contractor's exams can be a major task. This book shows you how to study, how questions are likely to be worded, and the kinds of choices usually given for answers. Includes sample questions from actual state, county, and city examinations, plus a sample exam to practice on. This book isn't a substitute for the study material that your testing board recommends, but it will help prepare you for the types of questions — and their correct answers — that are likely to appear on the actual exam. Knowing how to answer these questions, as well as what to expect from the exam, can greatly increase your chances of passing. **320 pages, 8½ x 11, $35.00**

Estimating With Microsoft *Excel*

Most builders estimate with *Excel* because it's easy to learn, quick to use, and can be customized to your style of estimating. Here you'll find step-by-step instructions on how to create your own customized automated spreadsheet estimating program for use with *Excel*. You'll learn how to use the magic of *Excel* to create detail sheets, cost breakdown summaries, and links. You'll put this all to use in estimating concrete, rebar, permit fees, and roofing. You can even create your own macros. Includes a CD-ROM that illustrates examples in the book and provides you with templates you can use to set up your own estimating system. **150 pages, 7 x 9, $39.95**

Electrician's Exam Preparation Guide

Need help in passing the apprentice, journeyman, or master electrician's exam? This is a book of questions and answers based on actual electrician's exams over the last few years. Almost a thousand multiple-choice questions — exactly the type you'll find on the exam — cover every area of electrical installation: electrical drawings, services and systems, transformers, capacitors, distribution equipment, branch circuits, feeders, calculations, measuring and testing, and more. It gives you the correct answer, an explanation, and where to find it in the latest *NEC*. Also tells how to apply for the test, how best to study, and what to expect on examination day. Includes a FREE CD-ROM with all the questions in the book in interactive test-yourself software that makes studying for the exam almost fun! Updated to the 2005 *NEC*. **352 pages, 8½ x 11, $39.50**

Contractor's Year-Round Tax Guide Revised

How to set up and run your construction business to minimize taxes: corporate tax strategy and how to use it to your advantage, and what you should be aware of in contracts with others. Covers tax shelters for builders, write-offs and investments that will reduce your taxes, accounting methods that are best for contractors, and what the I.R.S. allows and what it often questions. **208 pages, 8½ x 11, $26.50**

National Electrical Estimator

This year's prices for installation of all common electrical work: conduit, wire, boxes, fixtures, switches, outlets, loadcenters, panelboards, raceway, duct, signal systems, and more. Provides material costs, manhours per unit, and total installed cost. Explains what you should know to estimate each part of an electrical system. Includes a CD-ROM with an electronic version of the book with National Estimator, a stand-alone Windows™ estimating program, plus an interactive multimedia video that shows how to use the disk to compile construction cost estimates.
552 pages, 8½ x 11, $52.75. Revised annually

National Building Cost Manual

Square foot costs for residential, commercial, industrial, and farm buildings. Quickly work up a reliable budget estimate based on actual materials and design features, area, shape, wall height, number of floors, and support requirements. Includes all the important variables that can make any building unique from a cost standpoint.
240 pages, 8½ x 11, $28.00. Revised annually

Contractor's Plain-English Legal Guide

For today's contractors, legal problems are like snakes in the swamp — you might not see them, but you know they're there. This book tells you where the snakes are hiding and directs you to the safe path. With the directions in this easy-to-read handbook you're less likely to need a $200-an-hour lawyer. Includes simple directions for starting your business, writing contracts that cover just about any eventuality, collecting what's owed you, filing liens, protecting yourself from unethical subcontractors, and more. For about the price of 15 minutes in a lawyer's office, you'll have a guide that will make many of those visits unnecessary. Includes a CD-ROM with blank copies of all the forms and contracts in the book. **272 pages, 8½ x 11, $49.50**

National Home Improvement Estimator

Current labor and material prices for home improvement projects. Provides manhours for each job, recommended crew size, and the labor cost for the removal and installation work. Material prices are current, with location adjustment factors and free monthly updates on the Web. Gives step-by-step instructions for the work, with helpful diagrams, and home improvement shortcuts and tips from an expert. Includes a CD-ROM with an electronic version of the book, and National Estimator, a stand-alone Windows™ estimating program, plus an interactive multimedia tutorial that shows how to use the disk to compile home improvement cost estimates. **504 pages, 8½ x 11, $53.75. Revised annually**

Illustrated Guide to the International Plumbing & Fuel Gas Codes

A comprehensive guide to the *International Plumbing* and *Fuel Gas Codes* that explains the intricacies of the code in easy-to-understand language. Packed with plumbing isometrics and helpful illustrations, it makes clear the code requirements for the installation methods and materials for plumbing and fuel gas systems. Includes code tables for pipe sizing and fixture units, and code requirements for just about all areas of plumbing, from water supply and vents to sanitary drainage systems. Covers the principles and terminology of the codes, how the various systems work and are regulated, and code-compliance issues you'll likely encounter on the job. Each chapter has a set of self-test questions for anyone studying for the plumber's exam, and tells you where to look in the code for the details. Written by a former plumbing inspector, this guide has the help you need to install systems in compliance with the *IPC* and the *IFGC*. **312 pages, 8½ x 11, $37.00**

Markup & Profit: A Contractor's Guide

In order to succeed in a construction business, you have to be able to price your jobs to cover all labor, material and overhead expenses, and make a decent profit. The problem is knowing what markup to use. You don't want to lose jobs because you charge too much, and you don't want to work for free because you've charged too little. If you know how to calculate markup, you can apply it to your job costs to find the right sales price for your work. This book gives you tried and tested formulas, with step-by-step instructions and easy-to-follow examples, so you can easily figure the markup that's right for your business. Includes a CD-ROM with forms and checklists for your use. **320 pages, 8½ x 11, $32.50**

2003 International Residential Code

Replacing the *CABO One-* and *Two-Family Dwelling Code*, this book has the latest technological advances in building design and construction. Among the changes are provisions for steel framing and energy savings. Also contains mechanical, fuel gas and plumbing provisions that coordinate with the *International Mechanical Code* and *International Plumbing Code*. **578 pages, 8½ x 11, $62.00**
Also available: 2000 International Residential Code, $59.00
On interactive CD-ROM, $48.00

Profits in Buying & Renovating Homes

Step-by-step instructions for selecting, repairing, improving, and selling highly profitable "fixer-uppers." Shows which price ranges offer the highest profit-to-investment ratios, which neighborhoods offer the best return, practical directions for repairs, and tips on dealing with buyers, sellers, and real estate agents. Shows you how to determine your profit before you buy, what "bargains" to avoid, and how to make simple, profitable, inexpensive upgrades. **304 pages, 8½ x 11, $24.75**

Cost Records for Construction Estimating

How to organize and use cost information from jobs just completed to make more accurate estimates in the future. Explains how to keep the records you need to track costs for sitework, footings, foundations, framing, interior finish, siding and trim, masonry, and subcontract expense. Provides sample forms. **208 pages, 8½ x 11, $15.75**
